PHYSICAL MODELLING IN GEOTECHNICS

PROCEEDINGS OF THE 9TH INTERNATIONAL CONFERENCE ON PHYSICAL MODELLING IN GEOTECHNICS 2018 (ICPMG 2018), LONDON, UK, 17–20 JULY 2018

Physical Modelling in Geotechnics

Editors

Andrew McNamara, Sam Divall, Richard Goodey, Neil Taylor, Sarah Stallebrass & Jignasha Panchal
City, University of London, UK

VOLUME 1

CRC Press
Taylor & Francis Group
Boca Raton London New York Leiden

CRC Press is an imprint of the
Taylor & Francis Group, an **informa** business

A BALKEMA BOOK

CRC Press/Balkema is an imprint of the Taylor & Francis Group, an informa business

© 2018 Taylor & Francis Group, London, UK

Typeset by MPS Limited, Chennai, India

All rights reserved. No part of this publication or the information contained herein may be reproduced, stored in a retrieval system, or transmitted in any form or by any means, electronic, mechanical, by photocopying, recording or otherwise, without written prior permission from the publishers.

Although all care is taken to ensure integrity and the quality of this publication and the information herein, no responsibility is assumed by the publishers nor the author for any damage to the property or persons as a result of operation or use of this publication and/or the information contained herein.

Published by: CRC Press/Balkema
Schipholweg 107C, 2316 XC Leiden, The Netherlands
e-mail: Pub.NL@taylorandfrancis.com
www.crcpress.com – www.taylorandfrancis.com

ISBN: 978-1-138-55975-2 (Hbk set + USB)
ISBN volume 1: 978-1-138-34419-8 (Hbk)
ISBN volume 2: 978-1-138-34422-8 (Hbk)

ISBN: 978-0-203-71227-6 (eBook set)
ISBN volume 1: 978-0-429-43866-0 (eBook)
ISBN volume 2: 978-0-429-43864-6 (eBook)

Table of contents

Preface	XVII
International advisory board	XIX
Local organising committee	XXI
Manuscript reviewers	XXIII
Sponsors	XXV

VOLUME 1

Keynote and Themed lectures

Modelling tunnel behaviour under seismic actions: An integrated approach *E. Bilotta*	3
An example of effective mentoring for research centres *C.E. Bronner, D.W. Wilson, K. Ziotopoulou, K.M. Darby, A. Sturm, A.J. Raymond, R.W. Boulanger, J.T. DeJong, D.M. Moug & J.D. Bronner*	21
Geotechnical modelling for offshore renewables *C. Gaudin, C.D. O'Loughlin & B. Bienen*	33
Physical modelling applied to infrastructure development *R.J. Goodey*	43
The role of centrifuge modelling in capturing whole-life responses of geotechnical infrastructure to optimise design *S. Gourvenec*	51
Development of geotechnical centrifuges and facilities in China *Y.J. Hou*	77
Physical modelling of structural and biological soil reinforcement *J.A. Knappett*	87
Current and emerging physical modelling technologies *W.A. Take*	101

1. Sample preparation and characterisation

Investigation into 3D printing of granular media *O. Adamidis, S. Alber & I. Anastasopoulos*	113
Undrained shear strength profile of normally and overconsolidated kaolin clay *A. Arnold, W. Zhang & A. Askarinejad*	119
LEAP GWU 2017: Investigating different methods for verifying the relative density of a centrifuge model *R. Beber, S.S.C. Madabhushi, A. Dobrisan, S.K. Haigh & S.P.G. Madabhushi*	125
Centrifuge modelling of Continuous Compaction Control (CCC) *B. Caicedo & J. Escobar*	131
Shear wave velocity: Comparison between centrifuge and triaxial based measurements *G. Cui, C.M. Heron & A.M. Marshall*	137
Development of layered models for geotechnical centrifuge tests *S. Divall, S.E. Stallebrass, R.J. Goodey & E.P. Ritchie*	143

The influence of temperature on shear strength at a soil-structure interface *J. Parchment & P. Shepley*	149
Development of a 3D clay printer for the preparation of heterogeneous models *L.M. Pua, B. Caicedo, D. Castillo & S. Caro*	155

2. Engineered platforms

Centrifuge modelling utility pipe behaviour subject to vehicular loading *S.M. Bayton, T. Elmrom & J.A. Black*	163
Experimental model study on traffic loading induced earth pressure reduction using EPS geofoam *T.N. Dave & S.M. Dasaka*	169
Physical modelling of roads in expansive clay subjected to wetting-drying cycles *S. Laporte, G.A. Siemens & R.A. Beddoe*	175
Scaled physical modelling of ultra-thin continuously reinforced concrete pavement *M.S. Smit, E.P. Kearsley & S.W. Jacobsz*	179
The effect of relative stiffness on soil-structure interaction under vehicle loads *M.S. Smit, E.P. Kearsley & S.W. Jacobsz*	185
Plate bearing tests for working platforms *G. Tanghetti, R.J. Goodey, A.M. McNamara & H. Halai*	191
Geotechnical model tests on bearing capacity of working platforms for mobile construction machines and cranes *R. Worbes & C. Moormann*	197
1g physical modelling of the stoneblowing technique for the improvement of railway track maintenance *A.A. Zaytsev, A.A. Abrashitov & A.A. Sydrakov*	203

3. Physical/Numerical interface and comparisons

Millisecond interfacing of physical models with ABAQUS *S. Idinyang, A. Franza, C.M. Heron & A.M. Marshall*	209
Verification and validation of two-phase material point method simulation of pore water pressure rise and dissipation in earthquakes *T. Kiriyama, K. Fukutake & Y. Higo*	215
Centrifuge and numerical investigations of rotated box structures *T.A. Newson, O.S. Abuhajar & K.J.L. Stone*	221
Multibillion particle DEM to simulate centrifuge model tests of geomaterials *D. Nishiura, H. Sakaguchi & S. Yamamoto*	227
Trapdoor model test and DEM simulation associated with arching *M. Otsubo, R. Kuwano, U. Ali & H. Ebizuka*	233

4. Scaling

Variability of small scale model reinforced concrete and implications for geotechnical centrifuge testing *J.A. Knappett, M.J. Brown, L. Shields, A.H. Al-Defae & M. Loli*	241
Modelling experiments to investigate soil-water retention in geotechnical centrifuge *M. Mirshekari, M. Ghayoomi & A. Borghei*	247
Studies on the use of hydraulic gradient similitude method for determining permeability of soils *K.T. Mohan Gowda & B.V.S. Viswanadham*	253
A new insight into the behaviour of seepage flow in centrifuge modelling *W. Ovalle-Villamil & I. Sasanakul*	259

Applicability of the generalised scaling law to pile-inclined ground system 265
K. Sawada, K. Ueda & S. Iai

Permeability of sand with a methylcellulose solution 271
T. Tobita

5. Sensors

Investigation of an OFDR fibre Bragg system for use in geotechnical scale modelling 279
R.D. Beemer, M.J. Cassidy & C. Gaudin

Free fall cone tests in kaolin clay 285
A. Bezuijen, D.A. den Hamer, L. Vincke & K. Geirnaert

A new shared miniature cone penetrometer for centrifuge testing 293
T. Carey, A. Gavras, B. Kutter, S.K. Haigh, S.P.G. Madabhushi, M. Okamura, D.S. Kim, K. Ueda, W.Y. Hung, Y.G. Zhou, K. Liu, Y.M. Chen, M. Zeghal, T. Abdoun, S. Escoffier & M. Manzari

Shear wave velocity measurement in a large geotechnical laminar box using bender elements 299
J. Colletti, A. Tessari, K. Sett, W. Hoffman & J. Coleman

Low cost tensiometers for geotechnical applications 305
S.W. Jacobsz

A field model investigating pipeline leak detection using discrete fibre optic sensors 311
S.I. Jahnke, S.W. Jacobsz & E.P. Kearsley

Development of an instrumented model pile 317
A.B. Lundberg, W. Broere & J. Dijkstra

New method for full field measurement of pore water pressures 323
M. Ottolini, W. Broere & J. Dijkstra

Ambient pressure calibration for cone penetrometer test: Necessary? 329
Y. Wang, Y. Hu & M.S. Hossain

6. Modelling techniques

Development of a rainfall simulator in centrifuge using Modified Mariotte's principle 337
D. Bhattacherjee & B.V.S. Viswanadham

Development of model structural dampers for dynamic centrifuge testing 343
J. Boksmati, S.P.G. Madabhushi & N.I. Thusyanthan

Experimental evaluation of two-stage scaling in physical modelling of soil-foundation-structure systems 349
A. Borghei & M. Ghayoomi

Development of a window laminar strong box 355
S.C. Chian, C. Qin & Z. Zhang

Ground-borne vibrations from piles: Testing within a geotechnical centrifuge 359
G. Cui, C.M. Heron & A.M. Marshall

A new Stockwell mean square frequency methodology for analysing centrifuge data 365
J. Dafni & J. Wartman

Novel experimental device to simulate tsunami loading in a geotechnical centrifuge 371
M.C. Exton, S. Harry, H.B. Mason, H. Yeh & B.L. Kutter

A new apparatus to examine the role of seepage flow on internal instability of model soil 377
F. Gaber & E.T. Bowman

Centrifuge model test on the instability of an excavator descending a slope 383
T. Hori & S. Tamate

Transparent soils turn 25: Past, present, and future 389
M. Iskander

Application of 3D printing technology in geotechnical-physical modelling: Tentative experiment practice — 395
Q. Jiang, L.F. Li, M. Zhang & L.B. Song

Scaling of plant roots for geotechnical centrifuge tests using juvenile live roots or 3D printed analogues — 401
T. Liang, J.A. Knappett, G.J. Meijer, D. Muir Wood, A.G. Bengough, K.W. Loades & P.D. Hallett

Revisit of the empirical prediction methods for liquefaction-induced lateral spread by using the LEAP centrifuge model tests — 407
K. Liu, Y.G. Zhou, Y. She, P. Xia, Y.M. Chen, D.S. Ling & B. Huang

Physical modelling of atmospheric conditions during drying — 413
C. Lozada, B. Caicedo & L. Thorel

Centrifuge model tests on excavation in Shanghai clay using in-flight excavation tools — 419
X.F. Ma & J.W. Xu

Effect of root spacing on interpretation of blade penetration tests—full-scale physical modelling — 425
G.J. Meijer, J.A. Knappett, A.G. Bengough, K.W. Loades & B.C. Nicoll

Development of a centrifuge testing method for stability analyses of breakwater foundation under combined actions of earthquake and tsunami — 431
J. Miyamoto, K. Tsurugasaki, R. Hem, T. Matsuda & K. Maeda

Modelling of rocking structures in a centrifuge — 437
I. Pelekis, G.S.P. Madabhushi & M.J. DeJong

A new test setup for studying sand behaviour inside an immersed tunnel joint gap — 443
R. Rahadian, S. van der Woude, D. Wilschut, C.B.M. Blom & W. Broere

3D printing of masonry structures for centrifuge modelling — 449
S. Ritter, M.J. DeJong, G. Giardina & R.J. Mair

A mechanical displacement control model tunnel for simulating eccentric ground loss in the centrifuge — 455
G. Song, A.M. Marshall & C.M. Heron

Preliminary results of laboratory analysis of sand fluidisation — 461
F.S. Tehrani, A. Askarinejad & F. Schenkeveld

Rolling test in geotechnical centrifuge for ore liquefaction analysis — 465
L. Thorel, P. Audrain, A. Néel, A. Bretschneider, M. Blanc & F. Saboya

Design and performance of an electro-mechanical pile driving hammer for geo-centrifuge — 469
J.C.B. van Zeben, C. Azúa-González, M. Alvarez Grima, C. van 't Hof & A. Askarinejad

A new heating-cooling system for centrifuge testing of thermo-active geo-structures — 475
D. Vitali, A.K. Leung, R. Zhao & J.A. Knappett

Physical modelling of soil-structure interaction of tree root systems under lateral loads — 481
X. Zhang, J.A. Knappett, A.K. Leung & T. Liang

7. Facilities

A new environmental chamber for the HKUST centrifuge facility — 489
A. Archer & C.W.W. Ng

Upgrades to the NHRI – 400 g-tonne geotechnical centrifuge — 495
S.S. Chen, X.W. Gu, G.F. Ren, W.M. Zhang, N.X. Wang, G.M. Xu, W. Liu, J.Z. Hong & Y.B. Cheng

A new 240 g-tonne geotechnical centrifuge at the University of Western Australia — 501
C. Gaudin, C.D. O'Loughlin & J. Breen

Development of a rainfall simulator for climate modelling — 507
I.U. Khan, M. Al-Fergani & J.A. Black

The development of a small centrifuge for testing unsaturated soils — 513
K.A. Kwa & D.W. Airey

Full scale laminar box for 1-g physical modelling of liquefaction
S. Thevanayagam, Q. Huang, M.C. Constantinou, T. Abdoun & R. Dobry — 519

8. Education

Using small-scale seepage physical models to generate didactic material for soil mechanics classes
L.B. Becker, R.M. Linhares, F.S. Oliveira & F.L. Marques — 527

Centrifuge modelling in the undergraduate curriculum—a 5 year reflection
J.A. Black, S.M. Bayton, A. Cargill & A. Tatari — 533

Geotechnical centrifuge facility for teaching at City, University of London
S. Divall, S.E. Stallebrass, R.J. Goodey, R.N. Taylor & A.M. McNamara — 539

Development of a teaching centrifuge learning environment using mechanically stabilized earth walls
A.F. Tessari & J.A. Black — 545

9. Offshore

Development of a series of 2D backfill ploughing physical models for pipelines and cables
T. Bizzotto, M.J. Brown, A.J. Brennan, T. Powell & H. Chandler — 553

Capacity of vertical and horizontal plate anchors in sand under normal and shear loading
S.H. Chow, J. Le, M. Forsyth & C.D. O'Loughlin — 559

A novel experimental-numerical approach to model buried pipes subjected to reverse faulting
R.Y. Khaksar, M. Moradi & A. Ghalandarzadeh — 565

Wave-induced liquefaction and floatation of pipeline buried in sand beds
J. Miyamoto, K. Tsurugasaki & S. Sassa — 571

Surface pipeline buckling on clay: Demonstration
R. Phillips, J. Barrett & G. Piercey — 577

Centrifuge modelling for lateral pile-soil pressure on passive part of pile group with platform
G.F. Ren, G.M. Xu, X.W. Gu, Z.Y. Cai, B.X. Shi & A.Z. Chen — 583

Centrifuge model tests and circular slip analyses to evaluate reinforced composite-type breakwater stability against tsunami
H. Takahashi, S. Sassa, Y. Morikawa & K. Maruyama — 589

10. Offshore – shallow foundations

Centrifuge tests on the influence of vacuum on wave impact on a caisson
D.A. de Lange, A. Bezuijen & T. Tobita — 597

Physical modelling of active suction for offshore renewables
N. Fiumana, C. Gaudin, Y. Tian & C.D. O'Loughlin — 603

Cyclic behaviour of unit bucket for tripod foundation system under various loading characteristics via centrifuge
Y.H. Jeong, H.J. Park, D.S. Kim & J.H. Kim — 609

Physical modelling of reinstallation of a novel spudcan nearby existing footprint
M.J. Jun, Y.H. Kim, M.S. Hossain, M.J. Cassidy, Y. Hu & S.G. Park — 615

Reduction in soil penetration resistance for suction-assisted installation of bucket foundation in sand
A.K. Koteras & L.B. Ibsen — 623

Evaluation of seismic coefficient for gravity quay wall via centrifuge modelling
M.G. Lee, J.G. Ha, H.J. Park, D.S. Kim & S.B. Jo — 629

Sleeve effect on the post-consolidation extraction resistance of spudcan foundation in overconsolidated clay
Y.P. Li & J.Y. Shi — 635

Measuring the behaviour of dual row retaining walls in dry sands using centrifuge tests *S.S.C. Madabhushi & S.K. Haigh*	639
Verification of improvement plan for seismic retrofits of existing quay wall in small scale fishing port *K. Mikasa & K. Okabayashi*	645
Visualisation of mechanisms governing suction bucket installation in dense sand *R. Ragni, B. Bienen, S.A. Stanier, M.J. Cassidy & C.D. O'Loughlin*	651
Recent advances in tsunami-seabed-structure interaction from geotechnical and hydrodynamic perspectives: Role of overflow/seepage coupling *S. Sassa*	657
Evaluation of seismic behaviour of reinforced earth wall based on design practices and centrifuge model tests *Y. Sawamura, T. Shibata & M. Kimura*	663
Centrifuge tests investigating the effect of suction caisson installation in dense sand on the state of the soil plug *M. Stapelfeldt, B. Bienen & J. Grabe*	669
Centrifuge model tests on stabilisation countermeasures of a composite breakwater under tsunami actions *K. Tsurugasaki, J. Miyamoto, R. Hem, T. Iwamoto & H. Nakase*	675
Interaction between jack-up spudcan and adjacent piles with non-perfect pile cap *Y. Xie, C.F. Leung & Y.K. Chow*	681

11. Offshore – deep foundations

Centrifuge modelling of long term cyclic lateral loading on monopiles *S.M. Bayton, J.A. Black & R.T. Klinkvort*	689
Centrifuge modelling of screw piles for offshore wind energy foundations *C. Davidson, T. Al-Baghdadi, M.J. Brown, A. Brennan, J.A. Knappett, C. Augarde, W. Coombs, L. Wang, D.J. Richards, A. Blake & J. Ball*	695
General study on the axial capacity of piles of offshore wind turbines jacked in sand *I. El Haffar, M. Blanc & L. Thorel*	701
Dynamic load tests on large diameter open-ended piles in sand performed in the centrifuge *E. Heins, B. Bienen, M.F. Randolph & J. Grabe*	707
Centrifuge model tests on holding capacity of suction anchors in sandy deposits *K. Kita, T. Utsunomiya & K. Sekita*	713
A review of modelling effects in centrifuge monopile testing in sand *R.T. Klinkvort, J.A. Black, S.M. Bayton, S.K. Haigh, G.S.P. Madabhushi, M. Blanc, L. Thorel, V. Zania, B. Bienen & C. Gaudin*	719
Experimental modelling of the effects of scour on offshore wind turbine monopile foundations *R.O. Mayall, R.A. McAdam, B.W. Byrne, H.J. Burd, B.B. Sheil, P. Cassie & R.J.S. Whitehouse*	725
Centrifuge tests on the response of piles under cyclic lateral 1-way and 2-way loading *C. Niemann, O. Reul, Y. Tian, C.D. O'Loughlin & M.J. Cassidy*	731
Physical modelling of monopile foundations under variable cyclic lateral loading *I.A. Richards, B.W. Byrne & G.T. Houlsby*	737
Centrifuge model testing of fin piles in sand *S. Sayles, K.J.L. Stone, M. Diakoumi & D.J. Richards*	743
Dynamic behaviour evaluation of offshore wind turbine using geotechnical centrifuge tests *J.T. Seong, J.H. Kim & D.S. Kim*	749

An investigation on the performance of a self-installing monopiled GBS structure under lateral loading *K.J.L. Stone, A. Tillman & M. Vaziri*	755
Model tests on the lateral cyclic responses of a caisson-piles foundation under scour *C.R. Zhang, H.W. Tang & M.S. Huang*	761
Comparison of centrifuge model tests of tetrapod piled jacket foundation in saturated sand and clay *B. Zhu, K. Wen, L.J. Wang & Y.M. Chen*	767
Author index	773

VOLUME 2

12. Tunnel, shafts and pipelines

Study of the effects of explosion on a buried tunnel through centrifuge model tests *A. De & T.F. Zimmie*	779
Uplift resistance of a buried pipeline in silty soil on slopes *G.N. Eichhorn & S.K. Haigh*	785
Modelling the excavation of elliptical shafts in the geotechnical centrifuge *N.E. Faustin, M.Z.E.B. Elshafie & R.J. Mair*	791
Shaking table test to evaluate the effects of earthquake on internal force of Tabriz subway tunnel (Line 2) *M. Hajialilue-Bonab, M. Farrin & M. Movasat*	797
Effect of pipe defect size and maximum particle size of bedding material on associated internal erosion *S. Indiketiya, P. Jegatheesan, R. Pathmanathan & R. Kuwano*	803
Modelling cave mining in the geotechnical centrifuge *S.W. Jacobsz, E.P. Kearsley, D. Cumming-Potvin & J. Wesseloo*	809
Experimental modelling of infiltration of bentonite slurry in front of shield tunnel in saturated sand *T. Xu & A. Bezuijen*	815

13. Imaging

Flow visualisation in a geotechnical centrifuge under controlled seepage conditions *C.T.S. Beckett & A.B. Fourie*	823
A new procedure for tracking displacements of submerged sloping ground in centrifuge testing *T. Carey, N. Stone, B. Kutter & M. Hajialilue-Bonab*	829
Identification of soil stress-strain response from full field displacement measurements in plane strain model tests *J.A. Charles, C.C. Smith & J.A. Black*	835
Imaging of sand-pile interface submitted to a high number of loading cycles *J. Doreau-Malioche, G. Combe, J.B. Toni, G. Viggiani & M. Silva*	841
Image capture and motion tracking applications in geotechnical centrifuge modelling *P. Kokkali, T. Abdoun & A. Tessari*	847
A study on performance of three-dimensional imaging system for physical models *B.T. Le, S. Nadimi, R.J. Goodey & R.N. Taylor*	853
Visualisation of inter-granular pore fluid flow *L. Li, M. Iskander & M. Omidvar*	859

A two-dimensional laser-scanner system for geotechnical processes monitoring 865
M.D. Valencia-Galindo, L.N. Beltrán-Rodriguez, J.A. Sánchez-Peralta, J.S. Tituaña-Puente,
M.G. Trujillo-Vela, J.M. Larrahondo, L.F. Prada-Sarmiento & A.M. Ramos-Cañón

14. Seismic – dynamic

Dynamic behaviour of model pile in saturated sloping ground during shaking table tests 873
C.H. Chen, T.S. Ueng & C.H. Chen

Investigation on the aseismic performance of pile foundations in volcanic ash ground 879
T. Egawa, T. Yamanashi & K. Isobe

Effective parameters on the interaction between reverse fault rupture and shallow foundations:
Centrifuge modelling 885
A. Ghalandarzadeh & M. Ashtiani

Seismic amplification of clay ground and long-term consolidation after earthquake 891
Y. Hatanaka & K. Isobe

Evaluation of period-lengthening ratio (PLR) of single-degree-of-freedom structure via
dynamic centrifuge tests 897
K.W. Ko, J.G. Ha, H.J. Park & D.S. Kim

Centrifuge modelling of active seismic fault interaction with oil well casings 903
J. Le Cossec, K.J.L. Stone, C. Ryan & K. Dimitriadis

Centrifuge shaking table tests on composite caisson-piles foundation 909
F. Liang, Y. Jia, H. Zhang, H. Chen & M. Huang

Dynamic behaviour of three-hinge-type precast arch culverts with various patterns of
overburden in culvert longitudinal direction 915
Y. Miyazaki, Y. Sawamura, K. Kishida & M. Kimura

Dynamic centrifuge model tests on sliding base isolation systems leveraging buoyancy 921
N. Nigorikawa, Y. Asaka & M. Hasebe

Centrifugal model tests on static and seismic stability of landfills with high water level 929
B. Zhu, J.C. Li, L.J. Wang & Y.M. Chen

15. Seismic – liquefaction

Partial drainage during earthquake-induced liquefaction 937
O. Adamidis & G.S.P. Madabhushi

Centrifuge modelling of site response and liquefaction using a 2D laminar box and
biaxial dynamic base excitation 943
O. El Shafee, J. Lawler & T. Abdoun

Experimental simulation of the effect of preshaking on liquefaction of sandy soils 949
W. El-Sekelly, T. Abdoun, R. Dobry & S. Thevanayagam

Dynamic centrifuge testing to assess liquefaction potential 955
G. Fasano, E. Bilotta, A. Flora, V. Fioravante, D. Giretti, C.G. Lai & A.G. Özcebe

Experimental and computational study on effects of permeability on liquefaction 961
H. Funahara & N. Tomita

The importance of vertical accelerations in liquefied soils 967
F.E. Hughes & S.P.G. Madabhushi

The effects of waveform of input motions on soil liquefaction by centrifuge modelling 975
W.Y. Hung, T.W. Liao, L.M. Hu & J.X. Huang

Horizontal subgrade reaction of piles in liquefiable ground 981
S. Imamura

Experimental investigation of pore pressure and acceleration development in static
liquefaction induced failures in submerged slopes 987
A. Maghsoudloo, A. Askarinejad, R.R. de Jager, F. Molenkamp & M.A. Hicks

Centrifuge modelling of earthquake-induced liquefaction on footings built on improved ground 993
A.S.P.S. Marques, P.A.L.F. Coelho, S.K. Haigh & G.S.P. Madabhushi

Investigating the effect of layering on the formation of sand boils in 1 g shaking table tests 999
S. Miles, J. Still & M. Stringer

Centrifuge modelling of mitigation-soil-structure-interaction on layered liquefiable soil deposits with a silt cap 1005
B. Paramasivam, S. Dashti, A.B. Liel & J.C. Olarte

Centrifuge modelling of the effects of soil liquefiability on the seismic response of low-rise structures 1011
S. Qi & J.A. Knappett

Liquefaction behaviour focusing on pore water inflow into unsaturated surface layer 1017
Y. Takada, K. Ueda, S. Iai & T. Mikami

16. Dams and embankments

Performance of single piles in riverbank clay slopes subject to repetitive tidal cycles 1025
U. Ahmed, D.E.L. Ong & C.F. Leung

Load transfer mechanism of reinforced piled embankments 1031
M.S.S. Almeida, D.F. Fagundes, M.C.F. Almeida, D.A. Hartmann, R. Girout, L. Thorel & M. Blanc

Experiments for a coarse sand barrier as a measure against backwards erosion piping 1037
A. Bezuijen, E. Rosenbrand, V.M. van Beek & K. Vandenboer

Load transfer mechanism of piled embankments: Centrifuge tests versus analytical models 1043
M. Blanc, L. Thorel, R. Girout, M.S.S. Almeida & D.F. Fagundes

Physical model testing to evaluate erosion quantity and pattern 1049
M. Kamalzare & T.F. Zimmie

Physical modelling of large dams for seismic performance evaluation 1055
N.R. Kim & S.B. Jo

Centrifuge model tests on levees subjected to flooding 1061
R.K. Saran & B.V.S. Viswanadham

Centrifuge model test of vacuum consolidation on soft clay combined with embankment loading 1067
S. Shiraga, G. Hasegawa, Y. Sawamura & M. Kimura

17. Geohazards

Effects of viscosity in granular flows simulated in a centrifugal acceleration field 1075
M. Cabrera, P. Kailey, E.T. Bowman & W. Wu

Using pipe deflection to detect sinkhole development 1081
E.P. Kearsley, S.W. Jacobsz & H. Louw

Model tests to simulate formation and expansion of subsurface cavities 1087
R. Kuwano, R. Sera & Y. Ohara

Centrifuge modelling of a pipeline subjected to soil mass movements 1093
J.R.M.S. Oliveira, K.I. Rammah, P.C. Trejo, M.S.S. Almeida & M.C.F. Almeida

Effects of earthquake motion on subsurface cavities 1099
R. Sera, M. Ota & R. Kuwano

Preliminary study of debris flow impact force on a circular pillar 1105
A.L. Yifru, R.N. Pradhan, S. Nordal & V. Thakur

18. Slopes

Centrifuge modelling of earth slopes subjected to change in water content 1113
P. Aggarwal, R. Singla & A. Juneja

Centrifuge and numerical modelling of static liquefaction of fine sandy slopes — 1119
A. Askarinejad, W. Zhang, M. de Boorder & J. van der Zon

Modelling of MSW landfill slope failure — 1125
Y.J. Hou, X.D. Zhang, J.H. Liang, C.H. Jia, R. Peng & C. Wang

Effects of plant removal on slope hydrology and stability — 1131
V. Kamchoom & A.K. Leung

Centrifuge model test on deformation and failure of slopes under wetting-drying cycles — 1137
F. Luo & G. Zhang

Centrifuge model studies of the soil slope under freezing and thawing processes — 1143
C. Zhang, Z.Y. Cai, Y.H. Huang & G.M. Xu

An experimental and numerical study of pipe behaviour in triggered sandy slope failures — 1149
W. Zhang, Z. Gng & A. Askarinejad

19. Ground improvement

Investigation of nailed slope behaviour during excavation by Ng centrifuge physical model tests — 1157
A. Akoochakian, M. Moradi & A. Kavand

Relative contribution of drainage capacity of stone columns as a countermeasure against liquefaction — 1163
E. Apostolou, A.J. Brennan & J. Wehr

Observed deformations in geosynthetic-reinforced granular soils subjected to voids — 1169
T.S. da Silva & M.Z.E.B. Elshafie

Analytical design approach for the self-regulating interactive membrane foundation based on centrifuge-model tests and numerical simulations — 1175
O. Detert, D. König & T. Schanz

Earthquake-induced liquefaction mitigation under existing buildings using drains — 1181
S. García-Torres & G.S.P. Madabhushi

Deformation behaviour research of an artificial island by centrifuge modelling test — 1187
X.W. Gu, Z.Y. Cai, G.M. Xu & G.F. Ren

Effect of lateral confining condition of behaviour of confined-reinforced earth — 1193
H.M. Hung & J. Kuwano

An experimental study on the effects of enhanced drainage for liquefaction mitigation in dense urban environments — 1199
P.B. Kirkwood & S. Dashti

Influence of tamper shape on dynamic compaction of granular soil — 1205
S. Kundu & B.V.S. Viswanadham

Behaviour of geogrid reinforced soil walls with marginal backfills with and without chimney drain in a geotechnical centrifuge — 1211
J. Mamaghanian, H.R. Razeghi, B.V.S. Viswanadham & C.H.S.G. Manikumar

Centrifuge model tests on effect of inclined foundation on stability of column type deep mixing improved ground — 1217
S. Matsuda, M. Momoi & M. Kitazume

Large-scale physical model GRS walls: Evaluation of the combined effects of facing stiffness and toe resistance on performance — 1223
S.H. Mirmoradi & M. Ehrlich

Deep vibration compaction of sand using mini vibrator — 1229
S. Nagula, P. Mayanja & J. Grabe

Dynamic centrifuge tests on nailed slope with facing plates — 1235
S. Nakamoto, N. Iwasa & J. Takemura

Influence of slope inclination on the performance of slopes with and without soil-nails subjected to seepage: A centrifuge study V.M. Rotte & B.V.S. Viswanadham	1241
Performance of soil-nailed wall with three-dimensional geometry: Centrifuge study M. Sabermahani, M. Moradi & A. Pooresmaeili	1247
Behaviour of geogrid-reinforced aggregate layer overlaying poorly graded sand under cyclic loading A.A. Soe, J. Kuwano, I. Akram, T. Kogure & H. Kanai	1253
Physical modelling of compaction grouting injection using a transparent soil D. Takano, Y. Morikawa, Y. Miyata, H. Nonoyama & R.J. Bathurst	1259
Centrifuge modelling of remediation of liquefaction-induced pipeline uplift using model root systems K. Wang, A.J. Brennan, J.A. Knappett, S. Robinson & A.G. Bengough	1265
Comparative study of consolidation behaviour of differently-treated mature fine tailings specimens through centrifuge modelling G. Zambrano-Narvaez, Y. Wang & R.J. Chalaturnyk	1271
Physical modelling and monitoring of the subgrade on weak foundation and its reinforcing with geosynthetics A.A. Zaytsev, Y.K. Frolovsky, A.V. Gorlov, A.V. Petryaev & V.V. Ganchits	1277

20. Shallow foundations

Effect of spatial variability on the behaviour of shallow foundations: Centrifuge study L.X. Garzón, B. Caicedo, M. Sánchez-Silva & K.K. Phoon	1285
1g model tests of surface and embedded footings on unsaturated compacted sand A.J. Lutenegger & M.T. Adams	1291
Experimental study on the coupled effect of the vertical load and the horizontal load on the performance of piled beam-slab foundation L. Mu, M. Huang, X. Kang & Y. Zhang	1297
Determining shallow foundation stiffness in sand from centrifuge modelling A. Pearson & P. Shepley	1303
The effect of soil stiffness on the undrained bearing capacity of a footing on a layered clay deposit A. Salehi, Y. Hu, B.M. Lehane, V. Zania & S.L. Sovso	1309
Centrifuge investigation of the cyclic loading effect on the post-cyclic monotonic performance of a single-helix anchor in sand J.A. Schiavon, C.H.C. Tsuha & L. Thorel	1315
Bearing capacity of surface and embedded foundations on a slope: Centrifuge modelling D. Taeseri, L. Sakellariadis, R. Schindler & I. Anastasopoulos	1321

21. Deep foundations

Performance of piled raft with unequal pile lengths R.S. Bisht, A. Juneja, A. Tyagi & F.H. Lee	1329
Pile response during liquefaction-induced lateral spreading: 1-g shake table tests with different ground inclination A. Ebeido, A. Elgamal & M. Zayed	1335
Effect of the installation methods of piles in cohesionless soil on their axial capacity I. El Haffar, M. Blanc & L. Thorel	1341
Model testing of rotary jacked open ended tubular piles in saturated non-cohesive soil D. Frick, K.A. Schmoor, P. Gütz & M. Achmus	1347

Model tests on soil displacement effects for differently shaped piles — 1353
A.A. Ganiyu, A.S.A. Rashid, M.H. Osman & W.O. Ajagbe

Comparison of seismic behaviour of pile foundations in two different soft clay profiles — 1359
T.K. Garala & G.S.P. Madabhushi

Issues with centrifuge modelling of energy piles in soft clays — 1365
I. Ghaaowd, J. McCartney, X. Huang, F. Saboya & S. Tibana

Centrifuge modelling of non-displacement piles on a thin bearing layer overlying a clay layer — 1371
Y. Horii & T. Nagao

Rigid pile improvement under rigid slab or footing under cyclic loading — 1377
O. Jenck, F. Emeriault, C. Dos Santos Mendes, O. Yaba, J.B. Toni, G. Vian & M. Houda

Pull-out testing of steel reinforced earth systems: Modelling in view of soil dilation and boundary effects — 1383
M. Loli, I. Georgiou, A. Tsatsis, R. Kourkoulis & F. Gelagoti

Pile jetting in plane strain: Small-scale modelling of monopiles — 1389
S. Norris & P. Shepley

Influence of geometry on the bearing capacity of sheet piled foundations — 1395
J.P. Panchal, A.M. McNamara & R.J. Goodey

Kinematic interaction of piles under seismic loading — 1401
J. Pérez-Herreros, F. Cuira, S. Escoffier & P. Kotronis

Behaviour of piled raft foundation systems in soft soil with consolidation process — 1407
E. Rodríguez, R.P. Cunha & B. Caicedo

Displacement measurements of ground and piles in sand subjected to reverse faulting — 1413
C.F. Yao, S. Seki & J. Takemura

22. Walls and excavations

Centrifuge simulation of heave behaviour of deep basement slabs in overconsolidated clay — 1421
D.Y.K. Chan & S.P.G. Madabhushi

Soil movement mobilised with retaining wall rotation in loose sand — 1427
C. Deng & S.K. Haigh

Lateral pressure of granular mass during translative motion of wall — 1433
P. Koudelka

Deflection and failure of self-standing high stiffness steel pipe sheet pile walls embedded in soft rocks — 1439
V. Kunasegarm, S. Seki & J. Takemura

A new approach to modelling excavations in soft soils — 1445
J.P. Panchal, A.M. McNamara & S.E. Stallebrass

1g-modelling of limit load increase due to shear band enhancement — 1451
K.-F. Seitz & J. Grabe

Concave segmental retaining walls — 1457
D. Stathas, L. Xu, J.P. Wang, H.I. Ling & L. Li

A combined study of centrifuge and full scale models on detection of threat of failure in trench excavations — 1463
S. Tamate & T. Hori

Dynamic behaviour on pile foundation combined with soil-cement mixing walls using permanent pile — 1469
K. Watanabe, M. Arakawa & M. Mizumoto

Centrifuge modelling of 200,000 tonnage sheet-pile bulkheads with relief platform — 1475
G.M. Xu, G.F. Ren, X.W. Gu & Z.Y. Cai

Author index — 1481

Preface

The International Conference on Physical Modelling in Geotechnics is held under the auspices of Technical Committee 104 (*TC104: Physical Modelling in Geotechnics*) of the International Society for Soil Mechanics and Geotechnical Engineering (ISSMGE). Early workshops on physical modelling were held in Manchester, California and Tokyo in 1984 and, as the physical modelling community grew, the first international conference was held only 30 years ago in Paris in 1988. The possibilities offered by physical modelling became apparent around the world and the conference has developed into a quadrennial event that regularly attracts researchers from over 30 countries. The last meeting of the global community was in Perth, Western Australia; a veritable feast to sate the appetite of the hungry faithful, under the very capable leadership of Professor Christophe Gaudin. Regional conferences have also become established following the first Eurofuge held at City, University of London in 2008 followed by European regional conferences at TU Delft and IFSTTAR, Nantes and Asian regional conferences at IIT Bombay and Tongji University, Shanghai. These conferences bring together a community of great innovators; the most practical and capable engineers, in an exciting and specialist field.

TC104 selected London as the destination for the 9th International Conference (ICPMG 2018) which was held at City, University of London, in July 2018. The United Kingdom is a hotspot for physical modelling activity; centrifuges are established at Cambridge University, City, University of London, University of Dundee, University of Nottingham and University of Sheffield.

The conference coincided with the 4th Andrew Schofield Lecture, established by TC104 and named after Professor Andrew Schofield, the great pioneer of geotechnical centrifuge modelling. As the highest honour that can be bestowed upon a member of our community it is fitting that the lecture was delivered by Professor Neil Taylor of City, University of London and Secretary General of ISSMGE; a former doctoral student of Professor Schofield.

The conference programme was a physical modelling extravaganza divided into plenary and parallel sessions running over four days, 17th–20th July. Four keynote lectures were given in the areas of seismic behaviour, design optimisation, new facilities and environmental engineering representing significant areas of interest of the assembled audience. Themed lectures in the areas of education, new technology, urban infrastructure and offshore engineering addressed a key aim of TC104 in showcasing research opportunities to industry who attended a specific half day event. A total of 138 oral presentations were made from 230 papers submitted, originating from over 30 countries, and included in the conference proceedings in 22 chapters. All papers that were not presented orally were presented as posters. The conference gave delegates an opportunity to experience exciting and historic aspects of London that are normally inaccessible to those visiting the city. A welcome reception was held at the historic Skinners' Hall, home to one of the Great Twelve City livery companies and delegates enjoyed a sumptuous gala dinner at the spectacular Middle Temple Hall dating from 1573; one of the four Inns of Court exclusively entitled to call their members to the English Bar as barristers. A pleasant afternoon and evening was spent on a visit to Greenwich on the River Thames, home to the Meridian Line, the famous Cutty Sark, the Royal Observatory, the National Maritime Museum and the Old Royal Naval College.

Physical modelling has come of age and advances in all areas of technology, from digital imaging to computing, electronics and materials offer exciting opportunities to push boundaries well beyond the early experimental work. Visionary and adventurous physical modellers developed the basic techniques and important scaling laws that are the backbone of our work today. Such research made possible important contributions to the understanding of complex soil/structure interaction problems long before numerical modelling was capable of even attempting to establish such insight. Present day physical modellers are just as ambitious and adventurous as their forefathers and are anxious to build ever larger facilities and undertake increasingly complex experimental work. To this end, plans are underway for a 1000 g/tonne 'megafuge' capable of modelling the very largest of geotechnical structures. Physical modelling enjoys increasing popularity as a powerful means of exploring geotechnical problems. However, it rarely finds favour over numerical modelling in the eyes of industry where results of experimental studies are required soon after commissioning the work; regardless of accuracy and at minimal cost. Current work that focuses on exploring the interface between physical modelling and numerical modelling is a particularly exciting development and has the potential to yield important new knowledge applicable to both fields.

The organisation of a major international conference is a massive undertaking. My thanks go to the Local Organising Committee and the International Advisory Board and to everyone who participated in the very thorough review process. Particular thank are due to my colleagues, Sam Divall, Richard Goodey, Jignasha

Panchal, Sarah Stallebrass and Neil Taylor at City, University of London who rolled up their sleeves to help with all aspects of the conference; but notably in managing and editing the huge volume of poorly formatted papers. For anyone reading this far, please do not alter the template when writing your conference papers.

Andrew McNamara
Chair, Technical Committee 104 on Physical Modelling in Geotechnics, 2014 – 2018
International Society for Soil Mechanics and Geotechnical Engineering

International advisory board

Adam Bezuijen
Emilio Bilotta
Jonathan Black
Miguel Cabrera
Bernado Caicedo
Jonny Cheuk
Michael Davies
Jelke Dijkstra
Mohammed Elshafie
Vincenzo Fioravante
Christophe Gaudin
Susan Gourvenec
Stuart Haigh
SW Jacobsz
Ashish Juneja
Diethard König
Dong Soo Kim

Jonathan Knappett
Bruce Kutter
Jan Laue
Colin Leung
Xianfeng Ma
Alec Marshall
Tim Newson
Dominic Ek Leong Ong
Ryan Phillips
Kevin Stone
Andy Take
Jiro Takemura
Luc Thorel
David White
Daniel Wilson
Varvara Zania

Local organising committee

Andrew McNamara
Sam Divall
Neil Taylor
Richard Goodey
Sarah Stallebrass
Joana Fonseca
David White
Susan Gourvenec
Stuart Haigh
Mohammed Elshafie

Manuscript reviewers

O. Abuhajar
A. Ahmed
M. Alheib
I. Anastasopoulos
J. Barrett
A. Bezuijen
E. Bilotta
J.A. Black
M. Blanc
M. Bolton
A.J. Brennan
J. Breyl
L. Briancon
A. Broekman
M.J. Brown
M. Cabrera
Q. Cai
B. Caicedo-Hormaza
T. Carey
D. Chang
J. Cheuk
D. Chian
S.C. Chian
U. Cilingir
P.A.L.F. Coelho
G. Cui
T. da Silva
C. Dano
C. Davidson
M.C.R. Davies
A. Deeks
D. deLange
L. Deng
O. Detert
R. di Laora
T. Dias
J. Dijkstra
S. Divall
H. El Naggar
I. El-Haffar
G. Elia
S. Escoffier
V. Fioravante
J. Fonseca
T. Fujikawa
T. Gaspar
C. Gaudin
A. Gavras
L. Geldenhuys

R.J. Goodey
S.M. Gourvenec
G.J. Ha
J.G. Ha
S. Haigh
H. Halai
A. Hashemi
F. Heidenreich
H. Hong
K. Horikoshi
S.W. Jacobsz
A.J. Jebeli
A. Juneja
G. Kampas
T. Karoui
E. Kearsley
E.Y. Kencana
M.H. Khosravi
D.S. Kim
J.H. Kim
J.A. Knappett
D. Koenig
B. Kutter
R. Kuwano
L.Z. Lang
G. Lanzano
J. Laue
A. Lavasan
B.T. Le
S.W. Lee
F.H. Lee
C.F. Leung
L. Li
L.M. Li
T. Liang
H.I. Ling
M. Loli
C. Lozada
A. Lutenegger
F. Ma
A. Marshall
M. Masoudian
R. McAffee
A.M. McNamara
G.J. Meijer
M. Millen
H. Mitrani
A. Mochizuki
S. Nadimi

T. Newson
M. Okamura
D.E.L. Ong
J.P. Panchal
H.J. Park
J. Perez-Herreros
R. Phillips
G. Piercey
C. Purchase
M. Qarmout
S. Qi
M. Rasulo
S. Ravjee
A. Rawat
S. S.Chian
F. Saboya Junior
A. Sadrekarimi
M. Silva Illanes
G. Smit
S.E. Stallebrass
S.A. Stanier
K. Stone
A. Takahashi
A. Take
J. Takemura
G. Tanghetti
R.N. Taylor
L. Thorel
I. Thusyanthan
T. Tobita
K. Ueno
R. Uzuoka
R. Vandoorne
K. Wang
D.J. White
D. Wilson
K.S. Wong
H. Wu
Y. Xie
J. Yang
J. Yu
V. Zania
C. Zhang
G. Zhang
L. Zhang
Z. Zhang
B.L. Zheng
Y.G. Zhou
B. Zhu

Sponsors

PLATINUM SPONSOR – Actidyn http://www.actidyn.com/

ANDREW SCHOFIELD LECTURE RECEPTION SPONSOR – http://www.broadbent.co.uk/

CONFERENCE SPONSOR – Tekscan, Inc. – http://www.tekscan.com/

Keynote and Themed lectures

Modelling tunnel behaviour under seismic actions: An integrated approach

E. Bilotta
University of Napoli Federico II, Naples, Italy

ABSTRACT: This paper intends to describe the integration of physical and numerical modelling, focusing on tunnels under seismic actions. It shows how numerical calculations can be used in association with centrifuge testing to model different aspects of tunnel behaviour during earthquakes. The scope of the paper has been limited to a few aspects, mainly concerning the change of internal forces in the tunnel lining during shaking and the effect of soil liquefaction. The interaction between a tunnel and a building in a soil layer undergoing liquefaction has also been taken into account.

1 INTRODUCTION

The behaviour of tunnels under seismic action and their vulnerability to earthquakes is a topic that has received increasing attention in recent years. However, evidence of tunnel behaviour during natural events of ground shaking can be observed only after an earthquake occurs. The analysis of the problem based on post-earthquake reconnaissance only may give an incomplete picture of the problem.

The study of seismic vulnerability of tunnels is therefore a typical field where small scale physical modelling in a centrifuge finds a useful opportunity of application. In fact, artificial ground shaking can be produced in a centrifuge that simulates natural earthquakes in a ground layer surrounding a model tunnel. Hence, the complex interaction mechanism that arises between the tunnel structure and the surrounding soil during shaking can be reproduced in the model. Several studies have been based on centrifuge testing on reduced scale models of tunnels in sand (e.g. Cilingir & Madabhushi, 2011; Lanzano et al., 2012; Tsinidis et al. 2015, 2016a,b,c). They have provided experimental data on the changes of structural forces in a tunnel lining undergoing ground shaking. A few studies have also modelled in a centrifuge the effects on tunnels of earthquake-induced ground failure such as fault displacement (e.g. Baziar et al., 2014) or soil liquefaction (e.g. Chou et al., 2010; Chian et al., 2014).

On the other hand, numerical modelling has often served as a tool for analysing the problem or validating simplified analytical solutions (e.g. Kontoe et al., 2014). However, it is well acknowledged that when high quality centrifuge test data are available, they can also be used to validate the results of numerical modelling (Zeghal et al., 2014). For instance, for circular tunnels under seismic loading several numerical studies originated from a set of centrifuge tests specifically designed for that purpose and a comparison among experimental data and numerical results achieved using different constitutive models and numerical algorithms provided a deeper insight into the problem (Bilotta et al., 2014).

An integrated approach, including both physical and numerical modelling, also relying on an accurate soil characterization, appears therefore the most reliable tool to analyse boundary value problems involving dynamic conditions and complex soil behaviour. Such an approach, for instance, inspired validation exercises such as VELACS (Arulanandan & Scott, 1993) that has represented for many years a benchmark for the study of seismic-induced soil liquefaction. More recently the LEAP exercise has been launched that further implements the same idea (Kutter et al., 2017).

Large research projects such as the abovementioned concerning soil liquefaction require however a significant financial support. This can be provided from public funding agencies or private sponsors, probably focusing on broad and impacting research streams only. Less appealing problems, that receive lower attention, might be excluded from the benefit of such a combined approach. Repositories of the experimental data that are produced by different facilities for different purposes all over the world, play in this case a fundamental role.

This paper intends to describe the integration of physical and numerical modelling from the point of view of numerical modellers. Focusing on the dynamic behaviour of tunnels, and in particular on the internal forces in the tunnel lining, the use of the centrifuge results to calibrate a numerical model and extend the scope of application is shown in section 2 and section 3. In the former the results of centrifuge testing are boosted by including numerically the effect of a construction process for tunnelling and a more complex structural behaviour of the lining. In section 3 the back-analysis of the centrifuge test in dry sand is used in association with the results of cyclic simple shear testing in undrained conditions, for the calibration of a constitutive model suitable for modelling pore-water pressure build-up in undrained conditions. The effect on the lining of the excess pore-pressure arising during shaking is analysed. Finally, in section 4 the process is reversed from numerical analysis to centrifuge modelling. The calibration carried out in the previous sections is used to perform a preliminary analysis of tunnel-building interaction in liquefiable soil, in order to design a series of centrifuge tests.

2 INTERNAL FORCES IN A TUNNEL LINING

2.1 Background

Internal forces in the tunnel lining change during earthquakes. They can be calculated following several approaches (Hashash et al., 2001; Pitilakis & Tsinidis, 2014). Pseudo-static or uncoupled dynamic analyses are usually carried out in routine design. Full dynamic analysis, that is including dynamic soil-structure interaction, must be performed however, if the influence of the existing stress state around the tunnel has to be considered. Moreover, the latter may include the irreversible behaviour of soil that is likely to produce permanent ground deformation during shaking. Since the tunnel construction process may affect the static conditions before shaking, numerical analyses can include this aspect. Compared to plane strain, three-dimensional models permit the construction phases to be simulated in a more accurate fashion, including geometrical details of the lining that may affect its structural behaviour (for instance the segmental layout of precast lining). The effect of seismic waves propagating in any direction can be also analysed in a three-dimensional numerical model.

On the other hand, direct measurements of the effect of the complex interaction between a tunnel lining model and the surrounding soil during ground shaking can be achieved in centrifuge tests. This enables a large amount of experimental data to be collected and used for validation of numerical analyses.

As part of a research within the ReLUIS project funded by the Italian Civil Protection Department, a series of centrifuge tests were carried out at the Schofield Centre of the University of Cambridge on circular tunnel in dry sand, undergoing dynamic excitation (Lanzano et al., 2012). Internal forces (bending moments and hoop forces) in the tunnel lining were measured during shakings. Hence, experimental evidence was gained on a problem that had been previously explored via analytical solutions (mainly based on the elastic theory) and numerical modelling only.

Such tests, which are briefly recalled in the next section 2.2, were later used as an experimental benchmark for numerical modelling, aimed at extending the scope of the study. In fact, three-dimensional finite element analyses were performed that take into account the non-linear and irreversible soil behaviour. The tunnel excavation process, that is neglected in the centrifuge tests, was modelled to achieve a realistic state of stress effect before shaking. Moreover, the segmental layout

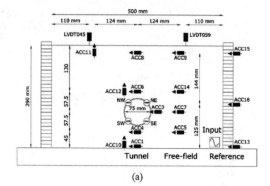

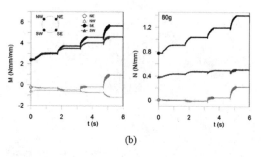

Figure 1. Model T3: (a) experimental layout; (b) measured time histories of bending moments and hoop forces (modified after Lanzano et al., 2012).

of a precast tunnel lining was modelled, although with a few simplifying assumptions (Fabozzi, 2017).

2.2 Experimental benchmark

The experimental benchmark used for validating the numerical model of a rather shallow tunnel (C/D = 2) in dense sand is the centrifuge model T3 (Figure 1), described in details by Lanzano et al. (2012).

In the model (Fig. 1a), an aluminium tube (diameter D = 75 mm, thickness t = 0.5 mm, cover C = 150 mm) representing a circular tunnel is embedded in a layer of dry Leighton Buzzard sand (fraction E), pluviated in the container at a relative density of 75% (Figure 1). The tube is instrumented with strain gauges in four positions along its transverse section (indicated as NE, NW, SW and SE in Fig.1a). After spin up at 80 g, a series of pseudo-harmonic signals of increasing amplitude and frequency was applied at the base of the model. The time histories of bending moment, M, and hoop forces, N measured during four subsequent dynamic events are shown in Fig. 1b at the model scale (Lanzano et al., 2012). It is worth noticing that permanent increments of internal forces arose in the tunnel lining after each events. These seem well correlated to the progressive densification of the sand layer that was observed in the experiments.

2.3 Numerical modelling

Numerical analyses were performed at prototype scale, using a scaling factor N=80. Hence, the corresponding

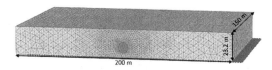

Figure 2. Numerical mesh.

Table 1. HS-small model parameters (Lanzano et al., 2016).

	sand
φ	38.6°
ψ	8.2°
c' (kPa)	0.01
E_{ref}^{50} (MPa)	18.6
E_{ref}^{oed} (MPa)	20.5
E_{ref}^{ur} (MPa)	62.2
$\gamma_{0.7}$	$0.60E^{-3}$
G_0^{ref} (MPa)	72.7
p_{ref} (kPa)	100
m	0.4

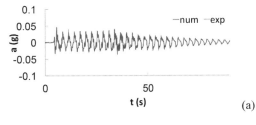

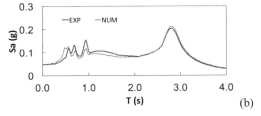

Figure 3. ACC9, experimental and computed (a) time history of acceleration and (b) response spectra.

tunnel diameter is assumed 6 m, the tunnel axis depth is 15 m and the lining thickness is comparable to that of a concrete lining about 0.06 m thick.

The numerical model has been implemented in the finite element code Plaxis 3D (Brinkgreve et al., 2016). The mesh is shown in Figure 2.

While the height of the model is 23.2 m, that is 80 times the relevant size at model scale, its width is larger than that and equal to 200 m, to minimise the influence of lateral boundaries. A reference section at the midspan of the tunnel was assumed to be compared to the experimental results. Hence, in order to guarantee plane strain conditions in the reference section, the size of the model along the axis of the tunnel was assumed as long as 150 m. The vertical sides of the mesh were fixed in the horizontal direction in static condition; viscous dashpots were applied during shaking.

The time history of acceleration recorded by the accelerometer ACC13 at the base of the centrifuge model (see Figure 1) was scaled up to prototype scale and band-pass filtered (15–130 Hz) in order to reduce the its high-frequency content. This signal (with nominal frequency 0.375 Hz and nominal amplitude 0.05 g) was applied as dynamic input at the base of the model.

The lining is an elastic plate (EA = 2.8·106 kN/m; EI = 3.7·102 kNm2/m) with a very smooth interface (the interface factor was assumed as Rint = 0.05).

The sand has been modelled using the Hardening Soil with small strain overlay constitutive model (Schanz et al., 1999; Benz et al., 2009), with the parameters shown in Table 1, derived by Lanzano et al. (2016).

This elastic-plastic with isotropic hardening soil model is able to reproduce the decay of shear stiffness with strain level from very small strain and the increase of hysteretic damping. The initial damping ratio at very small strain was modelled through a Rayleigh formulation (αR = 0.0668; βR = 0.704 10-3).

Figure 3a compares the time history of acceleration measured in the test by ACC9 with the corresponding computed results. In Figure 3b the corresponding response spectra at 5% of damping are shown. The dynamic response computed for the soil layer is close to the measurements, although there is evidence of a slight over-amplification of the signal at high frequencies, as observed also by Amorosi et al. (2014) in similar analyses.

Once validated against centrifuge results, the same 3D model was used to analyse the behaviour in the same sand of a different tunnel lining. This is a reinforced concrete lining with thickness t = 0.3 m (EA = 10.5E6 kN/m; EI = 78.75E3 kNm2/m; Rint = 0.7) and diameter D = 6 m.

A set of natural input signals was applied as time histories of acceleration at the base of the mesh. A few results of the study (Fabozzi, 2017) are presented in sections 2.4 and 2.5: the influence of the construction process on the seismic response of the tunnel is discussed in the former while the latter analyses the influence of the presence of joints in the segmental lining.

2.4 Pre-seismic conditions induced by tunnel construction

The influence of the construction process has been taken into account with reference to typical mechanized tunnelling with an earth pressure balance machine. Details of the procedure are described by Fabozzi & Bilotta (2016) and will not be discussed here. The seismic excitation was applied to the numerical model at the state of stress corresponding to the end of construction. Table 2 shows the main characteristics of the input signals applied as time history of acceleration at the base of the model. They are natural time histories of acceleration recorded on a rigid outcropping bedrock (soil type A according to EC8).

Table 2. Natural signals.

Earthquake event	Date	M_w	PGA
	-	-	g
Norcia	30/10/2016	6.5	0.78
Avej	22/06/2006	6.5	0.5
South Iceland (aftshck)	21/06/2000	6.4	0.36
Northridge	17/01/1994	6.7	0.68
Tirana	09/01/1988	5.9	0.33
Friuli	06/05/1976	6.5	0.35

Their mean response spectrum matches the Eurocode EC8–1 spectrum for ground type A (rock).

As an example, Figure 4a shows one of the time histories of acceleration applied at the base of the model. It is the record of the Norcia earthquake in Central Italy on 30/10/2016 (Mw = 6.5). In Figure 4b the corresponding normalized Fourier spectrum is shown.

In all the analyses that have been carried out, permanent changes of internal forces in the lining at the end of shaking were calculated. In some cases, they reach values as high as 30% of the maximum transient change during shaking. As an example, in Figure 5 a pair of time histories calculated for the input signal of the Norcia Earthquake (see Figure 4) are shown. They are the time histories of the increment of bending moment (a) and hoop force (b) calculated at the point NE of the reference central section of the tunnel lining.

The experimental evidence obtained by Lanzano et al. (2012) and shown in Fig. 1b are therefore confirmed in part by numerical modelling on a different lining and for different characteristics of ground shaking: permanent changes of internal forces are calculated at the end of shaking, as observed in the experiments, although they do not exceed the transient changes calculated during the event. It should also be remarked that, in order to capture such an effect a suitable elastic-plastic constitutive model for soil must be adopted, as in this case.

Figure 6 shows the distributions of internal forces in the tunnel lining calculated under static conditions prior to (continuous lines) and at the end of shaking (dashed line). Such distributions of bending moment (Fig. 6a), hoop force (Fig. 6b) and longitudinal force (Fig. 6c) were calculated in the transverse reference section, both after simulation of the tunnel construction process (black lines) and for an ideal "wished-in-place" tunnel (grey lines).

As one would expect, the stress change due to the excavation produces lower bending moments (Fig. 6a) and normal forces (Fig. 6b) in the tunnel lining, than in a wished-in-place tunnel. Furthermore, the latter is almost not loaded in longitudinal direction (Fig. 6c).

It is worth noting that, although the maximum values of pre-shaking internal forces (continuous lines) are quite different, such differences reduce after shaking (dashed lines). This implicitly means that the calculated permanent changes of internal forces depend on

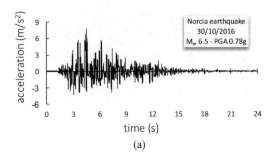

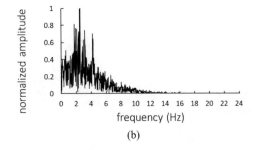

Figure 4. Norcia earthquake 2016 (M = 6.5): (a) time history; (b) Normalized Fourier spectrum.

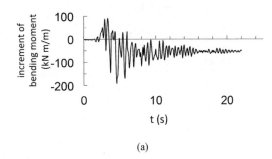

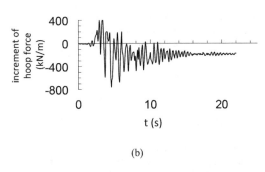

Figure 5. Time histories of the increment of internal forces in the point NE: (a) bending moment; (b) hoop force (Norcia earthquake).

the pre-seismic conditions: when the excavation process is modelled they are larger than in the case of the wished-in-place tunnel. The effect of the construction stages on the seismic behaviour of the tunnel lining is therefore evidenced by such numerical results.

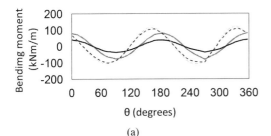

(a)

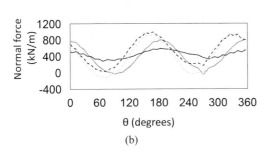

(b)

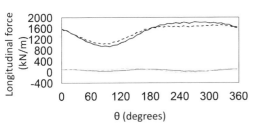

— Pre-shaking without excavation — Pre-shaking with excavation
⋯ Post-shaking without excavation - - Post-shaking with excavation

(c)

Figure 6. Distribution along the transverse reference section of (a) bending moment, (b) hoop force, (c) longitudinal force: static 'pre-shaking' (continuous lines) and 'post-shaking' (dashed lines), Norcia earthquake.

2.5 Segmental layout of the tunnel lining

Mechanised tunnelling in soft ground is usually associated with the use of a pre-cast concrete segmental lining to withstand external loads from interaction with the surrounding soil. Due to such a segmental layout the structural demand of the lining under static conditions is usually lower, because its flexural and axial stiffness is lower compared to a continuous lining of the same thickness.

The same numerical model that was described in section 2.4 was improved to introduce a segmental lining. The segments were modelled as elastic volumes of reinforced concrete with the same thickness as the continuous lining (EA = 10.5E6 kN/m; EI = 78.75E3 kNm2/m). Following Fabozzi (2017), the longitudinal joints between the segments were modelled as elastic-plastic elements (thickness = 0.30 m, width = 0.30 m): the values adopted for their mechanical parameters are

Table 3. Model parameters for the lining (Fabozzi, 2017).

	γ (kN/m^3)	E (GPa)	ν	c (kPa)	φ
segments	25	35	0.15	-	-
joints	25	6	0.15	9000	42

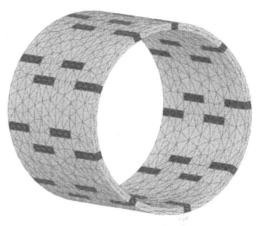

Figure 7. Detail of the numerical model of segmental lining

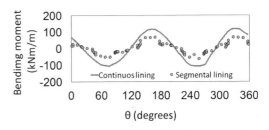

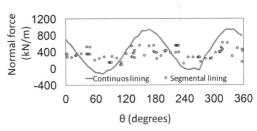

Figure 8. Distribution along the transverse reference section of (a) bending moment, (b) hoop force at the end of shaking: continuous vs. segmental lining (Norcia earthquake).

shown in Table 3. Interface elements with the same behaviour were assumed to represent the transverse joints between rings. Figure 7 shows details of the structural model. The excavation stage was not modelled.

A lower structural requirement for the segmental lining compared to the continuous lining is evident also at the end of shaking. In Figure 8, the distributions of bending moment (Fig. 8a) and hoop force (Fig. 8b) in the transverse section at the end of shaking are shown.

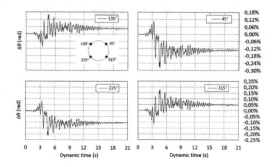

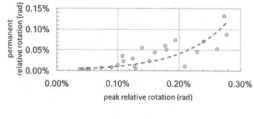

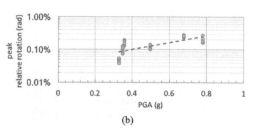

Figure 9. Time histories of relative rotation between segments during shaking (Norcia Earthquake).

The lower values of structural forces in the segments correspond to a larger deformability of the lining system at the joints, where relative displacements and rotation might be expected to occur.

Figure 9 shows the time histories of relative rotation between segments during shaking, calculated in the joints located at 45°, 135°, 225°, and 315° about the horizontal tunnel axis. At the end of shaking permanent relative rotations remain between segments. The magnitude of such permanent rotations is sometimes rather close to the peak values calculated during shaking. This result indicates a possible weakness of the segmental lining at the joints, where the rubber gaskets that guarantee water-tightness of real linings might be dislocated at the end of an earthquake.

The results in terms of relative rotations for the whole set of input signals shown in Table 2 are plotted in Figure 10. In Figure 10a a linear trend can be observed for the logarithm of the calculated peak relative rotation between segments versus the value of the peak ground acceleration of the input signal (PGA). It is worth noting (Figure 10b) that as far as the peak relative rotation increases (with increasing PGA), the permanent relative rotation increases faster. The ratio between the permanent and the peak rotation increases from as low as 10% until almost one half for the stronger earthquakes.

This further highlights the influence of the nonlinear behaviour of the surrounding soil on the value of permanent rotations experienced by the segmental lining at the end of shaking.

2.6 *Remarks*

The numerical analyses presented in this section were calibrated on a single benchmark centrifuge test and then extended to model more complex cases in terms of the geometry of the lining. The numerical model also allowed a straightforward application of natural input signals and a consideration of the influence of construction process on the seismic demand of the tunnel lining.

It is worth noting that an earthquake can hit a tunnel several years after construction, hence different "preseismic" conditions can be considered. Moreover, in earthquake-prone regions, the same tunnel may be

Figure 10. Peak relative rotation vs. peak ground acceleration (a) and permanent relative rotation vs. peak relative rotation (b), all input signals in Table 2.

subjected to sequences of seismic events, with variable intensity and effects. Hence, the influence of the "initial state" should be considered in the assessment of tunnel vulnerability.

Moreover, the numerical results from the segmental layout may create some concerns for tunnel linings in highly permeable soils, where an excessive rotation of joints may produce loss of water-tightness. This aspect may deserve attention in design and, at the same time, requires further experimental and numerical investigation.

3 TUNNELS IN LIQUEFIABLE SOIL

3.1 *Background*

Soil liquefaction may induce buoyancy of underground structures such as tanks, tunnels and pipelines. This is triggered when high excess water pressures develop, as those induced by strong motions. Several cases of uplift of underground tanks and pipelines have been observed in the past.

Although little evidence of liquefaction-induced damage to tunnels exists, physical modelling has shown that the high mobility of liquefied soil near surface would encourage floatation of very shallow or immersed tunnels. As a matter of fact, the uplift behaviour of underground structures caused by liquefaction has often been studied by physical models: for instance 1-g shaking table models of buried box structures, sewers and pipes and relevant possible mitigation measures (Koseki et al., 1997; Otsubo et al., 2014; Watanabe et al., 2016) or centrifuge models of tunnel of different shapes embedded in sand layers of different density, with several overburden and groundwater level (Yang et al. 2004; Chou et al, 2010; Chian and Madabushi, 2011; Chian & Madabhushi, 2012; Chian et al., 2014).

Experimental evidence has indicated that both the width of the underground structure and the depth of the liquefied layer have a large influence on the uplift displacement.

In general, physical modelling has been useful to collect important information and quantitative data on the phenomenon. In fact, although many numerical tools have been developed in the last decades to assess soil liquefaction, prediction of soil behaviour after liquefaction is still a challenging task. Hence, physical modelling has a further important role, that is to validate numerical models that can be used later for sensitivity analysis.

In this section, it is shown how starting from the back-analysis of the results of a centrifuge test on a model tunnel in dry sand undergoing shaking, the behaviour of the same tunnel in sand that has been saturated can be modelled numerically. The centrifuge model T4, described in detail by Lanzano et al. (2012), was used as an experimental benchmark. This centrifuge model has the same layout as model T3 in Figure 1, although the sand layer was looser (Dr = 40%).

The UBC3D-PLM constitutive model (Beaty & Byrne, 1998; Galavi et al., 2013) was used to represent the sand. It includes hardening plasticity and strain dependency of stiffness and damping. Hence it is able to capture the permanent deformation of the ground and changes in internal forces in the tunnel lining due to dynamic loading. Moreover, in undrained conditions it models the pore pressure build-up that may produce soil liquefaction and tunnel uplift.

The model is available in the 2D finite element code Plaxis (Brinkgreve et al., 2016) that has been used for the analyses. It was calibrated on the results of laboratory tests on the sand used in the centrifuge test along monotonic (Lanzano et al., 2016) and cyclic (Mele et al., 2018) stress paths.

3.2 Numerical analyses

A plane strain numerical model was defined in Plaxis 2D (Brinkgreve et al., 2016), at prototype scale. 'Tied degrees of freedom' between vertical sides were used as boundary conditions to simulate the laminar box behaviour during shaking. The nodes at the base of the finite element model were fixed in the vertical direction and a time history of acceleration was applied in the horizontal direction. The input signal applied at the base of the model is a pseudo-harmonic signal with nominal frequency 0.375 Hz and nominal amplitude 0.05 g at prototype scale. It was obtained after scaling up and filtering of the record of the base accelerometer in the centrifuge model ACC13 (see Figure 1).

3.3 Model calibration

The UBC3D-PLM is an elastoplastic constitutive model, which is a generalized formulation of the original UBCSAND model proposed for cyclic

Table 3. UBC3D-PLM model parameters.

	sand
φ'_{cv}	32°
φ'_p	35.5°
$c'(kPa)$	0.01
K^e_B	300
K^e_G	360
K^p_G	180
m_e	0.5
n_e	0.5
n_p	0.4
R_f	0.93
$N_{1,60}$	7.36
fac_{hard}	1.6
fac_{post}	1.0

loading by Beaty & Byrne (1998). The model uses isotropic hardening and a simplified kinematic hardening rule for primary and secondary yield surfaces respectively, in order to take into account the effect of soil densification and transition to the liquefied state during undrained cyclic loading.

The constitutive model is capable of modelling cyclic liquefaction for different stress paths (Galavi et al., 2013).

Table 3 reports the input parameters used in the UBC3D-PLM model. The calibration of the model mechanical parameters was performed by Colamarino et al. (2017) using the results of laboratory tests on the sand used in the centrifuge test along monotonic (Lanzano et al., 2016) and cyclic (Mele et al., 2018) stress paths.

3.4 Response of the model with dry sand

The numerical results are compared to the centrifuge test results in terms of time history of acceleration and the relevant response spectrum, 1.6 m below the ground surface (position of ACC9 in Figure 1) at prototype scale (Figure 11). For the sake of comparison, here and in the following figures, the experimental results of the centrifuge test are scaled up to prototype scale.

Other relevant comparisons between recorded data and simulation results are reported in terms of vertical displacements and bending moment in Figure 12a and b, respectively.

The main features of the experimental data are well-reproduced by the numerical predictions, both in terms of amplitude and frequency content, although an amplification larger than measured is calculated around 0.6 s, that is close to the natural period of the soil layer. Here, the agreement between the measured and the calculated amplitude achieved by using the UBC3D-PLM model is worse than by using HS-small model in similar conditions (see Figure 3). This might be in part due to the different amount of damping that the two constitutive models generate in stress-strain cycles.

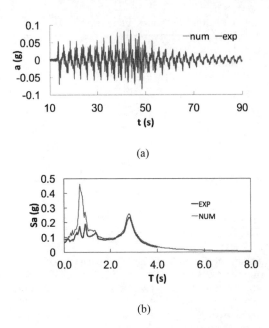

(a)

(b)

Figure 11. Simulated vs. experimental acceleration at 1.6 m under the surface (ACC9): time history of acceleration (a) and response spectra (b).

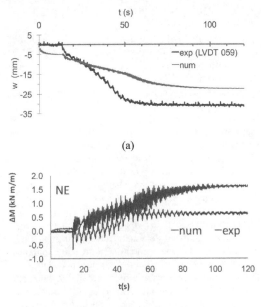

(a)

(b)

Figure 12. Simulated vs. experimental time histories of (a) settlement at the surface (LVDT 059) and (b) bending moment in the tunnel lining at position NE

Figure 12a shows that the numerical model computes settlement at ground surface since the very beginning of the analysis, before 10 s, that is when the amplitude of the input signal is still negligible. In the same time a slight increase of bending moment is calculated (Fig. 12b). On one hand this confirms the influence of sand densification (hence plastic volumetric deformation) on the permanent change of internal forces in the lining; on the other it also shows that the numerical model tends to overpredict the plastic volumetric strain during shaking. As a consequence, the residual value of bending moment that is calculated at the end of shaking is even larger than the experimental value, although the corresponding transient cyclic changes are very similar (Fig. 12b).

3.5 Response of numerical model in saturated sand

The same input motion was applied at the bottom of the mesh modelling the soil as completely saturated. In this condition, significant excess pore pressure developed in the soil layer above the tunnel, although full liquefaction was not triggered, due to the low amplitude of the input signal. The excess pore pressure ratio, r_u, defined as the ratio between the generated excess pore pressure and the initial effective vertical stress, did not exceed 0.77 (Figure 13).

Differences in the internal forces between dry and saturated conditions are shown in Figure 14.

This figure shows that the hoop force increased at both control points (NE and SE) in saturated sand compared to dry sand (Figure 14a, b), while bending moment increased in the upper part (NE) of the tunnel cross section and decreased in the lower part (SE).

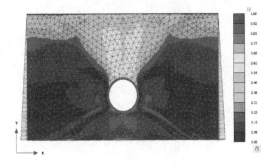

Figure 13. Excess pore pressure ratio at the end of shaking.

This indicates that the pore pressure build-up, associated with changes in effective stresses, affects the distribution of internal forces in the tunnel lining.

In general, a larger change of hoop force is induced in the lining during ground shaking if soil liquefaction approaches. The effect on bending moment depends on the position along the lining. However, for such a lining (the very flexible one used in the experiment) the values of bending moments are very low.

In order to evaluate the preliminary remarks that emerge on the basis of the comparison in Figure 14, the tunnel lining was changed, as in section 2.3, to a thicker reinforced concrete lining (EA = 10.5E6 kN/m; EI = 78.75E3 kNm^2/m).

Moreover, the numerical analyses were performed by applying at the base of the mesh the same signal 'EQ1' recorded in the centrifuge, two more signals obtained simply by scaling up 'EQ1' to twice ('2x')

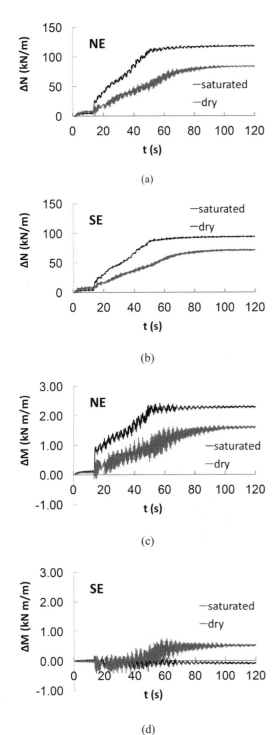

Figure 14. (a, b) Hoop force and (c, d) bending moment time histories along the tunnel on dry versus saturated soil conditions at (a, c) NE and (b, d) SE point.

and three times ('3x') its amplitude, and additionally the six natural signals shown in Table 2.

An overview of the analyses is given in Table 4.

The peak acceleration of the input signal ($a_{max,b}$), peak acceleration calculated at the ground surface ($a_{max,s}$), the maximum change of bending moment (ΔM) and hoop force (ΔN) in the tunnel lining at the end of shaking and the maximum uplift of the tunnel ($u_{v,max}$) are shown in the table. The last column of Table 4 also reports the average thickness of a continuous layer of soil (if any) where a value of the excess pore pressure ratio $r_u > 0.8$ was calculated.

In Figure 15 the time history of acceleration at the base of the model (grey line) is compared with that calculated at the surface (black line) for two cases from Table 4: 'Norcia' and 'Northridge' input. The achievement of liquefaction in the soil layer can be noticed in both cases.

Initially the signal is amplified (up to about 0.2 g, that is at about 2.5 s for 'Norcia' and 4 s for 'Northridge') at the surface compared to the base, then liquefaction occurs and the liquefied soil acts as an isolating layer: the amplitude of acceleration at the surface is lower than at the base from this point onwards. Figure 16 shows the distribution of the excess pore pressure ratio ru at the end of shaking in both cases. The shading has been limited to the range $0.8 \leq r_u \leq 1$. It can be observed that a continuous horizontal layer of soil near to the surface is very close to liquefaction if not liquefied. Moreover, the tunnel itself is partially interacting with liquefied soil, although deeper than the shallow liquefied horizontal layer (C/D=2).

In Figure 17 the ratio between peak acceleration at the surface and that at the base is plotted against the peak acceleration at the base, for all the input signals shown in Table 4. It can be noticed that in the cases where the peak acceleration of the input signal is lower than 0.2 g, such a ratio is higher than 1, indicating amplification, while for higher values of peak acceleration the ratio is lower than 1, indicating that de-amplification occurred.

In all cases de-amplification is caused by liquefaction occurring near the ground surface. The depth of the tunnel in the ground layer does not affect the dynamic response of the soil, as shown in the figure.

The effect of soil liquefaction on the tunnel lining is analysed by looking at the maximum changes of hoop force and bending moment at the end of shaking (Table 4 and Fig. 18a, b).

Figure 18 indicates that when the ground amplification prevails (for this ground conditions when amax,b < 0.2 g according to Figure 17) larger changes of internal forces arise for increasing amplitude of shaking. This trend is more evident for deeper tunnels (C/D=2). On the other hand, when liquefaction prevails (when $a_{max,b} > 0.2$ g), the change of internal forces is independent from the amplitude of the base acceleration.

Table 4. Overview of the analyses.

Input	$a_{max,b}$ g	$a_{max,s}$ g	ΔM kNm/m	ΔN kN/m	$u_{v,max}$ m	thickness $r_u > 0.8$ m
EQ1	0.054	0.114	43	126	0.004	-
2x (EQ1)	0.108	0.13	131	235	0.033	4
3x (EQ1)	0.162	0.197	132	210	0.238	7.5
Tirana	0.33	0.214	127	232	0.019	3
Friuli	0.35	0.25	137	240	0.026	3
South Iceland	0.36	0.315	139	234	0.087	-
Avej	0.5	0.339	189	246	0.061	-
Northridge	0.68	0.233	154	254	0.337	8
Norcia	0.78	0.26	150	233	0.128	7

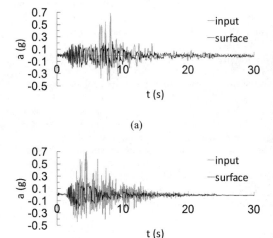

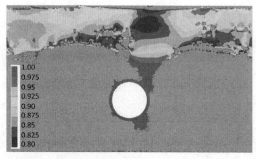

(a)

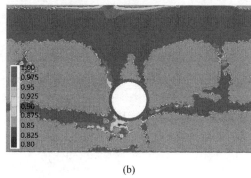

(b)

Figure 15. Time histories of acceleration at the base and at surface: input signal Norcia (a) and Northridge (b).

In terms of permanent displacements induced by soil liquefaction, it is worth noting that the calculations were performed by imposing undrained conditions. Pore-pressure build-up during shaking produces a very limited uplift of the tunnel, unless the soil liquefies and the liquefied ground interact with the tunnel, such as in the cases of Figure 16.

In Figure 19 the calculated uplift of the tunnel at the end of shaking is plotted as a function of the average thickness of a liquefied layer. This has been assumed to be a shallow continuous horizontal layer with $r_u > 0.8$ (see for instance the shaded areas in Figure 16). The trend in the figure shows that large amounts of liquefaction in the cover soil layer of the tunnel produces significant uplift of a shallow tunnel, although the cover upon diameter ratio is not too low (C/D=2). Similar trends were obtained for shallower tunnels (C/D = 0.5 and 1) using the pseudo-harmonic input signal only (that is 'EQ1', '2x', '3x' in Table 4) and are shown in Figure 20. Although the numerical results are limited, the effect of lower overburden can be read. The shallower the tunnel

Figure 16. Shadings of $r_u > 0.8$: input signal Norcia (a) and Northridge (b).

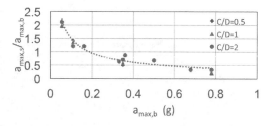

Figure 17. Ratio between peak acceleration at surface and at the base vs. peak acceleration at the base (all signals).

the larger the uplift associated with the mobility of the surrounding liquefied soil, as observed experimentally by Chian and Madabhushi (2011) in the centrifuge.

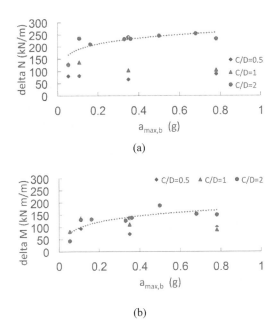

Figure 18. Maximum change of hoop force (a) and bending moment (b) in the lining at the end of shaking (all signals).

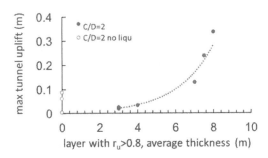

Figure 19. Maximum vertical displacement of the tunnel at the end of shaking (all signals, C/D=2).

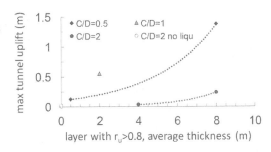

Figure 20. Maximum vertical displacement of the tunnel at the end of shaking ('EQ1', '2x', '3x')

3.6 Remarks

This section has shown how an advanced effective stress constitutive model, able to capture the cyclic behaviour of sand in both drained and undrained conditions, has been adapted to back-analyse a centrifuge test in dry sand in order to model afterwards a similar problem in saturated sand. The constitutive model has been calibrated using the results of laboratory tests carried out in monotonic and cyclic loading.

After comparing the numerical simulation in dry and saturated conditions, the numerical model has been used to extend the study to different conditions in terms of lining thickness, tunnel cover, input signal. This provided an insight into the behaviour of a shallow tunnel in a liquefiable sand layer. A form of limiting threshold to the change of internal forces induced by ground shaking in the tunnel lining has been observed in the numerical results once soil liquefaction occurs. At the same time, the influence of the overburden cover on the uplift induced at the tunnel in cases of extensive liquefaction has been discussed on the basis of the calculations.

Due to the lack of existing measurements for real cases, experimental campaigns using centrifuge modelling would be highly beneficial to corroborate or debate similar results.

4 TUNNEL-BUILDING INTERACTION IN LIQUEFIABLE SOIL

4.1 Background

Although uplift mechanisms for an underground structure experiencing soil liquefaction have been identified experimentally and numerically by several authors, the interaction of such mechanisms and the associated displacements of the underground structure with those induced in aboveground structures that may be founded nearby have not yet been investigated.

In urban areas shallow tunnels are likely to be close to the foundations of buildings and easily interact with them during earthquakes (i.e. Soil-Structure-Underground Structure-Interaction, SSUSI). Hence, the reciprocal influence of a tunnel and an adjacent building in the presence of soil liquefaction may be important.

Recent centrifuge testing on the behaviour of buildings founded in liquefiable ground layers has shown that smaller net excess pore pressures are generated within the liquefiable layer under a structure by increasing the contact pressure and height/width ratio of the building (Karimi & Dashti, 2016). Other studies have shown the reciprocal influence of adjacent buildings, affecting non-uniform settlement during liquefaction (Yasuda, 2014).

How the uplift mechanism of an adjacent underground facility is influenced by the presence of the building and how the floating of the underground structure can affect the tilt and settlement of the building are both aspects that deserve attention.

This problem appears rather important considering the rapid extension of the built environment, both above- and underground, to areas that may be subjected to risk of liquefaction. Hence an insight into such a problem may well contribute to increase the resilience of urban environment to natural hazards.

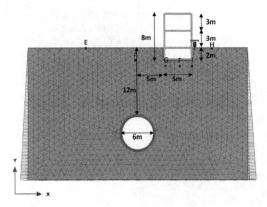

Figure 21. Numerical mesh (prototype scale).

The project STILUS, within the framework of the European funded network SERA (Seismology and Earthquake Engineering Research Infrastructure Alliance for Europe) intends to investigate this problem through a series of centrifuge tests.

In order to plan the centrifuge tests, a preliminary numerical study of tunnel-structure interaction in liquefiable soil was carried out as described in the following sections 4.2 and 4.3. A circular transverse section (modelling a bored tunnel) and a rectangular framed section (modelling a cut-and-cover tunnel) were taken into account at this preliminary stage, since they may be likely to occur in the urban environment.

4.2 Circular tunnel

The numerical calculations were carried out using the same layout as shown in Figure 1, that has been analysed in section 3. The numerical model and the input signal used were the same as described in that section, being the problem modelled using the UBC3D-PLM model for the soil (Galavi et al., 2013). A simple structure was added in the model mimicking a two-storey building as shown in Figure 21 (Colamarino, 2017).

The building consists of a two-floors (3 m high each) and a basement (2 m deep). The building rests to one side of the tunnel as shown in the figure.

The building frame was modelled using linear elastic beam elements. Two different material datasets were used, one for the basement (EI = 1.6×10^5, kNm^2/m, EA = 1.2×10^7 kN/m) and the other for the rest of the building (EI = 6.75×10^4 kNm^2/m, EA = 1.6×10^5 kN/m). The mass assigned per unit length to the beam elements takes into account also the presence of the floors and the walls.

The same set of input signals as in the previous section was used in order to compare the results to the "greenfield" conditions considered in that section. Figure 22 shows the highest values of excess pore pressure ratio ($r_u > 0.8$) calculated in undrained conditions at the end of the shaking for the three input signals 'EQ1' (Fig. 22a), '2x' (Fig. 22b) and '3x' (Fig. 22c).

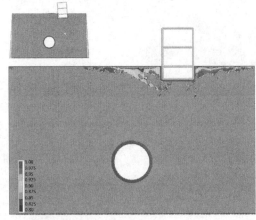

(a)

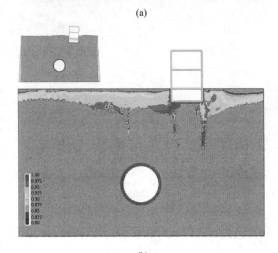

(b)

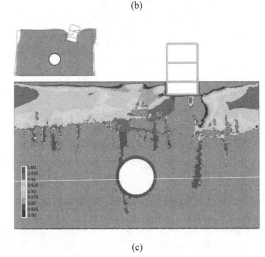

(c)

Figure 22. Excess pore pressure ratio distribution and mesh deformation (magnification 2) at the end of shakings (a) 'EQ1', (b) '2x' and (c) '3x.'

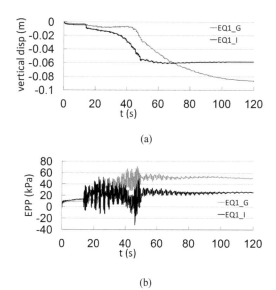

Figure 23. Time histories of settlement (a) and excess pore pressure (b) at the foundation level ('EQ1').

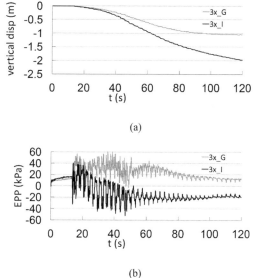

Figure 24. Time histories of settlement (a) and excess pore pressure (b) at the foundation level ('3x').

Insets in the same figure show the corresponding deformed configurations at the end of shaking.

As soon as the amplitude of the signal increases, larger areas of the sand layer are affected by liquefaction or are approaching it ($r_u > 0.8$). It is worth noticing that for 'EQ1' (Fig. 22a) the highest values of r_u are distributed in the area of maximum shear stresses around the building foundation. Instead, no evidence of liquefaction was observed in the results of the corresponding greenfield analysis in section 3 ($r_u < 0.8$, see Figure 13). The larger amplitude of the "2x" input signal (Fig. 22b) produces a continuous layer of shallow soil approaching liquefaction.

The influence of the building is still visible in this distribution but liquefaction does not affect the soil around the tunnel (C/D=2). When subjected to an even stronger shaking, a larger thickness of soil approached liquefaction (Fig. 22c). Compared to the corresponding greenfield analysis, the influence of the stresses induced by the building is evident both in terms of deviator and mean stress: calculated pore-pressure build-up are higher at the corners of the foundation due to initial higher shear stresses and lower towards the centre, where the mean stresses prevail. In this case liquefaction areas reached the tunnel below.

The building settles and tilts. Both settlement and tilt are influenced by the distribution of excess pore pressure around the foundation, that affects the degree of mobilization of shear strength in the foundation ground. However, when liquefaction reaches the tunnel depth, an increased uplift of the tunnel affects the building movements and the building starts to counter-rotate.

Figures 23 and 24 show the time histories of settlement and excess pore pressure calculated at point G and I (see Figure 21) with input signals "EQ1" and "3x". For the weaker "EQ1", the larger positive excess pore pressure that arises around point G (Fig. 23b) produces a larger settlement of the building on that side at the end of shaking (Fig. 23a).

On the other hand, for the stronger "3x", although negative excess pore pressure develop around point I (fig. 24b), the buildings settles more on the right side (fig. 24a), indicating an interaction with the uplift of the tunnel on the left side.

Tables 5a and 5b summarize the results achieved in the analyses with C/D=2. Positive tilt is assumed counter-clockwise.

Table 5a. Overview of the analyses with C/D=2.

Input	$a_{max,b}$ g	$a_{max,s}$ g	ΔM kNm/m	ΔN kN/m
EQ1	0.054	0.098	52	118
2x (EQ1)	0.108	0.139	146	250
3x (EQ1)	0.162	0.203	172	235
Northridge	0.68	0.249	242	301
Norcia	0.78	0.318	164	200

Table 5b. Overview of the analyses with C/D=2

Input	tunnel max uplift m	thickness $r_u > 0.8$ m	building max settlmt m	building max tilt rad
EQ1	0.003	-	0.23	0.032
2x (EQ1)	0.023	3	0.709	−0.026
3x (EQ1)	0.168	7	2	−0.142
Northridge	0.175	8	1.29	−0.017
Norcia	0.146	6.5	0.793	0.021

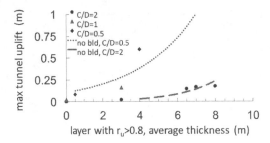

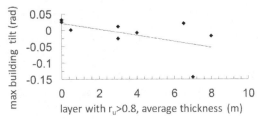

Figure 27. Building max tilt.

Figure 25. Tunnel max uplift in the analyses with building compared to corresponding trends calculated in analyses without.

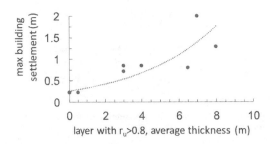

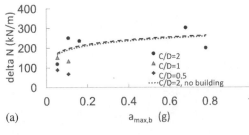

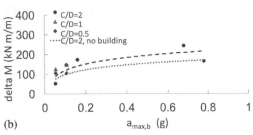

Figure 26. Building max settlement.

In Figure 25 the maximum value of uplift of the tunnel axis is plotted for different values of the ratio C/D as a function of the average thickness of a continuous horizontal layer of soil where the excess pore pressure ratio r_u is larger than 0.8. As in section 3, such a value has been assumed as a proxy for the effect of liquefaction in the ground layer. In the same figure two curves are shown that represent the trends calculated for C/D = 2 and C/D = 0.5 in the analyses without buildings (section 3).

Although with some scatter, the trends are the same in both sets of analyses (with and without buildings), indicating a minor effect of the presence of a building on the amount of tunnel uplift, providing that similar distributions of pore pressure build-up affect the soil surrounding the tunnel.

Trends of increasing building settlement and tilt can be observed in Figure 26 and 27. Very low values of average thickness of the layer with ru>0.8 (close to zero) indicate that liquefaction occurs only in the proximity of the foundation of the building (e.g. Fig. 22a). This corresponds to limited settlement, although non-negligible. Much larger settlement is calculated when liquefaction is approached in larger volumes of soil, as for instance in the cases shown in Fig. 22b-c.

Correspondingly, it might be noted that in Figure 27 there is a decreasing trend of tilt towards negative values as the average thickness of the layer with ru>0.8 increases. Hence, a larger pore-pressure build-up generally induces the foundation to rotate clockwise. This indicates the effect of the upheaval associated with the tunnel buoyancy on the left side of the building.

In Figure 28 the change of hoop force (a) and bending moment (b) in the tunnel lining at the end of

Figure 28. Maximum change of hoop force (a) and bending moment (b) in the lining at the end of shaking: all analyses in Table 5 compared to trend lines in Fig. 18 (no building).

shaking is plotted as a function of the peak acceleration of the input signal at the base of the model. Trend lines for the case C/D=2 are shown as dashed lines and compared with similar trend lines from Fig. 18 for the 'greenfield' cases, that is without the building (dotted lines). The change of hoop force N induced by pore-pressure build-up is independent of the presence of the building. On the contrary, the change of bending moment is generally larger than in the 'greenfield' case. This finds justification in the less uniform distribution of stresses induced around the tunnel by the presence of the building (compare for instance values of r_u: for 'greenfield' conditions in Fig. 16b with 'building" conditions in Fig. 22c).

4.3 Rectangular tunnel

In order to analyse a typical case of a cut-and-cover tunnel in an urban environment, a rectangular section has been assumed, as shown in Figure 29. The same liquefiable sand layer and the same building as in section 4.2 are modelled.

Table 6 shows an overview of the analyses that have been carried out. The legend for the input signals has been given in Table 4. A set of analyses with input signals of increasing amplitude was carried out

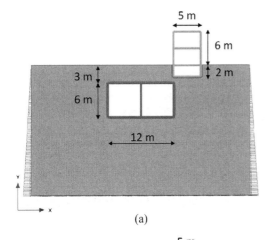

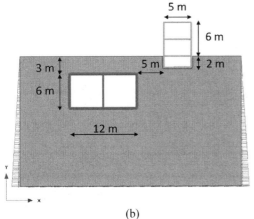

Figure 29. Models of rectangular tunnel with building: (a) building on the edge of the tunnel (d/C = 0), (b) building at a distance d = 5 m (d/C = 1.7).

Table 6a. Overview of 'greenfield' analyses without building.

input	tunnel max uplift m	thickness $r_u > 0.8$ m
EQ1	0.529	–
2x (EQ1)	0.998	2
3x (EQ1)	1.000	2
Northridge	1.300	12

Table 6b. Overview of the analyses without tunnel.

input	thickness $r_u > 0.8$ m	building max settlmt m	building max tilt rad
EQ1	3	0.572	0.058
2x (EQ1)	5	1.02	−0.009
3x (EQ1)	7	2.38	−0.1077
Northridge	7	1.2	−0.042

Table 6c. Overview of the analyses with tunnel and building.

d/C = 0

input	tunnel max uplift m	thickness $r_u > 0.8$ m	building max settlmt m	building max tilt rad
EQ1	0.095	1	0.23	−0.005
Northridge	0.421	12	1.32	−0.236

d/C = 1.7

input	tunnel max uplift m	thickness $r_u > 0.8$ m	building max settlmt m	building max tilt rad
EQ1	0.419	-	0.414	0.034
Northridge	1.340	8	1.72	−0.090

in 'greenfield' conditions (Table 6a), to study the effect on the dynamic response and the pore-pressure build-up of the presence of the tunnel. Similarly, a set of analyses was carried out for models with a building and without a tunnel (Table 6b). Finally, the tunnel-building interaction was analysed with two sets of numerical models with both structures (Table 6c). In the former the building was located on the edge of the tunnel, with a distance to cover ratio, d/C = 0 (Fig. 29a). In the latter, the building was located at a distance d = 5 m on the right side of the tunnel, corresponding to d/C = 1.7 (Fig. 29b).

The tunnel was very shallow, with a cover C = 3 m, compared to the depth of the basement (2 m).

The dynamic response of the soil layer in 'greenfield' conditions (no building) with and without this tunnel is shown in Figure 30. In the figure the ratio between the peak acceleration at the surface and at the base is plotted against the peak acceleration at the base. It can be noticed that the presence of the larger rectangular tunnel reduces the amplification at the ground surface compared to the circular tunnel. In all cases the dynamic amplification calculated in 'free-field' in the corresponding analyses without a tunnel is much larger.

The differences between the curves reduce as the peak ground acceleration increases, when de-amplification occurs due to soil liquefaction, as discussed in section 3.5.

The presence of the building affects the distribution of excess pore pressure, as shown in section 4.2. This is confirmed by the analyses without a tunnel as shown in Figure 31a. Here the continuity of the horizontal layer approaching liquefaction is broken below the building due to the higher effective mean stresses.

The presence of the tunnel further affects the distribution of excess pore pressure, as can be seen from the comparison between Fig. 31a and Fig. 31b.

It is worth noting that liquefaction is confined most in the free-field areas at the right side of the building and the left side of the tunnel. However isolated liquefied soil volumes are identified below the tunnel,

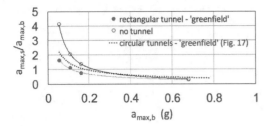

Figure 30. Ratio between peak acceleration at surface and at the base *vs.* peak acceleration at the base: with and without rectangular tunnel and comparison with trend for circular tunnel.

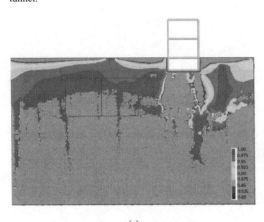

(a)

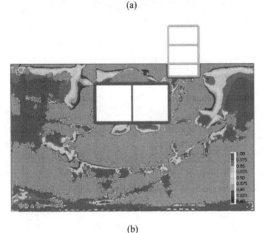

(b)

Figure 31. Excess pore pressure ratio distribution at the end of shaking 'Northridge': (a) 'no tunnel', (b) 'd/C=0'.

above the tunnel roof and between the tunnel and the building foundations. In the latter area the distribution of excess pore pressure depends on the relative distance d/C.

In Figure 32 the calculated values of building maximum settlement are plotted.

In the figure a trend of increasing settlement of the building with the average thickness of the layer approaching liquefaction is observed, although with some scatter. The presence of the tunnel reduces the

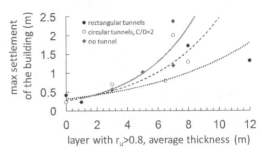

Figure 32. Max settlement of the building.

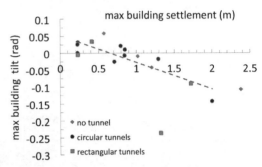

Figure 33. Max tilt vs. max settlement: analyses with building.

calculated settlement of the building. Such a reduction is more evident for the shallower and larger rectangular tunnels than for the deeper and smaller circular tunnels.

The calculated tilt and settlement of the building at the end of shaking are plotted one against each other in Figure 33. Despite some scatter, the plot shows a certain degree of correlation among the two quantities.

4.4 Remarks

This section has described a preliminary numerical study of tunnel-structure interaction in liquefiable soil. This study has been carried out to identify patterns of deformation that should be expected to occur in centrifuge tests to be carried out in a research project concerning such a problem.

The results have clearly shown how the distribution of excess pore pressure induced by shaking in undrained conditions is affected by the presence of the tunnel and of the building.

The relative distance between the two structures (here expressed in terms of ratio d/C of the horizontal distance between the tunnel wall and the building basement upon the tunnel cover) influences the solution both in terms of tunnel lining deformation and of building displacements. Consequence are observed in the distribution of internal forces in the tunnel lining and in the final configuration of the building.

A number of indications for implementing the physical modelling have been suggested by the results of the numerical analyses.

As far as the distribution of internal forces is concerned, since the analyses show that it is influenced by the presence of the building, it would be important to have a large number of measuring points along the tunnel lining, to get an experimental insight into this problem.

Furthermore, since building tilt is expected by the analyses, non-contact laser displacement transducers might be used in the centrifuge to measure such a tilt. They will be then associated to conventional transducers (LVDTs).

Moreover, the distribution of calculated ground movements induced by soil liquefaction may help to define areas where the addition of finer content (down to the nanoscale) may reduce the mobility of the soil. On the experimental side, this show the potential benefit of using digital imaging and particle image velocimetry in centrifuge tests, through a transparent side of the model container.

The numerical results also show that the tunnel uplift is driven by the increase of pore pressure below the tunnel invert and the concurrent reduction of effective stresses (and shear resistance) in the cover. Hence the safety factor against uplift is reduced. This will considered in the layout of the centrifuge tests by deploying transducers to measure pore pressures where the analyses show that a significant build-up may develop.

The calculated distribution of excess pore pressure induced around the tunnel and the building basement may also be useful to identify where mitigation techniques that may locally reduce pore-pressure build-up (for instance: drainage, densification, induced partial saturation) would be most effective against the effects of soil liquefaction and should be implemented in the tests.

Nevertheless, it should be pointed out that the results of the numerical predictions should be considered with care. Although the potential of the constitutive model used, and of other models of similar complexity, to predict pore pressure build-up and to identify the occurrence of soil liquefaction has been shown in several studies, their accuracy in predicting the deformation of soil approaching or experiencing liquefaction and large strain is still a matter of study. Hence the need to run tests on physical models, thus achieving an experimental assessment of the behaviour of the soil and interacting structures (tunnel and building) in such conditions.

5 CONCLUSIONS

This work has illustrated how numerical calculations can be used in association with centrifuge testing to model different aspects of tunnel behaviour during earthquakes. The scope of the paper has been limited to a few aspects, mainly concerning the change of internal forces in the tunnel lining during shaking and the effect of soil liquefaction. Tunnel-building interaction during shaking in liquefaction prone soil has also been investigated. However, those analysed are only examples of a larger number of applications where an integrated experimental-numerical modelling approach can be followed.

The point of view of this paper is on purpose slightly biased towards the numerical modellers that may benefit of centrifuge tests to calibrate their models. The use of centrifuge testing (and physical models in general) should be considered as complementing laboratory testing on single elements when it comes to study specific aspects of boundary value problems. Indeed the possibility of evaluating numerical models using well-defined and controlled experiments increase the reliability of any numerical study where advanced constitutive models are used.

On the other hand, experimental activities may benefit significantly from a preliminary numerical study that helps to define the scope of testing and the key aspects that the physical model should be able to reproduce. This permits efficient use of resources, possibly reducing the number of experiments, to focus effective efforts on the specified target.

The main achievements of this work are only partial and deserve further investigation. However, they help to show how a combined use of physical and numerical modelling is necessary to analyse earthquake-induced effects on tunnels and other similar subsystems of civil infrastructures.

ACKNOWLEDGEMENTS

The Author wishes to acknowledge the contribution of Dr Giovanni Lanzano and Dr Stefania Fabozzi within the framework of the ReLUIS project; Mr Gianluca Fasano, Ms Lucia Mele and Dr Anna Chiaradonna within the framework of the LIQUEFACT project (funded by EU under the grant agreement no. 700748); Mr Giuseppe Colamarino who carried out preliminary analyses for the STILUS project.

The centrifuge tests mentioned in the paper have been (for the ReLUIS project) and will be (for the STILUS project) carried out at the Schofield Centre (University of Cambridge, UK). The Author wishes to thank the staff of the Centre and its Director Professor Gopal Madabhushi.

REFERENCES

Amorosi, A., Boldini, D., Falcone, G. 2014. Numerical prediction of tunnel performance during centrifuge dynamic tests. Acta Geotechnica 9 (4): 581–596.

Arulanandan, K. & Scott, R.F. 1993. Verification of numerical procedures for the analysis of soil liquefaction problems. Proc. Int. Conf. on Verification of numerical procedures for the analysis of soil liquefaction problems (VELACS), UC Davis. Balkema, Rotterdam.

Baziar, M.H., Nabizadeh, A., Jung Lee, C., Yi Hung, W. 2014. Centrifuge modeling of interaction between reverse faulting and tunnel. Soil Dynamics and Earthquake Engineering, 65: 151–164.

Beaty, M. & Byrne, P. 1998. An effective stress model for predicting liquefaction behavior of sand. In P. Dakoulas,

M. Yegian & R.D. Holtz (Eds.), Geotechnical Earthquake Engineering and Soil Dynamics III, ASCE Geotechnical Special Publication, 75(1): 766–777.

Benz, T., Vermeer, P.A., Schwab, R. 2009. A small-strain overlay model. Int J Num Anal Meth Geomech 33(1): 25–44.

Bilotta, E., Lanzano, G., Madabhushi, S.P.G., Silvestri, F. 2014. A round robin on tunnels under seismic actions. Acta Geotechnica 9 (4): 563–579

Brinkgreve, R.B.J., Kumaeswamy, S. & Swolfs, W.M. 2016. PLAXIS 2016 User's manual. https://www.plaxis.com/kb-tag/manuals/

Chian, S.C. & Madabhushi ,S.P.G. 2011. Tunnel and pipeline floatation in liquefied soils. Proceedings of the 5th International Conference on Earthquake geotechnical Engineering, 5ICEGE, Santiago, Chile.

Chian, S.C., Madabhushi, S.P.G. 2012. Effect of buried depth and diameter on uplift of underground structures in liquefied soils. Soil. Dyn. Earthq. Eng. 41: 181–190.

Chian, S.C., Tokimatsu, K., Madabhushi, S.P.G. 2014. Soil liquefaction- induced uplift of underground structures: physical and numerical modeling. J. Geotech. Geoenviron. Eng. 140, 04014057.

Chou, J.C., Kutter, B.L., Travasarou, T., Chacko, J.M. 2010. Centrifuge modeling of seismically induced uplift for the BART transbay tube. J Geotech Geoenviron 137(8): 754–765.

Cilingir, U. & Madabhushi, S.P.G. 2011. Effect of depth on seismic response of circular tunnels. Can. Geotech. J., 48(1): 117–127.

Colamarino, G. 2017. Modelling the behaviour of a tunnel in liquefiable sand. Master Thesis, University of Napoli Federico II.

Colamarino, G., Fasano, G., Chiaradonna, A., Bilotta, E., Flora, A. 2017. Modelling the response of tunnel lining in liquefiable sand. Proc. of the 15th East Asia-Pacific Conference on Structural Engineering and Construction, Xi'an, Cina: 760–768.

Fabozzi, S. & Bilotta, E. 2016. Behaviour of a Segmental Tunnel Lining under Seismic Actions. Procedia Engineering, 158: 230–235.

Fabozzi, S. 2017, Behaviour of segmental tunnel lining under static and dynamic load. PhD Thesis, University of Napoli Federico II.

Galavi, V., Petalas, A., Brinkgreve, R.B.J. 2013. Finite Element Modelling of Seismic Liquefaction in Soils. Geotechnical Engineering Journal of the SEAGS & AGSSEA, 44(3): 55–64.

Hashash, Y.M.A., Hook, J.J., Schmidt B., Yao, J.I.-C. 2001. Seismic design and analysis of underground structures. Tunn and Undergr Space Technology 16: 247–293.

Karimi, Z. & Dashti, S., 2016. Seismic Performance of Shallow Founded Structures on Liquefiable Ground: Validation of Numerical Simulations Using Centrifuge Experiments. J. Geotech. Geoenviron. Eng., 142(6): 04016011

Kontoe, S., Avgerinos, V., Potts, D.M. 2014. Numerical validation of analytical solutions and their use for equivalent-linear seismic analysis of circular tunnels. Soil Dynamics and Earthquake Engineering, 66: 206–219.

Koseki, J., Matsuo, O., Koga, Y., 1997. Uplift behavior of underground structures caused by liquefaction of surrounding soil during earthquake. Soils Found. 37 (1): 97–108.

Kutter, B.L., Carey, T.J., Hashimoto, T., Zeghal, M., Abdoun, T., Kokkali, P., Madabhushi, G.S.P., Haigh, S.K., d'Arezzo, F.B., Madabhushi, S., Hung, W., Lee, C., Cheng, H., Iai, S., Tobita, T., Ashino, T., Ren, J., Zhou, Y., Chen, Y., Sun, Z. Manzari, M.T. 2017. LEAP-GWU-2015 experiment specifications, results, and comparisons. Soil Dyn Earthq Eng in press.

Lanzano, G., Bilotta, E., Russo, G., Silvestri, F., Madabhushi, S.P.G. 2012. Centrifuge modeling of seismic loading on tunnels in sand. Geotechnical Testing Journal 35(6): 854–869.

Lanzano, G., Visone, C., Bilotta, E., Santucci de Magistris, F. 2016. Experimental Assessment of the Stress–Strain Behaviour of Leighton Buzzard Sand for the Calibration of a Constitutive Model. Geotechnical and Geological Engineering, 34 (4): 991–1012.

Mele, L., Flora, A., Lirer, S., d'Onofrio, A., Bilotta, E. 2018. Effect of laponite addition on liquefaction resistance of sand. Proc. Geotechnical Earthquake Engineering and Soil Dynamics V 2018. Austin, Texas.

Otsubo, M., Towhata, I., Taeseri, D., Cauvin, B., Hayashida, T. 2014. Development of structural reinforcement of existing underground lifeline for mitigation of liquefaction damage. In: Geotechnics of Roads and Railways: Proceedings of the XV Danube-European conference on geotechnical engineering. Vol. 1, 119–125, Wien

Pitilakis, K. & Tsinidis, G. 2014. Performance and seismic design of underground structures, State-of-Art, Earthquake Geotechnical Engineering Design. Geotech Geol and Earthq Eng 28: 279–340.

Schanz, T., Vermeer, P.A., Bonnier, P.G. 1999. The Hardening Soil Model: formulation and verification. Plaxis symposium on beyond 2000 in computational geotechnics, Amsterdam: 281–296.

Tsinidis, G., Pitilakis, K., Anagnostopoulos, C. 2016a. Circular tunnels in sand: dynamic response and efficiency of seismic analysis methods at extreme lining flexibilities. Bulletin of Earthquake Engineering, 14 (10): 2903–2929.

Tsinidis, G., Pitilakis, K., Madabhushi, G. 2016b. On the dynamic response of square tunnels in sand. Engineering Structures, 125: 419–437.

Tsinidis, G., Pitilakis, K., Madabhushi, G., Heron, C. 2015. Dynamic response of flexible square tunnels: Centrifuge testing and validation of existing design methodologies. Geotechnique, 65 (5): 401–417.

Tsinidis, G., Rovithis, E., Pitilakis, K., Chazelas, J.L. 2016c. Seismic response of box-type tunnels in soft soil: Experimental and numerical investigation. Tunnelling and Underground Space Technology, 59: 199–214.

Watanabe, K., Sawada, R., Koseki, J., 2016. Uplift mechanism of open-cut tunnel in liquefied ground and simplified method to evaluate the stability against uplifting, Soils and Foundations 56 (3): 412–426

Yang, D., Naesgaard, E., Byrne, P. M., Adalier, K., Abdoun, T. 2004. Numerical Model Verification and Calibration of George Massey Tunnel Using Centrifuge Models, Can. Geotech. J., 41: 921–942.

Yasuda, S., 2014. Allowable settlement and inclination of houses defined after the 2011 Tohoku: Pacific Ocean Earthquake in Japan. Geological and Earthquake Engineering, 28: 141–157.

Zeghal, M., Manzari, M.T., Kutter, B.L., Abdoun, T. et al. 2014. LEAP: selected data for class C calibrations and class A validations. Proc. Fourth International Conference on Geotechnical Engineering for Disaster mitigation and Rehabilitation (4th GEDMAR), 16–18 September, 2014, Kyoto, Japan.

An example of effective mentoring for research centres

C.E. Bronner, D.W. Wilson, K. Ziotopoulou, K.M. Darby, A. Sturm,
A.J. Raymond, R.W. Boulanger & J.T. DeJong
University of California, Davis, California, USA

D.M. Moug
Portland State University, Portland, Oregon, USA

J.D. Bronner
GeoEngineers, Redmond, Washington, USA

ABSTRACT: Engineering centre research faculty and staff value the importance of performing educational outreach and mentoring graduate students. However, these activities are often less structured than research projects, which leads to variable and less effective results. The geotechnical group at the University of California, Davis (UC Davis), which includes research faculty and staff at the Center for Geotechnical Modeling and the Center for Bio-mediated and Bio-inspired Geotechnics, developed a Ladder Mentoring Model (LMM) for mentoring graduate students in academic environments to enrich graduate student development while minimizing additional demands on centre personnel. The LMM is a combination of several existing mentoring models and relies on six core principles where the outcome is students receiving guidance from a variety of mentors with different areas and levels of expertise or experience. This paper provides a brief overview of the UC Davis LMM and describes how it is integrated into three critical areas of graduate student development: technical training, professional skills, and educational outreach.

1 INTRODUCTION

Training graduate students is often a central objective for engineering research centres. Traditional models for training graduate students provide limited exposure to researchers other than faculty and staff related to their thesis project. There is often minimal development of non-research skills needed for successful academic careers, such as teaching, networking, and communication skills.

Centre personnel, however, have several other responsibilities including training visiting researchers on centre equipment, preparing for and performing experiments, maintaining centre equipment, and developing researchers. While centre research experiments are meticulously designed and orchestrated, a lack of structure often exists in mentoring and educational outreach activities. Despite recognizing the importance of these latter activities to prepare future engineers and scientists and broaden participation of underrepresented groups in STEM disciplines, centre researchers may feel burdened with other demands that produce timelier, more concrete results.

To improve graduate student mentoring and educational outreach effectiveness in research centres without excessive additional demands on personnel, a restructuring of these activities is needed. This paper presents a model for organizing mentoring and outreach activities to produce researchers with the technical expertise, networks of collaborators, ability to communicate to all audiences, and other professional skills that can help them achieve their career goals. After a brief overview of common mentoring practices, an overview of the UC Davis Ladder Mentoring Model (LMM) is presented along with its six core principles. The following three sections provide examples of how the LMM is applied with six core principles at UC Davis in three different areas: technical training, professional skills development, and educational outreach. The paper concludes with ideas for transferring and tailoring UC Davis's LMM model to other institutions.

2 MENTORING IN ACADEMIC ENVIRONMENTS

Table 1 summarizes the different types of mentoring models used in academic environments (Hanover 2014; Lee et al. 2015). The primary differences between the models include the distance in expertise between the mentor and mentee, the number of mentors, the combined breadth of expertise a mentee receives, and the amount of agency a mentee has in the mentoring process.

Table 1. Common mentoring paradigms used in academia (sources: Hanover Research 2014; Lee et al. 2015).

Mentoring Model	Example
Traditional one-on-one mentoring: Mentor seen as distributer of advice/help	Faculty advisor (mentor) guides graduate student (mentee) through the academic job search process
Peer mentoring: Mentoring between two or more individuals who are considered peers or have similar status	Graduate student (mentor) trains another graduate student (mentee) on how to set up a centrifuge test
Group/collective mentoring: Combination of traditional and peer mentoring	Faculty member (mentor) coaches their graduate group (mentees) on giving research presentations; students may also guide peers
Mutual mentoring: Mentoring relationships that include a wide variety of mentors and focus on specific areas of experience and expertise. Assumes that no single individual possesses all expertise that an individual needs	An assistant faculty member (mentee) mentored by a network of individuals (mentors) that may include peers, senior faculty, administrators, etc.
Reverse mentoring: The mentor in this role is often in the role of the mentee in other situations between these two individuals	Graduate student (mentor) guides a faculty member (mentee) through a new analytical approach
Mentoring up: Similar to a traditional mentoring model, however the mentee is proactive in determining the help they need and seeking it out	Graduate student (mentee) asks faculty advisor (mentor) for help on how to develop their professional network

2.1 UC Davis Ladder Mentoring Model

Geotechnical faculty at UC Davis encourage students to act as both mentees and mentors and to work in a collaborative environment. Often, students are mentored in research by near-peers who are just a few steps up the ladder from them (e.g., another graduate student who is one- or two-years ahead of them). Over time, the program has also developed structures that have integrated the LMM into the academic, professional development, and outreach training that graduate students receive. Through the LMM, graduate students obtain many of the benefits of traditional, peer, group, mutual, and reverse mentoring models, while practicing the pro-activeness from the mentoring up model.

Recently, the UC Davis team has started studying the LMM to evaluate its benefits and to share lessons learned with other institutions. It is posited that the model works due to the integration of the following six core principles into graduate student training in research, professional development, and educational outreach activities. Examples of how these principles are applied are provided in the next three sections.

1. Providing a sustainable structure with clear expectations
2. Tailoring mentoring to needs of the individual
3. Leveraging resources generously
4. Promoting an inclusive culture
5. Encouraging consistent assessment
6. Building networks that expand beyond the borders of the institution

The three organizations in Table 2 provide structure, vision, and resources for the sustainable implementation of the six LMM principles. The Center for Bio-mediated and Bio-inspired Geotechnics (CBBG) and Center for Geotechnical Modeling (CGM) are research centres, whereas the Geotechnical Graduate Student Society (GGSS) is a student organization.

Table 2. UC Davis geotechnical organizations.

Organization	Purpose
CBBG	Transforms geotechnical practice by developing technologies that leverage natural biogeochemical processes or leveraging principles/functions/forms from natural analogues (i.e., bio-inspired), resulting in more efficient and sustainable solutions
CGM	Provides access to geotechnical modelling facilities to enable major advances in the ability to predict and improve the performance of soil and soil-structure systems affected by natural hazards
GGSS	Promotes scholarship, service, leadership, and social events to foster collaboration within the UC Davis geotechnical group

*Abbreviations: CBBG = Center for Bio-mediated and Bio-inspired Geotechnics; CGM = Center for Geotechnical Model-ing; GGSS = Geotechnical Graduate Student Society

Many individuals in the UC Davis geotechnical group are connected to one or more of these organizations.

3 TECHNICAL TRAINING

The Center for Geotechnical Modeling (CGM) serves as a resource in the National Science Foundation's Natural Hazards Engineering Research Infrastructure program (NHERI). The facility hosts researchers from across the US and provides the technical training and oversight necessary to maintain a high standard of research quality. Currently 15 students are actively working across six projects at the CGM, including six non-UC Davis students. Typically, about 10 to 15 researchers per year will rotate through the testing facility for short durations.

New researchers start with varying skill levels, academic backgrounds, and hands-on mechanical

Table 3. Typical needs of different types of CGM researchers

Researcher Type	Typical Duration	Mentoring/ Training Need	Ability to Mentor
Undergraduate student from UC Davis	10 weeks to 2 years	Very high; transitioning to medium/high	Medium
Visiting undergraduate students	6 to 10 weeks	Very high	Low
UC Davis graduate students & post-docs	10 weeks to 6 years	Medium to high; transitioning to low or medium	High
Visiting graduate students & post-docs	2 to 6 week intervals over 1 to 3 years	Often high initially; transitioning to low or medium	High
Visiting research faculty	2 weeks to 1 year	Depends on experience	High

Figure 1. A typical experiment on the 9 m centrifuge at UC Davis includes 1500 kg of soil, over 100 sensors, in-flight characterization using cone penetrometers, and multiple simulated earthquake events. Experienced researchers may spend two months building, testing, and excavating such a model. New researchers learn through apprenticeships important centrifuge modelling techniques such as how to place soils, how to calibrate sensors, how to place and log sensors during model construction, how to design a test protocol, and how to manage their test schedule and facility resources, before attempting to lead an experiment.

expertise. Table 3 describes types of CGM researchers and their typical characteristics, including the amount of time they spend at the CGM.

The CGM follows an apprenticeship model to introduce new researchers to centrifuge testing. CGM staff train new users on methods directly through annual workshops and hands-on equipment training at the start of a researcher's time on site. However, new users can still be confused even after a lesson on what to do.

The apprenticeship model grew naturally from the mutual benefits gained by experienced users needing extra assistants and new users needing practice to support their training. Apprenticeship is formally integrated into current CGM operating protocols.

3.1 CGM apprenticeship model

At the CGM, researchers are responsible for their entire physical model test program (Fig. 1). New researchers must learn physical modelling techniques, sensor and data acquisition procedures, as well develop an engineering design of their research application. Researchers, acting as project managers, learn to supervise assistant researchers, productively direct staff, work with outside vendors, and manage non-personnel resources. Given the high cost of experiments on the 9 m centrifuge, both in terms of fees and consumed effort, projects cannot afford to let new researchers learn by failure in their first experiment. Thus, new researchers serve as apprentices to experienced researchers on other models/projects to learn how to run a centrifuge test.

The apprenticeship model requires new researchers to assist an experienced researcher during an experiment. The mentee is encouraged to participate in the experiment from beginning to end so that they can learn the entire process before becoming responsible for their own test. CGM staff still provide training on equipment, but focus primarily on personal and equipment safety. Apprentices "learn while doing" within a safe, supervised environment.

The apprenticeship model benefits both the mentee and the mentor. The mentee gains the experience and training required to design their future experiment. The mentor gains the advantage of having an extra set of hands and eyes. The CGM expects all researchers to serve as both mentees and mentors, so that all can gain experience and receive the benefit of outside help.

3.2 Role of CGM

The CGM has institutionalized the expectation for the apprenticeship model by incorporating the practice into facility use rates. Projects are charged a base fee for sending a "new lead researcher" to the CGM. New lead researchers require additional orientation, training, and interaction, which consumes effort of the CGM staff. Credits against this fee are given when the researcher has the tools to be self-sufficient in order to pass on the effort savings for the centre. For example, half the fee is returned if the new lead researcher has served a full apprenticeship at the CGM. Further credits are given for other forms of formal training such as attending the annual centrifuge users' workshop and taking courses in signal conditioning.

The CGM also has a fee for "basic researcher support" intended to recover costs of CGM staff providing the extra set of helping hands when a project only sends one researcher to perform a test. Credits are given if a project provides their own assistance, such as through mentoring other users.

The well-documented apprenticeship model together with the fee structure and credit incentives

Table 4. Centrifuge mentoring experience of Kathleen Darby.

Mentor or Mentee	Position and Affiliation*	Year	Role	Primary motivation in mentorship
R. Boulanger	Faculty	2014-2018	PhD Advisor	Lead research project
J. DeJong	Faculty	2014-2018	Mentor	Co-lead research project
D. Wilson	Faculty	2014-2018	Mentor	Train students on test methods
Jackee A.	GS	2014	Mentor	Transfer knowledge on NEEShub and data analysis
Mohammad K.	GS at VT	2014, 2017	Mentor	Gain assistance, train Kate and Jaclyn on test methods
Jaclyn B.	GS	2014, 2016	Peer Mentee	Co-apprentice under Mohammad. Co-lead 1 m centrifuge tests
Daniel C.	UG	2014	Peer Mentee	CGM UG employment. Experience research and assist researchers
Yunlong W.	VS from CEA	2015	Apprentice	Learn UC Davis test methods
Maggie E.	GS at OSU	2016	Apprentice	Learn 9 m test methods
Maddie H.	UG	2016	Mentee / Assistant	CGM UG employment. Experience research and assist researchers
Mohammad K.	Postdoc	2017	Assistant	Reciprocate assistance on test
Dexter H.	UG at MSU	2017	Mentee	NHERI REU to experience research
Gabby H.	GS (CBBG)	2017	Mentee / Apprentice	Gabby, Caitlyn, Alex, and Greg: Learn general 1m test methods and specific research protocols for their projects
Caitlyn H.	GS at ASU (CBBG)	2017	Mentee / Apprentice	
Alex S.	GS	2017	Mentee / Apprentice	
Greg S.	GS	2017	Mentee / Apprentice	
Jiarui C.	GS at UIUC	2018	Apprentice	Jiarui and Soham: Learn centrifuge testing methods (shared project)
Soham B.	GS at UV	2018	Apprentice	

* Institutional affiliation is UC Davis unless otherwise listed. Abbreviations: ASU = Arizona State University; MSU = Morgan State University; OSU = Oregon State University; UCD = UC Davis; UIUC = University of Illinois – Urbana-Champaign; UV = University of Vermont; VT = Virginia Tech; CEA = China Earthquake Authority; GS = graduate student; UG – undergraduate student; VS = visiting scholar; REU = Research Experience for Undergraduates.

have proven effective in getting 100% participation by project teams from UC Davis and near 100% by external users. External teams have an added burden of paying travel costs, which reduces their apprenticeship participation rate. When possible, external teams apprentice on the 1m centrifuge, where mentees can participate from beginning to end over a shorter time. The CGM has implemented parallel operating protocols across the 1m and 9 m centrifuge so that procedural training is consistent, which has improved the apprenticeship participation of external research teams.

The CGM use fees are located on the CGM website under the "information for users" area. https://cgm.engr.ucdavis.edu/information-for-users/

3.3 Connection to the LMM

The apprenticeship model for training researchers in centrifuge techniques aligns with the LMM framework and its six core principles as described below.

Providing a sustainable structure with clear expectations: The centrifuge test pricing incentives provide the primary structure for the success of the apprenticeship model. This structure offers users a price incentive to participate both as a mentee and as a mentor, and has helped the apprenticeship model of training to become "the norm" at the CGM.

Tailoring mentoring to needs of the individual: The model allows researchers to be paired with individuals who are their near-peers with respect to the experiment they will be performing. Researchers actively work with someone performing experiments using similar techniques to those they need to learn in addition to general training. As external researchers have additional housing costs, the CGM implemented parallel operating protocols for both the 9 m and 1 m centrifuge to allow researchers to train on either centrifuge. This flexibility reinforces the structure by making the program feasible for internal and external researchers.

The graduate students involved in mentoring develop advising skills, which is particularly important for those who plan to enter academia or serve in leadership roles. The high number of mentees a CGM graduate student mentors provides them more opportunity to develop their teaching style. Table 4 provides an example of doctoral student's mentoring experiences.

Leveraging resources generously: Leveraging of resources occurs between UC Davis and visiting centrifuge researchers. Through the apprenticeship program, a researcher is provided a necessary assistant at no cost, while another researcher receives training in centrifuge methods and a credit towards the cost of their centrifuge tests. The two projects benefit from reduced costs and the CGM staff can better utilize their expertise in centre operation and technical research advancement.

Promoting an inclusive culture: The apprenticeship model provides the opportunity to involve researchers from a broad range of backgrounds, abilities, expertise, and development levels. For example, the apprenticeship model allows for the inclusion of undergraduates in centrifuge research. Typically, undergraduates are not able to commit the time or flexible schedule needed to participate in centrifuge experiments. They can, however, offer valuable assistance as the third member of a centrifuge team while gaining valuable research experience. Provision of the primary assistance by the apprenticeship program produces more opportunities for undergraduates to work as an extra assistant when their schedule permits.

Encouraging consistent assessment: The CGM has a stated performance goal of developing its members for the future workforce. Objectives toward this goal include providing ladder mentoring toward the development of independent researchers, engaging researchers in education and outreach activities (EOT) (to be discussed later), and providing technical training on all facets of geotechnical centrifuge testing. Progress is assessed by tracking the percentage of teams with ladder-mentored lead or assistant researchers (target >90%, actual 10 of 11 since 2016), percentage of users engaged in EOT (target >50%, actual >75% since 2016), and through user satisfaction surveys (target > 90% of users satisfied or very satisfied with training, actual surveying has been informal to date). Our user surveys to date have indicated strong support for the apprenticeship model, but also a consistent desire for improved documentation.

The UC Davis geotechnical group is now working to improve and expand assessment of the ladder mentoring program across all activities in an effort to better quantify its impact on preparing its members for the twenty-first century workforce.

Building networks that expand beyond the borders of the institution: The CBBG and CGM both include participation by researchers across the US. These activities give users valuable opportunities to work with people from diverse institutions and academic backgrounds (Fig. 2). Anecdotal observations indicate that knowledge, beyond centrifuge testing skills, is being broadly disseminated and wide-reaching networks are being developed.

3.4 *Example: Experience of a graduate student*

To demonstrate the potential impact of the apprenticeship model, Table 4 highlights the centrifuge-related mentoring experiences of a graduate student participating in both the CGM and CBBG, Kathleen Darby. Her research included centrifuge tests over a period of five years. As Ms. Darby progressed through her graduate work, she worked with 17 different researchers (three faculty, one visiting scholar, one post-doc, nine graduate students, four undergraduate students) from eight different institutions covering a range of research roles, as described in the table.

Figure 2. Ladder mentoring in practice. Visiting PhD student Mohammad K. (VaTech) led an experiment looking at ground improvement using soil cement. He mentored three engineers during the test and benefited from the depth of support available for a complicated test. Dr Wang, a visiting scholar, gained experience in how to perform centrifuge testing that he would take back to his new centrifuge in China. Kate D., as an MS student, apprenticed during the experiment so that she could lead her own tests on the 1 m centrifuge and eventually the 9 m centrifuge as a PhD student. Daniel C. gained valuable research experience as an undergraduate and ultimately decided to further pursue his education as an MS student.

Due to the CGM's apprenticeship model, Ms. Darby's contact with researchers at several institutions allowed her to gain and distribute centrifuge-related skills beyond the boundaries of the CGM. She received mentorship from researchers within and outside of UC Davis, including initially serving as an apprentice under a visiting graduate student. As a graduate student, she mentored undergraduate and graduate students from seven different institutions, including several who apprenticed with her or were supervised by her.

4 PROFESSIONAL SKILLS DEVELOPMENT

The UC Davis geotechnical graduate program typically consists of about 30 graduate students and six full-time faculty, serving as their graduate advisors. A traditional mentoring system where knowledge transfer occurs only from faculty member to student would lead to limitations on mentoring in professional skills, such as restrictions based on faculty time constraints and variability based on an advisor's individual sense of importance for specific skills. Expansion of a mentoring system to include knowledge transfer between peers and research staff increases development of and feedback on professional skills.

At UC Davis, the Geotechnical Graduate Student Society (GGSS) provides an additional structure for graduate student professional development. The GGSS program actively fosters leadership, outreach, and mentorship skills in its members, making them better qualified and well-rounded to graduate to professional or academic careers. The organization's practices align with the LMM core principles and expand support originally provided through geotechnical faculty members and the CGM.

4.1 Geotechnical graduate student society

In 2007, the UC Davis geotechnical engineering faculty members guided the graduate students in initiating the GGSS to formalize and focus the activities used to develop the professional skills of graduate students. The goal of the GGSS is to promote scholarship, service, leadership, and social events for the geotechnical group at UC Davis. The intention is to foster community and collaboration, and provide opportunities to promote graduate student education and professional development.

The GGSS is governed by a board consisting of six officers: President, Treasurer, Seminar Coordinator, Social Events Coordinator, Field Trip Coordinator, and Outreach Coordinator. Each officer has clearly defined responsibilities and opportunities. For example, the seminar coordinator recruits and hosts seminar speakers, which allows them to develop a professional network that they can leverage for employment opportunities as they near graduation. The GGSS board is mentored by the faculty advisor.

Faculty members rotate the responsibility of GGSS faculty advisor so that the workload is fairly distributed. The faculty advisor provides historical context and advice to the students as they navigate their new roles. While the faculty advisor will always be a critical role, the GGSS board retains continuity of some members from year to year and draws on advice from past officers. The officers are usually established senior graduate students who in turn serve as mentors to junior officers and new GGSS members. New officers are elected in April and current officers end their terms the following June to ensure there is training time for new officers.

The GGSS organizes a variety of events including a weekly seminar series, field trips, educational outreach activities, and social outings, which diversifies the expertise and experiences to which graduate students are exposed. The largest GGSS event is the annual Round Table where about 80 geotechnical professionals from government and industry are invited to a full day of student presentations, poster sessions, panel discussions, and closing social. The goal of the Round Table is to foster connections between UC Davis researchers and leading professionals by providing opportunities for open conversations, exposure and feedback on current research, exchanges or collaborations, and connections among future colleagues.

4.2 Role of CGM

The CGM supports the goals of the GGSS by providing connections and institutional knowledge. Networking opportunities include interacting with visiting scholars at the CGM, utilizing the growing network of professional contacts when planning GGSS field trips and seminars, and connecting GGSS members with long-term educational contacts for outreach events.

The institutional history provided by the CGM was instrumental for the GGSS when developing its educational outreach program as it could build off the centre's previous experience and existing connections. Graduate students learned whom to contact and which activities had been the most successful.

4.3 Connection to the LLM

GGSS mentoring relationships strongly rely on characteristics of the mutual mentoring, peer mentoring, and mentoring up models. The GGSS structure relies on the six core principles in the LMM to provide effective professional skills development for graduate students.

Providing a sustainable structure with clear expectations: The GGSS provides a structure, outlined in its bylaws, with clear roles and responsibilities of officers. The election process and officer overlap period provide continuity for the organization and minimize the possibility of knowledge loss when students graduate.

Geotechnical faculty and current graduate students set a clear expectation that all graduate students in the group should be active participants in the GGSS. If students are not attending seminars, their faculty advisor is responsible for strongly encouraging their attendance, often through a reminder of the benefit they are missing out on. The importance of participation in the GGSS is highlighted from their first day on campus; prospective graduate student campus visits include attendance at a GGSS organized activities such as the Round Table event or a weekly seminar.

Tailoring mentoring to needs of the individual: Students in the UC Davis geotechnical group vary based on their experiences, career ambitions, and desired professional skills. The GGSS offers a variety of involvement levels, which requires students proactively decide how much they can or want to contribute and gain from the GGSS at a given time in their graduate study.

As a baseline, all students are expected to attend the weekly seminars and the Annual Round Table, which together provide essential exposure to professional practice and opportunities for networking. Note that all seminar speakers are taken to lunch by a group of two to four GGSS members, so all students have opportunities for establishing personal connections with various professionals during the year. In addition, GGSS members can participate in some combination of the field trips and outreach events held throughout the year, with that mix varying from year to year. For example, an MS/PhD student may only have time to participate in one or two outreach events in their first year (due to class workload), may participate more heavily for the next year or two, and then participate less frequently in the last year or two depending on other commitments or roles they assume. The same MS/PhD student may serve in an officer role (e.g., seminar coordinator) in their second or third year, followed by a second officer role (e.g., president) in the fourth or fifth year.

Additionally, the GGSS structure allows students to work on specific professional skills that they want to improve. For example, a student who has difficulty communicating their research to non-technical audiences may choose to participate in outreach activities to practice these skills. Another student who struggles in professional networking situations may become the seminar coordinator to hone these skills in a supportive environment.

Leveraging resources generously: Both the CGM and CBBG have responsibilities related to the professional development of graduate students. By these centres supporting the GGSS and encouraging their students to be active members, they leverage the enthusiasm of graduate students and provide a structured approach to professional development.

The GGSS, CGM, and CBBG also leverage resources for providing professional development. The Round Table event provides the majority of the funds for GGSS activities; the event's success is partially due to the reputation of the CGM and research faculty. CBBG resources (e.g., webinars) for supporting professional development of its students are often shared with other GGSS members. The CBBG also provides funding resources to support the outreach activities of the GGSS; these activities are further discussed in Section 5.

Promoting an inclusive culture: GGSS members actively recruit new graduate students as members. Their commitment to inclusivity is demonstrated by the policy in their bylaws that automatically makes any registered UC Davis geotechnical graduate student a voting member of the GGSS. The GGSS also has a practice of inviting visiting students and scholars to participate in GGSS activities as honorary members while they are in Davis.

The culture of inclusion is demonstrated through diverse leadership in the GGSS. Nationally, 20% of civil engineering graduate degrees are earned by women. Currently, 50% of the GGSS officers are women and 33% are underrepresented minorities. Three of the past seven presidents have been women. These statistics indicate that women and other underrepresented groups in engineering are supported and actively participating in the GGSS. GGSS students further stress the importance of inclusion by including presentations on topics such as inclusion in engineering education and impostor phenomenon in their seminar program.

Encouraging consistent assessment: After every large GGSS event, students host a debriefing session to identify strengths, weaknesses, and opportunities for improvement. Feedback on weekly seminars and social events is provided during quarterly GGSS board meetings. This consistent assessment followed by action to address concerns leads to ever-improving, high-quality events. For larger events, surveys are distributed to collect participant feedback and include their input in the debriefing meetings.

Building networks that expand beyond the borders of the institution: The GGSS members have helped expand the influence of the UC Davis geotechnical program beyond the institution's borders. Due to the GGSS's success at UC Davis, CBBG faculty and students used the GGSS as the model when designing the engineering research centre's Student Leadership Council (SLC). The SLC consists of graduate student and undergraduate student representatives from all CBBG partner institutions: Arizona State University, Georgia Institute of Technology, New Mexico State University, and UC Davis. To help establish similar expectations and a culture of inclusion in the SLC, UC Davis students, Michael Gomez and Alena Raymond, served as the president for the first and second years of the centre, respectively. Additional plans for expansion include collaborating with GGSS alumni now working at other universities to help establish a similar graduate student organization at their universities.

The professional network for UC Davis researchers has expanded through positive interactions of geotechnical professionals with students during seminars, field trips, professional and K-12 outreach activities, and the Round Table event. This reputation has helped a large percentage of students secure jobs before graduating; about 90% of master's students are hired by companies who attend the Round Table.

4.4 Example: Round Table event

The GGSS's Annual Round Table event foster connections between leading geotechnical professionals and UC Davis faculty and graduate students by providing opportunities for open conversations, exposure and feedback on current research, exchanges or collaborations, and connections among future colleagues. During the event, geotechnical graduate students present their research to professionals from industry, consulting firms, and government organizations through poster and oral presentations. The event also includes an industry panel discussion and social activities.

Round Table guests provide gifts that go to an account overseen by the civil and environmental engineering department, but controlled by the GGSS, and those funds support the GGSS activities throughout the year. These generous gifts reflect the fact the community has embraced the Round Table as an event they look forward to, they like to support the broader educational experience of graduate students, and they like the personal connections that lead to either hires or connections with future colleagues.

GGSS students plan and run all portions of the Round Table, which requires students to interact with professionals, plan out all logistics for the event, and develop an engaging program. Each year the GGSS President leads the event, however successful implementation requires a coordinated effort from all GGSS members. In their first year at UC Davis, students' participation at a minimum includes creating an abstract and poster presentation, informal conversations with professionals, and observations of their senior GGSS peers. By their second year, students will take on more responsibilities and may eventually lead the event or

Table 5. Different levels of GGSS member participation during Round Table.

Level of Involvement	Description of Mentoring	Role
First year graduate student	Mentoring focuses primarily on preparing individuals to present their research to a professional audience in a clear and engaging manner, including through their design of a research poster. Mentoring comes from faculty advisors and fellow GGSS students. Students make minimal contributions to larger planning efforts, mainly observing their peers.	Mentee
2+ years as graduate student	With respect to interactions with industry and poster preparation, students transition from mentee to mentor roles. Students receive mentoring from faculty advisors and fellow GGSS students on poster and/or oral presentations. Students make minimal contributions to larger planning efforts, mainly observing their peers.	Mentee & Mentor
GGSS Officer	Mentored by GGSS faculty advisor and provides mentoring to junior GGSS officers and members. Students contribute to larger planning efforts, such as program design and implementation and contacting professionals	Mentee & Mentor
GGSS President	Mentored by GGSS faculty advisor and provides mentoring to junior GGSS officers and members. Student is responsible for the event.	Mentee & Mentor

Table 6. Mentoring interactions initiated due to Round Table.

Mentoring Interactions at Round Table
Prior to Event
• GGSS past/senior officers mentor new officers on logistical processes involved, as well as how to handle moments of stress (near-peer mentoring)
• GGSS faculty advisor mentors GGSS president through check-in meetings and advising on logistics, especially those related to industry (traditional mentoring)
• GGSS senior members mentor new members on preparing research posters and how to interact with industry (near-peer mentoring)
• GGSS members give feedback to each other on their posters and presentations (peer mentoring)
During & Post Event
• Industry members and faculty members provide feedback and advice to graduate students on their research projects (form of mutual mentoring) – potentially forming new research contacts
• Faculty and GGSS members provide constructive feedback to each other on Round Table execution – strengths, weaknesses, opportunities (form of collective mentoring)

give one of the keynote presentations. Table 5 provides a potential Round Table path for GGSS members over their academic journey.

Table 6 lists examples of ladder mentoring interactions that occur during the preparation and implementation of the Round Table.

5 EDUCATIONAL OUTREACH

In addition to the technical training and professional development of graduate students, the mission of engineering research centres often includes providing service to the profession in the form of educational outreach activities. Despite good intentions, outreach activities are often ad hoc and their impact is seldom assessed. Funding agencies, such as the US National Science Foundation (NSF), are increasing the burden of evidence for demonstrating the impact of outreach efforts. Throughout its history, the CGM has and continues to provide hands-on tours of facilities to K to 12 students (US primary and secondary school levels, typical ages 5-17). Over time, these outreach events have added structure by rotating attendees through discrete stations, each led by a volunteer geotechnical graduate student. After its establishment, the GGSS took over the organization of outreach activities at the CGM. The post-activity assessment of outreach events includes discussion of what worked and what did not after each tour, but does not include assessment of activity learning outcomes.

In 2015, the UC Davis geotechnical group began transitioning to a more strategic approach to educational outreach due to three factors: the start of the CBBG, the hiring of a department faculty member with expertise in assessment, and the creation of a GGSS outreach officer position. One program in development is a graduate-level engineering education course in which students design educational activities to be implemented in annual outreach activities performed by the GGSS.

5.1 History of UC Davis geotechnical engineering outreach program

Before the GGSS began, CGM faculty, staff, and students developed relationships with local secondary schools and invited them on tours of CGM facilities (Fig. 3). They developed a series of modules for participants to rotate through. Modules are tailored to the needs of the participants, and more formal presentations on geotechnical earthquake engineering can be included. The most successful modules include significant physical interaction, while a tour of the 9 m centrifuge can impress students simply with its scale.

Current modules include a shake table where participants build structures with K'nex, a create your own earthquake station where participants jump on

Figure 3. Kathleen Darby (centre) leading an outreach module during a tour by middle school students during one of her experiments on the 1m centrifuge. Jaclyn B. (peer/mentee) and Mohammad K. (mentor/visiting graduate student) also participated in this tour event.

Table 7. Contributions to UC Davis geotechnical educational outreach.

Organization	Structure provided
CGM	Access to physical facility; institutional memory; technical support for demos
CBBG	Funded education-focused project; graduate course in engineering education; expectation of CBBG students to participate in two events per year
GGSS	Annual outreach coordinator; supply of volunteers
Department	Supporting tenure-track faculty hire in civil engineering education

an instrumented pad, a CGM module explaining the centrifuge and how it works, a CBBG module with bio-cemented sands, and a liquefaction module where users liquefy soil in a bucket to induce foundation failures.

With the creation of the GGSS, the students took over organizing the outreach events with the assistance of CGM staff. In 2014, the GGSS created an officer position for outreach coordinator. The result of these efforts was a time-efficient outreach system where new geotechnical students were trained on how to run different stations as they became involved in research. The participation in the activities provided opportunities for students to communicate technical topics to an audience with no or limited understanding of engineering. The direct interaction with K to 12 educators also exposes the graduate students to the curricular requirements of K to 12 education in the US.

The geotechnical group, however, did see a need for more intentional outreach that maximized impact without exhausting CGM staff and GGSS students. In 2015, the funding of the CBBG increased external demands for inclusive educational outreach and assessments of outreach efforts. This change coincided with the department hiring of a faculty member with an expertise in pedagogy and assessment.

Early steps have included the design of a two-course sequence for engineering graduate students in *Engineering Education Design (*discussed in section 5.4), intentional targeting of outreach activities to where they will have the most value, and developing tools for assessing outreach.

5.2 Role of CGM

As noted earlier, the CGM was the catalyst for early outreach efforts. Most connections with educators occurred organically. For example, one CGM development engineer, Tom Kohnke, initiated a now annual visit from a local high school where his daughter was attending. CGM personnel and students developed the first versions of the educational modules, and the facility attracted groups to UC Davis. The CGM currently support GGSS graduate students by providing access to the facility for tours and providing maintenance on outreach equipment (e.g., the shake table).

5.3 Connection to the LMM

Aligning the educational outreach program with the LMM maintains the sustainability of the program and trains graduate students to communicate their research to non-technical audiences.

Providing a sustainable structure with clear expectations: The structure for the outreach efforts are provided by the three geotechnical organizations and the UC Davis Civil and Environmental Engineering Department (Table 7). One of the most important factors is the expectation that graduate students participate in educational outreach, which allows more outreach to occur than if it were performed only by centre personnel.

Tailoring mentoring to needs of the individual: As with other GGSS activities, the level of involvement in educational outreach activities is flexible. Students with minimal interest may only participate in a couple of outreach events each year and receive basic training from more experienced GGSS members. However, students with a strong interest in outreach or teaching may enrol in the graduate course sequence and serve as GGSS outreach coordinator. More active students will have multiple mentors coaching them, including both geotechnical engineering faculty and a faculty member with expertise in engineering education.

Leveraging resources generously: For outreach programs and associated mentoring interactions to be sustainable, they must leverage funding, equipment, space, time, and expertise. The CGM and CBBG both contribute funding related to outreach activities. The CGM primarily funds equipment maintenance, some supplies, and contributes staff effort. The CBBG funds workshops, undergraduate assistants to help design and organize outreach events, and new module development, and provides faculty support.

Expertise is leveraged in the design of modules and training of graduate students. Modules depend on the technical expertise of the geotechnical graduate students and faculty and the engineering education expertise of an environmental engineering faculty member. By finding someone with an educational design and assessment background, the geotechnical group can more efficiently train their students and assess the impact of their activities. The CGM provides expertise and support in maintaining the equipment used for outreach and providing a facility for on-campus outreach activities.

Both the CGM and CBBG are required to perform educational outreach and contribute to broadening participation of underrepresented groups in geotechnical engineering. By working together and with the GGSS and pooling resources, different types of expertise are exchanged and activities are more strategically designed with respect to time and impact.

Promoting an inclusive culture: All three organizations are committed to an inclusive culture, both for participants in the outreach activities and for the graduate students, staff, and faculty involved.

Outreach activities typically are targeted at populations underrepresented in engineering, including students who are female, from an underrepresented minority or ethnicity, from low-income families, have a disability, or who would be the first in their family to go to a four-year university or graduate school. Examples of inclusive actions include partnering with schools where many students come from low-socioeconomic backgrounds and a one-week sustainable engineering academy designed for girls entering grades seven to nine.

Outreach activities are an opportunity for students to see role models with similar backgrounds to their own, and to envision themselves in similar roles. For example, in California, where approximately 50% of elementary students are Hispanic or Latino, it is important that some of our participating graduate students are Hispanic or Latino. The diverse group of geotechnical graduate students allows students to find someone who shares some characteristics with them. Currently 75% of UC Davis CBBG graduate students are female and 25% are Hispanic or Latino. The US averages for civil engineering graduate students are 24% and 12%, respectively (National Science Foundation 2017). Additionally, some of our outreach activities highlight the impact of less-known female civil engineers (e.g., Emily Roebling) to provide historical role models.

Recognizing and valuing the different areas of expertise needed for effective outreach, graduate students receive mentoring from each other, faculty, and secondary teachers in how to integrate the culture of inclusion into their educational modules. Examples of inclusive designs include designing flexible lesson components or challenges that can be increased or decreased in complexity and incorporating best practices for inclusive teaching in both the design and implementation of the module.

Encouraging consistent assessment: As with research experiments, assessment and evaluation are necessary to understand the results and make improvements. Assessment data has been collected from outreach participants through observations, surveys, and engineering assignments. For example, some of the modules ask participants to answer questions before and after the activity to determine if the learning outcomes are reached. In addition to these methods, assessment of secondary teacher feedback was collected through discussions on specific modules and on overcoming barriers to productive collaborations between the university and secondary schools. Graduate students are assessed in the engineering education course through reflection assignments and the process they use to design their educational module.

Through the assessment process there have been numerous lessons learned. Evaluation based on assessment data from the Sustainable Engineering Academy for Girls led to a modified recruitment plan, increasing the ages targeted, changing the duration from four to five days, adjusting the target number of participants to 15, and modifying educational modules for future implementations. The increased target age group was observed to be appropriate as students had the fundamental math skills desired for some activities (e.g., a Life Cycle Assessment activity). The older students also had a larger attention span and were all highly interested in science.

The recruitment strategy, based on conversations with middle school teachers, was modified in 2017 to have teachers nominate students for participation. Students came from five different schools and three different grades. Students were more racially and ethnically diverse than in 2016; 33.3% of students in the 2017 cohort were from underrepresented minorities and two of the students had disabilities.

As graduate students work with faculty in the assessment phase, they are mentored in the iterative process that is required when designing instructional activities. Graduate students also learn of the great impact of non-technical factors on the success of educational activities (e.g., the length of student's attention span, emotional needs of students, and preparing for sometimes random remarks/questions from students).

Building networks that expand beyond the borders of the institution: Through CBBG partner institutions, best practices and lessons learned are exchanged with respect to outreach design and implementation. That network also allows for an expanded library of educational modules.

While the CGM already had a network of secondary teachers, the revised outreach program has expanded the network and provided the teachers agency. They now are mentors and mentees in the overall LMM of the geotechnical group. By providing interactions during the academic year, hopefully these relationships will be strengthened and sustained. One mechanism for maintain relationships with secondary teachers is through the development of K to 12 educational modules.

Table 8. Summary of LLM Core Principles integration into the UC Davis geotechnical program.

Principle	Implementation in UC Davis Geotechnical Group
Providing a sustainable structure with clear expectations	Each program has multiple structures that provide clear roles or expectations
Tailoring mentoring to needs of the individual	Flexible options for participation depending on interests and needs of individuals
Leveraging resources generously	Financial, time, space, and expertise resources are leveraged
Promoting an inclusive culture	Common focus on increasing access to broaden participation
Encouraging consistent assessment	Assessment occurs in all activities and is increasing in rigor with time
Building networks that expand beyond the borders of the institution	Partners include other academic institutions and personnel, industry partners, secondary education teachers, etc.

5.4 Development of educational modules

A two-course sequence in *Engineering Education Design* was designed for graduate engineering students to offer guidance in intentional engineering educational design. The first course introduces students to engineering education topics (e.g., student learning outcomes (SLOs) and assessment, types of learning and communication styles, active learning strategies, project-based learning, and creating inclusive environments).

In the second course, students design educational outreach modules related to their research that target specific age groups, align SLOs with state or national education standards, and include SLO assessment strategies. After developing a draft of their modules, students pilot their designs for a sample target audience. In the past, pilot events have included a public outreach event and a one-week engineering academy for secondary school girls. The course has been offered twice, with plans to offer it annually.

It is necessary that students designing educational modules for elementary and secondary school levels receive feedback from teachers at these levels, as they are most knowledgeable on what would work and what is most important to cover in their classrooms. To provide this input, secondary school teachers participated in a one-week summer workshop in 2016 and 2017.

The workshop format included one to two graduate students teaching their educational modules each day followed by discussion on those modules. Other workshop activities introduced participants to the topics of engineering, civil engineering, geotechnical engineering, sustainability, and underrepresented groups in engineering (especially women). At the end of each day, facilitated discussions with teachers led to: 1) developing strategies for integrating workshop activities and content into lesson plans, 2) strategizing methods for involving underrepresented groups in outreach activities, 3) identifying potential partnerships between UC Davis and local schools, and 4) obtaining feedback for graduate students on the modules they presented.

The workshops achieved three main outcomes: 1) graduate students increased teachers' confidence to teach engineering in their classrooms, 2) teachers provided practical feedback on the modules designed by the graduate students, and 3) partnerships between teachers and the geotechnical group were nurtured. After incorporating feedback from teachers, graduate students revised their modules for future implementations.

Current modules are in the iterative revision and testing phase familiar to most engineers. Although some students have graduated, the modules remain part of the GGSS/CBBG/CGM library of activities. When the final educational modules are complete, they will be submitted to TeachEngineering (https://www.teachengineering.org/), a web-based digital library of standards-based engineering K to 12 curricula.

After evaluating activities from the past two years, an adjustment has been made to encourage more continuous interactions with secondary teachers (e.g., student visits to UC Davis, teachers attending some of the classes in the improved graduate student course, visits to science classes in the teachers' schools, teachers providing direct feedback on modules during the graduate course). One improvement implemented is a Google Form created in which teachers can submit requests for borrowing outreach equipment, touring UC Davis facilities including the CGM, and having undergraduate and graduate students visit their classrooms. There have also been improvements to assessing outreach activities and their impacts. For example, an online outreach form that the GGSS outreach coordinator fills out after each event maintains a record of all information needed for reporting to NSF and observations about the activity's implementation (e.g., features that could be improved).

6 SUMMARY AND FUTURE WORK

The Ladder Mentoring Model presented herein has provided a formal structuring of mentoring and outreach activities toward producing researchers with the technical expertise, networks of collaborators, ability to communicate to all audiences, and other professional skills that can help them achieve their career goals. While the specific mechanisms vary, the three different programs described herein address the six core principles of our LMM (Table 8).

The results in Table 8 are an initial effort to characterize the LMM at UC Davis. However, CGM and CBBG researchers continue to investigate impacts and perceptions of the LMM through surveys and interviews of current and past geotechnical engineering graduate students. The goal of these studies is to evaluate how and why the LMM model has been successful at UC Davis. Factors under investigation include quantifying mentoring interactions, understanding graduate student participation in program activities, and student perception of mentoring activities.

Future work will include piloting the LMM framework beyond UC Davis. GGSS alumni are now in faculty positions at other universities and we are making plans with them to pilot programs featuring the core principles at their institutions.

ACKNOWLEDGEMENTS

The work presented herein is the culmination of contributions from many participants over years of activity at UC Davis. We would particularly like to acknowledge the contributions of all current and past officers and members of the UC Davis Geotechnical Graduate Students Society, the teachers who bring their classes to UC Davis for outreach activities, whether one time or returning each year, and the broader engineering community who support the GGSS through activities such as the annual Round Table and weekly seminar series. We would also like to acknowledge the contributions of Professor Bruce Kutter, past director of the CGM and mentor to many centrifuge users and graduate students, and the entire geotechnical group at UC Davis.

The Center for Geotechnical Modeling is supported by the National Science Foundation as part of the Natural Hazards Engineering Research Infrastructure program through cooperative agreement CMMI-1520581.

The Center for Bio-mediated and Bio-inspired Geotechnics is supported by the National Science Foundation's Engineering Research Centre program under NSF cooperative agreement EEC-1449501.

REFERENCES

Hanover Research, 2014. Faculty Mentoring Models and Effective Practices. Washington, DC. Available at http://www.hanoverresearch.com/media/Faculty-Mentoring-Models-and-Effectives-Practices-Hanover-Research.pdf

Lee, S.P., McGee, R., Pfund, C. & Branchaw, J. 2015. "Mentoring Up": Learning to Manage Your Mentoring Relationships. In Wright, G. (ed.) The Mentoring Continuum: From Graduate School Through Tenure. Graduate School of Press of Syracuse University. Syracuse, NY Available at: http://graduateschool.syr.edu/wp-content/uploads/2017/03/Lee-et-al..pdf

National Science Foundation, National Center for Science and Engineering Statistics. 2017. *Women, Minorities, and Persons with Disabilities in Science and Engineering: 2017*. Special Report NSF 17-310. Arlington, VA. Available at www.nsf.gov/statistics/wmpd/.

Geotechnical modelling for offshore renewables

C. Gaudin, C.D. O'Loughlin & B. Bienen
Centre for Offshore Foundation Systems, The University of Western Australia, Perth, Australia

ABSTRACT: Centrifuge modelling has been used extensively over the last five decades to address offshore geotechnical challenges associated with oil and gas developments. In recent years, the development of offshore renewable energy devices and structures, including wind turbines and wave energy converters has increasingly mobilised the offshore geotechnical engineering community. This paper revisits the use of centrifuge modelling for offshore geotechnics in the light of the new challenges raised by offshore renewable energy developments. This is illustrated through some aspects of foundation loading regimes such as dynamic tensile loading and multidirectional loading over a large number of cycles, which are specific to offshore renewable energy applications. The emphasis is on the modelling techniques developed to address these challenges and the opportunities provided by centrifuge modelling.

1 INTRODUCTION

1.1 Historical background

Centrifuge modelling for offshore geotechnics has historically been driven by the needs and requirements of the oil and gas industry. The large size of offshore infrastructure, the emphasis on failure in design and the complexity of loading regimes have been key challenges to address, for which centrifuge modelling is particularly well suited. The modelling undertaken, focused first on phenomenological and site-specific studies (see the first use of centrifuge modelling in offshore geotechnics in Manchester University in 1973 and reported in Rowe & Craig, 1981) and developed progressively towards more general investigations to better understand soil-structure interaction, observe failure mechanisms or provide performance data.

The development of centrifuge modelling for offshore geotechnics has been well documented through the years, notably by Murff (1996), Martin (2001) and Gaudin et al. (2010). They present in particular a comprehensive review of the use of centrifuge modelling that can be categorised as:

- *Identification/Observation*: to develop an initial understanding of the engineering concern, to identify a particular failure mechanism such that an appropriate analytical solution can be developed, or to observe a particular mode of soil behaviour (e.g. is the response drained or undrained, does the soil flow or collapse?).
- *Validation*: of a technical solution (type and geometry of structure) or of a mode of behaviour upon which the design was based (i.e. mode of collapse).
- *Generation*: of performance data that can be used to calibrate numerical models, generate design charts, or understand the relative importance of particular parameters in the global geotechnical response.

Acceptance and awareness in the offshore oil and gas community (both industry and academic) of the benefits of centrifuge modelling has grown significantly over the past two decades. This is partly due to scientific and technical developments associated with motion control, instrumentation and data acquisition that has enabled more realistic and sophisticated modelling, but also due to an increasing need for performance data and understanding of offshore soil structure interaction. This awareness appears however to be limited within the offshore renewable energy community, evident by the very limited body of literature of offshore renewable energy studies involving centrifuge modelling that are driven by industry.

This paper revisits the use of centrifuge modelling for offshore geotechnics in the light of the new challenges raised by offshore renewable energy developments. This is illustrated using examples of foundation loading that are specific to offshore renewable energy applications, including dynamic tensile loading, and multidirectional loading over extremely high numbers of loading cycles. The emphasis of the paper is on the modelling techniques developed to address these challenges and the opportunities provided by centrifuge modelling, notably with respect to the economical constraints that are faced by the renewable energy industry.

1.2 The transition to renewables

In an era of escalating energy demand and climate change, securing the supply of low-emission energy is one of the major challenges of our generation. The world's oceans offer a largely untapped resource, with enormous potential for energy solutions. The most rapid development in offshore renewable energy has been in offshore wind (+2.2 GW in 2016, for a total of 14.4 GW, +3.5 GW in 2017), with most of the installed offshore wind capacity in European waters (+1.57 GW across 7 windfarms, +813 MW in Germany alone). The worldwide wind capacity reached 486 GW by the end of 2016, with a growth rate of 11.8 % (WWEA 2017). In the majority of these offshore developments monopile foundations are favoured (constituting 97% of the foundations for wind turbines installed in 2015) due to the shallow water depth (<30 m). The developments went hand in hand with ever increasing pile diameters, with 8 m diameters now relatively common, constantly redefining the boundaries of what is possible.

As the industry evolves, offshore wind farms will be sited further from the coast and in deeper waters (>50 m), requiring floating facilities that are moored to the seabed with anchors. The advantages of floating systems are (i) better energy resources can be tapped due to winds becoming higher and more consistent with distance offshore, (ii) larger wind turbines (8-10 MW) can be installed, thus increasing energy production, and (iii) maintenance costs can be potentially reduced, as turbines can be untethered and towed to shore. Similar trends are forecast for wave energy converters (WECs). The industry will need to transition from single or small-array demonstrator units (of moderate scale and power capacity) towards integrated arrays of larger, full-scale devices to realise commercially viable energy generation. This introduces a need to design multiple closely spaced foundations in water depths able to accommodate the typically larger draft of full scale WECs (up to depths of ~100 m).

Renewable energy generation from floating systems has been proven, and includes for example: (i) the Hywind spar floating wind turbine, which has been in operation in 198 m of water off the southwest coast of Norway since 2009, (ii) the WindFloat semi-submersible floating wind turbine that has been tested in 40-50 m water depths off the coast of Portugal since 2011, (iii) the Ocean Power Technology floating wave energy device, which has been tested off the coasts of Hawaii, USA and Scotland in water depths of up to 30 m since 2005, and (iv) the Perth Wave Energy Project from Carnegie Wave Energy, with three 240 kW WECs operating over 12 months offshore Garden Island in Western Australia in 2015. These small projects have aimed to demonstrate concept feasibility such that commercial developments can be expected in the coming decades.

1.3 The economic constraints

Previous offshore wind farm developments were subsidised, but now need to prove themselves to be competitive with other energy sources. This is indeed the case, with the cost of offshore wind power reported to be lower than that of nuclear power in the UK (BBC 2017). As these wind farms are large, with perhaps 200 turbines, even small improvements in design translate to large economic savings. Similarly, offshore floating renewables require reliable and economical anchoring systems that can perform in the type of seabed sediments encountered on the continental shelf, where floating renewables are expected to operate. Anchoring systems can contribute up to 22% of the total installed cost of an offshore wind turbine (Willow & Valpy 2011), and up to 30% of the total installed cost of a wave energy converter (Martinelli et al. 2012). This is one order of magnitude higher than for oil and gas structures (Kost et al. 2013), and contributes significantly to the high levelised cost of offshore wind and wave energy. The US Energy Information Administration forecasts a levelised cost of offshore wind energy of US$158/MWh (US$64.5/MWh for onshore wind) for plant entering service in 2022 (Energy Information Administration, 2017), while the cost of natural gas ranges from US$56/MWh to US$105/MWh. Without large commercial scale installations, the cost of wave energy is harder to forecast and varies significantly between the various types of converters (and capital cost) and between forecasters. For an array of 100 point absorbers, the cost has been estimated at around US$800/MWh (Neary et al. 2014), although a case study using an oscillating water column offshore Portugal estimated a cost as low as US$86/MWh (Castro-Santos et al. 2015).

A large volume of research is being undertaken to improve the efficiency of wind turbines and WECs, but considering the significant fraction of the capital cost they represent, savings in foundation engineering could potentially have a significant impact on the LCOE, provided a step wise improvement in technology and design is achieved (Gaudin et al. 2017).

Foundation design for offshore renewables is currently based on the knowledge and technology developed for and by the oil and gas industry over the last 50 years. This is reflected in the large number of rules and guidelines applicable to floating renewables, which overwhelmingly refer directly to oil and gas guidelines, such as API (2008) for mooring analysis for station keeping and API (2014) for foundation design. Neither guideline suggests adaptation for renewable energy.

A number of publications have emerged over the last few years (Stevenson et al. 2015; Knappet et al. 2015; Diaz et al. 2016), listing the existing anchoring solutions and design methodologies that would be suitable for floating renewables, highlighting particularities that require further investigation. Recently, an additional body of research has started to focus on the specific aspects of offshore renewables that are fundamentally different than oil and gas, such as changing loading direction (Rudolph et al. 2014) or ratcheting behaviour under very large number of cyclic loads (Houlsby et al. 2017). Foundation alternatives that could provide significant cost savings also start to be considered such as helical piles (Byrne & Houlsby 2015), or active suction caissons (Fiumana et al. 2018). These new developments will be required to generate the step change cost reduction in foundation engineering that is required to make offshore renewables economically viable at commercial scale.

Centrifuge modelling can play a significant role in assisting these developments, similar to how it has enabled some of the more significant advances in offshore geotechnics for oil and gas applications. New geotechnical challenges associated with offshore renewable energy applications often requires the development of new modelling techniques. This paper presents a snapshot of some of these developments, associated with:

1. Modelling dynamic tensile loading resulting from extreme storm loading on a floating wave energy converter.
2. Modelling very large number of loading cycles and multidirectional loading.

2 MODELLING DYNAMIC TENSILE LOADING

2.1 Motivation

Point absorbers are a category of wave energy converters that produce electricity by using the foundation as a reaction point. They are designed to resonate at the peak frequency of the energy in the wave spectra to ensure optimum power take-off (PTO), resulting in in magnitude tensile loads on the foundation, that are of the order of several MN under extreme (e.g. storm) loading. An approach to limit the loads on the foundation involves using a Coulomb-Damping PTO that caps the load to a maximum value, above which the PTO experiences continuous extension. The drawback of such an approach is the snatch load that occurs if the PTO reaches its maximum extension. Hydrodynamic analyses have demonstrated that this snatch load is dynamic in nature, such that the foundation experiences acceleration. Hence, an opportunity arises to design the foundation to satisfy two different loading states. The first is an operational loading state, where the foundation is designed to withstand the cyclic tensile loads due to movement of the WEC at the water line that translates (via the PTO) to loading on the foundation. Although this continuous tensile cyclic loading on foundations is relatively uncommon, and raises questions associated with drainage response (for coarse-grained seabeds) and the potential for ratcheting behaviour (e.g. Houlsby et al. 2017), from a modelling perspective it can be addressed using existing techniques. The second loading state is the extreme condition, associated with maximum extension of the PTO. This will result in a high magnitude but short duration tensile load on the foundation, such that geotechnical response is not only expected to be undrained, but may also include strain rate and inertia effects. Centrifuge tests can assist in identifying and quantifying these capacity components, although modelling such a loading event requires new modelling techniques.

2.2 Apparatus and preliminary results

An experimental arrangement designed to model the extreme loading condition described above is shown in

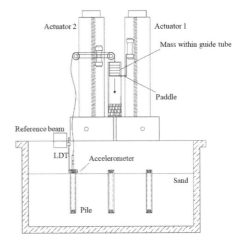

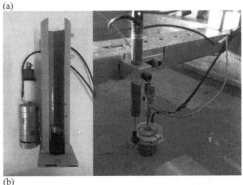

Figure 1. Experimental arrangement: (a) schematic representation, (b) falling mass and guide (left), instrumentation at the pile head (right).

Figure 1. Rather than using actuation systems to load the foundation, the foundation is loaded by allowing a mass to fall over a short distance in the elevated acceleration field of the centrifuge to generate a very short duration, high magnitude load.

Referring to Figure 1, a pile foundation is connected to a mass using a steel wire via two pulleys such that the dynamic tensile load is applied vertically to the pile. The mass falls within a slotted guide tube, with rubber foam at the base to absorb the impact if the tension in the steel wire does not arrest the fall. The mass is initially held in position using the 'paddle' located on the vertical axis of an actuator, where the slot in the guide tube allows access for the paddle. The pulley assembly is located on the vertical axis of a second actuator, which is adjusted to control the tension in the steel wire before the test. A load-cell, connected in series with the steel wire, is located just above the pile head, and a linear displacement transducer (LDT), located on an independent reference beam measures the pile head displacement. Two Microelectromechanical systems (MEMS) accelerometers, one on the falling mass and one on the pile, allow the acceleration and (through integration) the velocity of the mass and pile to be established.

The above experimental arrangement was adopted for tests conducted in dry and saturated silica sand conducted in the 1.8 m radius fixed beam centrifuge at The University of Western Australia. Details of the testing are due to be reported elsewhere, with snapshots presented here to illustrate some of the highlights from the tests.

A typical test programme using the falling mass loading system involves a monotonic test to establish the drained tensile capacity of the pile, followed by a series of dynamic tests to explore the pile response when loaded beyond the drained tensile capacity. Monotonic tests involve using the vertical axis of an actuator to displace the pile vertically slowly (via the steel loading wire) until a peak anchor capacity is measured. The dynamic tests would then select a mass such that the weight is a percentage of the drained capacity, but would result in a dynamic load that is considerably higher than the drained capacity. Each dynamic pile test could involve a single or multiple dynamic load events, depending on the pile response. Consecutive dynamic loads can be applied without stopping the centrifuge, by manipulating the paddle on the second actuator to raise and hold the mass, whilst raising the pulley assembly to control the initial tension in the steel wire.

Example test results are provided in Figure 2a and 2b for dry and saturated sand respectively. In each instance the weight of the falling mass (at the initial drop elevation) was approximately 50% of the drained tensile capacity. Figure 2a shows that in dry sand the dynamic pile capacity is approximately 50% higher than the drained monotonic capacity, and that the response in the dynamic test is much stiffer. As the sample is not saturated the additional resistance cannot be due to drainage, but must reflect an inertial component of resistance.

Figure 2b shows an equivalent comparison between monotonic and dynamic responses for a saturated sample (at the same relative density). In this instance the dynamic pile capacity is almost double the monotonic capacity, noting also that the monotonic capacity is lower than in the dry sample, reflecting the lower effective stress level in the saturated sample. As with the test in dry sand the pile response to dynamic loading is much stiffer than to monotonic loading, such that the pile displacements associated with these snatch loading events can be expected to be sufficiently low that the pile has sufficient residual capacity for additional operational or extreme loading events. The much higher ratio of dynamic to monotonic capacity for the saturated sample is due to the undrained response in the sand. This is to be expected, as the pile velocity reaches a maximum velocity, $v = 5$ m/s, such that the strain rate – approximated here as v/D, where D is the pile diameter – is 227 s^{-1}.

Returning to the test result from the dry sample, Figure 3 shows that the difference between the monotonic and dynamic resistance is close to the inertial resistance, calculated as the sum of the measured pile acceleration and the pile mass. This result suggests

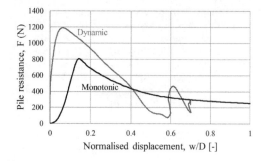

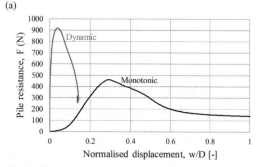

Figure 2. Example results from monotonic and dynamic tensile loading tests on a pile in sand: (a) dry sand, (b) saturated sand.

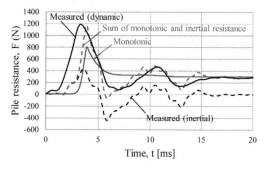

Figure 3. Interpretation of a dynamic tensile pile test in dry sand.

that for dry sand there are no other components of pile capacity, such as the pile impedance considered in pile driving. Consideration of the time duration of the dynamic load in these tests (~10 ms) relative to the likely time taken for a stress wave to travel along the pile and back (~0.06 s), supports this conclusion. An extension of the logic used in the interpretation of the tests in dry sand is that the dynamic resistance in saturated conditions is the sum of the undrained resistance plus an inertial component that is simply the product of the pile mass and acceleration.

These example results not only show that a pile in sand is capable of withstanding a short duration dynamic load, of a magnitude that is considerably in excess of the monotonic capacity, but also reveal how relatively simple measurements and permutations of test conditions reveal the components of capacity that are generated during dynamic loading, allowing for the development of appropriate prediction tools.

3 MODELLING CYCLIC MULTIDRECTIONAL LOADING

3.1 *High numbers of loading cycles*

While design for oil and gas structures typically involves consideration of the order of 10^3 loading cycles, the design of an offshore wind turbine generally requires the consideration of 10^6 to 10^7 load cycles undertaken a high frequency. This is not only to fulfil similitude requirements in terms of loading frequencies in the field and resulting drainage regimes, but also to minimise testing time while maximising the number of cycles. The drainage regime is important (Zhu 2018; Bienen et al. 2018) and can be controlled through pore fluid viscosity.

While it may not be economically feasible, nor indeed necessary, to perform long-term cyclic loading tests entirely in the centrifuge, it has been shown to be important to capture effects including installation history, stress level and drainage regime on the initial response under cyclic loading (Zhu 2018). This allows quantification of the initial rotation at field scale, which, when considered holistically with data from single gravity tests providing the accumulation trend over large numbers of load cycles enables evaluation of the long-term full-scale foundation response. An example from Zhu (2018) is provided in Figure 4 for monopod suction buckets in sand. The rate of accumulation in the single gravity and centrifuge tests is the same, although the low stress level in the single gravity tests results in a higher and incorrect initial rotation. Both the initial rotation and the rate of rotation accumulation with cycle number are important for the design of these dynamically sensitive structures, with strict limitations on non-verticality strictly enforced (DNV 2016).

3.2 *Accuracy and resolution of displacement measurements*

Offshore wind turbines are very sensitive to out-of-verticality, to the extent that rotation is typically limited to 0.5° over the design life of the structure (DNV 2016). This implies that in investigations of the foundation performance for offshore wind turbines, the expected displacements are very small. Further, precise knowledge of foundation stiffness is important as this affects the system stiffness, which is typically designed to fall in the narrow range between the blade and rotor forcing frequencies so as to avoid excitation of resonance. Experimental investigation therefore requires high accuracy and resolution of displacement measurements. Figure 5 illustrates the effects of average vertical stress and drainage regime (achieved through the use of different viscosity pore

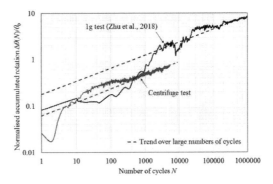

Figure 4. A combined approach of single gravity and centrifuge tests to predict the response of foundations subjected to large numbers of loading cycles (Zhu 2018).

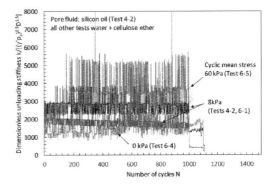

Figure 5. Unloading stiffnesses of suction buckets under vertical cyclic loading into tension (Bienen et al. 2018).

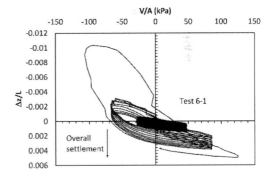

Figure 6. Displacement of suction bucket under vertical cyclic loading into tension (Bienen et al. 2018).

fluids) on the unloading stiffness of suction buckets in dense sand under vertical cyclic loading. While the displacement measurements (shown as an example for Test 6-1 in Fig. 5) are captured well, the deduced unloading stiffness (Fig. 6) shows some scatter, which is introduced by the division by very small differences in displacement.

Where investigations focus on foundation response under lateral cyclic loading, the displacement measurement technique is ideally non-contact as any

Figure 7. Centrifuge experimental arrangement that enables cyclic lateral loading with changing directionality (Rudolph et al. 2014).

resistance of the sensors may impact the measured displacements, in particular under low loading magnitudes.

3.3 Multidirectional loading

Loading of an offshore renewable energy device is expected to vary in cyclic loading magnitude, load eccentricity and even directionality. The latter has different origins, depending on the type of renewable energy installation. For offshore wind turbines, this relates closely to changes in metocean conditions over the design life of the structure.

The experimental apparatus hence needs to be sufficiently flexible to accommodate changes in loading characteristics, with an example shown in Figure 7 that enables cyclic lateral loading with changing directionality to be applied at an eccentricity above the soil. This is achieved by fixing a wheel at the top of the pile within which the pile can move freely in the vertical direction. A wire is connected to the wheel and to the vertical axis of an actuator. The vertical axis of the actuator is used to apply the load on the pile via the wire, while the horizontal axis modifies the direction of the application of the loading. The system enables either displacement or load control to be applied, while a set of two LDT sensors connected to the pile and separated by an angle of 90° provides information of the displacement history of the pile in the horizontal plane. The change in loading direction was found to significantly increase the monopile displacements (Rudolph et al. 2014), rendering consideration of uni-directional loading un-conservative.

Floating renewables are also subjected to multidirectional loading, but of a different type. One strategy currently considered for floating renewables (either wave energy converters of floating wind turbines) to significantly reduce the foundation costs involves sharing foundations across multiple devices assembled in an array (Karimirad et al. 2014). Different pattern of arrays can be considered as a function of the power output and shadowing effects between devices (see Child & Venugopal 2010), but in all cases, the number of foundations can be reduced significantly. For instance, for an array of 13 point absorbers each with three mooring lines, the honeycomb array pattern (Fig. 8a), allows the 13 devices to be anchored by 20 foundations (instead of 39).

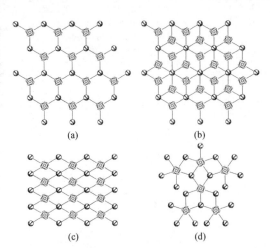

Figure 8. Array of wave energy converters (after Herduin et al. 2018).

A direct consequence of the foundation sharing strategy is the complexity of the loading regime on the foundation that can come from 2, 3 or 4 different directions (see Fig. 8). Depending on the wave spectra and period, and the spacing between the floating devices, the foundation can be subjected to loads coming from different directions that can be in phase (resulting in alternate loading along each of the mooring line) or out of phase (Herduin et al. 2018). This complex multi-directional loading mode, is fundamentally different from typical design considerations for floating oil and gas infrastructure, on which most of the design methodologies are based.

A first step in investigating the performance of foundations under multidirectional loading is to accurately and comprehensively define the load distribution and history acting on the foundation.

An initial development to define this load distribution and history (Herduin et al. 2016) used an analytical framework to establish the characteristics of the resultant load from multiple mooring lines, as a function of the individual load characteristics. The purpose is to characterise the variation of load magnitude and direction for a given array configuration, in order to define potential best and worst case loading scenarios from a foundation design perspective. Fig. 9 presents an example result that assumes an array of floating bodies assembled in a hexagonal configuration in constant water depth and subject to regular waves coming from a single direction. The wave series produces three harmonic loads of equal period and amplitude on the anchor from directions separated by 120°. Fig. 9 illustrates the variety of loading configurations applied to the anchor. As the phase difference θ between the three loads varies from 0 to π, resultant contours transform from a thin ellipse suggesting nearly bi-directional loading to a circle centred at the origin indicating constant amplitude and large variation in direction. This variety is further exacerbated when consideration is given to irregular waves and wave direction

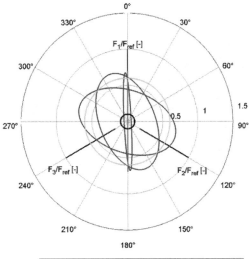

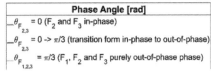

Figure 9. Variation of magnitude and direction of the load resultant from three harmonic loads of phase (after Gaudin et al. 2017).

with respect to the array configuration (Herduin et al. 2018). Fig. 9 is important from a geotechnical design perspective, considering that it is uncertain whether a resultant load of high magnitude and limited variation of direction is more detrimental than a resultant load of low magnitude varying over a wide range of directions. It is also important from a modelling perspective to define the requirements for a testing setup.

Indeed, the performance of traditional anchoring systems such as piles, skirted circular foundations or plate anchors, under multi-directional cyclic loading is poorly understood, but starts to receive attention. A few studies on multi-directional loading of offshore foundations suggested that a change in loading direction can increase plastic displacements and can reduce foundation performance as discussed in the previous section (Rudolph et al. 2014).

More recently, preliminary tests have been undertaken to better understand the performance of foundations under multidirectional alternate loading. Tests were performed on a suction caisson in sand with a setup that enabled two mooring lines separated by a planar angle ranging from 60° to 120° to be loaded alternatively (Fig. 10). The caisson was first installed at 1g (assuming wished in place conditions for this more fundamental study), and the two mooring lines were connected to two independent actuators that could apply either controlled displacement or loads to mimic any multi-directional loading scenario. Displacements are measured along the loading direction via encoders on the actuator, although this technique

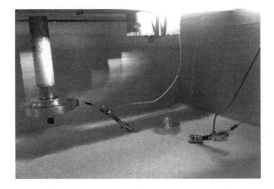

Figure 10. Multi-directional loading setup (after Herduin et al. 2016).

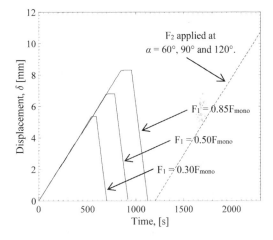

Figure 11. Multi-directional loading regime (after Herduin et al. 2018).

does not capture the whole displacement and rotation of the caisson in the six degrees of freedom. Further refinement, using PIV techniques have enabled measurements of the vertical horizontal and rotational displacements in one vertical plane (Gomez-Battista 2017) and developments are currently undertaken to expand the technique to the whole 6 degrees of freedom.

The loading regime applied in these preliminary tests is presented in Figure 11. A load at a fraction of the monotonic ultimate capacity was first applied in direction 1, and subsequently released. Immediately after a load was applied to failure along a direction 2, which is separated from direction 1 by a planar angle 30°, 90° or 120°.

Results are summarised in Figure 12 for initial loading of 30%, 50% and 85% of the monotonic capacity. They indicate a reduction in capacity of up to 10% for an initial loading higher than 50% of the maximum capacity and a loading direction of 120°. While this reduction may seem limited at first glance, it should be noted that this reduction occurs after one single episode of loading and that such a loading regime will be repeated multiple times over the life of the structure.

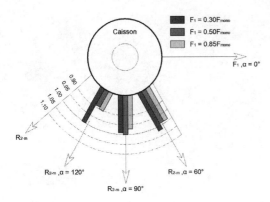

Figure 12. Change in caisson capacity under multi-directional loading (after Herduin et al. 2016).

The experimental approach is currently being improved to allow loading to be applied in three directions to mimic complex loading regimes with cycle numbers exceeding 10^5. In parallel, a macro-element model, capable of evaluating the change in capacity of the foundation under cyclic multi-directional loading, is being developed (see Gaudin et al. 2017). In this particular case, centrifuge modelling is used to provide insights into the behaviour of the foundation under multi-directional loading and to provide performance data to calibrate a theoretical model.

4 CONCLUSIONS

The paper presents a brief overview of some of the challenges faced by the offshore renewables community when designing foundations for fixed or floating structures. Because of the nature of these structures, the challenges are different than those faced by the oil and gas community over the last five decades and some of the standards and guidelines commonly used in design must be revisited.

Centrifuge modelling has an important role to play in addressing these challenges. Interestingly, the aims and objectives of centrifuge modelling remain identical; providing performance data, validating new concepts, identifying mechanisms, etc., but new modelling techniques are required to model new loading regimes, installation processes and serviceability constraints that are specific to offshore renewable energy applications. A few examples are provided, highlighting technological constraints and requirements and a snapshot of results demonstrate that centrifuge modelling is well positioned to assist the development of offshore renewables.

REFERENCES

API RP2SK 2008. *Design and Analysis of station keeping systems for floating structures*. American Petroleum Institute, 2005, Addendum 2008.

API 2GEO/ISO 19901-4, 2011. *Geotechnical and foundation design considerations, ANSI/API recommended practice*. American Petroleum Institute.

BBC, 2017. *Offshore wind power cheaper than new nuclear*. http://www.bbc.com/news/business-41220948.

Bienen, B., Klinkvort, R.T., O'Loughlin, C.D., Zhu, F., Byrne, B.W. 2018. Suction caissons in dense sand, Part II: Vertical cyclic loading into tension. *Géotechnique*, http://dx.doi.org/10.1680/jgeot.16.p.282.

Bienen, B., O'Loughlin, C.D., Zhu, F., 2017. Physical modelling of suction bucket installation and response under long-term cyclic loading. *Proc. 8th International Conference Offshore Site Investigation & Geotechnics "Smarter Solutions for Future Offshore Developments", Society of Underwater Technology*, London, UK, Vol. 1, pp. 524–531.

Byrne, B.W. & Houlsby, G.T., 2015. Helical piles: an innovative foundation design option for offshore wind turbines. *Philosophical Transactions of the Royal Society of London*, 373(2035).

Castro-Santos L., Garcia G.P., Estanqueiro A., Justino P.A.P.S., 2015. The levelized cost of energy (LCOE) of wave energy using GIS based analysis: The case study of Portugal. *Electrical Power and Energy Systems*, 65: 21–25.

Child B. F. M. & Venugopal V., 2010. Optimal configurations of wave energy device arrays. *Ocean Eng.*, 37(16): 1402–1417.

Diaz B.D., Rasulo M., Aubeny C.P., Landon M., Fontana C.M., Arwade S.R., DeGroot D.J., 2016. Multiline anchors for floating offshore wind towers. *Oceans 2016 MTS/IEEE Monterey*, 1–9.

DNV, 2016. *Support structures for wind turbines. Offshore standard (DNVGL-ST-0126)*. Oslo, Norway: DNV.

Energy Information Administration, 2017. *Levelized cost and levelized avoided cost of new generation resources*. Annual Energy Outlook.

Gaudin C., Cluckey E.C., Garnier J., Phillips R., 2010. New frontiers for centrifuge modelling in offshore geotechnics. *Proc. 2nd Intern. Symp. Frontiers in Offshore Geotechnics*, Perth, Australia: 155–188.

Gaudin, C., O'Loughlin, C.D., Duong, M.T., Herduin, M., Fiumana, N., Draper, S., Wolgamot, H., Zhao, L., Cassidy, M.J., 2017. New anchoring paradigms for floating renewables. *Proc.12th European Wave and Tidal Energy Conf.*, Cork, Ireland. Paper 666.

Global Wind Energy Council. www.gwec.net

Gomez-Batista D., 2017. *Cyclic loading of suction anchors for offshore floating renewables energy devices*. Master Thesis, The University of Western Australia.

Herduin M., Gaudin C., Zhao L., O'Loughlin C.D., Cassidy M.J., Hambleton J., 2016. Suction anchors for floating renewable energy devices. *Proc. 3rd Asian Wave and Tidal Energy Conference*, Singapore.

Herduin M., Gaudin C., Johanning L., 2018. Anchor sharing in sands. Centrifuge modelling and soil element testing to characterise multidirectional loadings. *Proc. 37th Intern. Conf. n Ocean, Offshore and Artic Engineering*, Madrid, Spain.

Houlsby G.T., Abadie C.N., Beuelelaers W., Byrne B.W., 2017. A model for nonlinear hysteric and ratcheting behaviour. *International Journal of Solids and Structures*. 120: 67–80.

Karimirad M., Koushan K., Weller S. D., Hardwick J., Johaning L., 2014. Applicability of offshore mooring and foundation technologies for marine renewable energy (MRE) device arrays. *Proc. 1st Int. Conf. on Renewable Energies Offshore*, Lisbon, Portugal.

Knappett J.A., Brown M.J., Aldaikh H., Patra S., O'Loughlin C.D., Chow S.H., Gaudin C., Lieng J.T., 2015. *A review of anchor technology for floating renewable energy devices and key design considerations*. Frontiers in Offshore Geotechnics III, 887–892.

Kost C., Mayer N.J., Thomsen J., Hartmann N., Senkpiel C., Philipps S., Nold S., Lude S., Saad N., Schlegl T., 2013. Levelized Cost of Electricity. PV and CPV in Comparison to Other Technologies. *Proc. 29th Intern. Photovoltaic Solar Energy Conference*, Amsterdam, The Netherlands.

Martin, C.M., 2001. Impact of centrifuge modelling on offshore foundation design. *Proc. Intern. Symp. Constitutive and Centrifuge Modelling; Two extremes*, Monte Verita, Switzerland, 135–154.

Martinelli L., Ruol P., Coretelazzo G., 2012. On mooring design of wave energy converters: The Seabreath application. *Proc. Intern. Conf. on Coastal Engineering*.

Murff, J.D., 1996. The geotechnical centrifuge in offshore engineering. Proc. *Offshore Technology Conf.*, OTC 8265.

Neary VS, Previsic M, Jepsen R.A., Lawson M.J., Yu Y.-H., Copping A.E., Fontaine A.A., Hallett K.C., 2014. *Methodology for Design and Economic Analysis of Marine Energy Conversion (MEC) Technologies*. Sandia Report. SAND2014-9040.

Rudolph C. Bienen B., Grabe J., 2014. Effect of variation of the loading direction on the displacement accumulation of large-diameter piles under cyclic lateral loading in sand. *Canadian Geotechnical Journal*. 51: 1196–1206.

Stapelfeldt, M, Bienen, B., Grabe, J., 2017. Advanced approaches for coupled deformation-seepage-analyses of suction caisson installation. *Proc. 36th International Conference on Ocean, Offshore & Arctic Engineering (OMAE)*, Trondheim, Norway, OMAE2017-61378.

Stevens R.F., Soosainathan L., Rahim A., Saue M., Gilbert R., Senanayake A.I., Gerkus H., Rendon E., Wang S.T., O'Connell D.P., 2015. Design procedures for marine renewable energy foundations. *Offshore Technology Conference*, 5: 3491–3507.

Weller S.D. & Johanning L., 2014. *Specific requirements for marine renewable energy foundation analysis*. Deliverable 4.2 of the DTOcean project.

Willow C. & Valpy B. 2011. *Offshore Wind Forecasts of future costs and benefits*.

Zhu F., 2018. *Suction caisson foundations for offshore wind energy installations in layered soils*. PhD Thesis, The University of Western Australia.

Physical modelling applied to infrastructure development

R.J. Goodey
City, University of London, London, UK

ABSTRACT: The physical modelling conference series has served as a primary means of sharing practise and disseminating current research in experimental geotechnics. Each conference highlights the trends, techniques and direction of current research. This paper summarises contributions to the 9th International Conference on Physical Modelling in Geotechnics from researchers broadly in the field of infrastructure development. This themed paper aims to identify innovative approaches to geotechnical problems, advances in experimental techniques and equipment in order to address new research questions and future trends in infrastructure research that might feature more significantly in future conferences. Some reflection on past conference proceedings is included with the hope that the community appreciates the scale of our achievements since the first conference in the series.

1 INTRODUCTION

Now in its ninth iteration, the International Conference on Physical Modelling in Geotechnics is the pre-eminent forum for the dissemination of research in all areas related to experimental geotechnics. From the early days, the conference series has grown and matured and this is reflected in the contributions submitted from researchers which have increased both in number as well as in complexity of topics addressed and range of techniques adopted.

The aim of this paper and the accompanying lecture is to highlight some of the contributions and advances made in the area of infrastructure development. Within this field there are many areas of interest and this is reflected in the high number of papers submitted to the conference. These papers detail work carried out using a wide variety of experimental techniques including large scale testing, centrifuge modelling and comparisons with field data.

The organisation of this paper follows the broad theme of the papers with the aim of identifying advances in the field as well as highlighting areas of future interest.

2 URBAN DEVELOPMENT

In urban areas there is a high demand to maximise available land and other resources. This has led to taller buildings with larger foundations and deeper basements. These types of structure can be difficult to construct in urban environments, there can be issues of noise during construction, interaction with existing infrastructure, as well as the need to ensure protection from earthquakes and other natural events. There are also problems of increasing subsurface congestion on new construction i.e. as buildings are redeveloped there are existing piles to consider as well as the need to avoid damage to existing infrastructure.

2.1 Driven piles

Conventional installation methods for driven piles (i.e. impact or vibration driving) cause undesirable noise and vibration in urban environments. One solution investigated by El Haffar et al. (2018) and Frick et al. (2018) is to use rotary jacked piles whose installation is much lower in both noise and vibration. These studies use coarse grained soils and investigate the influence of the installation parameters (forces and jacking stroke). Both studies produce broadly similar conclusions in that the installation forces and final capacity of the piles are strongly linked to the installation method adopted.

2.2 Deep basements

Deng & Haigh (2018) and Chan & Madabhushi (2018) both present preliminary work relating to deep basements. These papers investigate more efficient basement design (recognising that urban development now routinely incorporates deep basements) and both studies aim to investigate the underlying mechanisms. Deng & Haigh (2018) describe experimental work on soil movements behind a retaining wall. In this work, the wall movements are controlled and DIC (Digital Image Correlation, e.g. White et al. 2003) is used to monitor the soil response. This approach has been adopted as a more fundamental investigation of movements around excavations when compared with existing guidance (e.g. Clough & O'Rourke 1990) which is often based on empirical data and may not be universally applicable. Chan & Madabhushi (2018) also present work under development but here focussing on the heave behaviour of basement slabs founded on overconsolidated clay. The aim of the work is to study the influence of slab and basement stiffness on heave whereas previous work has focussed on specific cases or mitigation techniques. The results presented, whilst at an early stage, highlight not only the potential outcomes of the project, but also the complexity of the centrifuge modelling being carried out currently and the difficulties encountered.

3 ROADS AND PAVEMENTS

In the area of transportation infrastructure it is interesting to note that the majority of papers submitted to the conference are concerned with maintenance and prediction of long term performance. This is understandable given that many countries have well developed transport systems, elements of which may have originally been constructed more than a century ago.

3.1 Pipelines buried beneath roads

For ease of installation and maintenance, utility pipes are often buried beneath roads. This results in shallow pipelines which are subjected to significant cyclic loads from above. Bayton et al. (2018) report centrifuge tests of model pipelines subjected to simulated traffic loads. The motivation for the study is to minimise leakage from pipe networks with water supplies being highlighted. The work concentrates on the accumulation of bending moment within the pipe with repeated load cycles which could eventually result in damage to the pipe causing subsequent leakage. The effects of that leakage in the form of development of sinkholes are investigated in other papers (Kuwano et al. 2018, Indiketiya et al. 2018, Kearsley et al. 2018). Both Kuwano et al. (2018) and Indiketiya et al. (2018) use 1g testing and DIC to investigate the formation and propagation of a void above a simulated pipe with a defect. The experiments simulate the situation where soil is washed into the pipeline via the defect. It is interesting to note that in all cases very little movement is observed at the ground surface before the cavity collapses. The experimental arrangements in these papers are similar and in all cases soil below the water table is more prone to development of a cavity (as it is washed into the pipe via the defect). Indiketiya et al. (2018) conclude that cavity development is a function of the soil size when compared with the pipe defect but that it is difficult to identify a relationship between volume of soil lost and the size of the defect in the pipe. Kuwano et al. (2018) draw a similar conclusion with respect to the ratio of defect versus soil grain size but, due to the pipe defect having a fixed size in this study, do not comment on this aspect.

Laporte et al. (2018) report the development of apparatus to investigate the effect of wetting and drying cycles on expansive soils. Differential soil displacements can result in cracked pavements, damage to buried pipes and foundation movements. The experimental arrangement incorporates a model pavement and associated drainage ditch. This work shows significant differential movements between areas exposed directly to the elements and those shielded by the pavement surface. Again the work presented highlights the level of technical complexity that is being achieved in centrifuge modelling as well as the precise measurements that can be made via conventional instrumentation and DIC.

3.2 Pavement design

Two papers (Smit et al. 2018a,b) detail experimental investigations of Ultra-Thin Continuously Reinforced Concrete Pavement. This is proposed as a cost effective alternative to traditionally designed pavements and comprises a thin layer of heavily reinforced concrete. These papers highlight the problems associated with applying conventional design approaches to this innovative pavement design. The authors rightly highlight differences in design approach that would need to be accounted for given the observed difference in behaviour between this and traditional pavements.

4 PILES AND PILED FOUNDATIONS

Piled foundations are utilised in a wide range of applications; as foundations for medium to large size structures (both in isolation and as part of a piled raft), reinforcement under embankments or existing structures, to form walls and, more recently, as energy piles. This wide range of applications is reflected in the significant number of papers relating to pile performance. There is a particular focus on seismic and cyclic performance with the majority of the studies using centrifuge modelling techniques although 1g and shaking table tests are also utilised.

4.1 Piles under seismic action

A number of papers investigate the performance of piles under seismic loading. The experiments consider single piles (Yao et al. 2018, Ebeido et al. 2018, Chen et al. 2018, Pérez-Herreros et al. 2018) or small pile groups (Imamura 2018, Egawa et al. 2018). One paper (Garala & Madabhushi 2018) considered a comparison between a single pile and a small pile group (containing three piles). A range of soil conditions are used although sands and coarse grained soils are predominant. The majority of the papers investigate pile response in terms of bending moments and pile movements.

Many studies apply seismic loading either utilising input motion recorded during real earthquakes (e.g. Pérez-Herreros et al. 2018) or idealised sinusoidal motion, however Yao et al. (2018) have created an experimental apparatus to directly model the movement of a fault and used this to investigate the effect this has on piles close to and within the fault zone. This work also uses laser displacement transducers to measure the movement of the pile head and presents a good comparison between these measurements and those obtained from DIC. Significant bending moments and movements are observed in all piles, even those some distance from the fault zone. The authors conclude that, as in the real case, piles within the fault zone would most likely be completely sheared.

Egawa et al. (2018) present a series of tests on piles within layered soil models. The soils used are sand and volcanic ash and a variety of arrangements were tested (with respect to number and thickness of each layer). Despite the variation in test arrangements it was shown that large bending moments were consistently produced in the piles at around mid-height. It would be interesting to investigate whether this observation was repeated with different arrangements or sizes of pile. Pérez-Herreros et al. (2018) also use layered soil samples in their experiments on pile response, in this case the model is overconsolidated clay overlying dense sand. The model pile is predominantly embedded within the clay with its base just penetrating the sand layer beneath. Of great interest in this work is the observation that bending moments in the pile are strongly influenced by the amount of embedment into the dense sand layer. This obviously has great significance in areas where soft soils overly sands and end-bearing piles might be used.

4.2 Static behaviour of piles and piled foundations

Bisht et al. (2018) and Rodríguez et al. (2018) both present work investigating the performance of piled raft foundations. Bisht et al. (2018) show how the total load capacity is affected by the arrangement of piles the lengths of each pile within the group. A better understanding of how piles within the raft interact and contribute to the overall capacity could lead to more efficient designs. Rodríguez et al. (2018) concentrate on how the piled raft performs during changes of pore water pressures. These changes could arise from consolidation processes or by pumping from deep aquifers which is common in many cities. The authors present their results in terms of proportion of load carried by the raft or the piles. As pore water pressures decrease the proportion of load carried by the piles increases significantly, accompanied by a separation of the soil from the underside of the raft. This would have an impact on the design of both the piles and the raft although, as Bisht et al. (2018) point out, the contribution of the raft is often ignored in conventional design even though that implies poor economy.

Panchal et al. (2018) present a small study on a hybrid foundation system combining sheet piles with a pile cap. The aim is to produce a sustainable foundation design that can be used in already heavily developed urban areas. This type of foundation could be easily removed and recycled in the future which is not possible when dealing with bored, cast in-situ piles that are often found during site redevelopment. The results show a strong influence of geometry on the load capacity but that easily constructed square sheet pile groups could be a viable alternative to traditional bored piles.

4.3 Energy piles

Two papers highlight how physical modelling can be applied to new and emerging technologies. Energy piles combine structural requirements with the thermal performance of a ground source heat pump. The challenge is to assess the influence of the temperature changes on the structural performance. In a clay soil temperature changes will generally result in pore pressure variations due to the low permeability of the clay. The resulting change in effective stress is presumed to affect the interaction between soil and pile. Parchment & Shepley (2018) present a fundamental study of the influence of temperature on a soil-structure interface. A large number of direct shear tests between clay and a structural element (an aluminium block) are carried out. The study concludes that, for the rnage of temperatures that might be expected in a thermal pile, there is little effect in overconsolidated clay. Any effects seen relate to adhesion between pile and soil and may, in fact, be structurally beneficial. In normally consolidated clays the effect of heating is negligible.

Ghaaowd et al. (2018) present work on how heating affects the properties (undrained strength) of a clay sample and the resulting effect on pullout strength of a heated versus unheated pile. Significant increases in the pullout capacity are observed after heating. Only one (extended) cycle of heat was applied and it would be interesting to investigate how cycles of the type that would be expected in a thermal pile influenced the soil and pile behaviour.

5 SLOPES AND EMBANKMENTS

Slopes (both engineered and natural) present potential hazards primarily related to their long term performance and stability. A significant number of the papers submitted in this area deal with assessing and predicting the response of a slope to changes in pore water. The slopes studied are generally clay or clay dominated.

5.1 Slopes subjected to wetting and drying

Slopes and embankments are generally subjected to cyclic variation of wetting and drying. This could be due to seasonal variation, tidal variation or changes in reservoir level amongst other phenomena. Ahmed et al. (2018) present work investigating the movement of a slope subject to cyclic variation representative of tidal cycles. Similarly, Luo & Zhang (2018) simulated more extreme variations of wetting and drying more representative of the changing levels in a reservoir. Both of these studies adopted a similar, centrifuge based, approach and used DIC to monitor the movements. Movements are shown to accumulate with increasing numbers of cycles which obviously has implications for the long term stability of the slope. Ahmed et al. (2018) also carried out in-flight measurements of the soil strength within the slope and demonstrated that strength changed quite significantly with only a relatively small number of cycles. Again, this has implications for the long term stability of the slope.

Variations in water content can also be affected by vegetation on the slope. Vegetation is often cleared from embankments near roads and railway lines in order to reduce the potential for accidents and disruption. The effect of vegetation removal is investigated by Kamchoom & Leung (2018). These effects are twofold; firstly, removal of the vegetation would halt transpiration, potentially increasing pore water pressures in the embankment and secondly, the live roots act like reinforcement, the effectiveness of which will be reduced as the root decays. Kamchoom & Leung (2018) create a centrifuge model of a slope with artificial roots connected to a vacuum system. In this way, plant transpiration can be simulated and, by control of the suction from the artificial roots, plant removal can also be simulated. The results of these tests show that removal of plants from the upper portions of the slope has minimal effect on slope stability but stability is significantly compromised when vegetation is removed from the lower portion of the slope.

As well as fluctuations that might be interpreted as relatively easy to predict, if not account for, slopes and embankments are often subject to flooding. Saran and Viswanadham (2018) detail centrifuge tests on model levees which are subjected to flood events. Given that the potential for catastrophic failures to occur is well recognised and documented (e.g. Steedman & Sharp 2011) this work is significant and timely. Saran and Viswanadham (2018) perform centrifuge tests comparing the efficiency of horizontal and vertical (chimney) drainage layers within the levee. The experimental work is compared with numerical models. The experimental results suggest that the chimney drain increases the stability of the levee more effectively than the horizontal drain. This conclusion is not necessarily borne out by the numerical model which suggests that either drainage system results in a similar factor of safety against failure.

5.2 Embankments on soft soils

The problem of constructing an embankment over regions of soft soils is addressed by a number of papers. Founding an embankment on a soft underlying layer will generally result in long term settlements as the soft soil consolidates under the embankment load potentially damaging road and rail infrastructure. One solution to this problem is to improve the soft soil layer prior to embankment construction for which there are a number of approaches. Shiraga et al. (2018) detail centrifuge experiments on an embankment constructed using vacuum consolidation. Comparisons are made between embankment construction on the soft layer with and without the vacuum consolidation process. The results indicate that the vacuum consolidation process returns the pore water pressures in the ground to their original levels more quickly than simple embankment construction alone. Careful construction sequencing using this technique could reduce the possibility of having to undertake remedial works by ensuring settlements are mostly complete prior to installation of infrastructure on the embankment.

Another technique is to construct a piled embankment. The aim of this method is to reduce the load applied to the surface of the soft soil layer. This can be achieved by use of a geosynthetic such that the load spans between the piles beneath the embankment. Almeida et al. (2018) carried out multiple centrifuge experiments to investigate the influence on embankment performance of number geosynthetic layers, geosynthetic pretension, and pile size and arrangement. Their results concluded that use of a geosynthetic layer was extremely efficient in transferring load to the piles but there was no benefit to multiple layers and that the pretensioning effect was minimal. Blanc et al. (2018) also investigated the load transfer from the embankment (or granular mattress) to the piles below the embankment. In this work there was no geosynthetic reinforcement and only the pile spacing, size and height of embankment was varied. The work was carried out with the aim of validating previously published analytical models although the results suggest that each model is capable of representing some features of the system better than others.

6 TUNNELS AND PIPELINES

In previous conferences in this series, research and experimentation into tunnelling has been particularly well represented. It is interesting to note that for the 9th ICPMG the number of papers in this field is limited, perhaps indicating that researchers are adopting other techniques in this area.

6.1 Tunnelling

Tunnels are generally used in urban areas for mass transit systems. This is generally because of surface space constraints. Once constructed, tunnels can be subject to a variety of load conditions. Hajialilue-Bonab et al. (2018) describe 1g shaking table tests on an instrumented tunnel representative of a section of the Tabriz subway. Historically, underground structures have been subjected to lower levels of damage during earthquakes although the work presented here indicates that there may be a significant effect, particularly in strong ground shaking events. De & Zimmie (2018) investigated the effect of surface explosions on a tunnel. Measurements are presented on the basis of additional strains imparted to the tunnel lining during the event and some techniques for mitigation are investigated. The test arrangements are all shallow tunnels and it might be inferred that deeper cover (although not always possible) would result in more attenuation of the energy imparted by the explosion.

Xu & Bezuijen (2018) investigate the tunnelling construction process, specifically use of bentonite slurry to support the tunnel face during shield tunnelling. This is a 1g element testing study of how the bentonite filter cake develops as pore fluid infiltrates the soil surrounding the tunnel cavity. It was shown that different concentrations of bentonite affect the permeability of the surrounding soil with an almost direct relationship. The quantification of this would be useful information when designing tunnelling schemes through sandy strata.

6.2 Pipelines

Pipelines are used both onshore and offshore for the distribution of, for example, oil and gas. Extremely long networks are vulnerable to many hazards as they may cross earthquake zones, faults, slopes and many different soil conditions. Eichhorn & Haigh (2018) use the mini-drum centrifuge to examine the uplift resistance of pipes positioned parallel to the fall of a slope. Careful consideration is given to the scaling of the pipe model such that it is representative of a high pressure transmission pipeline that is currently in use. The experiments highlight deficiencies in the current methods used by industry.

The resistance to uplift is of critical importance during earthquakes or when large ground movements occur such as in a landslide. Wang et al. (2018) present a novel solution to the problem of uplift during earthquakes which is to strengthen the soil overlying the pipeline with vegetation. Model plant roots were created using a 3D printer and used to strengthen the soil in the upper layer of their experimental model. Results were presented in terms of the relative uplift of the pipeline with respect to the ground under the action of three earthquakes varying in intensity. Compared with a baseline case were there was no reinforcement, the introduction of roots to the soil did reduce pipeline uplift. The magnitude of reduction was related to the size of the roots. It was noted that uplift forces did not appear to be reduced and therefore the reduction was attributed to the increase in soil strength obtained from the root systems.

7 RETAINING WALLS

Retaining walls are found in a wide range of engineering projects including basement construction, retained embankments, and quays. There are also a wide variety of construction methods and materials. This diversity of application is represented by a number of papers investigating a range of topics including sheet piles, nailed walls and earth walls.

7.1 Earth walls

Walls constructed from earth offer advantages over other wall types such as low cost and ease of construction. To ensure stability, earth walls must contain some element of reinforcement and these systems are variously referred to as Geosynthetic Reinforced Soil Walls, Mechanically Stabilised Earth Walls and Reinforced Earth Walls. The performance of these types of walls relies on the interaction between the soil and the reinforcing elements. Mirmoradi & Ehrlich (2018) report large scale experiments on two Geosynthetic Reinforced Soil walls. Each wall was similar in design however one was faced with blocks and the other faced by wrapping the geosynthetic fabric around the soil. The wrap-faced wall was overall more flexible and transferred more of the applied surcharge load into the geosynthetic reinforcement. It is inferred that, given the good performance of both wall types, that the wrap-faced wall may be a preferred design solution on the basis of cost.

As stated earlier, the behaviour (and analysis) of these types of wall is dependent on the interaction between reinforcement and soil. Loli et al. (2018) have presented a large scale device for testing (and therefore characterising) reinforcement buried within soil. This device allows control of the overburden stress and is of a size sufficient to test many types of reinforcement. The performance of the device is compared with numerical modelling and the design choices justified on this basis. Whilst the focus of the paper is the device itself, it is clear that better understanding of the interaction between soil and reinforcement will enable more efficient and economical earth wall designs.

7.2 Soil-nailed walls

Two papers from the same group investigate walls reinforced with soil nails. Sabermahani et al. (2018) investigate the optimal arrangement of nails within an irregularly shaped excavation whereas Akoochakian et al. (2018) consider a more regular, rectangular excavation. The motivation for both these pieces of work relates to maximisation of available space during urban development. The work presented in both of these papers highlights that, even in relatively simple cases of regularly shaped excavations, the spacing of the nails, stiffness of the wall facing and presence of surcharge behind the wall all greatly influence the movements observed around the excavation.

8 SHALLOW FOUNDATIONS

As highlighted in the section described the papers submitted in the area of tunnels and pipelines, it is interesting to note how research activity changes over the years. In earlier iterations of the ICPMG there were many papers concerning the behaviour of shallow foundations but this number is very much reduced for the 9th ICPMG.

Qi & Knappett (2018) and Ghalandarzadeh & Ashtiani (2018) both present work relating to the response of shallow foundations under seismic loading. In the paper of Ghalandarzadeh & Ashtiani (2018) a similar approach to that taken by Yao et al. (2018) is adopted whereby an apparatus is developed that induces a predefined fault plane into the soil model and the response of the foundation to this is monitored. The footing load was generally maintained constant with embedment and distance from the fault plane being varied.

Results are presented in terms of footing rotation. In general, footings founded on the surface experience less rotation compared with footings that are initially embedded. The magnitude of the footing load appears to have little effect upon this result. As with Yao et al. (2018) the zone of influence of the fault is quite large so foundations are affected wherever they are placed within the experiment.

Qi & Knappett (2018) investigate the influence of soil permeability on shallow foundations (supporting a low-rise structure). The time histories of real earthquakes were applied sequentially. The results showed that, even if the potential for liquefaction could be identified, the prediction of structural damage is extremely difficult to estimate. In particular the effect of strong aftershocks seemed to place higher demands on the structure whilst not necessarily resulting in significant additional settlement or rotation.

9 GROUND IMPROVEMENT

There are many techniques available for ground improvement. The term generally refers to increasing soil strength but may also refer to the improvement of drainage. A number of papers submitted have considered ground improvement as a means to mitigating the effects of earthquakes. In particular, the use of drains (of various types) as mitigation against the effects of liquefaction is the subject of several papers. Paramasivam et al. (2018), García-Torres et al. (2018), Marques et al. (2018) and Kirkwood & Dashti (2018) all consider the use of vertical drains whilst Apostolou et al. (2018) consider the use of stone columns although their tests did not represent structural loads but rather, investigated dissipation of excess pore water pressures within the soil model. Finally, although not strictly speaking a ground improvement technique, Nigorikawa et al. (2018) describe a base isolation system as a mechanism to mitigate liquefaction effects.

All of these studies utilise centrifuge modelling techniques, highlighting the applicability of this method to studying earthquake related problems. The four papers that considered vertical drains underneath model structures (Paramasivam et al. 2018, García-Torres et al. 2018, Marques et al. 2018 and Kirkwood & Dashti 2018) all demonstrate a reduction in earthquake induced rotation and settlements when drains were used. There appeared to be a cost associated with this improvement however, in terms of the motion transferred to the superstructure. Both Kirkwood & Dashti (2018) and Paramasivam et al. (2018) saw increased accelerations of the structure when mitigation by drains was included. Additionally, in the tests of Kirkwood & Dashti (2018) an adjacent structure without drains was present which experienced an increase in rotations. The implication here is that this solution would either need to be applied to all structures in an area or that some sort of isolation or cut-off wall would be required to protect unmitigated structures.

10 SUMMARY AND CONCLUSION

Approximately fifty papers submitted to the 9th ICPMG have been reviewed in order reflect upon the contributions made, both in terms of experimental techniques being adopted and research questions currently being addressed. These contributions have been discussed with a view to identifying future trends and research questions whilst keeping in mind the progress that has been exhibited in the field over the entire conference series.

REFERENCES

Ahmed, U., Ong, D.E.L. & Leung, C.F. 2018. Performance of single piles in riverbank clay slopes subject to repetitive tidal cycles. Proceedings 9th Int. Conf. on Physical Modelling in Geotechnics, London.

Akoochakian, A., Moradi, M. & Kavand, A. 2018. Investigation of nailed slope behaviour during excavation by Ng centrifuge physical model tests. Proceedings 9th Int. Conf. on Physical Modelling in Geotechnics, London.

Almeida, M.S.S., Fagundes, D.F., Almeida, M.C.F., Hartmann, D.A., Girout, R., Thorel, L. & Blanc, M. 2018. Load transfer mechanism of reinforced piled embankments. Proceedings 9th Int. Conf. on Physical Modelling in Geotechnics, London.

Apostolou, E., Brennan, A.J. & Wehr, J. 2018. Relative Contribution of Drainage Capacity of Stone Columns as a Countermeasure against Liquefaction. Proceedings 9th Int. Conf. on Physical Modelling in Geotechnics, London.

Bayton, S.M., Elmrom, T. & Black, J.A. 2018. Centrifuge modelling utility pipe behaviour subject to vehicular loading. Proceedings 9th Int. Conf. on Physical Modelling in Geotechnics, London.

Bisht, R.S., Juneja, A., Tyagi, A. & Lee, F.H. 2018. Performance of piled raft with unequal pile lengths. Proceedings 9th Int. Conf. on Physical Modelling in Geotechnics, London.

Blanc, M., Thorel, L., Girout, R., Almeida, M.S.S. & Fagundes, D.F. 2018. Load transfer mechanism of piled embankments: centrifuge tests versus analytical models. Proceedings 9th Int. Conf. on Physical Modelling in Geotechnics, London.

Chan, D.Y.K. & Madabhushi, S.P.G. 2018. Centrifuge simulation of heave behaviour of deep basement slabs in over-consolidated clay. Proceedings 9th Int. Conf. on Physical Modelling in Geotechnics, London.

Chen, C.H., Ueng, T.S. & Chen, C.H. 2018. Dynamic behaviours of model pile in saturated sloping ground during shaking table tests. Proceedings 9th Int. Conf. on Physical Modelling in Geotechnics, London.

Clough, W. & O'Rourke, T.D. 1990. Construction induced movements of in situ walls. *Proceedings of design and performance of earth retaining structures, Ithaca, NY.* ASCE GSP 25: 430-470.

De, A. & Zimmie, T.F. 2018. Study of the effects of explosion on a buried tunnel through centrifuge model tests. Proceedings 9th Int. Conf. on Physical Modelling in Geotechnics, London.

Deng, C. & Haigh, S.K. 2018. Soil movement mobilised with retaining wall rotation in loose sand. Proceedings 9th Int. Conf. on Physical Modelling in Geotechnics, London.

Ebeido, A., Elgamal, A. & Zayed, M. 2018. Pile response during liquefaction-induced lateral spreading: 1g shake table tests with different ground inclination. Proceedings 9th Int. Conf. on Physical Modelling in Geotechnics, London.

Egawa, T., Yamanashi, T. & Isobe, K. 2018. Investigation on the aseismic performance of pile foundations in volcanic ash ground. Proceedings 9th Int. Conf. on Physical Modelling in Geotechnics, London.

Eichhorn, G.N. & Haigh, S.K. 2018. Uplift resistance of a buried pipeline in silty soil on slopes. Proceedings 9th Int. Conf. on Physical Modelling in Geotechnics, London.

El Haffar, I., Blanc, M. & Thorel, L. 2018. Effects of the installation methods of piles in cohesionless soil on their axial capacity. Proceedings 9th Int. Conf. on Physical Modelling in Geotechnics, London.

Frick, D., Schmoor, K.A., Gütz, P. & Achmus, M. 2018. Model testing of rotary jacked open ended tubular piles in saturated non-cohesive soil. Proceedings 9th Int. Conf. on Physical Modelling in Geotechnics, London.

Garala, T.K. & Madabhushi, G.S.P. 2018. Comparison of seismic behaviour of pile foundations in two different soft clay profiles. Proceedings 9th Int. Conf. on Physical Modelling in Geotechnics, London.

García-Torres, S. & Madabhushi, G.S.P. 2018. Earthquake-induced liquefaction mitigation under existing buildings using drains. Proceedings 9th Int. Conf. on Physical Modelling in Geotechnics, London.

Ghaaowd, I., McCartney, J., Huang, X., Saboya, F. & Tibana, S. 2018. Issues with centrifuge modelling of energy piles in soft clays. Proceedings 9th Int. Conf. on Physical Modelling in Geotechnics, London.

Ghalandarzadeh, A. & Ashtiani, M. 2018. Effective parameters on the interaction between reverse fault rupture and shallow foundations: centrifuge modelling. Proceedings 9th Int. Conf. on Physical Modelling in Geotechnics, London.

Hajialilue-Bonab, M., Farrin, M. & Movasat, M. 2018. Shaking table test to evaluate the effects of earthquake on internal force of Tabriz subway tunnel (Line 2). Proceedings 9th Int. Conf. on Physical Modelling in Geotechnics, London.

Imamura, S. 2018. Horizontal subgrade reaction of piles in liquefiable ground. Proceedings 9th Int. Conf. on Physical Modelling in Geotechnics, London.

Indiketiya, S., Jegatheesan, P., Pathmanathan, R. & Kuwano, R. 2018. Effect of pipe defect size and maximum particle size of bedding material on associated internal erosion. Proceedings 9th Int. Conf. on Physical Modelling in Geotechnics, London.

Kamchoom, V. & Leung, A.K. 2018. Effects of plant removal on slope hydrology and stability. Proceedings 9th Int. Conf. on Physical Modelling in Geotechnics, London.

Kearsley, E.P., Jacobsz, S.W. & Louw, H. 2018. Using pipe deflection to detect sinkhole development. Proceedings 9th Int. Conf. on Physical Modelling in Geotechnics, London.

Kirkwood, P.B. & Dashti, S. 2018. An experimental study on the effects of enhanced drainage for liquefaction mitigation in dense urban environments. Proceedings 9th Int. Conf. on Physical Modelling in Geotechnics, London.

Kuwano, R., Sera, R. & Ohara, Y. 2018. Model tests to simulate formation and expansion of subsurface cavities. Proceedings 9th Int. Conf. on Physical Modelling in Geotechnics, London.

Laporte, S., Siemens, G.A. & Beddoe, R.A. 2018. Physical modelling of roads in expansive clay subjected to wetting-drying cycles. Proceedings 9th Int. Conf. on Physical Modelling in Geotechnics, London.

Loli, M., Georgiou, I., Tsatsis, A., Kourkoulis, R. & Gelagoti, F. 2018. Pull-out testing of steel reinforced earth systems: modelling in view of soil dilation and boundary effects. Proceedings 9th Int. Conf. on Physical Modelling in Geotechnics, London.

Luo, F. & Zhang, G. 2018. Centrifuge model test on deformation and failure of slopes under wetting-drying cycles. Proceedings 9th Int. Conf. on Physical Modelling in Geotechnics, London.

Mirmoradi, S.H. & Ehrlich, M. 2018. Large-scale physical model GRS walls: evaluation of the combined effects of facing stiffness and toe resistance on performance. Proceedings 9th Int. Conf. on Physical Modelling in Geotechnics, London.

Nigorikawa, N., Asaka, Y. & Hasebe, M. 2018. Dynamic centrifuge model tests on sliding base isolation systems leveraging buoyancy. Proceedings 9th Int. Conf. on Physical Modelling in Geotechnics, London.

Panchal, J.P., McNamara, A.M. & Goodey, R.J. 2018. Influence of geometry on the bearing capacity of sheet piled foundations. Proceedings 9th Int. Conf. on Physical Modelling in Geotechnics, London.

Paramasivam, B., Dashti, S., Liel, A.B. & Olarte, J.C. 2018. Centrifuge modeling of mitigation-soil-structure-interaction on layered liquefiable soil deposits with a silt cap. Proceedings 9th Int. Conf. on Physical Modelling in Geotechnics, London.

Parchment, J. & Shepley, P. 2018. The influence of temperature on shear strength at a soil-structure interface. Proceedings 9th Int. Conf. on Physical Modelling in Geotechnics, London.

Pérez-Herreros, J., Cuira, F., Escoffier, S. & Kotronis, P. 2018. Kinematic interaction of piles under seismic loading. Proceedings 9th Int. Conf. on Physical Modelling in Geotechnics, London.

Qi, S. & Knappett, J.A. 2018. Centrifuge modelling of the effects of soil liquefiability on the seismic response of low-rise structures. Proceedings 9th Int. Conf. on Physical Modelling in Geotechnics, London.

Rodríguez, E., Cunha, R.P. & Caicedo, B. 2018. Behaviour of piled raft foundation systems in soft soil with consolidation process. Proceedings 9th Int. Conf. on Physical Modelling in Geotechnics, London.

Sabermahani, M., Moradi, M. & Pooresmaeili, A. 2018. Performance of soil-nailed wall with three-dimensional geometry: centrifuge study. Proceedings 9th Int. Conf. on Physical Modelling in Geotechnics, London.

Saran, R.K. & Viswanadham, B.V.S. 2018. Centrifuge model tests on levees subjected to flooding. Proceedings 9th Int. Conf. on Physical Modelling in Geotechnics, London.

Shiraga, S., Hasegawa, G., Sawamura, Y. & Kimura, M. 2018. Centrifuge model test of vacuum consolidation on soft clay combined with embankment loading. Proceedings 9th Int. Conf. on Physical Modelling in Geotechnics, London.

Smit, M.S., Kearsley, E.P. & Jacobsz, S.W. 2018a. Scaled physical modelling of ultra-thin continuously reinforced concrete pavement. Proceedings 9th Int. Conf. on Physical Modelling in Geotechnics, London.

Smit, M.S., Kearsley, E.P. & Jacobsz, S.W. 2018b. The effect of relative stiffness on soil-structure interaction under vehicle loads. Proceedings 9th Int. Conf. on Physical Modelling in Geotechnics, London.

Steedman, R.S. & Sharp, M.K. 2011. Physical modelling analysis of the New Orleans levee breaches. *Geotechnical Engineering*. 164(6): 353-372.

Wang, K., Brennan, A.J., Knappett, J.A., Bengough, A.G. & Robinson, S. 2018. Centrifuge modelling of remediation of liquefaction-induced pipeline uplift using model root systems. Proceedings 9th Int. Conf. on Physical Modelling in Geotechnics, London.

White, D.J., Take, W. & Bolton, M. 2003. Soil deformation measurement using Particle Image Velocimetry (PIV) and photogrammetry. *Géotechnique*. 53(7): 619-631.

Xu, T. & Bezuijen, A. 2018. Experimental modelling of infiltration of bentonite slurry in front of shield tunnel in saturated sand. Proceedings 9th Int. Conf. on Physical Modelling in Geotechnics, London.

Yao, C.F., Seki, S. & Takemura, J. 2018. Displacement measurements of ground and piles in sand subjected to reverse faulting. Proceedings 9th Int. Conf. on Physical Modelling in Geotechnics, London.

The role of centrifuge modelling in capturing whole-life responses of geotechnical infrastructure to optimise design

S. Gourvenec
University of Southampton, UK

ABSTRACT: Whole-life design relies on scrutinizing the geotechnical responses to whole of life loading sequences, through installation and operation or service, and partnering appropriate 'current' operational soil parameters with corresponding 'current' loading to optimize design outcomes. Whole-life design offers efficiencies over established design methods that are based on in situ soil parameters. In the current environmental and economic climate, established paradigms of design are being challenged to make way for enabling technologies to deliver projects of greater scale and complexity for less risk and cost. Whole-life design can be applied to a range of geotechnical boundary value problems - and can best be practically investigated in a centrifuge environment. This paper demonstrates the role of centrifuge modelling to identify governing mechanisms of whole-life response as a critical activity in the trajectory from design concept to implementation in engineering practice. The role of geotechnical centrifuge modelling in capturing whole-life response to optimize offshore foundation design is illustrated, although the overarching concepts put forward in the paper have much broader application.

1 INTRODUCTION

1.1 Whole-life response

Geotechnical centrifuge modelling has enabled field-scale infrastructure, soil stresses and geotechnical processes to be modelled realistically at small scale for over 5 decades (Roscoe 1970, Lyndon & Schofield 1970) – or a bit under 2 days at 100 g. Many hundreds of papers reporting significant insights into a range of field-scale geotechnical boundary value problems from centrifuge modelling are collected in the 8 sets of proceedings of the *International Conference on Physical Modelling in Geotechnics* and elsewhere in the literature.

A temporal spectrum can be used to describe how geotechnical centrifuge modelling can assist understanding of different geotechnical responses:

1. Short-term '*events*' (e.g undrained installation or failure of a geotechnical structure, or a change in load, water table level or other state);
2. Longer '*episodes*' that may be comprised of a series of events (e.g. construction or operational processes such as an excavation sequence or extreme weather event);
3. The '*whole life*' of a structure (e.g. a lifetime of weather episodes, such as freeze-thaw seasonal cycles or storms, or an operating life of changes in load level, such as tanks repeatedly filling and emptying, or thermal expansion loads from operation of equipment or facilities).

These temporal concepts are illustrated in Figure 1. Whole-life geotechnical response is the least investigated of the three classes of activity and is the focus of this keynote paper.

The whole-life concept partners whole-life loading sequences with whole-life soil responses to optimize geotechnical design outcomes. For example, this approach can lead to reduced foundation size where the soil strength rises through the design life, with knock on effects of reduced costs through

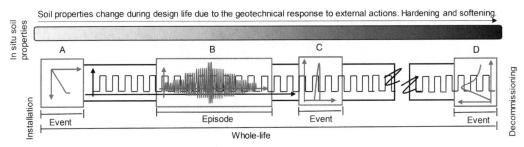

Figure 1. Temporal spectrum of geotechnical processes.

fabrication, transport and installation (Gourvenec et al. 2017a).

Consideration of the whole-life response requires identification of the dominant loading sequence that governs the geotechnical response of the soil throughout the design life. This inevitably leads to idealization of a field situation but is essential to obtain the effect of the whole-life loading on the whole-life capacity. In this way, the evolution of the soil strength, stiffness and other geotechnical parameters can be 'banked' where it is beneficial in meeting the design criteria throughout the whole life.

Events or episodes can be superimposed on a background of the whole-life response to enable greater scrutiny of specific activities or environmental influences. It is essential to understand current operative soil strength as well as the current position of a structure to inform predictions of the geotechnical response for an event during the life of the structure.

Questions such as "what is the soil strength and stiffness at the start and end of the episode 'B' in Figure 1" determine the (true) geotechnical stability of the structure during the episode and subsequently for future events, such as 'C'. Identifying the 'true', or current, operative shear strength and stiffness to inform that calculation enables a more realistic prediction of geotechnical resistance and optimized design, compared to assuming that the initial (in situ) properties apply throughout the life.

An example, topical in offshore engineering at present, is the need to predict retrieval loads to lift a structure from the seabed for decommissioning. Installation, self-weight consolidation, and a life time of operational loading and consolidation change the seabed state and strength, as well as the foundation position since installation. These conditions need to be predicted to assess the uplift resistance, which is likely to be (potentially much) greater than the resistance during installation (Small et al. 2015, Gourvenec & White 2017). This whole-life behaviour has clear implications in terms of crane requirement for vessels for planning decommissioning.

1.2 Offshore facility architecture and foundations

1.2.1 Field development architecture

Offshore developments for oil and gas are increasingly diverse in terms of architecture and support an increasingly diverse range of activities. Offshore structures range from single fixed platforms to fixed or floating hosts supporting a subsea development that may extend tens of kilometers from the host. Alternatively, subsea developments may be tied directly back to shore (Figure 2). Subsea developments comprise a network of flowlines (in-field pipelines) connected by structures to transport fluids to or from wells (Figure 3).

1.2.2 Fixed and floating platforms

Foundations for fixed offshore platforms are subjected to multi-directional loading derived from self-weight and cyclic lateral loads and moments dominated by

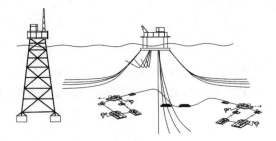

Figure 2. Examples of offshore structures; (L) single fixed platform and (R) floating host and subsea development.

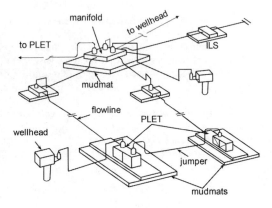

Figure 3. Examples of subsea architecture.

environmental forces such as wind, waves, current and in places sea ice. A permanent monotonic moment component may also arise from eccentricity of the supported superstructure relative to the foundation footprint. Foundations for fixed platforms are typically deep piled foundations or gravity bases.

Foundations, or anchors, for floating platforms are subject to vertical loads determined by the buoyancy of the floater and to multi-directional loading derived from environmental forces. One-way cyclic loading is typical for mooring anchors compared to the two-way cyclic loading seen by fixed platform foundations. Deep piled foundations, suction caissons and drag anchors are the most common anchoring systems for floating platforms although a number of novel and developmental anchors exist.

1.2.3 Subsea structures

Foundations for subsea structures supporting pipeline infrastructure are subjected to self-weight loading, but in contrast to fixed and floating platforms, the multi-directional horizontal loads, moment and torsion are dominated by installation and operational activities, rather than by environmental conditions.

Tie-in or 'metrology' loading occurs when pipelines and jumpers are connected to the subsea structure. Episodic monotonic loading occurs from thermal expansion and contraction of the attached pipelines and jumpers from start up and shut down cycles that form the operation of an offshore development.

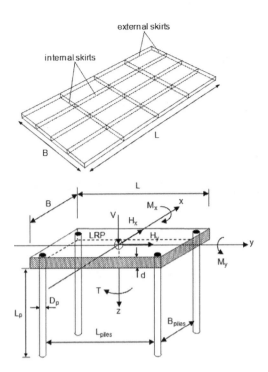

Figure 4. Options for enhancing capacity of subsea mudmats (upper) skirts and (lower) pinpiles.

Figure 5. Subsea pipeline end termination structure (Image from Subsea7).

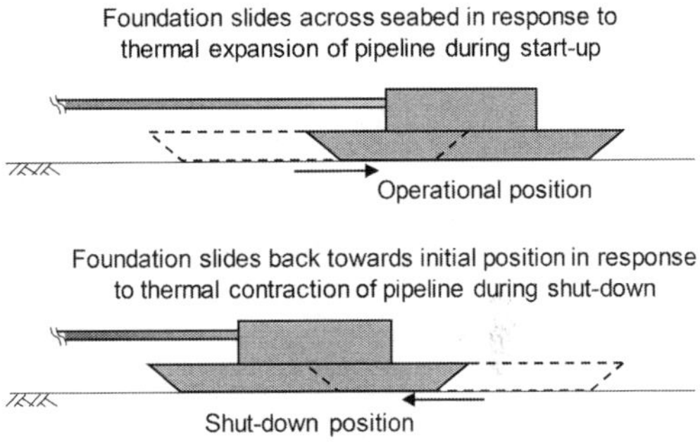

Figure 6. Concept of a tolerably mobile subsea foundation.

Pipelines expand as hot hydrocarbons pass through the pipelines during start up, remain expanded during an operational cycle, and contract when a well is shut down and the pipeline no longer contains hot hydrocarbons. Multi-directional lateral loads, moments and torsion are imposed to the foundation due to vertical and horizontal eccentricities of the pipeline and jumper connections to the orthogonal axes of the foundation.

Subsea structures are often supported on shallow foundations, or 'mudmats', due to the attractiveness of relatively straightforward self-weight installation, often performed from the pipe-laying vessel.

Increasingly, shallow foundations for subsea structures designed with traditional methods are too large for standard installation vessels. This is because of the more demanding operational requirements – capacity and stiffness – set by the supported structures and due to the softer seabeds found in deeper waters. This leads to increased project costs associated with heavier lift vessels.

Shallow foundations for mudmats can be augmented with skirts, caissons or pinpiles (Figure 4) that penetrate the seabed to increase resistance and reduce displacements (Dimmock et al. 2013, Feng et al. 2014, Hossain et al. 2015, Demel et al. 2016, Gourvenec et al. 2017b, Dunne & Martin 2017, Wallerand et al. 2017). However, these modifications lead to increased cost and risk in fabrication and installation (risk of failure to install) and cannot always deliver the reduction in mudmat footprint required. A photograph of a pipeline end termination structure on a skirted mudmat is shown in Figure 5.

1.2.4 *Challenging the design paradigm*

The basis of design for a mudmat is that it will spread the supported loads to the seabed with limited settlement and without geotechnical 'failure' – currently defined in practice by a required material factor on the mobilized soil strength (ISO 2016, API 2011). Traditional design methods for shallow foundations require a sufficiently large footprint to resist all applied loading and remain stationary in order to meet the basis of design – i.e. to avoid 'failure'.

To meet the demand for smaller subsea foundation footprints, the traditional design paradigm of static foundations has been challenged with concepts of 'tolerable mobility' or 'on-seabed sliding' (Cathie et al. 2008, Cocjin et al. 2014a, 2015, Deeks et al. 2014, Stuyts et al. 2015, Wallerand et al. 2015, Feng & Gourvenec 2016).

The concept of tolerable mobility is that the foundation is designed to move across the seabed to relieve the displacement-sensitive tie-in or operational loads in a way that is tolerable in relation to the function of the structure. However, foundations designed to slide across the seabed violate the current code definition of 'failure' (e.g. API 2011, ISO 2016). Nonetheless, sliding foundations have been designed and deployed for projects and centrifuge modelling has been a key element in making this possible (Client confidential).

Figure 6 illustrates the concept of a tolerably mobile mudmat for a pipeline structure. The mudmat rests on the seabed (i.e. without skirts) and is equipped with 'skis' to resist against overturning during sliding.

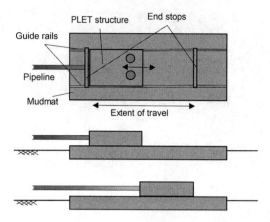

Figure 7. Concept of a static mudmat with a sliding mechanism.

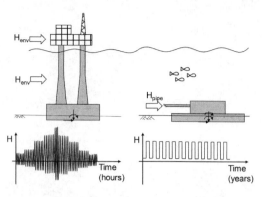

Figure 8. Comparison of loading sequences relevant to shallow foundations for (left) a fixed platform and (right) a subsea pipeline structure.

The foundation slides across the seabed in response to thermal expansion of the pipeline during start up and remains in the operational position while the well is producing, which may be for a few days to a few months before the next shutdown. The pipeline then slides back towards the initial position in response to thermal contraction of the pipeline on shut down. Shutdown is brief, typically a day or two. The process repeats episodically over the life of the structure leading to cycles of shearing (during sliding) and reconsolidation of the seabed at the operational and shut-down positions. The sliding foundation concept is similar to a snow sleigh or ski but is intended for only small distance of travel, of the order of a few meters.

It is worth noting that a 'sliding foundation' in this context is different to static subsea foundation equipped with a sliding mechanism (e.g. Jayson et al. 2008). In the latter case, a mechanical slider is mounted on the mudmat to absorb, to some extent, pipeline expansion and contraction movements (Figure 7). In contrast, a 'sliding foundation' is taken to mean a foundation that slides across the seabed.

Whole-life concepts are an additive tool that can be applied to reduce foundation footprints, whether designed to be static or tolerably mobile, and can yield particular efficiencies in subsea foundations.

1.3 Application of whole-life concepts to subsea foundations

Whole-life response can provide significant efficiencies to subsea foundation design outcomes due to the nature of loading, which is quite different to that for a fixed or floating platform.

A subsea mudmat may be set down some weeks or months in advance of a field becoming operational at which point the multi-directional operational loads are 'switched on'. The geotechnical foundation design can then rely on the enhanced shear strength of the seabed due to consolidation under self-weight of the foundation and structure for the operational load case. This is not so straightforward for platform foundations as the multi-directional loading is driven by environmental forces that are less predictable, which makes reliance on enhanced self-weight consolidated strength for the operational design load case more challenging. Nonetheless, the broadest concepts of whole-life response underpin the established practices of reliance on set-up in pile foundation and anchor design, active suction programs for gravity based platforms, staged installation processes (for GBS, embankments or artificial islands) and reuse of existing foundations. Moving beyond these examples of self-weight and post-installation consolidation, additional whole-life response benefits can be harnessed to optimize subsea foundation design.

Figure 8 illustrates schematically the load sequences transmitted to the shallow foundation for a fixed platform and a subsea pipeline structure. For the fixed platform, peak loading corresponds to extreme weather events that involve high amplitude and frequency cyclic loading. The duration of an extreme weather event, typically a few hours or days, prevents significant dissipation of excess pore pressure in fine grained seabeds. Excess pore pressures therefore accumulate, leading to a reduction in effective stress and undrained shear strength of the seabed, i.e. cyclic softening (e.g. Andersen 2015, Zografou et al. 2016) and a subsequent reduction in foundation capacity (e.g. Andersen et al. 1988, Xiao et al. 2016).

The whole-life loading sequence of a shallow foundation supporting a subsea structure will depend on the function of the structure and the environmental conditions, but in many cases will be dominated by operational activities, i.e. the thermal expansion and contraction of the attached pipelines and spools during start up and shut down operations. The duration of these operational activities are orders of magnitude longer than storm loading (months rather than days), such that excess pore pressures generated during the loading event may dissipate, even in fine grained seabeds, prior to the subsequent cycle.

Intervening reconsolidation between cycles of loading can lead to an increase in the shear strength of the seabed, i.e. cyclic hardening. Cyclic hardening has been demonstrated with in situ characterization tools (Hodder et al. 2010, Cocjin et al. 2014b) and for pipelines and foundations (Randolph et al. 2012, Cocjin et al. 2014a). Cyclic hardening of soft soil beneath a pipeline undergoing episodic axial movements is routinely relied on in design (White et al. 2015). In addition, pipeline expansion loads transferred to a subsea foundation will be relieved if the system is compliant (whether through spool deflection, sliders or on-bottom foundation sliding). This is in contrast to the environmental loading that dominates foundations for platform structures, for which the compliance of the structure does not lead to a significant reduction of the applied loading.

Periodic monotonic loading with intervening periods of consolidation, such as applied to a subsea pipeline structure, is well-suited to modelling faithfully in the centrifuge environment. In contrast, high frequency undrained cyclic loading such as relevant to storm loading of a foundation for a fixed platform, can incur issues of background consolidation during the cyclic load sequence. This arises because the high frequency of the cyclic loading cannot be faithfully scaled, and while a lower frequency of loading in the centrifuge does not prevent an undrained response for an individual cycle, significant consolidation can occur over the duration of an 'episode' in the centrifuge, that at field scale is an undrained activity.

Figure 9 illustrates potential gains from a 'lifetime' of periodic monotonic operational loading of a skirted mudmat. The foundation was subjected to 10 cycles of sustained horizontal load of $\pm 0.67 H_{cu}$ (where H_{cu} is the capacity after consolidation under self-weight. A separate test had determined that $H_{cu} = 2H_{uu}$, where H_{uu} is the in situ horizontal capacity.). The horizontal load did not fail the foundation, and the cyclic movements were less than 50 mm at prototype scale. The cycles included intervening periods of consolidation of varying durations. After 10 cycles the foundation was pushed sufficiently far to mobilise the post-cyclic undrained horizontal capacity (Gourvenec & White 2017). The greatest gains in capacity were observed for the longer periods of consolidation in between cycles of loading. Post-cyclic capacities of over 3 times the initial undrained capacity were observed. The measured capacities are well matched by predictions with a critical state theoretical framework (Gourvenec et al. 2014, Feng & Gourvenec 2015).

Figure 10 illustrates schematically a more complex scenario of operational loading than shown in Figure 9. This shows how a whole-life geotechnical response can be partnered with operational loading to reduce the required foundation footprint. The loading history shown is indicative of pipeline walking, which begins slowly at first and then stabilizes (Gourvenec et al. 2017a). The peak design load may therefore be partnered with an enhanced, cyclically hardened, seabed shear strength, enabling a smaller foundation footprint.

The whole-life approach to design (shown by the bold solid line), first captures the enhanced soil strength and hence geotechnical foundation resistance

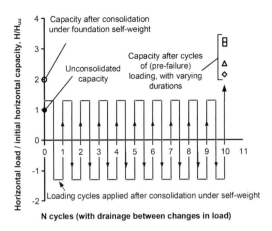

Figure 9. Gain in horizontal capacity of a skirted subsea foundation following cycles of pre-failure horizontal load with intervening consolidation.

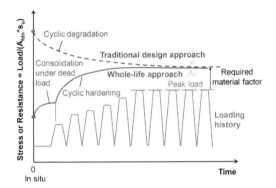

Figure 10. Comparison of whole-life response with traditional design approach applied to subsea foundation design.

from self-weight consolidation and then cyclic hardening in line with the operational loading and intervening consolidation. The foundation resistance, based on the cyclically hardening shear strength exceeds the required material factor at all stages of the life. In contrast, the traditional approach to design (shown by the bold dashed line), adopts an in situ soil strength that is adjusted down for cyclic softening, and hardening is neglected. For the same in situ soil strength, a much larger foundation would be required to ensure the material factor is met under peak loading using the traditional method relative to the whole-life response approach.

Designing a subsea foundation to slide across the seabed offers advantages compared with a static mudmat (i.e. designed to remain stationary on the seabed) with or without a mechanical sliding mechanism. A sliding foundation can realise reduction in footprint compared to a static foundation without a mechanical slider, through relief of applied displacement-sensitive loading, and is attractive over a foundation with a mechanical slider to overcome limitations of

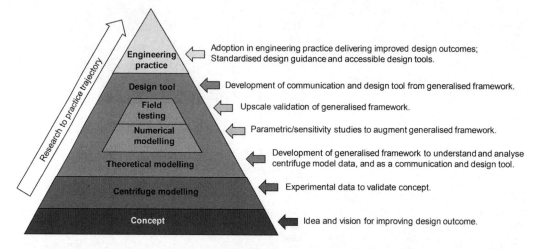

Figure 11. Pyramid of activities illustrating research trajectory from concept to engineering practice.

uni-directional sliding and reducing the risks of failure of moving mechanical parts on a subsea system.

1.4 Role of centrifuge modelling in optimizing geotechnical (offshore foundation) design

The time frames relevant to modelling a whole-life response, on which design guidance could be based, are impractical for laboratory floor or field testing. Tens of cycles of consolidation are required, which involve durations that are impractical to conduct in the field or in model tests at laboratory scale – centrifuge testing is necessary.

Figure 11 shows a pyramid of activities in the trajectory from a conceptual idea to implementation of a design method in engineering practice. A concept with a vision and idea for improving design outcome forms the starting block; centrifuge modelling then provides appropriately scaled experimental data to validate the concept is viable; a generalized theoretical framework is then developed to form the basis of understanding and interpreting the observed performance which, potentially augmented with numerical modelling and field testing, provides the basis for a predictive tool for communication and design; the design tool can then be adopted in engineering practice to deliver improved design outcomes, with the support of formalized design guidance.

The pyramid is applicable for the realization of any new geotechnical technology or methodology from concept to engineering practice – but critically for whole-life response since the time frames involved are impractical to investigate in environments other than a geotechnical centrifuge.

1.5 Structure of the paper

The remainder of this paper will illustrate the role of geotechnical centrifuge modelling in capturing whole-life response to optimize offshore subsea foundation design through consideration of a body of work on tolerably mobile, or 'sliding' foundations.

2 WHOLE-LIFE CENTRIFUGE MODELLING OF A TOLERABLY MOBILE FOUNDATION

2.1 Overview of test program

A program of centrifuge tests was carried out to investigate the response of a tolerably mobile subsea foundation on normally consolidated clay. The purpose of the tests was to provide cycle-by-cycle data on foundation performance, specifically sliding resistance, settlement and tilt, to demonstrate the concept and to provide validation data for a new analysis approach for design.

Each test consisted of 40 cycles of sliding with intervening consolidation representing a sequence of start up, pipeline operation and shut down in the field. A prescribed forward slide of the model foundation represented the foundation response to thermal expansion of the pipeline, a period of consolidation in the 'forward' position represented the period while the pipeline would be operational, and a backward slide, returning to the initial position, represented the foundation response to pipeline contraction during shut down.

A program of three foundation tests considered the effect on foundation response of operative vertical load of the foundation (i.e. buoyant self-weight relative to bearing capacity $q_{op}/q_{ult} = 0.3$ and 0.5) and duration of the consolidation period between sliding cycles ($t_{consol} = 3$ months and 1.5 years). The time histories for the tests carried out are illustrated in Figure 12. The longest continuous test phase lasted 70 years at prototype scale (2.5 days, or 61 hours, at 100g, of continuous centrifuge spinning).

The tests provided cycle-by-cycle data on foundation sliding resistance, settlement and tilt. In addition, characterization tests were performed with a T-bar penetrometer (Stewart & Randolph 1991, 1994), a pile penetrometer (Sahdi et al. 2015, Cocjin et al. 2014b), a piezocone (Randolph & Hope 2004) and a 'piezo-foundation' (Cocjin et al. 2014a, Vulpe &

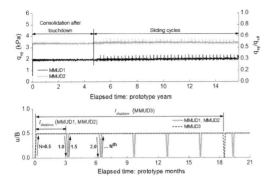

Figure 12. Time histories (prototype scale) of tests.

White 2014). These provided (i) shear strength profiles in undisturbed soil and through the foundation footprint following completion of each test; (ii) cyclic hardening and softening factors; (iii) a continuous representation of the change in shear strength across a lateral profile through the foundation footprint; (iv) and consolidation characteristics. Moisture content cores from undisturbed and disturbed sites were taken to provide vertical stress profiles and a comparison of void ratio change with measured foundation settlement. Full details of the centrifuge tests are reported in Cocjin et al. (2014a). Aspects of the tests are used here to illustrate the whole-life response.

2.2 Model

The foundation model represented a 2:1 rectangular plate with a rough base and smooth skis inclined at 30° from the horizontal. The model foundation had base plate dimensions of 50 mm × 100 mm, representing a 5 m × 10 m field-scale foundation.

The foundation model was fabricated from acetal (polyoxymethylene – POM), a stiff and light material that allowed the self-weight bearing pressure to be representative of field situations. Fine silica sand was glued to the base to provide a rough foundation-soil interface. Photographs of the model mudmat attached to the loading arm (described in the next section) are shown in Figure 13. A rough interface was chosen for the tests to encourage the most 'significant' response – i.e. the greatest possible excess pore pressure generation, cyclic hardening and settlements, to get the best quality data possible. The normally consolidated soil sample was selected for the same reason. A theoretical framework derived from the results from these tests (described later) enables insights into changes in interface roughness (as well as other input variables) on foundation performance.

2.3 Loading arm

The model mudmat was attached to a specially designed loading arm that enabled free rotation of the foundation about two axes and control of vertical load. The arm was connected to the foundation through a hinged joint sited 10 mm above the top of the base

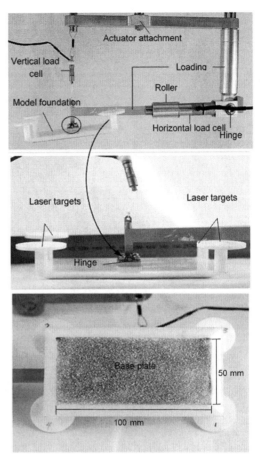

Figure 13. Centrifuge model of sliding foundation mounted on purpose-built loading arm.

plate ($0.1L$). The hinge enabled rotation about the short axis of the foundation, while a roller bearing within the loading arm enabled rotation about the long axis.

A vertically-oriented load cell was connected directly above the centre of the foundation through a padeye to measure the operative vertical foundation-seabed load (relative to the equilibrium state under the self-weight of the foundation and loading arm). An S-shaped axial load cell was positioned in-line with the loading arm between the roller bearing and the hinge at the end of the loading arm to measure the applied horizontal load.

2.4 Instrumentation

Vertical displacement at the four corners of the mudmat foundation were measured by four Keyence® laser displacement sensors (Model LB-70-11) mounted on a steel plate fixed to the actuator. The sensors tracked circular disk targets installed at each corner of the foundation, giving vertical displacements at these locations. The average of the four laser measurements was taken to provide the vertical displacement of the mudmat foundation. Rotation about the centre of the base of the mudmat was determined from the difference between pairs of vertical laser readings while the horizontal travel of the model foundation was measured using the horizontal displacement transducer of the actuator.

2.5 Centrifuge, control and data acquisition

The tests were carried out in a fixed beam centrifuge with a nominal radius of 1.8 m (Randolph et al. 1991) at the National Geotechnical Centrifuge Facility at the University of Western Australia. The tests reported in this paper were carried out at an acceleration level of 100 g. The soil sample was set up in a rectangular strong box with internal dimensions 390 mm by 650 mm by 300 mm high.

A loading actuator with vertical and horizontal axes of motion was used for load and displacement control (De Catania et al. 2010). Data acquisition used a high-speed Ethernet-based system with data streaming in real-time to a remote desktop (Gaudin et al. 2009).

2.6 Soil sample

A well characterised kaolin (Stewart 1992, Acosta-Martinez & Gourvenec 2006) was reconstituted from slurry with a water content twice the liquid limit (120%) and normally consolidated in flight at 100 g for 3.5 days. After consolidation was essentially complete, a thin layer of soil (up to 5 mm or 0.5 m prototype scale, depending on the location in the box) was scraped from the top of the sample to provide a smooth and level surface. After re-equilibration and swelling, the final height of the soil sample was 130 mm, equivalent to 13 m depth at 100 g.

2.7 Soil sample characterization

2.7.1 Soil characterization tools and methods

The intact soil sample was characterized in-flight with a laboratory scale T-bar penetrometer, a piezocone and a piezo-foundation to quantify strength and consolidation parameters. T-bar and novel pile penetrometer tests were carried out in the foundation footprints following testing to determine changes in strength through the whole-life activities. Core samples were also taken from intact and tested sites for determination of moisture content following the testing program. The extensive site investigation campaign underpinned understanding and analysis of the foundation test results. This is particularly the case for 'whole-life' testing in which spatially-varying changes in strength properties will occur during the simulation, and ought to be predicted within any back-analysis.

The tools used are shown in Figure 14 (with the exception of the pile penetrometer that was used only in the foundation footprint and is introduced later in the section on post-testing characterization). The miniature T-bar comprises a cylindrical bar, 20 mm long and 5 mm diameter, mounted on a strain gauged shaft to record resistance to penetration and extraction.

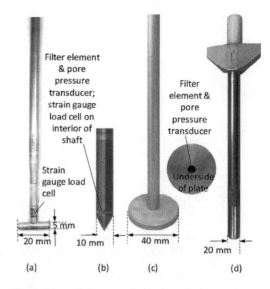

Figure 14. Laboratory scale in situ soil characterization tools (a) T-bar, (b) piezocone (c) piezo-foundation (d) corer.

The piezocone has a 10 mm diameter shaft and 60° cone, equipped with a filter element and pore pressure transducer at the shoulder ('u_2') position and internal integrated strain gauge load cell (similar to the T-bar) to measure penetration resistance. The piezo-foundation is a rigid circular foundation, 40 mm in diameter, instrumented with a central filter element and pore pressure transducer. Control of the bearing pressure, which is held constant during the dissipation test, is provided by a detachable load cell connected via a screw thread between the foundation shaft and actuator. The piston corer is inserted and extracted by hand and the contained core extruded and sliced in discs against the toe of the corer. Samples as thin as 5 mm can be reliably recovered to enable multiple layers in the near surface material to be collected for moisture content determination.

The information from the soil characterization tests was used to select the operative foundation bearing pressures for the sliding foundation tests, determine the degree of consolidation for the waiting times following touchdown and between cycles of sliding and assure that the rate of foundation sliding mobilized an undrained soil response.

2.7.2 Intact and cyclic strength

Figure 15 shows the undrained shear strength derived from the T-bar resistance in the intact sample following full consolidation and scraping of the sample. The linearly increasing with depth profile indicates that full consolidation was achieved, i.e. the sample is not under consolidated, which is indicated by a concave profile of resistance. The test was carried out at a rate to ensure undrained conditions and interpreted with a constant T-bar factor of 10.5 (Stewart & Randolph 1994) giving a mudline intercept undrained shear strength value $s_{um} \sim 0.5$ kPa and gradient of undrained shear strength

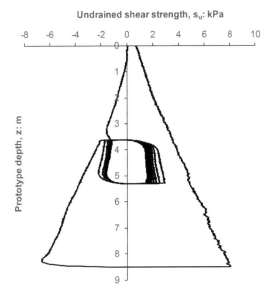

Figure 15. Undrained shear strength from cyclic T-bar test of intact in situ normally consolidated kaolin.

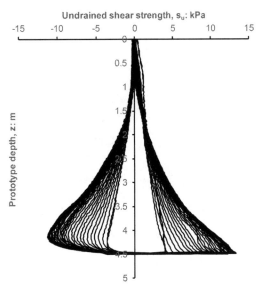

Figure 17. Undrained shear strength from cyclic T-bar test with intervening consolidation.

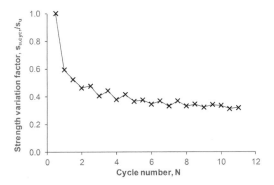

Figure 16. Strength degradation, or cyclic softening, derived from cyclic T-bar test.

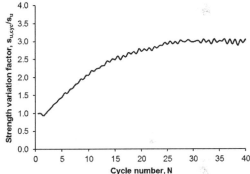

Figure 18. Strength regain, or cyclic hardening, derived from cyclic T-bar test with intervening consolidation.

profile $k \sim 0.9$ kPa/m. The continuous cyclic sequence at depth enables accurate zeroing of the monotonic test and quantifies the degradation of strength due to remoulding, i.e. cyclic softening. The strength reduction factor, defined by the strength during cycle N normalized by the initial intact strength, $s_{u,cyc}/s_u$, for the normally consolidated kaolin used in the study is shown in Figure 16. Strength degradation due to cyclic loading is conventionally described as a remoulding process (Randolph 2004), predominantly due to generation of positive excess pore pressure, but may also reflect a change in the mobilized strength ratio or a degradation of the soil structure (Randolph et al. 2007, White and Hodder 2010). The steady state resistance is indicative of the fully remoulded strength of the soil, which is ~ 0.4 of the intact strength for this soil, and is achieved after approximately 10 cycles. The reciprocal of the strength degradation factor is equivalent to the soil sensitivity, giving a value of $St \sim 2.4$ (Randolph 2004, White & Hodder 2010).

Cyclic T-bar tests over the full depth of penetration and permitting a period of consolidation between cycles were carried out to capture the cyclic hardening characteristics of the soil. The derived shear strength from a cyclic T-bar test with intervening consolidation is shown in Figure 17. The regain of shear strength taken at a depth of 4.0 m is shown in Figure 18 indicating an increase in shear strength of ~ 3 times the initial value. The ratio is similar to the relative penetration resistance during the first penetration of T-bar tests performed at drained and undrained rates in kaolin (House et al. 2001). This point will be revisited later in the interpretation of the foundation tests.

The contrasting behaviour between the two types of T-bar tests indicates that the gain in strength from reconsolidation can eclipse the loss of strength caused by remoulding. This behaviour is consistent with the critical state framework in which undrained failure of a contractile soil generates positive excess pore

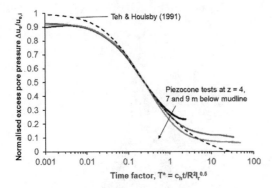

Figure 19. Piezocone dissipation test results.

pressures, which when allowed to dissipate lead to a reduction in moisture content and void ratio and increase in shear strength. The reduction in void ratio is evident in Figure 17 as the first resistance of T-bar is recorded deeper with each cycle. Local settlement near the touchdown point of the T-bar was also observed during testing. This mechanism of soil strength regain due to remoulding and reconsolidation is central to the whole-life response of seabed infrastructure and is linked later to the sliding foundation response.

2.7.3 Moisture content and vertical effective stress profile

The profile of in situ vertical effective stress with depth was derived from moisture content data obtained from core soil samples taken at different locations across the strongbox. The average effective unit weight over the depth of the sample was estimated to be $\gamma' = 5.7\,\mathrm{kN/m^3}$, which is typical for soft clays.

2.7.4 Consolidation characteristics

Horizontal and vertical coefficient of consolidation, c_h and c_v, were determined from piezocone dissipation tests and one-dimensional Rowe cell consolidation tests respectively, while a coefficient of consolidation relevant to shallow foundation loading was determined with a series of load-controlled 'piezo-foundation' tests. Photographs of the piezocone and piczo-foundation are shown in Figure 14 and the instruments are described briefly in Section 2.7.1.

Results from the piezocone tests carried out in the intact normally consolidated kaolin sample are shown in Figure 19. The tests were carried out at three different depths but collapse to a single line when plotted as excess pore pressure normalized by the initial excess pore pressure against dimensionless time factor $T*$. Here $T*$ is taken as $c_h t/R^2 I_r^{0.5}$ with R being the piezocone radius, 5 mm. I_r was taken as 88 following the method proposed by Mayne (2001) and the initial value of excess pore pressure was calculated following the method of Sully et al. (1999).

The coefficient of horizontal consolidation was determined from the piezocone tests with the theoretical solution based on T_{50}, the time for dissipation for 50% of the initial excess pore pressure (Teh &

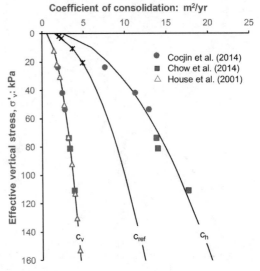

Figure 20. Coefficients of consolidation – c_v, c_h and c_{ref}.

Houlsby 1991). Values of the coefficient of vertical and horizontal consolidation are plotted against vertical effective stress in Figure 20 compared with other data from the public domain on the same material. The ratio of horizontal to vertical coefficient of consolidation is ~ 4.4, similar to although slightly higher than the theoretical prediction for normally consolidated kaolin of 3.5 (Mahmoodzadeh et al. 2015).

The consolidation response relevant to a shallow foundation was determined with the piezo-foundation, essentially a plate loading test with pore pressure measurement. Tests were carried out at two foundation bearing pressures relevant to the planned mudmat tests. The dissipation of excess pore pressure from the piezo-foundation tests are shown in Figure 21 as normalized excess pore pressure against dimensionless time factor $T = c_{ref} t/D^2$.

The reference coefficient of consolidation was determined by fitting with dissipation curves derived from numerical analyses for a rough circular foundation on an elastic half-space (Gourvenec & Randolph 2010) and a rectangular foundation on an elasto-pastic half space (unpublished but result from the analyses presented in Feng & Gourvenec 2015), in the same manner as determination of c_h from piezocone dissipation data. The resulting values of c_{ref} are shown on Figure 20 and can be seen to fall between values for vertical and horizontal coefficient of consolidation and can be given by $c_{ref} = 2.7 c_v$, which can be described by the power law $c_{ref} = 2.7((0.3 + 0.16\sigma_v')^{0.47})$.

2.8 Foundation test procedure

Each foundation test consisted of foundation touchdown followed by a period of consolidation, followed by 40 cycles of sliding with intervening consolidation where a cycle of sliding constituted a forward and backward slide. The period of consolidation following touchdown was determined to allow near full excess

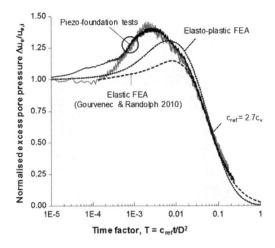

Figure 21. Piezo-foundation dissipation test results.

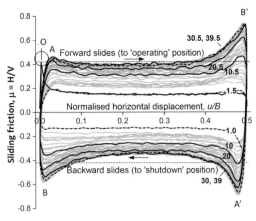

Figure 22. Sliding resistance with sliding distance.

pore pressure dissipation. In terms of the coefficient of consolidation relevant to foundation loading, c_{ref}, the time period corresponded to $T \sim 0.4$, which from the piezo-foundation tests (Figure 21) corresponds to > 95% dissipation of excess pore pressure.

Slides were displacement-controlled with a travel of 25 mm (0.25L, 0.5B) carried out at a rate of 1 mm/s, giving a one-way slide duration of 25 s, or 3 days at prototype scale. While the slide duration in the centrifuge tests was longer than expected in the field (where start up and shut down would last in the order or hours), negligible dissipation of the shear induced pore pressure would dissipate over the slide period, deduced from the piezo-foundation test results, and hence sliding was an undrained event.

Reconsolidation after the forward slide, representing the position of the foundation while the pipeline was operational, was taken as 3 months (780 s or 13 mins in the centrifuge) or 18 months (4680 s or 1 h 18 mins in the centrifuge). Results from the piezo-foundation test dissipation curve (Figure 21) indicates the consolidation periods enable approximately 10% and 70% of pore pressure dissipation respectively.

A pause of 8 s, or 1 day field scale, was modelled at the end of each cycle representing a brief shut down period. During the consolidation periods the foundation was prevented from moving horizontally but was free to settle under the operative vertical load.

2.9 Foundation test results

2.9.1 Cycle-by-cycle foundation sliding resistance

Figure 22 shows the evolution of interface sliding resistance over 40 cycles of sliding and reconsolidation described by the mobilized coefficient of sliding friction, $\mu = H/V$. Results shown relate to an operative vertical load of $0.3q_{ult}$ and period of consolidation t_{consol} of 3 months. Peak resistances are observed at the ends of the slide, but reach a steady residual value over the majority of the sliding movement, and increase with number of loading cycles.

Figure 23. Soil berm developed in centrifuge test as a result of cycles of sliding and reconsolidation.

Peak O reflects increased sliding resistance due to consolidation following set down and consolidation under self-weight at the start of the first slide. Peaks A and A' at the start of the forward and backward slides respectively reflect increased sliding resistance due to reconsolidation following a sliding event. The peak mobilized resistance at position A' is greater than at A since the period of post slide consolidation is greater following the forward slide and is therefore associated with a greater gain in sliding resistance. Peaks B and B' are associated with the interaction of the mudmat ski with soil berms that build up at the edges of the sliding footprint due to the ploughing action of the foundation during sliding and the consolidation settlement.

Figure 23 shows a photograph of the soil berm developed at the limit of the slide of the foundation footprint. The peaks at B' (at the end of the forward slide) are higher than at B (at the end of the backward slide), due to the larger consolidation period, and hence magnitude of consolidation settlements that take place at the end of the forward slide leading to a larger berm.

The evolution of the steady state sliding resistance with number of cycles is shown in Figure 24 for the half-way distance of each slide (i.e. at $u/B = 0.25$) for

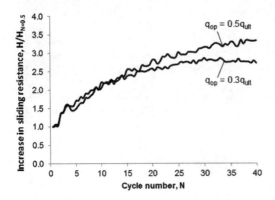

Figure 24. Evolution of sliding resistance with episodes of sliding and reconsolidation.

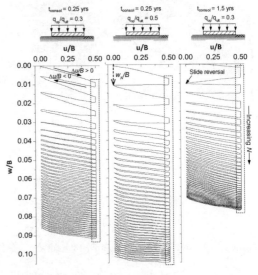

Figure 25. Foundation settlement and tilt with sliding distance.

the two tests with 3 months between slide events. It can be seen that the residual sliding resistance rises by a factor of 3 – 4 with continued cycles of shearing and reconsolidation. A similar rise in resistance is also observed in the response of the T-bar test with intervening consolidation (shown in Figure 17) and has also been observed for pipeline-soil interaction (Randolph et al. 2012, White et al. 2012, Yan et al. 2014).

The underlying mechanism is the same in the T-bar test, pipeline or sliding foundation test i.e. remoulding, or shearing, and reconsolidation to a critical state for the controlling stress. The long-term rise in sliding resistance tends to a drained value. This represents the state at which the soil beneath the foundation has undergone sufficient cycles of sliding, pore pressure generation and reconsolidation to reach the critical state and eliminate any tendency for contraction and further excess pore pressure generation (Cocjin et al. 2015). Depending on the duration of each cycle, this evolution may occur in a single cycle or progressively through multiple episodes of sliding and reconsolidation. Evolution of steady state sliding resistance can be unified for different degree of intervening consolidation by plotting against a normalized time factor, $T_{op} = c_{ref} t/B^2$, as opposed to cycle number, N (Cocjin et al. 2014a).

2.9.2 Cycle-by-cycle sliding foundation settlement

Figure 25 shows the full settlement history of the sliding foundation tests. It can be seen that settlement occurs during the undrained slide (indicated by the change in slope from left to right and right to left) and during consolidation when the foundation is at rest (indicated by the settlement at $u/B = 0.5$, operating position, and to a lesser extent at $u/B = 0$, shutdown position).

The settlement during the slide is initiated due to the settlement bowl sloping towards the operating position. The settlement bowl is developed in the early cycles due to the ploughing action of the foundation in the very soft soil and then sustained by the greater consolidation settlement that accumulates at the operating limit compared to the shutdown limit (since the period of consolidation at the end of the forward slide is much greater than at the end of the backward slide). After approximately 10 cycles, no further undrained shearing settlement is observed and the trajectory of the forward and backward slides are parallel. The inclination of the mudmat during sliding is less than 1° in any of the tests, which is relatively small compared to typical design tolerances.

Consolidation settlements are due to dissipation of shear induced excess pore pressure during sliding and greatest at the forward end of the slide, where the consolidation period is longest, and the rate of consolidation reduces with increasing number of slides.

Evolution of settlement with number of cycles of sliding is summarized in Figure 26 for each of the tests. It can be seen that the heavier foundation ($q_{op}/q_{ult} = 0.5$) is associated with higher settlement as would be expected. This is consistent with the higher sliding resistance mobilized (Figure 24) due to greater reduction in void ratio. Less intuitive is the observation that the prolonged intervening consolidation period reduces the settlements for a given number of cycles. The gain in strength due to consolidation causes the undrained settlement within each sliding movement to be reduced. Therefore, although the extended rest period between cycles allows a greater level of consolidation settlement, the undrained settlements incurred during each sliding movement are reduced.

Overall settlement at the end of the tests of 40 cycles of sliding and reconsolidation for the boundary conditions considered in this study reached between 7.5% and 11% of the foundation breadth (3.75% - 5.5% of the foundation length) depending on foundation weight and period of intervening consolidation.

2.10 Post-test soil characterisation

Characterisation of the soil within and around the footprints after the sliding foundation tests enabled

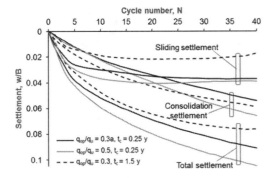

Figure 26. Evolution of foundation settlement and tilt with episodes of sliding and reconsolidation.

observations from the foundation tests to be partnered with changes in undrained shear strength and disturbance of the soil surface. Changes in strength in the foundation footprint were investigated at a discrete position with the T-bar and with cores for determining change in moisture content. A continuous lateral cross section of the foundation footprint area was investigated with a pile penetrometer.

2.10.1 *T-bar testing*

Figure 27 shows undrained shear strength profiles from data of T-bar tests carried out in (i) intact soil, (ii) through the footprint of a piezo-foundation under the operative vertical bearing pressure of the equivalent sliding foundation test, and (iii) through the footprint of a sliding foundation test following 40 cycles of sliding with intervening reconsolidation. The data relates to the same test as the results shown in Figure 22 – Figure 26. The T-bar test was carried out along the centreline relative to the short axis but offset by 0.25L relative to the centre line along the long axis to enable a pile penetrometer swipe to also be carried out through the footprint (as discussed in the next section).

The settlement of the footprint relative to the intact free-surface is clear. The slight difference in magnitude of settlement following the sliding foundation test indicated in Figure 27 compared to Figure 25 is due to the component of post-installation settlement (elastic and consolidation) prior to sliding, which for this case amounted to $w/B \sim 0.035$, or 0.175 m – and is in good agreement with the observed vertical settlement of the piezo-foundation test.

An increase in shear strength under foundation self-weight is apparent showing an increase above the intact value of up to $\sim 20\%$. A considerably greater increase in undrained shear strength is observed in the footprint following the sliding foundation test. The affected depth of influence, $z_{cons} \sim 0.3B$ with the greatest gain in strength near the mudline, more than doubling. This high strain shearing zone generates maximum excess pore pressure during foundation sliding that consequently translates to the considerable gains in strength following the dissipation of excess pore pressure during consolidation. At a depth of $\sim B/2$, the strength

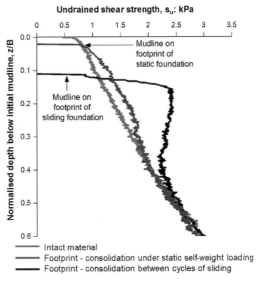

Figure 27. T-bar tests in an intact site, through a foundation footprint under self-weight consolidation only and through a foundation footprint following 40 cycles of sliding and reconsolidation.

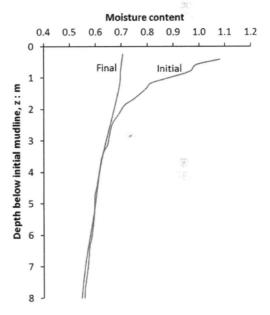

Figure 28. Moisture content in intact soil and in footprint of sliding foundation test after 40 cycles of sliding and intervening consolidation.

profiles in the disturbed sites converge with the intact profile.

2.10.2 *Moisture content cores*

Figure 28 shows the change in moisture content comparing a profile from core taken from an intact site with one from the footprint of a sliding foundation test following 40 cycles of sliding and intervening consolidation. Observed consolidation settlements were

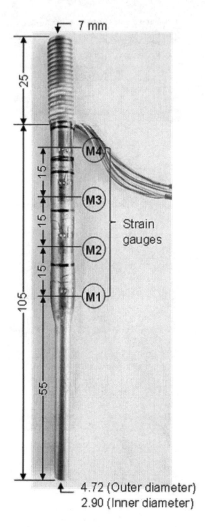

4.72 (Outer diameter)
2.90 (Inner diameter)

Figure 29. Pin pile penetrometer.

consistent with the reduction in void ratio determined from the moisture content data.

2.10.3 Pile penetrometer

The foundation footprint was also characterized with a novel pile penetrometer – so called because it resembles a tubular pile as shown in Figure 29. The pile penetrometer is driven vertically into the soil and then translated horizontally at fixed vertical displacement to determine the horizontal soil resistance (Sahdi et al. 2015, Cocjin et al. 2014b). Pairs of strain gauges are fixed on opposite sides of the device and are oriented to face the direction of motion. Connected via a bridge circuit, they record the bending moments, M_1 to M_4 at the specified elevations.

Figure 30 illustrates the conversion of the bending moment measurements on the pile penetrometer to an idealised profile of net pressure, q_h, acting on the pile.

The pile penetrometer translating horizontally in the soil acts as a cantilever beam subjected to a distributed horizontal soil resistance. The bending moment profile

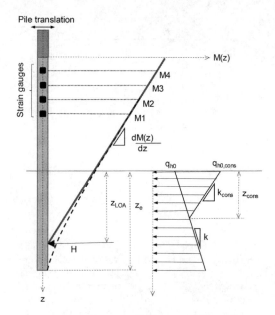

Figure 30. Interpretation of pile penetrometer.

in the strain gauges can be represented by an equivalent net horizontal load, H acting at a point along the pile at a depth, z_{LOA} from the soil mudline (Figure 30). H is equivalent to the total net pressure, q_h acting along the embedded depth (z_e) of the pile penetrometer. The magnitude of H is equal to the slope of the moment profile above the soil, $dM(z)/dz$, and z_{LOA} is the depth where the bending moment distribution projected from the recorded moment loads M_1 to M_4 reaches zero. A least squares method is employed to calculate the two-parameter best-fit moment line from the four measured moments. H and z_{LOA} are then used to calculate the distribution of net pressure, q_h, where the profile can be defined by two unknown parameters. For a linearly increasing profile the two unknowns are mudline value, q_{h0} and gradient with depth, k; and for a bi-linear 'consolidated' profile, the depth at which the strength profile returns to the intact value, z_{cons} and (decreasing) gradient with depth k_{cons}.

Soil strength is then derived by considering the failure mechanism of a pile under pure horizontal translation without rotation, where the undrained shear strength can be derived as $s_u = q_h/N_{pile}$ (Sahdi et al. 2015, Cocjin et al. 2014b). The horizontal bearing capacity factor, N_{pile} is assumed equivalent to the T-bar bearing factor, $N_{T-bar} = 10.5$ over the embedded depth of the pile penetrometer. This is consistent with the use of a constant T-bar factor, N_{T-bar} and neglect of the near-surface variation in N_{pile} and N_{T-bar} has minimal influence in this soft soil.

Comparison with T-bar profiles shows that shear strength profiles in the intact and disturbed soil is predicted well by the pile penetrometer.

The unique aspect of the pile penetrometer is the ability to provide a continuous profile of shear strength along a transverse section and determination

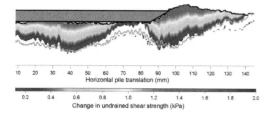

Figure 31. Change in undrained shear strength through foundation footprint following 40 cycles of sliding and intervening reconsolidation derived from pile penetrometer data.

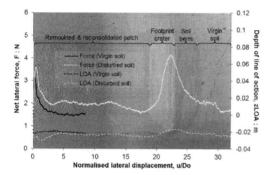

Figure 32. Determination of change in elevation of mudline from pile penetrometer data.

of changes in the elevation of the free surface. Figure 31 shows a contour map of variations in shear strength through the foundation footprint derived from the pile penetrometer. The data reveals a varying and localized change in undrained strength due to shearing and consolidation beneath the foundation and remoulding and consolidation of the material comprising the berm at the extremity of the foundation footprint. Figure 32 shows a topographical laser scan of the soil footprint with the net lateral force and depth of action from pile penetrometer traverse superimposed. The initial peak on the left side of the figure is the post-installation consolidated break-out resistance of the pile penetrometer at the start of the traverse while the second peak corresponds to the increased resistance as the berm is encountered.

3 GENERALIZED APPROACH TO PREDICTING WHOLE-LIFE FOUNDATION AND SEABED RESPONSE

3.1 Overview

Geotechnical centrifuge testing can provide data on a wide range of structure and soil response through instrumentation but rarely can provide a complete picture. Taking the centrifuge modelling program described above, the results provide information on the cycle-by-cycle resistance of the foundation and foundation displacements while information on the soil strength is only available for the intact sample and for the condition at the end of testing (via the post-test characterization) and no information is available on the distribution of shear stresses or excess pore pressure generated beneath the sliding foundation. Insight into the cycle-by-cycle changes in soil stress state and shear strength to partner with the observed foundation response requires theoretical or numerical modelling. Theoretical and numerical modelling are also needed to provide a tool for design – to scale to other combinations of soil strength, consolidation characteristics, foundation size and weight, movement history etc.

In this section, details of a generalized theoretical framework to capture the soil and foundation response to foundation sliding and intervening consolidation is presented (Cocjin et al. 2017a). Some preliminary results of numerical modelling are also shown in the subsequent section. Both the theoretical and numerical models offer opportunities to scrutinize the relationship between changes in soil state and foundation response to support guidance on whole-life design. Theoretical and numerical modelling also offer the opportunity for parametric or sensitivity studies, to explore the effect of changes in input variables on design output that would be impractical through centrifuge modelling.

3.2 Generalized theoretical framework

3.2.1 Conceptual basis of framework

A generalized theoretical model has been developed to capture the soil and foundation response to cycles of sliding and intervening consolidation, relevant to mud-mat foundations of subsea structures such as described in the preceding sections (Cocjin et al. 2017). Validation against the centrifuge model test results has shown that the framework captures the essential elements of the soil-structure interaction, which include (i) the changing soil strength from cycles of sliding and pore pressure generation; (ii) the regain in strength due to dissipation of excess pore pressures (consolidation); and (iii) the soil contraction and consequent settlement of the foundation caused by the consolidation process.

The combined effects on soil strength and infrastructure settlement of cycles of loading and consolidation have been captured by an effective stress framework based on critical state concepts. The basis of the model involves loading leading to generation of excess pore pressure, which when dissipated leads to a reduction in moisture content and void ratio with consequent compression of the soil skeleton, displacement of a supported structure and increase in shear strength of the soil and resistance to loading.

Frameworks based on these principles have been previously developed to analyze gains in foundation capacity from consolidation under (monotonic) self-weight loading (Gourvenec et al. 2014, Feng & Gourvenec 2015, Vulpe et al. 2016); changes in lateral break out resistance of pipelines as a result of self-weight or cyclic installation sequences (Chatterjee et al. 2014); and the effects of cyclic remoulding and consolidation of soil during a T-bar test and to vertical cyclic loading of a pipe element (White & Hodder

2010, Hodder et al. 2013). In these cases, significant components of vertical loading lead to development of a deep zone of influence. In the case of an installation moving horizontally on the soil surface, the shearing process is concentrated close to the surface and the associated generation of excess pore pressure varies with depth according to the distribution of mobilized shear stress (Randolph et al. 2012, Yan et al., 2014). As a result, a one-dimensional approach can be adopted as a practical approximation, in an extension of the well-established 'oedometer method' for prediction of foundation settlement (Skempton & Bjerrrum 1957).

3.2.2 Details of framework

The problem definition for the sliding foundation case is illustrated in Figure 33. An infinite half-space is considered, idealized as a one-dimensional column of soil elements. A constant vertical stress, σ_{op}, representing the submerged self-weight of the foundation and supported structure, and cycles of horizontal shear stress, τ_{op}, representing the effect of the sliding movement of the foundation, δu, are applied at the mudline. The stresses on an element of soil at depth, z below the mudline, σ_v and τ, are defined as a proportion of the mudline value by influence factors I_σ and I_τ according to stress distributions within an elastic half space defined by Poulos & Davis (1974).

Sliding is assumed to take place at an undrained rate with intervening periods of consolidation between sliding events. The soil responds in accordance with a simple critical state model defined by normal and shear stresses, void ratio and a critical state line (CSL) defined in the usual way. The critical state framework developed for the sliding foundation problem is defined in terms of vertical and horizontal stresses in the soil, σ_v and τ, since for the boundary conditions of a sliding foundation these are more convenient than the traditional mean principal effective stress, p', and deviatoric stress, q. The framework can be applied in a cycle-by-cycle manner solving for the response at each soil element in the column to determine the cumulative change in void ratio and the variation in shear stress and settlement at the soil surface. The critical state interpretation of the sliding foundation problem is illustrated schematically in Figure 34.

An initially normally consolidated (point A) or lightly over consolidated (point B) state is considered following consolidation under foundation self-weight. During sliding, positive excess pore pressure is generated, Δu_e, resulting in a decrease in the effective stress, σ'_v, and the stress path moves towards the critical state at constant voids ratio (assuming a rate of sliding to ensure an undrained soil response). Assuming a rough foundation-soil interface and that the ratio of applied shear stress to shear strength is lowest at the surface, the soil element at the mudline level will fail, so the stress state reaches the CSL (B to C). Elements of soil at depth will move towards but not reach the CSL (B to C') at least during the initial cycle. The current undrained shear strength is mobilized at the critical state (point C). During the subsequent period

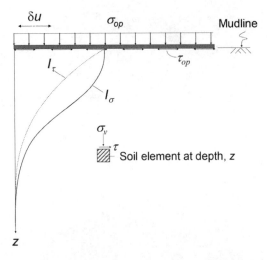

Figure 33. 1D idealization of sliding foundation problem.

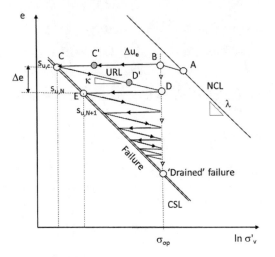

Figure 34. Critical state interpretation of sliding foundation.

of consolidation, the excess pore pressures dissipate either fully (C-D) or partially (C-D') and the vertical effective stress, σ'_v, returns towards the initial condition. The increase in σ'_v, follows the unload-reload line (URL) defined by the slope κ, causing a decrease in void ratio, Δe, and an accumulation of settlement at the soil surface. During the subsequent shearing cycle the soil element will fail at a higher σ'_v, (E) and subsequently mobilize a larger shear stress at failure.

Additional features of the framework include a curved normal compression line (NCL) in e, $\ln \sigma'_v$ space at very low stresses to account for the large changes in void ratio observed in the consolidation periods after the first few slides, following the concepts of the structured critical state model (Liu & Carter 2003); and a migrating critical state line (CSL) towards a limiting lower void ratio as a result of on-going cycles of shearing. The basic critical state parameters M, λ and κ can be derived from traditional

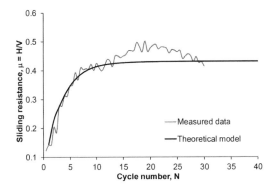

Figure 35. Comparison of predicted and observed sliding resistance.

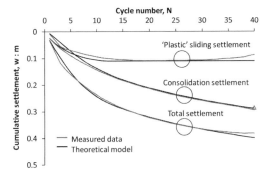

Figure 36. Comparison of predicted and observed settlements.

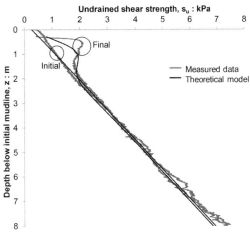

Figure 37. Comparison of predicted and observed undrained shear strength profile following 40 cycles of sliding and intervening consolidation.

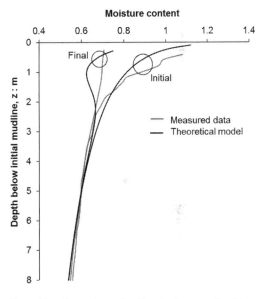

Figure 38. Comparison of predicted and measured moisture content in the intact state and following 40 cycles of sliding and intervening consolidation.

laboratory techniques, while soil parameter inputs specific to the framework can be derived from T-bar testing, a consolidation test (piezo-foundation, piezo-cone or Rowe cell) and moisture content cores. Full details of the formulation of the theoretical framework outlined in this paper are presented in Cocjin et al. (2017). The following section shows that key aspects of sliding foundation response can be captured by the existing framework.

3.2.3 Framework validation using centrifuge data

Figures 35–37 show the predicted response of a sliding foundation from the theoretical framework compared with observations from the centrifuge modelling. Clearly excellent agreement is achievable for the soil and foundation conditions modelled in terms of sliding resistance, settlement and change in soil strength. Further validation is given by the agreement between predicted consolidation settlement at the end of the foundation test (shown by the triangular data point in Figure 36) and the change in void ratio determined from the moisture content data from the footprint (Figure 38). The framework is currently being calibrated against centrifuge test results with the same model foundation in a natural carbonate soil.

3.2.4 Framework as a predictive tool

The framework is generalized such that it can be applied to other soil and foundation conditions provided the general soil response and boundary conditions are the same. The framework provides a powerful tool for first pass predictions of sliding foundation response.

The formulation of the theoretical framework is conducive to programming into an automated calculation tool enabling a quick and easy method for assessing sensitivity of input parameters on cycle-by-cycle foundation and soil response. The framework also enables scrutiny of soil response that cannot be observed or measured in the centrifuge environment, such as in a cycle-by-cycle change in soil strength, stress state and void ratio.

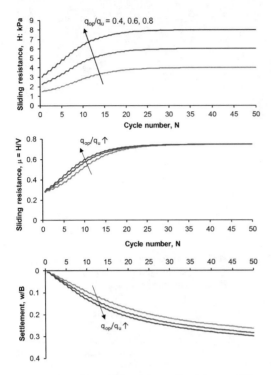

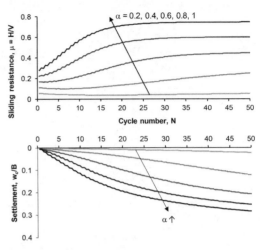

Figure 40. Effect of foundation interface roughness on sliding foundation response.

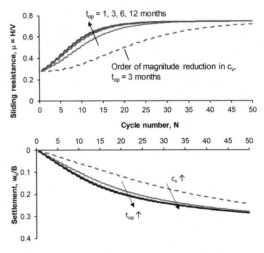

Figure 39. Effect of foundation self-weight on sliding foundation response.

Figures 39–41 show the effect of foundation self-weight, foundation-soil interface roughness and consolidation characteristics on the evolution of sliding resistance and foundation settlement for a hypothetical but realistic set of parameters, predicted by the theoretical framework described above. Figure 39 shows how increasing foundation self-weight leads to higher sliding resistance, due to consolidation of the underlying soil. This is contrast to the undrained sliding resistance based on the in situ mudline strength, which would give the same sliding resistance for all values of self-weight.

The steady state sliding resistance is directly related to the magnitude of self-weight and the results converge to a single line when sliding resistance is normalized by self-weight. The rate of increase of sliding resistance is weakly correlated with self-weight due to the dependency of consolidation coefficient on voids ratio and stress level. Increased self-weight also leads to increased foundation settlements as would be expected, and the framework allows these to be quantified for design purposes.

Figure 40 shows that increasing interface roughness leads to increased normalized sliding resistance and foundation settlement at constant self-weight as higher interface friction leads to mobilization of higher excess pore pressures, which on dissipating lead to greater reduction in void ratio and increase in strength. Figure 41 shows an apparently limited effect of consolidation time between cycles on sliding resistance and foundation settlement for the conditions considered for durations (1 – 12 months). A reduction in coefficient

Figure 41. Effect of consolidation period between slides and coefficient of consolidation on sliding foundation response.

of consolidation by an order of magnitude brings the behavior into a regime of partial drainage where c_h has an influence on the rate and magnitude of steady state sliding resistance and foundation settlement.

Figure 42 shows the evolution of undrained shear strength profile throughout a whole-life of surface sliding and intervening consolidation predicted by the theoretical framework. The representation also captures the reduction in mudline level in the foundation footprint.

As an indication of the usability of the framework, each of the parametric analyses, with results shown in Figures 39–41, took a matter of a minute or less to run.

3.3 Extension of theoretical framework

An extension of the same one-dimensional framework based on a kinematic hardening critical state model, the memory surface model (MSM), (Corti et al. 2016),

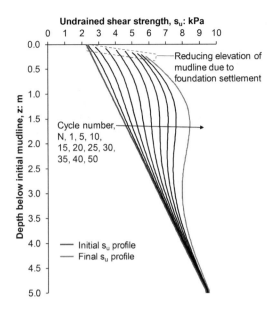

Figure 42. Cycle-by-cycle evolution of undrained shear strength and reduction in mudline level due to sliding and intervening consolidation.

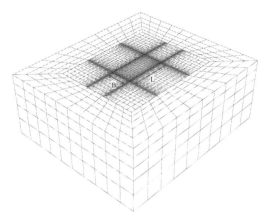

Figure 43. Finite element mesh for 3D small strain sliding foundation analyses.

has also been validated against the sliding foundation centrifuge test results outlined in this paper and presented fully in Cocjin et al. (2014a), (Corti et al. 2017). The boundary value problem was similarly defined (as shown in Figure 33) but the soil response was defined differently. In the MSM, the soil response is limited to drained conditions, calibration of the model parameters relies on drained element tests, which on clay can be challenging to achieve, and the framework cannot capture partial re-consolidation between sliding cycles. The approach is based on the validated assumption that a drained soil response is an appropriate proxy for alternating stages of undrained sliding and consolidation. The MSM framework was shown to capture the key elements of soil structure interaction for sliding with full drainage. The different frameworks show different ways to define the changing state of the soil in response to the whole-life loading sequences – either through a structure term and migrating CSL as in the Cocjin et al. (2017) approach or through kinematic hardening as in the Corti et al. (2017) approach.

4 NUMERICAL ANALYSIS

Finite element analysis also provides a powerful tool for parametric analysis of the sliding foundation boundary value problem and to scrutinize cycle-by-cycle soil response unachievable with centrifuge modelling. Numerical analysis has the benefit of being able to represent more complex boundary conditions than the theoretical model, but lacks in the rigour and efficiency of a theoretical model and is limited by available constitutive models in numerical software. Each approach (centrifuge modelling, theoretical modelling or numerical analysis) is simply a tool in a toolbox,

or block in the pyramid (Figure 11), no individual approach can provide all the information.

Numerical modelling of large amplitude sliding of a foundation requires large deformation finite element analysis (LDFEA) to capture the full picture of foundation and soil response. A code based on the LDFE remeshing and interpolation with small strain (RITSS) method (Hu & Randolph 1998) and constitutive model that captures the softening response and migrating CSL is in progress as part of the body of work described in this paper.

Insights into aspects of sliding foundation performance, rather than a holistic perspective, have been achieved with small strain finite element analysis with a standard critical state constitutive model for minimal computational cost. Three dimensional small strain FEA of sliding and reconsolidation of a surface foundation have been carried out to investigate the effect of partial drainage on sliding resistance and settlement (Figure 43, Feng & Gourvenec 2016). In the analyses, the foundation was translated sufficiently to mobilise the undrained strength of the soil at the interface, thus mobilizing relevant excess pore pressures, but did not model the full field-scale extent of a sliding foundation. The approach is akin to the one-dimensional theoretical model, in which the shear stress at the surface is sufficient to cause failure without capturing the full three-dimensional effects of the slide and sliding foundation footprint. A Modified Cam Clay (MCC) (Roscoe & Burland 1968) was adopted to represent the soil response, as implemented into the commercial finite element software package Abaqus. The results showed that prediction of the rate of rise in resistance and settlement as a function of partial drainage can be based on simple scaling between the zero and full drainage cases (Figure 44). Findings like this are useful in extending the scope of a theoretical framework.

5 FIELD TESTING

Referring to the pyramid in Figure 11, field testing is a potential and valuable activity on the journey from

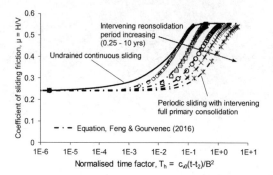

Figure 44. Evolution of sliding resistance under periodic shearing with partial intervening consolidation – results from 3D small strain finite element analyses.

concept to engineering practice to complement centrifuge modelling, theoretical modelling and numerical analysis. Practicality may preclude field testing, but upscaling centrifuge model tests to real field-scale conditions provides an additional layer of reassurance in the centrifuge tests and hence in the theoretical or numerical models validated against them. While complete whole-life sequences may be impractical in field tests, sufficient cycles to ensure the response is comparable with centrifuge test results may often be achievable.

As part of the activities of the OFFshore Hub (http://offshorehub.edu.au/), a program of sliding foundation field tests is planned at a soft clay site to further validate the existing whole-life framework. The field tests will be carried out at Onsøy, Norway, as part of the new Norwegian GeoTest Sites (NGTS) program (L'Heureux et al. 2016). The program of field tests will comprise development of an actuated rig to slide the foundation at constant vertical load with sliding resistance and settlement monitored during the test. The field tests will initially replicate the centrifuge tests presented in Cocjin et al. (2014a) – using many of the same actuation and data acquisition techniques, in scaled up form.

The field tests will inevitably model less cycles than in the centrifuge tests due to time constraints. However, the foundation will be sized to be representative of field scale rather than to field scale – around 1:5 scale so consolidation would be 25 times quicker than the equivalent prototype time in the centrifuge tests. Nonetheless, perhaps only $1/10^{th}$ of the life can practically be captured by field testing. The coefficient of consolidation of the natural clay may be higher than the kaolin, which would also 'speed up' the consolidation time between cycles of sliding.

The foundation field testing program at the Onsøy site will be complemented with field scale piezocone and T-bars, comparable with the centrifuge test program. The site is being additionally characterized as part of the NGTS project, providing greater insights and detail than would otherwise be available.

6 INDUSTRY CODES AND STANDARDS

Following development of a new design methodology, for example through the pyramid of activities illustrated in Figure 11, adoption by industry is greatly assisted if the method is formalized into a guidance note or recommended practice document for designers to use. Verification bodies such as Lloyd's Register, Bureau Veritas and Det Norske Veritas publish guidance notes that sit alongside international design standards and in cases provide guidance beyond that available in the International Standards.

Part of the activities of the OFFshore Hub include development of a guidance note with Lloyds Register on design of shallow foundations for subsea structures. The note will encapsulate the whole-life and tolerable mobility concepts and methods presented in this paper (amongst other subsea foundation related design solutions), which have been validated through centrifuge modelling. The resulting note will provide designers with a go-to document for selection and application of methodologies to optimize a subsea shallow foundation for the design scenario under consideration. Various other aspects of the design approaches beyond the codified approaches have relied on physical modelling for validation, notably mobilization of passive suctions to resist transient tensile load.

Guidance notes provide a pathway for seeking verification for designs based on approaches beyond the standard codified methods. In the future, the notion of 'geotechnical failure' could (and should) be challenged in International Standards (e.g. ISO 2016, API 2011) enabling alternative design methodologies – such as those embracing whole-life response and tolerable mobility - be included as routine established design practice. The evidence base to initiate those changes will require centrifuge modelling.

7 TOOLS FOR ENGINEERING PRACTICE

Development of a clear procedure, tools or software to make design methods accessible in engineering design practice are essential for widespread adoption in industry.

A calculation tool has been developed from the theoretical framework developed from the centrifuge tests, presented in this paper - for predicting foundation response to periodic cycles of sliding and intervening consolidation. The engine of the design tool is the theoretical model developed to aid understanding and interpretation of the centrifuge tests augmented with a user interface for input and output.

The tool is programmed in MATLAB requiring only a text input file to define the magnitude of the variables describing foundation geometry, soil conditions (stress state and constitutive model parameters) and operational loading characteristics (Figure 45). On running the script, the cycle-by-cycle response is calculated and displayed graphically in terms of shear, consolidation and total settlement; normalized sliding resistance; shear stress and shear strength profiles; and the stress state in e-ln σ'_v space (Figure 46).

Output of any of the calculated quantities can be written to file and exported for data analysis, to create figures such as those shown in Figure 39–Figure 42. The tool has also been compiled as an executable application requiring only MATLAB Runtime - a standalone set of shared libraries that enables the execution of compiled MATLAB applications without a licence. While the exe format enables more users, use is limited to editing the magnitude of variables in the input text file and observing the outcomes on the graphics page (shown in Figure 46); the code cannot be edited such that data cannot be written for exporting. The exe format is nonetheless a very useful tool for quick parametric studies to get a 'feel' for the sensitivity of the various inputs on the design output. The calculation tool for predicting foundation response to periodic cycles of sliding and intervening consolidation was developed as part of the activities of the OFFshore Hub and has been circulated to the industry partners of the project (Shell, Woodside, Lloyds Register and Bureau Veritas) under the name of SCAWL (Settlement and Capacity Analysis of Whole-Life response).

Similar calculation tools for other aspects of shallow foundation design are available in the public domain as web-based freeware www.webappsforengineers.com (Gourvenec et al. 2017b, Figure 47). The benefit of deployment of design calculation tools on a web-platform includes the assurance that users have the latest version of the code (rather than a host of local versions remaining in existence on the servers and laptops of individual users).

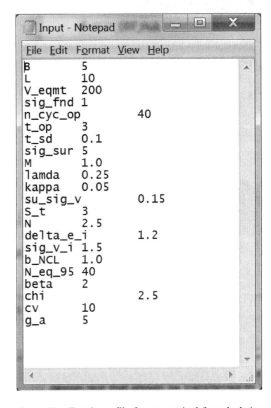

Figure 45. Text input file format required for calculation tool based on theoretical framework for prediction of foundation response to periodic cycles of sliding and intervening consolidation.

8 ENGINEERING IMPACT

Figure 48 demonstrates the improved engineering outcomes available by embracing whole-life design, illustrated through reduction in the area of a subsea mudmat. The conditions considered include a fixed foundation assuming (a) unconsolidated undrained capacity following Feng et al. (2014), (b) consolidated undrained capacity due to vertical self-weight preloading following Feng and Gourvenec (2015), (c)

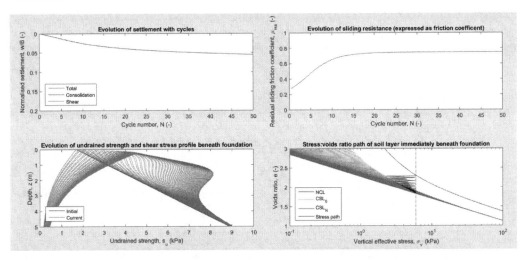

Figure 46. Screen capture of calculation tool output.

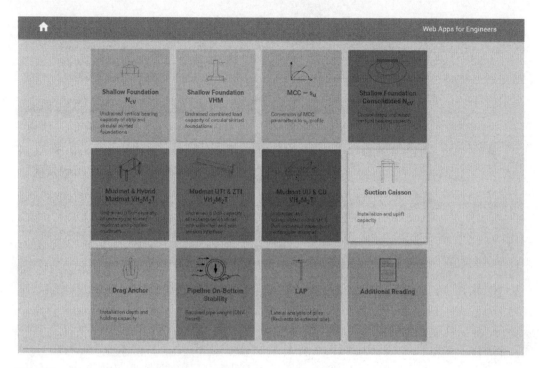

Figure 47. Screen capture of web app landing page www.webappsforengineers.com.

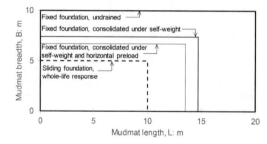

Figure 48. Beneficial design outcomes of whole-life response illustrated by reduction in required foundation footprint.

consolidated undrained capacity due to vertical self-weight and horizontal preloading following Feng and Gourvenec (2017); and a sliding foundation solution using a simplified variant of the theoretical framework presented in this paper (Cocjin et al. 2017) without CSL migration and a linear NCL in e-ln σ_v' space. Model inputs for the example are presented in Gourvenec et al. (2017a) and relate to a 'typical' deepwater seabed condition. The calculation tools for generating the results for the fixed foundation cases a and b are available through the freeware apps www.webappsforengineers.com.

Figure 48 shows that the plan area of the fixed foundation solutions are approximately halved if the effect of consolidation prior to operation are accounted for and that a sliding foundation concept further reduces the minimum required footprint. In the case of the sliding foundation, the size is limited not by capacity but by the settlement, which must be predicted and kept within limits to avoid overstressing the pipeline connections. Such reduction in foundation size can have significant cost benefits to a project.

9 CONCLUDING REMARKS

This paper has set out the role of geotechnical centrifuge modelling in the journey from design concept to engineering practice in capturing whole-life response to optimize geotechnical design outcomes. The principles of whole-life design have been illustrated through a body of work on a novel tolerably mobile sliding foundation concept although the concepts presented have broader application.

Whole-life design relies on partnering 'current' operational soil parameters with corresponding 'current' loading to optimize design outcomes. Changes in soil shear strength in response to whole of life loading sequences, through installation and operation or service, can be identified by geotechnical centrifuge modelling. Centrifuge modelling results can then be used to validate a generalized theoretical framework for use a predictive tool.

The role of centrifuge modelling as a tool for concept validation in the broader context of the development and adoption of a design methodology to improve geotechnical design outcomes has been set out with reference to a pyramid of activities including theoretical modelling, numerical analysis, field testing, formalization as design guidance and formulation of calculation tools for communication and dissemination and use in routine design.

The critical role of centrifuge modelling in validating a new design concept, underpinning the development of a design methodology, has been illustrated with the sliding foundation example. Centrifuge modelling is particularly critical for validation of a design method for whole-life response since the time frames of interest are impractical to model at 1g. Centrifuge modelling requirements for whole-life testing include capability for continuous spinning and detailed site investigation to characterize the intact stress, strength and consolidation properties of the soil and changes in soil state as a result of the whole-life activities.

Whole-life response is a concept that can be applied to a range of geotechnical activities. This paper has illustrated the application of whole-life principles to subsea shallow foundations, but the principles can be applied to prediction of a range of geotechnical boundary value problems including other foundation and anchoring systems, chain response, staged construction, embankment construction and in service performance; and to whole-life loading in its broadest sense, i.e. any external action, including the geotechnical response to humidity, wetting of unsaturated soils, rainfall, temperature, freeze-thaw, and climate change.

In the current environmental and economic climate, established paradigms of design must be challenged to make way for enabling technologies to deliver projects of greater scale and complexity for less risk and cost. Geotechnical centrifuge modelling has a critical role to play in development of new enabling design technologies and methodologies for improved design outcomes.

ACKNOWLEDGEMENTS

Sincere acknowledgement to Michael Cocjin, David White, Mark Randolph, Sam Stanier, Xiaowei Feng and Tianqiang Ja who all play an integral role in the sliding foundation story told in this paper, and to the entire technical team involved in the centrifuge modelling programs carried out at the NGCF at UWA.

The whole-life concepts outlined in this paper related to subsea foundations are being pursued as part of the Industrial Transformation Research Hub in Offshore Floating Facilities, supported by Shell, Woodside, Lloyds Register and Bureau Veritas and the Australian research council (ARC grant IH140100012).

REFERENCES

Acosta-Martinez, H.E. & Gourvenec, S.M. 2006. One-dimensional consolidation tests on kaolin clay. Research Report GEO: 06385 Centre for Offshore Foundations Systems. The University of Western Australia, Perth, Australia.

API 2011. Recommended Practice 2GEO Geotechnical and Foundation Design Considerations, 1st Ed. Washington, American Petroleum Institute.

Andersen, K.H. 2015. Cyclic soil parameters for offshore foundation design, Proc. 3rd Frontiers in Offshore Geotechnics, Oslo, 5-82.

Andersen, K.H., Kleven, A. & Heien, D., 1988. Cyclic soil data for design of gravity structures. Journal of Geotechnical Engineering, 114 (5), 517-539.

Cathie, D., Morgan, N. & Jaeck, C. 2008. Design of sliding foundations for subsea structures. Proc. BGA International Conference on Foundations. Dundee, Scotland, 24–27.

Chatterjee, S., Gourvenec, S. & White, D.J. 2014. Assessment of the consolidated breakout response of partially embedded seabed pipelines. Géotechnique, 64 (5): 391-399. http://dx.doi.org/10.1680/geot.13.P.215

Chow, S.H., O'Loughlin, C.D. & Randolph, M.F. 2014. Soil strength estimation and pore pressure dissipation for free-fall piezocone in soft clay. Géotechnique 64(10): 817-827. https://doi.org/10.1680/geot.14.P.107

Cocjin, M., Gourvenec, S., White, D.J. & Randolph, M.F. 2017. Theoretical framework for predicting the response of tolerably mobile subsea installations Géotechnique 67(7):608-620 http://dx.doi.org/10.1680/jgeot.16.P.137

Cocjin, M.L., Gourvenec, S.M., White, D.J. & Randolph, M.F. 2015. Effects of drainage on the response of a sliding subsea foundation. Proc. 3rd International Symposium on Frontiers in Offshore Geotechnics. Oslo, 777-782.

Cocjin, M., Gourvenec, S., White, D.J. & Randolph, M.F. 2014a. Tolerably mobile subsea foundations – observations of performance. Géotechnique 64(11): 895–909, http://dx.doi.org/10.1680/geot.14.P.098.

Cocjin, M., White, D.J. & Gourvenec, S. 2014b. Continuous characterisation of near-surface soil strength. Proc 32nd Intl. Conference on Ocean and Arctic Engineering (OMAE), San Francisco. OMAE2014-23469.

Colreavy, C., O'Loughlin, C. & Randolph, M.F. 2016. Estimating consolidation parameters from field piezoball tests Géotechnique, 66(4): 333-343. https://doi.org/10.1680/jgeot.15.P.106

Corti, R., Diambra, A., Muir Wood, D., Escribano, D.E. & Nash, D. 2016. Memory surface hardening model for granular soils under repeated loading conditions. J. Engineering Mechanics, ASCE 142(12) http://dx.doi.org/10.1061/(ASCE)EM.1943-7889.0001174

Corti, R., Gourvenec, S., Diambra, A. & Randolph, M.F. 2017. Application of a memory surface model to predict whole-life settlements of a sliding foundation, Computers in Geotechnics, 88:152–163 https://doi.org/10.1016/j.compgeo.2017.03.014

De Catania, S., Breen, J., Gaudin, C., & White, D.J. 2010. Development of a multiple-axis actuator control system. In Proc. Int. Conf. on Phys. Modelling in Geotechnics '10, Zurich (Springman S, Laue J and Seward L (eds)). Taylor & Francis Group, London, UK, pp. 325–330.

Deeks, A., Zhou, H., Krisdani, H., Bransby, M.F. & Watson, P. 2014. Design of direct on-seabed sliding foundation. Proc. 33rd Int. Conf. Ocean, Offshore and Arctic Engng, San Francisco, USA, V003T10A024, http://dx.doi.org/10.1115/OMAE2014-24393

Demel, J., Wallerand, R., Rebours, N., Cafi, M., Grelon, F., Mencarelli, G., Gioielli, P. & Olayera, O. 2016. Pipeline walking mitigation by Anchor In-Line Structures with the use of a Hybrid Subsea Foundation on Erha North Phase 2, Proc. Offshore Technology Conference, Houston, OTC-27295-MS

Dimmock, P., Clukey, E.C., Randolph, M.F., Gaudin, C. & Murff, J.D. 2013. Hybrid subsea foundations for subsea equipment. J. Geotech. Geoenvironmental Eng., ASCE, 129(12):2182-2192.

Dunne, H., Martin, C.M. & Wallerand, R. 2017. Limit analysis of hybrid subsea foundations under horizontal and torsional loading. Proc. 8th Int. Conf. on Offshore Site Investigation and Geotechnics (OSIG), London, UK. 2:818-825.

Feng, X. & Gourvenec, S. 2016. Whole-life cyclic shearing and reconsolidation around a mobile foundation. Géotechnique, 66(6):490-499, http://dx.doi.org/10.1680/jgeot.14-15.P.178

Feng, X. & Gourvenec, S. 2015. Consolidated undrained load-carrying capacity of mudmats under combined loading in six degrees-of-freedom. Géotechnique, 65(7): 563-575. http://dx.doi.org/10.1680/geot./14-P-090

Feng, X., Randolph, M.F., Gourvenec, S. & Wallerand, R. 2014. Design approach for rectangular mudmats under fully three dimensional loading, Géotechnique 64(1): 51-63. http://dx.doi.org/10.1680/geot.13.P.051

Gaudin, C., White, D.J., Boylan, N., Breen, J., Brown, T., De Catania, S., & Hortin, P. 2009. A wireless high-speed data acquisition system for geotechnical centrifuge model testing. Measurement Science and Technology 20(9): 095709. http://dx.doi.org/10.1088/0957-0233/20/9/095709

Gourvenec, S. & White, D.J. 2017. In situ decommissioning of subsea infrastructure, Proc. Conference of Offshore and Maritime Engineering; Decommissioning of Offshore Geotechnical Structures, Hamburg, Germany. Keynote 3-40 ISBN-13: 978-3-936310-40-5

Gourvenec, S., Feng, X., Randolph, M.F. & White, D.J. 2017. A toolbox approach for optimizing geotechnical design of subsea foundations – special session Proc. Offshore Technology Conference, Houston, OTC-27703-MS

Gourvenec, S. & Randolph, M.F. 2010. Consolidation beneath circular skirted foundations. International Journal of Geomechanics 10(1):22–29, http://dx.doi.org/10.1061/(ASCE)1532-3641(2010)10:1(22).

Gourvenec, S., Stanier, S.A., White, D.J., Morgan, N., Banimahd, M. & Chen, J. 2017a. Whole-life assessment of subsea shallow foundation capacity, Proc. 8th Int. Conf. on Offshore Site Investigation and Geotechnics (OSIG), London, UK. 2:1151-1160.

Gourvenec, S., Feng, X., Randolph, M.F. & White, D.J. 2017b. A toolbox approach for optimizing geotechnical design of subsea foundations – special session Proc. Offshore Technology Conference, Houston, OTC-27703-MS

Gourvenec, S., Vulpe, C. & Murthy, T. 2014. A method for predicting the consolidated undrained capacity of shallow foundations on clay. Géotechnique, 64(3):215–225, http://dx.doi.org/10.1680/geot. 13.P.101

Hodder, M.S., White, D.J. & Cassidy, M.J. 2010. Analysis of soil strength degradation during episodes of cyclic loading. Int. J. Geomechanics, 10(3), 117-123.

Hodder, M.S., White, D.J. & Cassidy, M.J. 2013. An effective stress framework for the variation in penetration resistance due to episodes of remoulding and reconsolidation. Géotechnique 63(1):30–43 http://dx.doi.org/10.1680/geot.9.P.145

Hossain, M.K., Shi, H., Abdalla, B. & Spari, M.K. 2015. Understanding hybrid subsea foundation design. Proc 33rd Intl. Conference on Ocean and Arctic Engineering (OMAE), Newfoundland, Canada. OMAE2015-42214

House, A., Olivera, J.R.M.S. & Randolph, M.F. 2001. Evaluating the coefficient of consolidation using penetration tests. Int. J. Physical Modelling in Geotechnics 1(3): 17–25.

Hu, Y. & Randolph, M.F. 1998. A practical numerical approach for large deformation problems in soil, Int. J. Numerical and Analytical Methods in Geomechanics, 22:(5): 327-350.

ISO 2016. ISO19901-4: Petroleum and natural gas industries specific requirements for Offshore Structures - Part 4: Geotechnical and foundation design considerations - 2nd Edition. Geneva, International Standards Organisation.

Jayson, D., Delaporte, P., Albert, J-P., Prevost, M.E., Bruton, D. & Sinclair, F. 2008. Greater Plutonio project – Subsea flowline design and performance. Proc. Offshore Flowline Technology Conference, Amsterdam.

L'Heureux, J.S., Degago, S., Instanes, A., Nordal, S. & Sinitsyn, A. 2016. The Norwegian GeoTest Sites (NGTS) project – An overview. Proc. GeoVancouver, 69th Canadian Geotechnical Conference.

Liu, M.D. & Carter, J.P. 2003. Volumetric Deformation of Natural Clays. International Journal of Geomechanics, 3(2): 236–252, http://dx.doi.org/10.1061/(ASCE)1532-3641(2003)3:2(236).

Lyndon, A. & Schofield, A.N. 1970. Centrifugal model test of a short-term failure in London Clay. Géotechnique 20(4): 440–442.

Mahmoodzadeh, H., Wang, D. & Randolph, M.F. 2015. Interpretation of piezoball dissipation testing in clay. Géotechnique, 65(10): 831-842. https://doi.org/10.1680/jgeot.14.P.213

Mayne, P.W. 2001. Stress-strain-strength-flow parameters from enhanced in-situ tests. Proc. Int. Conf. on In-Situ Measurement of Soil Prop. & Case Histories, In-Situ 2001. Bali, Indonesia, 27–47.

Poulos, H.G. & Davis, E.H. 1974. Elastic solutions for soil and rock mechanics. New York: John Wiley & Sons, Inc.

Randolph, M.F. 2004. Characterisation of soft sediments for offshore applications. Proc. 2nd Int. Conf. on Site Characterisation, Porto, Portugal, 209–231.

Randolph, M.F. & Hope, S. 2004. Effect of cone velocity on cone resistance and excess pore pressures', Proc. IS Osaka - Engineering Practice and Performance of Soft Deposits, Osaka, Japan, 147-152.

Randolph, M.F., Low, H.E. & Zhou, H. 2007. In situ testing for design of pipeline and anchoring systems, Proc. 6th International Conference on Offshore Site Investigation and Geotechnics, London, UK, 251-262.

Randolph, M.F., Yan, Y., & White, D.J. 2012. Modelling the axial soil resistance on deep-water pipelines. Géotechnique 62(9): 837–846. http://dx.doi.org/10.1680/geot.12.OG.010

Randolph, M.F., Jewell, R.J., Stone, K.J. & Brown, T.A. 1991. Establishing a new centrifuge facility. Proc. Int. Conf. on Centrifuge Modelling, Centrifuge '91, Boulder, 3–9.

Roscoe, K.H. 1970. The influence of strains in soil mechanics. Géotechnique 20(2):129–170.

Roscoe, K.H. & Burland, J.B. 1968. On the generalized stress-strain behaviour of wet clay. In Engineering Plasticity. (eds J., Heymen, and F. Leckie) Cambridge University Press, Cambridge, 535-609.

Sahdi, F., White, D.J., Gaudin, C., Randolph, M.F. & Boylan, N. 2015. Laboratory development of a vertically oriented penetrometer for shallow seabed characterization. Canadian Geotechnical Journal, 53(1), 93-102. https://doi.org/10.1139/cgj-2015-0165

Skempton, A.W. & Bjerrrum, I. 1957. A contribution to the settlement analysis of foundations on clays, Géotechnique, 7: 168-178.

Small, A., Cooke, G., Egborge, R. & Ejidike, A. 2015. Geotechnical aspects of North Sea Decommissioning,

Proc. International Ocean and Polar Engineering Conference, Kona, Hawaii, 1308 – 1314.

Stewart, D.P. 1992. Lateral loading of piled bridge abutments due to embankment construction. PhD thesis, The University of Western Australia.

Stewart, D.P. & Randolph, M.F. 1991. A new site investigation tool for the centrifuge. In Proceedings of international conference on centrifuge modelling, Centrifuge '91, Boulder. 531–538. Rotterdam, the Netherlands: Balkema.

Stewart, D.P. & Randolph, M.F. 1994. T-bar penetration testing in soft clay. Journal of Geotechnical and Geoenvironmental Engineering 120: 2230–2235. http://dx.doi.org/10.1061/(ASCE)0733-9410(1994)120:12(2230)

Stuyts, B., Wallerand, R. & Brown, N. 2015. A framework for the design of sliding mudmat foundations. Proc. 3rd Int. Symp. Frontiers in Offshore Geotechnics (ISFOG), Oslo. 2:807-818

Sully, J.P., Robertson, P.K., Campanella, R.G., & Woeller, D.J. 1999. An approach to evaluation of field CPTU dissipation data in overconsolidated fine-grained soils. Canadian Geotechnical Journal 36: 369–381. http://dx.doi.org/10.1139/t00-072

Teh, C.I. & Houlsby, G.T. 1991. Analytical study of the cone penetration test in clay. Géotechnique 41(1):17–34. http://dx.doi.org/10.1016/0148-9062(91)91308-E

Vulpe, C. & White, D.J. 2014. The effect of prior loading cycles on the vertical bearing capacity of clay: an experimental study. Int. J. Physical Modelling in Geotechnics 14(4): 88-98. http://dx.doi.org/10.1680/ijpmg.14.00013

Vulpe, C., Gourvenec, S. & Cornelius, A. 2016a. Effect of embedment on consolidated undrained capacity of skirted circular foundations in soft clay under planar loading. Canadian Geotechnical Journal. http://dx.doi.org/10.1139/cgj-2016-0265

White, D.J., Campbell, M.E., Boylan, N.P. & Bransby, M.F. 2012. A new framework for axial pipe-soil interaction illustrated by shear box tests on carbonate soils. Proc. Int. Conf. on Offshore Site Investigation and Geotechnics. SUT. 379-387.

White, D.J., & Hodder, M. 2010. A simple model for the effect on soil strength of episodes of remoulding and reconsolidation. Canadian Geotechnical Journal 47(7):821–826. http://dx.doi.org/doi:10.1139/T09-137

Wallerand, R., Stuyts, B., Blanc, M., Thorel, L. & Brown, N. 2015. A design framework for sliding foundations: Centrifuge testing and numerical modelling. Proc. Offshore Technology Conference, Houston, OTC-25978-MS

Wallerand, R., Cafi, M., Kay, D., Dimmock, P. & Randolph, M.F. 2017. Hybrid subsea foundations – from research to project application. Proc. Int. Conf. on Offshore Site Investigation and Geotechnics. SUT. 802-809.

White, D.J. & Hodder, M. 2010. A simple model for the effect of soil strength of episodes of remoulding and reconsolidation. Can. Geotech. Journal. 47(7):821–826, http://dx.doi.org/10.1139/T09-137.

Xiao, Z., Tian, Y. & Gourvenec, S.M. 2016. A practical method to evaluate failure envelopes of shallow foundation considering soil strain softening and rate effects., Applied Ocean Research, 59: 395-407, http://dx.doi.org/10.1016/j.apor.2016.06.015.

Yan, Y., White, D.J. & Randolph, M.F. 2014. Cyclic consolidation and axial friction for seabed pipelines. Géotechnique Lett. 4:165–169, http://dx.doi.org/10.1680/geolett.14.00032

Zografou, D., Boukpeti, N., Gourvenec, S. & O'Loughlin, C. 2016. Definition of failure in cyclic direct simple shear tests on normally consolidated kaolin clay. Proc. 5th Conf. Geotechnical and Geophysical Site Characterisation, Queensland, Australia, 1:583-588.

Development of geotechnical centrifuges and facilities in China

Y.J. Hou
China Institute of Water Resources and Hydropower Research, Beijing, China

ABSTRACT: While geotechnical centrifuges continued to grow in number and applications worldwide, centrifuge facilities started being developed in China in the 80s before reaching a flourishing period in recent years in number, size as well as simulation capacity. This paper introduces the past efforts put into geotechnical centrifuge development in China. The situation as it stands nowadays and plans for the future of Chinese Geotechnical Centrifuges are presented. Some of the auxiliary facilities for centrifuge model tests developed at the China Institute of Water Resource and Hydropower Research (IWHR) are also introduced.

1 INTRODUCTION

Geotechnical centrifuge modelling relies on a combination of technological developments in several fields of engineering, including mechanics, hydraulics, steel and materials, automation and controls, as well as the progress in civil engineering. Early in 1867, French engineer Edouard Phillips presented the idea of using a centrifuge to obtain similarity of stresses between models and prototypes and tried to apply his idea to bridge engineering. About 64 years later, the first appearance of a centrifuge was recorded at Columbia University (Bucky, 1931), used to perform some simple tests on mine roof structures built in rock. At almost the same time, USSR engineers Davidenkov and Pokrovskii (1936) developed a centrifuge of their own design and presented their research results at the First International Conference on Soil Mechanics and Foundation Engineering in 1936. But it was during the period from 1960 to 1990 that geotechnical centrifuge development really took off and that applications flourished due to the contributions of an increasing number of centrifuge researchers and enthusiasts, helped by the innovations in instrumentation and the rapid progress of computer science as well as the increasing capability of numerical analysis of that time.

Nowadays more and more geotechnical engineers realise the power and benefit of physical modelling on a centrifuge and apply it to solve a vast variety of geotechnical problems, such as those encountered in the design of high dams, slopes, deep foundations, piles and channels. Professors and researchers worldwide are attempting to discover and apply new mechanisms to geotechnical problem solving and to expand the theory of soil mechanics in the process, all with the help of centrifuge modelling. Appropriate tools are a prerequisite to the successful execution of any work. Therefore, improved centrifuge machines and facilities need to be developed involving less system errors, higher g levels where needed, larger and heavier payloads to allow bigger models to be embarked, and generally better centrifuge performances. Modern geotechnical engineering always requires the development of new appropriate geotechnical centrifuges capable of addressing the latest needs, incorporating the latest technological advances in centrifuge manufacturing.

After so many years of trial and error in geotechnical centrifuge design, two main designs have emerged: beam centrifuges and drum centrifuges rotating in a horizontal plane around a vertical axis. Beam centrifuges are more common and widespread as they allow large sized containers to be easily placed on their basket and are generally easier to operate. Researchers working on geostructures related to water in general, or the ocean in particular, will prefer to use drum centrifuges, in which the model is a continuous circumferential band of soil or water, as described by Schofield (1978). Originally most of the beam centrifuges were symmetrical, especially those with high rotation speeds, with the symmetrical basket used as counterweight. Most of the centrifuges in China adopt this type of design, which requires manual balancing before operating the centrifuge. As such large symmetrical centrifuges will require more labour and preparation time. But they also present the limitation of a hard to achieve real-time in-flight balancing in great part due to the deformation or collapse of the model inside the container in one of the baskets. Some of the advantages of unsymmetrical beam centrifuges are that they reduce power consumption and allow implementation of automatic in-flight balancing methods, such as the one provided by Actidyn Systèmes of France. However, the response time of the auto-balancing system may in some instances constitute a design challenge.

China started to set up geotechnical centrifuges since the 80's and has seen extremely rapid developments in recent years owing to economic growth and the government's emphasis on the importance of science and technology. This paper presents the past effort of geotechnical centrifuge development in China. Some of the associated facilities for centrifuge model tests developed in China Institute of Water Resource and Hydropower Research are also introduced.

2 DEVELOPMENT OF GEOTECHNICAL CENTRIFUGES IN CHINA

2.1 Period from 1983 to 1999

Following the growing international trend of geotechnical centrifuge research in the 50's and 60's, centrifuge feasibility studies were launched in China in 1960. These were unfortunately abruptly interrupted by the Cultural Revolution at that time. It was not until 1983 that the first 180 g-t geotechnical centrifuge (Fig. 1a) was set-up at the Yangzi River Scientific Research Institute. The radius at platform of the centrifuge was 3.0 m with a maximum carrying capacity of 600 kg, driven by a 450 kW DC motor capable of a maximum centrifuge acceleration of 300 g.

Initially a lot of original model tests were conducted with this 180 g-t centrifuge for projects related to dam construction, embankments, slopes, piles and soft foundations (Bao, 1991). Several young researchers from other institutions got their experiences of centrifuge modelling at this laboratory.

This centrifuge was decommissioned in 2002 and was later it was replaced with a modern 200 g-t modern centrifuge (Fig. 1b) in 2008 at the same location.

Several centrifuges (Table 1) were developed in China from 1983 to 1993. Two large centrifuges with a capacity exceeding 400 g-t were put in operation in 1991 and 1992 respectively. The largest centrifuge developed during this period was the 450 g-t Centrifuge at IWHR (Fig. 2), which passed acceptance testing in 1991 (Du et al., 1994). Figure 3 shows its old data acquisition system console in the main control room. Nowadays the same duties are performed by a single desktop computer. The centrifuge adopts a symmetrical arm design with a model basket at one end of the arm and a counterweight basket at the other. Its maximum radius is 5.03 m, maximum acceleration 300 g, effective payload 1.5 ton, motor power 700 kW, and the size of the model swing basket 1.5 m × 1.0 m × 1.2 m. The centrifuge still works very well after 27 years of operation, though after several upgrades of its drive and data acquisition systems.

Most Chinese engineers involved in geotechnical engineering were still not familiar with centrifuge model testing at the end of last century. Some research institutes and universities having centrifuge facilities were working hard to introduce the new physical modelling tools to related industries. Researchers at IWHR adopted centrifuge modelling mainly for the simulation of the behaviour of high dams. Most high dams in China with a height between 100 m to 260 m were tested on IWHR's centrifuge, such as Tianshenqiao concrete face slab rockfill dam (178 m in

Figure 1. (a) Old and (b) new looking of centrifuges in Yangzi River Scientific Research Institute.

height), Xiaolangdi inclined core earth dam (154 m), Nuozhadu core earth rockfill dam (261 m) etc. The centrifuge test results were mainly used to confirm the validity of the structural design or compare with those from numerical analysis.

Sometimes centrifuge testing may be the only solution to get appropriate results for construction. For example, the design of the 90m high cofferdam of the Three Gorges Project encountered the problem that most of its parts had to be constructed under 60 m deep river water by dropping soil and rocks from a boat. The density of those materials dropped under water had to be carefully defined. There was no experience in the past and no reliable applicable numerical method. Some 1 g model tests were performed in a channel pond with a water depth of 5∼6 m. After dropping the material into water to form a dam, the densities of the model dam were tested layer by layer, and the results showed that the maximum dry density of the dropped material was only 1.45 g/cm^3 on average, which means more gentle slopes were needed for the cofferdam and high cost seepage control measures had to be adopted.

The cofferdam material was weathered sand consisting of orthoclase, quartz, mica and other minerals. Its average specific gravity is 2.76, natural water content 5%–10%. Several centrifuge model tests were conducted with IWHR centrifuge, on the material with 40 cm in thickness and filled into the water with the same depth, with the designed scale ratio N = 150. The material was excavated from the construction site with a maximum particle size of 20 mm, so the original material can be directly applied for model preparation in a container with the size of 1.35 × 0.40 × 0.90 m (L × W × H).

All the models were tested at the acceleration of 150 g. LVDT and side markers were used to monitor

Table 1. Geotechnical centrifuges in China.

No.	Organisation	Max. radius (m)	Max. acc. (g)	Payload (kg)	Capacity (g-t)	Size of basket L×H×W (m³)	Year of operation
1	Yangzi River Scientific Research Institute	3	300	600	180	0.7×0.7×0.82	1983
2	Hehai University	3.4	250	100	25	0.9×0.2×0.35	1983
3	Nanjing Hydraulic Research Institute	2.1	250	200	50	0.7×0.35×0.5	1989
4	China Institute of Water Resources and Hydropower Research (IWHR)	5.03	300	1500	450	1.5×1.0×1.5	1991
5	Chengdu University of Technology	1.5	250	100	25	0.6×0.4×0.4	1991
6	Nanjing Hydraulic Research Institute	5	200	2000	400	1.2×1.2×1.1	1992
7	Tsinghua University	2	250	200	50	0.75×0.5×0.6	1993
8	Hong Kong University of Science and Technology	4	150	3000	400	1.5×1.5×1.0	2001
9	Southwest Jiaotong University	2.7	200	200	100	1.0×0.8×1.0	2003
10	Changan University	2.7	200	600	60	0.87×0.68×0.75	2005
11	Chongqing Jiaotong University	2.7	200	600	60	0.86×0.52×0.75	2006
12	Tongji University	3.5	200	1500	150	1.2×0.9×1.0	2007
13	Yangzi River Research Institute	3.7	200	1000	200	1.2×0.9×1.5	2008
14	Dalian University of Technology	0.7	600	750	450	Drum type	2009
15	Changsha University of Technology	3.5	150	1500	150	1.2×0.9×1.0	2009
16	Zhejiang University	5	150	2700	400	1.5×1.2×1.5	2010
17	Chengdu University of Technology	5	250	2000	500	1.4×1.5×1.5	2013
18	Tianjin Research Institute for Water Transport Engineering	5	250	2000	500	1.4×1.5×1.5	2015
19	Institute of Engineering Mechanics	5	100	3000	300	1.5×1.3×1.6	2017
20	Shanghai Jiaotong University	1.1	200	2000	400	Drum type	2018 Under Installation
21	Tianjin Port Engineering Design Consulting Company	4	200	1000	200	1.3×1.3×1.3	2018 Under Installation
22	Zhengzhou University	5	200	6000 at 100g	600	1.4×1.5×1.5	2018 Under Installation
23	China Institute of Water Resources and Hydropower Research (IWHR)	8.5	350	5000 at 200g	1000	2.0×2.0×2.0 2.0×1.3×1.5	2020 Under design
24	IWHR	4.5	1000	2000 at 200g	400	1.5×1.0×1.2 0.5×0.6×0.8	2020 Under design

Figure 2. IWHR 450 g-ton centrifuge4.

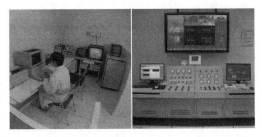

Figure 3. The old and new data acquisition system for IWHR centrifuge.

the model settlement and later for calculating the volume for density evaluation. There are mainly two types of weathered sands with different proportion of particle size greater than 5 mm (P5), namely P5=35% and P5=61%. The model with P5=35% had the dry density varying from 1.73 g/cm³ to 1.83 g/cm³, after three sample tests from model top to bottom. The average of the model dry density was 1.82 g/cm³ for the material P5=35%, and 1.85 g/cm³ for the coarse material P5=61%, much higher than the results from 1 g modelling. Centrifuge model tests provided a reliable density value for the cofferdam design, and this value was finally confirmed correct by in-situ test

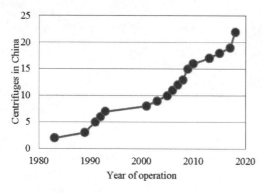

Figure 4. The tendency of geotechnical centrifuge development in China.

Figure 5. Symmetrical centrifuge (400 g-t) at Zhejiang University.

Figure 6. Unsymmetrical centrifuge (500 g-t) at Tianjin Research Institute for Water Transport Engineering.

in the cofferdam dam before it was removed (Li and Cheng, 2005).

2.2 Period from 2000 to 2017

During the nearly 7 years preceding 2000, centrifuge development in China stalled due, on one hand, to the unawareness of most engineers of the capabilities of centrifuge modelling and on the other hand to the economic condition of that time. The 21st Century saw a rapid increase in the number, size and simulation capacity of geotechnical centrifuges in China, as shown in Figure 4. More than 20 geotechnical centrifuges are now in operation in China with a capacity ranging from 50 g-t to 500 g-t (Table 1). Another large centrifuge with a 600 g-t capacity is under installation at Zhenzhou University at the time of this publication. Some even larger centrifuges are currently under design or planning, such as the 1000 g-t centrifuge under construction at IWHR and another huge multipurpose geotechnical centrifuge at Zhejiang University. Most of the newly built arm centrifuges are equipped with earthquake simulators, robots and other associated in-flight facilities for simulation and test.

The rapid increase of the number of centrifuges in China also required and triggered a large effort to be put into local design and manufacturing, an effort made by the General Research Institute (GRI) of the China Academy of Engineering Physics, which has designed and manufactured more than 40 centrifuges in different fields including a dozen geotechnical centrifuges. Figure 5 and Figure 6 show the centrifuges produced by GRI for Zhejiang University and Tianjin Institute for Water Transport Engineering.

The first drum centrifuge (Fig. 7) was set up at Dalian University of Technology, manufactured by Broadbent of the UK. The effective radius of the centrifuge is 0.6 m with appropriate acceleration of 514 g. The width of the model channel is 0.35 m, and the maximum capacity of the centrifuge is 450 g-t. The main advantage of a drum centrifuge resides in the circular model container, which provides a suitable environment for simulating marine pipelines, deep water slopes, or geotechnical structures under water forces or sea waves. A number of model tests related with marine engineering have been conducted with this drum centrifuge at Dalian University of Technology. Another drum centrifuge is currently being installed at Shanghai Jiaotong University, manufactured by Actidyn Systèmes of France. Including the centrifuges currently under construction, a wide variety of geotechnical centrifuges are now available in China for physical modelling, ranging mostly from 100 g-t to 1000 g-t in capacity as shown in Figure 8.

2.3 After 2018

The trend in China for the coming years is that of a continuous increase in centrifuge construction, with special emphasis on higher structures, complicated environment and higher accuracy. Industry specifications or standards in China have greatly contributed to the promotion of the application of centrifuge modelling in practical engineering problem solving, making it a conventional geotechnical test method.

Several universities and research institutes have the ambition to set-up their own centrifuges in the next five years. A mega centrifuge (Fig. 9) with a 1000 g-t capacity is being designed for IWHR, based on the need for simulating the behaviour of dams and slopes with a height around 300 m. It will be manufactured by Actidyn Systèmes of France and will be commissioned in 2020. The main technical parameters of this centrifuge are as follows:

– Maximum radius: 8.5 m
– Maximum acceleration: 350 g

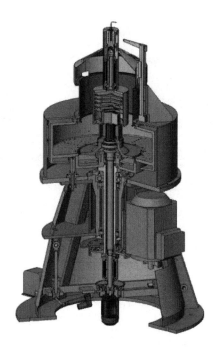

Figure 7. Drum centrifuge at Dalian University of Technology.

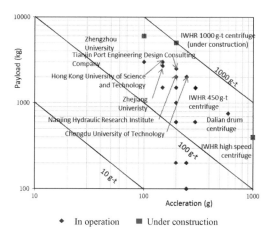

- In operation   ■ Under construction

Figure 8. The capacity distribution of centrifuges in China.

- 350 g-carrying capacity: 2500 kg
- 200 g-carrying capacity: 5000 kg
- Maximum capacity: 1000 g-t
- Size of 200 g swing basket: 2.0 m × 1.5 m × 2.0 m
- Size of 350 g swing basket: 2.0 m × 1.3 m × 1.5 m
- Continuous operation time: 48 h
- Maximum motor power: 2600 kW
- Accuracy of acceleration: 0.1 g, with dynamic balancing system
- Balanced earthquake simulator operating up to: 100 g

Geotechnical centrifuge studies related with many different kinds of research topics have been reported in the past decade owing to the growing number of research groups in centrifuge modelling. However, the number of researchers familiar with centrifuge modelling is no match for the rapid increase in the number of centrifuge laboratories. Some minor and not catastrophic incidents often occur in newly built laboratories. With the increasing number of centrifuges in China, it is important to stay vigilant and an emphasis on the safety of centrifuge operation is crucial to avoiding accidents. Sometimes for the newly built laboratories, test results have to be discarded due to improper operation or handling during modelling and testing. Some of the test results are biased due to lack of calibration and errors in analysis of the container, transducers, the shaft supporting the displacement sensors as well as of the influence on the model's behaviour of centrifuge acceleration and deceleration cycles. This situation has been improved nowadays after several years of centrifuge operation, with the accumulation of experience, understanding of simulation laws and modelling techniques, and the proper use of sophisticated equipment and monitoring instruments.

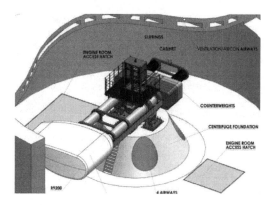

Figure 9. The 1000 g-t centrifuge at IWHR (under construction).

2.4 Development of a centrifuge laboratory

Centrifuge laboratory designs in China specifically emphasize the centrifuge chamber and its structure, and consider the chamber as an integral part of the centrifuge itself. Most of the large centrifuges adopt an underground layout for safety reasons, and leave more space in the hall above the chamber. Nowadays, foundation rigidity and deformation, thermal energy evacuation methods, fire protection facilities, water drainage and the convenience of model preparation, installation and testing are given the greatest attention and careful consideration. Cooling air is often used to stabilise the temperature and humidity inside the centrifuge chamber. Natural air during the cold season can also be circulated into the chamber for temperature control purposes. Therefore, it may be efficient to use both, an air cooling system as well as a natural air circulation system, introducing cool air from the centre of the centrifuge chamber's ceiling, to keep the temperature below 40°C in the chamber at all times.

When the outside temperature is lower than 25°C, a flow of natural air may be used for cooling the centrifuge chamber; above that, an air conditioned flow may be circulated. For some airtight or partially vacuumed centrifuge chambers, a cooling wall may be adopted by introducing cooling water inside the wall itself.

Besides the centrifuge chamber and offices, the layout of a centrifuge laboratory should also include the model preparation and materials processing areas, a power room, a main control room, rooms for soil tests and for transducer calibrations, rooms for material storage and discharge, rooms for compressed air and distilled water generation and one with hydraulic pumps, a room for electric facilities, etc. Enough space inside the laboratory should be provided to allow for freedom of movement, maintenance and replacement of equipment and parts. For an effective operation of a medium to large centrifuge laboratory, at least three dedicated experienced technicians are needed overseeing respectively the operation of the centrifuge, the data acquisition system and machinery.

2.5 Development of standard on centrifuge model tests

The rapid development of centrifuge labs in China requires more qualified engineers, operators and researchers in this field. To standardize the centrifuge operation, the first Specification on Centrifuge Model Tests was published in 1999 by the China Power Ministry (DL/T 5102-1999), and was revised in 2013 (DL/T 5102-2013) and published by the China Electric Power Press. This specification gives the general requirements on the centrifuge and test facilities, model preparation, simulation methods, test procedures, data acquisition and processing, as well as safety protection measures in centrifuge tests. It plays an important role in helping new researchers and students to use the centrifuge facilities for model tests. The basic requirements on the centrifuge design and manufacturing are also given in the specification for safety operation and maintenance.

The definition of some terms are provided, such as the maximum centrifuge radius which refers to the distance from the centrifuge axis to the upper surface of the basket platform, and the effective radius, used to calculate the effective centrifuge acceleration for data processing which refers to the distance from the axis of rotation to the point located in-flight at one third of the model height from the basket platform. Some other standards related with geotechnical model testing are also stipulated in different industries, such as transportation and water resources.

3 AUXILIARIES DEVELOPED IN IWHR FOR CENTRIFUGE MODEL TESTS

3.1 Coupled horizontal and vertical shaker

After the commissioning of IWHR's 450 g-t centrifuge (Hou and Han, 1994), all the tests performed with this

Figure 10. IWHR's combined horizontal and vertical shaker.

machine were static in nature until it was equipped with a combined testing horizontal and vertical shaker in 2004. This shaker (Fig. 10) aims at modelling dam deformation and stability during strong earthquakes in both horizontal and vertical directions (Hou, 2006). After its implementation, a dozen projects requiring dynamic centrifuge testing have been conducted, including the Milin Reservoir Project in Tibet, the evaluation of reinforcement effect of a core earth dam under strong earthquake, and the Zipingpu CFRD dam under the Wenchuan earthquake of 2008 (8.0 on the Richter Scale).

In spite of the force applied in the direction along the arm on the centrifuge axis due to the frequent vertical shaking during rotation, the IWHR centrifuge is still running safely without any detected mechanical damage. Typical vibration paths for combined horizontal and vertical shaking are illustrated in Figure 11, replicating the complex rocking effects real earthquakes have on structures.

The shaker was manufactured by ANCO of the US. It can generate a maximum 30 g acceleration horizontally and 20 g vertically, under centrifuge accelerations of up to 100 g. However, the capacity of IWHR's current centrifuge limits operation of this quake simulator to centrifuge accelerations below 80 g. The maximum payload is 415 kg and the maximum shaking frequency is 400 Hz. The internal dimensions of the laminar container are $810 \times 345 \times 500$ mm.

3.2 Controlled Tools System (CTS)

The Controlled Tooling System (CTS), also called robot, is a very useful tool in simulating construction in-flight. IWHR's CTS (Fig. 12) can be operated with CPT tools at centrifuge accelerations of up to 120 g and with an in-flight sand rainer operated below 90 g. The mobile body of the CTS moves on a strong frame sitting on a big container with inner dimensions of 1200 (length) × 800 (width) × 427 mm (height). The on-board three tool magazine allows the carrying of in-flight exchangeable tools, allowing a variety of complex in-flight operations such as pile driving, excavation and backfill simulations.

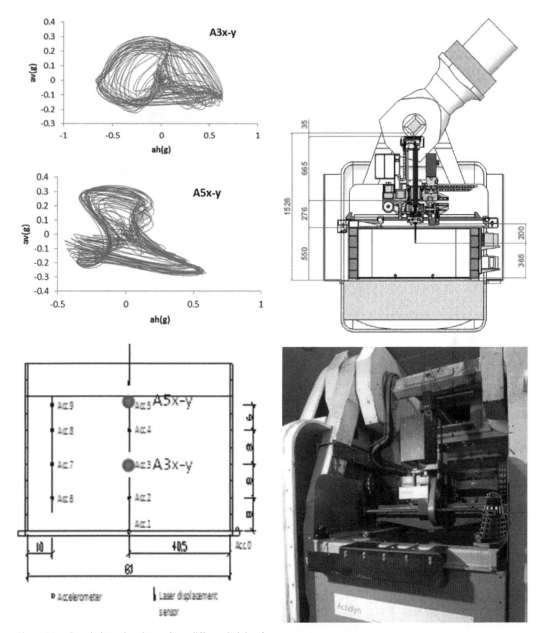

Figure 11. Coupled acceleration paths at different height of the model.

Figure 12. IWHR Controlled Tooling System.

The tool head has four degrees of freedom which allows mechanical actions to be applied and operations to be performed such as: developing torques, carrying water or other fluids to the model using the tool head or holding a laser displacement sensor for automated scanning measurements of the model surface topography.

IWHR's CTS was manufactured by Actidyn Systèmes of France and incorporates all of the features and performances of their former centrifuge robots. On board cameras and lights are also installed. Its main characteristics are summarized in Table 3 (Hou et al, 2012).

The CTS was adopted for measuring the cone penetration resistance over the model depth. Model foundations are made of medium sand and prepared with a sand-rainer. Figure 13 presents the centrifuge test results for different initial model densities, by using the CPT with the CTS under centrifuge acceleration of 50 g. It demonstrates that a higher initial model density usually generates more penetration resistance at the same depth of the sand foundation.

Table 2. Main features of IWHR's CTS.

Item	X	Y	Z	θ
Total displacement (mm)	700	450	400	±175
Velocity max (mm/s)	70	70	40	10°/s
Axis resolution (mm)	0.1	0.1	0.1	±1°
Force (kN)	1	1	±5	1 Nm

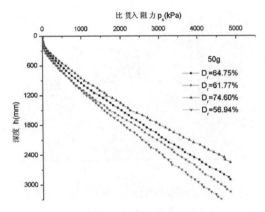

Figure 13. 50g CPT test results for different initial model densities.

Figure 14. Rotating container with its drive mechanism.

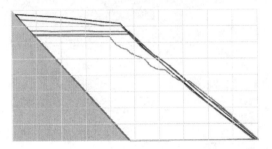

Figure 15. Deformation outlines of a MSW slope after test.

3.3 Rotating container

The large volume available in IWHR's centrifuge swing basket makes it capable of hosting a rotating container equipped with a driving mechanism (Fig. 14). Sometimes monitoring the slope failure in-flight during slope model testing can provide interesting and important information allowing the detection of the stability limit of a slope or its failure patterns. However, designing a model to be failed at a certain g level comes with its own intricacies, even when centrifuge acceleration is used as a parameter to be increased. The rotating container was developed to deal with such intricacies and to simulate the continuous change in slope gradients, so that the model can be failed at certain slope ratios at a given centrifuge acceleration. It can be operated under centrifuge accelerations of up to 50 g. The inner size of the rotating container is 568×342×308 mm. Its maximum inclination is 75° with different rates of rotation. The rotation rate of the container can be adjusted from 1°/min to 2°/min.

This container has been used to test the failure phenomenon of toppling slopes, soil slopes and slopes made of municipal solid waste (MSW). Figure 15 shows the deformation profile and failure patterns of a MSW slope at different rotating degrees at 40 g centrifuge accelerations, from which the tendency and mechanism of slope deformation can be analysed. The safety factor of the model slope may thus be estimated based on the test results recorded at the moment of slope failure.

3.4 A 3D optical displacement measurement system

The vertical displacement of a model surface can be measured by LVDT or laser displacement sensor. Only a limited number of points on the model surface can be detected. Sometimes it is important to know the displacement in an area on the model, such as the uneven settlement of concrete face slabs for a CFRD model dam. A 3D optical displacement measurement system (Fig. 16) was developed at IWHR, with the function of measuring displacements over an area of 240 mm×175 mm. It is composed of three high resolution cameras fixed at the bottom of a supporting frame, collecting pictures at a selectable rate from 0.1Hz to 1 Hz through a data acquisition system located near the centrifuge axis, and data transferred wirelessly to the main control room equipped with professional software G200. The system was manufactured by INNOWEP GmbH of Germany.

The camera supporting frame is made of carbon fibre featuring high strength and low deformations, and can be operated under centrifuge accelerations from 20 g to 200 g. The deformation and the layout of the frame are calibrated at different g levels and the calibration data serves as the basis of data pro-cessing by the software. At each given moment, three pictures are taken simultaneously to produce a topographical picture of the model surface. When pictures are taken at different moments or different g levels, the displacements at any point of the surface can be analysed in x,

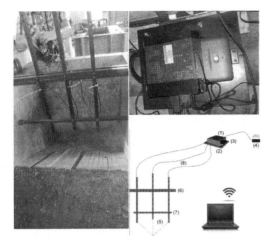

Figure 16. 3D optical displacement measurement system.

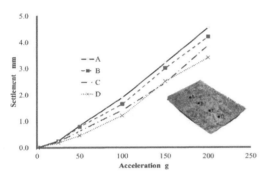

Figure 17. Model surface settlements at different points and g levels.

y and z directions, such as the settlement in z direction shown in Figure 17. The settlement of the model surface over the measured area at different times can also be analysed with the G200 software. The measurement accuracy is about 0.15 mm, depending on the quality of the preliminary calibration.

ACKNOWLEDGEMENTS

The author wants to express his thanks to those researchers and engineers who provided their centrifuge information for this paper. Sincere appreciation should give to all the designers and manufactures who involved in the construction of geotechnical centrifuges and related facilities in China. IWHR's new centrifuge and laboratory construction is supported by the Chinese government per Shuiguiji [2017] 69 notice.

REFERENCES

Baldi, G. Belloni, G. & Maggioni, W. 1988. The ISMES Geotechnical Centrifuge, *Proceedings of the International Conference on Geotechnical Centrifuge Modelling/Paris*: 45–48.

Bao, C.G. 1991. The state and prospect of geotechnical centrifuge model test in China, *Chinese Journal of Geotechnical Engineering*. Vol. 13(6):92–97.

Bucky, P.B. 1931. Use of models for the study of mining problems. Am. Inst. Min. Met. Eng., Tech. Pub. 425, 28.

China power industry standard. 2014. Specification for geotechnical centrifuge model test. DL/T 5102-2013, China Electric Power Press.

Craig, W.H. & Rowe, P.W. 1981. Operation of Geotechnical Centrifuge 1970 to 1979, *Geotechnical Testing Journal*, Vol. 4, 19–25

Craig, W.H. 1991. The future of geotechnical centrifuges. *Proc. ASCE Geotechnical Congress*, Vol. 2: 815–826.

Du, Y. & Zhu, S. 1994. LXJ-4-450 Geotechnical centrifuge in Beijing. *Centrifuge 94*, C.F. Leung, F.H. Lee & T.S. Tan (eds): 35–40.

Hou, Y.J. & Han, L.B. 1994. Performance of Large Geotechnical Centrifuge at IWHR, *1st International Symposium on Hydraulic Measurement*, Beijing.

Hou, Y.J., Zhang, X.D., Xu, Z.P., Liang, J.H. & Li, J.S. 2014. Performance of horizontal and vertical 2D shaker in IWHR centrifuge, *Physical Modelling in Geotechnics*, Gaudin & White (eds), ICPMG2014:207–214.

Hou, Y.J. 2006. Centrifuge shakers and dynamic testing, *Journal of China Institute of Water Resources and Hydropower Research* (in Chinese), 4(1):15–21.

Hou, Y.J., Wen, Y.F., Zhang, X.D., Sun, Q.L., Gauffre, C., Sabard, P. & Rames, D. 2012. Development of IWHR Centrifuge Controlled Tools System, *Proceedings of Asiafuge 2012*, Indian Institute of Technology Bombay, Mumbai, India, November 14–16

Ko, H-Y. 1988. The Colorado Centrifuge Facility, *Proceedings of the International Conference on Geotechnical Centrifuge Modelling/Paris*: 73–76.

Kutter, B.L., Li, X.S., Sluis, W. & Cheney, J.A. 1991. Performance and Instrumentation of the Large Centrifuge at Davis, *Proceedings of the International Conference on Geotechnical Centrifuge*: 19-28.

Li, Q.Y. & Cheng, Z.L. 2005. Analysis of the behaviour of stage II cofferdam of TGP, *Chinese Journal of Geotechnical Engineering*, Vol. 27 (4): 410-413.

Ng, C.W.W., Li, X.S., Van Laak, P.A. & Hou, Y.J. 2004. Centrifuge modelling of loose fill embankment subjected to uniaxial and biaxial earthquakes [J]. *Soil Dynamics and Earthquake Engineering*, Vol. 24(4): 305-318.

Pokrovsky, G.I. & Fedorov, I.S. 1936. Studies of Soil Pressure and Deformations by Means of a Centrifuge, *Proc. 1st. ICSMFE*, Vol.1.

Schofield, A.N. 1978. Use of centrifuge model testing to assess slope stability. *Canad. Geotech. J.*, 15, 14–31.

Schofield, A.N. 1980. Cambridge geotechnical centrifuge operations. *Géotechnique*, 20, 227–268.

Schofield, A.N., 1976. General Principles of Centrifuge Model Testing and a Review of Some Testing Facilities, Offshore Soil Mechanics: 328-339.

Physical modelling of structural and biological soil reinforcement

J.A. Knappett
School of Science and Engineering, University of Dundee, UK

ABSTRACT: This paper presents a number of different approaches that can be used to produce small scale models of soil reinforcing elements (here, piles and plant roots) for which similitude of relative soil-structure stiffness and soil-structure strength can be achieved simultaneously. This includes a discussion of the appropriate dimensionless groups that should be satisfied and a description of the modelling procedures. This is achieved via a series of worked examples of centrifuge model design for steel tubular piles, square reinforced concrete piles and plant roots, though the methods can in principle be applied to other types of reinforcement including retaining walls and soil nails. It is hoped that these will prove to be useful guidance in model design for those new to centrifuge modelling. The modelling procedures demonstrate how principles from materials science can be creatively applied to achieve simultaneous similitude of strength and stiffness, including (i) the use of heat-treatment processing of metal alloys; (ii) development of a micro reinforced concrete based on an understanding of size effect in brittle materials; and (iii) use of materials selection charts for identification of suitable analogue materials. The paper concludes with examples of the application of these procedures in assessing the resilience of reinforced slopes to earthquake ground motions using centrifuge modelling to determine whether vegetation (plant roots) can be used as a low-carbon alternative to conventional 'hard' engineering methods for such a problem.

1 INTRODUCTION

There are many geotechnical problems that involve soil-structure interaction. Scaled physical modelling, particularly using a geotechnical centrifuge, can be used to investigate such problems to (i) improve our understanding of the underlying phenomena for the development of analytical and numerical approaches for engineering design; (ii) provide data for validating such approaches; and (iii) simulate the behaviour of existing systems to evaluate performance against future actions. In some cases, e.g. where one of the performance requirements for the geotechnical system is that the structural elements must remain elastic, it is sufficient to model only the elastic soil-structure interaction (i.e. ensure similitude of relative soil-structure stiffness). This is often desirable when developing new designs and design methodologies (e.g. cases (i) and (ii) above), where ensuring that the structural elements do not fail prematurely before the soil will normally result in the optimal performance (e.g. minimum deformation) of the system. It is therefore not surprising that elastic similitude (only) has been used in the vast majority of physical modelling studies conducted to date.

However, in a World with ever-increasing pressures on resources and the environment it will become increasingly important to extend the life of existing infrastructure where possible. This will necessitate a growing requirement to model the behaviour of existing systems to assess their performance against actions beyond that which they were originally designed for. An example is the assessment of the performance of reinforcement schemes for slopes subjected to earthquakes or extreme rainfall events where an extension of the design life results in an increase in the size of the design event if the probability of failure is maintained due to an increase in exposure (an effect compounded by climate change in the case of rainfall events). In such cases, behaviour may transition into an inelastic range where it would be desirable to scale relative soil-structure strength to obtain an accurate model of the system's behaviour.

This paper will summarise approaches for producing reduced scale models of structural elements (both in steel and reinforced concrete) which can simultaneously scale stiffness and strength, and therefore simultaneously satisfy the appropriate dimensionless groups to ensure similitude in centrifuge tests. This will be discussed in the context of reinforcement of sloping ground against earthquake actions using a row of discretely-spaced vertical piles. The paper will also summarise an approach to producing representative physical analogues of plant roots using modern 3-D printing techniques (similarly ensuring similitude of both stiffness and strength simultaneously), which could be used as an alternative biological slope reinforcement with much lower cost and lower embodied carbon.

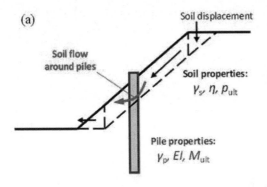

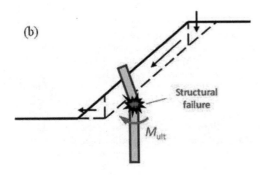

Figure 1. Failure mechanisms of a pile-reinforced slope: (a) soil fails first; (b) pile fails first.

2 SLOPE REINFORCEMENT USING PILES

The example problem of slope reinforcement has been selected as it represents a case where there are two competing failure modes at the ultimate limiting state, namely: (i) the pile is stronger than the soil so that the soil flows around the pile (a geotechnical failure); or (ii) the soil is stronger than the pile, so that the pile suffers a structural failure. These are shown in Figure 1.

In this section, appropriate dimensionless groups will be presented to achieve similitude of both relative soil-pile stiffness and relative soil-pile strength. These will be used to describe the requirements on the structural models that will be discussed in subsequent sections. In the remainder of the paper, dimensions are represented by M for mass, L for length and T for time.

2.1 Elastic soil-pile interaction

An appropriate dimensionless group that can be used to express relative soil-pile stiffness is:

$$\frac{\eta B L_p^4}{EI} \tag{1}$$

after Florres-Berrones & Whitman (1982), where η is the modulus of subgrade reaction (dimensions of $ML^{-2}T^{-2}$), B is the pile width or diameter and L_p a length of the pile (both with dimensions of L), and EI is the pile elastic bending stiffness (dimensions of ML^3T^{-2}).

In a 1:N (model:prototype) scale centrifuge test using the same soil in model and prototype, the soil stress-strain behaviour is scaled 1:1, so η scales as N:1, and EI therefore scales as 1:N^4 to ensure similitude. This is the conventional scaling law recovered in Wood (2004) or Madabhushi (2014), amongst others, and sets the requirement on the elastic bending stiffness of the model pile.

2.2 Soil-pile interaction at the ultimate limit state

Equation 1 represents the relative soil-pile lateral stiffness. To obtain the correct failure mechanism at the ultimate limit state, similitude in the relative soil-pile strength must be achieved to avoid a bias towards one of other mechanism. One possible dimensionless group expresses this in terms of the moment induced in the pile due to the maximum lateral soil pressure at soil failure (M_{sf}) and the structural moment capacity of the pile (M_{ult}):

$$\frac{M_{sf}}{M_{ult}} \tag{2}$$

The soil term in Equation 2, M_{sf}, can be approximated be considering the free body diagram shown in Figure 2. The maximum lateral earth pressure, p_{ult}, for a soil with a drained response (as an example) is:

$$p_{ult}(z) = 3K_p \cdot (\gamma_s' z) \cdot B \tag{3}$$

where γ_s' is the effective (buoyant) unit weight of the soil, z is the depth below the ground surface and K_p is the passive earth pressure coefficient. Therefore, by taking moments about the intersection of the slip plane with the pile when the slip plane is at a depth $z = L_a$ (active length):

$$M_{sf} = \frac{3K_p \gamma_s' B L_a^3}{6} \tag{4}$$

Combining Equations 2 and 4 suggests that a suitable dimensionless group for relative soil-pile strength is:

$$\frac{K_p \gamma_s' B L_a^3}{M_{ult}} \tag{5}$$

K_p has no dimensions, while unit weight has dimensions of $ML^{-2}T^{-2}$. Therefore, in a 1:N scale centrifuge test using the same soil in model and prototype, M_{ult} must scale by 1:N^3 for similitude.

3 MODEL PILE DESIGN – STEEL PILE

From Section 2, it has been shown that for complete similitude in the slope reinforcement problem, any model pile designed for a centrifuge test must have a bending stiffness scaled by 1:N^4 compared to the prototype and a moment capacity scaled by 1:N^3. This

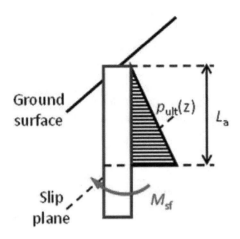

Figure 2. Free-body diagram for a pile subject to lateral pressure from plastically-yielding soil.

Table 1. Design of a 1:80 scale steel pile in 6000-series aluminium alloy.

Parameter (units)	Steel field pile	6063 prototype (T6 / T4 temper)	Prototype/ field
B (m)	0.496	0.496	1.00
t (m)	0.02	0.112	N/A
I (m^4)	0.00085	0.00270	3.18
E (GPa)	210	68	0.32
EI (MNm^2)	178	184	1.03
Z_p (m^3)	0.00453	0.01698	3.75
f_y (MPa)	355	250 / 90	0.70 / 0.25
M_{ult} (kNm)	1610	4246 / 1530	2.64 / 0.95

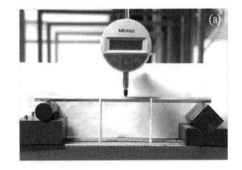

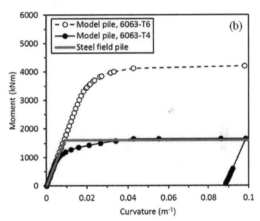

Figure 3. Structural behaviour of 1:80 model steel piles: (a) four-point bending test arrangement; (b) results at prototype scale.

section will present an example of the design of a model pile at a scale of 1:80 (N = 80) where the prototype is to represent an S355 steel tubular pile (i.e. yield stress $f_y = 355$ MPa) with outer diameter $B = 496$ mm and wall thickness $t = 20$ mm.

It is not practical in this case to produce a geometrically-scaled model pile in the same steel, as the resulting wall thickness would be 0.25 mm. This would be difficult to machine and small flaws in this process could result in significant localised weaknesses in the model pile wall. Indeed, it is common in centrifuge modelling to design a section with larger wall thickness in a material with a lower Young's Modulus so that EI can be appropriately scaled. Aluminium-alloy is a popular material as it is easy to machine and is corrosion resistant. Table 1 presents the results of using 6063-series aluminium alloy to produce the model pile (all values at prototype scale). 'T6' refers to the temper of the material (essentially a series of heat-treatment processes that have been applied during its production).

Table 1 demonstrates that a tube with outer diameter 6.2 mm and wall thickness of 1.4 mm can achieve representative B and EI (within 3% of the target value), satisfying Equation 1. The Young's Modulus of the 6063 alloy is approximately one third that of steel, so the second moment of area (I) has been increased by a factor of approximately three to compensate.

Increasing I, however, also increases the plastic section modulus Z_p (by a factor of nearly four) as these two parameters are both controlled by the geometry of the cross-section. This is problematic as f_y for 6063-T6 is only 70% of that of the steel and the plastic moment capacity is given by:

$$M_{ult} = f_y Z_p \quad (6)$$

Therefore, to simultaneously satisfy Equation 5, the yield stress of the aluminium-alloy needs to be reduced. This can be achieved by using an alloy of a different temper. The commonly available T6 temper is achieved through solution heat treatment (520°C for one hour followed by water quench) followed by aging (175°C for 8 hours). The ageing process increases f_y without changing E. A softer temper (T4) is achieved by applying the solution heat treatment without ageing. Table 1 shows that the lower value of $f_y = 90$ MPa for this un-aged material results in M_{ult} within 5% of the desired prototype value. The values of EI and M_{ult} in Table 1 were validated by conducting four-point bending tests on 200 mm long pile models in both 6063-T6 and 6063-T4 alloys, the results of which are shown in Figure 3.

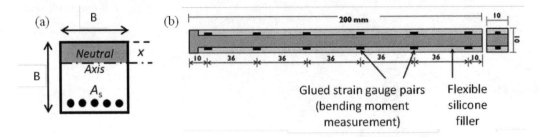

Figure 4. 1:50 elastic model of an RC pile: (a) singly-reinforced cracked concrete section (shaded zone in compression); (b) Aluminium-alloy composite section (dimensions in mm at model scale; additional instrumentation also shown).

It should be noted that T4 temper is not commonly available in the sizes of tube that were used to fabricate the model piles. However 6063-T6 machined to the correct dimensions can be annealed followed by heat-treatment to remove the effects of the aging and return the material to T4 temper (for the models presented in Figure 3 this was conducted by Beccles Heat Treatment, UK). It is desirable to do this as cold-working of the material and (a long) storage time will cause ageing of T4 towards the more stable T6 temper, so it is best to perform the heat treatment as the last process before testing.

4 MODEL PILE DESIGN – REINFORCED-CONCRETE PILE

If the piles to be modelled in the centrifuge are instead to represent reinforced concrete (RC) piles in the field, producing a model that can even just satisfy Equation 1 is complicated due to the composite nature of the material and the very different behaviour of the constituent materials (concrete and steel) in compression and tension (Knappett, 2008). It is first necessary to determine the effective (working) stiffness, as EI in a reinforced concrete beam is dependent on: (i) the overall section size (as in Section 3); (ii) the Young's Modulii of the concrete and steel; and (iii) the amount of steel reinforcement. In a beam-column (i.e. where there is combined bending and axial load), EI is also dependent on (iv) the axial load in the section (this will be discussed further in Section 4.3).

This section will present an example of the design of a model pile at a scale of 1:50 (N = 50) where the prototype is to represent a singly-reinforced square precast RC pile of size $B = 500$ mm, containing $A_s/A_g = 0.85\%$ steel (where A_s is the cross-sectional area of the steel reinforcement, and A_g is the gross area of the section, $A_g = B^2$). Being singly-reinforced, the pile will have the reinforcement all on the upslope side (to resist bending due to lateral earth pressure). The EI for a singly-reinforced rectangular RC section can be found using 'transformed area' theory after Kong and Evans (1987):

$$EI = E_c I_g \left[4\left(\frac{x}{B}\right)^3 + 12\left(\frac{E_s}{E_c}\right)\left(\frac{A_s}{A_g}\right)\left(1-\frac{x}{B}\right)^2 \right] \quad (7)$$

where E_c is the Young's Modulus of the concrete, I_g is the gross second moment of area ($=B^4/12$), x is the depth of the neutral axis (see Figure 4a) and E_s is the Young's Modulus of the steel. Equation 7 assumes that the loads on the beam are expected to induce moments greater than one third of the moment capacity, such that the concrete below the neutral axis that is in tension will be cracked and will therefore not contribute to the stiffness. This is indicated by the diagram in Figure 4a. The depth of the neutral axis is determined for the cracked section when considering the moment capacity of the section, which for a singly-reinforced section is given by:

$$\frac{x}{B} = -\left(\frac{E_s}{E_c}\right)\left(\frac{A_s}{A_g}\right) + \sqrt{\left(\frac{E_s}{E_c}\right)^2 \left(\frac{A_s}{A_g}\right)^2 + 2\left(\frac{E_s}{E_c}\right)\left(\frac{A_s}{A_g}\right)} \quad (8)$$

after Kong and Evans (1987). The moment capacity of a singly-reinforced section is given by:

$$M_{ult} = A_s f_y B \left[1 - \left(\frac{A_s}{A_g}\right)\left(\frac{f_y}{Kf_c}\right)\right] \quad (9)$$

where f_c is the concrete compressive strength and K is a stress-block factor ($K = 0.6$ is assumed herein). If the pile is circular, an approach to determine EI can be found in Knappett (2008).

4.1 Aluminium-alloy based ('elastic') model pile

Table 2 presents the results of using 6063-series aluminium alloy (as in Section 3) to produce a model pile that has the same EI as the RC section. This was achieved by machining a solid square rod to reduce the depth (d), while keeping the same width. Such a section does not have the correct external dimensions at prototype scale, but this can be remedied by using a flexible filler (e.g. silicone) with comparatively negligible stiffness to fill the gap (Figure 4b). Figure 5 shows the construction procedure of such piles, where the composite aluminium-silicone section is wrapped in a thin adhesive and waterproof aluminium tape (to prevent delamination of the filler). Fine sand was subsequently adhered to the surface of the pile using epoxy resin to mimic the rough soil-pile interface typical of concrete.

Table 2. Design of a 1:50 scale RC pile in 6000-series aluminium alloy.

Parameter (units)	RC field pile	6063 prototype (T6 / T4 temper)	Prototype/ field
B (m)	0.5	0.5	1.00
d (m)	0.5	0.26 (0.5*)	* with silicone
$E_{(c)}$ (GPa)	25	70	2.80
E_s/E_c (–)	8.4	N/A	N/A
A_s/A_g (–)	0.85%	N/A	N/A
EI (MNm^2)	47.7	51.3	1.08
f_c (MPa)	23.5	N/A	N/A
f_y (MPa)	460	250 / 90	0.54 / 0.20
M_{ult} (kNm)	230	3750 / 1350	16.30 / 5.87

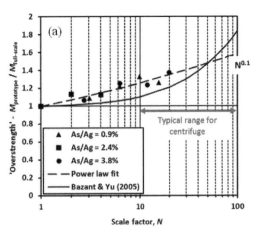

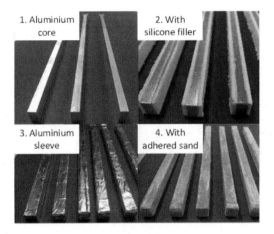

Figure 5. Construction process for aluminium-alloy based 1:50 model RC pile.

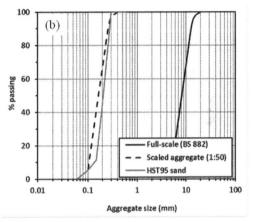

Figure 6. (a) Over-strength observed in previous bending tests of RC beams; (b) approach to geometric scaling of coarse aggregate.

While the elastic behaviour (EI) can be well approximated by this approach, the moment capacity for the model section at prototype scale (calculated using Equation 6) is greatly over-predicted compared to that of the field RC pile (calculated using Equation 9) irrespective of whether T6 or T4 temper materials are used. Therefore if both Equations 1 and 5 are to be satisfied simultaneously, a new modelling approach is required.

4.2 Scale model reinforced concrete

An alternative approach to the one outlined in Section 4.1 is to produce a purely geometrically scaled model, i.e. one where the Young's modulus and strength of all of the materials are scaled 1:1, with the concrete element modelled using a brittle model material and the reinforcement modelled using a ductile material. In principle, if this can be achieved then the behaviour of the model section will automatically display similar behaviour to the full scale section. If, as with the soil, the same concrete material is used in the model, the coarse aggregate size in the concrete will become increasingly larger relative to the size of the structural element as N is increased. As the tensile strength of concrete is controlled by the size of cracks/flaws, and their size is approximately proportional to the size of the aggregate (Bažant and Yu, 2005), use of the same material will tend to result in 'over-strength' in the small scale models. Results collated from previous tests on reinforced concrete beams at various scales with varying amounts of reinforcement are shown in Figure 6a (after Litle and Paparoni, 1966; Belgin and Sener, 2008).

The model of Bažant and Yu (2005) in Figure 6a suggests that the overstrength should be relatively small over conventional scales used for structural model testing (1 < N < 10), but will be unmanageable at scales commonly used in geotechnical centrifuge testing (N > 10). A fit to the test data suggests that the overstrength will be worse for N < 50. A solution to this is to scale down the particle size of the coarse aggregate proportionally as N increases. HST95 silica sand, commonly used at the University of Dundee for centrifuge modelling of sands (with $D_{10} = 0.15$ mm) represents a suitable material for a common scale of 1:50, as shown in Figure 6b.

This fine sand is mixed with plaster and water to produce a model gypsum mortar (model micro concrete) with representative compressive and tensile strength (assessed via the Modulus of rupture, f_r, i.e. the splitting strength of an unreinforced prism) of field concretes. This is summarised in Figure 7a. Different strengths of model concrete can be achieved by varying the water to plaster ratio, w/p (by mass) and also by using different types of plaster. Plaster comes in two principal forms, alpha-hemihydrate and beta-hemihydrate; the former is a stronger plaster typically used for producing dental moulds (Crystacal D™ from Lafarge Prestia, France was used herein), while the latter form is weaker and normally used in model making (Surgical plaster, also from Lafarge Prestia, France, was used herein). The range shown in Figure 7a can be used to represent weaker mass concretes up to structural grades of concrete. The error bars (based on testing 6 samples of each model micro-concrete in compression and 10 in rupture) indicate that the strength is variable; Figure 7b indicates that this variability is similar to that of well-produced concrete produced in the field. Approximate conversion between cylinder (f'_c) and cube (f_{cu}) compressive strengths can be achieved using:

$$f'_c \approx 0.90 f_{cu} - 6.26 \qquad (10)$$

after Mansur and Islam (2002). Further details about the model micro-concrete and the material testing conducted can be found in Knappett et al. (2011).

Reinforcement can be added to the model concrete by using steel wire to ensure 1:1 scaling of Young's modulus and tensile strength. Modelling conducted to date at the University of Dundee has used two materials – a grade 316 stainless steel cold-drawn wire with $f_y = 460$ MPa and a grade 304 stainless steel cold-drawn wire with $f_y = 380$ MPa. The use of stainless steel prevents corrosion which might split the small model sections (as oxidation products will not scale) but the smooth wires must therefore receive a coating of fine sand using epoxy resin (similar to the surface coating describe in Section 4.1 and shown in Figure 5). This allows a 'rough' bond with the fine sand in the model micro-concrete to approximate the effect of the ribbing on conventional deformed reinforcing bars.

To produce a model RC section, the modelling procedure shown in Figure 8 is followed. A formwork is firstly produced that is bolted together to allow it to be easily taken apart after casting to remove the model piles without damage. This contains small holes at either end that can be used to fix the longitudinal reinforcement and hold it in-place. Shear reinforcement may be added by threading the longitudinal wires through either (i) individually made stirrups (e.g. Loli et al., 2014); or (ii) through a continuous spiral wound around a rectangular former with marks to indicate the required spacing which is tied-off at either end (e.g. Al-Defae and Knappett, 2014). Once the reinforcing cage is fully-formed, the dry materials (sand and plaster) for the model concrete are measured-out and mixed, before the water is added. The plaster begins to harden rapidly, so the mixture is immediately poured into the formwork. The model elements can be de-moulded

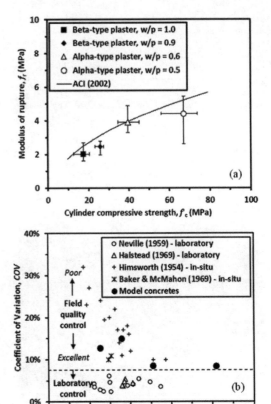

Figure 7. Basic material properties of model micro-concrete: (a) compressive and tensile strength; (b) variability.

after 24 hours (though care must be taken as they will be weak at this stage). The models are then left to cure for 28 days.

Figure 9 presents example results of four-point bending tests on 200 mm long singly-reinforced model concrete beams with different amounts of longitudinal reinforcement (represented by A_s/A_g, in grade 316 wire), both with and without shear reinforcement (a continuous spiral of grade 304 wire, representing 10 mm diameter links at pitch/spacing of 360 mm at prototype scale). So long as a suitable amount of shear reinforcement is included the beams can simultaneously scale the EI and M_{ult} (for the cracked section), and can therefore simultaneously satisfy Equations 1 and 5 for a reinforced concrete section, just as the heat-treated aluminium-alloy tubes achieved for the steel section in Section 3 (c.f. Figure 3b). The failure mechanism for such cases even exhibits tensile cracking and compressive spalling, as in full-scale RC beams; this is shown in Figure 10a. Furthermore if the shear reinforcement is removed a lower capacity is obtained in Figure 9 for both steel percentages, associated with a change in failure mechanism to flexural shear (see Figure 10b). This is consistent with the behaviour of RC beams with inadequate shear reinforcement at the

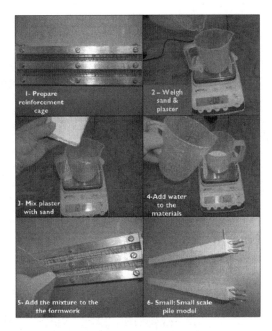

Figure 8. Casting procedure for model RC sections (pile shown).

shear-span ratio tested ($a/d = 5.3$ in Figure 10, e.g. Bažant and Yu, 2005).

4.3 Extension to combined axial load and bending

Having demonstrated that the model RC can replicate EI, M_{ult} and damage mechanisms in pure bending (which is most appropriate for the slope reinforcement problem), further testing has examined the behaviour of the composite material under combined axial load and bending moment. Achieving similitude under such conditions would make it suitable for use in piled foundation problems where lateral loads are also present (e.g. structural foundations, columns or bridge piers under earthquake loading).

Five doubly-reinforced concrete piles were fabricated as outlined in Section 4.2, with $A_s/A_g = 0.66\%$ (in total). These were tested as a vertical fixed-base cantilever with a static vertical load ($0 \leq P \leq 400$ kN) applied through a load hanger and a lateral worm screw to apply displacement-controlled lateral loading. The vertical loading resulted in instability of some of the tested sections once the moments had reached half of the expected values (due to P-Δ amplification as the lateral deflections, Δ, became larger). The moment-curvature behaviour up to this point was sufficiently developed to determine EI in all cases; however, only two of the columns reached a stable plastic failure for determination of M_{ult} (those at $P = 0$ kN and 150 kN). Test data for these cases is shown in Figure 11.

In Figure 11a there is evidence that increasing P increases both EI and M_{ult} of the model RC section. In terms of the strength, this is consistent with a typical RC interaction (P–M) diagram, where moment capacity increases with increasing axial load up to a

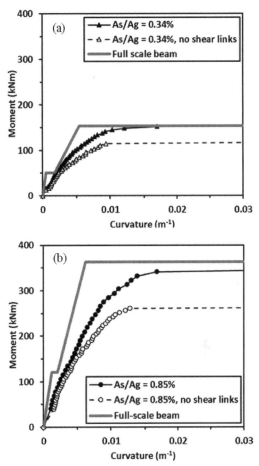

Figure 9. Structural behviour of 1:50 scale model RC piles from four-point bending tests (results at prototype scale).

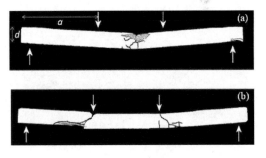

Figure 10. Damage to model RC beams in four-point bending tests ($A_s/A_g = 0.85\%$): (a) bending failure, beam with shear reinforcement; (b) flexural shear failure, beam without shear reinforcement.

maximum value before reducing to zero as the axial force approaches its limiting value (the 'squash' load, $P = f'_c A_g$). The interaction diagram for the representative field RC pile was here determined using BS8110 (1997) and part of this (for $P < 1$ MN) is shown in Figure 11b for comparison with the model RC test data. It is apparent that the model RC appears to capture the

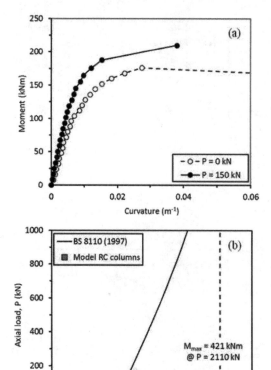

Figure 11. (a) Example behaviour of model columns/piles under combined axial load and bending; (b) interaction diagram.

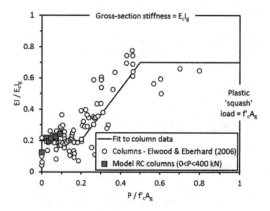

Figure 12. Effect of axial load on bending stiffness of reinforced concrete columns.

interaction behaviour at low values of P, though further testing is clearly required to confirm this result across a wider range of test conditions.

Figure 12 shows a comparison of the bending stiffness of the model RC sections tested as a function of axial load, against a database of full scale tests of RC columns (reported by Elwood and Eberhard, 2006).

This emphasises the difficulty in scaling the bending stiffness of RC in centrifuge soil-structure interaction models, as (i) $EI < E_c I_g$, as mentioned previously for singly-reinforced beams (see Equation 7); and (ii) EI is further affected non-linearly by the axial load P. It can be seen that the model RC datapoints are consistent with the behaviour of the full-scale RC sections. Further testing at higher values of P would be desirable to more fully understand the behaviour of the model RC for applications; however, it should be noted that $P = 400$ kN represents 15% of the elastic critical load of the model sections as tested, which would represent close to an upper limit for axial force carried by columns in conventional building design.

5 SLOPE REINFORCEMENT USING VEGETATION

As an alternative to piles (and other 'hard' engineering solutions), using the roots of plants to stabilise slopes offers a number of potential benefits including: (i) being cheaper; (ii) requiring minimum specialist equipment for installation; (iii) an ability to sequester carbon dioxide (i.e. actively remove it from the atmosphere) rather than requiring CO_2 to be produced during manufacturing (for the steel and concrete in the piles); (iv) offer aesthetic and acoustic benefits from the above-ground part of the vegetation. These benefits can only be realised however if the roots are able to meet a similar level of performance to more traditional solutions.

As with the pile-reinforced slope case discussed in Section 2, there are competing geotechnical and 'structural' failure modes when using vegetation as reinforcement. If the roots are strong enough, the most critical slip plane may deviate around the roots completely, normally moving into a deeper, less efficient position if the slope is uniformly planted across its face (Figure 13a). If the roots are weaker, the critical slip plane may pass through the rooted zone, with the roots failing either structurally in bending (which induces tensile strain within the roots) or in pull-out, depending on the relative soil-root strength (Figure 13b).

While the root soil interaction is elastic, the roots may be treated in the same way as piles such that the dimensionless group is essentially Equation 1. As it is rare and practically difficult to measure the bending stiffness of roots due to their high flexibility, it is useful to re-express Equation 1 idealising the root as a solid circular elastic element, i.e.

$$\frac{\eta L^4}{E_r D_r^3} \tag{11}$$

where E_r is the elastic stiffness (Young's Modulus) of the root and D_r is the root diameter. In a 1:N scale centrifuge test using the same soil in model and prototype, as η scales as N:1, E_r must therefore be scaled 1:1 to ensure similitude.

(a)

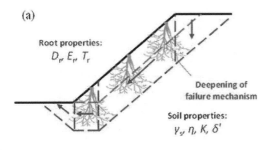

(b)

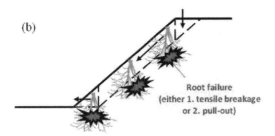

Figure 13. Failure mechanisms of a vegetation-reinforced slope: (a) soil fails around rooted zone; (b) roots fail.

Continuing to treat the roots as analogous to solid-circular rods/piles (so that $Z_p = D_r^3/6$), combining Equations 2, 5 and 6 results in the second dimensionless group to be satisfied, representing relative soil-structure strength in shear/bending, for studying the ultimate limiting state:

$$\frac{K_p \gamma_s' L^3}{T_r D_r^2} \quad (12)$$

where T_r is the root tensile strength (used in place of f_y in Equation 6 if the root behaviour is idealised as elastic-perfectly plastic, i.e. $f_y = T_r$).

Equation 12 is most relevant for those roots in the root system which are close to vertical/perpendicular to the shear plane. There may, however, also be roots which are close to horizontal/parallel to the shear plane. This roots will be subjected to predominantly axial, rather than lateral loads, which suggests a further dimensionless group:

$$\frac{P_{pullout}}{P_{ult}} \quad (13)$$

where $P_{pullout}$ is the root pullout strength and P_{ult} is the root axial strength:

$$P_{ult} = T_r \cdot \frac{\pi D_r^2}{4} \quad (14)$$

The pullout strength may be estimated from the surface area multiplied by the normal effective stress multiplied by the root-soil interface friction coefficient

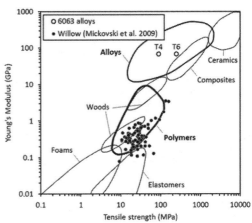

Figure 14. Identification of an analogue material for plant roots using a materials selection chart.

($=\tan \delta'$):

$$P_{pullout} = (\pi D_r L) \cdot (K \gamma_s' z) \cdot \tan \delta' \quad (15)$$

where K is an appropriate earth-pressure coefficient to convert vertical effective stress into normal effective stress and δ' is the interface friction angle. Combining Equations 13–15 suggests that a suitable dimensionless group is:

$$\frac{K \tan \delta' \gamma_s' z L}{T_r D_r} \quad (16)$$

In a 1:N scale centrifuge test using the same soil in model and prototype, Equations 11, 12 and 16 can be scaled simultaneously if E_r and T_r are both scaled 1:1. The simplest way to achieve this would be to identify an existing and available material that has similar stiffness and strength as plant roots. Figure 14 presents a materials selection chart in which the properties of some willow roots measured by Mickovski et al. (2009) are shown relative to various common engineering materials (plotted as regions by material type). From Figure 14 it can be seen that polymers appear to represent a suitable class of materials. This is highly advantageous as many polymers can now be formed using stereolithography (commonly known as 3-D printing) which allows the complex architecture of root systems to be fabricated easily.

6 MODEL PLANT ROOTS

Once a suitable material has been identified for 3-D printing, it is only necessary to produce a 3-D model within computer-aided design (CAD) software which has a geometry and architecture representative of a live-plant root system. This cannot always be a direct reduced scale version of every root as some roots at small scale will be of a size which is too small for the printer to print (e.g. 0.75 mm is the smallest (threshold) diameter that can reliably be printed using the machine described in the following paragraph). Different approaches may be taken to producing an idealised geometry and these are discussed through two examples in this section.

Figure 15 shows a deep-rooting system that will be used in the centrifuge testing application described in Section 7. Figure 15a shows printed root models at 1:10 and 1:30 scales, with lines indicating the prototype depths of potential shear planes in the centrifuge (when tested at N = 10 and 30, respectively). These were printed using a Stratesys Inc. *uPrint SE* printer. The roots were designed in an artificial structure based on the sampled data of root cross-sectional area (CSA) for roots of different size classes in Figure 15b. The bars in this figure represent the printed models, while the line represents the total root CSA at depth for an Oak tree (*Quercus alba*) as reported in Danjon et al. (2008). For simplicity, the model uses just four different root diameters. Further information about the detailed design process and the shear strength contribution that the model can make within rooted soil can be found in Liang et al. (2017).

Figure 16 shows a different model for a Pine (*Pinus pinaster*) root system which has a greater amount of lateral rooting material. In this case, each individual root was digitised from the uprooted tree (as reported in Danjon et al. 2005) and the digital data is shown in Figure 16a. Figure 16b shows an isometric view of the digital idealisation of this root system at 1:20 scale in a form suitable for 3-D printing (the two figures are at the same scale as indicated in Figure 16b). In this case, the digital data for the root centrelines was directly imported into the CAD software Solidworks (i.e. as a wire-frame model) before diameters were assigned. Due to the minimum threshold diameter for printing, some of the finer root material could not be included; to compensate, some of the larger roots were slightly thickened to produce a model with the same overall root volume.

In each case, the roots were printed using an acrylonitrile Butadiene Styrene (ABS) filament, which has been shown to have representative E_r and T_r of plant roots by Liang et al. (2015), thereby satisfying Equations 11, 12 and 16.

7 APPLICATION – SEISMIC RESILIENCE

This section will demonstrate how the different types of reinforcement described in Sections 4 and 6 can be used to study the relative effectiveness of using plant roots as a low-carbon alternative to (RC) discrete piling for improving slope performance at an ultimate limiting state, using physical modelling. The case considered is of a slope which is statically stable (with a factor of safety of ~1.6) but in which movement may be induced due to the additional action from earthquake ground motion. Existing centrifuge test data will be collated and compared in which all of the aforementioned stiffness- and strength-related

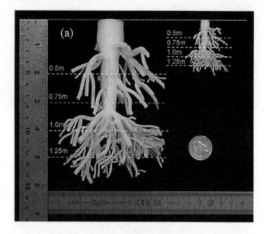
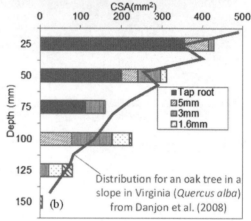

Figure 15. Artificially-created model root system produced using 3-D printing: (a) printed models at 1:10 and 1:30 scales; (b) root distribution data.

dimensionless groups have been satisfied for both types of reinforcement.

7.1 *Centrifuge testing*

A summary of the test arrangements considered can be found in Table 3. These consist of a series of pile reinforced cases (modelled at 1:50) where the piles are placed in a row midway between the crest and the toe at spacing, s, varying between $3.5 < s/B < 7.0$, and a vegetated case (modelled at 1:30) with 3-D printed analogue root clusters having the design described by Figure 15 placed at 1.4 m spacing (at prototype scale) in both horizontal directions (for a total of 36 clusters in the vegetated model).

The model layouts are shown in Figure 17. Both models were instrumented with accelerometers within the soil, but this paper will focus only on the settlement at the crest of the slope (where transportation infrastructure may be situated), which is an indicator of the stability of the slope (as a higher yield acceleration due to reinforcement will result in less slip).

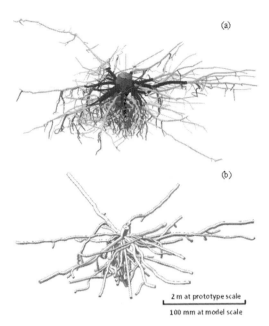

Figure 16. Model root system produced from digitised root data: (a) measured data; (b) idealised model at 1:20 scale.

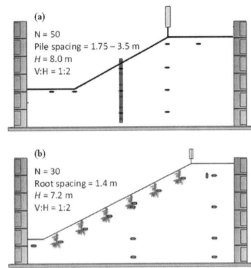

Figure 17. Centrifuge model layouts: (a) piled slopes; (b) vegetated slopes.

Table 3. Centrifuge test data summary (all values at prototype scale).

Test ID	Method	Spacing	EI (MNm^2)	M_{ult} (kNm)
AA01	Unreinforced	N/A	N/A	N/A
AA13	Elastic piles	7.0B	50.4	3750
AA14	Elastic piles	4.7B	50.4	3750
AA15	Elastic piles	3.5B	50.4	3750
AA09	RC piles	7.0B	48.9	230
AA10	RC piles	4.7B	48.9	230
AA11	RC piles	3.5B	48.9	230
AA18	Weak RC piles	7.0B	42.2	70
TL05	Unreinforced	N/A	N/A	N/A
TL06	Vegetated	1.4 m	N/A	N/A

In the piled cases, both 'elastic' (aluminium-alloy type) and model RC piles are considered. The elastic cases have the model pile construction and properties given in Table 2 and Figures 4 and 5 and are instrumented with strain gauges at the positions indicated in Figure 4. The model RC piles are singly reinforced with $A_s/A_g = 0.85\%$ on the upslope side and shear reinforcement representative (at prototype scale) of a 13 mm diameter spiral at a pitch of 275 mm. For the case of $s/B = 7.0$, an additional model RC case was also considered, in which the reinforcement was not roughened with sand to represent the use of non-deformed reinforcing bar. The two model RC sections have reasonably similar EI, but the M_{ult} for the weak section is only 30% of that of the conventional model RC pile. In this sense, the weak pile may represent a pile with only nominal reinforcement, while the conventional model RC section is one that has been specifically designed to have a capacity greater than the induced bending moments (see later). The design of such a pile could be achieved straightforwardly in practice using the modified Newmark sliding block procedure for piled slopes presented by Al-Defae and Knappett (2015), which can estimate the maximum bending moment in the (elastic) piles for a given earthquake ground motion series.

In both cases the soils were formed of dry sand at a relative density of 55% (to avoid liquefaction effects and as a model of a $c' - \varphi'$ soil). The slopes have an angle of 26.6° (vertical:horizontal = 1:2) and were formed within an equivalent shear beam container (ESB) to minimise dynamic wave reflection at the container boundaries. The slopes were of slightly different heights as indicated in Figure 17, though as the failure mechanism for the soil is very shallow and translational, they have the same static stability. They were also subjected to different earthquake time histories (using the QS67-2 servo-hydraulic earthquake simulator at the University of Dundee) and so results will be expressed in terms of the reduction in crest settlement compared to the unreinforced case. More detailed information on the modelling work, including details of the earthquake motions applied, may be found in Al-Defae and Knappett (2014) and Liang and Knappett (2017a).

7.2 *Comparative assessment of techniques*

A conventional stability analysis, conducted using available limit analysis methods could be undertaken to estimate the increase in the yield acceleration (k_{hy}) due to the presence of the reinforcement. Examples of how this may be done may be found in Al-Defae (2014) and Liang & Knappett (2017b). Such an analysis could be used to identify whether a design earthquake will trigger movement (e.g. for a return period of 475 years

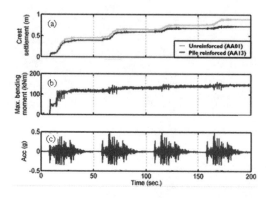

Figure 18. Measured response in elastic pile case for s/B= 7.0: (a) crest settlement; (b) maximum bending moment along the piles; (c) input ground motion.

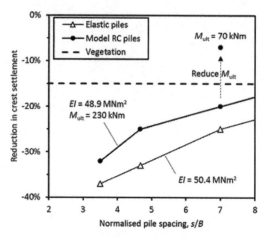

Figure 19. Comparative performance of piled and vegetated seismic slope reinforcement schemes.

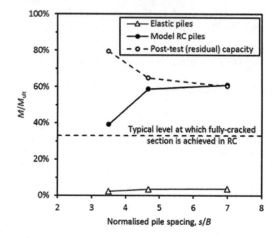

Figure 20. Proportion of pile bending capacity mobilised in pile-reinforced cases.

or 10% probability of exceedance with a 50 year design life). An increase in the yield acceleration (Δk_{hy}) by adding the reinforcement would allow a larger earthquake to be tolerated for the condition of no-slip (i.e. $k_{hy} \geq PGA_{design}$), which would extend the design life of the slope if the probability of exceedance of the design motion over the life of the slope was to be maintained.

However, it is becoming increasingly important to understand the resilience of engineering systems. In the case of a reinforced slope subjected to earthquakes, this involves considering the deformations which would occur if the design (no-slip) earthquake motion strength is exceeded, such that the implications for any supported infrastructure can be assessed. As both types of reinforcement considered here can model behaviour correctly into the inelastic range, centrifuge modelling can be used to investigate the performance when the earthquake motion is strong enough to induce slip. As an example, Figure 18 shows the response of the unreinforced and elastic pile reinforced slopes for $s/B = 7.0$. The earthquake motions used (from the 1999 Chi-Chi Earthquake in Taiwan with a peak ground acceleration of 0.41g) are strong enough to induce slip within the soil. This induces bending moments within the piles which in the elastic case are much less than M_{ult}.

Figure 19 compares the reduction in crest settlement (reinforced/unreinforced) for all of the test cases. Comparing vegetation to the RC piles, it can be seen that for these similar slope heights, vegetation can provide a similar benefit to widely spaced RC piles. Comparing specifically the case at $s/B = 7.0$, the vegetation is similar in performance to the 'designed' pile and better than the RC pile with the nominal reinforcement. There is also evidence (Liang and Knappett 2017a) that in smaller slopes vegetation becomes much more effective (−85% reduction has been reported for a slope 2.4 m high). It may therefore be the case that for some slope heights, vegetation can significantly outperform conventional reinforcement, rather than matching it (further pile-reinforced tests are required to confirm this).

It is further evident from Figure 19 that the model RC piles perform differently from the elastic piles, even though the induced moments appear to be well below the M_{ult} of 230 kNm (Figure 18). Figure 20 shows the maximum induced moments in the piles (assuming that the elastic pile measurements are representative also for the RC piles due to the similar EI) normalised by M_{ult} in each case. It can be seen that unlike the elastic piles which should be well within the truly elastic range ($M/M_{ult} < 5\%$) the RC piles mobilise a significant proportion of their capacity. The application of many cycles of loading at such high M/M_{ult} additionally results in a degradation in the residual capacity of the RC piles (measured from four-point bending tests of carefully exhumed piles from each test – three test piles per centrifuge test). This results in an even higher proportion of the instantaneous moment capacity being mobilised and explains the poorer performance of the RC piles compared to the elastic piles in Figure 19. This demonstrates the importance of developing models which can achieve similitude of stiffness and strength simultaneously, as the use of an elastic pile model would result in an overprediction of the resilience of the piled slope.

8 SUMMARY

Centrifuge modelling allows the constitutive behaviour of soils to be representative of full scale conditions when testing small scale models. However, for soil-structure interaction problems it is also important to appropriately model the behaviour of the structural elements. Dimensionless groups have here been developed to achieve this for piles and plant roots when used in soil reinforcement applications. Similar approaches could be developed for other types of reinforcement (e.g. retaining walls are similar to piles when loaded laterally, and soil nails are similar to plant roots, with structural and pull-out failure modes).

A range of possible techniques has subsequently been presented that can be used to achieve simultaneous similitude of both stiffness and strength to represent structural reinforcing elements made of steel or reinforced-concrete (RC), or plant roots. These are based upon various principles of material science, including metallurgical processes (heat treatment), behaviour of quasi-brittle materials (model RC) and the use of materials selection charts. This has been presented in the form of examples of model element design for centrifuge tests, and it is hoped that these worked examples will be useful to those learning centrifuge modelling.

Such models can be used with confidence when studying problems of soil-structure interaction at the ultimate limit state which is important for addressing topical issues such as assessing the resilience of old infrastructure against extreme events. An example of the application of the different modelling approaches was presented through an assessment of the potential for using vegetation as a low-carbon alternative to piles when used to increase the resilience of a slope to earthquake motions.

ACKNOWLEDGEMENTS

This paper is based on data collected by a number of researchers and students either formerly or currently at the University of Dundee, including: Dr Asad Al-Defae, Dr Teng Liang, Christine Reid, Patrick McDonnell and Kieran O'Reilly.

Thanks are also extended to Mark Truswell, Gary Callon, Colin Stark and Grant Kydd within the School of Science and Engineering for their assistance in fabricating the 3-D printed models, model formwork and beam/beam-column testing apparatus, and in running the centrifuge tests.

REFERENCES

Al-Defae, A.H. 2014. *Seismic performance of pile-reinforced slopes*. PhD Thesis, University of Dundee, UK.

Al-Defae, A.H. & Knappett, J.A. 2014. Centrifuge modelling of the seismic performance of pile-reinforced slopes. *J. Geotechnical & Geoenvironmental Engineering*, ASCE, 140(6): 04014014.

Al-Defae, A.H. & Knappett, J.A. 2015. Newmark sliding block model for pile-reinforced slopes under earthquake loading. *Soil Dynamics & Earthquake Engineering*, 75: 265-278.

American Concrete Institute (ACI) 2002. Building code requirements for structural concrete and commentary. *ACI 318-02*, American Concrete Institute, Farmington Hills, MI, USA.

Baker, W.M. & McMahon, T.F. 1969. Quality assurance in highway construction: Part 3 – Quality assurance of Portland Cement concrete. *Public Roads*, 35(8): 184-189.

Bažant, Z.P. & Yu, Q. 2005. Designing against size effect on shear strength of reinforced concrete beams without stirrups: II Verification and calibration. *Journal of Structural Engineering*, ASCE, 131(12): 1886-1897.

Belgin, C.M. & Sener, S. 2008. Size effect on failure of overreinforced concrete beams. *Engineering Fracture Mechanics*, 75: 2308-2319.

British Standards Institution (BSI) 1992. Specification for aggregates from natural sources for concrete. *BS 882:1992*, London, UK.

British Standards Institution (BSI) 1997. Structural use of concrete – code of practice for design and construction. *BS 8110-1:1997*, London, UK.

Danjon, F., Barker, D.H., Drexhage, M. & Stokes, A. 2008. Using three-dimensional plant root architecture in models of shallow slope stability. *Annals of Botany*, 101(8): 1281-1293.

Danjon, F., Fourcaud, T. & Bert, D. 2005. Root architecture and wind-firmness of mature Pinus Pinaster. *New Phytologist*, 168: 387-400.

Elwood, K.J. & Eberhard, M.O. 2006. Effective stiffness of reinforced concrete columns. *PEER Research Digest No. 2006-1*, Pacific Earthquake Engineering Research Centre, University of California, Berkeley.

Florres-Berrones, R. & Whitman, R.V. 1982. Seismic response of end-bearing piles. *Journal of the Geotechnical Engineering Division*, ASCE, 108(GT4): 554-569.

Halstead, P.E. 1969. The significance of concrete cube tests. *Magazine of Concrete Research*, 21(69): 187-194.

Himsworth, F.R. 1954. The variability of concrete and its effect on mix design. *Proceedings of the Institution of Civil Engineers, Part I*, 3(2): 163-200.

Knappett, J.A. 2008. Discussion: Design charts for seismic analysis of single piles in clay, by A. Tabesh & H. G. Poulos. *Proceedings of the Institution of Civil Engineers, Geotechnical Engineering*, 161(GE2): 115-116.

Knappett, J.A., Reid, C., Kinmond, S. & O'Reilly, K. 2011. Small scale modelling of reinforced concrete structural elements for use in a geotechnical centrifuge. *Journal of Structural Engineering, ASCE*, 137(11): 1263-1271.

Kong, F.K. & Evans, R.H. 1987. *Reinforced and prestressed concrete*, 3rd Edition. E&FN Spon, London, UK.

Liang, T. Knappett, J.A. & Duckett, N. 2015. Modelling the seismic performance of rooted slopes from individual root-soil interaction to global slope behavior. *Géotechnique*, 65(12): 995-1009.

Liang, T., Knappett, J.A., Bengough, A. G. & Ke, Y. X. 2017. Small-scale modelling of plant root systems using 3D printing, with applications to investigate the role of vegetation on earthquake-induced landslides. *Landslides*, 14: 1747-1765.

Liang, T. & Knappett, J.A. 2017a. Centrifuge modelling of the influence of slope height on the seismic performance of rooted slopes. *Géotechnique*, 67(10): 855-869.

Liang, T. & Knappett, J.A. 2017b. Newmark sliding block model for predicting the seismic performance of vegetated slopes. *Soil Dynamics & Earthquake Engineering*, 101: 27-40.

Litle, W.A. & Paparoni, M. 1966. Size effect in small scale models of reinforced concrete beams. *Journal of the American Concrete Institute*, 63(11): 1191-1204.

Loli, M., Knappett, J.A., Brown, M.J., Anastasopoulos, I. & Gazetas, G. 2014. Centrifuge modeling of rocking-isolated inelastic RC bridge piers. *Earthquake Engineering & Structural Dynamics*, 43: 2341-2359.

Madabhushi, S.P.G. 2014. *Centrifuge Modelling for Civil Engineers*. CRC Press, Boca Raton FL, USA.

Mansur, M.A. & Islam, M.M. 2002. Interpretation of concrete strength for non-standard specimens. *Journal of Materials in Civil Engineering*, ASCE, 14(2): 151-155.

Mickovski, S.B., Hallett, P.D., Bransby, M.F., Davies, M.C.R., Sonnenberg, R. & Bengough, A.G. 2009. Mechanical reinforcement of soil by willow roots: Impacts of root properties and root failure mechanism. *Soil Science Society of America Journal*, 73(4): 1276-1285.

Neville, A.M. 1959. The relation between standard deviation and mean strength of concrete test cubes. *Magazine of Concrete Research*, 11(32): 75-84.

Wood, D.M. 2004. *Geotechnical Modelling*. Spon Press, Abingdon, UK.

Current and emerging physical modelling technologies

W.A. Take
GeoEngineering Centre at Queen's-RMC, Queen's University, Canada

ABSTRACT: The physical modelling conference series has served as the primary mechanism for dissemination of physical modelling methods within the geotechnical community over the past several decades. Each set of proceedings charts the advances made in modelling methods from simple experiments to tests of increasing realism driven by increased sophistication of actuators and control systems, sensors, and understanding of scaling relationships. This manuscript reviews the contributions made to advance the state of the art of modelling methods contained within the proceedings of the 9th International Conference on Physical Modelling in Geotechnics 2018 to identify advances in measurement and control technologies, advances in modelling methods inspired by new research questions, and advances in boundary conditions, repeatability, and interpretation of scaling relationships.

1 INTRODUCTION

The physical modelling conference series, initiated in response to the introduction of the novel technology of geotechnical centrifuge modelling, has served as the primary mechanism for dissemination of physical modelling methods within the geotechnical community over the past several decades. Each set of proceedings charts the advances made in modelling methods and serves as a record of the historical evolution of methods from simple gravity turn-on failures of slopes or the response of foundations to loads imposed by simple 1D actuators, to experiments of increasing realism driven by increased sophistication of actuators and control systems, sensors, and understanding of scaling relationships.

The objective of this paper is to reflect on contributions made to advance the state of the art of modelling methods within the proceedings of the *9th International Conference on Physical Modelling in Geotechnics 2018* in the context of the wider literature to identify advances in measurement and control technologies, advances in modelling methods inspired by new research questions, and advances in boundary conditions, repeatability, and interpretation of scaling relationships.

2 ADVANCES IN MEASUREMENT AND CONTROL TECHNOLOGIES

2.1 *Image analysis*

The rapid increase in the spatial and temporal resolution of digital cameras and corresponding rate of advances in the accuracy of image analysis techniques such as digital image correlation (DIC) have had a transformative influence on the measurement of deformation fields in geotechnical physical models over the past two decades (e.g. White et al., 2001a,b; White et al., 2003; Take, 2003; Sadek et al, 2003; Rechenmacher and Finno, 2004; Take, 2015; Stanier et al., 2015; Chen et al., 2016).

High-speed cameras and advanced DIC analyses now permit the displacement-time history of dynamic systems to be measured with sufficient accuracy to generate time histories of velocity and acceleration (e.g. Murray et al., 2017). The paper of Kokkali et al. (2018) illustrates how a high-speed camera can be used to track the acceleration-time history of target markers on the surface of a liquefiable slope during earthquake simulations with excellent agreement with accelerometer data. Additional work by Carey et al. (2018) illustrate how a wave suppressing window installed below the water surface (i.e. similar to a glass bottom boat) eliminates the air-water interface and image distortion caused by surface water waves. Carey et al. (2018) go on to illustrate the effectiveness of this strategy can overcome this traditional obstacle to measure permanent displacements of submerged liquefying slopes. Work by Chian et al. (2018) aims to permit the cyclic deformation during earthquake simulation to be captured in profile through a design for a special transparent side walled laminar strong box.

The ability of DIC to measure full fields of deformation have inspired researchers develop new analytical techniques to mine this unique dataset. Of particular note is the work of Charles et al. (2018) who illustrate that optimisation can be used to reconstruct the soil stress-strain response from full field displacement measurements in plane strain model tests. In a different application, Gómez (2018) illustrate how DIC analysis of selected targets on a model suction anchor can be interpreted using rigid-body kinematics to calculate padeye displacements and rotations under load.

The experimental technique of X-ray tomography can yield 3D images of soil grains which can be analysed with 3D-DIC image analysis techniques to generate unique observations of internal deformations within non-transparent soils. Doreau-Malioche et al. (2018) present work conducted at Laboratoire 3SR, France which use these techniques to investigate the micro-scale mechanisms controlling the macroscopic behaviour of sand-pile interface during pile installation and cyclic loading.

Image analysis can also be used in hydraulic flow applications to infer degree of saturation (e.g. Peters et al., 2011; Sills et al., 2017) or used to follow tracer dyes. Beckett and Fourie (2018) evaluate the suitability of three tracer fluids (acrylic-resin ink; food-grade dye; and fluorescein) to infer flow paths without influencing the physics of the seepage flow and conclude that fluorescein is the most suited for centrifuge modelling.

2.2 Transparent soil

Refractive index matched soil skeleton-pore fluid systems, or transparent soils, have proved to be one of the most interesting physical modelling developments over the past twenty-five years. Iskander (2018) provides a comprehensive review of the historical development of this modelling technique, available transparent materials, and current and emerging applications of the technique to measure deformations internal to a soil model, interparticle fluid flow, degree of saturation, and heat flow. Further contributions by Takano et al. (2018) illustrate how the unique ability of this modelling technique to visualise internal deformations can be used to better understand the mechanism of compaction grouting to improve the liquefaction resistance of granular soils through densification and an increase in lateral earth pressure. At the pore-scale, Li et al. (2018) illustrates how transparent soils can be combined with DIC image analysis to measure the flow field in the pore space between particles, an exciting development for future research in internal erosion, piping, and pore-scale modelling. Of particular note in this study is the novel use of 3D printing to create up-scaled transparent particles representative of the particle geometry of Ottawa sand from Polyacrylamide hydrogel.

2.3 Embedded instrumentation

The ability to deploy dense networks of embedded sensors within highly controlled and characterised soil models is a significant advantage of physical modelling over field testing. Considerable recent advances have been made to embedded sensor technologies for the measurement of shear wave velocity, acceleration, and pore water pressure within the soil mass.

In order to better understand the propagation of railway-generated vibrations, Cui et al. (2018) have used stress-dependent shear wave velocity measurements from bender elements within a triaxial test to validate numerical processing techniques for the interpretation of shear wave velocity from buried airhammer and miniature accelerometers. At a larger scale, Colletti et al. (2018) demonstrate how 2D and 3D arrays of bender elements can be used in the large Geotechnical Laminar Box (GLB) at the University at Buffalo to characterise the changing shear wave velocity of the soil mass during dynamic testing. Dense arrays of instrumentation have also led to the development of new analysis methods. In particular, Dafni & Wartman (2018) describe the use of a new parameter, the Stockwell mean square frequency, to investigate the interaction of seismic waves with soil slopes.

Whereas accelerometers have been used for decades for dynamic testing, it is important to note that their use in physical modelling applications is not limited to these applications. Taking advantage of the availability of miniature micro electro-mechanical systems (MEMS) multi-axis accelerometers, Bizzotto et al. (2018) have developed a calibration methodology in which MEMS accelerometers can be used to monitor the forces acting on a pipe during backfilling indirectly through measurements of the acceleration of the pipe with known mass.

Advances in the measurement of pore water pressure within the geotechnical physical modelling community in recent years have focused on the identification of a replacement device for the Druck PDCR-81 miniature pore pressure sensor and the miniaturisation of sensors for the measurement of matric suctions for unsaturated applications (e.g. Take and Bolton, 2003). Jacobsz (2018) significantly advances these efforts with the development of a low cost sensor for the direct measurement of matric suction.

2.4 3D printed models, components, and soils

The technology of 3D printing has received considerable attention within the physical modelling community over the past four years. This technology permits the cost-effective inclusion of model structures with complex 3D geometries into geotechnical physical models. For example, Stathas et al. (2018) demonstrate the use of 3D printing to create miniature segmental retaining wall blocks whilst capturing their unique 3D geometry to capture their interlocking behavior during stacking in different configurations, Jiang et al. (2018) have used 3D printing to create tunnels and tunnel linings, Ritter et al. (2018) have used 3D printing of masonry structures for centrifuge modelling, and Liang et al. (2018) uses scaled 3D printed plant roots to capture the mechanical effects of root reinforcement in slopes. However, challenges remain with respect to the choice of printing material to adequately model the strength and stiffness of the structural member at reduced geometric scale, and with the inherent variation in material properties with respect to the orientation of the structural member with regards to the printing direction. These issues are investigated in depth by Ritter et al. (2018), and provide an excellent example of how these modelling challenges can be quantified and addressed in centrifuge models of soil-structure interaction using 3D printed components.

The use of 3D printing in geotechnical physical modelling can also be applied to the printing of soil grains and clay models. Adamidis et al (2018) investigate experimental methodologies for the generation of 3D printed particles with a geometry representative of a target reference material (e.g. type of scanning and required number of particles to be scanned) using different 3D printing technologies. Li et al. (2018) expand the scope of possible model grain types to include transparent particles. To investigate the influence of heterogeneity in clay soils, Pua et al. (2018) have developed a novel 3D clay printer to permit the construction of clay models exhibiting heterogeneity in three dimensions using up to eight different mixes of clay soils.

2.5 Fibre optic strain sensing

Fibre optic sensing systems have recently been gaining more widespread use in both geotechnical field monitoring and full scale geotechnical laboratory experiments. In particular, the ability to measure distributed strains at thousands of closely spaced locations is a revolutionary step-change in the physical modelling community's ability to observe soil-structure interaction problems as the location of maximum strain does not need to be known before placement of sensors. In their review of fibre optic sensing for geotechnical applications, Beemer et al. (2018) notes that three types of fibre optic strain measuring technologies have typically been used in the geotechnical literature: Brillouin optical time domain reflectometry (BOTDR), wavelength division multiplexing (WDM) of Fibre Braggs Grating (FBG), and optical frequency domain reflectometry (OFDR) of Rayleigh backscattering. Each of these techniques varies in operating principle, and as a result, the length of fibre that can be monitored, the density of strain measurements along the fibre (e.g. gauge spacing), resolution, and accuracy. The FBG technique can be used to measure strain at the discrete locations of the gratings. For example, Jahnke et al. (2018) has used this technique with gratings spaced at 1 m intervals to investigate the potential of using fibre optic strain measurements as a leak-detection system for water distribution systems. Alternatively, the Rayleigh backscatter technique can interrogate tens of meters of fibre, at 20 mm spacing, with 1 microstrain resolution without the need for gratings to be installed in the fibre. Using such a system, Ni et al. (2017) bonded a single fiber passing along the crown, invert, and springline of a PVC pipe to capture the response of the pipe to increments of normal fault displacements in a 8 m long, 2 m wide split-box.

The small gauge spacing of the Rayleigh backscatter technique (e.g. as small as 2.5–6 mm) raise the question whether this technique can be used in reduced scale models within a geotechnical centrifuge. To investigate this possibility, Beemer et al. (2018), tested whether an OFDR analyzer can measure the flexural strain experienced along a cantilever consisting of a 25 mm diameter, 600 mm long steel bar with increasing self-weight. The results of this study confirm this measurement technique has sufficient accuracy and precision for many geotechnical physical modelling applications and that further work is warranted to further refine the application of this sensing technology to centrifuge modelling.

2.6 Point cloud data

Detection and monitoring of ground surface change using point cloud data generated via terrestrial or airborne LiDAR/photogrammetry has seen widespread adoption within geotechnical field monitoring applications in the past ten years. The use of this technology within the geotechnical physical modelling; however, has not seen the same acceleration rate of use. The reasons for this are varied, but likely related to the high-cost of laser-scanning hardware and the perception that the data can primarily only be used to subtract point clouds to define maps of vertical ground surface change. Additional work is therefore required to identify more cost effective means of generating point cloud data (e.g. the use of structure from motion photogrammetry, 3D scanners and depth sensors) and to develop point cloud data analysis techniques to extract 3D displacement vector fields from the data.

In one of the few papers included in the proceedings that uses point cloud data, Valencia-Galindo et al. (2018) describe the implementation of a laboratory-scale LIDAR-type system installed on an environmental flume for geotechnical applications under 1-g conditions including the calibration of model rainfall systems, and capturing the morphology of a tailings flow deposit. In other work related to new point cloud analysis techniques, Berg et al. (2018) perform a large-scale physical model test of an 3.6 m high earth dam to illustrate how the surface roughness of point cloud can be used to track 3D displacement vectors and estimate the time to failure as the dam was brought to failure under high rainfall and elevated reservoir conditions. These two applications indicate that point cloud data has a very high potential for use by the physical modelling community.

2.7 Physical inclusion of structural components / hybrid modelling

Physical modelling of soil-structure interaction problems requires accurate reduced-scale representation of both the geotechnical and structural performance of the complete system. For complex structural responses, there is therefore considerable motivation to include this behavior within the geotechnical physical model. This is generally accomplished in one of two ways – to either include the structural complexity into the physical model itself, or to use a hybrid modelling approach in which physical and numerical modelling components running in parallel with the aim of achieving a boundary condition more representative of the complex structural response. Several contributions of note have been to both of these strategies are included within the proceedings.

Boksmati & Madabhushi (2018) discuss the development and testing of miniature viscous dampers capable of fitting into structural models for dynamic centrifuge testing. In this case, this physical representation of the more complex structural response is aimed at better understanding the geotechnical consequences of supplemental damping devices into structural frames to enhance the seismic capacity of structures.

For problems requiring the simultaneous modelling of the strength, stiffness, and ductility of structural elements, Knappett et al. (2018) describe the development of a geometrically scaled model of reinforced concrete through the combination of a quasi-brittle cementitious material (plaster-based mortar) with discrete steel wires.

Idinyang et al. (2018) demonstrate how hybrid modelling can be used to capture the soil-structure interaction response of an 8-storey concrete framed to tunnel-induced volume loss under its pile foundations within a centrifuge test. This application required the authors to successfully reduce the handshaking time (i.e. communication between the numerical and physical modelling domains) to 20 ms, thereby greatly expanding the range of applications of this method within the physical modelling community.

3 ADVANCES IN MODELLING METHODS INSPIRED BY NEW RESEARCH QUESTIONS

3.1 *Monopiles and other novel foundation systems*

Advances in physical modelling techniques are often inspired by new research questions. The investigation of the installation process (e.g. El Haffar et al., 2018) and resulting behavior of monopile foundations for offshore wind farms under cyclic lateral loads, for example, is one case in which a new research question has led to considerable advances in modelling methods. Of particular note is the work of the research team of Klinkvort et al. (2018) whose manuscript describes the combined efforts of five different physical modelling facilities to identify and quantify the physical modelling factors influencing the lateral resistance of monopile foundations in sand to reduce variability and improve design predictions. The need to impose variable long-term cyclic loading has inspired Richards et al. (2018) to develop an advanced cyclic loading system that is not restricted to fixed frequency uni-directional or basic multi-directional loading. This system uses two computer-controlled electric actuators to impose multi-directional loading, and six magnetostrictive displacement to track the pile's position in 3D space with high accuracy.

Other examples of advances in modelling methods reported in the proceedings to model the response of foundation systems include the work Schiavon et al. (2018) who investigate the cyclic loading behavior of helical anchors, and the work of Lundberg et al. (2018) who have created a novel instrumented pile with four horizontal contact stress sensors integrated into the model pile itself and an axial stress sensor located near the base of the pile.

3.2 *Energy piles*

The concept of thermally active pile foundations is another new research application that is inspiring new modelling methods. Ghaaowd et al. (2018) describe the development of a reduced-scale model energy pile consisting of a 25 mm-diameter, 255 mm-long split-shell cylinder instrumented with five temperature-compensated strain gages and thermocouples. The model energy pile also includes an internal electrical resistance heater in its core, and uses an annulus of fine sand to conduct heat to the outside of the pile. This pile is then used to assess the applicability of physical modelling to evaluate heat transfer, pore water pressure generation, volume change, and subsequent soil-structure interaction phenomena associated with energy pile operation in clay soils.

The contribution of Vitali et al. (2018) describes work performed at Dundee University to develop a new heating-cooling system for centrifuge testing of thermo-active geo-structures that can increase or decrease the internal temperature of model energy piles through circulation of heat carrier fluid under elevated gravity.

3.3 *Tunnelling and underground spaces*

The prediction of ground deformations and foundation response to volume loss generated by tunnelling is a research question in which physical modelling has made considerable impact to engineering practice. Continued modelling advances inspired by this research question include the work of Song et al. (2018) who present the development of an eccentric rigid boundary mechanical) model tunnel that has the ability to produce more realistic, non-uniform radial displacements around the tunnel lining (i.e. maximum soil displacements at the tunnel crown and no displacements at the tunnel invert) to investigate the resulting influence on the settlement trough geometry in comparison to that observed during radially imposed volume loss.

The work of Rahadian et al. (2018) investigates the long-term behaviour of tunnel joints for immersed tunnels using a physical model to investigate the hypothesis that a seasonal expansion and contraction of the tunnel elements allows sand to enter the joint gap between elements during winter, where it is compacted during summer, leading to an increasing amount of sand in the joint gap with time.

In a different application related to the stability of underground spaces, Jacobsz et al. (2018) describe the development of a weak artificial rock mass to investigate the scaling relationships associated with the physical modelling of cave mining processes (i.e. controlled undercutting of ore bodies in deep mines by drilling and blasting, allowing the rock mass to fracture under its own weight and horizontal in situ stresses).

3.4 Construction sequences and processes

Physical modelling of construction sequences and processes within centrifuge models has inspired the development of specialized modelling techniques. In particular, considerable efforts have been made to simulate the behaviour of retaining walls during excavation. One commonly used approach to achieve this objective is to use in-flight excavation tools to remove soil in controlled sequences. Such a strategy has been applied by Ma and Xu (2018) in an interesting study to simulate the construction sequence of multi-strutted excavations in soft Shanghai clay. The staged removal of heavy fluids or pressure boundary conditions has also been successfully used to model of excavations in soft soils. The work of Panchal et al. (2018) describes developments made at City, University of London to model to in-flight excavations in soft soils within a geotechnical centrifuge, using a novel arrangement of a latex air bag and support/guide structure to permit an isolation of base heave behaviour in the absence of lateral wall movement. For cases in which simulation of excavation requires the removal of a non-fluid pressure (i.e. differing horizontal and vertical pressures), Faustin et al. (2018) have devised an arrangement of stacked trays lifted vertically within a slotted cylinder to simulate the loads imposed on vertical shafts with elliptical plan geometries during excavation.

Considerable progress has also been reported in the development of modelling techniques for the investigation of construction processes such as pile installation, soil improvement, and compaction. Van Zeben et al. (2018) report the development of a reduced-scale model pile driver to investigate the process of pile driving within a geotechnical centrifuge. Of particular note is the design of the electro-mechanical lift system according to scaling laws to accurately recreate the blow rate (frequency) and blow energy (e.g. ram mass and stroke) of different prototype hammers. Advances related to soil improvement processes include the development of equipment to simulate dynamic compaction (Kundu & Viswanadham, 2018) and development of a $1/13.15^{th}$ scale mini-vibrator designed by Nagula et al (2018) with consideration of the appropriate 1g scaling laws to investigate the efficacy of the soil improvement technique of deep vibration compaction. In a different soil improvement strategy, Seitz & Grabe (2018) have developed a system to inject a water-cement mix into granular soils at controlled flow rate to investigate the efficiency of this rehabilitation technique for shallow foundation and retaining wall applications. Finally, Caicedo & Escobar (2018) present the development of a reduced-scale physical model of a vibratory roller compactor in which the drum's weight, diameter, and compaction effort were simulated according to scaling laws to permit testing at 10g within a geotechnical centrifuge. In particular, this development now enables the simulation of continuous compaction control in which the vertical acceleration of the drum is used to estimate the level of compaction.

4 ADVANCES IN BOUNDARY CONDITIONS, REPEATABILITY, AND INTERPRETATION OF SCALING RELATIONSHIPS

4.1 Boundary conditions in large length-scale problems

All physical modelling facilities have limits on the maximum size of models that can be tested. In certain cases, the length scale of the problem under investigation exceeds that of available facilities, requiring innovative approaches to the imposed boundary conditions in these special large length-scale problems. The work of Khaksar et al. (2018) investigates a novel approach to the large length-scale problem of stiff pipelines subjected to permanent vertical ground deformation in normal or reverse faulting. In the length-scale of typical centrifuge model strongboxes, the end displacement and rotation conditions imposed on the model pipeline are commonly either completely fixed or free – neither fully representative of field conditions. The work of Khaksar et al. (2018) provides a novel approach in which special end condition springs can be designed using numerical models to represent the behaviour of the length of pipe which was inevitably omitted from the physical model due to space limitations.

Physical modelling of the dynamic behavior of large earth dams (e.g. 100 m high) within a geotechnical centrifuge using traditional scaling approaches requires unrealistically large acceleration levels and performance envelopes for simulation of earthquake motions. In these cases Kim and Jo (2018) propose the use of a generalized scaling relationship (Iai et al., 2005) using a two-stage scaling strategy in which multiple correctly-scaled, but physically smaller prototypes can be used to infer the behavior of the full scale dam. Sawada et al. (2018) and Borghei & Ghayoomi (2018) further investigate generalized scaling relationships as applied to the behavior of pile foundations in liquefiable inclined ground and dynamic soil-structure interaction behavior of a simplified target prototype of 4-story concrete structure, respectively.

4.2 Land-climate boundary conditions

A growing awareness of the significant impact of climate on geotechnical structures has led to considerable recent developments in modelling technology to control land-climate boundary conditions within physical models. Whereas rainfall simulators have been used in the past to study infiltration-induced problems, these new developments seek to also control the factors influencing the rate of evaporation. Significant contributions towards this goal are presented by Archer & Ng (2018) and Lozada et al. (2018) who describe work to create new environmental chambers capable of controlling temperature, relative humidity, solar radiation, and wind for the Hong Kong University of Science and Technology centrifuge facility and the Universidad de Los Andes centrifuge facility, respectively. These new advances illustrate that independent control of temperature and relative humidity boundary conditions is now possible for geotechnical physical models (Archer & Ng, 2018) and that these boundary conditions constitute a significant advance in our ability to model the rate of evaporation from the soil surface (Lozada et al. 2018).

Additional contributions to the development of technology to impose land-climate boundary conditions relate to advances in the delivery of model rainfall systems, with Khan et al. (2018) and Bhattacherjee & Viswanadham (2018) providing details on independent advances relating to a higher degree of control on water supply to rainfall systems and methods of accounting for Coriolis acceleration. For the investigation of cold-regions phenomena, Zhang et al. (2018) describes the development of a system to impose freeze-thaw cycles on soil surfaces.

4.3 Characterization, variability, and quality control of soil models

The LEAP (Liquefaction Experiment and Analysis Project) is an international collaborative effort to validate modelling and simulation procedures of liquefaction problems. A key objective of the physical modelling associated with this project is to identify factors affecting the repeatability and variability of liquefaction experiments conducted within individual and between different facilities. These efforts are reported in three valuable contributions within the proceedings. Beber et al. (2018) describe work done as part of the project to investigate variability in void ratio of dry pluviated sand models through direct measurements of sand mass and volume during sand pouring and DIC image analysis of the deformation field during imposition of enhanced gravity within a geotechnical centrifuge. Carey et al. (2018) describe the development and validation of a novel miniature CPT device that is of a sufficiently reduced-cost to enable identical penetrometers to be used at eight different centrifuge facilities to assess quality control of similar experiments performed under the LEAP program. This strategy of using identical soil characterization equipment was shown to provide an excellent basis of comparison for experiments conducted at different facilities. Finally, the work of Tobita (2018) investigates possible sources of variability related to the use of viscous fluids in dynamic centrifuge modelling, to resolve conflicts in scaling of time between dynamic and diffusion events. The results of constant head permeability tests indicates that a commonly used viscous fluid (methylcellulose) may experience transient changes in permeability.

For the characterization of clay soils, Bezuijen et al. (2018) describes the development of a free fall cone (using accelerometer data to infer instanteous velocity and depth of penetration), and Wang et al. (2018) illustrates potential issues that may arise if an axial force is use to calibrate model CPT probes rather than an imposed pressure.

4.4 Scaling relationships of flows

Viscous fluids with viscosities larger than that of water are often used to overcome the inherent contradiction of time scaling between dynamic and diffusion phenomena associated with the technique of geotechnical centrifuge modelling (e.g. Tobita, 2018). However, additional consideration of scaling parameters is warranted for either high velocity pore fluid or granular/fluid flow systems such as debris flows. With regards to the former, Ovalle-Villamil & Sasanakul (2018) investigate the validity of Darcy flow assumptions under elevated seepage velocities associated with centrifuge testing. In particular, these researchers have used centrifuge permeability tests conducted at different acceleration levels in different gradations of sands to identify a critical Reynolds number as a threshold to identify the transition to non-Darcy's behavior. For the case of rapid flows of soil-fluid mixtures of varying viscosities, Cabrera et al. (2018) illustrate how Froude, Savage, Bagnold, and Stokes numbers can be used to assess scaling strategies and define flow regimes.

5 NEW PHYSICAL MODELLING FACILITIES

Descriptions of new and/or upgraded physical modelling facilities are important contributions as they record the evolution of the state-of-the-art of physical modelling technology within the geotechnical community. Contributions reported in the proceedings include descriptions of the following advanced facilities:

- A new 240 g-tonne geotechnical centrifuge at the University of Western Australia (Gaudin et al., 2018)
- Upgrades to NHRI – 400g-tonne geotechnical centrifuge (Chen et al., 2018)
- A full scale laminar box for 1-g physical modeling of liquefaction at the University at Buffalo (Thevanayagam et al., 2018)
- A 2D shaker for the RPI centrifuge (Shaker El Shafee et al., 2018)
- A small centrifuge for testing unsaturated soils (Kwa & Airey, 2018)

6 SUMMARY AND CONCLUSIONS

Nearly 80 conference manuscripts have been reviewed to reflect on the contributions made to advance the state of the art of modelling methods within the proceedings of the 9th International Conference on Physical Modelling in Geotechnics 2018. These contributions have been discussed in the context of the wider literature to identify advances in measurement and control technologies, advances in modelling methods inspired by new research questions, and advances in boundary conditions, repeatability, and interpretation of scaling relationships.

REFERENCES

Adamidis, O., Alber, S., & Anastasopoulos, I. 2018. Investigation into 3D printing of granular media. 9th Int. Conf. Phys. Modelling Geotechnics, London.

Archer, A. & Ng, C.W.W. 2018. A new environmental chamber for the HKUST centrifuge facility. 9th Int. Conf. Phys. Modelling Geotechnics, London.

Beber, R., Madabhushi, S.S.C, Dobrisan, A., Haigh, S.K., and Madabushi, S.P.G. 2018. LEAP GWU 2017: Investigating different methods for verifying the relative density of a centrifuge model. 9th Int. Conf. Phys. Modelling Geotechnics, London.

Beckett, C.T.S. & Fourie, A.B. 2018. Flow visualisation in a geotechnical centrifuge under controlled seepage conditions. 9th Int. Conf. Phys. Modelling Geotechnics, London.

Beemer, R.D., Cassidy, M.J. & Gaudin, C. 2018. Investigation of an OFDR Fibre Braggs System for use in Geotechnical Scale Modelling. 9th Int. Conf. Phys. Modelling Geotechnics, London.

Berg, N., Hori, T., and Take, W.A. 2018. Assessment of time to failure of an earth dam using DIC analysis of high temporal resolution point cloud data (submitted)

Bezuijen, A., den Hamer, D.A., Vincke, L., & Geirnaert, K. 2018. Free fall cone tests in Kaolin clay. 9th Int. Conf. Phys. Modelling Geotechnics, London.

Bhattacherjee, E. & Viswanadham, B.V.S. 2018. Development of a rainfall simulator in centrifuge using modified Mariotte's principle. 9th Int. Conf. Phys. Modelling Geotechnics, London.

Bizzotto, T, Brown, M.J., Breannan, A.J., Powell, T., and Chandler, H. 2018. Development of a series of 2D backfill ploughing physical models for pipelines and cables. 9th Int. Conf. Phys. Modelling Geotechnics, London.

Boksmati, J. & Madabhushi, S.P.G. 2018. Development of model structural dampers for dynamic centrifuge testing. 9th Int. Conf. Phys. Modelling Geotechnics, London.

Borghei, A., & Ghayoomi, M. 2018. Experimental evaluation of two-stage scaling in physical modeling of soil-foundation-structure systems. 9th Int. Conf. Phys. Modelling Geotechnics, London.

Cabrera, M., Kailey, P., Bowman, E.T., & Wu, W. 2018. Effects of viscosity in granular flows simulated in a centrifugal acceleration field. 9th Int. Conf. Phys. Modelling Geotechnics, London.

Caicedo, B. & Escobar, J. 2018. Centrifuge modelling of continuous compaction control (CCC). 9th Int. Conf. Phys. Modelling Geotechnics, London.

Carey, T., Stone, N, Kutter, B., and Hajialilue-Bonab, M. 2018a. A new procedure for tracking displacements of submerged sloping ground in centrifuge testing. 9th Int. Conf. Phys. Modelling Geotechnics, London.

Carey, T., Gavras, A., Kutter, B., Haigh, S.K., Madabhushi, S.P.G., Okamura, M., Kim, D.S., Ueda, K., Hung, W.Y., Zhou, Y-G., Liu, K., Chen, Y-M., Zeghal, M., & Manzari, M. 2018. A new shared miniature cone penetrometer for centrifuge testing. 9th Int. Conf. Phys. Modelling Geotechnics, London.

Charles, J.A., Smith, C.C., and Black, J.A. 2018. Identification of soil stress-strain response from full field displacement measurements in plane strain model tests. 9th Int. Conf. Phys. Modelling Geotechnics, London.

Chen, S.S., Gu, X.W., Ren, G.F., Zhang, W.M., Wang, N.X., Xu, G.M., Liu, W., Hong, J.Z., & Cheng, Y.B. 2018. Upgrades to NHRI – 400gt geotechnical centrifuge. 9th Int. Conf. Phys. Modelling Geotechnics, London.

Chen, Z., Li, K., Omidvar, M., & Iskander, M. (2016). Guidelines for DIC in geotechnical engineering research. International Journal of Physical Modelling in Geotechnics, 17(1), 3-22.

Chian, S.C., Qin, C., and Zhang, Z. 2018. Development of a window laminar strong box. 9th Int. Conf. Phys. Modelling Geotechnics, London.

Colletti, J, Tessari, A., Sett, K., Hoffman, W., and Coleman, J. 2018. Shear wave velocity measurements in a large geotechnical laminar box using bender elements. 9th Int. Conf. Phys. Modelling Geotechnics, London.

Cui, G, Heron, C.M., and Marshall, A.M. 2018. Shear wave velocity: comparison between centrifuge and triaxial based measurements. 9th Int. Conf. Phys. Modelling Geotechnics, London.

Dafni, J. & Wartman, J. 2018. A new Stockwell mean square frequency methodology for analyzing centrifuge data. 9th Int. Conf. Phys. Modelling Geotechnics, London.

Doreau-Malioche, J., Combe, G., Toni, J.B., Viggiani, G., and Silva, M. 2018. Imaging of sand-pile interface submitted to a high number of loading cycles. 9th Int. Conf. Phys. Modelling Geotechnics, London.

El Haffar, I., Blanc, M., & Thorel, L. 2018. General study on the axial capacity of piles of offshore wind turbines jacked in sand. 9th Int. Conf. Phys. Modelling Geotechnics, London.

El Shafee, O. Lawler, J., & Abdoun, T. 2018. Evaluation of RPI centrifuge 2D shaker and 2D soil response. 9th Int. Conf. Phys. Modelling Geotechnics, London.

Faustin, N.E., Elshafie, M.Z.E.B, & Mair, R.J. 2018. Modelling the excavation of elliptical shafts in the geotechnical centrifuge. 9th Int. Conf. Phys. Modelling Geotechnics, London.

Gaudin, C., O'Loughlin, C.D. & Breen, J. 2018. A new 240 g-tonne geotechnical centrifuge at the University of Western Australia. 9th Int. Conf. Phys. Modelling Geotechnics, London.

Ghaaowd, I., McCartney, J., Huang, X., Saboya, F., & Tibana, S. 2018. Issues with centrifuge modelling of energy piles in soft clays. 9th Int. Conf. Phys. Modelling Geotechnics, London.

Gómez, D.A. 2018. Image-based model displacement measurements system for centrifuge applications. 9th Int. Conf. Phys. Modelling Geotechnics, London.

Iai S, Tobita T. & Nakahara T. 2005. Generalised scaling relations for dynamic centrifuge tests. Géotechnique. 55(5): 355-362.

Idinyang, S., Franza, A., Heron, C.M., & Marshall, A.M. 2018. Millisecond interfacing of physical models with ABAQUS. 9th Int. Conf. Phys. Modelling Geotechnics, London.

Iskander, M. 2018. Transparent soils turns 25: past, present, and future. 9th Int. Conf. Phys. Modelling Geotechnics, London.

Jacobsz, S.W. 2018. Low cost tensiometers for geotechnical applications. 9th Int. Conf. Phys. Modelling Geotechnics, London.

Jacobsz, S.W., Kearsley, E.P., Cumming-Potvin, D., & Wesseloo, J. 2018. Modelling cave mining in the geotechnical centrifuge. 9th Int. Conf. Phys. Modelling Geotechnics, London.

Jahnke, S.I., Jacobsz, S.W., & Kearsley, E.P. 2018. A field model investigating pipeline leak detection using discrete fibre optic sensors. 9th Int. Conf. Phys. Modelling Geotechnics, London.

Jiang, Q., Li, L.F., & Song, L.B. 2018. Application of 3D Printing technology in geotechnical-physical modeling:

Tentative experiment practice. 9th Int. Conf. Phys. Modelling Geotechnics, London.

Khaksar, R.Y., Moradi, M., & Ghalandarzadeh, A. 2018. A novel experimental-numerical approach to model buried pipes subject to reverse faulting. 9th Int. Conf. Phys. Modelling Geotechnics, London.

Khan, I.U., Al-Fergani, M., & Black, J.A. 2018. Development of a rainfall simulator for climate modelling. 9th Int. Conf. Phys. Modelling Geotechnics, London.

Kim, N.R., & Jo, S.B. 2018. Physical modelling of large dams for seismic performance evaluation. 9th Int. Conf. Phys. Modelling Geotechnics, London.

Klinkvort, R.T., Black, J, Bayton, S., Haigh, S.K., Madabhushi, S.P.G., Blanc, M, Thorel, L., Zania, V., Bienen, B., and Gaudin, C. 2018. A review of modelling effects in centrifuge monopile testing in sand. 9th Int. Conf. Phys. Modelling Geotechnics, London.

Knappett, J.A., Brown, M.J., Shields, L., Al-Defae, A.H., & Loli, M. 2018. Variability of small scale model reinforced concrete and implications for geotechnical centrifuge testing. 9th Int. Conf. Phys. Modelling Geotechnics, London.

Kokkali, P., Abodun, T., and Tessari, A. 2018. Image capture and motion tracking applications in geotechnical centrifuge modeling. 9th Int. Conf. Phys. Modelling Geotechnics, London.

Kundu, S., & Viswanadham, B.V.S. 2018. Influence of tamper shape on dynamic compaction of granular soil. 9th Int. Conf. Phys. Modelling Geotechnics, London.

Kwa, K.A., & Airey, D.W. 2018. The development of a small centrifuge for testing unsaturated soils. 9th Int. Conf. Phys. Modelling Geotechnics, London.

Li, L., Omidvar, M., and Iskander, M. 2018. Visualization of inter-granular pore fluid flow. 9th Int. Conf. Phys. Modelling Geotechnics, London.

Liang, T, Knappett, J.A., Meijer, G.J., Muir Wood, D., Bengough, A.G., Loades, K.W., & Hallett, P.D. 2018. Scaling of plant roots for geotechnical centrifuge tests using juvenile live roots or 3D printed analogues. 9th Int. Conf. Phys. Modelling Geotechnics, London.

Lozada, C., Caicedo, B., & Thorel, L. 2018. Physical modelling of atmospheric conditions during drying. 9th Int. Conf. Phys. Modelling Geotechnics, London.

Lundberg, A.B., Broere, W., & Dijkstra, J. 2018. Development of an instrumented model pile. 9th Int. Conf. Phys. Modelling Geotechnics, London.

Ma, X.F, & Xu, J.W. 2018. Centrifuge model tests on excavation in Shanaghai clay using inflight excavation tools. 9th Int. Conf. Phys. Modelling Geotechnics, London.

Murray, C.A., Hoult, N.A., & Take, W.A. 2017. Dynamic measurements using digital image correlation. International Journal of Physical Modelling in Geotechnics, 17(1), 41-52.

Nagula, S., Mayanja, P, Grabe, J. 2018. Deep vibration compaction of sand using mini vibrator. 9th Int. Conf. Phys. Modelling Geotechnics, London.

Ni, P., Moore, I. D., & Take, W. A. 2017. Distributed fibre optic sensing of strains on buried full-scale PVC pipelines crossing a normal fault. Géotechnique, 68(1), 1-17.

Ovalle-Villamil, W. & Sasanakul, I. 2018. A new insight into the behavior of seepage flow in centrifuge modelling. 9th Int. Conf. Phys. Modelling Geotechnics, London.

Panchal, J.P., McNamara, A.M., & Stallebrass, S.E. 2018. A new approach to modelling excavations in soft soils. 9th Int. Conf. Phys. Modelling Geotechnics, London.

Pelekis, I., Madabhushi, S.P.G., and DeJong, M.J. 2018. Modelling of rocking structures in centrifuge tests. 9th Int. Conf. Phys. Modelling Geotechnics, London.

Peters, S. B., Siemens, G., & Take, W. A. (2011). Characterization of transparent soil for unsaturated applications. Geotechnical Testing Journal, 34(5), 445-456.

Pua, L.M., Caicedo, B., Castillo, D, & Caro, S. 2018. Development of a 3D clay printer for the preparation of heterogeneous models. 9th Int. Conf. Phys. Modelling Geotechnics, London.

Rahadian, R., van der Woude, S., Wilschut, D., Blom, C.B.M., & Broere, W. 2018. A new test setup for studying sand behavior inside an immersed tunnel joint gap. 9th Int. Conf. Phys. Modelling Geotechnics, London.

Rechenmacher, A. L., & Finno, R. J. (2004). Digital image correlation to evaluate shear banding in dilative sands. ASTM geotechnical testing journal, 27(1), 13-22.

Richards, I.A., Byrne, B.W, & Houlsby, G.T. 2018. Physical modelling of monopile foundations under variable cyclic lateral loading. 9th Int. Conf. Phys. Modelling Geotechnics, London.

Ritter, S, DeJong, M.J., Giardina., G., and Mair, R.J. 2018. 3D printing of masonry structures for centrifuge modelling. 9th Int. Conf. Phys. Modelling Geotechnics, London.

Sadek, S., Iskander, M. G., & Liu, J. (2003). Accuracy of digital image correlation for measuring deformations in transparent media. Journal of computing in civil engineering, 17(2), 88-96.

Sawada, K, Ueda, K, & Iai, S. 2018. Generalised scaling law for the liquefaction beahviour of pile-inclined ground system. 9th Int. Conf. Phys. Modelling Geotechnics, London.

Schiavon, J.A., Tshuha, C.H.C., & Thorel, L. 2018. Centrifuge investigation of the cyclic loading effect on the post-cyclic monotonic performance of a single-helix anchor in sand. 9th Int. Conf. Phys. Modelling Geotechnics, London.

Seitz, K-F. & Grabe, J. 2018. 1g-modeling of limit load increase due to shear band enhancement. 9th Int. Conf. Phys. Modelling Geotechnics, London.

Sills, L.A.K., Mumford, K.G., & Siemens, G.A. 2017. Quantification of Fluid Saturations in Transparent Porous Media. Vadose Zone Journal, 16(2).

Song, G., Marshall, A.M., & Heron, C.M. 2018. A mechanical displacement control model tunnel for simulating eccentric ground loss in the centrifuge. 9th Int. Conf. Phys. Modelling Geotechnics, London.

Stanier, S. A., Blaber, J., Take, W.A., & White, D.J. 2015. Improved image-based deformation measurement for geotechnical applications. Canadian Geotechnical Journal, 53(5), 727-739.

Stathas, D., Xu, L., Wang, J.P., Ling, H.I & Li, L. 2018. Concave segmental retaining walls. 9th Int. Conf. Phys. Modelling Geotechnics, London.

Takano, D., Morikawa, Y., Miyata, Y., Nonoyama, H., and Bathurst, R.J. 2018. Physical modeling of compaction grouting injection using a transparent soil. 9th Int. Conf. Phys. Modelling Geotechnics, London.

Take, W., 2003, The Influence of Seasonal Moisture Cycles on Clay Slopes, Ph.D. thesis, University of Cambridge, Cambridge, UK.

Take, W. A. 2015. Thirty-Sixth Canadian Geotechnical Colloquium: Advances in visualization of geotechnical processes through digital image correlation. Canadian Geotechnical Journal, 52(9), 1199-1220.

Take, W. A., & Bolton, M. D. 2003. Tensiometer saturation and the reliable measurement of soil suction. Géotechnique, 53(2), 159-172

Thevanayagam, S, Huang, Q, Constantinou, M.C., Abdoun, T., & Dobry, R. 2018. Full scale laminar box for 1g physical modelling of liquefaction. 9th Int. Conf. Phys. Modelling Geotechnics, London.

Tobita, T. 2018. Permeability of sand with a methylcellulose solution. 9th Int. Conf. Phys. Modelling Geotechnics, London.

Valencia-Galindo, M.D., Beltrán-Rodriguez, L.N., Sánchez-Peralta, J.A., Trujillo-Vela, M.G., Larrahondo, J.M., Prada-Sarmiento, L.F, & Ramos-Cañón, A.M. 2018. A two-dimensional laser-scanner system for geotechnical processes monitoring. 9th Int. Conf. Phys. Modelling Geotechnics, London.

van Zeben, J.C.B. Azúa-González, C., Alvarez Grima, M., van 't Hof, C., and Askarinejad, A. 2018. Design and performance of an electro-mechanical pile driving hammer for geo-centrifuge. 9th Int. Conf. Phys. Modelling Geotechnics, London.

Vitali, D, Leung, A., Zhao, R., & Knappett, J.A. 2018. A new heating-cooling system for centrifuge testing of thermo-active geo-structures. 9th Int. Conf. Phys. Modelling Geotechnics, London.

Wang, Y, Hu, Y., & Hossain, M.S. 2018. Ambient pressure calibration for cone prenetometer test: necessary? 9th Int. Conf. Phys. Modelling Geotechnics, London.

White D.J., Take W.A, Bolton M.D. 2001a. Measuring soil deformation in geotechnical models using digital images and PIV analysis. Proc. 10th Int. Conf. on Computer Methods and Advances in Geomechanics. Tucson, Arizona. pub. Balkema, Rotterdam pp 997-1002

White D. J., Take W.A, Bolton M.D. & Munachen S.E. 2001b. A deformation measuring system for geotechnical testing based on digital imaging, close-range photogrammetry, and PIV image analysis. Proc. 15th Int. Conf. on Soil Mech. and Geotech. Engng. Istanbul, Turkey. pp 539-542. pub. Balkema, Rotterdam

White, D.J., Take, W.A. and Bolton, M.D. 2003. Soil deformation measurement using particle image Velocimetry (PIV) and photogrammetry, Geotechnique, 53(7), pp. 619 – 631.

Zhang, C., Cai, Z.Y., Huang, Y.H., & Xu, G.M. 2018. Centrifuge model studies of the soil slope under freezing and thawing processes. 9th Int. Conf. Phys. Modelling Geotechnics, London.

1. Sample preparation and characterisation

Investigation into 3D printing of granular media

O. Adamidis, S. Alber & I. Anastasopoulos
ETH Zürich, Switzerland

ABSTRACT: Advances in additive manufacturing have recently motivated interest into 3D printing of geomaterials for an array of applications within geotechnical research. However, certain obstacles exist, which are investigated in this paper. Firstly, the geometry of the proposed particles has to be chosen, so that it is representative of an actual geomaterial. Here, Hostun sand was chosen as a reference material. It was found that CT scanning, microscopy, and morphology analysis were useful tools in assessing whether the proposed medium was representative of the reference material. Secondly, a specific 3D printing technology has to be chosen. Amongst the commercially available technologies investigated, PolyJet was found to be the most appropriate for the creation of small particles. Available materials, printing resolution, and support material removal were considered. Support material removal was found to be a limiting factor as very small particles can delaminate during this process. As a result, a particle diameter of 2 mm is proposed as the lower limit that can be reliably reproduced with the current level of PolyJet 3D printing.

1 INTRODUCTION

Advances in the field of 3D printing have sparked an interest for this methodology within geotechnical research. Interest stems from the unique opportunity provided by 3D printing for the creation of granular media with full and independent control of particle size, morphology, and material properties.

Significant promise exists in furthering our understanding of the influence of micro-scale particle characteristics on the macro-scale behaviour of granular media. For instance, Miskin and Jaeger (2013) focused on the role of particle shape on packing stiffness. They altered the shape of particles in the context of artificial evolution and used 3D-printing to recreate the computed shapes and verify the results of their simulations.

Another area of application is the potential synergy between 3D printing and Discrete Element Method (DEM) modelling. 3D printing could be used to recreate DEM generated particles and particle arrangements, offering a unique opportunity for meaningful validation of the numerical simulations. Kondo et al. (2017), performed such a validation by comparing the permeability of a granular medium, as calculated through numerical simulation using DEM and as measured for a 3D printed specimen which recreated the same particle arrangement.

Lastly, the use of 3D printing to produce standardised geomaterials for quality control and calibration of geotechnical testing devices has been proposed by Hanaor et al. (2015). By fully controlling the material properties and the size and morphology of particles, uniform and consistent specimens could be reliably reproduced.

Though the potential for the application of 3D-printing in geotechnical research is significant, several important issues remain. Firstly, the geometry of the particles that are to be 3D-printed has to be generated. Secondly, the appropriate 3D printing technology has to be chosen. Finally, quality assessment of the printed particles is necessary, in order to verify that the required geometries have been successfully reproduced. This last step is of increased importance since small particles deviate significantly from typical 3D printed objects, both in size and morphology.

2 DEFINING PARTICLE GEOMETRY

2.1 Choosing a reference material

Two avenues are available in defining the geometry of particles. The first is to choose simplified geometries in order to validate DEM simulations, as done by Kondo et al. (2017), who used spheres. The second is to produce particles that represent real geomaterials. In this case, geometries can be either directly obtained from a real material or stochastically generated through an algorithm, e.g. as proposed by Hanaor et al. (2015).

Here, the focus was on the capacity of 3D printing to recreate particles which include the small features that exist in real geomaterials. To that end, Hostun sand grains were chosen as reference particles to be replicated. Hostun sand is a poorly graded sand with $d_{50} = 0.34$ mm that has been widely used in geotechnical research (Mokni and Desrues 1999, for instance). It is an angular sand whose grains contain small features and are therefore appropriate for the pursued investigation.

Figure 1. The 65 CT scanned Hostun sand grains.

2.2 Micro Computed Tomography (CT) scanning

In order to capture the 3D geometry of Hostun sand grains, micro computed tomography (CT) was used. CT scanning has been used rather extensively in the geotechnical field, primarily to study the morphology of granular media and rocks (Desrues et al. 2006, for instance). Here, micro CT scanning was performed on 65 grains of Hostun sand. The number of grains chosen from different diameter size intervals corresponded to the relevant mass fraction calculated from the Particle Size Distribution (PSD) curve of the sand.

A μCT scanner ($\mu CT50$) of SCANCO Medical AG (Brüttiselllen, Switzerland) was used. Scanning was performed at the Institute for Biomechanics of ETH Zürich. In order to define the resolution for the scans, the approach suggested by Fonseca et al. (2012) was used. For Reigate Sand, they proposed a resolution of $5\,\mu m$, which corresponds to $0.018 \times d_{50}$. Extending this approach to Hostun sand, for which $0.018 \times d_{50} = 6\,\mu m$, it was decided that a resolution of $5\,\mu m$ would be sufficient. An overview of the relevant additional parameters to be set for CT scanning is given by Stauber and Müller (2008).

For the image processing of CT scans, the aspects that need to be considered include image filtering, segmentation, and component labelling (Stauber and Müller 2008). Image filtering reduces the noise within the image and hence enhances the visualisation of important features. During segmentation, images are discretized into two phases for further processing. Here, a threshold value was defined and voxels above it were set to white (object) while voxels below it were set to black (background). The last image processing step is component labelling, during which all small unconnected and undesired objects are removed. These small objects, which emerge because of image noise or improper sample preparation, are labelled by an algorithm and all particles below a certain size are eliminated. The sand grains as obtained after these processes are shown in Figure 1.

3 COMPARISON WITH REAL MEDIUM

In order to assess whether the 65 grains scanned were enough to produce a granular medium that is representative of Hostun sand, a morphology analysis was performed. The morphology characteristics of the proposed medium were compared to those of a sample of Hostun sand which was imaged using microscopy.

3.1 Morphology indices

There are numerous methods and definitions aiming to give a quantitative representation of particle morphology. Most of them rely on 2D images of particles (Wadell 1932, for instance). However, developments in 3D imaging technologies have increased the use of 3D-based morphology indices (Fonseca et al. 2012, Alshibli et al. 2015).

Indices related to the shape of a particle use the concepts of sphericity, aspect ratio, and convexity (Altuhafi et al. 2012). Sphericity indicates the degree of similarity of a particle to a sphere. This approach was introduced by Wadell (1932), who took the sphere as a standard of form. The aspect ratio is the ratio between the length of two axes of a particle. Convexity describes a particle's compactness. Roundness is another index related to the prominence and angularity of smaller features and is traditionally estimated through visual comparison with graphs (Powers 1953, Krumbein and Sloss 1951). Here, the numerical definitions of roundness proposed by Alshibli et al. (2015) were used. An overview of the morphology indices considered is given in Table 1.

3.2 Microscopy

In order to calculate the morphology indices for a large number of Hostun sand grains, which could then be compared to the 65 CT scanned grains, optical microscopy was used. A Leica Stereo microscope was used, provided by ScopeM, the Scientific Center for Optical and Electron Microscopy of ETH Zürich. A resolution of $7.14\,\mu m$ was used to capture 570 grains, an amount considered sufficient: a sample with more than 200 particles is expected to be reasonably representative of the medium (Bowman et al. 2001). The captured images were processed in a manner similar to the image processing described for the CT scans, once again involving the processes of image filtering, segmentation, and component labelling. This procedure is described in detail in Alber (2017).

In contrast to the 2D images captured via microscopy, the geometries obtained from CT scanning were in 3D. Comparing equivalent 2D and 3D morphology indices is not possible as they have different values for the same geomaterial (Alshibli et al. 2015). Therefore, it was chosen to compare the 2D morphology indices calculated from the microscope images to the 2D indices calculated from 2D projections of the CT scans. These 2D projections were chosen to be perpendicular to the shortest axis of the grains, since particles lying on a horizontal plane tend to have their smallest dimension vertically orientated (Cavarretta 2009). The grains lying on the horizontal surface of the microscope are hypothesised to have had a similar orientation.

Table 1. Overview of morphology indices.

Author	Formula	Name	Notation	Dimensions
Sphericity				
Wadell (1932)	$\dfrac{2\sqrt{A/\pi}}{d_{c,min}}$	Degree of sphericity	A : projected particle area $d_{c,min}$: diameter of minor circumscribed circle	2D
Alshibli et al. (2015)	$\dfrac{A}{A_S}$	Sphericity index	A : projected particle area A_S : area of circle with diameter the shortest particle diameter	*2D
Wadell (1932)	$\dfrac{SA_S}{SA}$	Degree of sphericity	SA : particle surface area SA_S : surface area of sphere with same volume	3D
Alshibli et al. (2015)	$\dfrac{V_P}{V_S}$	Sphericity index	V_P : particle volume V_S : volume of sphere with diameter the shortest particle diameter	3D
Aspect ratio				
Cavarretta (2009)	$\dfrac{d_{F,min}}{d_{F,max}}$	Aspect ratio	$d_{F,min}$: minimum Feret diameter $d_{F,max}$: maximum Feret diameter	2D
Fonseca et al. (2012)	$\dfrac{I}{L}$	Elongation index	I : intermediate principal particle axis L : longest principal particle axis	3D
Fonseca et al. (2012)	$\dfrac{S}{L}$	Flatness index	S : shortest principal particle axis L : longest principal particle axis	3D
Convexity				
Cavarretta (2009)	$\dfrac{A}{A_c}$	Solidity	A : projected particle area A_c : area of convex hull	2D
Fonseca et al. (2012)	$\dfrac{V_P}{V_{CH}}$	Convexity	V_P : particle volume V_{CH} : volume of convex hull	3D
Roundness				
Alshibli et al. (2015)	$\dfrac{0.78P}{\pi(L+S)/2}$	Roundness index	P : particle perimeter S : shortest principal particle axis L : longest principal particle axis	2D
Alshibli et al. (2015)	$\dfrac{SA}{4\pi\frac{(L+I+S)}{6}}$	Roundness index	SA : particle surface area S : shortest principal particle axis I : intermediate principal particle axis L : longest principal particle axis	3D

3.3 CT scans versus microscopy

Initially, the Particle Size Distribution (PSD) curve was examined. The PSD curve obtained for Hostun sand through sieving was compared to that calculated for the 570 particles captured through microscopoy, and to that of a medium consisting of the particles that were CT scanned, all appearing with the same frequency. This comparison is shown in Figure 2. For the microscopy PSD curve, the diameter depicted is the minimum Feret diameter multiplied by 0.91 (0.91 · $d_{F,min}$), as suggested by Altuhafi et al. (2012). The minimum Feret diameter is equal to the minimum distance between two parallel tangents to the particle perimeter. Further reduction of this value by the multiplier 0.91 is performed because particles were not restricted to pass through the sieve openings with their diameter $d_{F,min}$ normal to the sieve grid. The ordinate of Figure 2 for the microscopy curve depicts the percentage that is finer by area, rather than by mass. The curves calculated both through sieving and through microscopy are representative of Hostun sand and are very close to each other. The PSD curve for the CT scans was calculated in the same way as for the particles depicted through microscopy, using the 2D projections of the CT scans. Although the CT scans did not include the very small or the very large particles of the original particle size distribution curve, the depicted curves were deemed to be sufficiently similar. Therefore, it was decided that all 65 CT scans should appear with the same frequency in the proposed granular medium.

To further examine how representative such a granular medium is of Hostun sand, morphology indices were calculated. A favourable comparison was observed for the 2D sphericity index proposed by Alshibli et al. (2015) and presented in the histogram of Figure 3a, as well as for the aspect ratio, as defined by

Cavarretta (2009) and shown in Figure 3b. The definitions of these morphology indices are included in table 1. For the 2D roundness index proposed by Alshibli et al. (2015) and depicted in Figure 3c, a deviation was observed. The reason was that differences in perimeter were larger than in area, likely due to the particles examined with microscopy not lying on the horizontal plane with their shortest axis perpendicular to it, as was assumed. Particle solidity, as defined by Cavarretta (2009) was only slightly smaller for the CT scan projections than for the microscopy-captured particles (Fig. 3d). Overall, the morphology indices show that the proposed medium (CT scan projections) is similar to Hostun sand (microscopy). Even in terms of roundness, the small deviation that exists is likely an artefact of the assumption of particles lying perfectly perpendicular to their shortest principal axis.

4 3D PRINTING TECHNOLOGY SELECTION

The search for potential 3D printing technologies was based around those available within ETH Zürich. The technologies considered are shown in table 2.

The principal parameters considered included the existence of support material, the resolution, the available materials, and the time and cost involved. The existence of support material that had to be manually removed was prohibitive for 3D printing of large quantities of small particles. Regarding resolution, values that did not allow for the realisation of small features could not be accepted. Finally, it is important to note that quartz, out of which Hostun sand is made, is considerably harder than most materials available for 3D printing.

Stereolithography (SLA) is one of the oldest 3D printing technologies, where a UV laser beam hardens a liquid photopolymer resin layer by layer, creating the required object. While adequate resolutions are available, the existence of support structures which are made out of the same material as the object and which demand manual removal render this technology inappropriate for the proposed application of 3D printing particles.

PolyJet is a technology which is also uses liquid photopolymer and UV light. However, here the liquid resin is jetted and cured instantly. This allows multiple nozzles and materials to be used. Consequently, though support structures exist, they can be formed using a material different to that of the object. This material can be removed after printing, using a caustic solution. Resolutions for PolyJet are better than the other technologies investigated, with layer thickness being particularly fine. This technique is also rather affordable and quick. A drawback is that the photopolymers that are available in PolyJet printing are softer than quartz, of which Hostun sand consists, by up to two orders of magnitude.

Selective Laser Sintering (SLS) and Selective Laser Melting (SLM) are popular techniques for which a variety of materials exists. They work in a way similar to SLA, but use material powder instead of liquid polymer. In SLM, the powder is completely melted, while in SLS it is heated to the point that it can fuse on a molecular level. Both methods require no additional support material. The drawback of these techniques comes from the type of laser that is used, which has a bigger focus spot and thus limits the size of the smallest features that can be printed. Moreover, especially for metal powders, the cost is significantly higher than for PolyJet or SLA.

After comparing the above technologies, it was concluded that PolyJet is the most appropriate for small and highly detailed objects. It is on the high end of the examined technologies regarding resolution and it provides an avenue for easy support material removal. In the few cases reported in literature where 3D printing

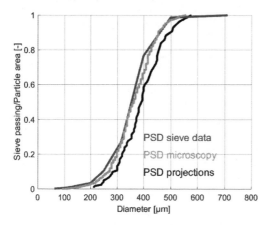

Figure 2. Particle size distribution curves, calculated through sieving, microscopy, and CT scan projection.

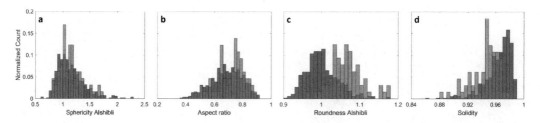

Figure 3. Normalised histograms for a. sphericity index (Alshibli et al. 2015), b. aspect ratio (Cavarretta 2009), c. roundness index (Alshibli et al. 2015), d. solidity (Cavarretta 2009), as calculated from microscopy images and from the projections of CT scans.

Table 2. Overview of investigated 3D printing technologies.

Techology	Material	Resolution	Support material	Cost
Stereolithography (SLA)	Photopolymer	25 μm (z) 140 μm (x, y)	Manual removal	Affordable
Polyjet	Photopolymer	16 μm (z) 40 μm (x, y)	Solution bath removal	Affordable
Selective Laser Sintering (SLS)	Metal, glass, ceramic, plastic powder	50–150 μm (z) 200–300 μm (x, y)	No support	Depends on material
Selective Laser Melting (SLM)	Metal powder	50–100 μm (z)	No support	Expensive

of geomaterial-like particles has been performed, Poly-Jet technology was also used (Miskin and Jaeger 2013, Hanaor et al. 2015). However, the available resolution of 40 μm in the x and y directions, although better than that of other technologies, still poses a limitation to printing the scanned particles in real scale. Considering that the diameter of some of the scanned grains was as low as 200 μm, important small features could not be captured in real scale with the available x and y resolution. Therefore, it was decided to scale up the available geometries, so that their morphology could be sufficiently reproduced. Three scaling factors were chosen: 4, 8, and 16.

5 PRINTING PROCESS

Printing was performed at the Department of Mechanical and Process Engineering of ETH Zürich. The printer used was the PolyJet Objet500 Connex3 of the company Stratasys. A printing resolution of 30 μm along the z axis and 40 μm along the x and y axes was used. The high temperature resistant material RGD525 was used for the prints, as it has the highest modulus of elasticity among the materials available for 3D printing with PolyJet technology (3.2–3.5 GPa). The support structures were created using the soluble support material SUP706. The support material was removed by submeging the prints in an aqueous solution bath, containing 2% NaOH and 1% NaSiO$_3$, which was then neutralised using vinegar and washed away under waterflow.

6 ASSESSMENT OF PRINTING QUALITY

Following printing, an assessment of quality was carried out. To this end, CT scans of the printed particles were performed, which were subsequently compared to the original CT scans of Hostun sand. For the scale factor of 16, 5 printed particles were scanned, for the scale factor of 8, 10 particles were scanned, and for the scaling factor of 4, 10 particles were scanned as well. All particles were randomly selected. Since the printed particles were larger than the originals, the scanning resolution could be modified accordingly. For the particles of scale factor 16, 8, and 4, a scanning resolution of 48.4 μm, 34.4 μm, and 20 μm was used respectively. Scanning was performed at the Institute for Biomechanics of ETH Zürich.

Table 3. Comparison between the original and the 3D printed particles: mean values of differences. 3D sphericity index, 3D roundness index, elongation, flatness, and convexity are included. See table 1 for definitions.

Property	scale 16	scale 8	scale 4	scale 4 new
Mean diameter	−3.2%	−6.3%	−9.5%	−3.3%
Area	−7.0%	−9.1%	+10.5%	+18.7%
Volume	−8.6%	−16.9%	−27.5%	−9.6%
Sphericity	+0.8%	−7.9%	−12.2%	+18.6%
Roundness	−1.0%	+3.3%	+33.8%	+25.2%
Elongation	+2.6%	−3.1%	−6.0%	−3.9%
Flatness	+1.0%	+4.6%	+5.8%	−8.8%
Convexity	+2.4%	+1.3%	+11.3%	+4.0%

Following CT scanning, geometrical properties and morphology indices were calculated for each particle and compared to those of the equivalent original particle of Hostun sand. The mean values of differences for each property considered are presented in table 3. Diameter, area, and volume were compared to the equivalent original particle, scaled up by the corresponding factor.

Morphology indices were close to the original grains for larger particles. As the scale factor decreased, deviations in morphology indices became larger. For example, roundness, which depends more on small features, deviated significantly for the particles of scale factor 4. Overall, morphology was adequately represented for scale factors 8 and 16.

Area, volume, and mean diameter deviated from the original particles. The printed particles were consistently smaller than expected. Considering the size of the grains, printing resolution could not be responsible for this deviation. It was hypothesised that the process of support material removal was culpable. The caustic solution used to dissolve the support material could have also affected the particles. This explains why the deviation in diameter and volume was larger for smaller grains, which were more severely affected by corrosion of their outer layers. This hypothesis was visually confirmed by comparing the same grain in all scales, as shown in Figure 4.

After this observation, new grains were printed for the scale factor of 4. These new prints were exposed to the caustic solution for 45 minutes, as opposed to several hours for previous prints, as is the common practice at the Institute for Biomechanics of ETH

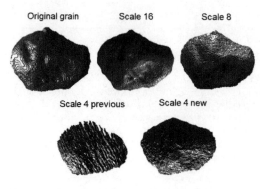

Figure 4. Comparison of printed particles at different scales with the original sand grain.

Zürich. Overall, the new particles required a much more labour intensive procedure for support material removal. The prints were initally broken up manually and most of the support material was removed under waterflow. Subsequently, the caustic solution was used. Finally, the particles were once again placed under waterflow and any remaining support material was manually removed. As seen in Figure 4 and table 3 the produced particles were a better representation of the original grains.

The outcome the new print of scale factor 4 highlights an additional limitation of PolyJet 3D printing: a fine line exists between support material removal using the proposed solution bath and particle delamination.

7 CONCLUSIONS

In this paper, the process of 3D printing particles that are representative of a geomaterial was investigated.

Initially, the choice of particle geometry was discussed. Hostun sand was chosen as a reference material. CT scanning was used to capture the geometries of 65 Hostun sand grains. The proportion of grains within different size intervals matched the Particle Size Distribution curve of the sand. The morphology characteristics of these particles were compared to those calculated for a sample of 570 grains of Hostun sand, subjected to microscopy. It was concluded that the 65 CT scanned grains adequately represented the morphology of the sand.

Afterwards, commercially available 3D printing technologies were assessed, in order to choose the most appropriate for printing of geomaterials. PolyJet was found to be the best option. It allowed for relatively easy support material removal and had the best printing resolution out of the technologies considered. However, the available resolution was still not fine enough to print the scanned sand grains in real scale. Consequently, it was decided to scale up the scanned geometries, by scale factors of 4, 8, and 16.

Post printing, some of the produced particles were subjected to CT scanning to assess their quality. It was found that the intended geometries were adequately reproduced for scale factors 8 and 16, following typical PolyJet 3D printing and support material removal

processes. For the scale factor of 4, the typically used process for support material removal, using a caustic solution bath, severely delaminated the particles and distorted their geometry.

Subsequently, a new set of particles was printed for the scaling factor of 4. These particles were submerged in the caustic solution for less than 1 h. Combined with manual removal of support material before and after submersion, better quality particles were attained. However, this process was too time consuming and labour intensive to be proposed for the production of large quantities of grains.

Overall, PolyJet is currently placed as the best option for 3D printing of geomaterials. However, it is proposed that particles with sizes above 2 mm are printed, corresponding roughly to scale factor 8 here. For such sizes geometrical and morphological characteristics are adequately reproduced using typical processes for printing and support material removal.

REFERENCES

Alber, S. 2017. *Investigation of 3D printed geomaterials*. Master's Thesis, ETH Zürich.

Alshibli, K., Druckrey, A., Al-Raoush, R., Weiskittel, T., & Lavrik, N. 2015. Quantifying morphology of sands using 3d imaging. *J. of Mat. in Civil Eng. 27 (10)*, 04014275.

Altuhafi, F., O'Sullivan, C., & Cavarretta, I. 2012. Analysis of an image-based method to quantify the size and shape of sand particles. *J. of Geotech. and Geoenv. Eng. 139*(8), 1290–1307.

Bowman, E. T., Soga, K., & Drummond, W. 2001. Particle shape characterisation using Fourier descriptor analysis. *Géotechnique 51*(6), 545–554.

Cavarretta, I. 2009. The influence of particle characteristics on the engineering behaviour of granular materials.

Desrues, J., Viggiani, G., & Bsuelle, P. 2006. *Advances in X-ray Tomography for Geomaterials*. ISTE Ltd.

Fonseca, J., O'Sullivan, C., Coop, M., & Lee, P. 2012. Noninvasive characterization of particle morphology of natural sands. *Soils and Found. 52 (4)*, 712–722.

Hanaor, D., Gan, Y., Revay, M., Airey, D., & Einav, I. 2015. 3d printable geomaterials. *Géotechnique 66 (4)*, 323–332.

Kondo, A., Matsumura, S., Mizutani, T., & Kohama, E. 2017. Reproduction of discrete element model by 3d printing and its experimental validation on permeability issue. In *Advances in Laboratory Testing and Modelling of Soils and Shales (ATMSS)*, pp. 517–524. Springer Int. Publ.

Krumbein, W. & Sloss, L. 1951. *Stratigraphy and Sedimentation*. W. H. Freeman & Co., San Francisco.

Miskin, M. & Jaeger, H. 2013. Adapting granular materials through artificial evolution. *Nature Mat. 12 (4)*, 326–331.

Mokni, M. & Desrues, J. 1999. Strain localization measurements in undrained plane-strain biaxial tests on Hostun RF sand. *441*(v), 419–441.

Powers, M. 1953. A new roundness scale for sedimentary particles. *J. of Sedim. Petrology 23*(2), 117–119.

Stauber, M. & Müller, R. 2008. Micro-Computed Tomography: A Method for the Non-Destructive Evaluation of the Three-Dimensional Structure of Biological Specimens. In *Osteoporosis, part of Methods in Molecular Biology series*, Volume 455, pp. 273–292.

Wadell, H. 1932. Volume, shape, and roundness of rock particles. *J. of Geology 40 (5)*, 443–451.

Undrained shear strength profile of normally and overconsolidated kaolin clay

A. Arnold
Lucerne University of Applied Sciences and Arts, Institute of Civil Engineering, Horw, Switzerland

W. Zhang & A. Askarinejad
Faculty of Civil Engineering and Geosciences, Delft University of Technology, Delft, The Netherlands

ABSTRACT: The profile of the undrained shear strength of kaolin clay samples has been determined using an automated shear vane device. The automated device can monitor and register the changes in the mobilised shear resistance as a function of shear strain, while merely peak undrained shear strengths of the sample can be determined using conventional vane shear tests. The clay samples were prepared using centrifuge consolidation of slurries at 30 g and 90 g. The results on the impact of different s_u-profiles to the load displacement curves and bearing capacities of shallow foundations which were tested at 30 g on normally consolidated and overconsolidated kaolin samples are shown. The reliability of field-vane-shear tests on borehole samples is also discussed.

1 MOTIVATION

1.1 Estimation of s_u-values in the field

Values of s_u in the field are sometimes estimated using pocket vane-shear devices (Serota & Jangle 1972; Brunori et al. 1989; Vahedifard et al. 2016; Gruchot & Zydroń 2016). Those vane shear tests are usually done on borehole samples. Having borehole samples at different depths a depth-dependent s_u-profile can be estimated. The question is how reliable those results from pocket vane-shear tests on borehole samples are?

1.2 Undrained analysis for shallow foundations

Shallow foundations on clayey soils are usually designed for undrained and drained conditions. The bearing capacity in the undrained case is mostly estimated by using the well-known approach of Prandtl (1920):

$$\sigma_f = 5.14 \cdot s_u \quad (1)$$

where σ_f = bearing capacity for undrained conditions; s_u = undrained shear strength.

This approach has the assumption that the undrained shear strength is a constant value for a given soil over a certain depth. The undrained shear strength of a certain clay soil is only dependent on the water content (for saturated soil) and will therefore in most cases not be constant with depth. Due to the overburden pressure and the decreasing water content with depth, s_u will increase with depth.

1.3 Effect of s_u-profiles to the bearing resistance of shallow foundations

Puzrin et al. (2011) show in their forensic analysis of the bearing capacity failure of the Transcona Grain Elevator, that the depth-dependent s_u-profile can have a huge impact to the bearing capacity of a foundation. In Equation 1 the width of the foundation does not play a role. But the size and therefore the depth of the failure mechanism are dependent on the foundation width. The influence of the s_u-profile can be of major interest for larger foundations. With other words: having a s_u-profile, Equation 1 is no longer independent of the foundation width due to an increasing s_u with depth.

2 LABORATORY TESTING OF UNDRAINED SHEAR STRENGTH ON KAOLIN CLAY

Results of laboratory vane shear tests as described for example in Knappett & Craig (2012) or Das & Sobhan (2016) are discussed in this contribution.

2.1 Preparation of the clay specimens

The Kaolin clay was mixed to clay slurry and then consolidated in a centrifuge at TU Delft (Askarinejad et al. 2017) with 30 g and 90 g respectively. The consolidation in a centrifuge has the advantage that s_u-profiles will be generated since the soil consolidates under its self-weight. The soil specimens were later used for foundation tests on normally consolidated soil (NC specimen; test-level 30 g, consolidation-level

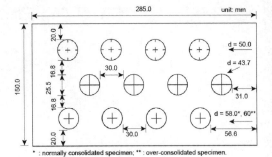

Figure 1. Top view of the conventional vane shear tests.

30 g) and on overconsolidated soil (OC specimen; test-level 30 g, consolidation-level 90 g, OCR = 3.0). The consolidation process in the centrifuge was monitored by tracking the settlement of the specimens.

2.2 Conventional vane shear tests

A hand-rotational vane test apparatus was used to conduct the conventional vane shear tests following BS 1377 standard (Standard 1990). Two springs with different rotational stiffness were selected based on the stiffness of the two samples, respectively. By turning the handle of the device at a constant angular velocity of one revolution per second, the spring will be deformed gradually which transfers the torsion force to the vane embedded in the soil. The torsion is increasing till the vane can rotate, which means the ultimate capacity of soil has been met. The torsion at the moment of failure (T) can be inferred using the pointer which shows the maximum rotation angle of the spring and then T can be determined through the calibrated deflection degree vs. torque relationship of the spring. The undrained shear strength of the soil, s_u, can be calculated based on Equation 2.

$$s_u = T/[\pi D^2(H/2+D/6)] \quad (2)$$

where, D and H are the diameter and height of the vane, respectively.

The vane used in this study has dimensions of 25.5 mm × 25.5 mm (D × H). NC and OC soil specimens had a height of 75.3 and 68.8 mm after consolidation, respectively. However, both specimens had a length of 285 mm and width of 150 mm. Three different test depths (d, from soil surface to the vane bottom) for each sample were defined and each embedded depth condition was repeated four times (Figure 1).

2.3 Automated vane shear test device

A rotation-controlled automated shear vane test apparatus (Figure 2) was used to measure the undrained shear strengths of the normally and over-consolidated specimens. The bearing capacity of a shallow foundation with a width of 50 mm (in model scale) was determined on these two samples. Only on one side of each specimen the vane shear tests were performed because of the cables of Pore Water Transduces (Figure 3). Three testing depths (d = 40.4 mm, 55.0 mm, and 74.1 mm) were defined for both specimens (Figure 3). Two types of vanes were used for these tests to check the effect of the size of vanes on the measured value of s_u: vane 1 has the dimensions of 25.5 mm in diameter (D) and height (H), whereas vane 2 has half that size, i.e. H = D = 12.75 mm.

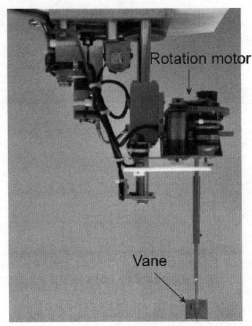

Figure 2. Automated vane test device.

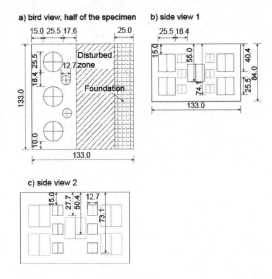

Figure 3. Automated vane shear tests design.

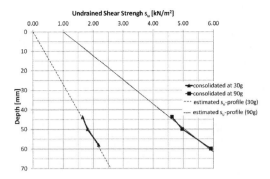

Figure 4. Measured undrained shear strengths in the two Kaolin clay specimens.

3 TESTS RESULTS

The vane shear tests were conducted at 1 g after consolidation in the centrifuge.

3.1 s_u-profiles gained from conventional vane shear tests

As one can see in Figure 4, in both clay specimens, three vane shear tests have been conducted in three different depths.

Due to the different g-levels in the consolidation-phase the clay specimens were consolidated with different weights. The bigger the overburden pressure, the smaller the specific volume and therefore the smaller the water content will become. This results in higher undrained shear strength in the soil, which was consolidated at 90 g.

3.2 Results obtained from the automated vane shear device

The change in undrained shear strength s_u during a vane shear test can be measured with the automated vane shear device. This gives more information about the development of the strength and allows estimation if the critical state or a peak state of s_u is measured with ordinary vane shear test devices.

Similar vane shear tests on synthetic clay soil (Laponite) have been done by Wallace & Rutherfort (2015) where peak values for the undrained shear strength could be shown as compared to the here presented results in Figure 5. It shows clearly, that the undrained shear strength s_u reaches a peak value at about 10° of vane-rotation. The critical value for s_u could not exactly be reached within a vane-rotation of 40°. It shows also a dependency of s_u on the depth of the investigated soil. A peak value for s_u was clearly expected since the Kaolin clay was consolidated at a much higher overburden pressure (30 times gravity) than it had during the vane shear tests. At a depth of 40.4 mm s_u drops from 1.5 kN/m² to around 0.75 kN/m² while at a depth of 74.1 mm s_u drops from 2.5 kN/m² to 1.5 kN/m².

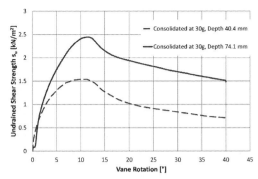

Figure 5. Undrained shear strength s_u in Kaolin clay consolidated at 30 g as a function of the shear vane rotation.

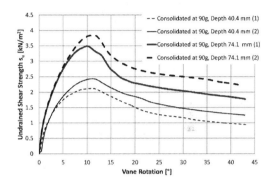

Figure 6. Undrained shear strength s_u in clay soil consolidated at 90 g as a function of the shear vane rotation.

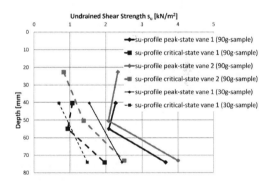

Figure 7. Profiles for s_u estimated from the automated shear-vane device.

Figure 6 shows the results from the tests conducted in the specimen consolidated at 90 g. One can see that there is a certain scatter in s_u determined at the same depth at different places. Here also the peak value is reached at a vane-rotation of around 10°. The values of s_u are generally bigger than in the specimen consolidated at 30 g which is feasible due to the bigger overburden pressure during the consolidation at 90 g.

Figure 7 shows profiles of peak- and critical states of s_u for different vanes (vane 1 and vane 2) and the two different soil specimens (consolidated at 30g and 90g). The results show differences in peak- and critical state,

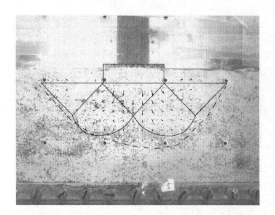
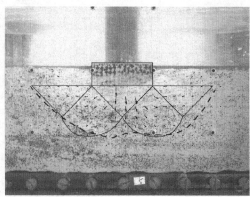

Figure 8. PIV image of the foundation (model width: 50 mm, prototype width: 1500 mm) test in the centrifuge at 30 g on normally consolidated clay at maximum load. Dashed line: Preliminary determined failure mechanism of the foundation at a model-load of 344.8 N (equal to $\sigma = 51.8$ kN/m^2). Continuous line: Failure mechanism after Prandtl (1920).

Figure 9. PIV image of the foundation (model width: 50 mm, prototype width: 1500 mm) test in the centrifuge at 30 g on overconsolidated clay. Dashed line: Preliminary determined failure mechanism of the foundation at a model-load of 344.8 N (equal to $\sigma = 51.8$ kN/m^2). Continuous line: Failure mechanism after Prandtl (1920).

in the use of the vane type and also in the g-level of clay consolidation. It should also be noted that the vane shear tests have been conducted after the foundation tests. Some disturbance of the soil due to the foundation tests cannot be excluded even though the distance between the vane shear tests and the disturbed zone from the foundation tests (compare Figure 3) should be big enough.

4 INTERPRETATION OF THE TEST RESULTS

The comparison of the s_u-profile in Figure 5 with the results of the automated shear vane device show, that the peak values in Figure 6 are in good agreement with the estimated s_u-profile for the specimen consolidated at 30 g. No good agreement is given from the tests at the specimens consolidated at 90 g. The automated shear vane tests show lower s_u-values even for the peak states. Figure 7 shows well the dependency of consolidation level and depth: lower consolidation level results in lower s_u-values as well as increasing depth gives higher s_u-values. Figure 7 shows as well that the vane-dimensions seem not to play an important role since the results of the two vanes used fit quite well.

5 UNDRAINED ANALYSIS FOR SHALLOW FOUNDATIONS WITH S_u-PROFILE

An ordinary dimensioning of a shallow foundation on clayey soil in undrained conditions will be done by using the approach after Prandtl (1920) as mentioned in Chapter 1.2. This approach takes into account constant undrained shear strength along all shear zones of the failure mechanism.

The undrained shear strength is usually not constant with depth as mentioned and discussed in relation to the vane shear tests in chapters 3.1, 3.2 and 4.

The increasing undrained shear strength will certainly have an effect on the shape of the occurring failure mechanism beneath the foundation.

Centrifuge tests on shallow foundations have been conducted to study the change of the shape of the failure mechanism depending on the overconsolidation ratio of the soil. Similar tests on shallow foundations are described by Madabushi (2014) after McMahon (2012). Preliminary results of these tests are given in figures 8 and 9. The shear bands are inferred by analysing the images taken from the transparent side of the boxes using Particle Image Velocimetry method (White et al. 2003; Askarinejad et al. 2015).

Kraehenbuehl (2018) shows in his Master thesis that taking into account an average s_u over the depth of the failure mechanism after Prandtl (1920) can lead to an overestimation of the ultimate limit state. More details will be published at later stage.

6 DISCUSSION & CONCLUSIONS

The test results confirm that overconsolidated soil shows a peak for s_u. This is a well-known fact (e.g. Muir Wood, 1990). Both soil specimens, consolidated at 30 g and 90 g show an overconsolidated behaviour, because the soil had enough time to swell at 1g after the consolidation phase in the centrifuge.

The test results show also that the ordinary shear vane device might show peak values but not s_u at the critical state; at least the comparison between the tests on the soil specimen consolidated at 30 g leads to this conclusion. Bjerrum (1973) showed that with increasing plasticity of the soil the s_u-value obtained from vane shear tests might be larger compared to other investigation methods of s_u.

If we now come back to the estimation of s_u in the field using pocket vane shear test devices on borehole samples, the question turns up whether a peak- or

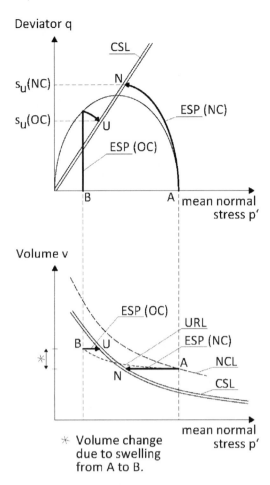

A: stress condition at a certain depth (initial condition in a borehole sample).

B: stress condition in the borehole sample after lifting and swelling process.

Figure 10. Swelling process in a borehole sample due to lifting and potential undrained stress paths (ESP) due to vane shear testing for normally (NC) and heavily (OC) overconsolidated soil after Roscoe & Burland (1968) and Schofield & Wroth (1968). CSL: critical state line. NCL: normal compression line. URL: unloading-reloading line.

a critical-state s_u-value is measured. A normally consolidated clayey borehole sample will behave overconsolidated on a vane shear test once the soil specimen has enough time to swell and to dissipate the excess pore water pressures, which arise when the specimens are taken out from a certain depth and therefore from a certain overburden pressure. The approximately measured peak-state s_u-value might be smaller than s_u under normally consolidated conditions (Figure 10, Schofield & Wroth, 1968).

The same behaviour should be given with an initially overconsolidated borehole sample. Due to the process of lifting and swelling the OC-borehole sample will behave even more overconsolidated and will also give a lower undrained shear strength at critical state. If s_u at peak state will be lower than s_u at critical state but with a smaller value of OCR cannot be answered – it depends on how much swelling will take place.

The drying process of a borehole sample will also have an effect on the undrained shear strength. Therefore, it seems to be of high importance to perform vane shear tests on freshly lifted borehole samples to make sure that the water content has not changed and therefore s_u will remain the same as it was before the borehole sample was lifted. Otherwise a wrong s_u might be used for the dimensioning of a shallow foundation for example.

The PIV-recordings of the foundation tests show at Figure 8 the maximum load and at Figure 9 the same load situation which is not the peak load for the foundation tested on overconsolidated soil. Both failure mechanisms are in rather good comparison to the Prandtl mechanism (1920).

The failure mechanism on the normally consolidated soil is deeper and the PIV-recordings show arrows which are orientated clearly downwards. This implies larger and more concentrated movements downwards which are feasible for normally consolidated soil. The failure mechanism on the overconsolidated soil is less deep but wider. The arrows below the foundation are not clearly orientated downwards but yet have a tendency to turn horizontally as the squeezing-mechanism seems to be dominant. If we compare the resulting downward movements at model scale of both tests at the same model-load of 344.8 N (normally consolidated: $\Delta s = 24$ mm; overconsolidated: $\Delta s = 16$ mm) our interpretation can be confirmed in the sense that overconsolidated soil does not tend to produce directly very large downward movements but will be subjected to a more pronounced lateral squeezing process.

If we compare these preliminary results with the results from the tests given in Madabushi (2014) after McMahon (2012) we can say that the distinctive lateral squeezing effect in the soil due to rising overconsolidation ratio OCR can also be observed in our PIV-analysis. The failure mechanism seems to get wider and less deep with rising OCR-level and with a more pronounced s_u-profile.

7 OUTLOOK

The automated vane shear device unfortunately did not work at 30 g. That's why only test results conducted at 1 g could be presented. In a next step centrifuge vane shear tests on normally- and overconsolidated clay soil will be conducted to get a better insight into the development of s_u. It would also be of great interest to conduct tests at borehole samples, both on freshly lifted and on swelled samples to find out how reliable those pocket vane shear tests are concerning the value of s_u.

The results of the foundation tests will be interpreted in more detail and will also be compared with analytic and numeric models to describe the dependency of the shape of the failure mechanism as a function of the existing s_u-profile in the soil.

ACKNOWLEDGEMENTS

We are most grateful to the Institute of Civil Engineering at the Lucerne University of Applied Sciences and Arts in Switzerland who funded the vane shear tests.

We are also grateful to the help of the centrifuge technicians Han de Visser, Ronald van Leeuwen, Kees van Beek and to the MSc-student Manuel Kraehenbuehl who conducted the vane shear tests in the lab. We also thank the MSc-student Philipp Baechler who helped with drawing Figure 10.

REFERENCES

Askarinejad, A., Beck, A. & Springman, S.M. 2015. Scaling law of static liquefaction mechanism in geo-centrifuge and corresponding hydro-mechanical characterisation of an unsaturated silty sand having a viscous pore fluid. *Canadian Geotechnical Journal*, 52: 1–13, doi:10.1139/cgj-2014-0237.

Askarinejad, A., Philia Boru Sitanggang, A. & Schenkeveld, F.M. 2017. Effect of pore fluid on the cyclic behaviour of laterally loaded offshore piles modelled in centrifuge. *In 19th International Conference on Soil Mechanics and Geotechnical Engineering*, Seoul.

Bjerrum, L. 1973. Problems of Soil Mechanics and Construction on Soft Clays. *Proceedings of the 8th International Conference on SMFE*, Moscow. 3: 111–159.

British Standard 1377, 1990. Methods of Test for Soils for Civil Engineering Purposes. *British Standard Institution*, London.

Brunori, F., Penzo, M. & Torri, D. 1989. Soil shear strength: its measurement and soil detachability. *Catena*, 16(1): 59–71.

Das, B.M., Sobhan, K. 2016. *Principles of Geotechnical Engineering*. Boston: Cengage Learning.

Gruchot, A. & Zydroń, T. 2016. Impact of a Test Method on the Undrained Shear Strength of a Chosen Fly Ash. *Journal of Ecological Engineering*, 17(4).

Knappett, J.A., Craig, R.F. 2012. *Craig's Soil Mechanics. Eighth Edition*. London: Spon Press, Taylor & Francis.

Kraehenbuehl, M. 2018. *Grundbruchwiderstand von Flachfundationen im undrainierten Zustand*. MSc thesis. Switzerland: Lucerne University of Applied Sciences and Arts.

Madabushi, G. 2014. *Centrifuge Modelling for Civil Engineers*. London: CRC Press, Taylor & Francis.

McMahon, B. 2012. *Deformation Mechanisms below Shallow Foundations*. PhD thesis, United Kingdom: University of Cambridge.

Muir Wood, D. 1990. *Soil Behaviour and Critical State Soil Mechanics*. Cambridge University Press.

Prandtl, L. 1920. Über die Härte plastischer Körper. Nachrichten von der Königlichen Gesellschaft der Wissenschaften zu Göttingen, Mathematisch-physikalische Klasse (1). 74–85.

Puzrin, A.M., Alonso, E.E., Pinyol, N.M. 2010. *Geomechanics of Failures*. Dordrecht: Springer.

Roscoe, K.H., Burland, J.B. 1968. On the generalized behaviour of "wet" clay. In: Heyman, J., Leckie, F.A. (eds). *Engineering Plasticity*. Cambridge University Press. 535–610.

Schofield, A.N., Wroth, C.P. 1968. *Critical State Soil Mechanics*. London: McGraw Hill.

Schofield, A. & Wroth, P. 1968. Critical state soil mechanics (Vol. 310). McGraw-Hill London (Pub).

Serota, S. & Jangle, A. 1972. A direct-reading pocket shear vane. Civil Engineering, 42(1).

Vahedifard, F., Howard, I.L., Badran, W.H., Carruth, W.D., Hamlehdari, M. & Jordan, B.D. 2016. Strength indices of high-moisture soils using handheld gauges. *Proceedings of the Institution of Civil Engineers-Ground Improvement*, 169(3): 167–181.

Wallace, J.F., Rutherford, C.J. 2015. Geotechnical Properties of LAPONITE RD®. *Geotechnical Testing Journal*. 38(5): 574–587.

White, D.J., Take, W.A. & Bolton, M.D. 2003. Soil deformation measurement using particle image velocimetry (PIV) and photogrammetry. *Géotechnique*, 53(7): 619–632.

LEAP GWU 2017: Investigating different methods for verifying the relative density of a centrifuge model

R. Beber
Department of Civil, Environmental & Mechanical Engineering, University of Trento, Trento, Italy

S.S.C. Madabhushi, A. Dobrisan, S.K. Haigh & S.P.G. Madabhushi
Schofield Centre, Department of Engineering, University of Cambridge, Cambridge, UK

ABSTRACT: The second phase of the LEAP project investigating the liquefaction of a 5° saturated sand slope has commenced, focused on the effects of relative density and input motion. Thus, reliably estimating the density and quantifying the associated uncertainty is cardinal. The sensitivity of the calculated relative density to small measurement errors on the sand mass and volume during model preparation is highlighted. Settlement during swing-up, quantified using particle image velocimetry, implies a small densification of the soil mass. Recognition of the stress dependent soil stiffness is required to model the measured settlement profile. Finally, in-flight miniature CPT tests are discussed for further analysis of density. Empirical correlations linking cone resistance to relative density appear to agree with measurements from the pouring stage. However, the dependence of cone resistance on other factors like the lateral stress state requires further investigation. Overall, the CPT provides instructive insights into key model properties.

1 INTRODUCTION

Following from the LEAP-GWU-2015 exercise, a second round of investigation to enrich experimental and numerical data on liquefaction problems has commenced. Kutter et al. (2017) presents the problem specifications and project guidelines. The problem being studied is that of the dynamic behaviour of a 5° saturated slope of F-65 Ottawa sand. In this second test series, the sensitivity of the liquefying slope's response to the relative density of the soil and the magnitude and frequency content of the ground motion are studied.

When compiling centrifuge data from different laboratories and comparing the results against numerical data the importance of estimating the critical model properties and quantifying the uncertainty is paramount.

This paper's focus is on characterising the relative density of the model slope through different, independent techniques. Traditional methods of determining the relative density are described which rely on measurements of the model mass and volume. The potential for changes in this density due to the centrifuge swing up is then touched upon, with image correlation techniques used to quantify the change. Finally, the use of an in-flight miniature cone penetration test (CPT) to check the relative density and small strain shear modulus is discussed.

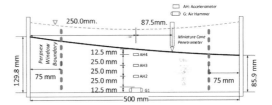

Figure 1. Model schematic.

2 EXPERIMENTAL SETUP

The desired prototype to be represented by the centrifuge model was a 5° saturated slope of F-65 Ottawa sand. The specified soil density for the test discussed was 1599 kg/m^3 or equivalently a target relative density of 50%. For the calculation, the minimum and maximum sand density for Ottawa sand were taken from Kutter et al. (2017). Figure 1 shows the corresponding centrifuge model schematic with the key instruments discussed in this paper illustrated. The earthquake motion is applied tangentially to the centrifuge motion. Hence, the model must have a logarithmic spiral to produce a prototype slope when the curvature of the radial centrifugal acceleration field is accounted for. The specification of these tests called for the use of a rigid model container in order to achieve numerically simple boundary conditions. The use of

Figure 2. Centrifuge model before swing-up with the PIV and CPT setup visible.

a window box featuring a perspex front facilitated imaging of the model cross section during the test.

Air pluviation was used for the preparation of the model slope. The instruments were placed at the required locations during the construction of the soil model. Once the model was completed, it was saturated under vacuum using 40 cSt methyl cellulose fluid to satisfy the dynamic scaling laws (Schofield 1981). Prior to saturation, the soil model was flushed with CO_2 gas in three cycles. The process removes the air from the soil model and ensures a high degree of saturation. The saturation was carried out using the CAM-SAT system which controls the rate of fluid mass flux into the base of the model (Stringer & Madabhushi 2009). The controlled mass flux of 0.5 kg/h prevents fluidisation of the soil bed during saturation whilst maintaining the required driving potential head. Following saturation, the model slope was created using a vertical cutting plate following guides which were machined and mounted at the top of the model container. The combination of guides and cutting plate allowed the logarithmic spiral geometry and a 1 : 40 slope correction to be realised. The second correction was due to the design of the Turner beam centrifuge resulting in the package base being held vertically up. The full slope geometry and modelling process has been previously detailed in Madabhushi et al. (2017).

A miniaturised cone penetrometer (6.40 mm diameter) was placed on the centrifuge package to obtain in-flight soil strength profiles. Furthermore an air-hammer (Ghosh & Madabhushi 2002) was installed at the base of the model. In conjunction with the accelerometers, this enabled measurements of shear wave velocity.

3 DENSITY ESTIMATION AT POURING

To attempt a reliable pour of the target density, an automatic sand pourer (Madabhushi et al. 2006) was used (figure 3). The machine is a spot pluviator. The flow rate is maintained by controlling the drop height and nozzle aperture. In general, models are poured and the average density back calculated from measurements of

Figure 3. Pouring setup with scale and laser in place with the automated sand pourer.

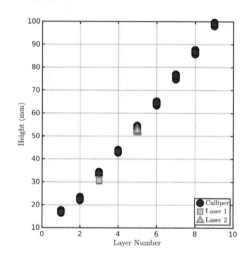

Figure 4. Calliper and laser height measured for each layer (laser layer 3 & 5).

the total height and poured mass. To refine this average, the model container was placed on a digital scale to allow continuous measurement of the model mass without disturbing the model. Measurements of the poured height were taken using a laser mounted on the sand hopper (figure 3) as well as a calliper which measured height with respect to a rigid reference relative to the model box. During the calibration and model pouring a set of 4 calliper measurements were taken for each layer of sand after it had been poured. Due

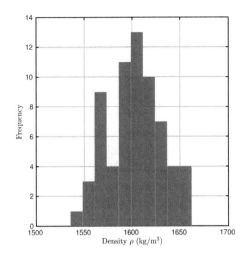

Figure 5. Density distribution.

Table 1. Sensitivity of ρ and I_d to measurement errors.

Parameter	±1% param. change		±5% param. change	
	ρ (kg/m³)	I_d (%)	ρ (kg/m³)	I_d (%)
Mass	1619–1587	55–44	1683–1523	77–20
Sand height	1587–1619	44–56	1527–1687	21–79

to constraints in the operation mode of the pourer, laser measurements were taken every four pours. Figure 4 highlights the consistency obtained in pouring similar sand layers. Moreover, good agreement was observed between the calliper and laser measurements. Hence these have been corroborated in a histogram of inferred density measurements from the height readings as shown in figure 5. Numerically the histogram translates to an average density of 1603 kg/m³ and standard deviation of 28 kg/m³, i.e. 1.7% of the mean value. However if converted to relative density, the uncertainty grows significantly. The result is an average I_d of 50% with standard deviation of 10%; a rise from 1.7% to 20% of the mean result. This highlights the sensitivity of I_d to pouring uncertainty and hence the need to understand it better. Part of the uncertainty is due to non-uniformities in flow-rate from the sand pourer. However, as table 1 shows, measurement errors play a significant part too.

In particular inaccuracies in measuring sand bed height can lead to large swings in inferred density. These swings can be as large as the standard deviation for height measurement errors within the range of accuracy of the measurement techniques used. Hence an assessment cannot be made on whether the spread in density shown in figure 5 is due to actual, physical non-uniformities in the pour or is just an image of measurement inaccuracy. Further experiments may tackle novel ways to measure sand height as well as methods to increase the confidence of the readings.

4 SWING-UP COMPRESSION

Particle image velocimetry was employed to investigate the soil displacements experienced by the model during swing-up. Specifically, it was desired to quantify the compression of the sand and the resulting effect on the voids ratio. This can be used to judge whether the density measurements and thus target voids ratio measured during pouring still retain their validity at high g.

The settlement field from the PIV is shown in figure 6. From this, the volumetric strains were obtained and were used to calculate changes in voids ratio. Across the model the average change in voids ratio during swing-up was found to be −0.012. For Ottawa sand this translates into a, small, 4% densification in terms of relative density.

The change in settlement across the height in response to the linearly increasing stress change from swing up can be used to characterise the soil stiffness. For this experiment the swing up induces a stress change of 40 kPa at the base of the model. For low stresses the instability of very loose sand configurations leads to behaviour better described by the load-unload compression index rather than the normal compression index. To rationalise the experimental results the predicted settlement profile based on an elastic soil model with compression index independent of stress level as well as a critical state-like model (1) were considered. Figure 7 shows the settlement profile with depth predicted by the two models for this swing-up scenario.

$$e = v_\kappa - \kappa \times \ln(\sigma'_v), \quad (1)$$

One can compare these to the measured PIV settlement shown in figure 8. The parabolic settlement derived from the elastic assumption is clearly inconsistent with the observed behaviour. This highlights the importance of accounting for the stress-dependent stiffness change even at relatively low stresses. By contrast, the critical state model predicts a linear settlement profile which is closer to the PIV result.

Nevertheless, figure 8 does not show constant strain with depth as implied by (1). This is true for the average settlement across the model and was checked by considering the settlement ± 1 standard deviation away as shown by the highlighted area in figure 8. A possible explanation might be the effect of the shear stress present in the sloping model, which may affect the volumetric strains generated. Overall, the tendency for volumetric straining and change in voids ratio due to swing up is comparatively small, but may be quantified using the PIV method.

5 INTERPRETATION OF CPT RESULTS

A cone penetration test (CPT) was carried out in-flight to obtain the sand strength profile along the depth of the model (figure 9).

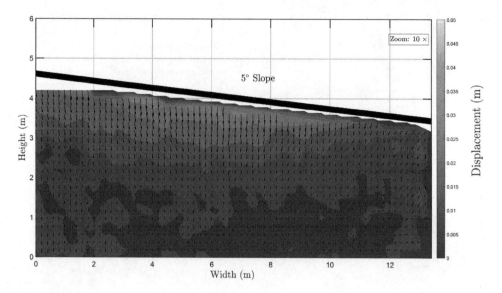

Figure 6. Swing-up PIV displacement that occur between 1 g and 40 g.

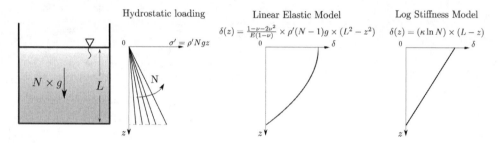

Figure 7. Model predictions.

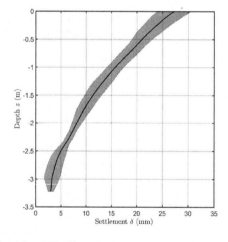

Figure 8. PIV settlement.

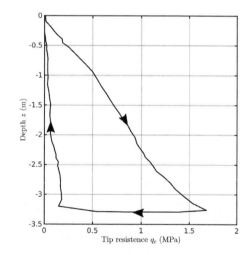

Figure 9. CPT strength profile.

Knowing the strength can lead to inferring important soil parameters, such as density and stiffness. Nonetheless, linking cone resistance to soil properties is hampered by the complexity of the soil deformation mechanism around an advancing cone tip. Presently no analytical solution has addressed this complexity, empirical correlations being the ones employed in practice. Jamiolkowski (1985), based on calibration chamber tests, proposed a link between relative density, tip resistance and vertical effective stress (2).

$$I_d(\%) = -98 + 66 \log_{10}\left(\frac{q_c}{\sqrt{\sigma'_{v0}}}\right), \quad (2)$$

where q_c and σ'_{v0} are in ton-force/m. Bolton et al. (1993) on the other hand focused on dimensional analysis as a tool to investigate CPT strength. They introduced a dimensionless parameter, the normalized cone resistance Q, to interpret CPT results (3).

$$Q = \frac{q_c - \sigma_{v0}}{\sigma'_{v0}}, \quad (3)$$

Through centrifuge tests on Fontainebleau sand, Bolton and Gui (1993) found the following correlation between I_d and Q:

$$I_d(\%) = 0.2831 \times Q + 32.964, \quad (4)$$

Investigations at Oxford in calibration chamber tests (Houlsby & Hitchman 1988) tried to separate the effects of vertical stress, horizontal stress and overconsolidation on cone resistance. They concluded that the effect of horizontal stress dominates. Houlsby (1998) changed the definition of the normalized cone resistance accordingly in the correlation he proposed (5).

$$\log_{10}\left(\frac{q_c - \sigma_{h0}}{\sigma'_{h0}}\right) = 1.51 + 0.0123 \times I_d(\%), \quad (5)$$

As shown by Gaudin et al. (2005), K_0 for normally consolidated sand in the centrifuge is well predicted by Jaky (1948):

$$K_0 = 1 - \sin\varphi'_p, \quad (6)$$

where the peak friction angle φ'_p was obtained from the LEAP database (Carey et al. 2016). This result can be used to estimate horizontal stress in (5). Fig. 10 shows I_d with depth as derived from (2), (4) & (5). Results are only plotted below a critical depth, below which surface effects are negligible. Figure 10 also includes the I_d estimate from the calliper measurement taken during pouring. For further comparison, the calliper I_d, adjusted for the swing-up densification as captured by the PIV, is also shown.

With the exception of (2), I_d predictions agree well between the different methods considered. This is especially true considering how sensitive I_d is to variability in voids ratio. (4) and (5) are particularly close, despite being derived through different test methodologies; centrifuge testing and calibration chamber respectively. It is worth remarking on some of the limitations of applying these correlations to the centrifuge data. Gaudin et al. (2005) have previously discussed the influence of miniature CPT's on the stress state in

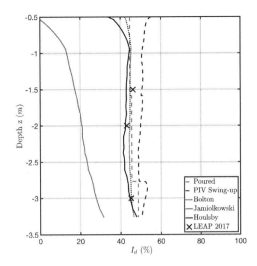

Figure 10. I_d predictions.

the centrifuge model, with measurable increases in the lateral stress recorded at the model boundaries. Given the average size of centrifuge packages, the presence of the probing instrument (CPT) may influences significantly the value it is probing (the perceived soil strength). This leads to some inaccuracy in using CPT field-derived correlations in the centrifuge. Hence, despite the good agreement shown in figure 10 further testing is required to validate the use of (4) or (5) in centrifuge CPT tests.

Towards this goal of further, repeatable testing, the LEAP 2017 exercise (Kutter et al. 2018) includes a series of 24 centrifuge tests across 9 research centres on centrifuge packages modelling the same prototype as the one presented in this paper. All centres employed the use of CPTs to characterise soil strength before and after the earthquake motion was fired. A compiled set of data as well as interpretation of the LEAP CPT results can be found in Kutter et al. (2018), including a unique relationship between q_c and I_d for this set of tests. As shown in figure 10 it compares favourably with the other results discussed.

Besides density, the sand stiffness, particularly the small strain shear modulus G_0 is a fundamental property in soil mechanics. Schnaid et al. (2004) proposed a range estimate of G_0 derived from CPT field data:

$$110\sqrt[3]{q_c\,\sigma'_{v0}\,p_a} \leq G_0 \leq 280\sqrt[3]{q_c\,\sigma'_{v0}\,p_a}, \quad (7)$$

where p_a is atmospheric pressure. Figure 11 shows the computed stiffness bound for the CPT experiment against G_0 data at two locations from an air-hammer test. The air-hammer point results are within the bounds. Nonetheless there is more than 100% difference between the lower and upper boundary. This highlights the inaccuracy of the method and the present limitations in deriving stiffness from CPT strength.

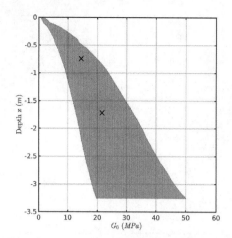

Figure 11. Small strain shear modulus predictions.

6 CONCLUSIONS

Overall it has been demonstrated that care must be taken when attempting to control the relative density of centrifuge models and in preparation of homogeneous soil deposits. The potential impact of measurement errors in model mass or height on the calculated soil density were discussed. The distinction between actual variations in the soil density and the confounding effect of measurement errors is highlighted.

Thus the option to verify the quality of the sand pouring using other instrumentation is useful. The PIV technique was used to demonstrate that the change in relative density during swing up was small but not necessarily negligible. In collaborative projects where target densities are prescribed for the final prototype scale, accounting for this change can become important.

The use of miniature CPTs to verify the relative density in flight was demonstrated. A number of empirical relations exist to convert the cone resistance directly into a relative density and were applied to the data collected in this test. Comparison between the variations with depth of the predicted and measured densities attest to the consistency of the pour in terms of a constant density across the model height. However, quantifying the relative density with the CPT is complicated by the potential dependence on the lateral stress state and the use of a rigid model container. Collaborative projects with large experimental data sets, such as LEAP, improve the confidence in experimentally derived correlations and can help mitigate against the uncertainty of measurements in individual tests.

REFERENCES

Bolton, M. D. & M. W. Gui (1993). The study of relative density and boundary effects for cone penetration tests in centrifuge. Technical report, University of Cambridge, Department of Engineering.

Bolton, M. D., M. W. Gui, & R. Phillips (1993). Review of miniature soil probes for model tests. In *Proceedings of the 11th Southeast Asian Geotechnical Conference*, pp. 85–90.

Carey, T. J., B. L. Kutter, M. T. Manzari, M. Zeghal, & A. Vasko (2016). Leap Soil Properties and Element Test Data. *DOI: http://doi.org/10.17603/DS2WC7W*.

Gaudin, C., F. Schnaid, & J. Garnier (2005). Sand characterization by combined centrifuge and laboratory tests. *International Journal of Physical Modelling in Geotechnics* 5(1), 42–56.

Ghosh, B. & S. P. G. Madabhushi (2002). An efficient tool for measuring shear wave velocity in the centrifuge. In *Proceedings of the International Conference on Physical Modelling in Geotechnics*, pp. 119–124. A.A. Balkema.

Houlsby, G. T. (1998). Advanced interpretation of field tests. In *Geotechnical Site Characterization*, Volume 1, pp. 99–112.

Houlsby, G. T. & R. Hitchman (1988). Calibration chamber tests of a cone penetrometer in sand. *Geotechnique 38*(1), 39–44.

Jaky, J. (1948). Pressure in silos. In *Proc. 2nd Int. Conf. Soil Mech.*, Volume 1, pp. 103–107.

Jamiolkowski, M. (1985). New developments in field and laboratory testing of soils. In *11th Int. Conf. on Soil Mechanics and Foundation Engineering*, San Francisco, pp. 57–153.

Kutter, B. L., T. J. Carey, T. Hashimoto, M. Zeghal, T. Abdoun, P. Kokkali, G. Madabhushi, S. K. Haigh, F. Burali d'Arezzo, S. Madabhushi, W.-Y. Hung, C.-J. Lee, H.-C. Cheng, S. Iai, T. Tobita, T. Ashino, J. Ren, Y.-G. Zhou, Y.-M. Chen, Z.-B. Sun, & M. T. Manzari (2017). LEAP-GWU-2015 experiment specifications, results, and comparisons. *International Journal of Soil Dynamics and Earthquake Engineering*, DOI: 10.1016/j.soildyn.2017.05.018.

Kutter, B. L., T. J. Carey, B. L. Zheng, A. Gavras, N. Stone, M. Zeghal, T. Abdoun, E. Korre, M. Manzari, S. P. G. Madabhushi, S. K. Haigh, S. S. C. Madabhushi, M. Okamura, A. N. Sjafuddin, S. Escoffier, D.-S. Kim, S.-N. Kim, J.-G. Ha, T. Tobita, & K. Hikaru (2018). Twenty-four centrifuge tests to quantify sensitivity of lateral spreading to dr and pga. In *5th Geotechnical Earthquake Engineering and Soil Dynamics Conference*.

Madabhushi, S. P. G., N. E. Houghton, & S. K. Haigh (2006). A new automatic sand pourer for model preparation at University of Cambridge. In *Proceedings of the 6th International Conference on Physical Modelling in Geotechnics*, pp. 217–222. Taylor & Francis Group, London, UK.

Madabhushi, S. S. C., S. K. Haigh, & S. P. G. Madabhushi (2017). LEAP-GWU-2015: Centrifuge and numerical modelling of slope liquefaction at the University of Cambridge. *International Journal of Soil Dynamics and Earthquake Engineering*, DOI: 10.1016/j.soildyn. 2016.11.009.

Schnaid, F., B. M. Lehane, & M. Fahey (2004). In situ test characterization of unusual geomaterials. In *Proc., 2nd Int. Conf. on Site Characterization*, Volume 1, Rotterdam, Netherlands, pp. 49–74.

Schofield, A. (1981). Dynamic and Earthquake Geotechnical Centrifuge Modelling. In *International Conferences on Recent Advances in Geotechnical Earthquake Engineering and Soil Dynamics*.

Stringer, M. E. & S. P. G. Madabhushi (2009). Novel Computer-Controlled Saturation of Dynamic Centrifuge Models Using High Viscosity Fluids. *Geotechnical Testing Journal 32*(6), 559–564.

Centrifuge modelling of Continuous Compaction Control (CCC)

B. Caicedo & J. Escobar
Universidad de Los Andes, Bogotá, D.C., Colombia

ABSTRACT: Continuous Compaction Control (CCC) is a process that uses the measurement of the drum's compactor vibrations (acceleration) to estimate the stiffness of the soil during compaction and then have a real time assessment of the soil compaction. This paper presents the development of a reduce-scale physical model of a vibratory roller compactor, which will be used to simulate continuous compaction control in a centrifuge. The model was conceived by scaling the drum's weight, diameter and the centrifugal force generated by a commercial vibratory roller compactor to be used in centrifuge at 10 g. Soil deformation analysis was also done during the test, by using particle image velocimetry (PIV) method, in order to analyse the soils' behaviour. Results show that the scale model reproduces the effects of vibratory roller compactors on soils and also how the force increases and displacements decrease as the soil is compacted in time.

1 INTRODUCTION

Static or dynamic compaction is a mechanical process used widely to improve soil's characteristics such as strength and stiffness in earthworks. When static or dynamic stresses are applied to the soil, at a higher value than its yield limit, the soil starts to experiment irreversible volumetric strains due to the rearrangement of the soil particles and by the expulsion of the air contained in it.

The study and the understanding of the effects of soil compaction have gone from Proctor's works, which related dry density and water content, to the relationship between water content, suction, density, stress history and the level of stress needed to generate an irreversible soil strain. Great advances in this topic, has been achieved in the last two decades.

Deciding which method of compaction is appropriate for a particular project will depend on its specific needs and the soil characteristics. However, dynamic compaction is one of the most common methods used in civil works, especially by the use of vibratory roller compactors. This is due to the following reasons: 1) a rapid succession of pressure waves that are scattered in all directions breaking the bond between the particles of the material, 2) the compaction depth reached with this kind of compaction and 3) the ease of covering large areas in relatively short periods of time, compared to other compaction methods.

However, one of the biggest issues that this compaction process has, is knowing the level of compaction reached at any time during the execution, which could avoid problems related to over compacted (softening) or sub compacted materials (low strength). These two cases affect directly the strength of the material and increase in budget and work schedule the project, due to the overpasses that the drum has done excessively or possible failure of structures supported on soils with not appropriate compaction level.

The solution for this problem has been studied since the 1970s, when Dr. Heinz Thurner, using a vibratory roller instrumented with accelerometers was able to correlate the acceleration frequencies of the drum with the stiffness level of the soil reached by the compaction (M. Mooney & Adam, 2007). Advances in technology have allowed to instrument vibratory roller compactors and to have a really good understanding of the relationship between drum's vibration and the level of compaction of the soil, opening the way to continuous compaction control technology (M. Mooney & Adam 2007).

An important number of compactor manufacturers companies have added this technology to their compactors and set their parameters for this technology (Vennapusa & White 2010). Likewise, several researchers and official entities have done large-scale tests on the field to evaluate the effectiveness and reliability of this technology (Camargo et al. 2006; C. L. Meehan et al. 2017; C. Meehan & Tehrani 2009; Petersen 2005). Numerical modelling and 2D and 3D FEM simulations have also been done (Herrera & Caicedo 2016).

Although several authors have studied the effects of compaction in soils in laboratory tests, no scale model tests of a real vibratory compactor have been done. This prototype will open the way to have a better understanding of the process of compaction and its effects on the soil, at a lower budget than the large-scale tests.

The purpose of this paper is to present a reduced scale model of a vibratory roller compactor to reproduce and study the continuous compaction control in a centrifuge machine, analysing forces and the displacement generated during the compaction process.

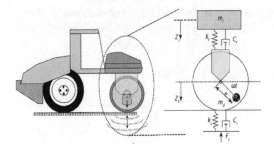

Figure 1. Two degrees of freedom system describing a vibratory roller, adapted from (M. Mooney & Adam, 2007; Pietzsch & Poppy, 1992).

2 THEORETICAL FRAMEWORK OF A VIBRATORY ROLLER COMPACTOR

An important issue to estimate the level of compaction trough continuous compaction control depends on the analysis of the interaction drum/soil. Several models were developed for this purpose (Mooney & Adam, 2007). The theoretical analysis, and the computer calculation capacity have been evolving and nowadays the models allow including several masses, and dampings (Pietzsch & Poppy 1992). As well, 2D and 3D FEM permit to use complex constitutive models of soil behaviour (Erdmann & Adam 2014). Furthermore, the use of commercial software such as ABAQUS among others, has let to perform analysis of soil-drum interactions (Capraru et al. 2014; Kim & Chun 2016).

However, field tests have allowed demonstrating that simpler lumped systems are good enough to simulate the compaction process and its results. Indeed, these lumped models permit to assess the level of compaction (Anderegg & Kaufmann, 2004).

Using the substructure method, Adam and Kopf, presented a theoretical analysis of the interaction between vibratory rollers and soils. Adam and Kopf describe a vibratory roller as the two degrees of freedom system shown in Figure 1. In this system, the drum moves as a result of the vibratory load and reacts in the soil and the frame of the compactor.

The set of two equations describing the movement is:

$$(m_d + m_e)\ddot{z}_1 + c_f(\dot{z}_1 + \dot{z}_2) + k_f(z_1 + z_2) = (m_d + m_e)g + m_e e \omega^2 \sin(\omega t) - F_s \quad (1)$$

$$m_f \ddot{z}_2 - c_f(\dot{z}_1 + \dot{z}_2) - k_s(z_1 + z_2) = m_f g \quad (2)$$

where m_d = mass of the drum; m_e = eccentric mass; m_f = mass of the frame; e = eccentricity of the mass; w = angular frequency of the rotating mass; F_s = reaction of the soil; z_1 = vertical displacement of the drum; z_2 = vertical displacement of the frame; k_f = stiffness of frame suspension and C_f = frame suspension damping. However, the connection between the drum and the frame of the compactor is conceived to produce negligible vibrations of the frame, $\ddot{z}_2 = 0$).

Table 1. Real and scale model compactors characteristics.

Drums Characteristics	CS78B Caterpillar	Scale Model 1:10
Weight at Drum with Cab (kg)	13440.0	14.10
Drum Diameter (mm)	1534.0	150.0
Drum Width (mm)	2134.0	200
Centrifugal Force Max/Min (kN)	322/166	2.14

Then, Equation 2 becomes:

$$c_f(\dot{z}_1 + \dot{z}_2) + k_s(z_1 + z_2) = -m_f g \quad (3)$$

As a result, only the static load of the frame is transmitted to the drum, and the two degrees of freedom system can be simplified to the single degree of freedom problem, given by the following equation:

$$(m_d + m_e)\ddot{z}_1 = (m_d + m_e + m_f)g + m_e e \omega^2 \sin(\omega t) - F_s \quad (4)$$

This problem can be solved for elastic or elastoplastic analysis, and the calculation of the model parameters will depend on the type of solution selected.

3 MATERIALS AND EQUIPMENT

The vibratory roller compactor selected to be represented by a scale model was the CS78B by Caterpillar (See properties Table 1), which was scaled 10 times in order to maintain the static and dynamic properties of the equipment during the test in the centrifugal machine, simulating the real effects caused on the soil by the compactor. The model is formed by 1) a steel drum and 2) a pneumatic turbine vibrator and it is instrumented by 1) two load cells, 2) an accelerometer and 3) a laser displacement sensor. These components describe and specify below. Figure 2a and b show their location in the scale model.

3.1 Compactor drum

The drum of the model was recreated by using half of a steel pipe to which a steel plate was welded in top of it. This drum has 15 cm in diameter, 20 cm in wide and 12.7 cm in thickness. The sides of the drum were cover with sheets of Teflon to avoid two possible situations:

- frictional forces between the sides of drum and the sides of the soil container during the test, due to the small gap between them (1 mm)
- as the test is performed, some soil particles close to the drum could rise up and get inside the drum.

3.2 Pneumatic turbine vibrator

A pneumatic turbine vibrator was used in the model to transmit vibrations to the compactor drum. The reference selected was BVS250 produced by VIBCO,

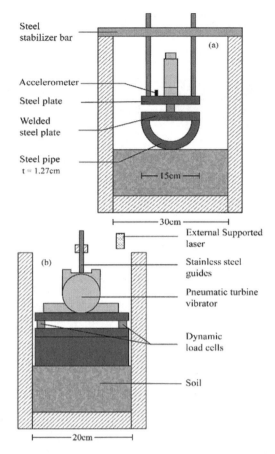

Figure 2. Location of the components a) Front and b) lateral view of the scale model.

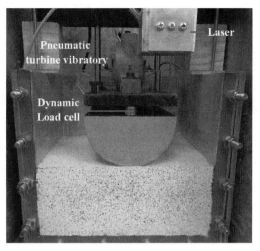

Figure 3. Front view photography of the scale model.

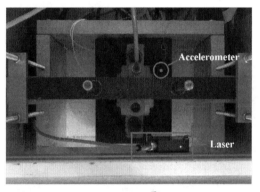

Figure 4. Top view photography of the scale model. Laser, accelerometer, and air intake location in the model is shown.

which generates a centrifugal force of 2136 N, value that is in the range that the real vibratory roller works, keeping scaling proportions. This element attached to a steel plate with screws, as it is shown in Figure 3 and Figure 4.

3.3 Load cells

Two piezoelectric dynamic load cells were placed as connectors between the drum and the plate to which the pneumatic turbine vibrator was attached. These load cells measure the soil response during compaction. The load cells capacity is 500 pounds each, equivalent to 5 volts. A calibration process was done for each cell, relating voltage and force.

3.4 Accelerometer

An accelerometer was installed in the steel plate where the turbine vibrator is attached (Figure 4), to measure the vertical acceleration of the drum in time, during the compaction process. The frequency analysis of this measurement permits to estimate the level of compaction, simulating the continuous compaction control on large-scale drum vibrators.

3.5 Laser

A laser with 5 cm measurement range was set up independently of the assembly made in the model and the soil container, to avoid the vibrations generated by the drum and transmitted to soil container at some point when the soil starts increasing its stiffness. Its ray was set up hitting the top of the plate that has to attach the pneumatic vibrator, following compactor's vertical path during the test. See Figure 4.

4 MATERIALS AND SPECIMEN

4.1 Material

Kaolin was the soil used for this test. Its properties are presented in Table 2. The soil was prepared according to (Tarantino & De Col, 2008), initially mixing dry soil manually while water was sprayed on it gradually, to minimize the formation of lumps, until the total amount of water, to obtain a 15% of water content. Afterwards, the soil was mixed mechanically, trying to homogenize the humidity present in the soil. A first

Table 2. Characteristics of the kaolin used in the investigation.

Liquid limit, w_L (%)	85
Plastic limit, w_P (%)	34
Specific Gravity (g/cm^3)	2.63

sieving stage was made at 1 mm to eliminate the presence of large aggregates, saving the material sieved in a hermetic plastic bucket. The large aggregates were broken up with the fingers or with a spatula so as not to lose the water present on them. This material was sieved again and poured in the bucket. As it is recommended in (Tarantino & De Col, 2008), the material was stored in a wet room for at least one week to guarantee good moisture equilibration and homogeneity.

4.2 Material pre-compaction

For the test, the material was poured in a box 30 cm long, 20 cm wide and 28 cm high, which has one of its long sides covered with an acrylic lid in order to see the soil behavior during compaction. The other three lids are made of aluminum.

A calibration test was performed at 1 g, showed that just by the weight of the compactor system, when it was set over the soil in a loose state, the soil suffered a considerable deformation. Because of this, a pre-compacting process was done at 30 kPa, applying an increasing charge all over the surface of the soil, by the use of an aluminum lid, until the charge was reached. The total settlement obtained during this process was 3.80 cm.

4.3 Markers for PIV analysis

Dyed sand was used as markers to do the PIV analysis due to lack of natural texture of the kaolin, for applying this technique (White et al., 2003; Y.D. Zhang & T.S. Tan, 2005) after the test. Markers were smaller than sieve # 30 and bigger than sieve # 50. These markers were added to the soil after pre-compaction, removing the acrylic lid and turning the box over, leaving this side horizontal. Sieve # 30 was used to spread the particles randomly creating an aleatory distribution over the soil. Particles smaller than Sieve # 50 were discarded as markers because they were attracted by the electrostatic force of the acrylic lid and at some point, they could stay with the acrylic and not moving with the soil during the compaction test.

5 COMPACTION TEST

5.1 Compaction stages

The compaction test had two stages. The first one corresponds to a static compaction, when the centrifugal machine goes from 1g to 10g. The second stage corresponds to the dynamic compaction, in which the vibratory compactor was turned on using the activation of an electro-valve, which allowed the air out from a storage tank. It was necessary to include this tank, due

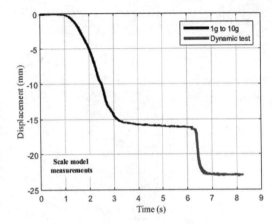

Figure 5. Total settlement reached during the static and dynamic compaction.

to the low pressure presented in the air network of the laboratory when it was open; pressure that was not enough to break the inertia of the vibratory equipment for compacting. The maximum pressure reached in the tank was 12 bars and kept the vibratory machine on approximately for 31 seconds.

A video was recorded during the complete test to generate images out of it, been able to do the PIV analysis and see the soil particles movement and determine their paths.

6 RESULTS

6.1 Displacement

As it was mentioned before, the test had to stages, corresponding to static compaction (1 g to 10 g) and dynamic compaction. Figure 5 shows the displacements measured during each of them, obtaining a settlement of 16.13 mm for the first one and 6.90 mm during the second one, for a total of 23.03 mm on the model.

The settlement caused by the drum's weight at the first stage was large and probably unacceptable as a construction specification, even though the soil was pre-compacted. It is important to mention that the layer thickness after pre-compaction was 12.2 cm, which is larger than a typical lift for compaction in real scale, in which a normal range would be from 40 to 60 cm for fine materials and for which the underlayer material should have greater stiffness.

For the second stage, the settlement was closer to the values which are frequently achieved in the field.

A Fourier Transform for displacement was done during the vibratory compaction stage, determining that the dominant frequency is 75 Hz approximately (Figure 6).

6.2 Dynamic load and displacement

Hysteretic cycles for dynamic load and displacement are shown in Figure 7 for seconds 2, 4, 6 and 10

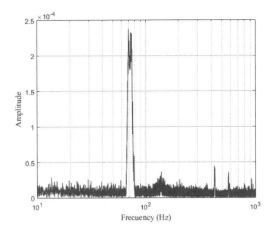

Figure 6. Fourier Transform for displacement during the vibratory compaction stage.

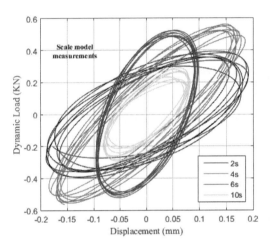

Figure 7. Hysteretic cycles for seconds 2, 4, 6 and 10.

during the compaction test. For the three first analyzed instants (2, 4, 6 s), as it is depicted in the graphic, the ellipses generated for this cycles rotate anti-clockwise indicating the increment of the dynamic force and the reduction of the displacement, which is an indicator of level of compaction that the soil reaches (Kenneally et al., 2015; M. A. Mooney & Rinehart, 2009). For the second 10 the ellipses are smaller, both in displacement and dynamic load.

This variation is directly related to the air supplying system, as it is depicted in Figure 8. When the air is released out of the tank, the air pressure starts decreasing as the time goes forward, until the pressure in the tank is lower than the minimum pressure need by the vibratory turbine to works properly.

6.3 PIV analysis

A basic PIV analysis was done for the static and dynamic compaction stages, using PIVlab

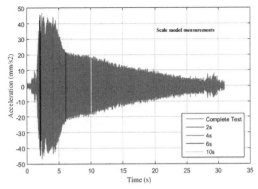

Figure 8. Drum's acceleration registered during the test. The acceleration for each second analyzed in Figure 7 is colored.

Figure 9. Vertical velocity component for the static stage.

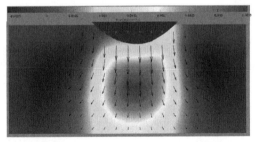

Figure 10. Simple strain rate for the dynamic stage.

(Thielicke & Stamhuis, 2014). The maximum single strain rate for the first stage was 2.5E–3 [1/s], while for stage two was 3.5E–3 [1/s]. The maximum vertical velocity components were 1.25E–4 m/s and 1.8E–4 m/s, for stage one and stage two, respectively.

Figure 9 shows that the highest speeds in the displacement of the soil occurred where the soil is in contact with the roller, in both stages, as expected. As it is seen in Figure 10, the trajectory vectors obtain by the PIV analysis, follow the typical path for this type of compaction.

Figure 11 shows the final state of the test, after static and dynamic compaction.

Figure 11. Final state of the test.

7 CONCLUSIONS

A reduce-scale physical model of a vibratory roller compactor was developed at the Universidad de los Andes in Bogotá, Colombia.

PIV analysis allowed to observe that the scale model reproduces the effects of vibratory roller compactors on soils as is referenced in the theory of this type of compaction.

The settlement caused by the drum's weight is unacceptable as a construction specification, even though the soil was pre-compacted. The settlement mesaurare on the second stage was closer to the values which are frequently achieved in the field.

Having a continuous vibratory system in time, will allow to obtain the hysteretic cycles at the end of the test, where it is expected to have higher forces and lower displacements due to the increment of soil stiffness.

Many tests have to conduct in order to correlate the drum's acceleration and the level of compaction, analyzing its frequencies and the soil parameters. Additionally tests with soils at different initial densities will help to compare the results obtained in the present investigation.

REFERENCES

Caicedo, B., Tristancho, J., Thorel, L. & Leroueil, S. 2014. Experimental and analytical framework for modelling soil compaction. Engineering Geology, 175, 22–34.

Camargo, F., Larsen, B., Chadbourn, B., Roberson, R. & Siekmeier, J. 2006. Intelligent compaction: A minnesota case history. 54th Annual University of Minnesota Geotechnical Conference, 17.

Capraru, C., Pistrol, J., Villwock, S., Völkel, W., Kopf, F. & Adam, D. 2014. Numerical Simulation of Soil Compaction with Oscillatory Rollers. XV Danube – European Conference on Geotechnical Engineering (DECGE 2014), (1).

Erdmann, P. & Adam, D. 2014. Numerical Simulation of Dynamic Soil Compaction with Vibratory Compaction Equipment. XV Danube – European Conference on Geotechnical Engineering (DECGE 2014), (119), 243–248.

Herrera, C. & Caicedo, B. 2016. Finite Difference Time Domain Simulations of Dynamic Response of Thin Multilayer Soil in Continuous Compaction Control. Procedia Engineering, 143 (July), 411–418.

Kenneally, B., Musimbi, O. M., Wang, J. & Mooney, M. A. 2015. Finite element analysis of vibratory roller response on layered soil systems. Computers and Geotechnics, 67, 73–82.

Kim, K. & Chun, S. 2016. Finite element analysis to simulate the effect of impact rollers for estimating the influence depth of soil compaction. KSCE Journal of Civil Engineering, 20(7), 2692–2701.

Meehan, C. L., Cacciola, D. V., Tehrani, F. S. & Baker, W. J. 2017. Assessing soil compaction using continuous compaction control and location-specific in situ tests. Automation in Construction, 73, 31–44.

Meehan, C. & Tehrani, F. S. 2009. An Investigation of Continuous Compaction Control Systems. Dct 204, 19716(July), 411.

Mooney, M. A. & Rinehart, R. V. 2009. In Situ Soil Response to Vibratory Loading and Its Relationship to Roller-Measured Soil Stiffness. Journal of Geotechnical and Geoenvironmental Engineering, 135(8), 1022–1031.

Mooney, M. & Adam, D. 2007. Vibratory roller integrated measurement of earthwork compaction: An overview. 7th FMGM 2007@ sField Measurements in Geomechanics, (Fmgm), 1–12.

Petersen, D. L. 2005. Continuous Compaction Control MnROAD Demonstration.

Pietzsch, D. & Poppy, W. (1992). Simulation of soil compaction with vibratory rollers. Journal of Terramechanics, 29(6), 585–597.

Tarantino, A. & De Col, E. 2008. Compaction behaviour of clay. Géotechnique, 58(3), 199–213.

Thielicke, W. & Stamhuis, E. J. (2014). PIVlab – Towards User-friendly, Affordable and Accurate Digital Particle Image Velocimetry in MATLAB. Journal of Open Research Software, 2.

Vennapusa, P. K. & White, D. J. 2010. A review of roller-integrated compaction monitoring technologies for earthworks, 1–26.

White, D. J., Take, W. A. & Bolton, M. D. 2003. Soil deformation measurement using particle image velocimetry (PIV) and photogrammetry. Géotechnique, 53(7), 619–631.

Zhang, Y.D., Tan, T.S. & C. F. L. 2005. Application of particle imaging velocimetry (PIV) in centrifuge testing of uniforme clay. International Journal of Physical Modelling in Geotechnics, 5(1), 15–26.

Shear wave velocity: Comparison between centrifuge and triaxial based measurements

G. Cui, C.M. Heron & A.M. Marshall
Nottingham Centre for Geomechanics, Department of Civil Engineering,
University of Nottingham, Nottinghamshire, UK

ABSTRACT: As part of a project investigating ground-borne vibration using a geotechnical centrifuge, which involves very small strain shear waves, the need for an accurate measurement of small strain shear modulus (G_{max}) arose. G_{max} can be back-calculated from measured shear wave velocities (v_s). Previous researchers have measured v_s in-flight by using an air-hammer device and sensitive accelerometers to detect the speed of the generated shear wave. However, issues relating to inaccurate position measurement of the accelerometers, background noise, and the recording of very small signals at sufficiently high frequency can cause errors in the measurement of v_s. This paper compares centrifuge air-hammer based measurements of v_s against those obtained from a bender element system integrated into a triaxial apparatus. The paper details the testing methodologies adopted and provides results which indicate that the two methods agree well with each other and with predicted values from literature. The paper highlights how the precise processing methodology adopted can affect the accuracy of the obtained values of v_s.

1 INTRODUCTION

With ever increasing urban population density and the corresponding need for efficient and expedient means of transport, rail-based public transportation systems are a popular choice (such as Crossrail and HS2 in the UK and the high-speed rail network in development in China). However, the noise and vibrations generated by such railways can cause problems in an urban setting. The need for accurate evaluations of the environmental impact of noise and vibrations from these transport systems has motivated recent research in this area (Kuo 2010, Nugent et al. 2012).

Yang et al. (2013) conducted centrifuge tests and numerical modelling to study the effect of the variation of soil parameters with depth on ground-borne vibrations. This work highlighted the sensitivity of ground-borne vibration behaviour to the precise selection of key soil parameters, such as shear modulus and damping. However, obtaining accurate values for these parameters, and how they vary with stress level, is challenging. This is particularly the case at low stress levels, near the ground surface, where parameter values tend to vary significantly.

A research project being conducted at the University of Nottingham is investigating ground-borne vibrations using a geotechnical centrifuge. This project focuses on the transmission of vibrations to, or from, a pile. Vibrations generated by a railway can propagate through a soil-pile-building pathway and affect the operation of sensitive equipment or cause disturbance to occupants (Kuo 2010). The centrifuge tests involve the vertical oscillation of a single pile embedded within a dry silica sand. An accurate evaluation of soil properties (e.g. dynamic shear modulus) over the pile depth is therefore required.

Dynamic shear modulus (G_{max}) is one of the most important parameters in soil dynamics and can be back-calculated by measuring shear wave velocity (v_s). Previous researchers have measured v_s in centrifuge models by using an air-hammer device and sensitive accelerometers to detect the speed of the generated shear waves. However, issues relating to inaccurate position measurement of the accelerometers, background noise, and the recording of very small signals at sufficiently high frequency can cause errors in the measurement of v_s.

The purpose of this paper is to demonstrate how air-hammer based measurements in a centrifuge compare against shear wave velocities measured from a bender element system integrated within a triaxial cell. Comparisons are also made against two empirical equations available from the literature. The paper provides an overview of the testing methodologies as well as details of data processing, which had an important impact on outcomes.

2 SHEAR MODULUS

Railway-induced vibration is small (0.1 to 1 cm/s^2) (Tsuno et al. 2005) and induce soil shear strains within the very small range (10^{-6} to 10^{-5}) (BS 2005). Within this range, shear modulus essentially remains constant (i.e. at G_{max}) (Rollins et al. 1998), which can be calculated from measured values of v_s using

$$G_{max} = \rho v_s^2 \quad (1)$$

where ρ is soil density. There are various factors that affect the magnitude of G_{max}, including confining stress, void ratio, over-consolidation ratio and geologic age (Dobry & Vucetic 1987); this study focuses solely on the effect of confining stress. G_{max} is commonly measured in the laboratory at the element scale using resonance column, torsion, and bender element tests (Shantz 2012). Several methods have been developed by centrifuge modellers to measure in-flight values of v_s, including bender elements (Brandenberg et al. 2006, Lee et al. 2012, Kim 2010, Lee et al. 2014) and air-hammers (Ghosh 2002, Arulnathan 2000).

Several empirical equations have been proposed to estimate G_{max} of dry sand, including Hardin and Richart (1963):

$$G_{max} = 7000 \frac{(2.17-e)^2}{1+e} \frac{(1+2K_0)^{0.5}}{3} (\sigma'_m)^{0.5} \quad (2)$$

and Hardin and Black (1968):

$$G_{max} = 6908 \frac{(2.17-e)^2}{1+e} (\sigma'_m)^{0.5} \quad (3)$$

where e is the void ratio of the soil model, K_0 is the at rest earth pressure coefficient, and σ'_m is the mean effective stress (kPa). These expressions are both applicable to the round-grained sand employed in the current study. Predictions obtained from these methods are compared against experimental data obtained from triaxial tests later in the paper.

3 MATERIAL PROPERTIES

HST95 Congleton sand was used for all experiments. It is a very fine and uniformly rounded silica sand. The basic parameters of Congleton sand are shown in Table 1 (Lauder 2010). Samples were prepared by air pluviation using flow rate and drop height calibrated to obtain a relative density of approximately 85% (± 2%) for both the centrifuge and bender element tests.

4 CENTRIFUGE MODELLING AND AIR-HAMMER TESTING

4.1 Centrifuge modelling

A cylindrical steel container with a diameter and depth of 500 mm was used for the centrifuge tests, which were performed at an acceleration of 60 g. The sand was poured to a depth of 440 mm (26.4 m at prototype scale). An undesirable problem caused by the rigid boundaries of the container is that vibration waves are reflected back into the model. To reduce this effect,

Table 1. Physical properties of Congleton sand (Lauder 2010).

Parameter	Value
Specific gravity (G_s)	2.63
D_{10}	0.090 mm
D_{30}	0.120 mm
D_{60}	0.170 mm
D_{90}	0.195 mm
Minimum dry unit weight ($\gamma_{d\ min}$)	14.6 kN/m^3
Maximum dry unit weight ($\gamma_{d\ max}$)	17.6 kN/m^3
Maximum void ratio (e_{max})	0.769
Minimum void ratio (e_{min})	0.467

a vibration absorbing material, Duxseal (Steedman et al. 1990, Cheney et al. 1990), was used to absorb the vibrations. The steel container was lined with Duxseal to a thickness of 19 mm along the sides and 34 mm on the bottom. A thicker layer was placed on the base to absorb vertically propagating waves from the end of the pile.

4.2 Air-hammer setup

A miniature air-hammer (AHA) was embedded in the sand to generate shear waves. As shown in Figure 1, the air-hammer is a 10 cm long, 6 mm diameter tube with a 20 mm long pellet inside. The pellet is made of 2 PTFE hollow cylinders, an steel cylinder and 2 bolts. To generate the shear waves within the soil, air is supplied through a tube to force the pellet to impact an end of the tube. Consequently, the tube moves slightly in the soil resulting in a small shear wave. A thin layer of sand was bonded to the surface of the tube to ensure shear waves are transferred to the soil effectively.

The progression of the shear wave is detected by a vertical array of four MEMS (Micro-Electro-Mechanical Systems) (type: ADXL001-70BEZ) accelerometers labelled AHA-M1 to M4 in Figure 1. The MEMS have a range of ±70 g and were mounted on a continual flexible plastic strip to reduce uncertainty in the inter-MEMS distances. The range of the MEMS is considerably greater than the measured accelerations generated from the shear wave, however MEMS with a smaller measurement range and a high frequency response are not commercially available. The sensitive axis of the MEMS was aligned with the horizontal particle motion direction of the shear wave. A high sampling rate of 100 kHz was used to record the acceleration time histories from the MEMS to create a sufficiently small sampling interval relative to the shear wave travel time, thereby minimising errors (Lee et al. 2014).

MEMS were installed at 60 mm (3.6 m prototype), 180 mm (10.8 m), 300 mm (18 m), and 420 mm (25.2 m) from the ground surface (shown in Figure 1) to measure the average shear wave velocity ($v_{s,AHA,1}$, $v_{s,AHA,2}$ and $v_{s,AHA,3}$) of each soil layer (L_1, L_2 and L_3). All dimensions regarding depths of MEMS and distances between soil layers, shown in Figure 1, are

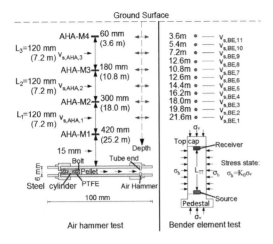

Figure 1. Air-hammer and bender element (BE) test setup (not to scale). Prototype depths/lengths shown in round brackets ( ).

specified in both model scale and prototype scale. The average shear wave velocity in a soil layer can be calculated by measuring the elapsed time for a shear wave to travel between two adjacent accelerometers (methods to do this are presented later in this paper).

5 TRIAXIAL BENDER ELEMENT TEST

In this study, a GDS Bender Element System (BES) was used to conduct the element scale shear wave velocity tests. A pair of bender elements, capable of producing separate shear and pressure waves, are incorporated into the top cap and base pedestal of a triaxial cell (Figure 1). The soil samples measured 50 mm in diameter and 100 mm in length. The wave velocity is measured in the BES as the time for a wave to the travel the distance L_{TT} between the source and receiver elements.

During the generation of shear waves in the bender element (BE) tests, three components appear. Of them, two propagate at shear wave velocity and the third travels at pressure wave velocity. The polarisation of the pressure wave is opposite to the input shear wave. If the source is too close to the receiver, the first arrival of the shear wave is masked by the pressure wave. This is referred to as the near field effect. The pressure wave attenuates quicker than the shear waves and its effect on readings can be minimised by increasing the ratio between the tip-to-tip distance L_{TT} and the wavelength λ of the faster shear waves (Camacho-Tauta et al. 2012). The length of the triaxial sample should therefore satisfy $2 < L_{TT}/\lambda < 9$, where $\lambda = v_s T$; the period T is an input parameter for the GDS BES. Using values of v_s obtained from air-hammer tests, T for the BE tests was estimated as 0.05×10^{-3} s to 0.1×10^{-3} s. However, it can be seen from Figure 3 that the near field effect still existed in the BE tests. A high speed data acquisition system with a sampling rate of 2000 kHz was used to obtain readings.

Two linear variable differential transformers (LVDTs) with a range of ±2.5 mm were installed to measure vertical deformation of the specimen. The measurement resolution using a 16 bit data acquisition is 0.1 μm, allowing the detection of very small axial strains (10^{-6}).

It has been found that σ'_v and σ'_h contribute equally to the value of the shear wave velocity, which implies that σ'_m in equations 2 and 3 can be taken as (Stokoe et al. 1985):

$$\sigma'_m = (\sigma'_v)^{0.5}(\sigma'_h)^{0.5} \tag{4}$$

For normally consolidated dry sand, the relationship between σ_v and σ_h satifies:

$$\sigma'_h = K_0 \sigma'_v \tag{5}$$

where $K_0 = 1 - \sin(\phi')$ (Jaky 1944), ϕ' is the angle of internal friction for dry sand. The ratio of horizontal to vertical stresses applied in the BE tests was 0.5 to match the anisotropic stress state in the centrifuge with an assumed $K_0 = 0.5$.

Wave velocities were determined at 11 levels of confining stress, corresponding to 11 depths within the prototype soil model, as illustrated in Figure 1. The tested confining stresses include those evaluated in the centrifuge to provide a basis for comparing air-hammer and BE data.

6 SIGNAL PROCESSING

The peak to peak and cross correlation methods are commonly used to determine wave travel time. The peak to peak method uses the time difference between the first two peaks as the travel time. Whereas, the cross-correlation method works by performing a pointwise multiplication to determine where two signals are best aligned. Two correlated discrete signals are denoted as $s[k]$ and $r[k]$ and the cross correlation function $T_{sr}[n]$ is defined as (Rabiner and Gold 1975):

$$T_{sr}[n] = \sum_{m=-\infty}^{\infty} s[m]r[m+n] \tag{6}$$

where the parameter n is an integer between positive and negative infinity. The time shift between two correlated signals is obtained at the maximum (or minimum if the signals are negatively correlated) of the cross correlation function. The cross correlation method avoids complications with picking characteristic points (i.e. first peak) from discrete signals, however the data length selected may affect results if more distorted data is involved, as discussed below.

Typical data from air-hammer and bender element tests are shown in Figure 2 and 3, respectively. The data from a depth of 3.6 m was chosen here as an example; data from triaxial bender element tests at

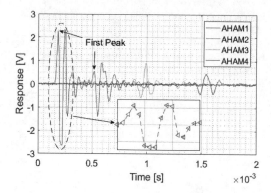

Figure 2. Typical data from a centrifuge air-hammer test.

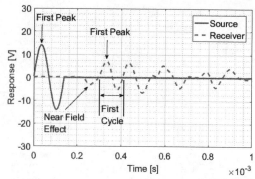

Figure 3. Typical data from a triaxial bender element test (depth = 3.6 m).

all depths had a similar pattern. background noise and reflected waves. Figure 2 shows that the determination of a first peak in each of the centrifuge MEMS data is not straightforward. The MEMS signals in the centrifuge are affected by several factors, including a relatively low resolution of data (9 points over one cycle, shown in Figure 2), reflecting waves from boundaries or elements buried in the soil (pile, piezoelectric accelerometers), decay of the wave signals, and electrical and mechanical noise. These issues are less pronounced for the first detected cycle (i.e. highest signal-to-noise ratio data). Despite this, the low resolution of data still poses a challenge for implementing the peak to peak method. The output of the cross correlation method is also be affected by these same signal quality issues. By including low signal-to-noise ratio data (i.e. the later detected cycles), the determined time difference between two signals can be impacted. It is therefore important to consider removing the distorted data in order to obtain a higher degree of correlation and hence a more accurate value for the travel time. This data length aspect will be discussed in the following section.

In contrast, the high resolution signals obtained from the BE tests (Figure 3) suffer to a lesser degree from these issues, and both the peak to peak and cross correlation methods prove to be reliable. Further results from these two methods are compared in the next section.

7 TEST RESULTS

7.1 Cross correlation method

To investigate the effect of data length on cross correlation results, the mean value of $v_{s,AHA,1}$ for the bottom soil layer (L_1) from 8 air-hammer tests using different data lengths was calculated. Table 2 shows results and indicates that a longer data length resulted in lower estimations of $v_{s,AHA,1}$. This outcome highlights the importance of considering data with the lowest possible signal-to-noise ratio.

In addition, three approaches were used to calculate v_s profile from the triaxial BE data: cross correlation

Table 2. Shear wave velocity from air-hammer testing using different data lengths.

Data length	Strikes	Mean $v_{s,AHA,1}$
1 cycle	8	293.7 m/s
1.5 cycles	8	281.1 m/s
2 cycles	8	264.8 m/s

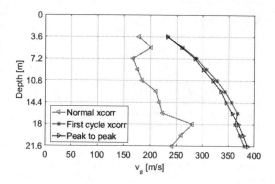

Figure 4. Shear wave velocity comparison using three approaches to analyse BE data.

using the entire data length (normal xcorr), cross correlation using the first cycle of the shear wave (first cycle xcorr), and the first peak to peak method (peak to peak), as shown in Figure 4. Results indicate that the peak to peak method agrees well with the first cycle xcorr method, but that using the full length of the data set results in a lower estimate of v_s as well as an unrealistic scattered pattern of v_s with depth.

7.2 Empirical curves versus BES data

Equations 2, 3, 4 and 5 can be manipulated to obtain expressions for v_s:

$$v_s = \sqrt{7000\frac{(2.17-e)^2}{(1+e)\rho}\frac{(1+2K_0)^{0.5}}{3}} \cdot (\rho g)^{0.25} \cdot K_0^{-0.125} \cdot h^{0.25} \quad (7)$$

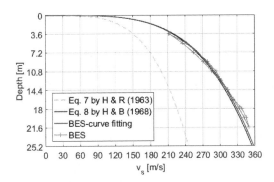

Figure 5. Shear wave velocities from BE tests.

$$v_s = \sqrt{6908\frac{(2.17-e)^2}{(1+e)\rho} \cdot (\rho g)^{0.25} \cdot K_0^{0.125} \cdot h^{0.25}} \quad (8)$$

where ρ is sand density (1739 kg/m³), e is 0.512, g is gravitational acceleration, and h is soil depth (see Figure 1). It can be seen from Figure 5 that Equation 8 agrees very well with the bender element test results, with differences less than 8 m/s. This indicates that Equation 3 can be used to calculate the shear modulus of dry Congleton sand. The difference between Equation 7 and BE data was significant, reaching more than 100 m/s at 20 m depth.

The trend of v_s with depth can be fitted with a power function in the form $v_s = ah^{0.25}$, where a is a fitting parameter. Figure 5 shows the result of fitting this curve to the BE data (BE-curve fitting); a value of $a = 159.6$ provided the best fit to the data. This curve is used for the analysis in the following section.

7.3 BE data versus air-hammer data

The measured shear wave velocities from air-hammer tests represent mean values across the depth between MEMS accelerometers (i.e. span lengths L_1, L_2 and L_3 from Figure 1). However, the BE test provides the shear wave velocity at a specific stress level, i.e. depth. Therefore, in order to make a direct comparison, the BE data spanning each of the depths (L_1, L_2, L_3) needed to be averaged. To do this, the $v_s = ah^{0.25}$ curve fitted to the BE data was used. The elapsed time t can be calculated by integrating $dh/(ah^{0.25})$ from h_1 to h_2 with respect to h:

$$t = \int_{h_1}^{h_2} \frac{dh}{a \cdot h^{0.25}} = \frac{h_2^{0.75} - h_1^{0.75}}{a \cdot 0.75} \quad (9)$$

Then the average velocity v_a from BE data can be obtained using:

$$v_a = \frac{h_2 - h_1}{t} = \frac{0.75a(h_2 - h_1)}{h_2^{0.75} - h_1^{0.75}} \quad (10)$$

where h_1 and h_2 correspond to the limits of the regions of L_1, L_2 and L_3.

Table 3. Shear wave velocities from air-hammer and BE.

Soil layer	Soil depth (m)	Air-hammer (m/s)	BE (m/s)	Difference
$L_3(v_{s,3})$	3.6-10.8	229.3	258.37	11.4%
$L_2(v_{s,2})$	10.8-18.0	304.13	310.64	2%
$L_1(v_{s,1})$	18.0-25.2	347.13	344.43	0.7%

Shear wave velocities of three soil layers from air-hammer and BE tests are given in Table 3. Shear wave velocities in Column 3 are measured from air hammer tests. Shear wave velocities in Column 4 are calculated using Equation 10. The data indicate that the middle (2% difference) and lower (0.7% difference) regions show good agreement. The larger difference of 11.4% for $v_{s,3}$ in the upper region is due to several factors: (1) the weak signal received by the upper MEMS, as shown in Figure 2; (2) the low stress levels in this shallow region, causing a relatively poor coupling between the soil and the accelerometer; (3) the influence of reflected pressure waves generated by the air-hammer at the air-hammer end; and (4) the low signal to noise ratio, making it hard to accurately distinguish the arrival of the shear wave.

Despite these issues, it can be said that the air-hammer is a good method to measure the in-flight shear wave velocity in the centrifuge in a simple, cheap, and efficient way. However, a greater level of scepticism should be applied to air-hammer results obtained near the ground surface.

8 CONCLUSIONS

An air-hammer device in the centrifuge and a bender element system (BES) integrated into a triaxial apparatus have been used to determine the shear wave velocity within a dry sand. Details of experimental methods, the data processing techniques, an evaluation of two empirical methods based on the BE data, and a comparison between the aim-hammer and BE shear wave velocity data have been presented. The main conclusions are as follows:

- The cross correlation method is significantly affected if more distorted data is included in the analysis; the data for the first detected shear wave cycle should be selected from the recorded signals and used to conduct the cross correlation analysis;
- The empirical relationship for shear wave velocity proposed by Hardin and Black (1968) fitted well to the BE data for the sand and test conditions applied.
- The air-hammer and BE data compared well, indicating that the air-hammer can be used as a reliable method for measuring shear wave velocity in the centrifuge. However, a higher level of scepticism should be applied to air-hammer measurements with distorted or weak signals, or those obtained near the ground surface.

REFERENCES

Arulnathan, R. (2000). New Tool for Shear Wave Velocity Measurements in Model Tests.

Brandenberg, S., S. Choi, B. Kutter, D. Wilson, & J. Santamarina (2006). A bender element system for measuring shear wave velocities in centrifuge models. In *6th International Conference on Physical Modeling in Geotechnics*, pp. 165–170.

BS (2005). Mechanical vibration Ground-borne noise and vibration arising from rail systems Part 1: General guidance. *BS ISO 148*, 54.

Camacho-Tauta, J., J. D. Jimenez Alvarez, & O. J. Reyez-Ortiz (2012). A procedure to calibrate and perform the bender element test. *Dyna 79*(176), 10–18.

Cheney, J. A., R. K. Brown, N. R. Dhat, & O. Y. Hor (1990). Modeling free-field conditions in centrifuge models. *Journal of Geotechnical Engineering 116*(9), 1347–1367.

Dobry, R. & M. Vucetic (1987). Dynamic properties and seismic response of soft clay deposits.

Ghosh, B ; Madabhushi, S. P. G. (2002). An efficient tool for measuring shear wave velocity in the centrifuge. In *Proceedings of the international conference on physical modelling in geotechnics*, pp. 119–124. AA Balkema.

Hardin, B. O. & W. L. Black (1968). Vibration modulus of normally consolidated clay. *ASCE – Proceedings – Journal of the Soil Mechanics and Foundations Division 94*(SM2, Part 1), 353–369.

Hardin, B. O. & J. F. E. Richart (1963). Elastic wave velocities in granular soils. *ASCE – Proceedings – Journal of the Soil Mechanics and Foundations Division 89*(SM1, Part 1), 33–65.

Jaky, J. (1944). The Coefficient of Earth Pressure at Rest. *Journal for Society of Hungarian Architects and Engineers* (October), 355–358.

Kim, N R ; Kim, D. S. (2010). Development of VS tomography testing system for geotechnical centrifuge experiments. *7th Physical modeling in geotechnics, 7th ICPMG, Springman 50*(1), 349–354.

Kuo, K. A. (2010). *Vibration from Underground Railways: Considering Piled Foundations and Twin Tunnels*. Ph. D. thesis, University of Cambridge.

Lauder, K. (2010). *The performance of pipeline ploughs*. Ph. D. thesis, Dundee University.

Lee, C.-J., W.-Y. Hung, C.-H. Tsai, T. Chen, Y. Tu, & C.-C. Huang (2014). Shear wave velocity measurements and soilpile system identifications in dynamic centrifuge tests. *Bulletin of earthquake engineering 12*(2), 717–734.

Lee, C.-J., C.-R. Wang, Y.-C. Wei, & W.-Y. Hung (2012). Evolution of the shear wave velocity during shaking modeled in centrifuge shaking table tests. *Bulletin of earthquake engineering 10*(2), 401–420.

Nugent, R. E., W. Village, & J. A. Zapfe (2012). Designing Vibration-Sensitive Facilities Near Rail Lines. *Sound and Vibration* (November), 13–16.

Rabiner, L. R. & B. Gold (1975). Theory and application of digital signal processing. *Englewood Cliffs, NJ, Prentice-Hall, Inc., 1975. 777 p. 1*.

Rollins, K. M., M. D. Evans, N. B. Diehl, & W. D. Daily III (1998). Shear modulus and damping relationships for gravels. *Journal of Geotechnical and Geoenvironmental Engineering 124*(5), 396–405.

Shantz, B. R. W. J. T. D. T. (2012). Guidelines for Estimation of Shear Wave Velocity Profiles. Technical report.

Steedman, R. S., S. P. G. Madabhushi, & U. of Cambridge. Engineering Department (1990). *Wave Propagation in Sand Medium*. CUED/D – soils TR. University of Cambridge, Department of Engineering.

Stokoe, K., S. Lee, & D. Knox (1985). Shear moduli measurements under true triaxial stresses. In *Advances in the art of testing soils under cyclic conditions*, pp. 166–185. ASCE.

Tsuno, K., W. Morimoto, K. Itoh, O. Murata, & O. Kusakabe (2005). Centrifugal modelling of subway-induced vibration. *International Journal of Physical Modelling in Geotechnics 5*(4), 15–26.

Yang, W., M. F. M. Hussein, A. M. Marshall, & C. Cox (2013). Centrifuge and numerical modelling of ground-borne vibration from surface sources. *Soil Dynamics and Earthquake Engineering 44*, 78–89.

Development of layered models for geotechnical centrifuge tests

S. Divall, S.E. Stallebrass, R.J. Goodey & E.P. Ritchie
City, University of London, London, UK

ABSTRACT: Centrifuge modelling is an established technique for investigating the ground response to complex and non-standard geotechnical events. These models are usually made from re-formed soil, allowing for comparisons with naturally occurring soil deposits. Clay models are formed by mixing clay powder and water into a slurry. This slurry is placed within a container and loaded to create a uniform stiff clay model. However, there is a fundamental disparity between this process and the deposition of natural soils, because natural soil is deposited in layers creating a unique structure. This structure is important for modelling true soil behaviour because some essential soil properties (such as permeability, stiffness and strength) are not identical in all directions. Currently, there are limited methods for creating layered soil samples. This paper describes the development of a new procedure for creating layered centrifuge models with structure – leading to potentially more representative models of naturally occurring ground.

1 INTRODUCTION

1.1 Background

The technique of geotechnical centrifuge modelling has enabled engineers and researchers to better understand geotechnical events and construction processes. One of main benefits has been the ability to observe mechanisms or patterns of movements in small-scale models that can be related to full-scale events. These small-scale models are often idealised homogeneous soil models which primarily consist of either sand or clay.

Mair (1979) developed a method for creating these homogenous samples for clay models by mixing kaolin powder with distilled water. Usual practice since then has been to prepare a slurry to a water content of 120% and place this slurry within a soil container known as a strongbox or strongtub (Grant 1998; McNamara 2001; Begaj 2009; Divall 2013; Le 2017). The slurry is subjected to a known stress history using hydraulic or pneumatic consolidation presses to arrive at a homogenous sample. Models representing overconsolidated soils often follow a period of swelling before the model preparation stage and further inflight consolidation. Studies concerned with short-term deformations have successfully replicated the stress-strain response of the prototype soil continuum (Grant 1998) by carefully considering the stress history and g-level.

However, reconstituted samples often cannot provide a realistic representation of natural soil structure, particularly the effect this has on permeability and inherent anisotropy. This is because the strength and stiffness of reconstituted soils are governed solely by their state (i.e. packing) and effective stresses (Atkinson et al.1990). The inability to physically model soil structure has led to considerable gaps in our knowledge regarding the deformations associated with geotechnical events in layered ground (Hird et al. 2006). Whilst the characteristics of layered soils have been fairly extensively documented (Burland 1990 and others), the effect of these differences on the ground response to geotechnical events is not well understood.

This paper gives details of the initial development of a novel procedure for creating a layered clay and sand soil sample for geotechnical centrifuge modelling, with representative effective stresses. The observations from an initial test including the data from a T-bar penetrometer are shown and a predictive framework for the layers is presented.

2 PREVIOUS PHYSICAL MODELLING OF LAYERED MODELS

Soil response is, in part, governed by the influence of structure (Leroueil & Vaugham 1990). Soil structure can be described as the combination of 'fabric' (or the arrangement of particles) and interparticle 'bonding' (Mitchell 1976).

Burland (1990) stated that the main aim of studies into the behaviour of natural sedimented soils were 'to bring the same unity and coherence that Critical State Soil Mechanics brought to reconstituted soils'. In some respects, this was also achieved by Cotecchia & Chandler (2000) who sought to better model the behaviour of fine-grained soils by introducing the term 'sensitivity'. Sensitivity is the ratio of the undisturbed to remoulded compressive strengths. Their study suggested that if data were normalised with respect to sensitivity there was a unique shape of the state boundary surface for all clays. The state boundary surface is defined in stress: volumetric space as the boundary of all possible states of soil.

Studies have attempted to clarify the shape of the state boundary surface by creating small samples, using sedimentation columns, to be used in element tests (Ward et al. 1959; Been & Sills 1981; Edge & Sills 1989; Mašín 2004). The latter used a specially fabricated sedimentation column laboratory apparatus to create triaxial samples. This sedimentation column was a 2 m high, 100 mm diameter, Poly(methyl methacrylate) or PMMA cylinder. A buoyant PMMA piston loaded the slurry and allowed for top and bottom drainage to accelerate the consolidation and sedimentation of the London Clay. London Clay was submerged in water and thoroughly mixed for 36 hours to an initial water content of 5800%. The samples were prepared using distilled water mixed with Saxa brand fine sea salt to 3.509% (similar to the conditions in the North Atlantic Ocean) to act as a flocculant. The slurry was left to sediment for approximately 3 days and the cycle was repeated 4 times to produce a 76 mm high triaxial sample. The result were samples that consisted of four layers and, like those of Been & Sills (1981), each layer had been separated according to size. Figure 1 (taken from Stallebrass et al. 2007) clearly shows the visual differences between a sedimented soil sample and a reconstituted soil sample.

Studies using layered samples in centrifuge tests have not applied this approach to create their models. Grant (1998) undertook a series of tests investigating movements around a tunnel in two-layered ground. These models consisted of a single Leighton Buzzard Sand layer overlying a Speswhite kaolin clay layer within a strongbox. The overall depth of the model was 225 mm with each of the clay layers ranging from 37.5 mm to 175 mm and the sand layers ranging from 67 mm to 187.5 mm. The clay layer was prepared in much the same way as Mair (1979) with the sand layer placed on top. Marshall et al. (2014) utilised a similar method for a study into non-displacement piles and pile groups within a strongtub. The bottom layer was Speswhite kaolin clay with Fraction E silica sand placed by air pluviation.

Muñoz & Caicedo (2014) modelled shallow tunnels in heterogeneous soils by omitting layers altogether. The study implemented a random field generator which dictated specific sections within a strongbox which could have various kaolin: bentonite proportions. This attempted to simulate the 2D variability of a clay model with inherent variability represented by soil with different Liquid Limits. Soils were placed with a 'caulking tool' in strips across the full depth of the strongbox. Samples were then consolidated in a consolidation press to a vertical effective stress of 50 kPa.

The aim of the new procedure is to establish a method for the creation of a sedimented sample (with 'sensitivity' as defined by Cotecchia & Chandler, 2000) within a strongbox. The soil bed created can be used during centrifuge testing to investigate the behaviour of geotechnical processes (such as piling or tunnelling). The combination of these two procedures would enable the effect of 'sensitivity' on ground response to be investigated and allow soil behaviour which is dominated by permeability, such as long-term tunnelling-induced ground movements, to be modelled experimentally.

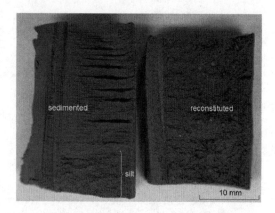

Figure 1. Sedimented vs reconstituted samples (Stallebrass et al. 2007).

3 EXPERIMENTAL WORK

3.1 Preparatory work

The clay sedimentation process started when a slurry containing disaggregated clay (see below) and Fraction E Leighton Buzzard sand was introduced into a strongbox containing distilled water. Prior to testing, an investigation was undertaken to determine i) the minimum water content of slurry that could be held in suspension and ii) whether the size of container affected the suspension of the slurry.

Figure 2a shows four sedimentation columns each with 500 ml of distilled water with 125 ml of slurry poured into the top. The 125 ml slurry introduced ranged in water content from 200% to 400%, such that the nominal water contents when the slurry and water are combined range from 1148% to 2153%. The minimum water content which was held in suspension was 250%. Figure 2 also shows three containers of different diameters and depths. The 250% water content slurry was placed in these demonstrating that regardless of container diameter, 125ml of slurry was held in suspension in 500ml water, which is equivalent to a water-content for the suspension of 1403%.

3.2 Mixing stage

Work undertaken by Phillips et al. (2014) established that when both natural and reconstituted clay cuttings are agitated in water they disaggregate into slurries containing a high proportion of silt sized agglomerations of clay particles or 'clay peds' rather than individual clay particles. Since many sedimented soils are deposited from eroded and transported material

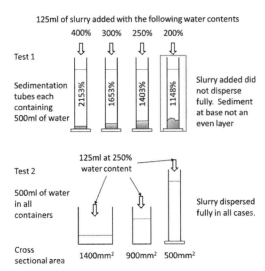

Figure 2. Test 1: Sedimentation columns with varying percentages of water-content slurries. Test 2: Sedimentation containers of various diameters.

it is likely that the majority of sedimented clays are formed from these silt sized clay peds and not the clay particles present in powdered clays such as Speswhite kaolin.

Consequently, before the sedimentation stages could be undertaken, an initial slurry of Speswhite kaolin clay powder (supplied by Imerys Minerals Ltd), was mixed with distilled water in a ribbon blade mixer to a water-content of 120%. This initial slurry was placed within a strongtub and subjected to a vertical stress of 350 kPa. The vertical stress was applied by a hydraulic consolidation press over approximately one week. This created a moderately stiff normally consolidated clay sample. The clay was removed from the strongtub and divided into 'cuttings' (of approximately 40–50 mm^3) and placed into the planetary mixer with more distilled water to a water-content of 1285%. This consisted of 500 g of clay, 200 g of sand and 9 litres of distilled water. It was then mixed for about 30 minutes until fully disaggregated.

3.3 Sedimentation stage

The silt-sized clay agglomerate and sand based slurry was poured into a second soil container (strongbox). The strongbox had been modified for this application with a PMMA window replacing the front face and with a porous plastic sheet silicone sealed to the bottom drainage plate. The slurry was subjected to acceleration on the centrifuge of 160g. This forced the larger soil particles and agglomerates to sediment first with the finer material sedimenting later.

This process created the layered soil structure that can be seen in standard sedimentation columns but across a soil container suitable for larger scale centrifuge model testing. The sedimentation of this first layer took approximately 1½ hours. The process could be observed through on-board USB cameras. The centrifuge was decelerated and pipette samples were taken of the remaining surface water on top to confirm that the sedimentation process was essentially complete. The water-content of the pipette samples was determined and it was found that the average water-content was 99.95%. This showed that the sedimentation process had completed within the time frame.

At this stage of testing a second slurry (identical to that previously described) was poured into the surface water. The centrifuge was then accelerated again until a second layer could be observed. This was repeated a third time to arrive at the final model height.

3.4 Testing

Once the third layer was sedimented a T-bar penetrometer (Gorasia, 2013) was bolted to the top of the strongbox at 1g (see Figure 3).

The undrained shear strength, S_u, of the model could be determined using the readings from the T-bar and the equation below originally from Stewart & Randolph (1991).

$$S_u = \frac{P}{N_b \cdot d} \quad (1)$$

where P = force per unit length acting of the bar; d = diameter of the bar (in this case 7 mm); and N_b = bar factor. Stewart & Randolph (1991) cited Randolph & Houlsby (1984) for a value of N_b as 10.5 for general use. Stewart & Randolph (1991) also state that the tool should be utilised in soft clay investigations. The T-bar was driven at 60 mm per minute. The aim was to record the resistance in at least two clay layers and their respective interfaces with the sand layers.

4 RESULTS

4.1 Visual inspection

Figure 4 shows the layers in the soil once the front PMMA window of the strongbox had been removed. The steel rule held against the layers shows the consistency and even distribution of the layers created. Moreover, it is possible to observe that the sand particles appear at the bottom of each layer, as expected, and there appears to be an interface between the largest clay agglomerates and smallest sand fines.

The soil on the top layer was very weak and had very little 'stand-up' time. This was assumed to be owing to the unloading process during deceleration of the centrifuge and the standing water above the soil as part of the sedimentation process. No drainage was allowed from the sample inflight and when the test was stopped, the sample swelled reabsorbing some of this standing water.

4.2 T-bar penetration data

The T-bar readings confirm that at 1g the clay layers had very low undrained strengths. Figure 5

Figure 3. T-bar penetrometer and frame used for determining the undrained shear strength with depth.

shows a standard undrained shear strength with depth profile.

There are three points of interest in this data. Firstly, the point at which the T-bar connects with the soil gives reading of 0.25kPa. Secondly, the readings increase dramatically when passing through the sand layer below and then drop to approximately 0.65 kPa. This layer is assumed to be the second clay stratum. This has been stressed by the weight of the layer above and therefore has a slightly higher undrained shear strength compared with the uppermost layer. Thirdly, the readings decrease for the intermediate clay layer before increasing once again for the sand layer.

This shows it is possible to identify the sand and clay layers and possibly their relative strengths. Although, caution should be applied here as sand derives its strength from angle of friction which cannot be identified on this plot.

4.3 Prediction of soil layers

To verify the approach, the depth of the layers was predicted using standard relationships for the compression of clays. These predictions should represent a lower bound to the thickness of the layers as they would not take account of 'sensitivity'.

The depth of each sublayer of sand or clay was calculated assuming the particles settled to a state which could be described by a normal compression line (in specific volume, v, and average stress, $\ln p'$, space). The height of the sand layers was determined by calculating the total volume using the weight of sand used

Figure 4. Three layers created through sedimentation (sand-Speswhite clay per layer).

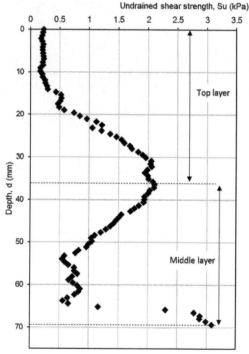

Figure 5. T-bar readings with depth for the layered soil model.

and average voids ratio values from Grant (1998). This volume was divided by the internal plan area of the strongbox to arrive at an average height.

The height of the clay sublayer was determined iteratively as the vertical stresses imposed at 160 times the earth's gravity are determined by the unit weight of the clay which changes with voids ratio, e. The voids ratio is in turn determined by the vertical stresses.

The overall height of the top sand and clay layer was computed to be 20.6 mm. It is difficult to make comparisons with Figure 4 because the soil has slumped forwards after the window was removed. However, given the relatively small measurements recorded this was considered a reasonable prediction of the height of the layers. The T bar readings indicate a combined height of sand and clay layer which is 36.4 mm. The second layer was predicted to be 19.2 mm and the T-bar indicates the depth of this sand and clay layer to be 30.1 mm.

5 CONCLUSIONS

It is possible to create a centrifuge model with a sedimented soil structure as defined by Mitchell (1976). This paper details the initial development of a novel procedure for creating a layered soil model within a geotechnical centrifuge. The results of an initial test are shown including a prediction of the height of the layers.

The one major shortcoming is that the method currently creates very weak soil samples. This would be overcome by creating more layers and allowing drainage after the layers have been created, to remove the surface water. This would help prevent the soil swelling and reabsorbing the standing water on top of the final layer. Further tests are required to characterise the layers to quantify the values of permeability in the vertical and horizontal directions.

ACKNOWLEDGEMENTS

This research was supported by the Pump Priming Award by City, University of London (project title: 'Layered Clay Samples for Geotechnical Centrifuge Modelling'). The authors wish to thank Eric Ritchie for his assistance during the testing and the supplying some of the photographs. Additional thanks should also go to members of the Research Centre for Multi-Scale Geotechnical Engineering at City, University of London for their support during this project.

REFERENCES

Atkinson, J.H., Richardson, D. & Stallebrass, S.E. 1990. Effects of recent stress history on the stiffness of overconsolidated soil. *Géotechnique*, 40(4): 531–540.

Been, K. & Sills, G.C. 1981. Self-weight consolidation of soft soils: an experiment and theoretical study. *Géotechnique*, 31(4): 519–535.

Begaj, L. 2009. Geotechnical centrifuge model testing for pile foundation re-use. PhD Thesis, City University London, UK.

Burland, J.B. 1990. On the compressibility and shear strength of natural clays. *Géotechnique*, 40(3): 329–378.

Cotecchia, F. & Chandler, J. 2000. A general framework for the mechanical behaviour of clays. *Géotechnique*, 50(4): 431–447.

Divall, S. 2013. Ground movements associated with twin-tunnel construction in clay. PhD Thesis, City University London, London, UK.

Edge, M.J. & Sills, G.C. 1989. The development of layered sediment beds in the laboratory as an illustration of possible field processes. *Quarterly Journal of Engineering Geology*. 22: 271–279.

Gorasia, R.J. 2013. Behaviour of ribbed piles in clay. PhD Thesis, City University London, London, UK.

Grant, R.J. 1998. Movements around a tunnel in two-layered ground. PhD Thesis, City University London, UK.

Hird, C.C., Emmett, K.B. & Davies, G. 2006. Piling in layered ground: risks to groundwater and archaeology. *Science Report* SC020074/SR. Environment Agency: Bristol.

Le, B.T. 2017. The effect of forepole reinforcement on tunneling-induced movements in clay. PhD Thesis, City, University of London, London, UK.

Leroueil, S. & Vaughan, P.R. 1990. The importance and congruent effects of structure in natural soils and weak rocks. *Géotechnique*, 40(3): 467–488.

Mair, R.J. 1979. Centrifugal Modelling of Tunnel Construction in Soft Clay. PhD Thesis, Cambridge University, Cambridge, UK.

Marshall, A.M., Cox, C.M., Salgado, R. & Prezzi, M. 2014. Centrifuge modelling of non-displacement piles and pile groups under lateral loading in layered soils. *Proceedings of the Eighth International Conference on Physical Modelling in Geotechnics* (ICPMG 2014), Perth, Australia, 2: 847–852.

Mašín, D. 2004. Laboratory and numerical modelling of natural clays. MPhil dissertation, City University London, London, UK.

McNamara, A.M. 2001. Influence of heave reducing piles on ground movements around excavations. PhD Thesis, City University London, UK.

Mitchell, J.K. 1967. *Fundamentals of soil behaviour*. New York: Wiley.

Muñoz, G. & Caicedo, B. 2014. Centrifugal modelling of shallow tunnels in heterogeneous soils. *Proceedings of the Eighth International Conference on Physical Modelling in Geotechnics* (ICPMG 2014), Perth, Australia, 2: 937–942.

Phillips, N.S., Stallebrass, S.E., Goodey, R.J. & Jefferis, S.A. 2014. Test development for the investigation of soil disaggregation during slurry tunnelling. *Proceedings of the Eighth International Conference on Physical Modelling in Geotechnics* (ICPMG 2014), Perth, Australia, 2: 979–984.

Stallebrass, S.E., Atkinson, J.H. & Mašín, D. 2007. Manufacture of samples of overconsolidated clay by laboratory sedimentation. *Géotechnique*, 57(2): 249–253.

Stewart, D.P. & Randolph, M.F. 1991. A new site investigation tool for the centrifuge. *Proceedings of the International Conference – Centrifuge 1991*, Colorado, USA, 531–538.

Ward, W.H., Samuels, S.G. & Butler, M.E. 1959. Further studies of the properties of London Clay. *Géotechnique*, 9(2): 33–58.

The influence of temperature on shear strength at a soil-structure interface

J. Parchment & P. Shepley
Department of Civil and Structural Engineering, University of Sheffield, Sheffield, UK

ABSTRACT: Thermal heat exchanger piles combine the requirements of a traditional pile foundation with the energy harvesting/dissipating potential of a ground source heat pump. Such installations are becoming increasingly popular as a means of reducing the energy demands of buildings. Behaviour of the soil-pile interface is central to predicting the shear response of floating piles; yet understanding of the effects of thermal loading on soil-structure interaction remains uncertain. In this paper, heated interface shear tests have conducted using a specially adapted direct shear apparatus; to mimic the behaviour of a thermal pile interface. Testing was undertaken at various over-consolidation ratios, normal stresses and temperatures. This investigation observed the interface friction angle to remain unaffected by heating until its thermal yield is reached and thermal consolidation takes place, at which point there is a reduction in the friction angle which remains when the soil is cooled. There is also an observed increase in apparent adhesion at the interface in response to heating which also remains after cooling.

1 INTRODUCTION

Thermal heat exchanger piles combine the structural requirements of a traditional structural pile with the energy harvesting/dissipating potential of a ground source heat pump. Such installations are becoming increasingly popular as a means of reducing the energy demands of buildings; yet understanding of the effects of thermal loading on soil-structure interaction remains uncertain.

Previously, this uncertainty has been addressed by using inflated factors of safety to ensure geotechnical performance. However, recent attempts to quantify the effects of thermal loading on pile behaviour have identified the primary mechanisms as: the differential axial strain between pile and soil, the effect of temperature on interface properties in cohesive soils and radial strain & confinement (GSHP Association 2012; Abuel-Naga et al. 2015). This work focuses on investigating the effect of thermal loading on shear behaviour at the soil-pile interface.

2 LITERATURE REVIEW

There are few investigations into the thermo-mechanical response of the soil-structure interface. As a result, much of the state of the art is derived from the thermo-mechanical response of soil bodies and the analysis of isothermal interface shear.

2.1 Soil response

Extensive research into the thermo-mechanical response of soils has shown that cohesive soils either contract or dilate, under thermal loading, dependent on the material's stress history. A normally consolidated soil is seen to undergo irreversible contraction upon heating, in a manner physically similar to creep (Plum et al. 1969; Sultan et al. 2002; Burghignoli et al. 2000; Cui et al. 2009). In contrast, over-consolidated soils initially undergo a reversible expansion upon heating (Towhata et al. 1993; Cekerevac & Laloui 2004); though, once subjected to a sufficiently high temperature, a thermal yield limit is reached and over-consolidated soils also begin to contract (Baldi et al. 1988; Towhata et al. 1993; Sultan et al. 2002). The extent of the volume change is also dependent on the stress history with more highly over-consolidated soils undergoing greater volume changes (Cekerevac & Laloui 2004). Previous studies have shown that OCR at which the soil transitions between contractive & dilative behaviour and the change in temperature at which over-consolidated soils begin to contract, vary between soil types and samples (Yavari 2014).

The characteristic response of cohesive soils subject to thermal loading is a combination of thermal softening and strain hardening. Soils that have experienced thermally induced strain hardening are considered to have undergone thermal consolidation.

Thermal consolidation has been observed to increase the peak friction angle of a soil, an expected response as thermal consolidation manifests as an apparent increase in pre-consolidation pressure (Kuntiwattanakul et al. 1995; Cekerevac 2003). However, critical state friction has been observed to increase, decrease or remain unaffected by thermal loading (Hamidi et al. 2015). This is an unexpected response as critical state friction is independent of stress history (Skempton 1970) suggesting that critical state theory may not be suitable in modelling thermo-mechanical response. In addition to stress history, it appears that this variation in critical state response is also dependent on soil type and experimental method (Hueckel & Baldi 1990; Hueckel et al. 2011; Laloui 2001).

2.2 Interface behaviour

There has been extensive research into the shear behaviour at the soil-structure interface. However, until recently all have been in isothermal conditions.

Surface roughness, soil composition, relative density, grain-size distribution, morphology of the soil particles, soil moisture, magnitude of normal stress and shearing rate have all been observed to effect interfacial shear behaviour (Chen et al. 2015). However, soil type, surface roughness and stress history are considered the most influential factors (DeJong & Westgate 2009). It is therefore reasonable to suggest that thermal loading may also influence interface behaviour.

2.3 Thermal interface behaviour

In thermal interface tests on normally consolidated Illite clay, (Di Donna et al. 2016) observed a small decrease in friction angle but an increase in adhesion, resulting in an overall increase in shear strength. The increase in adhesion is due to the increased mouldability of the clay deforming around the asperities at the interface surface.

The findings of (Di Donna et al. 2016) are further supported by (Yavari et al. 2016) in which kaolin samples at a range of over consolidation ratios were considered, but, the effects of thermal consolidation were eliminated by first heating all samples to a temperature higher than in testing. All test were therefore undertaken in the cooling phase of the heating-cooling cycle and places all responses within the elastic domain and a negligible variation in friction angle and adhesion were observed.

This work aims to provide a more complete understanding of the thermo-mechanical response of a soil by comparing the response at different over consolidation ratios and analysing different normal stresses in a lightly overconsolidated state. By investigating the thermo-mechanical response of a lightly overconsolidated soil it is possible to evaluate a greater range of results as both elastic and elasto-plastic thermal shear behaviour can be analysed by heating to levels below and at the thermal yield limit. As such the focus shall be on the thermo-mechanical response during the heating phase of a temperature cycle, though limited results on the cooling phase are available.

3 EXPERIMENTAL METHOD

3.1 Setup

The device used in this investigation was a direct shear apparatus accommodating a $60 \times 60 \times 20$ mm soil sample. A thermal interface was designed, shown

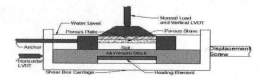

Figure 1. Schematic of apparatus setup.

Table 1. Properties of Speswhite kaolin.

Property	Value	
Liquid Limit, W_l	31.2	%
Plastic Limit, I_p	33.7	%
Specific Gravity, G_s	2.65	kg/m^3
Particle Diameter >50%, D_{50}	0.7	μm
Particle Diameter >95%, D_{95}	4	μm

in Figure 1, using an aluminium block containing a heating element with a temperature controller, data logger and thermocouples to replace the lower half of the shear box. The heating element was attached to the underside of the aluminium block directly below the shearing surface. The heating element was a 70 W Solid State PTC conductive heating element selected due to its slim profile and aluminium sleeve. Either side of the heating element two holes were drilled into which thermocouples were secured. The heating element and one thermocouple were attached to a self-regulating temperature controller, whilst the other was monitored through a data logger. Additional thermocouples were used to measure temperatures within the shear box, for calibration.

In traditional soil mechanics, the volumetric response of a cohesive soil is dilative or contractive, dependant on the overconsolidation ratio. At the pile-soil interface, this volumetric response is restricted by the surrounding soil. As such, testing under constant normal stiffness conditions would have been most accurate in replicating the behaviour at the pile-soil interface (Ooi & Carter 1987; Di Donna et al. 2016). However, due to the relatively low stiffness of cohesive soils and the small anticipated volumetric response, constant normal load was deemed suitable for this study (Di Donna 2014).

Prior to the main investigation, a series of calibration checks ensured the suitability of the system. The homogeneity of the temperature across the soil-solid interface was confirmed using the additional thermocouples. Further checks found that the thermally induced volume changes of the shear box apparatus and interface to have negligible effect on the results.

3.2 Materials

In this work, testing was undertaken on low-medium plasticity kaolin clay and at a kaolin-aluminium interface. Kaolin was selected due to its non-expansive and non-reactive microstructure.

Table 2. Selected stress histories.

Pre-consolidation Pressures (kPa)		Normal Effective stress (kPa)		
		50	100	150
Over-Consolidation Ratio	1	50	100	150
	2	100	200	300
	6	300	600	900

Table 3. List of tests conducted.

Test	Type	Thermal Load Path (°C)	OCR	Consolidation Pressure (kPa)
1	clay-clay	20	1	100
2	clay-clay	20	2	200
3–5	clay-clay	20	6	300/600/900
6	interface	20	1	100
7	interface	20	2	200
8–10	interface	20	6	300/600/900
11–13	interface	20-40	6	300/600/900
14	interface	20-60	1	100
15	interface	20-60	2	200
16–18	interface	20-60	6	300/600/900
19	interface	20-60-20	1	100
20	interface	20-60-20	2	200
21–23	interface	20-60-20	6	300/600/900

A smooth aluminium block, average roughness R_a–0.602 μm, was selected to represent the pile surface. Aluminium was chosen due to its high thermal conductivity, ensuring uniform heating of the soil samples. Aluminium is also not susceptible to rust, which would increase roughness during testing (Vaughan & Lemos 2000). Whilst not a material commonly used in the construction of thermal piles, the type of material used in interface tests has no significant effect on shear response (Potyondy 1961; Feligha & Hammoud 2015). Using a smooth surface avoided an artificial scenario in which turbulent shearing would occur allowing this study to focus on thermally induced interface behaviour.

3.3 Testing procedure

The properties of the clay and the influence of introducing an interface material were evaluated, before thermal loading was applied. All experiments were undertaken under fully saturated, drained conditions and failure was considered to be reached at 10% displacement (6 mm). Both the effects of thermal loading on clay at various normal stresses and a range of over-consolidation ratios were investigated and the selected tests are highlighted in Table 2.

Samples were pre-consolidated using a two-way drainage Rowe Cell, to ASTM D2435. Interrogation of the resultant consolidation curves lead to a drained

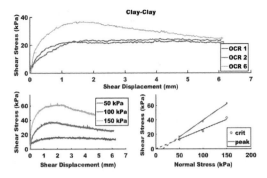

Figure 2. Clay-clay shear response – (a) OCR, (b) Normal.

shearing rate of 0.0377 mm/min being selected. A typical load path for the samples required saturation of the consolidated samples, before being loaded into the shear box with the desired normal stress applied for 30 minutes to allow vertical displacements to stabilise before either shearing (for tests at 20°C) or undergoing thermal loading before hearing.

During the thermal interface tests, the temperature was increased in 10°C increments every 15 minutes, until the desired temperature was reached. This temperature was maintained for 2.5 hours to allow thermally induced pore-water pressures to dissipate. For tests in the heated phase, shearing could begin. For cooling phase tests, the temperature was first reduced to the desired temperature at the same rate and temperature once again maintained for 2.5 hours before shearing.

4 RESULTS AND DISCUSSION

4.1 Clay-clay

Clay-clay shearing was performed at 20°C with resulting graphs shown in Figure 3. The observed peak strength at OCR 6 is evidence of an OC soil, though its contractive normal displacement suggests it is lightly OC. Comparison of the OCR 6 tests at a range of normal stresses has provided a critical state friction angle of 15° and peak of 24.7°. This critical state value is on the lower end of the range expected for Speswhite kaolin (Yavari 2014).

4.2 Response at a range of OCRs

4.2.1 Shear stress-shear displacement

As shown in Figure 4, introducing the aluminium interface there is a significant reduction in the observed peak strength, to values comparable to the critical state strength; in line with the observations of previous authors (Littleton 1976; Chen et al. 2015). However, it is noted that not only does a small peak remain, but a peak of similar magnitude is induced at OCR 2. The loss of peak strength at OCR 6 occurs as peak strengths are a result of the breaking strength of inter-particle

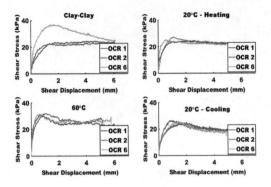

Figure 3. OCR shear stress response – (a) Clay-clay, (b, c & d) 20-60-20 Heating-cooling cycle.

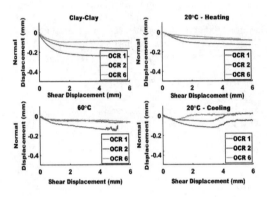

Figure 4. OCR shear displacement response – (a) Clay-clay, (b, c & d) 20-60-20 Heating-cooling cycle.

bonds as clay particles realign at the shear failure plane. Introduction of a structural interface reduced the need for this breakage as the two materials are not structurally bonded. As roughness decreases, this value of peak strength approaches that of the critical strength as there is less resistance at the interface and so particles are able to realign without significant bond breakage. The small increase in peak strength at OCR 2 and remaining presence at OCR 6 is due to the small degree of resistance still given by the roughness of the aluminium surface.

Throughout the heating-cooling cycle there is a clear agreement between the results at each OCR, suggesting that critical state shear strength remains independent of OCR under thermal loading. As a result, it is possible to draw conclusions on the behaviour of the samples at each OCR from one set of results. By comparing the results of OC samples further conclusions can be drawn from the peak strength response of the soil.

4.2.2 Normal displacement-shear displacement

As would be expected, Figure 5 shows that the normal displacement of the clay-clay sample is around twice that of the interface tests, as there is half the material at the interface.

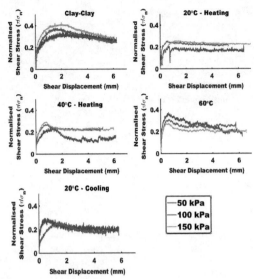

Figure 5. Normalised shear strength response at OCR 6 (a–d) 20-40-60-20 Thermal loading cycle.

After heating, the shear displacements remain relatively constant, suggesting no significant change in failure mechanism at any OCR during heating. However, the displacement behaviour during cooling is far more unusual. The samples dilate during shear, suggesting that thermal consolidation has taken place at the maximum temperature of 60°C. It is also noted that the magnitude of volume change appears to remain dependent on the OCR of the soil, with higher OCR samples dilating to a greater degree. When shearing after cooling to 20°C, the OCR 6 sample dilates early its shear displacement, as would be expected. However, the OCR 1 and OCR 2 samples contract until around 4 mm of shear displacement, before dilating. This highly unusual behaviour was consistently observed in repeat tests and further investigation is required understand the phenomenon.

4.3 Shear strength response

The shear strength analysis of this paper focuses on the results at OCR 6 to include peak strength behaviour whilst also characterising the thermo-mechanical response up the thermal yield limit

4.3.1 Shear strength-displacement relationship

The normalised shear-displacement curves in Figure 6 show critical state shear strengths to be consistent at different normal stresses. However, observed peak shear strengths increase with normal stress. Negative values returned for cohesion/adhesion are due to the side wall friction incident on the sample from the shear box ring. This created artificially lower cohesion/adhesion values, however, the observed trends remain valid in analysis.

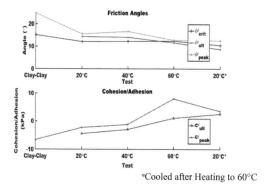

Figure 6. Friction angles and cohesion values.

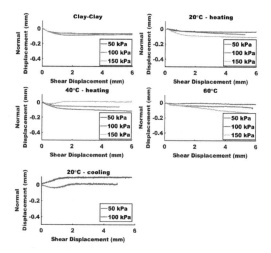

Figure 7. Displacement response at OCR 6 – (a) Clay-clay, (b-e) 20-40-60-20 Thermal loading cycle.

Heating affected both the friction angle and the adhesion observed at the interface. Heating to 40°C has minimal effect on the interfacial friction angle and also results in a small increase in adhesion. Further heating to 60°C causes a notable fall in the interfacial friction angle, but adhesion continues to rise. This drop in friction angle at 60°C is likely due to the thermal yield limit being surpassed for the given overconsolidation ratio, for which thermal consolidation occurs. This trend suggests that thermal consolidation at the soil-pile interface results in a drop in friction angle, whilst the continued increase in adhesion supports the notion that increasing temperature at the interface increases the mouldability of the clay, increasing adhesion. This is in contrast to the behaviour of clay bodies as found in literature, in which thermal consolidation causes the peak friction angle to increase.

During the cooling phase, the interface friction angle remains at levels comparable with that at the peak temperature (60°C), lower than those found at the equivalent temperature during heating. On the other hand, the observed adhesion is higher during the cooling than at the equivalent heating temperature; though slightly lower than the adhesion found at peak temperature. This further supports the notion of increased soil mouldability at raised temperature.

4.3.2 *Normal displacement-shear displacement*

The normal displacements observed during interfacial shear are much more varied than those observed in clay-clay shear. There was no appreciable trend in the spread of normal displacements.

During heating the normal displacement behaviour is consistently contractionary. However, during cooling there is consistent sample dilation during shear. This further supports the notion that the temperature increase to 60°C is large enough to cause the soil sample to reach its thermal yield limit, extending the elastic region.

Comparison of the friction angles compiled in Figure 7 show that in shearing clay against the structural material rather than at a clay-clay interface, there is a significant reduction in critical state friction angle. As previously discussed, this is evidence of a natural sliding failure at the interface as a result of an aluminium surface roughness below its critical value. As the interfacial friction angles at the interface are lower than the internal friction angle of the clay, the system should not be modelled using critical state theory and adhesion may be considered in ultimate strength analysis. However, caution is advised as calculation of adhesion using a linear function is not recommended due to the non-linear nature of the stress envelope at low stresses. The results of this investigation show that the reduction in friction angle is much greater at peak strength than ultimate, as expected, due to the lack particle bonding across the interface. There is also a marked increase in cohesion due to the mouldability of the clay around the asperities in the structural material at the interface.

5 CONCLUSION

Heating overconsolidated clay has negligible effect on the interface friction angle. However, once the thermal yield limit is reached, the friction angle falls, and if the clay is then cooled, the friction angle remains at the levels observed at maximum temperature. Adhesion at the interface increases with temperature and decreases, to a lesser degree, if the temperature is then reduced. This is possibly due to the increased mouldability of the clay at raised temperatures, but may be an artificial adhesion created by the non-linear stress envelope at low stresses.

The practical applications are that in the case of heavily overconsolidated clays, changes in friction angle are negligible for the typical temperature range in thermal pile systems. Further, the increased adhesion may act to further increase safety margins. For normally consolidated or lightly overconsolidated clays, heating causes a small reduction in friction angle if thermal consolidation occurs. Brief tests would be required to predict if the yield limit would be exceeded on a case by case basis.

A number of additional relationships and potential dependencies have been identified:

- Ultimate shear strength is independent of thermal loading conditions, unless the thermal yield limit is exceeded.
- Thermal consolidation at an interface may result in a delayed dilation response.
- The magnitude of dilation remains dependent on stress history following thermal consolidation.

REFERENCES

Abuel-Naga, H., Raouf, M., Raouf, A. & Nasser, A. 2015. Energy piles: current state of knowledge and design challenges. *Environmental Geotechnics*, 2(4), 195–210.

Baldi, G., Hueckel, T. & Pellegrini, R. 1988. Thermal volume changes of the mineral-water system in low-porosity clay soils. *Canadian Geotechnical Journal* 25(4), 807–825.

Burghignoli, A., Desideri, A. & Miliziano, S. 2000. A laboratory study on the thermomechanical behaviour of clayey soils. *Canadian Geotechnical Journal*, 37(4), 764–780.

Cekerevac, C. 2003. *Thermal effects on the mechanical behaviour of saturated clays: An experimental and constitutive study*. Swiss Federal Institute of Technology.

Cekerevac, C. & Laloui, L. 2004. Experimental study of thermal effects on the mechanical behaviour of a clay. *International Journal for Numerical and Analytical Methods in Geomechanics*, 28(3), 209–228.

Chen, X., Zhang, J., Xiao, Y. & Li, J. 2015. Effect of roughness on shear behavior of red clay – concrete interface in large-scale direct shear tests. *Canadian Geotechnical Journal*, 52(8), 1122–1135.

Cui, Y.-J., Le, T.T., Tang, A.M., Delage, P. & Li, X.L. 2009. Investigating the time-dependent behaviour of Boom clay under thermo-mechanical loading. *Géotechnique*, 59(4), 319–329.

DeJong, J.T. & Westgate, Z.J. 2009. Role of Initial State, Material Properties, and Confinement Condition on Local and Global Soil-Structure Interface Behavior. *Journal of Geotechnical and Geoenvironmental Engineering*, 135(11), 1646–1660.

Di Donna, A. 2014. *Thermo-mechanical aspects of energy piles*.

Di Donna, A. Di, Ferrari, A. & Laloui, L. 2016. Experimental investigations of the soil–concrete interface: physical mechanisms, cyclic mobilization and behaviour at different temperatures. *Canadian Geotechnical Journal*, 53(4), 659–672.

Feligha, M. & Hammoud, F. 2015. Experimental Investigation of Frictional Behavior Between Cohesive Soils and Solid Materials Using Direct Shear Apparatus. *Geotechnical and Geological Engineering*, 34(2), 567–578.

GSHP Association 2012. Thermal pile design, installation and materials standards. *Ground Source Heat Pump Association, Milton Keynes*, (1), 82.

Hamidi, A., Tourchi, S. & Khazaei, C. 2015. Thermomechanical Constitutive Model for Saturated Clays Based on Critical State Theory. *International Journal of Geomechanics*, 15(1).

Hueckel, T. & Baldi, G. 1990. Thermoplasticity of Saturated Clays: Experimental Constitutive Study. *Journal of Geotechnical Engineering*, 116(12), 1778–1796.

Hueckel, T., Francois, B. & Laloui, L. 2011. Temperature-dependent internal friction of clay in a cylindrical heat source problem. *Géotechnique*, 61(10), 831–844.

Kuntiwattanakul, P., Towhata, I., Ohishi, K. & Seko, I. 1995. Temperature Effects on undrained characteristics of clay. *Soils and Foundation*, 35(1), 147–162.

Laloui, L. 2001. Thermo-mechanical behaviour of soils. *Environmental Geomechanics*, 5(6), 809–843.

Littleton, I. 1976. An experimental study of the adhesion between clay and steel. *Journal of Terramechanics*, 13(3), 141–152.

Ooi, L.H. & Carter, J.P. 1987. A constant normal sltiffness direct shear device for static and cyclic loading. *Geotechnical Testing Journal*, 10(1), 3–12.

Plum, R.L. & Esrig, M.I. 1969. Some Temperature Effects on Soil Compressibility And Pore Water Pressure, *Highway Research Board Special Report*, 103, 231–242.

Potyondy, J.G. 1961. Skin Friction between Various Soils and Construction Materials. *Géotechnique*, 11(4), 339–353.

Skempton, A.W. 1970. First-time slides in over-consolidated clays. *Géotechnique*, 20(3), 320–324.

Sultan, N., Delage, P. & Cui, Y.J. 2002. Temperature effects on the volume change behaviour of Boom clay. *Engineering Geology*, 64(2–3), 135–145.

Towhata, I., Kuntiwattanaku, P., Seko, I. & Ohishi, K. 1993. Volume change of clays induced by heating as observed in consolidation tests. *Soils and Foundations*, 33(4), 170–183.

Vaughan, P.R. & Lemos, L.J.L. 2000. Clay–interface shear resistance. *Géotechnique*, 50(1), 55–64.

Yavari, N. 2014. *Aspects géotechniques des pieux de fondation énergétiques*. Universite Paris-Est.

Yavari, N., Tang, A.M., Pereira, J.M. & Hassen, G. 2016. Effect of temperature on the shear strength of soils and soil/structure interface. *Canadian Geotechnical Journal*, 53(7), 1186–1194.

Development of a 3D clay printer for the preparation of heterogeneous models

L.M. Pua, B. Caicedo, D. Castillo & S. Caro
Universidad de los Andes, Bogotá D.C, Colombia

ABSTRACT: The spatial distribution of soil influences the behavior of many geotechnical structures. The effect of spatial soil uncertainty has been modeled mainly by numerical approaches and some simplified experimental models. This article presents the development of a 3D clay printer that allows the construction of physical heterogeneous models. The 3D printer works with eight types of reconstituted clay soils, prepared with mixtures of kaolin and bentonite. To calibrate soils in terms of extrusion, indirect tests were carried out to determine the optimal water content at which soils have similar behavior. Although, the equipment can work with eight types of soil, the first prototype was made with four soils (from 0 to 33% of bentonite), due to the increase of viscosity of thixotropic clays. The heterogeneous layer shows that this new equipment can reproduce spatially random fields, and therefore it is a new tool that will allow the study of the effect of 3D variability in reduced scale models.

1 INTRODUCTION

Soils are natural materials formed from different geological processes; this results in a highly variable material. The spatial variability in soil behaviour leads to significant uncertainties that have influence in many geotechnical problems. Due to the challenges that soil variability represents in design, analysis and behaviour of geotechnical structures, various authors have studied this problem with different approaches.

The following studies are examples of geotechnical problems where soil variability can be an important variable that should be analysed in the design. Griffiths & Fenton (2001) studied the effect of spatial variability in the bearing capacity of undrained clay using numerical methods. They concluded, that bearing capacity decreases as the value of the coefficient of variation increases. Ni et al. (2016) analysed the influence of heterogeneity in slope stability using two types of clay soil. Their results show that the failure mechanism for heterogeneous soil differs considerably from the one obtained if one assumes homogeneous slope conditions.

Another example is the construction of tunnels in highly heterogeneous soils. However, methods for tunnel design commonly assume homogeneous soil conditions at tunnel face. Hu et al. (2012) developed a calculation method that allows two different soils at the face of the tunnel (clay and sand). Nevertheless, this approach does not consider spatial variability in the mechanical properties of a single type of soil.

Moreover, Chakrabortty et al. (2010) evaluated the effect of heterogeneity in the resistance to liquefaction for sand deposits. Their results suggest that excess pore water pressure generated during earthquakes in heterogeneous deposits is higher than in a homogeneous deposit, due to water migration from loose to dense zones. Another application is the analysis of soil-structure interaction. Spatial variability of soils is a big concern due to the generation of differential settlements, which is one of the main causes of structural damage, Breysee et al. (2005).

In summary, soil spatial variability has been widely analysed since it presents big challenges in geotechnical engineering. Nevertheless, the effect of soil variability has only been modelled by numerical approaches. Few efforts have been made to verify these results in the laboratory.

The first attempt to construct physical models with controlled variability based on 2D random field was by Garzón et al. (2015). Garzón et al. (2015), modelled soil spatial variability analytically using a discrete random field generation method, obtaining for each realization a probable distribution of different types of soil in the space. Afterward, they constructed physical models to match the realization. Models were handmade, prepared using nine types of reconstituted clay soils. After the preparation, models were subject to consolidation and the undrain strength was measured. Their results showed, that mechanical properties of soils depend not only on their intrinsic behaviour, but also on their spatial distribution.

Although previous work provides a method to construct heterogeneous models, it is limited to a 2D variable model since the models are hand-made. This article presents the development of a 3D clay printer that allows the construction of heterogeneous models with variability in all three dimensions. This new equipment facilitates experimental analysis of the effect of spatial variability of soils.

The article is organized as follows: Section 2 presents a description of the three-dimensional random field generation method. The experimental set-up, including the 3D printer and soil storage tub is presented in Section 3. Section 4 presents the materials and methods used, including a description of clay soils and the process of calibration through an indirect extrusion test. The first prototype of a printed clay layer accomplished is presented in Section 5. Finally, some conclusions and future work in the development of the equipment are present in Section 6.

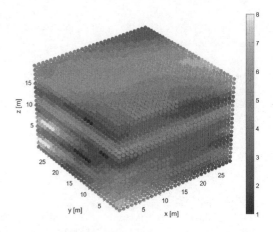

Figure 1. Schematic volume of heterogeneous soil generated with Discrete Fourier Transform

2 RANDOM FIELDS GENERATION BY DISCRETE FOURIER TRANSFORM

To represent soil variability in a natural deposit it is necessary to establish a probabilistic model. It has been widely accepted in engineering problems that random field theory is suitable to evaluate the variation of a parameter in space Vanmarcke (1982). A random field is a stochastic process indexed by a spatial variable Vanmarcke (1983); this is, an n-dimensional vector of random values that are correlated in space. It can be understood as a scalar function $X(\mathbf{t})$, where t is a vector of spatial coordinates.

Several numerical methods exist to create two- and three-dimensional random fields. Commonly used methods are: The Covariance Decomposition technique, the Turning Bands Method, the Fast and Discrete Fourier Transform and the Local Average subdivision, Fenton & Griffiths (2008). For this study the Discrete Fourier Transform (DFT) is used to generate the pattern that will be reproduced in the 3D clay printer. The DFT gives a field composed of random values that follow a normal-standard distribution.

For the Discrete Fourier Transform (DFT) method, the value of the random field at a given location, $X(\mathbf{t})$, is constructed as the summation of harmonic waves. These sinusoidal/cosine waves have random coefficients that control their amplitude, and are calculated at a set of frequencies with relative weight, determined by a Spectral Density Function (SPD). The scalar random field at each point can be computed as follows:

$$X(t_i, t_j, t_k) = \sum_m^M \sum_n^N \sum_p^P \begin{bmatrix} A_{m,n,p} \cos(\omega_m t_i + \omega_n t_j + \omega_p t_k) + \\ B_{m,n,p} \sin(\omega_m t_i + \omega_n t_j + \omega_p t_k) \end{bmatrix}$$
(1)

where $A_{m,n,p}$ and $B_{m,n,p}$ are the coefficients that contain the randomness of the field. These coefficients are calculated by using a combination of random values, the SPD, and the values of frequency ω, that are represented in the field (varying from a minimum to a maximum frequency, for each direction t). The same range and number of frequencies were used in every dimension, which means that for this study, $M = N = P$.

For the example of this paper, the generated field has 20 elements high and 30 elements wide. The correlation lengths of the soil field (i.e. the distance where correlation still exists between two elements along a given dimension) were defined as 10 units in the horizontal direction (x and y) and 3 units in the vertical direction. This parameter is essential to model the spatial variability of soil deposit Vanmarcke (1977). Although it is difficult to accurately measure these values in nature, the principle holds that as most soils are deposited in layers over great periods of time, a higher level of homogeneity is expected in the horizontal direction, while higher variability may appear when moving across the vertical direction.

The random field obtained is a set of continuum values, which are divided into eight intervals of equal size between the maximum and minimum values of the field. A transformation was applied to the random normal values, to create a field that contains only 8 values associated with soil types. The random fields obtain with the DTF method is show in Figure 1. The aim of this research is to print this pattern layer by layer with the 3D printer, to obtain a physically heterogeneous model that represents a random field.

3 EXPERIMENTAL SET-UP

3.1 Description of the 3D clay printer

3D printers work with additive manufacture, a process where an object is created by adding layers of material. The standard material that has been used with these machines is plastic. The equipment developed in the laboratory of geotechnical modelling of the University of los Andes, is a 3D printer that allows clay printing. Mainly two parts compose the system: the 3D printer and soil storage tubes.

3D printer is composed of a steel frame, three axes that allow movement in the x, y and z direction, a base plate to support the model, one multiplexing valve with an input for eight clay soils and an output for clay

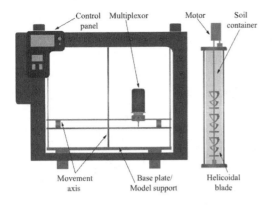

Figure 2. Experimental setup: 3D clay printer scheme at the left and soil storage tub scheme to the right.

extrusion. It has two control panels; one LCD screen to control the printer and a screen to control the multiplexor output. Figure 2 shows a scheme of the 3D printer. The equipment has the capacity to generate cubic models of 30 cm high in an area of 30×30 cm.

3.2 Soil storage tubs

The difference between standard 3D printers and this new equipment is the material used to build the model. Common printers work with plastic at high temperature, which is a stable material compared with clays. Therefore, soil storage tubs had to be designed to assure constant behavior of clays during the printing process.

Due to the high dependency of water content in clay behavior, it is necessary to avoid clay drying to assure continuous flow during printing process. Therefore, soil storage tubs are made with cylindrical acrylic walls sealed with two aluminum plates that are placed at the top and bottom of the storage tube.

Inside the storage tubes was a helicoidal blade that is connected to a motor that imposed a slow movement at constant speed, Figure 2. The heicoidal blade has two functions; firstly, is to remove voids within the soil mass, and secondly to prevent clay drying.

Connection between the soil storage tubs and the printer multiplexor was made with a hose of 6 mm diameter; the hose was placed at the bottom plate of the storage tub. To assure the clay flow, it was necessary to pressurize the soil storage tubs with a pressure valve installed at the top plate. To avoid pressure drops during the printing process, valves were connected to a pressure accumulator. The experimental set-up including the 3D printer and soil storage tubs is shown in Figure 3.

4 MATERIALS AND METHODS

4.1 Homogeneous soils

The 3D printer works with eight types of homogeneous clay soils prepared in the laboratory. The artificial soils

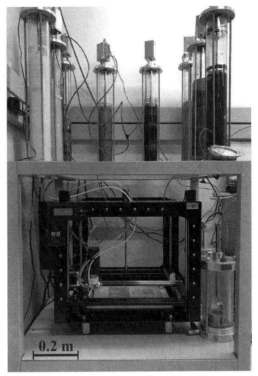

Figure 3. 3D clay printer equipment for the development of heterogeneous models.

Table 1. Definition of homogenous soils used in the 3D printer.

Soil	Kaolin	Bentonite	Colourant	Liquid limit (w_L)
	%	%	%	%
S1	100	0	0	87
S2	80	11	9	86
S3	68	23	9	121
S4	55	33	12	157
S5	44	44	12	189
S6	31	52	17	204
S7	21	63	16	241
S8	10	73	17	300

* Type of soils mineral colorant: a yellow, b combination of 2% and 7% blue, c combination of 4% red and 8% blue, d green, e red, f blue and g black. Garzón et al. (2015).

used are mixtures of kaolin, sodium bentonite and mineral colorants used for differentiation of soils. Relative proportions of soil preparation are shown in the Table 1. Soils used in this study were defined according to the characterization made by Garzón et al. (2015).

Garzón et al. (2015) constructed hand-made heterogeneous models using clay slurry prepared at a water content of 1.5 times the liquid limit. However, this is not a viable solution for the 3D printer to prepare the samples at this water content, because the flow pressure for each soil will be different.

The calibration of soils for the 3D printer was made by an indirect extrusion test. This test is designed to measure fine grained soil consistency limits Verástegui-Flores & Emidio (2014). However, the characteristic of this measurement allows us to evaluate the water content where soils have the same behavior in terms of extrusion pressure through a specific system.

4.2 Indirect extrusion test

Alternative methods to measure consistency limits in fine grain soils have been proposed due to the low reliability of commonly used tests. Verástegui-Flores & Emidio (2014) evaluated indirect extrusion tests as an alternative method to determine the liquid and plastic limits of fine grain soils.

Extrusion is a mechanical process where material is forced to flow through an orifice. The extrusion test determined the steady state extrusion pressure (PE) of a soil sample; this value is a function of water content of soil and geometry of the device. The test can be done directly if the direction of flow is the same as the direction of ram movement, or indirectly if the die orifice moves against the soil sample. The indirect extrusion test is preferred due to the low mobilization of frictional forces during the measurement Verástegui-Flores & Emidio (2014).

The curve of steady state extrusion pressure as function of water content is distinctive for a particular type of soil. Therefore, indirect extrusion tests are performed to evaluate the water content at which soils behave similar in terms of extrusion.

The indirect extrusion test set-up consists of a cylindrical steel container of 40 mm diameter where a soil sample of 10 mm high is located. The indirect extrusion device has an orifice of 5 mm diameter; where the soil will flow through it, see Figure 4. The indirect extrusion device moves against the soil sample at a constant rate of 4 mm/min that is imposed by a Digital Tritest 50 Load Frame machine. The extrusion load is measured with a load cell of 500 N capacity.

Indirect extrusion tests were performed on the eight homogeneous soils mentioned above. Soils were prepared at three water contents (w) higher than the liquid limit. The value of PE is defined from the curve of extrusion pressure as a function of displacement, where the extrusion pressure is constant (in a range of displacement of 4 to 7 mm). The behaviour obtained in the indirect extrusion test is show in the Figure 5, where results for the soil 2 are presented, similar trend is observed for the eight types of soil used in this study.

The value of steady state extrusion pressure at different water contents for each soil calibration curves are obtained. The curve of water content as a function of PE has a linear trend in a bi-logarithmic chart. Results are shown Figure 6. Finally, the water content appropriate to prepare soil for clay printing was determined as the water content associated with a value

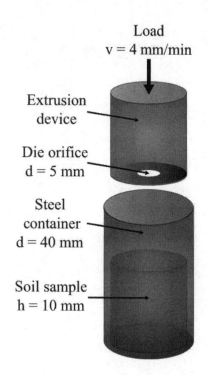

Figure 4. Experimental set-up of indirect extrusion test.

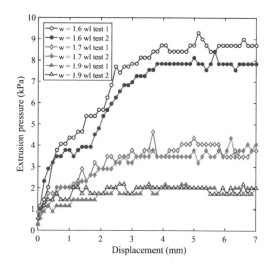

Figure 5. Extrusion curve of soil 2 (kaolin 80% and bentonite 11%).

of steady state extrusion pressure of 6kPa, because at this pressure the consistency of soils was appropriate to build the model. Results are shown in Table 2.

Finally, soils were prepared at the water content defined with the indirect extrusion test, and then were placed in soil storage tubs. To prevent soil drying during the process a continuous movement with the helicoidal blade was imposed.

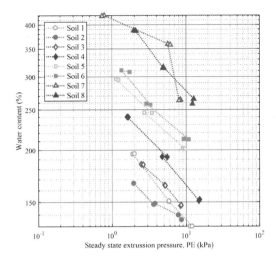

Figure 6. Extrusion pressure as a function of water content, the curve is presented for the eight homogeneous soils used in the 3D printer.

Table 2. Optimum water content for soil preparation determined through indirect extrusion test.

Soil	w %	w/w_L
1	153	1.8
2	142	1.7
3	159	1.3
4	188	1.2
5	220	1.2
6	235	1.2
7	309	1.3
8	316	1.1

*The proportion of the water content from the liquid limit (w/w_L) was calculated using the liquid limit reported Table 1.

5 RESULTS

The extrusion pressure depends on the geometry of the extrusion device; thus, it was necessary to determine the pressure in the printer system. Therefore, the extrusion pressure in soil storage tubs valves and infill speed (VI) were determined by printing squares of one homogeneous soil. The criterion used to define the optimal value for these variables was that the printed soil does not have voids in their structure. Results of soils 1 to 4 are presented in the Figure 7 and the printing properties are summarized in the Table 3.

Although the water content was determined with the indirect extrusion test to have the same extrusion pressure, results show that in the printer system the pressure is not the same. Still the difference does not inhibit the printer process. Therefore, one valve was used to regulate the pressure of soils 1 and 2, and a different one was used for soils 3 and 4.

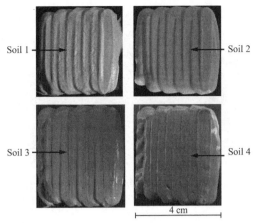

Figure 7. Individual printed soils.

Table 3. Printing properties of homogeneous soils from 1 to 4.

Soil	w %	P_E kPa	V_I mm/s
1	149	60	3
2	146	50	3
3	162	100	2
4	197	140	2

Even though the 3D clay printer can work with eight types of soils, just four types were printed. The reason is that soils have variable proportions of bentonite which is a thixotropic clay. Thixotropy is a property that exhibits non-Newtonian suspensions, where the viscosity decreases as the rate of shear strain increase. Viscous behaviour is generated due to particle redistribution as the shear speed increases. However, when the clay is at rest the initial structure is restored with time, increasing the viscosity Coussot (2014).

Thixotropy makes the clay printing process difficult; because when soils are at rest (or in a slow movement) viscosity increase significantly, and soils do not flow properly through the printer system. Despite soil storage tubs have a helicoidal blade that induces shear strain in the soil, their speed is considerably slower to reduce the viscosity of soils. For bentonite contents higher than 44% (soil 5) the effects of thixotropy became important, and preclude the printing process. Thereby, only soils 1 to 4 could be printed with the experimental set up proposed above. This phenomenon also explains the difference in extrusion pressure and speed infill obtained between soils with lower and higher content of bentonite (see Table 3).

With the printer parameters established for the soils 1 to 4, we proceeded to print a pattern based on layers of the 3D random field. Results are show in Figure 8. The dimensions of the layer printed are 20x20x1 cm, and printing time was 1 hour. Printed clay layers do

Figure 8. Printed pattern of heterogeneous soils.

not have voids within the homogenous soils. However, at the junction of different soils some voids are observed. Larger voids are present with the increase of bentonite content (in soils 3 and 4), because their increase in viscosity due to thixotropic effects. Heterogeneous models made by Garzón et al. (2015) shows that subjecting the model to vacuum can reduce these voids. The reduction of voids is crucial to evaluate the behaviour of the model, and to compare it with numerical models.

Results show that the 3D clay printer is new equipment that will allow the construction of physical models of heterogeneous soils variable in three dimensions. With this equipment physical modelling of soil uncertainty will be less time consuming, since the construction of the model is automated.

This new tool will give the possibility to validate and calibrate numerical models made to analyse the effect of spatial uncertainty in soils in a more efficient way.

6 CONCLUSIONS

Soil spatial uncertainty represents a challenge in civil engineering, since it can affect the behavior of geotechnical structures. Yet, it has modeled mainly by numerical approaches and some simplified experimental models. Within this context, this paper shows the development of a 3D clay printer. This equipment will allow the construction of heterogeneous models variable in three dimensions using eight types of clay soils.

The process of soil calibration though indirect extrusion testing allows the estimation of the optimal water content for the printing process. Results show that as bentonite content increases, the optimal water content for printing increases. Increase in bentonite content also made thixotropy effects more prominent. This property of non-Newtonian fluids makes the clay extrusion difficult, since the viscosity increases significantly at low shear rates. For this reason, the first prototype of heterogeneous models was made only with four types of clay soil, although the printer has the capacity to work with eight soils.

For future work, to construct models with eight clay soils, the speed of the helicoldial blade in the soil containers will be higher to reduce the soil viscosity. Additionally, to reduce the voids at the clay interface, models will be subject to vacuum during the printing process. With these improvements, the aim is to construct a multilayer heterogeneous model to evaluate its behavior in the centrifuge.

Despite remaining work with the equipment, the first prototype achieved show the potential of the 3D clay printer, as a new tool to construct heterogeneous models variable in three dimensions. With this equipment, the process of model construction will be less time consuming, and will allow the validation and calibration of numerical models.

REFERENCES

Breysse, D., Niandou, H., Elachachi, S., & Houy, L. 2005. A generic approach to soil–structure interaction considering the effects of soil heterogeneity. *Géotechnique* **55(2)**: 143–150.

Chakrabortty, P., Popescu, R., & Phillips, R. 2011. Liquefaction Study of Heterogeneous Sand: Centrifuge. *Geotechnical Testing Journal* **34(3)**: 227–237.

Coussot, P. 2014. *Rheophysics*. Switzerland: Springer Interna-tional Publishing.

Fenton, G. A & Griffiths, D. V. 2008. *Risk Assessment in Geotechnical Engineering*. New York: Wiley.

Garzón, L.X., Caicedo, B., Sánchez-Silva, M., & Phoon, K.K. 2015. Physical modelling of soil uncertainty. *International Journal of Physical Modelling in Geotechnics* **15(1)**: 19–34.

Griffiths, D. V., & Fenton, G. A. 2001. Bearing capacity of spatially random soil: the undrained clay Prandtl problem revisited. *Géotechnique* **51(4)**: 351–359.

Hu, X., Zhang, Z., & Kieffer, S. 2012. A real-life stability model for a large shield-driven tunnel in heterogeneous soft soils. *Frontiers of Structural and Civil Engineering* **6(2)**: 176–187.

Ni, P., Wang, S., Zhang, S., & Mei L. 2016. Response of heterogeneous slopes to increased surcharge load. *Computers and Geotechnics* **78**: 99–109.

Vanmarcke E. 1977. Probabilistic modeling of soil profiles. *Journal Geotechnical Engineering Division, ASCE* **103(11)**: 1227–1246.

Vanmarcke E. 1982. Developments in Random Field Modeling. *Nuclear Engineering and Design* **71**: 325–327.

Vanmarcke E. 1983. *Random Fields: Analysis and Synthesis*. Cambridge: The MIT Press.

Verástegui-Flores, R.D., & Di Emidio, G. 2014. Assessment of clay consistency through conventional methods and indirect extrusion tests. *Applied Clay Science* **101**: 632–636.

2. Engineered platforms

Centrifuge modelling utility pipe behaviour subject to vehicular loading

S.M. Bayton, T. Elmrom & J.A. Black
Centre for Energy and Infrastructure Ground Research (CEIGR), University of Sheffield, Sheffield, UK

ABSTRACT: Centrifuge model tests of buried flexible pipes in dry sand subjected to surface traffic loads are presented. Model pipe tests were performed at 25 gravities (25g) of a prototype pipe 355 mm in diameter. Pipe behaviour was observed for different burial depths of 0.5, 0.75, 1.0 and 1.5 m; and load eccentricities of 1 and 2 pipe diameters. Results show that pipes buried at shallower depths are subjected to significantly greater bending moments and corresponding shear stresses. When the load is applied at an eccentricity, if the pipe remains within the zone of stress influence, a comparable magnitude of maximum moment is anticipated. An initial series of 20 cycles were also carried out. Results indicate an amount of 'locked-in' bending moment upon unload; a phenomenon more evident for the shallower buried pipe.

1 INTRODUCTION

The integrity and performance of buried pipe infrastructure plays a significant role in modern society; providing the safe distribution of potable water, transportation of sewage and connection of communication services, amongst other applications. For this purpose, a diverse range of buried pipes are deployed such as large diameter concrete sewer pipes to small diameter plastic fresh water pipes. Municipal water distribution and wastewater infrastructure systems are of national importance and the aspect of water leakage, coupled with water scarcity, has become a serious problem in many countries. In England and Wales alone, over 3 billion litres of water is lost through leakage every day (DiscoverWater 2017), representing up to 30 per cent loss of the potable water from source to the consumers tap.

Typically, utility pipes are located beneath the surface of road networks to simplify installation, distribution and maintenance. Owing to the shallow burial depth of these critical infrastructure assets they are highly vulnerable from a number of factors that may influence their performance and deterioration; for example, the impact of heavy surface traffic loading and changeable burial conditions. A summary of the complexities surrounding buried utility pipes and the challenges faced by the water industry is illustrated in Figure 1.

Boussinesq (1883) developed a solution for the stress at any point in a homogeneous, elastic, and isotropic medium, caused by a point load applied on the surface of an infinitely large half-space. Based on these Boussinesq equations, Young & O'Reilly (1983) established a method for estimating the

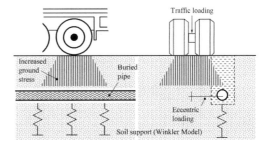

Figure 1. Buried pipe infrastructure and its challenges.

vertical stress on an underground pipe caused by traffic wheel loading. Various charts based on the Boussinesq method were developed, which are still in use in BS EN 1295-1 (BSI 1998). Other researchers, like Marston & Anderson (1913) and Burns & Richard (1964), introduced their theories about influence of loads on underground pipelines. Pocock et al. (1980) measured the bending strain developed in a shallow buried pipeline comprising eight cast iron pipes; while Taylor & Lawrence (1985) noticed that the response of a cast iron pipeline to heavy vehicles depended on structure of the pavement, backfill height, and pipe bedding.

Given the high leakages, combined with increased traffic loading, there is a high motivation to understand the behavior of buried shallow pipe networks to increase future resilience. At present however, there exists relatively little available literature with experimental results, in particular relating to these pressing challenges surrounding pipe-soil interaction. This is despite considerable uncertainties remaining.

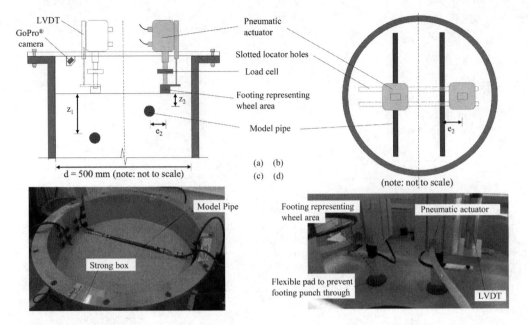

Figure 2. Test setup: schematic (a) elevation and; (b) plan view; (c) photo of pipe placement and; (d) actuator and LVDT setup.

This paper presents the results from a suite of small-scale buried pipe experiments performed in a geotechnical centrifuge at elevated gravity to simulate prototype stresses. The aim of the current research is to investigate the effect of pipe burial depth and load eccentricity on the behaviour of flexible HDPE pipes in sand with the view of providing an insight of the role of these external factors in the pipe-soil interaction.

2 CENTRIFUGE EXPERIMENTAL SETUP

A series of centrifuge model tests, at a centrifugal acceleration of 25 times Earth's gravity (25g) was conducted using the University of Sheffield's 4 m diameter 50g-tonne geotechnical beam centrifuge (Black et al., 2014). The centrifuge allows for the replication of prototype stress conditions to adequately capture the small-strain stiffness and dilation responses associated with the pipe-soil interaction problem.

A medium-fine grain sand commercially known as CH30 sand (see Table 1) was dry pluviated into a cylindrical strong box to an average relative density, R_d, over the test matrix of 81.0%. The model chamber had internal diameter and height of both 500 mm and provided a rigid boundary. The pluviation continued until reaching the desired level to the base of the buried pipe (i.e. burial depth). See Figure 2(a) for an illustrative diagram of the experimental setup; noting Figure 2(a) and (b) are not to scale.

The two acetal pipes were placed in parallel at a distance of 100 mm apart at the required burial depth. Further sand was pluviated on top until the desired cover above the respective pipes was achieved.

Table 1. CH30 sand properties.

Property	Value
Particle size, d_{10}	0.355 mm
d_{50}	0.450 mm
Specific gravity, G_s	2.67
Maximum void ratio, e_{max}	0.756
Minimum void ratio, e_{min}	0.508
Peak angle of shear, φ'_{peak}	37° (at $R_d = 80\%$)

Table 2. Flexible HDPE pipe properties.

Description	Prototype Dimension	Model Dimension
Gravity (g)	1	25
Material	HDPE	Acetal
Young's Modulus (E)	1.5 GPa	2.6 GPa
Diameter (D)	355 mm	14.2 mm
Thickness (t)	33.5 mm	2.1 mm
Flexural Stiffness (EI)	661 Nm²	3.91×10^{-3} Nm²
Scaling law of EI	1	$1/25^4$
Ratio of $EI_{model/proto}$	1	2.3

The model pipe material was selected to replicate prototype HDPE and can be considered to be flexible. Pipe and scaling properties are presented in Table 2. Using the bending stiffness scaling law of $1:N^4$, the model pipe's scaled bending stiffness provides an adequate representation of the prototype given that it is in the same order of magnitude and flexible behaviour should still be captured.

Eight pairs of strain gauges (KFG type) in half bridge configurations (four in the major axis from the mid-span along one half of the pipe, and four from the mid-span in the minor axis along the other half) were positioned on the pipe length at 30 mm intervals. This allowed the bending response captured to be mirrored at the mid-span given the load application was at this point. Bending moment calibration of the strain gauges was performed on each 90 degree orientation. The proportion of out-of-plane bending was recorded to be a maximum of just 7% of the respective in-plane bending. The pipe, and respective strain gauges, was then protected with a thin covering of electrical tape (with near zero stiffness ensuring no contribution to the stiffness of the pipe). Figure 2(c) presents the arrangement of the model pipes at different burial depths.

Additional scaling aspects considered were the sand-pipe particle interaction. In the present study, the pipe circumference, C, was 44.5 mm yielding a $C/d_{50} \approx 100$ and $D/d_{50} \approx 31$; conforming to the particle contact relationship outlined by Bolton et al (1999).

A bespoke loading frame was assembled from machined aluminium plate with specific locator slotted holes for the positioning of load actuation systems at any desired eccentricity across the width of the pipe (Figure 2(b)). The load application was achieved by the use of a 50 mm diameter dual acting pneumatic cylinder actuator, with a maximum load capacity of 1.3 kN. A positive pressure was supplied to the reverse action side during spin-up to prevent the piston from contacting the surface under increased self-weight. The opposing pressure line was then increased to apply load to the surface. A load cell was calibrated over the loading range and attached between the actuator and the model aluminium footing representing the lorry wheel load. The footing had dimensions of 10 mm × 20 mm to replicate a wheel contact area of 0.25 m × 0.5 m at prototype scale. The settlement of the footing was measured using one LVDT located directly on the footing. A flexible rubber mat was positioned on the sand surface to prevent the aluminium footing from punching directly through the sand since there is no restraint on the sand surface. Note, future tests have utilised sculpting plaster to model the contact road bending stiffness, which also plays an important role in dissipating the applied surface loading. Hence, it is likely that the observed load-displacement and bending moments reported herein reflect upper bound interaction levels.

3 TEST MATRIX

Six centrifuge tests were performed to study the effects of burial depth and load eccentricity on the pipe-soil behaviour. After the initial monotonic load, a further 20 load cycles were applied to observe the effects of repetitive traffic load. Table 3 presents the full test matrix.

Table 3. Experimental test matrix [prototype dimension in square brackets].

Test (#)	R_d (%)	Burial depth (z)		Load eccentricity (e)	
		(mm)	[m]	(mm)	[m]
1	80.1	60	[1.5]	0	[0]
2	79.6	40	[1.0]	0	[0]
3	84.8	30	[0.75]	0	[0]
4	81.9	20	[0.5]	0	[0]
5	80.1	40	[1.0]	14.2	[0.355]
6	79.6	40	[1.0]	28.4	[0.710]

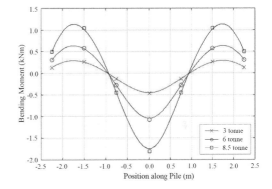

Figure 3. Evolution of bending moment with load for pipe embedded at $z = 1.0$ m.

The magnitude of the model load was selected to represent one half of a wheel axle of a 44 tonne articulated lorry at prototype scale (this equates to the largest permissible load on UK roads). This equals to a vertical load of 6 tonnes per wheel base. This is in accordance with monotonic and cyclic loading recommendations in AASHTO (2010).

4 RESULTS & DISCUSSION

Load was applied in stages and footing displacement and pipe bending moment were measured. Figure 3 presents the evolution of moment with load for the pipe buried at 1.0 m depth. All measurements are expressed in prototype dimensions. As expected, the greater the load on the footing, the greater the moment experienced within the pipe. It can also be seen that the point of zero moment remains in the same location throughout the load increments.

4.1 Effect of burial depth

For a footing loaded along the centreline and at mid-span of the pipe, settlement behaviour was observed for the range of pipe burial depths (sand cover of 0.5, 0.75 and 1.0 and 1.5 m) (Figure 4). It can clearly be

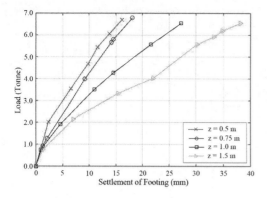

Figure 4. Variation in footing settlement with pipe burial depth.

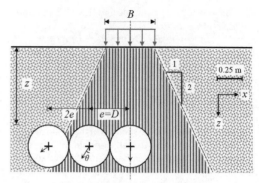

Figure 6. Schematic illustration of pipe behaviour under eccentric loading.

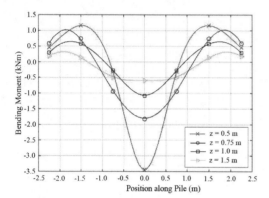

Figure 5. Variation in bending moment with pipe burial depth at applied load of 6 tonnes.

seen that a shallower pipe provides a stiffer footing settlement response signifying that the pipe is providing additional resistance to the increase in stress within the soil. Additionally, the initial small-strain stiffness appears to be very similar for each burial case suggesting that for low loads (up to 1 tonne) the soil close to the surface initially resists the additional stress and each test demonstrates the same behaviour in this part. Once the propagation of stress reaches the deeper soil, additional stiffening from the pipe comes into effect. During unloading, settlements are not fully recovered, most likely due to compaction and redistribution of sand grains.

Figure 5 presents the variation of bending moment with burial depth for these same surface loading of 6 tonnes. The moment is greatest for shallower burial depths. It can be seen that for the shallowest burial depth ($z = 0.5$ m), the bending moment peaks to a significant, sharp point directly below the load application. This is in contrast to the deeply buried pipe ($z = 1.5$ m), where the applied stress in the soil has radiated with depth and therefore when it reaches the deeper pipe, the area of influence is much greater, and the stress is much less. It can be seen that the bending response is much smoother, and there is a portion of almost equal bending across the middle 2 m section as the stress is more uniformly distributed in this region. The length of this section coincides very well with theoretical approximations of a 2V:1H stress distribution with depth.

A key pipe design parameter is to minimise the very large changes in bending over short distances as well as abrupt points of inflexion as these have large shear stresses associated with them. As shown, the selection of a greater burial depth facilitates this reduction in pipe stress and could aid with the longevity of pipe lifespan.

4.2 Effect of eccentricity

In addition to varying the burial depth of the pipe, the eccentricity of the applied footing load was increased to replicate variations in surface load conditions. Eccentricities of one and two pipe diameters were investigated, with loads again applied at the mid-span of the pipe. It is expected that due to the eccentric load, the pipe is subjected to additional lateral soil pressure and therefore experiences horizontal bending and displacement as well as vertical.

Figure 6 illustrates the test dimensions of the eccentrically loaded pipes. Overlaid on this figure is the dissipation of stress within the soil depth taken to be the classical 2V:1H zone of influence. As can be seen, the pipe at $e = 1D$ is well within the bounds of this zone. The pipe at $e = 2D$ skirts the edge of this zone. Table 4 presents the observations from these tests in terms of pipe major and minor axis and resultant bending moments as well as the corresponding resultant vector direction.

It can be seen that there is a clear distinction in pipe behaviour depending on whether the pipe is within or beyond the bounds of the zone of stress influence. For example, for the two pipes fully within this zone, the magnitude of the resultant bending moment is comparable, the difference being that for the

Table 4. Pipe bending magnitude and direction due to eccentric load.

Test (#)	Max Major Mom. (Nm)	Max Minor Mom. (Nm)	Resultant Mom. (Nm)	θ (°)
2	1080	3	1080	0.1
5	754	717	1040	43.5
6	256	317	407	51.0

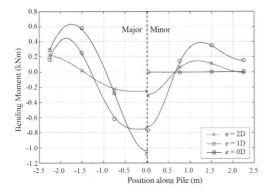

Figure 7. Variation in bending moment with load eccentricity at a burial depth of $z = 1.0$ m and applied load of 6 tonnes (N.B. D = external diameter of pipe).

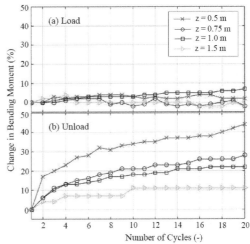

Figure 8. Evolution of (a) maximum (on load) and; (b) residual (on unload) bending moments in comparison to first cycle magnitude with number of cycles at 6 tonne load per cycle.

pipe eccentrically loaded at $e = $ 1D, the major bending moment has reduced and the minor moment has increased. This would suggest the same increase in resultant stress acting on the pipe here, but with different magnitudes of the lateral and vertical components. When the pipe is located right on the outer bound of the zone of stress influence there is a significant reduction in resultant bending moment, with both the major and minor moments reducing in parallel. The values do not reduce to zero, and the lateral stress is greater than the vertical at this location. Similar observations of bending moment reductions for eccentric loads have been made by Hosseini & Tafreshi (2010). The bending moment diagrams for each are presented in Figure 7.

Although loadings in the current study are statically applied, the observations of maximum and minimum bending moment occurring in the major and minor planes reflect the transitional stress states imposed on the buried pipe in serviceable conditions under transient loading associated with moving vehicles. Additional testing in this regard would prove highly valuable to understanding pipe longevity and deterioration owing to changes in burial confinement and relaxation that would occur with multiple transient loadings.

4.3 Effect of cyclic load

After the initial application of monotonic load, a series of 20 load cycles were applied to the footing to simulate the effect of repetitive traffic loading. The applied cyclic load simulated that of one single wheel of the articulated load and had a magnitude of 6 tonnes at prototype scale. It is admitted by the authors that 20 cycles is not a complete replication of the full cyclic loading over a lifetime pipe operation; however, it provided a preliminary insight into the cyclic behaviour with future testing in mind.

The change in bending moment (positive representing an increase in magnitude) on both the load and unload phase is shown in Figure 8. Upon each cyclic application of load, it can be seen (Figure 8a) that the maximum moment experienced within the pipe does not vary greatly with cycles. In contrast, when the load was removed the bending moment did not return to zero and residual 'locked-in' bending moment was observed (Figure 8b); increasing with cycle number. The phenomenon is clearly more pronounced for the shallower buried pipes. For the pipe buried at $z = 0.5$ m, the increase in residual bending moment is over 40% after 20 cycles which is a significant design consideration. This can be attributed to localised changes in sand grain distribution and density, and therefore new soil-structure interaction regimes on each cycle.

Upon spindown of the centrifuge, when the elevated stress field in the soil reduces, the 'locked-in' moments dissipated, confirming that this was indeed an increased confining stress within the soil restraining the pipe on unload. Over the lifetime of the pipe infrastructure, there is potential for significant 'locked-in' stresses to develop as several million cycles of load are applied to the pipe-soil configuration. This may have implications on pipe long-term performance. Strategically selected cyclic experiments investigating different loading conditions are ongoing at CEIGR in Sheffield to explore these effects.

5 CONCLUSION

A series of centrifuge modelling experiments has been carried out evaluating the behaviour of buried flexible pipes subjected to surface traffic loads, simulated by a metal footing, in dense sand. Comparison of pipes buried at different depths shows that shallower pipes are subjected to significantly greater bending moments. Footing settlement was also the lowest for the shallowest pipe suggesting increased stiffness of the system due to the pipe structure.

The buried pipes were also subjected to eccentric loads. When the pipe remained within the zone of stress influence from the eccentric load, the magnitude of the resultant bending moment remained comparable to that of the pipe loaded on the centreline. Once beyond this zone, the magnitude of resultant bending moment reduces significantly.

A series of 20 load cycles were also carried out. Upon each cyclic application of load, the maximum moment experienced within the pile remained reasonably constant. On unload on the other hand, a substantial amount of 'locked-in' bending moment was also observed, increasing with cycle number, owing to potential new soil-structure interaction regimes established. For the buried pipes at deeper depths, this phenomenon is much less pronounced suggesting less interaction.

ACKNOWLEDGMENTS

Funding support provided by the Engineering Physical Sciences Research Council (EPSRC) to establish the 4 m diameter beam centrifuge and Centre for Energy and Infrastructure Ground Research at the University of Sheffield (Grant No. EP/K040316/1) is gratefully acknowledged.

Thanks also to the support and expertise of the Department of Civil & Structural Engineering technical staff for in-house fabrication of the pipe and loading systems.

REFERENCES

AASHTO. 2007. AASHTO LRFD Bridge design specifications. 4th ed. *American Association of State and Highway Transportation Officials*, Washington, D.C.

Boussinesq, J. 1883. Application des potentials à l'étude de l'équilibre et du mouvement des solides élastiques. Gauthier-Villars, Paris.

Bolton, M.D., Gui, M.W., Garnier, J., Corte, J.F., Bagge, G., Laue, J. & Renzi, R. 1999. Centrifuge cone penetration tests in sand. *Géotechnique* 49(4): 543–552.

BSI. 1998. BS EN 1295-1: Structural design of buried pipelines under various conditions of loading. Part 1. General requirements. *British Standards Institution (BSI)*, London.

Burns, J.Q., & Richard, R.M. 1964. Attenuation of stresses for buried cylinders. *In Proceedings of the Symposium on Soil–Structure Interaction*, University of Arizona Engineering Research Laboratory, Tucson, Ariz., 8-11 June 1964. American Society for Testing and Materials, West Conshohocken, Pa.: 379–392.

Discover Water. 2017. [Date viewed: 04 October 2017]. http://discoverwater.co.uk/leaking-pipes

Hosseini, S.M.M. & Tafreshi, S.M., 2000. Soil-structure interaction of buried pipes under cyclic loading conditions. *WIT Transactions on the built environment*, 48

Marston, A. & Anderson, A.O. 1913. The theory of loads on pipes in ditches and tests of cement and clay drain tile and sewer pipes. *Bulletin 31. Iowa Engineering Experiment Station*, Ames, Iowa.

Pocock, R.G., Lawrence, G.J.L. & Taylor, M.F. 1980. Behaviour of a shallow buried pipeline under static and rolling wheel loads. *TRRL Laboratory Report* 954, Department of Transport.

Taylor, M.E. & Lawrence, G.J.L. 1985. Measuring the effects of traffic induced stresses on small diameter pipeline. *Pipe and Pipeline International*, 30(2): 15–19.

Young, O.C. & O'Reilly, M.P. 1983. A guide to design loadings for buried rigid pipes. *TRRL Laboratory Report*, Department of Transport.

Experimental model study on traffic loading induced earth pressure reduction using EPS geofoam

T.N. Dave
Institute of Infrastructure Technology Research and Management, Ahmedabad, India

S.M. Dasaka
Indian Institute of Technology Bombay, Mumbai, India

ABSTRACT: This paper presents results of 1-g experimental studies on the use of EPS geofoam to mitigate traffic loading induced earth pressure. Traffic loading was applied as combination of sustained static and cyclic loading using haversine waves for 10,000 cycles at 1 Hz frequency. Two different intensities, i) sustained static load of 30 kPa followed by cyclic load of 20 kPa and ii) sustained static load of 20 kPa followed by cyclic load of 50 kPa were applied. A load distribution system was devised and placed at the top centre of the backfill so as to apply 45° uniformly distributed load to backfill using servo hydraulic actuator. Three different densities of EPS geofoam, 10 kg/m^3 (10D), 12 kg/m^3 (12D) and 15 kg/m^3 (15D) were considered during the study, however results of 10D geofoam are discussed in this paper. It was observed that provision of EPS geofoam reduced earth pressures in the range of 39% to almost zero pressure for top half portion of the retaining wall and would be beneficial to reduce traffic loading induced earth pressures.

1 INTRODUCTION

Earth retaining structures are integral part of many infrastructure projects, and underground urban construction to retain soil on one of its sides. The lateral pressure acting on these structures due to backfill, surcharge load from adjacent structures and loads due to traffic and earthquake loads decides sectional dimensions of these structures. Techniques such as providing light weight backfill (Horvath, 2010), construction of relief shelves (Banerjee, 1977) were used in the past to reduce earth pressure on the retaining walls. However, structural behaviour and mechanism of stress transfer is questionable in case of relief shelves, while use light weight fill in retaining structures may cause excessive surface settlement due to the low stiffness of these fill materials (Horvath, 2010). Compressible geo-inclusion (also termed as geofoam) in the form of glass fibre (Rehnman & Broms 1972), cardboard (Edgar et al. 1989), glass wool (Purnanandam & Rajagopal 2008) etc. were also employed to reduce the earth pressure on retaining walls and to achieve overall economy in the construction of these walls. However, stress-strain behaviour of these materials is unpredictable and uncontrollable. In addition, these materials are either too compressible (glass-fibre) or biodegradable (card board). For any compressible inclusion, compressive stiffness (E/t, where E is modulus of elasticity of geofoam and t is thickness of geofoam) is the single most important behavioural characteristics influencing the reduction (Horvath 2010). For example, a geofoam with sufficiently lower stiffness and thickness t = 0.01 h (h is height of retaining wall) would provide active stress conditions in the backfill (Karpurapu & Bathurst 1992). Among the various materials used as compressible inclusion, provision of EPS geofoam inclusion at the retaining wall-backfill interface proved beneficial, because of its ready availability, ease in construction and most importantly predictable and controllable stress-strain behaviour.

1.1 Reduction of lateral thrust using EPS geofoam

The concept of reduction of seismic load induced earth pressure by providing geofoam inclusion (Zarnani & Bathurst 2007) showed that the peak lateral loads on the walls with compressible inclusions were reduced in the range of 30% to 60% compared to that of identical structure but with no compressible inclusion. It was pointed out that as backfill was densified during shaking, the geofoam was also compressed, which resulted in earth pressure reduction. The reduction of the lateral thrust and isolation efficiency of EPS geofoam is derived from its lower stiffness rather than from its low unit weight (Ertugrul & Trandafir 2011).

1.2 Earth pressure under traffic loading

Globalisation and urbanisation demand development of infrastructure facilities to meet the demand from heavy vehicular traffic efficiently, causing increased lateral thrust on retaining structures. Past studies

(Sherif & Mackey 1977) highlighted increase in the lateral pressure with increased cycles of traffic loading though the rate of increase of pressure decreased with number of cycles. Also, pressure distribution under cyclic loading was nonlinear and retaining structures undergo cumulative permanent deformations (Ashmawy & Bourdeau 1995). Further, the horizontal stress in the railway ballast in the loaded and unloaded state (at maximum and minimum cyclic loading) fluctuates between 20–60 kPa and eventually reaches 30 kPa (Aursudkij et al. 2009). Though few studies were conducted to consider the effect of traffic loading on retaining structures, the present state of understanding and design methods for structures subjected to cyclic loading is sparse.

1.3 Codal provision for traffic induced loading

Indian Standard code of practice considers a surcharge of 20 kPa due to live load along with surcharge load due to backfill as 15 kPa (IS 14458–Part 2: 1997) and Boussinesq's elastic solution to account for concentrated point loads and heavy line loads. Indian Road Congress code (IRC6-2014) suggests to consider minimum surcharge pressure (live load) of 20 kPa equivalent to 1.2 m height of fill material. As per trenching and shoring manual (Department of Transportation – California 2011), maximum surcharge loads due to vehicular traffic is restricted to 15 kPa. Code of practice for dead and imposed loads (Hong Kong Building Department 2011) recommends minimum surcharge of 20 kPa on earth retaining structures and to consider vehicular traffic minimum imposed loads in the range of 13.9 to 46.6 kPa, depending on type of vehicle. British design manual BD37/01 (Volume 1) suggests that in the absence of exact calculations, live surcharge may be assumed as: (i) HA loading (normal traffic) – 10 kPa (ii) HB loading (industrial vehicular loads) – 20 kPa (iii) RU loading (standard railway loading multiplied by appropriate dynamic factors) – 50 kPa and (iv) RL Loading (reduced loading for only passenger railway systems) – 30 kPa.

However, Shave et al. (2010) pointed out that uniform pressure method of BD37/01 [1] does not truly represent the earth pressure distribution on the wall due to vehicle loading and the pressures should be more concentrated towards the top of the wall. More realistic and critical average pressure values are suggested in PD6694–1, to consider a surcharge due to traffic loading during the design of structures. Table 1 presents the resulting average vertical pressure at the surface considering that vehicle weights uniformly distributed over the plan area of the vehicle.

1.4 Background and motivation

Review of previous studies revealed that very few studies are available related for realistic consideration of effect of traffic loading on retaining structures. Large variation is observed amongst the suggested methods to consider traffic loading on retaining wall among various standard codes of practices. Further, to the authors' knowledge experimental study on evaluation of earth pressure due to traffic loads and use of geofoam to mitigate the earth pressure is missing. Hence, this study is aimed at evaluation of earth pressure due to traffic loads and evaluation of EPS geofoam buffers to reduce earth pressure on a rigid non-yielding retaining wall.

Table 1. Approximate average pressure for vehicle loads (after Shave et al. 2010).

Vehicle Type	Pressure
Average pressure under one equivalent LM1 vehicle	36 kPa
Average pressure under rear pair of equivalent LM1 axles	98 kPa
Average pressure under trailer of the SV196 vehicle	56 kPa
Average pressure under trailer of SOV vehicle	66 kPa

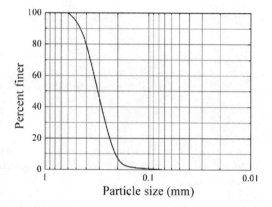

Figure 1. Particle size distribution curve.

2 EXPERIMENTAL PROGRAMME

2.1 Materials used in the study

In this study Indian Standard Sand of Grade III (Figure 1), classified as SP, as per Unified Soil Classification System (USCS), was used. The values of maximum void ratio (e_{max}) and minimum void ratio (e_{min}) were obtained as 0.817 and 0.549, respectively. EPS geofoam of density 10 kg/m^3 (10D) was used in this study. Stress-strain behaviour of EPS geofoam is presented in Figure 2.

The compressive strength (σ_c) and elastic modulus (E) of the geofoam obtained using strain controlled uniaxial compression test (as per ASTM D1621-10) were in line with standard values (ASTM D6817-11).

2.2 Experiment procedure

In the present study, experiments were carried out using an instrumented retaining wall as shown in Figure 3. A total of seven diaphragm type earth pressure

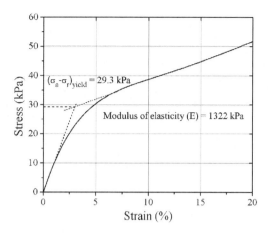

Figure 2. Stress-strain behaviour under uniaxial compression for 10 D EPS geofoam.

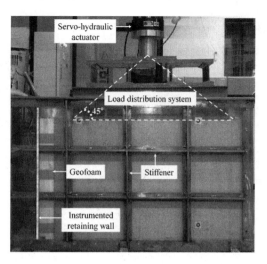

Figure 4. Pictorial view of experimental setup.

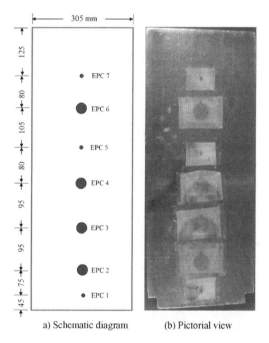

Figure 3. Model retaining wall with EPC (all dimensions in mm).

cells (EPC) were fixed flushed along the retaining wall surface at locations shown in the Figure 3. The wall was placed in a stainless steel test container of dimensions 1200 mm (length) × 310 (width) × 300 mm (depth), respectively. Three sides of the test tank consisted of 16 mm stainless steel plates perfectly welded while a longer side consisted of 25 mm thick Perspex sheet. Details of the strong box used in this study, method of sand bed preparation were discussed in Dave & Dasaka (2012). Sand beds were prepared to achieve backfill relative densities of 68% (dense) and 85% (very dense). A load distribution setup was placed at the centre of the backfill so as to apply uniformly distributed load at 45° (Figure 4). In the case of experiments with geofoam, an EPS sheet of thickness (t/H = 0.125) cut using hot wire cutter was pasted to retaining wall using duct tape to have proper contact between wall and EPS geofoam during the test.

In past, traffic loading was simulated in laboratory by application of successive haversine wave without rest period (Liu & Xiao 2010) or combined haversine (cyclic) and sustained static loading (AASHTO T 307). Most of the times the causes of dynamic loading on pavement are potholes, structural/alligator cracks, sudden acceleration/deceleration of vehicles. These defects are absent during laboratory testing hence the loading frequency does not play an important role (Liu & Xiao 2010). Also, the earth pressure is not sensitive to the loading frequency (Xue et al. 2011). Hence, in the present study traffic loading was simulated as a combination of sustained static and cyclic loading conditions and experiments were performed by applying cyclic loading in the form of haversine waves at a unified frequency of 1 Hz. The loading was applied using Shimadzu make servo-hydraulic actuator of maximum loading capacity 100 kN and maximum displacement capacity of ±50 mm, which was fixed upright on a reaction frame of 200 kN design capacity. Experiments were carried out without geofoam and with geofoam inclusion at the wall-backfill interface to study geofoam inclusion effect.

Loading confirming to AASHTO T 307 was applied on the backfill soil to simulate the traffic loading. Experiments were performed under two different sets of loading (1) Higher sustained load (30 kPa) followed by lower cyclic loading (20 kPa) (2) Lower sustained load (20 kPa) followed by higher cyclic loading (50 kPa). In both the sets, tests are performed at cyclic loading (haversine wave) of 1 Hz frequency for 10000 cycles, as shown in Figure (5, a-b). Details of experiments performed under traffic loading are presented in Table 2. During experiments, EPC 6 output was erratic and hence results of EPC 6 are

Table 2. Details of earth pressure cells used in present study.

	Haris EPC (EPC 2, 3, 4 & 6 in Figure 3)	TML PDA PA (EPC 1, 5 & 7 in Figure 3)
Pressure Range	0–2 kg/cm^2	0–2 kg/cm^2
Sensitivity	1.420 mV/V at Full Scale	+946 μV/V at Full Scale
Non-linearity & hysteresis	0.5% of Full Scale	0.5% of Full Scale
Dimensions	40 mm dia. × 10 mm thickness	6.5 mm dia. × 1 mm thickness

Table 3. List of experiments conducted.

Test no.	Test property	Loading magnitude Static	Loading magnitude Dynamic	R.D. of backfill
E1	Without geofoam	30 kPa	20 kPa	68%
E2	10D – 75 mm thick foam	30 kPa	20 kPa	68%

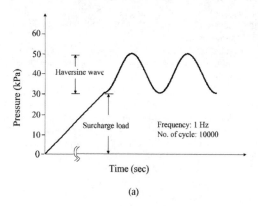

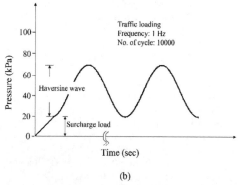

Figure 5. Schematic presentation of applied traffic loading; (a) Higher sustained – lower cyclic load and (b) Lower sustained – higher cyclic load.

excluded from presentation and remaining EPCs were monitored.

Magnitude and distribution of earth pressures and effect of EPS geofoam placement at interface of

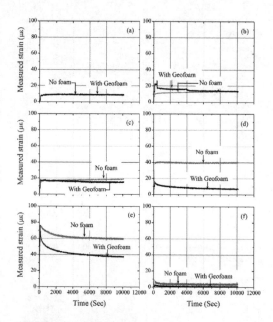

Figure 6. Measured earth pressure on retaining wall under traffic loading for tests E1 and E2 and comparison of output from (a) EPC 1, (b) EPC 2, (c) EPC 3, (d) EPC 4, (e) EPC 5, (f) EPC 7.

retaining wall and backfill when the backfill is subjected to traffic loading were evaluated. The study was carried out based on the dimensions of the available test facility. Though it is not possible to directly extrapolate the results from these study to understand behaviour of prototype walls, the results obtained are helpful in understanding the magnitude and distribution of earth pressure on rigid retaining wall subjected to traffic loading. Details of experiments performed under traffic loading are presented in Table 3.

3 RESULTS AND DISCUSSION

Experiments were performed to evaluate magnitude and distribution of earth pressure on model retaining wall under traffic loading. Magnitude of earth pressures on the retaining wall was monitored throughout the test. EPC (earth pressure cell) results obtained from experiments E1 and E2 are compared in Figure 6. Effectiveness of EPS geofoam inclusion to reduce earth pressures under traffic loading was also evaluated. Comparison of measured earth pressures on retaining wall for experiments E1 and E2, i.e. retaining wall without and with geofoam inclusions, is presented in Figure 6. Further, results of measured earth pressures at beginning of cyclic loading (at the end of ramp wave) were compared with measurements after 1000 cycles, 3000 cycles, 5000 cycles and 10000 cycles, at six locations (as EPC 6 stopped working) along the height of the retaining wall using earth pressure transducers.

Earth pressure results at each earth pressure cell from experiments with geofoam inclusion were

Table 4. Earth pressure at higher sustained load (Experiment E2/E1).

| | H | Normalised earth pressure | | | | |
		a	b	c	d	e
EPC 1	0.08	1.23	1.25	1.24	1.21	1.20
EPC 2	0.20	2.27	1.34	1.27	1.07	0.96
EPC 3	0.36	1.04	0.95	0.91	0.85	0.79
EPC 4	0.52	0.29	0.17	0.12	0.10	0.08
EPC 5	0.65	0.97	0.69	0.65	0.62	0.59
EPC 6	0.82	_*	_*	_*	_*	_*
EPC 7	0.96	0.32	0.32	0.27	0.23	0.24

H – Normalised height of earth pressure cell; a – Initial reading, b – After 1000 cycles, 3000 cycles; c – After 3000 cycles; d – After 5000 cycles; e – After 10000 cycles.

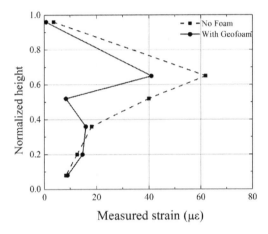

Figure 7. Earth pressure distribution under traffic loading.

normalised with respect to similar experiments without geofoam inclusion. Normalised earth pressures for all experiments are presented in Table 3. The position of earth pressure cell from bottom of the wall is normalised with respect to the total height of wall. As shown in Figure 6, the observed output from EPC was either remained steady (EPC 1 – EPC 2) or reduced (EPC 3 – EPC 7) with increase in number of cycles, which was contrary to previous studies (Sherif & Mackey 1977), wherein it was highlighted that earth pressure increased with number of cycles.

The magnitude of earth pressure in no-foam case was almost similar or even lower than with geofoam conditions for EPC 1 and EPC 2, located at lower heights on the retaining wall. However, comparison of outputs in no foam and with-geofoam conditions from EPC 3 to EPC 7, located at mid height to top of the wall, revealed reduction of earth pressure due to EPS geofoam inclusion. Provision of geofoam inclusion reduced earth pressures by 83%, 39% to almost zero pressure condition at locations corresponding to EPC 4, EPC 5 and EPC 7, respectively. The EPC output (in terms of measured strain) is plotted against normalised height as shown in Figure 7 to obtain earth pressure distribution on retaining wall.

Further, as can be observed from Table 3 and Figure 7, the measured output of a sensor in the presence of geofoam is normalised with respect to the measured output of same sensor for no-foam case. The normalised earth pressures reduced with increase in number of cycles, highlighting the effectiveness of geofoam in reducing earth pressures when subjected to traffic loading. Also, traffic loading effects are more pronounced in upper part of retaining wall, and the geofoam was found more effective in upper half. This effect is due to provision of geofoam, which might have compressed and helped to mitigate traffic loading induced the earth pressure. It proved placement of geofoam at retaining wall backfill interface effective in reducing traffic loading induced earth pressure.

As the calibration response is in general linear within the rated capacity of the sensor, measured earth pressures are directly proportional to the observed micro-strain of the sensors. In view of this, the nature of earth pressure variation along the depth is same as that of micro-strains. Similarly, the percentage reduction of earth pressure and micro-strains at any location due to use of geofoam is identical.

4 CONCLUSIONS

Experiments were conducted to evaluate magnitude and distribution of earth pressures under traffic loading. Following are main findings of the study:

- Increase in magnitude of earth pressure under traffic loading is more for the upper part of retaining wall and reduces with depth.
- Placing EPS geofoam as compressible inclusion between backfill and retaining wall can efficiently reduce earth pressure, specifically in the top upper half, where, increase in earth pressure under traffic loading is predominant.
- Earth pressures under traffic loading reduced with increase in number of cycles.
- The placement of geofoam was found more effective with increase in number of cycles.
- Geofoam of $t/H = 0.125$ used in the present study lead to reduction in the range of 39% to almost zero pressure for top half portion for the selected stiffness geofoam.

ACKNOWLEDGEMENT

Authors are thankful to staff of Advanced Geotechnical Engineering Laboratory, IIT Bombay.

REFERENCES

AASHTO, T. 307. 2007. Standard method of test for determining the resilient modulus of soils and aggregate materials, *American Society of State Highway Transportation Officials*, USA.

Ashmawy, A.K. & Bourdeau, P.L. 1995. Geosynthetic-reinforced soils under repeated loading: A review and comparative design study. *Journal of Geosynthetics International* 4(2): 643–678.

ASTM D1621-10. 2010. Standard Test Method for Compressive Properties of Rigid Cellular Plastics, ASTM Intl., West Conshohocken, PA.

ASTM D6817-11. 2011. Standard Specification for Rigid Cellular Polystyrene Geofoam, Annual Book of ASTM Standards, ASTM Intl., West Conshohocken, PA.

Aursudkij, B., McDowell, G.R. & Collop, A.C. 2009. Cyclic loading of railway ballast under triaxial conditions and in a railway test facility. *Granular Matter* 11: 391–401.

Banerjee, S. P. 1977. Soil behaviour and pressure on retaining structures with relief shelves. *Indian Highways*, 21–34.

BD 37/0. 2001. Design manual for roads and bridges, Summary of Corrections. 1(3) Part 14 – Loads for highway bridges, A/53.

Bouckovalas, G., Whitman, R.V., & Marr, V.A. 1984. Permanent displacement of sand with cyclic loading. *Journal of Geotechnical Engineering* 110(11): 1606–1623.

Boussinesq, J. 1885. Application des potentiels à l'Étude de l'Équilibre et due mouvement des solides élastiques. Gauthier-Villars, Paris.

Dave, T.N. & Dasaka, S.M. 2012. Assessment of portable traveling pluviator to prepare reconstituted sand specimens. *Geomechanics and Engineering – An International Journal* 4(2): 79–90.

Edgar, T.V., Puckett, J.A. & D'Spain, R.B. 1989. Effect of geotextiles on lateral pressures and deformation in highway embankments. *Geotextiles and Geomembranes* 8(4): 275–292.

Ertugrul, O.L. & Trandafir, A.C. 2011. Reduction of lateral earth forces acting on rigid nonyielding retaining walls by EPS geofoam inclusions. *Journal of Materials in Civil Engineering* 23(12): 1711–1718.

Geoguide I. 2000. Guide to retaining wall design. Geotechnical Engineering Off., Civil Engg. Dept., The Govt. of the Hong Kong, Special Administrative Region.

Horvath, J.S. 2010. Lateral pressure reduction on earth-retaining structures using geofoams: Correcting some misunderstandings. *Proc., ER2010: Earth Retention Conference* 3, ASCE, Reston, VA.

IRC: 6-2014. Standard specifications and code of practice for road bridges Section: II. *Loads and Stresses*, New Delhi.

IS: 14458-Part 2. 1997. Retaining wall for hill area-Guidelines, Design of retaining/breast walls. Bureau of Indian Standards, New Delhi.

IS: 4651-Part 2. 1989. Planning and Design of Ports and Harbours, Earth Pressure. Bureau of Indian standards, New Delhi.

IS: 875-Part 5. 1987. Design of loads (other than Earthquake) for Buildings and Structures. Bureau of Indian Standards, New Delhi.

Karpurapu, R. & Bathurst, R.J. 1992. Numerical investigation of controlled yielding of soil-retaining wall structures. *Geotextiles & Geomembranes* 11:115–131.

Liu, J., & Xiao, J. 2010. Experimental study on the stability of railroad silt subgrade with increasing train speed. *Journal of Geotechnical and Geoenvironmental Engineering* 136(6): 833–841.

PD 6694-Part 1. 2011. Recommendations for the Design of Structures Subject to Traffic Loading to BS EN 1997-1.

Purnanandam, K. & Rajagopal, K. 2008. Lateral earth pressure reduction due to controlled yielding technique. *Indian Geotechnical Journal*. 38(3): 317–333.

Rehnman, S.E., & Broms, B.B. 1972. Lateral pressures on basement walls: Results from full scale tests. *Proc., of 5th European conference on Soil Mechanics*, Madrid, Vol. 1:189–197.

Shave, J., Christie, T., Denton, S. & Kidd, A. 2010. Development of traffic surcharge models for highway structures. *Bridge Design to Eurocodes – UK Implementation*: 451–462.

Sherif, M. M. & Mackey, R. D. 1977. Pressures on retaining wall with repeated loading. *Journal of the Geotechnical Engineering Division* 103(11): 1341–1345.

Department of Transportation, 2011. *Trenching and Shoring Manual*. State of California, USA.

Xue, F., Ma, J. & Yan, L. 2011. Vibration Velocity, Earth Pressure, and Excess Pore Pressure Measured in Cyclic Dynamic Field Testing of Water-Rich Loess Tunnel for Passenger-Dedicated Line. Proc., GSP 217, Geo-Hunan 2011 *Advances in Unsaturated Soil and Geo-Environmental Engineering*, Hunan, China, June 9–11.

Zarnani, S. & Bathurst, R. 2007. Experimental investigation of EPS geofoam seismic buffers using shaking table tests. *Geosynthetics International* 14(3): 165–177.

Physical modelling of roads in expansive clay subjected to wetting-drying cycles

S. Laporte, G.A. Siemens & R.A. Beddoe
GeoEngineering Centre at Queen's-RMC, Royal Military College of Canada, Kingston, Canada

ABSTRACT: Expansive soils undergo excessive volumetric deformations during seasonal and annual wetting-drying cycles. Owners of infrastructure take extreme actions to mitigate expansive soil effects, however, the problem can be worsened rather than improved. Physical modelling in a centrifuge environment of expansive soil models allows for several wetting-drying cycles to be applied within a reasonable time period and for direct observation of the development of subsurface strains in response to wetting-drying cycles. This paper describes physical modelling of a road cross-section constructed in expansive soil and subjected to wetting-drying cycles utilising a new atmospheric chamber. This investigation examined the differential vertical subsurface displacement of the distinguishable features of the model including the impermeable road surface, adjacent drainage ditch, and far field. The results indicate centrifuge technology coupled with the atmospheric chamber's ability to control weather conditions can be used to measure the differential subsurface deformations of the soil-structure when exposed to cyclical wetting-drying events. These preliminary observations will provide the fundamental procedures and understanding to scope future investigations to aid government departments of transportation in optimising construction methods to increase sustainability of infrastructure in expansive soil areas.

1 INTRODUCTION

Infrastructure constructed in expansive soil is often subjected to cyclical swelling-shrinking movements responding to daily, seasonal, and annual wetting-drying periods. Differential soil displacements causes cracking in paved roads, foundation movements, and bursting of pipes. A mechanism of differential movements stressing a paved roadway constructed in expansive soil is illustrated in Figure 1. Pavement interrupts the natural soil-climate interaction below the impermeable road surface, creating a barrier for rain water infiltration and evaporation. Reduced soil-climate interaction below the roadway leads to reduction in vertical movements compared with the shoulder and ditch. Runoff from the road collects in the ditch, which experiences an increased magnitude of swelling and shrinking displacements compared to the soil below the centre of the roadway. Zornberg & Gupta (2009), showed field results where pavement differential displacements were shown to result in longitudinal pavement cracking.

The magnitude of the differential displacement is very difficult to measure in field studies. Montenegro et al. (2017) reports one of the few field measurements of a vertical displacement profile in expansive soil. Field observations of soil-structure-climate interactions require long time periods with a limited number of measurement points due to high cost and installation complexity. To use more complex computational models, which include application of weather data as a surface boundary condition adds additional cost and complexity.

Traditionally the swelling potential of expansive soil is studied by using an oedometer apparatus and introducing water to saturate the specimen (ASTM D4546 1996). Limited studies have been conducted measuring the swelling potential of expansive soil in a geotechnical centrifuge. Gadre & Chandrasekaran (1994) compared swelling potential results from tests conducted at 1g with tests done in the enhanced gravity field of a centrifuge with good agreement.

The challenges with field measurements and applicability of laboratory testing make physical modelling of the soil-structure-climate interactions attractive. Atmospheric chambers have been used to simulate climatic conditions to study seasonal effects of moisture changes for clay slopes by Take and Bolton (2011). Tristancho et al. (2012) reported an advanced chamber, which can apply a range of weather conditions to physical models.

The primary focus of this investigation is to measure the development of subsurface deformations of expansive soils below paved roads using physical modelling in a geotechnical centrifuge. The capabilities of the RMCC Geotechnical Centrifuge will be presented including payload, climate control, and digital image capture. Preliminary physical model results of a paved road constructed in expansive clay subject to cyclic wetting-drying cycles will be presented to demonstrate

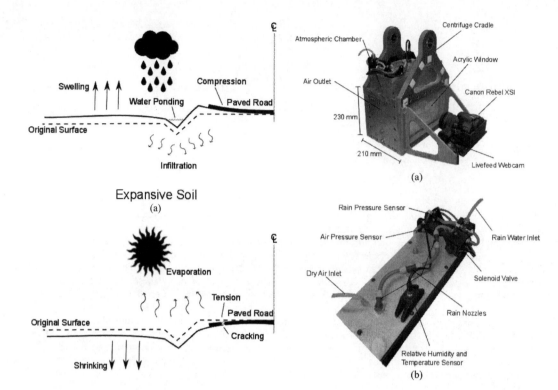

Figure 1. Schematic of soil-structure-climate interaction of a paved road constructed over expansive soil subjected to cyclical wetting and drying; (a) swelling of expansive clays during wetting cycle and (b) shrinking of expansive clays during drying cycle.

Figure 2. a) Centrifuge cradle with atmospheric chamber, b) Atmospheric chamber highlighting instrumentation type and location.

how physical modelling can represent the observed mechanism of damage to pavement surfaces.

2 METHODS AND MATERIALS

2.1 RMCC centrifuge

The 6G-ton geotechnical centrifuge at the Royal Military College of Canada has a diameter of 1.5 m, a top angular velocity of 638 RPM and can apply an acceleration of up to 300 g to two physical models simultaneously. The centrifuge cradles, shown in Figure 2a, are constructed from aluminium with acrylic viewing windows to allow collection of digital images. The internal dimensions of the cradles are $300 \times 100 \times 180$ mm (width × depth × height), a centrifuge model of these dimensions exposed to the maximum angular velocity produces a prototype model of $90 \times 30 \times 54$ m. The centrifuge includes an on-board high-speed 16 channel data acquisition system for recording applied atmospheric conditions.

Digital image data is collected using a Canon Rebel XSi DSLR camera with a 10–18 mm lens. Images are collected at 1 image per minute and the soil deformation is measured using geoPIV (White et al. 2003). A webcam is used to observe the model in real time and is accessed remotely.

2.2 Atmospheric chamber

The atmospheric chamber, Figure 2b, is constructed with 12.6 mm aluminium plates; creating an enclosed air volume above the soil structure in which variable climate conditions are controlled at the soil-structure-climate interface. Misting nozzles (Orbits cooling systems) that produce water droplets on the micron scale are used to simulate rain during wetting cycles. Dry air is injected to the chamber during drying periods and ports on the side of the cradle allow air venting. Chamber instrumentation measures air pressure (Motorola MPX2010GP), pressure in the rain control system (Motorola MPX2100GP) and relative humidity/temperature (VAISALA HMP110). Rain is controlled using a solenoid valve controlled in-flight.

2.3 Model preparation

The soil for this investigation is a mixture of 90% kaolinite clay and 10% bentonite clay by-mass. The combined material has liquid limit and plastic limit of 81.9% and 30.3% respectively resulting in a plasticity index of 51.6%, making it high-plasticity clay

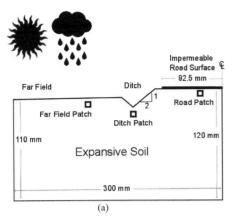

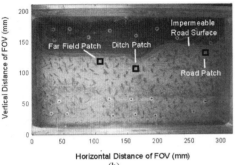

Figure 3. a) Dimensions of centrifuge model displaying locations of typical patches, b) Digital image of centrifuge model and typical patch locations.

(ASTM D2487, 2011). Kingswood (2017) showed this soil has similar properties as naturally occurring expansive soils.

To begin the sample construction process, the two constituents are carefully measure and mixed to create a homogenous mixture. The soil is then mixed with water to the desired initial gravimetric water content. The model is then compacted in an aluminium form with the same internal dimensions as the centrifuge cradle. The compaction procedure is conducted to replicate that of a standard proctor test with 600 kN-m/m^3 of applied energy. This method of model construction requires 6 lifts, applying 65 blows with the proctor hammer per lift. After the soil is compacted, the visible model face is lightly scored, and sand is applied to the surface to create texture for PIV analysis. The model is then trimmed to the desired geometry (Figure 3), and the impermeable road surface is applied using paraffin wax. After the wax cools and hardens the model is placed into the cradle and loaded into the centrifuge. Prior to testing the lighting and camera settings are set to ensure high quality image data.

2.4 Centrifuge model dimensions

The geometry of the model, shown in Figure 3A, is selected for a typical two lane paved road mirrored about its centreline. At 40 g, the 92.5 mm impermeable road surface is scaled to 3.7 m which is comparable to distance from centre line to shoulder of a paved road. The minimum height of the model, 110 m or 4.4 m at 40 g, is chosen to ensure the zone of influence of the shrinking and swelling cycles will not be impacted by the bottom boundary. The wax surface is used to ensure 'no flow' conditions below the pavement. No attempt is made to scale the pavement's structural properties.

2.5 Testing procedure

After setting up the camera, checking instrumentation and safety protocols are conducted; the centrifuge is initiated, brought to 40 g and the model equilibrates with the enhanced gravity environment. The model was exposed to 3 wetting/drying cycles, each 24 hours in duration, resulting in a 72 hour test.

Wetting cycles consist of 6 rain events one hour apart lasting 33 seconds each. Rain is imposed on to the model at a rate of 0.027 mm/s or 5.4 mm per wetting cycle. After the 6-hour raining cycle is completed, dry air is injected to the atmospheric chamber at 25 kPa from the air source. Vents in the atmospheric chamber maintain atmospheric pressure above the model surface. Dry air injection is shut off at the beginning of the next wetting cycle.

2.6 Data processing

Particle Image Velocimetry (PIV) is a routine method for measuring deformations in geotechnical models (White et al. 2003). The focus of this investigation is to capture the development of subsurface displacement below distinguishable features of the centrifuge model. geoPIV was used to track the movement of 3 representative patches (at 13 mm depth) from the image data collected during testing. The patches, shown in Figure 3b, are chosen to be display typical differential deformations of the expansive soil between the road, ditch, and in the far field. The patch locations included; the road patch below the impermeable road surface, the ditch patch below the drainage ditch which collected run-off, and the far field patch, located beneath the soil surface that was open to the variable weather conditions generated in the atmospheric chamber.

3 RESULTS

3.1 Test results

Results from a wetting-drying test are plotted in Figure 4 as prototype vertical displacement versus prototype time for three representative patches under the road, below the ditch and in the far field. The horizontal distance between adjacent patches is 4.2 m between the road and ditch patch and 2.7 m between the ditch and far field patch. Three 24-hour wetting-drying cycles were applied over 72 hours. Time is scaled at 40 g for a prototype time of 120 days. Rain events were

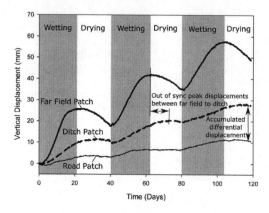

Figure 4. Prototype vertical displacement of PIV patches versus prototype time.

initiated at 0, 40 and 80 days prototype time. Drying cycles begin once water completely infiltrates the ground surface and ditch. Vertical displacement has also been scaled by 40g to prototype values.

The volumetric response of the three patches to the wetting-drying cycles is visually observable in Figure 4. The far field and road patch displacements plotted in Figure 4 are typical of those measured in adjacent locations at that depth. Each of the areas swell during wetting periods and shrink during drying periods. The magnitude of the vertical response is greatest in the far field patch and least below the road as expected. The peak vertical swelling deformation in the ditch patch is slightly out of sync with the far field owing to the water collecting in the ditch for a relatively longer period following the rain event. Over the three wetting-drying cycles the model experiences overall expansion, which is due to the soil being compacted to Proctor maximum for this test.

Differential displacements between the ditch and road accumulate over the three wetting-drying cycles. The differential displacements observed during the test illustrated the mechanism of damage to paved roadways shown in Figure 1. The far field and ditch patch move significantly more than the road patch both during wetting periods and drying periods. Below the impermeable road surface water can only infiltrate around the pavement structure. During the wetting cycle, only minor infiltration occurs resulting in attenuated vertical movements. The ditch and far field patches, being directly connected to the controlled atmospheric conditions, respond to the wetting-drying cycles at increased magnitudes.

The experimentally observed differential displacements agree with the pavement damaging mechanisms shown in Figure 1. Differential displacements accumulate over time and will damage pavement. The benefit of physical modelling is displayed in the directly observed subsurface deformations. Future analyses will consider development of soil strains at depth and differential displacements below the road surface. Future research will consider the effect of soil preparation conditions, paved shoulders, soil type, and climate change to aid governmental organisations to design and maintain sustainable and resilient infrastructure.

4 SUMMARY

This investigation was conducted as part of a larger research study to measure development of subsurface displacements due to cyclic wetting and drying periods. Physical modelling was conducted using a geotechnical centrifuge and the required variable weather conditions were controlled with an atmospheric chamber.

The experimental results display the response of typical subsurface locations to wetting and drying periods. Accumulated differential deformation correlates with what has been observed in the field. Physical modelling of the soil-structure-climate interaction using a centrifuge and atmospheric chamber can be used to accurately measure differential vertical subsurface displacements.

REFERENCES

ASTM D2487-11. 2011. Standard Practice for Classification of Soils for Engineering Purposes (Unified Soil Classification System), *Annual Book of ASTM Standards*, ASTM International, West Conshohocken, PA.

ASTM D4546. 1996. Standard Test Methods for One-Dimensional Swell or Settlement Potential of Cohesive Soils, *Annual Book of ASTM Standards*, Vol. 04-08, ASTM International, West Conshohocken, PA, pp. 693–699.

Boadbent, Inc. 2015. Broadbent Operating Manual, *Thomas Broadbent & Sons Ltd.*, Huddersfield, England.

Kingswood, J. 2017. Development and Implementation of The Royal Military College of Canada's Geotechnical Beam Centrifuge for Physical Modelling. MASc thesis in partial fulfillment of a Degree in Civil Engineering.

Madabhushi, G. 2014. *Centrifuge modelling for civil engineers*. Boca Raton, Fla.: CRC Press.

Montenegro, I.R., Whittle, A.J., & Germaine, J.T. 2017. Automated Station for Monitoring Seasonal Ground Movements in Expansive Clay. *In press. Proceedings of the 2nd Pan-American Conference in Unsaturated Soils*, Dallas, TX, 12–15 November 2017.

Take, W.A. 2015. Thirty-Sixth Canadian Geotechnical Colloquium: Advances in visualization of geotechnical processes through digital image correlation 1. *Canadian Geotechnical Journal*, 52(9): 1199–1220.

Tristancho, J., Caicedo, B., Thorel, L. & Obregón, N. 2012. Climatic Chamber With Centrifuge to Simulate Different Weather Conditions. *Geotechnical Testing Journal*, 35(1), 103620.

White, D.J., Take, W.A. & Bolton M.D. 2003. Soil deformation measurement using particle image velocimetry (PIV) and photogrammetry. *Geotechnique* 53(7): 619–632.

Zornberg, J.G., Plaisted, M.D., Armstrong, C.P. & Walker, T.M. 2013. Implementation of Centrifuge Testing for Swelling Properties of Highly Plastic Clays. No. FHWA/TX-13/5-6048-01-1. 2013.

Zornberg, J.G. & Gupta, R. 2009. Reinforcement of pavements over expansive clay subgrade, *Proceedings of the 17th International Conference on Soil Mechanics and Geotechnical Engineering*, 1.

Scaled physical modelling of ultra-thin continuously reinforced concrete pavement

M.S. Smit, E.P. Kearsley & S.W. Jacobsz
University of Pretoria, Pretoria, Gauteng, South Africa

ABSTRACT: Ultra-Thin Continuously Reinforced Concrete Pavement (UTCRCP) is an innovative pavement type researched and developed in South Africa. It consists of generic concrete pavement substructure overlain with thin, heavily reinforced concrete layer. Centrifuge modelling was used to study the response of UTCRCP to loading. This paper describes three model configurations used to study the response and reports the construction techniques used for tenth scale modelling of pavements. The pavement system was simplified by using compacted silica sand to construct the pavement layers supporting concrete surfacing. Cyclic loading was applied using line loads on a plane strain model, point loads on a three-dimensional model and a moving axle load on an additional three-dimensional model. Although similar behavioural patterns were observed for the different test setups, each setup provided unique information that can be used to understand better the behaviour of UTCRCP under cyclic wheel loads.

1 INTRODUCTION

Ultra-Thin Continuously Reinforced Concrete Pavement (UTCRCP) consist of a 50 mm thick ultra-thin high-performance fibre reinforced concrete layer containing steel mesh with a diameter of 5 mm and an aperture of 50 mm. The relatively high reinforcing content in the thin concrete layer provides it with significant flexural and tensile strength. UTCRCP has been utilised to rehabilitate existing asphalt pavements, similar to ultra-thin white-topping (Pereira et al. 2006), and overlay newly constructed pavements. The behaviour of UTCRCP under load is however not yet well understood.

In long-term pavement performance studies, pavements are monitored over extended periods of time. Accelerated Pavement Testing (APT) is popular because damage can be accumulated in a compressed timeframe (Metcalf 1998). Although some APT devices can be moved to different sites to test actual pavements (De Beer 1990, Van de Ven & De Fortier Smit 2000), they are also found in laboratories where controlled tests are conducted (Chan 1990, Juspi 2007). The fundamental feature is a loaded moving wheel. APT devices can run uni-directionally or bi-directionally and sometimes wander can be included (Donovan et al. 2016). Pavements can be tested at speeds up to 20 km/h and 800 passes per hour (CSIR 2017, Bowman & Haigh 2016, Dynatest 2017).

Accelerated testing of scaled pavements is often used because constructing full-scale prototypes is expensive. Models are rarely scaled smaller than 1:5 because material properties become difficult to replicate (Van de Ven & De Fortier Smit 2000, Bowman & Haigh 2016). Another problem with scaled modelling is that the stress in the model soil mass is not representative of that at full-scale. Centrifuge modelling can be used to ensure realistic soil stresses in pavement models and therefor scaled physical modelling and APT are currently being combined at the centrifuge facility of the University of Pretoria. This endeavour has made it possible to test a variety of pavement and loading configurations at relatively low cost in a limited period of time.

1.1 Background

The stress regime an element of material experiences as a result of a moving wheel load is complex and is affected by principal stress rotation (Brown 1996). Permanent strain accumulates faster if materials are subjected to principal stress rotation compared to simple monotonic load repetition. Additionally, bi-directional principal stress rotation causes permanent strain to accumulate even faster (Chan 1990).

Deflection profiles are often used to study the behaviour of pavements. They enable the back-calculation of the effective stiffness modulus of layers to estimate layer condition. Multi-Depth Deflectometers (MDDs) are used to measure deflection profiles. They consist of Linear Variable Displacement Transducers (LVDTs) that are retrofitted into the pavements at the desired depths (Scullion et al. 1988). Deflection bowls are measured when wheels pass, while permanent deformation is measured when the pavement is unloaded. Pavement behaviour has also been studied using destructive methods such as post-test trenching. More recently, Bowman & Haigh (2016) viewed the cross section of a pavement model subjected to APT through a window. APT models can be used to replicate the stress regimes caused by a moving wheel load and can thus be used to provide a representative prediction of the behaviour of different pavement configurations.

If there is no excessive wheel wander, the permanent deformation in pavements is essentially a plane-strain problem (Brown & Selig 1991). Plane-strain permanent deformation can be produced using line loads. It is reasonable to represent the effect of a wheel with a point load applied over a loading area matching the contact area of wheel. There is however a significant difference in load distribution underneath a point load and a line load. A line load that is exerting a contact stress equal to that of a point load would induce strain deeper into the supporting layers and result in greater deformation in the system.

1.2 Preliminary scaled modelling

Kearsley et al. (2014) tested a 1:10 scaled model of UTCRCP utilising a centrifuge to induce representative stress fields in the model. Care was taken to scale the pavement materials. The substructure system consisted of a 15 mm thick cement stabilised base, a 15 mm granular subbase and a 15 mm selected layer overlying 15 mm of backfill. The substructure was covered with a 5 mm High-Performance Concrete (HPC) layer. Moving wheel loads were applied by means of a two-axle cart. The two wheels on each axle, as well as the two axles, were spaced at 200 mm (to represent a typical prototype axle length of 2 m). The cart moved bi-directionally while loaded with weights to represent an 80 kN axle load at 10G. Stiff rubber wheels, with a contact area of approximately 22 mm by 33 mm, were used to induce a contact stress of approximately 560 kPa per wheel. The movement of the pavement system at the interface between layers was monitored by embedding LVDTs.

When a pavement, consisting of a number of layers of the same material, is subjected to a passing wheel for the first time, behaviour as indicated in Figure 1 (a) is expected at different depths below the concrete layer. All layers should deflect, with magnitude reducing with layer depth. When the load is removed, permanent or residual deformation, depending on load magnitude, but reducing with depth, should be visible. Figure 1 (b) shows the measured response of the model (Kearsley et al., 2014). At a glance, it is apparent that the response of the model pavement differed significantly from the idealised response. The response mimics the idealised response for the first passing load, until the load moves away from the measurement location. As expected, the magnitude of deflection reduces with depth, but the residual deformation underneath the cemented base and subbase is higher than under the concrete.

Figure 2 shows the deflection under load as measured underneath the concrete layer in the wheel path and on the centreline (between wheel paths). Positive deflection indicates downward movement. The relative deformation between the trough underneath the wheel and the centreline increases with every passing axle. Results presented in Figure 1 and Figure 2 were both

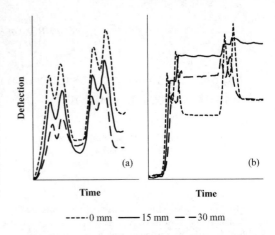

Figure 1. Multi-depth displacement of a) idealized model and b) the physical model.

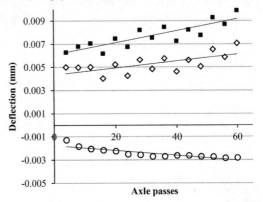

◊ Wheelpath ○ Centreline ■ Distance betwee trough and peak

Figure 2. Displacement along wheel path and centreline.

unexpected as these results contradict pavement design assumptions such as continuity between pavement layers and considering the load distribution underneath single wheels instead of axles (Brown 1996).

Experimental results are required to determine the size of the influence zone (depth and width) caused by a wheel load placed on a thin, flexible concrete surface. Research is currently underway to improve understanding of the interaction and load transfer between the HPC layer and the supporting soil layers that make up UTCRCP. The aim of this paper is to provide details of different centrifuge model configurations that were used to investigate different aspects of the complicated soil-structure interaction that takes place when a flexible concrete slab is placed on a multi-layer support system subjected to a moving wheel load.

2 MODEL CONFIGURATIONS

Results from the UTCRCP model tested by Kearsley et al. (2014) indicated that a need existed to study simplified pavement models. The multi-layer system was simplified to two- or three-layer systems of HPC slabs placed on layers of compacted and/or cemented silica sand. The load configuration was simplified to a single two-wheel axle load. Cyclic loading was applied. Pavement models were considered in three different ways by primarily applying the wheel axle load in different ways. First, the load was applied as two point loads in a three-dimensional model. Secondly, two line-loads were applied through strips, simplifying the model to a two-dimensional or plane-strain model. Thirdly, the pavement was once more considered as a three-dimensional model, where a moving axle was used to apply load in a test setup referred to as the centrifuge pavement tester. All scale models were tested at 10 G in the geotechnical centrifuge of the University of Pretoria (Jacobsz et al., 2014) to create representative stresses in the soil layers.

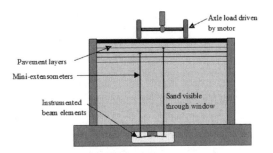

Figure 3. Cross-section of centrifuge pavement tester.

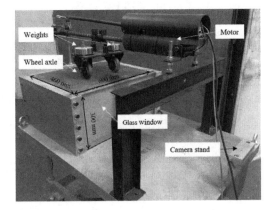

Figure 4. Testing setup of centrifuge pavement tester.

2.1 Point load model

The point load model was tested using an axle with two wheels spaced at 200 mm. A uniform pressure of 560 kPa was exerted under each wheel on the model surface to represent an equivalent standard axle load of 80 kN. The dimensions of the point load model were selected to match a full-scale model that would be loaded in the same manner. The model scale was 1:10 and the total depth of the layers was 60 mm with a plan area of 315 mm by 324 mm. The model height was limited to the depth of the layerworks, thus modelling support conditions of pavements constructed on solid rock.

A measurement system was incorporated into the base plate, underneath the model, to monitor deformation within the layerworks of the models. The system consisted of steel bending beams, measuring $50 \times 3.75 \times 0.62$ mm thick, with one end rigidly clamped between two aluminium blocks. The beams were instrumented with 120 Ω strain gauges and assembled in full Wheatstone Bridges. A needle, resting on the bending beams, was embedded in the model. When vertical movement occurred a small horizontal plate, on the top end of the needle, transferred the movement to the needle, causing the bending beams to deflect, creating mini-extensometers. During model preparation, the needle end plates were positioned at different depths in the model. The substructure was constructed around the needles, taking care to ensure that the surrounding material was compacted to the same degree as the rest of the model. Although the needles were smooth and slender they were prevented from deflecting by the confining pressure of the sand substructure. Vertical movement was measured underneath one of the wheels and on the centreline of the axle at depths above and below typical pavement layer levels: 0 mm (below the concrete layer), 15 mm (below the base course), 30 mm (below the subbase) and 45 mm (below a selected backfill layer).

2.2 Two-dimensional model

Loading strips were used to represent the wheel paths in which an axle would move. The strips were spaced 200 mm apart and loaded to a uniform pressure of 560 kPa. Although the stress distribution underneath a loading strip is not representative of rolling wheel load, it should give a qualitative indication of the expected response in a real pavement. A glass panel on the front of the model box made it possible to take images for Particle Image Velocimetry (PIV) as described by White et al. (2003) throughout the test. The model plan area measured 600 mm by 125 mm with a height of 160 mm. The increased model height was used to create boundary conditions representing UTCRCP on fill.

2.3 Centrifuge pavement tester

The centrifuge pavement tester (see Figure 3 & Figure 4) was constructed to test pavement scale models subjected to moving axle loads. The aluminium strongbox is 300 mm deep. Like the two-dimensional model, the increased model height creates a boundary condition representing UTCRCP constructed on fill. The path along which the wheel axle moves is 460 mm long and the width of the pavement is 600 mm. The front of the aluminium box is a 30 mm thick glass panel, making it possible to examine the cross-sectional pavement response using PIV.

Load was applied by moving a two-wheel motor-driven axle bi-directionally across the surface of the model at a speed of 9.94 mm per second. Weights were placed on the axle to induce the desired stress under the solid rubber tyre wheels (100 mm diameter, 22 mm wide under no load). At the end of each bi-directional cycle, the axle came to rest on the glass panel thus unloading the pavement model.

As with the point load model, mini-extensometer were incorporated into the model. The deflection was measured at four depths in one of the wheel paths and along the centreline between the wheel paths. The uppermost measurement location was directly underneath the concrete surfacing and the following three depths are 15 mm, 30 mm, and 60 mm from the top of the substructure.

3 MODEL CONSTRUCTION

For the construction of the ultra-thin concrete slab the concrete mix design was adjusted from Kearsley et al. (2014), decreasing the maximum fine aggregate size and increasing the superplasticizer dosage. Table 1 shows the mix design and properties of the concrete. To replicate the effect of the high reinforcing content,

Table 1. Mix design of scaled HPC.

	HPC
Material	Quantity (kg/m³)
Cement (CEM 42.5R)	450
Silica fume	50
Dolomite sand (<1.18 mm)	1850
Micro steel fibres	80
Superplastisizer*	4.4
Strength parameters	(MPa)
Compressive strength	104
Flexural strength	31
**Secant modulus of elasticity	42 200

*Percentage by mass of cementitious material
**$E_c = 10 f_{cube}^{0.31}$ (British Standards Institution, 2008)

micro steel fibres and wire mesh were used. Micro fibres replaced the hooked-ended steel fibres of the full-scale HPC. The fibres were 0.2 mm in diameter and 10 mm in length. The wire mesh had a diameter of 0.5 mm and an aperture of 5 mm by 5 mm, proportional to its full-scale counterpart. The wire mesh and the steel mesh had similar moduli of elasticity. As the model pavements was tested cyclically at low stresses, the matching initial stiffness rendered sufficiently representative flexural behaviour. To ensure bonding the mesh was treated with hydrochloric acid to increase its surface roughness.

Different approaches have been used for centrifuge modelling of UTCRCP at the University of Pretoria. Most often the slab was precast in a rigid smooth mould with the required plan area dimensions and a thickness of 5 mm. Roughness was created on the exposed surface. When constructing the model, the rough surface was placed on the substructure to ensure that sufficient friction developed at the interface. The mesh was positioned at mid-depth in the slab by compacting half the fresh concrete in the mould, placing the mesh and then adding the other half of the concrete.

When the slab was placed on the subsurface layers, the slab was lightly tamped to ensure proper contact with the underlying material. The model substructures were constructed by manually compacting sand into layers. It was attempted to keep the density uniform with depth. The subgrade was compacted in layers of 25 mm and the uppermost four layers each had a thickness of 15 mm (equivalent to the typical 150 mm layers at full scale). A target relative density of 100% was selected and the required mass was compacted into the calculated volume to obtain this density. Cement stabilized bases were constructed using the same method, mixing the required cement percentage and moisture content before compaction.

4 RESPONSE OF MODELS TO LOAD CYCLES

Figure 5 shows the permanent deformation measured in the wheel path for each model type.

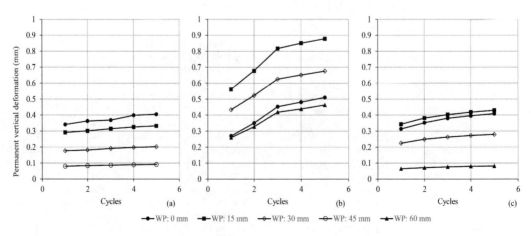

Figure 5. Permanent deformation of a) point load model, b) two-dimensional model and c) centrifuge pavement tester.

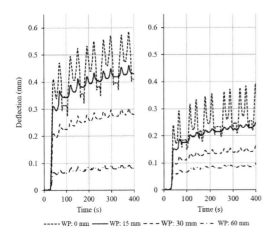

Figure 6. Multi-depth displacement of a) two-layer UTCRCP system and b) three-layer UTCRCP system with cemented base.

The deformations of four layers are reported with their depth away from the concrete layer as designation. The deformation reported was determined using PIV for the two-dimensional model, and mini-extensometers in the three-dimensional model and centrifuge pavement tester. The permanent deformations reported for the centrifuge pavement tester are for bi-directional cycles.

The permanent deformation measured beneath the concrete layer was less than beneath the layer below it for the two-dimensional model and the centrifuge pavement tester model. This was also seen in the results of Kearsley et al. (2014). These results indicate that the high strength reinforced concrete surface rebounds when the load is removed, while the base only shows limited recovery. This means that a void forms between the permanent rut in the base and the concrete layer.

It is worth noting that the model depth does not seem to have a significant effect on the deformation recorded, with the deformation of the 60 mm deep three-dimensional model loaded with a static axle load being similar to that of the 300 mm deep centrifuge pavement tester results. As expected the permanent deformation in the plane-strain model was more than in the other models. Although the rate of deterioration (increase in permanent deformation) is slightly higher for the moving wheel load than for the stationery wheel, the behavioural trend is similar.

As expected, the plane-strain load cycles resulted in significantly more deformation in all the layers of the pavement system. These results indicate that the use of plane-strain analysis to design concrete pavements could result in over-estimated deflections and thus over-conservative designs.

The multi-depth deflection in the wheel path of two models tested using the centrifuge pavement tester is presented in Figure 6. A significant advantage of using strain gauged bending beams to measure the movement of the pavement layers is the high rate of data acquisition in comparison with the number of images that can be obtained from a normal digital camera. These results indicate that the deflection of the UTCRCP model with a cemented base was significantly less than that of the model without it. For both models, the first load cycle resulted in significant permanent deformation and thereafter each additional cycle caused a smaller increase in permanent deformation. The stabilized base (WP: 15 mm) showed very little recovery when the load was removed from the pavement, while the dry sand recovered more. The deflection of the backfill layer (WP: 60 mm) in both models were approximately the same, while the increased stiffness of the stabilised base layer resulted in increased load-spreading ability and significantly reduced permanent deformation in the subbase.

These results indicate that the permanent deformation of UTCRCP can be reduced by the inclusion of a stabilized base layer. The fact that this layer does not seem to recover after load cycles, indicate that the stabilized layer does not behave as an elastic material, which probably means that the first load cycle results in cracking of the stabilized layer. Although more test results will be required before any meaningful conclusions can be drawn, these results show that the centrifuge pavement tester can be used to investigate the effect of relative pavement layer stiffness on pavement deterioration.

5 CONCLUSIONS

Based on the results from the three types of centrifuge models tested it can be concluded that all three types of models can give meaningful results that can be used to refine a design procedure for designing UTCRCP. The advantage of the plane-strain model is that image analysis gives a good indication of strain distribution through the whole cross section of the model and the contribution of each layer towards load distribution can be visually evaluated. The disadvantage of strip loading is that the strain distribution can be affected by the significantly larger loads placed on the pavement layers.

The advantage of the static wheel load is that it is a relatively simple test setup, but no visual observation is possible and the rate of deterioration is very slow, resulting in high numbers of cycles required to obtain meaningful results. Although further tests will be conducted to evaluate the centrifuge pavement tester, the initial test results indicate that the effect of a rolling wheel load can be simulated in a centrifuge and the composition of UTCRCP can be optimized without incurring the costs associated with full-scale field testing.

ACKNOWLEDGEMENTS

The authors would like to acknowledge the funding received from the South African Concrete Institute that made this research possible.

REFERENCES

Bowman, A. & Haigh, S.K., 2016. The Cambridge Airfield Pavement Tester. In *Physical Modelling in Geotechnics; Proc. of the 3rd European Conf., 1-3 June 2016.*, Nantes, France.

Brown, S.F., 1996. Soil mechanics in pavement engineering. *Géotechnique*, 46(3), pp. 383–426.

Brown, S.F. & Selig, E.T. 1991. The design of pavements and rail track foundations. In O'Reilly & Brown, (eds), *Cyclic loading of soils*. London: Blackie, pp. 249–305.

British Standards Institution. (2008). *Eurocode 2: design of concrete structures: British standard*. London, BSi.

Chan, F.W.K., 1990. *Permanent deformation resistance of granular layers in pavements*. Phd thesis. University of Nottingham.

CSIR, 2017. Heavy Vehicle Simulator. Available at: https://www.csir.co.za/heavy-vehicle-simulator [Accessed June 13, 2017].

De Beer, M., 1990. *Aspects of the design and behaviour of road structures incorporating lightly cementitious layers*. PhD thesis. University of Pretoria.

Donovan, P., Sarker, P. & Tutumluer, E., 2016. Rutting prediction in airport pavement granular base/subbase: A stress history based approach. *Transportation Geotechnics*, 9, pp. 139–160.

Dynatest, 2017. HVS Heavy Vehicle Simulator. Available at: http://www.dynatest.com/hvs [Accessed June 13, 2017].

Hambleton, J.P. & Drescher, A., 2008. Modeling wheel-induced rutting in soils: Indentation. *Journal of Terramechanics*, 45(11), 201–211.

Hambleton, J.P. & Drescher, A., 2009. Modeling wheel-induced rutting in soils: Rolling. *Journal of Terramechanics*, 46(2), 35–47.

Juspi, S., 2007. *Experimental validation of the shakedown concept for pavement analysis and design*. Phd thesis. University of Nottingham.

Jacobsz, S.W., Kearsley, E.P., Kock, J.H.L. The geotechnical centrifuge facility at the University of Pretoria. In Gaudin & White (eds) *Physical Modelling in Geotechnics; Proc. of the 8th int. conf., 14-17 January 2014*. Perth, Australia: Taylor & Francis Group, 1101–1106.

Kannemeyer, L., Perrie, B.D., Strauss, P.J. & Du Plessis, L., 2007. Ultra-Thin Continuously Reinforced Concrete Pavement research in South Africa. In *Proc. of int. conf. on concrete roads, 16-17 August 2007*. Midrand, South Africa: Cement & Concrete Institute 97–124.

Kearsley, E.P., Vd Steyn, W.J.M. & Jacobsz, S.W., 2014. Centrifuge modelling of Ultra-Thin Continuously Reinforced Concrete Pavements (UTCRCP). In Gaudin & White (eds) *Physical Modelling in Geotechnics; Proc. of the 8th int. conf., 14–17 January 2014*. Perth, Australia: Taylor & Francis Group, 1101–1106.

Metcalf, J.B., 1998. Accelerated pavement testing, a brief review directed towards asphalt interests. *Journal of the association of asphalt pavement technologists*, 67.

Pereira, D.D.S., Balbo, J.T. & Khazanovich, L., 2006. Theoretical and field evaluation of interaction between ultra-thin whitetopping and existing asphalt pavement. *International Journal of Pavement Engineering*, 7(4), pp. 251–260.

Scullion, T., Uzan, J., Yazdani J.I., & Chan, P., 1988. *Field evaluation of the multi-depth deflectometers*, Research Project Report 1123-2, Texas Transportation Institute, Texas A&M University, College Station, Texas.

Van de Ven, M.F.C. & De Fortier Smit, A., 2000. The Role of the MMLS Devices in APT. In *Proc. of the Southern African Transport Conference., 17-20 July 2000.*, Pretoria, South Africa: SATC, pp. 17–20.

White, D.J., Take, W.A. & Bolton, M.D., 2003. Soil deformation measurement using particle image velocimetry (PIV) and photogrammetry. *Géotechnique*, 53(7), pp. 619–631.

The effect of relative stiffness on soil-structure interaction under vehicle loads

M.S. Smit, E.P. Kearsley & S.W. Jacobsz
University of Pretoria, Pretoria, Gauteng, South Africa

ABSTRACT: Ultra-Thin Continuously Reinforced Concrete Pavement (UTCRCP) is designed using Conventional Concrete Pavement (CCP) design methodology. The primary structural component of this innovative pavement type consists of a heavily reinforced 50 mm high performance concrete layer. This alternative concrete material and geometry result in a flexible overlay that allows significant deflections. Centrifuge modelling of three multi-layer systems was conducted to investigate concrete overlay and substructure interaction. It was observed that non-vertical strains occur between the wheel paths, as well as cracking of the cement-stabilized layer and rutting in the wheel paths. These observations lead to the conclusion that CCP design methods should not be used to design UTCRCP.

1 INTRODUCTION

Ultra-Thin Continuously Reinforced Concrete Pavement (UTCRCP) is an innovative new pavement type that has been researched and developed in South Africa (Kannemeyer et al. 2007, Kearsley et al. 2014). It consists of a generic pavement substructure with a 50 mm high strength steel fibre concrete reinforced overlay, which is reinforced continuously with 5.6 mm diameter steel bar mesh at 50 mm longitudinal and transversal spacing. The generic pavement substructure can consist of a bound or unbound granular base and subbase, while the subgrade can be in either cut or fill.

Internationally, pavement systems similar to UTCRCP are referred to as ultra-thin whitetopping. Ultra-thin High-Performance Concrete (HPC) overlays can be used to rehabilitate existing pavements or to construct new pavements. With the bulk of South Africa's road network reaching the end of its design life and given severe budget constraints, the potential of UTCRCP to fulfil pavement repair strategy requirements has been recognised (Kannemeyer et al. 2007).

Research on ultra-thin high-performance concrete overlay systems has been based on accelerated pavement testing (APT) of full scale and scaled models. The performance is monitored using devices such as falling weight deflectometers, multi-depth deflectometers, joint deflection measuring devices and visual inspection techniques (Gerber 2011, Kannemeyer et al. 2007). Research has been focussed on the effect of substructure strength and overlay-substructure bond on design life (Isla et al. 2015, Kannemeyer et al. 2007, Gerber 2011).

The University of Pretoria has been involved in researching UTCRCP since 2006 and more recently there has been a move to investigate the behaviour of UTCRCP utilising scaled physical models at the institution's geotechnical centrifuge facility (Kearsley et al. 2014).

1.1 *Background*

Conventional Concrete Pavement (CCP) typically consist of a 300 mm Normal Strength Concrete (NSC) slab cast on a generic substructure. The principle cause of failure in CCP is cracking. Cracking is a result of environmental effects and traffic loading. Wide, uncontrollable cracks change the stress distribution in the concrete and substructure. Cracks also allow moisture ingress into the substructure which can results in loss of support (Huang 1993). Two types of CCP are Jointed Concrete Pavement (JCP) and Continuously Reinforced Concrete Pavement (CRCP). Thermally induced tensile stress is significant in concrete pavements and design techniques have been adopted to accommodate it. In JCP, thermally induced stresses are controlled by limiting the effective length of the concrete slabs by means of joints. These joints require regular maintenance and most failures in JCP are attributable to the joints.

The use of CRCP eliminates joints and incorporates longitudinal steel to control crack widths. Edge punch-out is the predominant distress type in CRCP (Huang 1993). It occurs when interlock at transverse cracks deteriorate and longitudinal cracks form due to the transient loading of passing vehicles. CCP are designed using rigid pavement design methodology. As with CRCP, UTCRCP eliminates problems related to joints by not using joints. It also mitigates edge punch-out through the use of high strength steel fibre reinforced concrete. This High-Performance Concrete (HPC) has superior post-cracking capacity and allows the reduction of the concrete layer thickness.

In South Africa UTCRCP is currently designed using rigid pavement methodology (Strauss et al. 2007, Gerber 2011). Load spreading occurs when the stresses at the wheel-pavement interface are spread or dissipated over an increasing area with depth. Rigid pavements are designed assuming that the flexural stiffness of the thick concrete layer ensures sufficient load spreading that the substructure properties are deemed to be unimportant as long there is not too much variation (Brown & Selig 1991, O'Flaherty 1967, Huang 1993). The traffic associated failure mechanism of rigid pavements is fatigue cracking. The structural models used to determine the critical tensile stresses at the bottom of the concrete slab for rigid pavements include spring foundations, multi-layer elastic foundations and elastic finite element methods. The tensile stress is used along with a distress model for concrete fatigue cracking to predict pavement life.

Load spreading through pavement layers is dependent on the stiffness of the layers. The stiffness is a function of the shape (thickness) of the layerworks, boundary conditions and the material's Young's modulus (often quantified by a secant modulus of elasticity) of the materials used in the layers (Clayton 2011). The effect of thickness and stress-strain response of bound materials on load spreading can be compared conceptually by looking at flexural rigidity of rectangular sections (Hibbeler 2008):

$$EI = E \frac{wt^3}{12} \quad (1)$$

where E = modulus of elasticity, w = layer width and t = layer thickness. The equation demonstrates the effect of layer thickness and indicates that the load spreading of the ultra-thin HPC layer will not be similar to that of thick pavements even though the material stiffness is greater. It is therefore proposed that UTCRCP should be designed incorporating aspects of flexible pavement design methodology for traffic associated failure mechanisms, which includes rutting in addition to fatigue cracking.

The relative stiffness of structural systems are considered differently for different design problems (Arnold et al. 2010, O'Flaherty 1967, Klar et al. 2005). Rigid pavement design uses the radius of relative stiffness which is calculated using Equation 2 (Westergaard 1926).

$$l = \sqrt[4]{\frac{Eh^3}{12(1-\mu^2)k}} \quad (2)$$

where E = modulus of elasticity of the concrete, h = thickness of the concrete slab, μ = Poisson ratio of the concrete and k = modulus of subgrade reaction. The radius of relative stiffness gives an indication of the extent of the zone of load influence around the wheel load (typically taken as up to 6l) (Westergaard 1926).

The modular ratio is used in flexible pavement design where the material stiffness of each layer is divided by the material stiffness of the layer underneath it (SANRAL 2013). Limiting the modular ratio leads to gradual load spreading through a pavement system and these pavements are termed balanced pavements.

The properties of pavement materials are improved by stabilization (O'Flaherty 1967). The most common types of stabilization are compaction and cement stabilization in addition to compaction. Different degrees of cementation are used. Soils that would otherwise be unacceptable for use are improved by introducing cement and the resulting material is intended to fragment and assume properties similar to that of the granular material. This is referred to as the equivalent granular state (SANRAL 2013, Visser 2017). Strongly cemented materials are not intended to become granular and crack in a more discrete fashion. The edges of the discretely cracked base cause distress in overlaying asphaltic layer which result in reflection cracking (Visser 2017). Introducing cement stabilized bases is an approach to improve load spreading from the ultra-thin concrete layer to the substructure. The flexible nature of the concrete layer may cause the base to crack and cause stress concentrations in adjacent layers.

2 EXPERIMENTAL SETUP

The objective of this study was to investigate the effect of relative stiffness on the response of pavement systems by considering simplified scaled physical models of two-layer and three-layer systems.

The 150 G-ton geotechnical centrifuge at the University of Pretoria was used for this study (Jacobsz et al. 2014). The stress-strain behaviour of soil is highly dependent on its stress conditions. Testing the 1:10 scaled models at 10 times the gravitational acceleration ensured that representative in-situ stress conditions occurred in the models.

2.1 Physical model

The scaling factor of 1:10 was influenced by the practicality of constructing the model. The models were constructed in a rectangular strongbox. Spacers were placed inside the strongbox to reduce the plan dimensions to 125 mm by 600 mm.

Three pavement configurations were investigated. The first was a CCP, the second a UTCRCP and the third represented a UTCRCP with a cement stabilized base. The substructure of all the models, except the third model, consisted of compacted dry cohesionless sand. A 100 mm subgrade was compacted first, followed by four 15 mm layers representing the layerworks.

For the cement-stabilized model the uppermost layer of compacted dry sand was replaced with sand stabilized by adding 10% water and 2.5% cement by mass. The CCP model had a 30 mm NSC slab placed on the substructure. To prevent shrinkage cracking the slab was reinforced with a wire mesh with a diameter of 0.5 mm and aperture of 5 mm. The UTCRCP models incorporated a 5 mm HPC slab placed on the substructure. The HPC consisted of high-strength micro-steel fibre reinforced concrete and included the same wire mesh.

2.2 Load application and data acquisition

For purposes of this study the movement of an axle with two wheels moving in the same wheel tracks was simulated using two load strips. The width and length of the load strips were 25 mm and 125 mm respectively. A stepper motor powered jack was used to apply the load and a load cell was used to measure it. The cyclic loading was applied manually, where load was increased until a pressure of 560 kPa was applied to the concrete through the strips. This simulated a standard axle load (SANRAL 2013). The pavement response was monitored using Particle Image Velocimetry (PIV). A camera was placed approximately 580 mm away from the glass panel. Photos were taken at six second intervals.

2.3 Material properties

The HPC mix design suggested by Kearsley et al. (2014) for UTCRCP was replicated. It was adjusted for NSC for the CCP model. Table 1 shows the mixtures used and the standard strength parameters measured (SANS 50196-1 2006). Although the material stiffness was not measured for the concrete, it was estimated using the relationship between compressive strength and secant modulus of elasticity.

The sand used was characterized by Archer (2014). It was classified as a poorly graded fine sand with a maximum dry density of 1669 kg/m^3. The substructures of the models were constructed by manually compacting the required mass into a specified volume to obtain a specific density. A target density of 1669 kg/m^3 was used for all the layers. The cement stabilized layer was constructed using the same method.

A secant modulus of elasticity of approximately 300 MPa for the dry sand and 1000 MPa for cemented sand was determined using triaxial testing, with locally attached LVDTs, at a confining stress of 100 kPa. An axial strain of 0.018% was imposed. These values will decrease as the axial strain increases and confining pressure decreases.

3 RESULTS AND DISCUSSION

3.1 Relative stiffness of pavement system

The relative stiffness of the pavement systems was considered using modular ratios and the radius of relative stiffness (Equation 2). The values in Table 2 were determined using the material properties mentioned earlier. The radius of relative stiffness should be between 570 mm and 2032 mm for a two-layer system to be adequately described by a slab-on-grade model (Gerber 2011). Typical modular ratio limits range from 2 to 9 for bound materials (SANRAL 2013). A flexible overlaying material tends to bend into its support,

Table 1. Mixture composition and strength parameters of scaled concrete.

	HPC	NSC
Material	Quantity (kg/m^3)	
Cement (CEM 42.5R)	450	350
Silica fume	50	N/A
Dolomite sand (<1.18 mm)	1850	2037
Micro steel fibres	80	N/A
Superplastisizer [a]	4.4	1
Strength parameters	(MPa)	
Compressive strength	104	57
Flexural strength	31	18
Secant modulus of elasticity [b]	42 200	35 000

a) Percentage by mass of cementitious material
b) $E_c = 10 f_{cube}^{0.31}$ (British Standards Institution 2008)

Table 2. Relative stiffness.

	CCP	UTCRCP	Cement-stabilized UTCRCP
Modular ratio	116.7	140.7	42.8 & 3.29[a]
Radius of relative stiffness (mm)[b]	1498	409	429[c]

a) Modular ratio of concrete to cemented sand and cemented sand to compacted sand
b) Modulus of subgrade reaction of 16 000 kN/m^3
c) Equivalent stiffness of concrete and cemented sand

increasing tensile stresses in itself and compressive strain in the supporting layer.

The radius of relative stiffness calculated for the CCP showed that a slab-on-grade model would be adequate for analyses. The modular ratio calculated for the CCP was very high at 116.7. The modular ratio of the UTCRCP model was higher than that of the CCP model, but the radius of relative stiffness indicated that the system did not behave as a slab-on-grade. The same was indicated for the UTCRCP with a cement-stabilized base. The modular ratio between the cement-stabilized base and the sand was acceptably low according to SANRAL (2013), but the concrete layer over the cement stabilized base was significantly higher than the upper limit of 9.

3.2 Pavement system response

Figure 1 illustrates the vertical displacement as load was applied cyclically and indicates the points of interest discussed in the following sections. It shows the first point of interest at point "a", which was the loaded pavement response during the first cycle. The deflection bowl width with depth, as well as displacement vector plots were used to interpret the pavement response. The points of interest were after the first cycle (indicated at point "b") and after five cycles (indicated at point "c") respectively. In these two cases the permanent deformation of the layers is reported as deflection bowls to examine the effect of repeated loading.

3.2.1 Loaded response to first cycle

Figure 2 shows the extent of deflection bowl width with depth for the UTCRCP models. It was estimated by fitting a Gaussian curve to the deflected shape at specific depths and calculating the distance from the maximum deflection (load application location) to where the deflection was 1×10^{-3} mm. The differential vertical deflection of the thick concrete layer in the CCP model was negligible and the estimated deflection bowl width therefore fell outside the model boundaries. At approximately 253 mm, the deflection bowl width over the ultra-thin concrete layer and the sand substructure was smaller than the radius of relative stiffness of 409 mm.

In the case of the stabilized model, the deflection bowl width just underneath the ultra-thin concrete layer was similar to that of the unstabilised UTCRCP model. This width did not increase significantly through the cemented base. It is likely that the base cracked during the first load cycle, and that the crack edge induced stress concentrations similar to reflection cracking. The high modular ratio between the concrete and cemented layer indicates that the cemented base would have been prone to cracking as it was forced to deflect with the flexible concrete layer, causing tensile stress at its base.

Figure 3 presents displacement vector plots of the first load cycle from the wheel centreline to the axle centreline. The displacement vectors were determined from the unloaded state to point "a" on Figure 1. Lateral load spreading through the thick concrete layer is confirmed by the displacement vectors all indicating vertical downward movement of the sand layers. The load spreading through the ultra-thin concrete surface to the compacted sand substructure contrasts with the CCP model. Immediately around the load location the vectors point directly downward, reducing in magnitude away from the load location. The horizontal vector component becomes greater away from the load location, with vectors pointing diagonally towards the axle centerline. The response of the substructure indicates lack of load spreading due to the low flexural rigidity of the ultra-thin concrete surface. The magnitude of displacement reduces with depth. The vector displacement plot for the cement-stabilized base UTCRCP model shows lateral load spreading by the stiff base. The response of this model is a combination of the two models presented earlier. The displacement of the base layer reduces marginally toward the axle centreline. The magnitude of displacement away from the wheel load reduces more rapidly than for the unstabilized UTCRCP model.

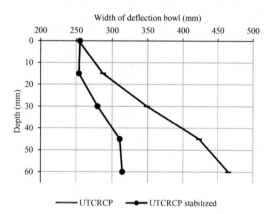

Figure 1. Model setup along with resilient and permanent displacement.

Figure 2. Horizontal influence zone of UTCRCP and UTCRCP stabilized.

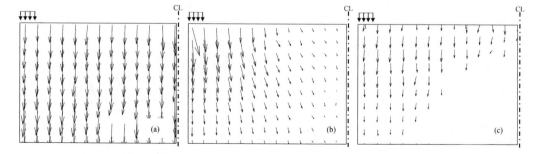

Figure 3. Displacement vector plots for a) CCP, b) UTCRCP unstabilized and c) UTCRCP stabilized.

3.2.2 Permanent deformation

The permanent deformation of the models was investigated by means of PIV by tracking the movement of rows of patches placed between pavement layers. Figure 4 illustrates the placement of patch rows. They are referred to by their depth from top of the pavement model (see Figure 4). Figure 5(a) to (c) show the permanent deformation of the models after one (point "b" in Figure 1) and five cycles (point "c" in Figure 1) respectively. The permanent deflection bowls determined from the patch rows reflect how stress was distributed through the pavement system.

Figure 5(a) shows minimal differential vertical permanent deformation. The high flexural rigidity of the thick concrete layer resulted in lateral load spreading so that the substructure probably experienced approximately the same vertical stress away from the point of load application. More permanent deformation accumulated toward the centre line. After five load cycles there was accumulation of permanent strain, but the shape remained approximately the same.

Figure 5(b) shows that there was a lack of lateral load spreading by the ultra-thin high performance concrete layer, resulting in differential vertical permanent deformation around the load location. The vertical difference between the trough and peak decreased with depth. After five cycles all the layers had settled. The vertical difference between troughs and peaks at 15 mm to 60 mm depths increased and the distance between them decreased as the layers compressed. This echoes the observations of the displacement vector plots (in Figure 3) and indicates non-vertical permanent strain in the layers.

The concrete layer of the UTCRCP model is represented by the top graph in Figure 5(b). If continuity was assumed between the concrete and base layer, it is anticipated that the permanent deformation at the surface would be more than at 15 mm because of its proximity to the load. In Figure 5(b) at 0 mm, less permanent deformation occurred than at 15 mm, indicating that there was some rebound of the ultra-thin high performance concrete layer as it lifted from the substructure, creating a gap. The permanently deformed shape of the concrete layer remained approximately the same after five load cycles, showing a smaller increase between its troughs and peaks than the other layers and confirming the hypothesis that the concrete layer rebounded.

The response of the cement-stabilized base UTCRCP fell between that of the CCP and UTCRCP models. Figure 5(c) show permanent deformation after the first cycle and minimal accumulation of settlement after five cycles. The bowl-shaped curves of 0 mm and 15 mm indicated that the poor load spreading from the ultra-thin HPC to the stabilized base caused damage which possibly resulted in the cement-stabilized layer cracking. The comparatively sharp slope change at 30 mm depth at offset 250 mm supports this observation. Nonetheless, the combination of the two layers spread the load such that differential vertical deformation was not pronounced in the layers toward the bottom of the pavement system. It is anticipated that the trends observed after five cycles for the three models will continue until the accumulation of permanent deformation reaches a plateau.

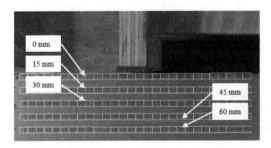

Figure 4. PIV patch arrangement for permanent deformation plots.

4 CONCLUSIONS

Scaled physical modelling was used to investigate the effect of relative layer stiffness on the response of

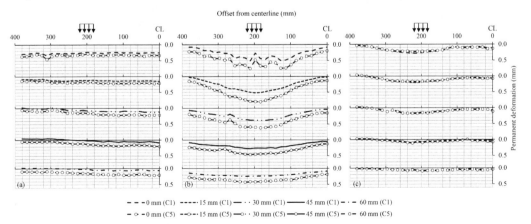

Figure 5. Permanent deformation after first and fifth cycle of a) CCP, b) UTCRCP unstabilized and c) UTCRCP stabilized.

multi-layer pavement systems. Differential permanent vertical deformation occurred in the unstabilized UTCRCP model. The ultra-thin concrete layer also rebounded off the substructure, forming a gap underneath. Although the combination of the ultra-thin concrete layer and cement-stabilized base reduced differential permanent vertical deformation, the base cracked during the first load cycle. This caused distress in substructure.

The response of UTCRCP should not be simplified to that of a slab-on-grade and should therefore not be designed using rigid pavement methodology. However, flexible pavement methodology will have to be adjusted if it is to be applied to UTCRCP. An alternative to the modular ratio as a measure of relative layer stiffness should be developed if UTCRCP is to be designed as a flexible pavement.

REFERENCES

Archer, A. 2014. *Using small-strain stiffness to predict the settlement of shallow foundations on sand*. Master's dissertation. University of Pretoria.

Arnold, A., Laue, J., Espinosa, T. & Springman, S.M. 2010. Centrifuge modelling of behaviour of flexible raft foundations on clay and sand. In Springman, Laue & Seward (eds), *Physical Modelling in Geotechnics; Proc. of the 7th int. conf., 28 June–1 July 2010*. London, England: Taylor & Francis Group: 679–684.

British Standards Institution. (2008). *Eurocode 2: design of concrete structures: British standard*. London, BSi.

Brown, S.F. & Selig, E.T., 1991. The design of pavement and rail track foundations. In O'Reilly & Brown, (eds), *Cyclic loading of soils*. London: Blackie: 249–305.

Clayton, C.R.I., 2011. Stiffness at small strain: research and practice. *Géotechnique*, 61(1): 5–37.

Gerber, J., 2011. *Characterization of cracks on Ultra-Thin Continuously Reinforced Concrete Pavements*. Master's dissertation. Stellenbosch University.

Hibbler, R.C., 2008. *Mechanics of Materials*, 8th ed., Pearson Prentice Hall.

Huang, Y.H., 1993. *Pavement Analysis and Design*, 2nd ed., Pearson Prentice Hall.

Isla, F. et al., 2015. Mechanical response of fiber reinforced concrete overlays over asphalt concrete substrate: Experimental results and numerical simulation. *Construction and Building Materials*, 93:1022–1033.

Jacobsz, S.W., Kearsley, E.P. & Kock, J.H.L., 2014. The geotechnical centrifuge facility at the University of Pretoria. In Gaudin & White (eds) *Physical Modelling in Geotechnics; Proc. of the 8th int. conf., 14–17 January 2014*. Perth, Australia: Taylor & Francis Group: 169–174

Kannemeyer, L., Perrie, B.D., Strauss, P.J. & Du Plessis, L., 2007. Ultra-Thin Continuously Reinforced Concrete Pavement research in South Africa. In *Proc. of int. conf. on concrete roads, 16–17 August 2007*. Midrand, South Africa: Cement & Concrete Institute: 97–124.

Kearsley, E.P., Vd Steyn, W.J.M. & Jacobsz, S.W., 2014. Centrifuge modelling of Ultra-Thin Continuously Reinforced Concrete Pavements (UTCRCP). In Gaudin & White (eds) *Physical Modelling in Geotechnics; Proc. of the 8th int. conf., 14–17 January 2014*. Perth, Australia: Taylor & Francis Group: 1101–1106.

Klar, A. Voster, T.E.B., Soga, K., & Mair, R.J., 2005. Soil-pipe Interaction due to tunneling: comparison between Winkler and elastic continuum solutions. *Géotechnique*, 55(6): 461–466.

O'Flaherty, C.A., 1967. *Highway Engineering Volume II*, 2nd ed., London: Edward Arnold Ltd.

SANRAL, 2013. Chapter 10: Pavement Enigneering. In *South African Pavement Engineering Manual, 1st ed*. South Africa: SANRAL Ltd.

Standards South African, 2006. SANS 50196-1:2006, 2006. Methods of testing cement. Part 1: *Determination of strength*. SANS

Strauss, P.J., Slavik, M., Kannemeyer, L., & Perrie, B.D., 2007. Updating cncPave: inclusion of ultra thin continuously reinforced concrete pavement (UTCRCP) in the mechanistic, empirical and risk based concrete pavement design method. In *Proc. of int. conf. on concrete roads, 16-17 August 2007*. Midrand, South Africa: Cement & Concrete Institute: 204–218.

Visser, A.T., 2017. Potential of South African road technology for application in China. *Journal of Traffic and Transportation Engineering (English Edition)*, 4(2): 113–117.

Westergaard, H.M., 1926. Stresses in concrete pavements computed by theoretic analysis. *Public Roads*, 7(2): 298–302.

Plate bearing tests for working platforms

G. Tanghetti, R.J. Goodey, A.M. McNamara & H. Halai
City, University of London, London, UK

ABSTRACT: During piling and other construction works, a working platform is often constructed across the site. These platforms comprise aggregate material placed and compacted to a designed thickness. Satisfactory performance of the platform may be confirmed by a plate bearing test. Current guidance given on plate bearing testing of granular soils suggests that the plate be at least five times the nominal size of the coarsest material. For a working platform this may be large and the reaction load required from plant and resources to carry out the bearing test may become excessively high. The aim of the research presented in this paper was to investigate the effect of particle to plate size ratios to establish if the use of a smaller plate would still allow a reliable test to be performed on site. Plate bearing tests were carried out in a centrifuge using a large, coarse grained limestone. The limestone was graded to a scale representation of 6F2 material, a commonly specified particle size distribution for working platforms. The size of plate was varied and the load displacement response recorded. The measured bearing capacity was correlated with the ratio of particle to plate size.

1 BACKGROUND

Plate bearing capacity tests represent a good method to investigate the behaviour of soils, especially the bearing capacity near the ground surface and the possible settlement under a certain load. The test is usually adopted when shallow foundations are to be used, or when temporary works requiring a working platform such as piling rigs or cranes are required on site. The standards applicable to this test are: the British Standard (BS) 1377 Part 9 and the American Society for Testing and Materials (ASTM) D1194. BS1377 refers to in situ plate bearing tests in this way:

"This method covers the determination of the vertical deformation and strength characteristics of soil in situ by assessing the force and amount of penetration with time when a rigid plate is made to penetrate the soil. Uses are to evaluate the ultimate bearing capacity, the shear strength and deformation parameters of the soil beneath the plate without entailing the effects of sample disturbance. The method may be carried out at the ground surface, in pits, trenches or adits, and at depth in the bottom of borehole"

It is common practice for the plate diameter in this test to vary, usually from 300 mm to 1000 mm. It is important to note that a bigger plate is often preferred, when available, in order to better mimic the actual conditions imposed by the foundation.

An important issue connected with the choice of plate size is the possible scale effect associated with testing soils where the ratio between plate diameter and maximum particle size is too small. With respect to this requirement, the standard BS1377 provides a specific indication of the minimum plate size which can be allowed in a plate bearing capacity test:
BS 1377-9:1990 (notes 4.1.2):

"When testing granular soil the plate diameter should exceed at least five times the nominal size of the coarsest material"

The implication of this limit provided by the standard is that the same response (in terms of stress-settlement) should be obtained for any plate size which is fulfilling this ratio value. There are, however, some difficulties in the interpretation and application of this guidance:

1. It is not completely clear what exactly is meant by "nominal size of the coarsest material". Dependent on interpretation is could relate to the maximum particle size, the D_{50} value or some other characteristic.
2. Working platforms are typically constructed using well graded sub-base granular material, such as one conforming to the 6F2 grading, which could be fresh aggregate or recycled demolition material (comprising concrete, brick and other materials). 6F2 grading is characterized by large particle size (up to 120 mm) and the large size of the particles would require a big plate diameter (up to 600 mm) in order to satisfy the limit proposed by the standard.

Assuming the "nominal size" refers to the maximum particle size of the material, the biggest problem related to plate bearing capacity tests remains, in this

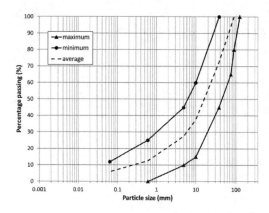

Figure 1. Grading curves representing minimum – maximum particle size values for 6F2 (solid lines) and the particle size distribution chosen to be scaled down for the test material (dashed line).

Table 1. Properties of test sample.

Property	Value
Minimum void ratio (e_{min})	0.332
Maximum void ratio (e_{max})	0.346
Specific gravity (G_s)	2.73
D_{50} (mm)	0.5
D_{max} (mm)	3.35
D_{min} (mm)	<0.18
γ, average value (kN/m^2)	20.05

application, the large size of the plate and the resulting high reaction load required to conduct the bearing test.

The main objectives of the research presented here is to understand if a smaller ratio between the diameter and particle size could be adopted during tests without changes in the results. In this way a cheaper procedure could be adopted for testing materials containing large particle sizes.

2 INTRODUCTION

A series of plate bearing capacity tests were carried out at City, University of London using the geotechnical centrifuge facility.

The tests were executed by the use of different plate sizes in order to verify if a scale effect can be associated with the use of plates with a diameter to maximum particle size ratio smaller than five (minimum value suggested in BS1377-9).

The material used for these tests is a grey Devonian limestone sourced from a quarry in Ashburton, Newton Abbot, UK. The limestone was graded with the intention of representing a scaled version of 6F2 material, commonly used for the construction of working platforms. Since the definition of 6F2 material covers a large range of particle size distributions, an average curve placed between minimum and maximum values of particle size distribution characterizing the 6F2 class (shown in Figure 1) was chosen as representative. The maximum particle size (90 mm) was then scaled down to a value of 3 mm, such that the acceleration level chosen to spin up the centrifuge model was equivalent to N = 30 g. The measured properties of the test material are summarised in Table 1.

The bearing capacity tests on this material were carried out using different plate diameters in order to verify the effect of diameter of the plate to maximum particle size ratio and therefore confirming or contradicting the indications presented in BS1377.

Assuming an acceleration level of 30 g, the plate diameters used (7.8 mm, 12 mm, 16.9 mm, 23.7 mm and 39.7 mm) represent prototype values of 234 mm, 360 mm, 507 mm, 711 mm and 1191 mm respectively. The corresponding B/D_{max} ratio (where B represents the plate diameter of the test and D_{max} the maximum particle size of the samples) was therefore equal to 2.3, 3.6, 5, 7.1 and 11.9 respectively, so that both higher and lower values of B/D_{max} ratio were tested to verify the effect of the ratio changes on the obtained stress/settlement curve for each test.

3 APPARATUS AND TESTING

3.1 Apparatus

The test was carried out in a circular centrifuge tub (working as container for the sample) with a loading frame above (whose function was to drive the plate into the soil at a constant rate of penetration equal to 1 mm/minute).

The test was driven for about twelve minutes such that the total penetration of the plate into the soil corresponded to approximately twelve millimetres, significantly further than might be expected in order to capture all features on the stress/settlement curve.

A large tub (having an internal height of 300 mm and a diameter of 420 mm) was chosen with the intention of avoiding boundary effects due to the proximity of the plate to the sides and base of the tub. The design chart presented by Ullah et al. (2017) provides a method to verify if the model geometry might be affected by boundary effects considering the ratio L_{BD}/D (where L_{BD} is defined as the distance measured from the centre of the plate to the inner edge of the tub and D is the diameter of the plate). It can be seen from the chart (Figure 2) that for uniform sand (D/H_s = 0, where H_s represents the thickness of sand) the minimum L_{BD}/D ratio allowed is equal to five. Considering that the maximum plate diameter used for the tests was equal to 39.7 mm, the diameter of the tub was considered large enough to prevent or reduce possible boundary effects.

The second important component of the testing apparatus, the loading frame, was used to push the plate into the sample and measure the force variation with increasing settlement. The frame consists of (Figure 3): a motor and screw jack assembly, a loading beam, a force plate and the test plate.

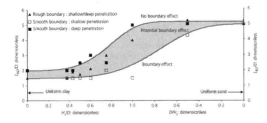

Figure 2. Centrifuge test design chart for estimating the safe normalized lateral boundary distance, Ullah et al. (2017).

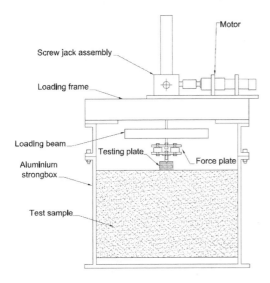

Figure 3. Test equipment and instrumentation.

The motor and screw jack drives the plate into the soil through the stiff loading beam, to which the force plate and the test plate were connected. The force plate is comprised of three load cells sandwiched between two stainless steel plates. Use of three loads cells in this arrangements prevented bending moments (which may arise from uneven seating of the plate on the test sample) to be eliminated. The total force acting on the plate was then calculated as sum of the readings from the three load cells.

Displacement of the plate was not measured directly but rather from knowledge of the precise speed of the jack and the time elapsed.

4 TEST SAMPLE PREPARATION

As a first step, the limestone was dry sieved using the method described in BS1377: Part 2 (1990). Once sieved, the different fractions were combined to create a particle size distribution corresponding to a scaled down (by a factor of 30) sample of 6F2 material.

The limestone was then placed into the tub, which was filled up in such a way that the height of sample was the same for each test (approximately 250 mm).

The sample was placed in around seven layers, each one comprising about 10 kg of material. The material was distributed inside the tub and each layer was accurately tamped before placing of the next one. The tamping operation was executed by hitting a heavy circular plate (placed on the soil surface) with a mallet. This led to some variance in compaction near the boundary which was corrected by manually tamping with a wooden block. This method of tamping gave a compact sample characterised by a low voids ratio of about 0.33. After the filling procedure, the distance (Δh) between the top of the tub and sample surface was measured in fourteen different positions in order to get an average height of the sample (calculated from the difference between internal height of the tub and average value of Δh). The height and diameter of the sample were used to evaluate its volume and, therefore, its voids ratio.

Preparation was completed by spinning the sample in the centrifuge for a short time (five minutes) with the intention of compacting the sample before starting the test. This further step ensuring that a repeatable, compact sample was obtained for each test. The height of the sample was checked again in order to ensure an accurate measure of voids ratio was obtained before testing.

4.1 Test procedure

After compaction the loading frame, instrumentation and test plate were mounted on the centrifuge model. The sample was spun in the centrifuge at an acceleration value of 30 g. During testing the motor and screw jack assembly pushed the plate into the sample at a constant rate of penetration, while the force plate measured the total force applied by the use of the three load cells.

Once the test was concluded it was possible to evaluate the settlement of the plate compared with the measured bearing capacity values. Therefore, for each test, a stress/settlement curve was generated. These curves were compared in order to evaluate the presence of any possible scale effect due to the use of small plate sizes.

5 TEST RESULTS

The results obtained from all tests are presented in Figure 4 as the variation of bearing stress (q) against the settlement of the plate (w).

From Figure 4 it can be observed that there is a general increase in plate bearing capacity with increase in plate size.

For a B/D_{max} ratio equal to 7.1 and 5, which considering the scale factor $N = 30$, represented prototype plates of 711 mm and 507 mm respectively, show a similar response. This is in accordance with the guidance from BS1377 Part 9 (1990), which suggests a plate diameter to nominal particle size ratio larger than five.

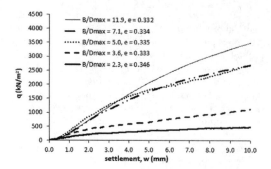

Figure 4. Bearing stress–settlement variation obtained from testing the same granular soil with different plate diameter sizes.

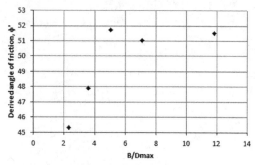

Figure 5. Angle of friction of platform material obtained from test results.

For a B/D_{max} ratio equal to 11.9 (representing a prototype plate of 1191 mm), the results showed higher values of bearing capacity compared with 7.1 and 5 ratios. The difference in stress value seems to increase with the settlement and may be related to a boundary effect due to a low L_{BD}/D ratio. Given that $L_{BD}/D = 5.3$ which is a value just larger than the minimum value of 5 indicated by Ullah et al. (2017), it is reasonable to consider the presence of an effect on results due to the proximity of the tub boundaries.

Tests conducted with a B/D_{max} of 2.3 and 3.6 (representing prototype plates diameter of 234 mm and 360 mm respectively) show significantly different results when compared with tests conducted at $B/D_{max} = 7.1$ and 5. They displayed lower bearing capacity values, apparently decreasing with the size of plate. For these tests the plate diameter to maximum particle size ratio was significantly lower than that recommended.

6 BACK CALCULATION OF FRICTION ANGLE

Plate bearing tests are often used to confirm the working platform design. The primary input into the design of the platform is the angle of friction of the granular material. The effective angle of friction can be back calculated from the results obtained using a simple bearing capacity formulation for a circular footing, shown in Equation 1 (Das, 2010):

$$q_{ult} = \sigma'_{zD} N_q + 0.3 \gamma' B N_\gamma \quad (1)$$

where q_{ult} = ultimate bearing capacity; σ'_{zD} = vertical effective stress at the depth the foundation is laid; γ' = effective unit weight; B = diameter of the foundation; N_q, N_γ = bearing capacity factors.

Figure 5 shows the angle of friction obtained by this method for each of the tests. It can be seen that once the plate diameter exceeds five times the maximum particle size the angle of friction is relatively constant at around 51.5°. At these high friction angles small variations in the value adopted would have a significant impact on the predicted capacity of any working platform design.

7 CONCLUSION AND FURTHER WORK

From the series of bearing capacity tests carried out using centrifuge modelling techniques it can be observed that plate diameter to maximum particle size ratio has an influence on results concerning the values of bearing capacity of the soil.

In particular, the results seem to confirm the validity of the BS1377 Part 9 (1990), which impose for plate tests on soils a plate diameter to nominal particle size exceeding five. This nominal particle size can be considered to be the maximum particle size in the material.

For plates corresponding to lower ratios the soil showed a different response manifesting a lower value of bearing capacity, which seems to decrease with reducing the plate diameter.

Another observation is related to the boundary effect which was found when testing the sample with the largest plate diameter. This phenomenon was observed for a test characterized by a boundary distance to plate diameter ratio equal to 5.3, very close to the lower value of 5 according to the chart presented by Ullah et al. (2017). It should, of course, be noted that this type of effect is unlikely to be present during full scale site testing.

Further tests could be carried out in future in order to investigate if a plate diameter to maximum particle size ratio between 3.6 and 5 could be used during testing without a change in results. This would be useful to understand if a slightly lower limit of ratio could be allowed without scale effects on results. This would then permit a smaller diameter (and thus cheaper equipment) to be used when testing granular soils having a large particle size.

REFERENCES

British Standards Institution. 1990c. British Standard Methods of Test for Soils for Civil Engineering Purposes: Part 2.

Classification Tests, BS 1377. *British Standards Institution, London.*

British Standards Institution. 1990b. British Standard Methods of Test for Soils for Civil Engineering Purposes: Part 9. In-Situ Tests, BS 1377. *British Standards Institution, London.*

Das, B., 2010. Principles of Foundation Engineering (7th edition). *Cengage, Stamford, USA.*

Highways Agency 2004. Manual of Contract Documents for Highway Works – Volume 1 Specification for Highway Works. *The Stationary Office, London.*

Halai, H., McNamara, A.M. & Stallebrass, S.E. 2012. Centrifuge modelling of plate bearing tests. *Second European Conference on Physical Modelling in Geotechnics, 23-24 April, Delft University of Technology, Netherlands.*

Ullah, S.N., Hu, Y., Stanier, S. & White, D. 2017. Lateral boundary effects in centrifuge foundation tests. *International Journal of Physical Modelling in Geotechnics* 17(3): 144–160.

Geotechnical model tests on bearing capacity of working platforms for mobile construction machines and cranes

R. Worbes & C. Moorman
Institute for Geotechnical Engineering, University of Stuttgart, Stuttgart, Germany

ABSTRACT: In a variety of construction projects, e.g. the construction of wind power plants and deep foundations, heavy working machines must be used on soft and low-bearing ground. In context of the research project "Bearing Layers for mobile Construction Machines and Cranes" the failure mechanism of geosynthetic reinforced multi-layered systems and the complex interaction between construction machines and supporting layers have been clarified by coupling model tests, field tests and numerical simulations. The performed 1g-model tests simulate the loading of construction machines on reinforced support layers subtended by a soft layer under static and cyclic loading conditions in the scale 1:3. This paper presents the results of model tests comparing the bearing and deformation behaviour of reinforced and an unreinforced support layers for working platforms.

1 INTRODUCTION

For the use of heavy mobile construction machines, e.g. drilling and trench wall units, rams, vehicle and crawler cranes, temporary working platforms are often created in the form of poured and compacted earth building materials, which are partially reinforced with geosynthetics. Particularly in the case of heavy working machines, the bearing capacity of the underlying subgrade is often insufficient to ensure a safe and usable installation considering all relevant operating and loading conditions. In a variety of construction projects, e.g. the construction of wind power plants and the production of deep foundations, heavy working machines must be used on soft and low-bearing ground. The use of working platforms and their correct dimensioning are therefore of fundamental importance for the durability of the construction machines and thus for work safety. In this context, there is a need for optimization, since the requirements of the construction machines are often not in line with the working platform design, and there are no commonly accepted technical regulations available for the design of temporary work platforms made of unreinforced and reinforced supporting layers. The aim of the research project, initiated by the Institute for Geotechnical Engineering at the University of Stuttgart, is the development of a design approach able to guarantee a safe and usable installation of mobile machinery under construction site conditions. Based on the acquired knowledge, a recommendation for the dimensioning, construction, testing and maintenance of temporary working platforms shall be derived, that should be able to optimize the working platform design both from technical and economical point of view. The research strategy is based on experimental and numerical investigations. The numerical simulation models are validated by measured data obtained from small-scale model tests and large-scaled field tests. Based on a numerical parameter study, the influence of geometric and geotechnical parameters is investigated to gain an improved understanding of the bearing and deformation behaviour of reinforced two-layer systems.

2 ANALYTICAL DESIGN APPROACHES FOR REINFORCED BEARING LAYERS

For the dimensioning of reinforced working platforms, various approaches are available for determining the bearing capacity of reinforced two-layer systems made of support and soft layer are available in literature. These are mainly based on the principle of load distribution or reduction on the underlying subgrade. In this case, the assumption is usually that the height of the supporting layer and the shear strength are chosen in such way, that a basic fracture failure occurs exclusively in the soft layer. After application of these approaches for load distribution due to the bearing layer, the bearing capacity of the subgrade is determined with the resulting stresses. In some of the design methods, the required bearing layer heights can be calculated directly.

2.1 Design approach after EBGEO

A widespread design approach is the EBGEO shown in Figure 1 for the dimensioning of reinforced foundation pads, which approximately corresponds to the system of temporary working platforms. In this approach, the ultimate bearing capacity is determined analogous to DIN 4017, with an additional vertical resistance increase due to the geogrid reinforcement, which is

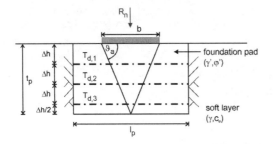

Figure 1. Design approach after EBGEO for foundation pads.

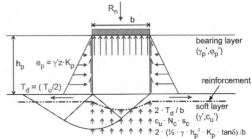

Figure 3. Design after BR 470 (2004/2007) based on Meyerhof's approach (1974) for reinforced support layers.

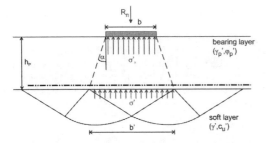

Figure 2. Design approach after Giroud & Nioray (1981).

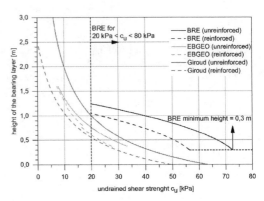

Figure 4. Comparison between the approaches for reinforced and unreinforced support layers for different ground condition.

dependent on the tensile strength, the pull-out resistance and the number of geogrid layers. Influence of inhomogeneity resulting from the foundation pad is considered with corrective factors. Disadvantage for the use for dimensioning working platforms is the limitation to a minimum number of two reinforcement layers with the same vertical distance between each layer.

2.2 Design approach after Giroud and Nioray

Figure 2 shows the approach according to Giroud & Noiray (1981), which applies a load distribution in the supporting layer and thus achieves a vertical stress reduction on the soft layer. This approach was originally developed as a semi-empirical evaluation method for unpaved roads reinforced with geosynthetics. The influence of the bearing behaviour of geosynthetics is considered by assuming an additional tensile membrane effect.

2.3 Design approach after MEYERHOF (BR 470)

A further possibility is the design approach by Meyerhof (1974) shown in Figure 3, where the support layer is considering the friction in the vertical fracture surface. The frictional force results here from the passive earth pressure, which is inclined by the wall friction angle δ and causes a reduction of the load in the subgrade directly below the load surface. The BR 470 from Directive of the British Research Establishment (2004/2007) is based on the design approach after Meyerhof and integrates it into a design method for self-propelled tracked working machines. This design method allows the dimensioning of unreinforced and reinforced working platforms. The bearing capacity of geosynthetics is only taken into account in a very simplified manner by considering an additional vertical reduction of the load on the subgrade due to the tensile membrane effect.

2.4 Comparison between the different design approaches

In the following section, the above-mentioned design methods for reinforced bearing layers are compared with the unreinforced ones in an example calculation. Similar to the example of Kleih et al. (2009), a large drilling plant with an additional load of 25.0 tons at the front and a chain width of 0.80 m is used for loading. In addition, the mast of the plant is tilted forward by 5° and the most unfavourable distortion of the superstructure is set by 12.5° to the driving direction. A coarse gravel with a friction angle of $\varphi' = 40°$ is assumed for the support layer. The required bearing layer heights in Figure 4 were calculated as a function of the undrained shear strength c_u for unreinforced and reinforced working platforms. Although the geosynthetic reinforcement of the bearing layers results in a considerable reduction in the thickness for all the design methods, significant differences in the required height of the support layer for different subgrade conditions are found depending on the applied design approach. It should be noted that the compared method according to BR470 is not permissible for undrained shear strengths less than $20\,kN/m^2$. In addition, it can be shown that results for higher shear strengths compared to EBGEO and Giroud's assessment approach are rather conservative.

3 EXPERIMENTAL CONCEPT

Using geotechnical model tests, it is possible to obtain deepened soil mechanic findings of the bearing and deformation behaviour of unreinforced and reinforced support layers over ground layers of low stiffness, under static and cyclic loading conditions. Aim of the experimental test concept hereby is the investigation of the failure form (fracture figure, perforation and slip surfaces) of unreinforced and reinforced two-layer systems. The serviceability states are represented by realistic load assumptions, frequencies and load cycle numbers of typical construction machines; the accumulation of deformations is considered. Furthermore, the influence of geogrid reinforcement and the geogrid behaviour is investigated during load transfer. The findings to be gained on this basis are the fundamentals for technically and economically optimized design approaches.

3.1 General testing setup

The geotechnical model tests are carried out as 1-g tests on a 1:3 scale. Figure 5 and 6 illustrate the geometry and the arrangement of the measuring sensors of the experiment. The basal area of the test field is 4.82 m × 2.72 m, in which two model tests can be carried out separately from each other. Each test field has the dimensions 2.41 m × 2.72 m. The subgrade, represented by a layer of loess loam (SC/CL classification according to USCS) with an undrained shear strength of $20\,kN/m^2$, has a height of 0.80 m. The shear strength cu of the soft layer can be varied from $10\,kN/m^2$ to $30\,kN/m^2$. The soil parameters of the densified loess loam are controlled by the moisture content and the undrained shear strength is measured by in-situ vane tests. Above the soft layer, the bearing layer is installed with a thickness of 0.20 m, which can be varied from 0.10 m to 0.30 m. For the installation of the bearing layer a well-graded sand-grit mixture with a grain size varying between 0 mm and 16 mm is used. The gravel mixture is incorporated with a proctor density of $D_{Pr} = 100\%$. The size of the load plate is 35 cm × 25 cm, and the vertical test load is applied with an eccentricity $e = 0.04 \times B = 1$ cm relative to the shorter foundation side to provide the direction of the basic fracture. Deformations on the surface are measured by potentiometric distance sensors at nine points. In the reinforcement test, the strains in the geogrid are additionally measured at seven points by strain gauges. In the unreinforced test, a nonwoven is used as a separating element between the soft layer and the

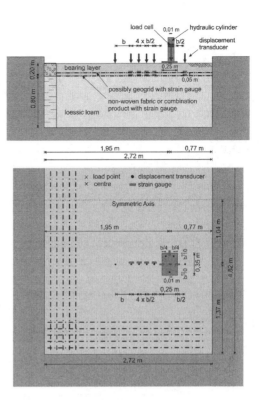

Figure 5. Setup and dimensions of the experimental test for the performed model tests.

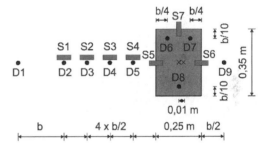

Figure 6. Arrangement of the potentiometric distance sensors and the strain-gauges on the geogrid around the load plate.

support layer. It can be assumed that that the nonwoven membrane has no reinforcement function. For the reinforcement of the second test a composite product made of a biaxial geogrid (laid, welded knots) with a maximum tensile strength of 30 kN/m and a nonwoven was used. This combination product has also been placed between the soft layer and the bearing layer fulfilling both the functions of reinforcing and separating.

3.2 Loading scheme

The loading scheme, shown in Figure 7 can be divided in three stages. The first stage is the monotonous loading, in which the load is initially increased to the

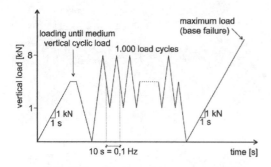

Figure 7. Load concept with three stages: initial loading, cyclic loading and maximum loading.

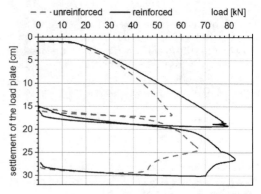

Figure 9. Comparison of the settlement of the load plate on thereinforced and unreinforced support layer for maximum loading.

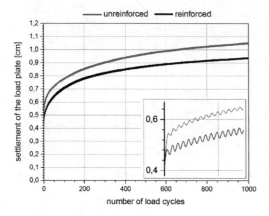

Figure 8. Comparison of the settlement of the load plate on the reinforced and unreinforced support layer for cyclic loading.

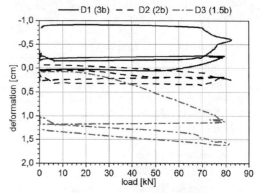

Figure 10. Deformation of the measuring points D1, D2 and D3 for the unreinforced support layer.

average cyclic load with a velocity of 0.1 kN/s. This causes plastic deformations prior to the cyclic loading stage and gives information about the initial stiffness. After that, the relief starts and the second stage, the cyclic loading, which simulates load effects under operating conditions, begins. In the second stage 1000 load cycles with a frequency of 0.1 Hz and an amplitude of 3.5 kN between 1 kN and 8 kN were applied. During the final stage, the load is increased up to a defined failure state, in order to obtain the bearing capacity.

4 TESTING RESULTS

In the following, the results of an unreinforced and a reinforced test are presented and compared to the analysis of the influence of the geogrid reinforcement.

4.1 Cyclic loading

Figure 8 shows the vertical settlement of the load plate during cyclic loading on the unreinforced and on the reinforced support layer. Obviously, the unreinforced bearing layer initially deforms more strongly, but for the cyclic load there is no substantial difference in the accumulation of permanent deformations. The settlement of the load plate has almost doubled in both systems, compared to the initial load, after about 1000 load changes. The reinforced system shows greater deformations intervals for each load cycle, and therefore a larger elastic deformation region.

4.2 Settlement and bearing capacity

Figure 9 shows the comparison of the load-settlement curves of the load plate for the unreinforced and the reinforced system in case of static load applied after the cyclic load. The stiffness of both two-layer systems is comparatively high up to almost 16 kN and there is almost no increase of the deformation. The main reason for this is the compression by the cyclic preload with a maximum load of 8 kN during stage two. At approximately 56 kN, the load plate is relieved due to the maximum press stroke in the unreinforced system. The relief shows high plastic deformations of 15 cm.

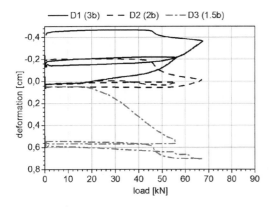

Figure 11. Deformation of the measuring points D1, D2 and D3 for the reinforced support layer.

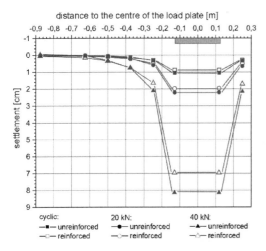

Figure 12. Comparison of the settlement of the load plate for the measuring points D1 to D9.

The plate is then loaded again and the bearing capacity is reached at 67.6 kN. The settlement of the plate increases from 24 cm to almost 30 cm, although the load decreases and the system fails. The reinforced system initially has approximately the same stiffness. The geogrid reinforcement requires sufficient deformation and settling to activate tensile forces. From about 30 kN the stiffness compared to the unreinforced system is significantly higher. The relief due to the maximum press stroke is at a load of approximately 80 kN. The plastic deformation of the bearing layer under the load plate is about 15 cm, which corresponds to the unreinforced system, although the load was about 45% higher. The maximum load is reached at 82.5 kN. The setting of the plate increases from 27 cm to 30 cm, although the load decreases. The maximum setting of both systems is almost identical. The bearing capacity of the reinforced system is 22% higher than the system with an unreinforced base layer.

Figure 8 shows the vertical displacements on the surface of the bearing layer at the measuring points D1 (3b), D2 (2b) and D3 (1.5b) for the unreinforced support system (location of the points according to Fig. 5b). Positive deformations mean lowering and negative deformations mean lifting. The measuring points D1 to D3 are the most distant to the lead plate and thus less influenced by the subsidence cavity around the load plate. Also, these points mark the transition point from settlement to heaving. At the beginning of the static load, small settlements are already present at the point D1 because of the first static and the cyclic preload. After a small increase of the load, the ground begins to rise up. The lifting on release is 0.2 cm. In the case of reloading, heaving increases to the maximum load bearing capacity of 67.6 kN. After reaching the basic failure load, further lifting occurs, although the load decreases. Point D2 shows almost no deformation up to the maximum load. Once the maximum load-bearing capacity is reached, the ground heaves by 0.2 cm. Point D3 continues from the beginning to the maximum vertical load and then heaves slightly. These slight lifts are probably due to elastic deformation caused by the relief of the soil.

The load-deformation envelopes at the measurement points D1 (3b), D2 (2b) and D3 (1.5b) for the reinforced system are shown in Figure 11. The displacements of the point D1 are affine to the unreinforced system, but the elevations are significantly higher. At the application of the vertical load, the ground settles at point D2. When reaching the maximum bearing capacity, the bearing layer starts lifting slightly. At the distance of 1.5b on point D3, the settlements are significantly greater than in the unreinforced system. The larger deformations for the reinforced system for this area can be explained by increasing the size of the subsidence cavity, due to the load spreading and the tensile membrane effect of the geogrid reinforcement. Figure 12 illustrates the vertical deformation along the measurement axis for the points D1 to D9 for the unreinforced and reinforced bearing layer. The settlement of the load plate is calculated from the results of the points D6 and D7. The deformations after the cyclic loading are approximately the same for the loading until 20 kN, although the deformation for the reinforced system is slightly lower. As the load is increased, the difference between both systems becomes more significant, due to the activation of the Geogrid reinforcement and the tensile membrane effect. The influence of this effect is locally limited to a range of about 1.5b around to the load plate, due to the punching of the load plate into the sub layer.

4.3 Strain-development

The measured strains in the geogrid reinforcement along the measuring axis of the second test are shown in Figure 13. It can be noticed that the strains in the geogrid are very small for lower loading conditions up to 20 kN (25% of the bearing capacity) and remain limited to a range of 1.5b distance around the centre

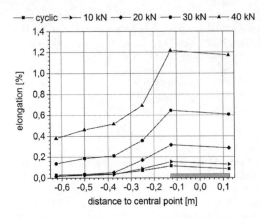

Figure 13. Measured strains on the geogrid reinforcement between the supporting layer and the subgrade.

of the load plate. At approximately 30 kN, a significant increase in the strain is achieved at all measuring points. With an increase in load to 40 kN, the strain on all measuring points increases disproportionately. The geogrid reinforcement at the peripheral measuring points S1 to S3 is activated much later, so that in contribution of these areas in transferring the load happens only at higher loads and thus larger deformations. The strains up to a load of 20 kN largely correspond to the lateral restraint forces due to load distribution, whereas the strain of 40 kN is additionally superimposed by the tensile membrane effect.

5 FAILURE MECHANISM

The failure mechanism consists of a combination of punching the bearing layer and base failure of the subgrade. In the case of an unreinforced system, base failure takes place after perforation of the supporting layer. Due to the low shear strength of the soft layer, there is no general shear failure and the system changes to a punching shear failure with increasing deformations. For the reinforced system both the perforation of the support layer and the punching shear failure take place long before the reinforcement fails. The final failure state, from which no further increase of the vertical load is possible, is marked by the failure of the geogrid. The geogrid used in the test rips under the edge of the load plate, due to the increased tensile forces in this area.

6 CONCLUSION

The comparison between a non-reinforced and a reinforced support system in the small-scale model tests shows that the maximum bearing capacity can be significantly increased with a geogrid reinforcement. The reinforcement improves the deformation behaviour, especially at higher loads due to the load spreading of the tensile membrane effect. The strain measurements show that geogrid-reinforcements clearly optimize the load and deformation behaviour. The bearing capacity in this case is increased about 22% and the settlements are reduced from 10% to 15%. For low stress conditions, both the unreinforced and reinforced system show a similar bearing behaviour. This depends on the minimum of deformation that is necessary for the activation of the geo-grid forces. The geogrid reinforcement also increases the elastic deformation region for unloading and reloading cases e.g. in cyclic loading conditions.

In context of the research project further model scale tests with various parameter setups are planned. Parameters to be examined are: the influence of the height of the support layer, the shear parameters of the soft layer and other influencing factors of the geosynthetics, like stiffness, arrangement and type of the geosynthetics.

REFERENCES

BRE – Building Research Establishment 2004/2007. *Working platforms for tracked plant: good practice guide to the design, installation, maintenance and repair of ground-supported working platforms (BR 470)*. IHS BRE Press, Bracknell, Berkshire, ISBN 186081 7009.

Deutsche Gesellschaft für Geotechnik / German Geotechnical Society 2010. *Empfehlungen für den Entwurf und die Berechnung von Erdkörpern mit Bewehrungen aus Geokunststoffen – EBGEO*, 2nd Ed., Berlin: Ernst & Sohn.

Deutsches Institut für Normung / German Institute for Standardization 2006. *DIN 4017:2006-03 "Baugrund – Berechnung des Grundbruchwiderstands von Flachgründungen"*.

Giroud, J.-P. & Noiray, L. 1981. Geotextile-Reinforced unpaved road design. *Journal of the Geotechnical Engineering Division* Vol. 107 (No. GT9), pp. 1233–1254.

Kleih, J. et al. 2009. Anforderungen an das Arbeitsplanum zur Gewährleistung der Standsicherheit von Spezialtiefbaugeräten. *BauPortal* 9/2009: 499–503.

Meyerhof, G. G. 1974: Ultimate bearing capacity of footings on sand layer overlying clay. *Canadian Geotechnical Journal* Vol. 11 (No. 2): 223–229.

1g physical modelling of the stoneblowing technique for the improvement of railway track maintenance

A.A. Zaytsev, A.A. Abrashitov & A.A. Sydrakov
Russian University of Transport RUT (MIIT), Moscow, Russia

ABSTRACT: Under cycle rail loading ballast is constantly degrading and thus requires periodical recovery. The commonly used method of rail track recovery – ballast tamping – possesses numerous shortcomings, which lead to high costs of track maintenance. Stoneblowing was suggested as an alternative technique of track level adjustment, free of tamping disadvantages. Unfortunately, application of this prospective technique still remains limited. This paper is focused on comparative study of tamping and stoneblowing techniques conducted by means of 1g physical modelling of recovered ballast. The difference in contamination levels and transverse resistance of railway track panel after different surfacing procedures. Conducted laboratory tests demonstrate that ballast, adjusted by stoneblowing, showed better mechanical characteristics in comparison to ballast adjusted by tamping.

1 INTRODUCTION

Tamping procedures comprise the adjustment of the sleepers' vertical and horizontal alignment and subsequent ballast packing beneath the sleepers. Due to numerous shortcomings, including ballast damage (Counter 2015), loosening and subsequent reversion of the ballast to its pre-maintenance state (Esveld 1989), application of this method leads to high track maintenance costs. Pneumatic ballast injection, or stoneblowing, is an alternative technique, originally proposed in the UK (Anderson et al. 2000). It is reported to be a more delicate technique that does not affect the existing well-compacted ballast and, thus, results in a better track quality (Claisse 1992, McMichael 2003). Unfortunately, this method has not found a wide application today. In this paper a comparative analysis of tamping and stoneblowing techniques is performed by means of 1g physical modelling of recovered ballast. In these experiments focus was on contamination and mechanical properties of the recovered ballast.

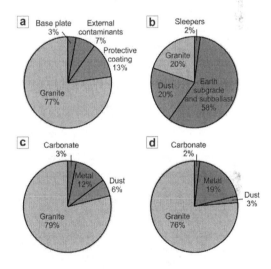

Figure 1. Petrographic studies of ballast from (a) UK railroads (adapted from Selig et al. 1996), (b) Australian railroads (adapted from Indraratna 2006), (c, d) Sample 1 and 2 respectively.

2 BALLAST CONTAMINATION SOURCE STUDY

Firstly, the main source of ballast contamination was investigated, since it can vary considerably (Fig. 1a, b). The process of ballast degradation is mainly dictated by contamination, thus, it is important to determine the nature of contaminants in ballast samples to develop correct new physical modeling tests.

For this purpose two ballast samples from railways with different operating and climatic environments were collected. These samples were used in all further experiments. Sample 1 was collected at the railway turnout of the Likhobory station on a site on the Small Ring Line of Moscow Railways, which is predominantly used for the passenger train operation. Sample 2 was collected on the Kovdor – Pinozero railroad section, where the main traffic is predominantly freight. The cone quartering method according the Russian State Standards (GOST P 54748 – 2011, GOST 25100-2011) were used to obtain representative samples of 8–10 kg mass for further petrographic study.

The results of the petrographic study are presented in (Fig. 1 c, d). The main source of contamination for both samples is granite, thus, the contamination

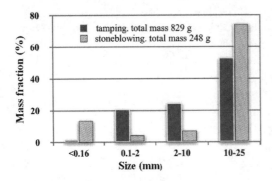

Figure 2. Comparison of size distributions of ballast particles after tamping and stoneblowing procedures.

is mainly caused by ballast degradation under high loads. The presence of metal in both samples may be explained by the fact that the first sample was collected on the railway turnout, whereas the second sample was collected in the vicinity of an iron ore mine. The comparison of the result of the petrographic study with the reference data for British railways shows a strong similarity between the two samples.

3 INFLUENCE OF DIFFERENT SURFACING TECHNIQUES ON CONTAMINATION

The influence of tamping and stoneblowing methods on overall ballast contamination were investigated in further experiments (Abrashitov et al. 2016).

The experimental setup included a universal testing machine EUS-40, Leipzig (frequency 5–25 Hz, maximum load 400 kN, box (300 × 600 × 400 mm) and metal press 240 × 250 mm). The experiments were conducted with Sample 1 (Likhobory station).

The tamping procedure and ballast settlement was modelled in the following way: the sample underwent dynamic loading for 5×10^5 cycles (12 Hz frequency), then ballast was poured out of the box and pressurized. After that it again underwent the same cyclic dynamic loading.

The stoneblowing procedure and ballast settlement were modelled in the following way: sample underwent dynamic loading for 10^6 cycles.

After load removal the sample was sieved through standard sieves.

The particle size distributions for the samples after loading in the two different regimes are presented in Figure 2. The obtained results demonstrate that in case of loading with repacking, the mass fraction of contaminants is shifted to smaller sizes. Small particles of degraded ballast penetrate into the space between large ballast stones and complicate further tamping procedures.

The presence of small particles (<0.16 mm) can lead to water accumulation in the ballast prism. It should also be emphasized that the overall contaminant fraction is 2–10 times higher in the case of the tamping compared to stoneblowing.

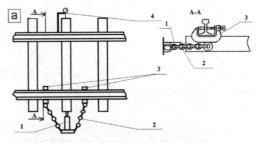

Figure 3. Schematic view of the experimental setup: 1 – hydrocylinder; 2 – metallic chain; 3 – rail clamp; 4 – bracket with dial gauge.

It is evident that if the rail track is maintained by methods that involve repacking (e.g. tamping) the ballast is degrading faster due to a higher extent of contamination by small granite particles.

4 INFLUENCE OF DIFFERENT SURFACING TECHNIQUES ON TRACK-BALLAST TRANSVERSE RESISTANCE

4.1 Contribution of different components to track-ballast lateral resistance

In the following experiments the contribution of different components (Vinogorov 2005), including resistance between the sleeper pad and ballast (1), resistance between the sleeper sides and ballast (2) and resistance of the ballast shoulder (3) were investigated.

In order to distinguish between different components of friction resistance were measured resistance of three samples:

– example 1: a fully covered sleeper (total lateral resistance)
– example 2: a sleeper shoulder of ballast prism (resistance between sleeper sides and ballast and resistance between sleeper pad and ballast)
– example 3: a sleeper without a shoulder of ballast prism and without a tie plate (resistance between sleeper pad and ballast)

The experiments were conducted at a 1:11 railroad switch setup used for training, situated on the territory of RUT (MIIT). The investigated sleeper was prepared according to the experimental scheme (Examples 1, 2 or 3) and instrumented as shown in Figure 3, with special devices such as: a hydrocylinder, a metallic chain, rail clamps and brackets with a dial gauge (Fig. 3). the threshold load values at which the sleeper began to slide were measured. The threshold load of was

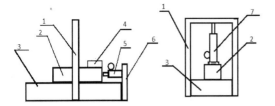

Figure 4. Schematic view of the experimental setup: 1 – frame; 2 – sleeper sample; 3 – box with ballast; 4 – areal vibrator; 5 – horizontal lifting jack with pressure sensor; 6 – fixed stop; 7 – vertical lifting jack with pressure sensor.

2.65 kN is for Example 1, 2.39 kN is for Example 2 and 1.41 kN is for Example 3. It may be concluded, that the main source of lateral ballast resistance is the resistance between the sleeper pad and ballast (52%). Thus, in further laboratory experiments only this factor was considered.

4.2 Modelling lateral resistance of railway track panel, reinstated by different maintenance techniques

The experimental setup used for these series of experiments was a vertically fixed with a firm metallic frame above a box (40×30×150 cm) filled with ballast gravel (Fig. 4). The following ballast materials were used for the experiment:

– ballast, collected at the site of the Likhobory station of Small Ring of Moscow Railways around a railway frog (with 30% contaminant fraction) i.e. a dirty sample;
– commercially available small-sized gravel (5–10 mm) as a stoneblowing material.

The sleeper sample of ShS-ARSÍ type (R-65 rail, ARS-4 fastening type) with lateral dimensions 28×65 cm was placed on the ballast surface (Fig. 4).

For the vertical load a hydraulic lifting-jack (maximum load 110 kN) was used, which pushed the frame and the rail apart, applying vertical pressure. For the horizontal shear load a hydrocylinder (maximum load 50 kN) was used, which pushed a fixed stop and the rail apart, applying horizontal pressure (Fig. 4). Both devices were equipped with pressure gauges, which measured pressure applied to the sample. The pressure gauge used for vertical load measurement was additionally calibrated with the help of the universal testing machine EUS-40.

The vibration of the rolling stock was simulated with an aerial vibrator IV-99B fixed on a sleeper sample. A frequency converter was used to reduce the vibration frequency to the 12 Hz required for the experiment.

Vertical loads of 0, 10, 30 and 50 kN were subsequently applied to each of the samples.

The maximum applied load 50 kN was chosen to provide stress $Pb = 0.320$ MPa under the press which corresponds to the maximum axle load in the Russian Federation (SP 119.13330, 2012).

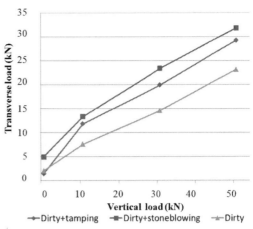

Figure 5. Dependency of transverse load from vertical load for different samples.

Experiments were conducted with the ballast materials listed above using different degree of packing. Ballast packing was performed by loading the sleeper sample with a vertical load of 49 kN at a constant frequency of 12 Hz with an amplitude of 1 mm. In total, 50,000 cycles were applied during the packing procedure.

During the shear tests the load was measured at a sleeper displacement of 0.4 mm. The same shift value was exploited in the early experiments of Bromberg (1966). Every measurement was repeated 5 times and the final results were averaged.

The stoneblowing technique was modelled in the following way:

– after reaching the required packing degree the sample sleeper was lifted by 40 mm above the ballast;
– gravel with a fine grain size was introduced under the lifted sample sleeper using a pneumatic injector and the resulting sample was packed with 50,000 additional loading cycles.

The tamping technique was modelled in the following way:

– after reaching the required packing degree the sample sleeper was lifted by 40 mm above the ballast and the tamping was performed by air drill in the chipping regime;
– the resulting sample was packed with 50,000 additional loading cycles.

The "dirty" sample, adjusted by the stoneblowing method shows a better transverse load at different vertical load values compared to the one adjusted by tamping (Fig. 5). Moreover, one can see that the lateral resistance without load is improved by a factor of 2.4, while for the transverse resistance for the sample adjusted by tamping without load, resistance did not change significantly.

The influence of the stoneblowing method on ballast settlement was investigated by means of the model system. The experiments were conducted with ballast from the site of the Likhobory station. It was demonstrated, that the stoneblowing method provides good results if applied two times. Even under a higher load the settlement level remains unchanged, though this level is higher than in the case of normal load.

5 CONCLUSIONS

The results of the petrographic study of the ballast contamination sources are presented the contamination is mainly caused by ballast degradation under high loads.

Comparative analysis was performed for stoneblowing and tamping techniques by 1 g physical modelling of recovered ballast. The focus on contamination levels and lateral resistance values and preliminary tests was performed to develop correct physical models. The results of the laboratory experiments clearly demonstrate that the ballast adjusted by stoneblowing possesses better mechanical characteristics in comparison to ballast adjusted by tamping. To prove fully the advantages of the stoneblowing technique, experiments are being carried out in the real-world environment.

The stoneblowing technique and the tamping technique were modelled for the transverse load investigation and after reaching the required packing degree the sample sleeper. It was demonstrated, that the stoneblowing method provides good results if applied two times. Even under a higher load the settlement level remains unchanged, though this level is higher than in the case of common load.

REFERENCES

Abrashitov, A.A, Zaytsev A.A., Semak A.A., Shavrin L.A. 2016. The estimation of pollution sources at the ballast layer from the granite crush stone and modeling of the destruction and abrasion of the ballast particles under the dynamic pressure. *Proceedings of the International Conference of the Projection, Construction and Maintenance of the Railway Track / Shahunyants Readings'XIII, Moscow, Russia, 31 March-01 April 2016* (in Russian).

Abrashitov, A. 2014. Stoneblowing by crushed rock ballast railway maintenance. *Put' i putevoe hozjajstvo (Railtrack and facilities)* 5: 11-14 (in Russian).

Anderson, W. F. & Key A. J. 2000. Model testing of a two layer railway track ballast. *Proc. Am. Soc. of Civil Engrs, J. of Geotechnical and Geoenvironmental Engng.* 126(4): 317-323.

Bromberg, E. 1966. *Stability of continuous welded rails* – Moscow, Transport publisher.

Claisse, P. 2003. Tests on a two-layered ballast system. *Proceedings of the Institution of Civil Engineers-Transport* 156(2): 93-102.

Counter, B. Franklind, A., Tannc, D. 2015. Refurbishment of ballasted track systems; the technical challenges of quality and decision support tools. *Construction and Building Materials.* 92: 51-57.

Esveld, C. 1989. Track Geometry and Vehicle Reactions. *Rail Engineering International Edition.* 4: 13.

Indraratna B., Salim, W., Khabbaz, H., Christie, D. 2006. Geotechnical properties of ballast and the role of geosynthetics in rail track stabilization *Sydney Geotechnical Consulting Division, Rail Corp (NSW).*

GOST P 54748. 2011. *Crushed stone from rocks for railway ballast. Specifications.*

GOST 25100. 2011. *Soils. Classification.*

McMichael, P and McNaughton, A. 2003. The Stoneblower-Delivering the Promise: Development, Testing and Operation of a New Track Maintenance System. *TRB Annual Meeting CD-ROM.*

Selig, E.T. 1994. *Track Geotechnology and Substructure Management.* Thomas Telford Services Ltd., London.

SP 119.13330. 2012 SNiP 32-01-95. Railway and gauge tracks 1520 mm.

Vinogorov, N.P. 2005. Stability of continuous welded rails *Put' i putevoe hozjajstvo (Railtrack and facilities)* 8 (in Russian).

3. Physical/Numerical interface and comparisons

Millisecond interfacing of physical models with ABAQUS

S. Idinyang
Nottingham Centre for Geomechanics, Faculty of Engineering, University of Nottingham, Nottingham, UK

A. Franza
Department of Engineering, University of Cambridge, UK

C.M. Heron & A.M. Marshall
Nottingham Centre for Geomechanics, Faculty of Engineering, University of Nottingham, Nottingham, UK

ABSTRACT: Hybrid modelling is an experimental methodology that integrates information from physical and numerical modelling components running in parallel with the aim of achieving a more realistic and accurate representation of a complex system. Interfacing physical and numerical models requires an efficient, reliable and fast data coupling method that enables the exchange of state updates (e.g. loads, displacements). This paper presents a hybrid analysis method for the study of tunnel-building interactions within a geotechnical centrifuge, referred to as the coupled centrifuge-numerical model (CCNM). The method incorporates a physical model in a geotechnical centrifuge and a numerical simulation using the finite element analysis software ABAQUS. Previously developed methods for communicating between ABAQUS and external programs used time-costly methods that delayed data transmission. The work presented here shows how the interfacing challenge can be overcome by adopting a network infrastructure within ABAQUS using FORTRAN subroutines. The new scheme can transfer data to and from ABAQUS in a 20 ms timescale, a vast improvement over previous hybrid testing applications This solution is beneficial to studies that need sub-second data interfacing between ABAQUS and an external interface.

1 INTRODUCTION

The capabilities of physical and numerical modelling applications have developed considerably in recent times, yet each modelling approach has its limitations. Physical modelling using a geotechnical centrifuge enables the replication of full-scale ground stresses within small scale models and has provided invaluable data related to many geotechnical problems. Sophisticated control systems have enabled the simulation of many complex geotechnical construction scenarios within the elevated gravity environment of a geotechnical centrifuge. However, it is very difficult to include detailed small-scale models of buried or connected entities (e.g. foundations, buildings, reinforcement elements) within a centrifuge that accurately replicate full-scale behaviour. Numerical models allow the simulation of complex material behaviour, geometric scenarios, and construction sequences. They can be effective predictive tools and are especially useful and efficient for conducting sensitivity analyses. The validity of numerical analyses in geomechanics relies on the ability of the applied constitutive model to accurately reflect real soil behaviour. Unfortunately, no numerical/constitutive model exists that can provide an accurate replication of the full complexity of soil behaviour. This is well illustrated by the inability of standard numerical modelling techniques to provide good predictions of greenfield tunnelling ground displacements (Franzius et al. 2005). On their own, physical and numerical modelling techniques have limitations regarding their accuracy and ability to model complex systems. Bringing together different modelling techniques offers a way to improve accuracy and the scope of application; this approach is referred to as hybrid modelling.

Hybrid modelling has been in use for over four decades. An early example of its application in geomechanics combined solutions from two types of numerical modelling methodologies to investigate soil-structure interaction (Gupta et al. 1980). Current hybrid testing applications generally combine solutions from physical and numerical domains. In some cases, results from the physical domain are obtained in an experiment and used to update numerical analyses. Online or pseudo-dynamic hybrid modelling indicates that the two problem domains are solved simultaneously, with each domain providing useful status data that is used to process or run the other domain analysis in real time. The term real-time is taken to mean a timescale where the concurrent interaction between the two systems is relevant and affects results.

Examples of hybrid testing in a geotechnical centrifuge are limited (Kong et al. 2015, Franza et al. 2016). Franza et al. (2016) reported a real-time system to investigate soil-structure interaction between tunnelling induced displacements and driven piles; the pseudo-dynamic hybrid test was labelled the coupled centrifuge-numerical model (CCNM). A tunnel, soil, and series of piles were modelled within a geotechnical centrifuge; pile displacements induced by tunnelling were measured and sent via a LabVIEW control program to a MATLAB model of a 2D elastic structural frame. The structural analysis then determined a set of updated pile loads based on the characteristics of the frame, which were sent back to LabVIEW and used to update the pile loads applied in the centrifuge. The overarching aim of the CCNM approach was to have a more accurate replication of tunnel-building interaction since pile loads applied in the centrifuge were dynamically adjusted according to the outcomes of the structural analysis. Franza et al. (2016) demonstrated the actuator and control systems as well as the integration capabilities of their application, however the implementation was limited to a 2D elastic structural analysis. Note that, for the linear elastic structure considered by Franza et al. (2016), the analysis could have been accomplished using a condensed stiffness matrix built prior to the hybrid tests, thereby negating the need for real-time communication between the numerical and physical models. However, the CCNM system was developed with the intention of extending its capabilities towards non-linear structural models.

This paper presents developments of the work presented by Franza et al. (2016) to extend the scope of analysis to include 3D structural models with the capability of including non-linear material behaviour. The developments mainly concern the communication protocols between the LabVIEW control system and the structural analysis, which is now performed within the finite element (FE) software package ABAQUS. The success of this implementation requires the integration of ABAQUS with LabVIEW to enable real-time transfer of pile displacements from the centrifuge model to ABAQUS and retrieval of computed reaction forces from ABAQUS to apply within the centrifuge. The paper consists of three sections. The first briefly presents the CCNM application and details of the physical and numerical models. The second presents the ABAQUS integration schemes that were tested, with advantages and disadvantages of each scheme discussed. Finally, some results are presented which illustrate the benefit of the CCNM application in the context of the analysis of the tunnel-building interaction problem.

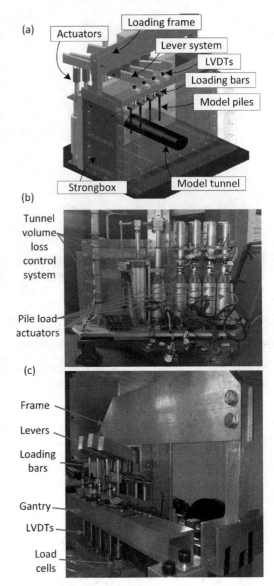

Figure 1. (a) Illustration of centrifuge model, (b) view of back of centrifuge model, and (c) view of lever and pile loading system.

2 CCNM APPLICATION

The CCNM application was developed using the University of Nottingham Centre for Geomechanics (NCG) centrifuge facility. The centrifuge model simulating the geotechnical domain consists of a tunnel, soil, and pile foundations (Figure 1); the numerical model simulates a slice of the overlying superstructure in the structural domain (Figure 2).

The centrifuge model includes a tunnel volume loss control system which enables the removal, in very small increments, of a volume of water from within a cylindrical flexible membrane that spans from the front to the back wall of the strongbox. This induces a ground loss around the tunnel, resulting in soil movements which are transferred through the soil to the piles. The piles are loaded independently using four MecVel ballscrew actuators (max 5 kN load and 100 mm stroke), as shown in Figure 1b. Due to space restrictions, the actuators were placed behind the box and a lever system was developed which transfers the actuator loads onto the piles via a series of loading bars (Figure 1c). A gantry with low-friction sleeves maintains the verticality of the loading bars and ensures that only vertical loads are transferred to the pile caps. The pile cap loads and displacements are measured using load cells and linear variable differential transformers (LVDTs), respectively.

The numerical analysis presented in this paper considered a slice through an 8-storey concrete framed building (Young's modulus $E = 30$ GPa, Poisson's ratio $v = 0.15$) with a height of 24 m, a width of 13.5 m, and supported by four piles spaced 4.5 m apart (see Figure 2 for additional geometric details). To match the constraint in the centrifuge test that only vertical loads are transferred to pile caps, the numerical analysis used pin joints at the pile-column connections. Franza et al. (2017) showed that the influence of the pile-structure connection is minimal when the tunnel axis is below pile tip level.

The CCNM application is relatively versatile and can accommodate different foundation types (e.g. displacement or non-displacement piles) and soils. In the implementation described here, displacement piles were used along with a relatively loose dry silica sand (relative density of 30%). All piles were initially loaded with the same magnitude of service load. The CCNM physical and numerical models are initiated and run in parallel, with pile displacement data (a function of the soil state and pile loading) being transferred from the physical to the numerical model, and pile load data (a function of the pile displacements and characteristics of the modelled structure) being transferred from the numerical to the physical model. A state change is initiated in the system by a small increment of tunnel volume loss (i.e. extraction of a small amount of water from the model tunnel), which triggers pile settlements and load redistribution amongst the piles. By updating the pile loads within the centrifuge, the CCNM application ensures that a realistic replication of the soil stresses around each pile is achieved, thereby increasing the accuracy of the experiment.

The CCNM application couples data at the boundary between the experimental and numerical models through rapid data exchange so that each model receives the most recent information from the other model. Figure 3 illustrates the overall architecture of the CCNM system operation. The user interface, application coordinator, numerical computation, and physical model represent action points that are all crucial to the success of the CCNM implementation. The CCNM (or application) coordinator is the central component of the application. It is written in LabVIEW and runs on a local PC within the centrifuge control room. The coordinator controls data flow between the numerical and physical models and operates the application based on set configurations from users through the user interface (UI). The UI allows users to interact with the system and updates users on system status, computational results and status, actuator system actions, pile loads/settlements, and tunnel volume loss.

3 CCNM IMPLEMENTATION

This section provides details of three CCNM implementations. IMP1 is based on the elastic 2D frame analysis presented in Franza et al. (2016); a brief summary is provided here. Due to the aim of incorporating more rigorous 3D structural models with the capability of including non-linear material behaviour, the FE software ABAQUS was integrated within the CCNM application. IMP2.1 was based on established methods

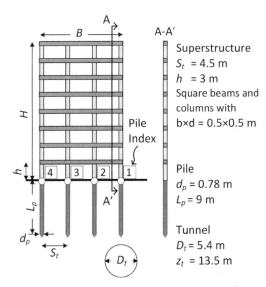

Figure 2. Illustration of structural scheme and soil-building geometry.

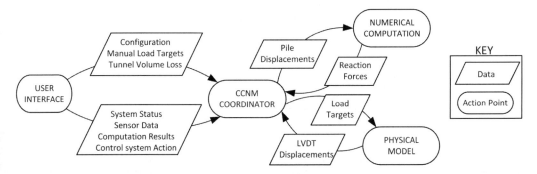

Figure 3. Architecture of CCNM application.

obtained from the literature. IMP2.2 represents a novel development for the integration of ABAQUS within a hybrid testing framework.

The following definitions are used to describe critical timings within the different CCNM implementations: $\Delta t_c =$ time required to exchange data; $\Delta t_p =$ time required to trigger the simulation program; and $\Delta t_r =$ time taken to compute the simulation.

3.1 MATLAB implementation (IMP1)

Integration of MATLAB with LabVIEW using the MathScript module embedded within LabVIEW is referred to as IMP1. The MathScript module calls MATLAB and transfers necessary functions and variables. MATLAB runs the numerical simulation and closes when completed, passing the outputs back to the MathScript module in LabVIEW. For the frame presented in Figure 4, the implementation took a total time of $\Delta t = \Delta t_p + \Delta t_c + \Delta t_r = 35$ ms. The simulation of the 2D frame in a standalone MATLAB application required $\Delta t_r = 8$ ms, hence $\Delta t_p + \Delta t_c = 27$ ms for IMP1. The data exchange between the Mathscript node and LabVIEW can be considered as negligible since data is passed in the computer memory, therefore $\Delta t_c = 0$ ms and $\Delta t_p = 27$ ms. This delay of 27 ms due to the program start/stop and pre-processes was deemed acceptable for the CCNM application.

4 ABAQUS INTEGRATION (IMP2)

Figure 4b shows the 3D solid model of the structural frame analysed in ABAQUS. The focus in this paper is to compare implementations, hence the ABAQUS model also considered a linear elastic structure with parameters equivalent to those used for the 2D frame. The challenge with integrating ABAQUS within the CCNM application was the need for fast and reliable data exchange between the application coordinator and the numerical model (i.e. the time used to send the boundary condition changes to ABAQUS and to retrieve the new reaction forces). Integration options considered were 1) ABAQUS input file scripting, and 2) FORTRAN socket connection. These were based on ABAQUS documentation and published methods for integrating ABAQUS with an external environment (Shen et al. 2010, Wang et al. 2006, Dassault-Systèmes 2014, Pan et al. 2016).

4.1 ABAQUS input file scripting (IMP2.1)

A common practice for using ABAQUS in hybrid testing is the modification of ABAQUS input files using a scripting program. The script may then go on to trigger the ABAQUS program to run a numerical simulation (using parameters contained in the input files) and extract results from an output file. These existing data exchange strategies are text based and involve time delays of up to 1 s, which is prohibitive for real-time data exchange operation. This is especially true when the numerical computation time is less than 1 s.

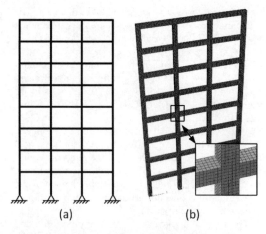

Figure 4. (a) 2D frame composed of beam elements, and (b) ABAQUS 3D finite element model of the frame.

Wang et al. (2006) developed a hybrid testing system using this strategy which comprised an earthquake simulation actuator system and ABAQUS model that was coupled via the external FEM (Finite Element Modelling) control program. A main application program, similar to the CCNM application coordinator, was developed in Visual Basic to do preliminary analysis, transmit target displacements to, and receive the reaction forces from, the FEM control program. The FEM control program was developed to generate input files for the FEM analytical steps, trigger the analysis, and extract the reaction forces from the analysis output file when the simulation was complete. This system is constrained by the waiting time required (known as a technical delay) before accessing the outputs stored in the data file. This delay made the integration method unsuitable for the CCNM application.

It was inferred that the externally developed FEM control program used by Wang et al. (2006) had limited integration capabilities that necessitated the technical delay. An alternative method, using ABAQUS scripting (implementation IMP2.1), was developed to try to eliminate the technical delay. The ABAQUS scripting system is able to inform on simulation completion which can instantly trigger an output file access method. ABAQUS scripting allows programmatic control of the ABAQUS Graphic User Interface (GUI) using the Application Programming Interface (API) developed in Python. A Python script for ABAQUS was developed to 1) edit the input file, 2) trigger the ABAQUS simulation, 3) wait for the simulation completion notice, and 4) access the output database.

Timing tests for this implementation showed that $\Delta t_p + \Delta t_r = 11.3$ s for the analysis of the 3D model shown in Figure 4b. The average simulation run time was $\Delta t_r \approx 312$ ms. The unexpectedly high Δt_p was due to the pre-processes that are called by ABAQUS before each job is executed. Unfortunately, this ≈ 11 s delay invalidated the use of IMP2.1 for the CCNM application. An improved ABAQUS integration method, discussed next, was therefore sought.

4.2 ABAQUS subroutines (IMP2.2)

Implementation IMP2.2 takes advantage of ABAQUS subroutines and its incremental modelling abilities. ABAQUS subroutines are called during computation to modify model properties and variables. In ABAQUS, an analysis step (i.e. a period of time over which the response of a model to a given set of loads and boundary conditions is determined) is broken down into a number of increments. This strategy has previously been implemented to: change model boundary conditions before each analysis (Lee et al. 2013, Hügel et al. 2008); access data (Ure et al. 2012); manage communication with a user database or access data (Ure et al. 2012). While Pan et al. (2016) was able to implement a Transmission Control Protocol/Internet Protocol (TCP/IP) to connect the FEM control software with the physical model, they were limited by the technical delay required before access to the data could be made. Their method also required the re-initialisation of ABAQUS and its pre-processes for every analysis.

The IMP2.2 CCNM application has, for the first time, extended the use of ABAQUS subroutines to use TCP/IP high speed data interfacing to connect ABAQUS with an external program. This implementation uses the incremental analysis capabilities of ABAQUS to obviate the need to restart the model for every analysis. A two-way data connection using network sockets was written into FORTRAN subroutines loaded into ABAQUS at the start of the simulation. Other utilised SUBROUTINES included DISP for boundary condition modification, UEXTERNALDB for programmatic interaction with the simulation, and URDFIL for access to simulation output at each increment. Figure 5 shows the procedure of the ABAQUS analysis in relation to the called subroutines. The network socket was made using the localhost network, a virtual network that resides on the PC. This TCP/IP connection can be extended to connect remote PCs and systems connected through a data network.

The CCNM application coordinator was also set up as a network socket to facilitate two way data exchange between the numerical and physical models. The coordinator was programmed to 1) act as the network socket server and accept incoming connections from clients such as the ABAQUS subroutine, 2) send boundary conditions based on pile displacements, 3) wait for reaction forces sent through the data connection, 4) pass these forces to the physical model, while retrieving the new pile displacements and 5) repeat steps 2-4 until the experiment was completed. The ABAQUS subroutine was set up to 1) connect to the network socket server, 2) update boundary conditions based on values received from the CCNM application coordinator, 3) extract the computed reaction forces at the end of each increment; and 4) send these forces to the application coordinator through the data connection. The data exchange between the coordinator and ABAQUS subroutines was done using 64bytes, which required about 1 ms.

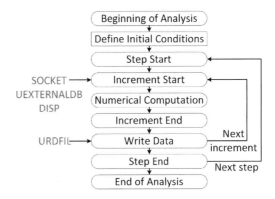

Figure 5. ABAQUS operation flow and subroutine calls.

IMP2.2 eliminates the need for starting ABAQUS processes between analyses where boundary conditions are changed. For a simulation time of $\Delta t_r \approx$ 312 ms (which depends on the computer used for the analysis), a total time between each analysis increment of $\Delta t = \Delta t_c + \Delta t_p + \Delta t_r = 330$ ms was achieved. The time required for the numerical computation was therefore the significant portion of the total time, with $\Delta t_c + \Delta t_p \approx 28$ ms. This is the very close to the timing achieved using the IMP1 MATLAB implementation, which is impressive given that IMP2.2 includes integration with a program running external to LabVIEW.

5 CCNM RESULTS AND PERFORMANCE EVALUATION

Data obtained from a centrifuge test using the IMP1 implementation was used to test the real-time ABAQUS integration in IMP2.2 and to evaluate the effect of longer simulation times (Δt_r). Measured pile displacements during the IMP1 centrifuge test were input to a 'virtual' IMP2 test (i.e. a centrifuge test was not done for IMP2.2) to calculate reaction forces. This virtual test ensured that any variability in results between implementations was due solely to the differences in the numerical models used (i.e. it removed the effect of variability between centrifuge experiments).

Figure 6 plots the change to pile load (ΔP) with tunnel volume loss from three tests: the CCNM IMP1 centrifuge test and two virtual (CCNMv) IMP2.2 tests with total times of $\Delta t \approx 300$ ms and $\Delta t \approx 30$ s. Space restrictions prevent an in-depth discussion of these results from a geotechnical/structural analysis perspective, however it should be said that this redistribution of loads has an important effect of the response of the pile foundation.

The IMP2.2 (30 s) analysis considers how the CCNM system would respond to a more computationally demanding (i.e. time consuming) ABAQUS analysis; an artificial delay was imposed in the communication of reaction forces obtained from ABAQUS. Figure 6 shows that the IMP2.2 results plot

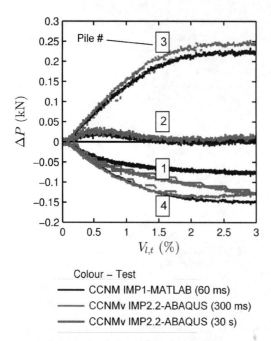

Figure 6. Target load variation for different CCNM implementations.

closely to the IMP1 data; the differences are due to the different outcomes of the structural analyses (which were expected given the different nature of the two analyses). The IMP2.2 (30 s) data is shown to track the IMP2.2 (300 ms) data but with a stepped trend. This indicates that, for the given application where the rate of tunnel volume loss can be set very low, IMP2.2 could effectively incorporate more complex and time consuming structural analyses. In fact, because the rate of tunnel volume loss can be set very low, IMP2.1 could also be effective for more time consuming structural analyses. However, if the experimental input event is required to occur relatively quickly then there is the need to reduce data transfer and numerical analysis times (e.g. tunnelling in undrained clay where the timing between volume loss increments is important to ensure undrained response).

6 CONCLUSIONS

This paper presented the development of the coupled centrifuge-numerical model (CCNM) pseudo-dynamic hybrid test application for analysis of tunnel-building interactions. The paper presented a novel implementation of ABAQUS finite element analyses within a hybrid test coordinated using LabVIEW. The use of ABAQUS subroutines was extended to incorporate a TCP/IP scheme that allows SOCKET connection between LabVIEW and ABAQUS, enabling millisecond communication of boundary condition changes to be made within ABAQUS without the need for time-consuming program start/stop sequences. Future work will involve CCNM IMP2.2 testing of non-linear structural systems and extension of the methodology to other applications.

ACKNOWLEDGEMENTS

This work was supported by the University of Nottingham and the Engineering and Physical Sciences Research Council (EPSRC) [EP/K023020/1, 1296878].

REFERENCES

Dassault-Systèmes (2014). ABAQUS 6.14 – ABAQUS scripting user's guide. *Providence, Rhode Island*.

Franza, A., S. Idinyang, C. Heron, A. Marshall, & A. Abdelatif (2016). Development of a coupled centrifuge-numerical model to study soil-structure interaction problems. In *Proceedings of the 3rd European Conference on Physical Modelling in Geotechnics (Eurofuge 2016)*, pp. 135–140.

Franza, A., A. M. Marshall, T. K. Haji, A. O. Abdelatif, S. Carbonari, & M. Morici (2017). Simplified elastic analysis of tunnel-piled structure interaction. *Tunneling and Underground Space Technology 61*, 104–121.

Franzius, J. N., D. M. Potts, & J. B. Burland (2005). The influence of soil anisotropy and Ko on ground surface movements resulting from tunnel excavation. *Geotechnique 55*(3), 189–199.

Gupta, S., T. Lin, J. Penzien, & C.-S. Yeh (1980). Hybrid modelling of soil-structure interaction. *Report to National Science Foundation* (UCB/EERC-80/09).

Hügel, H., S. Henke, & S. Kinzler (2008). High-performance ABAQUS simulations in soil mechanics. In *Proceedings of ABAQUS users conference*, pp. 192–205.

Kong, V., M. J. Cassidy, & C. Gaudin (2015). Development of a real-time hybrid testing method in a centrifuge. *International Journal of Physical Modelling in Geotechnics 15*(4), 169–190.

Lee, C.-H., P. J. Oomen, J. P. Rabbah, A. Yoganathan, R. C. Gorman, J. H. Gorman, R. Amini, & M. S. Sacks (2013). A high-fidelity and micro-anatomically accurate 3D finite element model for simulations of functional mitral valve. In *International Conference on Functional Imaging and Modeling of the Heart*, pp. 416–424. Springer.

Pan, P., T. Wang, & M. Nakashima (2016). *Development of Online Hybrid Testing*. Elsevier.

Shen, X., L. Cao, & R. Li (2010). Numerical simulation of sliding wear based on archard model. In *Mechanic Automation and Control Engineering (MACE), 2010 International Conference on*, pp. 325–329. IEEE.

Ure, J. M., H. Chen, & D. Tipping (2012). Development and implementation of the ABAQUS subroutines and plug-in for routine structural integrity assessment using the linear matching method. In *SIMULIA Community Conference 2012,(Formerly the ABAQUS Users Conference)*.

Wang, T., M. Nakashima, & P. Pan (2006). On-line hybrid test combining with general-purpose finite element software. *Earthquake Engineering & Structural Dynamics 35*(12), 1471–1488.

Verification and validation of two-phase material point method simulation of pore water pressure rise and dissipation in earthquakes

T. Kiriyama & K. Fukutake
Institute of Technology, Shimizu Corporation, Japan

Y. Higo
Kyoto University, Kyoto, Japan

ABSTRACT: Liquefaction of foundation ground during earthquakes can induce large deformations such as settlement of foundation structures, floating of underground structures, slope failures, and lateral spreading. Effective numerical simulation of these phenomena requires methods using specialized algorithms to account for the large deformations that geomaterials undergo. This work proposes a numerical simulator based on the particle-based Material Point Method (MPM), into which we introduce Biot's porous media theory. Discretizing the governing equation for a two-phase material according to the MPM framework, the Bowl model for liquefaction constitutive model is employed for dilatancy and the Ramberg-Osgood model for the nonlinearity of stress-strain relationship. The simulator is verified by comparing with an exact solution and validated by carrying out centrifuge model testing. This paper reports on the formulation, verification and validation of the newly developed simulator.

1 INTRODUCTION

There have been repeated reports of soil liquefaction leading to disastrous effects such as soil outflows from developed residential lots, collapse of embankments, lateral spreading and uplift/settlement of structures. These ground-related disasters involve large deformations of the ground. To predict such large deformations and quantify ground safety, recently developed particle-based numerical methods that can handle large deformations are used. The authors focus in particular on the material point method (MPM) (Sulsky et al. 1994) and extend it as a simulation method based on a two-phase formulation. The simulation is compared with the results obtained in experiments using a centrifuge, demonstrating that it is able to predict the behavior of liquefied ground, including the rise in pore water pressure during an earthquake and its later dissipation.

2 TWO-PHASE MATERIAL POINT METHOD

2.1 Biot's porous media theory for geomaterials

Biot's porous media theory (Biot 1941) is employed to model a geomaterial that is in two phases, solid and liquid. Specifically, the u-w formulation is employed, where u represents displacements of the solid phase and w represents displacements relative to the solid phase. The specific formulation is as follows. In the equations given below, superscripts s and f represent quantities in solid and liquid phases.

Principal of effective stress:

$$\sigma^f = -np\mathbf{I} \tag{1}$$

$$\sigma^s = \sigma' - (1-n)p\mathbf{I} \tag{2}$$

where σ^s, σ^f, p and n are the solid phase stress, liquid phase stress, pore water pressure and porosity, respectively.

Low of conservation of mass

$$\nabla \dot{\mathbf{w}} + \dot{\varepsilon}_v + \frac{n}{K^f}\dot{p} = 0 \tag{3}$$

where ε_v, K^f and p are the solid phase volumetric strain, bulk modulus of liquid phase and pore water pressure, respectively.

Equation of motion

$$\bar{\rho}^s \mathbf{a}^s = \nabla \cdot \sigma' - \nabla \cdot (1-n)p\mathbf{I} + \bar{\rho}^s \mathbf{b} + \frac{\bar{\rho}^f g}{k^f}\dot{\mathbf{w}} \tag{4}$$

$$\bar{\rho}^f \ddot{\mathbf{w}}^f = -\nabla \cdot np\mathbf{I} + \bar{\rho}^f(\mathbf{b} - \mathbf{a}^s) - \frac{\bar{\rho}^f g}{k^f}\dot{\mathbf{w}} \tag{5}$$

where $\mathbf{a}^s$, $\ddot{\mathbf{w}}^f$, $\mathbf{b}$, g and k^f are the solid phase acceleration, liquid phase acceleration, body force constant, gravitational acceleration and permeability of geomaterial, respectively.

$$\bar{\rho}^s = (1-n)\rho^s \tag{6}$$

$$\bar{\rho}^f = n\rho^f \tag{7}$$

where $\bar{\rho}^s$ and $\bar{\rho}^f$ are the surficial densities of the solid and liquid phase, which are related to the solid phase (ρ^s) and liquid phase (ρ^f) densities.

2.2 Nonlinear characteristics

To model the nonlinear characteristics of the geomaterial, the modified Ramberg-Osgood model for shear stress versus shear strain relationship and the Bowl model (Fukutake et al. 1990) for dilatancy characteristics are employed. The formulation of the Bowl model in two-phase theory is described below. Volumetric strain is expressed in Equation 8.

$$\varepsilon_v = \varepsilon_v^s + \varepsilon_v^c \tag{8}$$

where ε_v, ε_v^s and ε_v^c are the total volumetric strain, a component due to shear deformation and a component due to consolidation, respectively.

The shear deformation component is divided further into two sub components, a reversible component ε_Γ and an irreversible component ε_G, respectively.

$$\varepsilon_v^s = \varepsilon_\Gamma + \varepsilon_G \tag{9}$$

$$\varepsilon_\Gamma = A \cdot \Gamma^B \tag{10}$$

$$\varepsilon_G = \frac{G^*}{C + D \cdot G^*} \tag{11}$$

where A, B, C, D are parameters. Γ and G^* are physical quantities calculated using the equations below and called the resultant shear strain and accumulated shear strain, respectively.

$$\Gamma = \sqrt{\gamma_{xy}^2 + \gamma_{yz}^2 + \gamma_{zx}^2 + (\varepsilon_x - \varepsilon_y)^2 + (\varepsilon_y - \varepsilon_z)^2 + (\varepsilon_z - \varepsilon_x)^2} \tag{12}$$

$$G^* = \int \sqrt{d\gamma_{xy}^2 + d\gamma_{yz}^2 + d\gamma_{zx}^2 + (d\varepsilon_x - d\varepsilon_y)^2 + (d\varepsilon_y - d\varepsilon_z)^2 + (d\varepsilon_z - d\varepsilon_x)^2}\, ds \tag{13}$$

The consolidation component is described by the following equation.

$$\varepsilon_v^c = \frac{0.434 \cdot C_\alpha}{1 + e_0} \ln \sigma_m', \quad \alpha = \begin{cases} s & d\sigma_m' > 0 \\ c & d\sigma_m' < 0 \end{cases} \tag{14}$$

Next, the relationship between excess pore water pressure and volumetric strain in the Bowl model is described.

Firstly, excess pore water pressure is divided into the two components as described below.

$$\dot{p} = \dot{p}_{undrain} + \dot{p}_{drain} \tag{15}$$

where $p_{undrain}$ and p_{drain} are the undrained and drained components of excess pore water pressure. The former is a component that accumulates within the solid phase while the latter is a component that dissipates externally outside the solid phase. These undrained and drained components are described below.

$$\dot{p}_{undrain} = -d\sigma_m' = \frac{(1+e_0)\sigma_m'}{0.434 \cdot C_\alpha} d\varepsilon_v^s \tag{16}$$

$$\dot{p}_{drain} = -\frac{(1+e_0)\sigma_m'}{0.434 \cdot C_\alpha} \nabla \dot{w} \tag{17}$$

where σ_m' is the mean effective stress, expressing compression in the positive direction and tension in the negative. Finally the relationship between excess pore water pressure and volumetric strain in the Bowl model is described as below.

$$\dot{p} = \frac{(1+e_0)\sigma_m'}{0.434 \cdot C_\alpha}(-\nabla \dot{w} + d\varepsilon_v^s) \tag{18}$$

2.3 Discretization according to MPM framework

This section describes the discretization of the theoretical equations of porous media theory and geomaterial nonlinearity according to the MPM framework. In the equations given below, superscript k means the current step in the iterative numerical integration, and subscripts m and g represent quantities in the material point and in the grid point of the background mesh.

The equation of motion (Eq. 4, Eq. 5) and law of conservation of mass (Eq. 3, Eq. 16) are discretized below.

Equation of motion:

$$a_g^{s,k} = \frac{1}{m_g^{s,k}}\left(f_g^{s,int,k} + f_g^{s,ext,k} + f_g^{s,damp,k}\right) \tag{19}$$

$$\ddot{w}_g^{f,k} = \frac{1}{M_g^{s,k}}\left(f_g^{f,int,k} + f_g^{f,ext,k} + f_g^{f,damp,k}\right) \tag{20}$$

where f is the grid point force calculated from the stress/pressure exerted by surrounding material points. Superscripts *int*, *ext* and *damp* represent internal, external and damping forces. The various terms are described as follows.

$$f_g^{s,ext,k} = m_g^{s,k} \cdot b^k \tag{21}$$

$$f_g^{s,damp,k} = \sum_{i=1}^{n_p} \frac{m_{p,i}^{f,k} g}{k_p^f} \dot{w}_p^f S_{p,i}^k \tag{22}$$

$$f_g^{f,int,k} = n \sum_{i=1}^{n_p} V_{p,i} G_{p,i}^{T,k}\left(p_{p,i}^k - p_{p,i}^{ini}\right) \tag{23}$$

$$f_g^{f,ext,k} = m_g^{f,k} \cdot \left(b^k - a_g^{s,k}\right) \tag{24}$$

$$f_g^{f,damp,k} = -\sum_{i=1}^{n_p} \frac{m_{p,i}^{f,k} g}{k_p^f} \dot{w}_p^f S_{p,i}^k \tag{25}$$

Law of conservation of mass:

As a law of conservation, Eq. 3 is used for elastic response analysis.

$$\Delta p = -\frac{K^f}{n}\left(\Delta t \cdot \nabla \dot{w}_p^k - \Delta \varepsilon_v^k\right) \tag{26}$$

And Eq. 16 is used for nonlinear response analysis.

$$\Delta p = -\frac{(1+e_0)\sigma_m'}{0.434 \cdot C_\alpha}\left(\Delta t \cdot \nabla \dot{w}_p^k + \Delta \varepsilon_v^{s,k}\right) \tag{27}$$

Table 1. Relationship between theoretical solution and domains of problems for verification.

	Static	Dynamic
Small deformation	Terzaghi 1D consolidation theory	Simon 1D transient response
Large deformation	Mikasa, Gibson 1D consolidation theory	none

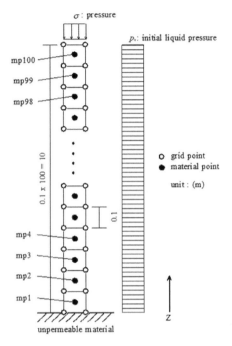

Figure 1. Analytical model and conditions for Terzarghi's 1D consolidation theory.

Table 2. Physical properties for Terzaghi's 1D consolidation simulation.

E (kPa)	ν	ρ (g/cm^3)	ρ_f (g/cm^3)	K_f (kPa)	n	k (m/s)
10000	0.3	2.2	1.0	2.2×10^6	0.2	0.0005

3 VERIFICATION BASED ON THEORETICAL SOLUTIONS

In order to verify the accuracy of the discretized equations and the implemented numerical simulator, two numerical calculations are performed. One is to obtain the static responses using Terzaghi's one-dimensional consolidation theory (Terzaghi 1924) and the other is to obtain the dynamic response using Simon's one-dimensional transient response method (Simon et al. 1984). Both are based on small deformation theory, while the scope of the MPM is large deformation problems.

Table 3. Numerical conditions for Terzaghi's 1D simulation.

dt (s)	Total steps	Analysis time (s)	σ (kPa)	P0 (kPa)
2.5×10^{-5}	1×10^8	2500	2000	2000

Table 4. Constants used to model external loads.

σ (kPa)	ω (s^{-1})	Δ (s)
1	62.83	0.024

Table 5. Material properties used for the simulations of Simon's 1D transient response.

E (kPa)	ν	ρ (g/cm^3)	ρ_f (g/cm^3)	K_f (kPa)	n	k (m/s)
3	0.2	0.306	0.2977	39.99	0.333	0.0142

Table 6. Numerical conditions for Simon's 1D simulation.

Dt (s)	Total steps	Analysis time (s)
0.0001	10000	1

For the verification of large deformation response, Mikasa's or Gibson's one-dimensional (1D) consolidation theory (Mikasa 1965, Gibson et al. 1981) is available for static response. On the other hand, there is no appropriate theory for dynamic response. Table 1 illustrates these theoretical solutions used for verification in each domain of interest. Where no theoretical solution is available for verification, another option is the method of manufactured solution (MMS). In this paper, however, verification is carried out only for small deformations.

3.1 One-dimensional consolidation theory

A one-dimensional analytical model (depth: 10 m, mesh size: 0.1 m) is prepared. The bottom boundary is set as fixed while the top boundary is free for both solid and liquid phases. The initial pressure P, which is equal to the surficial contact pressure σ, is applied to all material points in order to reduce initial numerical oscillation. One material point is set in each numerical grid. The analytical model is illustrated in Figure 1. The physical properties and numerical conditions are given in Table 2 and Table 3, respectively. Simulation results are shown in Figure 2, illustrating how the theoretical and numerical solutions for excess pore water pressure and degree of consolidation are in good agreement with each other. This demonstrates that the proposed formulation and the discretized equations are suitable for simulating static consolidation problems.

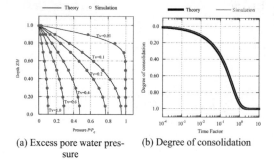

(a) Excess pore water pressure

(b) Degree of consolidation

Figure 2. Numerical solution compared with Terzaghi's 1D theory. Excess pore water pressure and degree of consolidation are both in good agreement.

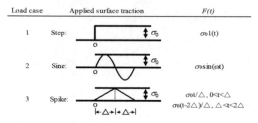

Figure 3. Step, sine and spike load functions are applied to top of 1D model.

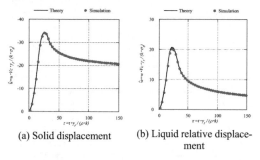

(a) Solid displacement

(b) Liquid relative displacement

Figure 4. Simulation result of Simon's 1D transient response for applied spike loading force. Numerical results and theoretical solution give good agreement with each other.

3.2 One-dimensional transient response

Simon et al. obtained a theoretical solution for the one-dimensional transient response of a two-phase material by employing a Laplace transformation. The same problems are solved here using the two-phase MPM in order to verify the applicability of the proposed method to transient/dynamic response analysis. A one-dimensional analytical model, which is an extension of Terzaghi's 1D column modified to have a depth of 100 m and a mesh size of 0.05 m, is used. Figure 3 and Tables 4–6 show the external forces, constants used to model external loading, material properties and numerical conditions for the simulations, respectively. The results are compared to the theoretical solution in Figure 4, where the two sets of results show good agreement with each other. This demonstrates that the proposed formulation is also applicable to the transient/dynamic response of two-phase problems.

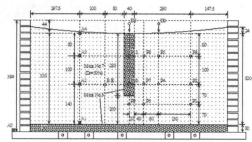

Figure 5. Liquefied ground is prepared with height of 32 cm in a shear sand box (H39.4 cm × W79.5 cm × D34.5 cm). No drain was installed in the cases of pure ground.

Table 7. Experimental cases. Maximum 100 Gal and 200 Gal random waveforms were used for base excitation following 30 Gal excitation.

Case number	Gravel drain	Base acceleration (Gal)
G-100	None	100
G-200	None	200
D-100	Installed	100
D-200	Installed	200

4 VALIDATION BASED ON CENTRIFUGAL TESTING

In the previous section, the applicability of the proposed method to both static and dynamic problems involving two-phase elastic materials was verified. In this section, applicability to geomaterials in particular is studied by investigating the rise and dissipation of excess pore water pressure in liquefied ground during earthquakes. Experiments are performed using a centrifuge and the results are compared with numerical simulations.

4.1 Experiment

A shear sand box (acceleration and rotation direction: 79.5 cm, acceleration orthogonal direction: 34.5 cm, height: 39.4 cm) is used for centrifugal testing. Silica No. 7 was used to represent the liquefied ground and silica No. 3 for the base ground and a gravel drain. After creating the base ground by tamping, the liquefied ground was created by air pluviation method. Liquefied ground both with and without a gravel drain was created; a casing was used to form the drain. piezo meters, accelerometers and bender-elements were installed as described in Figure 6. Base acceleration was applied by excitation with 100 Gal and 200 Gal random waveforms following low-level excitation with a 30 Gal random waveform. Centrifugal testing was performed under 30 gravities and silicon oil with a viscosity of 30 centistokes was used as the pore fluid. The experimental results are compared with numerical results (scaled to prototype values) in the discussion that follows. The experimental cases are listed in Table 7.

Table 8. Material properties and Bowl model parameters of silica No. 7 and No. 3 for numerical simulation. Each property is obtained by laboratory tests. Modified RO model and Bowl model parameters are obtained through a parameter fitting process.

Domain	v	ρ (g/cm^3)	k (m/s)	RO model			Bowl model							
				N	G_{0i}	$\gamma_{0.5i}$	h_{max}	A	B	C	D	X_1	$C_c/(1+e_0)$	$C_s/(1+e_0)$
Silica no. 7	0.33	1.84	3.5×10^{-4}	0.487	4235	5.77×10^{-6}	0.27	−0.2	1.4	5	40	0.13	0.007	0.006
Silica no. 3	0.33	1.99	5×10^{-3}	0.393	11815	5.983×10^{-5}	0.25	−3	1.4	25	50	0.3	0.0051	0.005

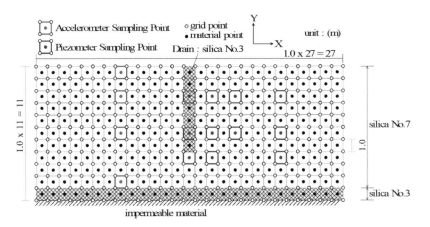

Figure 6. Numerical model for centrifugal experiments. A material point is placed in each mesh element (PPC: particle per cell = 1). Red and blue coloured cells correspond with accelerometers and piezo-meters, respectively.

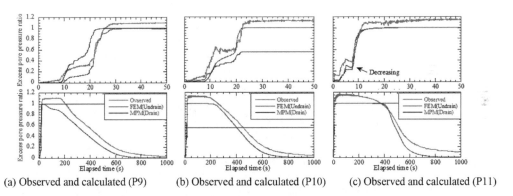

(a) Observed and calculated (P9) (b) Observed and calculated (P10) (c) Observed and calculated (P11)

Figure 7. Time history of excess pore water pressure ratio in case G-200. Elapsed time is from 0 to 50 (s) in upper figures and from 0 to 1000 (s) in lower figures. Experimental and calculated results show good agreement with each other.

4.2 Numerical simulation by two-phase MPM

Numerical simulations are performed with the following analytical conditions: particle per cell (PPC) = 1, grid size = 1 m, a cyclic lateral boundary, base excitation. The numerical model is described in Figure 6. Cells coloured in red and blue coincide with the locations of accelerometers and piezo meters in the centrifugal tests. Material properties, as obtained in laboratory tests, are listed in Table 8. Initial shear stiffness is calculated from V_s by bender element. Modified Ramberg-Osgood model and Bowl model parameters are fitted to a dynamic deformation test and liquefaction strength test.

4.3 Validation against experimental results

The observed excess pore water pressure ratios in the case of the ground-only responses (Case: G-200) are shown in Figure 7 alongside the numerical simulation results obtained by the proposed two-phase MPM. The figure also shows numerical results obtained by FEM for the single-phase (no-fluid flow) condition; these results confirm the effectiveness of the two-phase formulation.

During the pressure rise process, the inertial force arising from the earthquake is the main factor. According to Figure 7, all pressure ratios rise similarly during the earthquake (until an elapsed time of 180s), while

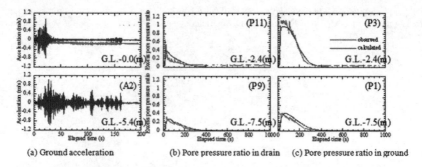

Figure 8. Time history of excess pore water pressure ratio in case D-100. Dissipation during earthquake due to drainage effect is confirmed. Piezo-meter in deep ground (P1) does not indicate liquefaction.

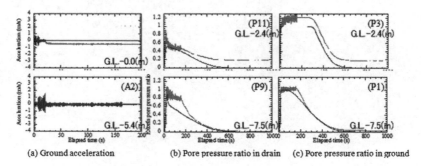

Figure 9. Time history of acceleration and excess pore water pressure ratio. (200 Gal).

the pore pressure ratio according to FEM at P10 (Figure 5, G.L. −5.4 m) does not reach 1.0, meaning that no liquefaction occurs in the single-phase condition. Because both the layers above and below P10 are fully liquefied, it seems the numerical results obtained by the single-phase formulation are incompatible with the experimental results and may lead to solutions on the dangerous side.

During the pressure dissipation process, seepage flow is the main factor. According to Figure 7, pore pressure ratios obtained with the two-phase formulation agree well with the observed results, meaning that the proposed two-phase formulation and the discretization procedure are quantitatively confirmed.

Figures 8 and 9 show the observed and calculated pore pressure ratios during/after an earthquake when a gravel drain is installed, focusing on the dissipation process (Cases: G-100 and G200). The observed and calculated results under different maximum base accelerations are in good agreement with each other, providing further confirmation of the proposed two-phase formulation and the discretization procedure.

The observed excess pore water pressure ratio exceeds 1.0 in shallow ground. Two factors give rise to this. One is a relative displacement between ground and pore fluid after the pore fluid flows out towards the ground surface. The other is a relative settlement of the piezo-meters with respect to the surrounding ground because of the reduced ground stiffness after liquefaction. In this paper, no modifications are applied to deal with these observations.

5 CONCLUSION

A modified formulation of the two-phase Material Point Method for ground liquefaction is proposed. The new formulation is verified by comparison with theoretical solutions and validated by comparison with the results of centrifugal experiments.

REFERENCES

Biot, M.A. 1941. General theory of three-dimensional consolidation, *Journal of Applied Physics*, 12(2): 155–164.

Fukutake, K., Otsuki, A., Sato, M. & Shamoto, Y. 1990. Analysis of satured dense sand-structure system and comparison with results from shaking table test, *Earthquake Engineering & Structural Dynamics*, 19(7): 977–992.

Gibson, R.E., Schiffman, R.L. and Cargill, K.W. 1981 The theory of one-dimensional consolidation of saturated clays. II. Finite nonlinear consolidation of thick homogeneous layers, *Canadian Geotechnical Journal*, Vol. 18, No. 2, pp. 280–293.

Mikasa, M. 1965. The consolidation of soft clay: a new consolidation theory and its application, *Japanese Society of Civil Engineers* (reprint from Civil Engineering in Japan 1965), in Japanese.

Simon, B.R., Zienkiewicz, O.C. & Paul, D.K. 1984. An analytical solution for the transient response of saturated porous elastic solids, *International Journal for Numerical and Analytical Methods in Geomechanics*, 8: 381–398.

Sulsky, D., Chen, Z. & Schreyer, H.L. 1994. A particle method for history-dependent materials, *Computer Methods in Applied Mechanics and Engineering*, 118: 179–196.

Terzaghi, K. 1924. Erdbaumechanik auf Bodenphysikalischer Grundlage.

Centrifuge and numerical investigations of rotated box structures

T.A. Newson & O.S. Abuhajar
Department of Civil and Environmental Engineering, Western University, Canada

K.J.L. Stone
School of Environment and Technology, University of Brighton, UK

ABSTRACT: Soil arching can occur due to the existence of buried structures inside a soil body. This arching can lead to an increase or decrease in soil pressures attracted to the buried structures, which cannot be accurately estimated by simply considering the self-weight forces generated by the prism of soil supported by the structure. The relative stiffness between the structure and the surrounding soil is the main factor that controls these soil pressures. This paper presents centrifuge and numerical results of two rotated box structures used to investigate the effect of different wall thicknesses on the contribution to the overall loads attracted to the structure. Results of comparative numerical analyses using PLAXIS are presented to aid the interpretation of the tests. Load reductions were found to occur for the most flexible portions of the structures (up to 20%) and both the individual flexibility of the members and the overall structure were found to be important.

1 INTRODUCTION

The loads attracted to buried structures, from both overburden and surcharge loads, are governed by the characteristics of the soil and geometry and stiffness of the structural components. In many instances, the redistribution of the free-field stresses as the result of the presence of a buried structure will lead to a decrease in loading over the deflecting or yielding areas of the structure and an increase over adjoining rigid or stationary parts. This transfer of load due to soil-structure interaction is known as 'arching'.

Whilst there have been numerous experimental and field studies to investigate stress distribution and arching (e.g. Lefebvre et al., 1976), the exact conditions required for this phenomenon to occur are still unclear and arching is often ignored in engineering design due to a lack of experience, and inclusion in codes of practice is rare. A range of problems such as underground conduits, tunnels, trapdoors, retaining walls and braced cuts can all experience significant arching action and theoretical analyses have been published on these subjects. These approaches have considered soil arching from both elastic and plastic soil states. In recent years, research has concentrated on scaled physical modelling of the behaviour of buried circular and square structures (e.g. Iglesia et al. 1999; Stone & Newson 2002; McGuigan 2010; Abuhajar et al. 2015a/b).

Buried box section culverts and conduits are commonly used around transportation infrastructure, e.g. to span highways. These are used to control water flow, storm runoff, divert municipal services, allow vehicular access and for other related activities.

AASHTO (2002) standard specifications for highway bridges takes some account of arching by changing vertical stresses over box culverts based on Marston-Spangler theory. However, this approach is quite conservative compared to more sophisticated numerical techniques, such as finite element analysis. For the case of relatively flexible buried structures, the soil-structure interaction is even more complicated, and the problem is difficult to solve theoretically or analytically.

Previous work by Abuhajar et al. (2009) investigated the differences between the loads attracted to buried flexible box structures orientated in different directions (i.e. with the sides parallel or at 45 degrees to the free surface). This study found that for the same structure the induced bending moments were higher, deflections smaller and attracted loads smaller for the rotated case. The principal aim of the current paper is to further investigate this problem using scaled modelling in the centrifuge and numerical analysis for two 45° rotated square box structures. These structures have different combinations of wall and slab thicknesses. The results are presented in the form of bending moments, deflections and vertical soil pressures, to enable a comparison of the contribution made by the structural elements to the loads attracted to the structure, related to the theoretical overburden loads.

2 EXPERIMENTAL TESTING

2.1 Model preparation

In general, underground box structures are constructed from short sections of reinforced concrete, which are

Figure 1. Model prior to embedment in sand.

joined together to form the desired cross-sections. Due to the problems of manufacturing concrete with micro-aggregates for scaled physical tests, model structures (shown in Figure 1) were made from an aluminum box section 400 mm long with an external dimension of 101.6 mm and wall thicknesses of 2.0 and 6.35 mm.

However, correct modeling of the structural deflection can be achieved in reduced scale models by ensuring the following relationship is maintained:

$$E_m/I_m = E_p/I_p/N^4 \qquad (1)$$

where, E = Young's modulus of the material, I = second moment of area per unit length of the material and N = scaling factor. The subscripts 'm' and 'p' refer to model and prototype respectively.

To enhance the response of the model, the walls of the aluminum section were machined to a thickness of only 2 mm. Miniature strain gauges were bonded at various locations on the structure to measure the internal and external strains. The outputs were used to determine the bending and axial strains at the location of each gauge pair.

2.2 Centrifuge tests

The centrifuge test procedure was as follows: firstly, a bed of sand was placed by air pluviation into a rectangular strongbox of dimensions 500 mm wide × 800 mm long × 600 mm high. The structure was placed by hand into the sand to the required depth and then the pluviation was completed. The strongbox was then mounted on the centrifuge platform and the structure placed on this layer. The strain gauges were connected to the data acquisition system, and after checking their initial gauge output, the data logger was zeroed. The centrifuge was then accelerated and held at the following acceleration levels, 5 g, 10 g, 20 g, 40 g, 60 g and 80 g. Data from the strain gauges was recorded continuously during the test. Congleton Sand was used for all of the tests. This is a uniform sand with a $d_{10} = 100\,\mu m$ and maximum and minimum densities of 1.78 and 1.51 g/cm^3 respectively. Air pluviation created soil beds with a relative density (D_r) of 35%. The

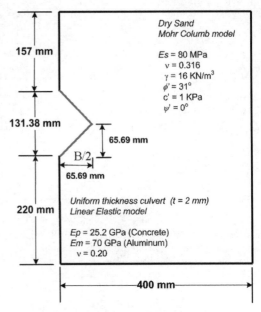

Figure 2. Diagram showing the orientation of the two structures and the properties used for analysis.

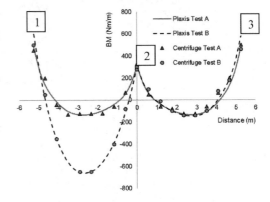

Figure 3. Comparison of bending moment diagrams at 80 g (prototype scale).

angle of internal friction of the sand at critical state is $\phi' = 31°$. In this paper, tests with two different box structures are presented: one with a uniform thickness of 2 mm (Test A) and the other with pairs of opposite sides with different thicknesses – top/bottom = 2 mm [thin] and sides = 6.35 mm [thick] (Test B). The boxes were orientated with their vertices rotated 45° into the vertical plane (as shown in Figure 2 below). The soil bed under the structure is 220 mm and the soil above is 157 mm, hence the normalised embedment (H/B) was equal to 1.14 for the modelled cases.

Figure 3 shows a comparison between the bending moment distributions for tests A and B for members '1-2' and '2-3' (shown as triangular and circular points) at 80 g. For member 2-3, the bending moment values are very close because of the similar wall thicknesses.

In contrast, for member '1-2', there is a clear difference in the values of the bending moments due the different thicknesses of the two structures. It is also seen from Figure 3 that the bending moment values at the upper vertex (point 2) are less than those at the mid-side vertices (points 1 and 3). For both members, the values of the bending moments are highest at the vertices and vary non-linearly along the members. Whilst these variations are approximately parabolic, they are asymmetric with the minima not coinciding with the centerline of the members.

3 NUMERICAL MODELLING

The numerical analysis results presented herein are for the 80 g test only. These tests were designed to investigate the interaction between the buried model and the surrounding soil related to the stiffness of the soil and deformation of the model. Two-dimensional plane strain finite element analysis was conducted using the package PLAXIS® to aid interpretation of the physical model. Drained soil conditions were assumed, and the modeling was carried out using Mohr-Coulomb elastic-perfectly plastic soil. The assumed material parameters were Young's modulus, $E = 80$ MPa, Poisson's ratio, $v' = 0.32$, effective unit weight, $\gamma' = 16$ kN/m³ (note the sand was dry), angle of internal friction, $\phi' = 31°$, cohesion angle, $c' = 1.0$ kPa and angle of dilation, $\psi' = 0°$. The structure material (assumed to be linear elastic) was assigned $E = 70$ GPa for aluminum and $v = 0.20$ and the test was modelled as a prototype at 80 g.

Due to the asymmetry of Test B, the whole soil-structure system was modeled. The domain was discretized using 15-noded triangular soil elements (with fourth order interpolation for displacements) and 5-noded beam culvert elements. Each node has three degrees of freedom per node: two translational and one rotational. Interface elements were used to provide for possible slippage and separation between the culvert and the surrounding soil. The 'roughness' of the interaction is defined by a strength reduction factor, R, which relates the interface strength to the soil strength parameters. A 33% reduction in soil strength was assumed ($R = 0.67$) at the interface between the culvert and the surrounding soil.

The modelled geometry and a typical finite element mesh used for the structures is shown in Figure 4. The boundary conditions used were to fix the domain sides in the horizontal direction (i.e. free 'y' movement), fix the bottom of the domain in both directions (x, y) and the upper surface was free to move in both directions. Figure 4 shows a deformed mesh (with the vertex numbering system used) and deflection of the model.

The numerical predictions of the bending moments for tests A and B are also shown in Figure 3. These generally show a very good match between the centrifuge model and numerical bending moments.

The corresponding deflections for the Test A structure are shown in Figure 5(a). The loaded structure

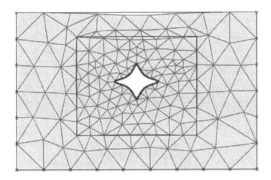

Figure 4. Geometry and finite element mesh for the model.

adopts the shape of a compressed hypo-trochoidal square, with the upper and lower vertices (points 1 and 4) moving inwards and the side vertices (points 2/3) moving outwards. The relative stiffness of the structure in the horizontal and vertical directions is approximately the same. This is similar to the behavior of circular culverts (e.g. Taleb & Moore 1999) with the crown settling and the shoulders moving outwards. The displacements vary non-linearly along the members of the structure and in common with the bending moments, the maxima do not coincide with the centerline of the members.

In contrast, the deflections of the members in Test B [Figure 5(b)] are relatively asymmetrical. The general pattern of behavior is the same with the upper and lower vertices moving inwards and the side vertices moving outwards. Again, the displacements vary non-linearly along the members of the structure. For the thin (2 mm) members the maxima do not coincide with the centerline of the members. For the thick (6.35 mm) members, the peak displacement occurs at the vertices. Interestingly the displacements at the four vertices suggest that the structure is rotating marginally in the anti-clockwise direction.

Figure 6 presents a comparison between the vertical soil pressures on members '1-2' and '2-3' of the two models at 80 g. In common with the bending moment diagrams in Figure 3, the results show very similar vertical soil pressures on the thin members (2-3) and greater variations for the thick/thin members (1-2). The highest pressures are seen at the vertices of the structure and the lowest minima for the zones with the peak deflections for the thinnest members (with about 20% of the pressure). For all of the members, the variation in vertical pressures along the structure is non-linear.

For the thick member (1-2) in Test B, the minima of the pressure is approximately 55% of the peak pressure. If the overall load attracted to the upper surface of the two models is assessed, the Test A structure carries 80% and Test B structure carries 96% of the weight of the overlying soil prism (from point 1 to 3). For Test B, 2/3 of this load is carried by the stiffer (thicker) member [1-2]. Whilst the asymmetric structure (Test B), conveys no advantage, the symmetrical flexible structure (Test A) sheds 20% of the load to the stiffer surrounding soil mass.

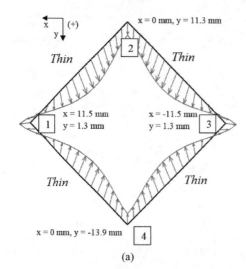

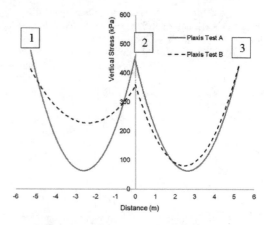

Figure 6. Comparison of vertical soil pressure diagrams for both structures at 80 g.

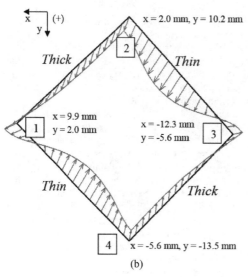

Figure 5. Member deflections for (a) Test A and (b) Test B structure at 80 g (prototype scale).

The incremental shear-strain contours for the two models are shown in Figure 7. Test A is shown in Figure 7(a) and indicates very little strain localisation along the middle 2/3 of the top of the structure. However, two shear strain localisations occur close to the side vertices (with peak magnitudes of 0.04%). Interestingly these originate from the points of contraflexure (BM = 0), approximately 4/5 of the way along the member and propagate towards the surface of the soil mass. On the underside of the structure the localisations are curved and span between the vertices on these members. For Test B [Figure 7(b)], only one localised shear strain zone was observed on the thinner side of the structure with a slightly higher peak strain of 0.12%. This has the same form as those shown in Figure 7(a) and again originates from the point of contraflexure on that member (2-3).

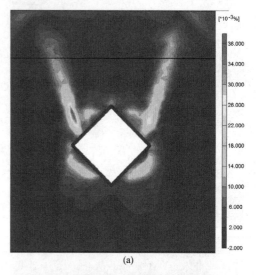

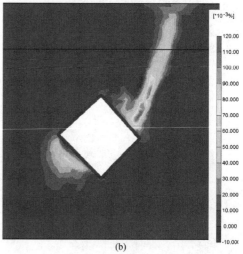

Figure 7. Incremental shear strain for (a) Test A and (b) Test B.

Although not shown in this paper, the horizontal effective stresses were also found using Plaxis. These indicate that significant portions of the box sides are subjected to low horizontal pressures (less than the active case in certain areas). The one exception is the side vertex of the boxes, which have pressures that begin to approach the passive pressure.

For the symmetrical rotated box (Test A), the two upper members behave like active trapdoors and the lower members like propped walls. The structure tends to reduce the total loads attracted to the top slabs, whilst concentrating the loads at the vertices (corners). This behaviour is further enhanced by the additional compression of the upper members acting together (in a similar manner to a loaded 'A-frame' structure). The shear strain diagram [Figure 7(a)] suggests that the material contained within the two localisations moves downwards with the structure and a discontinuity is beginning to form with stationary soil outside of the developing soil mechanism; this supports the observations made regarding the loads attracted to the model upper surfaces.

For the asymmetrical case (Test B), the more flexible portions of the structure are behaving in the same manner as Test A, but the localisations appear to be more intense and the developing soil mechanism much narrower, which affects the arching behavior and loads attracted to the structure.

4 CONCLUSIONS

This paper has described centrifuge and numerical modeling used to investigate soil-structure interaction of a buried rotated square box. The analyses have investigated the contributions of the form and location of the structural. The matches between the centrifuge and numerical predictions were found to be reasonable and this enabled further interpretation of the results to be conducted. The effect of member thicknesses showed their important contributions to the overall structural stiffness and the soil pressures attracted to the boxes. The shape of the soil pressure distributions was not uniform and were found to be approximately parabolic with high values at the edges and lower values at the center. Shear localisations were found to originate from the points of contraflexure on the more flexible portions of the upper surface of the structures, suggesting the initial formation of soil mechanisms. This lead to some arching and load transfer to the rigid portions of the soil mass and reductions in the overall loads on the structure. The more rigid portions of the structure also experienced much higher stresses and bending moments. For the asymmetric case (Test B), higher bending moments were found for the stiffer portions of the structure and quite significant increases in load were observed. Whilst it is unlikely this type of structure will be used in geotechnical practice, due to limitations of construction and usage, this study does provide information on fundamental aspects of arching phenomena around buried structures.

REFERENCES

Abuhajar, O., El Naggar, H. & Newson, T. 2015a. Static soil culvert interaction the effect of box culvert geometric configuration and soil properties. *Comput. Geotech.*, 69(6): 219–235.

Abuhajar, O., Newson, T., El Naggar, H. & Stone, K. 2009. Arching around buried square section structures 17th ISSMGE Conference, Alexandria, Egypt. 1–6.

Abuhajar, O., Newson, T. & El Naggar, H. 2015b. Scaled physical and numerical modelling of static soil pressures on box culverts. *Can. Geotech. J.*, 52(11): 1637–1648.

American Association of State Highway and Transportation Officials, Inc. (AASHTO) 2002. *AASHTO Standard Specifications for Highway Bridges*, Washington, D.C.

Iglesia, G.R., Einstein, H.H. & Whitman, R.V. 1999. Determination of Vertical Loading on Underground Structures Based on an Arching Evolution Concept, *Geotechnical Special Publication*, 90: 495–506.

Lefebvre, G., Laliberte, M., Lefebvre, L.M., Lafleur, J. & Fisher, C.L. 1976. Measurement of Soil Arching above a Large Diameter Flexible Culvert, *Can. Geot. J.*, 13: 58–71.

McGuigan, B. 2010. *Earth pressures and loads on induced trench culverts*. PhD Thesis. University of New Brunswick.

Stone, K.J.L. & Newson, T.A. 2002. Arching Effects in Soil-Structure Interaction, *4th Int. Conf. Physical Modelling in Geomechanics*, St. Johns, Newfoundland, 935–939.

Taleb, B. & Moore, I. 1999. Metal culvert response to earth loading: performance of two-dimensional analysis. *Transportation Research Record*. 1656: 25–36.

Multibillion particle DEM to simulate centrifuge model tests of geomaterials

D. Nishiura & H. Sakaguchi
Japan Agency for Marine-Earth Science and Technology, Kanagawa, Japan

S. Yamamoto
Obayashi Corporation, Tokyo, Japan

ABSTRACT: In this study, we developed centrifuge model simulations using the discrete element method (DEM) to quantitatively evaluate the earth pressure acting upon box culvert underground structures during the displacement of reverse faults directly beneath the structures. Furthermore, we investigated the reproducibility of ground deformation modes and the earth pressure upon underground structures through comparison with experimental results. We thus found that the earth pressure on the top slabs of underground structures is constant at the initial earth pressure, although the side wall earth pressure increases with fault displacement. Moreover, we found that the earth pressure acting upon underground structures is affected by the positional relationship between bottom slabs and fault lines. As a result, the full-scale three-dimensional DEM simulations proposed in this paper offer new possibilities for physical modelling studies of both ground deformation scenarios and earth pressure acting upon underground structures in geotechnical centrifuge tests.

1 INTRODUCTION

Ground surface fault displacement was observed for the first time during the Kumamoto Earthquake in April 2016, and reports of ground surface deformation attributed to inland, active fault earthquakes have increased in recent years. Overseas, cases of major damage to underground structures from large-scale ground deformation have been reported, typified by the 1999 Chi-Chi earthquake in Taiwan. Thus, there has recently been an increased emphasis on the design and construction of underground structures that take fault displacement into account, as well as an increasing number of studies using centrifuge model tests (Mason et al. 2013, Ulgen et al. 2015, Higuchi et al. 2017).

In addition, there have been many detailed studies on the interaction of soils, foundations, and underground structures using numerical analysis (Corigliano et al. 2011, Ha et al. 2014, Abate et al. 2015). Many of these studies focused on the evaluation of structural behaviour. By contrast, there have been very few cases (that have been quantifiably verified) in which external forces (attributed to earth pressure) acted upon underground structures experiencing fault displacement.

In most studies using numerical analysis, soils, foundations, and underground structures have been modelled using the finite element method (FEM) (Sharma & Bolton 1996, Sawamura et al. 2015). However, with FEM, soils are modelled in a variety of ways by using elastoplastic models with empirical parameters. This is done to reproduce the ground deformation and fault ruptures caused by friction and the mechanical and adhesion properties of the soil particles constituting the ground. Therefore, the constitutive law models of soils used, and the model parameters thereof, are selected to create reproducible experimental results, and thus FEM modelling is not possible without them. As a result, experiments are often reproduced using FEM analysis, and then the results are scaled up for analysis when it is not possible to measure physical quantities in experiments or conduct experiments at full scale.

On the other hand, the discrete element method (DEM) is a model that addresses the motion of individual particles and reproduces the viscoelastic plastic motion of soil by applying particle properties, such as particle size, mechanical properties, and friction, to individual particles (Cundall & Strack 1979). Thus, unlike FEM, it is possible to reproduce experimental results without prior centrifuge model tests simply through the proper use of the properties of particles constituting an actual soil with DEM.

That is, if it is possible to use DEM to determine particle properties by simulating the triaxial compression testing of sand particles constituting a soil, and then use these particle properties in a centrifuge model simulation with DEM, there is a strong possibility of supporting physical modelling with experimental results. However, calculations with DEM have previously been impossible owing to the large calculation loads, with particles in the range of hundreds of millions at sizes the same as the sand particles used in

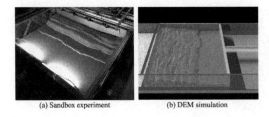

(a) Sandbox experiment   (b) DEM simulation

Figure 1. Snapshots of (a) sandbox experiment and (b) DEM simulation for fault slip formation process.

the centrifuge model tests. Thus, it was necessary to determine particle properties empirically to reproduce the centrifuge tests using DEM as well as FEM.

However, in recent years, there have been remarkable technological developments in computing environments and computer programs. As a result, it is presently possible to use 2.5 billion particles for the numerical analysis. Figure 1 shows the results of a sandbox experiment of fault formation, and the results of a DEM reproduction simulation using 2.5 billion particles. A fault structure closely resembling that of the sandbox experiment was observed using DEM.

Against this background, in this study, to evaluate the earth pressure acting upon box culvert underground structures during the reverse fault displacement of a fault directly beneath it, we developed the largest-scale DEM centrifuge model simulations in the world. We verified the reproducibility of the ground deformation modes and earth pressure acting upon the underground structures through comparison with experimental results obtained using geotechnical centrifuge tests. Thus, we provided new possibilities of the latest DEM simulations for supporting the physical modelling of geotechnical materials used in centrifuge tests.

2 METHODOLOGY OF CENTRIFUGE TEST

2.1 Analogue experiment (Higuchi et al. 2017)

In a previous study, Higuchi et al. designed a loaded soil tank for reproducing the reverse fault displacement of a model ground in a centrifugal field using a geotechnical centrifuge system. Figure 2 shows a schematic of the tank. The soil tank had a floor slab on the hanging wall side (length of 500 mm), gradually inclined at a 30° angle from the horizontal plane by a hydraulic jack so that reverse fault displacement acted upon the model ground.

A 1/50 scale model was used for the test, which was performed in a 50-G centrifugal environment.

Dry Gifu No. 7 silica sand was used as the soil material. The particle size distribution was as given in Table 1. Vibration compacting was used to create a model ground with a soil thickness of 300 mm, length of 1,545 mm, and depth of 1,000 mm (which would convert to a soil thickness of 15 m, length of 77.25 m, and depth of 50 m at full scale), aiming for a soil relative density of 90%.

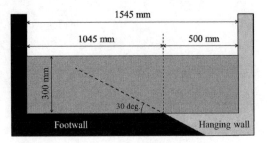

Figure 2. Schematic of 1/50 scale apparatus used in centrifuge test.

Table 1. Particle size distribution of Gifu No. 7 silica sand.

Median sieve diameter (mm)	Volume fraction (%)
0.075	0.9
0.089	4.0
0.122	30.8
0.163	30.8
0.217	30.8
0.326	2.7

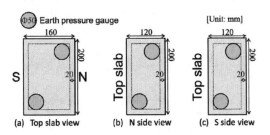

(a) Top slab view   (b) N side view   (c) S side view

Figure 3. Schematic of earth pressure gauge position on the box culvert.

For the underground structure model, a concrete box culvert was made with dimensions of length 160 mm, height 120 mm, and wall thickness 20 mm (which would convert to a length of 8.0 m, height of 6.0 m, and wall thickness of 1.0 m at full scale), and was placed within the soil so that it contacted the floor of the soil tank. The earth cover was 180 mm (which would covert to 9.0 m at full scale). As the effect of fault displacement on underground structures was the subject of investigation in this study, the underground structure model was placed along the depth direction of the model ground. Then, the underground structure model was divided into three sections in the soil tank depth direction (of widths 395 mm, 200 mm, and 395 mm), and the centre section was used as the measurement model.

The external force acting upon the underground structure model was measured using an earth pressure gauge, which was set on the side wall surface of the box culvert located at the centre section. The position of the earth pressure gauge was shown in Figure 3. Here, the side walls on the hanging wall side and on the footwall side are called the North (N) side wall and the South (S) side wall, respectively.

Then, fault displacement was applied to the model ground with a centrifugal force of 50 G for testing. As the soil contained dry sand, the displacement loading rate would have no effect, so the displacement was incrementally applied while confirming the ground deformation and underground structure motion scenarios. Here, the deformation of the ground surface was measured along the centreline in the soil tank depth direction by using a laser displacement meter.

2.2 Numerical simulation method

We used DEM by considering individual particle motion as a means of simulating a model ground. It is very difficult to perform simulations with DEM under the same conditions as actual experiments and phenomena because the computing load is high. Therefore, in this study, we applied a parallel computing method (Furuichi & Nishiura 2017) to DEM, and then attempted to perform simulations at the same scale as the experiments of Higuchi et al.

With the simulation system, we accounted for the computing load, reducing the depth to 2.0 mm, which is 1/500 of that in the experimental system. Accordingly, we set the friction coefficient between the particles and side wall to 0 to reduce the effects of the side wall. Otherwise, the system sizes were the same as during testing, as shown in Figure 2. In addition, the particle size was twice as large as actual Gifu No. 7 silica sand. That is, we used a length of 1,545 mm and width of 2 mm for the system size, and designed a ground with a thickness of 300 mm so that the relative density was the same as that of the experiment.

In addition, we mixed particles of four sizes (0.652 mm, 0.434 mm, 0.326 mm, and 0.244 mm) at volume fractions of 2.8%, 32.4%, 32.4%, and 32.4%, respectively. Then, we applied 50-G gravity as in the actual experiment, and moved the floor of the hanging wall side in a 30° direction from the horizontal plane. Here, we moved the floor at a constant vertical velocity of 0.03 m/s (fault direction velocity of 0.06 m/s).

Table 2 lists the particle properties used in DEM simulations for reproducing the centrifuge model tests using silica sand. The Young's modulus of the particles was set as a virtual value that is approximately 1/700 that of real silica grains by accounting for the computing load. If we used the real Young's modulus, the simulation could not be completed within the desired time because the discrete time used in time integration becomes smaller on increasing the Young's modulus. With these settings, the computing time was approximately 24 h using 768 nodes of super computer. In addition, we set the interparticle friction coefficient and the rolling friction coefficient of a single grain to the empirical values for silica sand because it is difficult to measure these properties for small nonspherical single grains. However, it was confirmed by the DEM simulation for triaxial compression tests (Sakaguchi 2004) that the internal friction angle of soil in the numerical model was almost the same as that in the experimental model.

2.3 Test cases

Table 3 lists the test cases. Case 1 is only soil without an underground structure model, and Case 2 is an underground structure model in which the bottom slab of the underground structure model is entirely on the hanging wall side. Case 3 is a fault line position that is directly beneath the bottom slab of an underground structure model. That is, 1/3 of the bottom slab of the underground structure model is on the hanging wall side, and the remaining 2/3 is on the footwall side.

3 RESULTS AND DISCUSSION

3.1 Ground deformation

Figure 4 shows the deformation of the model ground before and after fault displacement obtained from the DEM simulation. In all cases, the ground surface experienced bulging on the hanging wall side of the fault,

Table 2. Material properties of a DEM particle.

Property	Value	Unit
Density	2645	kg/m^3
Young's modulus	0.1	GPa
Poisson's ratio	0.2	–
Restitution coefficient	0.2	–
Friction coefficient	0.625	–
Rolling friction	0.05	–

Table 3. Conditions of underground structure used in centrifuge test.

Underground structure	Case 1	Case 2	Case 3
With or without	Without	With	With
Size (mm)	-	160 × 120, Side thickness 20	160 × 120, Side thickness 20
Placement position	-	Footwall / Hanging wall	Footwall / Hanging wall

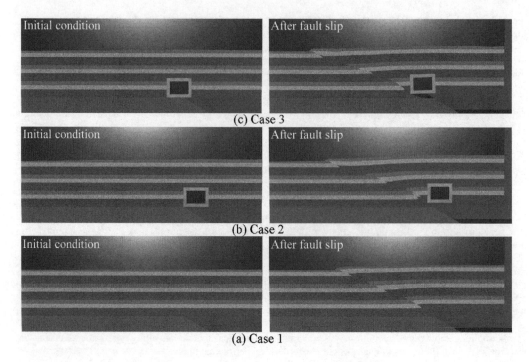

Figure 4. Deformation shapes of ground after fault slip obtained by DEM simulation in (a) Case 1, (b) Case 2, and (c) Case 3.

and a shear zone developed that reached the ground surface. The orientation of the shear zone broadly matched the displacement direction of the floor on the hanging wall side.

By contrast, in Case 3, in which the bottom slab of the underground structure model straddled the fault of a rock foundation, a shear zone developed in the soil from the lower left corner of the structure model. Compared to Case 1 and 2, parallel movement was seen in the shear zone on the footwall side. In addition, in Case 3, the underground structure was confirmed to have rotated slightly in the counter clockwise direction.

Next, Figure 5 shows the results of measuring the deformation of the ground surface in each case. These measurements were obtained from experiments and simulations using actual reduced values. In Case 1, which had only soil, and Case 3, in which an underground structure model was placed, the shape of the ground surface in the experiment and in the simulation largely matched. On the other hand, in Case 2, differences were observed in the results for the experiment and simulation.

In the experiment, regardless of the placement of the underground structure, the results for Case 2 were largely the same as those for Case 1, which had no underground structure. In the simulation, the structure appeared to affect the ground surface shape, and very likely, there were errors in the experimental results for Case 2. In addition, in Case 3, the position at which the resulting shear zone reached the ground surface had moved to the footwall side. When we assumed this was a shear zone from the lower left corner of the

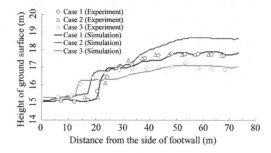

Figure 5. Shape profiles of ground surface in each case.

underground structure model, as shown in Figure 4, we found that it conformed to cases in which the position reached by the shear zone had moved to the footwall side.

3.2 *Earth pressure acting upon the underground structure*

We measured the external force of the earth pressure acting upon the underground structure model during fault slip. From this, we assumed the side wall of the structure on the hanging wall side to be the N side wall, and the side wall on the footwall side to be the S side wall. Figure 6 shows the relationship between the external force of the earth pressure from the soil acting upon the underground structure model and the amount

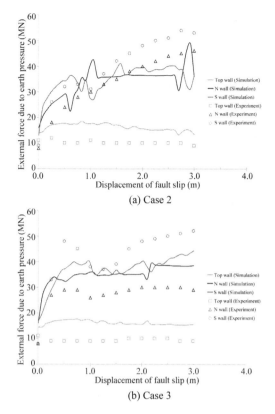

Figure 6. Comparison of earth pressure acting upon an underground structure in (a) Case 2 and (b) Case 3 between centrifuge experiment and simulation.

of slip in the fault direction of the hanging wall side floor.

Here, in the experiments, the external force of the earth pressure was computed by multiplying the surface area of the side wall surface (where each earth pressure gauge was placed) by the measured value of earth pressure gauges 50 mm in diameter. On the other hand, in the simulations, as the underground structures were modelled by rigidly connecting DEM particles, we measured the external force by summing only the force acting on the particles constituting the side wall of the structure.

Furthermore, for data reduction, we performed computations assuming the external force of the earth pressure was the acting force per block of the underground structure model (which would convert to a width of 10 m at full scale).

Quantitative differences were seen in the external force of the earth pressure between the experiment and simulation results. The placement method for the underground structure was considered a possible cause. It was conjectured that the ease of movement on the floor of an underground structure affects the earth pressure acting on its side wall. That is, the friction working between an underground structure and the floor needs to be properly established. In addition, although the earth pressure acting on the top slab also differed between the experiments and simulations, it is thought that the cause was the fact that the thickness and relative density of the earth cover of the underground structure did not match.

However, qualitative tendencies between the simulations and experiments closely resembled one another. We found that the external force of the earth pressure acting on the top slab was constant at the initial earth pressure regardless of the placement position of the underground structure model and the amount of fault displacement. Here, in the experiment, the external force of the earth pressure computed from the earth cover on the top slab was 11 MN.

On the other hand, although the external force of the earth pressure acting on the side wall of the underground structure increased along with the fault displacement, in Case 2 it momentarily levelled off at a fault slip of 0.5 m on the S side wall. Thereafter, it again gently rose from a fault displacement of close to 1.0 m. By contrast, on the N side wall, the external force continued to gradually increase.

On the S side wall in Case 3, the external force showed a peak at a fault slip of 0.5 m, and thereafter tended to increase after momentarily dropping. On the other hand, on the N side wall, the external force levelled off close to a fault displacement of 0.25 m, and thereafter was confirmed to no longer increase.

4 FUTURE DEVELOPMENTS FOR VIRTUAL CENTRIFUGE MODELLING TOOLS

Based on the above discussion, a large-scale DEM simulation can reproduce the results of centrifuge model tests with a certain degree of precision. In the future, we expect that the reliability of centrifuge model tests using this simulation could be improved by accurately setting the particle properties used in DEM through triaxial compression testing. In the DEM simulation for the triaxial compression tests, a membrane boundary surrounding test pieces was modelled by introducing membrane particles, which transmit hydrostatic pressure to the material particles constituting the test piece. We can perform a more accurate DEM simulation for centrifuge model tests using the particles with non-empirical friction properties, which are determined by the DEM simulation for triaxial compression tests as well as the procedure of an analogue experiment. We can call the DEM simulation consisting of the triaxial compression test and centrifuge model test a virtual laboratory for centrifuge model tests.

With the virtual laboratory, we can offer new information, which cannot be obtained through experiments. For example, we can reveal the effect of granular material properties on the spatial distributions of local stress and strain in the model ground. However, the experimental measurement is usually difficult because we need to install the instrument under the ground, which leads to the effect of artefacts due to the existence of the instrument as a foreign body. Although it is difficult in experiments to avoid the effect of artefacts, such as installed instruments, and initial heterogeneity in the model ground, we can evaluate these artefact effects using the virtual laboratory, which contributes to reasonable test planning during physical modelling. Moreover, with the virtual laboratory, we can adopt more difficult and complex conditions, similar to those encountered in practical situations in civil engineering work, when compared to experimentation. Even if some difficulties exist for the reproduction of structures and materials in experiments, they can be modelled using simulations, which contribute to physical modelling for obtaining new designs for effective construction and reinforcement of the foundation. Thus, we believe that the virtual laboratory could be useful for the physical modelling of geotechnical phenomena.

5 CONCLUSIONS

We developed a large-scale DEM simulation for reproducing the centrifuge analogue experiment of a model ground. We used a box culvert structure model placed on a rock foundation, backfilled underground, and simulated centrifuge model tests of the earth pressure acting on the structure when a fault directly beneath the structure experienced a reverse fault displacement. Then, we compared it against experimental results and investigated the reproducibility of ground deformation modes and the earth pressure on underground structures.

1. We found good reproducibility with the DEM simulation of ground surface deformation scenarios owing to fault development, whether it was only soil or also had an underground structure.
2. The earth pressure acting on underground structures appeared to qualitatively match between the simulations and experiments. On the top slabs, it was constant at the initial earth pressure. Although the side wall earth pressure increased with fault displacement, it levelled off with a constant amount of fault displacement.
3. At present, quantitative differences were seen in the earth pressure acting upon the underground structures, as it is not possible to fully reproduce underground structure placement methods, relative densities, layer thicknesses, or other experimental soil properties in simulations.

Although some quantitative differences between the results of experiment and simulation exist, in the future, they could be improved with continuous efforts toward the development of more accurate numerical modelling techniques. Then, we can investigate the detailed ground dynamics under ideal conditions without the effects of artefacts, which is difficult in experimental measurements. Thus, we believe that the simulation techniques will be useful to develop and verify the physical modelling of geotechnical phenomena.

REFERENCES

Abate, G., Massimino, M.R., & Maugeri, M. 2015. Numerical modelling of centrifuge tests on tunnel–soil systems. *Bulletin of Earthquake Engineering* 13: 1927–1951.

Corigliano, M., Scandella, L., Lai, C.G. & Paolucci, L. 2011. Seismic analysis of deep tunnels in near fault conditions: a case study in Southern Italy. *Bulletin of Earthquake Engineering* 9: 975–995.

Cundall, P.A. & Strack, O.D.L. 1979. A discrete numerical model for granular assemblies. *Geotechnique* 29: 47–65.

Furuichi, M. & Nishiura, D. 2017. Iterative load-balancing method with multigrid level relaxation for particle simulation with short-range interactions. *Computer Physics Communications* 219: 135–148.

Ha, J.G., Lee, S-H., Kim, D-S. & Choo, Y.W. 2014. Simulation of soil–foundation–structure interaction of Hualien large-scale seismic test using dynamic centrifuge test. *Soil Dynamics and Earthquake Engineering* 61: 176–187.

Higuchi, S., Kato, I., Sato, S., Itoh G. & Sato, Y. 2017. Experimental and numerical study on the characteristics of earth pressure acting on the box-shape underground structure subjected the strike slip fault displacement, *Journal of Japan Society of Civil Engineering Ser.A1* 73: I_19–I_31.

Mason, H.B., Trombetta, N.W., Chen, Z., Bray, J.D., Hutchinson, T.C. & Kutter, B.L. 2013. Seismic soil–foundation–structure interaction observed in geotechnical centrifuge experiments, *Soil Dynamics and Earthquake Engineering* 48: 162–174.

Sakaguchi, H. 2004. Virtual tri-axial compression/extension tests of dry sands using discrete element method. *Journal of the Faculty of Environmental Science and Technology, Okayama University* 2004: 127–131.

Sawamura, Y., Kishida, K. & Kimura, M. 2015. Centrifuge model test and FEM analysis of dynamic interactive behaviour between embankments and installed culverts in multiarch culvert embankments. *International Journal of Geomechanics* 15: 04014050.

Sharma, J.S. & Bolton, M.D. 1996. Finite element analysis of centrifuge tests on reinforced embankments on soft clay. *Computers and Geotechnics* 19: 1–22.

Ulgen, D., Saglam, S. & Ozkan, M.Y. 2015. Assessment of racking deformation of rectangular underground structures by centrifuge tests. *Géotechnique Letters* 5: 261–26.

Physical Modelling in Geotechnics – McNamara et al. (Eds)
© 2018 Taylor & Francis Group, London, ISBN 978-1-138-34419-8

Trapdoor model test and DEM simulation associated with arching

M. Otsubo, R. Kuwano & U. Ali
Institute of Industrial Science, The University of Tokyo, Tokyo, Japan

H. Ebizuka
The Tokyo Metropolitan Government, Tokyo, Japan

ABSTRACT: The estimation of earth pressure exerted on underground structures is not easy due to complex interactions between soil and structure. The distribution of earth pressure varies with differential settlement of soils around the structure. This contribution assesses evolution of earth pressures acting on an embedded structure caused by differential settlement. A soil box with five base platens, that can be lowered or uplifted separately, was developed to undertake trapdoor tests. A series of trapdoor tests was conducted to quantify earth pressures acting on the base platens while varying the burial depth and the density of Toyoura sand. The earth pressure was measured using load cells located inside each base platen. Similar test condition was considered in equivalent discrete element method (DEM) simulations in which spherical particles analogue to soil grains were used. The DEM results capture the deformation characteristics of the ground and the pressure distribution on the bottom platens observed in the model tests. The evolution of strong contact forces during lowering the trapdoor was analysed and development of an arch above the lowering trapdoor was confirmed.

1 INTRODUCTION

The soil-structure interaction is important to predict the earth pressure distribution around a buried structure. Earth structures such as box culvert (Fig. 1) can cause differential settlement of the ground due to the difference in stiffness between the structure and surrounding soils. Consequently, the earth pressure distribution around the box culvert becomes complex. The earth pressure acting on a box culvert can be increased by a factor of 1.6 according to the current design codes depending on the burial depth of the structure (Japan road association 1999). Marston & Anderson (1913) derived an analytical approach to estimate the earth pressure as a summation of weight of soil above the structure and frictional resistance along vertical shear planes. Their approach considers an ideal lateral boundary condition in which no boundary effect is considered. On the other hand, loosening of soil due to formation of an arch in the ground is also important (Chevalier et al. 2009), and this contribution assesses the soil deformation and pressure distribution associated with arching using both experimental and numerical approaches.

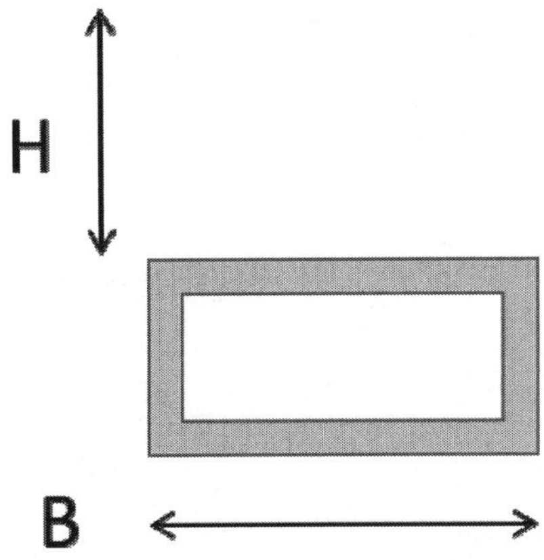

Figure 1. Schematic of a typical box culvert.

2 EXPERIMENTAL SETUP

Detailed specification of an experimental apparatus used in this study is explained in Ebizuka & Kuwano (2009), and is introduced briefly in this section. The trapdoor apparatus has dimensions of 700 mm × 293.6 mm × 555 mm (width × depth × height) (Fig. 2a). The base of the container is divided into seven parts, i.e. 100 mm × 7, where the five inner platens can be lifted or lowered individually (Fig. 2b). Each of the five platens contains five load cells that measure forces in the vertical and horizontal directions. In total, 25 load cells are installed with an

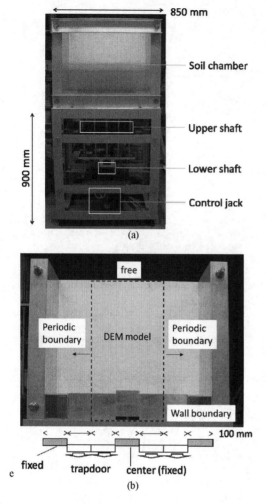

Figure 2. Trapdoor apparatus (a) soil container and loading system, (b) model for DEM simulations.

interval of 20 mm. Toyoura sand was used in the laboratory experiments, and its material properties are as follows: specific gravity $G_s = 2.62$, coefficient of uniformity $C_u = 1.61$, mean diameter $d_{50} = 0.203$ mm, maximum void ratio $e_{max} = 0.946$ and minimum void ratio $e_{min} = 0.637$. The model ground was prepared by means of dry pluviation at a fixed height. To visualise the ground deformation, coloured Toyoura sand was placed horizontally every 50 mm height close to the front wall. Referring to Table 1, two different densities of the model ground (e = 0.621–0.667 and 0.760–0.822) were prepared by changing the opening space of the pluviator, corresponding to relative densities $D_r \simeq 95\%$ (dense) and 40% (loose), respectively. The resultant e values varied slightly depending on the ground height. The ground height (H) tested were 100, 200, 300, 400 and 450 mm, giving H/B = 1.0, 2.0, 3.0, 4.0 and 4.5 where B = width of each platen (100 mm). In the experiments, the elevation of the centre platen was fixed while lowering the other four platens (Fig. 2b) with a constant velocity of 2×10^{-4} mm/s.

3 DISCRETE ELEMENT METHOD

The modified version of granular LAMMPS (Plimpton 1995) was used for three-dimensional DEM simulations of equivalent trapdoor tests, and the simulations were conducted using the Oakforest-PACS system in the Joint Centre for Advanced High-Performance Computing. Typical material properties of alkaline glass beads were used ($G_s = 2.5$, Young's modulus = 71.6 GPa and Poisson's ratio = 0.23) with a linear variation of particle size between 1.2 and 2.2 mm. Spherical particles were used and the Hertz-Mindlin contact model was adopted. Referring to Figure 2b a mixed boundary of lateral periodic and a rigid bottom wall was used. The same material and contact properties were applied for the contacts between the base wall and particles. Particles were generated randomly above the base wall having an initial void ratio about 3, and then gravitational forces were applied to the particles with an inter-particle friction coefficient of 0.05 and a damping. Particle kinetic energy was removed by applying an additional damping after the particles settled on the base wall. The dimensions of the model ground were 300 mm in width (X-direction) and 40 mm in depth (Y-direction), and the height (Z-direction) of the ground varied from 44 to 520 mm. The resultant e values were between 0.607 and 0.624 where a slightly denser packing was obtained for larger H (Table 1). The resultant H/B ratios were 0.44, 0.87, 1.7, 2.6, 3.5, 4.3 and 5.2 composed of 0.1, 0.2, 0.4, 0.6, 0.8, 1.0 and 1.2 million particles, respectively.

To simulate equivalent trapdoor tests as the laboratory, the base wall was divided into three parts, i.e. 100 mm × 3, in the X-direction and the earth pressure exerted on the base plates were measured with an interval of 20 mm. Two cylindrical elements having a radius of 1 μm were placed at the top convex corners of the centre plate to improve the contact detection. The inter-particle friction coefficient was increased to 0.35 before lowering the trapdoors. The trapdoors were lowered with constant velocities of 1×10^{-1} mm/s for test cases with H/B ≥ 3.5, and 5×10^{-2} mm/s for lower H/B cases to ensure quasi-static responses of the granular materials. No damping was applied during lowering the trapdoors.

4 RESULTS

4.1 Earth pressure distribution

The vertical earth pressure measured on the base wall at a large displacement of the trapdoor (p_{Zf}) normalised by the mean initial pressure (p_{Z0}) is illustrated in Figure 3 where displacements of 10 mm

Table 1. Test cases.

Test case	H mm	H/B	e	Density g/cm³	α_f
Lab 1	100	1.0	0.621	1.62 (dense)	1.33
Lab 2	200	2.0	0.637	1.60 (dense)	1.63
Lab 3	300	3.0	0.647	1.59 (dense)	2.22
Lab 4	400	4.0	0.647	1.59 (dense)	2.31
Lab 5	450	4.5	0.667	1.57 (dense)	2.44
Lab 6	100	1.0	0.760	1.49 (loose)	1.31
Lab 7	200	2.0	0.798	1.46 (loose)	1.59
Lab 8	300	3.0	0.818	1.44 (loose)	2.01
Lab 9	400	4.0	0.807	1.45 (loose)	2.33
Lab 10	450	4.5	0.822	1.44 (loose)	2.16
DEM 1	44.2	0.442	0.624	1.54 (dense)	1.21
DEM 2	87.3	0.873	0.617	1.55 (dense)	1.38
DEM 3	175	1.75	0.611	1.55 (dense)	1.72
DEM 4	261	2.61	0.610	1.55 (dense)	2.02
DEM 5	348	3.48	0.609	1.55 (dense)	2.17
DEM 6	434	4.34	0.608	1.55 (dense)	2.24
DEM 7	521	5.21	0.607	1.56 (dense)	2.28

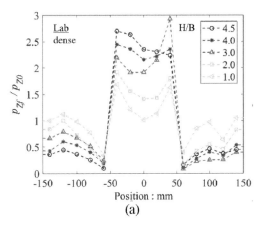

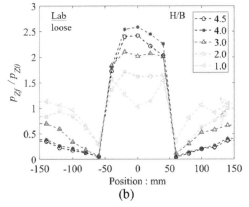

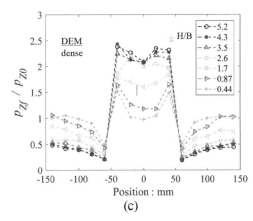

Figure 3. Normalised residual stress distribution on bottom platens (a) experiments for dense cases, (b) experiments for loose cases, (c) DEM simulations, for dense cases.

and 0.25 mm were considered for the experimental data and DEM data, respectively. The pressure ratio (p_{Zf}/p_{Z0}) at the centre plate increases with the H/B ratio, while the pressure ratio on the trapdoors decreases. Referring to the experimental data (Figs. 3a, b), the pressure distribution is affected by the packing density of the model ground, and Figure 3b shows the maximum pressures at the middle of the centre plate (position = 0 mm) in the loose cases except for the test case with H/B = 1. In contrast, the maximum pressures for dense model grounds are observed at the edges of the centre plate (position = −40 and 40 mm, Fig. 3a), which agrees with the DEM results for dense cases (Fig. 3c). For H/B ≤ 1.0, the pressure ratio remains close to 1.0, indicating no change in earth pressure at the middle part of the centre plate. For all the cases, the lowest pressures are evident on the trapdoors next to the centre plate (position = −60 and 60 mm). The maximum values of the pressure ratio are slightly higher for the experimental cases probably due to more angular shape of Toyoura sand grains. The minimum values of the pressure ratio are lower for the experimental cases.

The mean of the pressure ratio for the centre plate (α) is calculated as Equation 1, and the variations in α value with the lowering displacement of the trapdoors (δ) are illustrated in Figure 4.

$$\alpha = p_{z,center} / p_{z0} \quad (1)$$

where $p_{z,centre}$ is the mean vertical pressure exerted on the centre plate (−50 mm ≤ position ≤ 50 mm). For both experimental and DEM data, α increases with the H/B ratio where a clear difference in the peak strength between the dense and loose cases is observed for the experimental data (Fig. 4a), while the equivalent DEM dense samples do not exhibit a clear peak. The rate of increase in α with H/B is reduced with increasing H/B, i.e. the DEM results are similar between H/B = 4.3 and 5.2.

As considered above, the α values observed at $\delta = 10$ mm for the experimental data and $\delta = 0.25$ mm for the DEM data are defined as α_f and plotted against the H/B ratio in Figure 5. A linear relationship between

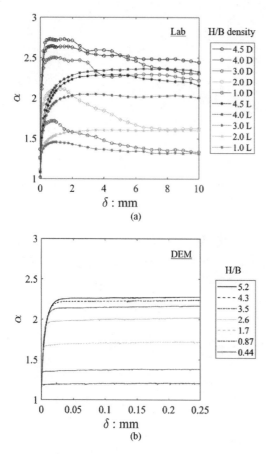

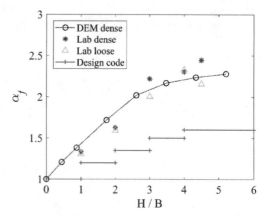

Figure 5. Relationship between residual pressure ratio (α_f) and depth-width ratio (H/B) of model ground.

Figure 4. Variation in α with lowering displacement of trapdoors (a) experiments, (b) DEM simulations.

α_f and H/B is marked for lower H/B < 3F where the DEM results of α_f values are slightly larger than that of experimental results, while a good match is observed between the experimental and DEM data for larger H/B. The maximum value of α_f may not exceed 3 even for larger H/B, and the reason is discussed below. Note that the current design codes underestimate α_f values considerably for the entire range of H/B.

4.2 Ground deformation

Experimental observations on deformation of dense model ground after lowering the trapdoors up to $\delta = 10$ mm for H/B = 2.0 and 4.0 are illustrated in Figure 6 where two shear planes develop symmetrically from both edges of the centre plate upwards. Ebizuka & Kuwano (2010) observed two types of shear planes: shear planes developed more vertically at around the peak strength and inner planes observed at a residual state. For lower H/B, the shear planes reach the ground surface and a non-uniform surface settlement is observed (Fig. 6a), while, for larger H/B, the shear planes do not reach surface and a uniform surface settlement is observed (Fig. 6b). This indicates that the upper side of model ground is not affected by the movement of the trapdoors at the base of the model ground.

Similar deformation characteristics are confirmed for DEM model grounds for H/B = 0.87, 2.6 and 4.3 (Figs. 7a, 8a, 9a) in which displacement vectors of individual particles at $\delta = 0.25$ mm along the middle slice of the ground is displayed (colour indicates the vertical displacement downwards). The shear planes develop more vertically along the XZ plane compared to the laboratory observation probably due to a lower shear strength of the spherical particles in DEM results (Stone & Muir Wood 1992). There seems to be a threshold value of H/B, depending on the material type, above which the surface settlement becomes uniform, and the threshold can be related to the formation of arching. The deformation characteristics are further discussed below.

4.3 Strong contact force

To discuss the particle-scale response associated with arching, strong contact forces (top 5% of contact force), connecting the centres of contacting particles are extracted from DEM data as illustrated for H/B = 0.87, 2.3 and 4.3 at $\delta = 0.25$ mm in Figures 7–9. As the contact force increases with the burial depth, the contact forces plotted in Figures 7–9 are normalised by the depth of the contact point from the ground surface.

Referring to H/B = 0.87, Figure 7b exhibits the strong contact forces observed before lowering the trapdoors, and a homogeneous distribution of strong contact forces is observed. After lowering the trapdoors, strong contact forces develop diagonally from the side parts of the centre (fixed) plate, and horizontal forces are marked above the trapdoors. In contrast, no obvious forces are confirmed above the middle part of the centre plate, which agrees with the earth pressure distribution as shown in Figure 3. Referring to Figure 7a for ground deformation, the particles above the middle part of the centre plate show negligible

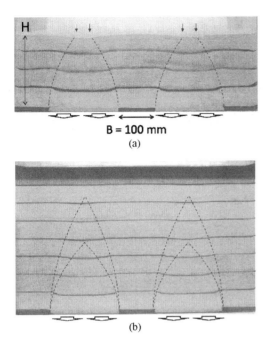

Figure 6. Residual deformation of model ground (a) H/B = 2.0, dense, (b) H/B = 4.0 dense.

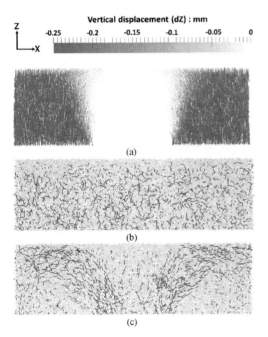

Figure 7. Particle-scale responses of DEM model ground for H/B = 0.87 (a) vertical displacement at $\delta = 0.25$ mm, (b) strong normal contact force at $\delta = 0$ mm, (c) strong normal contact force at $\delta = 0.25$ mm.

displacement even at the ground surface. The particles located above the trapdoors exhibit almost vertical settlement by the same amount of the settlement of trapdoors, i.e. $\delta = 0.25$ mm.

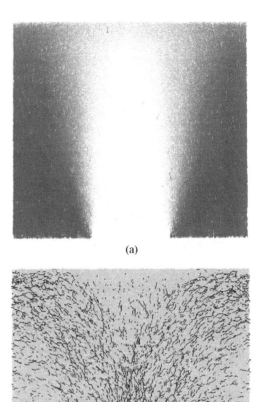

Figure 8. Particle-scale responses of DEM model ground at $\delta = 0.25$ mm for H/B = 2.6 (a) vertical displacement, (b) strong normal contact force.

For H/B = 2.6, strong contact forces are observed entirely on the centre plate, which agrees with Figure 3c. The direction of strong forces is rotated horizontally above the trapdoors (Fig. 8b). The movement of particles also indicates development of arching.

The evolution of strong contact forces for H/B = 4.3 (Fig. 9b) differs from that for lower H/B; the lower part of the ground shows a similar trend of that for H/B = 2.6 (Fig. 8b), while more homogeneous distribution of contact forces observed in the upper part of the ground is similar to that for the initial state of model ground (Fig. 7b). This indicates that all the masses of particles above the arch rest on the centre plate only, and further increment of H/B would not change the overall distribution of the contact forces. Thus, the maximum value of α_f depends on the configuration of the trapdoor test and this answers partially why α_f approaches 3 as discussed above. Additionally, the effect of particle characteristics on α_f for more general cases is being analysed.

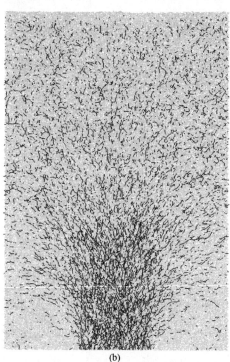

Figure 9. Particle-scale responses of DEM model ground at $\delta = 0.25$ mm for H/B = 4.3 (a) vertical displacement, (b) strong normal contact force.

5 CONCLUSIONS

This contribution assessed the earth pressure distribution and ground deformation around a buried earth structure subjected to differential settlement of surrounding soils. Laboratory model tests and equivalent DEM simulations of trapdoor test were conducted, and the following conclusions can be drawn.

1. The initial density of model ground affects the earth pressure distribution on base plates where the maximum pressures are observed at the middle or sides of the centre (fixed) plate for loose and dense cases, respectively.
2. The vertical pressure ratio measured on the centre plate increases with the burial depth (H) for a given width of an earth structure (B), which exceeds the consideration in the current design codes.
3. The maximum pressure ratio is 3 due to the periodic configuration of the trapdoor system considered in this study when arching is fully developed.
4. The overall earth pressure distribution on the base platens is unchanged after arching is fully developed despite further increase in H/B.
5. The direction of strong contact forces rotates horizontally above the trapdoors. Both ground settlement and contact force distribution become uniform above arches with larger H/B values.
6. DEM simulation results capture the overall trend of the laboratory experimental results, and more general cases of trapdoor configuration as well as material properties are being analysed.

REFERENCES

Chevalier, B., Combe, G. & Villard, P. 2009. Experimental and numerical study of the response of granular layer in the trap-door problem, *AIP Conference Proceedings*, 1145, 649–652.

Ebizuka, H. & Kuwano, R. 2009. Trap door tests for evaluation of stress distribution around a buried structure, *Proc. 8th International symposium on new technologies for urban safety of mega cities in Asia*, USMCA, Inchon, 245–250.

General guidelines for road earthworks – culvert work. 1999. Japan road association, ISBN 978-4889504101, Maruzen Print Co. Ltd. (in Japanese).

Kuwano, R. & Ebizuka, H. 2010. Trapdoor tests for the evaluation of earth pressure acting on a buried structure in an embankment, *Proc. 9th International symposium on new technologies for urban safety of mega cities in Asia*, USMCA, Kobe, 453–460

Marston, A & Anderson, A.O. 1913. The theory of loads on pipes in ditches and tests of cement and clay drain tile and sewer pipe, 31, Iowa State University Eng. Experiment Station.

Plimpton S. 1995. Fast parallel algorithms for short-range molecular-dynamics. *Journal of Computational Physics*, 117(1), 1–19.

Stone, K.J.L. & Muir Wood, D. 1992. Effects of dilatancy and particle size observed in model test on sand, *Soils and Foundations*, 32(4), 43–57.

4. Scaling

Variability of small scale model reinforced concrete and implications for geotechnical centrifuge testing

J.A. Knappett & M.J. Brown
School of Science and Engineering, University of Dundee, UK

L. Shields
Galliford Try, UK (formerly University of Dundee, UK)

A.H. Al-Defae
University of Wasit, Iraq (formerly University of Dundee, UK)

M. Loli
National Technical University of Athens, Greece

ABSTRACT: There are soil-structure interaction problems for which it is important to model both the relative soil-structure stiffness and strength. Examples from the earthquake engineering field include the design of resilient rocking-isolated foundations and the seismic stabilisation of slopes using piling. In both cases the aim is to ensure a preferred failure mode happens first in the soil instead of the structure i.e. controlled bearing failure of the foundation or soil yielding around piles. A recently developed model reinforced concrete for centrifuge testing can simulate stiffness and strength simultaneously, but suffers from variability in the material properties, as does the full-scale material. This paper presents a series of element tests on the variability of model reinforced concrete elements representative of large square monolithic bridge piers and slender square piles. Coefficients of variation for various material and element properties have been determined and shown to be similar to typical values for full-scale reinforced concrete elements obtained from the literature. It is also demonstrated that curing time beyond 28 days does not substantially affect strength and variability and that models of different absolute volume can be produced without inducing detrimental size effects. The results are used to discuss the selection of mean design strengths for model structural elements in centrifuge experiments using a quantitative statistical approach where there are competing structural and soil failure modes.

1 INTRODUCTION

While for many soil-structure interaction problems it is possible to identify a single non-dimensional parameter to ensure sufficient similitude when testing physical models in a geotechnical centrifuge, there also exist a number of important problems for which this is not the case. These are typically cases in which one of the possible failure modes is associated with the structural elements of the soil-structure system in question. In such cases it is important to achieve similitude of both the relative soil-structure stiffness and relative soil-structure strength. The former is important to ensure that the kinematic demands on the soil and structural elements are representative as the system approaches failure; the latter, to ensure that the appropriate ('weakest') failure mode is reached first, whether that is associated with the soil failing around the structure, or the structure failing before the soil.

Examples of such problems often occur in earthquake engineering, where high kinematic demands and failure may be unavoidable within design. These include assessing the performance of rocking-isolated structures (e.g. bridge piers on shallow foundations, Loli et al. 2014), pile reinforcement of slopes against seismic effects (e.g. Al-Defae & Knappett 2014) or damage to foundation pile groups due to liquefaction-induced lateral spreading (e.g. Stergiopoulou et al. 2016). In each of these aforementioned examples, the potentially damageable elements (piers or piles) were modelled using a novel micro reinforced concrete ('model RC') developed at the University of Dundee (Knappett et al. 2011) which can simultaneously achieve similitude of stiffness, strength and ductility at scaling factors suitable for centrifuge use.

This modelling approach involves combining a quasi-brittle cementitious material (plaster-based mortar) with discrete steel wires, as a geometrically scaled model of reinforced concrete. One of the implications of this, however, is that model RC elements will exhibit greater variability in key material properties (e.g. bending stiffness, EI and moment capacity, M_{ult})

Table 1. Model concrete cube compression testing results.

Mix*	t_{cure} (days)	n	$\bar{f}_{cu,100}$ (MPa)	COV
1	28	11	26.3	0.128
2	28	6	35.6	0.150
3	28	6	50.9	0.086

* as defined in Knappett et al. (2011)

Table 2. Model concrete modulus of rupture testing.

Size (mm)	t_{cure} (days)	n	$\bar{f}_r$ (MPa)	COV
10 × 10	28	10	2.02	0.149
30 × 30	variable	11	2.62	0.121

than equivalent 'elastic' models (made typically out of aluminium alloys, steel or plastics). If this variability is similar to that of field reinforced concrete, this would potentially represent another way in which the model RC is a closer analogue of field concrete.

This paper will address this issue of variability by presenting test data of both the variability in fundamental mechanical properties of the individual material components (e.g. compressive strength and tensile strength of the model concrete; yield strength of the model reinforcement) and of full reinforced concrete structural elements. This will be compared with extensive data from the literature for field reinforced concrete. The elements tested will be based on those used in recent geotechnical centrifuge testing programmes, and the results will be used to discuss the implications of variability on model design, using the example of a reinforced concrete bridge pier on a foundation designed to provide rocking-isolation under seismic actions.

2 TEST PROGRAMME

A series of both reinforced and unreinforced elements were cast using the materials introduced in Knappett et al. (2011). A total of 23 cube compression tests are reported herein, alongside 10 model reinforcement tension tests, 21 four-point bending tests on unreinforced model concrete beams (prisms) of different model sizes and 20 four-point bending tests on reinforced model RC elements of different size and reinforcement ratio, consisting of both singly-reinforced slender beams and stockier uniformly reinforced sections.

2.1 Model materials and element tests

Table 1 summarises the key compressive properties and variability of the model concrete materials, while Table 2 summarises the key tensile properties and variability. Compressive strengths ($f_{cu,100}$) were determined from crushing tests on 100 × 100 × 100 mm cubes; mean strength and coefficient of variation (COV) are shown in Table 1, across the n samples tested. All cubes were cured for $t_{cure} = 28$ days before testing.

Tensile strength was quantified by the modulus of rupture (f_r), determined from the maximum force (V_{ult}) applied at brittle failure of an unreinforced rectangular prism, loaded as a beam in a four-point bending test. The modulus of rupture represents the breaking strength at the edge of the beam on the tension side of the beam on the tension side, and can be considered to be the most representative tensile strength relating to the behaviour of reinforced elements in transverse bending.

Ten of the prisms were 10 mm × 10 mm in cross-section and 200 mm long and cured in air for 28 days before testing, while the remaining eleven were 30 mm × 30 mm in section and 200 mm long and cured for variable lengths of time before testing, between 14–122 days. By varying the curing length it will be possible to assess whether delays in a programme of centrifuge testing since curing would significantly affect the material properties of the model concrete.

Modulus of rupture is typically the parameter of a quasi-brittle material which is most sensitive to unwanted size effects, due to the likelihood of a larger crack existing within a specimen of larger volume. The prism dimensions considered here, while selected to match the later reinforced element tests (due to appropriate formwork being available for casting), demonstrate nearly one order of magnitude (a factor of nine times) difference in actual volume at model scale between the two sets of tests. This variation will therefore allow a fuller investigation of size effects than conducted in previous work (e.g. Knappett et al. 2011).

In the four-point bending tests, the distance (L) between supports was set at 180 mm (with the beam overhanging the supports at each end by 10 mm) and symmetrical vertical shearing loads were applied at 60 and 120 mm from one of the supports (i.e. a shear span of $a = 60$ mm). The modulus of rupture was then found using:

$$f_r = \frac{2V_{ult}L}{bd^2} \quad (1)$$

where $b =$ beam breadth and $d =$ beam depth.

All 21 prisms were cast using Mix 1 model concrete (see Table 1), and the basic test results are given in Table 2.

2.2 Model RC elements

Reinforced (model RC) elements were fabricated to different designs, representing (i) the 1:50 scale singly-reinforced piles used for slope stabilisation in the centrifuge tests reported by Al-Defae and Knappett (2014) and (ii) the 1:50 scale Eurocode 8 compliant square concrete bridge pier used in the centrifuge tests reported by Loli et al. (2014). These share the same exterior dimensions as the prisms cast for modulus of

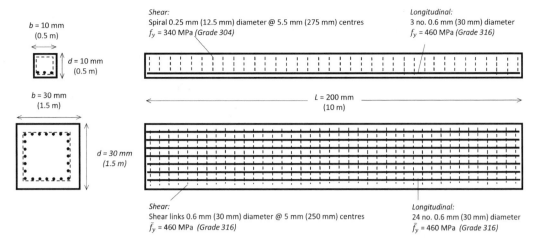

Figure 1. Reinforcement layout of model RC beams. Dimensions are model scale (prototype scale for 1:50 in brackets).

Table 3. Model RC element testing.

Type	Size (mm)	t_{cure} (days)	n	M_{pr} (kNm)
Pile	10 × 10	28	10	230
Pier	30 × 30	28	6	4500
Pier	30 × 30	variable	4	4500

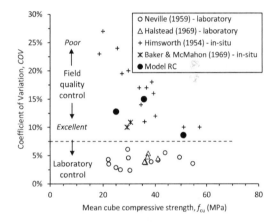

Figure 2. Variability of model concrete compressive strength compared to typical values for conventional concrete.

rupture testing described above, but containing model longitudinal and shear reinforcement modelled using drawn stainless steel wire (Grade 316 or 304) which was roughened for bond by gluing on a coating of fine sand (see Knappett et al., 2011 for further details). The reinforcement layouts are shown in Figure 1 and some key mechanical properties are summarised in Table 3.

M_{pr} in Table 3 represents the prototype Moment capacity that the sections simulate. The beams were tested in four-point bending using the same loading arrangement as for the modulus of rupture tests described above. It should be noted that the larger pier section was tested under zero axial load (though it would be used as a column in a centrifuge model). The model pier tests included some that were cured for longer periods than 28 days, to examine how any change in properties from curing (e.g. in f_r) may manifest in the global response of the reinforced element.

Ten sample lengths of Grade 316 stainless steel wire were tensile tested within an Instron 1196 load frame to evaluate the variability in the yield strength of the steel (f_y), which was evaluated at 0.2% strain for consistency with conventional definitions of this parameter for tests on steel reinforcing bar.

3 VARIABILITY OF MODEL RC ELEMENTS

3.1 Material components

Figure 2 shows a comparison of the compressive strength of the model concrete against data for conventional concrete collected from the literature (as detailed in the figure). It is clear that for the conventional data there is a significant difference in variability (as expressed by COV) between concrete mixed and placed in-situ, and that produced under laboratory conditions. The comments on quality control shown in the figure are as proposed by Walker (1955). It can be seen that the model concrete, prepared using the procedures outlined in Knappett et al. (2011), appears to have variability of compressive strength consistent with concrete cast in-situ with good quality control (model concrete data taken from Table 1).

Figure 3 shows a comparison of the yield strength of the model reinforcement compared to typical values for steel reinforcing bar taken from the literature. The dataset of Mirza & MacGregor (1979) covers tests on bars of diameter between 10–43 mm, produced in the UK, Canada and USA. The smaller dataset of Saputra et al. (2010) includes bars of between 25–32 mm diameter, produced in Indonesia, Japan and USA. It should be noted that the prototype scale diameter of the model wire tested (30 mm) falls within both datasets and has

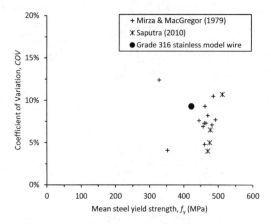

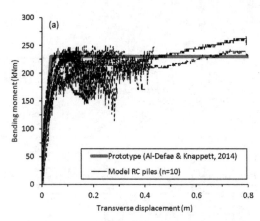

Figure 3. Variability of model reinforcement yield strength compared to typical values for conventional reinforcing bar.

a variability similar to that of conventional reinforcing bar. However, the test data indicate a mean yield strength which is lower ($\bar{f}_y = 422$ MPa) than the nominal value of 460 MPa shown in Figure 1 based on three preliminary tests.

3.2 Reinforced structural elements

Figure 4 shows the results of the bending tests for the model RC piles (Figure 4a) and piers (Figure 4b), to indicate the variability in the bending behaviour between nominally identical elements, both in terms of stiffness and strength. The data in this figure is shown at prototype scale (1:50). In each case, the prototype behaviour that the elements should reproduce is also shown. For the singly reinforced pile element, this was determined by hand calculation (see Knappett et al., 2011 for further details); for the pier, the uniform distribution of reinforcement complicated this and the prototype behaviour was instead determined using the numerical section analysis software USC_RC (Esmaeily & Xiao 2002).

There is a very limited amount of data in the literature relating to the variability of reinforced concrete elements, principally because of the wide variability in possible designs, and the time required to cast a sufficient number of elements for testing in the laboratory. However, MacGregor et al. (1983) present a numerical analysis in which simulations have been performed on a range of different types of structural elements (including beams) to determine the variability in the capacity, by accounting for the variability of the various constituent parts within the governing equations. Values of the ultimate moment capacity (M_{ult}) were taken from the data shown in Figure 4 and the COV of this data is shown in Figure 5, compared to the slender beams simulated by MacGregor et al. (1983). The parameter A_s in the figure is the total cross-sectional area of the longitudinal reinforcement (so that A_s/bd is the '% steel area' or reinforcement ratio).

Figure 5 shows that the model RC elements when cured for 28 days appear to exhibit similar or lower

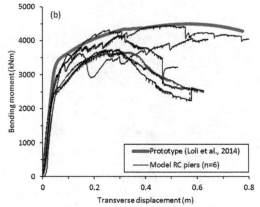

Figure 4. Four-point bending test results at prototype scale (1:50) for (a) piles; and (b) piers (only $t_{cure} = 28$ day cases shown).

variability than the conventional full-scale simulations. It is also demonstrated that if curing time was not controlled and the complete set of 10 pier tests was used to determine the COV for this case, the variability increases (as expected), but it is still reasonably representative of the conventional predictions. This suggests that if model RC elements were cast for a centrifuge test to be approximately 28 days old at the test time, but the test had to be postponed, it may not be necessary to recast new model RC elements.

This is a useful finding, given the amount of preparation, casting and curing time required to make such small and detailed model elements.

3.3 Effects of curing time and model volume

Figure 6 shows the effects of curing time and model volume (through comparison of the piles and piers) on the moment capacity of the model RC elements. M_{ult} has been normalised by the target prototype moment capacity (M_{pr}) so that the two sections with very different values can be compared on the same plot. Considering first the piles (only tested at 28 days) a very close replication of the target moment capacity is achieved, as noted previously by Knappett et al.

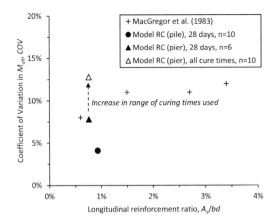

Figure 5. Variability of bending strength of model RC beams compared to simulated values for full-scale beams.

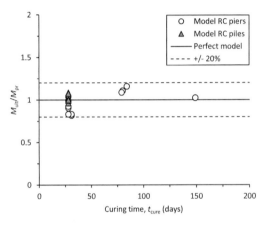

Figure 6. Effect of model volume and curing time on bending strength of model RC structural elements.

(2011). In comparison, the piers at $t_{cure} \approx 28$ days generally underpredict the target by up to 20% (though on average, around 12%). This is consistent with the pier section having a larger volume than the pile, which would suggest a slightly lower strength using the same materials based on fracture mechanics.

Previously (Knappett et al., 2011), it was suggested that the model RC could be used to produce prototypes of different sizes by varying the scaling factor. However, the results shown in Figure 6 suggest that models of different absolute size (here, by an order of magnitude in volume) can also be used without exhibiting significant over- or under-strength. This will provide greater flexibility in designing centrifuge models, particularly in cases where the scaling factor must by fixed by other considerations.

With increased curing, the underestimation in the piers reduces, and all of longer-term tests exhibit capacities which are slightly higher than the target. Along with Figure 5, this suggests that a longer wait between casting and testing will not only make the elements more variable in properties (the implication of which will be discussed in the following section), but also generally make them stronger. However, in both cases the effects are relatively small. Future testing of the pile section at different curing times would allow any size-dependency of the ageing effect to be determined.

4 IMPLICATIONS FOR MODEL DESIGN

It has already been demonstrated that the curing time and volume effects in the model RC are relatively small, providing flexibility in test scheduling and model design. However, the inherent variability of the model RC must also be taken into account when designing models which may induce structural failure. This will be demonstrated using the example of the rocking-isolated bridge pier experiments presented by Loli et al. (2014).

A soil-foundation-structure system such as a bridge pier will only be rocking isolated if the moment capacity of the foundation is lower than that of the structure. In Loli et al. (2014), two different foundations were considered to demonstrate the difference between rocking-isolated and conventional foundations. The former foundation was designed to have a deterministic moment capacity of 4.8 MNm, and the latter, 12.9 MNm. These values were confirmed by simulated push-overs in Loli et al. (2015). The moment capacity of the pier (as tested above) with the axial load of the bridge deck applied was 6.6 MNm. As this value is between the two footing moment capacities, this would appear to be suitable. However, this assumes that there is no possibility that due to material property variability in both the soil and the model RC pier, the rocking-isolated footing moment capacity is at one of its largest possible values and that this is higher than the moment capacity of the structure if this happens to be at one of its lowest possibilities (at which point it would cease to be rocking-isolated).

Figure 7 shows the probability density functions for the foundations and pier with the following assumptions: (i) the COV of the pier moment capacity is unchanged by axial load and only the mean capacity is increased; (ii) that the COV of the footing moment capacity is reflective of that of the friction angle used in its calculation. The value of COV selected is 10% after Schneider (1999); (iii) all moment capacities can be approximated by a normal distribution based on the deterministic calculations as the means.

The area beneath the overlapping parts of the footing and pier probability density function (pdf) curves (shaded zone in Fig. 7) represents the probability of the rocking isolation failing. For the case shown (using $COV = 7.8\%$ based on $t_{cure} = 28$ days), the probability is 7%. If there was greater variability, say due to a greater variability in curing time within the experimental programme, the likelihood of the test not working as designed would increase. Using the larger COV for the piers from earlier (12.8%), the probability of failure increases to 17%.

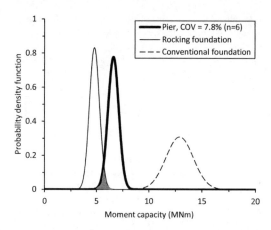

Figure 7. Determination of the probability of a rocking-isolated foundation not behaving as expected due to variability in the moment capacities.

This example demonstrates how the variability of model RC materials can be accounted for quantitatively in the design of centrifuge models including structural failure, to form a rational basis for decision making.

5 CONCLUSIONS

This paper has presented a study of the variability of a novel model RC material for use in centrifuge experiments which can simultaneously scale stiffness and strength. It has been shown that the variability of both the individual component properties and reinforced beams are similar to those of conventional field reinforced concrete. The effect of increased curing time increases variability and mean strengths, but the effect is small, suggesting that delays in a programme of centrifuge testing should not significantly affect the properties of the model RC elements. It has also been demonstrated how variability in the strengths associated with soil and structural failure modes can be quantitatively and statistically assessed in experimental design.

REFERENCES

Al-Defae, A. H. & Knappett, J. A. 2014. Centrifuge modelling of the seismic performance of pile-reinforced slopes. *J. Geotechnical & Geoenvironmental Engineering*, 140(6): 04014014.

Baker, W. M. & McMahon, T. F. 1969. Quality assurance in highway construction: Part 3 – Quality assurance of Portland Cement concrete. *Public Roads*, 35(8): 184–189.

Esmaeily, G. A. & Xiao, Y. 2002. Seismic behavior of bridge columns subjected to various loading patterns. *PEER Report 2002/15*, University of California, Berkeley.

Halstead, P. E. 1969. The significance of concrete cube tests. *Magazine of Concrete Research*, 21(69): 187–194.

Himsworth, F. R. 1954. The variability of concrete and its effect on mix design. *Proceedings of the Institution of Civil Engineers, Part I*, 3(2): 163–200.

Knappett, J. A., Reid, C., Kinmond, S. & O'Reilly, K. 2011. Small scale modelling of reinforced concrete structural elements for use in a geotechnical centrifuge. *Journal of Structural Engineering, ASCE*, 137(11): 1263–1271.

Loli, M., Knappett, J. A., Brown, M. J., Anastasopoulos, I. & Gazetas, G. 2014. Centrifuge modelling of rocking-isolated inelastic RC bridge piers. *Earthquake Engineering & Stuctural Dynamics*, 43(15): 2341–2359.

Loli, M., Knappett, J. A., Anastasopoulos, I. & Brown, M. J. 2015. Use of Ricker wavelet ground motions as an alternative to push-over testing. *Int. J. of Physical Modelling in Geotechnics*, 15(1): 44–55.

MacGregor, J. G., Mirza, S. A. & Ellingwood, B. 1983. Statistical analysis of resistance of reinforced and prestressed concrete members. *ACI Journal*, 80(3): 167–176.

Mirza, S. A. & MacGregor, J. G. 1979. Variability of mechanical properties of reinforcing bars. *ASCE J. of the Structural Division*, 105(ST5): 921–937.

Neville, A. M. 1959. The relation between standard deviation and mean strength of concrete test cubes. *Magazine of Concrete Research*, 11(32): 75–84.

Saputra, A., Limsuwan, E. and Ueda, T. 2010. Characteristics of material and fabrication for concrete structures in Indonesia. *Engineering Journal*, 14(4): 11–22.

Schneider, H. R. 1999. Determination of characteristic soil properties, *Proceedings of 12th European Conference on Soil Mechanics and Foundation Engineering*, Balkema, Rotterdam, 1: 273–281.

Stergiopoulou, E., Knappett, J. A. & Wotherspoon, L. 2016. Replication of piled bridge abutment damage mechanism due to lateral spreading in the centrifuge. *Proc. 3rd European Conf. on Physical Modelling in Geotechnics, EUROFUGE 2016*, Nantes, France, 1–3 June, 2016: 225–230.

Walker, S. 1955. Application of theory of probability to design of concrete for strength specifications. *14th Annual Meeting of National Ready Mixed Concrete Association*, Washington, 27 January, 1955.

Modelling experiments to investigate soil-water retention in geotechnical centrifuge

M. Mirshekari, M. Ghayoomi & A. Borghei
University of New Hampshire, Durham, New Hampshire, USA

ABSTRACT: Steady state flow and capillary rise from an identified water table are two common approaches to simulate unsaturated soils inside geotechnical centrifuge. However, they involve challenges with regards to introducing and monitoring unsaturated state, mapping the model results to prototype values, and accurate in-flight measurements. This paper presents planning, sensor calibration procedures, testing strategies, and preliminary results of a set of centrifuge experiments to evaluate the effect of g-level on soil-water retention; while both water content and suction were measured simultaneously throughout the tests. Considering the length scaling for capillary rise, negligible g-effect was observed on suction-water content relation. Steady state infiltration, however, resulted in suction-water content coordinates along the hysteresis curve depending on the sequence of hydraulic scenarios.

1 INTRODUCTION

Geotechnical centrifuge has been increasingly implemented to model or characterize hydraulic and mechanical behaviors of partially saturated soils. Different experimental approaches to control the level of saturation and conduct centrifuge tests on systems with unsaturated soils have been proposed that include experiments incorporating an unsaturated flow (e.g. Nimmo et al. 1987; Knight et al. 2000; McCartney & Zornberg 2010; Mirshekari & Ghayoomi 2017), tests with the vadose zone above an identified water table (e.g. Cooke & Mitchell 1991; Crançon et al. 2000; Depountis et al. 2001; Esposito 2000; Rezzoug et al. 2004), or those where a fine-grained soil was compacted with a certain moisture content prior to spin up (Deshpande & Muraleetharan, 1998). The challenges involved in centrifuge modeling of partially saturated soils comprise the experimental methods to control the degree of saturation throughout the test, in-flight measurements of degree of saturation as well as matric suction, the employment of 1-g Soil Water Retention Curve (SWRC) in higher gravitational fields, and mapping the scaled results from centrifuge tests to their corresponding prototype values.

Capillary rise similitude law was investigated through analytical derivations (Arulanandan et al., 1988; Lord, 1999; Rezzouget al. 2000) and, to some extent, by experimental simulations (Crançon et al., 2000; Depountis et al., 2001; Esposito, 2000; Knight et al. 2000; Rezzoug et al. 2004). The overall conclusion of these studies was that the modeled capillary rise is reduced by the factor of 1/N, similar to the length scaling factor. However, the experimental studies that addressed the capillary rise scaling law fell short in planning a testing scenario where the degree of saturation and matric suction are continuously measured during the centrifugation. Some of the shortcomings in the abovementioned testing programs included employment of optical measurements to determine the water-saturated zone, measurements of degree of saturation during the centrifuge stoppage times, absence of matric suction measurements, and assuming the validity of 1-g SWRC measurements at higher gravities (Cooke & Mitchell 1991; Crançon et al. 2000; Depountis et al. 2001; Esposito, 2000; Rezzoug et al. 2004). Although the centrifugation technique has been recommended for SWRC determination of fine-grained soils (as per ASTM D6836), the high-g SWRC measurements were only validated by means of indirect suction estimations using analytical formulations; i.e. the equation proposed by Gardner 1937 (Briggs & Mclane 1907; Khanzode et al. 2002; Reis et al. 2011; Russell & Richards 1939). Further, the degree of saturation was not continuously measured throughout the tests and only the values at stoppage times were reported.

Steady state infiltration technique was initially incorporated in in-flight permeameters with the purpose of characterizing hydraulic properties of unsaturated soils (Nimmo et al. 1987; Nimmo et al. 1992; Conca & Wright 1990; Zornberg & McCartney 2010). This technique was lately implemented in geotechnical centrifuges to study seismic response of partially saturated sands (Ghayoomi et al. 2011; Ghayoomi & McCartney 2011; Mirshekari & Ghayoomi 2017). McCartney and Zornberg (2010) studied SWRC and hydraulic conductivity of a low-plasticity clay in different g-levels and under different discharge velocities during consecutive drying and wetting processes in an in-flight permeameter. However, the effect of g-level on SWRC could not be clearly discovered as the specimen was rewetted successively. For the specific tested clay, the hysteresis was found to be negligible which, reportedly, could be either a result of continuous infiltration during the experiments, initial saturation level (i.e. not a fully saturated specimen), or the narrow range of moisture variations.

This paper aims to report the planning, calibration process, and preliminary results of centrifuge tests on partially saturated sand using two approaches: 1) where capillary rise generates a vadose zone above an identified water level (drainage experiments) and 2) during steady state infiltration. The objective is to introduce a systematic procedure to evaluate the influence of g-level on scaling, modeling, and interpretation of soil systems in partially saturated state through continuous measurement of volumetric water content (VWC, denoted by θ) and matric suction (ψ).

2 BACKGROUND

2.1 Capillary rise in centrifuge

Capillary rise in vadose zone above the groundwater table could be simplified using a "bundle of tubes" modeling the water paths throughout the soil layer (e.g. see Lord 1999). Similar to the capillary rise in a tube, water rises along different heights in unsaturated soils depending on the void sizes at various locations of soil layers. Considering the common shape of SWRC, the soil layer above the water table and below the air-entry suction head is almost saturated. This nearly-saturated zone is often referred as "capillary fringe" region; while, the suction head corresponding to the residual water content, is called the "capillary rise" height (Lu & Likos 2004).

Capillary rise in soils might be expressed as a function of surface tension between water and soil particles, void sizes between the soil granules, water density, and the gravity level. Therefore, in a higher gravitational field, if the soil/water surface tension is not affected by the g-level, the capillary rise is only a function of capillary rise in 1-g and the g-level; i.e. scaling factor of 1/N (Arulanandan et al. 1988; Lord 1999; Rezzoug et al. 2000). However, the surface tension between water and soil is a function of the shape of water menisci between granules in unsaturated soils, which could be influenced by the g-level (Schubert 1982). This phenomenon, although is hard to track in a scientific study, might lead to a different scaling factor.

The experimental programs, scheduled to study the capillary rise scaling factor, generally fall into two categories; drying and wetting experiments. Drying experiments involved draining an initially-saturated specimen in higher g-levels and monitoring the variations of capillary fringe or capillary rise heights (Cooke & Mitchell 1991; Esposito 2000). Wetting experiments, on the other hand, included setting up a water tank adjacent to the soil specimen, opening the water flow into the soil layer during the experiment, and monitoring the capillary characteristics throughout the tests at different g-levels (Crançon et al. 2000; Depountis et al. 2001; Rezzoug et al. 2004). The capillary rise (or fringe) height should be compared to drying or wetting paths of 1-g SWRC depending on the testing condition. Monitoring the capillary characteristics was carried out by means of optical distinction between wet and dry fronts (Burkhart et al. 2000; Depountis et al. 2001; Rezzoug et al. 2004), employing time domain reflectometry sensors to capture moisture content variations (Crançon et al. 2000), or simply weighing small soil samples during the centrifuge stoppage times (Cooke & Mitchell 1991; Esposito 2000). Among these studies only Burkhart et al. (2000) measured matric suction where the capillary rise heights estimated by optical measurements were about two times the ones measured by tensiometers.

2.2 Steady state infiltration in centrifuge

Steady state infiltration technique has been commonly used to generate uniform degree of saturation profiles in small permeameters (Nimmo et al. 1987; Nimmo & Akstin 1988; Nimmo et al. 1992; Conca & Wright 1990; Dell'Avanzi et al. 2004; Zornberg & McCartney 2010) as well as Geotechnical centrifuges (Ghayoomi & McCartney 2011; Mirshekari & Ghayoomi 2017). The concept is that a target water discharge is applied to an in-flight soil specimen while the water may freely drain out of the specimen. The infiltration can be applied using different approaches such as water sprinkling through several small porous stones, employing High Air Entry Value (HAEV) disks with different pore sizes on top of the soil sample, or spraying water onto the specimen. For the case of steady state infiltration in larger containers, the latter approach would be more practical. As a result of consistent infiltration and drainage in high-g, a uniform degree of saturation (or matric suction) profile would be achieved along a portion of specimen's length. The length of uniform degree of saturation profile is a function of g-level increasing as the g-level rises (Dell'Avanzi et al. 2004). Further, the degree of saturation in steady state infiltration could be varied by either changing the g-level or the discharge velocity (Dell'Avanzi et al. 2004; McCartney & Zornberg 2010).

The scaling factors of unsaturated flow problems were obtained through dimensional analysis or analyzing governing equations of unsaturated flow in prototype and model (Goodings 1982; Cargill & Ko 1983; Arulanandan et al. 1988; Cooke & Mitchell 1991; Goforth et al. 1991; Butterfield 1999; Lord 1999; Barry et al. 2001; Dell'Avanzi et al. 2004). By comparing the flow equations in prototype and model, Dell'Avanzi et al. (2004) concluded that when the ratio of the centrifuge arm to the length of specimen is more than 10, the matric suction profile would be scaled by the factor of unity. This so-called suction scaling factor becomes lower than one as the arm/length ratio decreases. Importantly, it should be noted that this scaling factor is only to be used for pure infiltration problems where the suction or degree of saturation profiles are compared with the length and discharge velocity scaled accordingly in the model. The appropriate scaling factor for measured matric suction values would be 1 considering their nature as a stress-type parameter. More detailed discussion on the suction scaling factor during steady state infiltration is available in Mirshekari et al. (2017).

3 EXPERIMENTAL PROGRAM

The 5 g-ton Geotechnical centrifuge with an arm radius of 1 m at the University of New Hampshire was used to conduct the target experiments in this study. The experimental procedure for steady state infiltration tests was similar to the one explained in Mirshekari and Ghayoomi (2017) where water was infiltrated using a set of nozzles and the drainage was conducted through a set of solenoid valves at the bottom of the container.

Employing the spraying system in steady state infiltration tests required using a relatively more permeable soil. Accordingly, F-75 Ottawa sand, a poorly graded material classified as SP as per USCS ($D_{50} = 0.2$ mm), was used in this study. The same soil was used in the capillary rise experiments since the suction head corresponding to the residual water content is high enough that the transition zone in SWRC could be captured with an acceptable resolution in higher g-levels.

The sand was dry pluviated in the container with a relative density of approximately 45%. The Soil Water Retention Curve (SWRC) of the soil was measured by the tensiometric technique. The specimens were instrumented with miniature tensiometers for matric suction and pore water pressure measurement, miniature pore pressure transducers for pore pressure measurement only, dielectric sensors for VWC measurement, and LVDTs for displacement measurement. In addition, during the capillary rise experiments, a plastic coverage was placed over the specimens to avoid surface evaporation. The instrumentations layouts for capillary rise and steady-state infiltration experiments are illustrated in Figure 1a and 1b, respectively.

The Druck PDCR81 miniature pore pressure transducers were used to measure positive pore water pressure and EPB-PW miniature tensiometers, from Measurement Specialties, were used to measure both pore pressure and matric suction in this research. To precisely measure water pressure/suction, both miniature tensiometers and pore pressure transducers must be completely saturated with water. In order to saturate the sensors, the concept introduced by Take and Bolton (2003) was used in this study. During the saturation step, the sensors were fixed in a sealed Plexiglass cell and an approximately −90 kPa vacuum was applied to the cell for 90 minutes. Thereafter, de-aired water was induced to the cell so that the sensors were soaked in water and the system was left on the vacuum for at least 12 hours so that any possible dissolved air bubble would be sucked out of the sensors.

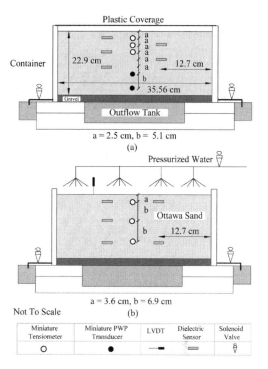

Figure 1. Schematics of instrumentation layout in (a) capillary rise tests and (b) steady state infiltration tests.

In order to calibrate the sensors, the Plexiglass cell was connected to a long burette where both cell and burette were completely filled with water. Afterwards, the water in burette was lowered sequentially and the head between water in burette and the diaphragm of sensors was measured and correlated with sensors' voltage output. The calibration of EPB sensors was performed for both positive and negative water pressures acting on the diaphragm where the negative pressure was obtained by adjusting the water level in burette lower than the diaphragm elevation. The calibration results demonstrated a negligible difference between the performance of sensors in positive and negative pressure (i.e. for each sensor an almost same linear calibration equation was obtained in pressure and suction). The calibration procedure was repeated for all the sensors prior to each test since the sensors' performance slightly varied over time. Prior to using the sensors in experimental programs, the cell was filled with Ottawa sand and the same steps of calibration with the burette were repeated to assure the accuracy of EPB sensors' measurements in soil. In addition, measurements of the EPB sensors were verified against a T5 Decagon lab tensiometer in a cylindrical container where a difference less than 0.5 kPa was observed between the miniature and lab tensiometers.

EC-5 Decagon dielectric sensors were used in this study to measure the VWC which correlate the dielectric constant of the soil with the volume of water. Since the resistivity of soils are also a function of mineralogy, texture, and salinity of the material, it is generally recommended to calibrate the sensors in the same testing conditions to increase the measurement accuracy (Cobos and Chambers 2010). Therefore, the EC-5 sensors were calibrated inside Ottawa sand mixed with water supplied from the same tank (as provided in the target centrifuge tests). The initial steps of the calibration procedure for the dielectric sensors consisted of uniformly mixing different amounts of water with sand, leaving the wet sand in a sealed plastic bag for one day to assure the moisture content homogeneity throughout the batch, and compacting the wet sand in three lifts inside a container to reach the relative density of almost 45%. Thereafter, the sensors were inserted into the soil specimen and the obtained VWC was recorded. After recording the VWC, several small samples of soil were obtained from different locations in the mold and oven-dried to measure an average moisture content of the specimen which, then, was linked to the VWC using the volume of container and total wet weight of soil.

This procedure was repeated for different VWCs and for each sensor. Since the accuracy of the dielectric sensors measurements in almost fully saturated condition was less than that in unsaturated condition, the data points with nearly-saturated condition were excluded when obtaining the linear calibration equation. To ensure that the horizontal alignment of the sensors in centrifuge tests does not affect the calibration results, the same procedure was repeated with horizontally-aligned sensors which demonstrated a negligible influence of sensors' alignment on the measured VWC. The calibration was accomplished in several molds (i.e. one compaction mold and two plastic molds with the diameters of 10 cm, 11 cm and 20 cm, respectively) to assure the insignificant influence of boundaries on the sensors' performance.

4 RESULTS AND DISCUSSION

4.1 Capillary rise tests

The set of tests planned to study the capillary rise similitude law and the influence of higher gravity on SWRC included experiments in several g-levels ranging from 5 g to 40 g. The captured results comprised of arrays of pore water pressure/suction time histories (with a sampling frequency of 10 Hz) and manually-recorded VWC values (at desired time intervals).

The pore water pressure/suction and VWC time histories during the 15-g experiment at the depth of 5 cm are shown in Figure 2a and 2b, respectively. When interpreting the time histories, it should be noted that the matric suction values did not exceed 8.2 kPa at the location of the EPB sensor. The matric suction values were also capped to values ranging from 7 to 15 kPa at the locations of other EPB sensors. This might have occurred because the water menisci shape at the porous stone/sand region did not permit water content to become lower than a certain value, locally.

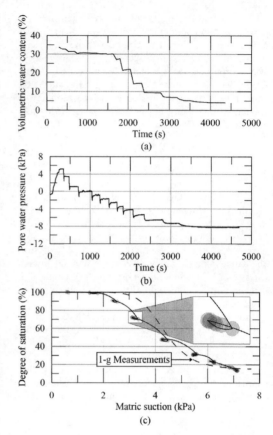

Figure 2. The results of 15-g experiment at the depth of 5 cm (a) VWC time history, (b) Pore water pressure/suction time history, (c) SWRC.

In addition, upon attempting to stabilize the water level at each sequence by closing the solenoid valve at the bottom, the values of pressure/suction showed a peak after which the pressure/suction was maintained constant. These peaks are due to small water level variations which could not be shown in the VWC time histories as the θ values were manually recorded at longer time intervals. The resulted SWRC for the 15-g experiment was obtained by graphing the degree of saturation versus their corresponding values of matric suction and is shown in Figure 2c along with 1-g SWRC curve from tensiometeric technique. It is noticeable that the small water table variations in the beginning of each sequence led to a small hysteresis effect which is reflected in the SWRC. These accumulative hysteresis effects shifted the obtained SWRC from the 1-g curve gradually as the test proceeded.

In order to study the capillary height scaling factor, degree of saturation profiles in each g-level and for different water levels were obtained. The degree of saturation profiles for the 15-g experiment for 4 water levels are illustrated in Figure 3. Results from the two pore water pressure transducers (Druck sensors) were employed to identify water table at each step. 1-g van Genuchten (VG) curves (van Genuchten 1980)

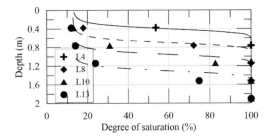

Figure 3. Degree of saturation profiles obtained in the drying experiment at 15 g.

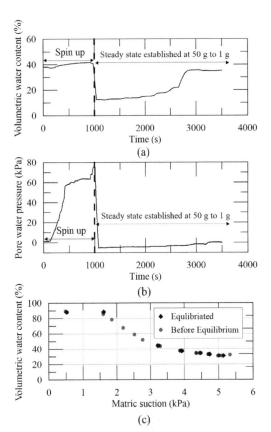

Figure 4. Sample time histories during steady state infiltration experiments for (a) VWC and (b) Pore water pressure/suction.

(obtained by fitting the VG equation to the results of 1-g tensiometeric tests), scaled by the factor of 1/N, initiating from the acquired water levels are demonstrated alongside the degree of saturation values in Figure 3. The agreement between the scaled VG curves and the measured degree of saturation values in higher gravity indicates that length scaling factor could be used with an acceptable precision for mapping the results of capillary rise. The same conclusion was reached by monitoring the results of experiments in other g-levels.

4.2 Steady state infiltration tests

The results of steady state infiltration experiments included matric suction and VWC time histories throughout the experiments under different discharge velocities where in each test the g-level was varied from 50 g to 1 g, gradually. The time histories of pore water pressure/matric suction and VWC during a test, with g-level varying from 50 g to 1 g, at the bottommost instrumented location are illustrated in Figure 4a and 4b, respectively. The time histories consist of two major parts of centrifuge spin up and steady state infiltration at target g-level.

During the spin up VWC slightly increased, which was due to the applied pressure in high-g resulting in higher VWC. The observed change in VWC was higher than the matric suction variations throughout the test as a result of relatively low suction level in Ottawa sand. The steady state condition, during the experiments under each discharge velocity, was initially reached at 50 g which led to the lowest degree of saturation throughout the test. Therefore, during each test under a constant water discharge the $\theta-\psi$ coordinate of the specimen shifted along the drying path of SWRC from the fully-saturated condition at 50 g. Then, the steady state infiltration was established for lower g-levels, which led to higher degree of saturation along the specimen. The increase in degree of saturation in lower g-levels led the $\theta-\psi$ coordinate of the specimen to shift along a scanning path starting from the degree of saturation at 50 g. By graphing VWC and matric suction values at times of equilibrated conditions in each g-level, the SWRC during steady conditions could be obtained.

5 SUMMARY AND CONCLUSIONS

This paper presented the testing procedure and preliminary results of a thorough study on the centrifuge modeling of unsaturated soils through two approaches of capillary rise and steady state infiltration. Dielectric sensors, pore water pressure transducers, and miniature tensiometers were successfully employed to measure volumetric water content, pore water pressure, and matric suction throughout the experiments. The calibration and validation procedures for dielectric sensors and miniature tensiometers were discussed. During the capillary rise experiments, water level was lowered gradually in different g-levels and the high-g SWRC and capillary rise scaling factor were studied using the obtained matric suction and VWC at each step. The results demonstrated the g-level independency of the SWRC, and also validated the length scaling of 1/N to be used for capillary rise scaling. The steady state infiltration tests comprised tests under several discharges where in each experiment the steady state condition was established in different g-levels. A significant hysteresis was observed during the infiltration tests under each discharge velocity when the steady state infiltration was established while lowering the g-level. More results and discussion on this subject would be available in Mirshekari et al. (2018).

REFERENCES

Arulanandan, K., Thompson, P., Kutter, B., Meegoda, N., Muraleetharan, K. & Yogachandran, C. 1988. Centrifuge modeling of transport processes for pollutants in soils. *Journal of Geotechnical Engineering*, 114(2), 185–205.

Barry, D., Lisle, I., Li, L., Prommer, H., Parlange, J., Sander, G. C. & Griffioen, J. 2001. Similitude applied to centrifugal scaling of unsaturated flow. *Water Resources Research*, 37(10), 2471–2479.

Burkhart, S., Davies, M., Depountis, N., Harris, C. & Williams, K. 2000. Scaling laws for infiltration and drainage tests using a geotechnical centrifuge. *Physical Modeling and Testing in Environmental Geotechnics*, 191–198.

Butterfield, R. 1999. Dimensional analysis for geotechnical engineers. Geotechnique, 49(3), 357–366.

Cargill, K. W. & Ko, H. 1983. Centrifugal modeling of transient water flow. *Journal of Geotechnical Engineering*, 109(4), 536–555.

Cobos, D. R. & Chambers, C. 2010. Calibrating ECH2o soil moisture sensors. Decagon Devices.

Conca, J. L. & Wright, J. 1990. Diffusion coefficients in gravel under unsaturated conditions. *Water Resources Research*, 26(5), 1055–1066.

Cooke, B. & Mitchell, R. 1991. Physical modelling of a dissolved contaminant in an unsaturated sand. *Canadian Geotechnical Journal*, 28(6), 829–833.

Crançon, C., Pili, E., Dutheil, S. & Gaudet, J. 2000. Modelling of capillary rise and water retention in centrifuge tests using time domain reflectometry. Paper presented at *the Proceeding of the International Symposium on Physical Modelling and Testing in Environmental Geotechnics*, 199–206.

Dell'Avanzi, E., Zornberg, J. G. & Cabral, A. R. 2004. Suction profiles and scale factors for unsaturated flow under increased gravitational field. *Soils and Foundations*, 44(3), 79–89.

Depountis, N., Harris, C. & Davies, M. 2001. An assessment of miniaturised electrical imaging equipment to monitor pollution plume evolution in scaled centrifuge modelling. *Engineering Geology*, 60(1), 83–94.

Esposito, G. 2000. Centrifuge simulation of light hydrocarbon spill in partially saturated dutch dune sand. Bulletin of Engineering Geology and the Environment, 58(2), 89–93.

Ghayoomi, M., McCartney, J. & Ko, H. 2011. Centrifuge test to assess the seismic compression of partially saturated sand layers. *ASTM Geotechnical Testing Journal*, 34(4), 321–331.

Goforth, G. F., Townsend, F. & Bloomquist, D. 1991. Saturated and unsaturated fluid flow in a centrifuge. Paper presented at the Proc. Centrifuge, 91, 497–502.

Goodings, D. 1982. Relationships for centrifugal modelling of seepage and surface flow effects on embankment dams. *Géotechnique*, 32(2), 149–152.

Knight, M., Cooke, A. & Mitchell, R. 2000. Scaling of movement and fate of contaminant releases in the vadose zone by centrifuge modeling. *Physical Modeling and Testing in Environmental Geotechnics*, 233–242.

Lord, A. 1999. Capillary flow in the geotechnical centrifuge.

Lu, N. & Likos, W. J. 2004. *Unsaturated soil mechanics*. Hoboken, New Jersey: John Wiley & Sons.

McCartney, J. S. & Zornberg, J. G. 2010. Centrifuge permeameter for unsaturated soils. II: Measurement of the hydraulic characteristics of an unsaturated clay. *Journal of Geotechnical and Geoenvironmental Engineering*, 136(8), 1064–1076.

Mirshekari, M. & Ghayoomi, M. 2017. Centrifuge tests to assess seismic site response of partially saturated sand layers. *Soil Dynamics and Earthquake Engineering*, 94, 254–265.

Mirshekari, M., Ghayoomi, M. & Borghei, A. 2018. A review on centrifuge modeling and scaling of unsaturated sands. *ASTM Geotechnical Testing Journal*, Accepted, In Press.

Muraleetharan, K. K. & Granger, K. K. 1999. The use of miniature pore pressure transducers in measuring matric suction in unsaturated soils. *Geotechnical Testing Journal*, 22 (3), 226–234.

Nimmo, J. R. & Akstin, K. C. 1988. Hydraulic conductivity of a sandy soil at low water content after compaction by various methods. *Soil Science Society of America Journal*, 52(2), 303–310.

Nimmo, J. R., Akstin, K. C. & Mello, K. A. 1992. Improved apparatus for measuring hydraulic conductivity at low water content. *Soil Science Society of America Journal*, 56(6), 1758–1761.

Nimmo, J. R., Rubin, J. & Hammermeister, D. 1987. Unsaturated flow in a centrifugal field: Measurement of hydraulic conductivity and testing of darcy's law. *Water Resource. Res*, 23(1), 124–134.

Rezzoug, A. 1 s., König, D. & Triantafyllidis, T. 2004. Scaling laws for centrifuge modeling of capillary rise in sandy soils. *Journal of Geotechnical and Geoenvironmental Engineering*, 130(6), 615–620.

Rezzoug, A., Konig, D. & Trantafylidis, T. 2000. Numerical analysis of scaling laws for capillary rise in soils. *Physical Modeling and Testing in Env. Geotechnics.*, 217–224.

Schubert, H. 1982. Kapillaritat in porosen feststoffsystemen. Berlin: Springer.

van Genuchten, M. T. (1980). "A closed-form equation for predicting the hydraulic conductivity of unsaturated soils." *Soil Sci.Soc.Am.J.*, 44(5), 892–898.

Zornberg, J. G. & McCartney, J. S. 2010. Centrifuge permeameter for unsaturated soils. I: Theoretical basis and experimental developments. *Journal of Geotechnical and Geoenvironmental Engineering*, 136(8), 1051–1063.

Studies on the use of hydraulic gradient similitude method for determining permeability of soils

K.T. Mohan Gowda & B.V.S. Viswanadham
Department of Civil Engineering, Indian Institute of Technology Bombay, Powai, Mumbai, India

ABSTRACT: The objective of the present study is to use the hydraulic gradient similitude method for assessing the permeability of soils with varying fines content and unit weights in the laboratory. For this purpose, a custom-designed hydraulic gradient similitude (HGS) test set-up was developed and used. In order to induce varying hydraulic head conditions, the Mariotte tube concept was adopted. This paper also discusses relevant scaling laws pertinent to the determination of permeability using HGS method. The developed HGS test setup was calibrated for varying pressure levels and instrumented using pore water pressure transducers. The model scaling laws implied by the hydraulic gradient similitude method are verified using modelling of models technique with experimental results. Three types of soils, namely, fine sand, fine sand with 10% fines, fine sand with 20% fines were adopted in the present study. The developed HGS test setup enables determination of the permeability of soils of equivalent thicknesses in the field.

1 INTRODUCTION

Physical model studies have been significantly helpful in the field of geotechnical engineering to understand and analyse the behaviour of the soil system. They include 1 g small scale tests, full-scale model testing (seldom adopted) and centrifuge model testing (popular). Simulation of actual stresses and stress-strain characteristics in physical soil models is necessary for understanding the actual behaviour of the prototype. Since in 1 g small scale models, only limited level of actual stresses can be achieved and full-scale testing is unreasonable, centrifuge model testing is used and accepted. But its widespread use is limited due to the high cost of the equipment. Another simple physical modelling technique is the hydraulic gradient similitude (HGS) method, wherein, the seepage forces are used to simulate the representative field stresses by maintaining required constant hydraulic gradient in small scale physical models (Zelikson 1969). Stresses are created in the soil model by using seepage forces, which increase the body forces in it, hence capable of achieving the stress representing the prototype level of stresses in the model. The major difference between the centrifuge modelling and this method is that in the former body force is due to the centrifugal acceleration whereas the latter uses seepage forces in the porous media. Higher the hydraulic gradient greater the depth of representative prototype for a soil model under consideration.

Zelikson (1969) used this approach to study the single pile penetration tests. Also, the experiment was conducted on a horizontal cylindrical anchor pulled by a wire at a constant depth. The resistance values obtained were compared with full-scale test results. Yan & Bryne (1989) studied small scale footing tests on the sand where a ceramic porous stone represented the footing. A linear pattern was observed when ultimate bearing pressure obtained was plotted with hydraulic gradient scale factor 'N' on a log scale and a similar trend was observed in centrifuge modelling results. Blackburn (1989) used a modified triaxial cell to study the uplift resistance of piles using this approach. Yan (1990) and Yan et al. (1991) developed a test setup and used this approach to study the behaviour of laterally loaded piles (single and group) under static and dynamic conditions in the sand. Birnbaum (1998) and Cohen et al. (2003) studied the capacity of driven piles and behaviour of driven piles under static loading and pull-out respectively. Avishur et al. (2006) studied the seismic response of piles using this method; wherein the model was subjected to gradient is attached to the shake table. Musso & Ferlisi (2009) used this approach to model the collapse behaviour of strip footing in the dense sand when vertical as well as eccentric loads are acting. Collapse mechanism is observed to depend on hydraulic gradient applied and load eccentricity. Leshchinsky et al. (2012) conducted eight pull-out tests for studying the uplift behaviour of piles in sand, wherein the total head values were measured at the boundaries of each soil layer before the uplift force was applied and calculated the hydraulic gradient associated with the layers. Yuan et al. (2016) studied the response and deformations of laterally loaded piles using this method. Small scale model studies using this method requires a container or a cell with facility to inlet and outlet the water, water tightness of the cell to maintain uniform gradient, provision to

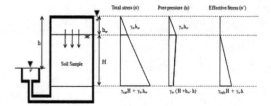

Figure 1. Pressure diagram for the representative soil model.

have steady state flow with continuous water supply, pressure measurement sensors like pore pressure transducers to monitor the hydraulic gradient uniformity and any other facility as per requirement. In the present study the scale factor for permeability is derived and experiments are conducted on three types of soils at different hydraulic gradients. The permeability values hence obtained are reported. Tests are conducted on two model heights to examine modelling of models concept.

2 SCALING LAWS

2.1 Hydraulic gradient (i) and scale factor (N)

It is now understood that in the HGS method prototype stresses in the model soil is achieved using seepage forces and reproduce the same mechanical properties in the model as that in the prototype. In the studies, it is shown that conservation of stresses and retention of material is taken care along with the increase of body forces. To obtain model similitude effective stress at each point in the small scale model and at the corresponding point in prototype should be the same. Let the scale factor with which the body forces are increased by N, where N is the ratio of prototype dimension to model dimension. Figure 1 shows the pressure diagram for the representative soil model of height H subjected to a head difference of h and derivation of the scale factor in terms of hydraulic gradient i is obtained as follows,

$$\sigma' = \gamma_{sub} H + \gamma_w h \qquad (1)$$

Hydraulic gradient,

$$i = h/H \qquad (2)$$

Using equations (1) and (2), the effective unit weight of the soil,

$$\frac{\sigma'}{H} = \gamma_m = \gamma_{sub} + i\gamma_w \qquad (3)$$

Here $i\gamma_w$ is the seepage pressure and can be seen that effective unit weight of the soil is increased by seepage pressure. Hence model unit weight has been increased by scale factor N times.

Therefore,

$$N = \frac{\gamma_m}{\gamma_p} = \frac{(\gamma_{sub} + i\gamma_w)_m}{(\gamma_{sub} + i\gamma_w)_p} \qquad (4, 5)$$

where γ_p is the effective unit weight of the prototype soil.

Equation (5) is further simplified in view of the practical consideration by the assumptions that the hydraulic gradient in the prototype is zero and the submerged unit weight is approximately equal to the unit weight of water.

$$N \approx \frac{\gamma_{sub}(1+i)}{\gamma_{sub}} \quad N \approx 1+i = 1+\frac{\Delta p}{\gamma_w H} \qquad (6, 7)$$

where Δp is the pressure difference between the the the top and bottom of the soil sample of height H.

Hence at homologous points in the model and prototype, the stresses developed will be same when a hydraulic gradient of i is achieved in the model scaled down by N times, which itself is the scaling law in the HGS approach;

$$\frac{H_m}{H_p} = \frac{1}{N} \qquad (8)$$

For example, if a prototype soil strata of 5 m in depth is to be represented in a small scale model of size 100 mm, then using HGS method, it is required to maintain a hydraulic gradient i of 49 in the small scale soil model. Based on equation (7), the scale factor $N = 50$.

2.2 Permeability

Using Darcy's law and steady-state seepage conditions, the scale factor for permeability is derived. Seepage force is given by,

$$F = i\gamma_w V \qquad (9)$$

where V is volume of the fluid phase (V = HA, A is c/s area of the soil model). Since unit weight of water is same, volume is scaled down by N3 and approximating, $i = N$,

$$\frac{F_m}{F_p} = N\frac{1}{N^3} = \frac{1}{N^2} \qquad (10)$$

Let K be the intrinsic permeability, then permeability k is given by,

$$k = \frac{K\rho_w g}{\mu_w} \qquad (11)$$

where, ρ_w, μ_w and g are the mass density and dynamic viscosity of water, and acceleration due to gravity respectively. Considering all these parameters (Eq. 11) to be equal in the model and the prototype scale factor for permeability obtained as,

$$\frac{k_m}{k_p} = 1 \quad (12)$$

And velocity of flow, ; where, n is the porosity and vs is the seepage velocity. Considering same porosity and using equation (12), scale factor for seepage velocity is,

$$\frac{(v_s)_m}{(v_s)_p} = \frac{k_m}{k_p} N \; ; \; \frac{(v_s)_m}{(v_s)_p} = N \quad (13, 14)$$

where m indicates model and p indicates prototype. The scale factors derived are used in the experimental study on determination of permeability using HGS method.

3 EXPERIMENTAL HGS TEST SET-UP

Figure 2 shows the schematic view of the developed HGS set-up. The model soil container has plan dimensions of 0.3 m by 0.2 m and height of 0.2 m internally and with the facility to inlet water, water collection tray to maintain the steady water table at the bottom and pore pressure transducers are placed to monitor the pore water pressures developed during various stages of the test. For maintaining the constant gradient Mariotte's principle is used and for inducing the higher gradients in the model soil, Mariotte tube is connected to an air compressor from which the water under pressure is introduced into the air tight container.

3.1 Mariotte's principle

The Mariotte's tube (Fig. 3) is a device that allows the movement of liquid at a constant pressure regardless of the level of liquid in the tube, but the liquid should remain above the bottom level of the tube which determines the exit pressure. Consider a cylinder filled with water to certain height (L) from the bottom outlet. The pressure above the water surface, A is atmospheric, a_o and the pressure at the outlet point, C is $a_o + \rho_w g L$, where ρ_w is the mass density of water and g is acceleration due to gravity. The pressure at D is also atmospheric, a_o. The pressure difference under which the flow is taking place is:

$$a_o + \rho_w g L - a_o = \rho_w g L \quad (15)$$

As the flow progresses the value of L varies and a constant head is not maintained. Therefore, a constant head difference h is created by inserting a thin tube, B to a certain level in the cylinder. The pressure at bottom of the tube inserted is atmospheric. It is known that at point D the pressure is atmospheric and now the flow is taking place under a head difference equal to:

$$a_o + \rho_w g h - a_o = \rho_w g h \quad (16)$$

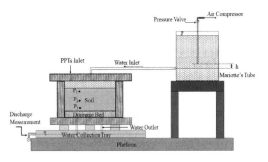

Figure 2. Schematic representation of experimental set-up.

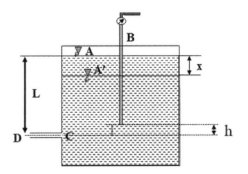

Figure 3. Schematic representation of the principle of Mariotte's tube.

where h is the constant head maintained. Thus, as long as the level of water inside the tube is above the bottom point of the tube inserted, the pressure at the exit level will remain constant at $\rho_w g h$. The air that enters the system bubbles up from the bottom of the tube to space at the top of the tube.

Rezzoug et al. (2004) used Mariotte's principle to study capillary rise in sand soils using centrifuge modelling. In their studies, it is assumed that air is behaving as a perfect gas medium with a pressure lower than atmospheric. With reference to Fig. 3, at initial position A ($P_o V_o$) when the flow has not yet started and at position A' ($P_1 V_1$) where the level of water is dropped by x, it can be written from the law of perfect gases as,

$$P_o V_o = P_1 V_1 = P_1(V_o + ax) \quad (17)$$

where $P_o = P_{atm}$ is the initial pressure; V_o is the initial volume; P_1 is the pressure at A'; V_1 is the volume at position A'; 'a' is the cross sectional area of the tank and 'x' is the drop in water level. Hence, this principle is employed to create a hydraulic gradient in the model soil to achieve the stresses homologous to the prototype by attaching an air compressor to it and supplying the water under desired pressures. In the present study the value of h is 0.08 m and the Mariotte's tube is filled leaving few centimeters of space at the top to release the built up pressure in case of requirement in the form of air pressure without spillage of water.

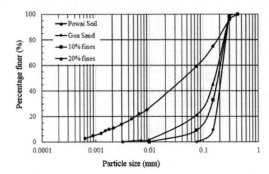

Figure 4. Particle size distribution curves of soils used in the study.

Table 1. Permeability values measured under different hydraulic gradients.

Model Height 0.094 m

Test legend	Prototype depth (m)	Scale factor (N)	Hydraulic gradient (i)	Permeability (m/s)
T1	Sand with RD = 40%			
T1a	1.24	12.79	11.76	3.33×10^{-05}
T1b	2.36	24.89	23.89	2.92×10^{-05}
T1c	3.6	38.29	37.29	2.48×10^{-05}
T2	Sand with RD = 50%			
T2a	1.24	12.79	11.76	2.93×10^{-05}
T2b	2.36	24.89	23.89	2.68×10^{-05}
T2c	3.6	38.29	37.29	1.99×10^{-05}
T3	Sand with RD = 60%			
T3a	1.24	12.79	11.76	2.64×10^{-05}
T3b	2.36	24.89	23.89	2.20×10^{-05}
T3c	3.6	38.29	37.29	1.78×10^{-05}
T4	Sand with RD = 85%			
T4a	1.24	12.79	11.76	2.43×10^{-05}
T4b	2.36	24.89	23.89	2.09×10^{-05}
T4c	3.6	38.29	37.29	1.53×10^{-05}
TF1	Sand with 10% fines ($\gamma_{dmax} = 17$ kN/m³; OMC = 13%)			
TF1a	1.24	12.79	11.76	5.12×10^{-06}
TF1b	2.36	24.89	23.89	3.96×10^{-06}
TF1c	3.6	38.29	37.29	4.44×10^{-06}
TF2	Sand with 20% fines ($\gamma_{dmax} = 17.4$ kN/m³; OMC = 14.6%)			
TF2a	1.24	12.79	11.76	2.24×10^{-07}
TF2b	2.36	24.89	23.89	2.05×10^{-07}
TF2c	3.6	38.29	37.29	2.43×10^{-07}

4 MATERIALS

The soils used in the study are pure sand (Goa Sand) without any fines and locally available Powai soil. The Powai soil is mixed with sand to obtain 10% and 20% fines soil blend in calculated proportions. Grain size distribution curves for Sand, Powai soil, 10% and 20% fines soil are presented in Fig. 4. As per USCS, Sand, Powai Soil, 10%, and 20% mixes are classified as SP, MI, SP-SM and SP-SM respectively.

5 TEST PROGRAMME

Tests are conducted on the Goa sand with relative densities 40%, 50%, 60% and 85%, and with fines content of 10% and 20% soil mixture. In this paper, results on two types of soils were used for formulating three soil types. They are (i) Goa sand and Goa sand blended with Powai soil (giving 10% fines and 20% fines). Two model heights (0.13 m and 0.094 m) were considered to perform modelling of models. Three hydraulic gradients are considered representing different prototype depths. The test series T1, T2, T3, and T4 for sand with relative densities and three gradients denoted by a, b and c. The inclusion of fines content is referred to as F1 and F2. Tables 1–2 summarize details of tests performed and discussed in this paper. The purpose of performing modelling of models is to validate derived scaling laws. This technique is also adopted in centrifuge model studies with the same intention, wherein, soil models are tested at different gravity levels. In the present study change in the model height implies change in the hydraulic gradient to be applied in order to arrive at the similar results.

In the preparation of a soil model, first, a drainage bed is prepared using an aggregate passing 12 mm and retained on 10 mm sieve, wrapped by non-woven geotextile which is helpful is avoiding the migration soil particles into the drain path. The drainage bed prepared is of very high permeability. Model soil was placed layer by layer to achieve the target dry unit weight throughout the model height H. The model thus prepared is kept for saturation under low gradient for a period depending on the type of soil. For saturation, water is supplied from bottom to top under low gradients. Further, water in the Mariotte's tube is pressurized using a compressor and then supplied through the inlet into the model soil. Pressure is monitored using pore pressure transducers. Water is collected at the bottom constant water table where the pressure is atmospheric. Discharge is measured from the tray outlet and is observed for the steady value. Permeability is calculated using $Q = kiA$, where Q is discharge measured, i is hydraulic gradient applied, A is the area of cross-section (0.06 m²) and k is the coefficient of permeability. The average permeability values measured at different hydraulic gradients for a specified model height and soil is representing corresponding prototype depth are presented in Tables 1–2.

6 RESULTS AND DISCUSSION

Figures 5–6 show the pore water pressure data for model height 0.094 m and 0.13 m respectively for all the hydraulic gradients. The pore water pressure transducer was placed 0.01 m below from the top surface of the sample, at center and 0.01 m above the bottom surface of the sample for both the model heights (P_1, P_2 and P_3 in Fig. 2). The pore water pressure data presented are for 10% fines soil sample.

Table 2. Permeability values measured under different hydraulic gradients.

Model Height 0.13 m

Test legend	Prototype depth (m)	Scale factor (N)	Hydraulic gradient (i)	Permeability (m/s)
Tm1	Sand with RD = 40%			
Tm1a	1.24	9.23	8.23	3.68×10^{-05}
Tm1b	2.36	18	17	3.33×10^{-05}
Tm1c	3.6	27.69	26.69	2.45×10^{-05}
Tm2	Sand with RD = 50%			
Tm2a	1.24	9.23	8.23	3.51×10^{-05}
Tm2b	2.36	18	17	3.18×10^{-05}
Tm2c	3.6	27.69	26.69	2.19×10^{-05}
Tm3	Sand with RD = 60%			
Tm3a	1.24	9.23	8.23	3.32×10^{-05}
Tm3b	2.36	18	17	2.93×10^{-05}
Tm3c	3.6	27.69	26.69	2.25×10^{-05}
Tm4	Sand with RD = 85%			
Tm4a	1.24	9.23	8.23	2.99×10^{-05}
Tm4b	2.36	18	17	2.75×10^{-05}
Tm4c	3.6	27.69	26.69	2.45×10^{-05}
TmF1	Sand with 10% fines ($\gamma_{dmax} = 17\,kN/m^3$; OMC = 13 %)			
TmF1a	1.24	9.23	8.23	2.35×10^{-06}
TmF1b	2.36	18	17	1.63×10^{-06}
TmF1c	3.6	27.69	26.69	2.02×10^{-06}
TmF2	Sand with 20% fines ($\gamma_{dmax} = 17.4\,kN/m^3$; OMC = 14.6%)			
TmF2a	2.73	21	20	2.37×10^{-07}
TmF2b	3.79	29.15	28.15	2.14×10^{-07}
TmF2c	5.1	39.29	38.23	2.18×10^{-07}

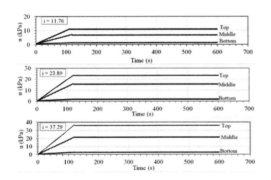

Figure 5. Pore-water pressure data of hydraulic gradients associated with sample height 0.094 m.

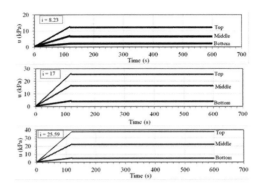

Figure 6. Pore-water pressure data of hydraulic gradients associated with sample height 0.13 m.

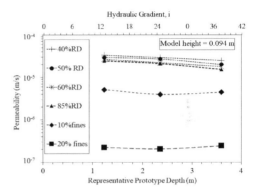

Figure 7. Variation of measured permeability with representative prototype depth.

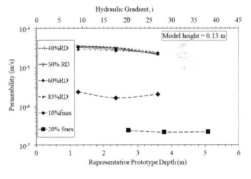

Figure 8. Variation of measured permeability with representative prototype depth.

This kind of constant variation in pressures were not observed in sand samples due to its high permeability. While performing HGS tests on sand samples, fluctuations in pore water pressure variations were observed continuously. This resulted in dropping of pressures unevenly.

Figures 7–8 present the measured permeability values with representative prototype depth for model heights 0.094 m and 0.13 m respectively. From the results, it can be noted that permeability decreases with an increase in gradient or scale factor or representative prototype depth. Also, for the same gradient permeability decreases with an increase in relative density, owing to the reduction in void spaces when density is increased. With an increase in gradient values, though the discharge values increased due to higher seepage forces, the permeability values decreased. Dunn (1983) also observed this behavior and, inferred that it is due seepage-induced consolidation and/or migration of fines creating local impermeable zones. In the case of 10% fines and 20% fines soil the permeability values decreased with an increment of the hydraulic

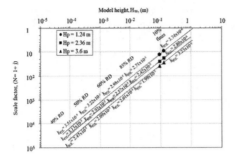

Figure 9. Scale factors used for two model heights to represent three prototype depths (shown in legends).

gradient but for the next increment, it increased by a certain amount though not significant. No outwash of fines was observed during tests and this is attributed to the presence of filter at the base (Fig. 2). Further, permeability values are measured for another model height that is 0.13 m, showed similar observations are presented in Table 4. Modelling of models by changing the sample height from 0.094 m to 0.13 m showed the same order of permeability values. With an increase in thickness of model soil height, marginal decrease in permeability values was observed. In sand, the hydraulic gradient was not uniform due to its high permeability but in the case of 10% fines and 20% fines soil, the uniformity of gradient was observed. This is found to be in good agreement with the remarks made by Leshchinsky et al. (2012).

From above observations, it is clear that for the uniform gradient to be attained, some percentage of fines are necessary. Samples for measuring water contents were collected at three different depths for all tests and water contents was observed to be mostly uniform and degree of saturation is close to 95% for all the samples. Figure 9 presents the scale factors used for both model heights considered for the study representing three prototype depths.

This indicates the versatality of the setup to vary the model height and gradients for the study. The permeability values presented in Fig. 9 are the average of two model heights representing same prototype depth. In the case of soil with 20% fines, test results could not be added in Figure 9 due to difficulty in maintaining same order of pressures, as that in other tests. Therefore, modelling of models could not be performed for this case. Though projections were made beyond model thickness of 1 mm, it is unrealistic.

7 CONCLUSIONS

In the present study HGS method is used to determine the permeability of soils by using a custom designed and developed HGS test setup. Relevant scale factors were derived and validated. An experimental test set-up using Mariotte's principle to maintain a constant gradient is designed and fabricated with facilities to suit the requirements from water tightness of the container to monitor pore-water pressure in the model.

A test programme was conducted on pure sand samples by varying relative density and by blending sand with Powai soil to obtain 10% and 20% fines soil. In order to study modelling of models two model heights and three hydraulic gradients representing three prototype depths were considered. Permeability values measured from both the model heights show very good aggrement with each other. With increase in density and inclusion of fines content reduced permeability values was observed. Also, for a given model height permeability values decreased with increase in hydraulic gradient. Uniform gradient was observed in soil with fines content. Unlike sand though constant gradient is maintained uniformity was not observed.

REFERENCES

Avishur, M., Klar, A. & Frydman, S. 2006. Hydraulic gradient models for study of seismic response of piles, *Proceedings of the 6th International Conference on Physical Modelling in Geotechnics*, Hong Kong, August 4–6, Ng et al. (eds.), Taylor and Francis, London, UK: 975–978.

Birnbaum, I. 1998. Use of models for the studying the capacity of driven piles in sand, M. Sc thesis, Technion-Israel Institute of Technology, Haifa, Israel.

Blackburn, R.A. 1989. An experimental study of the uplift resistance of piles, M.S. thesis, Univ. of Delware, Newark, DE.

Cohen, S., Klar, A. & Frydman, S. 2003. The behavior of driven piles in sand- hydraulic gradient and numerical models, *Int. J. Phys. Model. Geotech.* 3(4): 43–52.

Dunn, R.J. 1983. Hydraulic conductivity of soils in relation to subsurface movement of hazardous wastes, Doctoral Dissertation, University of California, Berkeley.

Leshchinsky, D., Vahedifard, F. & Meehan, C.,L. 2012. Application of a hydraulic gradient technique for modeling the uplift behavior of piles in sand, *Geotechnical Testing Journal, ASTM*, 35(3): 400–408.

Musso, A. & Ferlisi, S. 2009. Collapse of model strip footing on dense sand under vertical eccentric loads, *Geotech. Geol. Eng*, 27(2): 265–279.

Rezzoug, A., Konig, D. & Triantafyllidis, T. 2004 Scaling Laws for Centrifuge Modeling of Capillary Rise in Sandy Soils. *J.Geotech. Geoenviron. Eng., ASCE*, 130 (6): 615–620.

Yan, L. 1990. Hydraulic gradient similitude method for geotechnical modeling tests with emphasis on laterally loaded piles, Ph.D. Thesis, The University of British Columbia, Vancouver, BC, Canada.

Yan, L. & Byrne, P. M. 1989. Application of hydraulic gradient similitude method small-scale footing tests on sand, *Can. Geotech. Journal*, 26 (2): 246–259.

Yan, L., Byrne, P. M. & Dou, H. 1991. Model studies of dynamic pile response using hydraulic gradient shaking table tests, *Proceedings of the 6th Canadian Conference on Earthquake Engineering*, June 12–14, Sheikh & Uzumeri (eds), University of Toronto Press, Toronto, Canada: 335–345.

Yuan, B., Chen, R., Li, J., Wang, Y. & Chen, W. 2016. A hydraulic gradient similitude testing system for studying the responses of a laterally loaded pile and soil deformation. *Environmental Earth Science*. 75(1): 97–103.

Zelikson, A. 1969. Geotechnical Models hydraulic gradient using the similarity method, *Geotechnique*, 19(4): 495–505.

A new insight into the behaviour of seepage flow in centrifuge modelling

W. Ovalle-Villamil & I. Sasanakul
University of South Carolina, Columbia, South Carolina, USA

ABSTRACT: Application of centrifuge modelling for studying flow through soils is complex because the seepage velocity scales proportional to centrifuge gravity, resulting in greater potential for exceeding Darcy's flow conditions. Research efforts have been done to determine laminar flow limit based on Forchheimer's principles and the concept of critical Reynolds Number (R_{critic}). However, the interpretation of this limit remains ambiguous. This study provides new insights and establishes a connection between two different theoretical approaches. Centrifuge permeability tests were conducted at different gravitational levels for different gradation of silica sands. Results show that parameters of Forchheimer's regime remain constant regardless of the centrifuge gravity. Values of R_{critic} were found to be smaller than 1 for the finest sand, and between 3 and 10 for the coarser sands. These results indicate that reference values of R_{critic} are inaccurate for fine sands and very conservative for coarse sands.

1 INTRODUCTION

It is known that the centrifuge gravity affects the flow behaviour in a scaled model. If a prototype soil is used for a model subjected to N times Earth's gravity, it is to be expected that the flow velocity would be N times faster than the prototype condition. The increased flow velocity in a centrifuge model may be acceptable if the flow regime is known and the predominant behaviour of the model remains the same as the prototype condition. In other words, there is no conflict in similarities between prototype and model. Understanding flow behaviour of soils used for centrifuge modelling is important for the assessment of the flow regime that a model would experience under desired centrifuge conditions.

Majority of geotechnical engineering application involves Darcy's flow regime; thus it is important to ensure that Darcy's law remains valid in a centrifuge model. The critical Reynolds Number (R_{critic}), determined from the Moody diagram (Friction Factor versus Reynolds Number curve), has been used as a criterion to identify the transition where non-Darcy's behaviour would occur, and several researchers have proposed wide ranges of this parameter. However, inconsistency in values of R_{critic} depends upon several factors including different assumptions and expressions used to derive the Reynolds Number and difficulties to interpret the Moody diagram. The use of different definitions may result in significant error in subsequently calculating critical hydraulic gradients and flow velocities in a centrifuge model.

This study provides a review of formulations used to determine Reynolds Number and Friction Factor. The paper presents the results of centrifuge permeability testing of fine sands and the validation of R_{critic} values typically used in centrifuge modelling. Variation of critical values is assessed and discussed.

2 BACKGROUND

The relationship between pressure gradient ($\Delta P/\Delta L$) and velocity of flow (v) is used to describe the behaviour of flow through any media. This relationship represents the pressure drop ΔP in a system due to the flow of fluid at a given average velocity between two points separated by a distance ΔL. In porous media, a linear relationship indicates flow governed by viscous forces and is represented by Darcy's Law (Darcy 1856):

$$\frac{\Delta P}{\Delta L} = \frac{1}{\rho g}\frac{1}{k}v \quad (1)$$

where $\Delta P/\Delta L$ = pressure gradient; v = velocity of flow; ρ = density of fluid; g = gravitation acceleration; and k = Darcy's permeability. A non-linear relationship indicates flow governed by inertial forces. This behaviour is represented by the two-term, non-linear model proposed by Forchheimer (1901):

$$\frac{\Delta P}{\Delta L} = Av + Bv^2 \quad (2)$$

where A & B are Forchheimer coefficients.

Transition between Darcy's and Forchheimer's flow is analysed using the Moody Diagram (Moody 1944) relating the Darcy's Friction Factor (f) and the Reynolds Number (R_e). This method is widely used for studying flow through pipes, where Darcy's Friction Factor is a theoretical parameter that predicts the energy loss based on the velocity of the fluid and the resistance due to friction. Darcy-Weisbach equation defines Friction Factor as:

$$f = 2\frac{\Delta P d_p}{L_p \rho v_p^2} \quad (3)$$

where f = Friction Factor; d_p = diameter of pipe; L_p = length of pipe; v_p = average velocity of flow; and ρ = density of fluid.

Reynolds Number is a dimensionless parameter that compares inertial and viscous forces due to flow. This parameter is defined as:

$$R_e = \frac{\rho v_p d_p}{\mu} \quad (4)$$

where μ = dynamic viscosity of fluid.

The interpretation of Eqs. 3 and 4 for porous media uses the capillary model (Bear 2013) considering a medium as a set of capillaries similar to pipes. Therefore, it is assumed that the flow through an individual capillary is similar to the flow through a pipe, and f and R_e can be defined as:

$$f = \frac{\Delta P D}{\Delta L \rho v_s^2} = \frac{\Delta P D}{\Delta L \rho v^2} \quad (5)$$

and,

$$R_e = \frac{\rho v_s D}{\mu} = \frac{\rho v D}{\mu n} \quad (6)$$

where v_s = velocity of flow through an individual capillary; D = representative diameter of capillaries; n = porosity; and v = average velocity of flow through the medium obtained from Dupuit's relation ($v_s = v/n$). Goodings (1994) used Eqs. 5 and 6 to analyze non-Darcy's flow through different porous media assuming D as d_{50}.

Comiti & Renaud (1989), Comiti et al. (2000), and Khalifa et al. (2000), used a more precise approximation for the diameter of capillaries based on the concept of hydraulic radius (Trussell & Chang 1999; Richardson & Coulson 2002). Through this methodology, the diameter of capillaries is defined as:

$$d_c = \frac{2}{3}\frac{n}{(1-n)}d_{eff} \quad (7)$$

where d_c = diameter of capillaries; d_{eff} = effective diameter of particles (Carrier 2003); and $(1 - n)$ = fraction of medium occupied by solids. Furthermore, these works consider that the flow follows tortuous paths with length:

$$\Delta L_0 = \tau \Delta L \quad (8)$$

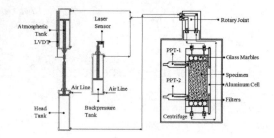

Figure 1. Experimental setup and test methodology.

where ΔL_0 = length of tortuous path; ΔL = length of porous media; and τ = tortuosity. Hence, Dupuit's relation is redefined as $v_s = \tau v/n$, leading to f and R_e given by:

$$R_e = \frac{2\rho \tau v d_{eff}}{3\mu(1-n)} \quad (9)$$

and:

$$f = \frac{4}{3}\frac{\Delta P}{\Delta L}\frac{n^3}{(1-n)}\frac{d_{eff}}{\rho \tau^3 v^2} \quad (10)$$

Due to the different expressions and interpretations used for f and R_e, several definitions for the upper limit of the Darcy's domain have been proposed. Goodings (1994) determined R_{critic} values between 3 and 11 for Ottawa Sand. Khalifa et al. (2000) and Comiti et al. (2000) defined the limit of Darcy's domain at an average R_{critic} between 4.3 and 4.9 for different sands. Salahi et al. (2015) estimated this limit at a markedly greater value of up to 30 for crushed angular materials.

2.1 *Experimental methodology*

A series of permeability tests, were conducted at centrifuge accelerations of 1 g, 10 g, 20 g and 30 g, using a customized setup assembled in a 1.3 m-radius geotechnical centrifuge located at the University of South Carolina. This setup was designed to allow precise increments of hydraulic gradients at different levels of gravity, inducing non-linear conditions of flow.

The system is composed of a specialized permeameter, three double-action (pneumatic) tanks, and two pressurized-air lines (Fig. 1). The permeameter allows testing cylindrical specimens with 15 cm of height and 7.5 cm of diameter. Two pressure sensors were connected at top and bottom of the specimens allowing continuous measurements of pressure loss in a length of 12.7 cm. In order to achieve a homogeneous distribution of flow to the specimen, two disks placed on top and bottom were filled with glass marbles with 0.5 cm of diameter. Particle migration was mitigated using No. 375, No. 200 and No. 150 filters.

The testing method consists of inducing pressure gradients from outside the centrifuge using pneumatic tanks. Top and bottom of the permeameter are connected to the tanks through the rotary joint of the

Table 1. Physical characteristic of porous media tested.

Specimen ID	Gradation	d_{eff} mm	n dimless
CS-U-10	Uniform	1.00	0.42
CS-U-05	Uniform	0.51	0.40
CS-W-05	Well-Graded	0.44	0.33
CSf-U-02	Uniform	0.16	0.41

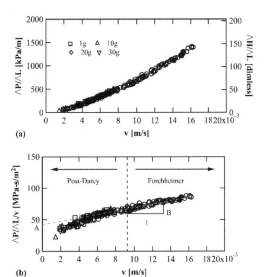

Figure 2. Specimen CSf-U-02: Evolution of (a) gradients and (b) normalized gradients, against the velocity of flow.

centrifuge. The top of the permeameter is connected to a head pressure tank while the bottom section is connected to a tank at atmospheric pressure. Pressure in the head tank is increased at a constant rate inducing flow through the specimen at different gradients. A Linear Variable Differential Transformer (LVDT) records the location of the pistons with time allowing measurements of induced velocity of flow.

Different gradations of silica sands from Columbia, South Carolina, were tested. Gradation, effective diameter (d_{eff}), and porosity (n) of each specimen are presented in Table 1.

Specimens were prepared using a dry compaction technique to reach the densest state possible and the lowest porosity to avoid additional changes in volume during centrifuge flights. For each specimen tested, layers with equal weight were carefully compacted using a rubber hammer allowing a homogeneous distribution of material.

Two techniques were used in order to produce a condition of full saturation of the specimens and the system during test. In first place, the air in the specimens was removed by flushing CO_2 at low pressure. Water was then flushed into the system in upward direction to remove the CO_2. The second stage of saturation consisted on pressurizing the system to compress the remaining air bubbles. As shown in Figure 1, the bottom section of the permeameter is also connected to an independent tank used to apply back pressure. This tank is connected to a pressurized airline and a laser sensor is attached to the piston. Back pressure was increased while the laser sensor recorded the change in volume of air reflected in the displacement of the piston. This process was repeated until the displacement stopped indicating the saturation of the specimen.

3 RESULTS AND DISCUSSION

This section presents the results and data analyses obtained from the centrifuge tests. Discussions of the transition to non-Darcy's domains and effect of centrifuge acceleration are included.

3.1 Centrifuge acceleration in non-Darcy flow

Results of hydraulic gradient against velocity of flow, and hydraulic gradient normalized by the velocity of flow (MacDonald et al. 1979) are presented in Figure 2 for specimen CSf-U-02. Figure 2a shows the non-linear relationship between gradients and velocity of flow confirming the occurrence of non-Darcy's flow. The non-linear behaviour is also observed in Figure 2b where the normalized hydraulic gradient increases as the flow velocity increases. According to Dukhan et al. (2014), the change in slope can be interpreted as the transition to Forchheimer domain where the initial slope reflects a post-Darcy condition but not a fully developed Forchheimer condition. The second slope represents a fully developed Forchheimer condition.

In Figure 2, it is observed that the relationship between gradients and velocity is very similar for each centrifuge gravitation acceleration. Effect of gravity is small or non-existent for the range of velocities tested. This is to be expected because velocity of flow increases in the centrifuge model but the relationship with the hydraulic gradient remains. Khalifa et al. (2000) indicated that despite some variations in the structure of the specimen due to compression during the centrifuge spinning, Darcy's law represented by the permeability k is not affected by gravity in a centrifuge environment. Results presented in Figure 2 lead to a similar conclusion for non-Darcy's domain represented by Forchheimer's Law (Eq. 2). Coefficients A and B used to describe non-linear flow are not affected by the centrifuge acceleration in the range of gravities tested (1 g – 30 g).

3.2 Friction Factor versus Reynolds Number

Evolution of Friction Factor as function of Reynolds Number is presented in Figure 3. Figure 3a shows the results obtained using Eqs. 5 and 6. Departure from viscous domain occurs at a Reynolds Number

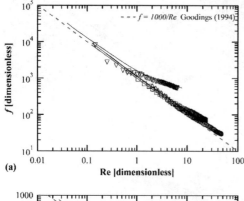

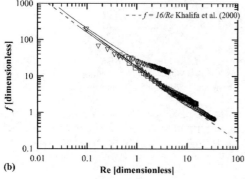

Figure 3. Friction Factor vs. Reynolds Number: (a) Goodings (1994), (b) Khalifa et al. (2000).

between 0.7 and 1 for the finest specimen CSf-U-02, and approximately between 3 and 10 for the other specimens. Similar results are seen in Figure 3b using Eqs. 9 and 10. In this figure, however, the magnitudes of Friction Factor experienced were nearly 100 times greater.

As presented by Goodings (1994), the range of validity of Darcy's domain could be located at Reynolds Numbers in a range from 3 to 11. Nevertheless, these values appear to fit certain porous media composed of particles with sizes of 0.5 mm and larger, regardless of the gradation. For finer sands, such as those usually used in centrifuge environments, this transition occurs at Reynolds Numbers lower than three. This indicates that there is a correlation between the particle size and the transition to non-Darcy's flow which occurs earlier in finer materials. This correlation is valid regardless of the formulation used to estimate the Reynolds Numbers, according to the results in Figure 3.

In contrast, Friction Factor is strongly dependent on the formulations. Despite the fact that this parameter is not an indicator of transition between regimes of flow, it is important to recognize that Friction Factor is a reference of shear stress due to seepage flow. Due to the differences seen among the results, this parameter should be estimated and used with caution and further research needs to be developed in order to fully understand this factor in porous media.

3.3 Assessment of critical Reynolds Number

Transition from Darcy's domain was found to occur at different Reynolds Numbers and varied with physical characteristics of the porous media. In absence of experimental data, the ranges of R_{critic} available in the literature have been used to approximate the range of validity of Darcy's law (Bear 2013). Nonetheless, since the flow behaviour represented by the relationship between hydraulic gradients and velocity of flow changes for each porous medium, the ranges of R_{critic} may result in wide ranges of gradients and velocities.

The evolution of hydraulic gradient against velocity of flow shown in Figure 4 is used to examine the variation of these two variables referencing to R_{critic}. This figure highlights the transition at the points corresponding to R_{critic} of 3 and 10, following Eqs. 5 and 6 (Goodings 1994), and R_{critic} of 4.9, following Eqs. 8 and 9 (Khalifa et al. 2000). Darcy's domain follows a straight line given by the inverse of Darcy's permeability. Transition is defined as the end of this linear relationship. Hence, a least square linear regression can be done for each R_{critic} and the coefficient of correlation, r^2, of each case can be used as an indicator of deviation from Darcy's Domain. As the r^2 deviates from 1, the behaviour of the flow is less linear. Similarly, r^2 values equal or close to 1 indicate Darcy's conditions of flow. Figure 4 shows the results obtained for each specimen tested in this study.

Figure 4a shows the flow behaviour for the coarser specimen CS-U-10. This specimen presents a fairly linear behaviour with a well-defined linear relationship between hydraulic gradients and velocity of flow. A clear transition from linear flow cannot be identified regardless of the decrease of r^2. For this specimen, Darcy's Law is valid to describe the flow corresponding to Reynolds Numbers of up to 10 calculated by using Eqs. 5 and 6. Transition to non-Darcy's flow for this specimen takes place at some value greater than 10. Figure 4b shows the flow behaviour of the specimen CS-U-05. The relationship is initially linear followed by a nonlinearity. A clear transition can be observed for R_{critic} within the values proposed by Goodings (1994). Curvature is more evident and the decrease of r^2 is more noticeable than with the coarser sand. Results for the well-graded specimen CS-W-05 are shown in Figure 4c and a similar behaviour with the specimen CS-U-05 is observed for Reynolds Number close to 10, using Eqs. 5 and 6, and 4.9, using Eqs. 8 and 9. Results for the specimen composed of the finest sand (CSf-U-02) are presented in Figure 4d. This specimen exhibits a well-defined, non-linear behaviour even for low velocities. Values of r^2 are noticeably low at the lowest value of R_{critic}. Results indicate that Darcy's domain is experienced at very low velocities of flow and transition occurs at Reynolds Number lower than 3.

Table 2. Critical values obtained for given R_{critic}.

ID	R_{critic}	v_c m/s	i_c m/m	r^2	Reference
CS-U-10	3.0	0.13	0.408	1.0000	Goodings (1994)
	4.9	0.29	0.928	0.9998	Khalifa et al. (2000)
	10	0.43	1.398	0.9997	Goodings (1994)
CS-U-05	3.0	0.24	2.554	0.9996	Goodings (1994)
	4.9	0.58	6.636	0.9979	Khalifa et al. (2000)
	10	0.80	9.567	0.9963	Goodings (1994)
CS-W-05	3.0	0.23	7.166	0.9999	Goodings (1994)
	4.9	0.76	25.24	0.9988	Khalifa et al. (2000)
	10	0.76	25.24	0.9988	Goodings (1994)
CSf-U-02	3.0	0.76	45.39	0.9658	Goodings (1994)
	4.9	1.85	187.0	0.9301	Khalifa et al. (2000)
	10	N/A			Goodings (1994)

The results presented in Figure 4 support the proportionality between the size of the particles in the porous medium and the magnitude of R_{critic}, using the root mean square approach. Critical values of hydraulic gradient and velocity of flow associated with the range of R_{critic} studied are presented in Table 2.

It is observed that the transition to non-Darcy's flow for the specimen CS-U-10 is expected to occur at an average hydraulic gradient of 1. At first glance, this critical value would represent a great disadvantage considering that this value is within the range of typical gradients in geotechnical applications. However, as noted in Figure 4a, the range of R_{critic} analysed in this study is conservative and real transition occurs at higher velocity of flow and hydraulic gradient.

Specimens with effective diameter of nearly 0.5 mm (CS-U-05 and CS-W-05) present similar critical velocities but higher critical hydraulic gradients are obtained for the well-graded material. Given the values of r^2 shown in Figures 4b and 4c, a critical Reynolds Number of 3 appears to be a better representative of the transition to non-Darcy's flow in both specimens.

Specimen CSf-U-02 presents the greater critical hydraulic gradients and velocities of flow corresponding to the reference values of R_{critic}. Nonetheless, according to Figure 4d, those values are not representative of the transition for this material. It is observed that transition occurs at much lower values of velocity and hydraulic gradient. This observation implies that the ranges of R_{critic} used as reference in this study are not valid for porous media composed of very fine-grained sands.

4 CONCLUSION

This study verifies that the effect of centrifuge acceleration is inexistent in the relationship between hydraulic gradient and velocity of flow, for accelerations of up to 30g. Therefore, critical values of Reynolds Number, hydraulic gradient, and velocity of flow are not affected by centrifuge gravity. Further research is required for higher levels of centrifuge gravity or different flow regimes.

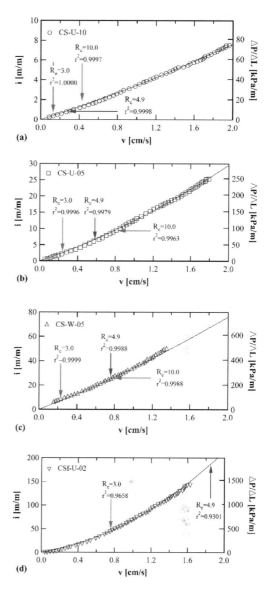

Figure 4. Hydraulic Gradient vs. Velocity of Flow: (a) CS-U-10, (b) CS-U-05, (c) CS-W-05, (d) CSf-U-02.

Critical values of Reynolds Number obtained using the Moody diagram ranged from 0.5 to 11. These results agree with the values reported by Goodings (1994), Comiti et al. (2000), Khalifa et al. (2000), among others. The transition to non-Darcy's flow was found to occur at Reynolds Number proportional to the size of the grains in the porous media. Although Reynolds Numbers obtained from different formulations were very similar, Friction Factors varied nearly two orders of magnitude between these formulations.

The critical Reynolds Numbers reported in the literature are fairly accurate for sands with effective diameter close to 0.5 mm and conservative for coarser materials. In the case of very fine sands, the transition to non-Darcy's domain occurs at lower values of Reynolds Numbers than the referenced values. This observation is critical for centrifuge modelling because the increased velocity of flow in a centrifuge model could exceed the critical magnitudes. Therefore, the typical range of critical Reynolds Number from 1 to 10 should be used with extreme caution as transition may occur at values 10 times smaller especially for very fine-grained sands.

REFERENCES

Bear, J. 2013. Dynamics of fluids in porous media. Courier Corporation.

Carman, P. C. 1956. Flow of gases through porous media. Academic press.

Carrier, W. D. 2003. Goodbye, Hazen; hello, Kozeny-Carman. Journal of Geotechnical and Geoenvironmental engineering, 129(11): 1054–1056.

Comiti, J., & Renaud, M. 1989. A new model for determining mean structure parameters of fixed beds from pressure drop measurements: application to beds packed with parallelepipedal particles. Chemical Engineering Science, 44(7), 1539–1545.

Comiti, J., Sabiri, N. E., & Montillet, A. 2000. Experimental characterization of flow regimes in various porous media—III: limit of Darcy's or creeping flow regime for Newtonian and purely viscous non-Newtonian fluids. Chemical Engineering Science, 55(15), 3057–3061.

Darcy, H. P. G. 1856. Dètermination des lois d'ècoulement de l'eau à travers le sable.

Dukhan, N., Bağcı, Ö., & Özdemir, M. 2014. Metal foam hydrodynamics: flow regimes from pre-Darcy to turbulent. International Journal of Heat and Mass Transfer, 77: 114–123.

Forchheimer, P. 1901. Wasserbewegung durch boden. Z. Ver. Deutsch. Ing., 45(1782): 1788.

Goodings, D. J. 1994. Implications of changes in seepage flow regimes for centrifuge models. In Centrifuge, 94: 393–398.

Khalifa, A., Garnier, J., Thomas, P., & Rault, G. 2000. Scaling laws of water flow in centrifuge models. In International Symposium on Physical Modelling and Testing in Environmental Geotechnics, 56: 207–216.

Khalifa, M. A., Wahyudi, I., & Thomas, P. 2000. A new device for measuring permeability under high gradients and sinusoidal gradients.

Macdonald, I. F., El-Sayed, M. S., Mow, K., & Dullien, F. A. L. 1979. Flow through porous media-the Ergun equation revisited. Industrial & Engineering Chemistry Fundamentals, 18(3): 199–208.

Moody, L. F. 1944. Friction factors for pipe flow. Trans. Asme., 66: 671–684.

Richardson, J. F., & Coulson, J. M. 2002. Chemical engineering: particle technology and separation processes. Elsevier Engineering Information, Incorporated.

Salahi M. B., Sedghi-Asl, M., & Parvizi, M. 2015. Nonlinear flow through a packed-column experiment. Journal of Hydrologic Engineering, 20(9).

Salem, H. S., & Chilingarian, G. V. 2000. Influence of porosity and direction of flow on tortuosity in unconsolidated porous media. Energy Sources, 22(3): 207–213.

Trussell, R. R., & Chang, M. 1999. Review of flow through porous media as applied to head loss in water filters. Journal of Hydraulic Research, 125(11): 998–1006.

Applicability of the generalised scaling law to pile-inclined ground system

K. Sawada
Graduate School of Engineering, Kyoto University, Kyoto, Japan

K. Ueda & S. Iai
Disaster Prevention Research Institute, Kyoto University, Kyoto, Japan

ABSTRACT: The generalized scaling law is based on the concept of two-stage scaling which overcomes limitations of the currently available facilities. A prototype can be scaled down into a physical model with a large scaling factor by multiplying a scaling factor (μ) for 1g tests with that (η) for centrifuge tests. In this study, the applicability to the dynamic behaviour of pile–inclined ground system in centrifuge tests is examined. Four cases of a combination of (μ, η) with keeping $\mu\eta = 100$ were experimented by strictly controlling the viscosity of pore fluid. Measured values in a model scale were converted into a prototype scale by applying the scaling law and the applicability can be verified if the values in the prototype scale coincide among the cases. As a result, the generalized scaling law can be applied to the dynamic behaviour of pile–inclined ground system in centrifuge model tests.

1 INTRODUCTION

The behaviour of soil-structure systems during earthquakes is generally non-linear. One of the ways to investigate the behaviour is a model test such as shaking table tests in a 1-g or centrifugal field. Shaking table tests with a large-scale model in a 1-g field can reproduce the real behaviour of ground, but it may cost too much money and time. So it is difficult to perform many experiments. Shaking table tests in a 1-g field with a small model can't reproduce the in-situ stress condition in ground exactly.

Centrifuge model tests can reproduce the confining pressure of real ground, but there are limitations of the centrifugal acceleration and thereby the reduction scale (prototype/physical model) depending on the performance of each centrifuge machine. To resolve such restrictions, Iai et al. (2005) proposed a scaling law by combining the scaling law for centrifuge testing with the one for 1-g dynamic-model testing (Iai 1989). It is called the "generalized scaling law" in dynamic centrifuge modelling. In the past studies, Nagaura et al. (2014) carried out a series of centrifuge model tests to examine the applicability of the law to the dynamic behaviour of dry and saturated lateral ground including pile foundation. In addition, the applicability of generalized scaling law has been studied by our colleagues (Wada et al. 2016, Bai et al. 2015, Tobita et al. 2009, 2011, 2014a, b, 2015, Tann et al. 2010). The objective of this study is to investigate the applicability of the generalized scaling law to the dynamic behaviour of pile-inclined ground system in centrifuge model tests. By strictly controlling the viscosity of viscous fluid, we especially give emphasis to the behaviour of bending moment, to which applicability was not be sufficiently confirmed in the past research (Wada et al. 2016).

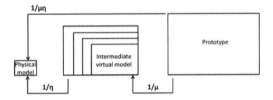

Figure 1. Concept of two-stage scaling, new modeling of models.

The generalized scaling law is based on the concept of two-stage scaling which can overcome limitations of the currently available facilities (Fig. 1).

In the first stage, a prototype is scaled down into an intermediate virtual model based on the scaling relation in a 1-g field with a scaling factor of μ (prototype/virtual model) [row (1) of Table 1]. In the second stage, this model is scaled down into a physical model using the conventional scaling relation in a centrifugal field with a scaling factor of η (virtual model/physical model) [row (2) of Table 1]. In this manner, a prototype can be scaled down into a physical model with a large scaling factor $\mu\eta$ (proto-type/physical model) [row (3) of Table 1].

2 CENTRIFUGE MODEL TEST

2.1 Test cases

Test cases are summarized in Table 2. Seven cases of a combination of (μ, η) with keeping $\mu\eta = 100$

Table 1. Scaling factors (prototype/model) in physical model tests.

	(1) for 1-g	(2) for centrifuge	(3) generalized scaling law
Length	μ	η	$\mu\eta$
Time	$\mu^{0.75}$	η	$\mu^{0.75}\eta$
Acceleration	1	$1/\eta$	$1/\eta$
Displacement	$\mu^{1.5}$	η	$\mu^{1.5}\eta$
Pore pressure	μ	1	μ
Viscosity	$\mu^{0.75}$	η	$\mu^{0.75}\eta$
Bending moment	μ^4	η^3	$\mu^4\eta^3$

Figure 2. Picture of the laminar shear box.

Table 2. Test cases.

Cases	m	n	1=un	Amplitude of acceleration (m/s²) prototype max	min
Case 1	2	50	100	3.08	−3.04
Case 2	3.3	30.3		3.11	−3.03
Case 3	4.72	21.2		3.15	−2.85
Case 4	7.63	13.1	100	2.96	−2.96
Case 5	4.72	21.2		3.14	−3.08
Case 6	3.3	30.3		3.39	−3.37
Case 7	4.72	21.2		3.22	−3.07
wada 50G	2	50		3.36	−3.10

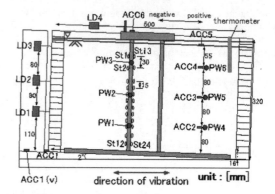

Figure 3. Cross-section view of the model (ACC: Accelerometer, LD: laser displacement sensor, PW: pore water pressure transducer, St: strain gauge).

were performed under saturated condition. In other words, without changing the actual size of the physical model, the virtual 1-g dimensions were varied. In the test series, centrifugal accelerations of 13.1, 21.2, 30.3, and 50 g were employed.

Tapered sinusoidal waves with the input acceleration amplitude of 3.0 m/s² and the frequency of 0.593 Hz in a prototype scale were horizontally given at the bottom of an experimental container (described later in Figure 2). Case 5 is a reproductive experiment of Case 3, and wada50G is quoted from the previous study (Wada et al, 2016). Due to the difference in the achieved acceleration amplitude, the cases are divided into two groups (see Section 3). Measured values (e.g. acceleration, bending moment of a pile) in the model scale were con-verted into the prototype scale by applying the generalized scaling law. The applicability can be verified if values in the prototype scale coincide among each group.

2.2 Experimental facilities and test-setup

The tests were performed in the geotechnical centrifuge of the Disaster Prevention Research Institute, Kyoto University, Japan. The centrifuge has an arm length of 2.5 m and is equipped with a shaking table that gives dynamic excitation to a model container in the tangential direction of flight.

The laminar shear box (Figure 2) with inner dimensions of 50 cm in width and 20 cm in depth (model scale) was filled up with Toyoura sand (described in the next section) to an intended height of 30.5 cm (model scale) corresponding to a 30.5-m-sand deposit in prototype scale. The laminar shear box has 20 layers which can move independently from each other due to linear bearings which support each layer.

This laminar shear box has the following advantages. First, it can reproduce the stress condition in the inclined ground more exactly than a rigid box. Second, it can reproduce the real vibration of the ground more exactly because it has many layers and the friction between each layer due to the overburden load from upper layers is significantly reduced.

The cross-section view of the model is shown in Figure 3. The angle of inclination of the ground was set to 2° in all cases by setting a sloping plate on the shaking table.

A pile structure is constituted of 4 piles of 30 cm height and a weight of 500 g in the model scale is in-stalled on the top of them as a footing foundation (Figure 4). The upper and bottom end of piles are welded to the weight and setting board respectively to reproduce the fixing condition. The pile is made of aluminium alloy and the Young's modulus (E) is 71.4 GPa (model scale). The section dimensions of the pile, which determine the second moment of area (I),

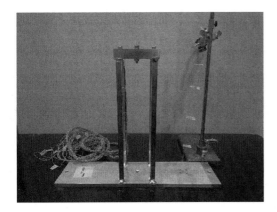

Figure 4. Side view of the model pile.

Table 3. Viscosity of pore water.

Cases	Saturation degree'FSt (%)	viscosity (cSt) prototype
Case 1	99.7	0.97
Case 2	99.7	1.02
Case 3	99.4	0.93
Case 4	99.7	0.98
Case 5	98.8	1.05
Case 6	98.4	0.98
Case 7	99.9	0.97
wada 50G	100	1.01

are decided so that the bending stiffness (EI) of the pile in the prototype scale coincides between all cases.

2.3 Material of the model ground

The model ground was made by Toyoura sand with air pluviation in all cases. The relative density was set to about 50%. The viscosity of the pore fluid has to be scaled with a factor of $\mu^{0.75}\eta$ relative to water according to the generalized scaling law as shown in Table 1. To produce water with the specific viscosity for each test, the methylcellulose solution (Metolose) was employed. Because the viscosity highly depends on temperature, the temperature in the ground during vibration was measured by a thermometer and controlled by air-conditioner. Table 3 shows the achieved viscosity of pore water.

3 RESULTS

Hereafter, unless otherwise noted, measured values are expressed in the prototype scale.

3.1 The first group (Case 1 to 5)

Plotted in Figure 5 are time histories of input accelerations in model and prototype scale. The amplitude and duration in the model scale are totally different among the cases, but they agree well in the prototype scale. The amplitude of the input acceleration is approximately $3.0\,m/s^2$ in the first group.

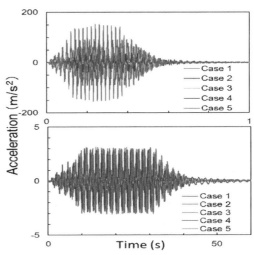

Figure 5. Time history of input acceleration (above: model scale, below: prototype scale).

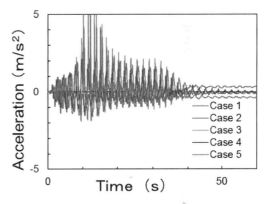

Figure 6. Time history of response acceleration.

Figure 6 shows the time history of response accelerations at ACC4. The response accelerations agree well among the cases including the amplitude increase around 12 s that may be due to positive dilatancy.

Plotted in Figure 7 are build-up and dissipation processes of excess pore water pressures in prototype scale. Liquefaction occurs in all cases because the excess pore water pressure reaches the initial effective stress (black line). Fairly similar build-up processes are obtained among the five cases. As shown in Figure 7, the excess pore water pressure in Case 2 dissipates slightly faster than others, but the four cases except for Case 2 are in good agreement.

The time histories of lateral displacement of the pile head and ground are shown in Figure 8. With regard to the pile head, very similar responses were observed. We can also observe the good agreement of ground displacements among the cases, although the residual dis-placement at LD3 in Case 5 was almost twice as large as those in other cases. This discrepancy may be because that identical experimental conditions were not accomplished among the cases due

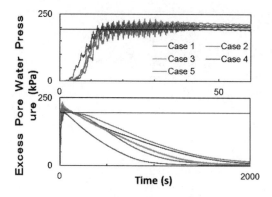

Figure 7. Excess pore water pressure (above: build-up process, below: dissipation process).

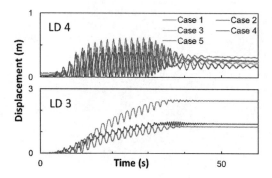

Figure 8. Time history of displacements (above: the pile head, below: ground).

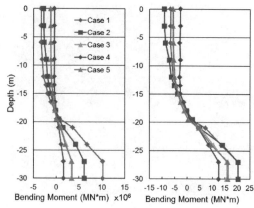

Figure 9. Residual bending moment profile of the pile (left: model scale, right: prototype).

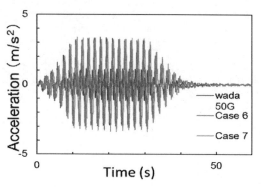

Figure 10. Time history of input acceleration.

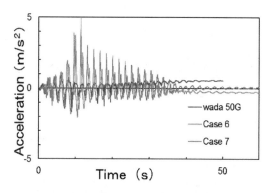

Figure 11. Time history of response acceleration.

to a change in the interlaminar friction, which was originally assumed almost zero, of the shear box.

Figure 9 depicts a comparison of vertical distributions of bending moment of the pile in model and prototype scale. The profiles in model scale display great variation among the cases, particularly in the vicinity of the bottom, but similar bending moments are obtained in prototype scale by applying the generalized scaling law.

3.2 The second group (Case6, 7, wada50G)

In this section, we compare the results of Cases 6 and 7, with data (hereafter Wada50G) quoted from the past study (Wada et al. 2016). Wada50G was conducted under the same experimental conditions as the two cases in prototype scale.

Plotted in Figure 10 are time histories of input accelerations in prototype scale. In each case almost the same input acceleration, of which amplitude is approximately 3.3 m/s^2, is given.

Figure 11 is the time history of response accelerations at ACC 4. Response acceleration agrees among all cases.

Plotted in Figure 12 are build-up and dissipation processes of excess pore water pressures in prototype scale. Liquefaction occurs in all cases because the excess pore water pressure reaches the initial effective stress (black line). Fairly similar build-up processes are obtained among the cases. Although the dissipation time of wada50G is earlier, in other cases it generally agrees.

The time histories of horizontal displacement of the pile head and ground are shown in Figure 13. The dis-placement amplitude, particularly the pile head amplitude, during shaking in Wada50G is relatively small and differs from those in other cases. However,

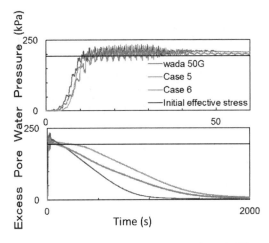

Figure 12. Excess pore water pressure (above: build-up process, below: dissipation process).

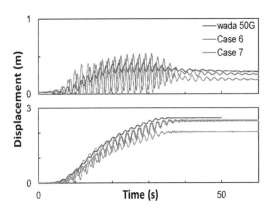

Figure 13. Time history of displacements (above: the pile head, below: ground).

the averaged components during shaking are quite similar among the cases and with regard to the residual displacement almost the same results are obtained.

Figure 14 depicts a comparison of vertical distributions of bending moments of the pile in model and proto-type scale. In model scale, discrepancy among the cases is observed, particularly in the vicinity of the bottom. However, the profiles in prototype scale show a very good agreement at all depths. In the past research, be-cause the viscosity was not strictly managed, sufficient applicability of the generalized scaling law has not been confirmed. However, in this study, the applicability of the generalized scaling law to the bending moment, in particular, was confirmed by firmly managing the viscosity.

4 CONCLUSION

In order to investigate the applicability of the generalized scaling law to pile-inclined saturated ground system, a series of dynamic centrifuge experiments was carried out based on the modelling-of-models

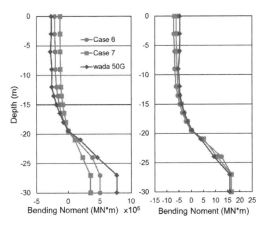

Figure 14. Residual bending moment profile of the pile (left: model scale, right: prototype).

technique and the results were compared in prototype scale. Primary conclusions of this study are summarized as follows:

- Response accelerations agreed very well among all cases at each measurement point (i.e. at the pile head and in the ground).
- Horizontal displacements of the ground had slight differences at an elevation (i.e. LD3) in the first group, but the pile head displacements were in good agreement.
- The build-up processes in excess pore water pressure were in good agreement, but a slight variation among cases was observed in the dissipation process.
- Bending moments of the pile also agreed well both during shaking (the results are not given due to limitations of space) and at the residual state.
- More precise control of the viscosity of pore fluid by keeping the ground temperature at the target level improved the applicability of the generalized scaling law compared to the past studies.

Thus, the generalized scaling law can be applied to the dynamic behaviour of pile–inclined ground system in centrifuge model tests by strictly controlling the viscosity of pore fluid.

REFERENCES

Bai, K., Tobita, T., Ueda, K. & Iai, S. 2016. New Modelling of Models for Centrifuge Model Testing. *Proceedings of the 2nd Asian Conference on Physical Modelling in Geotechnics* 146-151.

Iai, S. 1989. Similitude for shaking table tests on soil-structure-fluid model in 1g gravitational field. *Soils and Foundations* 29(1): 105-118.

Iai, S., Tobita, T. & Nakahara, T. 2005. Generalised scaling relations for dynamic centrifuge tests. *Geotechnique* 55(5): 355-362.

Nagaura, K. 2014. Applicability of the generalized scaling law to pile-ground system in centrifuge model tests. *Masters thesis* Kyoto University (in Japanese).

Tann, L. v. d., Tobita, T., & Iai, S. 2010. Applicability of two stage scaling in dynamic centrifuge tests on saturated sand deposits. *7th International Conference on Physical Modelling in Geotechnics (ICPMG 2010)* Springman, Laue & Seward (Eds), 191-196.

Tobita, T., Iai, S., & Noda, S. 2009. Study on generalized scaling law in centrifuge modeling with flat layered media. *Proceedings of the 17th International Conference on Soil Mechanics and Geotechnical Engineering (17th ICSMGE)* M. Hamza et al. (Eds.), Alexandria, Egypt, 664-667.

Tobita, T., Iai, S., von der Tann, L. & Yaoi, Y. 2011. Application of the generalised scaling law to saturated ground. *International Journal of Physical Modelling in Geotechnics* 11(4): 138- 155.

Tobita, T. & Iai, S. 2015. New modelling of models for dynamic behavior of a pile foundation. *The 15th Asian Regional Conference on Soil Mechanics and Geotechnical Engineering (15ARC)* Fukuoka, Japan.

Tobita, T., Escoffier, S., Chazelas, J.L. & Iai, S. 2014. Applicability of the generalized scaling law for fundamental physical properties. *Physical Modelling in Geotechnics* Gaudin & White (eds), Taylor & Francis Group, London, ISBN 978-1-138-00152- 7: 369-375.

Tobita, T., Escoffier, S., Chazelas, J. L., & Iai, S. 2014. Verification of the generalized scaling law for flat layered sand deposit. *Geotechnical Engineering Journal of the South Asian Geotechnical Society* 45(3): 32-39.

Wada, T., Tobita, T., Ueda, K., & Iai, S. 2016. New modelling of models for centrifuge model testing on the dynamic behavior of pile – inclined ground system. *Proceedings of the 2nd Asian Conference on Physical Modelling in Geotechnics* 140-145.

Permeability of sand with a methylcellulose solution

T. Tobita
Department of Civil, Environmental and Applied System Engineering, Kansai University, Japan

ABSTRACT: In dynamic centrifuge modelling, to resolve conflicts in scaling of time between dynamic and diffusion events, viscous fluids are used to reduce a model's permeability. In this study, permeability of Toyoura sand is measured with constant head permeability tests by varying the fluid viscosity, paying attention to transient changes in its permeability. Types of fluids used in the tests are purified water and methylcellulose (MC) solution. Results show that the permeability of sand with MC solution continuously decreased with the fluid passing, whereas it remained constant for purified water. It is found that the reduction rate of the permeability has two phases indicating possible clogging of MC fibres. By modelling of models of liquefaction model, the effect of permeability reduction is confirmed to be relatively minor. However, care should be taken if the duration of the diffusion process or settlements after liquefaction is of particular interest in testing a model with MC solutions.

1 INTRODUCTION

With recent developments of numerical analysis methods, demands on rigorous processes for validation and verification (V&V) of those simulation tools are increasing. Physical modelling has played a role to validate those tools for many years. Rigorous validation requires comparison to repeatable and reproducible (R&R) experiments. Among the efforts for producing high quality experimental data, LEAP (Liquefaction Experiment and Analysis Project), which is an international effort to validate modelling and simulation procedures for triggering and consequences of liquefaction, has been started since 2013 (Kutter et al. 2017, Tobita et al. 2017). LEAP is aiming at establishing a rigorous protocol for verification, calibration and validation of soil liquefaction models. To achieve the goal of LEAP, materials, such as substitutes of pore water in liquefaction models, should be thoroughly examined. Thus, in this study, characteristics of a type of viscous fluid which has been commonly used in centrifuge modelling is investigated with permeameter tests and physical modelling.

In centrifuge modelling, accelerations are scaled by N times that of the gravitational field, whereas lengths are reduced by N (Madabhushi 2014, Schofield 1980). Thus, from dimensional similarity, scaling for time is written as,

$$t_m = \frac{t_p}{N}, \tag{1}$$

where t represents time and m and p denote model and prototype, respectively.

Conversely, the diffusion process (i.e. the dissipation of pore pressure) is governed by consolidation phenomena, and its time is associated with the coefficient of consolidation. If the same soil is used in a model and a prototype, coefficient of consolidation c_v must be equivalent, i.e.

$$c_{v,m} = c_{v,p} \left(= \frac{k}{\rho g m_v} \right), \tag{2}$$

where k represents permeability, ρ is the density of pore fluid, g is acceleration due to gravity and m_v is the volume compressibility of the soil. From the equivalence of time factors between the model and the prototype, scaling for time of the diffusion phenomena is expressed as

$$t_m = \frac{d_m^2}{c_{v,m}} = \frac{(d_p/N)^2}{c_{v,p}} = \frac{t_p}{N^2}. \tag{3}$$

Thus, in the diffusion process, time is scaled by N^2, which differs from that of Equation 1 in which time is derived from the dimensional similarity.

In geotechnical centrifuge modelling, to compromise the above contradiction of time scaling between dynamic and diffusion phenomena, highly viscous fluids with viscosities larger than that of water are widely used as a pore fluid in the model ground. The rationale for this assumption is given by the following expression for permeability k,

$$k = \frac{K\rho g}{\mu} = \frac{Kg}{\nu}, \tag{4}$$

where K is the absolute or intrinsic permeability, μ is the dynamic viscosity of the pore fluid and ν $(=\mu/\rho)$ is the kinematic viscosity. According to Equation 4, by increasing kinematic viscosity N times, we can reduce the permeability of soil by one-Nth. Then, we have

$$t_m = \frac{d_m^2}{c_{v,m}} = \frac{(d_p/N)^2}{(c_{v,p}/N)} = \frac{t_p}{N}, \quad (5)$$

which is in agreement with Equation 1 (Stewart, et al. 1998).

On the basis of these observations, Ko (1994) used a viscous solution made of glycerine, whilst silicone oil has been widely used due to its chemical stability (Ko, 1994). More recently, methylcellulose (MC) solution has been widely adopted for use due to its convenience of handling (Dewoolkar, et al. 1999, Stewart, et al. 1998).

In experiments performed to validate the generalized scaling law (Iai, et al. 2005) Tobita, et al. (2011) modelled flat-layered liquefiable ground. They tested a 1/100 scale model under 5, 10, 50 and 70 g. The viscosity and other physical parameters were adjusted in accordance with the scaling law. What they found was a discrepancy in duration time of the excess pore water pressure dissipation among the models when low centrifugal acceleration (<10 g) was used. Based on these findings, they concluded that a major cause of the increase in the duration of dissipation was the very small value of elastic modulus due to low confining stress, which leads to a small coefficient of consolidation and hence results in an increase in duration for which the soil remains liquefied, e.g. Haigh, et al. (2012). They also suggested the possible transient change of permeability of the model ground due to the absorption of the MC polymer by sand grains; however, this has yet to be investigated and therefore is the focus of our present study.

2 CONSTANT HEAD PERMEABILITY TESTS

The permeability of Toyoura sand ($G_s = 2.636$, $e_{max} = 0.983$, $e_{min} = 0.609$, $D_{50} = 0.18$ mm, $U_c = 1.64$) was investigated with particular interest in its transient changes with a volume of fluid passing through the specimen. The constant head permeability test following (JIS A 1218:2009) JGS (2009) is carried out, in which permeability is computed as follows:

$$k_t = \frac{L}{h} \frac{Q}{A(t_2 - t_1)}. \quad (6)$$

Here, k_t (cm/s) is the permeability at T°C, L (cm) is the height of the specimen, A (cm^2) is the area of the specimen, h (cm) is the difference in water head, Q (cm^3) is the measured volume of the passing fluid, and $t_2 - t_1$ (s) is the duration for which Q is collected.

The soil specimen, at relative density (in this study, Dr = 40, 60 and 80%), was contained in a cylinder of cross-sectional area $A = 78$ cm^2 and height $L = 12.75$ cm with the air-pluviation method. It was rested on coarse wire mesh and deaired for 30 to 60 min. A steady vertical flow of liquid, under a constant head of $h = 2.89$ cm, was maintained through the soil. The water supplied to the specimen was purified with Amberlite ion exchange resins (Organo Co.) Weight of ejected water was measured with a digital scale (A&D Weighing, EJ-6100), which automatically recorded the weight in every 5 s for the case of pure water, every 10 s for viscous fluid of 20 mPa s, every 30 s for 44 mPa s, and every 60 s for 60 mPa s. The weight of fluid was divided by its density to get volume of ejected fluid. Then, a permeability was computed by Equation 6, and the correction was made by Equation 7 as described below.

The MC solution employed in this study is a Metolose type SM-100 produced by the Shin-Etsu Chemical Co. Ltd (Shin-Etsu Chemical Co., 2013). The SM-100 provides 100 mPa s of viscosity of a 2 wt. % aqueous solution at 20°C. In SM-100, some hydrogen atoms of the hydroxyl groups of cellulose are replaced by a methyl group ($-CH_3$). Another MC solution often used in centrifuge modelling is called hydroxypropyl methylcellulose (HPMC), in which some hydrogen atoms are replaced by the hydroxypropyl group ($-CH_2CHOHCH_3$). The major difference between MC and HPMC is their polymerization temperature on gel point, i.e. HPMC has a higher gelation point (70–80°C) than that of MC (50–60°C) (Nagura, et al. 1981). Thus, it is safe to assume that under normal operational temperature and general centrifuge model testing conditions (i.e. 10–30°C), soil containing viscous fluids of both types may exhibit similar behaviour. Note that the following observations/results may significantly vary with the type of methylcellulose employed for investigation.

2.1 Toyoura sand with purified water

A series of the constant permeability tests was conducted in a room temperature of 20 to 25°C. The reported coefficient of permeability in this study is adjusted to be the one at 15°C, as shown below (Eq. 7), by multiplying a correction factor, a ratio of dynamic viscosity at temperature of T°C, η_T, to the one at 15°C, η_{15} (after National Astronomical Observatory of Japan, 2011).

$$k_{15} = k_T \frac{\eta_T}{\eta_{15}} \quad (7)$$

Test specimens were prepared at void ratio $e = 0.833$ with relative density Dr = 40%, $e = 0.759$ for Dr = 60% and $e = 0.684$ for Dr = 80%. In what follows, the permeability of the specimen with purified water (1 mPa s) at Dr = 40% will be used as a reference and is denoted by $k_{15} = 9.05 \times 10^{-2}$ cm/s. To confirm the measurements, the tests with other target relative densities (Dr = 60 and 80%) were compared in Figure 1 together with the prediction by Taylor's equation for permeability (Taylor, 1948).

$$\tilde{k} = \frac{\gamma_w}{\mu} C_k \frac{e^3}{1+e} D_s^2 \times 10^4 \quad (8)$$

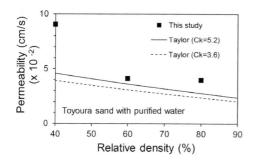

Figure 1. Measured and predicted permeability of Toyoura sand with purified water.

Here, γ_w (kN/m³) is unit weight of water, μ ($=1.138 \times 10^{-4}$ Pa s at 15°C) is the dynamic viscosity of water, C_k is shape factor, D_s ($=0.02$ cm) is diameter of a soil particle. Here the shape factor $C_k = 3.6$ is recommended for round river sand and $C_k = 5.2$ is for glass beads. In Figure 1, two predicted curves are plotted; one with the shape factor of $C_k = 3.6$, and the other is $C_k = 5.2$. Measured permeability shows fairly good agreements with the prediction with $C_k = 5.2$, except for the case of Dr $= 40\%$ in which the permeability is underestimated about 50%.

2.2 Toyoura sand with viscous fluid

To investigate transient changes of permeability with viscous fluid, the constant head permeability test was repeated but with MC solutions. As provided by the Metolose brochure (Shin-Etsu Chemical Co. 2013), firstly, a viscous fluid of 100 mPa · s having a viscosity 100 times that of water was made with 2% of the mass of the MC (SM-100) solution by mixing it with hot water (about 80°C).

Next, it is diluted with purified water until the desired kinematic viscosity was achieved at room temperature. The kinematic viscosity was measured using the Ubbelohde type viscometer (SHIBATA, SU-6709, measurable range 20–100 mPa s).

$$v = Ct \qquad (9)$$

where, v is the kinematic viscosity (mPa s) defined by $v = \mu/\rho$ where μ is the dynamic viscosity and ρ is the density of the fluid, t is the time for the liquid to pass through two calibrated marks on the viscometer, and C is the viscometer constant and is calibrated to be $C = 0.0827$ (mPa s) by the vender. In our institute, the viscometer constant was validated by using the standard fluid whose viscosity was 50 mPa s at 25°C.

The viscous fluids with kinematic viscosities of 20, 44, and 60 times that of water were prepared to see variation of the permeability of those fluids with the amount of water passing. Here, measured permeability was corrected by Equation 7 in the same way as the test with purified water.

Figure 2(a) compares variation of permeability of Toyoura sand with purified water and viscous fluids,

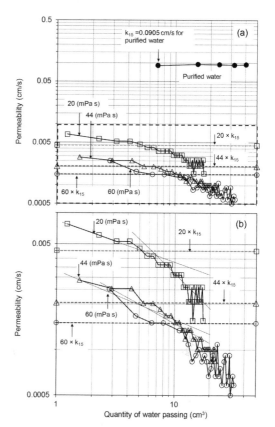

Figure 2. Variation of the permeability with quantity of passing water for Toyoura sand (Dr = 40%): (a) showing all curves including permeability of purified water, and (b) showing curves of viscous fluids only.

while Figure 2(b) shows curves of viscous fluids only to focus on the change of permeability of viscous fluids. Figure 2(a) shows that permeability with purified water, $k_{15} = 9.05 \times 10^{-2}$ cm/s, does not vary with the amount of passing water, while ones with viscous fluid show degradation soon after the water passing.

In both Figures 2(a) and 2(b), the target permeability of each viscous fluid is denoted with dotted horizontal lines. For example, the target permeability with viscous fluid whose kinematic viscosity is 20 mPa s is designated as a dotted line with "$20 \times k_{15}$." For 3 viscous fluids shown in Figure 2, the initial permeabilities of all of them are slightly larger than the expected one.

As shown in Figure 2(b), it seems that the larger is the kinematic viscosity, the larger is the rate of permeability degradation with the quantity of passing water. A closer look in Figure 2(b) reveals that these degradation curves constitute two phases, i.e., the initial and secondary degradation. For example, as indicated with thin line segments on the curve of 20 mPa s in Figure 2(b), the slope of the curve until about 7 cm³ of passing water is -5.43×10^{-4} cm/s/cm³, while after 7 cm³ the slope or degradation rate of permeability is

Table 1. Water content of Toyoura sand (Dr = 40%) taken from top and bottom of the testing cylinder.

Kinematic viscosity (mPa s)	Water content (%)		Difference (Top-Bottom)
	Top	Bottom	
20	26.69	26.01	7.16
44	39.96	32.80	3.68
60	40.16	30.27	9.89

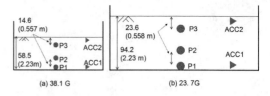

Figure 3. Model dimensions and sensor locations: (a) 38.1 g and (b) 23.7 g.

Figure 4. Side view of the models after shaking.

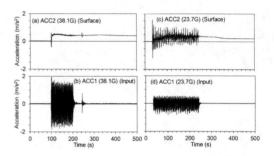

Figure 5. Time histories of input and response acceleration: (a) and (c) surface, and (b) and (d) input (prototype scale).

reduced to be -1.89×10^{-4} cm/s/cm^3. In other curves, the same trend is observed. The first slope of the curve may correspond to a rate of permeability degradation until complete coverage of pores near the surface in the sand specimen by MC fibres. That may be why the rate is significantly reduced after the break point where the rate change is observed.

Table 1 shows water contents of the sand taken from top and bottom of the testing cylinder after each test. For all cases, the water content of the top of the specimen is larger, and with the increase of the viscosity the difference of the water contents between top and bottom becomes larger.

From the observation above, it may be said that the permeability with MC solution drastically decreases with quantity of passing viscous fluid. This may be due to clogging of MC fibres in pores in sand grains.

3 CENTRIFUGE EXPERIMENTS WITH A LIQUEFACTION MODEL

Permeability degradation with the MC solution as a pore fluid poses a serious problem on scaling law of time in dynamic centrifuge modelling. The effect may be manifested by the time required for the excess pore pressure dissipation taking much longer than the expected in prototype.

To study the effects of degradation with quantity of passing water, centrifuge model tests with two different model scales were conducted in the geotechnical centrifuge in Kansai University, Japan, whose arm length is 1.5 m. To validate the scaling law of time in dynamic phenomena, the "modelling of models" technique was employed for a flat layered saturated sand deposit with viscous fluid of 38.1 mPa s and 23.7 mPa s. Deterioration of the viscous fluid was minimized by using purified water and reducing time for preparation until testing. The model ground was constructed with Toyoura sand by the water pluviation method and assumed to be fully saturated. Relative densities of the model ground before shaking in the case of 38.1 g and 23.7 g were, respectively, 52% and 50%, and those after the test was 56.5% and 61.9%, respectively.

Two accelerometers (ACC1: Input, ACC2: Surface) were placed at the specified depth of soil. 3 pore pressure transducers (P1 to P3) were glued on the backside wall of the sand box at the specified depth (Figure 3) so that they do not move with soil. In addition to the above mentioned two accelerometers, the third accelerometer (ACC3) was attached on the center line of longitudinal sidewall of the soil container (Figure 4) to measure the specified centrifugal acceleration at the 1/3 from the ground surface of the total thickness of the model ground, where the error due to the radial acceleration field on the confining stress is minimized.

Before shaking in the first test of 38.1 g, the centrifugal acceleration was monitored by ACC3 and was maintained with the acceleration required by the scaling law of time, 38.1 g. The thickness of the model ground was about 2.23 m in prototype scale. The model ground was initially made with a flat surface. However, due to the radial gravity, the shape of the ground is curved as shown in Figure 4.

The response and input accelerations are shown, respectively, in Figure 5(a) and (b). Peak amplitude of the input acceleration was about 1.5 m/s^2 and duration of shaking was about 100 s so that whole ground could be liquefied. As shown in Figure 5(a), due to liquefaction the sensor at the surface was rotated

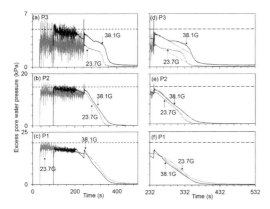

Figure 6. Time histories of excess pore water pressure: (a) P3, (b) P2, and (c) P1, and (d) – (f) for 232 s to 532 s. Dotted line indicates the initial effective vertical stress.

Table 2. Variation of pore pressure from 252 s to 352 s, and rate of dissipation.

			EPWP (kPa)		$(u_1-u_2)/(t_1-t_2)$ (kPa/s)	
			38.1G	23.7G		
		Time (s)	u_1	u_2	38.1G	23.7G
P3	t_1	252	4.29	2.76	−0.0264	−0.0252
	t_2	352	1.65	0.236		
P2	t_1	252	13.6	12.5	−0.115	−0.121
	t_2	352	2.05	0.32		
P1	t_1	252	15.2	14.5	−0.130	−0.112
	t_2	352	2.25	3.35		

and its amplitude was significantly reduced. In Figure 5(b), there is a miss-shot at around 240 to 260 s, which occurred by accident and might be caused by mechanical problem of the shaker.

Then, second test was conducted with the MC solution of 23.7 mPa s, by which the thickness of the model ground was determined to be (2.23 m/23.7 G=) 0.0941 m (Figure 3(b)) under the centrifugal acceleration of 23.7 g. The value of viscosity was taken arbitrary by diluting the MC solution with pure water. Figures 5(c) and (d) show response and input accelerations, respectively. The amplitude of input acceleration was about 0.7 m/s^2 and duration was 200 s, which is different from the one shown in Figure 5(b). Use of different input accelerations was intended to put more focus on the process of dissipation of excess pore pressure regardless of how the model ground was liquefied. As shown in Figure 5(c), acceleration at the surface shows periodically high amplitude which can be generally and typically observed when the sand is in the state of, so called, the cyclic mobility.

Figures 6(a) to 6(c) compares time histories of the excess pore water pressure. All those records, except for P3 in the case of 23.7 g, indicate that whole layer was liquefied. The record of P3 in the case of 23.7 g shows the ground was not fully liquefied; the effect of which is implied by the high spiky response due to the cyclic mobility shown in Figure 5(c).

Figures 6(d) to (f) compare dissipation phase of the excess pore pressure from 232 s to 532 s in prototype scale. Florin and Ivanov (1961) have given an expression for the velocity of the solidification front. If the sand grains are rigid and incompressible, the pore pressure in the solidified material would, in a uniform deposit, decay as a linear function of time to the steady-state hydrostatic value when the solidification is complete (Scott, 1986). This is verified in Figures 6(d) to (f) and Table 2. Rates of pore pressure dissipation between 252 s and 352 s are obtained as a slope of a linear segments shown with inclined dotted segments in Figures 6(d) to (f), which may well approximate the rate of dissipation. These rates at the same depths are nearly constant regardless of the variation of input acceleration.

In Figure 6(d), notice that the curves in the dissipation deviate from the linear segments in both cases of 38.1 g and 23.7 g (from 272 s to 232 s). This may indicate that, as pore pressure dissipates from the ground deeper than P3, ejected pore water may be accumulated near the ground surface because of the possible effect of clogging of MC solutions. If this is the case, similar observation could be made in previously conducted numerous experiments with liquefaction models. This will be investigated elsewhere.

By assuming that the settled volume of the ground after shaking is equal to the volume of fluid ejected from the ground, the quantity of passing water at surface level may be estimated. Based on the difference of the relative density before and after the shaking, which were computed from the volume change of the ground derived from the change of the ground height, the volume of the ejected water can be computed to be about 38 cm^3 and 99 cm^3, which can be converted to the ejected volume of fluid per unit area, that is 0.055 cm and 0.14 cm respectively for the cases of 38.1 g and 23.7 g. By multiplying area, $A = 78$ cm^2, of the sand specimen of the constant head permeability test, the equivalent amounts of volume of passing fluid can be obtained as 4.29 cm^3 and 11.3 cm^3. With these numbers and from Figure 3, for the case of 38.1 g in which the MC solution of 38.1 mPa s was used, the effect of the degradation of permeability is thought to be minor, while for the case of 23.1 g (23.1 mPa s), about 40% reduction in permeability is expected. Correlation between the results of this observation and the dips observed in Figure 6(d) should be investigated in future.

4 CONCLUSIONS

Constant head permeability tests were conducted with a particular focus on the transient changes of permeability with viscous fluids made of methylcellulose (MC) solution. Tested fluid viscosities were 20, 44 and

60 times that of purified water. Results of the constant head tests showed that the permeability tends to degrade with the increasing volume of passing fluid. The effect might be seriously affect the physical modelling in practice.

Results of the constant head tests showed that the derived curves of permeability versus the volume of passing fluid were comprised of two segments. On the first slope, permeability continuously decreased at an almost constant rate regardless of the viscosity of the fluid until specific volumes of viscous fluid passing were reached, at which point the curves reached a breakpoint and the second slope started. The first slope of the curve may correspond to a complete coverage of pores near the surface in the sand specimen by MC fibres. From the observation above, it may be concluded that the permeability with MC solution drastically decreases with quantity of viscous fluid passing. This may be due to clogging of MC fibres in pores in sand grains.

Then centrifuge model tests were conducted under two different g levels, 38.1 g and 23.7 g. These were determined by the viscosity of the MC solution of 38.1 mPa s and 23.7 mPa s as a pore fluid. A g level at 1/3 of ground thickness from the surface was monitored and it was maintained during the experiment so that the error of confining stress due to radial acceleration field can be minimized. Attention was given to the rate of pore pressure dissipation. What was found were that the curves of the near surface in the dissipation phase deviates from the linear segments in both cases of 38.1 g and 23.7 g. This may indicate that, as pore pressure dissipates from the ground, ejected pore water may be accumulated near the ground surface because of the possible effect of clogging of MC solutions.

By assuming that the settled volume of the ground after shaking was equal to the volume of fluid ejected from the ground, the quantity of passing water was estimated to be 4.29 cm^3 and 11.3 cm^3, respective for the cases of 38.1 g and 23.7 g. With these numbers and results of the constant head test results, for the case of 38.1 g in which the MC solution of 38.1 mPa s was used, the effect of the degradation of permeability would be minor, while for the case of 23.1 g (23.1 mPa s), about 40% of reduction in permeability was expected. However, it does not explain the observation of the same pattern of degradation in excess pore pressure at the surface. Thus, more detailed observation may be required.

We conclude that care should be taken if the duration of the diffusion process or settlements after liquefaction is of particular interest in testing a model with MC solutions. Duration times can be larger than that of a prototype, because of the possible clogging effect of the MC fibers. Note that the observations/results may significantly vary with the type of methylcellulose employed for the investigation.

ACKNOWLEDGEMENT

The author heartily thanks Mr. Kenta Yokoyama, a former undergraduate student at Kansai University, for his contribution to this study.

REFERENCES

Dewoolkar, M.M., Ko, H.-Y., Stadler, A.T. & Astaneh, S.M.F. 1999. A substitute pore fluid for seismic centrifuge modeling. Geotechnical Testing Journal, ASTM. 22(3):196–210.

Florin, V.A. & Ivanov, P.L. 1961. Liquefaction of sandy soils. Proc. 5th International Conf. on Soil Mechanics and Foundation Engineering, Paris. 1:107–111.

Haigh, S.K., Eadington, J. & Madabhushi, S.P.G. 2012. Permeability and stiffness of sands at very low effective stresses. Géotechnique. 62(1):69–75.

Iai, S., Tobita, T. & Nakahara, T. 2005. Generalized scaling relations for dynamic centrifuge tests. Géotechnique. 55(5):355–362.

JGS 2009. Japanese Geotechnical Society, Methods and Explanations of Laboratory Tests of Geomaterials.

Ko, H.-Y. 1994. Modeling seismic problems in centrifuges. Centrifuge 94, Leung, Lee & Tan (eds), Balkema, Rotterdam, ISBN 90 5410 352 3.3–12.

Kutter, B.L., Carey, T.J., Hashimoto, T., Zeghal, M., Abdoun, T., Kokkali, P., Madabhushi, G., Haigh, S., Hung, W.-Y., Lee, C.-J., Iai, S., Tobita, T., Zhou, Y.G. and Chen, Y. 2017. LEAP-GWU-2015 Experiment Specifications, Results, and Comparisons. Soil Dynamics and Earthquake Engineering (accepted).

Madabhushi, S.P.G. 2014. Centrifuge Modelling for Civil Engineers. CRC Press. Taylor & Francis Group, ISBN: 978-0-415-66824-8, 292pp.

Nagura, S., Nakamura, S. & Onda, Y. 1981. Temperature-viscosity relationships of aqueous solutions of cellulose ethers. Kobunshi Ronbunshu. 38(3):133–137.

National Astronomical Observatory of Japan 2011. Chronological Scientific Tables. Maruzen, 1054pp.

Schofield, A.N. 1980. Cambridge geotechnical centrifuge operations. Géotechnique. 30(3):227–268.

Scott, R.F. 1986. Solidification and consolidation of a liquefied sand column. Soils and Foundations, Japanese Geotechnical Society. 24(4):23–31.

Shin-Etsu Chemical Co., L. 2013. Metolose Brochure, Cellulose Dept., 6-1, Ohtemachi 2-chome, Chiyoda-ku, Tokyo, Japan.

Stewart, D.P., Chen, Y.-R. & Kutter, B.L. 1998. Experience with the use of Methylcellulose as a viscous pore fluid in centrifuge models. Geotechnical Testing Journal. 21(4):365–369.

Taylor, D.W. 1948. Fundamentals of soil mechanics. John Wiley & Sons, Inc, 700pp.

Tobita, T., Iai, S., von der Tann, L. & Yaoi, Y. 2011. Application of the generalised scaling law to saturated ground. International Journal of Physical Modelling in Geotechnics. 11(4):138–155.

Tobita, T., Ashino, T., Ren, J. and Iai, S. 2017. Effect of the radial gravity field on dynamic response of saturated sloping grounds in centrifuge model testing. Soil Dynamics and Earthquake Engineering (accepted).

5. Sensors

Investigation of an OFDR fibre Bragg system for use in geotechnical scale modelling

R.D. Beemer, M.J. Cassidy & C. Gaudin
Centre for Offshore Foundation Systems and ARC Centre of Excellence for Geotechnical Science and Engineering, The University of Western Australia, Perth, Australia

ABSTRACT: A common challenge in experimental research is making measurements without unintentionally omitting data at key locations or altering the mechanical properties of the object of interest. In recent years fibre optic sensing systems have been used to overcome these challenges in both geotechnical field monitoring and full scale geotechnical laboratory experiments. However, fibre optics have found limited use in small scale modelling, typically for measuring the deformation of geosynthetics. In this paper we present an investigation into a fibre optics strain measurement technology that combines optical frequency domain reflectometry (OFDR) and Fibre Braggs Gratings (FBG) and could be beneficial to small scale geotechnical modelling, including centrifuge modelling. Specifically, this device has the capability of measuring over 2,000 points of strain at spatial increments of 6.35 mm over a single data acquisition channel and can measure 3D shape of a fibre bundle in real-time. The technology is evaluated using cantilever deflections tests, with comparison strain gauge measurements and 3-dimensional shape with visual comparison to the bending of a thin flexible rod.

1 INTRODUCTION

A common challenge in experimental research is making measurements without (1) unintentionally omitting measurements at key locations and (2) altering the mechanical properties of the system and therefore its response. Geotechnical engineering has been particularly susceptible to both problems. Full-scale field experiments on infrastructure – such as mechanically stabilised earth walls or pipelines – are typically so large that economic considerations limit the number of sensing points. Reduced-scale models used in the geotechnical centrifuge, such as piles, risers or conductors, are so small that there are frequent concerns that the sensors in the model are altering the quantities being measured, with sensor wires often of the same order of magnitude as the model itself. Taking advantage of the rapid technological development in fibre optics geotechnical engineers have started using fibre optic sensing alleviate these issue.

In the geotechnical literature three types of fibre optic strain measuring technologies have been used spanning three different experiment scales, these are: Brillouin optical time domain reflectometry (BOTDR), wavelength division multiplexing (WDM) of Fibre Braggs Grating (FBG), and optical frequency domain reflectometry (OFDR) of Rayleigh backscattering. BOTDR is a distributed system used over large distances. It has sensing lengths up to 10 km, spatial resolution in the order of 1 m, and resolution in the order of 40 $\mu\varepsilon$ (Mohamad et al. 2010). It is typically used in field experiments, such as the measurement of retaining wall deformations at a construction site undertaken by Cambridge University (Mohamad et al. 2011).

FBG systems are discrete and tend to be used over short distances. Each FBG is etched (with a laser) into the fibre at a known location and serves as a gate that allows most light to pass, but reflects a specified wavelength back to the source. The reflected wavelength is highly dependent on the FBG's length and will vary as fibre is strained. Therefore, strain can be calculated from the measured change in wavelength of the reflected light. The difficulty in this technology actually arises from the interference caused by having multiple FBGs in a single fibre. The most common method to overcome this issue this is known as WDM. In these systems a fibre is manufactured such that each embedded FBGs reflect a unique wavelength, as long as the reflected wavelengths do not overlap, the system can read and identify each FBG. These systems can be very accurate, but have discrete and limited sensing points. Systems found in the literature (Silva et al. 2012; Corriea et al. 2016) can only sense 15-20 FBGs at once.

OFDR of Rayleigh backscatter system is also a distributed system that can be used to sense over intermediate distances, 30 m to 2 km. OFDR systems use a frequency dependent analysis of the sent and reflected light to sense both the variation in signal wavelength and the location of the reflection point (Duncan et al. 2007). This process can then be applied to Rayleigh backscatter, which is the result of light being reflected off imperfections in the fibre optic strand, to sense strain. OFDR can be used to measure the location of the imperfection and any deformation they undergo. This system is very accurate, and allows for measurement points 20 mm on centre (Simpson et al. 2015). However, spatial resolutions is highly dependent on the natural imperfection in the fibre and therefore each sensor is unique. Simpson et al. (2015) used this type of system to measure the deformation of large diameter pipes.

This paper presents a new OFDR FBG fibre optic sensing system that has not be used in geotechnical engineering to the best of the author's knowledge. It is capable of providing near continuous measurement of strain over small distances. As noted by its name, this systems uses OFDR methods to sense the location and deformation of FBGs along the length of a fibre. This system takes advantage of the OFDR technology to analysis FBGs that reflect identical wavelengths in a single fibre strand (Chan et al. 2015). Over short distance, <13 m, thousands of FBGs can be written back to back in the fibre resulting in a near continuous sensor (2048 FBGs, 6.3 mm on centre for the device – which is referred to as an interrogator – used in this paper). Since the FBGs are written into the fibre their locations are known and not dependent on natural variation (as required in Rayleigh scattering devices). Because OFRD systems can sense FBGs with identical reflective frequency they do not have the same interference issues of WDM FBG systems. Overall, this OFDR FBG system appear to be ideal for geotechnical scale modelling since they can provide high accuracy measurements at high spatial densities in a single fibre optic strand.

A major tool in geotechnical scale modelling is the geotechnical centrifuge. Any sensing technology meant for scale modelling should also be able operate in the centrifuge under macro-gravity. Though OFDR FBG systems have not been tested in the centrifuge WDM FBG systems have. They have been used in both drum and beam centrifuges to measure deformations in a geotextile (Kapogianni et al. 2010; Silva et al. 2012). In both experiments the fibre optic equipment performed efficiently and the interrogator was capable of surviving 6 g of lateral acceleration (Silva et al. 2012). This is important as it was reported by Corriea et al. (2016) that feeding the reflected signal from an FBG through a geotechnical centrifuge fibre optic rotary joint to an external interrogator (i.e. outside the centrifuge environment) induced a consistent low frequency (1-2.5 Hz) noise to the signal. Whilst this can be corrected, the system will be more accurate if this noise can be avoided, particularly during experiments involving cyclic loading (such as the steel catenary experiments described later).

2 EXPERIMENTS

2.1 Overview

The OFDR fibre optic system was evaluated by measuring the bending of a cantilevered round steel bar.

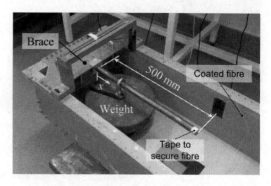

Figure 1. Cantilever bar experimental setup, with OFDR fibre bonded to the underside of the bar.

This statically determinate system was chosen for its approximation to a stiff pipeline and the ease at which tests could be conducted. A duplicate systems with half-bridge strain gauges was also tested for comparison.

The OFDR fibre optic system used in this experiment was the Sensuron Summit Elite (Senuron, 2016). This four channel interrogator is capable of sampling 2048 FBGs per channel, at a spatial resolution of 6.35 mm (0.25 in). The strain gauged configuration was monitored with the in-house DigiDAQ systems at the Centre for Offshore Foundation Systems at the University of Western Australia (Gaudin et al., 2009).

2.2 Experimental setup

The cantilever test was conducted on a 600 mm steel bar. The fixed end condition was provided by clamping 100 mm of the bar into an aluminium brace. The brace was bolted to a strong box typically used for geotechnical centrifuge testing. The final configuration result in a 25 mm diameter, 500 mm long cantilevered steel bar, elastic modulus taken as 200 GPa, (Fig. 1). Load was applied to the bar by hanging weights.

Two instrumented bars were tested. One had a single fibre optic strand bonded to its underside with cyanoacrylate. The fibre consisted of three portions: a 1 m sensing length, 195 μm in diameter, with FBGs spaced 6.35 mm on centre, a splice, and a 3 m long, 900 μm in diameter, coated extension cable (Fig. 2). The bar was oriented so the fibre was under it and opposite the applied load. The fibre was bonded from 0 mm to 450 mm on the x-coordinate (Fig. 1). The free 50 mm at the far end of the bar was used to secure the excess length of fibre.

The second bar had five sets of quarter-bridge gauges oriented on either side to create a half-bridge. They were bonded with cyanoacrylate, following the same process than the optic fibre. Cable ties were used to secure the wiring, which consisted of a set of three wired for each half-bridge. This type of half-bridge configuration is optimized the measurement of bending strain, but having gauges on top of the bar limited the number of load application points. Gauges were spaced at 70 mm on centre from 60 mm–340 mm on

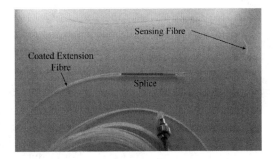

Figure 2. Fibre optic cable.

Figure 3. Cantilever bar experimental setup with strain gauges.

Table 1. Test Matrix.

Name	Increment mm	Weight kg	Repetition #
Light Fibre	20	5.6	5
Light Gauge	20*	5.6	5
Heavy Fibre	20	23.3	5
Heavy Gauge	20*	23.3	5

*Non-regular, all increments a multiple of 20 mm

the x-coordinate (Fig. 3). Unfortunately, due to technical issues the gauge at 340 mm was not recorded during testing.

2.3 Testing matrix

The cantilever bar experiment was designed to assess the precision and accuracy of the fibre system. Bending moment was applied to the bar via either a light or heavy weight (5.58 kg and 23.31 kg respectively). The moment was varied by moving the weight at 20 mm increments along the bar's length, 20 mm to 480 mm in the x-coordinate. Load increments were less regular, but always a multiple of 20 mm when testing the strain gauge bar due to the strain gauges and cable ties. These load combinations were selected to create magnitudes of strain typical of scale model experiments. An overview of the experiments is provided in Table 1.

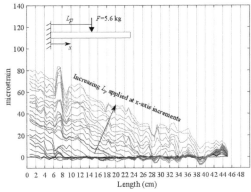

Figure 4. Incremental loading Light Fibre test.

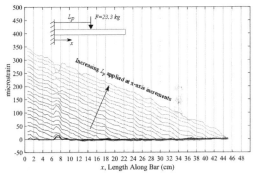

Figure 5. Incremental loading Heavy Fibre test.

3 RESULTS

3.1 Incremental loading

The results for a single repetition of the Light Fibre and Heavy Fibre have been plotted in Fig. 4 and Fig. 5, respectively. In the figures L_p is the distance to the load point and F is the applied load. In both cases the moment increments can be clearly seen and the strain measurements vary linearly along the length of the fibre as expected. There does appear to be a significant noise at low microstrains (i.e. below $80\,\mu\varepsilon$, Fig. 4). The load point can be clearly seen in the Heavy Fibre test, where each line intersects the x-axis (Fig. 5) because of the high spatial concentration of FBGs. Also of note is the spike in strain near 7 cm, it is believe to be an area of concentrated strain due to glue application process, as discussed below the fast setting cyanoacrylate did not appear to be an appropriate adhesive to use of the fibre.

3.2 Accuracy

The system accuracy was assessed for both the Light Fibre and Heavy Fibre tests at four different moment increments. Plots comparing measured strain from the OFDR fibre optic system to strain gauge output and theory are provide in Fig. 6 and Fig. 7. The strain gauges compare very well to theory; however, the fibre appears to under predict strain at large magnitudes.

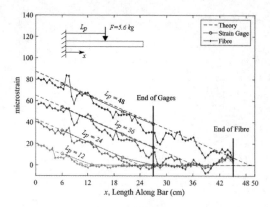

Figure 6. Comparison of OFDR fibre, strain gauge, and OFDR fibre for Light Fibre and Light Gauge tests.

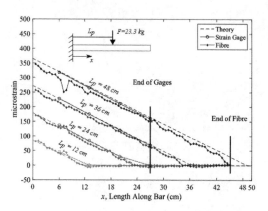

Figure 7. Comparison of OFDR fibre, strain gauge, and OFDR fibre for Heavy Fibre and Heavy Gauge tests.

Cantilever bar experiments using a similar OFDR FBG system conducted by Chan et al. (2015) also showed the fibre under-measuring large values of strain compared to gauges; however, it appears their error and noise was much lower. The mean error for all FBGs has been tabulated in Table 2, this is for a single repetition of each test. The results show a relatively consistent percent mean error with applied moment. It is arguable that the Heavy Fibre tests exhibit a larger percent error. The Heavy Fibre tests were conducted after the Light Fibre tests so it is possible that the adhesive was breaking down over time.

3.3 Precision and repeatability

The precision of the fibre systems under bending strain fibre was assessed from point by point statistics for the maximum moment ($L_p = 480$ mm) case over all five repetitions of the Light Fibre and Heavy Fibre tests (Fig. 8). It is observed that the mean standard deviation across all FBGs was 3.1 $\mu\varepsilon$ for F = 5.6 kg and $L_p = 480$ mm and 6.1 $\mu\varepsilon$ for F = 23.3 kg and $L_p = 480$ mm. In comparison the standard deviations for Light Gauge and Heavy Gauge cases were 0.4 $\mu\varepsilon$

Table 2. Accuracy of fibre measurements relative to theory.

Name	L_p mm	Moment N-m	Mean Error $\mu\varepsilon$	%
Light Fibre	12	6.57	−1.0	5.2
	24	13.14	−2.2	5.5
	36	19.71	−2.2	3.8
	48	26.28	−3.8	4.7
Heavy Fibre	12	27.44	−5.1	6.6
	24	54.88	−10.9	6.4
	36	82.32	−15.6	6.1
	48	109.76	−17.3	5.0

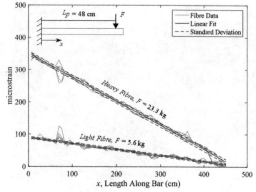

Figure 8. Variability in OFDR fibre measurements from five test repetitions.

and 0.9 $\mu\varepsilon$ respectively. It should be noted that these measures also include inherent variability in the testing technique; for example the user's ability to accurately place the weight at the targeted load point.

3.4 Deflections from numerical integration

Given the high spatial density of FBGs the OFDR fibre technology would be a useful technology for measuring the deflection in small scale models (e.g. pipelines, conductors, retaining walls and geomembranes). To that end deflections were computed by a solely numerical double integration of measured strain (steps: algebraic strain to moment, integration of moment to rotation, integration of rotation to deflection). This was done for a single repetition of $L_p = 480$ mm for each set of tests. The results are presented in Fig. 9. Overall this work very well, in-spite of the low signal to noise ratio in the measurements of strain. For the Heavy Fibre test, the maximum error relative to theory was 0.12 mm or 5.5%. While during Light Fibre test the maximum error was exceedingly small only 1.6×10^{-5} mm or 0.003%. In general the absolute error will increase along the x-coordinate with a numerical cumulative integration. The error at any point in the integration is the summation of error up to that point in the original function.

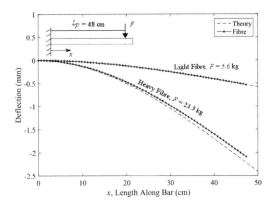

Figure 9. Measured deflections from OFDR fibre for a cantilever round bar.

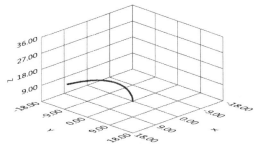

Figure 10. Example 3D Sensing output, axis units (cm).

4 DISCUSSION

4.1 Overview

Overall of performance of the OFDR FBG fibre appears acceptable. In these experiments, where the fibre was bonded to the steel bar with cyanoacrylate, the fibre was outperformed by strain gauges (though see discussion in section 4.2). That being said the system has two significant advantage. One: the bulk of the sensing system and its impact on the geotechnical model material properties is significantly reduced, Fig. 1 vs Fig. 3. This point is especially important for scale modelling where sensors and their wiring are disproportionately large compared to the models. Two: as shown in Fig. 9 the high spatial density of FBGS along the fibre's length allows for direct numerical double integration to calculate deflection. Under typical conditions, it is necessary to extrapolate data points between strain gauges in order to perform this type of analysis. As noted by Simpson et al (2015) minimizing extrapolation on flexible geotechnical structures is important to ensure areas of concentrated strain are not overlooked. This could have a major benefit in model tests of many soil structure interaction problems where accurate deflection measurements are critical.

4.2 Fibre bonding – cyanoacrylate

It should be noted that there is concern that the cyanoacrylate did not bond well with the fibre optic sensor. When the experiment was disassembled it was extremely easy to remove the fibre from the bar. If the cyanoacrylate was not properly bonded to the fibre it could explain the lower than expected measurements and potentially the relatively high noise to signal ratio (Fig. 6, Fig. 7 and Table 2). The geotechnical literature does not provide a systematic evaluation of how to bed bonding fibre optic sensors and a variety of methods have been used. Correia et al. (2016), for instance, used cyanoacrylate (though, did not compare their results to strain gauges or theory), Simpson et al. (2015) and Silva et al (2012) both used epoxy, and Kapogianni et al. (2010) used an optical adhesive (cures under UV light). Additionally, Chan et al. (2015) in the aerospace literature did not state what adhesive in their cantilever bar experiment. Given our experience in this set of experiments and the comparatively better results from Chan et al. (2015) we cannot recommend using cyanoacrylate for bonding optical fibre to scale models. We expect the fibre to perform better with a different adhesive, although this remains to be demonstrated.

4.3 3D shape sensing

An additional feature of OFDR FBG technology is the ability to measure 3D shape in real time. 3D shape can be measured in the same manner as 2D deflection was in the cantilever rod test (Fig. 9), however, three fibre strands are required. An example of real-time output from the 3D deflection of a 2.0 mm diameter, 700 mm long rod, with three fibres bonded about its circumference is provided in Fig. 10. Additionally the 3D shape of a flexible multi-core FBG fibre optic cable can be measured. This is of specific interest to offshore engineers since these 280 μm diameter fibre optic cables could be used to measure the real-time 3D shape of scale model anchor and mooring lines.

5 CONCLUSIONS

The OFDR FBG technology is an interesting new fibre optic technology that could be extremely useful in small scale geotechnical modelling given the high spatial density of FBG sensors in the fibre strand (6.35 mm on centre and over 2,000 on a single channel). Not only would this allow for significantly more measurement points on a model, but optic fibres are much less invasive than traditional strain gauges, which require multiple wires.

The following conclusions have been drawn from an initial assessment test of a point loaded cantilevered bar with a single fibre strand bonded with cyanoacrylate.

1. On a cursory examination the OFDR FBG systems appears to work well at large strain, greater than 80 μɛ, and had a strong response to small changes

in moment (Fig. 5). Below 80 $\mu\varepsilon$ it was not as strong (Fig. 4).
2. The mean error along the length of the bar appears relatively consistent with applied moment, Table 2
3. In spite of noise the system was fairly precise with measured standard deviation from repetitive testing being between 3.1 $\mu\varepsilon$ and 6.1 $\mu\varepsilon$.
4. The OFDR FBG technology worked extremely well for calculating deflections from numerical double integration of strain. This technology would be useful for monitoring the deformation during soil-structure interaction in small scale modelling.

Given the poor bonding between the fibre and steel bar using the cyanoacrylate adhesive these conclusions should be limited to this specific case. We believe the behaviour of the OFDR FBG measurement would be improved with the use of a better adhesive.

REFERENCES

Chan, H.M., Parker, A.R., Piazza, A. & Richards, W.L. 2015. Fiber-optic sensing system: Overview, development and deployment in flight at NASA. *2015 IEEE Avionics and Vehicle Fiber-Optics and Photonics Conference (AVFOP)*. IEEE, Santa Barbara, California, pp. 71–73

Correia, R., James, S.W., Marshall, A.M., Heron, C.M. & Korposh, S. 2016. Interrogation of fibre Bragg gratings through a fibre optic rotary joint on a geotechnical centrifuge. *Sixth European Workshop on Optical Fibre Sensors*. SPIE, Limerick, Ireland, pp. 1–4

Duncan, R.G., Froggatt, M.E., Jreger, S.T., Sceley, R.J., Gifford, D.K., Sang, A.K. & Wolfe, M.S. 2007. High-accuracy fiber-optic shape sensing. *Sensor systems and networks: phenomena, technology, and applications for NDE and health*. SPIE, Bellingham, WA

Gaudin, C., White, D.J., Boylan, N., Breen, J., Brown, T., De Cantania, S. & Hortin, P. 2009. A wireless high-speed data acquisition system for geotechnical centrifuge model testing. *Measurement Science and Technology*. 20(9): 1–11

Kapogianni, E., Sakellariou, M.G., Laue, J. & Springman, S.M. 2010. The use of optical fibre sensors in a geotechnical centrifuge for reinforced slopes. *Phys. Model. in Geotech*. CRC Press, Zurich, pp. 343–348

Mohamad, H., Benett, P.J., Soga, K., Mair, R.J. & Bowers, K. 2010. Behaviour of an old masonry tunnel due to tunnelling-induced ground settlement. *Géotechnique*. ICE 60(12): 927–938

Mohamad, H., Soga, K., Pellew, A. & Benett, P. 2011. Performance Monitoring of a Secant-Piled Wall Using Distributed Fiber Optic Strain Sensing. *J. of Geotech. and Geoenviron. Eng.*. ASCE 137(12): 1236–1243

Sensuron, 2016. *Summit Multi-Sensing Platform*. Austin, TX

Silva, T.S., Elshafie, M.Z.E.B. & Sun, T. 2012. Fibre optic instrumentation and calibration in the geotechnical centrifuge. *Proc. of 3rd Eur. Conf. on Phys. Model. in Geotech*. IFSTTAR, Nantes, France, pp. 129–134

Simpson, B., Hoult, N.A. & Moore, I.D. 2015. Distributed sensing of circumferential strain using fiber optics during full-scale buried pipe experiments. *J. of Pipeline Syst. Eng. and Pract*. ASCE 6(4)

Free fall cone tests in kaolin clay

A. Bezuijen
Ghent University, Ghent, Belgium
Deltares, Delft, The Netherlands

D.A. den Hamer
Ghent University, Ghent, Belgium

L. Vincke
MOW-Geotechniek, Ghent, Belgium

K. Geirnaert
dotOcean NV, Brugge, Belgium

ABSTRACT: Free fall cone tests have been performed in a model container filled with Kaolin clay with varying strength. The strength of the clay was varied by mixing it with different water contents. The free fall cone is equipped with accelerometers to determine the acceleration during the fall. One and two times integration of the measured acceleration gives the velocity and the displacement respectively. Measured penetration resistance is compared with penetration resistance measured with a standard CPT and a T-bar. Results showed good agreement when corrected for the difference in penetration rate.

1 INTRODUCTION

A free fall cone (FFC) is an interesting alternative for the traditional CPT or comparable penetration devices for soil exploration on deeper water. No rig is needed, just a small boat and a cone (Stark et al. 2009). It is especially suitable to investigate soft soil layers, since it will not penetrate very far in stiff layers. A complication is, however, that the penetration rate of a falling cone just below the sea bottom will be much higher than for a traditional cone that is pushed into the soil with 0.02 m/s. Rate effects will occur and it is necessary to correct the measurement results for these rate effects. These rate effects are also described in torpedo anchor tests O'Beirne et al. (2015 and 2016) and O'Loughlin (2013).

dotOcean NV has developed a FFC to explore the strength of the superficial layers of the sea bottom. The primary reason for that development was to investigate at what depth the sea bottom will cause so much penetration resistance that it hampers navigation. The research in this paper is performed to investigate whether it is possible to use the same FFC, called the Graviprobe, also for stronger soil layers and to what depths this is possible.

To simplify the experimental set-up, tests were performed under dry conditions. The FFC was dropped from a certain height into a sample of Kaolin clay.

This paper describes the FFC used, the sample preparation, the tests and the program used to interpret the results.

2 THE FFC USED, THE GRAVIPROBE

The FFC used is developed by dotOcean NV In the cone are two accelerometers that can measure acceleration to a maximum of 1.7 g and 70 g respectively. In this research, only the 70 g accelerometer is used. Furthermore, a data logger is built into the cone and the measured accelerations are stored with a sample rate of 2000 samples/s. The data logging is started just before the fall. Software tools are available to select the relevant data after a test. Only the acceleration is stored. Penetration force, velocity and displacement are derived from the measured acceleration and the weight of the FCC.

Two types of FFCs are developed, see Figure 1 and Figure 2, called "mud probe" and "soil probe respectively. When the mud probe falls into the soil, tip force and friction contribute to the deceleration, for the soil probe it is only the tip force and it had a slightly smaller tip. Consequently, the soil probe will penetrate deeper allowing exploration of the soil to a larger depth.

3 SOIL PREPARATION

A concrete model container was used for the Kaolin clay with a diameter of 0.8 m and 0.9 m height. In the container a 'big bag' and a polyethylene bag was applied before the soil sample was made. The big bag was for easy removal of the sample after the test. The polyethylene bag was to avoid water leakage from the

Figure 1. "Mud probe".

soil sample. Part of the soil sample and the container are shown in Figure 3.

The undrained shear strength of the clay depends on the amount of water added. The relation between the amount water added and the strength of the clay was determined in small scale model tests.

The standard way to make a relatively homogeneous clay sample is to prepare a clay slurry with a water content of 1 to 1.5 and to consolidate this sample to the required strength (Burland 1990). However, for the amount of clay necessary for these tests that would be too time consuming; therefore a different approach was followed. Starting from clay powder an amount of water was added and mixed with a high shear cement mixer. A large mixer is essential to prepare samples in this way. In this research we used a 100 litre horizontal counter current mixer normally used for concrete.

After mixing the clay was brought into the container. A vibration pestle was used to compact and homogenize the clay. The pestle worked on a circular 0.5 m diameter plate to avoid that the pestle sunk into the clay during operation. Using the pestle was essential to achieve a homogeneous clay sample see Figure 4.

It appeared that after a number of penetrations it was possible to recover the clay sample, i.e. to remove the penetration holes in the clay with the pestle. Preparation of a clay sample in this way without consolidation is less homogeneous than following the usual way, consolidating wet clay, described above but this was sufficient for this purpose, since we compare different

Figure 2. "Soil probe", Sketch and picture.

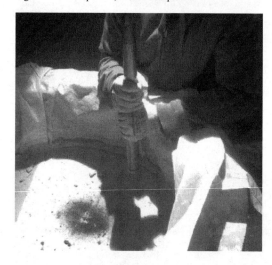

Figure 3. Part of the soil sample and the container with the bags just before a CPT test.

soil characterization methods and we can perform a number of tests to check the homogeneity.

Three different soil samples were made. Two samples had different types of kaolin clay see Table 1. The

Figure 4. Clay sample before and after the use of the pestle.

Table 1. Material properties of Rotoclay and Polwhite B.

Parameter		Rotoclay Sample A	Polwhite B Sample B-I
Water content	[%]	40.65	42.62
Dry solids	[%]	71.1	70.12
grain volume wt_assumed	[kg/m^3]	2650	2650
volume mass wet	[kg/m^3]	1746	1700
volume mass dry	[kg/m^3]	1242	1192
Pore number	[–]	1.34	1.223
Pore volume	[%]	53.15	55.01
Water content at saturation	[%]	42.8	46.14
Degree of saturation	[%]	94.98	92.36
Liquid Limit	[%]	49.99	58.09
Plastic Limit	[%]	39.90	41.34
W%_Real/ W% LL	[ratio]	81%	73%
Approx. strength	[kPa]	4–7	22–25

third soil sample was a layered sample. The bottom layer was the layer from the second sample – Polwhite B – again homogenized with the pestle. The upper layer prepared from Rotoclay clay, recovered from the first sample. The Rotoclay was mixed again and the amount of water was increased up to 45%. Thereby reducing the strength of this top layer to only a few kPa.

4 TESTS PERFORMED

In total 17 FFC tests have been performed. Falling height, kind of cone and soil sample varied.

Figure 5 shows a result of such a test. Before 57.09s the cone falls through air and is accelerating with 9.81 m/s^2, the acceleration of gravity. When it hits the clay the acceleration becomes negative and it decreases further when more of the cone comes into the clay. At around 57.20s the cone has come to a standstill. Integration of the data and 2 times integration results in the velocity and the displacement, see Figure 6 that shows the fall over 6 m in the air and in the clay (where the velocity decreases). The velocities and displacement in the clay are shown in a detail, Figure 7.

The FFC enters the clay in this test after a fall 5.4 m and then penetrates a bit more than 0.6 m into the clay.

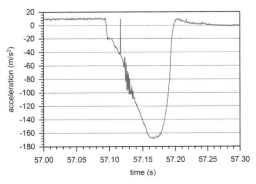

Figure 5. Result FFC test, measured acceleration.

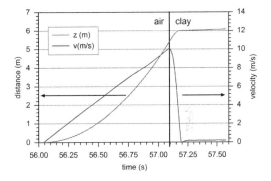

Figure 6. Measured velocity and displacement.

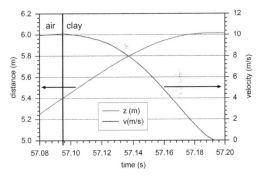

Figure 7. Measured velocity and displacement, detail part in clay.

5 OTHER TESTS

To be able to compare the soil strength measured with the FFC with a reference measurement, standard field equipment was used to measure the strength of the model soil in the container, see Figure 8. CPT, T-Bar and vane test were used. Since the field vane gave only one value in the relative small model container, the results of the CPT and T-Bar were used to compare with the results of the FFC. From the CPT the tip resistance and friction were used. The accuracy of a CPT is limited to 1 kPa. However, since the soil was relatively soft, also T-Bar cone penetration tests were

Figure 8. Field equipment used for CPT and T-Bar tests.

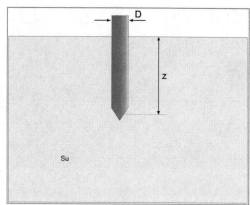

Figure 9. Definition sketch calculation model.

performed. The T-Bar has a 10 times larger surface area than the cone tip and therefore the measurements with the T-Bar are more accurate for low soil strengths.

Data were analysed assuming that for this soft clay the measured friction is S_u times the area of the friction sleeve. For the tip resistance of the CPT the relation used is:

$$S_u = \frac{q_c - \sigma_v}{12} \quad (1)$$

with q_c the tip resistance and σ_v the total vertical stress at penetration depth. For the T-Bar this relation is (Steward & Randolph, 1994):

$$S_u = \frac{q_T}{10.5} \quad (2)$$

where q_T is the resistance measured with the T-Bar.

6 EVALUATION OF THE FFC TESTS

6.1 Calculation model

A calculation method is derived to estimate S_u on the results of the FFC tests that is programmed in a spreadsheet. The model uses in principle only Newton's law:

$$F = m.a \quad (3)$$

with F the force in Newtons, m the mass and a the acceleration. In this calculation m is the mass of the cone, F the sum of the forces acting on the graviprobe. F has the following components:

$$F = W_g + F_T + F_f + F_b \quad (4)$$

where W_g the weight of the graviprobe, F_t the tip resistance of the tip when it penetrates into the clay, F_f the friction force between the shaft of the probe and the clay and F_b the buoyancy force that develops when the cone penetrates into the clay.

Assuming that the acceleration is positive according to gravity, the following relations can be derived for the forces, see also Figure 9:

$$F_T = -N_c f_d Su A_p \quad (5)$$

$$F_f = -f_f \sum_{i=1}^{n} Su_i . O_p \Delta z_i \quad (6)$$

$$F_b = -\gamma_c O_p z \quad (7)$$

where N_c is the bearing capacity factor (–), f_d the depth factor (–), f_f the friction factor (–), A_p the cross-sectional area of the graviprobe (m²), O_p circumference of the graviprobe (m), i is a counter (–), Δz_i is the thickness of small slice of clay defined as:

$$z = \sum_{i=1}^{n} \Delta z_i \quad (8)$$

with z the penetration depth (m).

The depth factor for undrained conditions is defined as (Brinch Hansen, 1970):

$$f_d = 1 + 0.4 \arctan\left(\frac{z}{D}\right) \quad (9)$$

with D the diameter of the cone.

The friction factor f_f is assumed to be 0.8 in the calculations.

S_u can be determined from the deceleration for the FFC. The FFC enters the clay with a velocity of several meters a second and slows down to a velocity of 0 m/s. Based on literature on dredging (Van der Schrieck, 2013), it is likely that the S_u depends on the penetration velocity. Normally it is a function of the strain rate and this is defined as:

$$S_u = S_{u,ref} \left(\frac{\dot{\gamma}}{\dot{\gamma}_{ref}}\right)^\beta \quad (10)$$

where $S_{u,ref}$ is the undrained shear strength at a reference strain rate (kPa), $\dot{\gamma}_{ref}$ the reference strain rate (1/s), $\dot{\gamma}$ the actual strain rate (1/s) and β exponent in the order of 0.1. This is a bit different from the relation proposed by Moavian et al. (2016), who suggested a logarithmic function. Biscontin & Pestana (2001) have shown that both functions are possible.

The shear rate is not known, but it will be a function of the penetration velocity and the diameter of the probe. At a given penetration velocity the shear rate will increase when the probe diameter decreases. The deformation pattern around a CPT will be comparable to the deformation pattern around a FFC. Therefore, Eq. 10 can be approximated to:

$$S_u = S_{u,ref} \left(\frac{d_{cpt} v_p}{d_p v_{cpt}} \right)^\beta \tag{11}$$

where v_{cpt} the velocity of a CPT (=0.02 m/s), v_p the velocity of the FCC, d_p the diameter of the FFC (0.05 m), d_{cpt} the diameter of a CPT cone (=0.036 m) and.

Eq. (12) has one problem: S_u becomes zero when $v_p = 0$. This is rather unrealistic. Therefore, it is possible to give a minimum value of S_u for which Eq. (12) can be used. In the model this minimum value was assumed to be 0.5 times the S_u measured with a CPT.

Since the various forces (F) are not constant, a numerical solution with a spreadsheet is made. For a given value of S_u as a function of depth we can write:

$$a = \frac{F}{m} = \frac{W_g + F_T + F_f + F_b}{m} \tag{12}$$

or after substitution with the various terms:

$$a_n = \frac{F_n}{m} = \frac{1}{m} \left\{ W_g - N_c f_{d,n} A_p S_{u,ref,n} \left(\frac{d_{cpt} v_p}{d_p v_{cpt}} \right)^\beta \right.$$
$$\left. - f_f \sum_{i=1}^n S_{u,ref,n} \left(\frac{d_{cpt} v_p}{d_p v_{cpt}} \right)^\beta \cdot O_p \Delta z_i - \gamma_c O_p z_n \right\} \tag{13}$$

The index n means that the parameter for the n step should be chosen, which is related to the penetration depth (Eq. (8)). If the velocity of the probe when it enters the clay is known, the acceleration can be calculated using Eq. (13) and with a known time step. The velocity can be calculated by numerical integration and the displacement by integration of the velocity to time.

To derive S_u from acceleration data, as measured by the FFC, Eq. (13) is rewritten and an approximation is made. The error is assumed to be small, if for the calculation of $S_{u,n}$ the friction along the pile is calculated until z_{n-1}. For Graviprobe FFC measurements described in this paper, this is allowed, since the sampling rate is 2048 samples/s. Displacement of the FFC

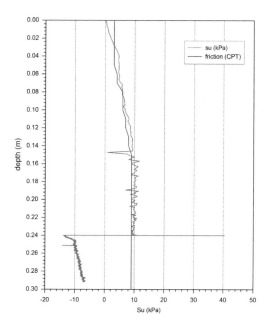

Figure 10. S_u determined from the measured friction (blue line) and from the FFC (red line). The falling depth was 0.24 m. The values below 0.24 m are the result of numerical drift in the first simulations.

between 2 measurements is never more than a few mm. With this approximation Equation 10 can be written as:

$$S_{u,ref,n} = \frac{\left(\frac{d_{cpt} v_p}{d_p v_{cpt}} \right)^{-\beta}}{N_c f_{d,n} A_p} \left\{ W_g - m a_n \right.$$
$$\left. - f_f \sum_{i=1}^{n-1} S_{u,ref,n-1} \left(\frac{d_{cpt} v_p}{d_p v_{cpt}} \right)^\beta O_p \Delta z_i - \gamma_c O_p z_n \right\} \tag{14}$$

In the spreadsheet $S_{u,ref,n}$ is determined using the measured acceleration. To determine the impact of the FFC, the integration starts when the graviprobe is released. Subsequently, Eq. (14) is used to determine $S_{u,ref,n}$ for each time step until the velocity of the FFC becomes zero. An example of the measured and calculated S_u profile is shown in Figure 10. The penetration depth of the FFC was 24 cm. Here the result is compared with the friction measured with a standard CPT and showed good agreement.

6.2 S_u values determined

The methodology described in the previous section was used to calculate the undrained shear strength from the measured deceleration. Table 2 presents the values used for the various parameters for evaluating the data from the mud probe.

The falling height and consequently the impact velocity were changed in the simulation depending on the real falling height that was measured with a laser device.

Table 2. Parameters used in simulation.

Parameter	Value	unit
acceleration of gravity	9.81	m/s^2
density water	1000	kg/m^3
density clay	1700	kg/m^3
diameter FFC	0.05	m
length FFC	1	m
mass FFC	17.49	kg
falling height	4.35	m
factor tip res. Shallow (N_c)	6.2	–
factor friction (f_f)	0.8	–
β	0.15	
Velos. cut off strain rate	0.5	–
impact velocity	9.24	m/s

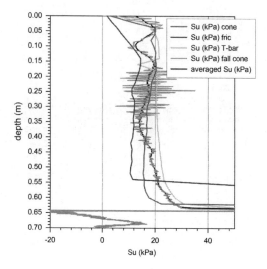

Figure 11. Example result of soil probe, compared with S_u determined from CPT tip (cone), CPT friction (fric) and T-Bar.

In the parameter set β depends on the type of clay. This parameter determines the deformation rate. It appeared that little information is available on this parameter and it is used as a fitting parameter to achieve agreement with the S_u values measured with the other devices. The value of the penetration velocity varies quite a lot, so a good fit also means that the model predicts the trend well. If the FFC is used on a regular base, it will be necessary to investigate the value of β for different clays.

For the soil probe, the diameter of the cone was 0.033 m, the height 0.032 m (because hardly any friction will develop along the rod between the tip and the actual cone) and the weight was 13.78 kg.

It appeared that result shown in Figure 10 shows exceptional good agreement. In other situations the agreement between the standard field determinations and the FFC was less, but still quite reasonable, as will be shown below.

The high sampling frequency of the fall cone causes quite some variation in the S_u value determined, but by

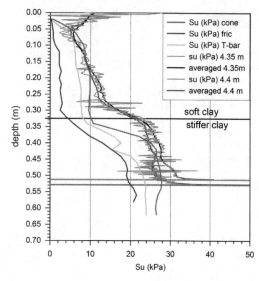

Figure 12. Measurements compared for two layer system, see also text. Averaged data are a moving average over 17 data points.

using a moving average over 17 measurement points (the number 17 is an arbitrary choice), this scatter can be removed and the resulting S_u is close to the value determined by the other methods.

Two tests were done with the mud cone in this sample with a slightly different falling height, see Figure 12. It appeared that the results are quite comparable. However, a remarkable deviation occurred in this two-layer system. The S_u determined in the weak layer agrees well with the S_u determined with the cone and is a bit higher than what is determined from the friction sleeve and the T-bar. The same is valid for the stiffer layer. However, the transition does not fit.

The FFC results show a transition to the stiffer clay layer at shallower depth than the CPT and T-bar results. It is all within 5 centimetres but the difference is remarkable since both the FFC tests results in more or less identical results and both the T-bar test and the CPT indicate a lower transition between the layers.

7 DISCUSSION

It appeared that there is a reasonable agreement between the undrained shear strength (S_u) determined with the standard field equipment and the FFC. This means that the FFC can be used as an alternative for CPTs or T-bar tests. However, even in the rather soft clay used in these tests (strength mostly between 10 and 20 kPa) the penetration depth is limited. The mud probe never fully penetrates until the bottom of the container (total depth of 1 meter), even when using a falling height of 5.4 m and a starting penetration velocity of 10.3 m/s. Figure 13 shows the penetration after a fall from 4.7 m.

Figure 13. Mud probe FFC penetration into the clay after a test with 4.7 m falling height.

A much heavier device will be necessary to penetrate to larger depth. Such a device will be less sensitive to determine the strength of soft upper layers, since they will hardly influence the acceleration of such a heavier device. A solution may by to use again 2 accelerometers just like in the Graviprobe tested here. One to measure small accelerations accurately and one for larger accelerations. The programme developed to evaluate these tests can also be used to design a FFC that can penetrate in clay layers of different strength. A sand layer of more than a few centimetres will in all cases be a 'show stopper' for this kind of probe.

8 CONCLUSIONS

The possibility to use a free fall cone for the determination of the undrained shear strength of soft clay was investigated. The fall cone was initially designed to measure the shear strength of mud instead of clay layers. From this laboratory study, we came to the following conclusions.

When correcting for the influence of the higher deformation rates compared to traditional penetration devices a reasonable to good agreement could be obtained with the undrained shear strength measured with the standard field equipment. To obtain this result it was necessary to fit the value of β, the parameter that determines the influence of the deformation rate. Additional research is needed in order to determine β for different types of clay and possibly as a function of the deformation rate.

It seems as if the detection of different layers can be done but determination of the actual position of these layers was not very accurate in the few tests we did. Further testing is required to determine the accuracy in depth localization.

REFERENCES

Biscontin, G. & Pestana, J. M. 2001. Influence of peripheralvelocity on vane shear strength of an artificial clay.ASTMGeotech. Testing J.24, No. 4, 423–429.

Brinch Hansen J. 1970. A revised and extended formula for the Bearing capacity. The Danish Geotechnical Institute, Bulletin No. 28.

Burland, J. B. 1990. On the compressibility and shear strength of natural clays. Geotechnique 40, No. 3, 329–378.

Moavenian, M.H. Nazem, M., Carter J.P., Randolph M.F. 2016 Numerical analysis of penetrometers free-falling into soil with shear strength increasing linearly with depth, Computers and Geotechnics 72 57–66.

O'Beirne, C.P., O'Loughlin, C.D., & Gaudin, C. 2015. Assessing the penetration resistance acting on a dynamically installed anchor in normally and over consolidated clay.

O'Beirne, C.P., O'Loughlin, C.D., Wang, D., & Gaudin, C. 2015. Capacity of Dynamically Installed Anchors as assessed through field testing and three dimensional large deformation finite element analyses. Canadian Geotechnical Journal, 52(2), 548–562.

O'Loughlin, C.D., Richardson, M.D., Randolph, M.F. & Gaudin, C. 2013. Penetration of dynamically installed anchors in clay. Géotechnique, 63(11), 909–919.

Stark N., H. Hanff & A. Kopf. 2009. Nimrod—A tool for rapidgeotechnical characterization of surface sediments. SeaTechnol. 50:10–14.

Stewart, D.P. & Randolph, M.F. 1994. T-bar penetration testing in soft clay. Journal of Geotechnical Engineering – ASCE, 120(12): 2230–2235.

Van der Schrieck G.L.M. 2013. Dredging Technology, Guest lecture notes CIE5300, Delft University of Technology/GLM van der Schrieck BV.

# A new shared miniature cone penetrometer for centrifuge testing

T. Carey, A. Gavras & B. Kutter
University of California Davis, Davis, California, USA

S.K. Haigh & S.P.G. Madabhushi
Cambridge University, Cambridge, UK

M. Okamura
Ehime University, Matsuyama, Japan

D.S. Kim
Korea Advanced Institute of Science and Technology, Yuseong, South Korea

K. Ueda
Disaster Prevention Research Institute, Kyoto University, Kyoto, Japan

W.Y. Hung
National Central University, Taoyuan City, Taiwan

Y.G. Zhou, K. Liu & Y.M. Chen
Department of Civil Engineering, Zhejiang University, Hangzhou, P. R. China

M. Zeghal & T. Abdoun
Rensselaer Polytechnic Institute, Troy, New York, USA

S. Escoffier
IFSTTAR, GERS, SV, Bouguenais, France

M. Manzari
The George Washington University, Washington, DC, USA

ABSTRACT: Cone penetrometers (CPTs) are commonly used for characterising the soil properties of centrifuge models; CPT data is useful for interpretation and quality control. This paper describes the development and design of a new robust CPT device for centrifuge testing. The new device consists of a 6 mm cone, an outer sleeve, and an inner rod that transmits cone tip forces to a load cell above the ground surface. The design eliminates the need for a custom submerged strain gauge bridge near the tip, significantly reducing cost. A direct comparison was performed between this CPT device and another similar device developed at the University of Cambridge. CPT's were manufactured using the new design and then shipped to eight different centrifuge facilities, for quality control of similar experiments performed for LEAP (Liquefaction Experiments and Analysis Projects). All the centrifuge tests simulated a 4 m deep deposit of soil, all consisting of Ottawa F-65 sand with relative densities ranging between about 45 to 80%. The results obtained have been extremely valuable as an independent assessment of the density calculated from mass and volume measurements at different laboratories.

1 INTRODUCTION

The ability of a cone penetrometer (CPT) to characterise the mechanical properties of geomaterials (Robertson and Cabal 2010) makes it an important tool for soil characterisation. Cone penetrometers have been used in centrifuge models by a number of researchers in the past (Bolton et al. 1999, Kim et al. 2015, Zhou et al. 2015 & Darby et al. 2016) for this reason.

The LEAP project, (Manzari et al. 2015) is an international collaboration to verify and validate numerical models that predict soil liquefaction. The current phase of the project, LEAP-UCD-2017, involves roughly 24 centrifuge experiments performed at 9 different research facilities. The LEAP tests were designed to determine the median response and the uncertainty of results. To assess the importance and influence of the uncertainty on the median response, it is also a goal of LEAP to quantify the sensitivity of test results to intended and unintended variations of input parameters. CPT results are especially valuable as an independent check of centrifuge model densities. To reduce variability due to differences in CPT equipment at each facility, it was decided to produce one economical design, fabricate the devices at one machine shop, and distribute them to the various centrifuge laboratories.

One of the challenges encountered following the previous phase of the LEAP project, LEAP-GWU-2015, was determining the achieved densities of the centrifuge models by mass and volume measurements (Kutter et al. 2016). Most researchers reported the achieved model density as the specified value and no independent checks were performed to evaluate the uncertainty of the mass and volume measurements. Therefore, it was considered critical for future LEAP exercises to have an independent check of model density. In flight CPTs were selected as a quality control check on prepared specimen density.

This paper describes the design, calibration, and provides a direct and cross comparison of the newly developed LEAP CPT device.

2 DESIGN

The new CPT, sketched in Figure 1, is 6 mm in diameter and is fabricated from stock stainless steel tubing and rod. This device measures tip forces using a load cell at the top of the cone, avoiding use of a costly custom submerged strain gauge near the tip. As shown in Figure 1, the inner rod is protected by a hollow sleeve and transmits tip forces to a load cell located in a rigid aluminium block. The yield stress of the Type 316 Stainless Steel inner rod is specified to be 200 MPa. Using an allowable yield stress of 130 MPa and considering the cross section of the inner rod is reduced by the presence of O-ring grooves, corresponds to a maximum tip force rating of 900 N (200 lbf). The relative density of a sand can be estimated with an expression proposed by Jamiolkowski et al. (1985):

$$D_r(\%) = 68\left[\log\left(\frac{q_c}{\sqrt{P_a \sigma_v}}\right) - 1\right] \quad (1)$$

where q_c is the cone tip resistance, P_a is atmospheric pressure, and σ_v is overburden vertical effective stress. For an overburden vertical effective stress of 100 kPa and $q_c = 900$ N/(area of the 6 mm diameter cone), it was estimated the cone could safely penetrate sand with relative density of 100%.

A guiding philosophy of the CPT design was to minimise use of components that require specialty machining. The completed manufacturing and assembly cost, including the load cell, is roughly $US 1,300. All the components of the device use 316 stainless steel, unless otherwise noted. The main components of the device are described herein.

2.1 Rod and sleeve

To avoid buckling, the unsupported length of the rod was reduced by a series of O-rings spaced at 100 mm; to minimise friction, the outer diameter of the O-rings were designed to be slightly smaller than the 5 mm inner diameter of the sleeve. The O-rings rest in grooves cut into the rod to prevent them from sliding while the rod is inserted into the sleeve. To align the rod in the centre of the sleeve, larger diameter "snug-fitting" O-rings are used at each end of the rod. To prevent sand and fluid from entering the gap between the rod and sleeve, a tip O-ring with a 4 mm diameter and 1 mm cross section is used. During final assembly, the inner cone rod is clamped to the slotted cone rod bolt by a M8 Jam nut, and is threaded into the load cell until 4 to 9 N of preload on the tip O-ring is achieved;

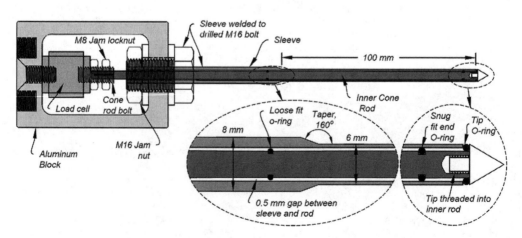

Figure 1. LEAP-UCD-2017 CPT design, illustrating cone tip, taper, and aluminium reaction block details.

the preload is easily measured by monitoring the load cell during assembly. Preloading the end O-ring to a specified small value ensures the gap between the cone shoulder and sleeve is closed and minimises potential inconsistencies of results caused by variable preloads.

The sleeve was manufactured from 8 mm outer diameter, 5 mm inner diameter tubing. Within 100 mm of the cone tip, the sleeve is machined down to 0.5 mm wall thickness, with an OD of 6 mm. 100 mm above the tip, the sleeve is tapered at 20 degrees to transition from 6 mm to 8 mm diameter. The location of the taper and the taper angle were selected to minimise increases in overburden stress at the cone tip from bearing loads produced at the transition. An abrupt 20 degree taper between the 6 mm and 8 mm diameters was chosen based on results by LeBlanc and Randolph (2008), who showed that resultant bearing loads on the tapered section would, perhaps counterintuitively, be larger on a gentler taper.

2.2 Reaction block and load cell

The aluminium block allows for simultaneous pushing of both the cone rod and sleeve. The block is 78 mm tall by 53 mm wide and 39 mm deep.

A M16 bolt connects the sleeve to the aluminium block. This is accomplished by thru drilling a bolt 25 mm in length, and welding the sleeve to the bolt. The bolt sleeve assembly is attached to the aluminium block using a jam nut. This connection is illustrated in Figure 1.

The load cell is attached to the block with a M8 bolt. The slotted cone rod bolt with a tapered jam nut clamps the cone rod to the load cell.

The load cell used for the CPT design is a 4500 N capacity, SML Mini Low Height S-Type Interface load cell. The SML line was selected for its small size and high capacity. To provide attachment for the completed device to an actuator, or an external load cell, four M6 threaded inserts are located atop the aluminium block.

3 CALIBRATION

Three calibration tests were performed prior to using the device on centrifuge models.

3.1 Friction between rod and sleeve

The first test measured the undesirable transfer of tip loads to the sleeve via the O-rings. A reference load cell was attached to the top of the aluminium block of the assembled device and measured total force while the cone tip was pushed into a block of plastic. The difference between the readings of the tip load and the external reference load cell is attributed to friction from the bracing O-rings. Under 425 N of axial load, the difference between the internal and external reference load cell was 13 N. This 3% difference is small and is accounted for by adjusting the calibration factor of the CPT load cell. In other words, the tip load measured by the CPT load cell should be multiplied by 1.03 to estimate the total tip load. This correction factor should be checked for each device in the calibration process.

3.2 Lateral force

The second test subjected the device to a 15 N lateral load at the tip, to determine if lateral loads would influence the load cell reading. Several cycles of lateral loading were applied, and no tip forces were measured. While lateral loads are expected to be small for a properly aligned device, this test shows unintended lateral loads should have negligible effect on the cone tip force reading.

3.3 Cyclic loading

The final test applied a sequence of several cycles of loading to determine the extent of the hysteresis of the completed device from friction of the O-rings. After five cycles of loading to 425 N, and back to zero load the peak difference between the internal tip and external reference load cell was measured at 13 N. The maximum and minimum widths between the loading and unloading paths of the hysteresis loops at 200 N of external loading is 1.3 N and 0.9 N respectively. The minimal change in the width of the hysteresis loops during successive loading and unloading suggests the friction contact between the sleeve and the O-rings is small and remains almost constant.

4 DIRECT COMPARISON

A direct comparison was performed with a CPT device developed at the University of Cambridge's Schofield Centre. Both CPT devices were pushed into the same tub of uniform sand, eliminating many sources of uncertain variability such as operator error during testing, placement method of sand, or different properties amongst sand batches. The container was filled with Hostun sand, dry pluviated to about 100% relative density to a depth that represented 14.5 m prototype at 50 g.

The University of Cambridge's CPT device is 6.35 mm in diameter and has a 60-degree cone tip. Similar to the new device, the Cambridge design uses an outer sleeve and has an inner rod that transmits tip forces to a load cell. As shown in Figure 4, the Cambridge CPT device uses a PTFE bushing behind the cone shoulder instead of the tip O-ring.

4.1 Results

Shown in Figure 2 are four cone resistances versus depth profiles for each device. In Figure 3 the cone tip load is isolated for depths of 2, 4, 6, 8, and 10 meters for each design. Overall, it can be concluded that the two devices produce comparable trends both in terms of stress magnitude and distinguishing characteristics. The device-to-device variability is similar to the profile-to-profile variability, but the average q_c of the LEAP CPT is about 5 to 10% larger than that from the Cambridge CPT.

The difference in average q_c values might be attributed to the different tip designs. It is also suspected that the details of the tip geometry can have significant effect on the tip resistance. In Figure 4 the PTFE bushing leaves a small (0.1 mm) gap before the sleeve, while the O-ring design results in a small ledge behind the cone shoulder.

5 CROSS COMPARISON

Twenty-four centrifuge experiments have been conducted so far at nine centrifuge facilities. Eight of these (UC Davis, Ehime, IFSTTAR, Kyoto, KAIST, NCU, RPI, and Zhejiang) used the LEAP-UCD-2017 CPT and Cambridge used their own, very similar design discussed above. Each facility has a custom rod and sleeve length due to unique container sizes, but all LEAP-UCD-2017 CPTs were manufactured at UC Davis to reduce variably of machinists and manufacturing tools.

5.1 LEAP-UCD-2017 experiment

The LEAP-UCD-2017 centrifuge experiment consists of a saturated Ottawa F-65 sand profile inclined with a five-degree slope in a rigid container. Each experiment used sand that was from the same batch that was shipped to UC Davis and then forwarded to participating facilities. Different facilities, with different box dimensions used different scale factors ($L^* = L_{model}/L_{prototype}$ between 1/20 and 1/50), but all the models represented 4 m deep (prototype scale) sand layers.

Shown in Figure 5 is the sensor layout for the LEAP-UCD-2017 experiment with accelerometers shown as triangles, and pore pressure transducers as circles.

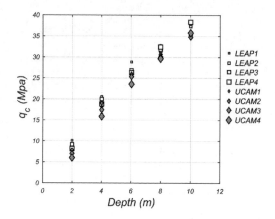

Figure 3. Cone tip resistance at 2, 4, 6, 8 and 10 m depths for the LEAP and Cambridge CPT.

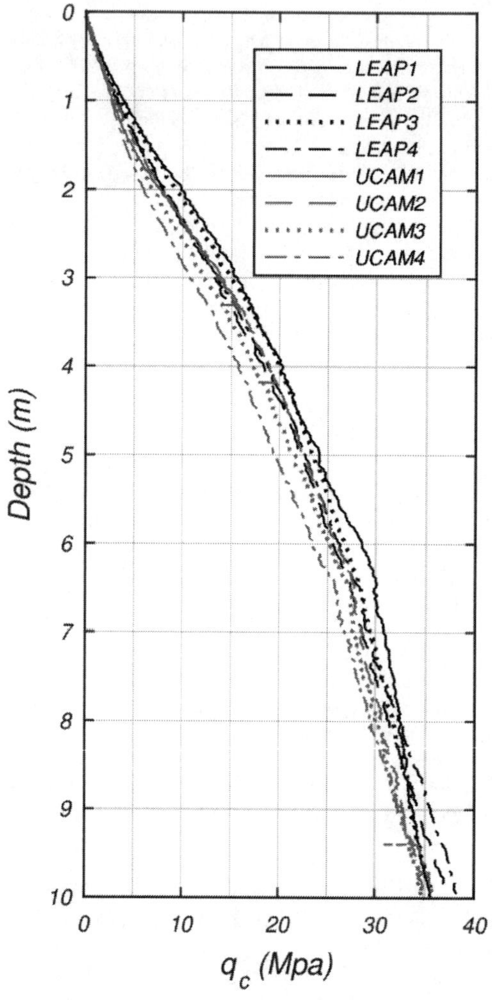

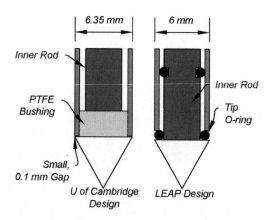

Figure 2. Stress profiles vs depth for the LEAP and University of Cambridge CPT.

Figure 4. Different in tip design between the Cambridge and LEAP CPT.

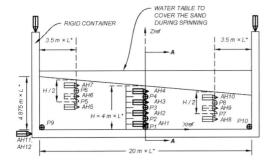

Figure 5. Sensor layout and test geometry for LEAP-UCD-2017.

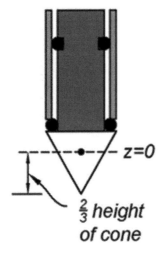

Figure 6. All CPT profiles were adjusted so zero depth coincided at 2/3 the height of the cone.

To ensure consistent drainage conditions for all penetration tests, the desired rate of cone penetration, in model scale, was specified for LEAP-UCD-2017 to depend on the pore fluid viscosity in the model:

$$V_{cpt,modelScale} = (100 mm/s)/\mu^* \qquad (2)$$

where $\mu^* = \mu_{model}/\mu_{water}$. Thus, if the test is done at 1/20 scale and according to scaling laws for dynamic testing ($\mu^* = 20$), the velocity of penetration should be 5 mm/s, model scale.

Kutter et al. (2016) presents a detailed discussion on the test geometry, sensor location and scaling laws for the LEAP-GWU-2015, which are similar to LEAP-UCD-2017. The detailed specifications for the LEAP-UCD-2017 centrifuge tests will be published prior to the ICPMG conference in London.

One goal of LEAP-UCD-2017 was to test models with a range of relative densities between about 50% (about 1599 kg/m^3) to 80% (1703 kg/m^3) to determine the sensitivity of the model response to relative density.

In centrifuge testing, the cone depth to diameter ratio is significant and hence a standard method for calculating the depth of cone penetration is important. Consistent with industry standards, the depth of zero penetration is adjusted to the 2/3 height of the cone tip, which is illustrated in Figure 6. In Figure 7 the recorded qc is given for depths of 1.5, 2, 2.5, and 3 m respectively versus the reported density of the model after the initial spin up. The solid lines in Figure 7 represent a linear regression, and the area between the dashed lines is the 95% confidence interval for the linear mean fit.

In Figure 8, the theoretical steady-state tip resistance, calculated from Idriss & Boulanger (2008), is compared with the q$_c$ data at 1.5, 2, 2.5, and 3 m depths.

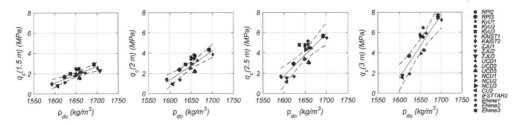

Figure 7. q$_c$ vs reported dry density at and confidence intervals for 1.5, 2, 2.5 and 3 m depths for LEAP-UCD-2017 experiments prior to destructive shaking.

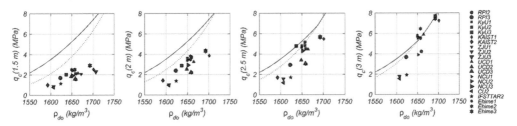

Figure 8. q$_c$ vs reported dry density at and theoretical steady-state tip resistance for 1.5, 2, 2.5 and 3 m depths for LEAP-UCD-2017 experiments prior to shaking.

The dashed line assumes the overburden correction factor (C_n) is unbounded, and the solid line assumes the correction factor is capped at 1.7. The C_{dq} factor for the curves is 0.65. At shallower depths, the theoretical cone tip resistance exceeds the recorded data significantly. One possible reason for this discrepancy is the lateral flexibility and limited distance to the walls of the model container. Another explanation could be that the model cones, being 120 to 250 mm in diameter prototype scale (depending on the g-level used in the LEAP experiments), were not sufficiently deep relative to their diameter to assume a deep failure mechanism.

6 CONCLUSION

A new, low cost, CPT device was developed. The device consists of an outer sleeve and an inner rod that transmitted cone tip forces to a load cell. A series of calibration tests were conducted, showing the device performed as expected. A direct comparison experiment was performed with the device at the University of Cambridge's Schofield Centre. Good agreement was observed with the Cambridge CPT and the LEAP-UCD-2017 CPT.

LEAP-UCD-2017 CPT's were used in similar models on eight different centrifuges. CPT results from all the LEAP facilities, especially at shallow depths, deviate from trends observed in large calibration chamber tests at 1 g. For tests at 3 m depth (Fig. 8(d)), agreement with correlations from calibration chamber tests is improved.

The use of a standard CPT on different centrifuges and standard methods of interpreting the depth of penetration are valuable, especially for comparing results of centrifuge tests performed at different facilities. It might be the case that tip resistance measurements using a standardised centrifuge CPT are more reliable for soil characterisation than direct density calculations based on mass and volume measurements.

ACKNOWLEDGMENTS

Funding for this work was provided by the National Science Foundation under the following CMMI grants: 1635307, 1635524 and 1635040, and the National Science Foundation of China (Nos. 51578501, 51778573). Thank you to Kate Darby for providing data during the initial design process. Design suggestions from professor Jason DeJong were very helpful. Barry Zheng's help during data processing was much appreciated. The authors would also like to thank Andy Cobb at UCD's BAE shop for providing construction recommendations and manufacturing each device.

REFERENCES

Bolton, M.D., Gui, M.W., Garnier, J., Corte, J.F., Bagge, G., Laue, J. & Renzi, R. 1999. Centrifuge cone penetration tests in sand. *Géotechnique*, 49(4): 543–552.

Darby, K.M., Bronner, J.D., Parra Bastidas, A.M., Boulanger, R.W. & DeJong, J.T., 2016. Effect of Shaking History on the Cone Penetration Resistance and Cyclic Strength of Saturated Sand. In *Geotechnical and Structural Engineering Congress 2016*:1460–1471.

Idriss, I.M. & Boulanger, R.W. 2008. *Soil liquefaction during earthquakes*. Earthquake Engineering Research Institute.

Jamiolkowski, M., Ladd, C.C., Germain, J.T. & Lancellotta, R. 1985. New development in field and laboratory testing of soils. In *Proc. 11th ICSMFE*. 1: 57–153

Kim, J.H., Kim, S.R., Lee, H.Y., Choo, Y.W., Kim, D.S. & Kim, D.J. 2014. Miniature cone tip resistance on silty sand in centrifuge model tests. In *ICPMG2014–Physical Modelling in Geotechnics: Proceedings of the 8th International Conference on Physical Modelling in Geotechnics* 2: 1301–1306. CRC Press, Boca Raton, FL, USA.

Kulhawy, F.H. & Mayne, P.H. 1990. *Manual on Estimating Soil Properties for Foundation Design,* Electric Power Research Institute, EPRI.

Kutter, B.L., Carey, T.J., Hashimoto, T., Zeghal, M., Abdoun, Madabhushi, S.P.G., Haigh, S.K., Burali d'Arezzo, F., Madabhushi, S.S.C., Hung, W.-Y., Lee, C.-J., Cheng, H.-C., Iai, S., Tobita, T., Ashino, T., Ren, J., Zhou, Y.-G., Chen, Y., Sun, Z.-B., & Manzari, M.T. 2016. LEAP-GWU-2015 Experiment Specifications, Results and Comparisons. *Soil Dynamics and Earthquake Engineering*.

LeBlanc, C. & Randolph, M.F. 2008. Interpretation of Piezocones in Silt, using Cavity Expansion and Critical State Methods, *12th International Conference of International Associate for Computer Method and Advances in Geomechanics*, 1–6 October, Goa, India, 822–829.

Manzari, M., B. Kutter, M. Zeghal, S. Iai, T. Tobita, S. Madabhushi, S.K. Haigh, L. Mejia, D. Gutierrez, R. Armstrong, M. Sharp, Y. Chen, & Y. Zhou. 2014. Leap projects: concept and challenges. In *Geotechnics for catastrophic flooding events*: 109–116. CRC Press.

Robertson, P.K. & Cabal K.L. 2010 Guide to cone penetration testing for geotechnical engineering. *Gregg Drilling and Testing Inc., USA*.

Zhou, Y.G., Liang, T., Chen, Y.M., Ling, D.S., Kong, L.G., Shamoto, Y. & Ishikawa, A. 2014, A two-dimensional miniature cone penetration test system for centrifuge modelling. In *ICPMG2014–Physical Modelling in Geotechnics: Proceedings of the 8th International Conference on Physical Modelling in Geotechnics* 2: 301–307. CRC Press, Boca Raton, FL, USA.

Shear wave velocity measurement in a large geotechnical laminar box using bender elements

J. Colletti, A. Tessari & K. Sett
University at Buffalo, Buffalo, New York, USA

W. Hoffman & J. Coleman
Idaho National Laboratories, Idaho Falls, Idaho, USA

ABSTRACT: Bender elements have been used by geotechnical engineers for decades in triaxial, resonant column, direct/simple shear, and centrifuge experiments. The primary use of a bender element (BE) transducer system is to generate and measure the propagation of elastic waves in soil to determine the compression and shear-wave velocities. These are valuable mechanical properties and may be correlated to other geotechnical indices as well as provide the initial small-strain value when determining modulus degradation curves. Bender elements have been implemented on a large field-type scale in the Geotechnical Laminar Box (GLB) at the University at Buffalo. Full-2D and 3D arrays of bender elements were used in the GLB to map the changing state of a soil model throughout two full-scale dynamic experimental programs. This paper will describe the equipment, model setup, methodologies, and a data set. We will also present a brief analysis of the results of the GLB-BE system as well as outline their potential use in uncertainty quantification and in future projects.

1 INTRODUCTION

Some of the cornerstone papers in geotechnical bender element (BE) hardware implementation include Shirley (1978), Dyvik and Madshus (1985), Rammah et al. (2006), and Brandenberg et al. (2006). These papers, and many others, have proven that bender elements are effective in traditional geotechnical testing apparatuses. Using bender elements in the University at Buffalo's (UB) Geotechnical Laminar Box (GLB) is the next phase in applying the capabilities of this technology to facilitate the quantification of uncertainty in the state of an experimental model, which will ultimately lead to better understanding of the fundamentals of large and full-scale geotechnical dynamic testing.

Large and full-scale geotechnical shaking is a useful testing methodology as it bypasses scaling issues, often seen at the centrifuge level, and boundary condition issues, typical of element-level laboratory testing, through the construction and testing of models at or close to field conditions. Given this, large-scale shaking can be used in conjunction with smaller-scale testing to maximize the financial and temporal efficiencies in reduced-scale testing with the realistic data sets produced at full-scale. Full-scale shake testing at UB is also proving to be a powerful tool when implementing real-time hybrid soil-foundation-superstructure interaction experimentation (Stefanaki et al. 2015: Colletti et al. 2017).

The primary benefit of using bender elements in the GLB is that they are non-destructive by design. The current method of using Seismic Cone Penetration Testing (SCPT) in the GLB is time consuming to operate and physically alters the 3D state of the soil. Further, since this method only provides a single vertical array of arrival times, the data is limited to a small zone in a large container. Uncertainty is not appropriately accounted for using this method of testing. Another method, currently in development at UB, is to freeze and extract undisturbed sand specimens such that soil microstructures can be correlated to macro-GLB behaviour. However, the same problems arise as with the SCPT method, as this type of testing is destructive and typically limited in use towards the end of an experimental plan.

To the authors' knowledge, the only reference of using bender elements in a large-scale geotechnical shaking test is Kobayashi et al. (2002), which examined the dynamic response of a bridge foundation. The exact setup and methodology of using bender elements are not described in detail, as it is not the focus of that particular paper.

This paper seeks to describe the physical components, electrical hardware, and software implementations necessary to rapidly obtain clear mechanical wave velocities in a large testing apparatus, such that it could be replicated in any laboratory or facility provided adequate resources.

2 OVERVIEW OF LAMINAR BOX AND BENDER ELEMENT SETUP

2.1 UB Laminar Box

The University at Buffalo (UB) Geotechnical Laminar Box (GLB) is a 2-D shaking laminar box apparatus, currently constrained to 1-D, allowing the construction and testing of large-scale geotechnical experiments. The GLB is used to test the dynamic response of soil-foundation models. The interior dimensions of the GLB are 6-m tall, 2.75-m wide and 5-m long (direction of shaking). To account for freeboard in the hydraulic filling method, the box may be filled with soil up to a height of 5 meters.

The initial testing height yields a soil-column volume of approximately 68 m^3. With this being the case, and given a specific gravity and minimum void ratio of the GLB material (Thevanayagam et al. 2002), the laminar box can contain up to 140 metric tons of saturated material in a typical test. The GLB is composed of a hydraulically actuated base and 40 laminate rings. The base rolls on ball bearings and the laminates are separated by ball bearings that do not constrain horizontal motion of the soil layers. The thickness of the shearing layers, i.e., height of one laminate, is 152 mm.

Ottawa F55 sand is placed into the GLB, by means of a dredging pump submerged in storage containers, via a hydraulic slurry process. The sand placement in the GLB is conducted with an extra meter of water, called freeboard, and is placed similar to natural alluvial deposition of sands in rivers, lakes, or man-made port islands (Thevanayagam et al. 2009). Since the material is placed hydraulically, it is generally loose, i.e. near the maximum void ratio, and weak prior to any base motions. The working volume is lined with a custom rubber bladder. This rubber liner allows for deformation while containing the soil material inside the GLB by preventing spillage of soil through the gaps between laminates.

Other in-situ characterization methods include the SCPT, embedded ShapeAccelArray (SAA) digital inclinometers, and pore-water pressure piezometers (PWPs).

2.2 Bender element setup

Bender elements were placed in strategic locations in the GLB to determine the changing state of the model over two different test plans. The goal was to create spatial arrays of transmitting actuators and receiving elements. Given this array, it is possible to infer, with a tomogram, the shear-wave velocity profile in the GLB. Investigation into the bender element idea was conducted in parallel with two-recent GLB projects, providing an opportunity implement and subsequently improve the BE system.

Geotechnical Laminar Box Test Plan 1 (TP1) consisted of a fully saturated soil-foundation model subjected to over 150 highly-plastic motions. The BE setup of TP1 was one of the first of its kind and it included 22 bender element transducers embedded

Figure 1. Photo of BEs in GLB prior to TP2.

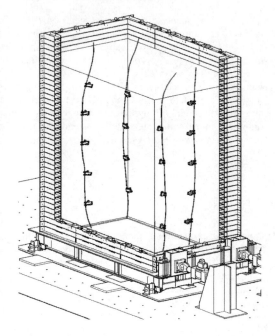

Figure 2. Section view of GLB and locations of BEs for TP2.

in the GLB soil column. They were placed in a 2-D arrangement, producing data primarily in the direction of shaking. The obtained data was clear and provided for a meaningful assessment of the state of the soil. For an initial treatment of the TP1 GLB-BE system see Colletti (2016).

A second experiment, Test Plan 2 (TP2) was conducted, in part to validate software developments, via a parametric study of over 140 nearly elastic, partially saturated, free-field motions. The TP2-BE setup consisted of an improved layout with 54 bender elements placed in a grid-like pattern in the GLB, as seen in Figures 1 and 2. The TP2-BE assemblies were designed with three orthogonal axes at both the source and receiving locations to allow for full 3-D compression and shear-wave component data sets. Figure 3 shows a sketch of the bender elements for both TP1 (bottom,

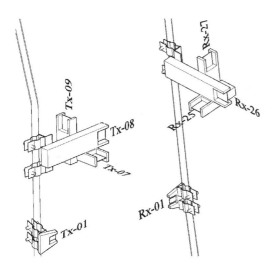

Figure 3. Rendering of bender element transmitter (Tx) and receiver (Rx) assemblies on SAAs.

Tx-01 and Rx-02) and TP2 (top, Tx-07, 08, 09 and Rx-25, 26 and 27).

During static steady-state, i.e. not during a GLB base motion and well after any input as determined by pore-pressure readings, transmitting bender elements were used to induce small-elastic body waves in the GLB soil column. This body wave moves out from the source location and triggers a receiving BE. The time-voltage history of the receiving bender elements gives valuable data on the state of the GLB model. Parameters of interest are compression and shear wave arrival times as well as the frequency content of the received signal. Further, with proper calibration of the bender elements, i.e. voltage-displacement correlation, these chips become active geophones for understanding in-situ dynamic characteristics, e.g. damping of the soil model during a GLB base motion. Future endeavours will attempt to combine all of this data to better monitor GLB testing and reduce uncertainty in material parameters.

3 DESIGN OF UB GLB-BE SYSTEM

3.1 Position via SAAs

ShapeAccelArrays have been used before to monitor the dynamic response of UB GLB models (Thevanayagam et al., 2009, Bennett et al., 2009). For both TP1 and TP2, the SAAs were used to keep track of the bender element and piezometer locations, as well as to measure soil response during GLB base motions.

The time-voltage histories of the BEs have little value without knowing, with good confidence, the position of the transmitted and received BE signals. By rigidly fixing the BE assemblies to the SAAs, the positions and orientations of the element tips are known at all times. This method of sensor location is not without limitation, as the SAA cannot register a rotation about an acceleration vector. However, the positions of nodes above the soil were verified during testing. This may become an issue in future testing programmes.

The SAA tips are fixed to known global coordinates at the bottom of the GLB. The SAAs and the bender elements are hung freely and cast in place as the GLB is hydraulically filled with the F55 sand. After filling has completed, this instrumentation is released from support and the SAAs and BEs are able to freely deform with the soil throughout the test plan.

3.2 BE assemblies

The transmitting and receiving transducer hardware was custom made at UB for use in the GLB. The basic design, for a BE transducer, is a piezoceramic chip fixed within a protective aluminium housing. The purpose of the housing is to protect the elements during the hydraulic filling process and to prevent damage to the chip during transport, installation, and storage. The aluminium housing is bolted near a node on an SAA so the location of the piezoceramic chip (BE) is known at all times.

The SAA data provides complete global locations and orientations of the bender elements. Tx and Rx are used here to abbreviate, respectively, transmitter and receiver. The bender elements are clustered in groups of three, shown in Figure 3, and may record shear waves, compression waves, or a combination of both depending on the orientation of the transmitting element. For example, in the cluster pair shown in the figure, the wave energy received via the transmitter-receiver pairs Tx-07/Rx-25 and Tx-08/Rx-27 is primarily shear while Tx-09/Rx-26 measures compression waves.

The chips are purchased from a third-party vendor and function under the typical physics of bender elements are explained, for example, in Shirley (1978), Dyvik and Madshus (1985), Rammah et al. (2006), and Brandenberg et al. (2006). The bender element chips are a composite of a brass-reinforced centre layer and two piezoceramic layers, all of which is manufactured with a protective nickel layer. Depending on the manufactured polarity of a chip, it will function as a transmitter or receiver requiring slightly different soldering procedures. All but two transducers in TP1 were of a smaller '303' size, all of the transducers in TP2 were of a larger '503' size, where the specifications the piezoceramic chips are shown in Table 1.

Signal conditioners (for TP2) and wiring are soldered to the BE chips, the chips and electronics are electrically potted in the aluminium casings and the wire is chased, out of the box, to a data-acquisition computer. Photos of the custom-made BE assemblies are shown in Figure 4.

The proof-of-concept design for TP1 consisted of smaller 303 chips individually located about a 2-D plane in the GLB. The 2-D setup included 12 transmitters and 10 receivers providing a possible 120 BE measurements per sample set. The chips used during TP2 were the larger 503 size and required a physically

Table 1. Dimensions and quantity of BE chips used in TP1 and TP2.

Test	Length mm	Width mm	Thickness mm	QTY	Type	Size
TP1	31.8	12.7	0.66	10	Trans.	303
TP1	31.8	12.7	0.66	10	Recvr.	303
TP1	63.5	31.8	0.66	2	Trans.	503
TP2	63.5	31.8	0.66	30	Recvr.	503
TP2	63.5	31.8	0.66	24	Trans.	503

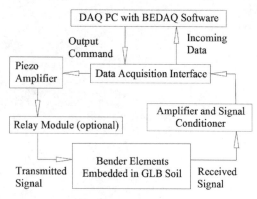

Figure 5. BE data acquisition equipment flow chart.

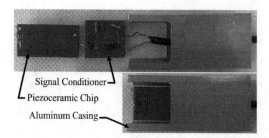

Figure 4. BE assembly circuitry (above). BE assembly epoxied and ready for service (below).

larger aluminium housing. The larger chips were found to have better transmission and reception capabilities in TP1 and therefore used exclusively in TP2. The 3-D setup of TP2 included 24 transmitters and 30 receivers providing a maximum of 720 conventional BE measurements.

The cantilevered length of chip that was in contact with the soil was 22.86 mm for TP1 and 38.10 mm for TP2. The width of a 503 chip is 2.5-times that of a 303 chip, thereby providing over four times the exposed cantilever area.

3.3 Data acquisition setup

Control of the bender elements is accomplished by a computer station at the top of the GLB and a visualization of the control hardware and software is presented as a conceptual flow chart in Figure 5.

Custom-made software, called BEDAQ, runs off a host 'DAQ PC' and controls the output signals while acquiring the received bender element signals. Transmitted signals are typically of sine shape and are fast stacked to average out random noise. Transmitted signals are piezo-amplified, as needed, to ensure that the body waves are perceivable at the receiving location. A relay module can be put in line to automatically cycle through all possible BE combinations.

Signal conditioning proved to be important while using the BEs in the GLB. During the TP1 project, received signals were amplified and conditioned individually at the control station with a stand-alone unit. There were limitations to this approach, primarily due to the long length of the cable between the data acquisition system and the bender element chip itself

acting as an antenna for noise. Given the low current produced by the receiving elements, the TP2 setup included amplification and conditioning at the source of measurement to increase signal quality. This was accomplished by custom building high-pass op-amp filter circuits and casting them in place with the receiving piezoceramic chips within the aluminium housing.

Amplifying a bender element is a controversial topic, as it may yield phase shifting and produce error in subsequent analyses. The operational amplifiers used in the amplification of the GLB bender elements were of significantly high slew-rate (>50 V/us), rapid settling time (<450 ns), gain-bandwidth product (>16 MHz), and had a total harmonic distortion with noise less than 0.00003% in the operating range. Further, to measure the effect of the amplifier and the system itself, bender elements were tested tip-to-tip with and without the amplifier. The total effect of the amplifier and filter is negligible in the design range for typical operation in the GLB. Those seeking to implement an amplifier should identify the frequency range of interest and select the amplifier and/or filter criteria accordingly.

4 BRIEF ANALYSIS OF RESULTS

4.1 Typical data sets and current analyses

Currently the analysis of data has been mostly geared towards looking at the shear-wave profiles through the duration of the test plans. This involves pairing SAA data with BE time-voltage data, interpreting received signals, and in verifying the results.

Full execution of TP1 resulted in extreme model consolidation that yielded a 533-mm settlement of the soil surface. This represents a roughly 10.5-percent decrease in the overall volume of the saturated GLB soil column. Naturally, this would imply that the soil has densified. The BE data from TP1 was used to monitor this densification throughout the project. Preliminary tomograms were generated and can be found in Colletti (2016).

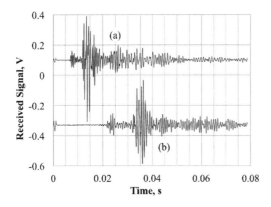

Figure 6. Typical shear wave data for TP2.

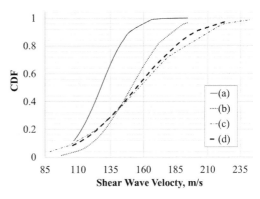

Figure 8. CDF of SWVs as function of layering.

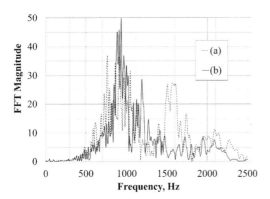

Figure 7. Frequency content of typical TP2 BE signals.

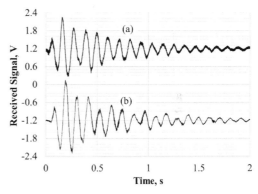

Figure 9. Signals acquired during a TP2 GLB base motion.

Typical BE measurements, for TP2, can be seen in Figure 6. Signal 6a is a shear wave measured along the short width of the box while signal 6b was measured along the longer direction of the GLB at about mid-height of the soil column. Signal interpretation, as proposed by Lee and Santamarina (2005), was used to determine arrival times of the UB GLB-BE system data.

Any biased choice of the arrival times will skew the shear-wave velocity profile. For this sample, using this basic approach, the shear-wave velocity along these two wave paths were measured as, respectively, 136 and 138 m/s.

The frequency analyses for signals 6a and 6b are shown Figure 7. These FFT plots simply verify that the BE system is working within reasonable accuracy because, for these signals, the transmitting frequency from the source bender element was set to 900 Hz.

The data was portioned into four layers and a representative shear wave velocity, measured during the experimental testing, of each layer was used as input parameters in a software validation exercise. The cumulative density function (CDF) of the four layers is shown in Figure 8. The lower velocities of curve 8a represent the shallowest layer of the GLB while curves 8b-c represent deeper layers.

4.2 Interpretation of dynamic data

During TP2, the BE system was also used to acquire dynamic data during GLB shaking. The signals shown in Figure 9 shown the voltage-time history of two receiver elements. Signal 9a represents in-situ soil motion in the direction of the moving laminates. Signal 9b represents soil motion in the direction of the based-induced vertically propagating shear wave. Using a simplified approach (Chopra, 2015), in-situ damping was calculated as 4 and 4.5% for curves 9a and 9b, respectively. Pending calibrations of the BEs will give further insight into the dynamic response of the GLB model motions as well as other elastic and plastic wave measurements.

5 CONCLUSIONS

This paper described the equipment, setup, methodologies, and a data set of the bender element system used in the large geotechnical laminar box at the University at Buffalo. A brief analysis of the results was presented and further analysis of the data is underway.

The goal of the first test and design was to initially implement a large-scale bender element system and determine feasibility. The second test and design improved the system by providing 3D data with a higher signal-to-noise ratio.

Future projects will look in to tightening the spread of the data. Additionally, full-waveform inversion techniques are pending in an effort to best characterize the laminar box soil models and changing boundary conditions. The authors envision that this system will be used to quantify uncertainty in the state of the model at any point during testing, yielding significant data for numerical analyses and in the explanation of experimental results.

REFERENCES

Bennett, V., Abdoun, T., Shantz, T., Jang, D. & Thevanayagam, S. 2009. Design and characterization of a compact array of MEMS accelerometers for geotechnical instrumentation. *Smart Structures and Systems*, 5, 663–79.

Brandenberg, S. J., Choi, S., Kutter, B. L., Wilson, D. W. & Santamarina, J. C. 2006. A bender element system for measuring shear wave velocities in centrifuge models. *6th International Conference on Physical Modelling in Geotechnics*. Hong Kong.

Chopra, A. K. 2015. *Dynamics of structures: theory and applications to earthquake engineering*, Upper Saddle River, NJ, Prentice Hall.

Colletti, J., Panthangi, S., Stefanaki, A., Tessari, A., Sivaselvan, M. & Whittaker, A. 2017. Large-Scale Hybrid Simulation of Soil-Foundation Structure-Interaction in a Geotechnical Laminar Box. *24th Conference on Structural Mechanics in Reactor Technology*. Busan.

Colletti, J. A. 2016. Analysis of Shear-Wave Velocities and Large-Scale Shaking Using Bender Elements. *CSEE*. Buffalo, NY, University at Buffalo.

Dyvik, R. & Madshus, C. 1985. Lab measurements of Gmax using bender elements. *Advances in the Art of Testing Soils Under Cyclic Conditions*. Detroit, ASCE.

Kobayashi, H., Tamura, K. & Tanimoto, S. 2002. Hybrid vibration experiments with a bridge foundation system model. *Soil Dynamics and Earthquake Engineering*, 22, 1135–1141.

Lee, J.-S. & Santamarina, J. C. 2005. Bender elements: Performance and signal interpretation. *Journal of Geotechnical and Geoenvironmental Engineering*, 131, 1063–1070.

Rammah, K. I., Ismail, M. A., and Fahey, M., 2006. Development of a Centrifuge Seismic Tomography System at UWA. *6th International Conference on Physical Modelling in Geotechnics*. Hong Kong.

Shirley, D. J. 1978. An improved shear wave transducer. *Journal of the Acoustical Society of America*, 63, 1643–5.

Stefanaki, A., Sivaselvan, M. V., Tessari, A. & Whittaker, A. 2015. Soil-Foundation-Structure Interaction Investigations using Hybrid Simulation. *23rd Conference on Structural Mechanics in Reactor Technology*. Manchester.

Thevanayagam, S., Kanagalingam, T., Reinhorn, A., Tharmendhira, R., Dobry, R., Pitman, M., Abdoun, T., Elgamal, A., Zeghal, M., Ecemis, N. & El Shamy, U. 2009. Laminar box system for 1-g physical modeling of liquefaction and lateral spreading. *Geotechnical Testing Journal*, 32, 438–449.

Thevanayagam, S., Shenthan, T., Mohan, S. & Liang, J. 2002. Undrained fragility of clean sands, silty sands, and sandy silts. *Journal of Geotechnical and Geoenvironmental Engineering*, 128, 849–859.

Low cost tensiometers for geotechnical applications

S.W. Jacobsz
University of Pretoria, Pretoria, South Africa

ABSTRACT: The paper describes a tensiometer that can be manufactured in most workshops and saturated in geotechnical laboratories at a cost of approximately US$27 (components only). Suctions of up to 500 kPa are routinely measured, with scope to increase the measuring range. The instruments are produced by fitting a porous ceramic filter of the desired air entry value to a low cost commercially available pressure sensor and sealing the assembly with an epoxy resin. The sensors are suitable for determining soil water retention curves using the drying method. They are also suitable for measuring positive pressures.

1 INTRODUCTION

Near-surface soils in many parts of the world occur in an unsaturated state. In order to model the behaviour of such soils, it is often necessary to apply principles of unsaturated soil mechanics, as indiscriminate application of saturated soil mechanics may be inappropriate. A crucial aspect controlling unsaturated soils behaviour is the pore water or matric suction. Matric suction contributes very significantly to the effective stress and hence strength of unsaturated soil as described by the extended Mohr-Coulomb failure criterion (Fredlund et al., 1978) or Bishop's effective stress parameter (Bishop, 1959). Matric suctions beyond 100 kPa has traditionally been difficult to measure reliably but suctions as large as 2 MPa are now routinely measured using high capacity tensiometers (Toll et al., 2013). Such tensiometers are however expensive, a factor that limit their use in laboratory and field applications.

This paper describes low cost miniature tensiometers developed at the University of Pretoria for routine use in the geotechnical laboratory. The tensiometers are of an informal construction, made using commercial pressure sensors, fitted with porous ceramic filters of the desire air entry value. The assembled sensors are sealed with a structural epoxy resin. Component cost per tensiometer amounts to approximately US$27 (based on 2017 prices). The measurement range of the tensiometer can be varied by choosing pressure sensors of the desired range and porous ceramics of the required air entry value. The sensors are easily assembled, even by undergraduate students, and are relatively simple to saturate. Suctions in excess of 500 kPa are routinely measured. Potential exists to extend this range significantly by using porous ceramics of higher air entry values and pressure sensors with matching ranges, although sensor saturation does become more difficult.

The tensiometers are typically used for the measurement of soil water retention curves using the drying-out-while-weighing method by Toker et al. (2004), for measuring pore pressures in physical models both at normal gravity and in the centrifuge and for teaching purposes. The sensors are also very well suited to measuring positive pore pressures.

This paper describes the tensiometer, its calibration and performance and presents results from suctions measurement in some geotechnical laboratory applications.

2 TENSIOMETER CONSTRUCTION

2.1 *Pressure sensor*

The tensiometers described in this papers are constructed using the MS54XX series of sensors from Measurement Specialties™. These are surface mountable miniature absolute pressure sensors and are available with full scale ranges of 1 bar, 7 bar, 12 bar and 70 bar respectively. The pressure sensors are small, measuring 6.2 × 6.4 mm (see Figure 1) and cheap, retailing for approximately US$20 (Nov 2016). The sensing element consists of a micro-machined silicon membrane with borosilicate glass wafer bonded under vacuum to the back side for reference pressure (Measurement Specialties, 2012). Pressure applied to the membrane is registered using implanted resistors operating by means of the piezo-resistive effect. The MS54XX sensors are typically used for altitude measurement and the measurement of atmospheric pressure in equipment ranging from the medical, industrial, consumer electronics and automotive industries. Typical applications include diving watches, barometers, electronic scales etc (Measurement Specialties, 2012).

The sensors operate over a temperature range of −40°C to 125°C. The active sensing element is housed

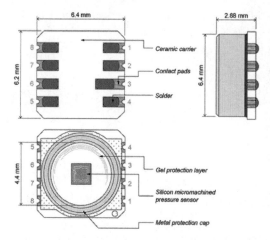

Figure 1. The MS54XX miniature pressure sensor (Measurement Specialties, 2012).

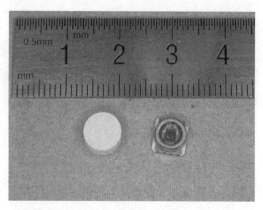

Figure 2. The ceramic disc and MS54XX miniature pressure sensor.

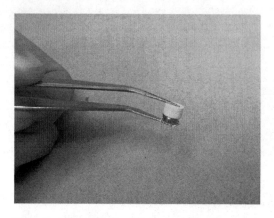

Figure 3. The ceramic disc glued to the MS54XX miniature pressure sensor.

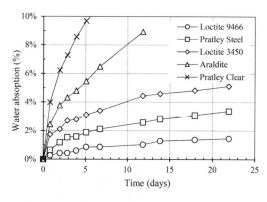

Figure 4. Water absorption ratios for various types of epoxy resins.

in a metal ring of anticorrosive alloy and is covered in silicone gel to protect against humidity and water while still allowing the sensor to register hydrostatic pressure. The sensor is reported to be stable in sea water (Measurement Specialties, 2012).

The sensors have a high output and support a direct current excitation voltage up to a maximum of 20 V. For most of the work described in this paper, sensors with a pressure range of 7 bar were used. The sensors were monitored using a Datataker DT80G data logger which provides current excitation of 2.5 mA, implying an excitation voltage of 8.25 V given the sensor's bridge resistance of approximately 3.3 kΩ.

2.2 Porous ceramic discs

Ceramic discs of 7 mm diameter and a thickness 4 mm were cut from larger discs to produce the tensiometers (see Figure 2). The diameter of the ceramics is such that it fits neatly over the alloy ring on the pressure sensor in which the active element of the sensor is mounted as shown in Figure 3.

The gap between the gel protection around the active sensing element and the attached ceramic disc is small and is estimated to have a volume of approximately 10 mm^3. This small volume suggests that the sensor may be suitable for measuring suctions, a requirement mentioned by Ridley & Burland (1993).

2.3 Structural epoxy resin

Experiments were carried out with a range of epoxy resins to find a product suitable for gluing the porous ceramic discs to the alloy ring of the sensor and to seal around the sensor. A product with a low water absorption was desired. Rough discs of several types of epoxy were formed on a glass plate and allowed to harden based on the respective manufacturers' recommendations. The discs were subsequently removed from the glass and immersed in water and weighed at regular intervals after drying with a paper towel to remove free water from the surface. Water absorption percentages over time are presented in Figure 4. The product with the best performance, i.e. the lowest water absorption ratio, was found to be Loctite Hysol 9466 which was subsequently used for the construction of the sensors.

A completed tensiometer, ready for saturation, is illustrated in Figure 5.

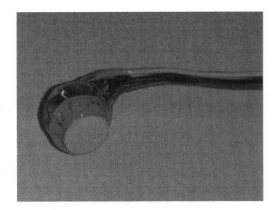

Figure 5. A completed tensiometer ready for saturation.

3 TENSIOMETER SATURATION PROCEDURE

The tensiometers were saturated in a triaxial cell fitted with a custom-built cell top with holes through which the tensiometer cables passed as shown in Figure 6. Sealing was accomplished by compressing an o-ring around the tensiometer cable using a bolt with a hole drilled through it to allow the cables to pass.

A saturation procedure as described by Take and Bolton (2003) was followed. Before saturation the tensiometers were oven dried at 60°C for at least 4 hours. After drying the tensiometer cables were threaded through the holes in the cell top and the sealing bolts were fastened. The cell top, with the tensiometers now fixed to it, were placed on the triaxial cell and the tensiometers were connected to a data logger. A vacuum was applied while monitoring the tensiometer response. Once the tensiometer readings had stabilized under vacuum, high quality deaired water, agitated under vacuum with a magnetic stirrer for at least 15 minutes, was allowed to fill the cell from the bottom, while keeping the vacuum pump running. The rising water level submerged the tensiometers and the vacuum pump was switched off just before the water level reached the top of the cell. After filling the cell with water, the air bleed was briefly opened until water emerged. This resulted in the cell pressure rising to the water head provided by the deaerator (about 1 m). The tensiometers, still registering vacuum, where then allowed to equilibrate to the pressure in the cell. Depending on the air entry value of the ceramic disc used, this could take from a few minutes to in excess of an hour (for 15 Bar ceramics). Tensiometers with ceramics that were not thoroughly pre-dried did not equilibrate under this low head.

Once the tensiometers had equilibrated to the cell pressure, the cell pressure was raised to approximately 700 kPa, the limit of the air pressure supply in the laboratory. The pressure was usually cycled from zero to 700 kPa in 100 kPa steps to test the tensiometer response. Initially the response was sluggish. Tensiometers were then typically left for 4 hours or more under a pressure of 700 kPa for trapped air to go into solution after which the pressure was removed

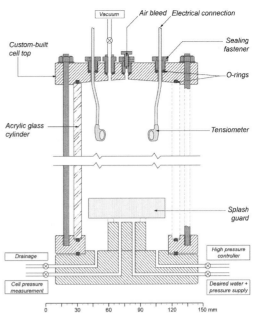

Figure 6. An illustration of the modified triaxial cell used for tensiometer saturation.

and a few pressure cycles repeated. The cell pressure was then cycled from zero to 100 kPa a number of times and the response assessed. A rapid response indicated satisfactory saturation. A slugging pore pressure response indicated that the pressurisation period had to be extended.

Due to the pressure available it was not possible to thoroughly saturated 15 Bar tensiometers. These tensiometers were typically only capable to measure suction of up to 600 kPa due to incomplete saturation.

4 TENSIOMETER CALIBRATION

Tensiometers were calibrated at positive pressure by raising the cell pressure in the triaxial cell while recording the sensors' response. The calibration curve was extrolated into the negative range and it was assumed to still be valid, a common practice (Toll et al., 2013). However, calibration in the negative range was checked by placing tensiometers surrounded in a low permeability clay in an oedometer. Known pressures were applied and complete consolidation was allowed to take place. The load was rapidly removed which resulted in a generation of a negative pore pressure theoretically equal in magnitude to the initially applied total stress because of perfectly undrained conditions. A similar approach was used by Ridley & Burland (1993), but in the triaxial apparatus.

The resulting calibration data is presented in Figure 7. It can be seen that the practice of extending the calibration curved from the positive range to the negative is reasonable, as only a slight deviation is evident for these tensiometers.

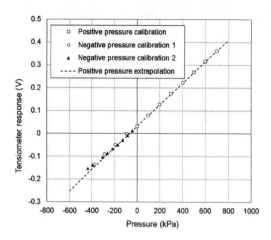

Figure 7. Tensiometer calibration data in the positive and negative ranges.

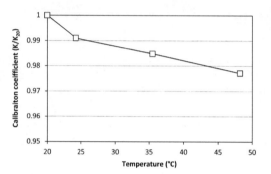

Figure 8. Effect of temperature on calibration data coefficient.

Temperature sensitivity was investigated by carrying out calibration cycles between zero and 700 kPa in the triaxial cell by filling it with water at different temperatures. Figure 8 illustrates that the calibration coefficient reduced in a somewhat non-linear fashion by approximately 2.5% as the temperature was increased from 20°C to 45°C. Given the fact that temperature in a laboratory environment is usually well controlled, the temperature sensitivity is considered satisfactory for practical suction measurement in most geotechnical applications.

It is important that piezometers and tensiometers should not be significantly affected by total stress changes. Total stress changes can have two potential effects. Type 1: Mechanical deformation of the sensors can potentially result in deformation of the sensing diaphragm, resulting in false pore pressure measurements. Type 2: Deformation of the sensor surround can result in compression of the water in the tensiometer reservoir, registering a positive pore pressure response which is the consequence of sensor compression and not a result of the process being monitored.

The MS54XX pressure sensor is illustrated in Figures 1 and 2. The sensing element is surrounded by

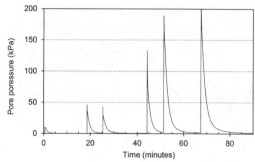

Figure 9. Pore pressure responses recorded in the oedometer to assess the sensor's total stress dependence.

a metal alloy ring and is covered by a layer of silicone gel. It is therefore well isolated from external mechanical strain. The total stress dependence of a tensiometer was examined by placing a saturated sensor in the oedometer in clay and loading the oedometer to a range of total stresses while checking that the measured pore pressure returns to a consistent zero value after consolidation. The result is presented in Figure 9. It can be seen that the registered pore pressure consistently returned to zero, indicating that deformation of the sensing element due to the applied total strain did not occur.

Due to the flexibility of the epoxy resin surround to seal around the sensor, the sensors are relatively compressible and will therefore register a positive pore pressure response upon compression. A more rigid metal surround will reduce this total stress dependence. Various schemes to improve sensor performance in this regard are presented by Take & Bolton (2003).

5 TENSIOMETER PERFORMANCE

5.1 Dry out tests

The capacity of a tensiometer to measure negative pore pressure can be assessed by allowing a thoroughly saturated tensiometer to dry in air while monitoring the registered pressure. As moisture evaporates from the ceramic disc, suctions are soon generated in the water reservoir inside the instrument. The suctions generated increase rapidly during drying until cavitation occurs or until the air entry value of the ceramic disc is reached. Figure 10 illustrates these two typical responses. In instances where cavitation occurs first, the registered suction drops rapidly to the cavitation pressure, the difference between the local atmospheric pressure and the liquid vapour pressure (Lu & Likos, 2004). The local atmospheric pressure depends on the elevation above sea level. The tests reported here were carried out in Pretoria at an elevation of approximately 1400 m. The atmospheric pressure measured at a local weather station was 86 kPa. The liquid vapour pressure is calculated from the relative humidity and the saturated vapour pressure, which at 25°C can have a

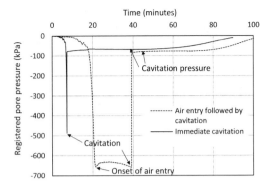

Figure 10. Tensiometer suction responses upon air drying.

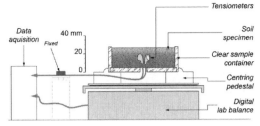

Figure 11. Two tensiometers installed in a soil sample mounted on a balance to determine the soil water retention curve.

maximum value of approximately 3 kPa, yielding a cavitation pressure of approximately 83 kPa. This correlates well with the measured cavitation pressures in Figure 10.

In instances where the air entry value of the ceramic disc is reached as water evaporates from the disc before cavitation occurs, the registered pressure will stabilise for some time at the air entry value. This is illustrated by the dotted curve in Figure 10. The initial erratic suction response of this tensiometer is through to be related to poor saturation, causing air voids trapped inside the ceramic to periodically expanding (resulting in a suction reduction) and then again becoming isolated by surface tension effects (once again allowing suction to increase) as drying progressed. As air was slowly drawn into the ceramic, the registered negative pore pressure slowly increased by another 100 kPa until cavitation occurred and the registered suction reduced to the cavitation pressure. Had air entry occurred into the tensiometer's water reservoir before cavitation, the registered pressure would have been expected to reduce to atmospheric pressure. This was not observed to occur with the tensiometers described here. As the tensiometers were allowed to dry further after cavitation, the registered pressure smoothly reduced over time to atmospheric pressure.

It was found that it was possible to resaturate tensiometers that had cavitated while they were still registering cavitation pressure by simply repressurising them for some time. However, once a tensiometer had reached atmospheric pressure, it was necessary to dry it thoroughly in an oven before attempting resaturation.

5.2 Measurement of soil water retention curves

Following the method by Toker et al. (2004), the tensiometers described here are routinely used for the determination of soil water retention curves of soil samples in the suction range from zero to approximately 500 kPa. (This range can be extended by using tensiometers fitted with ceramic discs of higher air entry values.) The method involves placing one or more tensiometers in the soil sample, which, in turn, is placed on a balance to allow continuous recording of the sample mass. This allows the gravimetric moisture content to be calculated at any time as the samples is allowed to dry out (see Fig. 11). The test is typically carried out from a state of complete saturation down to the moisture content at which the tensiometer cavitates. This test can be carried out rapidly compared to other methods for measuring soil water retention curves such as the pressure plate method.

Figure 12 illustrates soil moisture retention curves, plotting matric suction against gravimetric moisture content, for a gold tailings samples using three tensiometers and for a clayey soil sample using a single tensiometer. It can be seen that excellent correlation was obtained between the suction measurements from the three tensiometers in the same gold tailings sample up to the point where they cavitated. This indicates that matric suctions were uniformly distributed within the sample as it dried. For the clayey soil much greater suctions were registered during drying, reflecting the substantially finer particle size distribution. The method allows the retention curves to be observed in excellent detail over the low suction range.

The remainder of the soil water retention curves below cavitation moisture content can be measured using the filter paper method (see Hamblin, 1981 or Chandler & Gutierrez, 1986). To ensure consistency, these samples should be formed by drying out prepared samples to the required moisture content, similar to the way in which the tensiometer phases of the tests were carried out. It is good practice to allow some overlap between the two methods to allow comparison between the two sets of suctions measurements. Very good correlation between the two data sets is typically observed.

5.3 General

The tensiometers have a limited working life span but can generally be reused (allowed to cavitate, dry and resaturate) five or more times. It is believed that damage is caused to the silicone gel protecting the sensing element due to the pressure shock associated with cavitation. This limits the life of the pressure sensor. However, the low cost means that the sensors can be viewed as disposable and replaced.

The sensors have been tested for several week to record modest suctions (<100 kPa) satisfactory in a

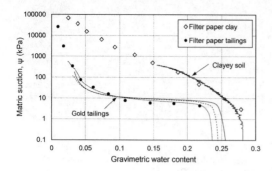

Figure 12. A soil water retention curve for gold tailings and a clayey soil measured using tensiometers and the filter paper method.

variety of granular materials. Long term performance under sustained suction exceeding 100 kPa has not been verified.

Due to the tendency of the epoxy resin to absorb a small amount of water over time, sensors should not be stored in water, but should be oven dried after use.

The tensiometers can be improved by designing a more robust surround to house the pressure sensor and porous ceramic disc. A more formal saturation system will be developed to suit the more robust tensiometer surround.

6 CONCLUSIONS

A low cost tensiometer that can be produced with relative ease in most workshops and saturated in most geotechnical laboratories was described. The tensiometer is capable of measuring suctions in excess of 500 kPa with potential to further improve the measuring range.

The sensors are suitable for measuring both positive and negative pressures. Calibration under positive pressure and extrapolation of the calibration curve into the negative range was found to be acceptable. Temperature and total stress dependence are acceptable for general laboratory applications. The low cost of the instrument places suction measurement within reach of emerging researchers and should prove advantageous to promote research in unsaturated soil behaviour.

ACKNOWLEDGEMENTS

The author wishes to thank Mr Paul le Roux who generated some of the graphics for this paper.

REFERENCES

Bishop, A.W. 1959. The principle of effective stress. *Teknisk Ukeblad I Samarbeide Med Teknikk*, Oslo, Norway, 106(39): 859–863.

Chander, R.J. and Gutierrez, C.I. 1986. The filter paper method of suction measurement. *Geotechnique* 36(2): 265–268.

Fredlund, D.G., Morgenstern, N.R. and Widger, R.A. 1978. Shear strength of unsaturated soils. *Canadian Geotechnical Journal* 15(3): 313–321.

Hamblin, A.P. 1981. Filter paper method for the routine measurement of field water potential. *Journal of Hydrology* 53: 355–360.

Lu, N. and Likos, W.J. 2004. *Unsaturated soil mechanics*. John Wiley & Sons.

Measurement Specialties. 2012. *MS54XX miniature SMD pressure sensor.* Product brochure DA54XX_020: 1–9.

Ridley, A.M. and Burland, J.B. 1993. A new instrument for the measurement of soil moisture suction. *Geotechnique* 43(2): 321–324.

Take, W.A. and Bolton, M.D. 2003. Tensiometer saturation and the reliable measurement of soil suction. *Geotechnique* 53 (2): 159–172.

Toker, N. K., Germaine, J. T., Sjoblom, K. J. & Culligan, P. J. 2004. A new technique for rapid measurement of continuous soil moisture characteristic curves. *Geotechnique* 54(3): 179–186.

Toll, D.G., Lourenço, S.D.N. and Mendes J. 2013. Advances in suction measurements using high suction tensiometers. *Engineering Geology* 165: 29–37.

A field model investigating pipeline leak detection using discrete fibre optic sensors

S.I. Jahnke, S.W. Jacobsz & E.P. Kearsley
University of Pretoria, Pretoria, South Africa

ABSTRACT: An estimated 26% of potable water is lost due to leaks from the water distribution network in the City of Tshwane (formerly Pretoria), the capital of South Africa. Leaks are often difficult to detect, with the consequence that they are not fixed. A need therefore exists for a detection system that would indicate a water leak. The potential exists to bury a fibre optic cable in the same trench with the pipe when constructing new pipelines. The paper investigates the use of fibre optic temperature and strain measurement as a means of leak detection on pipelines buried in an unsaturated soil profile with a deep water table. For the system to be successful it is necessary that leakage-induced temperature and strain changes must be distinguishable from naturally occurring changes. Results from tests using discrete Fibre Bragg Grating Sensors to evaluate the performance of such a system suggests that successful leak detection will be possible under the soil conditions tested.

1 INTRODUCTION

1.1 Background

The global problem of potable water losses during transmission was considered by investigating the use of temperature and strain measurements along a buried water pipeline to detect leaks. In the City of Tshwane in South Africa it is estimated that 26% of potable water can be classed as non-revenue water (NRW) (McKenzie et al. 2012). NRW can be classified in two categories, i.e. real losses, referred to as physical losses, and apparent losses, attributed to metering inaccuracies, theft and free basic water allocations to households. In older communities the percentage of NRW might be much higher as infrastructure is more likely to be in a state of disrepair (McKenzie et al. 2012). In an arid country such as South Africa with a fast growing population and a rapid trend to urbanisation, the loss in supply capacity from water leaks in major urban centres cannot be afforded.

NRW, especially physical losses due to leakage, does not only cause a financial loss to water suppliers and governing bodies, but also pose environmental and health risks. Contaminated surface water can, for example, enter a potable water pipeline through leaks during certain operational conditions when low pressures are present.

Another problem associated with water leaks from pipelines is that of sinkholes. Approximately 20% of the Gauteng province in South Africa is underlain by dolomite bedrock. Leaking water distribution networks can lead to the formation of sinkholes and subsidence in these areas. It has, in fact, been reported by Buttrick & Van Schalkwyk (1998) that 98.9% of all new sinkholes in the Tshwane area are triggered by leaking water pipes, either from the distribution network or from the waste water system. This poses a major risk to all types of infrastructure and inhabitants.

The major challenge faced by water authorities is the early detection of leaks before large amounts of water is lost. This is especially challenging when the leak does not surface, i.e. is not visible above ground. It has been reported that many leaks originating from potable water pipelines never surface and are often discharged via storm water systems, going undetected for long periods (McKenzie et al. 2012). A need therefore exists for leakage detection systems that can assist to identify leaks and enable remedial action to be taken before excessive amounts of water is lost.

In the last three decades fibre optic instrumentation has been developed to measure both temperature and strain with an unparalleled resolution, exceeding that of comparable conventional electro-mechanical sensors (Kreuzer, 2013). Two advantages of fibre optic sensors compared to conventional electro-mechanical sensors are the inert nature of the sensor as they are unaffected by electrical disturbances, and the long transmission lengths that are possible with fibre optic cables (National Instruments, 2013). Both discrete and distributed fibre optic sensors are available to measure strain, temperature and vibration.

The potential exists to use fibre optic cables extending along the length of pipelines as a means of leak detection. For this purpose distributed fibre optic sensing is preferred as strain and temperature can be measured continuously along the full length of a pipeline. Fibre optic cables are increasingly installed along major new pipelines for telemetry and pipeline performance data transmission. The potential exists to also utilise these cables as distributed leak detection sensors. Although distributed fibre optic readout units are costly, the cost may be offset by the savings that can be realised by early detection of water losses from water distribution pipelines. In addition, the cost is expected to reduce as the demand for the technology increases as it gains popularity.

The present study investigated the use of discrete fibre optic sensors to detect leaks in pipelines buried in unsaturated soil with a deep water table. Such conditions are typical in most Southern African cities. The use of discrete sensors for measuring temperature and strain limits the user to a finite number of monitoring locations but because of their excellent resolution, a detailed understanding of temperature and strain changes that typically occur in pipelines under normal operational conditions and when leaks occur could be gained. The study will be followed up by a study based on continuous strain measurement.

Discrete Fibre Bragg grating sensors (FBGS) comprise imperfections created in optic fibres, each with a unique wave length, at the desired spatial interval. This gives each Bragg sensor a unique identity in terms of a wavelength peak, typically in the range of 1500 nm to 1600 nm. The wavelength spacing is typically designed for either thermal or mechanical strain applications. For measuring pure thermal strain, the wavelength peaks are typically spaced at 1 nm, whereas for mechanical strain, the wavelength peaks are spaced at 5 nm, allowing for a greater strain range to be covered (Kreuzer, 2013).

A shift in the wavelength of light reflected from a particular sensor indicates a change in strain experienced by that sensor. This strain can be the result of mechanical or thermal strain or a combination of both. A compensation technique is required to account for changes in thermal strain when measuring mechanical strain. The thermal component can then be subtracted from the total strain component, leaving the mechanical strain. When pure thermal effects are investigated, the Bragg sensor needs to be strain-relieved and must not be affected by bending, tension, torsion or compression. In reality this is difficult to achieve (Kreuzer, 2013).

The success of a fibre optic leakage detection system depends on whether temperature and/or strain changes caused by a leak can be distinguished from natural temperature and strain changes in the ground and the temperature changes that naturally occur in water mains. It is hypothesised that by measuring temperature and strain profiles along the length of a pipeline at a specific time interval and comparing the readings to previous records, changes, possibly indicative of leaks, can be detected and appropriate and timeous remedial action taken. If this system proves successful, it is envisaged that new water distribution pipes can in future be fitted with fibre optic instrumentation which can be monitored periodically to detect leaks.

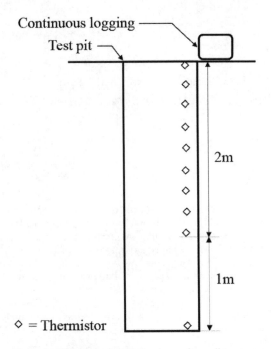

Figure 1. Thermistor installation to monitor soil temperature variation with depth.

2 EXPERIMENTAL SETUP

2.1 Ground temperature profiles and water mains temperature observation

Ground profile and water mains temperatures were investigated to study natural temperature cycles to understand the temperature variation that can be expected to occur in and around a water distribution system.

A set of thermistors were installed in the ground at an experimental facility of the University of Pretoria at intervals of 250 mm to a depth of 2 m, with another thermistor at a depth of 3 m. The excavation was backfilled with in-situ soil and compacted lightly with a tractor loader backhoe bucket. The ground conditions comprised of hillwash soil, consisting of slightly moist, red-brown, slightly clayey silty sands of medium dense to dense in situ consistency. The water table is deep and did not affect the site. The soils on site are unsaturated and the excavation required no support. The thermistor setup is shown in Figure 1. The thermistor chain was logged hourly to study ground temperature variation with depth over a two year period.

Water temperature in a water main was logged at a reservoir inlet chamber in a residential neighbourhood in Pretoria by means of a resistance temperature device (RTD) inserted into a 350 mm diameter live steel pipe. It was also logged at an hourly rate.

2.2 In-situ buried pipeline leakage experiment

A 12m long 110 mm diameter Class 9 un-plasticised polyvinyl chloride (uPVC) pipe was installed at the experimental facility of the University of Pretoria and connected to a municipal water connection. The pipe was bedded on a soil cradle of excavated material described above. Compaction around the pipe to 300 mm above the crown was done with hand tampers and mechanical tamping over the remainder of the trench to a density typically varying around 90% Mod AASHTO density.

The pipe was filled with municipal water to induce leaks at three points along the pipeline. The three discrete leak locations were spaced 2 m apart along the length of the pipe. The leak points comprised 4mm diameter polyurethane (PU) tubing tapped into the uPVC pipe, passing to the soil surface, through a ball valve to control the flowrate and to enable the leakage rate to be accurately measured, before returning into the ground to discharge immediately next to the pipe. This was necessary because it was desired to be able to leak water which has been in the pipe, and therefore at the same temperature as the surrounding soil, from the pipe into the ground. The ground surrounding the pipe was unsaturated before leaks were initiated. The temperature differential between stagnant water in the pipe and the surrounding soil can be expected to be less than that between circulating water and the ground. If a leak of stagnant water can be detected, the proposed detection system is likely to be successful as a leak on a pipeline with circulating water will be easier to detect because of the greater temperature differential.

Conventional thermistors and fibre optic Bragg sensors were used to monitor ground temperature and strain changes from daily and seasonal fluctuations, water flushing through the pipe and induced leaks. The thermistors were soldered to a multicore cable to form a chain of nine discrete measuring points at specific intervals laid close to the pipe as shown schematically in Figure 2. Three thermistor chains were installed at each of the three leak locations, one at the leak and one respectively 0.15 m upstream and downstream of the leak location. In total, 81 thermistors were laid around the pipeline. One thermistor was connected directly to the logger to record the ambient air temperature. Another thermistor was installed downstream of the water pipe where water leaves the system to monitor the temperature of water exiting the system.

The water network pressure and the flow rate entering the pipeline upstream of the installation was also monitored. A pressure transducer was used to monitor pressure within the pipeline. A gear-type Class C mechanical plastic flow meter was used to determine the volume of water entering the installation. The leakage rate was measured by disconnecting the tube after the valve and measuring the flow rate before joining the pipe again.

Sixteen FBGS, spaced 1 m apart along a single optic fibre cable, were used to instrument the pipe at discrete measuring locations. The FBGS had a 4 nm wavelength separation. Eight sensors were epoxied to the base along the length of the pipe. The fibre optic cable was looped around so that each strain measuring Bragg had a corresponding temperature measuring Bragg close to it. The section of the cable with the eight

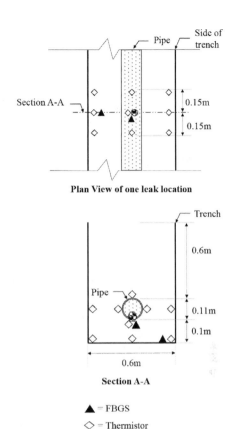

Figure 2. Instrumentation layout at a leak location around the experimental pipeline.

temperature measuring Braggs was threaded through in a 4 mm PU tube to isolate it from mechanical strain and was buried in the corner pipe trench, approximately 0.28 m from the pipe (see Figure 2). The reason for burying the cable in the corner of the trench was because that location was considered to offer the most protection to prevent cable damage on an actual pipeline project.

A HBM FS22DI BraggMeter fibre optic interrogator with a resolution of 5 pm and an absolute accuracy of ±10 pm (Kreuzer, 2013) was used to monitor the Braggs.

3 RESULTS AND DISCUSSION

3.1 Ground and water mains temperatures

The experimental observations commenced in January 2017. Figure 3 summarises the ground temperatures and daily rainfall data from January to August 2017. The left-hand axis represents ground temperatures at depths of 0.5 m, 1 m, 2 m and 3 m respectively. The right-hand axis gives daily rainfall depth measured over the same period. The reduction in ground temperature during the winter of 2017 is clearly evident. The figure illustrates that ground temperature changes

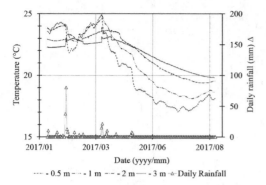

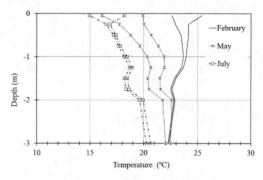

Figure 3. Soil temperature records at different depths compared to the rainfall record.

Figure 4. Monthly soil temperature with depth range.

reduced rapidly with depth. At a depth of 0.5 m a clear daily temperature variation can be observed when examining the temperature record more closely. The rapid reduction in temperature due to a major rainfall event can be observed in the month of February when a tropical cyclone caused significant rainfall.

Figure 4 presents a summary of the first and third quartile temperature values for each monitored depth for the months February (mid-summer), May (autumn) and July (mid-winter), giving an indication of temperature variation with depth. It is necessary that leakage-induced temperature changes must be distinguishable from these temperature variations for the proposed leakage detection system to be functional. The temperature variation at depth at mid-summer (February) and mid-winter was significantly smaller compared to the transition season (May) when temperature variation was at a maximum. The temperature stratification at a depth of 1.5 m and 1.75 m can probably be attributed to a change in backfill density and/or moisture content. Based on visual inspection the soil profile and the backfill soil appeared uniform. Slightly different compaction densities and backfill moisture contents might have resulted in slightly different thermally conductivities, possibly explaining the anomaly (Barry-Macaulay et al. 2013).

Figure 5 represents a brief period during which the temperature in water mains was observed. It was found that the water temperature followed the same trend as the daily ambient air temperature. However, the daily water temperature variation was less pronounced than the ambient air temperature.

The observed in-situ soil and water mains temperatures up to the end of August 2017 indicated that water temperature in the bulk supply pipeline was significantly different from the soil temperature at depths of 0.5 m and more. Figure 6 compares the first and third quartiles of the soil temperature with depth for July against the first and third quartiles of the temperatures observed in water mains. This comparison shows that, from a temperature differential perspective, there is potential for the proposed leakage detection system to be successful.

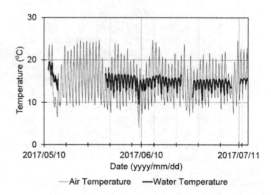

Figure 5. Air and water temperature at water mains.

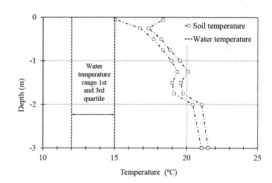

Figure 6. Water and soil temperature comparison for June 2017.

Water flowing through a pipe induces a temperature change in the soil surrounding the pipe, reducing the temperature differential between the pipeline water and the surrounding soil. This can potentially impact on the effectiveness of the proposed leak detection system requiring further investigation (Conway, 2010). However, it is proposed that the fibre optic cable measuring temperature near the pipeline be buried in the trench corner to minimise damage to the sensors during pipeline installation and repairs. This proposed location is a significant distance from the pipe and can therefore be assumed to be less affected by the thermal temperature gradient from the pipeline.

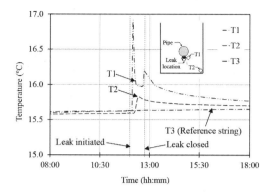

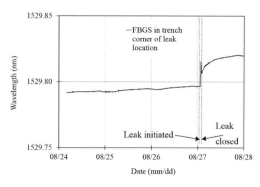

Figure 7. Temperature changes in the soil around the pipeline resulting from the induced leak (test date 27/08/2017).

Figure 8. FBGS wave length record at leak location in trench corner.

3.2 Results from pipeline leakage experiments

The installation of the in-situ experimental pipeline was carried out in July 2017. A period of two weeks was allowed for an initial settling in time. Thereafter, water from the municipal water connection was allowed to flow through the pipe and a leak test was conducted at one of the three leak locations. The temperature of the stagnant water within the buried pipe was lower than the water temperature entering the pipe from the municipal network, cooled down by the lower ground temperature in winter. As the municipal network connection was opened, the temperature of the water within the pipe gradually increased somewhat due to mixing of the water, but the test was largely carried out with stagnant water at a temperature close to that of the ground. If the smaller temperature differential can be detected by the measuring system, a larger temperature differential will be more readily detectable.

During the leak test the inlet flow rate from the municipal connection was 0.31 l/s and the internal pipe pressure approximately 180 kPa. The leak flow rate induced through a 4 mm PU pipe was 0.05 l/s. The balance of the flow (0.26 l/s) was allowed to drain from the pipe through the downstream valve.

Figure 7 presents the temperature recorded at the leak location in the soil. The thermistor in close contact with the pipe (T1) registered the most pronounced temperature spike with the temperature increasing from 15.5°C to 16.9°C. In the trench corner opposite the leak location (T2), the temperature change was less pronounced, only 0.2°C. A thermistor located 2 m upstream of the leak location (T3) in the pipe trench did not indicate any temperature changes and is shown as a control.

The shifts in FBGS wavelengths in close proximity to the leak location in the corner of the trench and the FBGS epoxied to the pipe were investigated. Wave lengths are reported instead of temperature because they were the result of both temperature-induced and mechanical strains. Figure 8 presents the wavelength record for the FBGS in the trench corner over a 3 day period leading up to the leak test, the duration of the leak test and 1 day after the leak test was completed. The figure indicates a rapid change in wavelength amounting to 0.024 nm at the time when the leak was imposed. This is equivalent to a change in temperature of 1°C, much more than the 0.2°C change in temperature experienced by the thermistor at the same location and indicates that the FBGS response also represented a strain effect. The strain effect in this case amounted to 24 $\mu\varepsilon$ and is considered to be the effect of wetting of the soil due to the initial unsaturated conditions around the optic fibre. This resulted in a change in the support condition around the optic fibre, causing the ground to undergo strain.

Figure 9 presents wavelength records from the FBGS epoxied to the pipe at the leak location and respectively 1 m downstream and upstream of the leak.

The initiation of the leak was detected by the FBGS upstream of the leak location which shows a reduction in strain due to a decrease in internal pipe pressure associated with opening of the valve to induce the leak. The end of the leak can also be detected by a corresponding increase in strain caused by the increase in the internal pipe pressure when the valve inducing the leak was closed. The pressure change was 196 kPa.

Initiation of the leak can also be detected from the strain record of the FBGS at the leak location, but it was less pronounced. This FBGS took some time to show the effect of the leak, but leak-induced strain eventually became clearly evident. These are indicative of a change in the support conditions around the pipe due to wetting of the soil, resulting in the pipe being strained.

The FBGS 1 m downstream also registered initiation of the leak. It exhibited more rapid strain than at the leak location itself and continued to strain even after the leak was closed and the connection to the water mains disconnected. The fact that it showed a more rapid response than the FBGS at the leak location suggest that the pipe was probably settling initially relatively uniformly at the leak location, causing the pipe to suffer flexural strains some distance away.

The fact that the behaviour of the downstream FBGS was not mirrored at the upstream FBGS is most likely attributable to variable support conditions around and

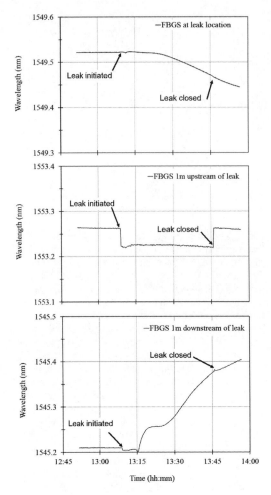

Figure 9. Response of FBGS epoxied to pipe surrounding the leak location (test date 27/08/2017).

underneath the pipe at the two locations. In practice it is impossible to provide perfectly uniform support conditions along the length of a pipeline and the extremely high resolution with which strains can be measured using the FBGS will clearly emphasis this fact.

The degree of bonding of the various FBGS to the pipe might have varied somewhat due to installation damage contributing to the variable responses of the various gauges to pressure fluctuation. Because the FBGS are sensitive to both strain and temperature, the leak could still be detected. This is testimony to the robustness of the system.

4 CONCLUSIONS

The following conclusions are presented:

- In order for a temperature or strain-based leakage detection system using fibre optic technology to be successful it is necessary that leakage-induced strain and temperature changes must be distinguishable from naturally occurring variations. The soil temperature variation at the experimental installation rapidly reduced with depth and was shown to be substantially different from the temperature variation occurring in the water mains monitored, suggesting that the proposed leakage detection system is likely to be successful.
- Tests in which stagnant water with a temperature nearly identical to that of the ground was leaked from a pipe, showed that the leak was easily identified from FBGS data.
- Water leakage from a pipeline results in both temperature and strain changes and because FBGS register a strain affected by both, it is substantially more sensitive that using temperature measurement alone. It is not necessary to distinguish between temperature and strain effects for the purposes of leak detection. A sudden change in FBGS measurements should be investigated as a potential leak.
- It is preferable not to attach the fibre optic sensor to the pipe because normal pressure fluctuation in the water networks results in significant strains which are likely to cause false alarms.

The success of FBGS to detect leaks on pipelines suggests that distributed strain/temperature sensing is also likely to be successful. A test programme examining the performance of distributed temperature and strain sensing as a means of leak detection should be carried out on pipelines which are part of an actual water network to examine system behaviour under the most realistic conditions. It is suspected that the leakage detection system described here will be less successful in saturated ground as leaks are likely to cause smaller temperature and strain variations.

REFERENCES

Barry-Macaulay, D. Bouazza, A. Singh, R.M. & Ranjith, P.G. 2013. Thermal conductivity of soils and rocks from the Melbourne (Australia) region. *Engineering Geology* 164: 131–38.

Buttrick, D. & Van Schalwyk, A. 1998. Hazard and risk assessment for sinkhole formation on dolo-mite. *Environmental Geology* 36 (1–2), November 1998, Springer-Verlag: 170–178.

Conway, M. 2010. *Heat Transfer in a Buried Pipe*. Masters Dissertation, University of Reading.

Kreuzer, M. 2013. Strain Measurement with Fibre Bragg Grating Sensors. Darmstadt, Germany: HBM.

McKenzie, R.S. Siqalaba, Z.N. & Wegelin, W.A. 2012. The State of Non-Revenue Water in South Africa. Pretoria: Water Research Commission. Report no. TT 522/12.

National Instruments. 2011. Products – Fibre optic sensing technology. Texas: USA. Available from: http://www.ni.com/white-paper/12953/en/ [Accessed 15 March 2017].

Development of an instrumented model pile

A.B. Lundberg
Geotechnical Department, ELU Konsult AB, Stockholm, Sweden

W. Broere
Geo-engineering Section, Delft University of Technology, Delft, The Netherlands

J. Dijkstra
Division of Geology & Geotechnics, Chalmers University of Technology, Gothenburg, Sweden

ABSTRACT: An instrumented model pile has been realized to study the displacement pile installation effects in sand in physical model tests. The system includes a model pile, instrumented with axial and horizontal contact stress sensors, and a corresponding calibration apparatus. The development of the instrumented model pile, including numerical analysis of the mechanical response during testing, and an optimization of the instrumentation to minimize thermal effects are described. The performance of this new model pile is demonstrated using calibration measurements and an example application in a physical model test at an elevated stress level in the geotechnical centrifuge.

1 INTRODUCTION

The axial bearing capacity of piles consists of the base resistance and the shaft resistance. For long piles in sand, the shaft resistance governs the total bearing capacity, (Klotz and Coop 2001, Randolph et al. 1994). In turn, the pile shaft resistance results from the sum of local shaft friction along the pile. In sands the local shaft friction is approximated in terms of Coulomb friction (Lehane et al. 1993), controlled by the effective horizontal contact stress at the pile-soil surface at the depth z and the interface friction. The interface friction angle is possible to obtain from laboratory tests, (Jardine et al. 1993), while the horizontal contact stress consists of the effective normal stress at the pile surface, i.e. the lateral stress in the subsoil. The horizontal contact stress is difficult to obtain in-situ, even from instrumented pile load tests, with some exceptions where highly controlled field test set-ups have been conceived, (Lehane et al. 1993, Jardine et al. 2013). The installation of displacement piles results in large disturbances of the initial soil state, i.e. the stress and density, due to the large soil deformations, (Poulos and Davis 1980, Jardine et al. 2013). The distribution of the horizontal contact stress $\sigma_h(z)$ along the pile shaft and consequently the pile bearing capacity is therefore highly influenced by the installation, resulting in significant differences in the load-displacement response of displacement piles compared to bored piles, where different installation effects have occurred, (Poulos and Davis 1980). Prediction methods for the axial pile bearing capacity have consequently relatively low accuracy, since the soil state after installation needs to be assessed based on the initial site investigation data, (Randolph et al. 1994). The bearing capacity and load-deformation response of a pile group is even more complicated to assess, due to the interaction between piles during installation, (Stuedlein and Gianella 2016). A more thorough comprehension of the pile installation process would therefore be of practical use.

Experimental research in which the normal stress $\sigma_n(z)$ acting on the pile is measured is required to arrive at an accurate description of the governing mechanisms of pile shaft friction. Due to the change in soil state during pile installation shown in (Lehane et al. 1993), local measurements are necessary to properly study the evolution of horizontal normal stress during the installation process. A measurement system has therefore been developed to locally measure soil horizontal normal stress $\sigma_n(z)$ on a continuous model pile surface. This system includes the design and instrumentation of an instrumented model pile, as well as a calibration system in which both normal stress sensors and axial stress sensors were calibrated. The design of the normal stress sensors was a compromise between the stress sensor sensitivity and minimization of influence from the axial load in the pile. The latter effect had been observed in other types of stress measurements as well, e.g. (Klotz and Coop 2001).

There are various methods to measure the local contact stress. These include surface sensors located on a structure (El Ganainy et al. 2013, Talesnick et al. 2014), and embedment cells surrounding the pile, (Foray et al. 1993, Jardine et al. 2013). Embedment cells require large soil models for instrumentation space, and are therefore not very suitable in the geotechnical centrifuge. The spatial accuracy of a small number of such cells is relatively low, as concluded by (Foray et al. 1993). Normal stress transducers mounted on the pile surface were therefore included in the instrumented model pile to measure the horizontal normal stress. Laboratory measurements show influence of grain size, stress distribution, hysteresis and sensor stiffness, (Talesnick et al. 2014). The soil grain size does not scale in centrifuge model tests (Garnier et al. 2007), therefore the soil grains are relatively large in relation to the stress sensor. Possible arching mechanisms redistribute the normal stress at the sensor, especially for external earth pressure cells mounted on the surface of a structure, (Foray et al. 1993). Installation of external earth pressure cells also result in a variation in pile surface roughness, e.g. as reported by (Klotz and Coop 2001). Especially, during cyclic shear loading, these effects redistribute the stress around the sensor (Randolph et al. 1994). A membrane strain gauge configuration directly embedded in the pile wall was therefore chosen for the current model pile to minimize the effect of stress redistribution. The membrane strain gauge consists of a flexible diaphragm on which a strain gauge bridge is mounted. In a novel configuration the flexible diaphragm is the pile surface itself, which deflects from the external soil pressure. As opposed to the null sensor (Talesnick et al. 2014), the membrane deflects under loading and the load is directly inferred from the membrane deflection rather than from a compensating fluid pressure. By omitting the null-sensor concept allows for extensive instrumentation on a relatively small model pile for the geotechnical centrifuge at the expense of allowing for movement in the soil that might affect the readings.

2 MECHANICAL DESIGN

The mechanical design of one half of the model pile is shown in Figure 2. The $125 \times 10 \times 10 \, mm^3$ instrumented pile was assembled from two identical milled steel sections. The model pile after assemblage is shown in Figure 1. The half bridge was mounted on the flexible membrane at cross-section A, shown to the right in Figure 2. The layout of the membrane was designed using analytical and numerical methods. The upper bound of the horizontal normal stresses, governing the membrane thickness, was estimated from available published data at 300–500 kPa (Boulon and Foray 1986, Foray et al. 1993, Lehane et al. 1993, Jardine et al. 2013). The membrane width was set to 8 mm, and the cross-section thickness h calculated from analytical formulas for a clamped circular plate (Reddy 2006):

$$\varepsilon_{rr} = \frac{3p(1-\nu^2)}{8Eh^2}(a - 3r^2) \quad (1)$$

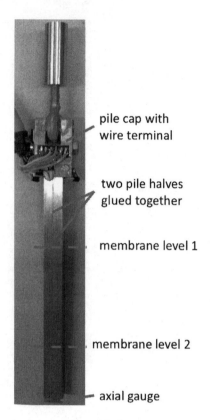

Figure 1. Image of the model pile.

Where E is the elastic modulus of the steel, p is the normal stress, h is the membrane thickness, a is the membrane radius, r is the position in the radial direction and ν is the Poisson's ratio. The maximum strain occurs in the middle of the membrane. This resulted in a membrane thickness of 0.3 mm for our geometry and material (stainless steel: Young's modulus E of 210 GPa and Poisson's ratio ν of 0.29).

Further analysis was carried out with the 3D-Finite Element program COMSOL Multiphysics version 4.3, (Comsol 2008). Ten-noded tetrahedral elements and triangular surface elements were used in the model to mesh the pile. The numerical model offered the possibility to model the influence of combined horizontal contact stress and pile axial load, which naturally would occur during the model tests. First the effect of variation in fabrication tolerance was studied by varying the membrane thickness to 0.27 mm, 0.29 mm, 0.31 mm, and 0.33 mm. The model pile was simulated loaded in the axial and in the horizontal normal direction at the pile surface, and the strain ε_{yy} perpendicular to the pile axis at the location of the stress sensors was obtained from the numerical model. The simulations were normalized with the strain level for the design dimension, $\varepsilon_{yy,30}$. The results demonstrate a +/−27% deviation in sensitivity for horizontal contact stress and a +/−8% sensitivity for axial load for each +/−10% successive change in the membrane thickness. After manufacturing of the membranes the tolerances in

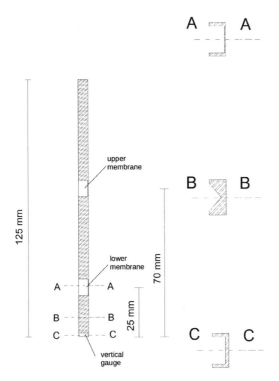

Figure 2. Details of model pile.

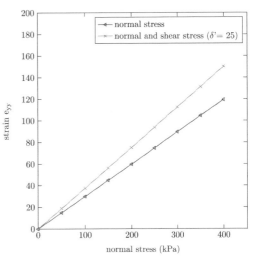

(a) The effect of shear stress on the membrane.

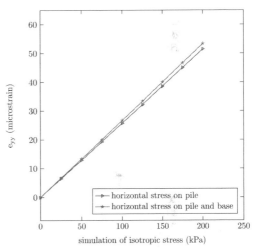

(b) Axial load effects.

Figure 3. The effect of additional shear stress and axial load in model pile on the contact stress gauge.

thickness were found to be < 5%. Therefore, the sensitivity of the sensor is within a reasonable bandwidth for the data acquisition equipment and easily accounted for in the calibration.

During installation the pile is subjected to both normal and shear strain at the pile shaft. The effect of shear stress on the membrane surface was simulated with the numerical model with a coefficient of friction $\mu = 0.466$, corresponding to an interface friction angle $\delta = 25°$, which is a reasonable assessment of common field and laboratory pile-soil conditions according to (Jardine et al. 1993). The simulation result is shown in Figure 3, which shows an increase in horizontal strain level ε_{yy} of around 25% at fully mobilized shear load.

Finally, a calibration of the horizontal normal stress sensor system in a pressure vessel was numerically simulated by application of an isotropic stress around the pile. The isotropic stress also resulted in axial stress component in the pile that loads the membranes axially. The extra horizontal strain ε_{yy} from combined axial and horizontal normal loading during isotropic calibration of the model pile is shown in Figure 3. The additional axial component leads to a higher horizontal strain level. This requires a correction of the measurements and resulting calibration factors for the horizontal contact stress sensors for the axial load. The numerical simulations lead to adopting a factor $R_{axial} = 0.963$, which was multiplied by the calibration measurements to retrieve the correct values of the calibration coefficients.

3 INSTRUMENTATION SYSTEM

3.1 Pile instrumentation

The model pile instrumentation system consisted of four horizontal contact stress sensors and one axial stress sensor located near the base of the pile. The horizontal contact stress sensors comprise a strain gauge half-bridge. The half strain gauge bridge consisted of two 120Ω strain gauges and two 120Ω 0.1% 5ppm/°C high precision SMD resistors. The strain gauge excitation voltage was 6 V, which was supplied by the in-flight amplifier system on the geotechnical centrifuge. The 120Ω strain gauges were of the

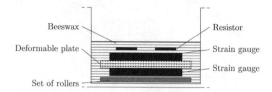

Figure 4. Detail of strain gauge configuration on membrane gauge (located at the bottom), not to scale. Rollers are 0.5 mm graphite pencil stifts.

type TML FLA 2-11, and had a tolerance of 0.1%, and a temperature compensation for steel of ($11\ 10^{-6}$ m/mK).

The two strain gauges were mounted on each side of a thin flexible plate, shown in Figure 4. The electrical configuration is such that only bending is picked up by the strain-gauges. Axial load in the plate is automatically compensated for. This plate could therefore be glued in the cavity to operate invariant of detrimental effects from axial loads. Also, shear stress in the membrane itself is compensated for by placing this 'bending-plate' at a distance from the membrane and only in contact to the membrane with small 0.5 mm graphite rods. As a result only the additional bending resulting from the axial loads are still influencing the measurements. The latter is compensated for by extensive calibration for load combinations supported by the numerical analysis of the pile, which has been previously discussed. The resistors, complementing the strain gauge bridge, were installed on top of the bending-plate with the strain gauges, before being covered with beeswax. The axial stress sensor consisted of two strain gauges placed opposite each other in a half Wheatstone bridge configuration and are mounted at opposite internal sides in the model pile. This electrical configuration doubles the sensitivity in axial direction and compensates for bending, however only in one direction, in this case the more compliant. The sensitivity of all the strain gauge bridges were optimized for an amplification of 1000x in the custom designed strain gauge amplifier such that no additional digital gain on the National instruments PCI-6220 acquisition card was required.

3.2 Excitation and data acquisition

Initial measurements showed significant heating of the instrumented model pile, which resulted in large deviations in the strain gauge output voltage V_{output}. This was the result of a reasonable low strain gauge resistance combined with a small heat dissipation area. Installing a switched strain gauge bridge supply successfully resolved the heating issues. The resulting output voltage V_{output} was subsequently processed to retrieve the valid measurements at the 6 V excitation level. The electric circuit, which was installed into a cable connected to the data acquisition system, consisted of a Diodes Inc ZXMP6A17G electronic switch, connected to a National Semiconductor LM555 timer.

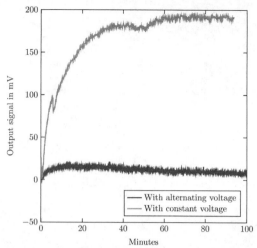

Figure 5. Effect of alternating bridge supply on thermal drift.

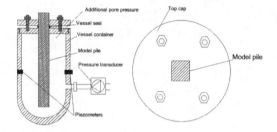

Figure 6. Vessel for the calibration of the horizontal stress gauges.

The duty cycle was 9% at 1.3 Hz frequency. Figure 5 shows the effectiveness of the system by plotting the evolution of V_{output} in time with and without the switched voltage regulation system.

4 CALIBRATION

4.1 Procedure

The instrumentation of the pile required a separate calibration procedure for the horizontal contact stress sensors and axial stress sensor. The horizontal contact stress sensor system was calibrated in a pressure vessel, providing an isotropic pressure at the stress sensor locations, shown in Figure 6. This vessel facilitates the calibration of the sensor without application of a pressure on the unsealed backside of the membrane via the pile head. The calibration vessel was sealed with a novel double layer cap top, in which a Teflon seal between model pile and the cap plates ensured a water tight seal upon compression in-between the top plates. A GDS standard pressure transducer controlled the pressure in the vessel. This vessel was equipped with two Druck PDCR-81 pore pressure transducers as a benchmark reading. During calibration the isotropic

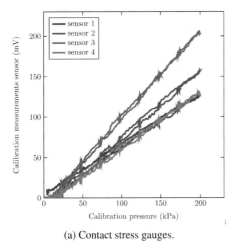

(a) Contact stress gauges.

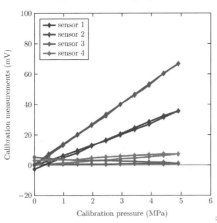

(b) Influence axial load on calibration of contact stress gauge.

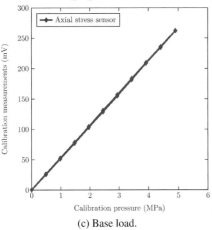

(c) Base load.

Figure 7. Calibration results for the horizontal stress gauges and pile base sensor.

pressure in the vessel was incremented with steps of 25 kPa up to 200 kPa and subsequently incrementally unloaded. The calibration was repeated four times to assess the stability and consistency of the system. The axial calibration was carried out with a lever system. The instrumented model pile was placed on a small steel sphere inserted into a drilled hole into the short end of the 5:1 lever for a moment-free load application. The vertical position of the model pile was adjusted for a horizontal lever position using a water level. During calibration the model pile was loaded to 5 MPa in 0.5 MPa increments, and then unloaded in 0.5 MPa increments.

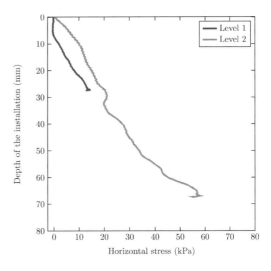

Figure 8. Pile installation test at 50-g using the newly developed model pile.

4.2 Results

A representative loading-unloading calibration loop in the pressure vessel for the horizontal contact stress sensors and for the calibration of the axial stress sensor using the lever system is shown in Figure 7. Additionally, Figure 7 shows the influence on the horizontal contact transducers from the axial load in the pile. The sensor oscillations at constant pressure level are the result from small pressure variations caused by minor leaks in the vessel and subsequent crude regulation of the GDS controller.

5 DEMONSTRATION TEST

The instrumented model pile was tested in a loose silica sand sample at 50-g acceleration level at the TU Delft centrifuge (Allersma 1994). Measurements of the upper and lower instrumentation levels are shown for continuous installation in Figure 8. The measurements were carried out through the same measurement setup as used for the calibration, and the installation test show a steadily increasing horizontal contact stress with some reduction for the higher level, confirming measurements reported by (Lehane et al. 1993, Jardine et al. 2013).

6 CONCLUSIONS

An instrumented model pile for axial and horizontal contact stress measurements and a complementary calibration system have been realized. A novel electric mechanical arrangement of the sensor compensates for all negative influence of the axial load in the horizontal contact stress readings. The additional bending of the membrane resulting from axial load, however, remains. The latter is compensated using extensive cross-calibration as well as numerical analysis of the system. The model pile is successfully tested in a physical model test performed in the TU Delft Geotechnical centrifuge.

ACKNOWLEDGMENTS

The research was funded by STW project number 10189. The support is gratefully acknowledged. The authors would like to thank Kees van Beek, Han de Visser and Ron van Leeuwen for their very kind and helpful assistance.

REFERENCES

Allersma, H. (1994). The university of delft geotechnical centrifuge. In *Proceedings International Conference Centrifuge 94*, pp. 47–52.

Boulon, M. & P. Foray (1986). Physical and numerical simulation of lateral shaft friction along offshore piles in sand. In *Proceedings of the 3rd International Conference on Numerical methods in Offshore piling, Nantes, France*, pp. 127–147.

Comsol (2008). Comsol multiphysics reference manual.

El Ganainy, H., A. Tessari, T. Abdoun, & I. Sasanakul (2013). Tactile pressure sensors in centrifuge testing. *Geotechnical testing journal 37*(1), 151–163.

Foray, P., J. Colliat, & J. Nauroy (1993). Bearing capacity of driven model piles in dense sands from calibration chamber tests. In *Offshore Technology Conference*, pp. OTC–7194–MS. Offshore Technology Conference.

Garnier, J., C. Gaudin, S. M. Springman, P. J. Culligan, D. J. Goodings, D. Konig, B. Kutter, R. Philips, M. Randolph, & L. Thorel (2007). Catalogue of scaling laws and similitude questions in geotechnical centrifuge modelling. *International Journal of Physical Modelling in Geotechnics 7*(3), 1–23.

Jardine, R., B. Lehane, & S. Everton (1993). Friction coefficients for piles in sands and silts. In *Offshore site investigation and foundation behaviour*, pp. 661–677. Springer.

Jardine, R., B. Zhu, P. Foray, & Z. Yang (2013). Interpretation of stress measurements made around closed-ended displacement piles in sand. *Géotechnique 63*(8), 613–627.

Klotz, E. & M. Coop (2001). An investigation of the effect of soil state on the capacity of driven piles in sands. *Géotechnique 51*(9), 733–751.

Lehane, B., R. Jardine, A. Bond, & R. Frank (1993). Mechanisms of shaft friction in sand from instrumented pile tests. *Journal of Geotechnical Engineering 119*(1), 19–35.

Poulos, H. G. & E. H. Davis (1980). *Pile foundation analysis and design*. Number Monograph.

Randolph, M., R. Dolwin, & R. Beck (1994). Design of driven piles in sand. *Géotechnique 44*(3), 427–448.

Reddy, J. N. (2006). *Theory and analysis of elastic plates and shells*. CRC press.

Stuedlein, A. W. & T. N. Gianella (2016). Effects of driving sequence and spacing on displacement-pile capacity. *Journal of Geotechnical and Geoenvironmental Engineering 143*(3), 06016026.

Talesnick, M., M. Ringel, & R. Avraham (2014). Measurement of contact soil pressure in physical modelling of soil-structure interaction. *International Journal of Physical Modelling in Geotechnics 14*(1), 3–12.

New method for full field measurement of pore water pressures

M. Ottolini
Technology Department, Heerema Marine Contractors SE, Leiden, The Netherlands

W. Broere
Geo-engineering Section, Delft University of Technology, Delft, The Netherlands

J. Dijkstra
Division of Geology & Geotechnics, Chalmers University of Technology, Gothenburg, Sweden

ABSTRACT: A cost effective method to measure pore water pressures in mixed granular media is described using 40 miniature MEMS pore pressure transducers. High accuracy in a single point is exchanged for lower accuracy full field measurements adjacent to the strongbox wall. The system is easily de-aired and calibrated due to the fact that the transducers are installed inside the strongbox wall. Additionally, the proof of concept test shows that the transducers are sufficiently accurate for problems with large pressure difference such as consolidation of clay while being subjected to elevated stress levels in the geotechnical centrifuge.

1 INTRODUCTION

In geotechnical engineering effort is made in quantifying the continuum stresses in granular media such as clays and sands. These exist out of three phases; particles in the solid phase leave pores in which water in the liquid phase and air in the gas phase is held. Experiments are often executed in a geotechnical centrifuge in order to satisfy similitude for the stress within these materials (Garnier et al. 2007). The acceleration of the centrifuge results into an increase of gravitational acceleration and therefore higher stress levels in the granular medium. These experiments require a strongbox to retain the soil. More conventional (miniature) pore pressure transducers, embedded in the soil or placed on the wall, are used where typically a high accuracy at a single point in space is required. Measurement of contact stress and pore pressures without influencing the soil state, especially the stiffness, around the sensor is not trivial (Talesnick 2005, Talesnick et al. 2014). Similar effects are observed for pore pressure transducers (Kutter, Sathialingam, & Herrmann 1990) or tensiometer design (Tarantino and Mongiovì 2002, Take and Bolton 2003), though most issues relate to the response time of the sensor for sensing dynamic events and stable operation at elevated stress levels and maintaining saturation in tensiometers. Traditionally, the Druck PDCR-81 is employed in physical model testing, however after its retirement alternatives are suggested (Stringer et al. 2014). Additional negative side effects of embedding transducers with their cables (Foray et al. , Jardine et al. 2013) in the specimen cannot be prevented. Also, the relatively high cost and/or large size of these transducers prohibit capturing the full spatial field. Recent developments in tactile sensing allow measurement of the contact load in a dense two-dimensional grid of points (Paikowsky and Hajduk 1997, Palmer et al. 2009). These promising sensors are, however, not easily converted for sensing pore water pressures in a soil sample and although the sensing element itself is inexpensive the data acquisition is not.

With the on-going developments in Micro Electro Mechanical System (MEMS) single chip solutions for sensing applications (Tanaka 2007, Tadigadapa and Mateti 2009) the instrumentation possibilities for geotechnical applications increase rapidly. MEMS accelerometers are already in use in the physical modelling community, e.g. Stringer et al. (2010). This paper, on the other hand, will introduce the use of cost effective MEMS pressure sensors (Eaton and Smith 1997) to acquire full field data of pore water pressures in soil samples adjacent to the wall in a strongbox. In principle the instrumentation is also applicable to measure the fluid phase in other mixed media.

2 EXPERIMENTAL SETUP

2.1 Design objectives

The MEMS pore pressure array (PPA) has primarily been developed for the TU Delft geotechnical centrifuge (Allersma 1994). This small beam centrifuge with a radius of 1.22 m has recently been re-equipped with modern data acquisition and camera facilities. The flight computer with wireless link, data acquisition and actuator control are all hosted on the central beam of the centrifuge. An embedded solution for data-acquisition near or in the sensor was preferred rather than developing a general robust miniature wireless data-acquisition system that interfaces conventional passive sensors (Gaudin et al. 2009), as the number of analog acquisition and amplifier channels is limited to 16. Furthermore, the setup should be autonomous from the centrifuge system, such that a similar setup could be quickly employed or cloned for instrumentation of 1-g physical model tests elsewhere in the laboratory and in long running autonomous in-situ tests. Finally, long-term stability and temperature compensation was deemed more important than dynamic response time as the primary motivation was to look into consolidation effects in clay resulting from installation of foundation elements.

2.2 Sensor selection and application

As opposed to MEMS accelerometers and gyroscopes, MEMS pressure transducers are more difficult to source, especially when considering the required (low) limit pressures and a sensing membrane that is insensitive to fluids. Finally, the Sensonor SP100 series transducers proved to be the best candidate for prototyping (Sensonor 2009). These pressure transducers are widely used for measurement of tire pressures in the automotive industry. Additionally to a sensing element these have a local embedded amplifier, 8 bit Analog to Digital Conversion (ADC), temperature and supply voltage sensing, and a digital communication interface (Serial Peripheral Interface: SPI). The triple stack glass-silicon-glass sensor element makes it suitable for wet environments such as saturated granular materials, as all instrumentation is encapsulated (Grelland 2001). Three different model types of the SP100 series have been used, the 1T (100 kPa), 2T (200 kPa) and 7T (700 kPa), to have a higher accuracy at the top part of the strongbox where lower pressures are expected. The stated resolution is 0.36 kPa for the 1T up to 1.25 kPa for the 7T model whilst the accuracy is 2% full scale. The latter is about 10 times worse than the PDCR-81 at 1/100th of the price. The surface mount packaging of the chip allows for a small centre-to-centre spacing. A hole in the top of the packaging allows for the fluid pressure to reach the sensing element. Isolation of the remainder of the chip from the soil and any harsh environment makes this solution more suitable for on wall measurements than embedment in the soil sample.

In the adopted application the MEMS transducers have been directly bonded to the aluminium wall of the strongbox. Contact between the water in the soil and the transducer is through a small channel in the wall with a 3 mm diameter at the transducer side and 5 mm diameter at the soil side. A sintered glass porous disc, of 5 mm diameter, prevents the soil from entering the channel. The channel and the porous disc are de-aired with silicon oil under a vacuum of -90 kPa to ensure a fast response of the transducer. Epoxy adhesive bonds the pressure transducers at the wall and

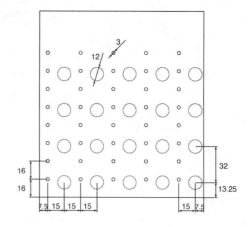

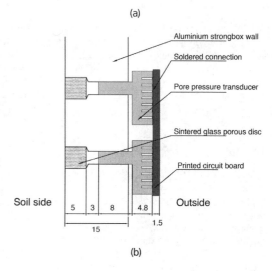

Figure 1. Design of pore pressure array (PPA); 1a: Mechanical drawing of PPA (measures in mm). 1b: Detailed mounting of MEMS package to sidewall.

ensures a watertight connection. The transducers have been spaced 30 mm horizontally and 16 mm vertically, in 8 rows of 5. A total of 40 transducers are placed to have sufficient field information. As a result the vertical spatial resolution is higher than the horizontal. The 14 pins sensor packages are soldered to a custom designed printed circuit board (PCB). The PCBs are a four-layer design consisting of a back and a front layer for placement of the electrical components and two inner planes serving as ground plane. Ten pore pressure transducers are combined on one sub-PCB and linked with a dedicated SPI BUS to the microcontroller. A technical drawing of the mechanical lay-out is shown in Figure 1.

2.3 Data acquisition

A functional block diagram of the electronic systems in the PPA is shown in Figure 2. The diagram is divided in three parts; the instrumentation on the wall itself,

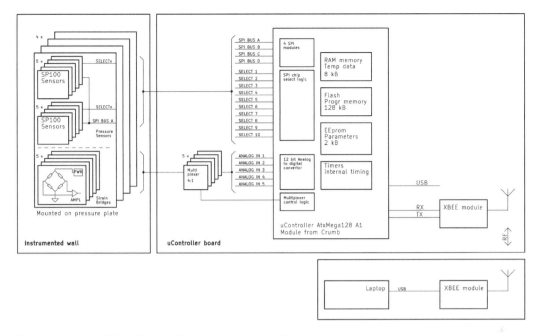

Figure 2. Functional block diagram of the electronic systems of the pore pressure array.

the data acquisition and transfer instrumentation and the data storage.

The data of each sub-PCB is connected through flat cables to a central CRUMBX microcontroller board (chip45 GmbH & Co. KG) that houses an Atmel AtxMega128 microcontroller. The availability of four SPI channels and 78 Programmable I/O Lines makes the microcontroller suitable for data acquisition of multiple channels on multiple PCBs. The flash drive of the CRUMBX is loaded with a control programme, written in C++, which handles the multiple channels retrieved. Ten I/O Lines are used as chip select lines. One transducer per PCB is selected and initialised simultaneously with the chip select line. The transducers are subsequently read and send to the microcontroller over a separate SPI bus, one for each PCB. This process is repeated until all pore pressure transducers are read. All the data is locally buffered until finishing reading the last sensor of the sequence. Subsequently, in sequences of 10 channels the data is wirelessly sent to a personal computer over a wireless connection through an XBee pro 60 mW wire antenna (Digi International Inc.). The latter acts as a wireless serial port at a baud rate of 115600 bps. Sampling of all transducers was set at a 1 second interval, though a real world upper sampling limit of 100 Hz can be reached. After reception the data is further handled in the custom written acquisition software developed at TU Delft, before being time stamped and saved to the hard drive.

The four sub-PCBs are equipped with the possibility to be used for strain gauge amplification with an Analog Devices AD8227 rail-to-rail output instrumentation amplifier and offset regulation of the strain gauges with a 500Ω potentiometer. On the data acquisition board five AD704 CMOS low voltage analog multiplexers are available to multiplex the data to the analog input channel of the microcontroller. The microcontroller is equipped with the necessary multiplexer control logic.

3 RESULTS

3.1 Calibration

The strongbox is modified to accommodate simultaneous saturation and calibration of all sensors. An aluminium lid equipped with two gas interconnectors seals the strongbox on the topside with a flat gasket inbetween the sidewalls and the lid. First the strongbox is filled with silicon oil and the air is evacuated under a vacuum to saturate the porous disc and cavity to the sensing membrane. Subsequently, the strongbox is filled with water and compressed air is used to calibrate the sensors in this temporary pressure vessel. The air compressor and regulator valve are too crude for accurate pressure readings, hence the second connector is attached to an analog precision manometer (accuracy <0.3 kPa) for measurement of the fluid pressure inside the strongbox. All the pore pressure transducers are simultaneously calibrated by applying pressure from 0 to 100 kPa in increments of 10 kPa. After stabilisation of the pressure, an average of 5 samples is taken from each transducer. A linear regression of the data is used to derive the calibration factor for further testing. Loading and reloading cycles have been applied to incorporate effects of hysteresis. The calibration is performed at room temperature (20°C ± 1°C) which is similar to the temperature in the

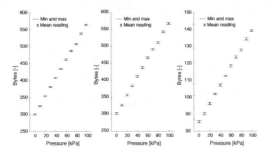

Figure 3. Calibration results for loading unloading loop of the MEMS pore pressure transducers with different pressure ranges (left: 100 kPa; middle: 200 kPa; right: 700 kPa. The mean, minimum and maximum readings of 10 sensors each have been plotted.

intended test. However, the readings can be corrected with the embedded temperature readings. Calibration results are shown in Figure 3 for the SP100-1T (left hand side), the SP100-2T (middle) and SP100-7T (right hand side) pressure is plotted against the raw sensor output in bytes. Minimum and maximum reading as well as the mean value are shown for 10 transducers of each type. The sensors are absolute pressure transducers, hence the intercept with the y-axis in the readings. Also, the limited calibration range for the 7T is due to the simultaneous calibration of all sensors with different maximum range. The calibration results indicate that in their new application the pressure sensors are operating within their rated specification of ± 2% full scale.

3.2 Proof of concept N-g test

The strongbox equipped with the PPA on one of the sidewalls has been used to monitor the pore pressure dissipation during consolidation of a kaolin soil specimen during self-weight consolidation in a geotechnical centrifuge. The standard procedure at TU Delft is that the kaolin clay powder is first mixed with de-aired water into slurry with high water content (>10 times the liquid limit). The strongbox used in the experiment has transparent windows made out of Plexiglas on the front and the back of the box and inner dimensions ($L \times W \times H$) $180 \times 155 \times 150$ mm^3 one side wall is equipped with the PPA. Before the slurry is poured in the strongbox, all walls are sprayed with PTFE spray to reduce wall friction. Subsequently the slurry is consolidated into a solid sample using the TU Delft beam centrifuge at an acceleration level of 100-g. In Figure 4 a typical result of the full field pore pressure evolution during consolidation is shown, for three distinct prototype times: 45 days, 449 days and 898 days (t_{45}, t_{449}, t_{898}). The vertical axis is exaggerated for clarity. The measurements clearly indicate dissipation of porewater pressure over time during self-weight consolidation. The difference pressure between sensors on the same depth stems from the eccentric placement of the strongbox in the centrifuge and the coriolis resulting from the rather short beam radius (Taylor 1995).

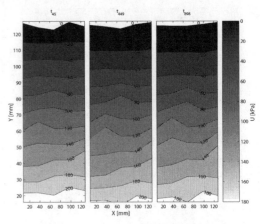

Figure 4. Full field measurement of dissipation of pore pressures due to self-weight consolidation of kaolin clay in a geotechnical centrifuge at 100-g after respectively 45 days (t_{45}), 449 days (t_{449}) and 898 days (t_{898}).

4 CONCLUSIONS

Only minor modifications are required to construct a cost effective pore pressure array (PPA) with a high number of sensors (40) using miniature MEMS pore pressure transducers. On-chip data acquisition allows for easy electrical interfacing to a micro-controller and subsequent wireless data transmission to any arbitrary receiver. High accuracy in a single point is exchanged for somewhat lower accuracy full field measurements adjacent to the strongbox wall. Due to the fact that the 40 pore pressure transducers are installed inside the wall they can be simultaneously de-aired and calibrated and additional detrimental effects from cables are prevented. The calibration of a large number of transducers corroborate that the sensors operate within their rated specifications (± 2% full scale) as well as that the proof of concept test shows that the sensors continue to function well at elevated stress levels in the geotechnical centrifuge. Unique full field data can be gathered on both deformations and pore pressure response in the sample when the PPA is combined with full field deformation measurements on an adjacent transparent wall.

ACKNOWLEDGMENTS

The authors would like to express their gratitude to C. van Beek, J.J. de Visser, R. van Leeuwen and M. van der Meer for their assistance during the design and construction of the setup.

REFERENCES

Allersma, H. (1994). The university of delft geotechnical centrifuge. In *Proceedings International Conference Centrifuge 94*, pp. 47–52.

Eaton, W. & J. Smith (1997). Micromachined pressure sensors: review and recent developments. *Smart Materials and Structures* 6(5), 530.

Foray, P., J. Colliat, & J. Nauroy. Bearing capacity of driven model piles in dense sands from calibration chamber tests. In *Offshore Technology Conference*, pp. OTC–7194–MS.

Garnier, J., C. Gaudin, S. M. Springman, P. J. Culligan, D. J. Goodings, D. Konig, B. Kutter, R. Philips, M. Randolph, & L. Thorel (2007). Catalogue of scaling laws and similitude questions in geotechnical centrifuge modelling. *International Journal of Physical Modelling in Geotechnics* 7(3), 1–23.

Gaudin, C., D. White, N. Boylan, J. Breen, T. Brown, S. De Catania, & P. Hortin (2009). A wireless high-speed data acquisition system for geotechnical centrifuge model testing. *Measurement Science and Technology* 20(9), 095709.

Grelland, R. (2001). Tyre pressure monitoring microsystems. *Advanced Microsystems for Automotive Applications 2001*, 245–251.

Jardine, R., B. Zhu, P. Foray, & Z. Yang (2013). Interpretation of stress measurements made around closed-ended displacement piles in sand. *Géotechnique* 63(8), 613–627.

Kutter, B., N. Sathialingam, & L. Herrmann (1990). Effects of arching on response time of miniature pore pressure transducer in clay. *Geotechnical Testing Journal* 13(3), 164–178.

Paikowsky, S. & E. Hajduk (1997). Calibration and use of grid-based tactile pressure sensors in granular material. *Geotechnical Testing Journal* 20(2), 218–241.

Palmer, M., T. ORourke, N. Olson, T. Abdoun, D. Ha, & M. ORourke (2009). Tactile pressure sensors for soil-structure interaction assessment. *Journal of geotechnical and geoenvironmental engineering* 135(11), 1638–1645.

Sensonor (2009). *Sp100 compensated pressure sensors*. datasheet.

Stringer, M., J. Allmond, C. Proto, D. Wilson, & B. Kutter (2014). Evaluating the response of new pore pressure transducers for use in dynamic centrifuge tests. In *Proceedings of the 8th International Conference on Physical Modelling in Geotechnics*, Volume 1, pp. 345–351.

Stringer, M., C. Heron, & S. Madabhushi (2010). Experience using mems-based accelerometers in dynamic testing. In *Proceedings of the 7th International Conference on Physical Modelling in Geotechnics*, Volume 1, pp. 389–394.

Tadigadapa, S. & K. Mateti (2009). Piezoelectric mems sensors: state-of-the-art and perspectives. *Measurement Science and technology* 20(9), 092001.

Take, W. & M. Bolton (2003). Tensiometer saturation and the reliable measurement of soil suction. *Géotechnique* 53(2), 159–172.

Talesnick, M. (2005). Measuring soil contact pressure on a solid boundary and quantifying soil arching. *Geotechnical Testing Journal* 28(2), 1–9.

Talesnick, M., M. Ringel, & R. Avraham (2014). Measurement of contact soil pressure in physical modelling of soil–structure interaction. *International Journal of Physical Modelling in Geotechnics* 14(1), 3–12.

Tanaka, M. (2007). An industrial and applied review of new mems devices features. *Microelectronic engineering* 84(5), 1341–1344.

Tarantino, A. & L. Mongiovì (2002). Design and construction of a tensiometer for direct measurement of matric suction. In *Proceedings of the 3nd International Conference on Unsaturated Soils, Recife, Brazil*, Volume 3, pp. 319–324.

Taylor, R. (1995). Centrifuges in modelling: principles and scale effects. *Geotechnical centrifuge technology*, 19–33.

Ambient pressure calibration for cone penetrometer test: Necessary?

Y. Wang & Y. Hu
School of Civil, Environmental and Mining Engineering, The University of Western Australia, Perth, Australia

M.S. Hossain
Centre for Offshore Foundation Systems, The University of Western Australia, Perth, Australia

ABSTRACT: The cone tip load cell needs to be calibrated before the cone penetration test (CPT) data in soil can be used to interpret the soil properties. The cone tip load cell is routinely calibrated through cone tip axial loading test. In this study, the CPTs were conducted in fully saturated layered clay deposits in a centrifuge. The net cone resistance was found to be negative by using the calibration factor obtained through axial loading test. This is illogical. Further CPT calibration was performed in water in the centrifuge under 150 times earth gravity. The CPT calibration factor under ambient pressure in water was found to be different from that obtained through axial loading test. By using both calibration factors from axial load test and under ambient pressure, the net cone resistance was recalculated. The interpreted soil undrained shear strength using the recalculated net cone resistance was consistent with that from a parallel ball penetrometer test. Since the soil overburden pressure acts on the cone tip as an ambient pressure, the cone tip load cell should be calibrated under ambient pressure to provide logical interpretation of CPT data.

1 INTRODUCTION

Due to the high cost of obtaining high quality soil samples offshore, field test becomes a popular choice to determine the geotechnical properties of seabed sediments. Cone penetrometer test (CPT) is the commonly adopted in-situ test, which is mainly due to its superior feature of providing a continuous resistance profile, compared to vane shear test. As shown in Figure 1, the cone penetrometer comprises a cone-shaped tip and a cylindrical shaft. The standard cone penetrometer used in the field has a base area of 10 cm^2 (diameter $D = 35.7$ mm), although the diameter can go up to 71.4 mm (40 cm^2) (Robertson and Cabal, 2015). There are also some miniature cones ($D = 7$ mm; 10 mm) that have been reported for laboratory test applications (Lee 2009).

The wide use of cone penetration test for soil characterisation in geotechnical practice rises the demand for accurate measurement, which relies on the accuracy of its sensors. The cone penetrometer generally consists of cone tip load cell and sleeve friction load cell. This study focuses on the former. The cone tip load cell is embedded in the shaft just above the shoulder of cone (as marked in Figure 1). The accuracy of the cone tip load cell and its calibration factor are crucial in terms of interpreting the corresponding soil strength. For the normal practice with a cone, zero-load error is a reliable indicator of output stability (ASTM 2012). Zero-load error refers to zero-load readings (or baseline readings; Mayne 2007) difference before and after a sounding. Recalibration is only necessary when the zero-load error drifts out of requirement (ASTM 2012; Robertson & Cabal 2015), and ASTM (2012) regulates this critical error as 2% full scale output.

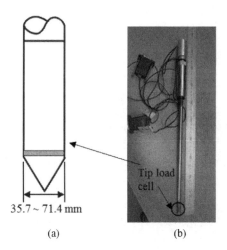

Figure 1. Cone penetrometer: (a) schematic figure; (b) cone penetrometer used in centrifuge tests.

To ensure sensible calculation using the measured data, in-house check-up for calibration factor of a load cell is routinely conducted. The recommended method for in-house calibration check of the cone tip load cell is through axial load test (Lunne et al. 1997), in which the cone penetrometer is loaded/unloaded against a reference load cell (Peuchen & Terwindt 2014). The

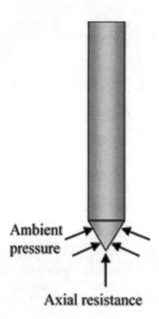

Figure 2. Cone penetrometer subjected to both ambient pressure and axial resistance when penetrating in soil.

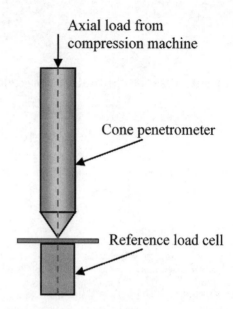

Figure 3. Axial load test at 1 g.

detail of the test can be found in Chen & Mayne (1994). Through loading/unloading, the voltage response of the tested cone tip load cell is recorded and the load on the cone is known from the reference load cell. The calibration factor can be obtained through best fitting the data, with the gradient expressed as,

$$C = \frac{q}{v} \quad (1)$$

where v is the voltage reading of the load cell and q is the corresponding pressure applied on the load cell, and the pressure-to-voltage ratio is called conversion or calibration factor (C). A qualified cone should prove good linear behaviour under loading with nonlinearity satisfy certain requirement (ASTM 2012).

The calibration factor obtained through the axial load test can only reflect the load cell behaviour under a tip resistance. However, when the cone penetrates into soil, it subjects to both tip resistance and ambient pressure (i.e. overburden pressure) (Figure 2). In this paper, firstly an in-house check-up for calibration factor is conducted through axial load test at 1g on the laboratory floor. Then the load gauge behaviour under ambient pressure is examined through a series of hydraulic pressure tests in a centrifuge. The calibration factors are compared, and both factors are used in the CPT data interpretation framework. Finally, the necessity of performing a calibration test under ambient pressure is demonstrated.

2 EXPERIMENT DETAILS

A miniature cone penetrometer was considered in this paper with a 60° cone tip and a 10 mm diameter. A tip load cell with 10 MPa capacity was equipped just above the cone shoulder (Figure 1b). In this section, firstly two types of calibration test were described—the axial load test that considers net tip resistance and hydraulic test that simulates ambient pressure. Then the CPT and ball tests in a layered clay sample are introduced.

2.1 Axial load test at 1g

In the axial load calibration test, the cone penetrometer concentrically placed on a high-resolution reference load cell was mounted on a compression machine, as shown in Figure 3. Compression loading and unloading was carried out in steps. In each step, sufficient time was allowed to stabilise the response readings, and the stabilised voltage (or bit from the cone tip load cell) and pressure (from the reference load cell) were recorded. Compression pressure was gradually increased up to 2865 kPa (force 224.9 N/total area of the cone 78.5 mm^2) prior to unloading.

2.2 Hydraulic pressure test at 150g

To test cone tip load cell behaviour under ambient pressure, the cone penetrometer was penetrated from the water surface to the designed depth (220 mm in model scale). To achieve a large water pressure, the hydraulic test was conducted under an enhanced gravity environment of 150 times the earth gravity (150 g) in the beam centrifuge at The University of Western Australia (Randolph et al. 1991). The experimental set-up is shown in Figure 4. The centrifuge strongbox with internal dimensions of 650 (length) × 390 (width) × 325 (depth) mm was filled with water. The cone penetrometer was penetrated in water at a constant velocity of 1.15 mm/s, which is the same as the

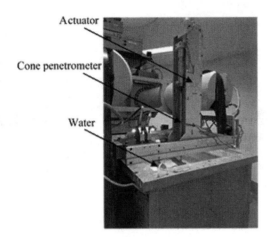

Figure 4. Centrifuge hydraulic pressure test in a centrifuge.

penetration velocity of the cone in the layered clay sample (described later). High-speed data acquisition system (10 Hz) was equipped with the centrifuge to allow in-flight data monitoring and collection. At the final depth of ~220 mm (~33 m in equivalent prototype scale), the maximum prototype hydraulic ambient pressure was 323 kPa.

2.3 Centrifuge penetrometer test on layered clays—cone and ball penetrometers

The same cone penetrometer just discussed was used to characterise a soft-over-stiff layered clay deposit at an acceleration level of 50 g. The soft and stiff clay layers were made from kaolin clay slurry pre-consolidated under the final pressures of 100 and 400 kPa respectively. The average effective unit weight (γ') of the soft and stiff clay layers was ~7.2 kN/m³. The thickness of the top soft clay layer and the bottom stiff clay layer was 9.5 m and 4 m respectively (prototype scale). A ball penetrometer test was also conducted in parallel to the CPT for comparison. The miniature ball penetrometer was 9 mm in dimeter. More information on the recently developed ball penetrometer can be found in Watson et al. (1998) and Newson et al. (1999). Ball penetrometer test was conducted at 1.25 mm/s and CPT at 1.15 mm/s to target an identical normalised velocity (nulling relative strain rate effect) and to ensure undrained conditions in the kaolin clay (Finnie & Randolph 1994).

3 RESULTS AND DISCUSSIONS

The following section discusses the calibration factors for the cone tip load cell derived from the axial load test at 1 g and the ambient pressure test at 150 g. The CPT data from the test in the layered clay are then interpreted using these calibration factors. The interpreted soil undrained shear strength from the CPT data is compared with that from the ball penetrometer test data.

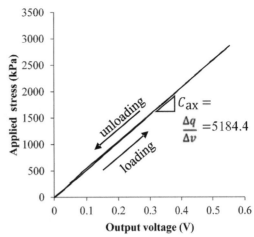

Figure 5. Stress-voltage response of cone tip load cell under axial load.

3.1 Calibration factor from axial load test

The recorded data from axial load test are shown in Figure 5. Applied pressure from the reference load cell is plotted as a function of voltage from the cone tip load cell. A perfectly linear response can be seen in both loading and unloading stage, with a negligible divergence between loading and unloading stage and minimal zero-load error (<0.1% full scale output). A linear fitting provides the calibration factor (C_{ax}, 'ax' stands for axial load) of 5184.4.

3.2 Processing of CPT data based on axial load calibration factor

The obtained calibration factor from axial load test was used to interpret the cone penetration test data in the layered clay. The soil undrained shear strength was interpreted according to Lunne et al. (2011),

$$S_{u\,(CPT)} = \frac{q_{net}}{N_{kt}} \quad (2)$$

$$q_{net} = q_t - \sigma_{v0} \quad (3)$$

where $S_{u\,(CPT)}$ is the undrained shear strength of clay from CPT, q_{net} is the net resistance from the cone tip, q_t is the total measured resistance from the cone tip, σ_{v0} (=$\gamma \times d$, γ is the total unit weight of the soil and d is the penetration depth of the cone shoulder) is the overburden pressure at corresponding depth. The bearing capacity factor of resistance factor of cone in clay, N_{kt}, was taken as 13.56 (Low et al. 2010).

The result from the ball penetrometer test can be processed to obtain the undrained shear strength as,

$$S_{u\,(ball)} = \frac{q_{ball}}{N_{ball}} \quad (4)$$

where q_{ball} is the resistance from the ball penetrometer test. The ball bearing capacity factor N_{ball} was

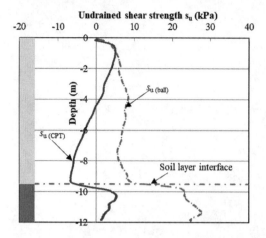

Figure 6. Comparison of undrained shear strength of layered clay from ball penetrometer test and cone penetrometer test.

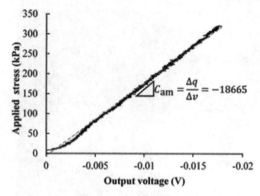

Figure 7. Stress-voltage response of cone tip load cell under ambient pressure from centrifuge hydraulic calibration test (150 g).

taken as 11.17 (Randolph et al. 2000). As the ball penetrometer penetration generally mobilises a full flow-round mechanism around the ball except for the shaft connection area, the soil overburden pressure has a minimal influence on the measured resistance (Randolph 2004).

The interpreted undrained shear strength profiles from the ball and cone penetrometer tests in soft-overstiff clay (with soft and stiff layer marked with light and dark shaded column separately) are shown in Figure 6. It is interesting to notice that a negative undrained shear strength profile between −4.5∼−9.6 m was obtained from the cone penetration data while an expected positive strength profile was obtained for the full penetration depths from the ball penetrometer data. The negative undrained shear strength in Figure 6 is resulted from the negative resistance calculated by Equation 2-2. A similar phenomena of negative resistance was reported by Boylan & Long (2006) from a field test on peat. The negative undrained shear strength is clearly illogical, leading to the further investigation on the cone calibration factor.

3.3 Processing of CPT data using both axial load calibration factor and hydraulic pressure calibration factor

The result from hydraulic pressure calibration is shown in Figure 7, plotting the ambient hydraulic pressure as a function of voltage response from the cone tip load cell. Surprisingly, the voltage-pressure response (Figure 7) shows a negative correlation between the voltage output under ambient pressure, which is the opposite to the one observed from the axial load test (see Figure 5). From the fitting curve in Figure 7, the calibration factor under ambient pressure (C_{am}, 'am' stands for ambient load) is −18665 kPa/V. It is worthwhile to note that, although the load range of the hydraulic test (323 kPa) is much less than that of the axial load test (2863 kPa), the linear response indicates a constant calibration factor. In the tested clay deposit, the maximum overburden pressure on the cone tip of 206 kPa also lies in this range.

To apply the cone tip load cell calibration factors in the CPT data interpretation, it should be noted that the measured cone tip resistance includes two parts: (1) the cone tip resistance from soil shearing during cone penetration; and (2) the cone tip resistance from the overburden pressure from the soil (see Equation 4). The recorded total voltage output can be seen as the contribution of these two components. Therefore, it can be derived from Equations 1 and 3 that,

$$v_t = \frac{q_{net}}{C_{ax}} + \frac{\sigma_{v0}}{C_{am}} \quad (5)$$

where v_t is the total voltage output. It should be noted that, if $C_{ax} = C_{am}$, the data interpretation becomes easier as $v_t = q_t/C_{ax}$ (or $q_t = C_{ax} \times v_t$). In this case, it is apparent that Equation 5 degenerates to Equation 6, which was used in the CPT data interpretation for Figure 6. However, in this case, different calibration factors were obtained from the axial load test and ambient pressure test. Thus, q_{net} should be calculated from Equation 6,

$$q_{net} = \left(v_t - \frac{\sigma_{v0}}{C_{am}}\right) \times C_{ax} \quad (6)$$

With the modified net tip resistance from Equation 6, a new strength profile can be interpreted. In this way, the undrained shear strength is modified (Equation 2) and compared with the ball penetrometer result. Figure 8 shows that the modified net tip resistance has produced a soil strength profile of the layered clay that is very close to that from the ball penetrometer test. This confirms that the modified net tip resistance can provide the logical interpretation of soil undrained shear strength. This exercise shows that failing to consider the difference in calibration factors for the two components may lead to interpretation error and even illogical result such as negative strength.

However, it is also worth noting that not all cone tip load cells show two different calibration factors

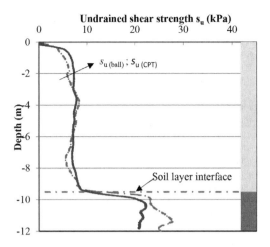

Figure 8. Modified CPT test results from considering ambient pressure calibration and its comparison with ball penetrometer results.

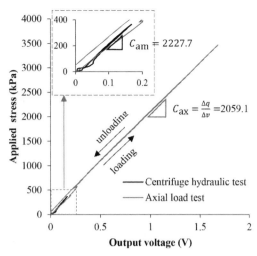

Figure 9. Axial load calibration test result and centrifuge hydraulic calibration test for another cone tip load cell.

under the axial load and the ambient pressure. Another group of parallel tests were conducted on the cone tip load cells, with the same size and the same type of cone (i.e. 10 mm in diameter and only equipped with tip load cell). Figure 9 depicts the results from both calibration tests. It can be seen that both calibration factors are consistent, with a difference being <8%. This difference of the calibration factors from 1 g axial load test and centrifuge hydraulic test might be due to the Poisson's strain in the load cell under ambient pressure and the increased self-weight of the cone tip under enhanced gravity (150 g) (Chow et al. 2017). Since the cone resistance profile was also measured under the influence of enhanced gravity, it is reasonable to adopt the calibration factor from centrifuge hydraulic test for subsequent interpretation. In this case, the negative net resistance would not happen because the overburden pressure would increase the output voltage in a similar ratio, so deducting it from the total tip resistance would not lead to a negative net resistance. Yet, the interpreted strength may not be accurate, as explained previously, emphasising the necessity of the ambient pressure test.

This study shows the importance of calibration under ambient pressure because calibration under pure axial pressure cannot always guarantee the load cell behaves in the same way under ambient pressure. Calibration of tip load cell under atmospheric condition, i.e. pressurised chamber test, has been reported by Peuchen & Terwindt (2014), which is able to reflect the ambient pressure effect on the load cell rather than axial load test. However, the pressurised chamber test is not suggested by the standards for cone tip load cell (ASTM 2012), and it is not commonly adopted in the in-house calibration check of the calibration factor. In fact, the pressurised triaxial apparatus are normally used to check for pore pressure transducer as well as net area ratio of cones (Mayne 2007). In addition, and critically, the hydraulic pressure test introduced in this paper can also examine the load cell behaviour under water e.g. potential water ingress and corresponding drifting/oscillation of response during actual testing, which could not be simulated with the air pressure chamber test.

4 SUMMARY AND CONCLUDING REMARKS

The total cone resistance measured by CPT is used to interpret soil undrained shear strength. The total cone resistance includes two parts: the shear resistance and the overburden pressure. The cone tip load cell needs to be calibrated using the axial load test (for the shear resistance) and the ambient pressure test (for the overburden pressure). When the calibration factors under the axial load test and under the ambient pressure test are the same (or similar), the interpretation can be straightforward. However, when the two calibration factors are different, both calibration should be used for interpreting the CPT data. The hydraulic test in centrifuge has been proved to work well to produce the cone tip load cell calibration factor under ambient pressure.

ACKNOWLEDGEMENTS

The research presented here was undertaken with support from the Australian Research Council (ARC) through the Discovery Grant DP140103997. The first author is the recipient of a University of Western Australia SIRF scholarship. The work forms part of the activities of the Centre for Offshore Foundation Systems (COFS), currently supported as a node of the Australian Research Council Centre of Excellence for Geotechnical Science and Engineering, through Centre of Excellence funding from the State Government of Western Australia and in partnership with The

Lloyd's Register Foundation. This support is gratefully acknowledged.

REFERENCES

ASTM. 2012. Standard Test Method for Performing Electronic Friction Cone and Piezocone Penetration Testing of Soils. West Conshohocken, PA: ASTM International.

Boylan, N. & Long, M. 2006. Characterisation of peat using full flow penetrometers. *Proceedings of the 4th International Conference on Soft Soil Engineering (ICSSE)*. Vancouver.

Chen, B.S. & Mayne, P.W. 1994. Profiling the overconsolidation ratio of clays by piezocone tests. *Report No. GIT-CEEGEO-94*, 1.

Chow, S., O'Loughlin, C., White, D. & Randolph, M. 2017. An extended interpretation of the free-fall piezocone test in clay. *Géotechnique*, 67, 1090–1103.

Finnie, I.M.S. & Randolph, M.F. 1994. Punch-through and liquefaction induced failure of shallow foundations on calcareous sediments. *Proceedings of International Conference on Behaviour of Offshore Structures*. Boston, MA.

Lee, K. K. 2009. Investigation of potential spudcan punch-through failure on sand overlying clay soils. Perth, The University of Western Australia.

Low, H.E., Lunne, T., Andersen, K.H., Sjursen, M.A., Li, X. & Randolph, M.F. 2010. Estimation of intact and remoulded undrained shear strengths from penetration tests in soft clays. *Géotechnique*, 60, 843–859.

Lunne, T., Andersen, K.H., Low, H.E., Randolph, M.F. & Sjursen, M. 2011. Guidelines for offshore in situ testing and interpretation in deepwater soft clays. *Canadian geotechnical journal*, 48, 543–556.

Lunne, T., Robertson, P.K. & Powell, J.J.M. 1997. *Cone penetration testing in geotechnical practice*, London, Blackie Academic & Professional.

Mayne, P.W. 2007. Cone Penetration Testing State-of-Practice. Atlanta, GA, Georgia Institute of Technology.

Newson, T.A., Watson, P.G. & Bransby, M.F. 1999. Undrained shear strength profiling using a spherical penetrometer. Crawley, Australia, Geomechanics Group, The University of Western Australia.

Peuchen, J. & Terwindt, J. 2014. Introduction to CPT accuracy. *Proceedings of 3rd International Symposium on Cone Penetration Testing*. Las Vegas, Nevada.

Randolph, M., Martin, C. & Hu, Y. 2000. Limiting resistance of a spherical penetrometer in cohesive material. *Géotechnique*, 50, 573–582.

Randolph, M. F. 2004. Characterisation of soft sediments for offshore applications. *Proceedings of 2nd International Conference on Site Characterization*. Porto.

Randolph, M.F., Jewell, R.J., Stone, K.J.L. & Brown, T.A. 1991. Establishing a new centrifuge facility. *Proceedings of the international conference on centrifuge modelling, Centrifuge'91*. Boulder, Colorado, Rotterdam, the Netherlands: Balkema.

Robertson, P.K. & Cabal, K.L. 2015. *Guide to cone penetration testing 6th Edition*, Gregg Drilling & Testing, Inc. 6th Edition, Signal Hill, California.

Watson, P., Newson, T. & Randolph, M. 1998. Strength profiling in soft offshore soils. *Proceedings of 1st International Conference on Site Characterisation*. Atlanta.

6. Modelling techniques

Development of a rainfall simulator in centrifuge using Modified Mariotte's principle

D. Bhattacherjee & B.V.S. Viswanadham
Department of Civil Engineering, Indian Institute of Technology Bombay, India

ABSTRACT: The paper presents design details of a simulator for replicating rainfall in the form of fine mist during centrifuge tests using specialized air atomizing nozzles. The relevant scaling laws involved in modelling rainfall at high gravities are discussed, with due consideration of unsaturation effects involved during rainwater infiltration. The importance of designing the pressurized water system in the form of Modified Mariotte's tube is highlighted, and the specialty of the developed actuator over previous air bladder systems used in literature is discussed in detail. Use of modified version of Mariotte's principle ensured that uniform rainfall intensity could be achieved over entire model surface area at high gravity, and the intensity could be regulated in-flight at any point of time during test to replicate all varieties of real-life natural hazards, ranging from lingering storms to intense cloudbursts. Further, measures adopted to minimize effects of Coriolis force on droplet trajectory are discussed, and the nozzle configuration selected to minimize chances of formation of 'blind zones' is presented.

1 INTRODUCTION

Recently, the frequency of rainfall-induced landslides has been increasing globally, coincident with the effects of climate change, causing innumerable deaths and severe damages to infrastructure. The annual statistical review report of the Centre for Research on the Epidemiology of Disasters reveals that, in the year 2016, almost 75% of the natural disasters that occurred globally may be attributed to rainfall, and the estimated financial losses (average) due to rain-triggered landslides exceed 92 million US$/annum, as reported by Dahal and Hasegawa (2008). The reason behind the instability may be attributed to the non-availability of good quality permeable granular material in recent times, which leads to the use of low-permeability soils having considerable fines content (alternatively termed as marginal soils) available locally at the construction site. During the process of rainwater infiltration into unsaturated marginal soils, the matric suction starts reducing rapidly with the progress of rainfall, leading to build-up of excess positive pore-water pressure and subsequent deformations, thereby triggering failures. Hence, there arise the necessity to investigate the behaviour of geotechnical soil structures (slopes, embankments, reinforced soil walls etc.), especially with low-permeable fill material under various rainfall intensities.

Small-scale physical modelling of engineered earth structures has been used in the past to provide insight into corresponding prototype behaviour. However, a limitation of scaled down models under normal gravity condition is that the stress levels are much smaller than that of prototype structures. Only full-scale physical models can include all these complexities, but they are expensive and time-consuming, and cannot be replicated with natural hazards like flooding and rainfall. In such situations, geotechnical centrifuge modelling can be used as an effective tool (Schofield, 1980, Taylor, 1995) to investigate the behaviour of earth structures as scaled-down models in a controlled environment. Drawing parallel to the above, the simulator under consideration was designed based on the principle of centrifuge modelling, in order to replicate accurately the effect of rainfall on various types of geotechnical structures, with special emphasis on the stability of natural and engineered soil slopes.

2 BASIC PRINCIPLE

2.1 Review of earlier set-ups

A review of existing set-ups developed by Craig et al. (1991), Tamate et al. (2010) and Ling and Ling (2012) reveal the fact that, majority of the researchers till date have simulated only a constant intensity of rainfall in-flight during centrifuge testing, in contrast to actual field conditions, where rainfall intensity is changing continuously with time. Moreover, measures taken to ensure uniform distribution of raindrops on model surface area at high gravities (with due consideration of Coriolis effects) are limited in the literature. Further, in most of the cases, the set-ups are deficient of a properly designed control system to regulate the intensity and duration of rainfall at the time of centrifuge testing. Another major concern arising out of the findings of earlier researchers is the erosion of model soil owing to large size of raindrops produced. The above-mentioned research gaps have prompted the necessity of designing a robust rainfall simulating system capable of addressing the existing drawbacks.

2.2 Scaling laws

The scaling laws involved in modelling rainfall at high gravities can be derived based on Dell'Avanzi et al. (2004) and Bhattacherjee and Viswanadham (2017). Let the notations p and m denote respectively the prototype and centrifuge model at $1/N$ scale, where the value of N corresponds to the geometric scale factor or gravity level. The scale factor for *suction* (N_ψ) under transient unsaturated rainfall conditions can be computed from Equation (1) as:

$$N_\psi = [Nz_m + N_v v_m K_p(z_p)] / [Nz_m \chi + v_m K_m(z_m)] \quad (1)$$

where, z_m represents the height of any specific point in the specimen from the base of the soil layer, χ is defined as the uniformity factor of the acceleration field, N_v represents the discharge velocity factor, v_m is the model discharge velocity, and $K_m(z_m)$ and $K_p(z_p)$ represent Gardner's k-function factor for the model and prototype respectively. As reported by Dell'Avanzi et al. (2004), if the ratio of centrifuge arm length (r_0) to model length (L_m) is sufficiently high (i.e, $r_0/L_m > 10$), then $\chi \sim 1$, and the suction scale factor (N_ψ) can be approximated as 1. In the present study, the r_0/L_m ratio is 17 for the centrifuge equipment and soil model dimension selected, which justifies adoption of a suction scale factor of 1 as reported in literature. Further, the *discharge velocities* in model and prototype are scaled as in Equation (2):

$$v_p = (1/N) v_m \quad (2)$$

Thus, for unsaturated flow, the discharge velocity (v) in the prototype scales by 1/N with respect to that in the model. Moreover, the *precipitation intensity* (r) and *duration* (t) are scaled in centrifuge as per Equation (3) and Equation (4) respectively:

$$r_m = N\, r_p \quad (3)$$

$$t_p = N^2\, t_m \quad (4)$$

Thus, a prototype rainfall intensity of 15 mm/h having 3 days duration in field replicates an intensity of 450 mm/h in model scale at 30 g, and corresponding rainfall duration is reduced to 5 minutes in centrifuge. Further, the droplet size should be reduced at higher gravities to ensure mist formation and occurrence of equivalent impact pressure on model ground surface as that experienced by prototype.

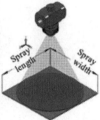

(a) A typical nozzle (b) Influence zone

Figure 1. Air-atomizing nozzle used in the study.

3 DESIGN DETAILS AND COMPONENTS

3.1 Air atomizing nozzles

The nozzles used in the present study (shown in Figure 1) are specialized full cone air-atomizing brass nozzles, capable of producing rainfall in the form of fine mist at high gravities by internal collision of pressurized air and pressurized water. In order to maintain a constant intensity of rainfall at high gravities, the air pressure (P_2) and water pressure (P_1) fed to the nozzles have to be kept constant. Accordingly, the water container used for supplying pressurized water to nozzles was designed to act like a Modified Mariotte's tube, capable of maintaining a constant head of water with progress of rainfall at high gravities, as discussed in subsequent sections.

3.2 Nozzle assembly attaching plate

The nozzle assembly attaching plate consists of slots along the entire length at fixed intervals. The nozzles are attached at the side slots, while central slots are used for housing nozzle water and air pipes, and for holding LVDTs.

3.3 Hanging arrangement

The nozzles are suspended at a certain height from the model soil surface by means of hanging rods, connected via brackets to the nozzle assembly plate through a nut-and-screw arrangement. The brackets are useful in securing the nozzles in proper position at high gravities, while the nut-and-screw connection serves as the point of rotation and facilitates adjustment in nozzle orientation, depending upon the inclination of model slope/soil wall to be tested. In addition, the nut-screw arrangement facilitates adjustment in the heights of nozzle hanging rods from the impact surface to arrive at various spray widths as per requirement.

3.4 Water container assembly with support system

3.4.1 Description

The water container assembly consists primarily of the overhead tank mounted on two supporting plates at the top and bottom, and is used to store the water required to create rainfall at high gravities. It is fabricated with mild steel and weighs 55 kg in empty condition, and 75 kg when filled to its full capacity (20 L). As the centrifuge equipment at Indian Institute of Technology Bombay has a maximum payload of 2.5 t [Chandrasekaran (2001)] at 100 g, it can be inferred that the weight imposed by the water tank at high gravity is a mere fraction of the load-carrying capacity of the centrifuge. Hence, from model preparation point of view, there is no limitation imposed by the weight of water tank on the weight of the soil model being tested. Below the bottom plate, a base plate stiffener is welded, and the entire assembly is supported by six metallic columns. The water container is provided with two-point openings at the top, one for venting air, and the other for supplying pressurized air from compressor. The main body of the container consists of a hollow pipe running upto 20 mm above the base, and has been designed on the principle of Mariotte's bottle for maintaining uniform intensity during rainfall. The application of Mariotte's principle in maintaining a constant head of water is well-known in literature, and was implemented by Thorel et al. (2002) during centrifuge testing. Rezzoug et al. (2004) also suggested the use of Mariotte's bottle for studies related to capillary rise in porous media at different levels of acceleration in centrifuge. However, the concept of Modified Mariotte's tube adopted in the present study for maintaining a constant water head during rainfall at high gravities is innovative, and has not been applied till date, to the best of the authors' knowledge.

3.4.2 Advantages over simple air-bladder systems

The air-atomizing nozzles used in the present study utilize a collision of pressurized water (P_1) and pressurized air (P_2) fed simultaneously at the two ends of each nozzle to produce an atomized spray. Thus, to simulate a uniform intensity of rainfall over the model soil surface, magnitudes of both water pressure (P_1) and air pressure (P_2) have to be maintained constant with the progress of rainfall. Although maintenance of constant air pressure is not a problem, but maintenance of constant water pressure is challenging due to the fact that, if a simple "air bladder" system (P_a) is used to pressurize the water (as adopted by previous researchers), the water pressure at nozzle tip (P_1) = P_a + Pressure due to existing height of water. As rainfall progresses, height of water reduces, and so does the magnitude of net water pressure (P_1) fed to the nozzle system, thereby creating variable rainfall intensity. The above problem can be overcome by either continuously changing the bladder air pressure (P_a) to compensate for the falling water head, or by designing the water container in the form of a Modified Mariotte's tube, as explained in Figures 2(a–c).

3.4.3 Illustration of Modified Mariotte's principle

Figure 2(a) represents a situation arising in the absence of Mariotte's bottle, where the water level, and hence water head drops continuously as a function of time (from h at $t = t_0$ to h' at $t = t_1$) with progress of rainfall. Hence, this is not advisable, as rainfall intensity changes with time. In the next case shown in Figure 2(b), a constant head (h_c) is being maintained with the help of Mariotte's bottle, as a result of which, the ordinate at the base of the pressure triangle (P_1) is equal to $h_c \rho g N$ at time intervals t_0 and t_1 with rainfall at high gravities. However, the case presented involves two major disadvantages, the primary one being that, the magnitude of water pressure supplied to the nozzles (P_1) in flight cannot be regulated, and is always equal to ($h_c \rho g N$), implying rainfall intensity during a particular test is always constant. Moreover, the magnitude of P_1 is not sufficient to replicate higher rainfall intensities at Ng due to limitations in the value of constant head (h_c). Hence, while designing the present rainfall simulator, a modified version of Mariotte's principle was considered [Figure 2 (c)], wherein pressure was applied externally through compressor (P_e) over and above normal atmospheric pressure (P_0). Thus, the magnitude of pressure (P_1) at the centre line of nozzles is equal to ($P_e + h_c \rho g N$) at any point of time during rainfall at Ng. In this case, it is possible to vary the intensity of rainfall in-flight condition by regulating the magnitude of P_e externally through valves in control room. In addition, by increasing or decreasing P_e, various prototype rainfall events can be simulated, from lingering storms to intense cloudbursts.

3.4.4 Activation of Mariotte's principle at Ng

As shown in Figure 3, when water is allowed to flow through the solenoid valve and nozzles at Ng (Case B), the initial water level (A-B in Case A) comes down to A'-B' (shown magnified in Case B), thereby increasing the volume A-B-C-D to A'-B'-C-D, and causing the pressure to drop in that region. This creates a pressure in-equilibrium inside the Mariotte's tube at the level of point E. Hence, in order to regain the equilibrium at the level of E, the water column inside the stand-pipe (Case A) comes down proportionally. The height at a particular instant of time is shown in Case B by the reduced water column, corresponding to which pressure at nozzle tip is indicated. Once the water inside the stand-pipe drains out completely, equilibrium is attained (Case C), such that constant atmospheric pressure (P_0) is attained within the Mariotte's tube at the level of point E, thereby indicating activation of Mariotte's principle.

3.4.5 Activation of Modified Mariotte system at Ng

If Case D is considered (before opening solenoid valve to create rainfall) where a Modified Mariotte's principle is applied, then Case A (just filled with water) is common to this process as well, where initial pressure in A-B-C-D is equal to atmospheric pressure (P_0). If an external gauge pressure P_e (over and above P_0) is applied at high gravity, then pressure at point E (just at the bottom tip of the stand-pipe) is equal to existing height of water + gauge pressure P_e. However, at the level of point E (to the left/right) inside the Modified Mariotte's water container, the pressure is equal to existing height of water (P_0 is not considered as we are dealing in terms of gauge pressure). To equilibrate pressure at the level of point E, water inside the

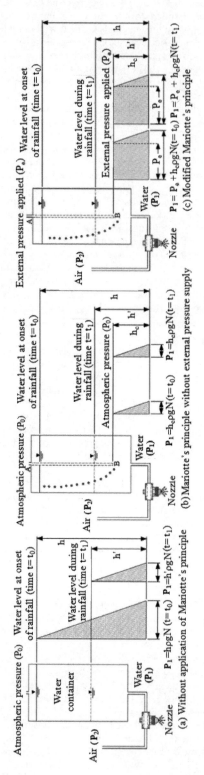

Figure 2. Design of pressurized water system based on Modified Mariotte's principle.

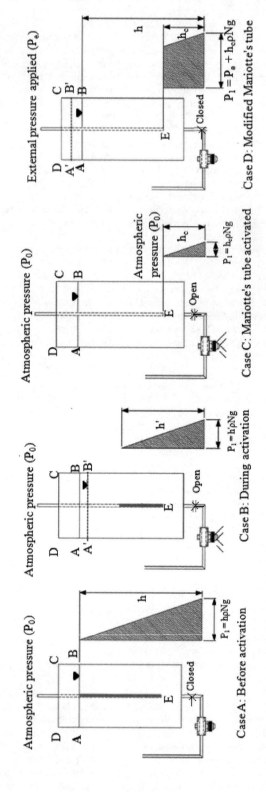

Figure 3. Activation of Mariotte's principle at high gravity.

340

stand-pipe drains out, thereby elevating slightly the level of A-B to A'-B' (shown magnified in Case D). This slight increase in level is not sufficient to equilibrate pressures at point E, and subsequently is followed by air bubbling to compensate for further pressure differences, until pressure in Modified Mariotte's water container at level E becomes equated to P_e, ensuring activation of the Modified Mariotte's principle, as presented in Figure 3. Although the activation process described above is almost instantaneous, as a precautionary measure, a waiting time of about 5 min was allowed after setting the external gauge pressure (P_e) at high gravity before opening the solenoid valve to create rainfall.

3.5 Additional components

In addition to the above components, a solenoid valve was installed in the path of water flowing from the water container assembly to water distribution network system connected to the nozzles for controlling the flow of water during tests. The valve can be operated from the control room, by means of which, rainfall can be started/stopped at any point of time during centrifuge tests. In addition, a seepage tank was used during centrifuge tests to maintain an initial water table in model slopes/soil walls before the onset of rainfall. Lastly, the excess water flowing as run-off during tests was collected in a container placed on the swing basket of centrifuge.

4 SALIENT FEATURES AND MERITS

4.1 Advantages over existing set-ups

The developed rainfall simulator has salient features and advantages over other simulators developed till date, which are summarized herein. Firstly, it is capable of producing rainfall in the form of fine mist, thereby reducing chances of washing away of soil particles due to erosion, as reported by earlier researchers. As a consequence, impact pressure on ground surface in centrifuge is inferred to be reduced to approximately the same level as that of prototype, and the phenomena of failure of earth structures due to rainwater percolation inside unsaturated soils is properly replicated. Secondly, the water container is designed to act like a modified Mariotte's tube when filled with pressurized water at high gravities, on account of which, the simulator can produce rainfall at a uniform rate in-flight condition, and the rainfall intensity can be changed at any point of time during test by changing magnitude of external pressure applied (P_e). This is a notable advantage of the developed rainfall simulator compared to other test set-ups. Further, by increasing the water pressure (P_1) relative to the air pressure (P_2) fed to the nozzles within a short period of time, it is possible to create a short spell of very high intensity rainfall, thereby simulating cloudburst conditions due to major rainfall. On the other hand, by increasing P_2 relative to P_1 over a relatively long duration, it is possible to study the response of soil structures subjected to long term medium/low intensity rainfall, which is a typical pattern of antecedent rainfall. Replication of antecedent rainfall patterns are of utmost importance in case of unsaturated soils, as reported by Lee et al. (2011) and Rahimi et al. (2011), as it controls the rate of decrease in factor of safety by creating an initial wetting condition in the soil.

4.2 Measures adopted to nullify Coriolis effects

A major error that arises in centrifugal rotational field is due to Coriolis force, which generates a motion in a direction opposite to that of rotation of centrifuge. As a result, the droplets tend to get drifted away from the impact surface, and lesser number of raindrops reach the target as desired. While designing the present simulator, side slots were provided at regular intervals along the entire length of the plate arrangement fabricated for housing the nozzle hanging rods, in order to enable horizontal shift in position of the nozzles with nut-screw arrangement, depending upon direction of centrifugal rotation. As another precautionary measure, the maximum g level was restricted to 30 g at the time of centrifuge model tests on earth structures, on account of the observations reported by Caicedo et al. (2015), which states that Coriolis effects becomes more pronounced with increase in centrifugal acceleration, especially above 50 g.

4.3 Prevention of 'blind zones'

The nozzles used in the present case cover a circular area, which is approximated to the equivalent square impact area (of 100 mm by 100 mm), as presented in Figure 1. Let us consider Case P shown in Figure 4, where influence area of a typical nozzle is shown. In this case, area of circular spray zone $= [\pi(100^2)/4]$ mm^2, whereas, area of equivalent square impact area $= 100 \times 100$ mm^2. Hence, ratio of area covered in practice on account of rectangular model geometry $= [\pi(100^2)/4]/(100 \times 100) = 80\%$, which is quite satisfactory. Thus, chances of formation of zones receiving no rainfall (referred to as blind spots) are much reduced. Moreover, due to the effect of Coriolis force at high gravities, the effective spray width tends to be somewhat elliptical, as shown in Case P [which, however remains to be validated]. This, together with the mist formation at high gravity is assumed to increase, to some extent, the spread of the nozzles, and ensures adequate distribution of rainfall over the impact surface. Thereby, chances of zones (a, b, c, d) receiving rainwater during centrifuge testing are increased, as shown in Figure 4(a).

A second option is the layout shown in Case Q, [Figure 4(b)], where overlapping of spray occurs from adjacent nozzles, and an area is covered twice in the process. However, in such cases, the approximation in area covered on account of rectangular model geometry as opposed to observed circular spray

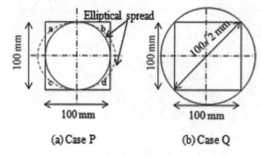

Figure 4. Equivalent influence area of nozzles.

pattern $= [\pi(100\sqrt{2})^2/4]/[100 \times 100] = 157\%$, which is erroneous, and unacceptable. Hence, in the present study, the latter setting was not considered, and the former arrangement was opted for. However, as explained in previous sections, as the height of the nozzles from the impact surface can be adjusted in the developed rainfall simulator, it is possible to vary the effective spray-width (if required) and create a certain zone of overlapping (in between two extreme cases depicted in Case P and Case Q so that chances of formation of 'blind zones' are still further reduced using the developed rainfall simulator.

5 CONCLUSIONS

In the present paper, design and development of a rainfall simulator for inducing rainfall at high gravity environment in a geotechnical centrifuge is presented. Detailed discussions have been conducted on the design of the pressurized water system in the form of Mariotte's tube for ensuring uniform rainfall intensity at higher gravity levels. Further, the use of modified version of Mariotte's principle ensured that the intensity simulated could be regulated in-flight at any point of time to replicate all varieties of real-life natural hazards. Further, measures adopted to minimize effects of Coriolis force and to minimize chances of formation of 'blind zones' are presented. The developed rainfall simulator can be used to investigate the stability of a wide range of geotechnical structures subjected to rainfall of various intensities and duration, including landfill slopes, ash-pond and tailing dam slopes, geosynthetic reinforced soil walls and slopes, soil nailed slopes and anchored slopes.

REFERENCES

Bhattacherjee, D., & Viswanadham, B.V.S., 2018, Design and Performance of an In-flight Rainfall Simulator in a Geotechnical Centrifuge, *Geotechnical Testing Journal*, ASTM, 41(1), 72–91.

Caicedo, B., Tristancho, J., & Thorel, L., 2015, Mathematical and physical modelling of rainfall in centrifuge, *International Journal of Physical Modelling in Geotechnics*, 15(3), 150–164.

Chandrasekaran, V. S., 2001, Numerical and Centrifuge Modelling in Soil Structure Interaction, *Indian Geotechnical. Journal*, 31(1), 30–59.

Craig, W.H., Bujang, B.K.H., & Merrifield, C.M., 1991, Simulation of climatic conditions in centrifuge model tests, *Geotechnical Testing Journal*, ASTM, 14(4), 406–412.

Dahal, R. K., & Hasegawa, S., 2008, Representative rainfall thresholds for landslides in the Nepal Himalaya, *Geomorphology*, 100(3–4), 429–443.

Dell'Avanzi, E., Zornberg, J. G., & Cabral, A. R., 2004, Suction Profiles and Scale Factors for Unsaturated Flow Under Increased Gravitational Field, *Soils and Foundations*, 44(3), 79–89.

Lee, L.M., Kassim, A., & Gofar, N., 2011, Performances of two instrumented laboratory models for the study of rainfall infiltration into unsaturated soils, *Engineering Geology*, 117(1-2), 78–89.

Ling, H., & Ling, H.I., 2012, Centrifuge Model Simulations of Rainfall-Induced Slope Instability, *Journal of Geotechnical and Geoenvironmental Engineering*, ASCE, 138(9), 1151–1157.

Rahimi, A., Rahardjo, H., & Leong, E.C., 2011, Effect of antecedent rainfall patterns on rainfall-induced slope failure, *Journal of Geotechnical and Geoenvironmental Engineering*, ASCE, 137(5), 483–491.

Rezzoug, A., König, D., & Triantafyllidis, T., 2004, Scaling Laws for centrifuge modelling of capillary rise in sandy soils, *Journal of Geotechnical and Geoenvironmental Engineering*, ASCE, 130(6), 615–620.

Schofield, A. N., 1980, Cambridge geotechnical operations, *Geotechnique*, 30(3), 227–268.

Tamate, S., Suemasa, N., & Katada, T., 2010, Simulating shallow failure in slopes due to heavy precipitation, *7th ICPMG, Switzerland*, S. Springman, J. Laue, and L. Seward (Eds.), Taylor and Francis group (Pubs.), Vol 2, pp. 1143–1149.

Taylor, R. N., 1995, Centrifuges in modelling: principles and scale effects, Geotechnical Centrifuge Technology, R. N. Taylor, (Eds.), Blackie Academic and Professional, Glasgow, UK.

Thorel, L., Favraud, C., & Garnier, J., 2002, Mariotte bottle in a centrifuge: a device for constant water table level, *International Journal of Physical Modelling in Geotechnics*, 2(1), 23–26.

Development of model structural dampers for dynamic centrifuge testing

J. Boksmati & S.P.G. Madabhushi
Schofield Centre, Department of Engineering, University of Cambridge, UK

N.I. Thusyanthan
Chief Engineer, Atkins, UK

ABSTRACT: Supplemental energy dissipation devices are often selected based on the fixed base response of structures. There is limited knowledge regarding the actual seismic performance of these buildings with regard to soil-structure interaction. This paper discusses the development and testing of miniature viscous dampers capable of fitting into structural models for dynamic centrifuge testing. An electromagnetic shaker was utilised to test the miniature dampers under high frequency small stroke conditions, similar to those expected during shaking in the centrifuge. Three damping fluids have been investigated; H68 oil, H32 oil, and water. Different piston stroke frequencies were applied for each of the fluids being tested. The miniature viscous dampers were successful in dissipating energies at small strokes. However, the damper setup seems to possess some inherent stiffness in addition to its viscous behaviour. Cavitation and entrapped air plays a critical role in limiting the effect of fluid viscosity on the damping behaviour for the small strokes and high frequencies being applied.

1 INTRODUCTION

Project design engineers are usually tasked with developing earthquake mitigation plans to accomplish specific seismic performance objectives set by the client. The plan essentially highlights how the structure will provide the necessary capacity to meet the predicted seismic demand; this is often an interplay between the structure's kinetic energy, elastic deformations, inelastic deformations, and material hysteretic damping. Critical infrastructure assets and other key projects often have very stringent performance objectives. In many cases, clients demand operational continuity post-earthquakes, and request maintenance to be limited to non-structural components in their assets.

One traditional design approach would be to dissipate the seismic energy through elastic deformations within the structural frame. However, this usually results in uneconomical designs in which the structural member sizes are governed by a low probability event rather than more frequent actions. There is an increasing trend in industry to incorporate supplemental damping devices into structural frames as means of enhancing the seismic capacity of structures. Kasai (2016) conveniently classifies these supplemental devices as being either hysteretic (displacement-based), viscous (velocity-based), or visco-elastic. Other forms of supplemental damping, such as base isolation and tuned-mass dampers, have also been investigated and implemented in industry.

Unfortunately, the bulk of the research in the field of supplemental damping has been primarily driven

Figure 1. New education building retrofit at the University of Canterbury, New Zealand. (Victor Seismic 2017).

by structural engineering. There is very limited input from geotechnical engineers about how soil flexibility influences the performance of structures equipped with these devices.

Recent centrifuge work by Jabary & Madabhushi (2015) highlighted potential drawbacks in setting the tuned mass damper frequency to the fixed-base natural frequency of structures. Jabary & Madabhushi (2015) concluded that a tuned-mass damper tuned to the soil-structure natural frequency outperforms that which is tuned to the fixed-base natural frequency. This finding raises questions about the performance of the other damping devices under real soil-structure conditions.

Among the several damping technologies listed earlier, viscous dampers have proven to be quite a popular option in projects requiring better storey-drift control and energy dissipation. Their dependency on velocity implies that the damping forces generated are out-of-phase with the applied seismic accelerations, hence, minimising any influence the additional damping forces may have on the structural members (Lee & Taylor 2001). In fact, their dependence on velocity allows them to mitigate both floor accelerations and storey shears simultaneously (Symans & Constantinou 1998). A wealth of knowledge about the performance of viscously damped buildings originates from intensive research programs, such as Kasai et al. (2010) and Chang et al. (2008), who investigated scaled down prototypes founded on rigid foundations. A review of current literature has shown very limited efforts towards understanding the effects of soil flexibility on the overall viscously damped structural response. Numerical and analytical evidence by Li et al. (2015), Zhou et al. (2012), and Zhao et al. (2017) has hinted towards a drop in damper efficiencies with softer ground. Dynamic centrifuge testing of model scale sway frames with these miniature dampers can provide vital insight into how seismic performance of real buildings could be improved by such devices. This paper will focus on the development and testing of miniature dampers under high frequency small stroke conditions using a Ling-200 series electromagnetic shaker.

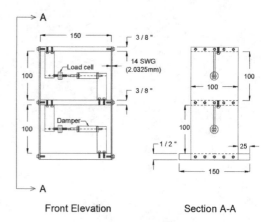

Figure 2. Cross-sectional view of the centrifuge model frames. All dimensions are in mm unless otherwise noted.

Figure 3. Damped and undamped models fully assembled with miniature dampers in place.

2 PHYSICAL MODELLING

2.1 Model structures

Two identical model scale structures were constructed for the purpose of this investigation. The frames were fabricated using 6082-T6 aluminium alloy plates assembled to represent a two degree of freedom system. Figures 2 and 3 illustrate the dimensions and cross-section of the model frames. One model has been fitted with side plates to hold miniature dampers in place, while the other frame was used as an undamped control reference. The stiffness and mass distribution in the physical models were proportioned to achieve fixed base natural frequencies of 52 Hz and 119 Hz model scale.

At the intended centrifuge acceleration of 50 g, the models would represent a fixed-base prototype with natural frequencies of 1.0 Hz and 2.38 Hz respectively. The choice of replicating a moderately flexible prototype was necessary to ensure practical inter-storey drift magnitudes during centrifuge shaking. The floor displacements at model scale had to be large enough to trigger proper viscosity induced damping behaviour from the miniature dampers. Otherwise, for very small damper piston movements (i.e. stroke), friction at the piston-seal interfaces would dominate the damping mechanism. In an attempt to further promote damper piston movements, the miniature dampers were positioned horizontally in-line with the floor displacements. The stocky aspect ratio of the model frames and the large raft foundations helped promote inter-storey shear deformations and damper activation by lowering the centre of gravity to minimise possible rocking during shaking.

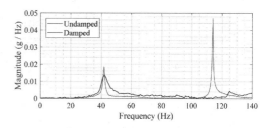

Figure 4. FFT of normalised roof accelerations for both structures during impact testing.

Impact testing was conducted on the two model frames to determine their as-built natural frequencies. The models were clamped at their base and subjected to small lateral impulses using a hammer. Figure 4 presents the results of a Fast-Fourier Transform conducted on normalised roof acceleration traces for both models. Data from each frame was normalised relative to the peak accelerations to account for the different

Table 1. Dynamic properties of the centrifuge models.

	Undamped	Damped
Mode 1	41.7 Hz	42 Hz
Mode 2	114 Hz	–
Damping ratio, ξ		
Mode 1	1.06%	3.7%

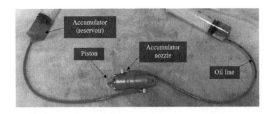

Figure 5. Actual damper setup.

magnitudes of hits applied during impact. The structures exhibited a fundamental frequency of 41.7 Hz and 42 Hz for the undamped and damped frames respectively. This marked a 29% approximate drop in stiffness when comparing that of the fabricated models to that which has been assumed in design. This drop in stiffness is attributed to the partial fixity provided by the single row of bolts used for the floor-to-wall connection detail.

Table 1 summarises the dynamic characteristics of the two structural frames. The damped model is approximately 2% stiffer than the undamped model. This minor variation in stiffness is the result of different tightening torques applied to the frame bolts during assembly. The additional connections fitted to support the miniature dampers in the damped model increased the frame's hysteretic damping. This increase in damping ratio suppressed the occurrence of a second mode as shown in Figure 4.

At a centrifuge acceleration of 50 g, the 42 Hz damped model represents a prototype building with 0.84 Hz fundamental frequency. Using the equation provided in Eurocode 8 (BS EN 1998-1) for predicting structural natural periods, and assuming a steel structural frame with $C_t = 0.085$, the model frames represent a 34m tall building (9-storey structure).

$$T_0 = C_t H^{0.75} \quad (1)$$

where T_0 = fundamental period of structure, C_t = structural frame constant, and H = effective height.

2.2 Miniature dampers

Model-scale hydraulic cylinders were procured from a supplier who specialises in manufacturing small-scale components for R/C (remote control) applications. The cylinders were slightly modified from their initial configuration to better simulate the internal components of passive dampers. Essentially, an O-ring seal originally installed around the piston head has been removed to create an annular gap between the piston head and damper body. This annular gap effectively acts as an orifice as the piston head moves in and out of the damper. Figure 5 shows a schematic representation of the damper setup. The actual setup can be seen in Figure 6.

The modular construction of the cylinders allowed access to all the internal components of the damper, and provided flexibility for testing different damping fluids. During the development phase, three types

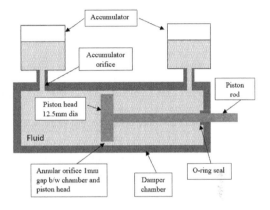

Figure 6. Schematic representation of the damper setup.

Table 2. Fluid properties of the different fluids tested.

	H68 Oil	H32 Oil	Water
Density (kg/m³)	865	857	1000
Kinematic viscosity at 20°C	220 cSt	87 cSt	1 cSt

of fluids were investigated; H68 hydraulic oil, H32 hydraulic oil, and water. Table 2 gives the basic fluid properties at room temperature for the three fluids.

3 DAMPER CHARACTERISATION

3.1 Test setup

Prior to testing the dampers in the centrifuge, it was important to understand, and where possible, replicate the conditions that these miniature dampers will experience during a dynamic centrifuge test. The first challenge was that these miniature dampers had to function under very small strokes (in the order of a few millimetres). The second challenge was that the strokes would be occurring at frequencies higher than what is typically experienced by prototype dampers in real structures. As illustrated earlier, the damped model frame has a fundamental fixed-base frequency of 42 Hz (model-scale). Hence, any base input motion applied in the centrifuge would trigger damper strokes

Figure 7. Experimental setup for testing miniature dampers at high frequency, small strokes.

at frequencies close to that range. A special setup, utilising a Ling-200 series shaker, was used to characterise damper behaviour under small stroke, high frequency inputs. The damper was clamped by a stand on one side, and connected to the shaker on the other. A function generator was used to drive the mechanical shaker at different frequencies. The experimental setup was instrumented with a tension-compression load cell to record the damping forces generated during vibration. A MEMS accelerometer was positioned on the damper piston to record piston acceleration traces with time. From this data, and through band-pass filtering and integration, piston velocities and stroke magnitudes were derived.

3.2 Results and discussions

3.2.1 Ideal viscous behaviour

Force-velocity and force-displacement plots are essential to characterising any damper performance. Ideally, viscous dampers are velocity dependant devices, with a force-velocity relationship governed by the equation:

$$F = cV^\alpha \quad (2)$$

where F = damping force, c = damping coefficient, V = velocity, and α = velocity exponent.

The velocity exponent is a function of piston head valving and orifice design in a damper (Duflot & Taylor 2008). Figure 8 shows the typical viscous damper characteristics for different velocity exponents. Linearity in the force-velocity relationship tends to simplify the selection and design process of dampers in a structure (Lee & Taylor 2001). However, most practical applications favour dampers with velocity exponents less than 1 (Lee & Taylor 2001). These nonlinear dampers generate greater damping forces at lower velocities compared to linear dampers (Hwang, 2002). Moreover, they exhibit a force cap with increasing velocity, which helps protect structural members from excessive forces during intensive shaking. The hysteresis in the force-displacement plot represents energy dissipated per cycle of piston movement.

For ideal viscous dampers which are purely velocity dependant, force displacement loops are perfect ellipses. The smoothness of the ellipse is a function of the velocity exponent.

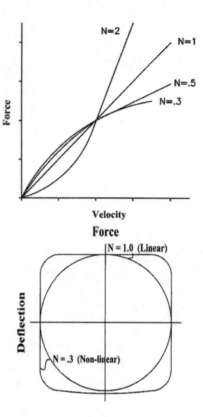

Figure 8. Variation of force-velocity and force-displacement response of viscous dampers with different velocity exponents (Adapted from Lee and Taylor, 2001).

3.2.2 Effect of input frequency

The shaker was set at four different frequencies for each of the fluids being investigated to examine the effect of input frequency on the damping behaviour.

The force traces have been filtered to remove very high frequency spikes for a clearer illustration of the force-velocity and force-displacement patterns observed. These spikes in force are attributed to contact between the moving coil assembly driving the damper, and the permanent magnet in the shaker. A low-pass filter was applied in MatLab setting the frequency cut-offs at 200 Hz, 250 Hz, 300 Hz, and 350 Hz for the 20 Hz, 30 Hz, 40 Hz, and 50 Hz frequencies being investigated.

Unlike the ideal behaviour presented earlier, the force-velocity plots for H68 oil shown in Figure 9 display considerable hysteresis in velocity. This implies that a portion of the input energy going into the damper gets stored rather than fully dissipated (i.e. the damper possesses some stiffness). At the 20 Hz and 30 Hz frequencies, there was a clear separation between the acceleration branch and deceleration branch of the compression stroke. Interestingly, this separation was not as excessive during the rebound portion of the cycle.

The very high frequency strokes applied by the shaker caused local cavitation and air bubbles to form

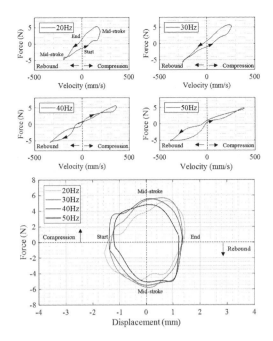

Figure 9. Force-velocity and force-displacement for H68 oil.

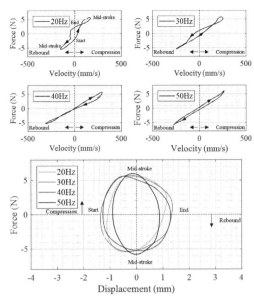

Figure 10. Force-velocity and force-displacement for water.

in the oil at the vicinity of the piston head. At 40 Hz and 50 Hz, the problem of oil cavitation becomes worse and the force-velocity traces become highly irregular for both compression and rebound strokes. This is mainly induced by the inconsistent viscous behaviour of foamed oil flowing around the piston head. A noticeable drop in gradient of the force-velocity plots, hence damping efficiency, can be observed as frequency increases from 20 Hz to 50 Hz. Symans & Constantinou (1998) observed the same trend and attributed this to speed incompatibility between the accumulator nozzle discharging oil to the accumulator and the dynamic piston displacements at high frequencies. This lag accentuates the effect of fluid compressibility in the overall damper response. Foamed oil is more compressible that de-aired oil, which explains the substantial hysteresis observed in velocity. Jiao et al. (2017) go a step further and associate the drop in damper efficiency to the shear thinning properties of the damping fluid.

3.2.3 Effect of fluid viscosity

Comparing the force-velocity plots for H68 oil and water in Figure 10, both damping fluids exhibit hysteresis in velocity. However, the lag between damping force and piston velocity was noticeably smaller for water than for the H68 hydraulic oil. This highlights the adverse effects that cavitation can have on the damping response in oils.

Basically, the more viscous H68 oil was not capable of flowing fast enough through the narrow orifice around the piston to match the high frequency strokes applied. This initiated local cavitation in the oil and resulted in a rather chaotic and unpredictable response from the foamed oil. Less viscous water flows around the piston head with lower resistance which limited the initiation of cavitation. The net result was a more predictable and smoother force-velocity relationship.

4 CENTRIFUGE TEST RESULTS

Testing the miniature dampers at 1 g using the Ling-200 shaker has proven that the devices are indeed capable of dissipating work at high frequency small-stroke inputs. Nevertheless, the damping efficiency of the devices seems to be hindered by cavitation and foaming of oil. In an attempt to evaluate the model damper performance under enhanced g-level, the damped and undamped structural frames presented in Figure 3 were embedded into dry Hostun sand (HN-31 80% relative density) and simultaneously tested at 50 g. The damped model was equipped with dampers filled with H68 hydraulic oil. Roof acceleration results for 0.6 Hz (30 Hz model scale) sinusoidal input motion of 0.12 g magnitude are presented in Figure 11.

The model dampers have successfully reduced peak roof accelerations in the damped frame compared to its undamped counterpart. A substantial reduction in the free vibration phase can also be observed when comparing the damped and undamped frame responses at the end of the base excitation.

Damper force-velocity plots during two steady state cycles are presented in Figure 12. Damping coefficients in the centrifuge (i.e. slope of the force-velocity plots) was much higher than its 30 Hz counterpart at 1 g. It is believed that the enhanced g-level increased the oil pressure difference generated across the piston

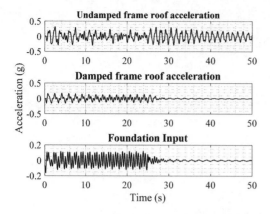

Figure 11. Prototype structural accelerations for model frames embedded in dense sand test at 50 g experiencing 0.6 Hz (30 Hz model scale) sinusoidal input motion.

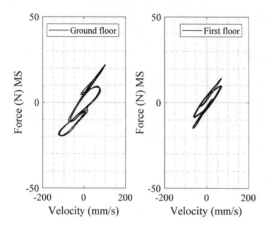

Figure 12. Force-velocity plots for damper on the first floor and on the ground floor during steady-state cycles. Force in model scale.

head in the dampers during piston motion. Hysteresis in velocity was still observed.

5 CONCLUSIONS

An electromagnetic shaker was utilised to determine the damping characteristics of three different damping fluids (H68 oil, H32 oil, water) at four different frequencies (20, 30, 40, and 50 Hz). As frequency increased, an overall decrease in the damping coefficient was observed for the hydraulic oils tested. This was attributed to the formation of air bubbles and foaming of the oil at high frequencies. Despite the force-velocity lag which has been recorded for all the fluids tested, the miniature dampers have been quite successful in dissipating energy at very small strokes and high frequencies. The model dampers were tested in the centrifuge as part of a two-degree of freedom frame embedded in dense dry sand. The enhanced g-levels resulted in higher damping coefficients than at 1 g. However, lag between damping forces and velocity was still observed.

REFERENCES

BS EN 1998-1:2004 Eurocode 8: Design of structures for earthquake resistance.

Chang, K., Lin, Y. & Chen, C. 2008. Shaking Table Study on Displacement-Based Design for Seismic Retrofit of Existing Buildings Using Nonlinear Viscous Dampers, 134(April), 671–681.

Duflot, P. & Taylor, D. 2008. Experience and Practical Considerations in the Design of Viscous Dampers, in Footbridge 2008: Third International Conference.

Hwang, J. 2002. Seismic Design of Structures with Viscous Dampers, International Training Programs for Seismic Design of Building Structures. Taiwan.

Jabary, R.N. & Madabhushi, S.P.G. 2015. Tuned mass damper effects on the response of multi-storied structures observed in geotechnical centrifuge tests. Soil Dynamics and Earthquake Engineering, 77: 373–380.

Jiao, S., Tian, J., Zheng, H. & Hua, H. 2017. Modeling of a hydraulic damper with shear thinning fluid for damping mechanism analysis, JVC/Journal of Vibration and Control, 23(20): 3365–3376.

Kasai, K., Ito, H., Ooki, Y., Hikino, T., Kajiwara, K., Motoyui, S., Ozaki, H. & Ishii, M. 2010. Full-scale shake table tests of 5-story steel building with various dampers, in 7th International Conference on Urban Earthquake Engineering (7CUEE) & 5th International Conference on Earthquake Engineering (5ICEE). Tokyo, 11–22.

Kasai, K. 2016. Performance of Seismic Protective Systems for Super-Tall Buildings and Their Contents, International Journal of High-Rise Buildings, 5(3):155–165.

Lee, D. & Taylor, D. P. 2001. Viscous damper development and future trends, Structural Design of Tall Buildings, 10(5), pp. 311–320. doi: 10.1002/tal.188.

Li, P., Yang, J. & Lu, X. 2015. Study on Influence Effect of Vicious Dampers Considering Soil-Structure Dynamic Interaction, in 6th International Conference on Advances in Experimental Structural Engineering. Illinois.

Symans, M.D. & Constantinou, M.C. (1998) Passive Fluid Viscous Damping Systems for Seismic Energy Dissipation, ISET Journal of Earthquake Technology, 185–206.

Victor Seismic, 2017. Case Studies. [Online] Available at: https://www.victorseismic.co.nz/case-studies-0

Zhao, X., Wang, S., Du, D. & Lie, W. 2017. Optimal Design of Viscoelastic Dampers in Frame Structures considering Soil-Structure Interaction Effect, Shock and Vibration, Volume 17.

Zhou, Y., Guo, Y. & Yong, Z. 2012. Influence of Soil-Structure Interaction Effects on the Performance of Viscous Energy Dissipation Systems, 15th World Conference on Earthquake Engineering (15WCEE).

Experimental evaluation of two-stage scaling in physical modelling of soil-foundation-structure systems

A. Borghei & M. Ghayoomi
University of New Hampshire, Durham, New Hampshire, USA

ABSTRACT: Geotechnical centrifuges have been successfully used to study Soil-Structure Interaction (SSI) problems. There is growing demand to conduct SSI tests for large prototype systems. Consequently, these experiments should be done either in larger centrifuge facilities or in higher centripetal acceleration fields. However, earthquake actuators which can produce seismic load with very high frequencies are not currently available. An innovative solution would be developing new scaling factors between a prototype and a physical model, so that dynamic experiments with high scaling factors can be done in lower acceleration fields. In this study, the capability of the "two-stage scaling" method is experimentally evaluated. Two models representing the same prototype were designed based on the conventional centrifuge scaling factors and the proposed method. Dynamic responses of the models in the prototype scale were approximately similar. The method was more successful to capture the inertial interaction effects than the kinematic interaction effects.

1 INTRODUCTION

During an earthquake, it is well known that the motion at a foundation of a structure, called the Foundation Motion (FM) is different from the motion far from the structure, which is usually named the Free-Filed Motion (FFM). Terms such as Soil-Structure Interaction (SSI) or Soil-Foundation-Structure Interaction (SFSI) are commonly used in the literature to study the mechanics and extent of this phenomenon (for example, Ghayoomi & Dashti 2015).

Shaking table tests both at 1-g and in geotechnical centrifuges have been extensively used to study SSI. Scaling factors implemented in these tests are generally different, for example, the strain scale factor is unity in centrifuge tests, while it is not unity in 1-g tests. Thus, it is necessary to make assumptions about the nonlinear behavior of soils to approximately estimate the strain scale factor in 1-g tests. Consequently, capturing the true soil nonlinear response is more difficult in 1-g shaking table tests, which makes geotechnical centrifuges more attractive for SSI studies.

The peak acceleration developed by in-flight servo-hydraulic actuators inside a geotechnical centrifuge is a function of the payload mass shaken by the table, the performance of the servo-valves, the volume of the accumulators, and pressure of the hydraulic fluid (Scott 1994). Generally, as the frequency of the motion increases, the amplitude of the motion decreases. This constrains dynamic centrifuge testing to be done with a scaling factor (prototype/ model) ranging from 10 to 100 g. Thus, this would limit the size of models that can fit inside the centrifuge platforms. To date, there is demand from the engineering community for physical modeling of large prototype systems that involve SSI. The scale factor needed to bring dimensions of these large prototypes down to an appropriate size and can be spun by current centrifuge facilities is between 100 to 1000 (Iai et al. 2005). One solution is to build larger centrifuges; however, building and conducting experiments with these facilities are relatively expensive. Another solution is to develop new similitude laws which combine standard scaling factors from traditional 1-g and centrifuge tests to come up with physical models that can capture the response of large structural models. Iai et al. (2005) theoretically developed generalized scaling factors for dynamic centrifuge tests by combing scaling factors of 1-g tests and centrifuge tests, which is called the "two-stage scaling" method. The method not only enables researchers to test models with high scaling factors in low centripetal acceleration fields but also can be used for testing large prototypes in small centrifuges with lower payload capacities. In this paper, a procedure is designed to experimentally evaluate the capability of the two-stage method in SSI analysis of a single degree of freedom structural model on dry sand.

A target prototype structure with significant SSI influence was selected, then, it was scaled down based on 1) regular centrifuge scaling factors and 2) the two-stage scaling factors to design the Conventional Centrifuge Model (CCM), and the Two-stage Centrifuge Model (TCM), respectively. Both experimental models were characterized so they represent the same dynamic system according to pre-defined criteria. Physical properties of the two models such as the first fixed-based natural frequency, the damping ratio, mass, and their dimensions were measured. After preparing the soil inside a laminar container,

Table 1. Scaling Laws (after Iai et al. 2005).

Quantity	General scaling factors	Scaling laws		
		Partitioned scaling factors		
		1-g	Centrifuge	Two-stage scaling
Length	λ	μ	η	$\mu\eta$
Density	λ_ρ	1	1	1
Strain	λ_ε	$\mu^{1/2}$	1	$\mu^{1/2}$
Acceleration	λ_g	1	$1/\eta$	$1/\eta$
Time (dynamic)	$(\lambda\lambda_\varepsilon/\lambda_g)^{1/2}$	$\mu^{3/4}$	η	$\mu^{3/4}\eta$
Frequency	$(\lambda\lambda_\varepsilon/\lambda_g)^{-1/2}$	$\mu^{-3/4}$	$1/\eta$	$\mu^{-3/4}/\eta$
Displacement	$\lambda\lambda_\varepsilon$	$\mu^{3/2}$	η	$\mu^{3/2}\eta$
Velocity	$(\lambda\lambda_\varepsilon\lambda_g)^{1/2}$	$\mu^{3/4}$	1	$\mu^{3/4}$
Stress	$\lambda\lambda_\rho\lambda_g$	μ	1	μ
Stiffness	$\lambda\lambda_\rho\lambda_g/\lambda_\varepsilon$	$\mu^{1/2}$	1	$\mu^{1/2}$
Axial Force	$\lambda^3\lambda_\rho\lambda_g$	μ^3	η^2	$\mu^3\eta^2$

TCM was placed on the soil surface. Centrifuge was spun up to 25-g, and the seismic motion was applied. Then, the centrifuge was stopped, the structure was replaced by CCM, centrifuge was spun up to 50-g, and the earthquake motion was applied.

2 BACKGROUND

2.1 Generalized scaling factors

Considering scaling factors for length (λ), density (λ_ρ), strain (λ_ε), and centrifugal acceleration (λ_g) as independent factors and using the similitude laws, Iai et al. (2005) determined scaling factors of other physical and mechanical parameters; listed in the second column of Table 1.

2.2 Scaling factors of centrifuge tests

The scaling factors for centrifuge tests can be developed from the generalized scaling factors, by replacing notation λ with η. If the acceleration is consistently scaled with the geometry scaling ($\eta_g = 1/\eta$) and the same soil with the same destiny is used for a test ($\eta_\rho = 1$), and the strain scaling factor is considered to be unity ($\eta_\varepsilon = 1$), the generalized scaling factors can be converted to the conventional centrifuge scaling factors; as shown in the fourth column of Table 1. These scaling factors are consistent with the ones suggested by other researchers (Lord 1987; Wood 2004; Iai et al. 2005; Towhata 2008; Garnier et al. 2007; Madabhushi 2014).

2.3 Scaling factors of 1-g tests

The scaling factors for 1-g model tests can also be derived from the generalized scaling factors by replacing notation λ with μ. When the same soil with the same density is used for a test ($\mu_\rho = 1$), the scale factor of strain (μ_ε) equals to $\mu(V_m/V_P)^2$; where V_m and V_P are shear wave velocity of the soil in the model and the prototype, respectively. By measuring these two velocities, μ_ε can be calculated for 1-g model tests. However, when they are not available, μ_ε for sandy soils can be estimated based on assumption that the shear modulus at small strain is proportional to the square root of the confining pressure. For this condition, μ_ε equals the square root of the length scale factor ($\mu_\varepsilon = \mu^{1/2}$). Scaling factors of 1-g model tests in the mentioned conditions are showed in the third column of Table 1.

2.4 Two-stage scaling method

The two-stage scaling method is developed by combing the scaling factors for 1-g tests and centrifuge tests, and it includes two stages. First, scaling factors associated with 1-g test (μ; prototype/virtual model scale factor) are used to scale down the prototype to an intermediate virtual model. Then, the intermediate virtual model is scaled down to a physical model based on centrifuge scaling factors (η; virtual model/physical model scale factor). The main goal is to break down a large scaling factor ($\lambda = \mu\eta$; prototype/physical model scale factor) into the smaller scaling factors (μ and η). Therefore, the specimen would be spun with a centrifugal acceleration field correspond to η instead of λ. Theoretically, any combination of μ and η which produces λ can be chosen. However, since 1-g scaling factors are approximate values, η is suggested to be selected based on the maximum centrifugal field where the in-flight earthquake actuator can produce acceptable seismic motions. Two-stage scaling factors are shown in the fifth column of Table 1. It can be easily seen that each scale factor in the two-stage scaling method is the product of scaling factors in 1-g and centrifuge (Iai et al. 2005).

3 DESIGN AND CONSTRUCTION OF PHYSICAL MODELS

Dynamic centrifuge experiments were designed and conducted using the 5 g-ton (1 m radius) geotechnical centrifuge facility at the University of New Hampshire (Ghayoomi & Wadsworth 2014). The soil was prepared

Table 2. Scaling factors used for design of the physical models in the present study.

Quantity	CCM $\lambda = 50$	TCM Partitioned scaling factors		Two-stage scaling $\lambda = \mu\eta\ 50=2*25$
		1-g $\mu = 2$	Centrifuge $\eta = 25$	
Length	50	2	25	50
Density	1	1	1	1
Strain	1	1.414	1	1.414
Acceleration	0.02	1	0.04	0.04
Time (dynamic)	50	1.682	25	42.045
Frequency	0.02	0.595	0.04	0.024
Displacement	50	2.828	25	70.711
Velocity	1	1.682	1	1.682
Stress	1	2	1	2
Stiffness	1	1.414	1	1.414
Axial Force	2500	8	625	5000

in a laminar container not only to allow the soil to move in the shear beam mode but also to minimize undesirable boundary effects. The width, length, and depth of the laminar container are 17.78 cm, 35.56 cm, and 24.13 cm, respectively. One directional seismic motion was applied to the bottom of the laminar container in the tangential direction of flight.

F-75 silica sand was used in this study. The soil was classified as poorly graded sand (SP) with D_{50} of 0.19 mm according to the Unified Soil Classification System (USCS) (ASTM D2487-11 2011). Maximum and minimum void ratios of the soil are 0.8 and 0.49, respectively (Khosravi et al. 2010). The oven-dry soil was poured into a hopper suspended from a crane above the laminar container. The length of the orifice of the hopper is 26.67 cm while the width can be changed to adjust the sand flow. The drop height and the flow rate through the nozzle were calibrated to deposit the sand at a relative density approximately 60%. The hopper was moved horizontally to cover the area of the container. After raining 2.54 cm of the sand, the hopper was lifted 2.54 cm. The achieved relative density in this test was 62.6%, and the dry density of the soil was 1650.2 kg/m³.

In this study, a simplified target prototype of 4-story concrete structure with a 1 m thick surface mat foundation was modeled. By considering the size of the laminar container, a mat footing with a width of 3.5 m was selected as the foundation. The bearing pressure was estimated as 145 kPa by considering the dead and live loads. The first natural frequency of the prototype structure was selected to be 3.36 Hz. A length scaling factor (λ) of 50 is selected to model the system. Table 2 shows scaling factors used to design the physical models for CCM and TCM.

The bearing pressure, the first fixed-base natural frequency, the ratio of the mass of the foundation to the mass of the superstructure, and the dimensions of the foundation were chosen as the criteria to design and construct single-degree-of-freedom oscillator-foundation physical models. Abaqus Finite Element package was used to estimate the first natural frequencies of the physical models. Moments and axial forces in the columns were estimated when the physical models would be excited by target seismic loads. These forces and moments were used to make sure the columns would remain in elastic range. Width and thickness of the columns were changed until the natural frequencies would be in an acceptable range, and the columns would stay elastic during seismic events. To make sure the physical models would not fall when shaken; the factor of safety against overturning was controlled. Also, the bearing capacity and the settlement of the physical models were estimated to ensure that a bearing capacity failure or excessive settlement would not happen, when the models are spun in the centrifuge.

Available overhead space in the centrifuge limits the heights of the physical models, and the soil depth to 9.65 cm and 19.05 cm, in the model scale, respectively. The small-strain shear modulus (G_{max}) was estimated by the equations suggested by Seed and Idriss (1970), then, the shear wave velocity was estimated in depth by using the soil density. Since the shear wave velocity changes with depth, the average shear wave velocity of the prototype soil was estimated by using the harmonic mean as 194.5 m/s. The first natural frequency of the soil was estimated as 5.1 Hz. The soil depth for the TCM model was calculated in the way that the soil would have the same first natural frequency as the first natural frequency of the soil for the CCM model. It can be shown that to satisfy the mentioned condition, the soil depth in both models should be equal. As a result, the soil depth in both experiments was chosen as 19.05 cm, which is approximately equal to 9.53 m at the prototype scale; shown in Figure 1.

Accelerometers and the Linear Variable Differential Transformers (LVDT), as shown in the Figure 1, were used to measure acceleration and displacement at different locations of the experiment. Figure 1 shows the instrumentation layout of the centrifuge model tests for both CCM and TCM structures. The schematic drawing and the photograph of the models also demonstrated in Figure 2.

The fixed-base first natural frequency of the physical models was measured by conducting impact tests.

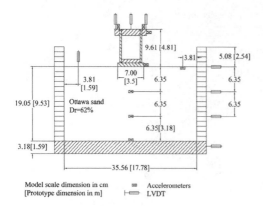

Figure 1. Elevation view of centrifuge model tests layout with instruments in CCM and TCM experiments. Dimensions of the structural physical models vary between the two experiments.

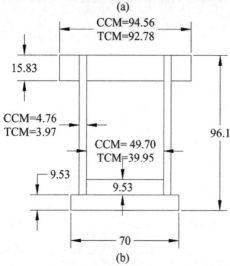

Figure 2. The physical model used in centrifuge experiment; a) Photograph of the models; b) schematic drawing (Dimensions are in the model scale and in mm).

The bottom part of the foundation in each physical model was replaced by a large steel plate tightened to a 1-g shake table. Then, a modal hammer was used to tap the mass and acceleration time histories of the

Table 3. Properties of physical models in the prototype scale.

Properties	CCM	TCM	Difference (%)
Foundation width(m)	3.50	3.51	0.27
foundation thickness (m)	0.96	0.96	0.00
Base pressure (kg)	143.39	142.58	0.56
f (Hz)	3.37	3.23	4.15
m_0/m	0.50	0.50	0.57
h/r	1.51	1.51	0.27
σ	19.35	20.19	4.33
γ	1.98	1.97	0.37

damped free vibration were recorded. The first natural frequency of CCM and TCM in the model scale is measured as 168.3 Hz, and 135.9 Hz, respectively; also, these frequencies are reported in the prototype scale in Table 3. Furthermore, the logarithmic decrement method was used to estimate damping ratios of CCM and TCM as 0.32% (standard deviation of 0.18%), and 0.85% (standard deviation of 0.19%) (Chopra 1995).

Veletsos & Nair (1975) defined dimensionless parameters mainly controlling the interaction between the structure and the soil. These parameters are the dimensionless wave parameter, $\sigma = V_s/(fh)$; the relative mass density of the structure and the supporting soil, $\gamma = m/(\rho\pi r^2 h)$; the ratio of the height of the structure to the equivalent radius of the foundation, h/r; the ratio of the dominant frequency of the excitation to fixed-based natural frequency of the system f_e/f; the damping capacity of the soil; the ratio of the foundation mass to the mass of the superstructure, m_0/m; the damping of the structure in fixed based conditions; and the Poisson's ratio of the soil. Where, $h =$ the distance from the foundation to the centroid of the inertia forces of the fundamental mode of vibration of the fixed-base structure; $\rho =$ the soil density; $r =$ equivalent radius of the foundation; and $V_s =$ average shear wave velocity of the soil. Some of these parameters, shown in Table 3, are estimated and compared for the two physical models. In addition, important properties of the models are also compared in Table 3. The comparison shows very low to zero difference between the two models.

4 RESULTS AND DISCUSSION

The 1994 Northridge earthquake motion recorded at Newhall- W Pico Canyon Rd. station, scaled to 0.1 g PGA (Peak Ground Acceleration), and was picked as the desired motion. Then, the desired acceleration time history was scaled for the physical modeling experiments based on the acceleration and time scaling factors for CCM and TCM experiments according to Table 2. Following the procedure by Mason et al. (2010), the shake table was calibrated to achieve a similar acceleration time history and frequency response. For example, Figure 3 shows the achieved motion at the bottom of the soil layer in the CCM experiment in prototype scale.

Transfer functions are commonly used to study inertia and kinematic interactions. Theoretically, the

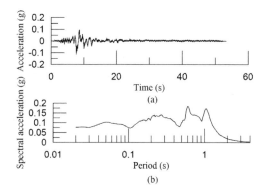

Figure 3. Achieved acceleration at the bottom of the soil layer during CCM test: a) acceleration time history; b) 5%-damped acceleration response spectra.

transfer function is the ratio of the Fourier amplitude of the output signal to that of the input signal. Often, transmissibility functions (ratios of power and cross spectra density functions of the input and output signals) are used to achieve clearer and smoother transfer functions (Kim & Stewart 2003, Mikami et al. 2008, Lalanne 2010). In order to compare SSI in the two models, Structural Horizontal Motion (SHM) to FM, Foundation Rocking Motion (FRM) to FM, and FM to FFM transfer functions were plotted and analyzed in prototype scale to capture both inertial and kinematic interaction.

Figure 4a shows SHM/FM transfer functions for both experiments in the horizontal direction. According to the figure, the flexible base natural frequency of CCM and TCM are 2.33 Hz and 2.11 Hz, respectively (difference 9.44%). The ratio of the flexible-base first-mode period to the fixed-base period of the structures, $\tilde{T}/T$, for CCM and TCM are 1.45 and 1.53, respectively (difference 5.23%). The overall trend of the transfer functions is similar. However, since the achieved motion (input at the base) in the two experiments are not identical, the amplitude of the transfer functions is expected to be different; further tests are underway to address this problem. Rocking time history of the two models was estimated by using the two accelerometers attached to the top of the model, see Figure 1. Foundation Rocking Motion transfer function (FRM/FM) is shown in Figure 4b. The rocking natural frequency of CCM and TCM are 2.30 Hz and 2.11 Hz, respectively (difference 8.26%).

By comparing the lateral and rocking component of FM with FFM, the extent of kinematic interaction can be evaluated for the models. It has been observed that FM in surface foundations has smaller translational component compared to FFM, however torsional and rocking motions are introduced into FM. Veletsos et al. (1997) analytically developed transfer functions for surface-supported, rigid, massless rectangular foundations. The soil was assumed as an elastic half-space, and the bond between the foundation and the soil was assumed rigid. The transfer functions were developed for Foundation Input Motion (FIM) which is a motion

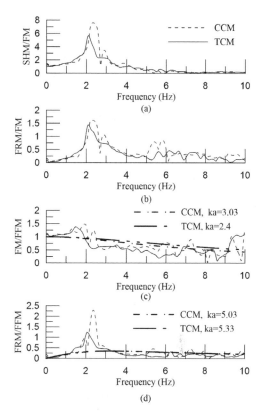

Figure 4. Comparing transfer functions of the two experiment: a) structural translational transfer function, SHM/FM; b) Foundation rocking transfer function, FRM/FM; c) foundation translational transfer function, FM/FFM; d) Foundation rocking transfer function, FRM/FFM.

that would be occurred at the foundation if the foundation were rigid, and both the superstructure and the foundation were massless. FIM was mathematically modeled by including two terms; i.e. the incoherence of the motion and the wave passage effect. In the current study, the wave passage effect was negligible because seismic motions were applied horizontally to the bottom of the laminar container, and the motions propagated vertically. Veletsos et al. the proposed analytical transfer function between FIM and FFM based on Equation 1.

$$s(r_1, r_2, \omega) = e^{-\left(\left(\frac{\omega}{v_s}\right)k_a|r_1 - r_2|\right)^2} s_g(\omega) \quad (1)$$

where r_1 and r_2 = the position vectors of the two arbitrary points; $s_g(\omega)$ = a local Power Spectral Density (PSD) function of FFM; $s(r_1, r_2, \omega)$ = a cross PSD function of FIM; ω = circular frequency; and k_a = the incoherence parameter.

Figure 4c shows the FM/FFM transfer functions for the two models. Nonlinear regression analyses were performed to estimate the coherence parameter, k_a, by matching the experimental and analytical transfer functions. When the coherence between the two signals for frequencies below 10 Hz is more than 0.8, the transfer function is used in the regression. For lateral components of FM, k_a was estimated to be 3.03 and 2.4 for CCM and TCM, respectively (difference 20.79%). Kinematic interaction is very sensitive to input motion, thus, a larger error was observed. Moreover, for the torsional component of FM shown in Figure 4d, k_a was approximated as 5.03 and 5.33 (difference 5.62%).

5 SUMMARY AND CONCLUSIONS

In this study, dynamic centrifuge experiments were performed to experimentally evaluate the performance of the two-stage scaling method to be used in Soil-Structure-Interaction (SSI) analyses. A 1DOF structure built on dry sands with a significant SSI effect was selected as the target prototype. Parameters of the prototype system such as the first fixed-based natural frequency, the bearing pressure, dimensions of the foundation, and the ratio of the mass the foundation to the mass of the superstructure were considered to design and build two physical models based on the Conventional Centrifuge Model (CCM), and the Two-stage Centrifuge Model (TCM). Dynamic centrifuge tests were conducted on both models. The two physical models approximately have the same response in the prototype scale. One of the reasons for the difference in the dynamic responses is that motions generated at the base of the container in the two experiments were slightly different in the porotype scale. According to the results, the two-stage scaling method was more successful in capturing the inertia interaction effects than the kinematic interaction effects. However, further tests introducing different motions with higher intensity, different frequency content, and duration are ongoing and the results will be presented in future publications.

REFERENCES

ASTM D2487-11. 2011. *Standard practice for classification of soils for engineering purposes (Unified Soil Classification System)*. West Conshohocken, PA: ASTM International.

Chopra, A. K. 1995. *Dynamics of structures: theory and applications to earthquake engineering.*, Englewood Cliffs, NJ: Prentice Hall.

Garnier, J., Gaudin, C., Springman, S. M., Culligan, P. J., Goodings, D., Konig, D., Kutter, B., Phillips, R.,

Randolph, M. F., & Thorel, L. 2007. Catalogue of scaling laws and similitude questions in geotechnical centrifuge modelling. *International Journal of Physical Modelling in Geotechnics,* 7(3), 1-23.

Ghayoomi, M., & Dashti, S. 2015. Effect of ground motion characteristics on seismic soil-foundation-structure interaction. *Earthquake Spectra,* 31(3), 1789-1812.

Ghayoomi, M., & Wadsworth, S. 2014. Renovation and reoperation of a geotechnical centrifuge at the University of New Hampshire. *ICPMG2014–Physical Modelling in Geotechnics: Proceedings of the 8th International Conference on Physical Modelling in Geotechnics 2014 (ICPMG2014), Perth, Australia, 14-17 January 2014,* CRC Press.

Iai, S., Tobita, T., & Nakahara, T. 2005. Generalised scaling relations for dynamic centrifuge tests. *Géotechnique,* 55(5), 355-362.

Khosravi, A., Ghayoomi, M., McCartney, J., & Ko, H. 2010. Impact of Effective Stress on the Dynamic Shear Modulus of Unsaturated Sand. *In GeoFlorida 2010: Advances in Analysis, Modeling & Design,* 410-419.

Kim, S., & Stewart, J. P. 2003. Kinematic Soil-Structure Interaction from Strong Motion Recordings. *Journal of Geotechnical and Geoenvironmental Engineering,* 129(4), 323-335.

Lalanne, C. 2010. *Mechanical Vibration and Shock Analysis, Random Vibration.* Hoboken, NJ: John Wiley and Sons.

Lord, A. E. 1987. Geosynthetic/soil studies using a geotechnical centrifuge. *Geotextiles and Geomembranes,* 6(1), 133-156.

Madabhushi, G. 2014. *Centrifuge Modelling for Civil Engineers.* London: CRC Press.

Mason, H. B., Kutter, B. L., Bray, J. D., Wilson, D. W., & Choy, B. Y. 2010. Earthquake motion selection and calibration for use in a geotechnical centrifuge. *In 7th int. Conf. on Physical Modeling in Geotechnics.*361-366, Leiden, Netherlands: CRC Press, .

Mikami, A., Stewart, J. P., & Kamiyama, M. 2008. Effects of time series analysis protocols on transfer functions calculated from earthquake accelerograms. *Soil Dynamics and Earthquake Engineering,* 28(9), 695-706.

Scott, R. F. 1994. Review of progress in dynamic geotechnical centrifuge research. *Dynamic Geotechnical Testing II,* ASTM STP 1213.

Seed, H. B., & Idriss, I. M. 1970. Soil moduli and damping factors for dynamic response analyses. Earthquake engineering research center, Berkeley, California.

Towhata, I. 2008. *Geotechnical earthquake engineering.* Berlin: Springer.

Veletsos, A. S., Prasad, A. M., & Wu, W. H. 1997. Transfer function for rigid rectangular foundation. *Earthquake Engineering & Structural Dynamics,* 26(1), 5-17.

Veletsos, A. S., & Nair, V. V. 1975. Seismic interaction of structures on hysteretic foundations. *Journal of the Structural Division,* 101(1), 109-129.

Wood, D. M. 2004. *Geotechnical Modelling.*, Florence: CRC Press.

Development of a window laminar strong box

S.C. Chian, C. Qin & Z. Zhang
National University of Singapore, Singapore

ABSTRACT: In physical modelling, laminar strong boxes are commonly used to allow for deformation comparable to the soil which is subject to shaking at the base of the model. This is achieved with a series of rings capable of sliding over one another. This therefore permits the development of stresses and strains associated with the upward propagation of shear waves and achieve the semi-infinite extent of prototype geotechnical structures, allowing dissipation of energy without significant reflection of stress waves due to rigid boundaries. Unfortunately, due to the requirement of the boundary to be made into separate rings, it is often unable to accommodate a window panel for Particle Image Velocimetry (PIV) analysis, which has become very popular amongst physical modellers. A unique window laminar box was hence developed to achieve both objectives. The design philosophy and construction details, as well as results from dynamic centrifuge tests are presented in this paper.

1 INTRODUCTION

Geotechnical centrifuge modelling is a technique to replicate geotechnical structures in the field by subjecting small scale models to high centrifugal accelerations. Centrifuge modelling allows testing to be conducted in a controlled environment as compared to full scale testing which is costly and sometimes unfeasible such as in the case of earthquake loading. These benefits have led to the increasing popularity of geotechnical centrifuge modelling with over 100 geotechnical centrifuges currently operating worldwide.

Soil is a highly non-linear material. It is therefore essential to replicate identical stress and strain conditions as in the prototype scale in laboratory tests. Geotechnical centrifuge modelling achieves these conditions with the use of high centrifugal acceleration to scale up the model. A scaled model is made to correspond to the prototype model under a pre-determined centrifuge g-level. As a result, a 1:N model experiences the same stress-strain conditions as the prototype model when subjected to a centrifugal acceleration of $N \times g$ (Schofield 1980, Schofield 1981). A set of scaling laws are used to interpret other centrifuge testing parameters in prototype scale.

Dynamic centrifuge modelling involves extending the above principles to dynamic loadings such as earthquake, blast and waves. Similar to the scaling of centrifugal acceleration, a 1:N scale model subjected to $N \times g$ of centrifugal acceleration with a given lateral acceleration of $N \times a$ will exhibit identical behaviour under a horizontal earthquake acceleration of a at prototype scale. However, dynamic time is different to the seepage time scale and often corrected with a pore fluid of N times of the viscosity of the actual pore fluid. This is particularly important for sand model which has high permeability.

2 REQUIREMENTS OF A CENTRIFUGE MODEL CONTAINER

While centrifuge modelling is now commonly accepted as a useful tool for study of geotechnical problems. The reliability of output of the centrifuge test is dependent heavily on the simulation of appropriate boundary conditions. This is particularly true for dynamic centrifuge tests where the degradation of soil stiffness and shear strength with increasing strain should ideally be reproduced by appropriate physical boundaries for the centrifuge model.

The soil in the field has semi-infinite boundary unlike a centrifuge model test which has a limited plan area dimension. When a soil column is subjected to lateral shaking, the amplitude of displacement of the column varies with depth similar to a shear beam. Hence, the design of a centrifuge model strong box should replicate such realistic deformation so as to obtain accurate and true presentation of the behaviour of similar soil in the field (Fig. 1).

In order to address the above challenge, the boundary of a strong box is usually treated with: 1) mouldable materials (e.g. Duxseal, plasticine or sponge), 2) multiple rings attached with rubber (e.g. equivalent shear beam (ESB) box), or 3) multiple rings sitting on rollers (e.g. laminar box).

In the case of the side boundaries treated with mouldable materials, the benefit is its ability to reduce reflection of waves as well as allowing the soil to deform with the shaking (Steedman & Madabhushi, 1991). Although this method allows a rigid box to be transformed into a "flexible" boundary box, it suffers from the uncertainty of the actual boundary conditions offered by the mouldable material.

The equivalent shear beam (ESB) box involves designing the shear stiffness of the walls of the box to match the shear stiffness of the soil. In this case, the

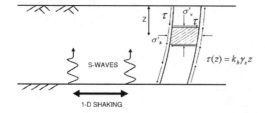

Figure 1. Soil with infinite lateral extent subjected to base shaking at the bedrock (Bhattacharya et al. 2011).

walls of the box are deemed to deform similar to the soil when subjected to shaking. The shortcoming is its inability to follow the changing stiffness of softening soil especially in cases of soil liquefaction.

The design principle of a laminar box is to minimise lateral stiffness of the box so as to allow the soil to deform freely according to the shaking. Water tightness is ensured with a bag wrapped around the internal boundary of the laminar box. Since the laminar box is most capable of allowing free movement of the model, the issue of changing shear stiffness of soil and flexible boundary conditions are addressed and hence often the most desirable design in most dynamic centrifuge laboratories worldwide especially for liquefaction models. Unfortunately, given the multiple sliding rings design of most laminar boxes, Particle Imaging Velocimetry (PIV) is not possible since a clear sectional view of the soil is not available. In order to depict the deformation of liquefying soil in the centrifuge, a rigid box with window opening treated with mouldable wall boundaries is often adopted (Chian et al. 2014, Cilingir 2009).

In this paper, the development of a laminar box with window opening is discussed so as to allow the increasingly popular PIV technique to be applied with ease.

3 DESIGN OF THE NOVEL WINDOW LAMINAR BOX

Figures 2 and 3 show the photograph and sketch of the window laminar box designed at the Centre for Soft Ground Engineering, National University of Singapore. The front panel was made up of a thick transparent Perspex supported by stainless steel brackets bolted to the base plate. This permits easy dismantling of the front panel for cutting of slope surface, boring a tunnel opening and sprinkling of coloured particles for better clarity of deformation of soil in photographs for PIV analysis.

In order to incorporate the laminar nature of the wall boundaries, the conventional rectangular shaped rings were designed as C-shaped rings so as to permit viewing of soil through the front panel while still allowing the soil to displace freely when subject to one-dimensional shaking. These C-shaped rings were designed to be more robust than most closed-looped rectangular rings given the potentially large lateral forces acting on the cantilever ends of the C-shaped rings at 100g centrifuge level.

Figure 2. Isometric view of window laminar box.

Another challenge faced during the design is the uneven sliding resistance of the roller bearings on these C-shaped rings. In addition, the width of the C-shaped rings would have to be excessively large to permit lateral displacement of 3% for a 390 mm height soil model, which makes the use of bearings uneconomical. Hence, bearings were substituted with a smooth slot system (see Fig. 3) that allows large lateral displacements to develop which is particularly useful for lateral spreading tests.

Other challenges include the need to: 1) ensure water tightness between the membrane bag holding the soil and the Perspex, 2) provide drainage opening at the base of the model for saturation under vacuum, and 3) permit consolidation of slurry clays for preparation of model prior to dynamic centrifuge test. These concerns were correspondingly addressed with 1) the fabrication of a groove on the Perspex to insert the membrane and a stiff metal piece to achieve water tightness requirements, 2) a drilled opening hole with casing through the Perspex to minimise concentration of stress onto the opening, and 3) design of height extension that can be sealed onto the uppermost ring of the laminar box.

4 PERFORMANCE OF WINDOW LAMINAR BOX

In order to validate the suitability of the window laminar box, a centrifuge test is presented herein. Figure 4 shows the instrumentation layout of the centrifuge model. The soil used in this model is saturated Kaolin clay with specific gravity, liquid and plastic limits of 2.60, 60 and 42 respectively.

Figures 5 and 6 show the acceleration time histories and corresponding lateral displacements after double integration respectively. It can be observed that the soil deformation follows that of a shear beam as depicted in Figure 1. In addition, these experimental displacements correspond closely with theoretical calculated lateral displacements adopted from Hardin & Black (1968) and Hardin & Drnevich (1972).

Lastly, a PIV analysis is carried out based on the photographs taken during the centrifuge. As observed

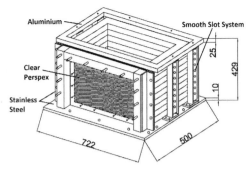

(a) Full assembly

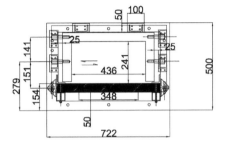

(b) Plan view

Figure 3. Design drawings of components of the window laminar box.

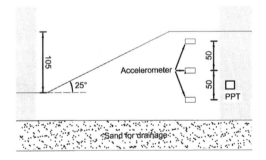

Figure 4. A 1:50 model layout of slope with instrumentation.

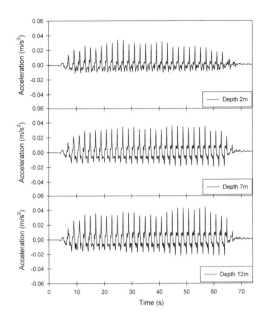

Figure 5. Acceleration time histories measured at different depths of soil in centrifuge test.

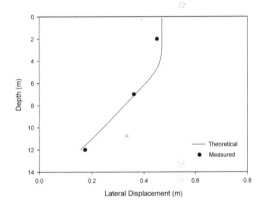

Figure 6. Corresponding lateral displacement of soil in centrifuge test.

Figure 7. PIV analysis of slope collapse in centrifuge test.

in Figure 7, the collapse of the slope can be captured with ease with the convenience of the Perspex window, while meeting semi-infinite boundary condition alike to the field.

In view of the success of the ability of the window laminar box to represent a reliable lateral soil deformation and perform PIV technique concurrently, similar design of such centrifuge strong boxes may be adopted in future centrifuge tests so as to maximise the research output of each test.

5 CONCLUSION

A novel window laminar box was developed at the Centre for Soft Ground Engineering, National University of Singapore. The purpose is to enable the viewing of the cross-section of the centrifuge model while allowing natural movement of the soil subjected to shaking. C-shaped sliding rings supported with slots were adopted in contrast to typical closed-loop rings supported with roller bearings. These changes permit greater lateral displacements to be achieved, while allowing viewing of soil deformation from a Perspex window along the length of the box.

Results presented from a typical centrifuge test showed that the acceleration time histories and lateral deformation across the soil depth resembles that of a shear beam. Lateral displacements also corresponded well with theoretical estimates which confirmed the suitability of design of the window laminar box to be used for dynamic geotechnical centrifuge testing.

REFERENCES

Bhattacharya, S., Lombardi, D., Dihoru, L., Dietz, M.S., Crewe, A.J. & Taylor C.A. 2011. Model container design for soil-structure interaction studies. In Michael N. Fardis & Zoran T. Rakiecevic (ed.), *Role of seismic testing facilities in performance-based earthquake engineering.* Dordrecht: Springer.

Chian, S.C., Tokimatsu, K. & Madabhushi, S.P.G. 2014. Soil liquefaction-induced uplift of underground structures: physical and numerical modeling. *Journal of Geotechnical and Geoenvironmental Engineering,* 140(10): 04014057.

Cilingir, U. 2009. *Seismic response of tunnels*, PhD Thesis, Cambridge University.

Hardin, B.O. & Black, W. 1968. Vibration modulus of normally consolidated clay. *Journal of Soil Mechanics and Foundations*, 94(2): 353-369.

Hardin, B.O. & Drnevich, V.P. 1972. Shear modulus and damping in soils: design equations and curves. *Journal of Soil Mechanics and Foundations*, 98(7): 667-692.

Steedman, R.S. & Madabhushi, S.P.G. 1991. Wave propagation in sand medium. *Proc. 4th International Conference on Seismic Zonation, Stanford, 26-29 August 1991.* California: EERI.

Ground-borne vibrations from piles: Testing within a geotechnical centrifuge

G. Cui, C.M. Heron & A.M. Marshall
Nottingham Centre for Geomechanics, Department of Civil Engineering,
University of Nottingham, Nottinghamshire, UK

ABSTRACT: Vibrations propagating from railways can affect the living environment of individuals residing nearby and have a negative impact on structures and the equipment housed within. Centrifuge testing of ground-borne vibration (GBV) problems can provide valuable insights but involves various challenges, mainly associated with the very high frequencies and small amplitudes of the vibrations. This paper presents new developments for testing GBV problems on the Nottingham Centre for Geotechnics centrifuge. The propagation of vibrations generated from the vertical oscillation of a single pile is studied, which, for example, simulates a train running on a viaduct with piled foundations. The vertical oscillations of the pile are imposed by an electro-magnetic shaker with the dynamic load and acceleration of the pile being monitored. Sensitive piezoelectric accelerometers are used to monitor the transmission of vibrations through the soil body. The challenges faced during such tests include: low signal-to-noise ratios, physical vibrations from wind buffeting, ambient vibrations, and choice/control of appropriate input signal parameters. This paper illustrates how these issues were overcome to obtain high quality and high resolution signals such that the transmission of the imposed vibrations could be monitored.

1 INTRODUCTION

1.1 Ground-borne vibration

Increasing population and traffic congestion in major cities have led to the rapid expansion of railway lines, including trams, light railways, surface railways and underground railways. However, one disadvantage is that vibrations travelling from railway lines can affect sensitive buildings (Kuo 2010) or sensitive equipment housed in laboratories (Nugent et al. 2012). These vibrations can generate noticeable vibration, usually at frequency of 2 to 80 Hz, and re-radiated noise usually in the range of 30 to 250 Hz (Thompson 2009). In recent years, issues related to ground-borne vibrations have gained increasing attention due to the expansion of railways and the decreased tolerance to disturbance, particularly in residential properties.

1.2 Challenges associated with modelling GBV

Previous research conducted using a geotechnical centrifuge focused on vibration sources from surface and underground railways (Tsuno et al. 2005, Yang 2012). These studies utilized an electro-magnetic shaker to generate the simulated railway-induced vibration. Itoh et al. (2002) developed a ball dropping system to impose impulse vibration at the ground surface and also used a shaker to study the effect of periodic vibration in the centrifuge. In this study, the propagation of vibrations generated from the vertical oscillation of a single pile (about 0.5 m in diameter and 17 m in length, at prototype scale) is investigated. The outcomes can be related to scenarios such as trains running on piled structures, such as a viaduct. Critical to modelling any soil-structure-interaction problem is the replication of correct interface characteristics and soil behaviour. Therefore the Nottingham Centre for Geomechanics 4 m diameter geotechnical centrifuge was used to provide a full scale stress profile. Vertical oscillations of the pile were imposed by an electro-magnetic shaker. Dynamic load and acceleration of the pile were monitored and sensitive piezoelectric accelerometers, denoted as 'Piezos' in this paper, were used to measure vibrations within the soil body.

One of the challenges with such experiments is low signal-to-noise ratios. Railway-induced vibrations are small, with field monitoring indicating an acceleration amplitude of about 0.1 to 1 cm/s^2 (Tsuno et al. 2005). The corresponding velocity amplitude is about 0.1 to 0.5 cm/s depending on the distance from track, train speed and train type (Connolly et al. 2014). Yang (2012) investigated the acceleration noise level induced by ambient electrical noise and physical vibrations in the Nottingham centrifuge. It was found that the centrifuge induced acceleration amplitude was significant, at about 1 to 5 cm/s^2 and up to 20 cm/s^2 at frequencies less than 1200 Hz (20 Hz in prototype scale). However, Yang did not extend the investigation to examine the effect of shaker induced noise and vibrations on the measured signals. This challenge can be overcome by imposing higher amplitude vibrations through the shaker, however care must be taken that the imposed vibrations and soil shear strains are applicable to the prototype scenario. Preliminary tests with the adopted shaker indicated that a 500 to 1000 mV input excitation amplitude was appropriate for the test configuration, detailed in Section 2. The detected acceleration levels (0.5 to 8 cm/s^2) within the soil were similar to the field scale monitoring records, as was the shear strain levels. Therefore, all the results presented in this paper were obtained from tests performed within this range.

An additional challenge relates to the selection of an appropriate input signal type to ensure the soil-pile system can reach a steady state level of response. However, with the large range of frequencies being investigated, combined with the potential for very large data files, input signals need to be kept as short as possible to allow for efficient testing. In this study, two types of sinusoidal based vibration signals (sine chirp signal and single frequency signal (Avitabile 2001)) were used. The coherence function between the input signal and the output measurements is used to examine whether the model reached a steady state, evaluate the quality of measurements, and explore which type of input signal provides the best model response within an acceptable testing time.

The final significant challenge faced with the centrifuge tests was related to the effect of wind buffeting on the shaker armature and upper portion of the pile, which protruded above the top of the centrifuge container.

This paper discusses the challenges mentioned above and illustrates how they were overcome to obtain high quality and high resolution signals such that the transmission of the imposed vibrations could be extracted from the recorded data. Results from selected Piezos (detailed in Section 2.2) are presented in this paper to illustrate the effects of those challenges; data from other Piezos within the soil (not presented in the paper) showed similar outcomes at a given depth.

2 GBV EXPERIMENTAL SETUP

2.1 Centrifuge modelling

Centrifuge tests were conducted at an acceleration of 60 g. A steel cylindrical tub measuring 500 mm in diameter and 500 mm in height was used to contain the soil. To reduce the reflections generated by the rigid boundaries of the container, a vibration absorbing material (Duxseal, as adopted by Cheney et al. (1990)) was moulded to the sides and bottom of the container.

The soil was a fine, uniformly rounded dry sand referred to as Congleton (HST95 silica sand). Material properties are given in Table 1 (Lauder 2010). The sand was dry pluviated to achieve a relative density of 85 ± 2% (accuracy determined from repeated calibration pours).

Table 1. Physical properties of Congleton Sand (Lauder 2010).

Parameter	Value
Specific Gravity (G_s)	2.63
Minimum dry unit weight ($\gamma_{d,min}$)	14.6 kN/m^3
Maximum dry unit weight ($\gamma_{d,max}$)	17.6 kN/m^3
Maximum void ratio (e_{max})	0.769
Minimum void ratio (e_{min})	0.467
D_{10}	0.090 mm
D_{30}	0.120 mm
D_{60}	0.170 mm
D_{90}	0.195 mm

2.2 Model configuration

The model layout, locations of the instrumentation, and a schematic of the equipment are provided in Figure 1. The dead load (simulating the superstructure) was applied using 76.2 mm diameter brass discs which were located above a static load cell (SLC), providing a load of about 120 to 440 N at model scale (432 to 1584 kN at prototype scale). These brass discs screw onto a central steel rod and were held in place by a lock nut. This setup allowed the mass to be changed easily between flights with minimal disturbance to the pile. A 500 N capacity static load cell (type: Richmond Industries 210 Series inline load cell) was employed to measure the dead load being applied to the top of the pile.

An electromagnetic shaker (type: LDS V201) was used to apply the dynamic load to the 8 mm diameter aluminum model pile with an embedded length (L) of 280 mm. The small shaker can provide a peak force of 17.8 N across a frequency range of 5 Hz to 13 kHz. The shaker was controlled (via the centrifuge electric slip rings) by a signal generator, the signal of which was increased to the required level by a power amplifier. A dynamic load cell (type: ICP 221B02) was connected to the shaker to measure the dynamic load being applied during tests. The dynamic load cell (DLC) is able to measure frequencies from 10 Hz to 15 kHz.

The Piezos in Figure 1 are labelled according to their orientation and location: VP denotes vertical, HP is horizontal, NP is 'near pile' (50 mm from pile), and DP is 'distant to pile' (135 mm from pile). To evaluate the progression of the shear waves propagating horizontally from the pile, six 5 g Piezos were buried in the soil and oriented to measure vertical accelerations (VPNP3, VPDP3, VPNP2, VPDP2, VPNP1, VPDP1). Two 5 g Piezos (HPNP3, HPNP1) were also used to measure horizontal vibrations (which can be caused by sideways rocking of the pile during the vertical loading by the shaker). 5 g Piezos were used as a balance had to be found between minimizing transducer size while still being able to record across the broad band of frequencies involved in the tests. Ideally a smaller range of instrument would have been used in the soil however compromises with regard to size and dynamic response were deemed unacceptable. It was found that

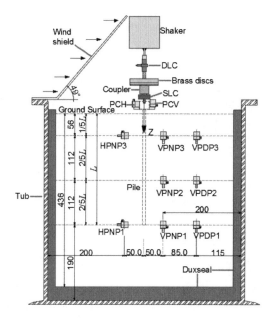

Figure 1. Schematic view of model layout in the centrifuge (dimensions in mm).

at the pile-cap, 10 g Piezos were required. Therefore, two 10 g Piezos were attached to the pile cap; one vertically and one horizontally (PCV & PCH). PCH was mounted horizontally to evaluate the rocking of the pile as mentioned above.

The Piezos (type: DJB Instruments A/120) have a linear response between 0 Hz and 7 kHz and deviate by up to 10% between 7–10 kHz. Therefore, care should be taken when using the instruments to monitor any higher harmonics experienced during the tests. As discussed in the introductory section of this paper, the frequency range of interest is 1 Hz to 120 Hz in prototype scale (60 Hz to 7200 Hz in model scale).

2.3 Data acquisition system

A National Instruments Ethernet Rio expansion chassis (NI-9149) forms the backbone of the data acquisition system on the University of Nottingham centrifuge. For this testing, the piezoelectric accelerometers and dynamic load cell, which are IEPE (integrated electronic piezoelectric) instruments, required the use of NI-9234 modules which are capable of a synchronous sampling frequency of 51.2 kHz. The voltage output instruments (static load cell and displacement transducers), despite not being high-frequency instruments, were connected to a NI-9220 module which is capable of a synchronous sampling frequency of 100 kHz. This high speed module was needed for recording MEMS accelerometers which were used in the tests to assist with the measurement of the dynamic, in-flight, properties of the soil (see other Cui et al paper within these proceedings). Both data acquisition modules allowed for the recording of frequencies well in excess of the Nyquist frequency (3.5 times for the Piezos).

3 EXPERIMENTAL CHALLENGES

3.1 Low signal-to-noise ratio

Electrical noise and physical vibrations come from the shaker (in addition to the intended oscillation of the pile), the centrifuge drive system, the associated on board electronics, and other equipment located in the laboratories surrounding the centrifuge. Noise and vibrations can distort the real vibration level caused by soil-pile interaction, referred to as the 'real signal' in this paper, and can lead to low signal-to-noise ratios. Electrical noise cannot be entirely eliminated, however an in-house developed junction box system used to power and amplify the instrument signals (voltage output instruments) has significantly reduced electrical noise to a background level of less than 2 mV. Careful cleaning of the piezoelectric cable connectors and routing of the cables also assists with reducing noise in the IEPE instruments. Physical vibrations generated by the centrifuge and shaker vary but isolation systems can be used to reduce their effect. The centrifuge and the shaker induced noise and vibrations were investigated and methods developed to reduce the interference they cause.

3.1.1 Centrifuge induced interference

To reduce the vibrations from the centrifuge operation, a 6 mm thick rubber mat was placed between the centrifuge platform and the container. A noise and vibration test was conducted after the model was accelerated to 60 g without applying harmonic excitation with the shaker. The measured data was transferred from the time domain to the frequency domain by using a discrete fast Fourier transform. Figure 2 shows the measured accelerations from 2 vertical Piezos (VPNP1 and VPNP3) in the soil, denoted as 'soil Piezos', and indicates that the main frequency range of the centrifuge induced noise and vibrations is in the low frequency range (less than 600 Hz), and occurs mainly between 550 Hz and 600 Hz (an improvement compared to the significant noise levels up to 1200 Hz measured by Yang 2012).

3.1.2 Shaker induced noise and vibrations

In the previous centrifuge tests carried out by Tsuno et al. (2005) and Yang (2012), the shaker was directly connected to the model container. However, in addition to the intended vibrations being transmitted to the pile, the shaker can transmit vibrations through its support which then pass through the model container and into the soil hence affecting the real signals. To reduce these undesirable effects, a supporting system (see Figure 3) was designed to isolate the container/soil from the supports for the shaker. The system consists of a steel (for stiffness) beam and two aluminum (to minimize weight) rectangular columns, which are mounted on the centrifuge platform instead of the model container.

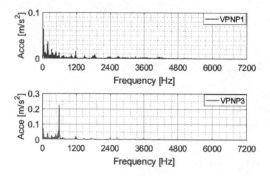

Figure 2. Vibration and electrical noise from centrifuge operation.

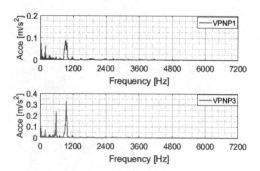

Figure 4. Frequency range of background noises and vibrations from centrifuge and shaker.

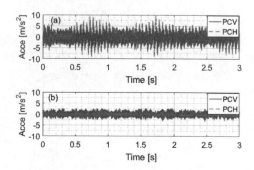

Figure 5. Wind effect on PCV and PCH: (a) without wind shield and (b) with wind shield.

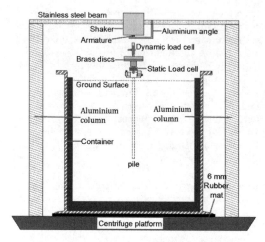

Figure 3. Supporting system in the centrifuge tests.

Another noise and vibration test was conducted to estimate the shaker induced vibration level unrelated to the soil-pile interaction. This was achieved by disconnecting the shaker from the pile. The movable component of the shaker (armature) was fixed by an aluminum angle, see Figure 3, mounted to the steel beam. Therefore, all the recorded signals were influenced by the electrical noise or the physical vibrations from the centrifuge and the shaker. A 20 second sweep signal with a frequency range of 0.2-7200 Hz was used. The input voltage amplitude adopted was 500 mV based on preliminary tests, as discussed in Section 1.2. The measured signals from VPNP1 and VPNP3 are illustrated in Figure 4. It can be seen that soil Piezos are still significantly affected within some frequency ranges. This is likely to be due to electrical noise as opposed to physical noise given the measures taken to avoid the transmission of unwanted mechanical vibrations into the soil. Figure 2 and Figure 4 suggest there are two frequency ranges (less than 1000 Hz) which could affect the soil-pile interaction induced vibrations: 550 Hz to 600 Hz caused by the centrifuge and 900 Hz to 1000 Hz caused by the shaker.

3.1.3 Wind effect

Another issue is related to the wind generated during the centrifuge tests. The centrifuge rotates at 3 cycles per second (approximately 38 m/s) at the specified centrifugal acceleration, thus generating undesirable wind buffeting. Two exposed accelerometers on the top of the pile, PCV and PCH were vulnerable to the wind effects. The wind effect was examined by recording the accelerations with and without a wind shield, as shown in Figure 1. A 6 mm thick aluminum plate with a projection area onto the windward plane of $22 \times 70 \text{ cm}^2$ (height × width) was used. The wind shield was higher than the shaker top and wider than the model container so that the two Piezos were protected, to a degree, from the wind. The results, shown in Figure 5, indicate that the wind had an adverse effect on the response of PCV and PCH.

3.2 Selection of shaker signal parameters

As discussed in Section 1, a trade-off between the quality of measurements and testing time prompted the investigation of two types of input signals. The coherence function was used to evaluate the quality of measurements (De Silva 2007). The coherence gives a value between zero and one in frequency domain, where zero indicates no correlation between the outputs and inputs and a value of one indicates the outputs are 100% aligned with the inputs. In this case the value from the dynamic load cell was used as the input signal data stream while the data from the Piezos was taken as

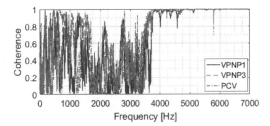

Figure 6. Coherence values for VPNP1, VPNP3, and PCV (Test 1: broad-band sweep signal).

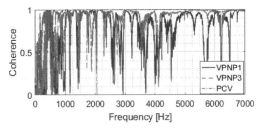

Figure 7. Coherence at VPNP1, VPNP3 and PCV (Test 2: narrower-band sweep signal).

the output. All the tests presented in this section were carried out under the same dead load (719.8 g) and with the same amplitude of 1000 mV for the shaker so that only one influence factor, type of input signal, is being investigated.

3.2.1 Sine chirp signal

A sine chirp signal is a sweep signal in which the frequency varies across the whole frequency range (i.e. 0.2 Hz to 7200 Hz) with a short duration (e.g. 5 s). The advantage of a sine chirp signal is the short testing time, which minimizes data file size and allows a wider range of parameters to be investigated dung one centrifuge flight. However, the rate of change of frequency is very rapid, especially for the low frequencies. Within an extremely short period, the model response may not reach a steady state. Two forms of sine chirp sweep signals were assessed: broad-band and narrower-band:

Broad-band sweep signal

In Test 1, a 5 second sine chirp signal (0.2 Hz-7 kHz) was used to excite the model. The coherence values between the dynamic force and accelerations in the soil (VPNP1 and VPNP3) and on the top of the pile (PCV) are shown in Figure 6.

It was found that the coherence was generally low and extremely variable in the frequency range of 0.2 to 4000 Hz. This is related to the fact that the signal generator breaks the total frequency range (0.2 to 7000 Hz) into 2048 steps, giving a frequency spacing of 3.42 Hz. Each frequency component only lasts 0.00244 seconds (5/2048=0.00244) which is shorter than the period of frequency components below 410 Hz (1/410 = 0.00244). Figure 6 indicates that the coherence was larger than 0.9 at most frequencies above 4000 Hz. Given that the period of 4000 Hz is 0.00025 seconds, and 0.00244 seconds is about 10 times the period of 4000 Hz, this suggests that each frequency component needs at least ten periods to allow the pile-soil system to reach a steady-state response.

Narrower-band sweep signal

A narrower-band sweep signal has a narrower frequency range (e.g. 0.2 Hz to 1200 Hz) and a longer duration (e.g. 20 s) compared to the broad-band sweep signal, therefore the rate of change of frequency is slower. However, the model response may still not reach a steady state at low frequencies. In Test 2, the frequency range of 0.2 to 7000 Hz was divided into 7, approximately equal, sections.

The resulting coherence is given in Figure 7. It can be seen that coherence values were larger than 0.9 at most frequencies. However, there were still many localised reductions in coherence, especially for VPNP1. This is due to the vibration level generated through soil-pile interaction decreasing with depth and signal-to-noise ratio being relatively low.

3.2.2 Single frequency signal

The second input type considered was a single frequency sinusoidal signal (e.g. 200 Hz) with a shorter duration (e.g. 0.1 s). This type of signal can guarantee the steady state response of the model is measured because the duration of each single frequency can be controlled to last long enough so that soil has adequate time to reach a steady state response. Some preliminary tests were performed using a duration of 0.1 s for each frequency employed, detailed below, to verify the steady state. However, the estimated total time (around 1100 s) was much longer than that for the sine chirp signals (5 s and 140 s respectively), explained in detail below.

In this study, the scaling factor for frequency is $N = 60$, so 6 Hz (0.1 Hz in prototype) was chosen as the frequency spacing. A total of 1101 input frequencies: 600, 606, 612 ... 7200 Hz (10, 10.1, 10.2, 10.3 ... 120 Hz in prototype scale) were used one by one to excite the model in Test 3. The recorded duration of each single frequency input signal was around 1 s, however only about 0.4 s of data corresponded to the action of the dynamic load. This is due to the control system taking approximately 0.3 s to send the command from the signal generator to the shaker and for the system to respond, as shown in Figure 8. The 0.4 s of useful data was extracted from the data stream for subsequent analysis. Results from the sine chirp signal indicated that to obtain a good response, the duration of each excitation frequency component was about 10 times the period. In this study, 0.4 s corresponds to 10 times the period at 25 Hz and it is hence at least 10 times the period of every frequency used. Therefore, steady state model response should have been achieved at each frequency level. Coherence function values from Test 3 are shown in Figure 9. It can be seen that coherence values at selected locations were quite close to 1 across most frequencies. Therefore,

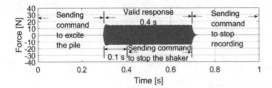

Figure 8. An example of single frequency input signal (dynamic load in Test 3: single frequency signal).

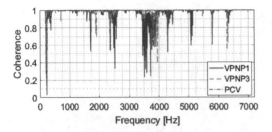

Figure 9. Coherence at VPNP1 VPNP3 and PCV (Test 3: single frequency signal).

the single frequency signal with a good control over frequency, amplitude and duration should be adopted in the future to obtain high quality and high resolution signals despite the drawbacks of long test durations and management of large data files.

It should be noted that in this study the objective is to understand how the pile-soil system responds at specific frequencies to subsequently allow for the calibration of a numerical model using the centrifuge test data. Although the vibrations emanating from a train will vary rapidly in terms of amplitude and frequency, this transient behaviour can only be investigated once the simpler response to a single frequency input is fully understood and can be captured within the numerical analysis.

4 CONCLUSIONS

This paper presents new developments for testing ground-borne vibration generated from the vertical oscillation of a single pile on the Nottingham Centre for Geotechnics centrifuge. The real signals caused by the soil-pile interaction can be affected by electrical noise and physical vibrations generated from the centrifuge and the shaker, wind buffeting, and the characteristics of the input signal. This paper investigated challenges related to these issues and proposed approaches to overcome them in order to obtain high quality measurements.

Electric noise and physical vibrations from the shaker and the centrifuge system cannot be considered negligible when undertaking ground-borne vibration studies. Despite two methods being employed to isolate the vibrations, their effect within some frequency ranges were still evident: 550 Hz to 600 Hz from the centrifuge and 900 Hz to 1000 Hz from the shaker. Wind buffeting created by the centrifuge can affect the pile, accelerometers and wires, thus distorting the measurements, particularly from two exposed accelerometers on the top of the pile. The use of a wind shield was shown to reduce the effect of wind on measurements. An evaluation of input signals based on coherence values showed that the single frequency input signal was preferable to the sine-chirp signal. The drawback of the single frequency input tests is a longer test duration, however tests were achieved with an acceptable time frame. The quality of the data obtained using the single frequency input signal justified the additional time and effort required.

REFERENCES

Avitabile, P. (2001). Experimental Modal Analysis. *Sound and Vibration* (January), 1–11.

BS (2005). Mechanical vibration Ground-borne noise and vibration arising from rail systems Part 1: General guidance. *BS ISO 148*, 54.

Cheney, J. A., R. K. Brown, N. R. Dhat, & O. Y. Hor (1990). Modeling free-field conditions in centrifuge models. *Journal of Geotechnical Engineering 116*(9), 1347–1367.

Connolly, D. P., P. Alves Costa, G. Kouroussis, P. Galvin, P. K. Woodward, & O. Laghrouche (2015). Large scale international testing of railway ground vibrations across Europe. *Soil Dynamics and Earthquake Engineering 71*, 1–12.

Connolly, D. P., G. Kouroussis, P. K. Woodward, P. A. Costa, O. Verlinden, & M. C. Forde (2014). Field testing and analysis of high speed rail vibrations. *Soil Dynamics and Earthquake Engineering 67*, 102–118.

De Silva, C. W. (2007). *Vibration: fundamentals and practice / Clarence W. De Silva*. Boca Raton, Fla.: Boca Raton, Fla.: CRC/Taylor & Francis.

Itoh, K., M. Koda, L. K. I, & O. Kusakabe (2002). Centrifugal simulation of wave propagation using a multiple ball dropping system. *International Journal of Physical Modelling in Geotechnics 2*, 33–51.

Kuo, K. A. (2010). *Vibration from Underground Railways: Considering Piled Foundations and Twin Tunnels*. Ph. D. thesis.

Lauder, K. (2010). *The performance of pipeline ploughs*. Ph. D. thesis.

Nugent, R. E., W. Village, & J. A. Zapfe (2012). Designing Vibration-Sensitive Facilities Near Rail Lines. *Sound and Vibration* (November), 13–16.

Thompson, D. D. (2009). *Railway noise and vibration: mechanisms, modelling and means of control / David Thompson with contributions from Chris Jones and Pierre-Etienne Gautier*. Oxford: Oxford: Elsevier.

Thusyanthan, N. I. & S. P. G. Madabhushi (2003). Experimental study of vibrations in underground structures. *Proceedings of the Institution of Civil Engineers: Geotechnical Engineering 156*(2), 75–81.

Tsuno, K., W. Morimoto, K. Itoh, O. Murata, & O. Kusakabe (2005). Centrifugal modelling of subway-induced vibration. *International Journal of Physical Modelling in Geotechnics 5*(4), 15–26.

Yang, W. (2012). *Physical and numerical modelling of railway induced ground-borne vibration / by Wenbo Yang*. Ph. D. thesis.

A new Stockwell mean square frequency methodology for analysing centrifuge data

J. Dafni & J. Wartman
University of Washington, Seattle, Washington, USA

ABSTRACT: Dense arrays of accelerometers were used to record the dynamic response of stepped soil slopes in an equivalent shear beam container at the Center for Geotechnical Modeling at the Univ. of California, Davis. Data from these arrays was utilized in performing time and time-frequency domain analysis. The propagation and interaction of seismic waves with the soil slopes was interpreted using a new parameter, deemed the Stockwell mean square frequency. The Stockwell mean square frequency consists of calculating the mean square frequency of the S transform at each time step for a ground motion time history. Dominant frequencies of the ground motion were tracked in time, allowing frequencies amplified by the presence of the slope to be identified.

1 INTRODUCTION

A comprehensive geotechnical centrifuge study was performed using dense arrays of accelerometers to investigate the dynamic response of stepped sandy slopes. The aim of this study was to better understand the topographic modification of ground motion, or "topographic effects." Topographic effects have been studied extensively using numerical and field-based methods, with only a few smaller-scale centrifuge tests previously performed (Brennan & Madabhushi 2009, Ozkahriman et al. 2007, Yu et al. 2008). However, due to quantitative differences between numerical and field studies (Geli et al. 1988, Pagliaroli et al. 2011, Massa et al. 2014), topographic effects are still not fully understood. Therefore, new methods of data collection and interpretation should be explored. In addition to using a unique data collection method (the centrifuge) for this study, alternative time- and time-frequency-based analysis methods were also utilized.

Data were analyzed in the time-frequency domain using variations of the S transform (Stockwell 1996), the result of which is often deemed the Stockwell spectrum. The mean square frequency (MSF), or inverse of the mean square period (Rathje 1998), of the Stockwell transform, was then used to track dominant frequencies in time. Other parameters, such as the Hilbert transform (Liu 2012) and variations of cumulative Arias Intensity (AI), or Husid parameter (Husid 1969) without normalization by the total ground motion AI, were also used to perform time domain analysis. Additionally, the data was visualized by creating several versions of videos with the data. However, this paper focuses on a new Stockwell mean square frequency methodology for time-frequency domain analysis of geotechnical centrifuge data.

2 BACKGROUND

To better understand the analysis results presented in the next section, the Stockwell spectrum, MSF, common terminology used to describe topographic effects, and an overview of the centrifuge study will be briefly described here.

2.1 Stockwell Spectrum (S Transform)

The Stockwell spectrum was first presented by Stockwell et al. (1996) as an improved means for studying a time signal in the time-frequency domain. The S transform extends the ideas of the continuous wavelet transform and uses a localized Gaussian window that scales with the time signal frequency. It is also directly related to the commonly used Fourier spectrum, in that the average of the local spectra over time is equal to the Fourier spectrum.

2.2 Mean square frequency

The MSF is typically based on the Fourier spectrum and is calculated as follows:

$$MSF = \frac{\Sigma_i C_i^2 f_i}{\Sigma_i C_i^2} \text{ for } 0.25 \leq f_i \leq 20 Hz \quad (1)$$

where C_i are the Fourier amplitudes of the time history and f_i are the corresponding frequency for i discrete points. The calculation is like that used to define a geometric centroid. Thus, the MSF is a scalar representation of a centroid-like frequency for a given ground motion.

The MSF defined above provides a singular frequency measure for an entire ground motion. While this can provide insight into the ground motion frequency content, it is not sufficient for understanding when, or how frequency content may relate to the observed ground motion, such as that associated with topographic effects. Applying this same calculation to the Stockwell spectrum at each discrete ground motion time step, however, allows dominant frequencies to be tracked in time. The Stockwell MSF is not a standard measure, in that, to the author's knowledge, it has not been previously published. However, it proved useful in this study and could be applied to other ground motion studies as well.

For this study, the Stockwell MSF was calculated using an upper cutoff frequency of 12 Hz, instead of 20 Hz. The Stockwell MSF was also calculated based on the Stockwell spectrum of the ground motion velocity time history, rather than the acceleration time history traditionally used for the Fourier spectrum. Velocity time histories were calculated by integrating acceleration time histories. Different variations may be appropriate for other studies.

2.3 Description of centrifuge study

The centrifuge study consisted of testing six model configurations, including three stepped slopes, two dam-like configurations and a flat ground model, which served as an experimental baseline. The slopes were constructed in an equivalent shear beam container using dry Nevada sand, which was vibrated to 100% relative density, except for one of the dam-like configurations, which used cemented sand. The stepped slopes were constructed at slope inclinations of 20, 25, and 30 degrees. The results presented in this paper focus on the 30-degree slope.

Testing was performed at two gravity accelerations (g levels), 55 g and 27.5 g, allowing a modeling-of-models exercise to be performed (Dafni 2014). The same suite of ground motions was introduced to each model configuration. These ground motions consisted of idealized (e.g., sine wave motion, Ricker wavelets, and frequency sweeps), as well as non-idealized or realistic ground motions (e.g., earthquake motions). An example of an idealized and non-idealized ground motion, introduced at 55 g will be presented in the next section. Further details of the centrifuge study can be found in Dafni (2017).

2.4 Topographic effects

Topographic effects are often analyzed by comparing the dynamic response at and around the crest of slopes to the free field response at some distance behind the crest of the slope. In this way, the amount of amplification or deamplification resulting from the slope geometry as compared to typical 1-D site conditions can be interpreted. This method of comparison can be employed by defining a topographic factor for different ground motion intensity measures (Ashford & Sitar 1997).

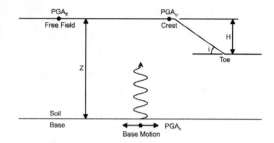

Figure 1. Reference diagram for a stepped slope with properties of interest defined.

Using peak ground acceleration (PGA) as an example, the topographic factor (TF) can be written as

$$TF_{PGA} = \frac{PGA_s - PGA_{ff}}{PGA_{ff}} \quad (2)$$

where PGA_s is the PGA at any point on the slope (at the crest $PGA_s = PGA_{cr}$) and PGA_{ff} is the PGA for free field (1-D) site conditions at some point behind the slope crest (Fig. 1). By removing the influence of the free field response, the topographic factor provides a method for determining the percent change in ground motion amplitude associated with topographic effects. This factor can be applied to many ground motion parameters, including in the time-frequency domain, using the Stockwell spectrum.

The topographic factor accounts for ground motion amplitude; however, frequency content is also affected by topographic features. Much in the same way that a resonant site frequency (f_s) can be identified for a 1-D soil column, a topographic frequency (f_t) can be defined for a given slope (Ashford et al. 1997). The site frequency can be written as

$$f_s = \frac{V_s}{4Z} \quad (3)$$

where V_s is the average shear wave velocity of the soil profile and Z is the depth of the soil column from surface to bedrock. The topographic frequency can be defined as

$$f_t = \frac{V_s}{5H} \quad (4)$$

where H is the slope height. Thus, maximum amplification for the site may be expected at the site frequency, while maximum amplification at the crest may occur at the site or topographic frequency.

3 SAMPLE RESULTS

3.1 Sine wave motion

An example of an idealized ground motion recorded at the base, crest, and free field (Fig. 1) is presented in

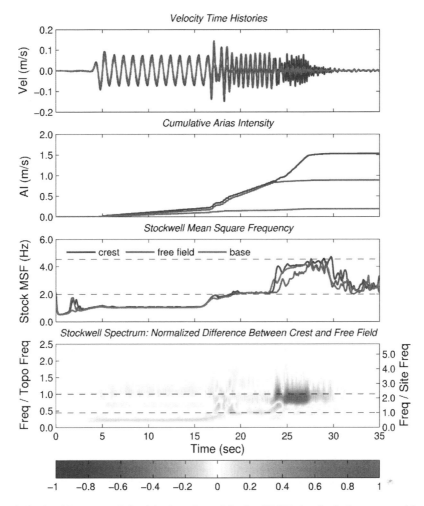

Figure 2. Velocity time history, cumulative Arias Intensity, and Stockwell MSF plots for the base, crest and free field; and the difference between the crest and free field Stockwell amplitude, normalized by the maximum difference (i.e., a value of 1 is equivalent to maximum amplification at the crest, −1 maximum deamplification, and 0 represents no difference between the crest and free field Stockwell amplitude), for 1 Hz, 2 Hz, and 4 Hz sine wave motion.

Figure 2. The base motion was introduced as 12 cycles of sine wave motion at central frequencies of 1, 2 and 4 Hz. The latter two frequencies roughly coincide with the site and topographic frequency (Equations 3 and 4) of 2 Hz and 4.5 Hz, respectively.

Differences in the ground motions at the base, crest and free field are well illustrated by the velocity time histories in the top plot of Figure 2. However, these differences can be quantified and better visualized using the cumulative AI, Stockwell MSF, and normalized Stockwell plot in the next three plots. Additionally, differences not detected in the velocity time histories can be observed.

The central frequencies of the base motion are well defined in the Stockwell MSF plot. At 1 and 2 Hz, and the transition between, the MSF of the free field and crest track well with the base motion. That is, the frequency of the sine wave motion at the crest and free field matches the forcing frequency. As the forcing frequency transitions to 4 Hz, the frequency content at the three locations begins to deviate. The MSF at the crest quickly moves to 4 Hz and then trends upwards towards the topographic frequency (4.5 Hz), even after the base MSF (which transitions more slowly and then maintains an MSF of 4 Hz) begins to decrease. This decrease in MSF at the base begins around 27.5 seconds and coincides with the transition towards an attempt at 6 Hz sine wave motion (which could not be achieved). In the free field, the MSF transitions upward from 2 Hz (the site frequency) more slowly and doesn't match the base MSF until it peaks and begins to drop off.

Inspection of the cumulative AI plot reveals similar patterns, but considering the build-up of ground motion energy, at the different central frequencies. That is, the behavior of the free field and crest is similar

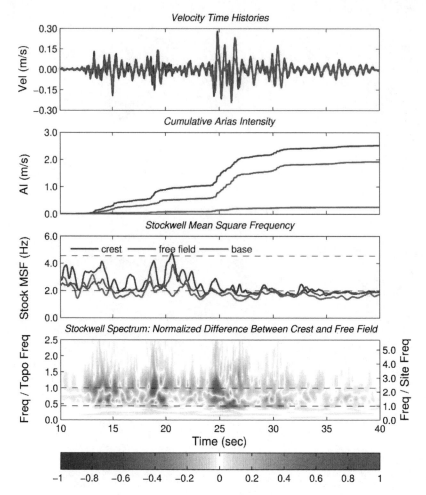

Figure 3. Velocity time history, cumulative Arias Intensity, and Stockwell MSF plots for the base, crest and free field; and the difference between the crest and free field Stockwell amplitude, normalized by the maximum difference (i.e., a value of 1 is equivalent to maximum amplification at the crest, −1 maximum deamplification, and 0 represents no difference between the crest and free field Stockwell amplitude), for the Chi Chi earthquake motion.

at 1 and 2 Hz, but deviates at 4 Hz. Energy at the base builds up gradually with time across all frequencies. At 1 Hz, energy builds up slightly faster than the base at the crest and free field. The rate of energy build-up then increases at the crest and free field for 2 Hz motion (the site frequency). At 4 Hz, the energy at the crest continues to build at an even faster rate, while the energy build-up in the free field levels out. The crest and base motion then both start leveling out where 6 Hz motion was attempted. Similar amplitude patterns can also be seen in the velocity time history plot.

The frequency and amplitude patterns of the crest and free field, relative to each other, is also illustrated in the normalized Stockwell plot in Figure 2. The normalized Stockwell plot consists of the difference between the crest and free field Stockwell spectrum at a discrete time, normalized by the maximum difference for the entire motion. It is similar to the TF (Equation 2) but takes a difference instead of a ratio.

A value of 1, and −1 represent maximum amplification, and deamplification, respectively, of the crest over the free field, in terms of the Stockwell amplitude. A value of 0 means the Stockwell amplitude at crest and free field are equal. This plot reveals that the spectral amplitude at the crest is slightly amplified at 1 Hz (0.5 times the site frequency), is either slightly deamplified or not amplified at 2 Hz (site frequency), and reaches maximum amplification at 4 Hz (near the topographic frequency of 4.5 Hz), when compared to the free field.

The behavior at the crest and free field deviate as ground motion shifts to 4 Hz. The response in the free field lags as the transition to 4Hz ground motion occurs at the base. This is likely a residual effect of the tendency towards resonance at the site frequency, and the low amplitude motion at 4Hz, which wasn't amplified in the free field. In other words, it took time for energy at the site frequency to dampen, and the competing energy at 4 Hz was minimal, which delayed the shift in MSF to 4 Hz.

At the crest, however, the transition towards the topographic frequency occurred simultaneously with the shift in the base motion, despite similar resonant conditions at the site frequency. Any residual effects from the resonant mode at the site frequency were suppressed by excitation near the topographic frequency.

3.2 Earthquake motion

The plots in Figure 2 provide insight into the ground response for an idealized motion. However, using these analysis methods, insight can also be gained for earthquake motions. Figure 3 presents these same plots for the Chi-Chi earthquake motion.

Despite an increase in the complexity of the ground motion, similar patterns to those seen with the sine wave motions are revealed. The Stockwell MSF gravitates towards the site and topographic frequencies at the free field and slope crest locations, respectively (marked by horizontal dashed black lines in the third plot in Figure 3). Deviations between the crest and free field Stockwell MSF coincide with differences in overall ground motion amplitude, as illustrated by the cumulative AI plot (second plot in Figure 3), where the response at the crest is shown to be amplified compared to the free field. Inspection of the normalized Stockwell plot (fourth plots in Figure 3) illustrates that differences in amplitude can be attributed to amplification of ground motion near the topographic frequency at the crest. Note that the dashed lines in the normalized Stockwell plots coincide with the site and topographic frequencies.

Additional complexities are also highlighted by the plots in Figure 3. Increases in the Stockwell MSF at the crest typically follow increases in the Stockwell frequency of the base motion. This indicates that the input (base) amplitude at frequencies closer to topographic influences the level of amplification observed at the crest. The Stockwell MSF also highlights that lower frequency energy can be present while greater levels of amplification at the crest occur. For instance, amplification of the crest over the free field is greatest around 18.5 and 24.5 seconds, despite a higher Stockwell MSF at 18.5 seconds.

These observations highlight that for complex ground motion, differences in the ground response can vary dramatically. However, using the tools illustrated in Figures 2 and 3 and studying ground motions in the time and time-frequency domain can provide significant insight into the ground response. These tools can be used to not only observe what occurs but also start to understand why they occur.

4 CONCLUSION

Data collected in a series of geotechnical centrifuge physical model experiments was successfully used to study the ground response of sandy stepped slopes. Analysis of the data was accomplished time and time-frequency domain analysis. The methods utilized in this study have proven to be powerful for interpreting the mechanisms and causes of topographic effects. Several of the methods applied have not been used in other studies of topographic effects to the knowledge of the author. In particular, the Stockwell spectrum, and the newly introduced Stockwell MSF which can be used to track ground motion frequency content in time; and the cumulative AI which can be used to track changes in ground motion amplitude energy.

To understand the complexities of topographic effects, it is important to explore new methods for collecting and analyzing data. New methods, such as those presented here, should continue to be explored. These analysis methods may also be useful for understanding other ground motion phenomenon, such as liquefaction, or the initiation of co-seismic landslides, among others.

More results using these analysis methods, and more focused on the analysis of topographic effects will be presented in future publications.

ACKNOWLEDGMENTS

Financial support for this research was provided by the U.S. National Science Foundation under grant 0936543. The centrifuge experiments were performed at the Center for Geotechnical Modeling at the Univ. of California, Davis, a George E. Brown Jr. Network for Earthquake Engineering Simulation (NEES) facility.

REFERENCES

Ashford, S.A. & Sitar, N. 1997. Analysis of topographic amplification of inclined shear waves in a steep coastal bluff. *Bull. Seismol. Soc. Am.* 87(3): 692–700.

Ashford, S.A., Sitar, N., Lysmer, J. & Deng, N. 1997. Topographic effects on the seismic response of steep slopes. *Bull. Seismol. Soc. Am.* 87(3): 701–709.

Brennan, A.J. & Madabhushi, S.P.G. 2009. Amplification of seismic accelerations at slope crests. *Can. Geotech. J.* 46: 585–594.

Dafni, J. 2017. Experimental investigation of the topographic modification of earthquake ground motion (doctoral dissertation). *University of Washington*. ID No. 17212.

Dafni, J. & Wartman, J. 2014. Centrifuge modeling of dynamic response in slopes. *8th International Conference on Physical Modeling in Geotechnics*. Paper No. 159.

Geli, L., Bard, P. & Jullien, B. 1988. The effect of topography on earthquake ground motion: a review and new results. *Bull. Seismol. Soc. Am.* 78(1): 42–63.

Husid, L.R. 1969. Caracteristicas de terremotos, analisis general. *Revista del IDIEM 8, Santiago del Chile*: 21–42.

Liu, Y.W. 2012. Hilbert transform and applications. *Fourier Transform Applications*. InTech. Ch.12: 291–300.

Massa, M., Barani, S. & Lovati, S. 2014. Overview of topographic effects based on experimental observations:

meaning, causes and possible interpretations. *Geophysical Journal International* 197: 1537–1550.

Ozkahriman, F., Nasim, A. & Wartman, J. 2007. Topographic effects in a centrifuge model experiment. *4th International Conference on Earthquake Geotechnical Engineering*. Paper No. 1262.

Pagliaroli, A., Lanzo, G., & D'Elia, B. 2011. Numerical evaluation of topographic effects at the Nicastro Ridge in Southern Italy. *Journal of Earthquake Engineering* 15: 404–432.

Rathje, E.M., Abrahamson N.A., & Bray, J.D. 1998. Simplified frequency content estimates of earthquake ground motions. *Journal of Geotechnical and Environmental Engineering*

Stockwell, R.G., Mansinha, L., & Lowe, R.P. 1996. Localization of the complex spectrum: the S transform. *IEEE Transactions on Signal Processing* 44(4):998–1001.

Yu, Y., Deng, L., Sun, X. & Lu, H. 2008. Centrifuge modeling of a dry sandy slope response to earthquake loading. *Bull. Earthquake Eng.* 6: 447–461.

Novel experimental device to simulate tsunami loading in a geotechnical centrifuge

M.C. Exton, S. Harry, H.B. Mason & H. Yeh
Oregon State University, Corvallis, Oregon, USA

B.L. Kutter
University of California, Davis, USA

ABSTRACT: Soil scour from tsunami hazards causes substantial damage to coastal infrastructure (e.g., the damage caused by the 2011 Great East Japan Tsunami). Simulating tsunamis in a laboratory setting is necessary to further understanding of tsunami-induced soil scour, but the simulations are challenging because dynamic similitude cannot be achieved for small-scale models. The ability to control the body and viscous forces in a centrifuge environment considerably reduces the mismatch in dynamic similitude. A novel centrifuge "Tsunami Maker," designed specifically for exploring the basic physics of soil response to tsunami-like loadings, is presented. The apparatus was constructed and calibrated for use on the 9.1-m geotechnical centrifuge at the Center for Geotechnical Modelling at the University of California, Davis. The Tsunami Maker contains removable soil specimen containers to increase system modularity. Tsunami flooding is created by lifting a gate and releasing water from a reservoir, and tsunami drawdown is achieved by lifting another gate to drain the flooded water. Measurements of the flow characteristics indicate that the Tsunami Maker can produce tsunami-like loading.

1 INTRODUCTION

Tsunamis are rare, extreme events and cause significant damage to coastal infrastructure. The sustained inundation flow damages infrastructure above the ground surface and it also greatly alters the behaviour and strength of the soil surrounding the infrastructure and its foundation. Figure 1 shows the consequences of the 2011 Great East Japan Tsunami in Kirikiri, Japan. A seawall designed for tsunami protection experienced significant scour. The scour was observed below the front and rear faces of the seawall. During the tsunami drawdown, several seawall components were pushed seaward and scattered on the beach. The structural components of the seawall were largely undamaged; therefore, the observed failure was likely caused by soil instability.

Due to the severity of damage, and infrequency and complexity of tsunami phenomenon, it is necessary to study tsunami-induced soil scour in a laboratory setting. Tsunami scour has been studied experimentally using water-wave tanks, which results in substantial mismatch in the Reynolds and Froude numbers as well as in the soil response. The ability to control the body force and the viscous force in a geotechnical centrifuge environment considerably reduces the mismatch in dynamic similitude.

Few researchers study the effect of tsunamis using a geotechnical centrifuge, and the ones who do only consider the runup stage. For instance, Takahashi et al. (2014) used a beam centrifuge to model seepage under

Figure 1. Displaced seawall components in Kiriki, Japan following the 2011 Great East Japan Tsunami.

breakwater caissons during tsunami runup. They used a water-supply tank to suddenly discharge water and maintain a head difference between the seaward and landward sides of the breakwater with discharge holes in the side of the model container. Similarly, Sassa et al. (2016) modelled a breakwater system with the addition of a water supply tank that uses the "mariotte bottle" concept. This mariotte bottle system allows for a constant flow rate by prescribing the location of the air-open interface in the sealed tank using an inlet tube, thus maintaining a constant pressure head in the water supply. Iai (2015) also used a geotechnical centrifuge to model the effect of tsunami inundation on breakwaters as well as pile-founded buildings and was able to achieve horizontal flow through the use of a standpipe-type supply tank on the seaward side of the breakwater.

Here, we describe a newly constructed Tsunami Maker, which is used with the large centrifuge at the University of California, Davis. The Tsunami Maker is the first centrifuge tsunami apparatus capable of simulating both inundation and drawdown stages in a single spin.

2 CENTRIFUGE WAVEMAKER DESIGN

2.1 *Structural design and operation*

The Tsunami Maker was designed for use in the 9.1m geotechnical centrifuge at the University of California, Davis Center for Geotechnical Modelling. This centrifuge is capable of spinning up to 75 g and a payload capacity of 265 g-tons. The large capacity provides sufficient space for the multi-component tsunami apparatus system.

Schematic drawings of the Tsunami Maker are depicted in Figure 2 and photographs of the apparatus are shown in Figure 3. A fluid reservoir is raised above the floor of the container. The reservoir floor (and thus the released tsunami) is level with the surface of the soil specimen. Pneumatically driven gates on either end of the reservoir control tsunami runup and drawdown.

The reservoir gates are sealed with rubber O-rings to prevent liquid leaking into the soil model. When the reservoir is full, the resulting hydrostatic pressure applies pressure to the O-rings, aiding the seal.

The discharge gates are lifted with 102 mm bore pneumatic cylinders mounted on steel frames with an operational pressure of 8.6 bar. Each cylinder is activated with a solenoid valve. Removable bump stops are attached to the cylinder frames to cushion the impact of the lifting gate. Gate 1 is opened first to release the liquid from the reservoir and apply the tsunami runup load. Following runup, tsunami drawdown is initiated by opening Gat e 2 and draining the flooded liquid from the soil specimen. The flooded liquid is drained into the chamber underneath the reservoir.

Two crossbars above the soil specimen boxes (see Figure 3) serve as support for instruments as well as the lighting panels, and 127 mm diameter holes in the sidewalls serve as the view windows for high-speed, high-resolution cameras. The holes also ensure that the water level in the Tsunami Maker is kept under the designed maximum water level.

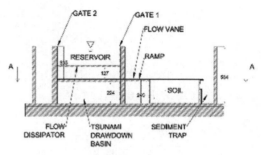

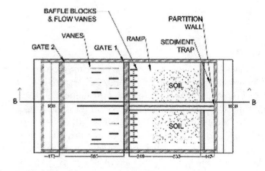

Figure 2. Schematic drawing of the Tsunami Maker in elevation view (top) and plan view (bottom). Dimensions are in millimetres. Gate 1 opens to initiate the runup stage followed by Gate 2 for the drawdown stage. No gate opening devise is shown for brevity.

Figure 3. Photographs of the Tsunami Maker from the side (top); showing the reservoir with the flow dissipater (bottom left); and a plan view of the ramps with baffle blocks and flow vanes, and soil specimen compartments separated by a partition.

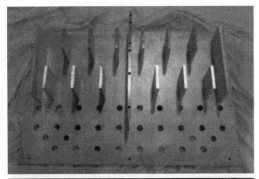

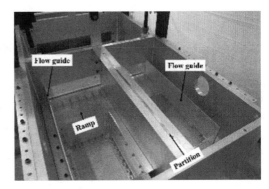

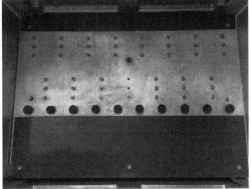

Figure 4. Photographs of the flow dissipater plate from the bottom view (top), and top view (bottom). The dissipater plate is placed in the reservoir to reduce the runup flow velocity. The orifices in the dissipater plate were adjusted with rubber plugs and PVC sheet to reduce the flow velocity.

Figure 5. Ramps with flow vanes and baffle blocks to dissipate energy before the tsunami bore reaches the soil specimens. The soil specimens are separated by the partition. Flow guides prevent flow expansion and maintain uniform flow from the reservoir to the soil specimens.

2.2 Tsunami wave design

The Tsunami Maker was designed with representative tsunami conditions in mind. An inundation height of approximately 4 m and an average celerity of approximately 5 m/s was used to represent "typical" tsunami inundation conditions based on previously observed tsunami events (see for example Sanuki et al. (2013) who estimated the flows for the 2011 Japan Tsunami in Fukushima). The preceding tsunami conditions dictated the height of the gate openings and the amount of liquid used. The amount of liquid was determined by conservation of mass throughout the experiment. Based on the location of the liquid at each stage (pre-tsunami, runup, and drawdown compartments) the sizes of the reservoir, tsunami runup and drawdown basin were determined.

To reduce the flow velocity during the tsunami runup phase, a dissipater plate was installed horizontally in the reservoir immediately above the gate openings as seen in Figure 4, top. Flow vanes wer connected to the dissipater plate to disturb and reregulate flow as well as reduce the Coriolis effects. Note that fine-tuning of the discharge rate can be performed by filling the orifice holes as shown in Figure 4, bottom. The dissipater plate and vanes are supported by threaded rods in the reservoir floor. Controlling the flow discharge with the dissipater plate is sufficiently effective and structurally simpler than a more sophisticated control device (e.g. mariotte tanks used by Sassa et al. 2016).

Immediately outside of the runup gate, ramps were installed to allow reregulation of the runup flow onto the soil specimens. The ramps, shown in Figure 5, are 254 mm long with small baffle blocks (no more than 6.4 mm tall) to dissipate the thinnest runup flow during the first stage of gate opening (n.b., at this stage, the reservoir head is highest and gate opening is smallest), and flow vanes to straighten this flow.

The specimen partition allows two soil specimens to be tested simultaneously with a single experimental run (Fig. 5). The specimen partition is 19 mm thick and secured at the centreline of the Tsunami Maker with a frame made of 38 mm thick angle iron. The resulting discontinuity in the flow path is a source of flow separation, and a vane was installed in the reservoir to minimise the flow separation effects (see the long vane along the centreline shown in Figure 4, top). As shown in Figure 5, flow guides were installed at the outside edges of the specimen containers to prevent flow expansion and maintain uniform and straight flow conditions on the specimen.

2.3 Soil container design

The two specimen boxes are each 533 mm long, 364 mm wide, and 249 mm deep. The boxes are composed of 9.5 mm thick aluminium plates bolted together using 38 mm thick angle iron at the inside seams; the boxes are sealed with a marine grade sealant. To reduce deflection, stiffening members at the top edge of the specimen boxes and at mid-height are installed outside of the boxes. The specimen boxes are prevented from lateral movement by adjustable brackets on the floor of the Tsunami Maker apparatus.

3 VALIDATION EXPERIMENT SETUP

The goal of the validation experiment was to test the designed loading scenario (40 g) for the Tsunami

Maker. The water levels in the container were monitored to ensure no leakage from the entire apparatus container and within the isolated compartments of the container. Accelerometers were placed on the pneumatic cylinder frames to estimate the gate opening time and absolute pressure transducers were placed along the specimen boxes parallel to the direction of the flow to measure the propagating bore height. Video cameras (GoPro) were used to monitor the splash height on the end wall, gate opening, and the wave profile. Analogue cameras were used to monitor the experiment and water levels in the container. A high-speed, high-resolution camera (Photron Mini AX) was used with a reference grid to determine the bore speed. The gate opening, activated by the solenoid valve, is synchronised with the high-speed camera.

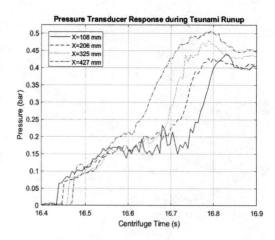

Figure 6. Pressure transducer response illustrating the measured wave height at positions along the direction of wave propagation with respect to time ($X = 0$ mm is at the edge of the soil specimen box that first encounters the tsunami bore). The bore speed of 8.5 m/s is determined from the initial response of the sensors. Note that the data, originally sampled at 5000 Hz, are subsampled to one in every 30 points for visual clarity.

4 RESULTS

At a centrifugal acceleration of 40 g, the runup gate opens in 0.09 seconds: this was determined from the GoPro footage. The pressure transducer readings during the runup stage are shown in Figure 6 and indicate a wave speed of 8.5 m/s. This was computed by using the known pressure transducer locations and the time difference of the initial response of each sensor. During runup, the transducers respond in order of increasing distance in the direction of tsunami wave propagation ($X = 0$ mm is at the edge of the soil specimen box that first encounters the tsunami bore). Note that the order of pressure transducer response data shown in Figure 6 reverses at a time of approximately 16.48 seconds (e.g., the transducer located at $X = 427$ mm responds before the transducer at $X = 325$ mm). This is due to the tsunami bore completing the travel distance across the soil specimen surface and reflecting from the back wall of the Tsunami Maker.

After a prescribed four-minute waiting period, the drawdown gate (Gate 2 in Figure 2) opens in approximately 0.09 seconds. Note that the waiting period is adjustable and represents the inundation time. The pressure transducer readings for the drawdown stage are shown in Figure 7. The drawdown occurs more slowly than runup with only small differences in the measured wave height recorded in the pressure transducers. Figure 7 presents an illistrtion of the overall drawdown response. The approximate total drain time is three seconds.

The celerity of the wave (which scales by one using conventional centrifuge scaling laws) can also be determined by using the high-speed camera footage captured at 4000 frames per second. Figure 8 shows two frames taken during runup (frame numbers 748 and 866). Each square on checkerboard grid on the right of the image represents 25.4 mm. The time difference between the two frames is 0.0295 seconds, and the centreline of the bore travels about 254 mm, which corresponds to celerity of approximately 8.6 m/s, confirming the readings from the pressure transducers in Figure 6.

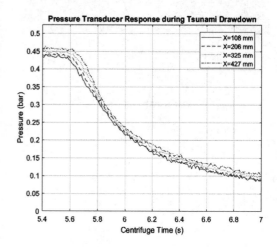

Figure 7. Pressure transducer response of the measured wave height during tsunami drawdown. Complete drawdown occurs much slower than runup and its total duration is not shown in this figure. Note that the data, originally sampled at 5000 Hz, are subsampled to one in every 50 points for visual clarity.

5 CONCLUSIONS AND FUTURE WORK

The Tsunami Maker performed well for the pilot test. All structural components withstood the designed loading scenario. The discharge gates opened in the anticipated time and the use of pressure transducers reasonably measure the propagating bore speed. The flow velocity is confirmed using the high-speed camera footage. Though the flow velocity is higher than expected, the flow speed is not unreasonable for tsunami runup.

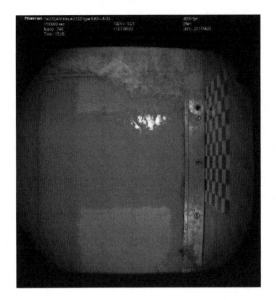 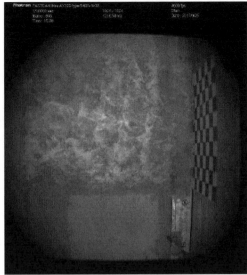

Figure 8. Frame 748 (left) and 866 (right) taken from tsunami runup video and used to determine the bore speed. Footage was taken at 4000 frames per second with an exposure time of 1/10000 seconds and a resolution of 1024×1024.

It is emphasised that the Tsunami Maker was designed and constructed with modification capabilities. Further tuning and adjustments are planned because exact flow predictions under centrifuge conditions are formidable. In the forthcoming experiments, improvements will be made regarding the flow velocity, and the ramp design.

ACKNOWLEDGEMENTS

The authors would like to thank Lars Pederson and Greyson Termini for their contributions to the design and machining of the Tsunami Maker apparatus, and Dan Wilson and the staff members at the Davis Center for Geotechnical Modeling for their support during the experiments. We would also like to thank the National Science Foundation for supporting this project (award number CMMI-1538211).

REFERENCES

Iai, S. 2015. Combined failure mechanism of breakwatersand buildings subject to Tsunami during 2011 East Japan earthquake. Geotechnics for Catastrophic Flooding Events; Proc. ISSMGE Geotechnical Engineering for Disaster Mitigation and Rehabilitation, Kyoto, Japan, 16-18 September 2014. London: Taylor & Francis.

Sanuki, H., Tajima, Y., Yeh, H., & Sato, S. 2013. Dynamics of tsunami flooding to river basin, Proc. Coastal Dynamics, 2013, Bordeaux, France.

Sassa, S., Takahashi, H., Morikawa, Y., & Takano, D. 2016. Effect of overflow and seepage coupling on tsunami-induced instability of caisson breakwaters. *Coastal Engineering* 117: 157-165.

Takahashi, H., Sassa, S., Morikawa, Y., Takano, D. & Maruyama, K., 2014. Stability of caisson-type breakwater foundation under tsunami-induced seepage. *Soils and Foundations* 54(4): 789-805.

A new apparatus to examine the role of seepage flow on internal instability of model soil

F. Gaber & E.T. Bowman
Department of Civil and Structural Engineering, University of Sheffield, UK

ABSTRACT: Nearly 50% of embankment dam stability problems are due to internal instability of the soil of homogeneous dams, or dam filters in the case of engineered dams. This instability can manifest as seepage induced internal erosion of fine particles through the coarser soil matrix. This study discusses the design of a new triaxial permeameter capable of investigating the influence of seepage on internal erosion of a model soil under controlled stress conditions. Two triaxial tests are reported here, using model soil of a specific gap graded particle size distribution, prepared to the same initial conditions of density and stress. The first test was conducted as a control, to examine the strength and stiffness under non-eroded conditions. In the second test, downward seepage was applied to the sample prior to testing in order to erode fine particles from the matrix. The gap-graded model soil samples were used in two distinguishable colours (corresponding to coarse and fine) to ensure that the eroded particles seen were from the original fines content and not a product of crushing in the coarse matrix.

1 INTRODUCTION

As well as providing and storing water and regulating flows, hydroelectric dams function as a sustainable and clean source of energy and energy storage. However, many large embankment dams were constructed before engineers had a sound understanding of filter design. The most frequent cause of failure in many distressed embankment dams involves internal erosion of the dam materials (Bonelli 2013).

Internal erosion is a very complex phenomenon, involving a multitude of factors, from the fines content of the soil, to more subtle factors like the stress state within the soil. According to Moffat & Fannin (2006), constraints on the internal erosion of dams can be separated into different categories: geometric constraints, and hydro-mechanical constraints (see Figure 1, after Garner & Fannin (2010)). The geometric constraints include the fabric, porosity and particle size distribution of the soil, while hydromechanical constraints encompass effects due hydraulic gradient or velocity of seepage flow, and critical stress conditions. In general, the geometric constraints will determine whether particles can theoretically move within the soil matrix. This is influenced by whether the pores of the primary matrix are sufficiently large for the fine particles to move through. The hydromechanical constraints will then determine the flow velocity under a critical stress condition required to initiate erosion of fine particles.

Internal erosion can cause a variety of problems for an embankment dam, such as loss of fine particles, change in the soil skeleton, change in porosity, settlement, differential movement, and even structural failure. Settlement, in particular, is problematic for an embankment dam as it decreases the maximum water level it can hold and thus makes the dam more prone to overtopping. Settlement which occurs locally in only one region of the dam, such as differential settlement, can pose additional structural stability issues. For this reason, the geotechnical aspects of the soil need to be better understood so as to enable the design and maintenance of a wide variety of embankment dams which can fulfil local needs safely.

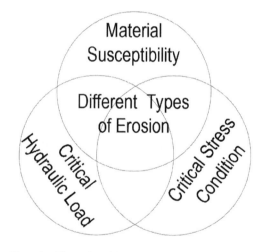

Figure 1. Illustration of the factors affecting internal erosion modified after Garner & Fannin (2010).

A large share of currently existent embankment dams are susceptible to internal instability (Head 1986; Bonelli 2013; Chang et al. 2012). It is necessary to

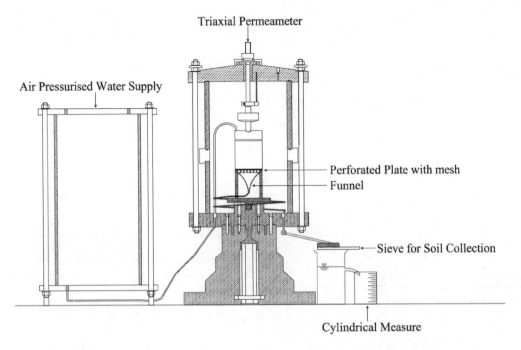

Figure 2. Design of erosion triaxial permeameter.

understand the conditions under which these dams may actually become unstable in order to prevent failure incidents that can lead to the destruction of nearby infrastructure or loss of lives (DeutschesTalsperren-Komitee 2007). In order to develop a framework capable of dealing with the phenomena that embankment dams are exposed to during their lifetime, it is important to develop a means of analysing the soil under laboratory conditions which mirrors as closely as possible the conditions to be found in situ. Development of Triaxial Permeameter.

Here, a physical model of seepage induced instability has been developed at a small scale by studying the effect of a controlled flow rate of seepage through a model soil under controlled stress conditions. The triaxial permeameter (Figure 2) enables the influence of seepage on the shear strength of the sample to be examined, as well as the quantity of eroded soil over a chosen duration and the change of permeability throughout the erosion process.

A Bishop and Wesley stress path cell has been adapted to create the permeameter. An air pressurised water supply allows a high flow rate to be achieved to simulate the seepage flow through a section of soil. In place of a pore stone, a Perspex perforated plate and mesh (Figure 3) are placed at the soil bottom to support the body of the sample while allowing the eroded fine particles to pass through to the soil collection point.

A funnel and wide transparent tube guide the eroded soil through from the base of the sample to exit external to the pressurised permeameter. A sieve enables further investigation of the phenomenon by noting the quantity of the eroded fine particles for each duration of time. Furthermore, the flow rate is determined periodically

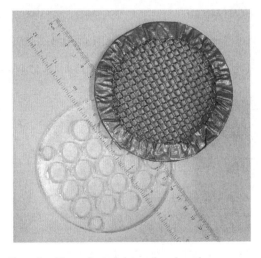

Figure 3. The perforated plate and mesh used.

via measuring cylinder to study the change in permeability throughout the internal erosion process. For the purpose of this paper, we concentrate on the effect of the seepage flow on the strength of the model soil.

2 MODEL SOIL MATERIAL

2.1 *Particle size distribution*

During the construction of a dam, soil segregation often occurs, hence the original broadly-graded soils may become gap-graded soils at local points of the dam. The potential for internal erosion may become

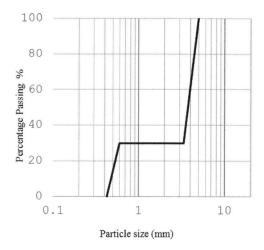

Figure 4. Particle size distribution.

higher at this point. This was one of the motivations to develop a method to assess gap-graded soils.

The apparatus presented here can be used with different types of gap-graded soil by simply changing the size of the mesh on the perforated plate to block the coarse particles and allow the fine particles through to be collected in the soil collection system. The "soil" used in the experiments reported here is a cohesionless model soil (see Figure 4). A large number of existing dams have filters made of cohesionless soils, and therefore this category is of importance in investigating internal erosion.

The soil is internally (and generally arbitrarily) divided into coarse and fine particles. For the chosen soil here, the finer fraction constitutes 30% of the mass and the gap ratio (D'_0/d'_{100} where D' and d' represent the coarse and fine fraction respectively) is 5.58. The sample is unstable according to the three most common methods used in practice for investigating whether a filter is stable or unstable based on geometric constraints (Kenney & Lau, Kezdi and Burenkova methods).

The Kenney and Lau method appears to be best suited for widely graded soils, where it has been found to have better predictive power than the other two methods for instability (Li & Fannin 2008; Moffat 2005). They categorise the granular material as being composed of three different elements: the primary matrix, which is responsible for carrying stresses and which is generally stable and not prone to movement, the loose particles present which are prone to being transported in the soil skeleton, and the constrictions generated by the presence of different size particles which can prevent the aforementioned particle transportation (Kenney & Lau, 1985). This makes it evident that cohesionless filters which are gap-graded, such that there is an absence of sufficient particles of the size required to block travelling paths of fines in between the much larger coarse particles, are more likely to be internally unstable.

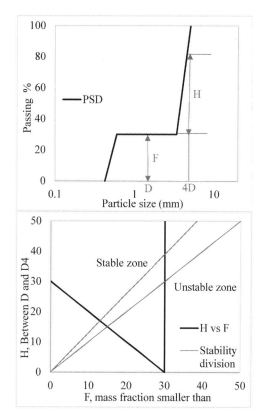

Figure 5. Kenny and Lau criteria check.

Theoretically, Kenny and Lau propose that a soil may be assessed by ensuring that for all particle diameters D chosen, corresponding to points on the PSD curve with a cumulative fines content by mass smaller than 20% for a widely-graded material, the fines content F at that diameter must be smaller than or equal to the mass fraction H that exists between diameter D and 4D (Figure 5). Initially, it was thought that H had to be larger than 1.3 F (Kenney & Lau, 1985), however, this proved to be overly conservative, and it was readjusted to H having to be bigger than 1F (Kenney & Lau, 1986). This can be simplified mathematically into H/F > 1 (previously 1.3). From Figure 5, it is evident that part of the soil passes through the unstable region, suggesting it is unstable over these sizes.

While the Kenney and Lau method appears to be the best predictive method for internal stability, the actual behaviour may be more difficult to predict. Whether erosion actually takes place in those soils will be determined by the hydromechanical constraints imposed on them, such as the critical hydraulic gradient or seepage velocity and stress conditions.

2.2 *Sample preparation*

Model soil samples of aquarium sand and gravel with a specific gravity, Gs, of 2.716 were used in two distinguishable colours to ensure that the eroded particles

Figure 6. Sample with two distinguished colours.

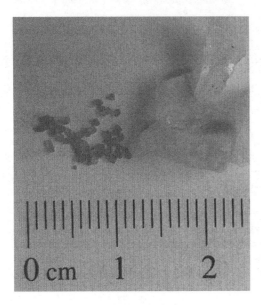

Figure 7. The model soil material in large scale.

Table 1. Maximum and minimum dry density and void ratio of soil mixture tested.

ρ_{min}	ρ_{max}	e_{max}	e_{min}
1.255	1.633	0.992	0.531

one another (S_{max}) (Shire et al. 2014). The fine particle composition is important because it determines how loads are transmitted through the soil. At fines compositions lower than S_f, fines are generally held completely within voids between coarse particles and offer little to no support in carrying the loads. At fines compositions above S_f (especially above S_{max}), fine particles dominate the primary matrix and contribute more along with coarse particles to load transmission (Skempton & Brogan 1995; Shire et al. 2014). These latter stressed matrices offer greater resistance to suffusion, as particles which are under stress are more difficult to remove by seepage flow than unstressed particles.

In this particular sample where the fines content is intermediate to these extremes, both types of particles (coarse and fine) carry loads, hence, eroding of some of the fine particles is expected to change the shear resistance of the sample.

3 TESTING PROCEDURE

For these preliminary tests, samples of 100 mm in diameter and 150 mm in height have been prepared with 0.854 initial void ratio and 30% relative density for testing via moist tamping using the undercompaction method proposed by Ladd (1978). The method helps to ensure an even distribution of particles through the sample and is very efficient in the compaction of sand, avoiding the issue of having the asymmetrical compaction throughout the sample (bottom layers having a higher relative density in comparison to top layers) moreover, the method prevents soil segregation during preparation. Two consolidated drained triaxial tests have been undertaken on the gap-graded material with and without erosion to assess the effect of seepage on the shear resistance of the sample.

For each test, a low cell pressure of 15 kPa was applied to prevent the collapse of the sample. The saturation process was verified using a B-check, reaching a value of B (ratio of $\Delta u/\Delta \sigma'$) greater or equal to 0.92. The sample was then isotropically consolidated to a radial effective stress of 100 kPa. When the desired radial effective stress was reached and there was no more volume change with time, the consolidation process was considered to be complete. The shearing phase of the test then commenced at a rate of 0.05 mm/min until a maximum axial strain of 12%. Local strain transducers mounted directly on the sample were used to record the axial displacement and hence axial strain throughout the test. The shearing phase generally lasted 6-7 hours.

Prior to shearing in one of the tests, the sample was exposed to erosion by applying seepage flow.

are from the original fines content and not a product of crushing within the course matrix (see Figure 6). The coarse particles have relatively flat faces and angular edges (Figure 7). The maximum and minimum dry densities for the particle size distribution (Table 1) are determined after Head (1996).

Skempton & Brogan (1994) defined a critical fine particle content (S_f) at, or above which the internal voids between the coarse particles are filled by fine particles. S_f was calculated to be 24% for densely packed specimens and 29% for loosely packed specimens based on the permeameter seepage studies performed. The other important fine particle content limit is 35%, which represents the fine particle composition when coarse particles are completely separated from

Figure 8. The eroded fines after exposing the sample to seepage flow.

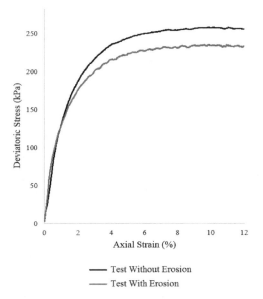

Figure 9. Illustration of the stress-strain relationship, at 100 kPa effective radial stress.

The pressure in the air pressurised water supply was increased 5 kPa every 5 minutes until reaching the critical skeleton-deformation hydraulic gradient which, when achieved, leads to very rapid and significant loss of fines content (see Figure 8) and a sudden change in the matrix of the soil. Figure 9 below compares the experimental data on stress-strain relationship without and with erosion on the gap-graded sample.

4 RESULTS AND DISCUSSION

There was a significant reduction in the shear strength of the sample of 13.5% after erosion occurred. This may have occured as the eroded finer particles were partially in contact with the coarse particles, providing some support to the soil matrix; hence losing that contribution the shear strength of the soil has been reduced.

Interestingly, there was also a slight increase in the shear stiffness at the beginning of the shearing test that might be due to the initial migration of the fine particles to the contact points of the coarser particles, as recently suggested by Akihiro (personal communication). Further work is required to investigate this possibility.

5 CONCLUSION

A new permeameter apparatus, capable of investigating the influence of seepage under controlled triaxial stress path conditions has been developed. The aim of the research is to investigate the stability of susceptible gap-graded soils under complex triaxial stress states undergoing internal erosion. In this way, the test in question can physically model a zone of soil within a larger geotechnical structure undergoing local stress and hydraulic gradient conditions.

The results of two preliminary tests under isotropic mean effective stress and downward seepage are presented. Model soil samples were used in two distinguishable colours to ensure that the particles eroded by seepage were from the original fines content and not a product of crushing in the coarse matrix. The results show a clear decrease of the shear strength and an increase in stiffness of the tested gap-graded sample after internal erosion of the fines has occurred.

The tests reported here show that the triaxial permeameter can be used to identify the effect of seepage flow on the stability of a gap-graded sample. It can be used also to quantify the eroded fine particles over a chosen duration and to investigate the change of permeability throughout the erosion process, which will be the focus of future work.

ACKNOWLEDGEMENTS

The first author is supported by an Engineering and Physical Sciences Research Council (EPSRC) DTA Scholarship, while the experimental research is being supported through EPSRC grant EP/P010423/1 Particle-scale investigation of seepage induced geotechnical instability.

REFERENCES

Akihiro, T. 2017. Personal communication.
Bonelli, S., 2013. Erosion in Geomechanics Applied to Dams and Levees. s.l.:ISTE Ltd and John Wiley & Sons Inc.
Bonelli, S., Brivois, O. & Lachouette, D., 2007. The scaling law of piping erosion. Grenoble, s.n.
Brown, A. J. & Bridle, R. C., 2008. British Dams Website. [Online] Available at: http://www.britishdams.org/2008 conf/papers/Internal%20erosion/P33%20Brown%20Final.pdf [Accessed 02 08 2016].

Chang, D., 2012. Internal Erosion and Overtopping Erosion of Earth Dams and Landslide Dams. PhD Thesis.

Chang, D., 2012. Internal Erosion and Overtopping Erosion of Earth Dams and Landslide Dams. PhD Thesis.

Chang, D. S. & Zhang, L. M., 2011. A Stress-controlled Erosion Apparatus for Studying Internal Erosion in Soils. Geotechnical Testing Journal, 34(6).

Chang, D. S., Zhang, L. & Xu, T. H., 2012. Laboratory Investigation of Initiation and Development of Internal Erosion in Soils under Complex Stress States. Paris, s.n., pp. 895-902.

Chang, D. & Zhang, L., 2013. Extended internal stability criteria for soils under seepage. Soils and Foundations, 53(4), pp. 569-583.

Crawford-Flett, K. A., 2008. An improved hydromechanical understanding of seepage-induced instability phenomena in soil. PhD Thesis.

Deutsches Talsperrenkomitee, 2007. Assessment of the Risk of Internal Erosion of Water Retaining Structures: Dams, Dykes and Levees. Freising, Germany, Deutsches Talsperrenkomitee.

Deutsches Talsperrenkomitee, 2007. Assessment of the Risk of Internal Erosion of Water Retaining Structures: Dams, Dykes and Levees. Freising, Germany, Deutsches Talsperrenkomitee.

Fannin, J., 2008. Karl Terzaghi: From Theory to Practice in Geotechnical Filter Design. Journal of Geotechnical and Geoenvironmental Engineering, 134(3), pp. 267-267.

Head, K. H., 1993. Manual of Soil Laboratory Testing. 2nd Edition ed. London: Pentech Press.

Igwe, O., Fukuoka, H. & Sassa, K., 2012. The Effect of Relative Density and Confining Stress on Shear Properties of Sands with Varying Grading. Geotechnical and Geological Engineering, 30(5), pp. 1207-1229.

Ke, L. & Takahashi, A., 2014. Experimental investigations on suffusion characteristics and its mechanical consequences on saturated cohesionless soil. Soils and Foundations, 54(4), pp. 713-730.

Kenney, T. C. & Lau, D., 1985. Internal Stability of Granular Filters. Can. Geotech. Journal, Issue 22, pp. 215-225.

Kenney, T. C. & Lau, D., 1986. Internal stability of granular filters: Reply. C. Geotechnical Journal, Issue 23, pp. 420-423.

Kezdi, A., 1979. Soil physics – selected topics. Amsterdam: Elsevier Scientific Publishing Company.

Ladd, R. S., 1978. Preparing Test Specimens Using Undercompaction. Geotechnical Testing Journal, 1(1), pp. 16-23.

Li, M., 2008. Seepage Induced Instability in Widely Graded Soils. PhD Thesis.

Li, M., 2008. Seepage Induced Instability in Widely Graded Soils. PhD Thesis.

Li, M. & Fannin, R. J., 2008. Comparison of two criteria for internal stability of granular soil. Canadian Geotechnical Journal, 49(9), pp. 1303–1309.

Li, M. & Fannin, R. J., 2012. A theoretical envelope for internal instability of cohesionless soil. Geotechnique, 62(1), pp. 77–80.

Luo, Y.-l. et al., 2013. Hydro-mechanical experiments on suffusion under long-term large hydraulic heads. Natural Hazards, 65(3), pp. 1361–1377.

Moffat, R., 2005. Experiments on the Internal Stability of Widely Graded Cohesionless Soils. Thesis.

Moffat, R. A. & Fannin, R. J., 2006. A Large Permeameter for Study of Internal Stability in Cohesionless Soils. Geotechnical Testing Journal, 29(4).

Moffat, R. & Fannin, R. J., 2011. A Hydromechanical Relationship Governing Internal stability of Cohesionless Soil. Can. Geotech, I(48), pp. 413–424.

Rees, S., 2013. Introduction to Triaxial Testing. [Online] Available at: http://www.gdsinstruments.com/__assets__/pagepdf/000037/Part%201%20Introduction%20to%20triaxial%20testing.pdf [Accessed 25 08 2016].

Shire, T., O'Sullivan, C., Hanley, K. J. & Fannin, R. J., 2014. Fabric and Effective Stress Distribution in Internally Unstable Soils. J. Geotech Geoenviron. Eng..

Skempton, A. W. & Brogan, J. M., 1994. Experiments on Piping in Sandy Gravels. Geotechnique, 44(3), pp. 449-460.

U.S. Dep., 2015. Best Practices. [Online] Available at: http://www.usbr.gov/ssle/damsafety/risk/BestPractices/Chapters/IV-4-20150617.pdf [Accessed 01 08 2016].

Wan, C. F. & Fell, R., 2008. Assessing the Potential of Internal Instability and Suffusion in Embankment Dams and Their Foundations. Journal of Geotechnical and Geoenvironmental Engineering, 134(3), pp. 401-407.

Centrifuge model test on the instability of an excavator descending a slope

T. Hori & S. Tamate
National Institute of Occupational Safety and Health, Tokyo, Japan

ABSTRACT: Several accidents occur each year wherein an excavator topples whilst traversing on a slope. Analysis of past accidents showed that toppling accidents occurred even on slopes with a smaller angle than the stability of the machinery. In this study a 1/10th scale model of an excavator was developed to clarify the mechanisms that render such machinery unstable whilst moving on a slope in centrifuge model tests were performed. The results of the centrifuge model tests reproduced the toppling phenomenon of the excavator model on slopes with an inclination angle smaller than the stability angle of the model excavator. It was clarified that the toppling phenomenon that occur while an excavator is moving on a slope can be roughly explained by approximate values based on the law of conservation of energy.

1 INTRODUCTION

In Japan, nearly eighty workers are killed every year owing to labour accidents caused by construction machinery. In particular, labour accidents caused by excavators are the most common among construction machinery. Most of the accidents involve workers being struck while the machinery drives backwards, but there are also many accidents where the machinery topples over while moving on slopes. Some cases of toppling occurred when the machinery traversed a slope with an inclination angle smaller than the stability angle onf the model excavator. There are very few studies aiming to clarify the toppling phenomenon during movement on such slopes, and there is a lack of sufficient knowledge on the subject. In this study, a 1/10 scale model of an excavator was constructed to determine the necessary traveling conditions to prevent toppling accidents while moving on slopes, and centrifuge model tests were performed. We assessed the risk factors during movement on a slope and utilized an evaluation method based on the law of conservation of energy.

2 MODELLING

2.1 Details of centrifuge model tests

In this section, the scaling law concerning the interaction between the ground and the machinery is described. The rotary motion of the excavator model is composed of four kinds of forces: inertial force, damping force, subgrade reaction force and moment (Kagawa, 1978).

An inertial force is a product of the inertial mass and the angular acceleration; the ratio of inertial force of the model and the prototype is shown in the following equation:

$$r_i = \frac{I_m \cdot \ddot{\theta}_m}{I_p \cdot \ddot{\theta}_p} \tag{1}$$

where I is the inertial mass and $\ddot{\theta}$ is the angular acceleration. The subscripts 'm' and 'p' denote 'model' and 'prototype' respectively. The inclination angle and the density of material of the model are the same as those of the prototype. Therefore, Equation 1 can be rewritten as Equation 2:

$$r_i = \frac{I_m \cdot \ddot{\theta}_m}{I_p \cdot \ddot{\theta}_p} = \frac{\rho_m l_m^5 (\theta_m/T_m^2)}{\rho_p l_p^5 (\theta_p/T_p^2)} = \frac{1}{n^5} \times \left(\frac{T_p}{T_m}\right)^2 \tag{2}$$

The damping force is a product of the damping factor and the angular velocity, and the ratio of force for the model and the prototype as shown in Equation 3:

$$r_c = \frac{c_m l_m A_m \dot{\theta}_m}{c_p l_p A_p \dot{\theta}_p} = \frac{1}{n^3} \times \frac{c_m}{c_p} \times \frac{T_p}{T_m} \tag{3}$$

where l is the length, A is the surface area, and θ is the inclination angle.

The subgrade reaction force ratio is shown in Equation 4:

$$r_e = \frac{k_m \varepsilon_m l_m A_m \theta_m}{k_p \varepsilon_p l_p A_p \theta_p} \tag{4}$$

where k is the elastic modulus and ε is the strain. These are the same in the prototype as in the model. Therefore, the force ratio can be expressed as Equation 5:

$$r_e = \frac{1}{n^3} \tag{5}$$

Table 1. Comparison of the specifications of the prototype and model.

	Excavator			
	Prototype		Model (1/10 scale)	
Centre of gravity	Horizontal x(m) 0.24	Vertical y (m) 0.81	Horizontal x (m) 0.01	Vertical y (mm) 0.06
Gross weight	46.16 kN		38.6 N (10 g in 385.8 N)	
Contact pressure	40.4 kPa		40.2 kPa (10 g)	
Stability	forward (deg) 42.2	backward (deg) 57.3	forward (deg) 49.4	backward (deg) 56.3

The moment ratio is shown in Equation 6:

$$r_m = \frac{m_m l_m g_m}{m_p l_p g_p} = \frac{1}{n^4} \times \frac{g_m}{g_p} \quad (6)$$

The condition in which all ratios of the moments shown in Equations 2, 3, 5, and 6 are equal is shown in Equation 7:

$$r_i = r_c = r_e = r_m = \frac{1}{n^3} \quad (7)$$

The following equations were used to reproduce the necessary prototype conditions:

$$\frac{T_m}{T_p} = \frac{1}{n} \quad (8)$$

$$\frac{c_m}{c_p} = n \quad (9)$$

$$\frac{g_m}{g_p} = n \quad (10)$$

$$\frac{v_m}{v_p} = 1 \quad (11)$$

The velocity of the model was the same as that of the prototype.

2.2 Outline of the excavator models

Most of the toppling accidents involving excavators were caused by a small machinery with a bucket capacity less than 0.2 m³. Therefore, this study focused on small machinery.

The positions of the motor, decelerator, and battery of the model were designed relative to the prototype machinery in terms of the centre of gravity.

Figure 1 shows the excavator model, and Table 1 shows a comparison between the specifications of the prototype and the model. Comparing the stability of the actual machine with the model, the forward stability angle of the actual machine was 42.2° while the forward stability angle of the model was 49.4°. Which means that, the model was more stable than the actual machinery.

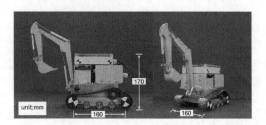

Figure 1. Excavator model.

Although the actual machine could not be reproduced perfectly with respect to mechanical characteristics such as rigidity and deflection of the model axle and crawler track, it was presumed that these effects could be considered negligible.

2.3 Modelling of the ground

Two ground types were modelled. The model grounds were made using either expanded Poly-Ethylene (EPE) or "Kanto loam" (a kind of cohesive volcanic soil). EPE is a homogeneous chemical material; it was used for ensuring the reproducibility of the experiment. The model ground using Kanto loam was made using a centrifuge apparatus to ensure equal compaction pressure. Weight was placed so that the compaction pressure was 50 kPa. After being compacted at a centrifugal acceleration of 50 g, the material was cut into the shape of a slope. The ultimate bearing capacity of the model ground produced by compaction was about 250 kPa. Because the acting pressure of the excavator model (40.2 kPa) was smaller than the elastic limit of the ground, the influence of ground damping was considered to be small.

3 TEST PROGRAM

3.1 Test apparatus and test conditions

All the tests described here were conducted on the NIIS Mark II centrifuge (Horii et al. 2006). The model is operated by a wireless controller. The transmitter for operating the model is separated from the controller, the controller and the transmitter are connected via a slip ring.

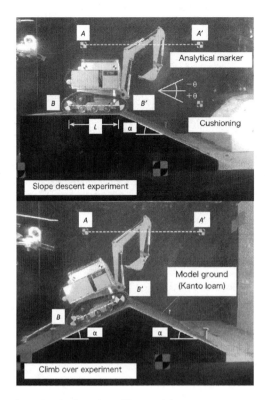

Figure 2. Outline of centrifuge model tests.

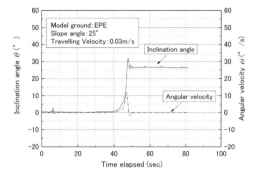

Figure 3. Relationship between the inclination angle θ and the time history of angular velocity ω.

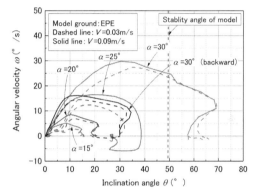

Figure 4. Relationship between the inclination angle and the angular velocity.

Figure 2 shows the outline of the centrifuge model test. In this study, two kinds of experiments were conducted: one involving descending the slope, and the other one experiment involving climbing over an embankment simulating surplus soil. These experiments are named "slope descent experiment" and "climb over experiment", respectively. In the slope descent experiment, the model starts at the top of the embankment and descends the slope. In the climb over experiment, the model climbs over the embankment simulating the soil. The inclination of the slope of the model ground was set at 15°, 20°, 25° and 30°.

The model was operated after the centrifugal acceleration was increased to 10 g. The velocities of the model considered were 0.03 m/s and 0.09 m/s, because the model had a maximum velocity of 0.15 m/s.

The behaviour of the model was recorded with a high-speed camera. The resolution of the high-speed camera was 512 × 412 pixels, and the shooting speed was 500 fps. The oscillation of the model was measured by image analysis of the captured movie. In the analysis, the angle formed by the line A-A' (connecting the two target markers installed on the container wall) and the line B-B' (mounted on the axle portion of the model) was defined as the inclination angle, θ. As for the polarity of θ, the direction in which the model was inclined forward was considered +. (as shown in Figure 2)

Moreover, for the slope descent experiment, we experimented with the following motions: descending while moving forward and descending backwards, this to investigate the influence on the instability arising from different directions of travel.

4 EXPERIMENT RESULTS

4.1 Slope descending experiment

The inclination angle, θ, and the angular velocity, ω, were determined by image analysis. Figure 3 shows the inclination angle, θ, and the time history of angular velocity, ω, for a test condition where the inclination of the slope of EPE ground was 25°. The value of the inclination angle was almost zero while moving on the crest (t = 0 to 38 s). The values of θ and ω of the model fluctuated by a large amount when the model passed over the top of the slope and began its travel on the slope (t = 39 s). Subsequently, while moving on the slope (t = 50 s), ω became almost zero and θ also fluctuated minimally. Figure 4 shows the relationship between θ and ω. The dashed line in the figure represents $V = 0.03$ m/s and the solid line shows the results for $V = 0.09$ m/s. In this figure, when comparing the angular velocities at the same inclination angle, the larger the angular velocity, the greater the rotational kinetic energy, which leads to more instability. As the slope angle increases, the angular velocity

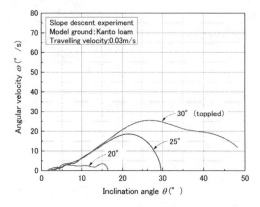

Figure 5. Results of the slope descent experiment (Kanto loam ground).

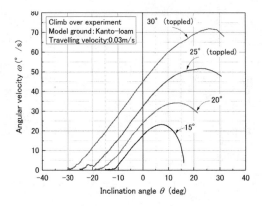

Figure 6. Results of the climb over experiment (Kanto loam ground).

also increases and it can be confirmed that it is destabilised. At a slope angle of 30°, the model toppled when descending the slope during forward travel. The forward stability of the model was 49.4 degrees, and this confirmed that the model topples even on a smaller angle than the stability angle of the model excavator (slope angle < angle of forward stability). On the other hand, on the same 30° slope, when descending by backward travel, it was possible for the model to descend the slope without toppling over. The reason for this discrepancy is that the centre of gravity of the model is located slightly ahead of the axis of rotation, so the stability in front and rear directions are different.

In the case of descending the slope by forward travel, the load acts concentrically on the front of the crawler. From these results, it was found that it is safer to descend a slope by backward travel than forward travel.

4.2 Comparison of "slope descent experiment" and "climb over experiment"

Figures 5 and 6 show the results of the "slope descent experiment" and "climb over experiment" on the Kanto loam ground. The ω in the climb over experiment had a larger value than in the slope descent experiment, indicating that it is more unstable. Moreover, conditions in which the model toppled over were found in three cases: a slope angle of 30° in the slope descent experiment, and slope angles of 25° and 30° in the climb over experiment. For the slope descent experiment at 30° and the climb over experiment at 15°, although the maximum value of the ω was nearly equal, the model toppled during the slope descent experiment (30°), but did not topple during the climb over experiment (15°). As described above, even if the maximum angular velocity (rotational kinetic energy) was the same, there were differences in the results. It was considered that the inclination angle of the model and the slope angle were related when the maximum angular velocity occurred. In other words, the stability is different for a model placed on the slope angle of 30° and the model set on a 15° angle, and thus, the amount of energy required for toppling is different. The discussion of the toppling conditions of the machine will be described later. From the above results, it was found that the model became unstable at smaller slope angles when climbing over surplus soil, as compared to the case of descending the slope, and there was a risk of toppling even on a relatively gentle slope.

4.3 Investigation of toppling phenomenon of the excavator based on energy equilibrium

Herein, we consider the calculation of the approximate value of ω during both slope descent and climb over. Figure 7 shows the movement of the model's centre of gravity in the centrifuge model test. The displacement magnitude of the centre of gravity in the slope descent experiment is expressed by Equation 12, and in the climb over experiment by Equation 13.

$$\Delta h_1 = h - h \cdot \cos\alpha \quad (12)$$

$$\Delta h_2 = h/\cos\alpha + c \cdot \cos\alpha - h \cdot \cos\alpha \\ = 2 \cdot h\sin\alpha \cdot \tan\alpha \quad (13)$$

In the slope descent experiment, assuming that the displacement magnitude of the centre of gravity (amount of potential energy) is converted into rotational kinetic energy based on the law of conservation of energy, the expression in Equation 14 is obtained.

$$m \cdot g \cdot \Delta h_1 = 1/2 \cdot m \cdot h^2 \cdot \omega_1^2 \quad (14)$$

Where, m is the model's mass, g is the gravitational acceleration, and h is the radius of rotation.

From Equation 14, the expression describing the relationship between the displacement magnitude of the centre of gravity and the angular velocity is shown in Equation 15.

$$\omega_1 = \sqrt{2 \cdot g \cdot \frac{\Delta h_1}{h^2}} \quad (15)$$

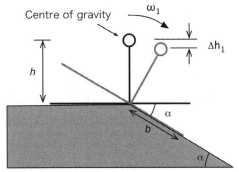

(a) Slope descent experiment

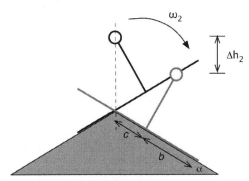

(b) climb over experiment

Figure 7. Modelling of the movement of the model's centre of gravity in the centrifuge model test.

Similarly, expressions for the climb over experiments are expressed by Equations 16 and 17.

$$m \cdot g \cdot \Delta h_2 = 1/2 \cdot m \cdot (h/\cos\alpha)^2 \cdot \omega_2^2 \tag{16}$$

$$\omega_2 = \sqrt{2 \cdot g \cdot \frac{\Delta h_2}{h^2} \cdot \cos^2\alpha} \tag{17}$$

From Equations 15 and 17, the approximate value of the angular velocity at the time of slope descent and climb over can be obtained.

In this study, the value obtained by dividing the maximum angular velocity, (ω_{max}, measured by the experiment) by the calculated approximate value was defined as the angular velocity ratio, $R\omega$. $R\omega$ is expressed by the following equation:

$$R_\omega = \omega_{max} / \omega_{cal} \tag{18}$$

Figure 8 shows the relationship between the slope angle and $R\omega$. $R\omega$ was approximately less than 1. Therefore, Δh was calculated assuming that the model rotated around the fulcrum at approximate values.

However, in fact, it continued to travel forward while the model was rotating.

As a result, it was considered that Δh became larger than the approximate value.

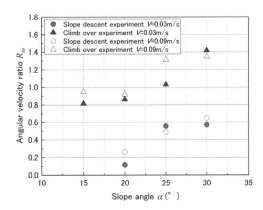

Figure 8. Relationship between slope angle α and angular velocity ratio R_ω.

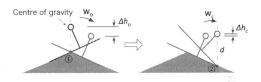

Figure 9. Movement of the centre of gravity when the model topples over.

4.4 Investigation of toppling conditions

Figure 9 shows a conceptual diagram of the movement of the centre of gravity when the model topples over. In a machine that uses a crawler as its travelling equipment, the model starts to rotate (with 1 at the fulcrum) when the centre of gravity of the model passes the top of the slope or the top of the embankment. The rotational kinetic energy generated at this time corresponds to the variation in the potential energy. That is, the kinetic energy required for the model to topple, Δh_c, is expressed by the following equation:

$$\Delta h_c = d - (b \cdot \sin\alpha + h \cdot \cos\alpha) \tag{19}$$

From the law of conservation of energy, the expression for the relationship between the potential energy and the rotational energy can be expressed by Equation 20.

$$E_c = m \cdot g \cdot h_c = 1/2 \cdot m \cdot d^2 \cdot \omega_c^2 \tag{20}$$

where, d is the turning radius at the time of toppling.

Table 2 lists the slope angle, (Δh_c) and corresponding critical angular velocity (ω_c).

When the slope angle is large, Δh_c and the angular velocity are small, and it can be confirmed that the energy required for toppling is also small. The kinetic energy generated at the time of descent or climb over (point ①) is defined as E_0, and the kinetic energy required for toppling is defined as I (point ②). The value obtained by dividing E_0 by E_c was defined as the degree of toppling risk, D_t.

Table 2. Slope angle, α, Δh_c and critical angular velocity, ω_c.

slope angle α (°)	Δh_c (mm)	Critical angular velocity ω_c (°/s)
15	13.7	18.5
20	9.9	15.7
25	6.6	12.9
30	4.0	10.0

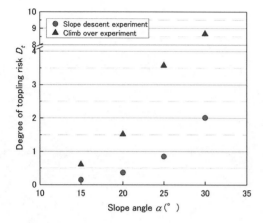

Figure 10. Relationship between slope angle α and degree of toppling risk D_t.

Equation 21 shows the expression for this relationship. If D_t is greater than 1, it indicates that the model will topple.

$$D_t = E_o/E_c \qquad (21)$$

Figure 10 shows the results of a theoretical analysis of D_t for each experiment. These results confirm that when climbing over surplus soil, the machinery was more unstable compared to when it was descending the slope, and it was found that the machine topples over even under conditions with a gentler slope. Moreover, D_t for the slope descent experiment with a slope of 30° is 2.01, which indicates 'topple over'; however, D_t for the climb over experiment on the slope of 15° is 0.61, which means 'will not topple over'. This prediction agrees with the experimental results.

In the climb over experiment with a slope of 20°, D_t was 1.5, but the model did not topple over in the experiment. The reason is considered to be that the model ground subsided when the model was located at the top of the embankment, and Δh became smaller than the approximate value.

As a result, during the experiment, E_0 became smaller and D_t became less than 1 (which means, the model did not topple over).

In all other cases beside this condition, there was good agreement between The theoretical analysis and experimental results.

Therefore, it was found that the analytically obtained topple risk, D_t, assuming that there is no settlement of the crawler, is a good indicator of the safety of toppling risks.

5 CONCLUSIONS

This study examined instability in excavators caused by descending slopes and climbing over embankments. The conclusions of this study are summarised as follows:

(1) It was possible to reproduce a toppling accident of an excavator during the centrifuge model test in which the accident occurred while moving on a slope with a smaller angle than the stability angle of the model excavator.
(2) As the inclination of the slope increased, the machinery became unstable, but it was determined that the absolute value significantly differed relative to the direction of travel of the machinery. In other words, it was found that it is possible to descend the slope more safely when moving backward rather than when moving forward.
(3) The theoretical conditions leading to toppling of the machinery were derived and compared with the index of toppling risk, D_t. When compared with the slope descent experiment, the D_t of the climb over experiment was larger by a factor of more than two, and it became clear that the risk of falling was high in this case.

REFERENCES

Horii, N., Itoh, K., Toyosawa, Y.& Tamate, S. 2006. Development of the NIIS Mark-II geotechnical centrifuge, *Proceedings of the 6th international conference on physical modeling in geotechnics*, (1): 141–146.

Kagawa, T. 1978. On the similitude in model vibration tests of earth structures, Proceedings of Japan society of civil engineering, 275 : 69–77.

Transparent soils turn 25: Past, present, and future

M. Iskander
Civil & Urban Engineering Department, New York University, Brooklyn, New York, USA

ABSTRACT: Transparent soils debuted in 1993. Since their introduction, they have established their utility as a useful tool for physical modelling of (1) soil-structure interaction, (2) saturated-unsaturated hydraulic behavior, and (3) thermal processes in soils. This paper traces the history of the development of transparent soil surrogates and provides a brief literature review to aid new users interested in adopting transparent soils for use in laboratory 1g bench-scale and centrifuge tests to investigate a variety of geotechnical applications. The paper also looks to the future to anticipate where further developments in these materials and complementary technologies may be expected.

1 INTRODUCTION

Transparent soils are made by refractive index (RI) matching of the pore fluid and solids representing the soil skeleton (Iskander 2010). Development of optical impedance matching methods occurred independently in a number of disciplines including physics, chemistry, agronomy, and geotechnical engineering.

In geotechnical engineering, early attempts to probe the internal behavior of soils using the visible wave light spectrum involved the use of crushed glass or photo-elastic disks to represent soil grains along with a variety of unstable matching fluids (Allersma 1982, Desrsues et al. 1991, and Konagai et al. 1992). These pioneering studies did not document the geotechnical behavior of the soil surrogates, and as a result these techniques were not widely adopted.

Modern transparent soils trace back their history to Mannheimer (1990) development of a transparent slurry for study of non-Newtonian flow. The geotechnical properties of core samples (Fig. 1) consolidated from Mannheimer's slurry were first established by Mannheimer and Oswald (1993) and Iskander et al. (1994). These two publications were the forerunners of the current interest in physical modelling with transparent soils.

Geotechnical engineering research is distinct from that of many other disciplines by the need for larger test models, so the development to a very large extent has been characterized by the need to reduce the cost of the constituent materials through the use of inexpensive industrial substances and/or the use of refuse materials, rather than pure scientific-grade materials typically employed in other disciplines.

The use of a soil surrogate raises some concerns about how well the surrogate represents natural soils. For that reason, early developments in transparent soils have been concerned with correlating the behavior of discovered surrogates to those of natural soils.

Modelling with transparent soils has evolved over the past 25 years. Early models have been rather rudimentary; however, with time more sophisticated experiments have been conducted. A large number of surrogates was also introduced. Similarly, tools to reconstruct model behavior from digital images are becoming increasingly more powerful. Transparent soils are now considered a useful tool for physical modelling of soil-structure interaction mechanisms,

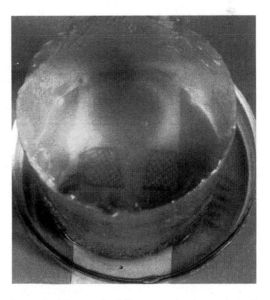

Figure 1. Photograph of first generation transparent soil consolidated from a slurry of amorphous silica and mineral oil blend.

saturated-unsaturated hydraulic behavior, and thermal processes in soils (Iskander et al. 2015a, b, Ganiyua et al. 2016).

2 AVAILABLE SURROGATES

2.1 Precipitated silica

The first family of transparent soils was made of precipitated amorphous silica powder commonly used as filler to manufacture paint, cosmetics, and paper. A variety of pore fluids can be used, offering a plethora of advantages and drawbacks (Zhao et al. 2010), but fluids made of mineral oil and solvent have been the most popular. Two pioneering studies were conducted using this material. The first investigated flow properties around prefabricated vertical (wick) drains (PVDs) (Welker et al. 1999). The second investigated deformations due to pile penetration (Gill and Lehane 2001).

Amorphous silica is suitable for modelling strength, permeability, and consolidation behavior of low plasticity clays (Iskander et al. 2002; Liu et al. 2003). The material was well received, especially by British and Australian researchers.

2.2 Silica gel

Silica gel, which is a commonly used desiccant, can be used to model the static and dynamic behavior of sand (Iskander et al. 2003, Zhao and Ge 2014). It has the same RI as precipitated silica, and thus the two can be combined in the same model to represent a layered system of sand and clay (Sadek et al. 2002). The main drawback of silica gel is that the particles are relatively soft, and also prone to crushing. This has led to the material being overtaken by fused quartz, in subsequent years.

Silica gel and precipitated silica have the same RI, and can thus be combined in the same model using the same pore fluids to model layered stratigraphy.

2.3 Aquabeads

A transparent water-based hydrogel polymer, renamed Aquabeads for convenience, has been used as a surrogate (Lo et al. 2010). Aquabeads absorbs 200 times its own weight in water. Aquabeads particles can be easily crushed, allowing the formation of custom grain-size distributions (Fernandez Serrano et al. 2011). The material is suitable mainly for study of flow in soils and modelling weak marine sediments. Use of Aquabeads has been limited, because it is available from only one manufacturer, with inconsistent availability.

2.4 Fused quartz/silica

At this time, fused quartz is the most popular surrogate for modelling sand. Fused quartz and fused silica are non-crystalline forms of silicon dioxide. Crushed fused quartz employed as a soil surrogate is often a byproduct of industrial processes for the manufacture of semiconductors, solar cells, telescope and microscope lenses, telecommunication equipment and glass chemical containers. The static and dynamic properties of the developed materials are similar to natural angular sand (Ezzein and Bathurst 2011, Guzman et al. 2014; Cao et al. 2011).) A variety of pore fluids can be used including mineral oil blends, pre-manufactured sucrose solutions and sodium thiosulfate-treated sodium-iodide (STSI) (Carvalho et al. 2015, Guzman and Iskander 2013). The material has been especially well received in China where fused quartz is readily available.

2.5 Laponite RD

Laponite is a fine particulate powder made of lithium sodium magnesium silicate. When hydrated with water, it turns into a transparent colloid known as Laponite RD, commonly used to manufacture personal care products. Laponite RD is suitable for modelling very soft clay (Wallace and Rutherford, 2015, Beemer et al. 2016). The material has been employed by Chini et al. (2015) to visualize the shear surfaces formed by several different laboratory tests. It is presently being used to visualize the penetration mechanisms occurring during the penetration of model Torpedo Anchors, for support of offshore structures, as shown in Fig. 2.

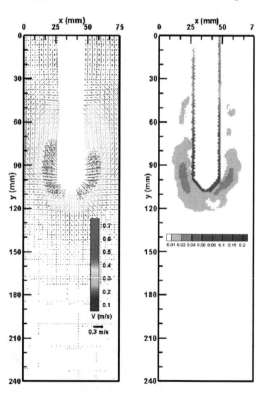

Figure 2. Deformation patters and shear strains developed during penetration of a finless torpedo anchor in Laponite RD representing soft Clay (Initial impact velocity = 5.5 m/sec).

3 ANALYTICAL TOOLS

3.1 Displacement analysis

At this time the gold standard for displacement analysis of transparent soil is Digital Image Correlation (DIC). A method for slicing transparent soil using laser light sheets was first introduced by Sadek et al. (2003). Sometimes transparent soils are seeded with dyed particles in order to assist with image correlation under difficult lighting conditions (Hover et al. 2013, Chen et al. 2014). To date analyses are performed on gray-scale images, using an advanced form of DIC known as adaptive digital image correlation (Liu and Iskander 2004, Omidvar et al. 2014). Several commercial packages have been used, as well as the open source packages GeoPIV (White et al. 2003) and Magic Geo (Iskander et al. 2016). Several papers offer advice on performing DIC analyses including Stanier et al. (2012), Chen et al. 2017 and Black and Take (2015). An example of DIC analysis on Laponite representing soft marine clay being penetrated by a torpedo anchor at an impact velocity of approximately 5.5 m/sec is presented in Fig. 2, where displacements and sheer strains are obtained, non-intrusively.

3.2 Flow analysis

Three-dimensional mapping of non-aqueous phase liquid (NAPL) distribution within saturated porous media is an important issue in bench-scale geo-environmental studies. Most analyses involve correlating gray-scale intensity with NAPL Concentration.

A recent development debuting in this conference is the ability to visualize inter-particulate flow using Particle Image Velocemetry (PIV). PIV is mathematically similar to DIC. The method involves a combination of (1) transparent soils with sizes that are 20-times those of natural soil, (2) seeding the pore fluid using florescent particles, (3) laser illumination, and (4) high-speed and high-resolution photography of the flow (Li et al. 2018). Combined with particle detection and tracking algorithms, the method allows for study of coupled flow problems in geomechanics as shown in Fig. 3.

3.3 Unsaturated-soil analysis

Transparent soils appear white when completely dry and transparent when fully saturated. In between these states, the intensity of the partially saturated transparent soil surrogates has been found to correlate with the degree of saturation. Peters et al. 2009, 2011 and Siemens et al. 2013, 2014 have employed this property in order to investigate the hydraulic properties of partially saturated granular media.

3.4 Thermal analysis

Optical transparency of index-matched granular media is a function of temperature. Black and Tatari (2015) and Siemens et al. (2015), took advantage of this property to visualize heat flow in transparent soils.

4 APPLICATIONS

Transparent soils have been used in a large number of applications. A complete listing is not possible here. The following are examples, of what is possible. For more details, readers are encouraged to explore the *Transparent Soil Wiki* (Iskander 2017).

4.1 Soil-structure interaction

Modelling of the interactions between soils and foundations has been a popular application of transparent soils. To date a wide range of foundation problems has been investigated including (a) pile penetration in clays (Lehane and Gill, 2004; Ni et al. 2010; Hird et al. 2011, Liu and Iskander 2010, Kong et al. 2015), (b) helical anchors (Stanier et al. 2013), (c) specialized expanded base piling (Qi et al. 2017), as well as conventional, (d) shallow foundations (Iskander and Liu 2010).

A variety of surrogates have also been used to investigate the behavior of offshore foundations including (i) plate anchors in amorphous silica (Song et al. 2009, Wang et al., 2011); (ii) drag anchors in Laponite (Beemer and Aubeny 2012); and (iii) torpedo anchors in fused quartz and Laponite (Omidvar and Iskander 2017).

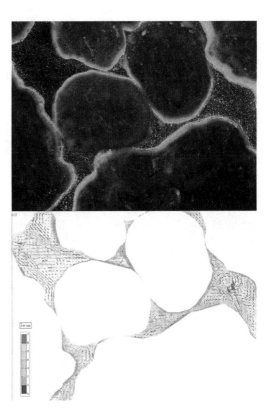

Figure 3. Inter-particulate flow visualization. (a) Image captured by high-speed camera (Top) and, (b) Flow patterns obtain from PIV (bottom).

A few studies employed transparent soils to investigate ground improvement mechanisms. For example, McKelvey (2004) modeled vibrated stone columns in soft clay using a surrogate made of Amorphous silica powder. Similarly, Sui et al. (2015) and Gao et al. (2015) modeled grout propagation in sand using fused quartz.

Several recent studies have explored soil-geogrid interactions using fused quartz surrogates representing sand including Ezzein and Bathurst 2014, Bathurst and Ezzein 2015, 2016, and Ferreira and Zornberg 2015.

Both silica gel and fused quartz have been used as surrogates to visualize tunneling induced ground movement in sand and to investigate the required support pressure (Sun & Liu 2014, Ahmed & Iskander 2011a, b).

Modelling of fast phenomena, such as pile driving and installation of torpedo anchors, using transparent soils presents a number of challenges in terms of image acquisition, providing adequate lighting, and having sufficient contrast within the captured images. These challenges have been overcome, and Fused quartz has been successfully employed for modelling rapid projectile penetration into sand (Chen et 2014; Omidvar et al. 2015, 2016; Iskander et al. 2015a).

4.2 Flow and surfactant flushing

Amorphous silica powder has been used to study flow near PVDs (Welker et al. 2000), and in 2D tank tests (Liu et al. 2005). Aquabeads have also been employed to visualize surfactant flushing (Lo et al. 2010, Tabe 2015). In a recent development, Kashuk & Iskander (2015) employed color space analysis of orthogonal images acquired at model boundaries to reconstruct the NAPL volumes at various stages of contamination and remediation using a previously published calibration model Kashuk et al. (2014).

4.3 K-12 outreach

Aquabeads is particularly suited for use in educational settings and has been employed by Suescun-Florez et al. (2013) for STEM outreach programs in elementary schools. Laponite is water based, like Aquabeads, and is therefore believed to similarly suited for K-12 environments.

5 EXPECTED FUTURE DEVELOPMENTS

The following ideas can in the foreseeable future lead to significant gains in understanding of geotechnical problems through studies that employ transparent soil surrogates.

5.1 Centrifuge testing

Several studies have employed transparent soils in a centrifuge to qualitatively investigate the behavior of offshore foundations. Black (2015) demonstrated a system that permits laser illumination of transparent soil models in a centrifuge. Adoption of similar systems should make transparent soils more popular for investigation of properly scaled soil-structure interaction problems.

5.2 New materials

Efforts are underway in China to develop a transparent porous rock. The author also believes that development of a surrogate suitable for modelling the behavior of fat plastic clays is much needed.

5.3 Color image analyses

Kashuk and Iskander (2014) observed that chromatic families of color space are less sensitive to optical noise compared to luminance (gray-scale) components, which permitted significantly improved flow field analyses (Kashuk et al. 2015). It is therefore expected that color image analyses will in the foreseeable future become routine for image analysis of both flow and soil-structure interaction problems.

5.4 Visualization of particle scale phenomena

Sanvitale and Bowman (2012) were able to resolve individual soil particles to image granular flows. Li et al. (2018) were able to visualize inter-particulate flow (Fig. 3). Together, these technologies permit investigating coupled flow and deformation problems.

6 CONCLUSIONS

The author hopes that this review will encourage geotechnical engineering students and researchers to explore the transparent materials and technologies described in this paper. A wealth of information and examples is available at the *Transparent Soil Wiki* to aid readers in their research.

REFERENCES

Ahmed, M. & Iskander, M. 2011a. Analysis of tunneling-induced ground movements using transparent soil models. ASCE J. of Geotechnical and Geoenvironmental Engineering, 137(5): 525–535.

Ahmed, M. & Iskander, M. 2011b. Evaluation of tunnel face stability by transparent soil models. Tunneling and Underground Space Technology, 27(1): 101–110.

Allersma, H. 1982. Photo-elastic stress analysis and strains in simple shear. Proc. IUTAM Symposium on Deformation and Failure of Granular Materials, Netherlands, 345–53.

Bathurst, R.J. & Ezzein, F.M. 2015. Geogrid and soil displacement observations during pullout using a transparent granular soil. ASTM Geotechnical Testing J., 38(5).

Bathurst, R.J. & Ezzein, F.M. 2016. Geogrid pullout load–strain behaviour and modelling using a transparent granular soil. Geosynthetics International, 23(4), pp. 271–286.

Beemer R.D. & Aubeny, C. 2012. Digital image processing of drag embedment anchors in translucent silicate gel. Proc. GeoManitoba, CGS, Canada.

Beemer, R.D., Shaughnessy, E., Ewert, K.R., Boardman, N., Biscontin, G., Aubeny, C.P. & Grajales, F.J. 2016. The Use of Sodium Pyrophosphate to Improve a Translucent Clay Simulate. Proc. Geo-Chicago, pp. 83–93.

Black, J. 2015. Centrifuge modelling with transparent soil and laser aided imaging. ASTM Geotechnical Testing J., 38(5).

Black, J. & Tatari, A. 2015. Transparent soil to model thermal processes: A thermal pile example. ASTM Geotechnical Testing J., 38(5).

Black J.A. & Take W.A. 2015. Quantification of optical clarity of transparent soil using the modulation transfer function. Geotechnical Testing J., 38(5).

Cao, Z., Liu, J. & Liu, H. 2011. Transparent fused silica to model natural sand. Proc. Pan American Conference, Toronto, Canada.

Carvalho, T., Suescun, E., Omidvar, M. & Iskander, M. 2015. A water based transparent soil for modeling mechanical response of saturated sand. ASTM Geotechnical Testing J., 38(5), doi:10.1520/GTJ20140278

Chen, Z., Omidvar, M., Iskander, M. & Bless, S. 2014. Modelling of projectile penetration into transparent sand. Int. J. of Physical Modelling in Geotechnics, 14(3): 68–79.

Chen, Z., Li, K., Omidvar, M. & Iskander, M. 2017. "Guidelines for DIC in geotechnical engineering research," Int. J. of Physical Modelling in Geotechnics, Vol.17, No.1, pp. 3–22, doi: 10.1680/jphmg.15.00040, ICE.

Chini, C., Wallace, J., Rutherford, C. & Peschel, C.J. 2015. Shearing failure visualization using digital image correlation and particle image velocimetry in soft clay using a transparent soil. ASTM Geotechnical Testing J., 38(5).

Desrsues, J., Mokni, M. & Mazerolle, F. 1991. Tomodensitometry and localization in sands. Proc. of X ECSMFE: Deformation of Soils and Displacements of Structures, 61–65.

Ezzein, F.M. & Bathurst, R.J. 2011. A transparent sand for geotechnical laboratory modeling. ASTM Geotechnical Testing J., 34(6): 1–12.

Ezzein, F.M. & Bathurst, R.J. 2014. A new approach to evaluate soil-geosynthetic interaction using a novel pullout test apparatus and transparent granular soil. Geotextiles and Geomembranes, 42(3): 246–255.

Fernandez Serrano, R., Iskander, M. & Tabe, K. 2011. 3D Contaminant flow imaging in transparent granular porous media. Geotechnique Letters, 1(3): 71–78.

Ferreira, J. & Zornberg, J. 2015. A transparent pullout testing device for 3-D evaluation of soil-geogrid interaction. ASTM Geotechnical Testing J., 38(5).

Gao, Y., Sui, W. & Liu, J. 2015. Visualization of chemical grout permeation in transparent soil.

Gill, D. & Lehane, B. 2001. An optical technique for investigating soil displacement patterns. ASTM Geotechnical Testing J., 24(3): 324–29.

Ganiyua, A., Rashid A. & Osman M. 2016. Utilisation of transparent synthetic soil surrogates in geotechnical physical models: A review. J. Rock Mech. & Geotech. Engr. 8(4): 568–576.

Guzman, I. & Iskander, M. 2013. Geotechnical properties of sucrose-saturated fused quartz for use in physical modeling. ASTM Geotechnical Testing J., 36(3): 448–454.

Guzman, I., Iskander, M., Suescun, E. & Omidvar, M. 2014. A transparent aqueous-saturated sand-surrogate for use in physical modeling. Acta Geotechnica, 9(2): 187–206.

Hird, C., Ni, Q. & Guymer, I. 2011. Physical modelling of deformations around piling augers in clay. Geotechnique, 61(11): 993–999.

Hover, E.D., Ni, Q. & Guymer, I. 2013. Investigation of centerline strain path during tube penetration using transparent soil and particle image velocimetry. Geotechnique Letters, 3(2): 37–41.

Iskander, M., Lai, J., Oswald, C. & Mannheimer, R. 1994. Development of a transparent material to model the geotechnical properties of soil. ASTM Geotechnical Testing J., 17(4): 425–33.

Iskander, M., Liu, J. & Sadek, S. 2002. Transparent amorphous silica to model clay. ASCE J. of Geotechnical and Geoenvironmental Engineering, 128(3): 262–273.

Iskander, M., Sadek S. & Liu, J. 2003. Optical measurement of deformation using transparent silica gel to model sand. Int. J. Physical Modeling in Geotechnics, 2(4): 13–26.

Iskander, M. 2010. Modeling with Transparent Soils, Visualizing Soil Structure Interaction and Multi Phase Flow, Non-Intrusively, Springer.

Iskander, M. & Liu, J. 2010. Spatial deformation measurement using transparent soil. ASTM Geotechnical Testing J., 33(4):1–7.

Iskander, M., Bless, S. & Omidvar, M. 2015a. Rapid Penetration into Granular Media. Visualizing the Fundamental Physics of Rapid Earth Penetration, Elsevier, 458p.

Iskander, M., Bathurst, R. & Omidvar, M. 2015b. Past, present, and future of transparent soils, Geotechnical Testing J., Vol. 38, No. 5, pp. 1–17, ASTM.

Iskander, M, Chen, Z. & Omidvar, M. 2016. Magic Geo: Multi Scale Analyses for Granular Image Correlation Software. wp.nyu.edu/magicgeo

Iskander, M. 2017. Transparent Soil Wiki, website, wp.nyu.edu/ts

Kashuk, S. & Iskander, M. 2014. Evaluation of color space information for the visualization of contamination plumes. J. of Visualization, DOI 10.1007/s12650-014-0232-3, Springer.

Kashuk, S., Mercurio, R. & Iskander, M. 2014. Visualization of dyed NAPL concentration in transparent porous media using color space components. J. of Contaminant Hydrology, v. 162–163, pp. 1–16.

Kashuk, S., Mercurio, R. & Iskander, M. 2015. Methodology for optical imaging of 3D NAPL migration in transparent porous media. ASTM Geotechnical Testing J., 38(5), doi: 10.1520/GTJ20140153.

Kashuk, S. & Iskander, M. 2015. Reconstruction of three dimensional convex zones using images at model boundaries. Computers & Geosciences, Vol. 78, May, pp. 96–109, doi:10.1016/j.cageo.2015.02.008. Elsevier.

Konagai, K., Tamura, C., Rangelow, P. & Matsuhima, T. 1992. Laser-aided tomography: a tool for visualization of changes in the fabric of granular assemblage. Proc. JSCE No: 455 I-21.

Kong, G., Cao, Z., Zhou, H. & Sun, X. 2015. Analysis of piles under oblique pullout load using transparent soil models. ASTM Geotechnical Testing J., 38(5).

Lehane, B. & Gil, D. 2004. Displacement fields induced by penetrometer installation in an artificial soil. International J. of Physical Modeling in Geotechnics, 4, No. 1, 25–36.

Li, L, Omidvar M. & Iskander, M. 2018. Visualization of inter-granular pore fluid flow, Proc. 9th ICPMG

Liu, J. & Iskander, M. 2010. Modelling capacity of transparent soil. Canadian Geotechnical J., 47(4): 451–60.

Liu, J. & Iskander, M. 2004. Adaptive cross correlation for imaging displacements in soils. ASCE J. of Computing in Civil Engineering, 18(1): 46–57.

Liu, J., Iskander, M. & Sadek, S. 2003. Consolidation and permeability of transparent amorphous silica. ASTM Geotechnical Testing J., 26(4): 390–401.

Liu, J., Iskander, M., Tabe, K. & Kostarelos, K. 2005. Flow visualization using transparent synthetic soils. Proc. 16th ICSMGE, Osaka, Japan, 2411–2414.

Lo, H.C., Tabe, K., Iskander, M. & Yoon, S.H. 2010. A transparent water-based polymer for simulating multiphase flow. ASTM Geotechnical Testing J., 33(1): 1–13.

Mannheimer, R. 1990. Slurries you can see through. Technology Today, March Issue, p. 2., Southwest Research Institute, San Antonio, Texas.

Mannheimer, R. & Oswald, C. 1993. Development of transparent porous media with permeabilities of soils and reservoir materials. Ground Water, 31(5): 781–788.

McKelvey, D., Sivakumar, V., Bell, A. & Graham, J. 2004. Modelling vibrated stone columns in soft clay. Geotechnical Engineering, 157(3): 137–149.

Ni, Q., Hird, C. & Guymer, I. 2010. Physical modelling of pile penetration in clay using transparent soil and particle image velocimetry. Geotechnique, 60(2): 121–132.

Omidvar, M., Chen, Z. & Iskander, M. 2014. Image based Lagrangian analysis of granular kinematics. ASCE J. of Computing in Civil Engineering, Vol. 29, No. 6.

Omidvar, M., Malioche, J.D., Chen, Z., Iskander, M. & Bless, S. 2015. Visualizing kinematics of dynamic penetration in granular media using transparent soils. ASTM Geotechnical Testing J., 38(5).

Omidvar, M., Iskander, M. & Bless S. 2016. Soil-projectile interactions during low velocity penetration, Int. J. of Impact Engineering, Vol. 93, pp. 211–221, Elsevier.

Omidvar, M. and Iskander M. 2017. Soil deformations during finless torpedo anchor installation Proc. Geotechnical Frontiers 2017, ASCE.

Peters, S.B., Siemens, G.A. & Take, W.A. 2011. Characterization of transparent soil for unsaturated applications. ASTM Geotechnical Testing J., 34(5): 1–11.

Peters, S.B., Siemens, G.A., Take, W.A. and Ezzein, F. 2009. A transparent medium to provide a visual interpretation of saturated/unsaturated hydraulic behavior. Proc. GeoHalifax, Canada, CGS, 59–66.

Sadek, S., Iskander, M. & Liu J. 2003. Accuracy of digital image correlation for measuring deformations in transparent media. ASCE J. of Computing in Civil Engineering, 17(2): 88–96.

Sadek, S., Iskander, M. & Liu, J. 2002. Geotechnical properties of transparent silica. Canadian Geotechnical J., 39: 111–124.

Sanvitale, N. & Bowman, E.T. 2012. Internal Imaging of Saturated Granular Free-Surface Flows, Int. J. Physical. Modelling in Geotech., Vol. 12, No. 4, pp. 129–142.

Siemens, G., Mumford, K. & Kucharczuk, D. 2015. Characterization of variably transparent soil for use in heat transport experiments. ASTM Geotechnical Testing J., 38(5), doi:10.1520/GTJ20140218.

Siemens, G.A., Peters, S.B. & Take, W.A. 2013. Comparison of confined and unconfined infiltration in transparent porous media. Water Resources Research. 49: 851–863.

Siemens, G.A., Take, W.A. & Peters, S.B. 2014. Physical and numerical modeling of infiltration including consideration of the pore air phase. Canadian Geotechnical J., 51: 1475–1487.

Song, Z., Hu, Y., O'Loughlin, C. & Randolph, M.F. 2009. Loss in anchor embedment during plate anchor keying in clay. ASCE J. of Geotechnical and Geoenvironmental Engineering, 135(10): 1475–85.

Stanier, S.A., Black, J. & Hird, C.C. 2013. Modelling helical screw piles in soft clay and design implications. Geotechnical Engineering, 167(5): 447–460 ICE.

Stanier, S.A., Black, J. & Hird, C.C. 2012. Enhancing accuracy and precision of transparent synthetic soil modelling. Int. J. Physical Modelling in Geotechnics, 12(4): 162–175.

Suescun-Florez, E., Iskander, M., Kapila, V. & Cain, R. 2013. Geotechnical engineering in US elementary schools. European J. of Engineering Education, 38(3): 300–315.

Sui, W., Qu, H., & Gao, Y. 2015. Modeling of grout propagation in transparent replica of rock fractures. ASTM Geotechnical Testing J., 38(5).

Sun, J. & J. Liu 2014. Visualization of tunnelling-induced ground movement in transparent sand, Tunnelling and Underground Space Technology 40 (2014): 236–240.

Tabe, K. 2015. "Transparent Aquabeads to Model LNAPL Ganglia Migration Through Surfactant Flushing," Geotechnical Testing J., Vol. 38, No. 5, pp. 787–804.

Qi, C., Iskander, M. & Omidvar, M. 2017. "Soil deformations during casing jacking and extraction of expanded-shoe piles, using model tests," Geotechnical & Geological Engineering, Vol. 35, No. 2, pp. 809–826.

Wang, D., Hu, Y. & Randolph, M. 2011. Keying of rectangular plate anchors in normally consolidated clays. ASCE J. of Geotechnical and Geoenvironmental Engineering, 137(12): 1244–1253.

Welker, A., Bowders, J. & Gilbert, R. 1999. Applied research using a transparent material with hydraulic properties similar to soil. ASTM Geotechnical Testing J., 22(3): 266–270.

Welker, A., Bowders, J. & Gilbert, R. 2000. Using a reduced equivalent diameter for a prefabricated vertical drain to account for smear. Geosynthetics International, 7(1): 47–57.

Wallace, J. & Rutherford, C. 2015. Geotechnical properties of Laponite RD. ASTM Geotechnical Testing J., 38(5) doi:10.1520/GTJ20140211.

White, D.J., Take, W.A. & Bolton, M.D. 2003. Soil deformation measurement using particle image velocimetry (PIV) and photogrammetry. Geotechnique, 53(7): 619–631.

Zhao, H., Ge., L. & Luna, R. 2010. Low viscosity pore fluid to manufacture transparent soil. ASTM Geotechnical Testing J., 33(6): 1–6.

Zhao, H. & Ge, L. 2014. Investigation on the shear moduli and damping ratios of silica gel. Granular Matter, 16:4449–456.

Application of 3D printing technology in geotechnical-physical modelling: Tentative experiment practice

Q. Jiang, L.F. Li & M. Zhang
State Key Laboratory of Geomechanics and Geotechnical Engineering Institute of Rock and Soil Mechanics, Chinese Academy of Sciences, Wuhan, China

L.B. Song
College of Resource and Civil Engineering, Northeastern University, Shenyang, Liaoning, China

ABSTRACT: For developing the physical modelling for rock mechanics, tentative experiment investigation of 3DP technology in producing models had been carried out in this paper. These tests of 3DP models, including specimens with small cracks and holes, tunnel body, supporting structures, had been printed with the GP material and PLA materials. Practical investigation indicated that the 3DP technology can produce the models with high consistency and less artificial errors in a time-saving. What's more, the loading tests indicated that not only the small specimens with cracks and holes but also large tunnel models had the similar deformation and failure patterns to the mechanical response of rock material and rock engineering in general. This investigation lets us firmly believe that further development of 3DP technology, integrating with other technologies, will eventually greatly promote the development of rock/rock mass mechanics research.

1 INTRODUCTION

The existence of joints, fractures, flaws etc is the basic characters of the rock mass distinguishing from other material due to their significant influence by these inherent defects (Jaeger 1971; Brown 1981; Goodman 1989; Ghazvinian et al. 2010; Barton 2011). Thus, complicated mechanical behaviour of geotechnical engineering is often investigated through the way of physical modelling (Wong & Einstein 2009; McNamara 2009; Chantachot et al. 2016; Zhang et al. 2017). The physical test in laboratory not only can objectively expose the interactional relationship between rock medium and the artificial engineering structures, but can directly observe and measure the deformational and stress responses induced by the artificial excavation, mining and added load (Manzella & Labiouse 2008; Yavari et al. 2014; Chen et al. 2016; Montrasio et al. 2016; Zhang et al. 2017).

During the physical modelling for geotechnical engineering, the geometric similarity in geological structure is a key issue. As a "bottleneck" problem, how to model the complicated joints, vast flaws and complicated artificial structures always challenges the preparation of physical specimens, such as time-consuming processing, inevitable artificial errors, difficulties in close defect etc. (Jiang C et al. 2016; Jiang Q et al. 2016a) Therefore, it is very importance to efficiently and uniformly prepare a physical model similar to the engineering rock mass or in line with research requirements.

Current, three-dimensional printing (3DP) technology, which can quickly and easily produce a complex-structure 3D entity, has attracted wide attention of experts and scholars in various fields. At present, 3DP technology has been well used in biomedicine (Hong et al. 2001), building manufacturing (Henke et al. 2013), industrial manufacturing (Ma et al. 2014), aerospace (Joshi et al. 2015), food industry (Godoi et al. 2016), etc. With the continuous development and update of 3DP technology and print materials, rock-like materials have gradually been applied in preparing model for rock mechanics research, typical works including Ju (et al. 2014), Jiang C (2016), Jiang Q (et al. 2016a; 2016b), Head & Vanorio (2016), Fereshtenejad & Song (2016). These researches fully demonstrated the potential and advantages of 3DP technology in experimental study of rock mechanics. However, most of these studies focus on 3DP rock model, and the feasibility of 3DP technology in physical models test is still lack.

Here, the 3DP has been introduced to produce the several kinds of models for physical test. The small specimens with cracks and holes, the large tunnel models, the rockbolts and linings had been printed with the gypsum powder (GP) or polylactic acid (PLA) materials. Corresponding loading tests for these models showed that their deformation and failure patterns were similar to the mechanical response of rock material and rock engineering. These tentative experiments indicated that the 3DP would have a great application prospect in experimental rock mechanics.

Table 1. The characteristic parameters of 3DP model for test.

Model type	Model material	Model size	Virtual digital model	Physical model
Cylinder containing holes	Plaster powder	Diameter=5cm Height=10cm Hole's diameter=0.3cm		
Rectangle containing cracks	Plaster powder	Edge=5cm Height=10cm		
Intact cylinder	PLA	Diameter=3.5cm Height=7cm		
3D tunnel	Plaster powder	Model: Height=Width=19.6cm, Thickness=2.5cm Tunnel: Height=3.5cm, Width=4cm		
Rockbolt and lining	PLA	Rockbolt: Diameter=0.3cm, Height=2.4cm Lining: Width=3.9cm, Height=3.4cm		

2 EXPERIMENTAL SCHEME

2.1 Physical modelling method

The 3DP technology, which bases on a virtual digital model, is a quickly additive manufacturing or additive layer manufacturing in fact. Generally, the general printing process of 3DP for a physical model includes three basic steps:

1. Model construction. A desired virtual model should first be constructed by a designer using 3D computer-aided way, such as AutoCAD software or CT scanning technology.
2. Slice of digital model. Each 3D digital model need be divided into horizontal layer by layer for additive manufacturing before printing by professional software, such as Make Ware.
3. Model printing. Basic printing parameters, such as filled ratio, printing precision, and layer thickness, should be set after the operating software of the printer has read the virtual model.

In our experiment, the filled ratio was set as 100% for printing the solid model; the printing precision and layer thickness for the FDM printer with PLA material were set as 0.15 and 0.2mm, respectively. The characteristics of printed 3DP models, including specimens and physical models, as listed as Table 1.

2.2 Printing principle for physical model

The applied machine for producing above models were belong to the plaster-based 3D printer (PP) and fused deposition modelling printer (FDM), since printing principle is generally classified according the printing material.

The basic principle of PP printers is that the binder cartridge to pave a layer of powder material on the work platform first, and then the binder cartridge injects binder to the location of the solid part according to the designed section profile information under the control of the computer, so that the powder particles bond together. After bonding of a layer of material is completed, the work platform of build chamber is lowered by one-layer thickness, while the feed material chamber is elevated by a certain height, pushing out a number of powders. The powders are further pushed to the moulding cylinder by the powder roll, to be paved and compacted (Figure 1a). Typically a PP type printer, of which the layer thickness is 0.1 mm, the print resolution is 600×540 dpi, the print accuracy is 0.1 mm (Figure 1b). The powder particle used is powder-gypsum.

The principle of FDM printer is to heat and melt filamentous hot melt material in the printing process. The print head selectively coats the material on the work platform according to the designed section profile

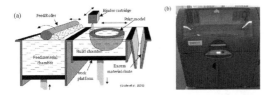

Figure 1. (a) PP printing principle and (b) a commercial machine.

Figure 2. (a) FDM printing principle and (b) corresponding PLA printing material.

information, forming a layer of cross-section after rapid cooling. After a layer of moulding is completed, the machine work platform is lowered by certain height (i.e., layer thickness), followed by moulding of the next layer, until the formation of the entire entity print model. In this paper, the used FDM printer is shown in Figure 2a, whose layer thickness is 0.05–0.5 mm, and print accuracy is 0.1 mm–0.2 mm. The used hot melting material is PLA material with a diameter of 1.75 mm in general (Figure 2b).

3 COMPRESSIVE MECHANICAL BEHAVIOURS OF 3DP MODELS

The loading test for the 3DP models was finished on the RMT-150C experiment system. In our test, the loading ratio of RMT-150C was set as 0.001–0.005 mm/s under the axial strain control condition.

3.1 Test result of cylinder containing holes

Since many holes inside the natural rock mass will affect the mechanical performances, we firstly produced some specimens with many small ball hole by 3D printer, and then investigate their mechanical performance under axial compression test.

The typical stress-strain curves of 3DP specimens showed that their compressive deformations before peal strength were coincide with each other, and their peak strengths were also almost equal to each other (Figure 3). What's more, the stress-strain curves of 3DP specimens passed the peak strengths were different to each other due to the failure randomness. Obviously, the strain behaviours were similar to the general rock (Jiang et al, 2016), which indicated the advantages of 3DP technology in produce the rock-material in high homogeneous way.

Investigation of the break 3DP specimens indicated that their failure included two types, i.e. tensile failure and shearing failure, as shown in Figure 4. In some

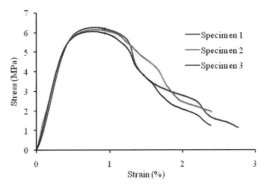

Figure 3. Axial stress-strain curves of 3DP cylinders with holes.

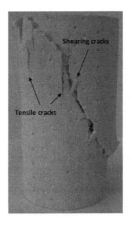 

Figure 4. Failure performances of 3DP cylinders with holes.

specimens, there were a main shearing crack and some small tensile cracks, but many large tensile cracks in other specimens can also be observed. In these failure model, the promoting effect of holes can also been observed.

3.2 Test result of rectangle specimens containing cracks

Existence of many cracks is a typical phenomenon in rock. Thus, specimens with one crack or two cracks had been printed for observing their cracking performance (Seeing Table 1). These experiment had been finished also on the RMT-150C system, and its axial loading rate was set as 0.005 mm/s.

The testing results indicated that the 3DP specimens with one crack and two cracks performed the same stress-strain behaviour, embodied as crack closure stage, elastic stage, crack initiation and coalescence, macro-scale cracking extension stage and strain soften stage (Figure 5), as discussion about general rock (Martin 1997, Cai & Kaiser 2014). What's more, the experimental result indicated that the stress-strain curves of three specimens were similar to each other also indicated that the 3DP technology can produce the specimens with small discreteness.

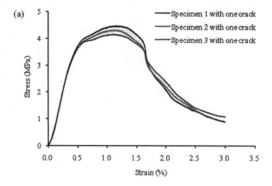

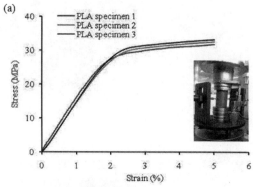

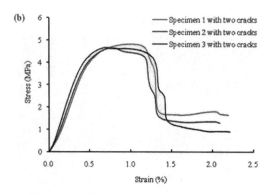

Figure 5. Axial stress-strain curves of 3DP rectangle specimens (a) Specimens with one crack and (b) Specimens with two cracks.

Figure 7. Experimental results of the PLA specimens (a. Compressive stress-strain curves, b. Deformation pattern).

3DP technology and corresponding material can be used for physical modelling in geotechnics.

3.3 Test result of PLA cylinder specimens

The PLA cylinder was a resin material with typical plastic property. Compression test for the PLA cylinders showed that they performed as elasto-plastic behaviour (Figure 7a), which is similar to the steel material. A further observation for the compressed PLA specimens found that there was no any obvious cracks or shearing zone on its surface, which was also in according with steel material (Figure 7b). This test indicated that the printed PLA model cannot be used to simulate the rock material, but can be used to replace the supporting structures in physical modelling, such as rockbolt and lining.

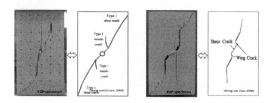

Figure 6. Experimental results of the specimens with pre-existed cracks.

What's more, observation for the break specimens also indicated that the new tensile crack of specimens with one crack firstly initialised at the pre-exist crack's tip and developed in a tensile-shearing way, but the new tensile crack of specimens with two crack firstly appeared at the zone between two pre-exist cracks and then grew into a large shearing crack (Figure 6). The experimental phenomena were also similar to the Wong's summarisation for rock (Wong & Chau 1998; Wong & Einstein 2009). This test for rectangle specimens containing cracks also indicated that the 3DP technology can be used to study the crack extension of rock material.

Above tests for the 3DP specimens with holes and cracks showed that their mechanical characteristics, including deformation, strength and failure pattern, was similar to natural rock, which indicated that the

4 COMPRESSION TEST FOR PHYSICAL TUNNEL MODELS

For investigating the applicability of 3DP model in physical modelling for rock engineering, two tunnel models with no support and two tunnel models with rockbolt and liner support had been printed by 3D printer. Then, these physical models had been place on a transparent PMMA holder for axial compression (Figure 8) with the loading rate 0.005 mm/s.

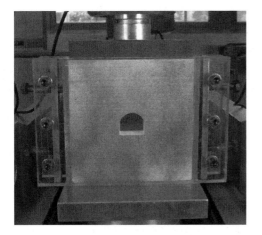

Figure 8. Compression tests for 3DP models with tunnel.

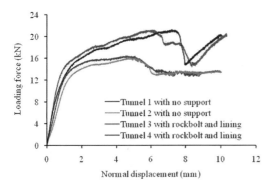

Figure 9. Experimental curves between loading force and normal displacement.

Compression loading curves of force and displacement indicated that: (i) The same type of tunnel models showed the similar loading curves, which meant the 3DP physical models can overcome most of specimens' discreteness induced by the artificial production; (ii) The comparison test for two type models with "rockbolt + lining" or no support shown that the reinforced support for tunnels can obviously improve its loading capability, about 29.75% improvement (Figure 9).

Analysis for the failure form of 3DP models also indicated that: (i) Many serious failure zone of 3DP tunnel with no support could be observed on the sidewall and the crown, which was in according to some test results in laboratorial tunnel model and the actual in-situ failure models of tunnels (Figure 10a); (ii) The failure degree of supporting tunnels was relatively less slight, although their loading force was larger than that for no-support tunnels and the their loaded normal displacements were bigger (Figure 10b), which proved that the reinforce support for tunnel model can distinctly improved its loading capability.

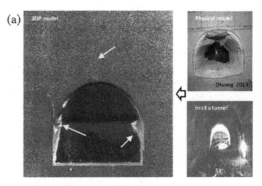

Figure 10. Typical failure forms of 3DP models after compression.

5 CONCLUSIONS

Tentative experiment investigation of 3DP technology in physical modelling for rock engineering had been carried out. The testing result indicated that not only the small specimens with cracks and holes, but also large tunnel models had the similar deformation and failure patterns to the mechanical response of rock material and engineering in general.

These tests for 3DP models, including small specimens with cracks and holes, the large tunnel models, the rockbolt and lining printed with the GP material and PLA materials, indicated that the 3DP technology can produce the models with high consistency and less artificial errors in a time-saving way.

It needs note that there are still some problems to be overcome before its full application in geotechnics. Obviously, we firmly believe that the further development of 3DP technology, integrating with other technologies, will eventually greatly promote the development of rock/rock mass mechanics research.

ACKNOWLEDGEMENTS

The authors gratefully acknowledge the financial support from the State Key Research Development Program of China (Grant No. 2016YFC0600707) and the National Natural Science Foundation of China (Grant No. 51779251 and 51379202).

REFERENCES

Barton, N.R. 2011. From empiricism, through theory, to problem solving in rock engineering. In: *Harmonising Rock Engineering and the Environment*, Qian & Zhou (eds), Taylor & Francis Group, London, ISBN 978-0-415-80444-8.

Brown, E.T. 1981. Rock characterization, testing and monitoring – ISRM suggested methods. *Pergamon Press*, Oxford.

Cai, M. & Kaiser, P. 2014. In-situ Rock Spalling Strength near Excavation Boundaries. *Rock Mech Rock Eng*, 47(2): 659–675.

Chantachot, T., Kongkitkul, W., Youwai, S. & Jongpradist, P. 2016. Behaviours of geosynthetic-reinforced asphalt pavements investigated by laboratory physical model tests on a pavement structure. *Trans Geotec*, 8: 103–118.

Fereshtenejad, S. & Song, J.J. 2016. Fundamental study on applicability of powder-based 3D printer for physical modeling in rock mechanics. *Rock Mech Rock Eng*, 49(6): 2065–2074.

Ghazvinian, A.H., Taghichian, A., Hashemi, M., Marashi, S.A. 2010. The shear behavior of bedding planes of weakness between two different rock types with high strength difference. *Rock Mech Rock Eng*, 43: 69–87.

Godoi F.C., Prakash, S. & Bhandari, B.R. 2016. 3d printing technologies applied for food design: Status and prospects. *J Food Eng*, 179: 44–54.

Goodman, R.E. 1989. Introduction to Rock Mechanics, 2nd Ed., *John Wiley and Sons*, New York.

Grote, D.L., Park, S.W. & Zhou, M. 2001. Dynamic behavior of concrete at high strain rates and pressures: I. experimental characterization. *Int J Impact Eng*, 25(9): 869–886.

Head, D. & Vanorio, T. 2016. Effects of changes in rock microstructures on permeability: 3-D printing investigation. *Geophys Res Lett*, 43(14): 7494–7502.

Henke, K., & Treml, S. 2013. Wood based bulk material in 3D printing processes for applications in construction. *Holz Roh Werkst*, 71(1): 139–141.

Hong, S.B., Eliaz, N., Sachs E.M., Allen, S.M. & Latanision, R.M. 2001. Corrosion behavior of advanced titanium-based alloys made by three-dimensional printing (3DP TM) for biomedical applications. *Corros Sci*, 43(9): 1781–1791.

Huang F., Zhu H., Xu, Q. 2013. The effect of weak interlayer on the failure pattern of rock mass around tunnel–Scaled model tests and numerical analysis. *Tunn Undergr Sp Tech*, 35: 207–218.

Jaeger J.C. 1971. Friction of rocks and stability of rock slopes. *Geotechnique*, 21: 97–134.

Jiang, C., Zhao, G.F. & Zhu, J., 2016. Investigation of dynamic crack coalescence using a gypsum-like 3D printing material. *Rock Mech Rock Eng*, 49(10): 3983–3998.

Jiang, Q., Feng, X. & Gong, Y. 2016a. Reverse modelling of natural rock joints using 3D scanning and 3D printing. *Comp Geotec*, 73: 210–220.

Jiang, Q., Feng, X. & Song, L. 2016b. Modeling rock specimens through 3D printing: Tentative experiments and prospects. *Acta Mech Sinica*, 32(1): 101–111.

Jiang, Q., Zhong, S., Cui, J., et al. 2016. Statistical Characterization of the Mechanical Parameters of Intact Rock Under Triaxial Compression: An Experimental Proof of the Jinping Marble. *Rock Mech Rock Eng*, 49: 4631–4646.

Joshi, S.C. & Sheikh, A.A. 2015. 3D printing in aerospace and its long-term sustainability. *Virtual and Physical Prototyping*, 10(4): 175–185.

Ju, Y., Xie, H. & Zheng Z. 2014. Visualization of the complex structure and stress field inside rock by means of 3d printing technology. *Chinese Sci Bull*, 59(36): 5354–5365.

Ma, W.L., Tao, F.H. & Jia, C.Z. 2014. Applications of 3D Printing Technology in the Mechanical Manufacturing. In Applied Mechanics and Material, *Trans Tech Publications*, 644: 4964–4966.

Manzella, I. & Labiouse, V. 2008. Qualitative analysis of rock avalanches propagation by means of physical modelling of non-constrained gravel flows. *Rock Mech. Rock Eng.* 41 (1): 133–151.

Martin, C.D. 1997. Seventeenth Canadian geotechnical colloquium: the effect of cohesion loss and stress path on brittle rock strength. *Can Geotech J*, 34(5): 698–725.

McNamara, A.M; Goodey R.J. & Taylor R.N. 2009. Apparatus for centrifuge modelling of top down basement construction with heave reducing piles. *Int J Phys Model Geo*, 9(1): 1–14.

Montrasio, L., Schiliròl, L. & Terrone, I.A. 2016. Physical and numerical modelling of shallow landslides. *Landslides* 13: 873–883.

Muller, L. 1966. The progressive failure in jointed media. *Proc. of ISRM Cong.*, Lisbon: 679–686.

Wong, L.N.Y. & Einstein H.H. 2009. Systematic evaluation of cracking behavior in specimens containing single flaws under uniaxial compression. *Int J Rock Mech Min Sci*, 46(2): 239–249.

Wong, L.N.Y. & Einstein, H.H. Systematic evaluation of cracking behavior in specimens containing single flaws under uniaxial compression. *Int J Rock Mech Min Sci*, 2009, 46(2):239–249.

Wong, R.H.C. & Chau, K.T. 1998. Crack coalescence in a rock-like material containing two cracks. *Int J Rock Mech Min Sci*, 35(2): 147–164.

Yavari, N., Tang, A.M. & Pereira J.M. 2014. Experimental study on the mechanical behaviour of a heat exchanger pile using physical modelling. *Acta Geotech*, 9: 385–398.

Zhang, Q.Y. Duan, K. & Jiao, Y.Y. 2017. Physical model test and numerical simulation for the stability analysis of deep gas storage cavern group located in bedded rock salt formation. *Int J Rock Mech Min Sci*, 94: 43–54.

Zhang, Q.Y., Duan, K., Jiao, Y.Y. & Xiang, W. 2017. Physical model test and numerical simulation for the stability analysis of deep gas storage cavern group located in bedded rock salt formation. *Int J Rock Mech Min Sci*, 94: 43–54.

Scaling of plant roots for geotechnical centrifuge tests using juvenile live roots or 3D printed analogues

T. Liang, J.A. Knappett, G.J. Meijer & D. Muir Wood
University of Dundee, Dundee, UK

A.G. Bengough
University of Dundee, Dundee, UK
The James Hutton Institute, Dundee, UK

K.W. Loades
The James Hutton Institute, Dundee, UK

P.D. Hallett
University of Aberdeen, Aberdeen, UK

ABSTRACT: Geotechnical centrifuge modelling of vegetated slopes requires appropriately scaled plant roots. Recent studies have independently suggested that juvenile live plants or 3D printing to fabricate root analogues could potentially produce representative prototype model root systems. This paper presents a critical comparison of juvenile versus 3D printed approaches in terms of their representation of root mechanical properties, root morphology and distribution of the additional shear strength generated by the roots with depth. For the 3D printing technique, Acrylonitrile Butadiene Styrene (ABS) plastic material was used, while for live plants, three species (Willow, Gorse and Festulolium grass), corresponding to distinct plant group functional types (tree, shrub and grass), were considered. The tensile strength and Young's modulus of the 'roots' were collected from uniaxial tension tests and shear strength data of rooted soil samples was collected in direct shear. The prototype root characteristics as modelled were then compared with published results for field grown species and the benefits and challenges of using these two modelling approaches is discussed. Finally, some recommendations on realistically modelling plant root systems in centrifuge tests are given.

1 INTRODUCTION

Vegetation as a low-cost, carbon-neutral natural alternative to conventional ground reinforcement techniques, has been recognised in geotechnical and ecological engineering practice to prevent shallow landslides and erosion (Stokes et al., 2009, 2014). However, they are rarely incorporated explicitly within geotechnical design, principally due to perceived issues of unpredictability in location and variability in biomechanical properties of the roots and hydrological properties of the soil-root composite.

The University of Dundee has had a long-running collaboration with the James Hutton Institute on this issue and has provided some new insights into the design and implementation of projects to mitigate slope instability. These include element scale investigation into root mechanical and hydrological effects (e.g. Mickovski et al., 2009; Loades et al., 2010, 2013; Boldrin et al., 2017a, b; Leung et al., 2017), large-scale investigation into the global slope performance and detection of the failure mechanism of vegetated slopes under rainfall or earthquakes using either centrifuge modelling or numerical modelling approaches (e.g. Sonnenberg et al., 2010, 2011; Liang et al., 2015, 2017a; Liang and Knappett, 2017a), development of analytical models for predicting the deformation response of vegetated slopes (Liang and Knappett, 2017b), and development of rapid in situ testing technique for determining rooted soil properties (Meijer et al., 2016, 2017).

Among these studies, geotechnical centrifuge modelling can provide relatively low cost testing while maintaining a high level of fidelity. It can be used to investigate the global performance of vegetated slopes under known boundary conditions and identify deformation and failure mechanisms of vegetated slopes through image analysis techniques such as particle image velocimetry (PIV) (White et al., 2003). However, in using a geotechnical centrifuge to investigate in detail the engineering performance of vegetated slopes, correct scaling of plant root systems is a substantial challenge (Liang et al., 2017b).

Recent studies have independently shown that using juvenile plants or 3D printing techniques could potentially produce prototype root systems that are highly representative of corresponding mature root systems both in terms of root mechanical properties and root morphology. These methods offer different advantages as an approach for scaling root systems. For example, 3D printed root analogues have good repeatability of architecture and mechanical properties and can be easily and quickly produced. Live plants, on the other hand, can provide highly representative root-soil interaction properties and also more correct stress-strain response. However, many challenges and uncertainties still exist for the use of both types of model roots.

The aim of this paper is to compare these two types of modelling approaches in terms of their representation of root mechanical properties, root morphology and distribution of the additional shear strength generated by the roots with depth based on databases collected at the University of Dundee. Insights and recommendation are made based on these comparisons for better selection of root analogues that may be applied to a wider range of practical problems.

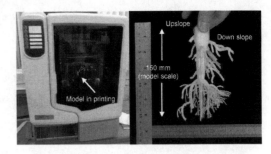

Figure 1. 1:10 geometrically-scaled ABS plastic root model from 3D printer.

2 ROOT MODEL

2.1 3D printed root analogues

The root analogues discussed in this paper are a 1:10 geometrically-scaled tree root cluster consisting of a tap-root system (see Fig. 1), the root architecture used as a template was based on the tap-root system of a white oak tree located at the Warnell School for Forestry and Natural Resources, University of Georgia (Danjon et al., 2008). The root analogue was fabricated using a Stratesys Inc. uPrint SE ABS rapid prototyper (known more commonly as a 3D printer) following the procedures outlined in Liang et al. (2014). Further details relating to the design and fabrication process for this model can be found in Liang et.al (2017a).

2.2 Juvenile live plants

The juvenile live plants discussed in this paper represent 1:15 geometrically-scaled model roots. Three species, *Salix viminalis* (Willow, variety Tora), *Ulex europaeus* L.(Gorsc) and *Lolium perenne* × *Festuca pratensis* hybrid (Festulolium grass), which correspond to distinct plant functional groups (tree, shrub and grass, respectively) with contrasting root systems were selected following a preliminary assessment of suitable species for use in slope engineering applications.

These were cultivated for approximately two or three months (two month for Willow and Festulolium grass, three months for Gorse, due to slower growth) in 150 mm diameter tubes under controlled lighting and temperature (16 h daylight per day under controlled temperature of 27.25 ± 0.38°C (Mean ± SE), and 8 h night per day at a temperature of 22.15 ± 0.13°C). Water was supplied every two days using a watering can. The amount of water supply was decided on the basis of maintaining soil field capacity (5 kPa suction),

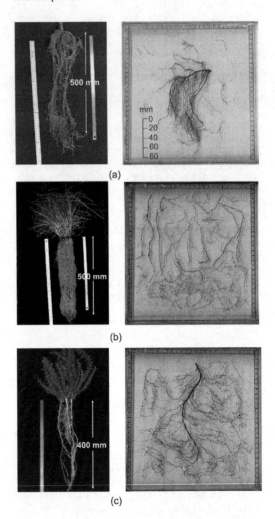

Figure 2. The model juvenile live plants used in this study, showing maximum rooting depth and scanned root architecture (at model scale) : (a) Willow; (b) Festulolium grass; (c) Gorse.

which corresponds with a gravimetric water content of $0.25\,\text{gg}^{-1}$.

The resulting root systems obtained from growth are shown in Fig. 2. After the desired growing time,

Table 1. Scaling laws for centrifuge testing related to this study (After Taylor, 1995; Muir Wood, 2003).

Parameter	Scaling law: Model/Prototype	Dimensions*
Root diameter	1/N	L
Rooting depth	1/N	L
Tensile Strength	1	M/LT²
Shear strength	1	M/LT2
Young's Modulus	1	M/LT²

* L = length; M = mass; T = time.

the three species developed significantly different root morphologies. Festulolium grass had a typical fibrous root system (Fig. 2b); Gorse had a tap root system that consisted of a thick tap root with numerous secondary roots less than 0.5 mm in diameter were attached (Fig. 2c); and Willow developed a root system with numerous branches of different diameters (Fig. 2a).

3 METHODOLOGY

Here only a brief description of how the parameters were obtained will be presented and full details about the testing setup can be found in Liang et al. (2017a,b) for 3D printed root analogues and juvenile live plants, respectively.

The biomechanical properties (specifically, strength and stiffness) of model roots were determined from uniaxial tensile tests and three-point bending tests. The shear strength of rooted and fallow samples were obtained from a custom-designed large direct shear apparatus (Liang et al., 2015; Mickovski et al., 2009). The additional shear strength provided by roots was taken as the maximum difference in shear resistance between the rooted and fallow soil samples divided by the 'working' shear plane area (e.g. Zone of Rapid Taper (ZRT), Danjon and Reubens, 2008). The centrifuge scaling laws related to this paper are shown in Table 1.

4 RESULTS AND DISCUSSION

4.1 Bio-mechanical properties of roots

The measured values (Mean ± SE) of tensile strength and Young's Modulus for 3D printed root analogues and juvenile live plants within different diameter ranges were scaled up according to centrifuge scaling laws (see Table 1) and plotted against the upper and lower bounds of mature root data collated from the literature (Fig. 3). Specifically, data for tensile strength of mature field roots were collected from 40 species of trees, 12 species of shrubs and 21 species of grasses/herbs (see Mao et al., 2012). While data Young's Modulus of mature field roots were collected from 6 species of trees, 5 species of shrubs and 2 species of grasses/herbs (Operstein and Frydman,

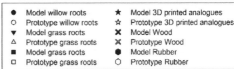

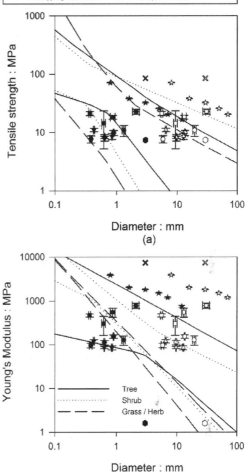

Figure 3. Comparison of root biomechanical properties between model roots (juvenile live plants (Mean ± SE) and root analogues) and mature plants collected from the literature (Root tensile strength data, n = 40, 12 and 21 for trees, shrubs and grasses/herbs, respectively; Root Young's Modulus data, n = 6, 5 and 2 for trees, shrubs and grasses/herbs, respectively): (a) Tensile strength; (b) Young's Modulus.

2000; Van Beek et al., 2005; Mickovski et al., 2009; Fan and Su, 2008; Teerawattanasuk et al., 2014).

Compared with juvenile live plants, 3D printed ABS root analogues are stronger and stiffer in terms of modelling root biomechanical properties. However, it is still a great improvement compared with previous analogue materials (e.g. wood) used in previous studies. It should be noted that the mechanical properties for the 3D printed root analogues shown in Fig. 3 were collected from straight rod samples with individual layering of material aligned parallel to the axis of the root analogue. This scenario may represent an ultimate material mechanical condition. However, when a root cluster with complicated root morphology is printed (like shown in Fig. 1), individual fibres within one root segment will not always be aligned to be parallel to the root axis for each root. As a result, the mean values of tensile strength and Young's Modulus within different diameter ranges in a complex 3D architecture are expected to be lower than the ones shown, as obtained from uniaxial tensile tests or three/four points bending tests. To identify such assumptions, further material characterisation tests are required on cylindrical samples with individual layers aligned in different directions relative to the root axis.

However, such a potential lowering of strength has been indirectly observed in terms of a lower overall additional shear strength provided by roots through a comparison of the 3D root cluster in Fig. 1 and a group of straight rods with aligned layers having the same root distribution across a certain shear plane; further details can be found in Liang et al (2017a).

4.2 Root morphology

Through attentive selection of plant species and growing time, use of live plants in the centrifuge can potentially simulate many types of root morphology in the field. This is also true of 3D printing. However, it should be noted here that, both the 3D printing technique and juvenile live plants approach have a certain threshold of root diameter which can be modelled. For example, the minimum root diameter within the uPrint printer is 0.75 mm (Liang et al., 2017a); as a result, a large amount of very fine or fine roots will not be included in prototype root models. For juvenile plants, the minimum root diameter observed was less than 0.1 mm. Such a drawback should be given particular attention when modelling vegetated slope problems in which fine roots play a major role on root mechanical effect, such as for grassed areas. It is possible that the fine material in the 3D printed case could be simulated by the addition of a quantity of fibres surrounding the ABS model.

4.3 Shear strength of root-reinforced soil

The shear strength measured for 3D printed root analogues and juvenile live plants in the direct shear apparatus is compared in Fig. 4 to root contributions from large in-situ shear box tests conducted for some common species collected from field from a database collated by Liang et al. (2017b). It should be noted here that the shear tests conducted on juvenile live plants could not consider the variation of soil confining stress due to the limitations of the testing apparatus. As a result, the original measured values, only represented the lower bound values (as shown in Fig. 4) considering the effect of soil confining stress on the shear strength increase of rooted soil (Duckett, 2013; Liang et al., 2017a). The upper bound values of root contribution to soil shear strength for juvenile live plants

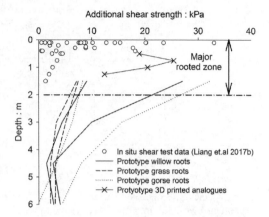

Figure 4. Comparison between the increased shear strength provided by the juvenile plants and 3D printed ABS plastic root analogues at prototype scale and root reinforcement data collected from the literature.

shown in Fig. 4 was derived using Wu and Waldron's model (WWM, Wu et al., 1979) through assuming all roots were mobilized and broken simultaneously as Waldron (1977) did:

$$c_r = 1.15 \times T_r \times RAR \qquad (1)$$

Where c_r is additional shear strength provided by roots, T_r is average tensile strength, and RAR represents root density across each shear plane, defined as the ratio between root cross sectional area and the 'working' cross-sectional area of shear plane (e.g. ZRT). It should be noted here that the derived upper bound values represent the ultimate shear strength root can provide, and the actual root contributions are generally than such values (e.g. Pollen and Simon, 2005; Docker and Hubble, 2008; Bischetti et al., 2009; Loades et al., 2010; Mao et al., 2012).

Fig. 4 clearly demonstrates that in situ direct shear tests on field plants were generally performed on very shallow shear planes (less than 0.2 m deep) due to the limitations of available shear apparatus, which highlights the benefit of using the centrifuge modelling approach for vegetated slope problems, where much more representative stresses can be simulated. It should also be noted that due to the large size of test apparatus required, field tests considered only small elements of rooted soil, while the scaled tests were able to test the root system for a complete plant or tree. It is not surprising therefore that the model systems generally provide higher amounts of reinforcement compared to the field, as the root systems are able to redistribute stresses internally via their interconnected architecture. In terms of the magnitude of root reinforcement, both 3D printed root analogues and juvenile live plants could provide a reasonable magnitude of rooted strength within the major rooted zone (down to 2 m below ground level) compared with the values reported in the literature for direct in situ shear tests, which generally between 2–20 kPa (Norris et al., 2008; Bischetti et al., 2009).

However, compared with the successful control of rooting depth for printed root analogues, the prototype rooting depth of live plants (even for such a short growing period) reached deep within the soil (to 6 m in Fig. 4), leading to a different slope response compared with the field conditions, where roots are mainly concentrated in the top 2 m of soil (Jackson et al., 1996). However, as indicated by Liang et al. (2017b), although juvenile live plants penetrated deeper than the field conditions at prototype scale, the main effective contributions of roots to soil strength are still located in the shallow layer. The reason for this is because the root contribution in the deeper soil layers (>3 m) are relatively small (less than 30%) compare with the fallow soil strength at the same depth. In other words, using juvenile plant roots to scale root reinforcement under field conditions may not be perfect, but it still can provide a representative and informative mechanical model, including the major root reinforcement at the surface and strongly reducing shear strength with increasing depth.

5 FUTURE INSIGHT

Using juvenile plants or 3D printing technique as reported in this paper is currently limited to use in modelling root mechanical reinforcement, and hydrological effects (chiefly transpiration) have not been taken into consideration. Some trials, which combine both mechanical root reinforcement and evapotranspiration have been reported by Ng et al. (2014, 2016).

The idea of applying external suction through a vacuum system on live poles (Ng et al., 2016) or high air-entry value (AEV) porous filters which are made of cellulose acetate (Ng et al., 2014) appears to be effective in modelling water uptake behaviour. Unfortunately, these models are currently based on very simple root geometry, mainly straight rods, occasionally with some highly simplified branching patterns. Considering the influence of root morphology on root mechanical reinforcement (e.g. Ghestem et al., 2014) and root water uptake (e.g. Boldrin et al., 2017a), such a model may not provide a reliable simulation of slope response.

However, the concept of using external suction to simulate root water uptake behaviour may be combined with the 3D printing technique through fabricating hollow root analogues with more realistic root morphology. In terms of fabricating a porous structure, 3D printing can easily achieve this, however, finding a printable material with a high air entry value is a challenge.

In contrast with this, whether live plants can still maintain evapotranspiration behaviour during enhanced gravity is still uncertain. Even if this was possible, the time scaling is likely to mean that it would be difficult to model water uptake representatively at prototype scale.

6 SUMMARY AND CONCLUSIONS

This paper presented a critical comparison of juvenile live plants versus 3D printed root analogues in terms of their representation of root mechanical properties, root morphology and distribution of the additional shear strength generated by the roots with depth. The results suggest that both approaches are imperfect but can still provide a representative and informative mechanical model. Specifically, juvenile plants provide more representative root mechanical properties, and also more correct stress-strain response, including the maximum strain and stress localisation. They are also ideal for use in tests where the ground conditions would prohibit placement of a 3D printed model (e.g. in compacted or cohesive soils). However, these properties are biologically variable and so this method may not be ideal for cases when multiple centrifuge tests must be compared which are to have the same rooted soil properties.

In contrast with this, 3D printed analogues have much better control in modelling the distribution of the additional shear strength generated by the roots with depth, even though the analogues are stiffer than live roots. Combined with their repeatability, this approach is better suited to testing programmes where comparative tests must be undertaken with directly comparable rooting conditions (e.g. when slope height or loading conditions are variables) in granular media.

Further developments that can more realistically model coupled root mechanical and hydrological effects are required.

ACKNOWLEDGEMENTS

This research was funded by the Engineering and Physical Sciences Research Council (EPSRC, EP/M020355/1; a collaboration between the Universities of Dundee, Southampton, Aberdeen, Durham and The James Hutton Institute. The authors thank Professor Mike Humphreys (IBERS, Aberystwyth University) and Scotia seeds for providing seeds used in this study. The James Hutton Institute receives funding from the Scottish Government (Rural & Environmental Services & Analytical Services Division).

REFERENCES

Bischetti, G.B., Chiaradia, E.A., Epis, T. & Morlotti, E. 2009. Root cohesion of forest species in the Italian Alps. Plant Soil 324(1), 71–89.

Boldrin, D., Leung, A.K. & Bengough, A.G. 2017a. Correlating hydrologic reinforcement of vegetated soil with plant traits during establishment of woody perennials. Plant Soil 1–15. doi:10.1007/s11104-017-3211-3

Boldrin, D., Leung, A.K. & Bengough, A.G. 2017b. Root biomechanical properties during establishment of woody perennials. Ecol. Eng. 109(Part B), 196–206.

Danjon, F., Barker, D.H., Drexhage, M. & Stokes, A. 2008. Using three-dimensional plant root architecture in models of shallow-slope stability. Ann. Bot. 101(8), 1281–1293.

Danjon, F. & Reubens, B. 2008. Assessing and analyzing 3D architecture of woody root systems, a review of methods and applications in tree and soil stability, resource acquisition and allocation. Plant Soil 303(1-2), 1–34.

Docker, B.B. & Hubble, T.C.T. 2008. Quantifying root-reinforcement of river bank soils by four Australian tree species. Geomorphology 100(3-4), 401–418.

Duckett, N. 2013. Development of Improved Predictive Tools for Mechanical Soil-Root Interaction. PhD thesis, University of Dundee, UK.

Fan, C.C. & Su, C.F. 2008. Role of roots in the shear strength of root-reinforced soils with high moisture content. Ecol. Eng. 33(2), 157–166.

Ghestem, M., Veylon, G., Bernard, A., Vanel, Q. & Stokes, A. 2014. Influence of plant root system morphology and architectural traits on soil shear resistance. Plant Soil 377(1-2), 43–61.

Jackson, R.B., Canadell, J., Ehleringer, J.R., Mooney, H. A., Sala, O.E. & Schulze, E.D. 1996. A global analysis of root distributions for terrestrial biomes. Oecologia 108(3), 389–411.

Leung, A.K., Boldrin, D., Liang, T., Wu, Z.Y., Kamchoom, V. & Bengough, A.G. 2017. Plant age effects on soil infiltration rate during early plant establishment. Géotechnique. doi:10.1680/jgeot.17.t.037.

Liang T., Knappett J.A. & Bengough A.G. 2014. Scale modelling of plant root systems using 3-D printing. In ICPMG2014–Physical Modelling in Geotechnics, Perth, Australia, 14-17 January 2014, pp: 361–366.

Liang, T., Knappett, J.A. & Duckett, N. 2015. Modelling the seismic performance of rooted slopes from individual root–soil interaction to global slope behaviour. Géotechnique 65(12), 995–1009.

Liang, T., Knappett, J.A., Bengough, A.G. & Ke, Y.X. 2017a. Small scale modelling of plant root systems using 3-D printing, with applications to investigate the role of vegetation on earthquake induced landslides. Landslides 14(5), 1747–1765.

Liang, T., Bengough, A.G., Knappett, J.A., Muirwood, D., Loades, K.W., Hallett, P.D., Boldrin, D., Leung, A.K. & Meijer, G.J. 2017b. Scaling of the reinforcement of soil slopes by living plants in a geotechnical centrifuge. Ecol. Eng. 109(Part B), 207–227.

Liang, T. & Knappett, J.A. 2017a. Centrifuge modelling of the influence of slope height on the seismic performance of rooted slopes. Géotechnique 67(10), 855–869.

Liang, T. & Knappett, J.A. 2017b. Newmark sliding block model for predicting the seismic performance of vegetated slopes. Soil Dyn. Earthq. Eng. 101, 27–40.

Loades, K.W., Bengough, A.G., Bransby, M.F. & Hallett, P.D. 2013. Biomechanics of nodal, seminal and lateral roots of barley: Effects of diameter, waterlogging and mechanical impedance. Plant Soil 370(1-2), 407–418.

Loades, K.W., Bengough, A.G., Bransby, M.F. & Hallett, P.D. 2010. Planting density influence on fibrous root reinforcement of soils. Ecol. Eng. 36(3), 276–284.

Mao, Z., Saint-André, L., Genet, M., Mine, F.X., Jourdan, C., Rey, H., Courbaud, B. & Stokes, A. 2012. Engineering ecological protection against landslides in diverse mountain forests: Choosing cohesion models. Ecol. Eng. 45, 55–69.

Meijer, G.J., Bengough, A.G., Knappett, J.A., Loades, K.W. & Nicoll, B.C. 2016. New in situ techniques for measuring the properties of root-reinforced soil – laboratory evaluation. Géotechnique 66(1), 27–40.

Meijer, G.J., Bengough, G., Knappett, J., Loades, K. & Nicoll, B. 2017. In situ root identification through blade penetrometer testing – part 2: field testing. Géotechnique. doi:http://dx.doi.org/10.1680/jgeot.16.P.204.

Mickovski, S.B., Hallett, P.D., Bransby, M.F., Davies, M.C.R., Sonnenberg, R. & Bengough, A.G. 2009. Mechanical Reinforcement of Soil by Willow Roots: Impacts of Root Properties and Root Failure Mechanism. Soil Sci. Soc. Am. J. 73(4), 1276–1285.

Muir Wood, D. 2003. Geotechnical modelling. London and New York: CRC Press.

Ng, C.W.W., Leung, A.K., Kamchoom, V. & Garg, A. 2014. A Novel Root System for Simulating Transpiration-Induced Soil Suction in Centrifuge. Geotech. Test. J. 37(5), 1–16.

Ng, C.W.W., Leung, A.K., Yu, R. & Kamchoom, V. 2016. Hydrological Effects of Live Poles on Transient Seepage in an Unsaturated Soil Slope: Centrifuge and Numerical Study. J. Geotech. Geoenvironmental Eng. 4016106.

Norris, J.E., Stokes, A., Mickovski, S.B., Cammeraat, E., Van Beek, R., Nicoll, B.C. & Achim, A. 2008. Slope stability and erosion control: Ecotechnological solutions. Springer, Netherlands.

Operstein, V. & Frydman, S. 2000. The influence of vegetation on soil strength. Proc. Ice-gr. Improv. 4(2), 81–89.

Pollen, N. & Simon, A. 2005. Estimating the mechanical effects of riparian vegetation on stream bank stability using a fiber bundle model. Water Resour. Res. 41(7), 1–11.

Sonnenberg, R., Bransby, M.F., Bengough, A.G., Hallett, P.D. & Davies, M.C.R. 2011. Centrifuge modelling of soil slopes containing model plant roots. Can. Geotech. J. 49(1), 1–17.

Sonnenberg, R., Bransby, M.F., Hallett, P.D., Bengough, A.G., Mickovski, S.B. & Davies, M.C.R. 2010. Centrifuge modelling of soil slopes reinforced with vegetation. Can. Geotech. J. 47(12), 1415–1430.

Stokes, A., Atger, C., Bengough, A.G., Fourcaud, T. & Sidle, R.C. 2009. Desirable Plant root traits for protecting natural and engineered slopes against landslides. Plant Soil 324(1), 1–30.

Stokes, A., Douglas, G.B., Fourcaud, T., Giadrossich, F., Gillies, C., Hubble, T., Kim, J.H., Loades, K.W., Mao, Z., McIvor, I.R., Mickovski, S.B., Mitchell, S., Osman, N., Phillips, C., Poesen, J., Polster, D., Preti, F., Raymond, P., Rey, F., Schwarz, M. & Walker, L.R. 2014. Ecological mitigation of hillslope instability: Ten key issues facing researchers and practitioners. Plant Soil 377(1-2), 1–23.

Taylor, R.N. 2003. Geotechnical centrifuge technology. London, UK: CRC Press,

Teerawattanasuk, C., Maneecharoen, J., Bergado, D.T., Voottipruex, P., Lam & L.G. 2014. Root strength measurements of Vetiver and Ruzi grasses. Lowl. Technol. Int. 16(2), 71–80.

Van Beek, L.P.H., Wint, J., Cammeraat, L.H. & Edwards, J.P. 2005. Observation and simulation of root reinforcement on abandoned mediterranean slopes. Plant Soil 278(1-2), 55–74.

Waldron, L.J. 1977. The Shear Resistance of Root-Permeated Homogeneous and Stratified Soil. Soil Sci. Soc. Am. J. 41(5), 843–849.

White, D.J., Take, W.A. & Bolton, M.D. 2003. Soil deformation measurement using particle image velocimetry (PIV) and photogrammetry. Géotechnique 53(7), 619–631.

Wu, T.H., McKinnell III, W.P. & Swanston, D.N. 1979. Strength of tree roots and landslides on Prince of Wales Island, Alaska. Can. Geotech. J. 16(1), 19–33.

Revisit of the empirical prediction methods for liquefaction-induced lateral spread by using the LEAP centrifuge model tests

K. Liu, Y.G. Zhou, Y. She, P. Xia, Y.M. Chen, D.S. Ling & B. Huang
Key Laboratory of Soft Soils and Geoenvironmental Engineering, Institute of Geotechnical Engineering, Zhejiang University, Hangzhou, P.R. China

ABSTRACT: Large lateral spread, which causes considerable damage to structures and infrastructure systems, was widely observed during past earthquakes. Empirical and semi-empirical methods of liquefaction induced lateral spread are reviewed and analyzed based on LEAP centrifuge model tests. The maximum shear strain and accumulated shear strain are obtained from the shear strain time history, which derived from the acceleration records at different depths of the model ground. The methods proposed by Zhang et al. (2004) and Shamoto et al. (1998) are selected as the representative ones to estimate the lateral spread. The rationality and accuracy of the predictions are discussed. For lateral displacement less than 0.2 m, a good correlation between the measured and the estimated values was found for both existing methods, and a similar relationship was found between the measured lateral displacements and the LDI expressed in terms of the accumulated shear strain. However, for larger flow type lateral spread associated with strain localization, these methods will considerably underestimate the measured value and other type of analysis is desired.

1 INTRODUCTION

1.1 General introduction

Liquefaction induced lateral spread is a significant cause of damage to structures and lifelines in past earthquakes, which especially happens near a waterfront or mildly sloping deposit. Such as during the 1964 Niigata earthquake, the 1983 Nihonkai-Chuba earthquake, the 1976 Tangshan earthquake and 1995 Hyogoken-Nambu earthquake (Shamoto et al. 1998; Balakrishnan et al. 1999; Hamada et al. 1992). Lateral spreads are the pervasive types of liquefaction-induced ground failures for gentle slopes or for nearly level (or gently inclined) ground with a free face (e.g., river banks, road cuts) (Zhang et al. 2004).

Several methods have been proposed to estimate the lateral displacements, including numerical models, element tests, model tests at 1g and Ng, and field investigations. Hamada et al. (1987), Youd et al. (2002), Zhang et al. (2004), Shamoto et al. (1998), have proposed several methods for estimating the lateral spread. During these studies, centrifuge model tests were used in validating the rationality and accuracy of the above methods (Sharp et al. 2002).

Hamada et al. (1987) proposed an empirical equation which can calculate the magnitude of the displacement using: $D_h = 0.75 H^{1/2} \times \theta^{1/3}$; where H is the thickness of liquefiable layers, and θ is ground slope. It should be noticed that Hamada thought that lateral spread is caused by soil flow induced by liquefaction rather than by the effect of shear deformation of the ground. In his equation, soil type and seismic characteristics are not taken into account. Empirical equations were developed by Youd et al. (2002) from the multi-linear regression of a large case history database, which were suitable for predicting in-situ lateral spread for seismic characteristics, topography, and soil properties are considered. But multi-linear regression is empirical. Zhang et al. (2004) developed a semi-empirical approach to estimate the liquefaction induced lateral spread using SPT and CPT data. And the lateral displacement index (LDI) was introduced, which is obtained by integrating the maximum cyclic shear strain with respect to the depth. Also a correction term for topography, $F(S) = S + 0.2$, is proposed based on regression of field case histories. Then the lateral spread (LD) is given by

$$LDI = \int_0^{Z\max} \gamma_{\max} dz \qquad (1)$$

$$LD = (S + 0.2) \times LDI \text{ for } (0.2 < S < 3.5) \qquad (2)$$

Based on post-liquefaction stress-strain constitutive analysis, Shamoto et al. (1998) proposed a new method for predicting post-liquefaction ground settlement and lateral spread characterizing ground geometry for level ground with or without a free face. As for level ground without a free face, the expression for predicting lateral spread is,

$$D_h = C_h \times \int_0^H (\gamma_r)_{\max} dz \qquad (3)$$

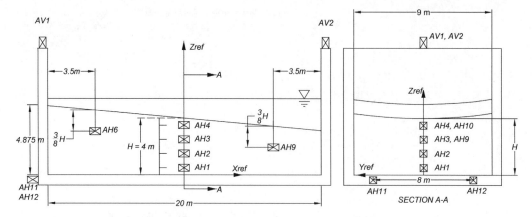

Figure 1. Configuration of accelerometer of Zhejiang University for LEAP (in prototype scale).

$$(\gamma_r)_{max} = (e_0 - e^*_{min})/(1+e_0) \times R_0^* / M_{cs.o} \times \gamma_{max}^m \quad (4)$$

$$e^*_{min} = e_{max} - 1.3 \times (e_{max} - e_{min}) \quad (5)$$

in which, C_h is coefficient for lateral spread derived from statistical analysis of sites data ($C_h = 0.16$ is advised by Shamoto et al. (1998)); $(\gamma_r)_{max}$ is maximum residual shear strain; e_0 is initial void raito; e_{max} and e_{min} are maximum and minimum of void raito respectively; $R_0^* = 2$, is a constant independent of type of soils; M_{cs} is the critical deviator-isotropic stress ratio, and $M_{cs.o}$ is different from M_{cs} and follow the relation $M_{cs.o} = a \times M_{cs}$ where $a = 0.142$; γ_{max} is maximum shear strain. Lateral spread is associated with the residual shear strain in essence.

It is obvious that Zhang et al. (2004) (hereafter abbreviated as Z method) and Shamoto et al. (1998) (hereafter abbreviated as S method) took the maximum shear strain to characterize the residual shear strain. The maximum shear strain was considered as the primary factor controlling the liquefaction induced lateral spread of a soil layer.

In this paper, these two methods are used to estimate the lateral spread, the predictions are compared with the measured lateral spread of centrifuge tests in LEAP-2015 and LEAP-2017. Results and discussions are presented, and the LDI associated with accumulated shear strain is also proposed for estimating lateral spread.

2 LEAP SPECIFICATION

LEAP (Liquefaction Experiments and Analysis Projects) is an international effort to formalize the process and provide data needed for validation of numerical models designed to predict liquefaction phenomena (Kutter et al. 2017). In LEAP-2015 and LEAP-2017, centrifuge experiments were conducted at 6 centrifuge facilities (Cambridge University, Kyoto University, University of California Davis, National Central University, Rensselaer Polytechnic Institute, and Zhejiang University).

The instrumentations are shown in Figure 1. It represents a prototype 5-degree, 4 m depth at midpoint, 20 m length, uniform medium dense sand slope deposit of Ottawa F-65. The specified ground motion is a ramped 1 Hz sine wave base motion. Surface markers manufactured by nylon zip ties were used for tracing surface displacement. Other details are available in the literature (e.g., Kutter et al. 2017).

3 EVALUATION OF SHEAR STRAIN HISTORY

3.1 Processing the acceleration records

It has been widely accepted to determine the dynamic displacements by integration of acceleration records. The acceleration records must be band-filtered prior to integration to eliminate unwanted components. It is very important to filter data at high frequency to eliminate noise and at low frequency to eliminate drift errors. In addition, the noisey acceleration signal prior to and after the shaking event are analyzed to determine the cut-off frequencies.

The average acceleration data of AH11 and AH12 is taken as acceleration record at the depth of 4 m in prototype scale. Second order nonlinear interpolation was used to obtain the surface acceleration record (i.e., $z = 0$ m in prototype) by Eq. (6):

$$a_{i-1} = a_i \frac{(z_{i-1}-z_{i+1})(z_{i-1}-z_{i+2})}{(z_i-z_{i+1})(z_i-z_{i+2})} + a_{i+1} \frac{(z_{i-1}-z_{i+2})(z_{i-1}-z_i)}{(z_{i+1}-z_{i+2})(z_{i+1}-z_i)}$$
$$+ a_{i+2} \frac{(z_{i-1}-z_i)(z_{i-1}-z_{i+1})}{(z_{i+2}-z_i)(z_{i+2}-z_{i+1})} \quad (6)$$

in which, $a_i = a_i(z_i, t)$ is the acceleration record at the depth of z_i.

As shown in Figure 1, the accelerometers of AH1-AH4 located at the central line of the model ground with depths of 3.5 m, 2.5 m, 1.5 m and 0.5 m respectively, to minimize their sensitivity to boundary effects

Table 1. Specification of abbreviation in figures in section 4.

Centrifuge tests for LEAP	Density kg/cm³	Abbreviation
Test at CU in 2015	1620 ± 20	CU
Test at RPI in 2015	1650	RPI
Test at UCD in 2015	1652 ± 10	UCD
Test at KU in 2015	1652	KU
Motion 2 at ZJU in 2015	1644 ± 54	ZJU-2
Motion 4 at ZJU in 2015	1658 ± 54	ZJU-4
1st model at ZJU for motion 1 in 2017	1672	ZJU1-1
1st model at ZJU for motion 2 in 2017	1679	ZJU1-2
1st model at ZJU for motion 3 in 2017	1686	ZJU1-3
2nd model at ZJU for motion 1 in 2017	1606	ZJU2-1
2nd model at ZJU for motion 2 in 2017	1618	ZJU2-2
2nd model at ZJU for motion 3 in 2017	1624	ZJU2-3
3rd model at ZJU for motion 1 in 2017	1706	ZJU3-1
3rd model at ZJU for motion 2 in 2017	1713	ZJU3-2
3rd model at ZJU for motion 3 in 2017	1718	ZJU3-3

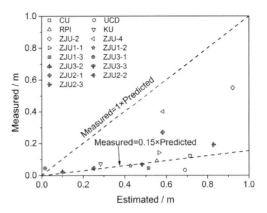

Figure 2. Measured lateral displacements versus predicted in Z method.

from the side walls of the container. The shear strain time histories are derived from accelerometers located at different depths in five centrifuge tests in LEAP-2015 and three centrifuge tests conducted by Zhejiang University in LEAP-2017.

3.2 Processing the shear strain time history

The calculation of shear strain at the depth z_i, as detailed by Zeghal et al. (1999), adopted second order nonlinear expression in the form Eq. (7).

$$\gamma(z_i) = \left[(u_{i+1} - u_i) \frac{(z_i - z_{i-1})}{(z_{i+1} - z_i)} + (u_i - u_{i-1}) \frac{(z_{i+1} - z_i)}{(z_i - z_{i-1})} \right] / (z_{i+1} - z_{i-1}) \quad (7)$$

Thus, six acceleration records are taken into Eq. (7) to obtain four shear strain records corresponding to AH1-AH4.

4 RESULTS AND ANALYSIS

4.1 General discussion of the procedures and the tests data

The measured lateral spread at the surface are compared with that of the existing empirical correlations applicable for model test. The Z method and S method are included in comparison. For simplicity and field applications, the correlation between the maximum cyclic shear strain and the factor of safety proposed by Ishihara et al. (1992) is adopted by Z method to estimate the maximum shear strain, and S method took the maximum shear strain related to the cyclic stress ratio induced by an earthquake and the standard penetration test blow count of a layer proposed by Tokimatsu et al. (1983) as the basis of calculation.

To estimate the lateral spread in centrifuge models and to be consistent with the above procedures, the maximum shear strain at different sand layers are determined from dynamic displacement of accelerometers by means of the second order nonlinear interpolation. It should be noted that the coefficients used in (3), (4), and (5) are not the proposed values of S method but derived from physical property tests and cyclic triaxial tests of Ottawa F-65 sand in LEAP. The measured lateral spreads for each shaking in LEAP are average values, which are partly quoted from the literature (Kutter et al. 2017).

For the sake of clarity, abbreviations are adopted for different dynamic centrifuge tests. And the initial density of each model test is also listed in Table 1. Note that the change of model slope angle is very small during multiple shaking events.

4.2 Results and analysis of Z method and S method

Figure 2 shows the prediction of Z method versus the measured displacements. It is obvious that the prediction from Z method is much greater than the measured displacements except for ZJU3-1, which agrees well with the prediction. It seems that the measured displacements are approximately 15% of the prediction of Z method. The measured displacements of ZJU-2 and ZJU-4 are greater than that of other tests and are 50% of the prediction. The Z method is recommended for use of mildly sloping ground (e.g., $0.2 < S < 3.5$), not for models with steeper slope such as the LEAP model, which represents a 5-degree deposit corresponding to $S = 8.7$, and the minimum ground slope (S) is 7.3 even after several shaking events in these LEAP tests.

It should be noted that the slope correction term in Z method is just a one order linear function of the slope, and may not be applicable to the cases of steeper slope, which leads to disagreement between the measurements and the predictions in this work. Further

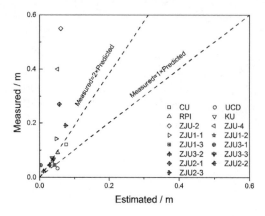

Figure 3. Measured lateral displacements versus predicted in S method for level ground.

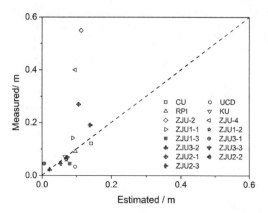

Figure 5. Measured lateral displacements versus predicted in S method with $C_h = 0.424$.

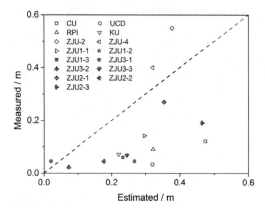

Figure 4. Measured lateral displacements versus predicted in S method for level ground near waterfront.

study is desired to improve the ground slope correction term.

Figures 3–5 show the prediction of the S method versus the measured displacements. It can be seen that the lateral spread from S method for level ground without a free face is much smaller than the measured displacements, and the measurements are very close to 2 times of the prediction except for ZJU1-3, whose measurement is very close to the prediction. For level ground without a free face, $C_h = 0.16$ in (3) is recommended by S method, and $C_h = 1$ for ground near waterfront to estimate the lateral spread. $C_h = 1$ for gently sloping ground is tried and Figure 4 shows the results. It seems that the predictions are somewhat greater than that of measured, which implies taht $C_h = 1$ is not applicable for mildly sloping ground to estimate the lateral spread. Figure 5 shows the results when C_h equals to 0.3, and the prediction are in good agreement with the measurements.

It should be mentioned that the correlation between the maximum shear strain and the factor of safety adopted in Z method was derived from cyclic simple shear testing, in which the soil specimen is isotropically consolidated before shearing, and the maximum shear strain related to the cyclic stress ratio and standard penetration test blow count of a layer proposed by Tokimatsu et al. (1983) is only suitable for level ground. In other words, the maximum shear strain adopted by Z method and S method did not take the effect of deviator stress into account. However, there is initial static shear stress in sloping ground. Hence it is reasonable for the Z method to take the ground slope correction into consideration to estimate the lateral spread. Therefore the maximum shear strain in the Z method and S method reflects both the effects of seismic loading and soil properties. The maximum shear strain used in this work, derived from acceleration records in gently sloping ground, reflects the effects of initial static shear stress in sloping ground. So the procedure in this work is reasonable for S method, and relating LDI with lateral spread is more acceptable for Z method.

4.3 Liquefaction induced lateral spread with LDI associated with accumulated shear strain

It has been mentioned above that Z method and S method took the maximum shear strain during shaking to characterize the residual shear strain. The maximum shear strain could be correlated to the residual shear strain as it reflects the combined effect of loading characteristics and soil properties. On the other hand, Sento et al. (2004) found that the accumulated shear strain developed during the undrained cyclic shear process is more versatile than the maximum shear strain. The accumulated shear strain γ_{acm} is defined as:

$$\gamma_{acm} = \int_0^t |\dot{\gamma}(t)| dt \qquad (8)$$

where $\dot{\gamma}(t)$ denotes the shear strain rate at time of t. The LDI proposed by Z method has been adopted

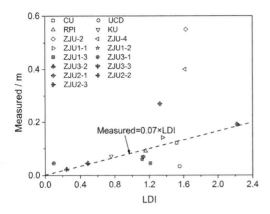

Figure 6. Measured lateral displacements versus LDI associated with γ_{acm}.

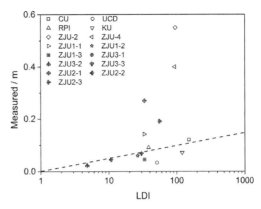

Figure 7. Measured lateral displacements versus LDI associated with $\log(\gamma_a)$.

in this work for estimating the lateral spread. LDI is defined as:

$$LDI = \int_0^Z \gamma_{acm} dz \qquad (9)$$

Figure 6 shows the correlation between the updated LDI and the measured lateral spread. It is seen that measured displacements are in good linear correlation with LDI calculated with γ_{acm}, which indicates that γ_{acm} is a good index for characterizing the residual shear strain as well.

For further study, the authors propose a new form of the accumulated shear strain, γ_a, which is defined as:

$$\gamma_a = \int_0^t \Delta\gamma dt \qquad (10)$$

where $\gamma = 0$ for $\gamma(t) = <\gamma_{th}$; $\gamma = |\gamma(t) - \gamma_{th}|$ for $\gamma(t) > \gamma_{th}$; γ_{th} denotes the threshold shear strain of the specific soil. The above definition is based on that there will be irreversible shear strain generating and accumulating as to residual shear strain when the shear strain is larger than the threshold shear strain. Similarly LDI could be defined as:

$$LDI = \int_0^Z \gamma_a dz \qquad (11)$$

For simplicity, the threshold shear strain here is taken as 0.01% (Dobry et al. 1982). Figure 7 shows the measured displacements versus LDI associated with $\log(\gamma_a)$, which indicates a good linear relationship between the measured displacements and LDI defined by γ_a.

4.4 Some preliminary findings

It should be noted that all the above mentioned methods, including Z method and S method, show a strong linear relationship between the prediction and the measurements when measurements are less than 0.2 m (i.e., 5% of average shear strain in the whole profile), but for the cases of larger lateral spread, such as ZJU-2, ZJU4 and ZJU2-1, the datasets depart from the expected linear line. The same situation happens when correlating the measured lateral spread with LDI associated with accumulated shear strain defined by Sento et al. (2004) and the present authors.

It should be noted that all above methods are based on the hypothesis of consecutive shear deformation of continuum. Lateral spread will be triggered by flow failure instead of limited shearing when strain localization appear in sand layers, such as the cases of ZJU-2, ZJU-4 and ZJU2-1, where shear bands were founded in relatively deep sand layers during excavation after spin down (Zhou et al. 2017). It is interesting that such finding is consistent with that of Hamada et al. (1987).

Kutter et al. (2017) pointed that the limited sample size along with the limited accuracy of the manual measuring tools would produce uncertainties on the order of 0.03 m in the measurements in LEAP-2015 (1 mm model scale corresponds to 0.02–0.05 m prototype scale depending on the scale factor adopted), so does the tests at Zhejiang University in LEAP-2017. Therefore the deviation between the measurements and the prediction when the measurements are less than 0.2 m is acceptable.

5 CONCLUSIONS

The methods proposed by Zhang et al. (2004), Shamoto et al. (1998) for predicting lateral spread are used to estimate lateral spread in LEAP centrifuge tests, and the rationality and accuracy of the results were compared and discussed. Both maximum shear strain and accumulated shear strain are considered to characterize residual shear strain. The conclusions are as follows.

The prediction of Z method is much larger than the measured lateral displacement, and the measured is approximately 15% of the prediction. Z method seems not applicable for ground of large sloping angle (i.e., more than 5 degrees). One possible improvement might be adjusting the slope correction term in

Z method, which could be expressed as a nonlinear function of slope instead of a linear one.

The prediction of S method for level ground without a free face is smaller than the measured displacements, and the measured lateral displacements approximately double the prediction. The coefficient of C_h is taken as unity as an attempt to estimate lateral spread and the results are poor, which demonstrates $C_h = 1$ is not applicable for mildly sloping ground. And $C_h = 0.3$ is recommend for sloping ground to estimate the lateral spread in this case.

The accumulated shear strain proposed by Sento is adopted to characterize the residual shear strain. And the measured lateral spread are in good correlation with LDI calculated with γ_{acm}, which indicates that γ_{acm} is a good index for residual shear strain as well. A new definition of accumulated shear strain (γ_a) is proposed based on threshold shear strain. The measured lateral spread and LDI associated with $\log(\gamma_a)$ is also in good correlation. Accumulated shear strain is potentially more suitable for liquefaction-induced deformations with long shaking duration.

The predictions of Z method and S method are literally linear with measured lateral spread when the measurements are less than 0.2 m in prototype, but will considerably underestimate the measured value when the lateral spread is larger than 0.2 m, which was probably caused by sand flow as shear strain localization was formed during strong shaking. It is the same when correlating measured lateral spread with LDI associated with accumulated shear strain defined by Sento et al. (2004) and the present study.

ACKNOWLEDGEMENTS

This study is partly supported by the National Natural Science Foundation of China (Nos. 51578501, 51778573), the National Program for Special Support of Top-Notch Young Professionals (2013), the Zhejiang Provincial Natural Science Foundation of China (No. LR15E080001) and the National Basic Research Program of China (973 Project) (No. 2014CB047005). These supports are gratefully appreciated.

REFERENCES

Balakrishnan, A. & Kutter, B.L. 1999. Settlement, sliding, and liquefaction remediation of layered soil. Journal of geotechnical and geoenvironmental engineering 125(11): 968–978.

Dobry, R., Ladd, R. S., Yokel, F. Y., Chung, R. M. & Powell, D. 1982. Prediction of pore water pressure buildup and liquefaction of sands during earthquakes by the cyclic strain method (Vol. 138). Gaithersburg, MD: National Bureau of Standards.

Hamada, M., Towhata, I. & Yasuda, S., et al. 1987. Study on permanent ground displacement induced by seismic liquefaction [J]. Computers and Geotechnics 4(4): 197–220.

Hamada, M. & O'Rourke, T.D. 1992. Case Studies of Liquefaction and Lifeline Performance during Past Earthquakes, Vol. 1, Japanese Case Studies Technical Rep. NCEER-92-0001, National Center for Earthquake Engineering Research, Buffalo, N.Y., February.

Ishihara, K. & Yoshimine, M. 1992. Evaluation of settlements in sand deposits following liquefaction during earthquakes. Soils and foundations 32(1): 173–188.

Kutter, B.L., Carey, T.J., & Hashimoto, T., et al. 2017. LEAP-GWU-2015 experiment specifications, results, and comparisons [J]. Soil Dynamics and Earthquake Engineering. https://doi.org/10.1016/j.soildyn.2017.05.018

Sento, N., Kazama, M., & Uzuoka, R., et al. 2004. Liquefaction-induced volumetric change during reconsolidation of sandy soil subjected to undrained cyclic loading histories. In Proc. International Conference on Cyclic Behaviour of Soils and Liquefaction Phenomena, Bochum, Germany. pp. 199–206.

Shamoto, Y., Zhang, J.M., & Tokimatsu, K. 1998. New charts for predicting large residual post-liquefaction ground deformation. Soil dynamics and earthquake engineering 17(7): 427–438.

Sharp, M.K., & Dobry, R. 2002. Sliding block analysis of lateral spreading based on centrifuge results. International Journal of Physical Modelling in Geotechnics 2(2): 13–32.

Tokimatsu, K., & Yoshimi, Y. 1983. Empirical correlation of soil liquefaction based on SPT N-value and fines content. Soils and Foundations 23(4): 56–74.

Youd, T.L., Hansen, C.M., & Bartlett, S.F. 2002. Revised multilinear regression equations for prediction of lateral spread displacement. Journal of Geotechnical and Geoenvironmental Engineering 128(12): 1007–1017.

Zeghal, M., Elgamal, A.W., & Zeng, X., et al. 1999. Mechanism of liquefaction response in sand-silt dynamic centrifuge tests. Soil Dynamics and Earthquake Engineering 18(1): 71–85.

Zhang, G., Robertson, P.K., & Brachman, R.W.I. 2004. Estimating liquefaction-induced lateral displacements using the standard penetration test or cone penetration test. Journal of Geotechnical and Geoenvironmental Engineering 130(8): 861–871.

Zhou, Y.G., Sun, Z.B., & Chen, Y.M. 2017. Zhejiang University benchmark centrifuge test for LEAP-GWU-2015 and liquefaction responses of a sloping ground. Soil Dynamics and Earthquake Engineering. https://doi.org/10.1016/j.soildyn.2017.03.010

Physical modelling of atmospheric conditions during drying

C. Lozada
Civil Engineering Department, Escuela Colombiana de Ingeniería Julio Garavito, Bogotá, Colombia

B. Caicedo
Civil and Environmental Engineering Department, Universidad de Los Andes, Bogotá, Colombia

L. Thorel
IFSTTAR, Department GERS, Geomaterials and Modelling in Geotechnics Laboratory, Bouguenais, France

ABSTRACT: Atmospheric conditions during drying have been modeled in a climatic chamber to evaluate the effect of desiccation in soil layers. Different environmental variables such as relative humidity, wind velocity, radiation, and temperature have been measured. The climatic chamber was designed and calibrated to simulate properly each environmental variable and the heat transfer mechanisms during desiccation process. A potential evaporation tests was performed measuring the weight of water and actual evaporation tests were performed in a soil layer of Speswhite clay, in these tests, potential and actual evaporation rate were determined. This paper presents the evolution of the environmental variables and the soil properties such as soil temperature, water content, and suction during the evaporation process. As a result, a comparison between potential and actual evaporation rate indicates the large effect of suction on the actual evaporation rate.

1 INTRODUCTION

Water evaporates during drying periods depending on the environmental conditions such as solar radiation, temperature, relative humidity and wind velocity. Desiccation from the soil surface may generate geotechnical problems such as differential settlements in foundations, damages in roads and a general change in design conditions. Different authors have reported and study damages caused by this phenomenon (Bozozuk 1962; Perpich et al. 1965; Yesiller et al. 2000).

The study of the potential evaporation rate (evaporation of water) and the actual evaporation rate (evaporation of soil) is essential to understand the conditions that influence the desiccation process.

Different empirical approaches have been obtained to assess potential evaporation (Dalton 1802; Bowen 1926; Penman 1948). On the other hand, the actual evaporation rate is a more complex process that depends on climatic conditions on the soil surface and the soil properties, mainly soil suction (Gray et al. 1970; Wilson et al. 1994; Blight 2002; Song et al. 2014).

This paper presents the results of evaporation tests performed in water and clay. The experimental work was achieved using the climatic chamber of the Universidad de Los Andes in Bogotá, Colombia for modelling sun irradiance, temperature, relative humidity, and wind velocity. This chamber was conceived to work at 1×G to simulate an artificial atmosphere adequately and was used as a suitable tool to study the phenomenon of actual evaporation.

2 THE CLIMATIC CHAMBER FOR STUDYING DRYING PROCESS

This paper presents the study of the evaporation process of thin clay layers using the climatic chamber of the Universidad de Los Andes. This chamber allows the user to impose temperature, relative humidity, solar irradiance, and wind velocity through different mechanisms as a part of a closed circuit of air (infrared lamps, heaters, impulsion fans, extraction fans and cold plate coolers) (Fig. 1). The operation of the chamber consists in:

1) Two cold plate coolers condensate water to obtain a specific relative humidity in the chamber. A condensate water reservoir is located following the plates.
2) After the air is cooled, it passes through two air heaters to impose the temperature in the chamber.
3) To propel the air from the heaters to the chamber and to simulate wind velocity above the soil samples, a set of six fans were installed.

Figure 1. Climatic chamber Universidad de los Andes.

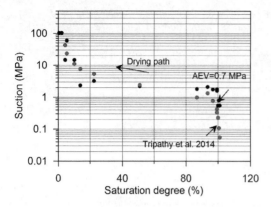

Figure 2. Soil water retention curve Speswhite kaolin clay.

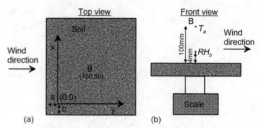

Figure 3. Position for the measurement of the climatic variables. a) top view b) front view. Distance a = 27 mm and b = 25 mm.

Table 1. Tests specifications.

Test	Duration min	Radiation W/m²	Wind Velocity m/s	RH RH %	Water temp. °C	Air temp °C
T1w	180.0	1040	1.74	34.1	27.3	40.9
T1	211.3	1040	1.74	23.0	37.4	42.0
T2	255.0	0	1.44	37.6	26.6	28.0
T3	327.2	0	1.36	47.2	26.1	27.0

4) After the air passes above the soil sample, it is propelled to the cold plate coolers through a set of four fans. Afterwards, the circuit starts again.

The climatic chamber includes many instruments of measurement (ie. temperature, wind velocity, and relative humidity sensors). The operation ranges of the chamber are relative humidity [18% - 40%] and temperature [23°C - 38°C]. A detailed description of the chamber is presented by Lozada et al (2016).

3 EXPERIMENTAL SETUP

3.1 Soil properties

The clay used in this investigation is Speswhite kaolin clay. The values of liquid limit, plastic limit, and specific gravity are 55, 32.3, and 2.65, respectively (Thorel et al. 2011).

The total suction of the Speswhite kaolin clay was measured using a dewpoint potentiometer, WP4 (Delage et al. 2008). The soil was dried at the laboratory conditions from a water content of 1.5 liquid limit (same initial condition as the models) and each point of measurement was stabilized during 72 hours in a sealed recipient. The volume change was measured to calculate the saturation degree for each suction measurement. The obtained water retention curve is shown in Figure 2 and is similar to the curve obtained by Tripathy (2014) for the Speswhite clay.

3.2 Physical modelling of environmental variables during drying

Three desiccation tests were performed in a layer of soil with 1 mm in thickness ($T1$, $T2$, and $T3$) and one test was performed in water ($T1_w$). The procedure for the desiccation tests involves the following steps:

1) The environmental conditions in the climatic chamber were stabilized for 30 minutes.
2) A container filled with water or soil was put into the chamber, the soil models were prepared with a water content of 1.5 liquid limit, and the different environmental variables were measured. The temperature of the surface was recorded with an infrared camera, the relative humidity was measured with a sensor at 4 mm from the surface and the air temperature was measured with an RTD temperature sensor at 100 mm from the surface.
3) To compute the potential or the actual evaporation rate, an electronic scale was placed at the bottom of the container to measure the weight of the sample.

Figure 3a shows the position B where the measurements of temperature, T_a, and relative humidity, RH_b were taken and Figure 3b depicts the lateral view of the container and the position of the temperature and the relative humidity sensor.

General specifications and the environmental variables obtained at the end of the tests are shown in Table 1.

4 EVAPORATION OF WATER SURFACE

4.1 Environmental variables. Temperature and relative humidity

Various mechanisms of the climatic chamber control temperature and relative humidity: heaters, cold plate coolers, extraction and impulsion fans, and infrared lamps. The evolution of relative humidity and temperature during the test $T1_w$ is shown in Figure 4. Air and water temperature increase with time and relative humidity decreases as temperature increases. The temperature of the water is lower than the temperature of air due to the latent heat of evaporation during the phase change (liquid to vapour).

The coupling of the environmental variables leads to a specific evaporation rate of water. The evaporated water with time is shown in Figure 5. The potential evaporation rate PE is calculated as the first derivate of the curve of the evaporated water in function of time. In this test, the potential evaporation rate is a constant value of 14.7 mm/day.

5 DRYING OF SOIL LAYER

5.1 Environmental variables. Temperature and relative humidity

The variation of soil temperature, T_s, and relative humidity, RH, of the tests performed in clay are depicted in Figure 6. Actual evaporation is a more complex process, which depends on the environmental conditions and also on the availability of water at the surface.

The tests $T1$ has the highest temperature and the lowest relative humidity due to the imposition of radiation with the infrared lamps at the soil surface. The test $T3$ has the highest relative humidity and the lowest temperature because of the absence of radiation and the lowest wind velocity.

5.2 Actual evaporation rate

Figure 7 shows the evaporated water with time for the tests performed in clay. In these tests, the evaporation rate is constant at the beginning of the test and then, decreases with time as water evaporates from the soil surface.

The water content of the soil layers is depicted in Figure 8. This figure shows a decrease in water content with time until reaching a constant water content, close to the residual value, indicating the end of the evaporation process.

The equation that fit the Water Characteristic Curve of the Speswhite kaolin according to Van Genuchten (1980) and Fredlund and Xing (1994) is:

$$w = w_r + \frac{w_s - w_r}{\{\ln[e + (\psi/a)^n]\}^m} \quad (1)$$

where w is the water content, w_s is the gravimetric water content at which air starts to enter in the soil, taken as 0.5, w_r is the residual gravimetric water content taken as 0.04, e is the Euler's number, ψ is the total

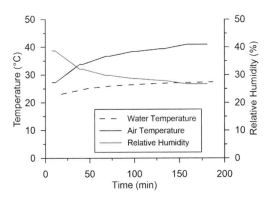

Figure 4. Evolution of the environmental variables with time. Test $T1w$.

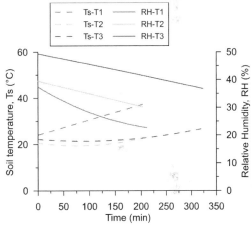

Figure 6. Evolution of the environmental variables with time. Test $T1$, $T2$, and $T3$.

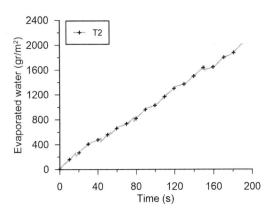

Figure 5. Evaporated water. Test $T1_w$.

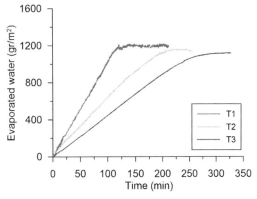

Figure 7. Evaporated water. Test $T1$, $T2$, and $T3$.

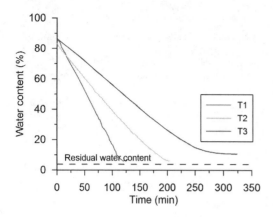

Figure 8. Water content. Test $T1$, $T2$, and $T3$.

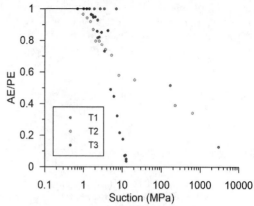

Figure 10. AE/PE vs suction. Test $T1$, $T2$, and $T3$.

– In the test $T1$ the ratio AE/PE seems to remain constant until complete desiccation, having higher suction values only at the end of the test, after 100 minutes (also shown in Fig. 8).
– In the tests $T2$ and $T3$, for values of AE/PE lower than 0.7, the test $T2$ presents a higher suction than the test $T3$. This is due to the lowest desiccation velocity in $T3$.

6 CONCLUSIONS

The climatic chamber of the Universidad de Los Andes was used to perform evaporation tests in soil layers. Thanks to this device, potential and actual evaporation process with different environmental conditions were studied. The principal findings of experimental results are:

– Potential evaporation rate is controlled by the coupling of all environmental variables such as the temperature of the air, solar radiation, wind velocity, and relative humidity.
– The results of the Actual Evaporation tests indicate that soil suction is the main physical restriction on evaporation process.

Further research is needed in the effect of suction on Actual Evaporation rate including the influence of thickness and material type.

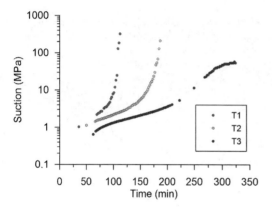

Figure 9. Soil suction. Test $T1$, $T2$, and $T3$.

suction [kPa] and the fitting parameters are $a = 1.40$, $n = 3.1$, and $m = 1.0$.

Figure 9 shows the suction values for each test using equation 1. This figure shows that suction value increases when reaches the AEV=0.7 corresponding to a water content of $w = 50$ %. Suction values of the test $T1$ increase faster as in the test $T2$ and $T3$ because of the extreme environmental conditions. Test $T3$ presents the smallest increase in suction with time due to the applied environmental conditions, mainly absence of solar radiation.

Similar to potential evaporation rate PE, the actual evaporation rate AE is calculated as the first derivate of the curve of the evaporated water in function of time. According to the experimental results and presented also by Wilson et al (1994), the actual evaporation rate is constant if the suction is lower than AEV. Under this condition, AE is found as 14.64 mm/day, 9.53 mm/day and 6.44 mm/day for the test $T1$, $T2$, and $T3$, respectively. The evaporation rate of the test $T1$ is similar to the test $T1_w$, 14.7 mm/day, due to the similar environmental variables in the chamber.

Figure 10 presents the evolution of the ratio AE/PE with suction in the three tests performed in clay. According to the results, one can state next remarks:

REFERENCES

Blight, G.E. 2002. Measuring evaporation from soil surfaces for environmental and geotechnical purposes. *Water SA*, 28(4), pp.381–394.

Bowen, I.S. 1926. The ratio of heat losses by conduction and by evaporation from any water surface. *Physical review*, 27(6), pp.779–787.

Bozozuk, M. 1962. Soil shrinkage damages shallow foundations at Ottawa, Canada. Engineering Journal, 45(7), pp.33–37.

Dalton, J. 1802. Experimental essays on the constitution of mixed gases; on the force of steam or vapor from water and other liquids in different temperatures, both in a Torricellian vacuum and in air; on evaporation and on the expansion of gases by heat. *In Proceedings of Manchester Literary and Philosophica Society.* pp. 535–602.

Delage, P., Romero, E. & Tarantino, A. 2008. Recent developments in the techniques of controlling and measuring suction in unsaturated soils. *In Keynote Lecture, Proc. 1st Eur. Conf. on Unsaturated Soils* pp. 33–52.

Fredlund, D.G & Xing, A. 1994. Equations for the soil-water characteristic curve. *Canadian Geotechnical Journal* 31(3): 521–532.

Gray, D.M., Norum, D.I. & Wigham, J.M. 1970. Infiltration and physics of flow of water through porous media *D. Gray, ed., Ottawa: Canadian National Committee of the International Hydrological Decade.*

Lozada, C., Caicedo, B. & Thorel, L. 2016. Improved climatic chamber for desiccation simulation. *In proceedings of the 3rd European Conference on Unsaturated Soils – "E-UNSAT 2016". E3S Web of Conferences* (Vol. 9, p. 13002). EDP Sciences.

Penman, H. 1948. Natural Evaporation from Open Water, Bare Soil and Grass. *In Proceedings of the Royal Society of London. Series A, Mathematical and Physical Sciences.* pp. 120–145.

Perpich, W.M., Lukas, R.G. & Baker, Jr, C.N. 1965. Desiccation of soil by trees related to foundation settlement. *Canadian Geotechnical Journal*, 2(1), pp.23–39.

Song, W.K., Cui, Y.J., Tang, A.M. & Ding, W.Q. 2013. Development of a Large-Scale Environmental Chamber for Investigating Soil Water Evaporation. *Geotechnical Testing Journal*, 36(6), pp.847–857.

Thorel, L., Ferber, V., Caicedo, B & Khokhar, I.M. 2011. Physical modelling of wetting-induced collapse in embankment base. *Géotechnique* 61(5): 409–420.

Tripathy, S., Tadza, M.Y.M. & Thomas, H.R. 2014. Soil-water characteristic curves of clays. *Canadian Geotechnical Journal* 51(8), pp.869–883.

Van Genuchten, M.T. 1980. A closed-form equation for predicting the hydraulic conductivity of unsaturated soils. *Soil science society of America journal* 44(5): 892–898.

Wilson, G.W., Fredlund, D.G & Barbour, S.L. 1994. Coupled soil-atmosphere modelling for soil evaporation. *Canadian Geotechnical Journal* 31(2): 151–161.

Yesiller, N., Miller, C.J., Inci, G. & Yaldo, K. 2000. Desiccation and cracking behavior of three compacted landfill liner soils. Engineering Geology, 57(1), pp.105–121.

Centrifuge model tests on excavation in Shanghai clay using in-flight excavation tools

X.F. Ma & J.W. Xu
MOE Key Laboratory of Geotechnical and Underground Engineering, Department of Geotechnical Engineering, Tongji University, Shanghai, China

ABSTRACT: With the development of urbanization, more and more infrastructures have been constructed in Shanghai, and excavations in the field tend to be deeper and closer to adjacent structures. As Shanghai clay is very soft and highly compressible with high water content and low strength, ground deformation and wall deflection are of prime concerns to engineers in deep excavations constructed in Shanghai. A series of small-scale centrifuge model tests using reconstituted Shanghai soft clay have been carried out to simulate the construction sequence of multi-strutted deep excavation. A new testing device has been applied to simulate the whole process of excavation, in which a scraper moves freely in both horizontal and vertical directions to excavate the soils and the bracing system is modelled by multiple pairs of struts powered by air pressure. Ground surface settlement, wall displacement and lateral earth pressure were monitored during the excavation.

1 INTRODUCTION

The safety of clay excavation in many cities like Shanghai has always been one of the main concerns in the development of underground space, whether it be metro station, deep basement or other underground utilities. Besides extensive numerical analysis involving computed constitutive models, physical modelling has always been performed as an alternative to investigate the behaviour of excavation. Geotechnical centrifuge modelling has been particularly applied because self-weight stress in the field can be replicated in centrifuge tests by increasing the centrifugal force acting on the model.

The early centrifuge model test on excavation was carried out by Lyndon & Schofield (1970) to study the behaviour of London clay excavation, which was modelled in 1-g level before increasing the centrifugal acceleration to the designated value. This simulation method has been used extensively by researchers (Liu 1999; Ma et al. 2009; Liang et al. 2012). The drawback of this method is that the stress condition induced in the model is different from that on site (Zhang et al. 2014). Heavy fluid discharge method has also been used to model the excavation, by which liquids with the same density as the soil were filled into the rubber bags and released to simulate excavation at a high g level (Lade et al. 1981; Kusakabe 1982; Bolton & Powrie 1987; Powrie & Kantartzi 1996; Richards & Powrie 1998; Leung et al. 2006; Elshafie et al. 2013). Such a method cannot realize the equivalent stress field in the soil base as the prototype because the lateral pressure coefficients of the soil and the liquids are significantly different and the passive earth pressures in the excavation area are difficult to simulate (Lee et al. 2010; Zhang & Yan 2014).

Several devices have been developed to simulate excavation in centrifuge model tests using different approaches (Kimura 1997; Takemura et al. 1999; Ng et al. 2001). However, these devices cannot simultaneously simulate the soil excavation and the supporting structure installation and the costs to develop these devices are high. Lam et al. (2012, 2014) developed a new apparatus for modelling excavations, which mainly consisted of an inflight excavator, a cylinder support system and a gate system. This apparatus effectively simulated the field construction sequence of a multi-propped retaining wall during centrifuge flight using Kaolin clay as model ground. The excavation was narrow and shallow with a final excavation depth of 9.6 m and width of 3.6 m at prototype.

Developing new modelling technique and simulating the in situ excavation has become major problems in the model tests. In order to investigate the performance of excavation and its effect on the surrounding environment in Shanghai clay, a series of inflight excavation and prop systems have manufactured based on the idea of Lam et al. (2012, 2014). Centrifuge model tests on Shanghai clay excavation have been performed using the inflight excavation tools in the geotechnical centrifuge at Tongji University. The wall displacement, ground settlement behind the wall and lateral earth pressure of the excavation are also discussed in this study.

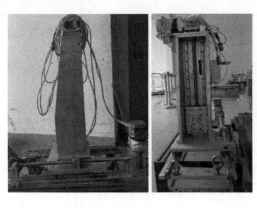

Figure 1. Inflight excavation tools.

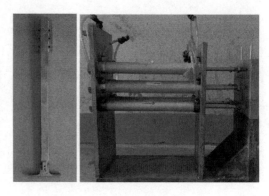

Figure 2. Scraper and struts system.

2 EXPERIMENTAL EQUIPMENT

2.1 Centrifuge facility

The geotechnical centrifuge used in the model tests belongs to Tongji University in China. It has an effective radius of 3.0 m, a maximum payload of 2 tons (at 75 g), and a maximum centrifugal acceleration of 200 g. The geotechnical centrifuge facility is composed of a power system, a rotational system, a data acquisition system, a control system, and a video surveillance system. The strong box designed for the centrifuge has an internal dimension of 0.9 m × 0.7 m × 0.7 m (length × width × height). The tests presented in this paper were carried out at an acceleration level of 100 g. Consequently, the dimensions of the prototype were scaled down by a factor of 1/100, while the stress and strain were the same as the prototype.

2.2 Inflight excavator and struts system

As has been developed by Lam et al. (2012), a T-shaped scraper controlled by a two-axis servo actuator is employed to excavate the clay, the actuator can move freely in both horizontal and vertical directions at 100g and the bracing system is modelled by multiple pairs of struts powered by air pressure. Three pairs of 40-mm-high blocks serve as a temporary wall retaining the clay designated to be removed. The blocks are supported by three pairs of struts in order to ensure full consolidation before excavation.

At the first stage of excavation, the upper pair of struts retreat before the scraper cut into the soil and scrape 2 mm-thickness soil to the reserved areas. As the scraper continues to move downwards and excavate soil, the unrestrained blocks would consequently fall down, as a result, the first level of struts is allowed to advance and reach to the retaining wall.

The following excavation repeats the prior steps until the soil in the excavated area is all scraped to the designated space and three pairs of struts meet the model wall.

2.3 Model preparation

The remoulded and reconsolidated Shanghai clay was used in the centrifuge model tests. The soil sample was consolidated inflight in the container. The disturbed clay was sunbaked, pulverized, and screened before being poured into the predesigned mixing unit. To increase the degrees of saturation of the remoulded clay, evacuating device is employed. The screened clay powder was stirred to the predefined water content by the mixing unit in a vacuum environment. De-aired clay slurry was then slowly placed into the container and transferred into the centrifuge swing platform after the instrumentation was placed in the designated positon. The angle of internal friction and cohesion of the clay were 10° and 11 kPa respectively.

The retaining wall was made by an 8-mm-thick and 260-mm-high aluminium alloy plate with a stiffness (EI) of 2700 MNm^2/m at prototype. The wall was installed at a depth of 260 mm before inflight excavation.

2.4 Measuring instrumentation

In the tests, the clay slurry was placed into the strongbox in layers, and in this process, the earth pressure transducers and strain gauges glued on the surfaces of model wall were placed in the predesigned location. After the clay reached the predesigned height, the strongbox was transferred and then fixed to the centrifugal basket.

In order to analyse the deformation mechanism and failure mode of the excavation during flight, a digital camera was inserted into an aluminium block, which was connected to a rigid frame and provided a stable mount for the camera. The camera was mounted approximately 300 mm from the Perspex side wall with LED strip lights attached around to facilitate imaging.

To receive a photo series over the whole test, it is necessary to have a remote control software for the camera and thereby a connection to a computer within the centrifuge. Therefore, a netbook was tied firmly to the centrifuge axis before centrifuge flight. The netbook was operated by a computer in the control room via Ethernet connection over slip rings.

Table 1. Description of test procedure.

Test No.	Excavation depth/mm	Excavation width/mm	Wall condition
1	120	120	cantilever
2	120	120	3-level strutted

Several earth pressure transducers and strain gauges were buried into the clay during test preparation and run orthogonal to the direction of principal soil movement. Transducers embedded within the soil model were also miniature with dimension of 8 mm, which reduced the effect of acting as ground anchors, and these transducers were rugged enough to resist not only their self-weight but also mechanical handling during placement.

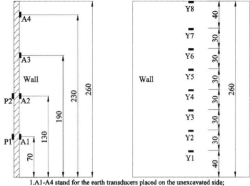

1. A1–A4 stand for the earth transducers placed on the unexcavated side;
2. P1 and P2 stand for the earth transducers placed on the unexcavated side;
3. Y1–Y8 stand for strain gauges; 4. Unit: mm.

Figure 3. Lateral earth pressure transducers and strain gauges.

3 CENTRIFUGE MODEL TEST PROGRAM

3.1 Test procedure

Two centrifuge model tests were carried out in this study to simulate the excavation in clay using inflight excavator. Test 1 and 2 investigated the performance of untrusted excavation and excavation supported by multi-strutted retaining wall respectively. The description of test procedure is shown in table 1.

The subjects of analysis were mainly the wall displacement, ground settlement, and the relationship between them. The measured lateral earth pressures developed during the excavation were also discussed and compared with theoretical Rankine earth pressures. The final excavation depth and wall height were 12 m and 26 m in the two tests. In test 1, the excavation was supported by a cantilever retaining wall, and in test 2 the excavation was supported by a three-level-strutted retaining wall.

3.2 Layout of test setup

The placement of lateral earth pressure transducers and strain gauges are shown in Figure 3. The transducers on both the excavated and unexcavated sides were placed with the same space of 60 mm with each other. 16 gauges were pasted on two faces of the model wall in order to obtain enough strain data. The dimensions of transducers and gauges were quite small and would not affect the accuracy of measurement during inflight excavation.

Figure 4 shows the diagram of model package in test 2. The inflight excavation was recorded by a high-definition camera mounted in front of the Perspex window with LED strips attached around, and the excavation rate of the scraper was reasonable compared with the construction rate in site. The three-layered blocks, two at one layer, were supported by the struts and serve as a boundary to ensure clay consolidation before excavation. The temporary blocks fell into the

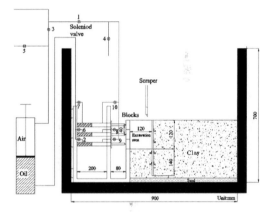

Figure 4. Diagram of model package.

spare space as the excavation progresses. The solenoid valves and the tank containing air and oil could control the movement of the struts, thus facilitate the scraper' excavating and support the retaining wall promptly.

4 TEST RESULTS

4.1 Wall displacement and ground settlement

Figure 5 shows the wall displacement and ground settlement at three main construction stages in test 1. Without support, the movement of the retaining wall in the first two stages was similar to rotating about a fixed point near the wall toe and the magnitude escalates as the excavation progresses. When the excavation height reached to final excavation depth (12 m), it seemed that the wall displacement is too much and the profile of the cantilever wall deflection displays a spandrel type. As to the ground settlement, clay right behind the wall presented the maximum value and the impact of excavation on ground beyond final excavation depth began to reduce rapidly, the range of influence area was less than 2 times the final excavation depth.

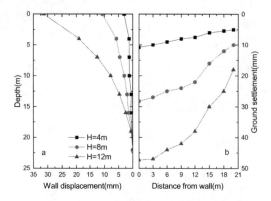

Figure 5. Wall displacement (a) and ground settlement (b) during excavation in test 1.

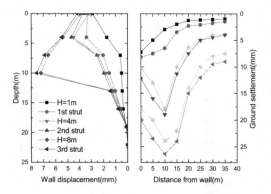

Figure 6. Wall displacement (a) and ground settlement (b) during excavation in test 2.

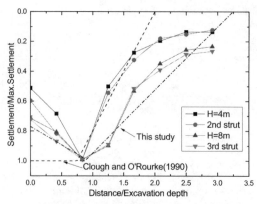

Figure 7. Normalized wall displacement versus distance during excavation in test 2.

Figure 6 shows the wall displacement and ground settlement at six main construction stages in test 2. According to the wall displacement, prior to the first pairs of struts, a cantilever pattern was characteristic of the wall displacement with a maximum displacement of 29 mm. The struts restrained the wall displacement and when the excavation became singly-strutted, the maximum value added slightly to 34 mm and the maximum displacement occurred closely above the base. As the scraper continued to excavate the soil, the mode of wall displacement remained the same as the one when the wall supported by the first pairs of struts supports the wall and the maximum value reaches to 47 mm. However, the wall performance after the second pair of struts was quite different compared with the previous ones. To be specific, the incremental displacement below 5 m was increasingly enlarging until the depth of 10 m. Although the incremental displacement at the depth of 5 m was only 1 mm, the maximum increment, at the depth of 10 m, rise to 5.5 mm at the end of excavation. One thing worth noticing was that the maximum displacement of the three-level-strutted retaining wall occurred at the depth of 10 m, close to the final excavation depth of 12 m.

The ground settlement curve is spandrel and the maximum settlement occurs near the retaining wall at the first stage. This occurrence was attributed to the fact that the wall behaved like a cantilever at this stage. This cantilever type abided by the graphical form proposed by Peck, R.B. (1969), which was the summary of the field observations of ground settlement around several excavations. The strutted excavation showed a mitigation of ground settlement and a change of pattern from spandrel type to concave one. The characteristics of concave settlement profile became apparent in the following construction stages, as shown in the later four settlement curves. The lowest point of the concave settlement curves was located about 10 m behind the wall and did not change with the excavation depth. According to filed observations on Shanghai excavation cases (Liu and Hou 1997; Xu 2007), the maximum ground settlement usually falls between 0.5 and 0.7 times the final excavation and occurs at the same distance from the retaining wall. The maximum ground settlement in this study occurred at 0.8 times the final excavation depth. In practice, the axial force of struts is adjusted simultaneously with reference of the wall displacement in order to avoid large ground settlement. The retaining wall in this test, however, was not supported promptly by the struts after excavation, was thus likely to cause greater ground settlement. Note from Figure 6 (a) that ground settlement extended to a considerable distance from wall, which tended to remain the steady at a distance of more than 30 m.

Figure 7 presents the relationship between normalized ground settlement (ground settlement divided by the maximum settlement) and normalized distance (distance from the retaining wall divided by the excavation depth). The occurrence of the maximum settlement during excavation coincided with that of the experiential curve proposed by Clough and O'Rourke (1990), whereas the influence area expanded as the excavation proceeds. The method introduced by Clough and O'Rourke (1990) provided a conservative estimation of ground settlement behind the retaining wall, compared with the normalized settlement caused by the latter stages of Excavation. In this study, the ground settlement right behind the wall was less than 0.8 times the maximum settlement and

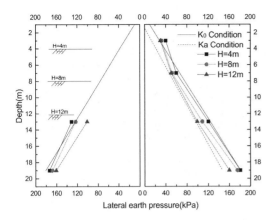

Figure 8. Measured lateral earth pressure and Rankine earth pressure in test 2.

the influence area extended to 3.2 times the final excavation depth.

4.2 Lateral earth pressure

Figure 8 compares the measured lateral earth pressures and the theoretical Rankine active pressures and at-rest (K_0) earth pressures for excavation depth of 4 m, 8 m and 12 m in test 2.

The lateral earth pressures on the unexcavated side approached the limited value as the excavation progressed, but were less than the Rankine active pressures, especially for soil at deeper level. This was due to two aspects. For one, the strutted retaining wall was restrained to restrain large lateral wall movement. For another, the Rankine solution assumes that no friction exists between the soil and wall, which accounts for the overestimation of the theoretical lateral earth pressure applied on the unexcavated side. The measured lateral earth pressures on the excavation side were much smaller than the Rankine earth pressures and decreased throughout the excavation.

5 CONCLUSIONS

Behaviour of unstrutted and strutted clay excavation is investigated by 1/100 scale centrifuge model tests. Main conclusions drawn in this study are:

1. The inflight excavation tools and struts system have been applied effectively in the centrifuge model tests on Shanghai clay excavation, the inflight excavation tests in this study simulates the sequences of in situ excavation process and the available monitored data could be used to assess the performance of excavation.
2. The deformation mechanism of unstrutted excavation is different from that of strutted excavation. In test 1, the displacement of the cantilever wall continues to develop as excavation progresses and ends at 32 mm, which has surpassed the allowable value. The maximum ground settlement behind the wall occurs near the wall and increases as the wall incline to the excavated area. In test 2, the maximum wall displacement occurs at a deeper level as excavation progresses and is much smaller compared with unstrutted condition. The ground settlement has been controlled by the struts-supported retaining wall, and the maximum point is 10 m from the wall and does not change with the excavation height.
3. The normalized ground settlement profile is larger than the one proposed by Clough and O'Rourke (1990). This is attributed to the fact that the wall is not strutted after every stage of excavation and the clay is low in strength. However, the maximum settlement occurs at 0.8 times the final excavation depth, almost the same position as the experiential method by Clough & O'Rourke (1990).
4. The lateral earth pressure on the unexcavated side is less than the Rankine active pressure at deeper level. This behaviour is perhaps due to the assumption of the Rankine solution that no friction exists between the soil and wall.

REFERENCES

Bolton, M.D. & Powrie, W. 1987. The collapse of diaphragm walls retaining clay. *Geotechnique* 37(3): 335–353.

Clough, G.W. & O'Rourke, T.D. 1990. Construction-induced movements of in situ walls. *Proceedings, Design and Performance of Earth Retaining Structure*: 439–470.

Elshafie, M.Z.E.B., Choy, C.K.C., & Mair, R.J., 2013. Centrifuge modeling of deep excavations and their interaction with adjacent buildings. *Geotech. Test. J.* 36(5): 1–12.

Kimura, T., Takemura, J., Hirooka, A., et al. 1997. Excavation in soft clay using an in-flight excavator. *Proceedings of the International Conference on Geotechnical Centrifuge Modelling*: 649–654.

Kusakabe, O. 1982. Stability of excavations in soft clay. Ph.D. Thesis, University of Cambridge.

Lade, P.V, Jessberger, H.L, Kakowski, M. & Jordan, P. 1981. Modelling of deep shafts in centrifuge tests. *Proceedings of 10th International Conference. on Soil Mechanics and Foundation Engineering*, Vol. 1, 683–692.

Lam, S.Y, Haigh, S.K, Elshafie, M.Z.E.B., & Bolton, M.D. 2012. A new apparatus for modelling excavations. *International Journal of Physical Modelling in Geotechnics* 2012: 12(1): 24–38.

Lam, S.Y., Haigh, S.K., & Botlon, M.D. 2014. Understanding ground deformation mechanisms for multi-propped excavation in soft clay. *Soils and Foundations* 2014: 54(3): 296–312.

Lee, F.H., Almeida, M.S.S. & Indraratna, B. 2010. Physical modelling of soft ground problems, *Proceedings of 7th International Conference on Physical Modelling in Geotechnics*: 45–66.

Liang, F.Y., Chu, F., Song, Z., et al. 2012. Centrifugal model test research on deformation behaviors of deep foundation pit adjacent to metro stations. *Rock and Soil Mechanics* 33(3): 657–664.

Liu, J.H. & Hou, X.Y. 1997. Excavation engineering handbook, Chinese Construction Industry Press, Beijing, P.R. China.

Liu, J.Y. 1999. Centrifugal model test in soft soil foundation pit. Ph. D. Thesis. Tongji University.

Lyndon, A., Schofield, A.N. 1970. Centrifuge model test of short term failure in London clay. *Geotechnique* 20(4): 440–442.

Ma X.F., Zhang H.H., & Zhu W.J., et al. 2009. Centrifuge model tests on deformation of ultra-deep foundation pits in soft ground. *Chinese Journal of Geotechnical Engineering* 31(9): 1371–1377.

Ng, C.W.W., Van Laak, P.A., Zhang, L.M., Tang, W.H., Li, X.S., & Xu, G.M. 2001. Key Features of the HKUST Geotechnical Centrifuge. *Proceedings of the International Symposium on Geotechnical Centrifuge Modelling and Networking: Focusing on the Use and Application in the Pan-Pacific Region*: 66–69.

Peck, R.B. 1969. Deep excavation and tunneling in soft ground. *Proceedings of 7th International Conference on Soil Mechanics and Foundation Engineering, Mexico City, State-of-the-Art-Volume*: 225–290.

Powrie, W. & Kantartzi, C. 1996. Ground response during diaphragm wall installation in clay, Centrifuge Model Tests. *Geotechnique* 46(4): 725–739.

Richards, D.J. & Powrie, W. 1998. Centrifuge model tests on doubly propped embedded retaining walls in over consolidated kaolin clay. *Geotechnique* 48(6): 833–846.

Takemura, J., Kondoh, M., Esaki, T., Kouda, M., & Kusakabe, O. 1999. Centrifuge model tests on double propped wall excavation in soft clay. *Soils Found.*, 39(3): 75–87.

Xu Z.H. 2007. Deformation behavior of deep excavations supported by permanent structure in shanghai soft deposit. Ph.D. Thesis, Shanghai Jiao Tong University.

Zhang, G., Li, M., & Wang, L.P. 2014. Analysis of the effect of the loading path on the failure behaviour of slopes. *KSCE J. Civ. Eng.* 18(7): 2080–2084.

Effect of root spacing on interpretation of blade penetration tests—full-scale physical modelling

G.J. Meijer & J.A. Knappett
Division of Civil Engineering, University of Dundee, Dundee, UK

A.G. Bengough
Division of Civil Engineering, University of Dundee, Dundee, UK
James Hutton Institute, Invergowrie, Dundee, UK

K.W. Loades
James Hutton Institute, Invergowrie, Dundee, UK

B.C. Nicoll
Forest Research, Northern Research Station, Roslin, Midlothian, UK

ABSTRACT: The spatial distribution of plant roots is an important parameter when the stability of vegetated slopes is to be assessed. Previous studies in both laboratory and field conditions have shown that a penetrometer adapted with a blade-shaped tip can be used to detect roots from sudden drops in penetrometer resistance. Such drops can be related to root properties including diameter, stiffness and strength using simple Winkler foundation models, thereby providing a field instrument for rapid quantification of root properties and distribution. While this approach has proved useful for measuring single widely-spaced roots, it has not previously been determined how the penetrometer response changes as a result of roots being in close proximity. Therefore in this study 1-g physical modelling (at 1:1 scale) was conducted to study the effect of vertical root spacing using horizontal, straight 3D-printed root analogues. Results show that when roots are closely spaced, there is significant interaction between them, resulting in higher apparent root displacements to failure and an increased amount of energy being dissipated. This preliminary work shows that the interpretive models used to analyse the penetrometer trace require further development to account for root-soil-root interactions in densely rooted soil.

1 INTRODUCTION

Plant roots can reinforce soil through mechanical action (similar to fibre-reinforced soil) and hydrological effects (by reducing the soil water content and increasing soil matric suction through water uptake by the plant) (e.g. Coppin and Richards 1990). To make quantitative predictions for either of these effects, the spatial distribution of roots is an important parameter. It is however difficult to obtain this information without time-consuming root sampling techniques.

To address this problem, Meijer et al. (2016) proposed new field measurement techniques. One of these, the so-called 'blade penetrometer', can be used to infer root depths and root properties from the penetrometer depth–resistance curve. The penetrometer tip shape was enhanced with a thin blade. This greatly increases the chance a root will be hit per unit of tip area, making the penetrometer more sensitive to identifying roots (Meijer et al. 2016). Roots will be visible in the depth–resistance trace as peaks: from the moment a root is hit the penetrometer resistance will increase, up to the point at which the root fails, visible as a sudden decrease in penetrometer resistance (Figure 1).

Interpretative methods were developed to infer root properties such as root diameter from the characteristics of these root reinforcement peaks, based on the assumption that the root either fails in pure bending or pure tension. These properties can then in turn be used to calculate mechanical root-reinforcement using existing models (e.g. Meijer et al. 2017a).

This methodology was successfully tested in the laboratory using dry sand and 3-D printed Acrylonitrile Butadiene Styrene (ABS) root analogues (Meijer et al. 2017a). It was shown that for these root analogues the model based on root failure in bending worked best. The best predictions were made using the magnitude of the sudden decrease in penetrometer resistance associated with root analogue failure, rather than the displacement required to reach failure. Testing in real soils with live vegetation confirmed these results (Figure 1), although for thinner roots an interpretative model based on root failure in tension worked better (Meijer et al. 2017b).

The laboratory investigations by Meijer et al. (2017a) only considered individual roots. In reality however roots may be closely spaced. When the displacement required to reach root failure is larger than

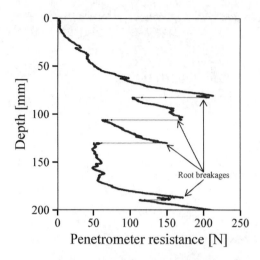

Figure 1. Field depth–penetrometer resistance trace measured in a clayey silt soil rooted with oak trees (Meijer et al. 2017b). Depth and resistance were sampled at 100 Hz (measurement points in green).

the root spacing, multiple roots might be loaded by the penetrometer at the same time, making interpretation of the results much more difficult. Furthermore, root–soil–root interaction might occur.

To investigate these effects, a preliminary physical modelling study was performed using ABS root analogues in dry medium dense and dense sand investigating the effect of closely spaced roots on the penetrometer depth–resistance response.

2 METHODS

2.1 Laboratory experiments

A series of blade penetrometer tests was performed in the laboratory, using closely spaced root analogues in dry sand. All tests were performed at 1-g at a 1:1 scale. To ensure accurate scaling, the following conditions should match the conditions in typical field conditions:

1. Root mechanical properties
2. Root diameter, length and spacing
3. Soil stress level

Similar to Meijer et al. (2017a), Acrylonitrile Butadiene Styrene (ABS) plastic was used as root analogue material. This material has comparable mechanical characteristics to plant roots (Liang et al. 2015, Meijer et al. 2016), see Figure 2. The bending strength and stiffness has been measured in 3-point bending (Meijer et al. 2017a). Peak strength and stiffness reduces with increasing diameter. This is fitted using the following curve:

$$\sigma_b = \alpha_\sigma \left(\frac{d}{d_{ref}}\right)^{\beta_\sigma} \quad (1)$$

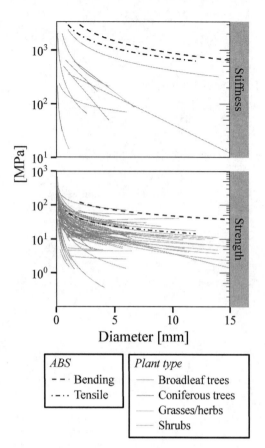

Figure 2. ABS tensile and bending properties compared to tensile strength and Young's modulus for real plant roots reported in the literature (see Meijer (2016) for sources). The ABS stiffness is plotted as the secant stiffness at 90% strength.

where σ_b is the peak bending strength, d the diameter and $d_{ref} = 1$ mm a reference diameter. $\alpha_{\sigma,b} = 180$ MPa and $\beta_{\sigma,b} = -0.577$. A similar relation was used for the bending stiffness E_b:

$$E_b = \alpha_E \left(\frac{d}{d_{ref}}\right)^{\beta_E} \quad (2)$$

where $\alpha_{E,b} = 4937$ MPa and $\beta_{E,b} = -0.749$. Stiffness E_b is defined as the secant stiffness at 90% strength rather then the Young's modulus, as it gives a better approximation of the non-linear stress–strain curve when a linear elastic material model is used in the interpretation (Meijer et al. 2017a).

The soil material used was dry Congleton silica sand (HST95) with relative densities of $I_d = 50\%$ (medium dense) or $I_d = 80\%$ (dense). The critical state friction angle is $\phi_{cv} = 32°$ and peak friction angles are $\phi' = 39°$ and $\phi' = 45°$ respectively. Dry unit weight γ' was 16.0 and 16.9 kNm^{-3} respectively (Lauder 2010).

Root analogues were printed in segments of 200 mm using a rapid prototyper ('3D-printer'). Root analogue diameters were 2 or 4 mm. The tested root length was

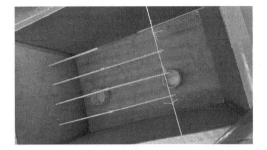

Figure 3. Box prior to pluviation.

$L = 400$ mm, obtained by gluing together two root analogue segments using epoxy resin and a printed ABS coupler with a length of 15 mm and an external diameter 3 mm larger than that of the analogue. The root analogue diameters tested are near the commonly used threshold between 'thin' and 'thick' roots of $d = 2$ mm (e.g. Achat et al. 2008), making them representative of an 'average' root. The root length is on the short end of the root length/root diameter ratio compared to values measured for Norway spruce roots (Giadrossich et al. 2013):

$$L = 390d^{0.56} \quad (3)$$

A plastic box lined with 10 mm thick adhered wooden panels with internal dimensions 530× 330 × 310 mm (length × width× height) was used. Prior to sand pluviation, roots were glued into pre-drilled holes in the wooden panel. Roots were suspended by wires (cut prior to commencing a test) to prevent significant deformations due to self-weight (Figure 3).

For every test, two horizontal root analogues were used. The first root analogue was located at $z = 150$ mm below the soil surface and the second analogue at a distance of $3d$ below this. (centre-to-centre spacing $s/d = 3$). These depths correspond with typical rooting depths in the field were most roots grow near the surface because of availability of water, oxygen and nutrients. Typically, over 50% of plant roots can be found in the top 300 mm of the soil (Jackson et al. 1996). Effective soil stresses in experiments are therefore representative of those found in the field (assuming suction levels are small). The adopted root spacing can be expressed in terms of root area ratio (RAR, i.e. the percentage of soil cross-sectional area covered by root):

$$RAR \approx \frac{\frac{\pi}{4}d^2}{ws} \quad (4)$$

where $w = 30$ mm is the width of the penetrometer. For $d = 2$ mm root analogues, $RAR = 1.75\%$ while for $d = 4$ mm $RAR = 3.49\%$, which is high compared to the field where typically $RAR < 1\%$ (e.g. Bischetti et al. 2005. Field values are however average values, so locally RAR might be higher. Thus, $s/d = 3$ can be seen as an upper limit for realistic root spacing.

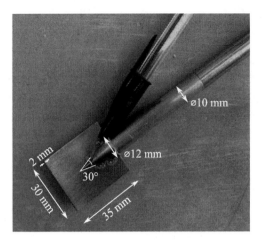

Figure 4. Blade penetrometer. Ballpoint pen for scale.

Subsequent to root placement the box was filled with dry sand to a height of 300 mm using a slot pluviator. Multiple penetrometer tests were conducted per box. The lateral spacing between roots was at least 65 mm ($s/d > 16$) to minimise interaction between tests.

2.2 Equipment and test programme

The blade penetrometer was constructed by welding a blade to a standard agricultural penetrometer (⌀12 mm, 30° tip). The shaft was thinner (⌀10 mm) to minimise shaft friction (Figure 4).

Tests were performed using a universal testing machine (Instron 5980). The extension rate was 300 mm min^{-1}, similar to Meijer et al. (2016). Both force and displacement were logged at 20 Hz.

Roots were loaded by the penetrometer at a distance of 300 mm from the point where the root was anchored in the side of the box. Tests were performed using both root diameters (2 and 4 mm) and both soil densities (50% and 80%), totalling 4 tests.

Additional blade penetrometer tests were performed in areas of the container that contained no roots. The results of these 'fallow' tests were subtracted from the results of the rooted test to find the contribution of the root analogues to penetrometer resistance.

Root reinforcements measured in rooted tests were compared to the results obtained for single roots as measured by Meijer et al. (2017a).

2.3 Interpretative methods

Meijer et al. (2017a) derived an interpretative method for roots loaded in bending by a penetrometer based on Euler-Bernoulli beam theory. The maximum force a root can sustain before breaking in bending when loaded perpendicularly by a penetrometer (F_u) can be estimated by:

$$F_u = \xi_F d^2 \sigma_b^{0.5} p_u^{0.5} \quad (5)$$

where $\xi_F = 1.02$, d is the root diameter [mm], σ_b the maximum root strength in bending [MPa] and p_u the soil resistance against lateral root displacement [MPa]. The latter is expressed as $p_u = p/d$, where p is the maximum mobilised lateral soil resistance according to p-y theory (e.g. Reese and Van Impe 2011). The corresponding lateral root displacement to failure u_u is given by:

$$u_u = \xi_u d \sigma_b^2 E_b^{-1} p_u^{-1} \quad (6)$$

where $\xi_u = 0.098$ and E_b is the root stiffness in bending. The force-displacement behaviour is given by:

$$F(u) = \xi_F \xi_u^{-0.25} u^{0.25} d^{1.75} E_b^{0.25} p_u^{0.75} \quad (7)$$

Integrating Equation 2.3 between $u = 0$ and $u = u_u$ gives the total amount of work (W_u) dissipated by the root before failure occurs:

$$W_u = 0.8 \xi_F \xi_u d^3 \sigma_b^{2.5} E_b^{-1} p_u^{-0.5} \quad (8)$$

Equations 2.3–2.3 can also be used for roots reinforcing soil in direct shear. When they cross the shear plane perpendicularly, $\xi_F = 0.89$ and $\xi_u = 1.05$. Thus the results from a blade penetrometer test can be directly used to estimate the root behaviour in direct shear loading.

p was estimated using a p-y model for piles in dry sand, ignoring surface wedge formation near the surface (Reese and Van Impe 2011):

$$p = A_s dz \gamma' \left[K_a \left(\tan^8 \beta - 1 \right) + K_0 \tan \phi' \tan^4 \beta \right] \quad (9)$$

where z is the depth, $K_a = 0.4$ and $K_0 = 1 - \sin \phi'$ coefficients of lateral earth pressure, A_s a dimensionless model constant and $\beta = 45° + \phi'/2$. It can be readily seen from Equations 2.3–2.3 that the behaviour of the root does not only depend on root properties but also on the soil resistance. At $z = 150$ mm depth, $p_u \approx 0.19$ ($I_d = 50\%$) or $p_u \approx 0.47$ MPa ($I_d = 80\%$).

3 RESULTS

The penetrometer force–displacement behaviour proved difficult to interpret, mainly due to post-failure root analogue effects. After root failure, during subsequent penetrometer displacement both broken ends might have got stuck on the shoulder of the cone, causing additional peaks and troughs in the trace. When dealing with multiple roots, the behaviour of subsequently loaded roots got obscured by these artefacts. To address this problem, sudden drops in force that corresponded with root analogue failure were identified by audio observations. Failure coincided with a clear 'snapping' noise.

The depth–root reinforcement traces are shown in Figures 5–8, both for tests with two closely spaced roots and tests containing only a single root at $z \approx 150$ mm. Both the experimental and model results

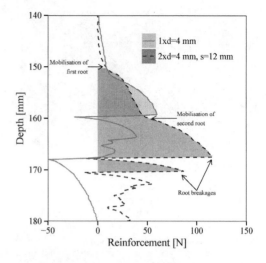

Figure 5. Root reinforcement traces for a single and two $d = 4$ mm roots in $I_d = 50\%$ sand. 's' is root centre-to-centre distance. Shaded areas indicate root action.

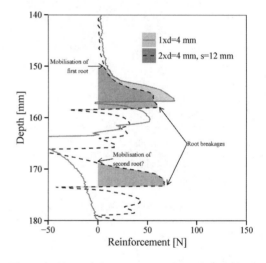

Figure 6. Root reinforcement traces for a single and two $d = 4$ mm roots in $I_d = 80\%$ sand. 's' is root centre-to-centre distance. Shaded areas indicate root action.

for single roots shows an increase in F_u with diameter and soil relative density. In contrast, u_u increases with diameter but is inversely correlated to relative density. The only exception is the $d = 2$ mm root analogue in dense soil, for which u_u is much larger than expected compared to other tests.

Increasing the relative density has a distinct effect on the effect of closely spaced $d = 4$ mm root analogues (Figure 6). In the $I_d = 80\%$ case $s \approx 2-3u_u$, resulting in what appears to be independent behaviour of the two analogues. When $I_d = 50\%$ however, $s \approx u_u$, resulting in interaction between the two root analogues (Figure 5). The upper root analogue in this case displaced much further before reaching failure compared to a test containing only a single analogue. Similar interaction effects were observed for two $d = 2$ mm

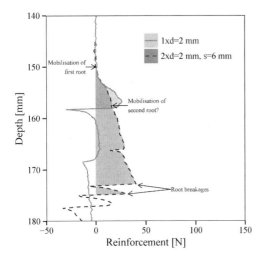
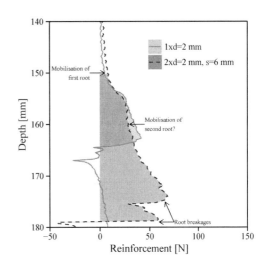

Figure 7. Root reinforcement traces for a single and two $d = 2$ mm roots in $I_d = 50\%$ sand. 's' is root centre-to-centre distance. Shaded areas indicate root action.

Figure 8. Root reinforcement traces for a single and two $d = 2$ mm roots in $I_d = 80\%$ sand. 's' is root centre-to-centre distance. Shaded areas indicate root action.

Table 1. Summary of experimental results.

Diameter [mm]	I_d [%]	Measured					Predicted/Measured			
		F_u [N]	u_u [mm]	W_u [Nmm]	$2W_{u,single}/W_{u,double}$ [–]		F_u/F_u [–]	u_u/u_u [–]	W_u/W_u [–]	f_m [–]
2	50	26	8.3	105			0.75	0.62	0.76	
2	80	41	14.5	372			0.75	0.14	0.14	
4	50	60	9.8	372			1.07	0.79	1.06	
4	80	77	7.0	294			1.31	0.45	0.86	
2 & 2	50			526	0.40					0.16
2 & 2	80			1092	0.68					0.47
4 & 4	50			1187	0.63					0.39
4 & 4	80			544	1.08					1.17

roots in $I_d = 50\%$ (Figure 7) or $I_d = 80\%$ sand (Figure 8). Breakage of both roots occurred almost at similar depths and at much larger displacements compared to tests on a single root.

The total amount of work dissipated by two closely spaced roots is higher then the sum of two individual roots. This effect is more pronounced for thinner roots and in soil with lower relative densities (Table 1).

4 DISCUSSION

Since root properties σ_b and E_b are constant, according to Equation 2.3 only a decrease in p_u can explain the increased amount of work dissipated by closely spaced roots. The same holds for the increase in lateral root displacement u_u. An explanation for this reduction in p_u can be found by looking into the mobilisation mechanism of the soil resistance. It is hypothesised p_u will reduce on the first load once the second root analogue is mobilised since the first root moves through a zone 'shaded' by the second root (Figure 9).

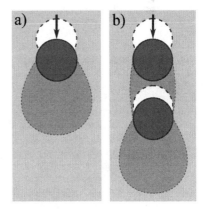

Figure 9. Schematic representation of mobilisation of soil resistance in the case of two roots.

Such a reduction in p_u bears similarities to reductions in lateral pile capacity for piles in pile groups compared to single piles. This reduction can be taken into account by reducing the lateral resistance by a factor f_m, a so-called p-multiplier (e.g. Brown et al. 1988).

McVay et al. (1995) conducted centrifuge tests on a 3×3 pile configurations in dry sand, spaced at $s/d = 3$. For loose sand ($I_d = 33\%$) they found $f_m = 0.65$, 0.45 and 0.35 for piles in the leading row, middle row and back row respectively. For medium dense sand ($I_d = 55\%$), these values were 0.8, 0.4 and 0.3 respectively. This shows that the more a pile is 'shielded' by piles ahead of it, the lower the lateral soil resistance will be, an effect that will be more pronounced in looser soil.

To explain the amount of work dissipated in the testing described in this paper, average p-multipliers required are either higher ($d = 4$ mm, $I_d = 80\%$) or lower (all other tests) compared to these literature values (see Table 1, last column), despite similar s/d values as used by McVay et al. (1995). One explanation for the lower values is that literature values for f_m are typically derived from tests with relative small relative displacements between piles. In the case of McVay et al. (1995), piles were connected at the top to model a superstructure, resulting in s/d not significantly changing during the test. However, in the tests described here the first root hit will slowly move towards the second root. Therefore s/d will reduce with increasing penetrometer displacement before breakage, resulting in gradually decreasing values of f_m. This reduction might be smaller when the ratio between u_u and s is small since the root will have failed before the first root reaches the non-displaced position of the second root. This might (partially) explain why f_m values measured for the test in $I_d = 80\%$ using $d = 4$ mm root analogues were higher compared to the other tests.

5 CONCLUSIONS

It was shown that root–soil–root interaction has a significant effect on the behaviour of root analogues loaded by a blade penetrometer. Due to a reduction in soil resistance caused by close spacing, root analogues displaced further and dissipated more energy compared to tests only containing a single root. This effect was particularly pronounced for the first root hit. These effects will be stronger when the displacement required to failure is large compared to the root spacing (large u_u/s), for example in weaker soils or for strong, flexible roots. However, not enough test data was available to quantify the magnitude of the soil resistance reduction factor f_m (p-multiplier) as a function of soil properties, root properties and root spacing.

As a result of root–soil–root interaction, closely spaced roots will not necessarily appear as discrete separate peaks in the depth-penetrometer resistance trace. The experiments showed that they might break at a similar point in time. In these cases, inferring root properties from the amount of energy dissipated by displacing roots might be more feasible.

The existing penetrometer method is currently most suitable for identifying more widely spaced structural roots. Physical modelling indicated that development of an interpretative model should look at root-soil-root interaction when roots are closely spaced.

REFERENCES

Achat, D. L., M. R. Bakker, & P. Trichet (2008). Rooting patterns and fine root biomass of *Pinus pinaster* assessed by trench wall and core methods. *Journal of Forest Research* 13(3), 165–175.

Bischetti, G. B., E. A. Chiaradia, T. Simonato, B. Speziali, B. Vitali, P. Vullo, & A. Zocco (2005). Root strength and root area ratio of forest species in Lombardy (northern Italy). *Plant and Soil* 278(1–2), 11–22.

Brown, D. A., C. Morrison, & L. C. Reese (1988). Lateral load behaviour of pile groups in sand. *ASCE Journal of Geotechnical Engineering* 114(11), 1261–1276.

Coppin, N. & I. Richards (1990). *Use of vegetation in civil engineering, CIRIA book 10*. Kent: Butterworths.

Giadrossich, F., M. Schwarz, D. Cohen, F. Preti, & D. Or (2013). Mechanical interactions between neighbouring roots during pullout tests. *Plant and Soil* 367(1–2), 391–406.

Jackson, R. B., J. Canadell, J. R. Ehleringer, H. A. Mooney, O. E. Sala, & E. D. Schulze (1996). A global analysis of root distribution for terrestial biomes. *Oecologica* 108(3), 389–411.

Lauder, K. (2010). *The performance of pipeline ploughs*. Ph. D. thesis, University of Dundee.

Liang, T., J. Knappett, & N. Duckett (2015). Modelling the seismic performance of rooted slopes from individual root–soil interaction to global slope behaviour. *Géotechnique* 65(12), 995–1009.

McVay, M., R. Casper, & T.-I. Shang (1995). Lateral response of three-row groups in loose to dense ssand at 3d and 5d pile spacing. *ASCE Journal of Geotechnical Engineering* 121(5), 436–441.

Meijer, G. J. (2016). *New methods for in situ measurement of mechanical root-reinforcement on slopes*. Ph. D. thesis, University of Dundee.

Meijer, G. J., A. G. Bengough, J. A. Knappett, K. W. Loades, & B. C. Nicoll (2016). New in-site techniques for measuring the properties of root-reinforced soil – laboratory evaluation. *Géotechnique* 66(1), 27–40.

Meijer, G. J., A. G. Bengough, J. A. Knappett, K. W. Loades, & B. C. Nicoll (2017a). In situ root identification through blade penetrometer testing – part 1: interpretative models and laboratory testing. *Géotechnique*. Published ahead of print.

Meijer, G. J., A. G. Bengough, J. A. Knappett, K. W. Loades, & B. C. Nicoll (2017b). In situ root identification through blade penetrometer testing – part 2: field testing. *Géotechnique*. Published ahead of print.

Reese, L. C. & W. F. Van Impe (2011). *Single piles and pile groups under lateral loading, 2nd edition*. Leiden, The Netherlands: CRC.

Development of a centrifuge testing method for stability analyses of breakwater foundation under combined actions of earthquake and tsunami

J. Miyamoto, K. Tsurugasaki & R. Hem
Technical Research Institute, Naruo, Toyo Construction, Nishinomiya, Japan

T. Matsuda
Toyohashi University of Technology, Toyohashi, Japan

K. Maeda
Nagoya Institute of Technology, Nagoya, Japan

ABSTRACT: A centrifuge testing method for combined actions of earthquake and tsunami was developed in a drum centrifuge. This method was made possible by installing a shaking table device to the channel of drum centrifuge for the tsunami test. This paper describes the experimental method developed and shows the experimental results of a combined failure of breakwaters subjected to earthquake ground motion and tsunami actions. Specifically, two experimental cases were performed on composite breakwaters on loose sand beds under a centrifuge acceleration of 30 g: Case 1 where earthquake and tsunami separately attack to the breakwater model; Case 2 where the earthquake and tsunami simultaneously attack to the breakwater model. The principal experimental results may be summarized as follows: (1) tsunami overflow easily occurs due to the significant settlement of the caisson induced by ground motion; (2) even if the level of tsunami was moderate, the breakwater easily collapsed when the ground motion acted during a tsunami overflow.

1 INTRODUCTION

In the 2011 off the Pacific coast of Tohoku Earthquake, the coastal area of the Tohoku region of Japan was severely damaged by the huge earthquake and tsunami (Sugano et al. 2014). In Japan, there are concerns about serious damage of coastal ports and industrial complexes due to strong earthquakes and tsunamis that may occur in the future. In order to design and examine countermeasures against earthquakes and tsunamis of port structures, it is important to grasp what kinds of damages are caused by a combined force of earthquake and tsunami. Damages due to combined forces of earthquake and tsunami have been studied by several experiments and analyzes (Matsuda et al. 2016; Miyake & Sawada 2010; Tobita & Iai 2014). The authors conducted experiments on interaction between tsunamis and breakwaters using the drum centrifuge and investigated the instability mechanism of the breakwater foundation during the tsunami (Miyamoto et al. 2015; Tsurugasaki et al. 2016). However, because shaking test cannot be performed in the drum centrifuge, we could not investigate the damage mechanism of breakwaters due to a combined force of ground motion and tsunami.

In this research, we installed a shaking device in the drum centrifuge tsunami channel to develop a testing method of the combined force of earthquake ground motion and tsunami. The developed experiment method under combined force of ground motion and tsunami is applied to investigate the stability of the composite breakwater. Experiments were conducted by assuming two scenarios. The first scenario is a scenario where a tsunami comes after an earthquake, the second scenario is a scenario where an earthquake occurs when a tsunami is attacking. In other words, it is a scenario where earthquakes and tsunami act simultaneously.

The structure of this paper is as follows. First of all, we will explain about the development of the experiment method with combined ground motion and tsunami. Next, the experimental conditions are described. Finally, we will describe the behaviour of the composite breakwater with the two scenarios with combined force of ground motion and tsunami.

2 DEVELOPMENT OF A CENTRIFUGE TESTING METHOD FOR EARTHQUAKE AND TSUNAMI TESTS

Figure 1 shows the water channel in the drum centrifuge. Both the shaking device and the tsunami generator are installed in the cylindrical channel. Figure 2 shows the detailed mechanism of shaking system. In this mechanism, rollers (free bearings) are mounted

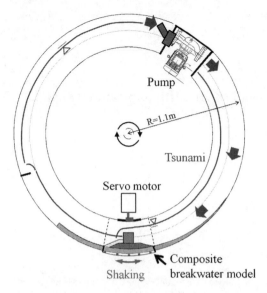

Figure 1. Cross section of water channel installed in the drum centrifuge for earthquake and tsunami tests.

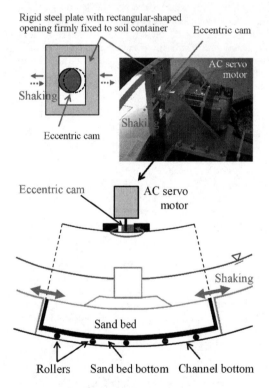

Figure 2. Shaking system by AC servo motor and eccentric cam in the tsunami testing channel.

on the bottom of the soil container, and the soil container is shaken by the AC servo motor and the eccentric cam, so that shaking motion can be applied to the breakwater model. Tsunami is generated by a circulation pump. Since the mechanism of the generation

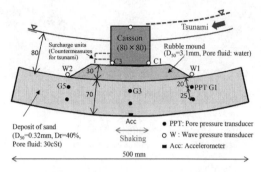

Figure 3. Cross section of a composite breakwater model on the loose sand bed.

of ground motion and tsunamis are independent, the way of combining forces of earthquakes and tsunamis is free and various scenarios can be performed. However, the waveform of the seismic motion is limited to only sine wave.

It should be notified that the shaking motion would cause an upward pumping effect between the sides of the sand bed container and the channel bottom. However, the moving of the sand bed was in a small range (amplitude 2 mm), so this effect was considered to be small and it was included in tsunami flow which is measured near the caisson (C1).

The experiment was conducted under a centrifuge acceleration of 30 g. The cross section of the experimental model is shown in Figure 3. The loose sand bed of relative density of 40% (Dr = 40%) was made by using silica sand No. 6 ($D_{50} = 0.32$ mm). The gravel with an average particle grain size of 3.1 mm ($D_{50} = 3.1$ mm) was used as mound material. A 30cSt methylcellulose solution is used as a pore fluid of the foundation ground, so that the time similarity rules related to seismic ground motion, tsunami propagation, and soil consolidation are matched (see Table 1). With reference to Takahashi et al. 2014, water was used for external fluid and pore fluid in the gravel material such as mound. Although, the density of the methylcellulose solution is very slightly larger than that of water, during the experiment, there is a possibility that these two fluids may be slightly mixed at the contact surface.

As countermeasures for tsunami, two layers of surcharge units (mass = 15g/unit) made with gravels of $D_{50} = 30$ mm, are installed behind the breakwater caisson, so that it will not collapse only by a medium-scale tsunami.

Two scenarios were conducted for the combined force experiments. In Case 1, just after shaking, a medium-scale tsunami attacks the breakwater model. In Case 2, while medium-scale tsunami is attacking the breakwater model, the breakwater was shaken. Here, a medium-scale tsunami (tsunami level is the same as the caisson top level) is a tsunami with the extent to which a slight overflow occurs. In the experiments carried out separately, we confirmed that the breakwater was remained stable under an action of this medium-scale

Table 1. Similarity ratio in this study.

		Prototype	Present study
Experimental condition	Centrifugal acceleration	1	N
	Model size	1	1/N
	Mound Grain size	1	1/N
	Dynamic viscosity of pore fluid	1	1
	Sand Bed Grain size	1	1
	Dynamic viscosity of pore fluid	1	N
Ground motion (earthquake)	Acceleration	1	N
	Velocity	1	1
	Time	1	1/N
Tsunami	Pressure	1	1
	Velocity	1	1
	Time	1	1/N
Seepage in mound (Turbulent)	Mean flow velocity	1	1
	Seepage time	1	1/N
Seepage in sand bed (Laminar)	Mean flow velocity	1	1
	Seepage time (consolidation time)	1	1/N

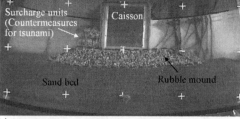

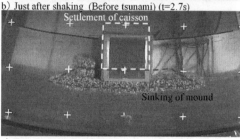

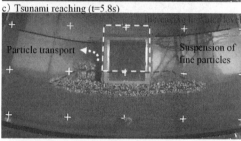

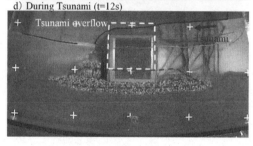

Figure 4. Observed behaviour of breakwater under the actions of (a)–(b) ground motion and (c)–(d) tsunami (Case 1).

tsunami alone (Case 0). The acceleration amplitude of the ground motion was about 0.2 g. The frequency of 50 Hz and a total of 34 cycles for Case 1 and 75 cycles for Case 2 of ground motion were input. Regarding the boundary conditions in model, the boundary of the sand bed is the sand bad container, and there is no boundary for exterior fluid.

3 TEST RESULTS

3.1 Tsunami after earthquake

The appearance of the breakwater behaviour due to the ground motion and tsunami is shown in Figure 4. The caisson subsides with the liquefaction of free field due to the ground motion (Fig. 4b). The caisson settlement occurs almost symmetrically in the same way as in the previous studies (Iai et al. 1998; Kim & Sekiguchi 2002). After shaking, when a tsunami is generated, the water level seaward side starts to rise (Fig. 4c). At this time, it is possible to observe the soil particles floating on the free field at the seaward side and at the vicinity of the backside of caisson. These are thought to be related to the excess pore pressure remained just before the start of rising water level. Despite the fact that the tsunami is medium size, the caisson has greatly subsided with liquefaction at the time of the shaking, so a significant overflow has occurred (Fig. 4d).

In addition, the embankment unit at the back of the caisson, as a countermeasure against the tsunami, is detached from the caisson by ground motion (Fig. 4b). Under such conditions, the possibility that the caisson will collapse is high due to the continuation of the subsequent tsunami overflow. It is necessary to examine the countermeasures in consideration of both the earthquake and tsunami.

The pore pressure response in the ground is shown in Figure 5. This figure also shows water pressure difference between the front and behind of caisson due to the tsunami (Fig. 5a) and the input acceleration (Fig. 5b). After shaking, the excess pore water pressure in the ground is increased in free field (G1, G5) (Figs 5c, e). The elevated level of excess pore water pressure has reached the initial effective overburden pressure

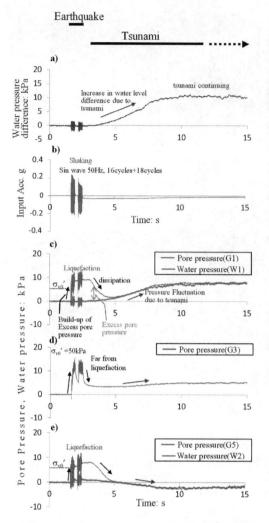

Figure 5. Time history of (a) water difference between the harbour side and the seaward side, (b) input acceleration, and (c, d, e) pore pressure responses and water pressure responses (Case 1).

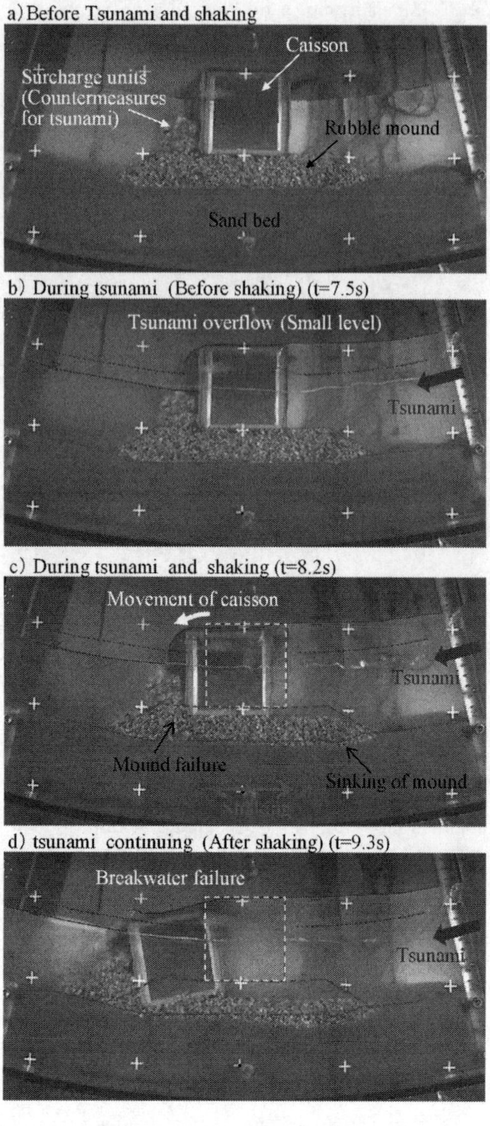

Figure 6. Observed behaviour of breakwater under the actions of tsunami and ground motion (Case 2).

level σ_{vo}', indicating the occurrence of liquefaction. Although the excess pore pressure rises in the ground directly under the caisson (G3), it does not reach liquefaction (Fig. 5d). The tsunami will come after this. After shaking, the excess pore pressure begins to dissipate, then the water level at the front of breakwater rises due to the arrival of the tsunami (Fig. 5c). In this experiment, when the rise of the water level by the tsunami started, it was noted that a slight excess pore water pressure remained.

3.2 *Earthquake during tsunami*

Figure 6 shows the behaviour of caissons and mounds when an earthquake occurred during the tsunami attack. As shown in Figure 6b, if the tsunami is as medium-scale as of this experiment level, the caisson hardly moves when it receive the action from only tsunami. Interestingly, when the action of the ground motion is added to tsunami action, the caisson, mound, and the ground may be readily unstable (Figs 6c, d). It is shown that the ground at the seaward side is liquefied and the mound sinks into the ground and the caisson is greatly displaced and collapsed due to collapsing the mound at the harbour side. From this study, it was found that the caisson collapsed if it is acted simultaneously by the earthquake and tsunami even for a medium-scale tsunami.

The development process of the caisson displacement is shown in Figure 7. In this figure, the vertical displacement and the horizontal displacement of the lower end of the caisson are shown to increase. While

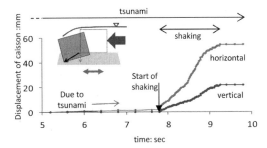

Figure 7. Time histories of development of displacement of caisson during tsunami and earthquake (Case 2).

the tsunami is acting, it remains a slight horizontal displacement. However, when seismic motion is added, it can be seen that horizontal displacement sharply increases with vertical displacement (subsidence). As described above, it was also confirmed by another experiment that the displacement of the caisson does not develop when only keeping the tsunami acting (Case 0).

The time history of pore water pressure in the ground is shown in Figure 8. The figure also shows the water pressure difference at the seaward side due to the tsunami (Fig. 8a) and the input acceleration (Fig. 8b). First, attention is paid to the behaviour of the ground at the seaward side (Fig. 8c). When the water level rises due to the tsunami, pore water pressure in the ground is also increased. At this time, after the start of shaking, the excess pore water pressure suddenly rises and leads to liquefaction. Next, we will look at the behaviour of the ground just below the caisson (Fig. 8d). The increase of excess pore water pressure caused by ground motion does not reach the level of about 40% of the initial effective overburden pressure. In the ground at the harbour side, the pore water pressure drops due to the lowering of the water level by the tsunami (Fig. 8e). At this time, when the ground motion is applied, the excess pore water pressure increases to become close to the liquefaction state. When we closely look at this figure, cyclic mobility was observed. When the ground motion is applied while the tsunami is acting, it is important that the pore water pressure behaviour is different between the ground at the seaward side (Fig. 8c) and the ground at the harbour side (Fig. 8e).

4 CONCLUSIONS

In this research, we developed an earthquake and tsunami experiment method in the drum centrifuge water channel. This experimental method was applied to the evaluation of stability of breakwater foundation. The main results obtained are summarized as follows.

1) When a medium-scale tsunami comes after the earthquake occurs (Case 1), the caisson subsides significantly with the liquefaction of the free field due to the ground motion. Despite the fact that the tsunami after the earthquake is medium size,

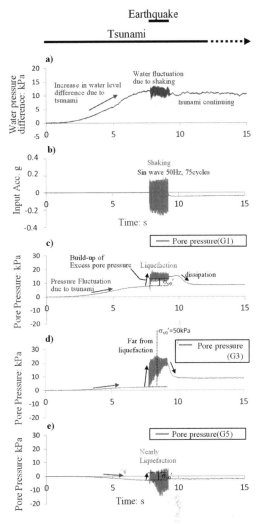

Figure 8. Time history of (a) water difference between the harbour side and the seaward side, (b) input acceleration, and (c, d, e) pore pressure responses (Case 2).

since the caisson was already largely settled by the ground motion, tsunami significantly overflows the caisson. It was shown that it is important to consider countermeasures considering both earthquake and tsunami.

2) Even if the caisson is stable due to the effect of only the tsunami, but in case the earthquake and tsunami come simultaneously (Case 2), it was obtained that the caisson collapsed very easily. When the earthquake and tsunami come simultaneously, it was shown that stability of the breakwater was significantly declined.

REFERENCES

Iai, S., Ichii, K., Liu, H. & Morita, T. 1998. Effective stress analyses of port structures, *Special Issue of Soils and foundations*. 97–114.

Kim, H. & Sekiguchi, H. 2002. Plastic deformation behaviour of composite breakwaters under earthquake shaking, *Physical Modelling in Geotechnics, ICPMG2002*, 587–592.

Matsuda T., Maeda, K., Miyake, M., Miyamoto, J., Sumida, H. & Tsurugasaki, K. 2016. Instability of a Caisson-Type Breakwater Induced by an Earthquake-Tsunami Event, International Journal of Geomechanics, ASCE, Vol. 16-5, (doi: 10.1061/(ASCE) GM.1943-5622.0000619).

Miyake, M. & Sawada, Y. 2010. Development of centrifuge modeling for tsunami and earthquake, Marine Voice 21 Summer 2010, Vol. 270, 20–23. (in Japanese)

Miyamoto, J., Miyake, M., Tsurugasaki, K., Sumida, H., Sawada, Y., Maeda, K. & Matsuda, T. 2015. Instability of Breakwater Foundation during Tsunami in a Drum Centrifuge, *Proceedings of the 25th International Offshore and Polar Engineering Conference, ISOPE2015*, 846–850.

Sugano, T, Nozu, A., Kohama, E. Shimosako, K & Kikuchi, Y. 2014. Damage to coastal structures, *Soils and Foundations*, 54-4, 883–901.

Takahashi H., Sassa, S., Morikawa Y., Takano, D. & Maruyama, K. 2014. Stability of caisson-type breakwater foundation under tsunami-induced seepage. *Soils and Foundations*, 54-4, 798–805.

Tobita, T. & Iai, S. 2014. Combined failure mechanism of geotechnical structures *Physical Modelling in Geotechnics, ICPMG2014*, 99–111.

Tsurugasaki, K., Miyamoto, J., Hem, R., Nakase, H. & Iwamoto T. 2016. Collapse Mechanism of Composite Breakwater under Continuous Tsunami Overflow and its Countermeasure. *Proceedings of the 26th International Offshore and Polar Engineering Conference, ISOPE2016*, 754–760.

Modelling of rocking structures in a centrifuge

I. Pelekis, G.S.P. Madabhushi & M.J. DeJong
Department of Engineering, University of Cambridge, UK

ABSTRACT: Structural rocking and foundation rocking are two design strategies to provide seismic base isolation. Two building models designed to exhibit each rocking behaviour respectively, were tested simultaneously in a centrifuge under earthquake loading. The instrumentation, detailed in this paper, allowed soil and structural deformations to be derived from measured accelerations, and the force demand of the superstructure (e.g. storey drift) to be derived from measured strains and accelerations. Importantly, these measurements enabled direct comparison of the two rocking systems, quantifying relative benefits. In selected earthquakes, foundation rocking caused larger dynamic differential settlements while structural rocking led to larger rocking rotations.

1 INTRODUCTION

1.1 Soil and structures tests with centrifuge

Centrifuge modelling has become increasingly popular for replicating soil-structure interaction during dynamic events (blasts, earthquakes) or quasi-static events (deep excavations, tunneling etc.). As a result, the modelling challenges are shifting towards the mimicking of complex prototype scale structures (Madabhushi 2015). In light of this, new practices and materials are employed to create small scale models matching the properties of prototype structures, such as 3D printing (see Ritter et al. 2017). Monitoring the structural response is also of interest, with visual means (high-speed cameras) and miniature sensors (MEMS accelerometers, strain gauges) typically used in centrifuge tests. This paper presents response parameters from specific earthquake centrifuge tests on soil-building systems, obtained by using MEMS and piezoelectric accelerometers for building and soil deformations, respectively, and strain gauges for axially loaded members. The tests are part of a wider centrifuge campaign to assess the performance of buildings allowed to rock above their foundation level with rotational motion of their superstructure, or alternatively below their foundation level with rocking of their footings on the soil.

1.2 Structural and foundation rocking

Since the 1960's the rocking performance of slender structures (structural rocking) has been studied extensively. Previous studies have focused on small scale models with forced and free rocking on a shaking table, or generally on a strong floor (Housner 1963, Makris & Vassiliou 2014, Acikgoz & DeJong 2016, Acikgoz et al. 2016) with a few exceptions referring to flexible supports for soil simulation (Psycharis & Jennings 1985, Palmeri & Makris 2008, Ma & Butterworth 2012). Overall, structural rocking can prevent earthquake energy from exciting vibrations in the structure, but internal load amplification due to impact at the pivot points might develop, thus exciting higher structural modes (Acikgoz et al. 2016).

On the other hand, testing the response of rocking foundations on soil is a challenge as this would require a massive shake table to enable realistic stressing of the soil across its depth. Therefore, in contrast with structural rocking, foundation rocking has typically been studied with centrifuges assuming rigid body structural response (Gajan & Kutter 2008, Loli et al. 2014). In some cases, rocking foundations, which behave as energy dissipaters, have been combined with plastic hinges in the superstructure's beams, with a kinematic mechanism similar to the typical strong column-weak beam objective dictated by modern seismic codes (Mason et al. 2010, Deng et al. 2011, Gelagoti et al. 2012, Liu et al. 2015). However, energy dissipation from soil alone can be sufficient to achieve a high damping ratio, thus structural plastic hinges might not be necessary (Heron et al. 2015).

Although both structural and foundation rocking have the potential for effective seismic base isolation, they have never been directly compared in experimental testing. This paper details the first application of structural rocking in centrifuge testing, and in parallel, provides comparison with the traditional foundation rocking approach, thus setting the two philosophies side to side in the same test.

2 SELECTED EARTHQUAKE TESTS IN CENTRIFUGE

To assess the benefits of foundation and structural rocking, two building models were designed as identical, with their only difference being the type of support release at the foundation (Figure 1). While model RA

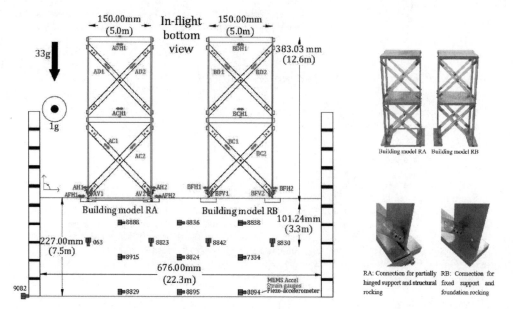

Figure 1. Cross-sectional view of the centrifuge model (left) and detail of the connectivity for building models RA and RB (right).

(Rocking Above the foundation) steps on its footings and rotates about a discrete pivot point, model RB (Rocking Below the foundation) rotates about a pivot point as set by the contact of its footing with the soil (Pelekis et al. 2017). Model RA is also designed to have mounted on its footing an elastoplastic element (fuse), to provide additional energy dissipation to the rocking response.

Results from four tests, each involving a sequence of different earthquakes, are described in this paper. More tests were also run but are not discussed here. Two tests involved dense sand and two tests involved loose sand (Hostun HN31, Table 1). The centrifuge model, consisting of the models RA and RB on dense or loose sand and the instrumentation, are shown in Figure 1. For the first dense or loose sand test (Table 1), the stiffening fuses were included on model RA. For the second test on each sand type (not presented here), the fuses were removed. Six ground motions were applied during each test, including real earthquake time histories, but only the results from a 6-cycle sinusoidal excitation with a driving frequency of 0.91 Hz prototype scale (Table 1) are shown here. Note that the fixed base natural frequency of the model structures was 1.5 Hz.

Figure 2 shows the response of the servo-shaker (Madabhushi et al. 2012) used to fire earthquakes of the same input at the bottom of the centrifuge box, when testing with dense sand (top) and loose sand (bottom).

The response near the free surface (sensor 8836 in Figure 1) is also shown and overall a very similar input acceleration signal was produced for the two tests. Therefore, for the load and deformation response

Table 1. Selected earthquake tests and sand details.

Test	Relative density (%)	Excitation frequency	
		Model Hz	Prototype Hz
DENSE SAND TEST1 EQ4	96	30	0.91
LOOSE SAND TEST1 EQ4	58	30	0.91

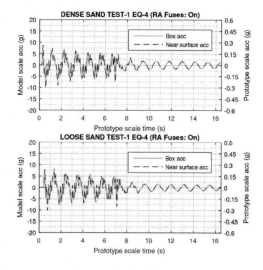

Figure 2. Input acceleration signals.

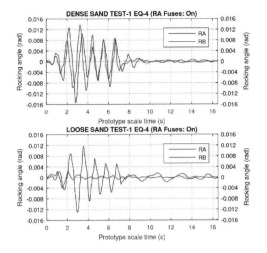

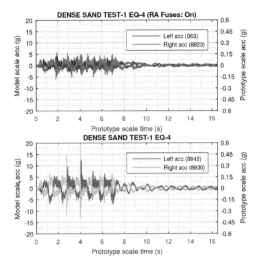

Figure 3. Dynamic rotations of the two building models.

Figure 4. Vertical acc. Below model RA-top and RB-bottom.

parameters compared below, the input is considered practically the same.

3 INDIRECT DEFORMATION RESPONSE

To assess the structural deformation of the two building models and their settlements, typically direct means, such as LVDTs or PIV are used. In this case a set of MEMS accelerometers was used to monitor vertical and horizontal (lateral) accelerations of the building models. Piezoelectric accelerometers were used for monitoring the vertical movement below the footings of each building model, as well as laterally at different layers (Fig. 1).

The angle of rocking θ experienced by the two buildings can be approximated by using the vertical accelerometers placed at the column bases (AV1 and AV2). For example, for building RA:

$$\theta = \iint \ddot{\theta} dtdt = \iint \frac{AV1 - AV2}{2B} dtdt \quad (1)$$

where B is the half-width of the building. Figure 3 shows the response of the two buildings. The signals had their integration offsets filtered out, thus any cumulative rotation was not captured and only the dynamic rotation is shown. It is observed that in both loose and dense sand the buildings behaved initially very similarly. However, in loose sand building RB had its rotational response suppressed, whereas building RA responded nearly the same in dense and loose sand.

Figure 4 shows the acceleration response of the vertical accelerometers in the soil from the dense sand test. The response has an out of phase profile and to assess the extent to which the different types of rocking influence the soil at that depth, double integration of the vertical accelerometers was used to obtain approximately the corresponding dynamic vertical

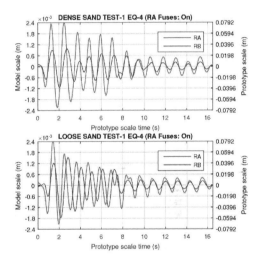

Figure 5. Differential displacement as obtained from the vertical accelerometers in the soil.

displacements and finally the dynamic differential settlements (Fig. 5).

Building model RB was seen to have its dynamic differential settlement attenuated with loose sand, a trend which followed the suppressed rotations in Figure 3. Similarly, the dynamic differential settlements of model RA mirrored its rocking response in both loose and dense sand tests. Overall, model RB appeared to cause larger deformations in the soil than model RA based on Figure 5.

To obtain the interstory drift ratios, the MEMS accelerometers placed at the building slabs and column ends were used (Fig. 1). When large rocking develops, the rotational acceleration of the building needs to be

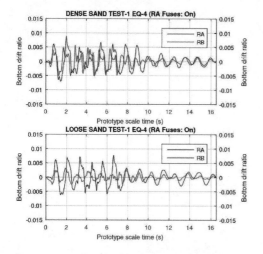
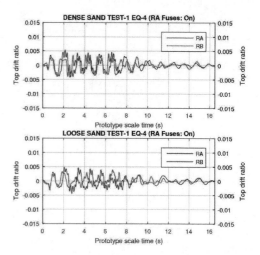

Figure 6. Bottom storey drift ratio from model RA and RB.

Figure 7. Top storey drift ratios from models RA and RB.

considered (Acikgoz 2014), therefore the interstory drifts are obtained by the following equation:

$$u_{r,n+1} - u_{r,n} = \iint (MEMS_{n+1} - MEMS_n) - (H_{n+1} - H_n)\ddot{\theta} dt dt \quad (2)$$

where $n = 0, 1$, with the elevations H_n considered as the column bottom ends ($n = 0$), the bottom slab ($n = 1$), and the top slab ($n = 2$). The interstorey drift ratios are defined as the interstorey drift over the storey height. Figures 6-7 show the bottom storey drift ratios and the top storey drift ratios, respectively. The response of model RA is very similar for both loose sand and dense sand. In contrast, model RB experienced significantly smaller interstorey deformations for loose sand than for dense sand.

For the dense sand test, drift ratios for RA and RB were relatively similar, though slightly higher for RA structures. On the contrary, for loose sand, RB structures clearly experienced smaller drifts. The results indicate that the loose sand significantly reduced the rocking rotation (i.e. uplift) of the structure, probably due to reduced bearing capacity causing local failure at the foundation edges during rotation. Additionally, although the loose sand did not significantly reduce the propagation of the input motion through soil (Figure 2), it did effectively provide a softer soil beneath the building, which might have led to reduced drifts once rocking did not occur. Meanwhile, model RA deformed more, both in terms of storey deformations and angle of rocking, in exchange for smaller differential soil displacements.

4 DIRECT LOAD RESPONSE

The design of the building models ensured that the bracing elements did not experience buckling or material failure and that their connections remained intact throughout the centrifuge experiments. Additionally, a low stiffness was necessary to tune the natural frequency of the buildings (60 metric tonnes prototype total mass), so that the latter represent prototypes with 3-4 storeys. To meet these requirements, the polyester PETG was chosen with tubular sections. PETG is a derivative from polyethylene terephthalate (PET), produced after the partial addition of cyclohexane-1.4 – dimethanol and is typically used for plastic sheet or where high clarity parts are required (Focke et al. 2009, Chen et al. 2011).

Table 2. Bracing cross sections with nominal dimensions.

Story	Outside diameter mm	Inside diameter mm
TOP	11.3	10.7
BOTTOM	14.5	13.4

To assess the capacity and stiffness performance of the bracing members with their connections (see Table 2 for cross sections) before the centrifuge tests and thus verify their design, a series of slow cyclic tests was performed on specimens, involving both a varying amplitude protocol (FEMA 461 2007) and a steady amplitude protocol. Although the former was developed for non-structural elements, it also applies to drift sensitive structural members (Krawinkler 2009), such as this bracing system. Figure 8 shows a typical cyclic testing of the bottom bracing elements, with safety factored design resistances and expected design actions in model scale and with no stiffness degradation being evident.

Next, the bracing elements of one side of the building models were instrumented with strain gauges to monitor their axial load. The calibration was carried while the bracing elements were removed from the building models. This instrumentation, combined with a very small column bending stiffness, allowed the direct measurement of storey shear, although in terms of axial compression and tension. For the top storeys,

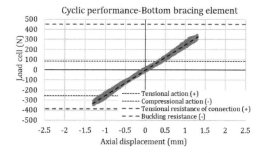

Figure 8. Performance of a bottom bracing member in slow cyclic tests (+. Tension, −. Compression).

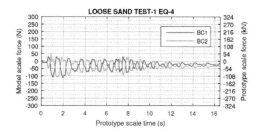

Figure 9. Bottom bracing earthquake response of model RB.

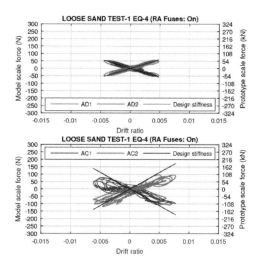

Figure 10. Top bracing earthquake response (top) and bottom bracing earthquake response (bottom), model RA.

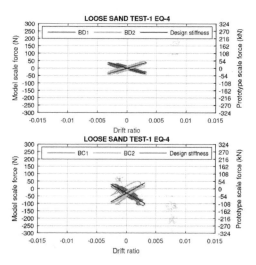

Figure 11. Top bracing earthquake response (top) and bottom bracing earthquake response (bottom), model RB.

the prototype design lateral stiffness is 7.9 MN/m with 85% provided by the bracing members. Regarding the bottom storeys, the same figures account for 15.9 MN/m and 98% for RA, while for RB 16.5 MN/m and 85%. Figure 9 shows as an example the time-history response of the bottom bracings of model RB for the loose sand test. Generally, it was observed that offset values of loading occurred at the beginning and end of a given earthquake for both models across all tests. These offsets were taken into account sequentially for a following earthquake. Their existence was the result of a new state of equilibrium for the soil-structures. Since the time-history response is well within the boundaries set by the linear elastic performance under cyclic testing (Fig. 8), it was concluded that these offsets were not due to plastic deformation.

The bracing load response with the offset was plotted against the obtained interstorey drift ratios from the MEMS accelerometers (Figs 10-11). The design axial stiffness verified in the slow cyclic tests is also included in the graphs. Generally, the centrifuge response of model RB matched the design stiffness for both storey bracings. Regarding model RA, there was a mismatch of stiffness in the bottom bracing only, due to the cross sections being smaller than the nominal dimensions and this trend was found in other tests too, yet the response remained linear elastic. Overall, any non-linear effects from the superstructures of the models RA and RB were minimal. Therefore, any type of rocking and soil deformation were the only potential non-linear phenomena to develop and be observed clearly, as planned from the initial design of the building models.

5 CONCLUSIONS

The soil-structure interaction of two building models with different footing-column connections was investigated with reference to low frequency input motions in a centrifuge. The assessment was based on miniature accelerometers and strain gauges; residual or cumulative deformations were not measured. The linear elastic design of the two models, representative of structural and foundation rocking, was validated, thus allowing their relative differences to manifest during centrifuge testing. Structural rocking led to larger rocking rotations and interstorey deformations, with these response parameters being similar across loose and dense sand tests. Foundation rocking led to smaller rocking rotations and interstorey deformations as opposed to structural rocking, and even smaller in loose sand. However, foundation rocking also led to greater dynamic differential settlement beneath the footings. Finally, a wider series of input motions is required to assess the rocking performances across different driving frequencies and real earthquake records and this is part of a larger centrifuge campaign.

REFERENCES

Acikgoz, M.S. 2014. *Seismic assessment of flexible rocking structures*, Ph.D Thesis, University of Cambridge, UK.

Acikgoz, S. & DeJong, M.J. 2016. Analytical modelling of multi-mass flexible rocking structures. *Earthquake Engineering & Structural Dynamics*, 45(13): 2103–2122.

Acikgoz, S., Ma, Q., Palermo, A., and DeJong, M.J. 2016. Experimental Identification of the Dynamic Characteristics of a Flexible Rocking Structure. *Journal of Earthquake Engineering*, 20(8): 1199–1221.

Chen, L., Zhang, X., Li, H., Li, B., Wang, K., Zhang, Q., and Fu, Q. 2011. Superior tensile extensibility of PETG/PC amorphous blends induced via uniaxial stretching. *Chinese Journal of Polymer Science*, 29(1): 125–132.

Deng, L., Kutter, B.L., and Kunnath, S.K. 2011. Centrifuge modeling of bridge systems designed for rocking foundations. *Journal of Geotechnical and Geoenvironmental Engineering*, 138(3): 335–344.

FEMA 461. 2007. *Interim testing protocols for determining the seismic performance characteristics of structural and nonstructural components*. Redwood City, California.

Focke, W.W., Joseph, S., Grimbeek, J., Summers, G.J., and Kretzschmar, B. 2009. Mechanical properties of ternary blends of ABS + HIPS + PETG. *Polymer-Plastics Technology and Engineering*, 48(8): 814–820.

Gajan, S. and Kutter, B.L. 2008. Capacity, settlement, and energy dissipation of shallow footings subjected to rocking. *Journal of Geotechnical and Geoenvironmental Engineering*, 134(8): 1129–1141.

Gelagoti, F., Kourkoulis, R., Anastasopoulos, I., and Gazetas, G. 2012. Rocking-isolated frame structures: Margins of safety against toppling collapse and simplified design approach. *Soil Dynamics and Earthquake Engineering*, 32(1): 87–102.

Heron, C.M., Haigh, S.K., and Madabhushi, S.P.G. 2015. A new macro-element model encapsulating the dynamic moment–rotation behaviour of raft foundations. *Géotechnique*, 65(5): 442–451.

Housner, G.W. 1963. The behaviour of inverted pendulum structures during earthquakes. *Bulletin of Seismological Society of America*, 53(2): 403–417.

Krawinkler, H. 2009. Loading histories for cyclic tests in support of performance assessment of structural components. *In: Third international conference in experimental earthquake engineering*. San Francisco.

Liu, W., Hutchinson, T.C., Gavras, A.G., Kutter, B.L., and Hakhamaneshi, M. 2015. Seismic Behaviour of Frame-Wall-Rocking Foundation Systems. I: Test Program and Slow Cyclic Results. *Journal of Structural Engineering*, 141(12): 4015059-1-4015059-12.

Loli, M., Knappett, J.A., Brown, M.J., Anastasopoulos, I., and Gazetas, G. 2014. Centrifuge modeling of rocking-isolated inelastic RC bridge piers. *Earthquake Engineering & Structural Dynamics*, 43(15): 2341–2359.

Ma, Q.T. and Butterworth, J.W. 2012. Simplified expressions for modelling rigid rocking structures on two-spring foundations. *Bulletin of the New Zealand Society for Earthquake Engineering*, 45(1): 31–39.

Madabhushi, S.P.G. 2015. Session report: physical modelling in geotechnical earthquake engineering. *International Journal of Physical Modelling in Geotechnics*, 15(2): 91–97.

Madabhushi, S.P.G., Haigh, S.K., Houghton, N.E., and Gould, E. 2012. Development of a servo-hydraulic earthquake actuator for the Cambridge Turner beam centrifuge. *International Journal of Physical Modelling in Geotechnics*, 12(2): 77–88.

Makris, N. and Vassiliou, M.F. 2014. Are Some Top-Heavy Structures More Stable? *Journal of Structural Engineering*, 140(5): 6014001-1-06014001-5.

Mason, H.B., Trombetta, N.W., Gille, N.W., Lund, J.N., Zupan, J.D., Puangnak, H., Choy, B.Y., Chen, Z., Bolisetti, C., Bray, J.D., Hutchinson, T.C., Fiegel, G.L., Kutter, B.L., and Whittaker, A.S. 2010. *Seismic performance assessment in dense urban environments: centrifuge data report for HBM02*. Davis, California.

Palmeri, A. and Makris, N. 2008. Response analysis of rigid structures rocking on viscoelastic foundation. *Earthquake Engineering and Structural Dynamics*, 37(7): 1039–1063.

Pelekis, I., Madabhushi, G.S.P., and DeJong, M.J. 2017. A Centrifuge Investigation of two different Soil-Structure Systems with Rocking and Sliding on Dense Sand. *In: M. Papadrakakis and M. Fragiadakis, eds. 6th ECCOMAS Thematic Conference on Computational Methods in Structural Dynamics and Earthquake Engineering*. Rhodes Island, Greece, 15–17.

Psycharis, I.N. and Jennings, C. 1985. Upthrow of objects due to horizontal impulse excitation. *Bulletin of Seismological Society of America*, 75(2): 543–561.

Ritter, S., Giardina, G., DeJong, M.J., and Mair, R.J. 2017. Centrifuge modelling of building response to tunnel excavation. *International Journal of Physical Modelling in Geotechnics*.

A new test setup for studying sand behaviour inside an immersed tunnel joint gap

R. Rahadian
Delft University of Technology, Delft, The Netherlands

S. van der Woude
Van Hattum en Blankevoort, Woerden, The Netherlands

D. Wilschut
Municipality of Rotterdam, The Netherlands

C.B.M. Blom & W. Broere
Delft University of Technology, Delft, The Netherlands

ABSTRACT: During inspections of several immersed tunnels in the Netherlands, damage of immersion joints has been observed. In some cases the Gina seal has moved inwards from its original location, and in other cases a permanent elongation of the entire tunnel structure has been measured. For both cases it has been hypothesised that a seasonal expansion and contraction of the tunnel elements allows sand to enter the joint gap between elements during winter, where it is compacted during summer, leading to an increasing amount of sand in the joint gap over the years. In order to study this mechanism and assess its impact, a 1:3 scale model joint gap has been designed and constructed. This setup can simulate expansion and contraction cycles of the joint and measure stresses in the joint gap and deformations of the Gina seal. First test results are presented here and show that compaction of the sand entering the joint gap indeed occurs and leads to the observed large inwards deformations of the Gina seals.

1 INTRODUCTION

Immersed tunnels are constructed from several elements which are immersed next to each other and joined together by making an immersion joint, which forms the temporary and permanent watertight connection between elements (Lunniss and Baber 2013). The most common immersed tunnel design uses thick-walled concrete shells for the tunnel construction and a double rubber seal to waterproof the immersion joint. The roughly trapezoid shaped Gina seal consists of 2 main parts, the stiff base and the soft nose that will compress once the elements are lined up to one another. It is bolted to the concrete shell of the tunnel on one side of the joint and compressed during the immersion procedure. According to the immersed tunnel design guidelines this serves as a temporary seal (COB Commissie T202 2015). The Omega seal is a curved rubber strip placed around the inner side of the immersed joint after the water between the bulkheads have been drained. It functions as the secondary measure against leakage and is intended as the permanent seal. See Figure 1 for a sketch of the immersion procedure and Figure 2 for a detail of the joint and seals.

The Gina gasket compresses due to the water pressure at the far bulkhead of the element during

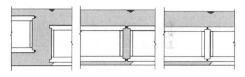

Figure 1. Construction procedure of immersion joint.

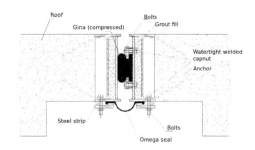

Figure 2. Gina and Omega seal solution in an immersed tunnel joint.

immersion, but a gap between the end frames remains, as shown in Figure 2. The gap allows for water and soil to enter after the soil cover has been backfilled on top of the tunnel.

During regular inspections of several immersed tunnels in the Netherlands damage has been observed at immersion joints, and in some occasions an increasing elongation of the entire tunnel has been observed over the years. It has also been observed that the tunnel elements expand during summer and contract during winter, resulting in a seasonal contraction-expansion of the immersion joint (Berkhout 2015). These observations have led to the hypothesis that during expansion soil enters into the gap and is compressed during joint contraction. The seasonal contraction-expansion may lead to failure of the soil mass in the joint, densifying the soil in the joint gap and applying an increasing pressure on the outside of the Gina seal.

The Gina seal is not intended to withstand high loads on the outside. In one occasion it was observed that the Gina gaskets punched through the inner side of the tunnel joint and damaged the Omega gaskets. This not only implies an increased pressure on the outside of the Gina, but also failure of the bolts by which it is attached to the tunnel structure, and a continued inflow of soil into the joint gap as the Gina gradually moves downward. As finite element simulations were unable to prove the exact failure mechanism occurring in the joint gap and reliably estimate the pressure exerted by the soil on the Gina, a scale model of the joint gap has been designed to study the behaviour of the Gina during seasonal compression-expansion cycles.

2 EXPERIMENTAL SETUP

2.1 Design considerations and test setup

The test setup is intended as a physical model of the outer side of the immersed joint gap. It is divided into 2 main parts:

- the top part (a) models the soil column outside of the joint gap. The soil column is put inside a tubular drum container with an opening at the bottom that connects it to the bottom part. The stress conditions in the drum should be controlled to resemble the overburden above the tunnel.
- the bottom part (b) models the immersed joint gap, which is rectangular shaped. This section houses the Gina profile pushed against the tunnel structure and a mechanical jack that allows it to move in a cyclic motion, to simulate expansion-contraction throughout the seasons.

The entire setup is contained in a rectangular metal frame with two rings bolted to the sides around mid height at the center of gravity. These rings allow the entire setup to be lifted by a crane and flipped over.

After considering scaling effects and the expected forces in the model setup, a model scale of 1:3 has been chosen. A side view sketch of the equipment is shown in Figure 3, which shows the soil drum containing soil and an inflatable bladder that can be filled with air to simulate a vertical overburden pressure, the fixed wall on the left to which the Gina profile is attached and the movable wall on the right. The latter is attached to a

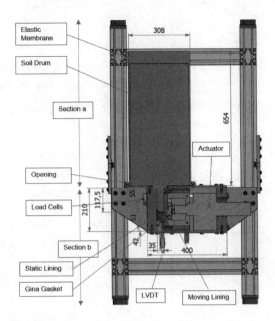

Figure 3. Cross-section of the test equipment.

mechanical jack (Madler 50 kN), which is used to close the joint gap and compress the Gina in its initial position and simulate the seasonal contraction-expansion. In addition two load cells (Burster type 8523-100) are mounted on the fixed wall with modified 10 mm diameter load plates positioned flush with the joint wall and in contact with the soil. A LVDT (ELE 10) is placed on the bottom side of the Gina seal.

The top edges of the lining are slightly curves, to mimic the actual tunnel structure. This small detail might further facilitate the densification process as the soil easily intrudes into the gap when the joint expands.

In addition, there are two 7 mm diameter openings in the opposing side walls of the joint gap (perpendicular to this cross-section) at the center height of the load cells, which allow a pocket penetrometer access to the soil in the gap.

2.2 Model Gina profile

The model Gina gasket (Figure 4) used in this experiment is supplied by Trelleborg, the manufactrer of most Gina seals used in immersion projects. The gasket is molded out of a similar type of material compared to the original gasket, albeit with lower elastic modulus to compensate for the limitations of the actuator. The elastic modulus of the gasket is at 1.0 MPa, whereas the a Gina profile made in the regular Sh-50 quality would have a 2.2 MPa elastic modulus.

However, as the Gina gasket is made out of similar rubber material, the Poissons ratio of the model gasket is comparable to the prototype at $v = 0.498$. The high Poissons ratio implies that the gasket is nearly incompressible, and any axial compression is translated into lateral extension.

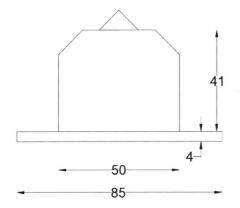

Figure 4. Model Gina seal.

Figure 5. Modified pocket penetrometer.

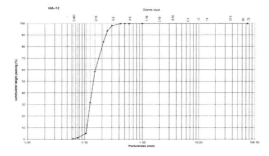

Figure 6. Grain size distribution.

2.3 Modified penetrometer

Various authors have correlated penetrometer resistance to soil density. A commonly used relationship is Kulhawy & Mayne (1990), which takes into account the overconsolidation ratio of the soil. The relative density is derived from

$$D_r^2 = \frac{Q_{cn}}{305 Q_C Q_{OCR}} \qquad (1)$$

where

$$Q_{cn} = \frac{\left(\frac{q_c}{p_a}\right)}{\sqrt{\left(\frac{\sigma_v'}{p_a}\right)}} \quad ; \quad Q_{OCR} = OCR^{0.18} \qquad (2)$$

and Q_C is a compressibility factor, which is 0.91 for high, 1.0 for medium and 1.09 for low compressible soils.

Jamiolkowski (2001) modified the relationship from Kulhawy & Mayne (1990) to

$$D_r(\%) = 26.8 \ln Q_{cn} - b_x \qquad (3)$$

where b_x takes values of 52.2, 67.5 and 82.5 for high, medium and low compressible soils respectively.

Schmertmann (1976) obtained a different correlation

$$D_r = \frac{100}{C_2} \ln \frac{q_c}{C_0 (\sigma_{v0}')^{C_1}} \qquad (4)$$

where suggested values for a number of normally consolidated sand are $C_0 = 0.050$, $C_1 = 0.700$ and $C_2 = 2.91$ (Sandrekarimi 2016).

These relations are based on results from cone penetrometer tests, but the size of the model joint does not allow the use of a full-scale 10 cm^2 CPT cone. Instead a scaled penetrometer will be used, based on a modified Eijkelkamp 500 kPa pocket penetrometer. This device has a rod diameter of 0.25" which is small enough to be used in the joint gap. However, in its unmodified state, the effective penetration depth is only 10 mm. Therefore, the rod is extended to provide a 70 mm effective penetration depth, as shown in Figure 5.

The pocket penetrometer only records the cone resistance q_c, as the sleeve area is considered to be too small to contribute to friction, and the device itself usually lack the means of measuring friction resistance. Due to the extended probe length, some friction resistance is may contribute to the readings, but this is not corrected for.

A further point is that the penetrometer is normally used in a vertical penetration direction. In this setup, however, the penetrometer will be inserted horizontally. Research on horizontal cone penetration conducted by Broere & van Tol (1998) and Broere (2001) produced a relationship between vertical and horizontal cone resistances, indicating that depending on the sand density, a deviation between horizontal and vertical CPT readings might occur. Based on cavity expansion theory, a ration of $q_{c,hor}/q_{c,ver}$ up to 1.5 might occur for medium dense sand with $K = 0.5$. This correction for the horizontal orientation has been included in the derivation of relative densities in Table 2.

2.4 Sand properties

The sand used in the tests is a mixture of Geba Weiss (a fine sand with D_{50} of 100 μm) and medium sized fractions of Maas river sand (sieve openings between 125 and 250 μm) in a ratio of 80/20, to obtain a distribution closely resembling field conditions, see Figure 6.

The resulting sand has a dry volumetric weight γ_d of 17 kN/m^3 and minimum and maximum void ratio have been determined as 0.548 and 0.929 respectively (Elmi Anaraki 2008).

2.5 Test procedure

The experimental procedure consist of a preparation and execution stage. During the preparation phase the

Table 1. Summary of test series.

Test	Test Code	Overburden (kPa)	Stroke (mm)	No. of Cycles
Calibration Tests				
Zero cycle	TC1-XX	68	0	0
Stepwise penetrometer	TC2-XX	68	3.5	50
Sandless test	TC3-YY-XX	0	Varies	25
Main Series				
Reference (1.5 mm)	TVS15-XX	68	1.5	25
0.5 mm stroke	TVS05-XX	68	0.5	75
1.0 mm stroke	TVS10-XX	68	1.0	50
2.5 mm stroke	TVS25-XX	68	2.5	20
3.5 mm stroke	TVS35-XX	68	3.5	10

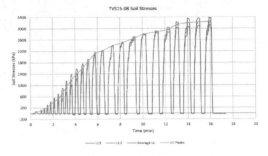

Figure 7. Soil stresses measured in TVS15-08.

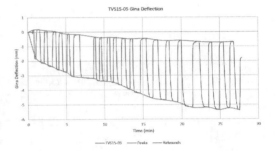

Figure 8. Gina gasket deflection during TVS15-05.

entire setup is initially in a flipped position, with the soil drum on the bottom side. The air bladder in the drum is not pressurized. The entire setup is rattled while suspended to ensure no soil remains in the joint gap and the soil in the drum is in a loose state. After this the joint gap is closed to a 35 mm gap size. Next, the entire setup is gently flipped over, so the soil drum is in top position, and lowered to the ground. Finally, the air bladder is inflated to the desired pressure of 58 kPa. Including the weight of the soil column this results in a 68 kPa overburden pressure at the top of the joint gap.

Prior to starting the actual test, a penetrometer reading is taken at one side of the joint. The cyclic deformation of the joint is started by a half-range outward (expansion) of the joint, followed by the desired number of full range compression-expansion cycles. Normally, at the end of the full series of loading-unloading cycles, a second penetrometer reading is taken, at the other side, the side not previously disturbed by the first penetrometer reading.

In addition to this regular test schedule, a number of tests have been performed for calibration purposes. These include tests where the penetrometer readings were taken at both sides, without performing any loading-unloading cycles at all. These tests serve to show the resistance of the newly loosened sand after each test setup reset.

In a different calibration test, penetrometer readings were taken after each 5 loading-unloading cycles. During each penetrometer measurement, the probe is inserted into the soil to a depth of 7 cm. This action might locally disturb the soil inside the joint gap and provide false readings if subsequent tests are made at the same side without resetting the test setup.

Finally, a series of tests has been performed without soil in the setup. These tests serve to investigate the deflection of the Gina during subsequent compression-extension cycles, in order to verify that without a soil load on the outside, the Gina expands downwards during compression due to its near incompressibility, and fully bounced back during unloading.

The expansion and contraction cycle of the joint gaps depends on the tunnel material, the length of each element and the seasonal temperature variation. The prototype tunnel joints contract and expand a maximum of 10 mm throughout the years. In the model would this translates to max. 3.5 mm deformation cycles. However, as the capacity of the screw jack is limited, the reference stroke length used will be 1.5 mm for 50 cycles. In test variations 0.5, 1.0, 1.5, 2.5 and 3.5 mm will be used, although it proved to be necessary to limit the number of loading cycles at larger stroke lengths due to the limitations of the screw jack. This results in the test configurations listed in Table 1.

3 RESULTS

Graphs for some test results are shown below. Figure 7 shows the load cell measurements for TVS15-08. The results from both load cell, indicated by LC1 and LC2 are averaged and a line connecting the subsequent peak stresses at the end of each compression stage. The increase of the stress in the joint after each cycle is visible. The rate of stress increase drops off in later cycles, but no clear limit is reached during the first 27 cycles.

Figure 8 shows the downward deflection and rebound of the Gina during TVS15-05. A line is included that connects the peak downward deformation at the end of each compression cycle, as well as a line connection the remaining deformation after rebound at the end of each extension cycle. It is clear that the Gina gasket moves increasingly downwards at the end of each compression cycle, and does not fully rebound to its initial position after relaxation.

Figure 9 shows the results for sandless control test TC3-15-01, which has the same 1.5 mm stroke length

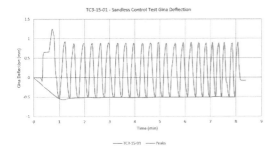

Figure 9. Gina gasket deflection during TC3-15-01 (1.5 mm deflection sandless control test).

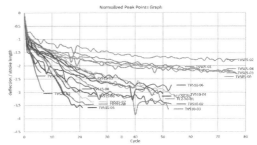

Figure 11. Normalized peak deflection points for all tests.

Table 2. Relative densties derived from test results.

Test	No. of cycles	D_r (%) Start[1]	End[1]	End[2]
TVS05-01	56	4.7	29.3	43.2
TVS05-02	77	3.5	23.9	37.8
TVS05-03	77	4.7	27.1	41.0
TVS05-04	74	0	26.3	40.2
TVS05-05	77	0	23.0	36.9
TVS10-02	53	0	25.0	38.9
TVS10-03	52	4.7	22.4	36.3
TVS10-04	54	0	23.4	37.3
TVS10-05	47	0	24.6	38.5
TVS10-06	53	0	24.4	38.3
TVS15-01	5			
TVS15-02	20			
TVS15-03	22			
TVS15-04	20	0.3	30.4	44.3
TVS15-05	21	0	27.1	41.0
TVS15-06	26	0	18.3	32.2
TVS15-07	25	0	17.4	31.3
TVS15-08	25	0	14.0	27.9
TVS25-01	7	0	17.4	31.3
TVS25-02	7	0	14.0	27.9

[1] Lower estimate: corrected for horizontal test direction.
[2] Upper estimate: not corrected for stress conditions.

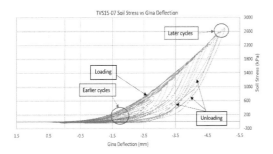

Figure 10. Soil stresses vs. Gina deflection for TVS15-07.

as TVS15-05. It can be observed that without the influence of the sand, both the gasket deflection peak and rebound points remains relatively constant throughout the test despite a too large unloading stroke in the first cycle and small inaccuracies in subsequent strokes. Comparing the gasket deflection values of both graphs shows that the sand has a significant effect on the overall gasket deflection.

Figure 10 plots the averaged soil stress (measured in the horizontal direction) versus the vertical deflection of the Gina gasket over the entire loading cycle. It can be observed that during the loading part of every cycle, the system follows a certain gradient. However, hysteretic behaviour is present during unloading, where the stress relief is much more rapid compared to the rebound of the Gina gasket. The hysteretic behaviour is more apparent at later cycles and at higher gasket deflections, with more pronounced lagging of gasket rebound compared to the stress decrease.

Figure 11 shows the peak deflection point graphs for every configuration normalized with their respective stroke lengths. It can be observed that the deflection results for equal stroke lengths are similar. This shows the reliability of the test setup, and signifies that the soil in the joint gap demonstrates similar behaviour at similar load conditions.

However, observing results for different stroke lengths reveals that the gradient of the peak points lines differ from each other. The curve becomes steeper as the stroke length increases. This suggests that the soil undergoes escalating plastic deformations, which allows more soil to enter the joint gap, increasing the soil density in the gap and propagating the effect.

Figure 12 visualises the penetrometer test results shown in Table 2. Results with similar stroke lengths were grouped and at rend line was drawn. Similar to the stress increase discussed above, the trend line results also show an increase in gradient with increasing stroke lengths. This signifies a greater rate of densification at higher strains. Even for the highest penetrometer resistances plotted here, the derived relative density remains at 30%, which would be categorized as a medium low to medium dense sand. This is, however, after the final relaxation cycle, which still means that overall significant and permanent densification has taken place. Also, the results of TVS05 and TVS10 tests show that the cone resistance values continue to increase well after the 50th cycle. Results from a limited number of penetrometer readings at minimal joint width, or maximum compaction, are not reported here, as all readings exceeded the 500 kPa limit of the pocket penetrometer, indicating effectively a relative density above 80% during maximum compaction.

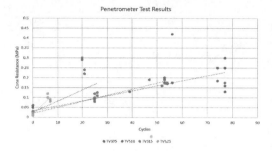

Figure 12. Penetrometer test results.

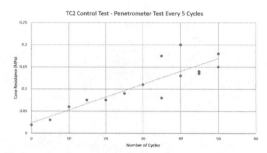

Figure 13. Stepwsise penetrometer tests (TC2-XX) results.

Figure 13 plots the TC2 test results with a trend line drawn through. It is clear that cone resistance, and thereby relative density, steadily increases with each cycle.

4 CONCLUSIONS

The results of a series of compression-extension experiments of a model immersed tunnel joint gap show that the cyclic, seasonal, movement of the tunnel segments significantly increases the soil stresses in the joint. This increase in soil stresses occurs due to overlying sand entering the joint gap, where it is subsequently compacted. Penetrometer test results confirm that the sand in the joint gap increases in cone resistance, and therefore density, after subsequent loading and unloading cycle.

During each deformation cycle, the soil stresses in the joint are observed to increase. Also a cyclic deformation of the Gina gasket towards the inside of the tunnel is observed from LVDT readings, showing that these increasing soil stresses are also exerted onto the Gina gasket. However, during unloading the Gina gasket does not fully rebound, resulting in a gradual inward movement of the Gina, that can lead to its ultimate failure.

ACKNOWLEDGMENTS

The authors would like to express their gratitude to S. van Herk, C. van Beek, R. van Leeuwen and J.J. de Visser for their assistance during the design and construction of the setup, to J. van Stee and Trelleborg Netherlands BV for manufacturing a custom scaled Gina profile, and to the Municipality of Rotterdam for financial support of the first Author during this project.

REFERENCES

Berkhout, B. (2015). Instandhouding zinkvoegen. Technical report, COB.

Broere, W. (2001). *Tunnel Face Stability & New CPT Applications*. Ph. D. thesis, Delft University of Technology, Delft.

Broere, W. & A. van Tol (1998). Horizontal cone penetration testing. In P. Robertson and P. Mayne (Eds.), *Geotechnical Site Characterization, Proc. ISC'98*, pp. 989–994. Balkema.

COB Commissie T202 (2015). Handboek tunnelbouw. Technical report, COB.

Elmi Anaraki, K. (2008). Hypoplasticity investigated: Parameter determination and numerical simulation. Master's thesis.

Jamiolkowski, M. (2001). Evaluation of relative density and shear strength of sand from cpt and dmt. In D. Lo Presti and M. Manassero (Eds.), *LADD Symposium*.

Kulhawy, F. & P. Mayne (1990). Manual on estimating soil properties for foundation design. Technical Report EPRI-EL-6800, Electric Power Research Inst., Palo Alto, CA (USA).

Lunniss, R. & J. Baber (2013). *Immersed Tunnels*. CRC Press.

Sandrekarimi, A. (2016). Estimating relative density of sand with cone penetration test. *Ground Improvement 169*(4), 253–263.

Schmertmann, J. (1976). An updated correlation between relative density d_r and fugro-type electric cone bearing, q_c. Technical Report DACW 39-76, M 6646 WES, Vicksburg, Miss. (USA).

3D printing of masonry structures for centrifuge modelling

S. Ritter & M.J. DeJong
Department of Engineering, University of Cambridge, Cambridge, UK

G. Giardina
Department of Architecture and Civil Engineering, University of Bath, Bath, UK

R.J. Mair
Department of Engineering, University of Cambridge, Cambridge, UK

ABSTRACT: Centrifuge modelling necessitates large scale factors due to space and payload limitations. Hence, replicating details of a prototype is difficult. This is particularly true for masonry buildings with highly non-linear material properties and building features that affect the structural behaviour. This paper discusses powder based 3D printing to replicate masonry structures in centrifuge models. Four-point-bending tests determined the mechanical properties of the 3D printed material. Results reveal a variation in material properties with position and orientation of the 3D printed object in the print bed. After restricting the position of the model in the print bed, repeatable material properties with lower stiffness and higher strength than typical masonry were observed. However, building layout and window opening percentage could be adjusted to create building models with overall bending and axial stiffness typically obtained in the field. These improved 3D printed scale models were subsequently used in centrifuge tests exploring building response to tunnel subsidence. Results show that powder based 3D printed models provide a level of detail not previously simulated in the centrifuge, unlocking new information regarding this soil–structure interaction problem.

1 INTRODUCTION

In the last 20 years, centrifuge modelling research has been performed to study the response of building models to tunnelling works (Taylor and Grant 1998, Caporaletti et al. 2005, Farrell 2010, Taylor and Yip 2001). Although these investigations highlighted important mechanisms governing this soil–structure interaction problem, much of the research has been limited by simple structural models. This limitation is mainly caused by the necessity to employ large scale factors, often between 1:20 and 1:100, when studying soil–structure interaction phenomena in a geotechnical centrifuge (Knappett et al. 2011).

Recent developments in rapid prototyping (RP) technologies have opened the door to an array of applications in civil engineering research (DeJong and Vibert 2012, Feng et al. 2015, Liang et al. 2014, Liang et al. 2015). Specifically, detailed small-scale models that replicate structurally important features of the prototype can be fabricated. This is crucial when studying the response of surface structures to tunnelling-induced ground movements.

Appropriate mechanical properties of the material used to replicate a prototype structure are vital to realistically model the deformation and strength behaviour of structures in contact with the soil (Knappett et al. 2010). Current procedures to assess building response to tunnel excavation are based on relations between the soil and structure stiffness (Potts and Addenbrooke 1997, Franzius et al. 2006, Son and Cording 2005, Goh and Mair 2011). Consequently, the main objective of this research is to realistically model the global stiffness of surface structures subject to tunnelling-induced ground movement. In the light of latest generations of tunnel boring machines, which often induce very small soil displacements and distortions of nearby structures, the initial building stiffness is a key parameter governing this tunnel-soil–structure interaction problem. Nevertheless, as pointed out by Knappett et al. (2011) the scaling of strength properties is crucial when studying collapse mechanisms of soil–structure interaction phenomena. This is particularly relevant for historic masonry structures where cracking could occur at small differential displacements.

This work discusses the 3D printing of 2-storey surface structures on shallow foundations at a scale of 1:75, which are subsequently exposed to tunnelling-induced ground displacements at $75g$. The aim of these 3D printed structures is to model realistic building characteristics including front, rear, end and intermediate walls, window openings and strip foundations while realistically modelling the axial, EA, and bending stiffness, EI of the building. Therefore, the used approach carefully balances the Young's modulus, the moment of inertia and the area of the reduced

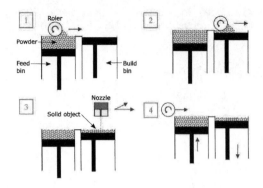

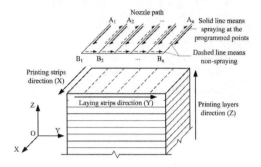

Figure 1. Overview of the 3DP procedure: (1 and 2) roller places a layer of powder, (3) inkjet head prints binder on powder and (4) steps 2 and 3 are repeated until entire 3D structure is created and loose power is removed (adapted from Butscher et al., 2011).

Figure 2. Coordinate system adopted for the 3D printing procedure (after Feng et al., 2015).

scale model. The used 3D printing technique is first introduced, after which the preparation of the 3D printed building models for the centrifuge model tests is described. Then, the mechanical properties of the 3D printed material are discussed and initial observations of an experimental programme to improve the strength properties of the 3D printed material are presented. Finally, example results of series of centrifuge model tests exemplify the value of using 3D printed models to study building response to tunnelling.

2 3D PRINTING PROCEDURE

The surface structures were printed on a Z Corporation Zprinter350. This rapid prototyping technology can be classified as a Three-Dimensional Printing (3DP) technique using a cementitious powder (Feng et al. 2015). Figure 1 depicts the main steps involved in the 3DP process.

Within this research, the following 3DP process was adopted: Firstly, the 3D printed building and model and specimen were first created using computer aided design (CAD) software (i.e. AutoCAD 2015). Secondly, the CAD model was saved as a standard triangulation language (STL) file, which converts the solid sections of the CAD model into numerous thin digital layers. Thirdly, the printing process was started by transmitting the STL file to the 3D printer and subsequently the 3D printer prints each layer atop another. Finally, after the printing process was finished and sufficient initial curing time was allowed, the printed components were removed from the powder bed and an air nozzle was used to remove the remaining powder from the printed parts.

For this research the 3D printed building models had to be printed in parts; the process is described below. Each part of the building model required about 7 hours of printing. Additional time was necessary for initial curing (about 3 hours for the 3D printed objects created herein) and removal of the powder (about 1.5 hours). Depending on the different building configurations used, the time to 3D print an entire structural model was between 3 to 4 days.

2.1 Coordinate system

Previous research (Asadi-Eydivand et al. 2016, Feng et al. 2015, Gharaie et al. 2013) identified that the printing direction affects the material properties of the 3DP material. To study these orthotropic material characteristics the coordinate system defined by Feng et al. (2015) and depicted in Figure 2 is adopted. The X axis is the direction in which the nozzle moves when it drops binder in the build chamber. They Y axis is perpendicular to X and both the X and Y axis define the nozzle path for one layer of the 3DP object. The Z axis is in vertical direction perpendicular to the layers of the 3DP structure and the feed and build bins (Figure 1) move in Z direction.

2.2 Material composition

The used powder and binder was the Visijet PXL Core powder and the Visijet PXL Clear binder supplied by 3D Systems. Based on safety data sheets (3D Systems 2013) the main component (80-90%) of the powder is calcium sulphate hemihydrate ($CaSO_4 \frac{1}{2} H_2O$), which is also called *plaster of Paris* (Butscher et al. 2011). The remaining components are not specified but the previous generation of powder (zp150; Z Corporation 2009) consisted of vinyl polymer (<20%) and carbohydrate (<10%). Data sheets for the binder indicate that the binder is a mixture of primarily water and a humectant (0-1% 2-pyrrolidone, C_4H_7NO). Properties of the binder are very similar to water as was identified by Asadi-Eydivand et al. (2016). From the components of the powder and the binder it can be followed that the binder dissolves the calcium sulphate cements and the polymer to form a solid structure while the carbohydrate acts as a filler. The main binder/powder setting reaction can be written as

$$CaSO_4 \frac{1}{2} H_2O + 1\frac{1}{2} H_2O \rightarrow CaSO_4 2H_2O \qquad (1)$$

where the calcium sulphate hemihydrate reacts with water to form gypsum ($CaSO_4$ $2H_2O$, calcium sulphate dihydrate). The polymer reaction remains a company secret of 3D Systems. However, an X-ray diffraction phase analysis performed by Asadi-Eydivand et al. (2016) identified that the zp150 powder before and after printing consisted of $CaSO_4 \frac{1}{2} H_2O$ and $CaSO_4$ $2H_2O$, respectively. This suggests that the main reaction causing the solid 3DP object can be attributed to the hydration of calcium sulphate hemihydrate leading to the crystallization of gypsum.

2.3 Microstructure effects

The 3DP material is characterised by a distinct orthotropic behaviour that is related to the orientation of the structure in the print bed (Asadi-Eydivand et al. 2016, Chan 2013, Feng et al. 2015, Gharaie et al. 2013). This observation can be related to the printing procedure that results in a characteristic layered microstructure.

The following observations were made: (i) In XZ plane distinct layers were identified which represent the vertical layers of the 3DP procedure. (ii) In XY plane further layers are visible which is caused by the nozzle pattern printing distinct strips in X direction (Figure 2). Interestingly, these patterns were not observed at all XY planes which might be related to the position of the XY plane with respect to the layers in Z direction. Specifically, XY surfaces facing towards the nozzle were very smooth. (iii) The YZ plane was not characterised by a layered structure and had a similar roughness than the XZ surface.

These observations imply that the print orientation has a significant effect on the 3D printed material properties. Chan (2013) studied the mechanical properties of specimen printed with the equivalent printer used for this work and a previous generation of powder and binder. This previous work shows that the 3DP material is weakest when loaded in the XY plane. Equivalent findings were reported by Feng et al. (2015) when testing 3DP specimen in tension and bending. They related their observations to the lower strength between layers (in Z direction) compared to strips (in X direction). Moreover, Chan (2013) identified a certain area within the print bed that results in consistent material properties of the 3DP material. These findings of previous researchers were considered when designing the building models.

3 BUILDING MODELS

To make use of the lower interlayer bond strength pointed out above, the building model was printed so that the facade walls were perpendicular to the XY plane of the 3D printer, as shown in Figure 3. The dashed line surrounding the building model indicates the area of consistent material properties identified by Chan (2013). Due to the size of the print space (250×200×150 mm) and the required orientation in the print bed, the printing models were printed in

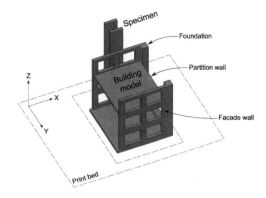

Figure 3. Building model and specimens in print bed.

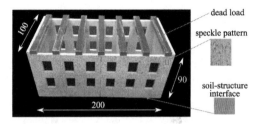

Figure 4. Building model (dimensions in mm).

two or three parts for the structures with $L = 200$ mm and 260 mm, respectively, and subsequently glued together. In every print job, specimens were also fabricated (Figure 3) and subsequently tested to derive the 3D printed material properties.

To finalise the building models after 3DP, the following steps were carried out: (i) connecting the individually printed building parts with Araldite standard glue, (ii) colouring the front facade of the building window with a so-called speckle pattern which enabled to track building displacements with DIC, (iii) attaching brass dead load bars to the top of the front and rear facades of the building model to replicate a vertical stress of 100 kPa beneath the strips of the front and rear facades, (iv) attaching a black sheet of paper to the back of the front facade and (v) installing MEMS accelerometers on the structure models.

A complete building model is shown in Figure 4 and indicates that the 3DP procedure enabled to obtain complex building models with realistic building layout (e.g. front, rear, end and intermediate walls) and building features such as strip foundations, window openings and a rough-soil structure interface by printing an uneven foundation base. As stated above, replicating these structural details at 1/75th scale was the primary aim of adopting the 3DP technique. The dead load bars and the speckle pattern can also be seen in Figure 4.

4 MATERIAL TESTING

Four-point bending tests were conducted to determine the material properties of the 3D printed material.

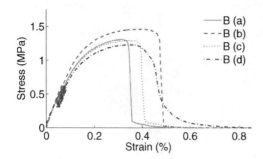

Figure 5. Stress-strain curve of the 3D printed material for model B (see Table 1).

Therefore, two specimens (20 mm × 4 mm × 125 mm) were printed in every print job.

4.1 Test procedure

The four-point-bending test followed the test procedure described in ASTM D 790M-86 II, Procedure A ASTM Standards (1986). The specimens were tested flatwise with a support span of 99 mm and a load span of 33 mm. The position of the specimen in the material tests was defined so that the load was applied in the XY plane to test the weak interlayer bond of the 3DP material. A cross-head motion of 4.5 mm/min was applied and a laser extensometer was used to monitor the mid-span deflection of the samples. After a load threshold of 2 N was reached three unloading and loading cycles between 2 N and 1 N were performed using the identical cross-head motion rate of 4.5 mm/min. The test data was acquired using a sampling frequency of 10 Hz.

5 RESULTS

Figure 5 shows representative stress-strain curves for the 3DP samples of a building model, which was subsequently tested in a centrifuge. The stress-strain curves of the samples associated with a single centrifuge test showed similar initial response (Figure 5). This indicates a consistent Young's modulus value. By contrast, the samples typically fractured at considerably different strain values. This is likely related to 3DP defects causing a weaker bond between layers of the 3DP material. Nevertheless, it is evident from Figure 5 that the 3DP material exhibits a softening behaviour typical of brittle materials. This implies that the 3DP material can be used to experimentally model cracking damage.

Table 1 summarises the mechanical properties (according to ASTM D 790M-86 II, Procedure A (ASTM Standards 1986)) of printed building models of different layout, which is further detailed elsewhere (Ritter et al. 2017). Additionally, typical material properties of masonry are presented for reference. The 3D printed material has a lower density than masonry while the Young's modulus values are in the range of historic masonry. The derived flexural strength of the

Table 1. Mechanical properties of the 3DP building models (A-F), where ρ is the density, f_t the flexural strength, E the Young's modulus and ϵ_{ult} the ultimate strain to failure.

Model	ρ (kg/m^3)	f_t (MPa)	E (GPa)	ϵ_{ult} (%)
A	1293	1.362	0.893	0.298
B	1278	1.311	0.800	0.357
C	1261	1.130	0.727	0.282
D	1272	0.934	0.516	0.352
E	1280	1.139	0.690	0.309
F	1247	1.702	1.039	0.246
Mean	1272	1.263	777.7	0.307
Masonry*	1900	0.1–0.9	1.0–9.0	0.05**

*Masonry properties follow Giardina et al. (2015)
**Strain at onset of cracking (Burland and Wroth 1974).

Table 2. Effect of different curing temperatures on 3DP material properties.

T (°C)	f_t (MPa)	E_{3DP} (MPa)	ϵ_{ult} (%)
room temp.	1.08	568.8	0.33
70°C	1.42	863.5	0.25
110°C	0.78	355.0	0.33

3D printed material and the ultimate strain to failure are higher than typical values for masonry. As a result of the greater ϵ_{ult}, cracking is expected to initiate at greater building distortions than in real structures.

For this reason, an initial investigation into the influence of the temperature during curing of the 3DP was conducted to provide further guidance for future research aiming to replicate more realistic masonry properties. Twelve specimens (identical to above) were printed and exposed to the curing temperatures presented in Table 2. After 24 hours of curing, the specimens were stored in a desiccator. Seven days after the 3DP they were tested in four-point-bending using the identical procedure described above.

An increase of the curing temperature to 70°C results in stronger, stiffer and more brittle material properties than curing at room temperature (Table 2). By contrast a curing at 110°C reduced the strength and stiffness properties compared to curing at room temperature or at 70°C. This is likely due to evaporation of the binder that reduces the calcium sulphate hemihydrate reaction (Equation 1). The data in Table 2 suggests that a curing at 70°C could result in 3D printed materials that are more similar to masonry. Further experimental investigations are recommended to optimize the properties of the 3D printed material.

6 APPLICATION IN CENTRIFUGE MODEL TESTING

To estimate the plane-strain global building stiffness of the 3DP building models, the contribution of each

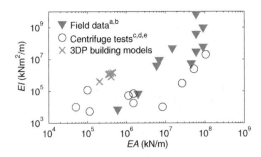

Figure 6. Global stiffness of the 3D printed building models compared to field data ([a]Mair and Taylor (2001), [b]Dimmock and Mair (2008)) and previous centrifuge tests ([c]Taylor and Grant (1998), [d]Taylor and Yip (2001), [e]Farrell (2010)).

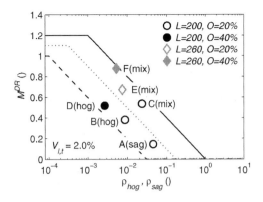

Figure 7. Observed modification factors for the deflection ratio in Goh and Mair (2011) design chart. (L building length in mm; O facade opening percentage).

structural element (i.e. wall and foundation) according to Melis and Rodriguez Ortiz (2001) was considered. Plane-strain stiffness values for each structural member were estimated by dividing the total width of each member perpendicular to bending by the entire building width (Giardina et al. 2017). A stiffness reduction due to window openings was considered for the bending (Melis and Rodriguez Ortiz 2001) and axial stiffness (Pickhaver et al. 2010). Figure 6 indicates that the obtained overall stiffness values of the 3D printed buildings are close to the range of case studies and previous research. Differences between the field data and the 3D printed buildings can be related to different methods of estimating the global building stiffness, as detailed in Ritter (2017).

Results of a series of centrifuge model tests (Ritter et al. 2017), investigating the influence of building characteristics on the ground and building response to tunnelling, are next adopted to demonstrate the value of using 3D printed building models. Figure 7 plots observed modification factors for the deflection ratio, M^{DR}, which relate the deflected shape of a building to the theoretical greenfield (i.e. no structure present) equivalent (Potts and Addenbrooke 1997), along relative stiffness expressions proposed by Goh and Mair (2011). All tests fell within the upper and lower design envelopes identified by Goh and Mair (2011) and the data in Figure 7 reveal interesting trends: (i) The building-to-tunnel position significantly influences the building response. Structures placed in the hogging/sagging transition zone ('mix', test C) experienced significant distortions while identical structures (tests A and B) located predominantly in either sagging or hogging showed notably fewer distortions. (ii) Increasing the building length, L, towards the tunnel (tests E and F) further increased building deflections. (iii) Doubling the opening percentage, O, caused substantially greater DR values (compare test B with test D or test E with test F).

With respect to practical implementation, this data might suggest that the upper design envelope (solid line in Figure 7 may be applicable for structures placed in the hogging/sagging transition zone. By contrast, the mean design envelope (dotted line in Figure 7) may be used for structures in the hogging or sagging region, though further data is required to confirm this suggestion.

7 CONCLUSIONS

This paper has considered the application of 3DP to fabricate small-scale building models that were subsequently subjected to tunnel excavation in a geotechnical centrifuge. The applied powder-based 3DP technique was introduced, after which details of the building models were discussed. The mechanical properties of the 3D printed material were derived by four-point-bending tests. From this testing programme it was observed that the 3D printed material exhibits brittle material properties similar to that of masonry and hence can be employed to study cracking damage. The stiffness of the 3D printed material is comparable to historic masonry but the 3D printed material is notably stronger than masonry. A preliminary investigation into different curing temperatures suggests that strength properties in better agreement with masonry could potentially be achieved.

Based on the data from the four-point-bending tests, a careful balancing of the building layout and the facade openings resulted in building models with global stiffness values in the range of typical case studies. When applied to centrifuge model tests, results revealed that building characteristic notably affect tunnelling-induced structural distortions. This demonstrates that the application of 3D printed building models enables the controlled investigation of structural details in centrifuge model testing.

ACKNOWLEDGEMENTS

Financial support provided by Crossrail and EPSRC grant EP/K018221/1 and the help of the technicians of the Schofield Centre are gratefully acknowledged. The associated research data is available at https://doi.org/10.17863/CAM.20400.

REFERENCES

3D Systems (2013). *Safety Data Sheet VisiJet PXL Core*. 3D Systems, Inc., USA.

Asadi-Eydivand, M., M. Solati-Hashjin, A. Farzad, & N. A. A. Osman (2016). Effect of technical parameters on porous structure and strength of 3D printed calcium sulfate prototypes. *Robotics and Computer-Integrated Manufacturing 37*, 57–67.

ASTM Standards (1986). *D790M-86 II, Standard Test Methods for Flexural Properties of Unreinforced and Reinforced Plastics and Electrical Insulating*. ASTM Int., West Conshohocken, PA.

Burland, J. B. & C. P. Wroth (1974). Settlement of buildings and associated damage - SOA review. In *Proc. Conf. Settl. of str., Cambridge*, pp. 611–654.

Butscher, A., M. Bohner, S. Hofmann, L. Gauckler, & R. Müller (2011). Structural and material approaches to bone tissue eng. in powder-based three-dimensional printing. *Acta Biomaterialia 7*(3), 907–920.

Caporaletti, P., A. Burghignoli, & R. N. Taylor (2005). Centrifuge study of tunnel movements and their interaction with structures. In *Geotech. Aspects of Underg. Const. in Soft Ground (K.J. Bakker, A. Bezuijen, W. Broere and E.A. Kwast eds.), Amsterdam, The Netherlands, 15-17 June*, pp. 99–105.

Chan, D. (2013). Investigation of tunnelling-induced building settlements using 3d printed models. Fourth-year undergraduate project, University of Cambridge.

DeJong, M. J. & C. Vibert (2012). Seismic response of stone masonry spires: Computational and experimental modeling. *Eng. Structures 40*, 566–574.

Dimmock, P. S. & R. J. Mair (2008). Effect of building stiffness on tunnelling-induced ground movement. *Tunnelling and Underground Space Techn. 23*(4), 438–450.

Farrell, R. P. (2010). *Tunnelling in sands and the response of buildings*. Ph. D. thesis, University of Cambridge.

Farrell, R. P., R. J. Mair, A. Sciotti, A. Pigorini, & M. Ricci (2011). The response of buildings to tunnelling: a case study. In *Geotech. Aspects of Underg. Const. in Soft Ground (G.M.B. Viggiani ed.), Rome, Italy*.

Feng, P., X. Meng, J.-F. Chen, & L. Ye (2015). Mechanical properties of structures 3D printed with cementitious powders. *Const. and Build. Mat. 93*, 486–497.

Franzius, J. N., D. M. Potts, & J. B. Burland (2006). The response of surface structures to tunnel construction. *Proc. ICE-Geotech. Eng. 159*(1), 3–17.

Gharaie, S. H., Y. Morsi, & S. Masood (2013). Tensile Properties of Processed 3D Printer ZP150 Powder Material. In *Advanced Materials Research*, Volume 699, pp. 813–816.

Giardina, G., M. J. DeJong, B. Chalmers, B. Ormond, & R. J. Mair (2017). A comparison of current analytical methods for predicting soil–structure interaction due to tunnelling. *in review*.

Giardina, G., M. A. N. Hendriks, & J. G. Rots (2015). Sensitivity study on tunnelling induced damage to a masonry façade. *Eng. Structures 89*, 111–129.

Goh, K. H. & R. J. Mair (2011). Building damage assessment for deep excavations in Singapore and the influence of building stiffness. *Geot. Eng. J. SEAGS & AGSSEA 42(3)*, 1–12.

Knappett, J. A., C. Reid, S. Kinmond, & K. O'Reilly (2011). Small-scale modeling of reinforced concrete structural elements for use in a geotechnical centrifuge. *Journal of Structural Eng. 137*(11), 1263–1271.

Liang, T., J. A. Knappett, & A. G. Bengough (2014). Scale modelling of plant root systems using 3-D printing. In *ICPMG2014 (Gaudin & White eds.), Perth, Australia*.

Liang, T., J. A. Knappett, & N. Duckett (2015). Modelling the seismic performance of rooted slopes from individual root–soil interaction to global slope behaviour. *Géotechnique 65*(12), 995–1009.

Mair, R. J. & R. N. Taylor (2001). Settlement predictions for Neptune, Murdoch, and Clegg Houses and adjacent masonry walls. *Building Response to Tunnelling: Case Studies from Construction of the Jubilee Line Extension, London, Vol. 1 CIRIA SP200*, 217–228.

Melis, M. J. & J. M. Rodriguez Ortiz (2001). Consideration of the stiffness of buildings in the estimation of subsidence damage by EPB tunnelling in the Madrid Subway. In *Proc. Int. Conf. Response of buildings to excavation-induced ground movements, Imperial College, London, CIRIA SP201*, pp. 387–394.

Pickhaver, J. A., H. J. Burd, & G. T. Houlsby (2010). An equivalent beam method to model masonry buildings in 3d finite element analysis. *Computers & structures 88*, 1049–1063.

Potts, D. M. & T. I. Addenbrooke (1997). A structure's influence on tunnelling-induced ground movements. *Proc. ICE-Geotech. Eng. 125*(2), 109–125.

Ritter, S. (2017). *Experiments in tunnel–soil–structure interaction*. Ph. D. thesis, University of Cambridge.

Ritter, S., M. J. DeJong, G. Giardina, & R. J. Mair (2017). Influence of building characteristics on tunnelling-induced ground movements. *Géotechnique 67*(10), 926–937.

Son, M. & E. J. Cording (2005). Estimation of building damage due to excavation-induced ground movements. *Journal of Geot. and Geoenv. Eng. 131*(2), 162–177.

Taylor, R. N. & R. J. Grant (1998). Centrifuge modelling of the influence of surface structures on tunnelling induced ground movements. In *Tunnels and Metropolises (A. Negro Jr & A.A. Ferreira eds.), World Tunnel Congress '98, São Paulo, Brazil*, pp. 261–266.

Taylor, R. N. & D. L. F. Yip (2001). Centrifuge modelling of the effect of a structure on tunnel-induced ground movements. In *Proc. Int. Conf. Response of buildings to excavation-induced ground movements, Imperial College, London, CIRIA SP201*, pp. 401–432.

Z Corporation (2009). Zprinter 350/ zprinter 450 quick start guide. zcentral.zcorp.com, 09/25/2009.

A mechanical displacement control model tunnel for simulating eccentric ground loss in the centrifuge

G. Song, A.M. Marshall & C.M. Heron
Nottingham Centre for Geomechanics, Faculty of Engineering, University of Nottingham, Nottingham, UK

ABSTRACT: Understanding the ground response to tunnelling is important for evaluating its effects on existing structures and buried infrastructure. The simulation of tunnel ground loss in soil using a geotechnical centrifuge is a convenient and widely accepted approach to analysing ground movements induced by tunnelling. Tunnel volume loss in the centrifuge is commonly simulated as a plane strain condition in which the internal tunnel pressure is released by extracting fluid from inside a flexible membrane (i.e. pressure control). There are relatively few cases in which mechanical models have been developed to simulate the tunnelling problem, where the tunnel volume loss is induced by an inwards radial movement of the model tunnel lining (i.e. displacement control). Existing mechanical model tunnels impose concentric movements of the tunnel lining towards the tunnel centre. This paper presents the development of an eccentric rigid boundary mechanical (RBM) model tunnel that has the ability to produce non-uniform radial displacements around the tunnel lining, causing maximum soil displacements at the tunnel crown and no displacements at the tunnel invert. The paper provides the detailed mechanical design of the model tunnel and results from three centrifuge tests in sand with tunnel cover to depth ratios of 1.3, 2.0 and 2.4. Greenfield tunnelling settlement trough data obtained using the newly developed RBM model tunnel are compared against previously published data obtained using a flexible membrane model tunnel.

1 INTRODUCTION

Tunnel excavation will inevitably result in soil disturbance that could cause damage to nearby structures, such as pile groups, foundations, and pipelines. It is important to understand the ground response to tunnelling in order to evaluate their potential to cause damage to these important civil engineering assets. Centrifuge testing has been widely used as a tool to investigate soil movements caused by tunnelling (Mair and Taylor 1999). The simulation of tunnel excavation considering construction sequence and tunnel face pressure release within a centrifuge environment (increased acceleration field) is a major challenge. To simplify the simulation, tunnel excavation is generally considered as a plane strain condition. Early tunnel volume loss simulations were achieved by decreasing internal air pressure within a tunnel shaped flexible membrane (Atkinson et al. 1975; Potts 1976; Hagiwara et al. 1999). This technique makes the measurement of the tunnel volume loss (a measure of the amount of displacement occurring around the tunnel lining) difficult due to the compressibility of air. To allow a better measurement of tunnel volume loss, the air can be replaced by water or oil, as done by Loganathan et al. (2000), Vorster et al. (2005) and Jacobsz (2003) using a flexible membrane sealed onto a shaft located centrally within the membrane. Around the same time, a rigid boundary mechanical (RBM) model tunnel was developed by Katoh et al. (1998), in which concentric tunnel volume loss displacements were imposed around the tunnel circumference (i.e. displacement control – DC). A concentric displacement profile does not necessarily replicate soil displacements around shallow tunnels (Loganathan & Poulos 1998), where an eccentric shape is typical (maximum displacements at the tunnel crown and minimal displacements at the invert). Marshall (2009) developed an eccentric flexible membrane (FM) model tunnel by moving the central shaft within the membrane close to the model tunnel invert, thereby ensuring that minimal displacements occurred near the invert.

There is no definitive choice regarding what type of model tunnel should be used to simulate tunnelling in the centrifuge; each method has limitations compared to reality. However, it is important to understand the consequences of the type of model tunnel used. It was highlighted by Marshall and Franza (2016) and Boonsiri and Takemura (2015) that the different tunnel boundary condition (FM versus DC) was the likely cause of discrepancies between greenfield tunnelling ground displacements (i.e. just the tunnel with no other structures within the soil) obtained using the two tunnel modelling approaches. A thorough investigation of the effect of the tunnel modelling technique (FM versus DC) and configuration (eccentric versus concentric) has never been conducted. This paper presents the design and testing of a newly developed eccentric DC RBM model tunnel. The data obtained using this model tunnel will be used alongside data from

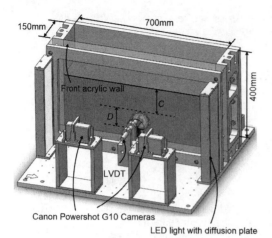

Figure 1. Centrifuge model strongbox layout.

other model tunnels and numerical analysis results in a future paper to develop a better understanding of the consequences of using the different types of model tunnels.

In this paper, a detailed description of the newly developed eccentric DC RBM model tunnel is provided, as well as an overview of the experimental setup adopted for testing. Soil displacement data are provided from greenfield centrifuge tests in sand with certain tunnel cover to diameter (C/D) ratios. The characteristics of the settlement trough obtained using the DC RBM model tunnel are compared with FM model tunnel data given in Marshall et al. (2012) and Boonsiri & Takemura (2015) to evaluate the effect of tunnel type (FM versus DC) on settlement trough shape.

2 EXPERIMENTAL SYSTEM & SET UP

The centrifuge tests presented in this paper were conducted on the University of Nottingham Centre for Geomechanics (NCG) 2 m radius, 50 g-tonne geotechnical centrifuge at an acceleration of 80 times gravity (i.e. 80 g). Figure 1 shows the strongbox layout, which is 150 mm wide, 700 mm long, and 400 mm high (internal dimensions). An 80 mm thick transparent acrylic wall is mounted at the front of the box to allow the acquisition of images for the analysis of soil displacements. The model tunnel has a diameter of 90 mm and was located 75 mm above the strongbox base. Leighton Buzzard Fraction E sand with a D_{50} of 0.12 mm was used for the tests. The sand has a specific gravity G_s of 2.65. The maximum (e_{max}) and minimum (e_{min}) void ratios of the sand are 1.01 and 0.61, respectively (Franza 2016). Three centrifuge tests were undertaken with C/D ratios of 1.3, 2.0, and 2.4.

2.1 Model tunnel

Figure 2 illustrates the newly developed eccentric RBM model tunnel. The eccentric tunnel volume loss shape is achieved by controlling six segments that move towards the tunnel centerline at different radial displacements; a maximum displacement of 1.4 mm at the tunnel crown, 0.9 mm at the spring line, and zero displacement at the tunnel invert can be reached, resulting in a maximum tunnel volume loss of approximately 3.5%.

A single bi-directional screw shaft is placed at the tunnel centerline, fixed by two ball screw supports located on either side (front and back) of the strongbox. Two flange nuts are secured on the screw shaft in opposite directions. Each flange nut is connected with a hexagonal wedge shaped shaft (see Figure 2). The hexagonal wedge shaped shaft consists of six faces with different taper angles, which are varied from 4° at the tunnel crown to 0° at the tunnel invert. Six arch shaped tunnel segments are connected to the two opposing hexagonal wedge shaped shafts via two linear guide rails to generate the tunnel circumference. Each segment is equipped with two linear carriages, secured to the opposing wedge shaped shafts. The linear guide rail with linear carriage enables the relative movement between the six segments and the two shafts along the taper angle. A stepper motor linked with a 10:1 ratio gearbox drives the screw shaft via a flexible coupler, which permits some degree of misalignment between the screw and gearbox shafts. Tunnel volume loss is achieved by rotating the bi-directional screw shaft, causing the two hexagonal wedge shaped shafts to move apart from each other, and the six segments to move towards the tunnel centerline. Due to the differences in taper angle, the non-uniform tunnel boundary displacement profile is formed.

Figure 3 shows the positions of the six tunnel segments before and after tunnel volume loss. Tunnel volume loss is calculated by measuring the radial displacement of each segment, which is related to the horizontal displacement of the wedge shaped shafts and the taper angle (radial displacement=horizontal displacement × tan(taper angle)). The displacement of the wedge shaped shaft can be estimated from the number of signals sent to the stepper motor (i.e. rotation of screw shaft). During each volume loss stage (equivalent to 0.18% of tunnel volume loss), the screw shaft (with a 1/8 screw pitch) was rotated by 450°, corresponding to 0.625 mm of horizontal displacement of the wedge shaped shaft. The displacement of the wedge shaped shaft was further confirmed using a Linear Variable Differential Transformer (LVDT) mounted in front of the strongbox (see Figure 1). Figure 4 shows the LVDT measurement and stepper motor estimation of wedge shaped shaft horizontal displacement during several volume loss stages. The stepper motor estimation of displacement is shown to be a close match to the LVDT readings under the tested conditions. The maximum horizontal displacement of the wedge shaped shafts is around 20 mm.

Initially (before volume loss), a 2 mm clearance gap exists between adjacent tunnel segments (Figure 3), which decreases with volume loss. To generate a smoother tunnel surface, flexible copper strips were

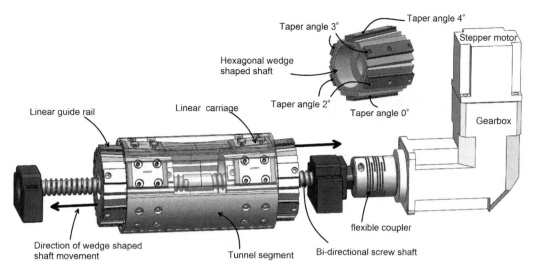

Figure 2. Eccentric rigid boundary mechanical model tunnel.

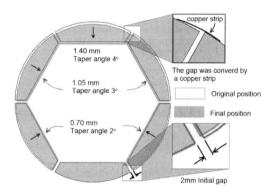

Figure 3. Eccentric mechanical tunnel deformed shape.

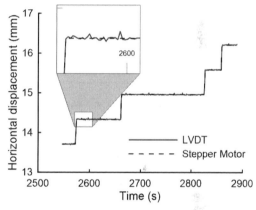

Figure 4. Cross check the volume loss measurement.

used to cover the gaps along the tunnel length. The tunnel was subsequently sealed within two latex membranes during tests to protect it from soil intrusion.

2.2 Sample preparation

The sand model was prepared by dry air pluviation. The tunnel was secured within the strongbox and the whole box was placed on its side (with the back wall on the floor). This orientation allowed sand pouring in the direction of the tunnel axis to ensure a uniform sample was obtained (consistent with the methodology used by Jacobsz (2003), Vorster et al. (2005), and Marshall (2009)). Sand pouring was then conducted using a hopper with a controlled pouring rate and drop height calibrated to obtain a relative density (I_d) of approximately 90%. After sand pouring, a thin layer of dyed sand was placed uniformly on the surface of the sample (adjacent to the acrylic front wall) to improve the texture of the soil for image analysis. The front acrylic wall was then bolted to the strongbox and the strongbox was rotated slowly to the upright position.

The tunnel volume control gearbox, stepper motor and LED light panels, as well as the LVDT used to measure the wedge shaped shaft movement were then installed.

2.3 Instrumentation

Surface and subsurface soil displacements were measured by analysing images taken at each tunnel volume loss step. Two 14.7 mega-pixel Canon Powershot G10 cameras were used to capture images during the tests. A uniform illumination of the front acrylic wall was generated using two LED light panels equipped with diffusion plates placed on either side of the strongbox. The soil displacements were measured using the GeoPIV-RG (Stanier et al. 2015) image analysis technique. To evaluate the accuracy of the image analysis applied in these tests, an image of the tested sand sample was taken under the centrifuge test lighting environment. A region of the image containing soil only was then shifted in both x and y directions by 10 pixels to generate a second image. By analysing the

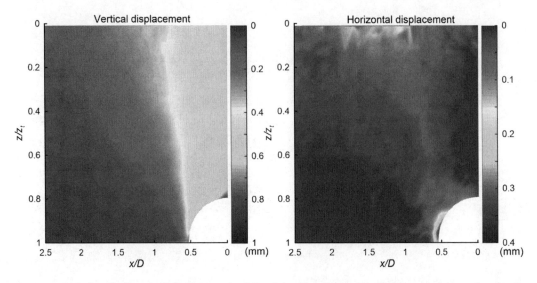

Figure 5. Vertical and horizontal soil displacement for $C/D = 2.4$ at $V_{l,t} = 2\%$ (positive displacements downwards and to the right).

object space coordinates of the original and shifted images using GeoPIV-RG, a horizontal and vertical standard deviation of 0.0019 mm and 0.0011 mm, respectively, were obtained. These standard deviation values are similar to those quoted by Marshall and Mair (2011), who examined PIV precision by analysing seven images taken at elevated gravity when no soil displacement was expected.

It is worth noting that the soil displacement data presented in this paper were all obtained at the acrylic wall interface. The interface friction between the soil and the acrylic wall will affect soil displacements, however similar centrifuge tests have shown that the boundary friction did not have a significant effect on settlement trough shape (Marshall et al. 2012).

2.4 Testing procedure

After mounting on the centrifuge cradle, the model was spun to 80 g in stages of 10 g. The model was then put through four stabilisation cycles (going from 80 g to 10 g and back to 80 g). It has been found that, for this soil, four stabilisation cycles provides a consistent measurement of surface displacement from 10 to 80 g. The stabilisation cycles are done to help achieve consistency between tests by reducing localised high-stress zones ('hung-up' particles), thereby achieving a more uniformly stressed soil profile. The tunnel volume loss process was then started and images taken from the two cameras at every volume loss stage.

3 RESULTS

3.1 Soil deformation mechanism

Figure 5 shows the vertical and horizontal soil displacements from the eccentric RBM centrifuge test for $C/D = 2.4$ at a tunnel volume loss of $V_{l,t} = 2.0\%$. From the plot of vertical displacements, a localised maximum vertical settlement zone can be identified close to the tunnel crown. The vertical settlements are mainly concentrated in a chimney-shaped zone directly above the tunnel, with a reasonably well defined shear band discernible that propagates up from the sides of the tunnel. The vertical band of shear becomes less well defined nearer the surface where vertical displacements spread out. A zone of relatively large horizontal displacements is noted close to the tunnel springline. Horizontal displacements within the region directly above the tunnel are very small, as is expected due to the symmetry of the problem. Horizontal displacements within the main soil body are relatively small, but a zone of larger displacements near the surface was generated.

3.2 Settlement trough curve fitting

Curves are generally fitted to settlement trough data in order to quantify the characteristics of the soil displacements and evaluate how they change with influential parameters (e.g. depth, soil density, tunnel volume loss). The standard approach is to use the simple 2-parameter Gaussian curve: $S_v = S_{max} \exp(-x^2/(2i^2))$, where S_{max} is maximum settlement and i is the horizontal distance from the tunnel centerline to the inflexion point of the curve. The standard Gaussian curve generally provides a good fit to settlement trough data, especially for tunnels in clay where, for undrained conditions, the volume of the soil is constrained and settlement trough shape seems to be mainly insensitive to the properties of the soil and the magnitude of tunnel volume loss. The Gaussian curve does not, however, always provide a good fit to settlement trough data, especially for tunnels in sands. This is due to the complex contractive/dilatant response of the soil

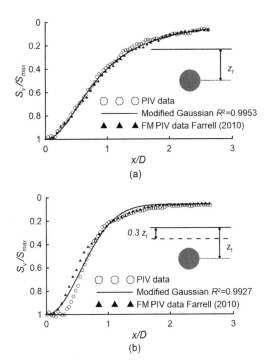

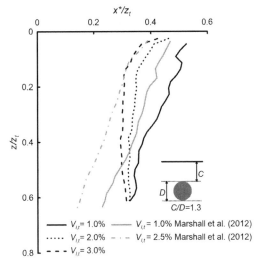

Figure 7. Variation of x^* with depth for $C/D = 1.3$.

Figure 6. Curves fitted to (a) surface and (b) subsurface ($z/z_t = 0.3$) settlement data from test with $C/D = 1.3$ at $V_{l,t} = 3.0\%$.

during tunnel volume loss, which is a function of the soil properties, the confining stress, and the magnitude of shear strain experienced by the soil (which is proportional to tunnel volume loss) (Marshall et al. 2012). A modified Gaussian curve was proposed by Vorster et al. (2005), which has an additional parameter (α) compared to the standard Gaussian curve but maintains the meaning of i as the horizontal distance to the inflexion point. The modified Gaussian curve has been shown to give a better fit than the standard Gaussian curve for settlement trough data above tunnels in sands (Vorster et al. 2005, Marshall et al. 2012) and is therefore used within this paper.

Figure 6 shows surface and subsurface (at $z/z_t = 0.3$) settlements from the centrifuge test with $C/D = 1.3$ at $V_{l,t} = 3\%$ as well as the fitted modified Gaussian curves. The goodness of fit of the curves to the displacement data is quantified based on the coefficient of determination, R^2. Data from an equivalent FM test (tunnel diameter $D = 150$ mm) at the same tunnel volume loss (Marshall et al. 2012) are also provided.

There are some qualitative differences between the shape of the settlement trough data from the DC RBM tunnel and FM model tunnel. The settlement trough shape of the DC RBM tunnel, especially at depth, is relatively flat immediately above the tunnel. This makes it difficult to achieve a good fit even with the 3-parameter modified Gaussian curve (especially near the tunnel centreline). This observation can also be made by comparing the DC model tunnel data in Boonsiri and Takemura (2015) with the FM data in Marshall et al. (2012).

3.3 Difference between FM and DC model tunnels

To study the differences in settlement trough shape characteristics between the eccentric RBM and the flexible membrane model tunnels, the settlement trough width parameter x^* is used (Marshall et al. 2012). This parameter gives the offset from the tunnel centreline to the point on the fitted settlement curve where $S_v/S_{max} = \exp(-0.5) \approx 0.606$. For a Gaussian curve, $x^* = i$, hence the use of x^* allows a qualitative comparison between data obtained when using Gaussian and modified Gaussian curves. Figure 7 compares x^* against depth for the RBM model tunnel and the equivalent FM model tunnel data from Marshall et al. (2012). Both model tunnels show a decrease in x^* with depth, indicating that the settlement trough shape becomes narrower towards the tunnel. With an increase in tunnel volume loss, the value of x^* decreases. For a given volume loss, the data from the eccentric RBM model tunnel shows higher x^* values compared with the FM model tunnel. At a tunnel volume loss of 1%, the gradient of x^* with depth for the two model tunnels is similar. As volume loss increases, the gradient of x^* remains about the same for the FM model tunnel, whereas for the eccentric RBM model tunnel the gradient increases. It is clear that the different model tunnels have an important effect on the resulting shape of the greenfield settlement trough.

4 CONCLUSIONS

Details of a newly developed eccentric rigid boundary mechanical (RBM) model tunnel were presented, along with an overview of the centrifuge experiment testing procedure. Surface/subsurface soil displacement data from three greenfield tests were presented and the outcomes of curve fitting were provided. Data of the settlement trough width (x^*) from the eccentric RBM model tunnel were compared with those from a flexible membrane (FM) model tunnel. Results showed that the settlement trough shape is generally wider for the eccentric RBM model tunnel than the FM model tunnel for the same volume loss. In addition, it was shown that the rate at which the settlement trough narrows with depth changes with tunnel volume loss for the eccentric RBM model tunnel, whereas it stays relatively constant for the FM model tunnel. The results presented in this paper should help researchers conducting centrifuge testing of tunnelling related problems to better understand the implications of their choice of model tunnel.

REFERENCES

Atkinson, J., E. Brown, & D. Potts (1975). Collapse of shallow unlined tunnels in dense sand. *Tunnels and Tunneling* 7(3), 81–87.

Boonsiri, I. & J. Takemura (2015). Observation of ground movement with existing pile groups due to tunneling in sand using centrifuge modelling. *Geotechnical and Geological Engineering* 33(3), 621–640.

Elshafie, M., C. Choy, & R. Mair (2013). Centrifuge modeling of deep excavations and their interaction with adjacent buildings.

Franza, A. (2016). *Tunnelling and its effects on piles and piled structures*. Ph. D. thesis, University of Nottingham.

Hagiwara, T., R. J. Grant, M. Calvello, & R. N. Taylor (1999). The effect of overlying strata on the distribution of ground movements induced by tunnelling in clay. *Soils and Foundations* 39(3), 63–73.

Jacobsz, S. W. (2003). *The effects of tunnelling on piled foundations*. Ph. D. thesis, University of Cambridge.

Katoh, Y., M. Miyake, & M. Wada (1998). Ground deformation around shield tunnel. In *Centrifuge*, Volume 98, pp. 733–738.

Loganathan, N. & H. Poulos (1998). Analytical prediction for tunneling-induced ground movements in clays. *Journal of Geotechnical and Geoenvironmental Engineering*.

Loganathan, N., H. Poulos, & D. Stewart (2000). Centrifuge model testing of tunnelling-induced ground and pile deformations. *Geotechnique* 50(3), 283–294.

Mair, R. & R. Taylor (1999). Theme lecture: Bored tunneling in the urban environment. *of XIV ICSMFE [131]*, 2353–2385.

Marshall, A. (2009). *Tunnelling in sand and its effect on pipelines and piles*. Ph. D. thesis, University of Cambridge.

Marshall, A., R. Farrell, A. Klar, & R. Mair (2012). Tunnels in sands: the effect of size, depth and volume loss on greenfield displacements. *Géotechnique* 62(5), 385–399.

Marshall, A. M. & A. Franza (2016). Discussion of "observation of ground movement with existing pile groups due to tunneling in sand using centrifuge modelling" by ittichai boonsiri and jiro takemura. *Geotechnical and Geological Engineering*, 1–5.

Marshall, A. M. & R. J. Mair (2011). Tunneling beneath driven or jacked end-bearing piles in sand. *Canadian Geotechnical Journal* 48(12), 1757–1771.

Potts, D. M. (1976). *Behaviour of Lined and Unlined Tunnels in Sand*. Ph. D. thesis, University of Cambridge.

Ritter, S., G. Giardina, M. J. DeJong, & R. J. Mair (2017). Centrifuge modelling of building response to tunnel excavation. *International Journal of Physical Modelling in Geotechnics*, 1–16.

Stanier, S. & D. White (2013). Improved image-based deformation measurement in the centrifuge environment.

Stanier, S. A., J. Blaber, W. A. Take, & D. White (2015). Improved image-based deformation measurement for geotechnical applications. *Canadian Geotechnical Journal* 53(5), 727–739.

Vorster, T., A. Klar, K. Soga, & R. Mair (2005). Estimating the effects of tunneling on existing pipelines. *Journal of Geotechnical and Geoenvironmental Engineering* 131(11), 1399–1410.

Zhou, B. (2015). *Tunnelling-induced ground displacements in sand*. Ph. D. thesis, University of Nottingham.

Preliminary results of laboratory analysis of sand fluidisation

F.S. Tehrani
Deltares, Delft, The Netherlands

A. Askarinejad
Delft University of Technology, Delft, The Netherlands

F. Schenkeveld
Deltares, Delft, The Netherlands

ABSTRACT: This paper reviews the preliminary results of an experimental study on fluidization of sand in the laboratory scale. The study aims to investigate the hydro-mechanical process that takes place when fluidization is initiated and evolves in sand. To that end, small-scale laboratory experiments are performed on fully saturated samples of silica sand with different initial relative densities. The sand samples are prepared inside a transparent cubic container with Plexiglas walls. A nozzle is embedded at the bottom corner of the container to discharge an upward axisymmetric flow into the sand sample. The discharge is generated through the controlled application of hydraulic gradient in the sample. Two high resolution digital cameras are utilized to take sequential images from the soil at two transparent sides of the container during the discharge process. The Digital Image Correlation technique (DIC) is used to analyse the images for quantification of soil displacement before and at the onset of fluidization. The results show that the soil failure at the onset of fluidization is highly dependent upon the initial state of the tested sand samples such that as the relative density of soil increases, a greater zone is influenced due to fluidization.

1 INTRODUCTION

Fluidization of geomaterials is a phenomenon in which granular geo-material is transformed into a fluid-like state. This can be driven by several reasons such as earthquake loading (liquefaction) and downstream outward flow (piping). Fluidization of coarse-grained soil, e.g. sand and gravel, can take place when water is pumped into the soil body (e.g. De Jager et al. 2017). This type of fluidization is very common when a buried water pipeline is damaged (van Zyl et al. 2013) or when water pressure leakage takes place through sheet pile walls (Cashman & Preene 2001).

The previous experimental studies on water injection through an orifice in coarse-grained soils suggest that when the injection rate is low, fluidization can barely occur (e.g. Alsaydalani & Clayton 2014). It was shown that at low rates of water injection in soil, Darcy's law holds where flow is linearly proportional to the applied head difference. However, at higher injection rates, more complex relationship may form between head loss and flow rate (Niven 2002). Once the flow rate exceeds certain threshold, fluidization starts to happen at the vicinity of the water discharge zone. Such flow rate, when the flow is directed upward, may be predicted using the equation proposed by Ergun (1952). After the fluidization is initiated, a limited fluidized zone around the zone of water discharge is developed which gradually progresses, provided that a sufficiently high flow rate is injected, until it reaches the surface of the soil (van Zyl et al. 2013, Alsaydalani & Clayton 2014).

In this paper, we study internal fluidization of sand, due to water discharge inside the soil body. Similar study has been done by Alsaydalani & Clayton (2014) with the difference that in their work only dense granular materials were taken into account, whereas in the present study the attention is paid on the effect of sand relative density (loose, medium dense and dense) on the response of soil body before and at the onset of fluidization.

2 PROPERTIES OF EXPERIMENTS

2.1 Test setup

The test setup consists of a transparent box with the side length of 180 mm and height of 194 mm. The side walls are made up of Plexiglas with a thickness of 15 mm. To model the axisymmetric problem a nozzle, with the internal diameter of 5 mm, is glued to the bottom corner of the box where it is connected to a water tank (reservoir) that is used to adjust the applied head

Figure 1. Test setup for the sand fluidization study. 1: transparent test box containing the soil specimen, 2: digital timer, 3: Digital SLR cameras, and 4: nozzle.

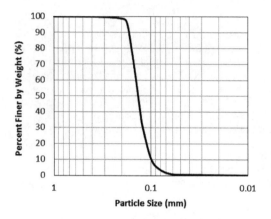

Figure 2. Grain size distribution of Baskarp sand.

Table 1. Basic characteristics of Baskarp sand.

D_{50} (mm)	C_c (-)	C_u (-)	e_{max} (-)	e_{min} (-)	G_s (-)
0.14	1.04	1.43	0.89	0.55	2.65

pressure. Two Digital SLR cameras of type Cannon EOS 750D with the focal length of 24 mm are facing the perpendicular sides of the box in order to capture images during the water discharge. The distance of the cameras from the sides of the box is 300 mm. Next to every transparent wall, a digital timer is installed to record the time stamps of events as the test progresses. Figure 1 shows the test setup.

2.2 Material and sample preparation

Baskarp sand is used as the test material. Figure 2 and Table 1 show the grain size distribution and basic properties of Baskarp sand, respectively.

Sample preparations are performed using wet pluviation accompanied by side tapping until the desired relative density is achieved.

2.3 Testing procedure

Tests are performed on samples in loose ($D_R = 31\%$), medium dense ($D_R = 60\%$) and dense ($D_R = 88\%$)

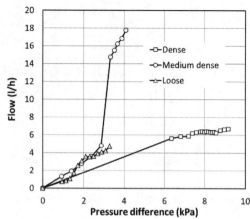

Figure 3. Relationship between the increasing upstream hydraulic pressure and the discharge flow rate for different relative densities of Baskarp sand.

states. When necessary, tests were repeated to verify the repeatability of the results. All tests are performed under head-controlled conditions where the applied head is increased in small steps until fluidization of the sand sample takes place. For the tests on medium dense and dense sands, the head increment is equal to +5 cm (increased at every 5 minutes), whereas for the test on the loose sand the head increment is equal to +2 cm (increased at every 5 minutes). The flow rate is measured at every head increment. In order to visualize the displacement and deformation of soil before and during fluidization, sequential images of sand sample are taken at the frame rate of 2 fps (frame per second). These images are later analysed using the Digital Image Correction (DIC) technique. Commercial software VIC-2D (Correlated Solutions 2009) is used for the analysis of the images.

3 RESULTS AND DISCUSSION

Two sets of results are presented in this section: a) effect of change of hydraulic head on discharge flow rate and b) soil displacement at the onset of fluidization.

Figure 3 shows the variation of flow rate with increase in the hydraulic head of the input flow for loose ($D_R = 31\%$), medium dense ($D_R = 60\%$) and dense ($D_R = 88\%$) sand samples.

It can be seen in Figure 3 that the pressure head required for fluidization of sand samples increases as the relative density of the samples increases. However, it is observed that the increase in the flow rate as the hydraulic head increases is not proportional to the relative density of the sand samples. The results of this study shows that, before the onset of fluidization, the flow rate in the medium dense sand sample exceeds that of loose and dense sand samples as the hydraulic head increases. To ensure the validity of this finding,

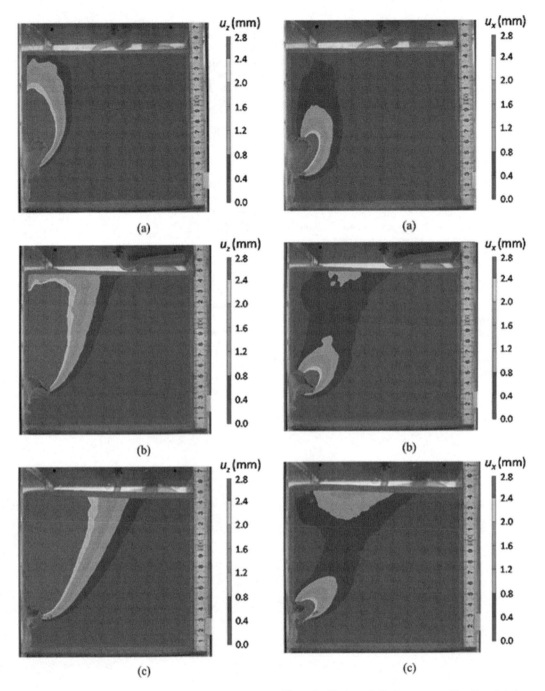

Figure 4. Vertical displacement profile of sand during fluidization: (a) loose sand ($D_R = 31\%$), (b) medium dense sand ($D_R = 60\%$), and (c) dense sand ($D_R = 88\%$).

Figure 5. Horizontal displacement profile of sand during fluidization: (a) loose sand ($D_R = 31\%$), (b) medium dense sand ($D_R = 60\%$), and (c) dense sand ($D_R = 88\%$).

the test on the medium dense sand sample was repeated and a similar outcome was observed.

Figure 4 and Figure 5 show, respectively, the vertical and horizontal displacement profiles of loose, medium dense and dense sand samples before the fluidized zone reaches to the soil surface. The results show that the extent of the zone that is influenced by the soil fluidization increases as the relative density of the sand sample increases. Similar finding is reported in the numerical study of Martinelli et al. (2017). As shown in Figure 4 and Figure 5, the displacement profile is mainly vertical on top of the nozzle and the zone with negligible horizontal displacement is extended as the density of the sand samples increases. These are important findings which can be useful in design of buried water pipelines. This finding may suggest that poorly compacted trench material can limit the extent of damaged zone when a water pipeline is ruptured. However, further investigation is needed to draw a firm conclusion.

4 SUMMARY AND CONCLUSION

In this paper, we showed the preliminary observations of an experimental study on fluidization of sand. Small-scale laboratory experiments were carried out on fully saturated sand samples at loose, medium dense and dense sates. The sand samples were prepared inside a transparent test box. An upward axisymmetric head-controlled water discharge was applied at the bottom of the sand samples through a nozzle embedded at the bottom corner of the container. Two high resolution digital cameras were utilized to take sequential images from the soil at two transparent sides of the container during the discharge process. The flow-head measurements and the image-based analyses of the captured images showed that the soil failure at the onset of fluidization was highly dependent upon the initial state of the tested sand samples. The hydraulic measurements showed that the hydraulic response of the medium dense sand sample was quite different than the loose and dense sand samples. Also the image-based analysis results indicated that as the relative density of soil increased, a greater zone was influenced by soil fluidization. Such findings can be useful in the risk assessment and design of buried pipelines.

REFERENCES

Alsaydalani, M.O.A., & Clayton, C.R.I. 2013. Internal fluidization in granular soils. Journal of Geotechnical and Geoenvironmental Engineering, 140(3), 04013024.

Cashman, P.M., & Preene, M. 2001. Groundwater lowering in construction: A practical guide, Spon, London.

Correlated Solutions 2009. Vic-2D. Columbia, SC, USA: Correlated Solutions.

De Jager, R.R., Maghsoudloo, A., Askarinejad, A. & Molenkamp, F. 2017. Preliminary results of instrumented laboratory flow slides. Procedia Engineering, 175, pp. 212–219.

Martinelli, M., Tehrani, F.S., & Galavi, V. 2017. Analysis of crater development around damaged pipelines using the material point method. Procedia Engineering, 175, 204–211.

Niven, R.K. 2002. "Physical insights into the Ergun and Wen & Yu equations for fluid flow in packed and fluidized beds." Chem. Eng. Sci., 57(3), 527–534.

van Zyl, J.E., Alsaydalani, M.O.A., Clayton, C.R.I., Bird, T., & Dennis, A. 2013. "Soil fluidization outside leaks in water distribution pipes – Preliminary observations." Proc. Inst. Civ. Eng. Water Manage., 166(10), 546–555.

Rolling test in geotechnical centrifuge for ore liquefaction analysis

L. Thorel, Ph. Audrain, A. Néel, A. Bretschneider & M. Blanc
IFSTTAR, GERS Deptartment, Geomaterials & Models in Geotechnics Laboratory, Bouguenais, France

F. Saboya
UENF, Brazil

ABSTRACT: To study the development of liquefaction in ore cargo during maritime transportation, a new Rolling Test device has been designed to support similar stresses to those observed in a vessel. It can be used in an $80 \times g$ macrogravity field in the 5.5 m radius IFSTTAR geo-centrifuge. Its main characteristics and benchmark test results are presented.

1 INTRODUCTION

Combination of cyclic loading, presence of fine particles and variable moisture content within a bulk carrier's ore cargo can result in liquefaction causing the vessel to list or capsize and possibly loss of human life. Several accidents with vessels have been attributed to ore liquefaction, mostly carrying iron, bauxite and nickel (IMO, 1998, IMO 2012). Three elements may generate such a catastrophic event: the cargo properties, the ship design and the sea conditions. In order to investigate the origin of cargo liquefaction during transportation, a new device has been developed at IFSTTAR in the framework of the Franco-German European LiquefAction project.

From a geotechnical point of view liquefaction is a hazardous phenomenon that consists in a change of the soil behavior from "friction" to "liquid". This phenomenon is related to the presence of interstitial water, which under specific loading condition, can generate overpressures on the soil grains, up to a level sufficient to undermine the friction resistance or, in other words, the shear strength. The liquefaction phenomenon occurs "rapidly", it means that the overpressures generated (e.g. by compression) cannot be dissipated due to a very rapid solicitation on medium-low permeability soil. Liquefaction accidents are often observed during earthquakes and may, for instance, involve building foundations, slopes and earth embankments.

Ore cargo liquefaction is a complex and still not fully understood phenomenon. It is not necessarily similar to seismic liquefaction even if analogous effects can be observed. It could be related to different hypotheses ascribable to cyclic loading, fluid migration, and soil initial state. Of course, the type of material, loading conditions and water content are the main parameters that influence the triggering of this phenomenon. To identify the risk of cargo liquefaction, several tests can be found in the literature (ClassNK, 2012, IMSBC 2013): flow table test (derived from ASTM C230), penetration test, weight penetration test and rolling test. The latter consists in a $0.3 \times 0.3 \times 0.3$ m Perspex cubic box, which is rotated around a horizontal axis located in the middle of the box base. The Rolling period simulated is 10s, the maximum rotation angle of rolling is $\pm 25°$, and the test duration is limited to 5 min. All those tests are focused on the identification of the Transportable Moisture Limit (TML) of the ore cargo. The TML is the maximum Moisture Content (MC) of the ore cargo for which there is no risk of "flow".

Atkinson & Taylor (1988) have developed a small scale model for studying in centrifuge the stability of iron ore concentrate. The box was 0.35 m large and 0.16 m high. The tests were performed at $100 \times g$, using a silicon fluid, but the device allowed only to reach a roll angle of 40° in 20s at the model scale. GBWG (2017) mention centrifuge tests on bauxite performed at $50 \times g$ using a rolling table (with a container size of $0.6 \times 0.2 \times 0.45$ m), applying a maximum angle of 25° during 20 cycles at a frequency of 5Hz.

Liquefaction in ore cargoes is still an open problem; no observations being possible during the shipment. In this sense physical modelling is a useful tool to observe the evolution of the material during shipment by artificially reproducing conditions similar to those occurring on the oceans.

2 EXPERIMENTAL SETUP

The Rolling Test Device presented here has been designed to be used in the IFSTTAR's geotechnical centrifuge (Fig. 1), the target is the reproduction of similar movement, stresses and pressures on the sample inside the vessel.

Figure 1. 200×g-tons & 5.5 m radius IFSTTAR's geo-centrifuge.

Figure 3. Detail of the IFSTTAR's Rolling box.

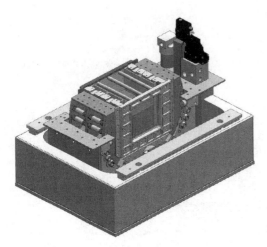

Figure 2. Scheme of the device.

2.1 Rolling test device

The device has been developed in order to simulate a rolling movement of a ship that transports possible iron ore cargo.

2.1.1 Rolling box

The box itself (Figure 2 & 3) is designed similarly to the Rolling Test Equipment suggested by ClassNK (2012), but it has been reinforced in order to support the stresses induced by the macrogravity field. A transparent face allows the observation of phenomena occurring inside the box during the movement. Inside the box, the two other lateral walls (port and starboard sides) have been designed with special features that allow different boundary conditions (rough, smooth, drained...).

The box is fixed in a rotating cradle (radius = 400 mm), placed on an assembly including rolls and hydraulic rotary actuator. A cog-wheel, linked to the actuator, moves a rack and pinion fixed on the cradle. The axis of rotation of the box is perpendicular to the centrifuge rotation axis.

The elevation of the box (Fig.4) may be adjusted in order to simulate the rolling movement in different cases: when the centre of gravity is higher than the axis of rotation (light ship) or when the centre of gravity

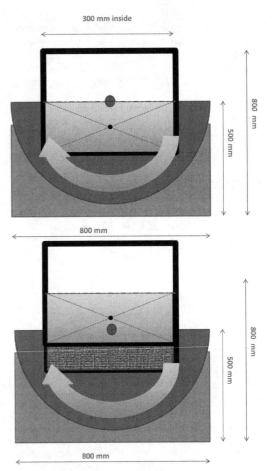

Figure 4. The axis of rotation (fixed) may be relatively more (top) or less (bottom) elevated than the cargo's center of gravity. The red arrow represents 112.5 mm maximum.

is lower than the axis of rotation (heavy cargo). The elevation is fixed before the test by adjusting wedges of different thicknesses.

The stresses induced in the cargo are similar to those existing in a vessel, as the maximum height of cargo is 0.3 m in the box, which corresponds to 24 m at a centrifuge acceleration of 80×g.

2.1.2 Hydraulic rotary actuator

The rotary actuator has been selected in order to apply the required torque in the worst work condition, when the centre of gravity is less elevated than the centre of rotation and taking into account the required frequency.

The model selected is a Parker HTR45 hydraulic rotary actuator, which allows a torque of 2000 N·m and a maximum service pressure of 80 bar. It contains an oil volume 1.8 l, which requires an oil debit of 21 l/min at 0.1 Hz, or 105 l/min at 0.5 Hz. Those performances require of course an adequate hydraulic power supplies by a high-pressure hydraulic pump with a flow of more than 100 l/min. Due to hydraulic constraints, the frequency cannot be scaled. So it has been chosen to reproduce the same range of frequency than in situ.

2.1.3 Control-command

The macrogravity field in the centrifuge basket precludes any human intervention. All on-board equipment is remotely guided from the control room.

The movement applied to the Rolling box is controlled by a servo-controller manufactured by MOOG. A control-loop was created with the rotation sensor, which is an absolute single turn encoder with a precision of 21bits. The software associated with the controller allows a real time control of the movement applied to the Rolling box. The movement could be a sine signal, or other signal as required.

2.1.4 Instrumentation

The data acquisition system HBM Spider enables conditioning and digitizing measurements by means of synchronous 8-channel modules that may be linked. Any type of sensor may be conditioned: full bridge, half-bridge, voltage source, temperature probe…The sampling frequency reaches up to 1.2 kHz.

Pore pressure measurements are necessary to evaluate the overpressure generated by the mechanical solicitations and to compare to the effective stress for liquefaction analysis. The sensors used classically are Druck or Measurement sensors with a range of 700 kPa.

Earth pressure sensors Kyowa (200 and 500 kPa) will be installed on the walls of the box and at the bottom.

A digital camera will be installed in front of the glass of the container. This one will turn with the rolling box to observe the movement of the cargo. This full HD color camera allows observation and measurement of the phenomenon. If measurement needs more precision, a higher definition camera will be installed.

Small size B&K IEPE accelerometers could be installed in the cargo.

Table 1. Performances of the centrifuge Rolling test device.

Max. g-level	80
Maximum angular velocity [°/s]	60
Maximum rotation angle of Rolling [°]	±25
Mass of the box (empty) [kg]	140
Maximum mass of material in the box [kg]	40
Maximum moving mass [kg]	400
Total mass of the device (empty) [kg]	1485*

*Including centrifuge container

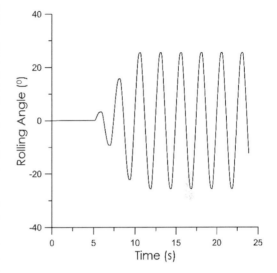

Figure 5. Rolling test frequency response.

Roll angle can be measure with the rotation sensor installed to control the Rolling box.

Pressure sensors are installed on the hydraulic inputs of the actuator to verify the approximate torque issued by the system.

2.1.5 Performances

The performances have been selected to simulate one degree of freedom of a rigid vessel rolling movement during shipment. Thanks to the centrifuge technique, the stresses and pressures are similar to the ones encountered in the vessel. The technical characteristics are presented in Table 1.

A first proof test has been performed up to 80×g. The box, filled with water, has been tested at the lowest elevation under a frequency of 0.1 Hz. The movements follow a sine signal and pressures in the system were in line with the expectations (Figure 5). A second series of proof tests have been realized with an empty box and with dry sand inside. The Figure 6 shows the envelope curve that links the maximum frequency to the maximum roll amplitude. All the combinations below this curve are available.

2.1.6 Scaling laws

To simulate the movement of the boat on the sea, the dynamic phenomenon should be scaled. This is not possible for this device, so it is assumed that the wave

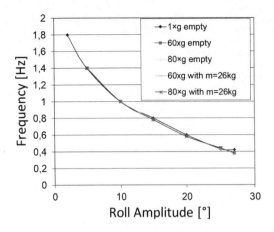

Figure 6. Envelope response curve of the Rolling test (model scale).

frequency (and the wave period) are not scaled. The scaling factors, in a macrogravity field of N, are then $\ell* = 1/N$ for length, $\sigma* = 1$ for stresses, $t* = 1$ for time and $f* = 1$ for frequency. This is not perfect, but it permits the observation of phenomenon.

An assumption is that, as the movement is quite slow, there is an equilibrium at each instant. So the scaling on dynamic effect would not play any role.

Concerning the flow of water, if it follows the Darcy law, and diffusion effect on time, that is scaled as $t* = 1/N^2$. For water, the Darcy velocity becomes $v* = \ell*/t* = N$. This is too high to avoid turbulent flow or other phenomenon like erosion, so the technique will be to increase the fluid viscosity by $\eta* = N$.

This parameter will be studied in the first series of tests.

3 CONCLUSIONS

A new device has been developed to simulate Rolling test for studying the liquefaction hazards of ore cargo. Designed for centrifuge testing at $80 \times g$, it has been successfully tested under those conditions in the framework of approval testing. In the future, the first tests with ore cargo will concern iron concentrate and lateritic-nickel ore. The objectives are: 1) to observe; 2) to understand; 3) to simulate liquefaction of ore cargo and 4) to test counter-measures on small-scale models to avoid this phenomenon.

ACKNOWLEDGMENTS

This research has been supported by the programme ERANet MarTec (Maritime Technologies) via the French Ministry of Ecology, Sustainable Development and Energy, General Direction of Infrastructures, Transports and Sea (DGITM/SAGS/EP1). They are greatly acknowledged.

REFERENCES

ASTM C230/C230M – 14 Standard Specification for Flow Table for Use in Tests of Hydraulic Cement. 6p.

Atkinson J.H. & Taylor R.N. 1994. Moisture migration and stability of iron concentrate cargoes. Centrifuge 94, Leung, Lee & Tan (eds). Balkema, Rotterdam. pp. 417–422.

Class NK, 2012 Guideline for the safe carriage of Nickel Ore. 188p.

Corté J.F., 1989. Essais sur modèles réduits en géotechnique. Rapport général, session 11. XII ICSMFE Rio, vol. 4, pp. 2553–2571.

GBWG (Global Bauxite Working Group) 2017. Report on research into the behavior of bauxite during shipping. 149p. International Maritime Solid Bulk Cargoes Code 2013.

ISSMGE web site: http://www.issmge.org/committees/technical-committees/fundamentals/physical-modelling

IMO (1998). International Maritime Organization. Code of safe practice for solid bulk cargoes. London: International Maritime Organization.

IMO (2012). International Maritime Organization. International maritime solid bulk cargoes code. London: International Maritime Organization.

Philips E., 1869. De l'équilibre des solides élastiques semblables. Comptes rendus hebdomadaires des séances de l'Académie des Sciences, vol. 69, série 1, 75–79.

Design and performance of an electro-mechanical pile driving hammer for geo-centrifuge

J.C.B. van Zeben & C. Azúa-González
Delft University of Technology, Delft, The Netherlands

M. Alvarez Grima & C. van 't Hof
Royal IHC, Kinderdijk, The Netherlands

A. Askarinejad
Delft University of Technology, Delft, The Netherlands

ABSTRACT: Pore pressure development along with other soil processes occur during pile driving. There is still limited knowledge about these processes and especially pore pressure development. To that aim an experimental research has been launched to investigate soil response mechanisms during offshore pile driving. A new model pile driving device for the geo-centrifuge has been designed and constructed for the purpose of this research. The model pile driving device uses an electro-mechanical lift system to drive open ended steel piles. The design of the hammer focuses mainly on accurately recreating the blow rate and the blow energy according to the corresponding scaling laws. Each of the three major characteristics of hammering, i.e. frequency, ram mass, and stroke, can be adjusted to accommodate different prototype hammers. The applied energy to the pile is measured through a custom-made load sensor integrated into the pile cap and passive coil measuring the impact speed. The model pile driving hammer is designed to work between 30g and 50g. This paper summarizes the design and performance of the pile driving device.

1 INTRODUCTION

The installation of large open ended steel piles triggers several soil processes along the pile. One of these is a change in pore pressure (Iskander, 2010). This pore pressure development can have an effect on the drivability of the pile and efficiency with which it is installed. IHC MTI together with the Delft University of Technology (DUT) are researching this topic. Alongside a numerical study on the process, an experimental study in the centrifuge of DUT was initiated.

For this experimental study a new centrifuge pile driving hammer was developed. The need to model the large blow energy of an offshore pile driving hammer and achieve a high blow rate to properly model the pore pressure development during driving were central to the design of the pile driving hammer.

The DUT centrifuge is a beam centrifuge with a diameter of 2.6 meters and capable of accelerating a weight of 300 N up to 300g (Allersma 1994). The carriages can hold models up to 300 by 400 by 550 mm. The pile driving hammer is designed to drive piles at 30g and 50g with a continuous blow rate of 10 to 30 Hz or a single blow. Model piles of 41.4 mm outer diameter and a wall thickness of 2.6 mm can be driven up to 120 mm into a fine GEBA sand sample (De Jager et al, 2017, Maghsoudloo et al, 2017). The soil sample is saturated with a viscous fluid in order to model pore fluid response correctly.

This paper describes the development of the pile driving system and the performance. Test results of the first test series in dry sand are also presented and discussed.

2 PILE DRIVING HAMMER

In the past a small pile driving device was created for the DUT centrifuge to hammer piles with a diameter of 15 mm using a pneumatic system. This system has been decommissioned. Moreover the system was also not powerful enough to drive larger diameter piles at a high blow rate. Therefore a new pile driving device needed to be designed to accommodate the current research's need for a system that could drive large diameter open ended piles with the required blow rate and blow energy. A literature review of current hammer devices, such as the pneumatic system described by De Nicola & Randolph (1994) and other systems as described by Levacher et al (2008), showed that the requirements needed were quite high especially regarding ram mass, fall height and blow rate.

An important research goal was the capability to accurately model pore pressure development resulting from the pile driving process. To model this, the blow rate had to be scaled correctly. The prototype pile driving hammers were the IHC Hydrohammer S series; specifically the Hydrohammer S200, S280 and S500 with a blow rate of 45 blows per minute. The model pile driving device was designed for 50g operations. An acceleration of 50g proved optimal as fall time, fall height and model dimensions were still within a range that was achievable within the centrifuge facility while being able to model large diameter piles.

At 50g a model blow rate of 35 blows per second is required. The prototype hammers are commonly used as offshore pile driving hammers with a ram mass larger than those of onshore hammers. The prototype hammers deliver a blow energy between 200 and 500 kJ. To achieve the correct scaled energy per blow a relatively heavy ram mass was needed. Scaling laws showed that a ram mass between 110 and 240 g, was needed to be able to achieve the blow energy needed at a maximum impact speed of 6.3 m/s. This combination of a heavy ram mass and blow rate of 35 Hz provided the main design requirements of the pile driving hammer. In order to fulfil this requirement an electrical mechanical lift system was chosen.

The basis of the lift mechanism is a flywheel that picks up the ram mass at its lowest point and releases it at a predefined point. The flywheel is powered by a 1.2 kW electric brushless motor. A set of gears provides a ratio of 4.55 between the motor and the flywheel. The gear system ensures that the motor can operate at maximum efficiency while keeping the flywheel at the correct speed for pile driving. Figure 1 provides a schematic overview of the pile driving device.

A modular system was created for the ram mass consisting of a bearing house and a top cap to achieve the correct mass. The bearing house fixes the ram mass to the ram mass guiding beam. This beam ensures the position of the ram mass in the horizontal plane while accommodating vertical movement with limited friction. Three different top caps were made so to allow the ram mass' mass to change between 110 g, 140 g and 220 g. The modular system makes it possible to change the blow energy in between tests and also accommodate different masses for future research without the necessity of major alternations. The fall height is determined by several factors and can be varied from 40 to 55 mm resulting in an impact speed on the anvil between 4 and 6.3 m/s at 50g.

The entire motor housing is connected with a sliding bearing to the load frame to allow it to follow the pile as it is driven. This ensures that distance from the pile anvil, and thus fall height, is kept constant. The motor housing is connected with a vibration free pinhole connection on the pile anvil. As the motor housing – including the flywheel, gears and electric motor – weighs 3 kg, a stiff connection would put a disproportional load onto the pile. Therefore a counterweight was connected to the back of the motor housing via a cable and pulley system.

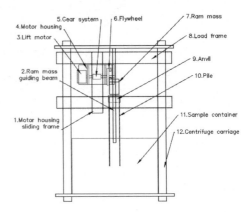

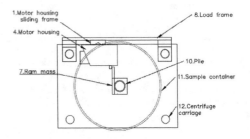

Figure 1. Schematic overview of pile driving device.

As the pile penetrates the soil, the motor housing follows it downwards. The g-level is dependent on the distance from the axis of the centrifuge. As the motor housing moves away from the centrifuge axis, the counterweight moves towards the axis, resulting in a change in the balance between these two components. To counteract this, a spring is connected to the motor housing that comes under tension as the motor housing moves down. This spring compensates the unbalance resulting from the changing radial distance from the centre of rotation. The system is balanced in such a way that the motor housing is always pressing very slightly on the pile and has a resulting force downwards. The extra load from lifting the ram mass is also transferred to the pile through the connection between motor housing and pile anvil. In practice, the load of the hammer and the resulting force from lifting the ram mass are also transferred into the pile and not into the piling rig.

To perform single blow tests a retractable pin is located on the motor housing. This pin can block the ram mass in the highest falling position. The pin can be retracted by a servo motor thus releasing the ram mass. This is also part of the multiple blow system as the electric motor first spins up to the desired RPM without the extra load of lifting the ram mass. After the desired RPM has been reached the ram mass is released into the trajectory of the flywheel and continuous pile driving commences.

When the ram mass is picked up by the flywheel there is a momentarily drop in angular velocity of the wheel as the electric motor needs to deliver extra power. This effect is only present at the first few blows after which the extra energy in the following blows is captured by the momentum of the flywheel.

2.1 Model pile

The model pile has an outer and an internal diameter of 41.4 and 36.2 mm, respectively, with a length of 190 mm. The pile can be embedded up to 120 mm. Height restrictions within the carriage of the centrifuge prevented a longer pile. The pile is restricted from horizontal movement by a Teflon bearing ring embedded in a PVC beam that sits on top of the sample container. The bearing ring does allow for vertical and rotational movement.

The pile cap is on top of the pile. The pile cap has several functions. Besides being the anvil for the ram mass; first, within the anvil four strain gauges are connected in a full Wheatstone bridge configuration. This sensor measures the incoming blow energy of each blow. It also picks up reflections coming back from the pile. Secondly, on top of the anvil is a PVC ring with 400 revolutions of 0.1 mm copper wire. The copper wire acts as a passive coil. On the ram mass four small magnets are attached. As the ram mass moves through the coil a small electrical current is produced. From calibration tests it was deduced that the peak created during the movement through the coil gives a high quality measurement of the speed just before impact without interfering with the free fall of the ram mass. Lastly, the anvil is the connection point for the draw wire displacement sensor that is attached to the frame of the centrifuge. This measures the absolute displacement of the pile during centrifuge operation and pile driving. All of these sensors have a data acquisition rate of 10 kHz.

2.2 Performance of pile driving hammer

The key design requirement for the pile driving hammer was the capability of operating at frequencies up to 35 Hz. In an initial system concept test, a secondary setup was used outside the centrifuge to better control and inspect the performance. The motor housing, ram mass and ram mass guiding beam were placed in a frame outside the centrifuge. Attached to the ram mass was a long spring that simulated the increased force and acceleration that would be present during centrifuge flight. The system performed well in this setup and blow frequencies varying from 10 Hz to 30 Hz were achieved. Thereafter two continuous blow rate centrifuge tests were performed, one at 20g and one at 50g. Sand samples were prepared with a high relativity density of 90%. For the initial tests the blow frequency was set at 10 Hz and the lightest ram mass of 110 g was used. A continuous blow rate was maintained for 6 seconds resulting in a total of 60 blows.

The main focus of these tests was to check the performance of the lifting mechanism and the sensors installed under increased g-level loading. The

Figure 2. Photograph of pile driving hammer in centrifuge.

performance achieved by the pile driving device was similar to the performance measured during 1g tests. Some minor adaptations were made to the counterweight and spring system and to the reaction time of the electric motor to optimize the pile driving process. Figure 2 shows the complete setup installed in the centrifuge.

2.3 Sample preparation

Three new sample containers were constructed. To mitigate boundary effects the largest possible containers were used. Three circular containers with a diameter of 315 mm, a wall thickness of 10 mm and a bottom plate of 20 mm were constructed out of HDPE plastic. They have a free internal height of 175 mm.

In order to scale the pore pressure development correctly the permeability of the soil has been reduced by using a viscous fluid (Askarinejad et al, 2014 & 2017). The viscous fluid used was developed by Deltares (Allard & Schenkeveld, 1994).

For sample preparation the filling method described by Rietdijk et al (2010) was used. This method involves three main steps. First, a water saturated sand is pumped into the container and drizzled down submerged. The sand is then compacted to the desired density using shockwave compaction. Finally the water level is lowered to just below soil level and viscous fluid is slowly pumped on top of the sample. From the bottom the water is slowly pumped out at a pump rate that allows the complete substitution of water by viscous fluid as pore fluid. A drainage layer with a pump outlet is present at the bottom of each container. This drainage layer consists of 10 mm high permeable rubber mats which on top have 4 layers of mesh. From

Table 1. GEBA sand properties (Krapfenbauer, 2016).

Property	Symbol	Value	Unit
Specific gravity	G_s	2.65	[–]
Average grain size	D_{50}	0.103	[mm]
Lower 10% grain size	D_{10}	0.90	[mm]
Lower 60% grain size	D_{60}	0.110	[mm]
Maximum void ratio	e_{max}	0.934	[–]
Minimum void ratio	e_{min}	0.596	[–]
Coefficient of curvature	C_c	1.47	[–]
Coefficient of uniformity	C_u	1.22	[–]
Angularity		Rounded-Subangular	[–]
Critical state friction angle	φ_{cs}	30 ± 1	[°]

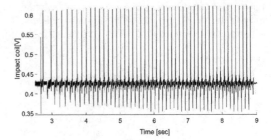

Figure 3. Passive coil measurements.

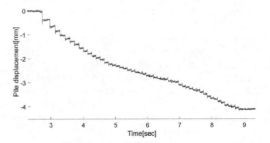

Figure 4. Pile displacement, multiple blow, 50g.

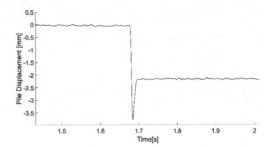

Figure 5. Pile displacement, single blow, 50g.

top to bottom the mesh size is 3 mm, 50 micron, 150 micron and 3 mm. The 3 mm mesh is used to fixate and protect the finer mesh from the sand.

For the tests GEBA sand was used. This is highly uniform sand that has been extensively tested before at DUT and ETH Zurich (Krapfenbauer, 2016).

GEBA sand has a narrow grain size distribution between 80 and 200 μm and consists of 99% silica. The sand was chosen because of the small grain size. Two important factors in scaling the soil structure interaction are the wall thickness, t, over D_{50} ratio and the development of the shear band along the pile. According to De Blaeij (2013) and De Nicola & Randolph (1997), the t/D_{50} should be at least 10. With GEBA sand and the chosen pile configuration this is easily achieved.

3 TEST RESULTS

The following test results are from a multiple blow test and the single blow test both performed at 50g. For the multiple blow test the 110 g ram mass was used and for the single blow test the heaviest ram mass of 240 g was used.

Figure 3 shows the impact speed measurement from 52 separate blows during 5.5 seconds. The y-axis displays the voltage measured over the coil. This is not the actual impact speed displayed; the signal is converted into impact speed by subtracting the high peak and the low peak directly following it and applying a calibration factor. The lift action from the pile driving can also be seen in this measurement from the sharp low peaks in between the high peaks. Analysis of the measurements gave an average impact speed of 4.9 m/s with a deviation of less than ± 5%. Analysis of the data sets from the multiple blow, single blow and the calibration tests show that the measurement indicates a reproducible and reliable measurement of the impact speed. The single blow impact speed was measured at 4.75 m/s.

The pile displacement sensor data for the multiple blow test is displayed in Figure 4. The total displacement after the 52 blows is 4.2 mm.

Figure 5 shows the displacement of the pile from the single blow with the heaviest ram mass of 240 g at 50g. A moving average filter has been put over the data to remove some of the noise. The total displacement from this single blow was 2.1 mm. Closer examination of the displacement measurement show the same trend of a small rebound of the pile anvil system after each blow. Pile displacement was numerically simulated beforehand by (Azúa-González et al, 2018). Comparison between the numerical simulation and experimental data showed similar behavior in both quantitative and qualitatively aspects. The difference in penetration between the lightest and heaviest ram mass was also similar to the expected difference.

Measurement from the load cell showed a constant repetitive force going into the pile during multiple blow testing. The measurements show minor distortions as the impact surfaces of the anvil and ram mass are not always completely perpendicular. This influences the way the load is measured by the strain gauges. Secondly owing to the speed with which the impact waves propagate through the pile system it is difficult to exactly quantify the load going into the pile. At a measuring frequency of 10 kHz the wave was still too quick to be measured accurately. A slight optimization of the measuring frequency is maybe possible and will be investigated in future tests. Also placement of strain gauges on the pile wall is an option to further investigate the energy transferred into the pile.

4 CONCLUSIONS

A pile driving device was successfully constructed for the DUT Centrifuge facility. Main design aspects regarding blow rate and ram mass were achieved. Single blow and multiple blow rates reaching from 10 Hz to 30 Hz can be achieved and maintained. A new sensor system to monitor pile behavior during installation in combination with high rate acquisition system has proven to be operational and adequate to measure key aspects of the pile driving process.

Initial tests show impact speed and pile displacement are within the design specifications. Pile displacement and soil structure interaction compare well to previously numerical simulations. Further testing will be done to simulate the pore pressure development under different blow rates.

ACKNOWLEDGEMENTS

This project was partly funded by Royal IHC. Additional support was provided by the Geotechnical Engineering Department of DUT. Sincere appreciation goes out to all the lab technicians involved with the project and in special to J.J. de Visser and C. van Beek for their continued support.

REFERENCES

Allard, M. & Schenkeveld, F. 1994, The Delft geotechnics model pore fluid for centrifuge tests, in *'Centrifuge'*, Vol. 94, Balkema, Rotterdam, pp. 133–138.

Allersma, H. 1994, *The university of delft geotechnical centrifuge*, in 'Centrifuge', Vol. 94, Balkema, Rotterdam, pp. 47–52.

Askarinjad, A., Beck, A. & Springman, S. M. 2014. Scaling law of static liquefaction mechanism in geocentrifuge and corresponding hydromechanical characterization of an unsaturated silty sand having a viscous pore fluid. *Canadian Geotechnical Journal*, 52, pp. 708–720.

Askarinejad, A., Philia Boru Sitanggang, A. & Schenkeveld, F. M. 2017 Effect of pore fluid on the cyclic behavior of laterally loaded offshore piles modelled in centrifuge. *19th International Conference on Soil Mechanics and Geotechnical Engineering*, 2017 Seoul, South Korea. pp. 905–910.

Azúa-Gonzalez, C., Van Zeben, J.C.B., Alvarez Grima, M., Van 't Hof, C., Askarinejad, A. 2017. *Dynamic FE analysis of Soft Boundary Effects (SBE) on impact pile driving response in centrifuge tests*, in 9th Int. Conf. on Physical Modelling in Geotechnics [under revision].

De Blaeij, T. 2013. On the modelling of installation effects on laterally cyclic loaded monopoles. *MSc thesis*, TU Delft.

De Nicola, A., Randolph, M.F. 1994. Development of a miniature pile driving actuator. in *'Centrifuge'*, Vol. 94, Balkema, Rotterdam, pp. 473–478.

De Nicola, A. & Randolph, M. 1997. The plugging behaviour of driven and jacked piles in sand. *Geotechnique*, 47 (4).

Iskander, M., 2010. *Behavior of Pipe Piles in Sand*. Brooklyn: Springer.

Krapfenbauer, C. 2016. Experimental investigation of static liquefaction in submarine slopes. *MSc thesis*, ETH Zurich.

Levancher, D., Morice, Y., Favraud, C., Thorel, L. 2008. A review of pile drivers for testing in centrifuge. *Xèmes Journées Nationales Génie Côtier*, Okt 2008, pp. 573–583.

Maghsoudloo, A., Galavi, V., Hicks, M.A., Askarinejad, A., 2017, Finite element simulation of static liquefaction of submerged sand slopes using a multilaminate model. *In of Proceedings of the 19th International Conference on Soil Mechanics and Geotechnical Engineering*, Seoul, Korea.

Richard R. de Jager, Arash Maghsoudloo, Amin Askarinejad, Frans Molenkamp 2017, Preliminary results of instrumented laboratory flow slides, in *Procedia Engineering*, Volume 175, pp. 212–219, ISSN 1877-7058.

Rietdijk, J., Schenkeveld, F.M., Schaminée, P.E.L., Bezuijen, A., 2010. The drizzle method for sand sample preparation. *Physical modelling in Geotechnics – Proceedings of the 7th international conference on Physical modelling in Geotechnics*, pp. 267–272.

A new heating-cooling system for centrifuge testing of thermo-active geo-structures

D. Vitali, A.K. Leung, R. Zhao & J.A. Knappett
University of Dundee, Dundee, Scotland, UK

ABSTRACT: Centrifuge modelling has been considered as an effective means for investigating the energy and geotechnical performance of thermo-active geo-structures. A major challenge to correctly model (i) soil-structure heat transfer and (ii) thermo-mechanical behaviour of the geo-structure in a centrifuge is to design a system that could deliver sufficient heat energy (i.e. in terms of flowrate and temperature) under enhanced gravity conditions. This paper reports a new and robust heating/cooling system developed for these purposes and evaluates its performance. The proof heating tests performed up to 50-g suggest that when an appropriate pipe configuration is designed, the heating system is capable of producing a water flowrate up to 13.5 ml/s, which is sufficient to generate a turbulent flow regime within the water circulation pipe, hence maximising the convective heat transfer mechanism. The heating system has been successfully applied to deliver a controllable amount of heat energy, simultaneously, to multiple thermo-active piles in a row for warming up the surrounding soil. With proper thermal insulation of the pipework of the system, temperature loss between the target value at the pipe inlet and the one registered at the entrance of the model structure could be less than 2°C. An idea for extending the system to lower the temperature below ambient is also presented.

1 INTRODUCTION

Thermo-active foundations are sustainable technology that do not only provide structural support to the superstructure but also exploit clean and renewable geothermal energy for satisfying the energy demand in the structure. Typically, this kind of foundation is made of reinforced concrete (RC), where high-density polyethylene (HPDE) pipe loops are embedded and attached along the steel reinforcement. The pipes are to circulate heat-carrier fluid so that the thermal energy could be exchanged between the supported structure and the soil surrounding the foundations. The cyclic thermal loading applied to the soil introduces thermal and mechanical stresses and strains to the RC foundation. Improved understanding of the complex soil-structure interaction associated with the combined thermo-mechanical loadings is required to inform engineering design.

Centrifuge modelling has been increasingly used to study the thermo-mechanical behaviour of thermo-active geo-structures such as a pile and its interaction with the surrounding soil under combined mechanical and thermal loadings. Compared to field experiments, testing small-scale models is relatively cheap and can be performed under much more controlled testing and boundary conditions. Hence, this approach can be used to isolate undesirable and uncontrolled field conditions (e.g. soil heterogeneity and groundwater flow), exclusively focusing on the complex soil-structure interaction involved. This is important for benchmarking numerical and analytical tools. The analysis performed by Rotta Loria et al. (2015) has shown that centrifuge model tests are capable of capturing fundamental aspects and provide highly consistent and comparable responses of thermo-active piles.

As far as the authors are aware, there are two active centrifuge facilities testing the geotechnical performance of thermo-active geo-structures, namely the University of Colorado Boulder (Stewart 2012, Stewart & McCartney 2014) and the Hong Kong University of Science and Technology (Ng et al. 2014, Shi et al. 2016). Both systems adopt a closed circuit system in which a circulating fluid transfers heat to/from the model energy geo-structure.

The system developed by Stewart (2012) uses a Julabo F25-ME heating/cooling water bath circulator that is set outside of the centrifuge chamber for controlling the temperature of ethylene glycol. The fluid is directed to a model structure through the centrifuge slip rings using a pump placed outside the centrifuge. The setup requires a pre-heating phase prior to centrifuge spinning so that all the components including the slip rings within which the fluid is circulating can reach a constant temperature for minimising heat losses. Thus, this system requires two slip rings for fluid circulation, which could limit inflight activities for small- to medium-size centrifuge facilities that have only a limited number of slip rings. Normally, inflight activities such as hydraulic jacking, earthquake simulation, application of rainfall would require at least one slip ring for operation.

The system developed in Hong Kong, on the contrary, avoids the use of multiple slip rings by introducing two independent components: one for the heating of the glycol-water solution and one for its cooling. The heating is achieved by using an electric heating element that is submerged in an aluminium tank filled with the circulating fluid. The cooling, on the other hand, is obtained through the Peltier effect to bring the fluid temperature below the ambient. The two components are arranged on the centrifuge together with a pump that allows the fluid to be delivered to the model foundations (Shi et al. 2016). The system requires a considerable amount of space to mount the two components onto a centrifuge and is therefore less suitable to be adopted by small- to medium-size beam centrifuges.

This paper details the development of a new heating/cooling system that can be mounted at the geotechnical beam centrifuge facility at the University of Dundee. Due to the limit of space, only the heating component of the system is discussed here. The working principle of the system is elaborated and its performance under different gravity levels is evaluated.

2 SYSTEM DESIGN

The design objective is to satisfy the following three conditions, namely (1) the need to flexibly both increase and decrease the internal temperature of model geo-structures through the circulation of heat-carrier fluid – to mimic the process occurring in the prototype; (2) the ability to operate under high gravity conditions; and (3) to avoid/minimise the use of dedicated centrifuge slip rings. Objective (1) is crucial for allowing the study of not only the geotechnical performance of the model, (i.e. thermo-mechanical responses of structure) but also energy performance (i.e. heat transfer efficiency) which has not been fully taken into account in previous centrifuge modelling studies.

Figure 1 shows the schematic and an overview of the design. The entire system is mounted on the centrifuge arm, so no slip ring is required. The system consists of an aluminium tank, a magnetic centrifugal pump, two nylon manifolds, a solenoid valve, a manual regulator valve and a series of sensors (i.e. k-type thermocouples and a flow meter) for monitoring the system efficiency. The tank can store up to 5.5 l of water. The water can be heated by a 1.0 kW electric heating element, whose temperature is controlled via a temperature controller and a submergible thermocouple. The temperature controller is used to maintain a desired temperature when it is deviated from the target value at the heating rod location by 1.5°C. Once the target water temperature is reached, the pump can be switched on via an electromechanical relay to circulate the hot water into the nylon pipes fixed along the centrifuge arm.

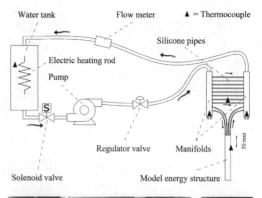

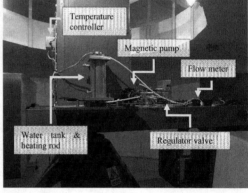

Figure 1. Schematic diagram (top) and overview (bottom) of the heating system mounted on the Dundee beam centrifuge.

A multi-port manifold is placed at the end of the nylon pipe for connecting with ten flexible silicone pipes which have an internal diameter of 1.5 mm each. These silicone pipes can be embedded in any model structure and are used to deliver the hot water to the structure. The pipes coming out from the model structure are then connected to a second identical manifold and finally to the top of the water tank through a nylon pipe. This forms a closed circuit.

A solenoid valve is placed between the water tank and the pump to isolate the water reservoir from the rest of the circuit in case any unexpected leakage happens along the circuit during a centrifuge test.

A regulator is installed behind the pump. The regulator, after calibration, can be used to adjust the water flowrate before the start of a centrifuge test. At the same time, a flow meter is placed behind the nylon manifolds both to verify whether the applied flowrate remains constant during the test and to check for any pipe leakage.

3 PROOF TESTING

Three centrifuge tests were conducted to evaluate the performance and functionality of the heating system under different g-levels ranging from 5- to 50-g. Test

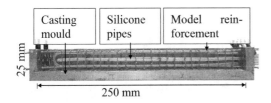

Figure 2. Detail of the reinforcement cage and of the silicone pipes embedded in the model thermo-active piles.

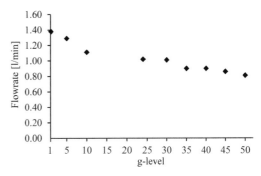

Figure 3. Steady-state flowrate (model scale) at different g-levels.

1 aimed to evaluate any influence of g-level on the flowrate when circulating the hot water, while Tests 2 and 3 were to quantify the heat losses in the system.

In each test, a model thermo-active pile made of a new type of model concrete developed by Vitali et al. (2016) was connected to the manifolds. The model concrete has similar thermal properties (thermal conductivity and coefficient of thermal expansion) of concrete at prototype scale, so that the energy performance (i.e. temperature distribution and heat transfer efficiency) of the model pile could be evaluated. The model concrete scales the bending stiffness of the pile. The model reinforced concrete (RC) pile (Fig. 2) has longitudinal reinforcements from which hang two U-shaped silicone pipes for circulating the heat-carrier water, similar to what is similarly found in prototype RC thermo-active piles (e.g. Brandl 2006, Bourne-Webb et al. 2009). The model pile also has shear reinforcement links which are square in shape. Further details about the thermo-mechanical properties of the model concrete and the model pile made of this material may be found in Vitali et al. (2016) and Minto et al. (2016), respectively. Pile surface temperature was measured by a k-type thermocouple (see sensor location in Figure 1).

3.1 Effects of g-level on water flowrate

Prior to centrifuge spinning, the temperature controller was set to 50°C and the heating element was activated. The regulator valve positioned after the magnetic pump was left fully open. The centrifuge was then spun up to 5-g, and then the pump was switched on to allow the hot water to circulate within the pipes. At steady state (when the flow meter measurement did not show any observable change for 4 min), the flowrate was recorded. Subsequently, the g-level was raised to 10-g and, again once a steady state was reached, the flowrate was recorded. This test procedure was repeated at 24-g, 30-g, 35-g, 40-g, 45-g and finally 50-g.

The water flowrates recorded under steady-state conditions at different g-levels are shown in Figure 3. There is an evident reduction of flowrate as the g-level increases, especially between 1-g and 10-g. The drop thereafter is less significant and seems to stabilise at a rate of ∼0.9 l/min. The observed drop is attributed to the reduction of the efficiency of the magnetic centrifugal pump as the g-level increases. According to the number of pipes connected to the manifold and the pipe geometry, the observed range of flowrate represents a laminar flow regime. However, it is worth-noting that transitional and turbulent regimes, if desired, can also be achieved by reducing the number of pipes to four and two, respectively. The test results also suggests that the heating system is capable of circulating hot water with a reasonable flowrate for thermo-active piles with a wide range of pile size in prototype (from 125 × 125 mm at 5-g to 1.25 × 1.25 m at 50-g).

3.2 Temperature distribution and heat efficiency

Two tests, namely Test 2 and Test 3, were performed at 24-g to investigate the heat losses occurring along the system when nylon pipes were thermally insulated (Test 2) or not (Test 3). In these tests both the heating rod and the magnetic pump were switched on before spinning (i.e. at t = 0 min of Figure 4).

In both tests, the temperature was monitored at five different locations (Fig. 1): three of them measuring the temperature of the water circulating within the system (i.e. within the water tank and the two manifolds) and two recording the temperature of the external surface of both the silicone pipes and the model piles.

Figure 4 shows the temperature data obtained from the five different control points for Test 2. A progressive increase in temperature was recorded by all the thermocouples after the heating element was switched on and the pump was activated. Although the target temperature was kept constant at 50°C at the heating rod location, the temperature of the water in the tank dropped slightly by 3–4°C (i.e. down to 46–47°C) as expected due to the heat exchange with the cooler air inside the centrifuge room during spinning (room temperature between 22 and 23.5°C). Apparently, the temperature recorded in the water tank was oscillating by no more than 2.2°C. This is an indication of the feedback system associated with the temperature controller, which tries to maintain the temperature at 50°C by switching on and off the heating rod intermittently. The temperature drop at the pile surface at elapsed time of about 10 min is because of the reduction of pump efficiency when the g-level rose from 1 to 24 g (refer to Figure 3).

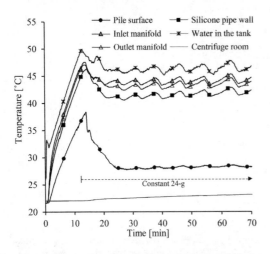

Figure 4. Temperature recorded during Test 2. Both the heating element and the pump were switched on at t = 0 min. The spinning was started at t = 2 min. Thermocouple positions are shown in Figure 1.

As the heat energy travels from the water tank to the inlet manifold at steady state, the mean temperature drop is less than 2.5°C when thermal insulation is applied to nylon pipes. A further drop of temperature from 44.9°C at the inlet manifold to 28°C at the surface of the model piles (i.e. reduced by 16.9°C) was recorded. This reduction was attributed both to the thermal resistance of the material of which the model pile is made (i.e. the internal silicone pipes and the model concrete) and to the heat lost due the wind generated by the centrifuge spinning. The thermal conductivity of the silicone pipes and model concrete are 0.12 and 0.73 W/(m·°C), respectively (Vitali et al. 2016). For the given pipe thickness (0.75 mm) and cover thickness between the pipe and the pile surface (4 mm), it may be estimated by Fourier's law (i.e. assuming steady state heat conduction mechanism and neglecting superposition effects due to the double U-shape pipe configuration) that the pile surface temperature should be 30.9°C. This means that no more than 17% of the heat loss was attributed to convective mechanisms occurring on the model surface due to the wind introduced by centrifuge spinning. In fact, during a centrifuge test, a wind cover may be applied to the centrifuge model box to minimise this kind of heat loss.

4 AN EXAMPLE OF SYSTEM APPLICATION

To further evaluate its performance for engineering applications, the heating system was connected to a row of three model thermo-active piles embedded in an unsaturated compacted soil. This centrifuge test (Test 4) aims to evaluate the temperature distribution in the piles and the surrounding soil through the circulation of hot water in the pile row.

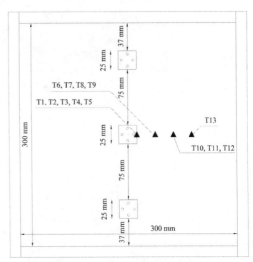

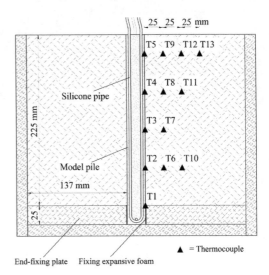

Figure 5. Plan view (top) and elevation view (bottom) of the centrifuge test setup in Test 4. Note that only results from T2, T4, T8 and T11 are presented in this paper.

A g-level of 24 was selected for this test, and the regulator valve was partially closed to control a steady-state flowrate of 2.22 ml/s in each one of the six silicone pipes connected to the manifolds. Details of the test setup are given in Figure 5.

The soil used was A50 silica flour, which is an artificial crushed soil with a quartz content higher than 99.5%. Some soil properties are summarised in Table 1. The soil was compacted to a target dry density of 1.58 g/cm^3 and an initial water content of 18% (by mass). During compaction, dummy aluminium piles were used to prevent any structural damage to the model RC pile. After soil compaction, the dummy piles were then replaced by the model RC ones.

The target temperature set at the heating rod was 53°C, while the temperature recorded at the inlet

Table 1. Properties of A50 silica flour.

Index Properties	
d_{10} [mm]	0.004
d_{50} [mm]	0.04
d_{90} [mm]	0.1
c' [kPa]	9
φ' (plane strain conditions) [°]	38°
Thermal properties	
Coefficient of thermal expansion [10^{-6} °C^{-1}]	0.55
Thermal conductivity [W/(m °C)]	1.38
Van Genuchten (1980)'s hydraulic parameters	
K_s [mm/s]	$7.8 \cdot 10^{-4}$
θ_r [cm^3/cm^3]	0.0295
θ_r [cm^3/cm^3]	0.2844
α [cm^{-1}]	0.0022
n [-]	1.3313
l [-]	0.5

Note: K_s is saturate permeability; θ_r is residual water content (by volume); θ_s is saturated water content; α, n and l are fitting parameters that control the shape of the water retention curve.

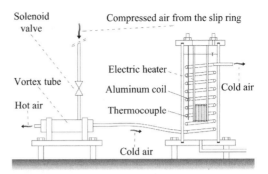

Figure 7. Schematic of the proposed cooling system.

The soil temperature recorded at horizontal distances of 25 and 50 mm from the pile surface shows that steady-state condition was not reached even after 165 minutes of spinning, corresponding to a prototype time interval of 66 days at 24-g.

5 ONGOING SYSTEM UPGRADE

The next phase of the system development is to incorporate a cooling component for applying and controlling the temperature of the heat-carrier fluid below ambient. The ultimate goal is to develop a holistic system that can flexibly apply cyclic thermal loadings (i.e. controlled heating/cooling cycles) to a model geo-structure during a centrifuge test.

Figure 7 shows the proposed modification. Cooling of the heat-carrier water can be achieved by using a Ranque-Hilsch vortex tube. Compressed air coming from a centrifuge slip ring will be injected into the tube. A vortex would be generated inside the tube to separate the compressed air into hot and cold streams and they will be ejected at the two ends of the tube. The stream temperature depends on the entering pressure and temperature of the compressed air. For example, compressed air with a pressure of 0.4 MPa and a temperature of 20°C would produce a cold stream as low as 6°C.

The cold stream will be then directed into an aluminium coil mounted inside the water tank. Hence the passage of the cold stream would cool the surrounding water (when the heating rod is switched off). Preliminary proof testing at 1-g shows that with proper thermal insulation of the water tank and the continuous circulation of the cold stream, the temperature of the heat-carrier water dropped to 7°C.

In order to apply cyclic heating/cooling loads, the temperature controller will be replaced by a remote controlled one, which will switch the electric heating rod on and off following a prescribed thermal test path. When the heating rod is switched off, the relay-controlled solenoid valve placed between the slip ring and the vortex tube (Fig. 7) will be opened. Subsequently, the compressed air will be injected into the vortex tube and the cooling procedure starts.

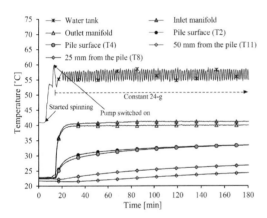

Figure 6. Temperatures recorded at different location during the 24-g test. Thermocouple positions are shown in Figure 5.

manifold is 40.6°C (Fig. 6). Compared to Test 2, the greater heat loss in this test is because no thermal insulation was applied to the nylon pipes. It can be seen that the rate of increase of the pile surface temperature was relatively slower than in the manifolds. This is because of the equivalent thermal resistance of the model pile, which has a value of 1.64°C/W. At steady state, the two thermocouples attached on the pile surface both showed 33.2°C, meaning that there was 8°C of temperature lost through the model material. This smaller reduction of temperature (compared to the 16.9°C drop recorded in Test 2) is expected because the pile, in this case, was in contact with soil material, instead of air. Due to the pile embedment, convective heat loss at the pile surface was minimised.

The performance of the proposed modification of the system is currently under evaluation through a further series of 1-g and high-g centrifuge tests.

6 CONCLUSIONS

A new and robust heating system for centrifuge testing of thermo-active geo-structures was developed. The system is compact in size and does not require any slip rings because all its main components are placed on the side of the central centrifuge cabinet, leaving only the pipes and some sensors running along the centrifuge arm.

The performance of the heating system was evaluated at various g-levels between 1-g and 50-g using the University of Dundee beam centrifuge. The flowrate of the heat-carrier water circulating into, and from, a small-scale model thermo-active pile was g-level dependent. The steady-state flowrate reduces from 1.38 to 0.81 l/min as g-level increased from 1-g to 50-g. Nonetheless, the system is capable of reproducing any water flow regime in the 1.5 mm silicone pipes by controlling the number of pipes connected to the manifolds, in order to achieve any desirable condition that could maximise the heat transfer between the different elements of a centrifuge model (model structures and surrounding soil).

With a proper thermal insulation of the pipework of the system, the temperature drop between the target value (in the inlet manifold) and the one registered at the entrance of the model structure is less than 2°C (i.e. equivalent to 5% when the target temperature in a model structure is set at 40°C). As the heat energy diffuses through the model structure, heat losses into the air through the convection mechanism due to the centrifuge spinning are minimal. This kind of heat lost could be further eliminated by applying an aerodynamic cover on the model box.

The robust heating system has been successfully applied to deliver controllable amount of heat energy to a row of thermo-active piles for heating up the surrounding soil. Ongoing work is underway to further expand the capability of the system for controlling the temperature of the heat-carrier water below the ambient, hence applying heating/cooling cycles.

ACKNOWLEDGMENTS

The authors would like to acknowledge the studentship and the research cost supported by Energy Technology Partnership (ETP), Scottish Road Research Board (SRRB) through Transport Scotland, and the EPSRC Doctoral Training Award.

REFERENCES

Bourne-Webb, P., Amatya, B., Soga, K., Amis, T., Davidson, C. & Payne, P. 2009. Energy pile test at Lambeth College, London: geotechnical and thermodynamic aspects of pile response to heat cycles". *Géotechnique*, 59(3): 237–248.

Brandl, H. 2006. Energy foundations and other thermo-active ground structures. *Géotechnique*, 56(2): 81–122.

Minto, A., Leung, A.K., Vitali, D. & Knappett, J.A. 2016. Thermomechanical properties of a new small-scale reinforced concrete thermo-active pile for centrifuge testing. In *Energy Geotechnics: Proceedings of the 1st International Conference on Energy Geotechnics, ICEGT 2016, Kiel, 29–31 August 2016*. CRC Press.

Ng, C.W.W., Shi, C., Gunawan, A. & Laloui, L. 2014. Centrifuge modelling of energy piles subjected to heating and cooling cycles in clay. *Géotechnique Letters*, 4(4): 310–316.

Rotta Loria, A.F., Donna, A.D. & Laloui, L. 2015. Numerical study on the suitability of centrifuge testing for capturing the thermal-induced mechanical behavior of energy piles. *Journal of Geotechnical and Geoenvironmental Engineering, ASCE*, 141(10): 04015042.

Shi, C., PA, V.L., Gunawan, A. & Ng, C.W. 2016. Development of a heating and cooling system for centrifuge modelling of energy piles at HKUST. *Japanese Geotechnical Society Special Publication*, 2(74): 2559–2564.

Stewart, M.A. 2012. *Centrifuge modeling of strain distributions in energy foundations* (Doctoral dissertation, University of Colorado at Boulder).

Stewart, M.A. & McCartney, J.S. 2014. Centrifuge modeling of soil-structure interaction in energy foundations. *Journal of Geotechnical and Geoenvironmental Engineering, ASCE*, 140(4): 04013044.

van Genuchten, M. Th. 1980. A closed-form equation for predicting the hydraulic conductivity of unsaturated soils. *Soil Science Society of America Journal* 44(5): 892–898.

Vitali, D., Leung, A.K., Minto, A. & Knappett, J.A. 2016. A new model concrete for reduced-scale model tests of energy geo-structures. In *Geo-Chicago 2016: Geotechnics for Sustainable Energy*, 185–194.

Physical modelling of soil-structure interaction of tree root systems under lateral loads

X. Zhang, J.A. Knappett, A.K. Leung & T. Liang
University of Dundee, Dundee, UK

ABSTRACT: This paper presents some initial physical modelling of root-soil interaction for trees under lateral loading, such as from wind or debris flows. 3-D printing of interconnected systems of ABS plastic root analogues, which are mechanically representative of tree roots, is used to produce idealised root architectures with varying root depth, spread and root area ratio (RAR, i.e. amount of roots). These models are installed within a granular soil with a wooden model trunk attached and laterally loaded to measure the push-over behaviour. From this data, the moment capacity and rotational stiffness of the model root systems are evaluated. The results suggest that (i) the effect of the central tap root on strength and stiffness is low, compared to the lateral and sinker roots, but otherwise proportional to RAR; and (ii) that development of analytical predictive models may be able to adapt existing procedures for conventional geotechnical systems, such as pile groups.

1 INTRODUCTION

1.1 Background

Understanding 'windthrow' of trees (their behaviour under strong lateral wind loads) has long been of interest in forestry, where it is important for understanding how much of a crop may be damaged by bad weather and how selective planting may be used to protect stands of vulnerable trees (Ulanova, 2000; Gardiner & Quine, 2000; Mickovski et al., 2005). It is also of interest in civil engineering, where windthrow of trees on sloping ground may be a trigger for landsliding, or where fallen trees may disrupt transportation services, such as occurred in the UK during Storm Doris in February 2017, where a fallen tree brought down overhead power lines on the West Coast Mainline. While research has previously been conducted into the subaerial aspects of this problem (i.e. fluid-structure interaction, e.g. Langre, 2008), little is known about how to model the interaction between the tree and the ground through the root system, which acts as a foundation of highly complex geometry. This is important as changing ground conditions (e.g. due to rainfall accompanying high winds) may soften and weaken the ground and increase the vulnerability of a tree to overturning.

1.2 Aims and objectives

In this paper, two contrasting 1:10 scale root systems were designed. One type is narrow and deep while the other is wide and shallow. Different numbers, sizes and distributions of roots were also considered to construct root system with different root area ratio (RAR).

3-D printing techniques were employed to fabricate these models using Acrylonitrile Butadiene Styrene (ABS) plastic material, which has mechanical properties highly similar to tree roots and has been used by other researchers to create root models for studying slope stabilisation by vegetation (e.g. Liang et al., 2015; Meijer et al., 2016; Liang & Knappett, 2017). The models were subsequently embedded into soil with model tree trunks connected to the top. A linear actuator was then used to conduct lateral load tests close to the bottom of the trunk to measure the rotational stiffness and moment capacity of the root systems acting as foundations, and including the second-order P-Δ effects from the trunk.

2 ROOT MODELLING

2.1 Trunk, stump and rooted zone geometry

Root area ratio (RAR) is an important characteristic when describing a root system. It refers to the proportion of total root cross sectional area within the zone of rapid taper (ZRT), a zone that includes most of structural roots with a radius conventionally defined as approximately 2.17 times bigger than the diameter at breast height (DBH; Danjon et al., 2005). In order to quantify the influence of RAR, three RARs are considered here, namely 0.5%, 2.0% and 4.5%, and this is believed to cover a wide range of species (Mao et al., 2012).

The thick main stem of a tree, which is also called the trunk, transfers lateral loading directly to the root system via the stump. According to research conducted by Danjon et al., (2008), the diameter of the stump has

Table 1. Root numbers for the deep and narrow root system.

RAR	Diameter of individual root: mm				
	12 (tap)	5	3	1.6	0.8
0.57%	0	1	2	4	6
2.01%	1	1	2	4	6
4.47%	1	4	13	25	34

been observed in the field to be approximately 250 mm for some tree species in slopes. This value was adopted in this study, resulting in a stump 25 mm in diameter at 1:10 scale. To simplify the design of the tree model and subsequent analysis, a circular trunk was designed having the same diameter, constant with height, such that DBH was 25 mm. The length of the trunk was 750 mm, thereby representing a tree 7.5 m tall.

For the selected DBH, the ZRT dimensions of the short and wide root system would exceed the threshold of the available 3-D printer. In order to make the models printable, the ZRT was set to twice the DBH, for both the deep and shallow model systems.

2.2 Root system

Liang et al. (2015, 2017) reported a prototype tree root system that was constructed based on the statistical data of tree root characteristics available in the literature. The system was shown to provide similar direct shear behavior to real root systems. This study bases on the same prototype root system to develop model root analogues for studying active push-over loading. The analogues were made of vertical circular rods of constant diameter (i.e. no tapering with distance from the stump). Five distinct types of root analogues with diameter 12 mm (just for the tap root below the stump), 5 mm, 3 mm, 1.6 mm and 0.8 mm were used within the design of the model root systems in this paper, which is consistent with previous work (Liang et al., 2015; 2017). The root architecture was idealized as a perforated circular disc at the ground surface, consisting of lateral 'spokes' and circumferential rings, mechanically connecting the set of vertical root analogues to the stump.

A range of RAR values between 0.5% and 4.5% was selected. It should be noted that for $RAR = 0.5\%$, the central tap root (immediately beneath the stump) was not included because in this case, the overall total root cross sectional area was smaller than the cross sectional area of a single tap root. Finally, three values of RAR (0.57%, 2.01% and 4.47%) were selected to use. It should be noted that the only difference between the models of $RAR = 0.5\%$ and $RAR = 2.0\%$ was only the addition of a single tap root at the centre of the root system. The number of each size of root analogue in each model system is shown in Table 1.

The cumulative root fraction (percentage of the total amount of roots within a certain distance from the stump) decreases sharply with increasing depth below the ground surface and for trees, shrubs and grasses,

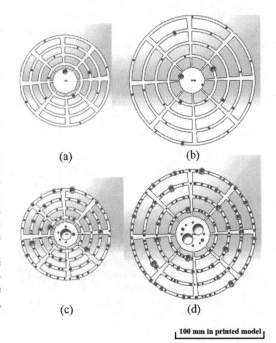

Figure 1. Root system constructed in Solidworks: (a) & (c) represent a deep and narrow root system with $RAR = 0.57\%$ and 4.47% respectively; (b) & (d) represent a shallow and wide root system with $RAR = 0.57\%$ and 4.47% respectively.

at least 90% of the cumulative root fraction reaches to 1 m depth (Jackson et al., 1996). So the deepest and narrowest model root system was designed to be 100 mm deep at model scale. The shallower, wider root system represents the case of the roots being half as deep, but with twice the plan area, without changing the RAR (i.e. having twice the number of each size of root analogue from Table 1, each being half the length).

The computer-aided design software Solidworks was used to construct the root system models and generate an input file for the 3-D printer. The vertical root analogues were distributed across five concentric rings forming the surface disc. The thickness of the rings and of the spokes connecting them was 3 mm, which was consistent with Sonnenberg et al., (2010) and the width of rings were designed to also be 3 mm. The spokes represent stiff lateral branches off the main tap root (see Fig. 1). The width of the eight uniformly distributed spokes ranges from 0 mm at the central point to 6 mm at edge, such that the average width was also 3 mm.

In this study, individual roots and perforated discs were printed separately, with the discs incorporating printed holes, 2 mm in depth, for subsequent attachment of the root analogues. In addition, a coupler with 25 mm internal diameter, 10 mm in height and 1 mm in thickness was designed to attach the trunk to the stump.

2.3 Model fabrication and assembly

The model roots were fabricated from Acrylonitrile Butadiene Styrene (ABS) plastic which was chosen as

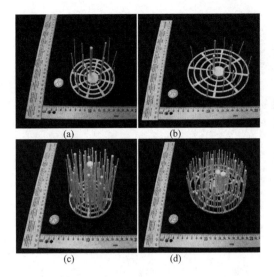

Figure 2. 3-D printed root system model: (a) & (c) show the deep and narrow root system with $RAR = 0.57\%$ and 4.47% respectively; (b) & (d) show the shallow and wide root system with $RAR = 0.57\%$ and 4.47% respectively.

its properties are similar to those of real roots (Liang et al., 2015; 2017). All of the models were fabricated using a Stratesys Inc. uPrint SE ABS rapid prototyper (3-D printer) at the University of Dundee. The fabrication process, which involves progressive layering of the ABS plastic was exploited to print the root analogues with the layering running along the vertical root axis, to approximate the fibrous structure of real roots (Liang et al., 2014).

Epoxy resin adhesive was used to connect the root analogues with the disc. This required some abrasion of the printed holes to ensure the root analogues would fit snugly. Some assembled model root systems are shown in Figure 2.

3 TEST METHODOLOGY

A 900 mm × 450 mm × 450 mm box, which is large enough to avoid boundary effects, was used to contain HST95 Congleton silica sand. This material was pluviated in air to a relative density of approximately 50% for each test. Root system models were subsequently pushed into the soil after pluviation, this being similar to root growth in reality.

Lateral force was applied using a GDS Force Actuator (GDSFA), which is a general purpose electrically driven uniaxial loading system with feedback control and continuous displays of force and displacement. This paper reports only push-over tests to large displacement (90 mm), which were conducted under displacement control at a constant speed of 6 mm/min. Force and displacement readings were logged at a sampling frequency of 1 Hz.

The loading was applied towards the bottom of trunk, 140 mm above the ground surface. A trolley

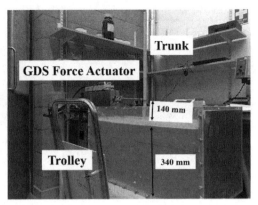

Figure 3. Schematic test setup.

was used to raise the model container to this height. The wheels of the trolley were locked during testing to ensure there was no displacement of the reference frame. The geometry of the test setup is shown in Figure 3. The thickness of box base was 30 mm and the soil 350 mm in depth.

4 RESULTS AND DISCUSSION

4.1 Push-over curves

Figure 4 shows the direct results of pushover curves measured for the different types of root system. It can be seen that, in general, the push-over force experienced a rapid increase in the initial stage of the test. However, after reaching a certain value, the push-over force decreases at a similar rate with initial increase. For the deep and narrow root system, the behavior differs between smaller and larger displacements, the reason for which will be discussed later. The push-over curves exhibit some noise. A possible reason is that when roots are pushed to move, no matter horizontally or vertically, root-soil interface friction generated may not be uniform in dry sand because of the shape of the sand particles in contact with the roots However, further experiments are needed to confirm this.

4.2 Root system moment capacity

The variation of moment capacity with RAR is shown in Figure 5. It can be seen that in general, moment capacity grows with the increase of RAR from 2.01% to 4.47%, while no obvious difference can be observed when RAR changes from 0.57% to 2.01%. It is important to state at this stage that the only difference between the root systems with $RAR = 0.57\%$ and $RAR = 2.01\%$ is the presence of the central tap root (two central adjacent tap roots in the case of the shallow and wide root system). In this case, it is reasonable to say that the tap root(s) makes few contributions to the moment capacity, while the smaller diameter roots spaced further away from the trunk have a more significant effect on the moment capacity of the root system.

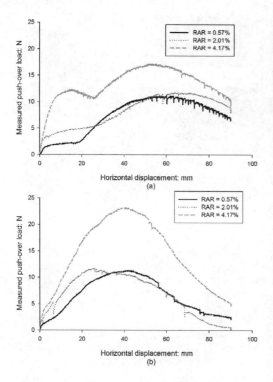

Figure 4. Push-over curve for (a) deep and narrow root system and (b) shallow and wide root system, with different values of RAR.

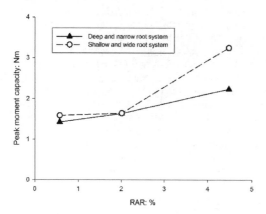

Figure 5. Moment capacity for different types of root systems with different root area ratio.

It is also evident from Figure 5 that the shallow and wide root system, which has the same amount of root material and the same surface area of the root analogues, has a much larger resistance to overturning at higher RAR. This is consistent with the moment capacity being the sum of the individual root pull-out capacities in each case multiplied by the radius from the trunk. At lower RAR (<2%) this effect is not apparent, possibly because the increased moment capacity due to the increased radius is offset by the

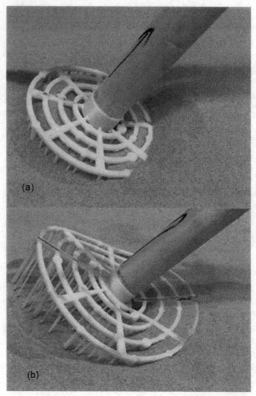

Figure 6. Different failure mechanisms observed in the tests (a): overturning of the whole root system; (b) breakage of the lateral branches close to trunk.

lower average effective confining stress along the analogues, given their reduced length and the low number of roots in these models.

In the case of $RAR = 4.47\%$, the more than doubling in root material from the $RAR = 2.01\%$ case, offsets the reduction (by approximately half) in the confining stress on its own, explaining the increased effectiveness of this case. This is not true in increasing RAR to 2.01% from the 0.57% case, as all of the increase is concentrated in the large tap roots, the pull-out resistances of which are not mobilised fully due to their position immediately beneath the trunk.

4.3 Failure mechanism

Two types of failure modes were observed in the tests presented here: (i) overturning of the whole system involving extensive pull-out of the vertical root analogues on the 'windward' side (Fig. 6(a)); and (ii) breakage of the lateral branches close to the trunk (Fig. 6(b)). These two failure mechanisms could occur either separately or simultaneously.

For shallow and wide root systems, as the horizontal displacement increased, the lateral branches on the 'leeward' side broke, with sinker roots on windward side only partially pulled out from the soil (i.e. mechanism (ii); Fig. 6(b)).

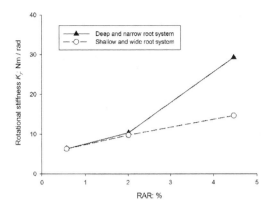

Figure 7. Rotational stiffness for different root systems with different root area ratio.

For deep and narrow root systems, both mechanisms occurred during the tests, resulting in the distinctive flattening or drop in the push-over curve at smaller displacement (Fig. 4(a)). The whole system therefore overturns (i.e., root pull-out; mechanism (i), Fig. 6(a)) at lower displacement, and exhibits the first peak resistance shown in Fig. 4(a) (i.e., around 2 N, 4 N and 12 N for increasing RAR). As the horizontal displacement increases further, the resistance provided by the lateral branches is increasingly mobilised, causing the second higher peak in Fig. 4(a). The subsequent drop of push-over load is because of the breakage of the lateral branches (mechanism (ii); Fig. 6(b), similar to what is observed in the shallow and wide root system.

For all tests, no single vertical root was broken during the pushover process. In other words, lateral roots are more vulnerable to be damaged during push-over compared to sinker roots.

4.4 Root system stiffness

The following equation was used to calculate the total stiffness of root system, to account for the flexibility in the 140 mm of trunk between the load point and the soil surface:

$$K_{total} = \frac{1}{h^2(\frac{h}{3EI} + \frac{1}{K_r})} \quad (1)$$

where h refers to the distance between loading point and bottom of trunk, E is the flexural modulus of trunk, I is for second moment of area of trunk and K_r is the rotational stiffness of the root system. This equation is derived from considering the lateral force-displacement behavior of a linear elastic cantilever with a rotational spring at the point of fixity.

The properties of the trunk were obtained from conducting four-point bending tests on $n = 6$ samples. The mean value of E was 0.502 MPa, the coefficient of variation was 11.8% and the mean value of EI was therefore 86.188 Nm^2.

Using Equation (1) it was found that the displacement caused by trunk bending was 10^3 times smaller than that triggered by rotation and could be ignored. As a result, $K_{total} \approx K_r$ and was determined by linear fitting of the lateral moment-rotation curves, up to moments half of the maximum moment capacity. The value of K_r for each case is shown in Figure 7. It can be seen that the rotational stiffness of deep and narrow root system is larger than that of shallow and wide one, which is contrary to the observation made earlier for moment capacity. This is because up to half of the maximum moment capacity, the lateral roots have not broken in either test – if these control the initial lateral stiffness, then the reduced length of these roots in the deep and narrow case provides greater stiffness against bending.

Figure 7 also suggests that the lateral surface roots make a larger contribution to rotational stiffness than the tap root.

5 CONCLUSIONS

All of the factors considered (e.g. root type and root area ratio) in this study were observed to influence the stiffness and capacity of tree root systems in soil under lateral loadings. Deep and narrow root systems show higher rotational stiffness at larger RAR while shallow and wide root systems show larger moment capacity. So when using trees to mitigate against different types of hazards, different species should be considered, depending on whether the design requires that the trees should remain stable (e.g. trees on sloping ground or next to transport infrastructure) or whether they are sacrificial (e.g. as exterior protection for large stands of commercial forest).

It has also been shown that stiffness is largely controlled by the near-surface lateral roots bending, while moment capacity is controlled by pull-out of vertical sinker roots. This suggests that existing analytical models for foundations (e.g. soil-structure interaction for flexible rafts/beams or overturning resistance of large pile groups) may be adapted for analysing tree root systems under lateral loading.

REFERENCES

Danjon F, Fourcaud T and Bert D, 2005. Root architecture and wind-firmness of mature Pinus pinaster. New Phytologist 168(2): 387–400.

Danjon F, Barker DH, Drexhage M and Stokes A, 2008. Using three-dimensional plant root architecture in models of shallow-slope stability. Annals of Botany 101(8): 1281–1293.

Gardiner BA and Quine CP, 2000. Management of forests to reduce the risk of abiotic damage – a review with particular reference to the effects of strong winds. Forest Ecology and Management 135: 261–277.

Jackson RB, Canadell J, Ehleringer JR, et al., 1996. A global analysis of root distributions for terrestrial biomes. Oecologia 108(3): 389–411.

Langre E De, 2008. Effects of Wind on Plants. Annual Review of Fluid Mechanics 40: 141–168.

Liang T, Knappett JA and Bengough AG, 2014. Scale modelling of plant root systems using 3-D printing. In ICPMG2014–Physical Modelling in Geotechnics, Perth, Australia, 14-17 January 2014. pp: 361–366.

Liang T, Knappett JA and Duckett N, 2015. Modelling the seismic performance of rooted slopes from individual root–soil interaction to global slope behaviour. Géotechnique 65(12): 995–1009.

Liang T and Knappett JA, 2017. Centrifuge modelling of the influence of slope height on the seismic performance of rooted slopes. Géotechnique 67(10): 855–869.

Liang T, Knappett JA, Bengough AG and Ke YX, 2017. Small scale modelling of plant root systems using 3-D printing, with applications to investigate the role of vegetation on earthquake induced landslides. Landslides 14(5): 1747–1765.

Meijer GJ, Bengough AG, Knappett JA, et al., 2016. New in situ techniques for measuring the properties of root-reinforced soil – laboratory evaluation. Géotechnique 66(1): 27–40.

Mickovski SB, Stokes A and Beek LPH Van, 2005. A decision support tool for windthrow hazard assessment and prevention. Forest Ecology and Management 216: 64–76.

Sonnenberg R, Bransby MF, Bengough AG, et al., 2011. Centrifuge modelling of soil slopes containing model plant roots. Canadian Geotechnical Journal 49(1): 1–17.

Ulanova NG, 2000. The effects of windthrow on forests at different spatial scales: a review. Forest Ecology and Management 135: 155–167.

7. Facilities

A new environmental chamber for the HKUST centrifuge facility

A. Archer & C.W.W. Ng
Department of Civil and Environmental Engineering, The Hong Kong University of Science and Technology, Hong Kong SAR, China

ABSTRACT: The advent of climate change has influenced weather patterns which in turn adversely affects geotechnical infrastructure. This has resulted in the use of environmental chambers coupled with centrifuge modelling to study the effect of climate change on geotechnical works. This paper presents a new environmental chamber developed at the Hong Kong University of Science and Technology Centrifuge facility. The aim of the new centrifuge environmental chamber (CEC) is to simulate different climate conditions by decoupling the influence of different climate variables. The design and construction of the new centrifuge environmental chamber is discussed. The CEC has the ability to control temperature, relative humidity, wind, radiation and rainfall.

1 INTRODUCTION

The Intergovernmental Panel on Climate Change (IPCC 2013) concludes that climate change will result in increased temperatures as well as changes in the water cycle, i.e. precipitation and relative humidity (RH). This has prompted an increase in studies over the past decade which aims to understand the impact of these changes on geotechnical works, more importantly the long term and seasonal performance (Davies 2011, Mendes 2011, Toll et al. 2012, Pritchard et al. 2014).

Figure 1 shows a schematic of the soil-atmosphere boundary in the context of a typical embankment problem. The climate variables that influence the hydraulic response and therefore effective stress changes include: temperature, relative humidity, solar radiation, wind and precipitation. The first four of these variables are driving factors for evaporation. Kilsby et al. (2009) points out that to fully assess the impact of climate change require a comprehensive understanding of the soil-atmosphere boundary. Our fundamental understanding of the soil-atmosphere boundary relies on investigating both coupled as well as de-coupled behaviour of climate variables.

In recent years environmental chambers coupled with centrifuge modelling have been used to study the effect of climate variables on embankments and slopes (e.g. Take & Bolton 2002, Tristnancho et al. 2012, Castiblanco et al. 2016). Centrifuge modelling has proven to be an effective modelling tool for these studies due to the advantage of reproducing the correct stress states and also the benefit of modelling long term soil response in a single test.

This paper introduces a newly developed centrifuge environmental chamber (CEC), with the aim to independently controlling climate variables in-flight. The design and construction of the new environmental chamber is discussed. Results of simulated atmospheric paths are presented to illustrate the capability of the new chamber.

2 CURRENT STATE-OF-THE-ART

Various geotechnical facilities have developed tools to control climate conditions at the soil-atmosphere interface (Askarinejad et al. 2012, Hudacsek et al. 2009, Matziaris et al. 2015, Take & Bolton 2002, Tristancho et al. 2012). Take & Bolton (2002) presented a climate chamber able to induce wetting-drying cycles by means of rainfall and dry air circulation. In their tests RH was used as reference and not explicitly controlled.

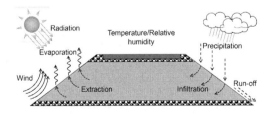

Figure 1. Climate variables and processes associated with the soil-atmospheric boundary in a typical embankment.

Castiblanco et al. (2016) developed a climate chamber that is able to control RH for the purpose of studying desiccation processes and crack formations in soil. The vapour equilibrium technique is used to control RH. This however limits the design to a fixed RH for a single test with no in-flight control.

Tristancho et al. (2012) presented the first chamber capable of simulating various climate variables including temperature, RH, rainfall, wind and radiation. Their system employs heating and cooling elements through a convection controlled setup. They demonstrated successful use of these elements at high-g conditions and the ability of a convection control system to produce the desired climate conditions. Their results however show only one particular atmospheric path with no separate control of relative humidity and temperature. Since soil responds differently to changes in temperature and RH it may be beneficial to have a system capable of independently controlling these variables.

The success and versatility of the convection system makes this the most suitable choice for the environmental chamber design. In addition, to correctly model climate conditions the variables shown in Figure 1 should be considered, i.e. temperature, RH, radiation, wind and rainfall.

3 SCALING LAWS

Simulating climate characteristics requires scaled relationships for the processes at the soil-atmosphere interface. In addition, scaling laws associated with unsaturated soil are required. Three key processes at the soil-atmosphere interface are: heat exchange, evaporation and infiltration. These involve heat and mass transport driven by diffusion processes. In unsaturated soils the time for diffusion have the same scaling factor of N^2, with N being the gravitational acceleration. Flow of mass and heat occurs N times faster in the centrifuge model.

Tristancho et al. (2012) conducted a comprehensive investigation on the scaling laws for heat exchange. Adopting a scale factor of N between model and prototype sensible heat, scale relationships for irradiance, thermal emission, convection, and latent- and sensible heat were derived.

With regard to evaporation, a general form of Dalton's equation (Dalton 1834) was used by Tristancho et al. (2012) to derive scaling laws. It was established that vapour pressure deficit (VPD) scales N times, assuming a scaling factor of unity for wind velocity. VPD is a function of relative humidity and soil surface temperature, showing the importance of controlling these variables.

Infiltration, due to rainfall, involves scale relationships for duration, intensity and frequency.

Considering the scale relationships for time and water flow, rainfall intensity, duration and frequency adopt scale factors of N, $1/N^2$ and N^2, respectively.

Relevant scaling laws associated with the soil-atmosphere boundary are presented in Table 1.

Table 1. Summary of relevant scaling laws (adapted from Tristancho et al. 2012).

Parameter	Unit	Scaling factor
Length	L	1/N
Density	M/L^3	1
Mass	M	$1/N^3$
Time (diffusion)	T	$1/N^2$
Seepage velocity	L/T	N
Irradiance	I	$1/N(\alpha_p/\alpha_m)$[†]
Rain duration	T	$1/N^2$
Rain intensity	L/T	N
Rate of evaporation	E	N
Vapour pressure deficit	D*	N[††]
Wind velocity	V_w	1[††]

[†] α is the mean absorption coefficient.
[††] Other combinations of wind velocity and vapour pressure deficit are possible to satisfy the scaling law concerning the rate of mass evaporated per unit surface.
*$D = e_s - e_a$, where e_s is the saturation vapour pressure at the soil surface and e_a is the vapour pressure in the air.

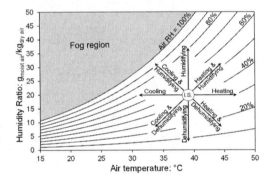

Figure 2. Psychrometric chart showing typical atmospheric paths. (I.S. refers to the initial air state).

4 NEW CENTRIFUGE ENVIRONMENTAL CHAMBER DESIGN

4.1 Design considerations

The study of psychometrics was adopted as a theoretical basis for the design of the CEC. Figure 2 shows a psychrometric chart as well as various atmospheric paths. From an arbitrary initial air state, one of the paths can be adopted by inducing changes in the air state (i.e. temperature, RH, etc.). These processes generally involve fluctuations in air humidity (i.e. water vapour in the air) and temperature.

Unsaturated soil behavior is primarily influenced by changes in soil suction. The new CEC allow soil suctions to be manipulated by controlling RH as described by Lord Kelvin's thermodynamic equation (Sposito 1981):

$$\psi = -\frac{RT}{\upsilon_{w0}\omega_v}\ln(RH) \qquad (1)$$

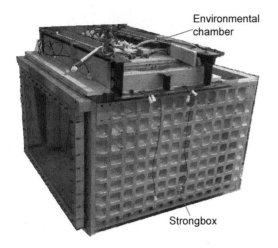

Figure 3. Overall view of environmental chamber on strongbox.

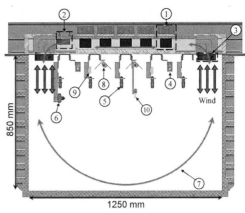

1. Cooling unit
2. Heating unit
3. Fans
4. UV lights
5. Rain nozzles
6. Humidity nozzle
7. Circulation (Both directions)
8. IR lights
9. Temperature & RH sensors
10. Anemometer

Figure 4. Cross-sectional elevation drawing of the environmental chamber.

where ψ is total suction (kPa); R is the universal gas constant [J/(mol*K)]; T is the absolute temperature (K); υ_{w0} is the specific volume of water (m^3/kg); ω_v is the molecular mass of water vapour (g/mol); $RH = \bar{u}_v/\bar{u}_{v0}$, $\bar{u}_v$ is the partial pressure of water vapour (kPa); $\bar{u}_{v0}$ is the saturation pressure of water vapour over a flat surface of pure water (kPa).

From the information presented in the previous sections, the climate variables considered in this design are temperature, RH, rainfall, wind and radiation.

This will allow infiltration and evaporation processes to be simulated. In addition, by allowing independent control of the variables more diverse climate conditions will be permitted.

4.2 Overall design

Figure 3 shows the complete CEC developed for the Hong Kong University of Science and Technology centrifuge facility (Ng et al. 2001). Figure 4 shows a cross-sectional elevation drawing of the complete CEC system. The CEC is mounted on top of a strongbox with model space dimensions of 1250 mm × 900 mm × 800 mm (length × width × height). The EC comprise of two channels in which cooling and heating units are placed. Variable speed axial fans are used to circulate air through the channels. Below the channels infrared (IR), visible and ultraviolet (UV) lights are located to simulate radiation. An adiabatic model space is insured by means of thermally insulated rigid foam board installed inside the model container. The foam board has a thermal resistance of ±3500 (cm°C)/W.

4.2.1 Relative humidity and temperature

Relative humidity and temperature is altered by imposing water and heat flux in accordance with the psychrometric processes. This is achieved by cooling and heating units placed in the airflow channels which are used in combination, depending on the desired air state.

The cooling units consist of thermoelectric coolers (TEC), i.e. Peltier plates, attached to cold sinks. The TECs cool the cold sinks, which in turn cool the air flowing past it. Twenty 128 W TECs were used resulting in 2560 W of cooling power. The heating units comprise of finned strip aluminium heaters. The heaters have a temperature output of ±200°C. In addition to changing the air temperature, RH changes are also induced by the cooling and heating units. When the cooling units are cooled to below the air dew point, condensation will occur, extracting moisture from the air. This results in a decrease in RH as well as a decrease in temperature. The heaters will dry the air, also lowering the RH during an increase in temperature. An increase in RH is possible by allowing condensation and, by means of the airflow, allowing the condensed water to flow into the model space. This process can be time consuming and requires some computation work to achieve a desired RH value.

A key feature of the CEC is the addition of the humidity nozzle (B1/4J+SUF1 from Spray Systems Inc.). The nozzle allows more control of the RH variable without additional computational effort or unwanted changes in the temperature. The humidity nozzle, with a drop size of 6 μm, allows water vapour to be introduced to the artificial atmosphere. Assuming linear scaling dimensions, the droplet size equates to vapour size particles associated with humidity at prototype scale. This permits a change in the RH, allowing only vapour transfer with the soil and no mass flux. Current designs employ the nozzles used for rainfall to change the relative humidity (Take & Bolton 2002). Rain nozzles typically presented in the literature have a mean drop diameter ranging from 30 μm to 60 μm, which are equivalent to light- and heavy rain at prototype scaled. Adopting the conventional scaling factor for linear dimensions, Figure 5 shows the variation of

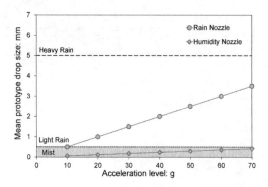

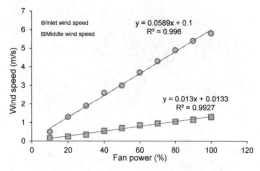

Figure 5. Drop size variation with centrifuge acceleration.

Figure 6. Measured model wind speed as the inlet and middle of the model space.

drop size with g-level for the different nozzles as well as typical drop sizes for rain (Mason & Andrews 1960). This highlights the importance of the humidity nozzle to allow additional relative humidity control. The CEC can simulate temperature and RH ranges from 15°C to 50°C and from 30% to 100%, respectively.

4.2.2 Rainfall

The CEC system employs two nozzle types, LNN 1/4J-1.5 and LNN 1/4J-12 from Spray Systems Inc., connected to a pressured water tank. The flow rates of the nozzles range between 1.33–3 cm^3/s (LNN 1/4J-1.5) and 11–24 cm^3/s (LNN 1/4J-12) for water pressures between 200–1000 kPa. The nozzles produce a fine mist and depending on the experimental setup, they are geometrically adjusted to ensure a uniform distribution during rainfall application. Rainfall intensity is controlled by varying the pressure of the water tank, thus changing the flow rate through the nozzles. Rainfall duration is a function of the nozzle discharge time. For the CEC prototype rainfall intensities up to 100 mm/h can be simulated, depending on the g-level and nozzle setup.

4.2.3 Wind

Axial fans with variable speeds are used. The fans have a dual purpose of allowing air circulation as well as inducing a wind boundary condition. The variable speed allows for different wind speed to be simulated. In addition, the fans are installed to allow air circulation in opposite directions (i.e. fans at one end will push air through the channels and out at the opposite end). This facilitates a more efficient system as cold- or hot air can directly be blown into the model space, dependent on whether heating or cooling is employed. By changing the power supplied to the fans different wind speed are obtained. Figure 6 shows the measured wind speeds for different percentages of power applied to the fans. Wind speeds were measured with an anemometer at the inlets to the model space as well as in the middle of the model space. The wind speed at the inlet is much higher than in the middle.

This shows that depending on the model size, different wind speeds may be induced at different locations. This should be a consideration when planning a model test setup.

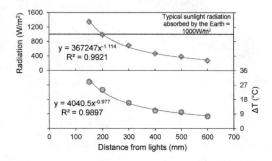

Figure 7. Measured model radiation and increase in temperature due to IR lights.

4.2.4 Radiation

Radiation is replicated with three infrared (IR), four ultraviolet (UV) and visible lights (LED lights). The major driving factor for radiation is the IR lights. Three 500 W IR lights are installed. The irradiance output of the lights was measured with a pyranometer. Figure 7 shows the measured radiation at various distances from the lights. As radiation results in heating the temperature difference (ΔT) from ambient (22°C) was also measured. Higher radiation and temperatures were measured at closer to the lights with a non-linear trend further away.

4.2.5 Instrumentation

Numerous sensors are installed in the CEC; employed as performance indicators as well as measurement devices. Temperature sensors (DS18B20, accuracy ±0.5°C) are installed on heating and cooling elements, as well as the inlet and outlet of the EC channels. Relative humidity and temperature sensors (Sensirion SHT25, 1.8% RH accuracy) are used inside the model space. A pyranometer (Campbell Scienititic CS300) is mounted inside the model space for solar radiation measurements. An anemometer (Lutron AM-4204) is installed in the model space for wind speed measurements. Single point laser sensors (Wenglor YT44MGV80, 0.2 mm resolution) are used to capture displacement measurements. A summary of the instruments employed in the CEC is presented in Table 2. In-house designed software is used for data acquisition as well as to control the CEC. This allows for

Table 2. Summary of measurement instruments used in the CEC.

Sensor	Unit	Range	Accuracy
RH & Temperature	% & °C	RH: 0–100 Temp.: −10–55	RH = ±2% Temp. = ±0.3°C
Temperature sensor	°C	0–100	±0.75°C
N-type thermocouple	°C	−50–250	±1.5°C
Pore-water pressure transducer	kPa	−70–3000	±1 kPa
Pyranometer	W/m^2	0–1750	±5% for daily total radiation
Anemometer	m/s	0.2–20	±1%
Single point laser	mm	0–200	1%

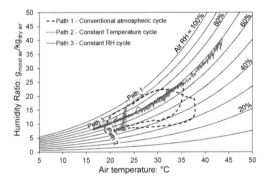

Figure 8. CEC performance test results plotted on a typical psychrometric chart.

continuous atmospheric cycles to be simulated without cessation of the centrifuge.

5 PERFORMANCE TEST

To assess the performance of the environmental chamber three separate atmospheric paths were induced. One of the main objectives of the new CEC is to have independent control of temperature and RH. Figure 8 shows three atmospheric paths simulated atmospheric paths during a centrifuge test carried out at 40 g. The paths are plotted on a psychrometric chart. Path 1 is a typical atmospheric cycle with no definite RH or temperature control. As expected the RH decreases during temperature increase as the air dries out. During cooling the air needs less moisture to become saturated, hence an increase in RH as observed. Path 2 shows RH cycles while maintaining a constant temperature. RH varied between ±45%–90% while the temperature was maintained with a variation of roughly 1°C. Path 3 shows temperature cycles while maintaining a constant RH. The temperature varied from 18°C–45°C with the RH roughly maintained at ~70%. Undulations in this path are due to the addition of water vapour in order to keep the RH constant. The atmospheric paths presented show the ability of the EC to separately control RH and temperature as boundary conditions.

6 CONCLUSIONS

The paper presented details of a new CEC for use during centrifuge tests. The CEC has the ability to control temperature, RH, wind, radiation and rainfall. Isolating theses climate variables in the centrifuge were previously not possible. A humidity nozzle was introduced which permits control of the RH boundary condition. This allows for the independent control of the temperature and RH boundary conditions. Results presented show that it is possible to maintain a constant temperature while varying the RH and vice versa. This permits studies where the influence of climate variables can be de-coupled, allowing for more fundamental studies on the impact of climate variables.

ACKNOWLEDGEMENTS

The authors would like to acknowledge the contribution from research grants T22-603/15N and C6012-15GF supported by the Research Grants Council (RGC) of the Hong Kong Special Administrative Region (HKSAR). The first author is grateful for the support of the Hong Kong Ph.D. Fellowship Scheme (HKPFS) provided by the RGC of the HKSAR.

REFERENCES

Askarinejad, A., Laue, J., Iten, M., Zweidler, A., Bleiker, E., Buschor, H. & Springman, S.M. 2012. Physical modelling of rainfall induced landslides under controlled climatic conditions. In *Eurofuge 2012*, Delft, The Netherlands, April 23-24, 2012, Delft University of Technology and Deltares, Published on CD only.

Castiblanco, P., Lozada, C., Caicedo, B. & Thorel, L. 2016. A new climatic chamber adapted to the mini-centrifuge for simulating soil drying. In *Eurofuge 2016, 3rd European conference on Physical Modelling in Geotechnics*, June 2016, Nantes, France, 111–115.

Dalton, J. 1834. *Meteorological observations and essays*. Second Edition, Harrison & Crosfield, London.

Davies, O. 2011. *Numerical analysis of the effects of climate change on slope stability*. PhD thesis, School of Civil Engineering and Geosciences, Newcastle University.

Hudacsek, P., Bransby, M.F., Hallett, P.D. & Bengough, A.G. 2009. Centrifuge modelling of climatic effects on clay embankments. *Proceedings of the ICE-Engineering Sustainability* 162, No. 2, 91–100.

IPCC 2013. *Climate Change 2013: The Physical Science Basis. Contribution of Working Group I to the Fifth Assessment Report of the Intergovernmental Panel on Climate Change*. Eds.: Stocker, T.F., D. Qin, G.-K. Plattner, M. Tignor, S.K. Allen, J. Boschung, A. Nauels, Y. Xia, V. Bex & P.M. Midgley. Cambridge University Press, Cambridge, United Kingdom and New York, NY, USA, pp. 1535.

Kilsby, C., Glendinning, S., Hughes, P.N., Parkin, G. & Bransby, M.F. 2009. Climate-change impacts on long-term performance of slopes. *Proceedings of the ICE-Engineering Sustainability* 162, No. 2, 59–66.

Mason, B.J. & Andrews, J.B. 1960. Drop-size distributions from various types of rain. *Quarterly Journal of the Royal Meteorological Society* 86, No. 369, 346–353.

Matziaris, V., Marshall, A.M. & Yu, H.S. 2015. Centrifuge Model Tests of Rainfall-Induced Landslides. In *Recent Advances in Modeling Landslides and Debris Flows*, W. Wu, Ed., Springer International Publishing, 73–83.

Mendes, J. 2011. *Assessment of the impact of climate change on an instrumented embankment: an unsaturated soil mechanics approach*. PhD thesis, Durham University.

Ng, C.W.W., Van Laak, P.A., Tang, W.H., Li, X.S. & Zhang, L.M. 2001. The Hong Kong geotechnical centrifuge. In *Proceedings of the 3rd International Conference on Soft Soil Engineering*, 225–230.

Pritchard, O.G., Hallett, S.H. & Farewell, T.S. 2014. Soil impacts on UK infrastructure: current and future climate. *Proceedings of the ICE-Engineering Sustainability* 167, No. 4, 170–184.

Sposito, G. 1981. *The Thermodynamics of Soil Solutions*. London: Oxford Clarendon Press.

Take, W.A. & Bolton, M.D. 2002. An atmospheric chamber for the investigation of the effect of seasonal moisture changes on clay slopes. In *Proceedings of the International Conference on Physical Modeling in Geotechnics*, R. Phillips; P.J. Guo and R. Popescu, Eds., St. John's, Newfoundland, Canada, Balkema, Rotterdam, July 10–12, 2002, 765–770.

Toll, D.G., Mendes, J., Hughes, P.N., Glendinning, S. & Gallipoli, D. 2012. Climate change and the role of unsaturated soil mechanics. *Geotechnical Engineering Journal of the SEAGS & AGSSEA* 43, No. 1, 76–82.

Tristancho, J., Caicedo, B., Thorel, L. & Obregon, N. 2012. Climatic Chamber with Centrifuge to Simulate Different Weather Conditions. *ASTM Geotechnical Testing Journal* 35, No. 1, 159–171.

Upgrades to the NHRI – 400 g-tonne geotechnical centrifuge

S.S. Chen, X.W. Gu, G.F. Ren, W.M. Zhang, N.X. Wang & G.M. Xu
Nanjing Hydraulic Research Institute, China

W. Liu, J.Z. Hong & Y.B. Cheng
China Academy of Engineering Physics, China

ABSTRACT: In Nanjing Hydraulic Research Institute (NHRI), the NHRI – 400gt centrifuge had been operating for more than twenty years since 1991. In 2016 and 2017, the machine was thoroughly upgraded without changing its basic engineering. Upgrades were made to the mechanical system, electrical system and data-acquisition system. A torque motor and a matched drive control system were employed to drive the machine directly, rather than the traditional DC motor – reducer drive system. Also, the new data-acquisition system includes 90 static data channels and 38 dynamic data channels. The brand new NHRI - 400gt geotechnical centrifuge is expected to operate for decades more.

1 INTRODUCTION

1.1 Establishment of the machine

From 1984 to 1986, Nanjing Hydraulic Research Institute carried out technical and economic evaluations and investigated many centrifuges in the former Soviet Union, United Kingdom, United States, France, Nederland, Denmark, German, Italy and Japan (Zhu, 1986).

In 1987, the Ministry of Water Resources and the Ministry of Communications agreed to jointly invest in a project to construct the NHRI – 400gt geotechnical centrifuge laboratory. In the same year, the 602 Research Institute won the bidding and signed a contract with NHRI for technical development of the machine. The facility was first put into use in 1991 (Dou & Jing, 1994).

1.2 Features

The main features for the NHRI – 400gt geotechnical centrifuge are as follows:

(1) Maximum acceleration is 200 g at 5.0 m radius. Maximum payload is 2 t at 200 g. The capacity is 400 gt.
(2) Largest radius is 5.5 m from rotation center to the platform surface.
(3) Clear swinging basket dimensions are 1.2 × 1.2 × 1.1 m³.

There are also some other important design and construction features as follows:

(4) The original driving system consisted of DC motor, coupling, reducer and driving shaft. The speed governing system employed a digital-analog double enclosed ring, including a current regulator and a speed regulator.
(5) An unbalance monitoring system was installed which can auto-balance an in-flight unbalanced force up to 200 kN, by controlling the movement of water through valves in special tanks installed in the arm. The unbalanced force was measured by a pair of force sensors installed on the arm. The system would stop the centrifuge if the unbalanced force reached its capacity.
(6) A braking system was used during decelerating.
(7) A slip ring of 100 channels and an electric power ring with 10 channels were installed.
(8) An air/water conveyance joint (maximum pressure 20 MPa), a water transfer joint (maximum discharge 380 L/min) and two oil joints (maximum pressure 20 MPa and maximum discharge 30 L/min) were commissioned.

The original monitoring system and data acquisition system were useful but have been superseded by technology now.

1.3 Operation and maintenance

From 1991 to 2015, upgrades of sensors, monitoring system and data acquisition system had been performed from time to time.

Particularly, major maintenance works were carried out twice besides daily operation and maintenance.

(1) In 2000, the speed governing system was upgraded to a direct current governor system using an automatic control program with PLC modules and parts. With a touch panel, the centrifuge was able to be operated in a more accurate way.

(2) From 2007 to 2009, mechanical system had been checked and maintained thoroughly. The major works were as follows:

The DC motor was repaired. The badly worn reducer was replaced by a new one. The gear coupling with elastic pin between motor and reducer was changed with a new bearing box.

The old lubricating system was dismantled and two new lubricating systems were built, one for the reducer and one for the drive shaft.

The braking system was abandoned due to its limited contribution during the decelerating process.

The auto-balance system had failed for long period and was left unused. The old force sensors were replaced by two new ones. The unbalanced force was measured and used to modify the counterweight equation.

An instrument cabinet was designed for installment, energy supply, and control of auxiliary devices such as shaking table, robotic manipulator, loading device and so on.

1.4 Troubles

In 2015, the centrifuge began to fail further and more frequently. The main troubles were as follows:

(1) The speed governor system failed several times since 2007. First, the battery on PLC CPU ran out and the control program was lost. Then, the touch panel, PLC CPU, one of the cables, mother board and some parts were changed one after another. Gradually, the old PLC modules and parts were out of date and ceased to be produced. It became more and more difficult to find replacements.
(2) The swinging basket was worn and rusted badly. The platform surface was no longer flat. Model container was unable to keep stable. The basket deformed greatly under high g-level.
(3) The data-acquisition system was unable to meet the growing needs of research work. The channels were far from enough and data transmission speed was too low.
(4) The switch cabinet was too old to be used.
(5) Force sensors failed.

2 UPGRADE PLAN

2.1 Preliminary work

The centrifuge was investigated by experts from NHRI and China Academy of Engineering Physics in 2015. Then an application was submitted to the Ministry of Water Resources, with a detailed upgrade plan.

Soon in 2015, the upgrade project was approved by the Ministry of Water Resources. With financial support, the project started at the beginning of 2016.

2.2 Upgrade aims

The aims of upgrade works are as follows.

(1) To recover the capacity of the NHRI – 400 gt centrifuge.
(2) To build a more efficient drive system and matched electrical speed governing system.
(3) To upgrade the data-acquisition system.
(4) To improve the centrifuge's compatibility for further upgrade works of various auxiliary devices, especially for a new shaking table.

2.3 Design of upgrade work

According to the aims, the upgrade project was designed in the following detailed procedures.

(1) Check the machine thoroughly. Try to figure out whether there are additional hidden issues. Refine the upgrade design.
(2) Remove the old parts which are to be replaced or upgraded. These parts include old drive system, speed governing system, lubricating system, principle axis, joints, swinging basket, instrument cabinet, unbalanced force sensors and switch cabinet.
(3) Install and debug a new switch cabinet for energy supply.
(4) Assemble the new hollow principle axis, arm, new swinging basket and unbalanced force sensors to the axis. The new accumulators and joints for a new shaking table are also mounted in the arm.
(5) Install instrument cabinet. Setup and debug the new data-acquisition system.
(6) Install the new driving system, lubricating system and new.
(7) Setup the new speed governing system. Debug the machine under higher and higher g-level gradually.
(8) In the centrifuge containment, make a lighting system on the ceiling, reinforce the inner wall, paint the machine and so on.

3 MECHANICAL SYSTEM

3.1 Driving system

The most characteristic upgrade of the NHRI – 400gt centrifuge is the new direct drive system.

Due to the tight budget, it is impossible to change the laboratory basic engineering to make upgrade work easier. A new system with a torque motor was designed.

As shown in Figure 1 & 2, the direct drive system with the new type motor needs no coupling or reducer compared with a traditional design. The simple structure has improved air flow conditions in the tunnel. During the rainy season, it is easy to keep moisture from the motor.

Compared with the old one (Fig. 3), a new hollow principle axis (Fig. 4) is manufactured. The torque motor is also hollow. In this way, setup of new joints for air, water and oil conveyance is now convenient.

These joints are designed not only for precipitation simulation, but also for the new shaking table and other auxiliary devices' load or energy supply.

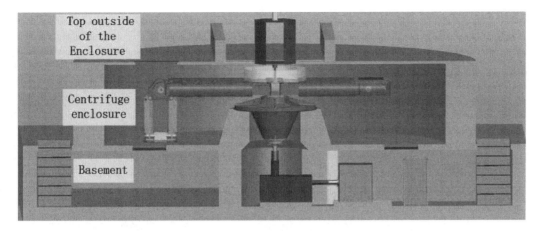

Figure 1. Layout of the NHRI – 400gt centrifuge before upgrades.

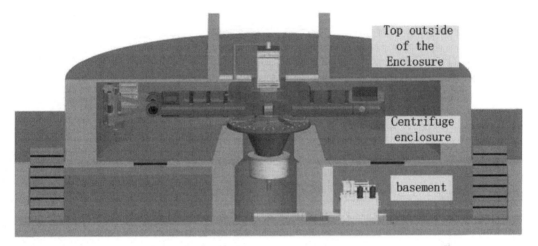

Figure 2. Layout of the NHRI – 400gt centrifuge after upgrades.

3.2 *Lubricating and cooling system*

According to the new drive system, new lubricating and cooling system is designed. Good operation of this system is the basic requirement in the electrical speed governing system.

Temperature sensors are set in several key parts and the data are shown for the electrical speed governing system. Normally, the lubricating oil should be kept cool. For long-term operation, water circulation can be used for extra cooling.

3.3 *Swinging basket*

The old basket was connected to the arm with a link axis (Fig. 5). This made the model placement and removing very inconvenient. The old structure (Fig. 6) is a composite system and is in poor condition after so many years. The material under the platform is severely rusted. The old basket was abandoned.

As shown in Figure 7, the new basket has a monolithic construction with tension plates and platform. The plates and platform are assembled in two axes. The new basket structure is suitable for static and dynamic tests both.

The clear dimensions are $1.4 \times 1.2 \times 1.2$ m^3 of the new swinging basket, providing slightly larger room for models.

3.4 *Rotating arm*

As shown in Figure 8, the rotating arm consists of tension straps, constraint plates, protecting coat and a counterweigh box.

The water tanks of broken auto-balance system are removed. Then four unbalanced force transducers are installed symmetrically near the principle axis.

Accumulators and related pipe system are installed for a new shaking table. There is still enough room for joints and pipes installation.

The old link axis is changed to two shorter axes (Fig. 5).

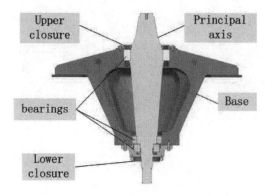

Figure 3. The old main transmission system.

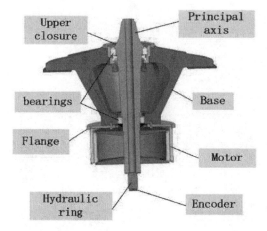

Figure 4. The new main transmission system.

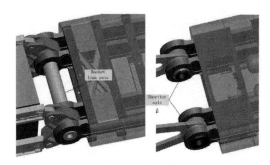

Figure 5. Basket mounting axis before and after upgrades.

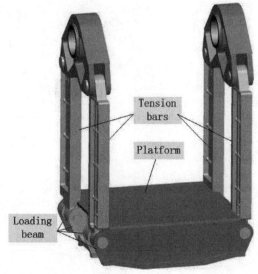

Figure 6. The old swinging basket.

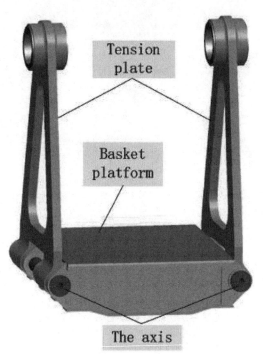

Figure 7. The new swinging basket.

3.5 *Instrument cabinet*

A new instrument cabinet is installed on the arm (Fig. 1). The cabinet (Fig. 9) is divided into three layers. The upper layer is for static data-acquisition modules. The middle layer is for installation of dynamic data-acquisition system. The lower part is for the controlling parts of auxiliary devices such as the shaking table, robotic manipulator, loading device and so on.

4 ELECTRICAL SYSTEM

4.1 *Speed governing system*

The new direct drive system needs a matched speed governing system.

The new speed governing system consists of a drive control system, a central control system and an upper computer.

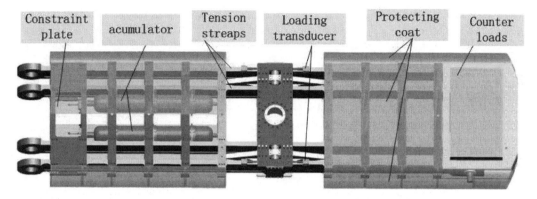

Figure 8. The new rotating arm.

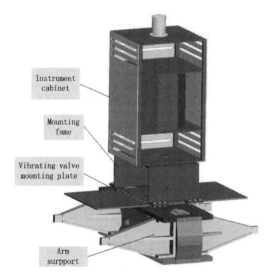

Figure 9. The instrument cabinet and its mounting.

The drive control system is based on a full-digital AC inverter of the latest type. The central control system is based on a PLC. They are integrated in control cabinets.

There is a touch panel on the main PLC cabinet. Centrifuge operation can be controlled either through the upper computer or by touch panel.

The system can only be started when all safety requirements are satisfied, such as lubrication, pressure and temperatures, counterweight balance, door closing and so on.

4.2 Safety monitoring system

Safety monitoring system includes 8 video channels, 5 for the monitoring of the basket (3 channels), centrifuge containment and lubricating system, 3 for further use.

The 3 video channels for the basket are not only for safety concerns, but also for high quality pictures and videos record of tests. Recorded pictures and videos are transferred through fiber slip ring.

5 DATA-AQUSITION SYSTEM

5.1 Slip ring system

The upgraded centrifuge is equipped with a fiber slip ring, signal collector ring and power collector ring.

The fiber slip ring has 4 channels. The signal collector ring has 60 channels (2 A, 220 V). The power collector ring has 20 channels (30 A, 380 V).

5.2 Static & dynamic data-acquisition system

The upgraded data-acquisition system includes a static system and a dynamic one.

The static data-acquisition system consists of IMP modules, DC supply and upper computer. The system has 90 channels, 70 for strain and 20 for voltage signal. Strain signal range is 0–$10000\,\mu\varepsilon$. Voltage signal range is 0–10 V. The system gives aviation plug interfaces for different transducers. The data sampling frequency is up to 1 Hz.

The dynamic data-acquisition system includes 16 channels for strain and 22 channels for voltage signal. Strain signal range is 0–$25000\,\mu\varepsilon$. Voltage signal range is 0–10 V. The data sampling frequency is up to 5 kHz.

6 OTHER UPGRADES

Some other upgrades are as follows:

(1) Two circular light belts are installed on the ceiling of the centrifuge containment. The illumination is improved and pictures of models are of better quality.
(2) Inner wall, ceiling and floor of centrifuge containment are reinforced and painted. The reinforced floor is water-proof for dam breach simulations.
(3) The machine will be painted to a brand new appearance.

Figure 10. Effect diagram of upgraded NHRI – 400 gt centrifuge.

(4) Preparations are made for further upgrade of the shaking table, dam breach simulation system.

7 SUMMARY

The NHRI – 400gt centrifuge had been operating for more than twenty years since 1991. In 2016 and 2017, the machine was thoroughly upgraded without changing its basic engineering. Upgrades were made to the mechanical system, electrical system and data-acquisition system.

The debugging of the upgraded centrifuge has been done. The machine's capacity has been fully recovered. Several tests have been performed and good results achieved.

The brand new NHRI - 400gt geotechnical centrifuge (Fig. 10) is expected to operate for decades more. Also, it is about to be equipped with more new facilities.

REFERENCES

Y. Dou & P. Jing. 1994. Development of NHRI – 400gt geotechnical centrifuge. Leung, Lee & Tan (eds), *Centrifuge* 94: 69–74. Rotterdam: Balkema.

W. X. Zhu. 1986. Centrifuge modeling for geotechnical engineering in the world. *Chinese Journal of Geotechnical Engineering* 8(2): 82–95.

A new 240 g-tonne geotechnical centrifuge at the University of Western Australia

C. Gaudin, C.D. O'Loughlin & J. Breen
Centre for Offshore Foundation Systems, The University of Western Australia, Perth, Australia

ABSTRACT: A new 5 m radius, 240 g-tonne geotechnical centrifuge was commissioned at the University of Western Australia (UWA) in September 2016. The centrifuge is a C72 model manufactured by Actidyn Systems, capable of a maximum centrifuge acceleration of 130g. The centrifuge can accommodate samples up to 1 m^3 and provides additional capacity and capability to the centrifuge modelling facilities at UWA, which includes a beam and drum centrifuge that have been operating at UWA since 1989 and 2001, respectively. The paper provides details on the centrifuge and its ancillary equipment, and also on the capabilities of the National Geotechnical Centrifuge Facility that now hosts the three centrifuges in a new 700 m^2 laboratory.

1 INTRODUCTION

The Centre for Offshore Foundations (COFS) at The University of Western Australia (UWA) has been running two geotechnical centrifuges for over two decades. The 1.8 m radius, 40 g-tonne beam centrifuge was commissioned in 1989 (Randolph et al, 1991) and has totalled over 100,000 hours of rotation in its 28 years of operation. The 0.6 m radius 290 g-tonne drum centrifuge was established in 2001 (Stewart et al., 1998) to complement the existing beam centrifuge and to allow for additional modelling capability in modelling the seabed response due to wave action and parametric studies of foundation response (Stewart et al., 1998). Both centrifuges have been used extensively, each undertaking 25 to 35 research projects a year (~270 days of spinning per year) on offshore geotechnics, including shallow and deep foundation performance, pipeline-soil interaction, anchor performance, seabed characterisation and submarine landslides.

In 2012, the decision was made to commission a third centrifuge to complement the existing facilities. This decision was motivated by (i) the need to service a growing group of researchers approaching 25 academics and over 40 PhD students, (ii) the significant usage requirements from the local and international offshore oil and gas industry, and (iii) a strategic decision to extend the research scope to include onshore geotechnical engineering and outreach to other research groups in Australia to give the facility a national focus.

After reviewing the existing capabilities and expected future needs, the decision was made to acquire a large beam centrifuge. The larger package complements the two existing centrifuges, notably enabling larger and better instrumented models to be used, while allowing an easy integration of the existing on-board equipment developed in-house for the smaller centrifuge.

The new centrifuge was purchased from Actidyn Systems and funded partly by a research grant from the Australian Research Council, in collaboration with the University of Newcastle, the University of Wollongong, Monash University, Queensland University and the University of Adelaide, and partly by internal funding from COFS and UWA.

Associated with acquiring the new centrifuge was the development of a new centrifuge laboratory located within a new Indian Ocean Marine Research Centre located on the UWA campus. The new centrifuge was commissioned – and the two existing centrifuges recommissioned – in this new laboratory in September 2016 together with the mechanical and electronic workshop facilities. Collectively these facilities form the National Geotechnical Centrifuge Facility (NGCF – www.ngcf.edu.au).

2 THE C72 CENTRIFUGE

2.1 Centrifuge specifications

The new centrifuge is a C72-2 model manufactured by Actidyn Systems and rated as a 240 g-tonne fixed beam centrifuge (Fig. 1). It has a radius of 5 m and a platform of 1.2 × 1.2 × 1.2 m that can accommodate a payload of 2400 kg at 100 g (133.7 rpm) or 1400 kg at the maximum acceleration of 130 g (152.5 rpm). The centrifuge is similar to that commissioned by the Korea Advanced Institute of Science and Technology (Kim et al., 2006).

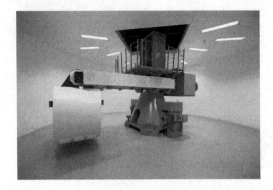

Figure 1. The new C72 beam centrifuge at UWA.

Two key features of this new generation of centrifuges are the drive assembly and the automatic balancing system. The centrifuge is powered by 2 AC 110 kW variable speed induction motors and a belt that rotates the boom, which is supported by a set of tapered roller bearings housed in a pyramidal steel and reinforced concrete casting. The motors are controlled by a Rockwell Variable Frequency Drive that controls the motor speed and acceleration. This power arrangement is an upgrade of the previous motor/gearbox/brake arrangement, and allows for easier maintenance and reduces the manufacturing cost.

The automatic balancing system is a two-step system that is undertaken in-flight. Coarse balancing is executed during the starting phase, at 5 g, by moving the 16 t counterweight. Fine balancing is executed at accelerations above 10 g by moving steel cylinders located within the hollow arms of the centrifuge. This enables continuous adjustment of the balance during testing should a change in payload occur (e.g. from fluid evaporation). The hydraulic power pack and programmable logic controller (PLC) that run the automatic unbalance system are located within the top cabinet above the centrifuge arms where they experience low centrifuge acceleration.

2.2 Control and services

The centrifuge is controlled by a stand-alone computer running custom-built control software that displays the centrifuge parameters in real-time, including centrifuge acceleration (as a function of time), rotation speed, unbalance, temperature and humidity. The operator has control of the radius, the target acceleration, the maximum centrifuge acceleration and the rate of acceleration. The latter, which can be varied from 0.002 g/min to 8 g/min, is important to avoid segregation of soft soil samples that are consolidated from a slurry in-flight. The interface also controls access to the centrifuge chamber via interlocks on the door, and features an alarm and automatic stop system that triggers at threshold unbalance levels.

Supply services to the centrifuge are via an electrical power rings, a hydraulic rotary joint (HRJ) and a fibre optic rotary joint (FORJ). The HRJ is located

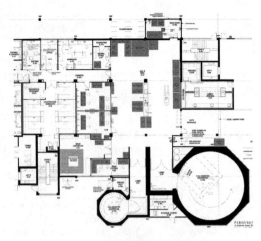

Figure 2. Layout of the centrifuge laboratory.

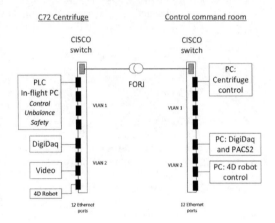

Figure 3. Networks architecture between the centrifuge and the control command room, via a Fibre Optic Rotary Joint (FORJ).

underneath the centrifuge, within the conical pedestal and permits hydraulic and pneumatic connection (up to a pressure of 20 MPa) to the payload. It is used mainly to maintain a constant head of water above the sample, and to action pneumatic on-board devices such as the pneumatic hammer used for pile driving (de Nicola and Randolph, 1994). The FORJ is located above the centrifuge cabinet and permits video and data transmission to the control command room at a rate of 1 gigabits per second (1 Gbps). Fibre to copper conversion is also implemented, providing 12 network Ethernet ports (see Fig. 3).

2.3 Laboratory layout

The NGCF is located in a brand new 700 m^2 laboratory on the ground floor of the IOMRC building (Fig. 2). The laboratory includes a mechanical and electronic workshop, the three geotechnical centrifuges and their dedicated control command rooms, a large soil preparation and storage area (including a soil quarantine room) that includes the soil sample preparation equipment (which are common to each centrifuge) and five 1 g mock-up/testing stations.

The layout of the laboratory was designed to host the nine technical staff and to facilitate the flow of staff, centrifuge equipment and samples between the different areas. The 1 g mock-up/testing stations allows physical modelling activities at 1 g, but importantly trialling of the experiments at 1 g before testing in the centrifuge. Each 1 g test station mimics a particular centrifuge environment and has the same data acquisition and actuation control as on the associated centrifuge, such that there is minimal requirement for fine-tuning the experimental arrangement after moving the model to the centrifuge, which allows for a quick turnaround between projects.

3 ON-BOARD EQUIPMENT

3.1 Data acquisition system

The centrifuge setup uses the data acquisition system developed at UWA, DigiDaq (Gaudin et al. 2009), which has been implemented in a modified form factor and is located on a platform within the shroud of the centrifuge basket. DigiDaq is a self-contained data acquisition system that features a C8051 microchip and 256k of solid state storage to undertake amplification, filtering and conversation from analog to digital of the electric signal, at a real-time sampling rate of up to 333 Hz. In parallel, DigiDaq caters for short dynamics events by short-burst sampling at up to 1 MHz, electrically storing the data in the solid state memory for downloading via Ethernet after the test. The DigiDaq software is a custom-built labview interface that enables data from a particular sensor to be plotted in real-time either against time, or against data from a different sensor (e.g. force versus displacement), over several window displays.

Data transfer is via the FORJ that connects at each end using CISCO 3560 switches. The switches are also used to connect the PLC, the 4D Robot (see Section 3.3) and any other on-board device to the control command room (Fig. 3). One specific feature of the architecture developed is the two independent networks that separates the critical control system of the centrifuge to the user network that hosts DigiDaq and PACS2 (see Section 3.2).

3.2 2D actuator and motion control

The 2D Actuator developed at UWA has been modeled on the very successful smaller version designed for the C661 centrifuge. It enables motion along the vertical and one horizontal axis along a stroke of 700 and 600 mm respectively (Table 1). The redesign of the 2D actuator was the opportunity to improve its capabilities, bringing the capacity and velocity of each axis to 12 kN and 13.5 mm/s. (against 7 kN and 3 mm/s in the small beam actuators). The enhancement in velocity

Table 1. 2D actuator specification.

	x	z
Displacement	600 mm	700 mm
Velocity	13.5 mm/s	13.5 mm/s
Force	12 kN	12 kN

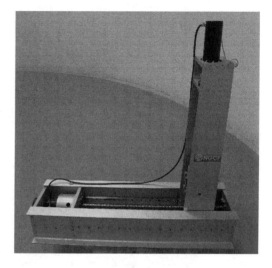

Figure 4. 2D actuator for the C72 centrifuge.

was motivated by the need to apply high frequency cyclic loading to model environmental loading.

Control of the 2D actuator is achieved using an updated version of a multiple axis actuator control system, the Package Actuator Control System or PACS, developed in-house at the University of Western Australia (De Catania et al., 2010). This update, labelled PACS2, was inspired by the original PACS system but was rewritten entirely in Labview. It now runs on the NI 9030 CompactRIO controller, a powerful platform that provides dedicated real-time closed-loop control.

Each PACS2 system can simultaneously and independently control four axes of motion (e.g. enabling the simultaneous control of two 2D actuators) in either load or displacement control. Additionally, actuators can be controlled either directly by the operators, who can prescribes targeted load or displacement values, or load or displacement rates, or run a pre-programmed sequence. The sequence can take the form of an ASCII file, or can be generated by a powerful waveform editor, where highly complex motion or load sequences can be easily assembled by the operator.

The graphical user interface (Fig. 5) connects to the motion controller over the network and allows for remote operation of the actuators from the centrifuge control room. Considerable effort was put into designing an intuitive and clean user interface that still provides a comprehensive view of the status and operation of the actuators.

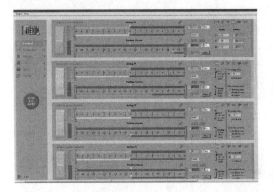

Figure 5. The PACS 2 user interface showing four axes with either load (orange gauge) or displacement (blue gauge) control.

Figure 6. The 4D robot sitting on the centrifuge basket.

Table 2. 4D robot specification.

	X	Y	Z	θz
Displacement	1.0 m	0.65 m	0.5 mm	+/−175°
Velocity	50 mm/s	50 mm/s	50 mm/s	10° sec
Force	1 kN	1 kN	5 kN	5 Nm

3.3 4D robot

The 2D actuator is a versatile and powerful device that can be easily and readily setup on the centrifuge to model a wide range of geotechnical problems (Fig. 6). There are however problems that specifically requires motion in the free dimensions of space. This is notably the case of 3D excavation, steel catenary riser soil interaction, or multidirectional loading of foundations for offshore renewables, the latter two being extensively investigated at COFS.

To complement the 2D actuator, a 4D robot was purchased from Actidyn Systems. The 4D Robot is a general purpose devices that features four axes of motion along X, Y, Z and θz, driven by electric servo actuators, and a tool changing systems with four magazines. The robot was developed based on the initial design used at IFSTTAR (former LCPC, Derkx et al. 1988), and is identical to that used at KAIST (Kim et al. 2006). Its capabilities are summarised in Table 2.

The robot has been designed to facilitate in flight tool/instrument change. Up to four tools can be loaded onto the robot prior to commencement of a test, negating the need to stop the machine between testing sequences. The tool head of the robot features two user controllable hydraulic ports, two pneumatic or water ports, multiple power and data lines, and a built-in 20 kN load cell. The electric and data connection enables instruments and models to be energised and instrumented, while the hydraulic and pneumatic ports allows for mechanical actuation (like for a pneumatic hammer for instance).

Control is managed by an onboard Rockwell Automation RSLogix 5000 PLC located in the central tower of the centrifuge, and a user interface that sits on a dedicated computer in the control command room. The user interface provides supervision of the system, user input of control parameters, and sequence control for all four axes. It also features an anti-collision system that prevents the robot head to enter areas defined by the user.

Another feature of the system is the ability to extract data for integration into peripheral systems. In the present case, displacements and rotation along the four axes, and load cell readings are extracted and integrated into DigiDaq in real time, to be combined with data coming from the centrifuge (e.g. acceleration, unbalance, temperature, etc.) and from any instrumentation in the model.

4 SOIL SAMPLES RECONSTITUTION

4.1 Strongboxes

The C72 centrifuge can accommodate two types of strongboxes, a square container 1 × 1 m in plane view and 0.5 m high, or a cylindrical container, 0.895 m diameter and 0.35 m high. In both cases, two containers can be stacked one onto another to prepare samples up to 1 m and 0.70 m high, respectively.

Both the square and cylindrical strongboxes features drainage systems at the bottom that allows for two way consolidation and free hydraulic connection between the bottom of the sample and the free water on top of the sample. The ability to partially or fully close the bottom drainage offers versatile drainage configurations to suits various types of modelling.

Typically, square strongboxes are used to reconstitute sand samples, as they are more appropriate for reconstitution by raining (see Section 4.3). When saturated samples are required, saturation can be achieved from the bottom of the strongbox to guarantee expulsion of the air, at a flow rate slow enough to prevent piping. The square strongbox is structurally designed to exhibit minimal horizontal deformations under pressure of the contained soil, so to maintain a K0 state in the sample. It is accordingly heavy, weighing 490.8 kg. When two containers are stacked, the weight increases to 894.2 kg, so a full sample of dry sand 1 m high would typically reach about 2,400 kg, i.e. the maximum payload at 100 g.

Figure 7. 500 kPa electric consolidation press.

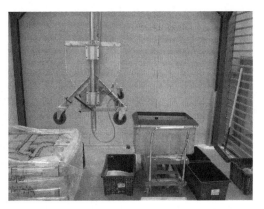

Figure 8. Automatic sand rainer.

Cylindrical containers are preferred for clay and silt samples that are consolidated under the electric press (see Section 4.2). The full 700 mm high container weighs 225 kg, so a normally consolidated clay sample would typically weigh about 790 kg and could spin at the maximum acceleration of 130 g.

4.2 Electric consolidation press

Consolidation of clay samples for the small C661 centrifuge is typically undertaken in-flight as the centrifuge can operate remotely and spin 24/7. While it is intended for the large C72 centrifuge to also operate continuously and consolidate samples in-flight, there is still a need for a versatile consolidation press that can be used to prepare normally and over consolidated samples for both centrifuges.

The large consolidation frame (Fig. 7) was designed and assembled by the NGCF team to consolidate samples in either a C72 circular container or a C661 rectangular container. The press is based on a 350 kN screw jack coupled to a brushless DC motor, and a 350 kN load cell that provides closed loop feedback to a National Instruments CompactRIO controller. The piston has an overall travel of 1000 mm and has a capacity of 500 kPa. The modular design also allows mounting of a square piston suitable for consolidating C661 samples.

User control is facilitated by a built in Touch screen located on the main control of the machine. Control for the machine is based on the same platform that that used for the 2D actuators, hence complex loading sequences can be created, which enable the user to automate part or all of the consolidation process. All components of the machine are an integral part of the frame, making it easy to move within the laboratory.

4.3 Automatic sand rainer

Sand samples are prepared by raining (or pluviation), a technique that is used in many different centers and have proven to generate homogeneous and repeatable sample conditions. Similarly to the electric press, the sand rainer has been designed and assembled by the NGCF team. It is capable of preparing samples for both beam centrifuges, preferably in rectangular or square strongboxes. The whole rainer is housed in a sealed enclosure for dust containment and can be accessed via two transparent roller shutters. Dust control is managed by an extraction system on the top of the enclosure.

Samples density is controlled by the thickness of the sand curtain, and the horizontal velocity of the hopper. The maximum horizontal travel of the hopper is 1600 mm with a velocity of up to 300 mm/s. The hopper has a capacity of 300 kg with a vertical travel of 1600 mm.

All controls and parameters are set using a Labview based graphic user interface on a PC located outside the sand rainer enclosure. An Ultra sonic displacement sensor monitors the drop height between the hopper and the sand surface, and automatically adjusts the hopper vertical displacement between passes to maintain a constant drop eight throughout the process.

Following reconstitution, samples are transported to the 1 g mock-up/testing stations or to the centrifuges (see Fig. 2). The laboratory layout has been designed to minimise transport distances between the centrifuge and the different stations and to avoid any disturbance that might affect the density of soil samples.

4.4 Soil characterization tools

Along with the commissioning of the centrifuge, a number of soil characterisation tools were manufactured, including a 20 MPa, 10 mm diameter and 500 mm long piezo-cone penetrometers (also featuring sleeve friction measurement) and a 440 N, 500 mm long T-bar penetrometer that can accommodate bars 5×20 or 10×40 mm These are up-scaled version of the penetrometers developed over the years for the

C661 beam centrifuge. They feature a versatile clamping arrangement, so they can be fitted either in the 2D actuator or in the 4D robot.

ACKNOWLEDGEMENTS

The National Geotechnical Centrifuge Facility was established with funding from the Australian Research Council (LE120100011), the University of Western Australia, the Centre for Offshore Foundation Systems, Newcastle University, the University of Wollongong, Monash University, Queensland University and the University of Adelaide. The support of these institutions, and the support and continuous enthusiasm of the eight technical staff of the NGCF are gratefully acknowledged.

REFERENCES

De Catania, S., Breen, J., Gaudin, C. & White, D.J. 2010. Development of a multiple axis actuator control system. *Proceedings of the 7th International Conference on Physical Modelling in Geotechnics*, Zurich, Switzerland, 1: 325–330.

De Nicola, A. & Randolph, M. F. 1994. Development of a miniature pile driving actuator. *Centrifuge 94*, 473–478.

Derkx, F., Merliot, E., Cottineau, L.M. & Garnier J. 1998. On-board remote controlled centrifuge robot. *Centrifuge 98*, 97–102.

Gaudin, C., White, D.J., Boylan, N., Breen, J., Brown, T.A., De Catania, S. & Hortin P. 2009. A wireless data acquisition system for centrifuge model testing. *Measurement Science and Technology*. 20: 095709. DOI:10.1088/0957-0233/20/9/095709.

Kim, D.S., Cho, G.C. & Kim, N.R. 2006. Development of KOCED geotechnical centrifuge facility at KAIST. *Proceedings of the 9th International Conference on Physical Modelling in Geotechnics*, Hong-Kong, 1: 147–150.

Randolph, M.F., Jewell, R.J., Stone K.J.L. & Brown T.A. 1991. Establishing a new centrifuge facility. *Centrifuge 91*, 3–9.

Stewart, D.P., Boyle, R.S. & Randolph M.F. 1998. Experience with a new drum centrifuge. *Centrifuge 98*, 35–40.

Development of a rainfall simulator for climate modelling

I.U. Khan, M. Al-Fergani & J.A. Black
Department of Civil and Structural Engineering, Centre for Energy & Infrastructure Ground Research (CEIGR), University of Sheffield, Sheffield, UK

ABSTRACT: This paper outlines the design and development of a rainfall simulator for the centrifuge for the purpose of evaluating infrastructure resilience in changing climate conditions. Pressure losses in the water supply system are evaluated and related to nozzle pressure and flow rate at both 1 g and elevated gravity up to 100 g. Coriolis effects are investigated and quantified with increasing gravity accounting for effects of air flow through the experimental package on the University of Sheffield 50 gT beam centrifuge. The rainfall simulation capabilities are demonstrated on a slope boundary problem to validate the developed system. Slope deformation is captured using digital image correlation techniques which confirm triggering of slope movements with applied rainfall events.

1 INTRODUCTION

Simulating rainfall within the centrifuge is now a mainstream technique for the purpose of evaluating the impact of infiltration and cyclic pore pressures in engineered earth structures. The rain droplets are influenced by parameters such as coriolis force and artificial windage around the model (Caicedo & Thorel 2014, Hudacsek & Bransby 2008, Take & Bolton 2002). Due to the presence of these forces, it is difficult to ensure a uniform distribution of rainfall. Hence it is necessary to estimate the effect of parameters such as the centrifuge acceleration, nozzle position, injection pressure and wind velocity (Caicedo & Thorel 2014).

Considerable effort has been made to unearth the answers of rainfall/climate-induced slope failure via full-scale monitoring and instrumentation, however; this process is protracted and may not yield effective understanding due to the isolation of the test case (Brackley & Sanders 1992). Modelling this type of problem under controlled laboratory conditions, at full scale, is also not truly feasible owing to the magnitude and time taken to evaluate performance over many years (Ling et al. 2009). Small scale modelling at elevated gravity in a centrifuge is attractive, compared to 1g counterparts, as prototype self-weight body forces that drive downward slope movements are simulated.

Previous literature on slope instability triggered by climate conditions has identified various factors such as the soil properties, rainfall intensity, initial water table location, and slope geometry to contribute to the instability of slopes during rainfall (Kim et al. 2004, Yoshida et al. 1991). The aspect of rainfall simulation is further complicated owing to coriolis effects that affect the rainfall trajectory and how this advances to the model surface. To provide an adequate distribution over the soil surface, the coriolis effect and other properties previously highlighted should be calibrated.

To minimize the coriolis effect and to prevent surface erosion due to rainfall impact, the vertical distance between the surface of the model and spraying nozzles has to be carefully considered. Askarinejad et al. (2012) determined a stand-off distance between 70 to 100 mm was suitable to minimise coriolis effects, but noted this was dependant on the rainfall intensity and nozzle type. Caicedo et al. (2015) proposed a theoretical rainfall framework of coriolis incorporating factors such droplet size, initial velocity of droplet, wind velocity, rate of evaporation and relative velocity between air and droplet. Experimental observations validated the predictions.

In this paper, the aspect of rainfall generation and dispersion is evaluated for the University of Sheffield 50 gT 2 m radius beam centrifuge. This investigation supports ongoing work to develop an environmental climate chamber for modelling infrastructure resilience. The water supply system for the rainfall simulator is discussed in detail. The coriolis effect is studied based on the experimental tests at various g levels. A slope test has been conducted to analyze and validate the functionality of the developed system. The previous contributions identified, offer a brief summary of some of the major technical and scientific contributions within centrifuge rainfall/climate simulation and form a valuable database to support the endeavours described in this paper.

2 SYSTEM DESCRIPTION

A schematic block diagram of water supply system to the centrifuge is shown in Figure 1. Two air supply lines pass to the centrifuge via a 4 port rotatory

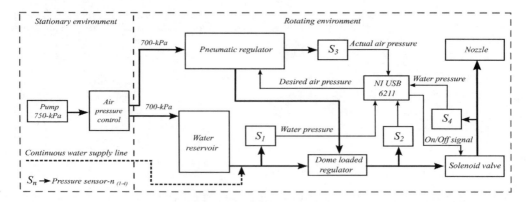

Figure 1. Centrifuge rainfall simulator block diagram.

union interface from the external lab air ring main. One air supply line is connected to an air-water reservoir which provides the initial positive water pressure head to drive the rainfall system. Alternatively, this pressure head can originate by passing water directly over the central axis and allowing centrifugal force to generate water head at the package. The later method is suitable for longer test runs where fluid volume requirements exceed that of the local water reservoir.

The second air supply line is connected to a pneumatic regulator (ITV0050) that is used to step down and regulate the air pressure provided to a dome loaded valve (IR4001) to control the rainfall sprinkler nozzle. The pressure for both the air supply lines is set to 700 kPa. The pressurized water line from the reservoir is connected to the input of the dome loaded valve. The output of the dome loaded valve is connected to a solenoid valve and two pressure sensors, S_1 and S_2, monitoring the input and output pressures respectively. The output of the solenoid valve is connected to the rainfall nozzle and a further pressure sensor (S_4) measures the precise nozzle supply. The water pressure at the nozzle is controlled by regulating the output water pressure at the dome loaded valve. The dome loaded valve offers variable ratio of input to out pressure; however, in the current setup it is calibrated to yield 1 to 1 for air input to water output pressure. The air pressure input supply to the dome valve is regulated using a pneumatic regulator with a pressure sensor (S_3) and monitored using a Labview Virtual Instrument (VI) with integrated hardware USB-6211 data acquisition system. The air pressure is controlled via a PID (Proportional Integral Derivative) control loop with a continuous feedback from the pressure sensor (S_3). All the sensors are monitored online and the control signals are sent to the pneumatic regulator and solenoid valve using the user VI. Figure 2 shows the front and the top view of the experimental setup.

3 CENTRIFUGE WATER SUPPLY SYSTEM

It is very important to monitor the changes in the water pressure as this will affect the flow rate of the rainfall. These affects also vary with increasing gravity

(a) Setup

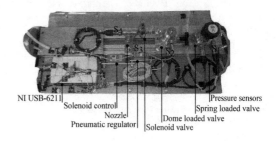

(b) Top plate

Figure 2. Experimental setup.

thus several calibration tests were conducted to understand the water pressure supply system and quantify head losses. The relationship between the flow rate and the applied pressure for the nozzle used is given by Equation (1):

$$Q = 4 \times 10^{-8} \, P_n^{0.5023} \tag{1}$$

where Q is the nozzle flow rate and P_n is nozzle pressure. Tests at four different gravity levels were considered, 25, 50, 75 and 100 g. A time history trace of a tests at 100 g, demonstrating the pressure response of the water supply system starting from stationary (0–3 min), spinning (3–29 min) and rainfall simulation (14–20 min) is presented in Figure 3.

The air pressure to the dome loaded valve is set to 300 kPa and the air pressure to the air-water interface

Table 1. Centrifuge water pressure.

Gravity (m/s²)	P_1 (e) (kPa)	P_2 (t) (kPa)	P_3 (e) (kPa)	P_4 (t) (kPa)	P_5 (e) (kPa)
1 g	697	2.25	697	0.39	0
25 g	646	45.1	637	7.90	14
50 g	590	101	571	17.9	32
75 g	536	158	507	27.8	52
100 g	489	214	447	37.7	80

$P_{1 \rightarrow 4}$ = water pressure.
(e) = experimental, (t) = theoretical.

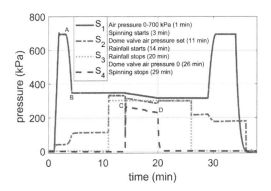

Figure 3. Water pressures at 100 g in centrifuge water supply system starting from stationary (0–3 min), spinning (3–29 min) and rainfall simulation (14–20 min).

is set to 700 kPa. At 1, 25, 50, 75 and 100 g, the water pressure read by sensor S_1 at the input of the dome loaded valve is 697 kPa prior to spinning. At t = 3 min, the centrifuge starts spinning and after steady state, the water pressure at S_1 has reduced (Table 1. P_1(e)). The magnitude of pressure reduction increased with higher gravity levels up to a maximum of 489 kPa at 100 g. The pressure loss due to a height difference between the water reservoir position (resting on the centrifuge platform at a radius of 2 m) and the rainfall system top plate on top of the centrifuge plane strain box which is at a radius of 1.56 m. The theoretical pressure loss is given by Equation (2):

$$P_{loss} = \rho N g h. \quad (2)$$

where ρ is the density of water, N is a multiple of earth gravity g and h is the height difference between the water reservoir and the top plate (0.23 m). The P_2 (t) in Table 1 shows pressure loss calculated using Equation 2. It can be seen that the pressure drop at S_1 from the input pressure of 700 kPa (shown as P_1 (e) in Table 1) and the pressure calculated in P_2(t) are similar in magnitude. This confirms that the location of the water reservoir is an important consideration in the pressure distribution system as the available head to drive the rainfall simulator can be significantly reduced.

Another interesting observation during the rainfall event that commenced at t = 14 min, for 6 minutes, was a constant reduction in supply pressure to the dome valve demonstrated in Table 1 (P_3(e)). This was due to the water column height reduction in the reservoir as rainfall continued. As expected the magnitude of the pressure reduction increased with increasing levels of applied gravity. Crucially, this reduction in system pressure also affected the water pressure supplied to the nozzle. P_4(t) in Table 1, shows the calculated pressure loss due to the decreasing water column height (i.e. 0.041 m in this test case) which reflects the pressure loss observed during the rainfall event. While this supply pressure loss is unavoidable when supplying water via a closed reservoir system, a critical factor is whether these changes in pressure are sufficient to affect the characteristics of the rainfall produced and the discharge intensity. Hence, as demonstrated by Table 1, it would be necessary to refill the water reservoir in flight after each rainfall event to minimise the pressure losses and adverse effects on the nozzle pressure. To emphasise this aspect Figure 3 shows results of an additional test at 100 g where the water reservoir was only half filled. The initial decrease in the water pressure (from point A to B) once 100g was achieved was 142 kPa greater than a previous 100 g test shown in Table 1, in which the reservoir was full. In this test the nozzle pressure (S_4) at the start of the rainfall was 263 kPa (Point C) and dropped to 230 kPa (Point D) after the 6 minute rainfall event. Clearly this is not desirable as the supply pressure fell below that of the specified nozzle pressure of 300 kPa which affect the output flow rate.

The use of water reservoir in the centrifuge environment can create problems related to water pressure. An alternate to mitigate this problem would be the provision of a continuous water supply line. However, if the use of a water reservoir is unavoidable then it is recommended to be placed close to the nozzle position (i.e. as close to the top of the package) in order to minimise pressure losses associated with the height difference and reducing head. Another important observation from all the above mentioned results is an additional increase in the output water pressure of the dome loaded valve shown in Table 1 (P_5(e)) by S_2 at the steady state. This additional increase went up at higher g levels and is attributed to the internal mechanical spring assembly that controls the output orifice. Under elevated gravity, the spring compresses inside the valve due to increasing self-weight forces. However, once calibrated and the magnitude known, this can be adjusted to ensure the desired pressure is achieve by reducing the control pressure by a known offset amount.

4 CORIOLIS EFFECT

Coriolis is one of the most important cited phenomena in the centrifuge when dealing with rainfall simulations. In order to study the coriolis effect different experimental tests have been performed at different gravity levels. In all the tests, nozzle pressure is regulated to 400 kPa using the pneumatic regulator. At

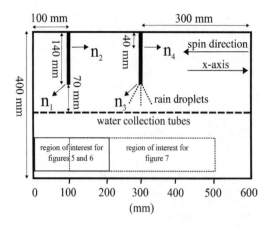

Figure 4. Nozzle positions in the package for coriolis tests.

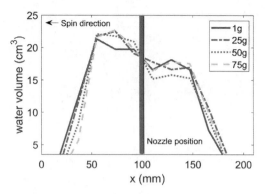

Figure 5. Water volume distribution at different g levels with top plate and nozzle position at n_1.

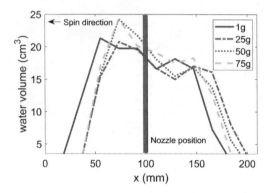

Figure 6. Water volume distribution at different g levels without top plate and nozzle position at n_1.

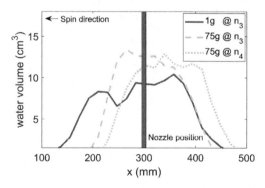

Figure 7. Water volume distribution at 1 g & 75 g without top plate and nozzle in the center of package.

this pressure, the nozzle flow rate is 8.05×10^{-7} m^3/s. The rainfall event lasts for 5 minutes. For the first three set of tests, the distance between the tip of the nozzle and the water collection tubes is set to 70 mm. In the fourth set of tests, this distance is increased to 140 mm. The centrifuge rotation is clockwise and the displacement of droplets in all the tests is measured with respect to the droplets position at 1 g.

Figure 4 shows four different positions of the nozzle (i.e. n_1, n_2, n_3 and n_4). The first set of tests are performed (at 1, 25, 50 and 75 g) with the top plate on to remove the air flow windage effects and the nozzle position at n_1. The results are shown in Figure 5. With each higher g level there is a 3 mm displacement in rainfall distribution towards the nozzle position. In the second set of tests, the top plate has been removed with the nozzle position at n_1 and the results are shown in Figure 6. At all the g levels a greater displacement of rainfall distribution was observed up to 19 mm from the original position. The coriolis effect is very small without air interference in the package. In the second set of tests, the air flow is generating increased drag force on the water droplets, resulting in greater translation of the distribution position. Note, it is observed that there is a non-symmetrical distribution from the nozzle. This was checked and is believed to be due to manufacturing tolerances.

In the third set of tests, to check the effect of the air on water droplets, the nozzle has been moved from position n_1 to n_2 without the top plate. At 75 g with the nozzle at n_1, there is a displacement of 19 mm along the x-axis. At 75 g with the nozzle at n_2, there is a displacement of 22 mm along the same axis, which is not significant for the change in vertical position. In the fourth set of tests the nozzle position is moved between n_3 and n_4 without the top plate. The results are shown in Figure 7. At 75 g with the nozzle at n_3 there is a displacement of 63 mm long x axis. At 75 g with the nozzle at n_4, there is a displacement of 87 mm along the same axis. The effect of air is minimal at both n_1 and n_2 because of being very close to the package side wall. Therefore, the displacement of water droplets at n_1, n_2 is almost similar. However, the effect of air at n_3 and n_4 significantly influence the distribution position of the rainfall. Hence it is clear that windage substantially adds to the coriolis effect observed and this should be minimised. For both cases, the vertical nozzle position has some influence on the coriolis due to the increased free fall distance, although this is small in comparison to windage effects.

Figure 8 shows an increase in the water volume and a reduction in water collection area, with higher gravity levels (i.e. from 1 g to 75 g). This phenomenon

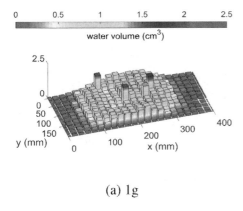

(a) 1g

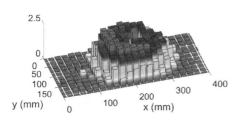

(b) 75g

Figure 8. Water volume distribution at 1 g and 75 g without top plate and nozzle in the center of package (n_3).

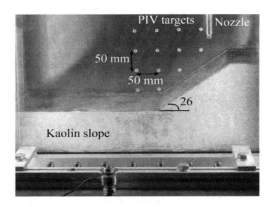

Figure 9. Slope prepared from Kaolin clay, with 40% of water content.

is seen in all of the above tests. At 1 g the water volume collected is 140×10^{-6} cm^3, whereas at 25 g is 171×10^{-6} cm^3, at 50 g is 183×10^{-6} cm^3 and at 75 g is 191×10^{-6} cm^3. There is 300×10^{-6} cm^3 additional water volume collected at 25 g with respect to 1g; however after 25 g, at every higher gravity the additional water volume collected is 10×10^{-6} cm^3. There are two factors behind the increase of water volume, gravitational force and nozzle pressure. The nozzle pressure has increased at a fixed rate at each higher gravity level. The reason for this increase is spring compression inside the dome valve due to increasing self weight forces at every higher gravity level. The flow rate of the nozzle is dependant on the water pressure. This increase in the flow rate is adding 10×10^{-6} cm^3 at higher gravities. An additional increase of 20×10^{-6} cm^3 at 25 g is because of the gravitational force. The increase in gravitational force is not allowing the water droplet to move further away from the nozzle, as the water loss at 1g was seen on the front and back wall of the package. Beyond 25 g the effect of increased gravitational force reduces the water collection area.

5 SLOPE TEST

The previous sections outlined the initial development of a rainfall simulation system that will be incorporated within a climate control chamber to investigate infrastructure resilience. The impact of temporal changes in precipitation on the performance of engineered slopes and earthworks pose increased risk to linear infrastructure systems. To demonstrate the functionality of the rainfall system and the associated movements induced by rainfall, a trial validation case was considered for an engineered slope. A model slope of height 120 mm and slope angle of 26 degrees was tested at 25g, representative of a 3 m prototype (Figure 9). The slope was prepared from Kaolin clay, prepared at a moisture content of 40% and was compacted into the centrifuge payload in 50 mm thick layers using a 2.5 kg hammer. The clay block was trimmed to the desired slope profile with the aid of side cutting templates and a wire saw. Trimmings of the removed soil confirmed uniform moisture conditions of the placed clay layers. In order to observe the impact of precipitation, the validation test case was configured as a plane strain model whereby the side of the slope was exposed and visible through the Perspex viewing window. To enable Digital Image Correlation (DIC) analysis, texture was applied to the front face of the slope model during preparation.

Due to the increase in self-weight stresses imposed by elevated gravity, an initial spin up time of 30 minutes was allowed at 25 g for the slope to bed in. This ensured slope movement due to the imposed gravity during spin up occurred prior to commencing the rainfall stage. In addition this also allowed equilibrium of internal pore water pressures, although note these were not monitored. A continuous rainfall event was implemented in order to trigger the onset of slope movement. Note the ability for individual seasonal conditions (i.e. summer and winter) at various rainfall intensities has been incorporated into the control system development for future more complex rainfall conditions. The nozzle pressure was regulated to 400 kPa. After the initial equilibrium phase, rainfall was simulated for 20 minutes at a flow rate of 8.05×10^{-7} m^3/s. The volume of rainfall is equivalent to a prototype rainfall of 2800 mm per year, which corresponds to the most extreme rainfall predictions for the UK.

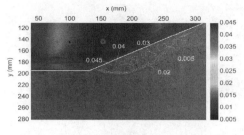

(a) Resultant displacement contour

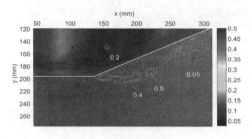

(b) Shear strain contour

Figure 10. Contour plots for the slope test using Digital Image Correlation (DIC).

Figure 10(a) outlines the resultant displacements that occurred at the end of the 20 minute rainfall event across the slope. The effect of the applied rainfall is clear whereby movement of the slope is observed parallel with the slope surface, commencing at the toe and propagating backwards to the crest. The maximum resultant displacement captured is 0.045 mm [0.001 m in prototype] at the interface of the slope and toe position, with displacements generally reducing with slope elevation. This location of maximum displacement also corresponds with the region of highest shear strain (Figure 10b) which can be seen extending from the toe into the slope. These observations confirm that the simulated rainfall event triggered the onset of slope movement. These observations are consistent with rainfall triggered movements in other tests reported by Take and Bolton (2002), corresponding to where the shear stresses and pore pressures are highest. Other positive aspects to be drawn from the preliminary trail test is that no preferential drainage paths along the clay-window boundary were observed, confirming that the application of silicone grease along this edge was sufficient to mitigate water ingress.

Furthermore, on completion of the test, no adverse surface erosion effects were observed, confirming that the selection of nozzle pressure, height and dispersion angle were suitable to model the rainfall event. After the test some surface cracking across the width of the slope between the front and back side walls was detected; which may offer some insight as to the deeper movement observed within the slope. At this stage it is not clear whether these cracks occurred during the initial spin up phase or as a consequence of rainfall induced movements; however, nevertheless they would have accelerated infiltration of water in the slope at increased depths which could give rise to the observed movements within the slope. Further discussion is beyond the scope of this paper and continued work relating to climate simulation is ongoing to reveal greater insight of infrastructure resilience in changing environments.

6 CONCLUSIONS

In this paper, characteristics of a rainfall simulator have been considered for the centrifuge environment. The water supply system for the rainfall simulator has been analyzed in detail. It has been concluded from the results that the use of a water reservoir in the centrifuge can pose problems related to water pressure losses. To avoid the pressure decrease during the rainfall events, a continuous water supply line should be used. The effect of coriolis has been evaluated, nozzle position and angular velocity are shown to influence the windage component for coriolis consideration. A slope test has been conducted to validate the developed system. Preliminary results indicated that the rainfall has induced slope instability. Further development of the system will allow the investigation of infrastructure resilience with more complex scenarios of the rainfall events.

REFERENCES

Askarinejad, A., J. Laue, A. Zweidler, M. Iten, E. Bleiker, H. Buschor, & S. Springman (2012). Physical modelling of rainfall induced landslides under controlled climatic conditions. *Proceedings of 2nd European Centrifuge Physical Modeling in Geotechnics*, 23–24.

Brackley, I. & P. Sanders (1992). In situ measurement of total natural horizontal stresses in an expansive clay. *Geotechnique 42*(3), 443–451.

Caicedo, B. & L. Thorel (2014). Centrifuge modelling of unsaturated soils. *Journal of Geo-Engineering Sciences 2*(1-2), 83–103.

Caicedo, B., J. Tristancho, & L. Thorel (2015). Mathematical and physical modelling of rainfall in centrifuge. *International Journal of Physical Modelling in Geotechnics 3*(15), 1–15.

Hudacsek, P. & M. Bransby (2008). Centrifuge modelling of embankments subject to seasonal moisture changes. In *Proc. Conf. Advances in Transportation Geotechnics*.

Kim, J., S. Jeong, S. Park, & J. Sharma (2004). Influence of rainfall-induced wetting on the stability of slopes in weathered soils. *Engineering Geology 75*(3), 251–262.

Ling, H. I., M.-H. Wu, D. Leshchinsky, & B. Leshchinsky (2009). Centrifuge modeling of slope instability. *Journal of Geotechnical and Geoenvironmental Engineering 135*(6), 758–767.

Take, W. & M. Bolton (2002). An atmospheric chamber for the investigation of the effect of seasonal moisture changes on clay slopes. In *Proc., Int. Conf. on Physical Modeling in Geotechnics*, pp. 765–770. Balkema Rotterdam, The Netherlands.

Yoshida, Y., J. Kuwano, & R. Kuwano (1991). Rain-induced slope failures caused by reduction in soil strength. *Soils and foundations 31*(4), 187–193.

The development of a small centrifuge for testing unsaturated soils

K.A. Kwa & D.W. Airey
University of Sydney, School of Civil Engineering, Australia

ABSTRACT: Geotechnical centrifuges of various sizes have been used to obtain the hydraulic properties and model the behaviour of unsaturated soils. Centrifuges that are capable of taking inflight and real time pressure and flow measurements tend to be large, complex and expensive to operate. Whereas smaller centrifuges are simpler and have much smaller preparation and testing times, but have little to no real time or inflight measuring systems because of space and weight restrictions. In this study, a Beckman Coulter J6-M1 centrifuge with a diameter of 280 mm and a swinging bucket assembly was modified so that the pore water pressure in unsaturated soil samples could be measured and recorded inflight. Miniature pore pressure transducers were embedded into the soil sample and were also connected to an on board power source and wireless data acquisition system. Battery powered wireless cameras were mounted onto the specimen holder so that the sample could be visually monitored in real time. Results are presented to demonstrate the ability of the system to monitor dynamic moisture changes in the soil and show how these vary with initial degree of saturation. The advantages and limitations of small centrifuges for these studies are discussed.

1 INTRODUCTION

1.1 Background

Small centrifuges have been developed and used in the past to model slope stability problems and the soil response under shallow foundations with various loading geometries (Allersma, 1991, 1994a, 1994b) but small centrifuges have not been widely used in geotechnical engineering due to difficulties with space and weight restrictions for inflight monitoring systems. However, recently, small commercial medical centrifuges with test-tube or swinging bucket assemblies have become available that can be purchased at relatively low cost and modified to test small samples of soil. Small, wireless monitoring equipment that are also relatively inexpensive, can also be purchased and installed to observe and measure soil deformations in model slope stability, retaining wall and shallow foundation problems with varying geometries. Generally, these centrifuges are simpler in design due to space restrictions and carry small scale, low weight samples, making the centrifuge easier to assemble and operate and in particular, have been used for teaching purposes (Airey & Barker, 2010, Black & Clarke, 2012).

Small commercial medical centrifuges have also been modified and used as an alternative method for measuring the soil water characteristic curves (SWCC) and the hydraulic conductivity of various unsaturated fine grained soils. Using a centrifuge considerably reduces the time required to obtain a soil's SWCC as equilibrium is established more quickly due to the higher flow rates that develop in the sample under high gravity fields (Khanzode et al., 1999, 2002). The results from using a centrifuge were found to be consistent with conventional unsaturated soil testing techniques (Reis et al., 2012) and analytical flow solutions (Dell'Avanzi et al., 2004, Parkes, 2007, Zornberg & McCartney, 2010a, b). However, the small centrifuges, unlike the larger, more complex centrifuges, generally have no on board in-flight monitoring systems. During SWCC testing, small centrifuges often need to be stopped and the samples weighed to check whether constant water content has been reached for a specific g-level. Centrifuge studies have also been performed to investigate the hydraulic conductivities of compacted unsaturated soil under transient centrifugal flow conditions (Singh & Gupta 2000, Singh & Kuriyan 2002). However, these centrifuges also had no inflight, live monitoring systems installed and the movement of water through the samples during testing was not measured. The centrifuge also had to be stopped and the samples removed and sliced to obtain a moisture content profile. The need to stop the centrifuge can be avoided if the flow of water is monitored live during centrifugation and there are some small centrifuge set ups that can achieve this. An unsaturated flow apparatus (UFA) modification for a small centrifuge has been developed and is commercially available, however, it has been designed to predict the transport of reactive solutes through clays with low hydraulic conductivities (Timms & Hendry, 2008).

A small geotechnical centrifuge at the University of Sydney was developed through modifying a commercially available Beckman-Coulter J6-M1 centrifuge

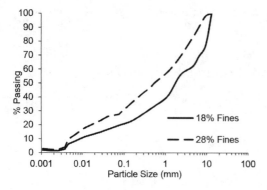

Figure 1. Grading curves.

Table 1. Permeabilities of materials.

Material	Permeability (m/s)
18% Fines	2.7×10^{-7}
28% Fines	3.0×10^{-8}

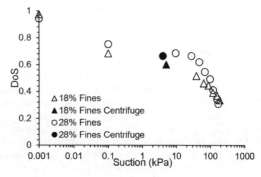

Figure 2. Soil water characteristic curves (SWCC).

with a swinging bucket assembly. This centrifuge was originally modified for teaching purposes and used for demonstrating soil behaviour in modelling slope stability and retaining wall problems (Airey & Barker, 2010). This paper describes further modifications to contain soil samples with on board, wireless cameras that can observe the dynamic changes in moisture throughout the sample and inflight instrumentation that measures the pore water pressures at the base of the sample live during centrifugation. Some preliminary tests, which have been performed as part of a project exploring the performance of unsaturated bulk ore cargoes in ships, are presented and discussed. The development of the small centrifuge capable of monitoring the movement of water in a soil sample at increased g-levels is intended to help in understanding the unsaturated behaviour of the cargo, which can be up to 16.5 m deep within the hold, during transportation.

1.2 Material

The gradings of the materials used in this study, shown in Figure 1, are similar to iron ore fines, a cargo that is prone to liquefaction during transportation, as it is transported moist, in a relatively loose state and can be subjected to a number of significant loading cycles from ship vibrations and wave rocking motions (TWG, 2013).

The materials used in this study are comprised of basalt aggregates ranging in size between 9.5 mm to 150 μm and feldspar fines, less than 75 μm in size, with fines contents of 18 and 28%. Saturated triaxial testing has been performed to investigate how the grading, and in particular the fines content present in these materials, affects their liquefaction response (Kwa & Airey, 2017). The permeabilities and soil water characteristic curves (SWCCs) of these materials at their tested densities are summarised in Table 1 and Figure 2, respectively.

2 CENTRIFUGE SPECIFICATIONS

The J6-M1 Beckman-Coulter centrifuge has a diameter of 280 mm, four swinging hangers and is designed to take a load of 6 kg at 4000 rpm. A picture of the centrifuge bowl with a soil sample container and counter weight ready for a test is shown Figure 3.

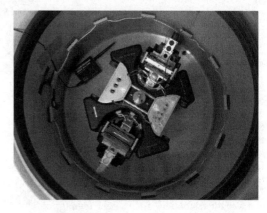

Figure 3. Centrifuge bowl with soil sample and counterweight.

A soil sample with an embedded pore pressure transducer at the base of the sample and camera mounted to the outside of the soil container has been constructed to fit into one of the swinging micro-plate carriers that has plan dimensions of 130 mm × 85 mm as shown by the photograph in Figure 4. To provide a counterweight, an identical soil container with a mounted dummy camera was placed in the other micro-plate container and filled the same soil. The swinging micro-plate carriers that came with the centrifuge are manufactured to slide into position in a sample bay on the rotor piece and therefore the micro-plates and rotor piece were not modified. The Perspex soil containers have inner dimensions of 75 mm × 38 mm in plan and 75 mm deep.

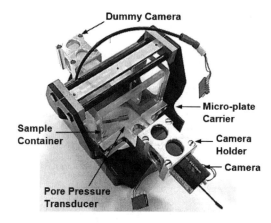

Figure 4. Photograph of soil container in micro-plate with on board instrumentation.

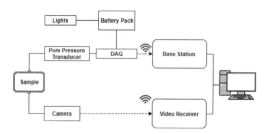

Figure 5. Schematic of on board monitoring system.

A schematic of the monitoring set up for a sample is shown in Figure 5. The CMOS, battery powered, wireless cameras that were mounted onto the specimen holders have dimensions of 20 mm × 20 mm × 63 mm and a resolution of 628 × 582 pixels. Their video data was transmitted to a receiver on the base of the centrifuge lid which was connected to a PCIe video capture card in the computer. The centrifuge has safety features that only allow operation with the lid closed and lighting has to be provided to obtain high quality video. This was achieved through mounting three 12V LEDs on the sample carrier bays of the rotor and thirteen around the bowl of the centrifuge so that the LEDs sufficiently illuminated the micro-plate carrier and sample during centrifugation.

The embedded miniature pore pressure transducers were connected to a wireless Lord Microstrain (LM) 2 channel data acquisition (DAQ) system, which transmits data to a base station placed on the bottom of the centrifuge bowl and was connected to the computer via USB. A small hole was cut in the centrifuge lid to allow cabling between the wireless camera and LM base stations and the computer. The wireless LM DAQ and lights were powered by on board batteries placed in an aluminium container manufactured to sit in an unused sample bay on the rotor. This container was balanced by another aluminium container in the opposite unused sample bay which contained the LM DAQ system, also shown in Figure 3.

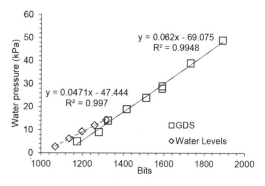

Figure 6. Calibration lines.

3 EXPERIMENT

3.1 Calibration

The miniature pressure transducer was calibrated using a GDS pressure/volume controller and the calibration line relating the pressure transducer's readings in bits and water pressure is shown in Figure 6.

The pressure transducers were set at a hardware gain of 80, resulting in a sensitivity of approximately 0.062 kPa/bit and a maximum pressure reading of 248 kPa at a full scale output of 4000 bits which was sufficient for the speeds used in this study. Due to time constraints, the centrifuge rotor was only balanced by ensuring equal weights of both micro-plate holders with the soil specimens in place. As a result the centrifuge could only test samples with attached instrumentation at a maximum speed of 500 rev/min. An apparatus that is capable of balancing the rotor with all the instrumentation attached is currently under construction, and should enable tests to be performed at speeds up to 4000 rev/min (approximately 2000 g).

To further check whether the hardware gain and therefore, sensitivity of the pressure transducer was set appropriately for testing in the centrifuge, sample containers were filled with water at constantly increasing water levels to a maximum of approximately 50 mm which was the target height for the soil samples. The transducer readings for the different water levels were plotted against theoretical water pressures p for an angular velocity ω of 450 rev/min, calculated by using Equation 1 where ρ is the density of water, h is the water level before centrifugation and gravity, g was calculated by integrating throughout the depth of the water.

$$p = \rho h \omega^2 \int_{r_1}^{r_2} r \, dr \qquad (1)$$

The radius r was taken from the centre of rotation to the top of the water (r_1) down to the middle of the pressure transducer (r_2), located at the bottom of the sample. It can be seen from Figure 6 that Equation 1 leads to an overestimation of the water pressures. This is believed to be because no allowance has been made for the water surface becoming curved at elevated acceleration levels which results in the height of water in the centre of the container reducing. Nevertheless, these values from the centrifuge calibration provide an indication of the saturated water height in the model tests.

3.2 Sample testing

Samples containing 18 and 28% fines were mixed with water at various moisture contents to achieve degrees of saturation ranging between 60 to 90% when prepared at the target void ratios of 0.35 and 0.4 for each material respectively. These void ratios were representative of the low relative densities that result from the compaction levels associated with the loading process of ore cargoes into bulk carriers (TWG, 2013). Once the sample was prepared to a target height of 50 mm the container was assembled in the micro-plate holder and balanced together with the counterweight. Then the micro-plate holders containing the sample and the dummy sample were slotted into place on the rotor and the on board instrumentation was connected to the DAQ system and the monitoring software was started. The lights in the centrifuge were switched on and once the centrifuge lid was closed, the centrifuge was spun up to 450 rev/min for up to 5 minutes which was sufficient time for all the samples to reach equilibrium as can be seen from the typical pore water pressure responses in the samples, recorded and displayed live during centrifugation and summarised in Figure 7. Once the test was finished and the centrifuge stopped, the samples were disassembled and results were saved in the computer for further analysis.

4 ANALYSIS

As expected from the typical pore water pressure responses shown in Figure 7, the unsaturated samples initially had negative pore water pressures after assembly, however, once the centrifuge was started, the pore water pressures rapidly increased in samples that were prepared at degrees of saturation 65 and 70% for samples containing 18 and 28% fines, respectively. The increase in pore water pressures was most noticeable when the centrifuge reached a speed at which the samples started to lift from a near vertical to a horizontal position and therefore, forces in the sample became closer to being orientated parallel to the centripetal force. Samples with lower degrees of saturation of 60 and 65% in samples containing 18 and 28% fines respectively, did not appear to compress significantly and there was no evidence of moisture redistribution from the pore water measurements taken at the base of the samples as the suctions measured did not change despite the stress increase resulting from the centrifuge rotation.

Samples that had higher initial degrees of saturation reached larger maximum pore water pressures as can be seen from the peaks in the pore pressure responses in Figure 7. The final pore pressures suggest that the soil has become fully saturated in most of the tests with initial degrees of saturation exceeding 65%. The

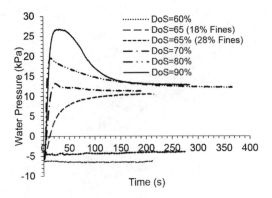

Figure 7. Typical pore water pressure responses.

increase in the degree of saturation at the base of the specimen is in part due to the compression that accompanies the stress increase, and in part due to the flow of water. With high initial degrees of saturation a peak in the water pressures occurs that is greater than the final equilibrium value. These peaks are believed to occur because of the rapid centrifuge acceleration. As a result, the soil at the base quickly becomes saturated and then subsequent total stress increases are reflected in generation of excess pore pressures which do not have time to dissipate. In addition when the centrifuge was started, it initially exceeded the set speed of 450 rev/min by a maximum of 20 rev/min and then stabilised to 450 rev/min after a few seconds. After the pore water pressures reached their maximum value, consolidation occurred and from the rate of pore water pressure dissipation of the test with 90% degree of saturation, a coefficient of consolidation, c_v of 3×10^{-5} m^2/s can be approximated which was found to be consistent with existing data from triaxial tests performed on these materials. Peaks were not observed in the pore water pressure response of samples at lower degrees of saturation less than and equal to 65% as at lower moisture contents, water did not flow as easily through these samples.

An increasing amount of time was required to achieve equilibrium in the samples with higher degrees of saturation. Equilibrium was achieved once the water pressure readings in the samples stabilised as can be seen in the levelling off of the water pressure response curves in Figure 7. Figure 8 summarises the final pore water pressures that occurred in samples containing 18 and 28% fines when equilibrium was reached during centrifugation.

As expected, the pore water pressures increased as the degree of saturation increased and higher pore water pressures were measured in samples containing 18% fines at degrees of saturation less than 80% because this material was more permeable than the material containing 28%, fines. There were no significant differences in measured pore water pressures between the two materials at degrees of saturation above 80%. The measured water pressures and observed water levels in the unsaturated samples

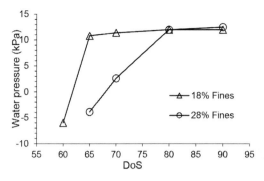

Figure 8. Summary of pore water pressures in samples at all tested degrees of saturation.

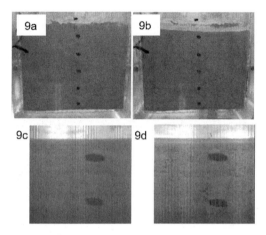

Figure 9. Before and after photos taken outside the centrifuge test. Start and end photos taken inside the centrifuge.

were also consistent with the water pressures measured when the container was filled only with water at different water levels. The corresponding water levels for various water pressures were always less than the maximum theoretical water level and were also consistent with the water levels observed in samples containing degrees of saturation greater than 80% as these samples typically consolidated, became saturated and formed a thin layer of water on top during centrifugation as shown in the before and after testing as shown in Figures 9a, b, c, d. No significant changes were observed in the samples containing lower degrees of saturation during centrifugation.

The suctions measured in both materials when tested at degrees of saturation of 60 and 65% were consistent with SWCC data previously obtained as can be seen by the data points shaded in black in Figure 2. Samples that were tested in the centrifuge at elevated g-levels reached equilibrium much more quickly (less than 5 minutes) than when tested in the pore pressure plate apparatus. However, further modifications to the sample container would be required to measure and allow drainage of water through a porous disk into a region at the bottom of the container if this small centrifuge were to be used as an alternative method to obtain a soil's SWCC.

The main benefit of testing samples in a small centrifuge was that the total time required for testing was significantly less than if a large centrifuge was used. In this study, only 30 to 45 minutes was required for one person to assemble, test and disassemble a sample. Therefore, a variety of tests on materials with different gradings could be performed within a day. However, a challenge with small centrifuge testing is in understanding how the results are affected by the large variation of gravity within the samples. Therefore, different angular velocities and rates of acceleration will be used in future tests and more pore pressure transducers will also be embedded at different levels in the sample container to monitor the movement of water and air within the sample. Unfortunately, the J6-M1 centrifuge does not provide a computer interface to allow monitoring of the angular velocity with time and it is proposed to mount accelerometers to the centrifuge's rotor to measure the acceleration and therefore, the angular velocity during centrifugation. The quality of the samples' video footage captured live during testing needs to be improved to avoid the streaking seen in the video frames shown in Figure 9d, and to allow accurate displacement measurement. The poor quality is believed to be a result of intermittent and weak signals when the camera is located furthest away from the receiver, and also due to insufficient lighting. To prevent this in future tests, the receiver will be moved and placed in the middle of the centrifuge lid to stop fluctuations in signal strength and more lights will be added to the base of the bowl of the centrifuge.

5 CONCLUSION

A small centrifuge has been modified to test soil samples with on board instrumentation that accurately monitors the pore water pressures and also have a wireless camera that takes live video of samples during centrifugation. A number of tests were performed during this study on soil samples similar in grading to iron ore fines, containing 18 and 28% fines. Results were validated with tests performed on water and the pore water pressure results were also found to be consistent with the soils' permeabilities and SWCCs. Tests were relatively quick due to the simple, light weight set up and also because equilibrium within the samples was reached within 5 minutes of centrifugation. The miniature pore pressure transducers, wireless DAQ system, wireless camera and video capture cards were purchased at a relatively low cost of approximately $4000. A remaining challenge is in placing and balancing the instrumentation in a confined, small space so that tests can be performed up to the maximum possible speed of 4000 rev/min. Future modifications to the centrifuge apparatus include installing more pore pressure transducers to monitor the flow of air and water throughout the sample, improving the camera signal and lighting, monitoring the angular velocity of the centrifuge with respect to time during centrifugation and properly balancing the rotor with all on board instrumentation to enable testing at higher g-levels.

ACKNOWLEDGEMENTS

The authors also acknowledge the support provided by the Australian Research Council Discovery Scheme (grant DP150103083) and the technical ingenuity of Ross Barker that has made the tests possible.

REFERENCES

Airey, D.W. & Barker, R. 2010. Development of a low cost teaching centrifuge. *Proc. 7th International Conference on Physical Modelling in Geotechnics*: 205–210.

Allersma, H.G.B. 1991. Using image processing in centrifuge research. *Centrifuge 9, Balkerna, Rotterdam*: 551–558.

Allersma, H.G.B. 1994a. Development of miniature equipment for a small geotechnical centrifuge. *Transportation Research Record 1432*: 99: 105.

Allersma H.G.B. 1994b. The Unversity of Delft geotechnical centrifuge. *Proc. Int. Conf. Centrifuge 94*: 47–52.

Black, J.A. & Clarke, S.D. 2012. The development of a small-scale geotechnical teaching centrifuge. *Enhancing Engineering Higher Education, Royal Academy of Engineering*: 37–41.

Dell'Avanzi, E., Zornberg, J.G. & Cabral, A.R. 2004. Suction profiles and scale factors for unsaturated flow under increased gravitational field. *Soils and Found*. 44:3: 79–89.

Khanzode, R.M., Fredlund, D.G. & Vanapalli, S.K. 1999. An alternative method for the measurement of soil-water characteristic curves for fine grained soils. *Proc. 52nd Can. Geotech. Conference*: 623–230.

Khanzode, R.M., Vanapalli, S.K. & Fredlund, D.G. 2002. Measurement of soil-water characteristic curves for fine grained soils using small-scale centrifuge. *Can. Geotech. J.* 39: 1209–1217.

Kwa, K.A. & Airey, D.W. 2017. The effect of fines on the liquefaction behaviour in well graded materials. *Can. Geotech. J*.

McCartney, J.S. & Zornberg, J.G. 2010. Centrifuge Permeameter for unsaturated soils. II: Measurement of hydraulic characteristics of an unsaturated clay. *Journal of Geotech. and Environ. Eng.*: 136(8): 1064–1076.

Parkes, J.M. 2007. Investigation of infiltration and drainage flow processes in unsaturated soil using a centrifuge permeameter. *PhD. Thesis*. University of California, Berkley.

Reis, R.M., Sterck, W.N., Ribeiro, A.B., Dell'Avanzi, E., Saboya, F., Tibana, S., Marciano, C.R. & Sobrinho, R.R., 2011. Determination of the soil-water retention curve and the hydraulic conductivity function using a small centrifuge. *Geotech. Testing J.* 34:5: 1–10.

Singh, D.N. & Gupta, A.K. 2000. Modelling hydraulic conductivity in a small centrifuge . *Can. Geotech. J.*: 37: 1150–1155.

Singh, D.N. & Kuriyan, J. 2002. Estimation of hydraulic conductivity of unsaturated soils using a geotechnical centrifuge. *Can. Geotech. J.*: 39: 684–694.

Timms, W.A & Hendry, M.J. 2008. Long-Term Reactive Solute Transport in an Aquitard Using a Centrifuge Model. *Ground Water*: 4:46: 616–628.

Technical Working Group (TWG). (2013). TWG 3rd report TWG 4th report.

Zornberg, J.G & McCartney, J.S. 2010. Centrifuge Permeameter for unsaturated soils. I: Theoretical basis and experiment developments. *J. of Geotech. and Environ. Eng.*: 136(8): 1051–1063.

Full scale laminar box for 1-g physical modelling of liquefaction

S. Thevanayagam, Q. Huang & M.C. Constantinou
University at Buffalo, Buffalo, New York, USA

T. Abdoun & R. Dobry
Rensselaer Polytechnic Institute, Troy, New York, USA

ABSTRACT: Brief details of a 1-g full scale laminar box system capable of simulating the liquefaction phenomenon of saturated loose soil deposit up to 6 m deep are presented. The internal dimensions of the box are 5 m in length and 2.74 m in width. The system consists of a two dimensional laminar box made of 40 rectangular laminates stacked on top of each other supported by high capacity and very low friction bearings, a base shaker sitting on the strong floor, two high speed actuators mounted on the rigid blocks, a dense instrumentation array and a close loop hydraulic filling system for building loose sand deposit. Intricate details of the system are presented and its capabilities are illustrated using a recent shake test leading up to liquefaction.

1 INTRODUCTION

Liquefaction of loose saturated soil is one of the common phenomena during earthquakes, causing severe damage to waterfront buildings and other pile-supported structures (McCulloch and Bonilla 1970; Hamada et al. 1986; Hamada and O'Rourke 1992; Youd 1993; Ishihara et al. 1996; Tokimatsu 1999; Dobry and Abdoun 2001; Koyamada et al. 2005). Efforts have been made to study the liquefaction phenomena and its associated effects on the pile foundations through a number of centrifuge tests (e.g. Dobry et al. 1995; Abdoun et al. 2003; Haigh and Madabhushi 2005; Gonzalez 2008). The challenges in centrifuge tests include difficulties in satisfying the most relevant scaling laws for the desired experiment and in implementing dense instrument array to measure the soil response. Recently, large-scale geotechnical shaking facilities have been developed to test larger specimens, in the attempt of simulating the soil response and the soil-structure interaction of 'real world' when subjected to seismic shaking. He et al. (2006) studied the effect of liquefaction-induced lateral spreading on the piles in a 1.8 m high soil deposit using a rigid wall soil box with the approximate dimension of 5 m long, 1.2 m wide and 2.1 m deep. Thevanayagam et al. (2009) and Dobry et al. (2011) investigated the phenomena of liquefaction and lateral spreading in loose saturated sand deposits of approximately 5 m in depth using a full scale laminar box system with 5.0 m in length and 2.75 m in width. A number of large scale shake tests have also been performed to study the soil-pile-structure interaction in a rigid wall box (16 m × 4 m and 5 m deep) as well as laminar boxes (8 m diameter × 6.5 m deep and 12 m × 2.5 m and 6 m deep) (Suzuki et al. 2005; Tamura and Tokimatsu 2005; Suzuki et al. 2008; Motamed et al. 2009; Motamed et al. 2013). Fox et. al (2014) introduced a rigid rectangular box (10.1 m × 4.6 ~ 5.8 m wide and 7.6 m deep) capable of a wide variety of applications such as large-scale dynamic rocking tests of shallow foundations. Large soil boxes facilitate more realistic materials and sample preparation methods as well.

This paper describes the features of a new full scale laminar box system at University at Buffalo. This system is capable of simulating the liquefaction of saturated loose sands under level ground conditions. It includes a dense array of instrumentation to record detailed data on the progression of liquefaction phenomenon. A recent liquefaction experiment and results are presented to illustrate the functionality of the system. Detailed analyses of the liquefaction experiments are presented elsewhere (Dobry et al. 2018).

2 LAMINAR BOX SHAKING SYSTEM

The recent full scale laminar box shaking system built at University at Buffalo (UB) has the capability to test up to approximately 80 m^3 of sand simulating a soil deposit up to a maximum of 5~6 m deep. It consists of three parts: a laminar box, a base shaker actuators, and a safety-bracing frame system.

2.1 Laminar box

Figure 1 schematically shows the inner details of the newly developed laminar box. It is a stack-ring system consisting of 40 steel rectangular-shaped laminates

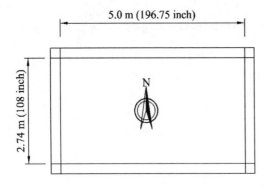

(a) Plan view

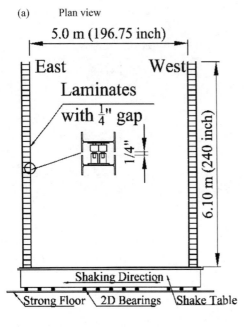

(b) Cross-section

Figure 1. Schematic details of laminar box.

Figure 2. Snapshot of laminar box.

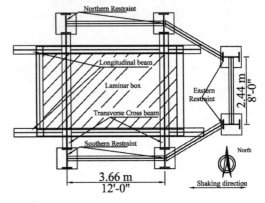

Figure 3. Bracing system.

(rings) (L0-L39). A snapshot of the fully assembled box is shown in Figure 2. The laminates are made of steel I-beams, typically each 0.26 m wide and 0.15 m high. The laminates are separated and supported by a number of distributed high load capacity and very low friction ($\mu < 0.75\%$) ball bearings (#41163, General Bearings Corporation). The inner dimensions of the box is 5.00 m × 2.74 m and up to 6.1 m deep when all 40 laminates (L0-L39) are assembled. The ball bearings are mounted inside the top channel of each I-beam laminate-ring. The under side of the I-beams, where the ball bearings mounted on the I-beam laminate below makes contact with the laminate above, are lined with hardened flat and smooth steel plates. When the laminates are stacked vertically together, the laminates sit on each other resting on ball bearings. A small gap of typically 6.35 mm is provided between the flanges of any two adjacent I-beam laminates by adjusting the height of ball bearings, to prevent any direct horizontal contact interference between any adjacent laminates during horizontal shaking/sliding of the laminates. For safety, internal inter-laminate displacement (drift) restrainers are also placed in the immediate vicinity of the bearing support units to prevent excessive horizontal displacements, should the specimens experience large displacements beyond preset limits anticipated for the experiments to avoid overstressing the laminates and damage to the box.

The total weight of the top 39 laminates (L1-L39) including the inter-laminate bearings and displacement restrainers is about 393 kN excluding any soil weight. The ratio of the laminate weight to the saturated soil weight is about 25% (for $\gamma_{sat} = 19.2$ kN/m³).

2.2 *Base shaker, actuators and safety braces*

The laminar base ring (L0) is bolted to the base shaker placed on the horizontal strong floor of the laboratory. Base shaker is a home-made shaking table supported on inverted high capacity ball bearings (#41163) supported on a horizontal hardened steel plate resting on the strong floor under the base shaker, with future capability for 2-D horizontal shaking.

Two computer-controlled MTS actuators are employed to provide sinusoidal or any preset scaled earthquake shaking history at the base shaker, to be

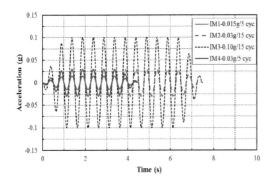

Figure 4. Planned input base motions.

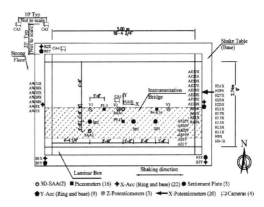

Figure 5. Instrumentation plan (plan view only)

decided on a case-by-case basis depending on limitations of tolerable total and interlaminate displacements for the box. Each actuator is mounted to 'rigid' reaction blocks mounted on the strong floor or strong wall and is tied to the base shaker. The location of the actuator-base shaker tie location could be adjusted, if needed, to minimize any small unwanted moments the actuators may produce at the shaking base of the shaker. Further details of the base shaker and actuators are presented in Bethapudi (2008) and Thevanayagam et al. (2009).

A safety bracing frame system attached to the strong floor is installed on three sides of the laminar box, with optional spring loaded bracings, with vertical motion restrainer beams on top of the laminate L39 (Figures 2 and 3).

3 INITIAL LIQUEFACTION EXPERIMENTS

3.1 Experiments and input base motions (IM)

A set of initial liquefaction experiments were conducted, partly to evaluate the laminar box system, and for liquefaction research. It involved 8 series of shaking experiments, totaling 58 different shaking events (El-Sekelly 2014, Dobry et al. 2018). Figure 4 shows four planned sinusoidal input base motions (IM1 through IM4) with frequency of 2 Hz. IM1 is a small-magnitude sinusoidal motion with 5 cycles at peak base acceleration (a_{max}) of 0.015 g. The purpose of IM1 was to shake the soil at small strain levels without causing rise in pore water pressures. IM2 and IM3 are with 15 cycles but at different peak base accelerations of 0.03 g and 0.10 g, respectively. IM4 is with 5 cycles at $a_{max} = 0.03$ g. Only one experiment with shaking using IM3 is discussed herein, for brevity.

3.2 Sand filling and instrumentation

The soil box was lined with an impermeable 40mil thick EPDM rubber liner bladder with top opening, and another protective EPDM curtain of the same thickness between the bladder and the metal-interior of the box to prevent abrasive damage to the liner. Loose sand was deposited inside the box by hydraulic filling technique developed at UB (Thevanayagam et al. 2009)

using Ottawa F#55, and the sand deposit inside the box was heavily instrumented. F#55 is a clean, fine, uniform sand having a specific gravity of 2.67 and maximum and minimum densities of 1720 kg/m^3 and 1475 kg/m^3, respectively.

Prior to saturated sand filling, an instrumentation bridge was placed above the box on the safety-bracing frame, together with a pipe-grid placed at the bottom of the laminar box, to facilitate instrumentation positioning. Table 1 summarizes the instrumentation sensors installed in this test. Figure 5 shows the locations of the sensors. A total of 16 Piezometers (P) and 2 high-band-width ShapeAccelArray (SAA, Measurand Inc.) sensors were placed inside the laminar box before the sand placement.

Following instrumentation placement inside the box, sand was transferred using a large diameter hose in the form of a slurry and was pumped into the laminar box container. The sand grains were allowed to settle in quiet water, simulating many alluvial and artificial hydraulic field deposits. The target relative density was 35~40%. Based on density measurements using steel buckets placed in the soil box during deposition, the density ranged from 29 to 39% with an average of 36%. CPT tests were also done immediately after sand filling, and are reported in Dobry et al. (2018).

After sand filling, three ground surface settlement plates were placed on the top surface of the sand and connected to vertical potentiometers (V) hung from the instrumentation bridge above the box. A total of 25 accelerometers (AE, AW) and 19 horizontal displacement sensors (H) were placed on the outside walls of selected laminates. Six accelerometers (B) and one horizontal displacement sensor (H) were placed on the base shaker. Four digital cameras were used to monitor critical components or portions of the laminar box from above and from different angles looking at the box from ground level (base shaker level). All sensors were synchronized to a common clock and data was recorded using a common data acquisition system at 200 or 256 Hz frequency. The SAA data were recorded separately but synchronized to the same common clock.

Table 1. Instrumentation summary.

Instruments	Quantity	Model
○ 3D-SAA (SAA)	2	Measurand Inc.
■ P-Piezometers (P)	16	GE-PDCR81 and Kulite XCL-11-250-50A
◆ X-Acc (Ring and base)(AE,AW,B)	22	Honeywell Sensotec 10G
○ Y-Acc (Ring and base) (AE,B)	9	
← X-Potentiometers (H)	20	MTS-Tempsonics-011004050207
⊗ Z-Potentiometers (V)	3	Space Age Control Inc.
● Settlement Plate (SP)	3	PN-62-55-7171
Cameras (CA)	4	CA2 CA3 CA4: Go Pro

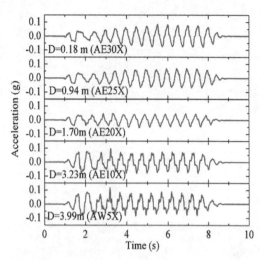

Figure 7. x-acceleration.

4.1 Accelerations

Figure 6 presents the recorded base acceleration at different locations on the base shaker for both x-(longitudinal) and y-(transverse) directions for the shaking of interest ($a_{max} = 0.10$ g/15 cycles). The measured x-direction acceleration in Figure 6 was very close to the planned input base motions (Figure 4). The ratio of the peak base acceleration in y-direction to that in x-direction is about 9%, indicating that the targeted one dimensional shaking was fairly achieved. Figure 7 shows the x-acceleration histories of the laminates at different depths D = 0.18, 0.94, 1.70, 3.23, and 3.99 m below ground surface.

4.2 Pore pressures

Figure 8 shows the shaking-induced normalized excess pore pressure ratio ($r_u = \Delta u/\sigma_{vo}'$) response at depths D = 0.36, 2.03, 3.56, and 4.32 m, respectively, where Δu = excess pore pressure and σ_{vo}' = initial effective vertical stress. It is clear that liquefaction did occur in the soil at all depths in the laminar box. The r_u at each depth reached unity very quickly after the first three shaking cycles.

4.3 Horizontal displacements

Figure 9 shows the peak horizontal inter-laminate relative displacement between any two adjacent laminates (called drift) and the peak absolute base displacement in both positive and negative longitudinal directions. The absolute displacements were well within the tolerable limits for the laminar box. The drift, that could cause overstressing of the laminates and damage to the box, was within tolerable limits.

Further details and analyses of other instrumentation data are not presented herein due to page-length limitations. They are presented in El-Sekelly (2014) and Dobry et al. (2018).

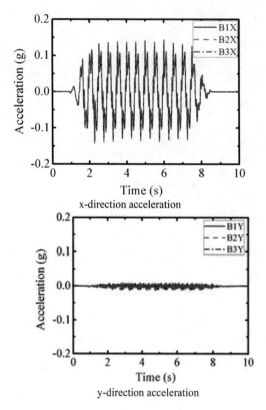

Figure 6. Recorded input base motion.

4 TEST RESULTS

Although a total of 58 individual shakings were done on this specimen, only the results for the third shaking Test T3 (15 cycles at peak base accelerations of 0.10 g, IM3) is presented herein. The soil was subjected to two very small and preliminary pre-test-shakings of 5 cycles at peak base acceleration 0.015 g (IM1) and 5 cycles at 0.03 g (IM4)) prior to Test T3. No significant deformations or excess pore pressures were observed during those two pre-shakings.

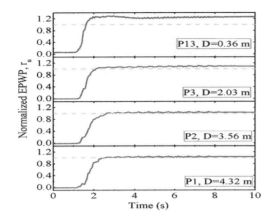

Figure 8. Normalized pore pressure – Test T3.

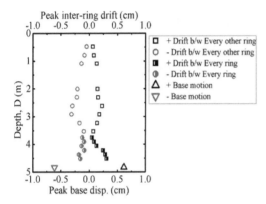

Figure 9. Horizontal peak inter-laminate displacement.

5 CONCLUSIONS

Brief details of a recent soil laminar box system are presented, illustrating that the system is capable of simulating the liquefaction phenomenon in a level ground with the soil deposit of approximately 5 to 6 m in depth. It was possible to prepare hydraulically-placed high quality loose saturated sand deposits that are susceptible to liquefaction. A dense instrumentation with accelerometers, piezometers, potentiometers, high speed video cameras and ShapeAccelArray (SAA) sensors was employed to examine the response of soil during shaking. Liquefaction was observed at all depths during shaking of a saturated sand at an average 36% relative density. The laminar box performed well and the interlaminate relative displacements (drift) were well within tolerable limits. Further analyses of the data are presented elsewhere (Dobry et al. 2018, El-Sekelly (2014)).

ACKNOWLEDGMENTS

Authors wish to thank the NSF-supported UB-NEES facility, SEESL Directors A. M. Reinhorn, A. Filiatrault, and A. S. Whittaker, and the technical staff Mark Pitman, Robert Staniszewski, and Chris Budden for the technical help during this study and in conducting the laboratory tests.

REFERENCES

Abdoun, T., Dobry, R., O'Rourke, T. D. & Goh, S. H. 2003. Pile Response to Lateral Spreads: Centrifuge Modeling, J.Geotech. Geoenviron. Eng., Vol. 129, No. 10, pp. 869–878.

Bethapudi, R. 2008. Liquefaction Induced Lateral Spreading in Large-Scale Shake Testing. Master's thesis, State University of New York at Buffalo, Buffalo, NY.

Dobry, R., Taboada, V. & Liu, L. 1995. Centrifuge Modeling of Liquefaction Effects During Earthquakes, Proceedings of the First International Conference on Earthq. Geotech. Eng. (ISTokyo), Keynote Lecture, Tokyo, Japan, A. A. Balkema, Rotterdam/Brookfield, pp. 1291–1324.

Dobry, R. & Abdoun, T. 2001. Recent Studies on Seismic Centrifuge Modeling of Liquefaction and Its Effect on Deep Foundation, Proceedings of the Fourth International Conference on Recent Advances in Geotech. Earthq. Eng. and Soil Dyn.,Vol. 2, San Diego, CA, March 26–31, S. Prakash, Ed., University of Missouri-Rolla Continuing Education, Missouri.

Dobry, R., Thevanayagam, S., Medina, C., Bethapudi, R., Elgamal, A., Bennett, V., Abdoun, T., Zeghal, M., El Shamy, U. & Mercado, V. M. 2011. Mechanics of lateral spreading observed in a full-scale shake test, J.Geotech. Geoenviron. Eng., Vol. 137 No. 2, pp. 115-129.

Dobry, R., Thevanayagam, S., El-Sekelly, W., Abdoun, T. & Huang, Q. 2018 Large-scale modeling of effects of pre-shaking on liquefaction resistance, shear wave velocity and CPT point resistance of clean sand, in preparation for ASCE J. Geotechnical Engineering.

El-Sekelly, W. 2014 The effect of preshaking history on the liquefaction resistance of granular soil deposits. PhD dissertation, Rensselaer Polytechnic Institute, Troy, NY.

Fox., P.J., Sander, A.C., Elgamal, A., Greco, P., Isaacs, D., Stone, M. & Wong, S. 2014. Large Soil Confinement Box for Seismic Performance Testing of Geo-Structures, Geotechnical Testing Journal, 38 (1), 72-84).

Gonzalez, M. A. 2008. Centrifuge Modeling of Pile Foundation Response to Liquefaction and Lateral Spreading: Study of Sand Permeability and Compressibility Effects Using Scaled Sand Technique, Ph.D. Thesis, Dept. of Civil and Environmental Engineering, Rensselaer Polytechnic Institute, Troy, NY.

Haigh, S. K. & Madabhushi, S. P. G. 2005. The Effect of Pile Flexibility on Pile-Loading in Lateral Spreading Slopes, Geotech. Spec. Publ., Vol. 145, R. W. Boulanger and K. Tokimatsu, Eds., pp. 24–37.

Hamada, M. & O'Rourke, T. D. 1992. Case Studies of Liquefaction and Lifeline Performance During Past Earthquakes. Vol. 1: Japanese Case Studies, Tech. Report No. NCEER-92-0001, NCEER, SUNY-Buffalo, Buffalo, NY.

Hamada, M., Yasuda, S., Isoyama, R. & Emoto, K. 1986. Study on Liquefaction Induced Permanent Ground Displacements, Research Report, Assn. for Development of Earthquake Prediction, Japan, November.

He, L., Elgamal, A., Abdoun, T., Abe, A., Dobry, R., Meneses, J., Sato, M. & Tokimatsu, K. 2006. Lateral Load on Piles Due to Liquefaction-Induced Lateral Spreading During 1G Shake Table Experiments, Proceedings of the Eighth U.S. National Conference on Earthq. Eng., Paper No. 881, San Francisco, CA, EERI, Oakland CA.

Ishihara, K., Yasuda, S., & Nagase, H. 1996. Soil Characteristics and Ground Damage, Soils Found., Special Issue: January, pp. 109–118.

Koyamada, K., Miyamato, Y. & Tokimatsu, K. 2005. Field Investigation and Analysis Study of Damage Pile Foundation During the 2003 Tokachi-Oki Earthquake, Geotech. Spec. Publ., Vol. 145, R. Boulanger and K. Tokimatsu, Eds., pp. 97–108.

Motamed, R., Towhata, I., Honda, T., Yasuda, S., Tabata, K., & Nakazawa, H. 2009. Behavior of Pile Group behind A Sheet Pile Quay Wall Subjected to Liquefaction-induced Large Ground Deformation Observed in Shaking Test in E-Defense Project, Soils and Foundations, Vol. 49 No. 3, pp. 459-475.

Motamed, R., Towhata, I., Honda, T., Tabata, K. & Abe, A. 2013. Pile group response to liquefaction-induced lateral spreading: E-defense large shake table test, Soil Dynamics and Earthquake Engineering, Vol. 51, pp. 35-46.

McCulloch, D. S. & Bonilla, M. G., 1970. Effects of the Earthquake of March 27, 1964 on the Alaska Railroad, Professional Paper No. 545-D, U.S. Geological Survey, Reston, VA, http://pubs.er.usgs.gov/usgspubs/pp/pp545D.

Suzuki, H., Tokimatsu, K., Sato, M. & Abe, A. 2005. Factor Affecting Horizontal Subgrade Reaction of Piles During Soil Liquefaction and Lateral Spreading," Geotech. Spec. Publ.,Vol. 145, R. Boulanger and K. Tokimatsu, Eds., pp. 1–10.

Suzuki, H., Tokimatsu, K., Sato, M. & Tabata, K. 2008. Soil-pile-structure Interaction in Liquefiable Ground Through Multi-dimensional Shaking Table Tests Using E-Defense Facility. Proceedings, 14th World Conference on Earthquake Engineering, Beijing, China, 8 p.

Tamura, S. & Tokimatsu, K. 2005. Seismic Earth Pressure Acting on Embedded Footing Based on Large-Scale Shaking Table Tests, Geotech. Spec. Publ., Vol. 145, R. W. Boulanger and K.Tokimatsu, Eds., pp. 83–96.

Thevanayagam, S., Shenthan, T. & Kanagalingam, T. 2003. Role of Intergranular Contacts on Mechanisms Causing Liquefaction and Slope Failures in Silty Sands, Research Report No. 240.6 T338r 2003, U.S. Geological Survey, Department of Interior, USA.

Thevanayagam, S., Kanagalingam, T., Reinhorn, A., Tharmendhira, R., Dobry, R., Pitman, M., Abdoun, T., Elgamal, A., Zeghal, M., Ecemis, N. & El Shamy, U. 2009. Laminar Box System for 1-g Physical Modeling of Liquefaction and Lateral Spreading, Geotechnical Testing Journal, Vol.32, No. 5, pp. 1-12.

Tokimatsu, K., 1999. Performance of Pile Foundations in Laterally Spreading Soils, Proceedings of the Second International Conference on Earthquake Geotechnical Eng., Vol. 3, Lisbon, Portugal, P. S. E. Pinto, Ed., pp. 957–964.

Youd, T. L. 1993. Liquefaction-Induced Damage to Bridges, Transportation Research Record. No. 1411, Transportation Research Board and the National Research Council, Washington, D.C., pp. 35–41.

8. Education

Using small-scale seepage physical models to generate didactic material for soil mechanics classes

L.B. Becker, R.M. Linhares, F.S. Oliveira & F.L. Marques
Department of Civil Engineering, Federal University of Rio de Janeiro, Rio de Janeiro, Brazil

ABSTRACT: Physical models of small-scale geotechnical problems are thought-provoking didactic tools that facilitate the understanding of some physical phenomena and theoretical concepts. In a glass wall tank measuring 150 × 50 × 10 cm, classical two-dimensional seepage problems, such as the flow under sheet pile wall and through earth dams with and without drainage blanket, were reproduced on reduced model tests. Medium sand was used. The total seepage discharges were measured as well as the piezometric heads in certain points of the models. Dyes were applied to reveal the flow lines. The simple and well defined boundary conditions helped introducing methods for tracing flow nets by hand as well through numerical methods, such as the finite element method (FEM). FEM results were compared to the physical model measurements and the flow nets obtained by FEM were superimposed on model photos, enhancing the learning process. More complex phenomena, such as the flow in the unsaturated zone, could also be visualised. This article presents the use of these reduced models as a didactic tool in undergraduate and post postgraduate courses in Civil Engineering at UFRJ.

1 INTRODUCTION

The use of physical models while teaching geotechnical engineering allows students to experiment, interpret data and work as a team, improving the process of learning and motivating students (Wartman, 2006). Several universities worldwide are employing physical modelling as a didactic tool to better explain basic and complex concepts of geotechnical engineering. Some examples include the notions of bearing capacity, lateral earth pressure, slope stability, and flow through porous media (Craig, 1989; Mitchell, 1998; Dewoolkar et al., 2003). However, the production of such models can be difficult, because great care and effective techniques are necessary to create reliable physical models that can be successfully compared to analytic and numerical results. When inappropriate modelling techniques are used, problems such as segregation, anisotropy, erosion and concentrated flow can generate misleading results.

On this study, three different classical geotechnical seepage problems were reproduced as small-scale physical models performed in an Armfield S1 Drainage and Seepage Tankglass, measuring 150 × 55 × 10 cm: (i) flow under sheet pile wall; (ii) flow through homogeneous earth dam; (iii) flow through homogeneous earth dam with drainage blanket. All tests were monitored, photographed and then compared to the results of numerical analysis performed on a Finite Element Method (FEM) software, SeepW (GeoStudio 2012).

The resulting data from this work was used to create a courseware composed by a report, photographs and videos. This material is used both in theoretical and computer laboratory classes of undergraduate and post-graduate students at Federal University of Rio de Janeiro (UFRJ) and Alberto Luiz Coimbra Institute for Graduate Studies and Research in Engineering (COPPE/UFRJ).

Table 1. Sand characteristics.

Parameters	Medium sand
D_{10} (mm)	0.43
D_{50} (mm)	0.68
D_{60} (mm)	0.70
D_{60}/D_{10}	1.60
e_{min}	0.39
e_{max}	0.67

2 EXPERIMENTAL PROGRAMME

2.1 Grading and permeability of the soil employed on the models

A medium poorly graded sand from Pontal Beach (Rio de Janeiro, Brazil) was used in this study. The characteristics are shown in Table 1 and the particle size distribution curve is shown in Figure 1.

The following rational equation (Eq. 1) deduced by Taylor (1948), shows the main factors which affect permeability of saturated sands:

$$k = D_s^2 \frac{\gamma_w}{\mu} \frac{e^3}{1+e} C \qquad (1)$$

where k = coefficient of permeability; D_s = diameter of a sphere of equivalent "volume-surface area" ratio of all grains of soil together; γ_w = specific weight of the fluid; μ = viscosity of the fluid; e = void ration (dimensionless); and C = shape factor depending on soil structure.

In an attempt to evaluate (Eq.1) correlation in this medium sand, constant head permeability tests were performed with different void ratios and the results are shown in Figure 2.

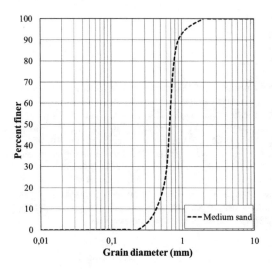

Figure 1. Particle size distribution curve.

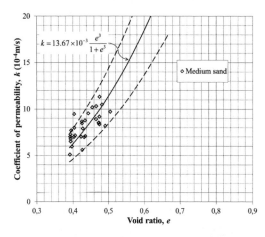

Figure 2. Relationships between the coefficient of permeability and the void ratio.

2.2 Construction sequence and test procedures

Before modelling, the tank was filled with water to assure soil saturation, and the model's geometry was drawn in the tank's glass wall. After that, dried soil was poured through a funnel in the water filled tank, following the drawn geometry of the model and compacting each 2.5 cm thick layer. A poorly graded sand was used to avoid soil segregation during modelling process and the consequent induction of heterogeneity in the physical model.

After the modelling process was finished, it was imposed to the model a constant difference of water head controlling the upstream and the downstream water levels by means of two overflow pipes. The overflowed water through these pipes was sent to a reservoir under the tank and was continuously pumped to the upstream side of the model.

Sufficient time was given so the stationary flow condition could take place. Then the flow lines were made visible using green or fluorescent dyes inserted in the soil surface through needles fixed in the glass wall. The needles' position and the dye flow rate needed to be carefully controlled to prevent excess dye dispersion and leaking. When the dyed flow lines arrived at the downstream surface of the model photographs were taken from a fixed position what made it easy to compare similar physical models.

The model flow rate was measured at the downstream overflow pipe and the hydraulic head was assessed on the piezometers. The soil weight was measured after disassembling the model. As the model volume was also known, the void ratio could be calculated (Table 2). This enabled an estimation of the soil permeability using Figure 2.

3 SEEPAGE ANALYSIS

3.1 Handmade flow nets and numerical analysis

Given the geometry of each model (Figure 3) and the respective coefficient of permeability (Table 2), before carrying out the numerical analyses it is suggested that students sketch the flow nets by hand and calculate the expected flow rates through the equation:

$$q_t = \frac{N_F}{N_D} \cdot k \cdot \Delta h \qquad (2)$$

where q_t = flow rate; N_F = number of flow paths; N_D = number of equipotential drops; k = coefficient

Table 2. Expected ranges of coefficient of permeability of medium sand for each model.

Test	Void Ratio	Water temp. (°C)	Viscosity correction factor	Coefficient of permeability × 10^{-4} (m/s)		
				Minimum	Mean	Maximum
A	0.44	25	1.13	6.8	9.1	11.5
B	0.45	25	1.13	7.2	9.7	12.2
C-WL1	0.44	26	1.15	6.9	9.3	11.7
C-WL2	0.44	24	1.10	6.6	8.9	11.2

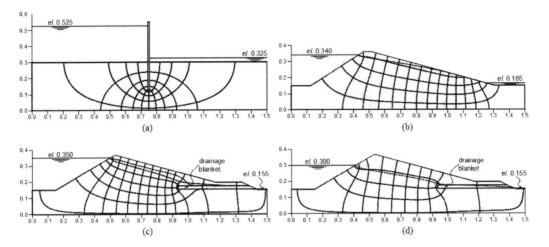

Figure 3. Calculated flow nets: (a) Test A: Flow under sheet pile wall; (b) Test B: Flow through a homogeneous earth dam; Test C-WL1: Flow through an earth dam with a drainage blanket – (c) Water Level 1; (d) Water Level 2 (Units: m).

Table 3. Calculation of the total flow rate and comparison with the measured values.

Test	$k \times 10^{-4}$ (m/s)	Δh (m)	Flow net		Calculated flow rate (cm³/s)		Measured flow rate (cm³/s)
			N_F	N_D	By hand	Numerical Analysis	
A	9.1	0.200	4.3	10	8.22	8.09	21.30
B	9.7	0.175	3.6	14	4.58	4.42	11.97
C-WL1	9.3	0.195	5.0	13	7.32	6.93	19.11
C-WL2	8.9	0.145	3.0	10	4.07	4.28	12.81

of permeability and Δh = total head dissipated within the flow net.

Numerical seepage analyses by means the Finite Element Method were performed assuming soil isotropy and the mean values of permeability shown in Table 2. Sufficiently refined meshes of triangular linear elements and quadrangular bilinear elements were used. The flow nets and the total flow rates obtained are shown on Figures 2a-d and on Table 3.

3.2 Comparison between results of the numerical analyses and of the reduced models

The flow rates obtained on the numerical analysis are compared to the measured values in Table 3.

The flow lines were also drawn with the software on the same positions of the physical model's dye needles. Both images were superimposed, as shown in Figure 3.

In order to obtain a better agreement between the numerical and the experimental flow lines of Test B, it was necessary to divide the sand domain into three regions, each one with a different coefficient of permeability. This can be attributed to the difficulty to densify the sand on the slopes, due to the lack of confinement. Also, to obtain a good agreement between the flow lines of C Tests, the coefficient of permeability adopted to the drainage blanket was 45 times that of the sand.

After this, the coefficient of permeability on each model was adjusted to obtain the calculated total flow rate equal to the measured, resulting on the values shown on Table 4.

Table 4. Coefficients of permeability of the bi-dimensional models compared to that expected based on the constant head permeability tests.

Test	Adjusted "k" $\times 10^{-4}$ (m/s)	Expected "k" $\times 10^{-4}$ (m/s)		
		Min.	Mean	Max.
A	21	6.8	9.1	12
B	26	7.2	9.7	12
C-WL1	22	6.9	9.3	12
C-WL2	22	6.6	8.9	11

3.3 Analysis of results

Overall, there is a good agreement between the dyed flow paths and the numerical ones. It is visible that the main difference is due to dye dispersion along the paths.

However, as can be can be observed in Table 3, the measured flow rates are 2.6 to 3.0 times the ones calculated on the numerical analysis using the estimated permeability. There are some possible causes for this discrepancy but the main one seems to be the inaccuracy in the permeability measurement by constant head tests.

While the reduced models were moulded submerged, guarantying saturation of the sand, the specimens of the constant head permeability tests were moulded with dried sand in order to ease void ratio determination. The samples were subsequently

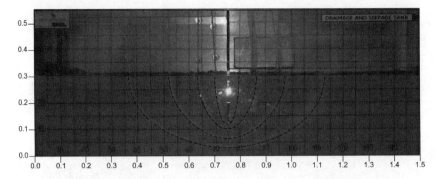

Figure 3a. Test A - photograph with the flow lines of the numerical analysis superimposed on the dye.

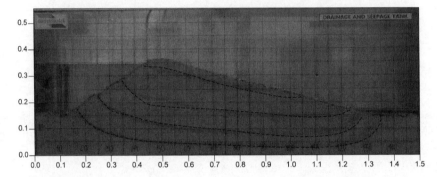

Figure 3b. Test B - photograph with the flow lines of the numerical analysis superimposed on the dye.

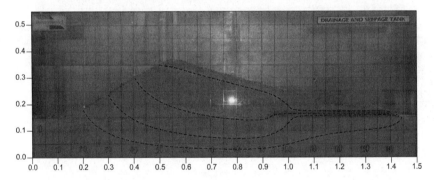

Figure 3c. Test C-WL1 - photograph with the flow lines of the numerical analysis superimposed on the dye.

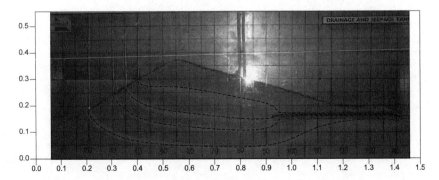

Figure 3d. Test C-WL2 - photograph with the flow lines of the numerical analysis superimposed on the dye.

saturated and care was taken to avoid the formation of air bubbles in the specimens. However, it was impossible to eliminate all the bubbles, resulting in reduced permeability of the specimens.

The Hazen equation ($k = D_{10}^2$, with D_{10} in mm and k in cm/s) yields $k = 18 \times 10^{-4}$ m/s. After the corrections for temperature, k varies from 19.8 to 20.7×10^{-4} m/s. These values are very close to the adjusted ones in Table 4, highlighting the detrimental effects of poor saturation.

Unsaturated flow above the water level was observed in some models, especially in Test C-WL2. However, it was not considered an important cause of discrepancy because, for the Test A, where the soil was completely saturated, the ratio between the measured and the calculated flow rates were similar to that corresponding to the other tests with unsaturated zones.

4 IMPROVEMENT IN THE LEARNING PROCESS

The photographs and measurements can be used to enhance the learning process in many ways.

Students can compare their handmade flow nets with the actual results. The photographs and piezometric head measured in the models can also help the students who perform numerical analysis by setting a unique benchmark for their flow nets and pore pressure calculations.

Test A can be used to show how an increase in the embedded length of the sheet pile wall lengthens the flow paths and, consequently, reduces the flow rate and the exit gradient.

The Test B allow to show the effect of pipping when the water table intersects the downstream slope without any superficial protection, and how this effect is avoided when a gravel layer is disposed over the sand slope.

Tests C show how the drainage blanket improves the downstream slope stability and avoid the pipping by moving away the water table from it.

Comparison between tests C-WL1 and C-WL2 shows how the upstream water level elevation can affect the pore pressures distribution in the dam and, consequently, its stability.

The water flow in the unsaturated zone can also be visualised in tests B and C.

5 SUMMARY AND CONCLUSIONS

On this study, three different classical geotechnical seepage problems were reproduced as small-scale physical models: (i) flow under sheet pile wall; (ii) flow through homogeneous earth dam; (iii) flow through homogeneous earth dam with drainage blanket. All tests were monitored, photographed and then compared to the results of numerical analysis performed on a Finite Element Method (FEM) software.

Four flow models showed good agreement with the results of finite element analysis due to the careful assembly described herein. The construction without these precautions leads to physical models that cannot be compared to the analytical and numerical models.

Dye lines were used to show flow paths in the physical models. The photographs of the physical models were superimposed on the finite element results to demonstrate the good agreement between theory and reality.

The photographs and measurements can be used to enhance the learning process, for the students can compare their predictions with actual results.

Some important effects of seepage on safety and stability of earth structures can be exposed in a very didactic way.

REFERENCES

Craig, W.H. 1989. Use of a centrifuge in geotechnical engineering education. *Geotechnical Testing Journal* 12(4): 288–291.

Dewoolkar, M.M., Goddery, T. & Znidarcic, D. 2003. Centrifuge modeling for undergraduate geotechnical engineering instruction. *Canadian Geotechnical Journal* 26(2): 201–209.

Mitchell, R.J. 1998. *The eleventh annual R.M. Hardy keynote address, 1997: Centrifugation in geoenvironmental practice and education*. Canadian Geotechnical Journal 35(4): 630–640.

Taylor, D.W. 1948. Fundamentals of Soil Mechanics. New York: John Wiley and Sons.

Wartman, J. 2006. Geotechnical Physical Modeling for Education: Learning Theory Approach. *Journal of Professional Issues in Engineering Education and Practice* 134:4(288).

ized physical models at high gravitational acceleration to reproduce in-situ self-weight stresses.

Centrifuge modelling in the undergraduate curriculum—a 5 year reflection

J.A. Black, S.M. Bayton & A. Cargill
Centre for Energy and Infrastructure Ground Research (CEIGR), University of Sheffield, Sheffield, UK

A. Tatari
University of Portsmouth, UK

ABSTRACT: A summary overview of 5 years working with a small-scale educational centrifuge at the University of Sheffield is presented. Various geotechnical design problems have been successfully performed throughout this period, including slope stability, bearing capacity failure and tunnelling. The opportunity to experimentally explore the theoretical content taught in lectures has had a positive impact on student learning in the undergraduate curriculum. The authors advocate there is an immediate need for greater adoption of experimental based observation/demonstration, either conducted at 1 g or Ng, to be embedded within the geotechnical undergraduate curriculum to enrich and deepen the student learning experience of geotechnical system performance.

1 INTRODUCTION

Civil Engineering undergraduate students often struggle to design with soil as an engineering material compared to steel or concrete owing to the variable nature of soil. Within the field of Geotechnical Engineering Burland (2006) describes the '*Soil Mechanics Triangle*' which identifies the interdependencies embedded within geotechnical design. Competency extends beyond understanding the soil material itself, but also requires knowledge of complex theory and analysis methods; many of which have evolved from empiricism, observation and experience, which undergraduate students lack in the embryonic stage of their career. The formative years of a degree programme typically focus purely on theoretical aspects and are void of opportunity to embed experience of actual geotechnical design in practice and learn through observation. Hence, active experimental, observation and reflection pedagogy described by Kolb (1984) via model testing, leading to the establishment of '*well-winnowed experience*' as referred by Burland (2006), represents an exciting opportunity to enhance comprehension and understanding of the design process in undergraduate students.

Laboratory based demonstrations form a valuable learning tool as they provide an opportunity to explore design scenarios, challenge and reinforce theory taught in lectures. Typically these demonstrations are limited to element tests used to assess soil properties such as compressibility and strength. While beneficial, these tests fail to provide any insight of how actual full-scale structures perform; for example, rotational instability of an embankment slope. Without observing failure of structures of their own design, students will not truly fully appreciate the impact of their design assumptions, design philosophy/concept and appreciate the consequence of inadequate design.

1.1 Physical modelling in education

Reduced-scale physical models at 1 g can provide a basic insight of geotechnical performance with respect to indicative behavior, i.e. mode of failure. Quantifiable observations derived from small-scale model tests are limited as realistic prototype self-weight stresses are not preserved. Similitude of stress can be achieved by testing models in a high gravitational acceleration field produced by a centrifuge; hence, the stress and strain distributions in the model will reflect the field situation.

Craig (1989) was one of the first to formally discuss physical modelling for geotechnical engineering education. He described a modelling initiative that began in the mid-1970s at the University of Manchester where experiments were performed using an inexpensive "teaching centrifuge". Mitchell (1994), Collins et al. (1997), Newson et al. (2002) and Dewoolar et al. (2003) also demonstrated the applicability of centrifuge modelling for instructional purposes to illustrate concepts of slope stability, retaining walls, foundations, tunnel stability, and lateral earth pressure theory (Wartman 2006). A summary of several educational centrifuge facilities is reported in Table 1.

A small-scale educational centrifuge has been developed by the lead author at the University of Sheffield, and been in continuous operation since 2012. The educational centrifuge used to support a number of taught modules and dissertation projects. One specific module is the final year '*Advanced*

Table 1. Existing educational centrifuge facilities.

Reference	Gravity (g)	Radius (m)	Model size (mm L × H × W)
Newson et al. (2002)	400	0.325	80 × 80 × 80
Craig (1989)	500	0.30	125 × 70 × 25
Dewoolar et al. (2003)	400	0.61	223 × 165 × 25
Caicedo (2000)	500	0.565	140 × 120 × 70

Geotechnics: CIV4501' elective course. This seeks to enhance students understanding of geotechnical design through enquiry and problem based learning to promote critical/lateral thinking and reflective practice. This is achieved through the integration of advanced geotechnical theory relating to constitutive models to describe soil behaviour, small-scale physical model centrifuge tests, self-learning laboratories and complementary analytical and numerical analysis methods. The purpose of this paper is to highlight a number of projects that have been successfully completed during the last 5 years of operation to demonstrate the value this facility offers undergraduate students.

2 UNIVERSITY OF SHEFFIELD CENTRIFUGE

A small-scale state-of-the-art beam centrifuge 1 m diameter was designed and is capable of rotating a payload up to 20 kg at 100 gravities (100 g), referred to as UoS 2 gT, and is shown in Figure 1. The maximum sample size that can be tested is 160 mm (L) × 100 mm (H) × 80 mm (W) which represents prototype dimensions of 16 m × 10 m × 8 m at 100 g. This is sufficient to test a diverse range of reduced scale engineering structures such as slopes, retaining walls and foundations, while providing stress conditions that realistically duplicate prototype behaviour. The centrifuge is equipped with electrical power slip rings, dual port hydraulic rotating fluid union enabling the delivery of air and water in-flight, digital image capture, signal acquisition, onboard PC and real-time wireless data communication/transfer. Images of samples captured in-flight enable real-time observations of displacement and failure mechanisms. Detailed information about the centrifuge design and development is reported in Black et al. (2014). In the 5 years of operation from 09/2012 to 10/2017, the centrifuge has conducted excess of 500 tests and directly impacted on approximately 300 students via taught modules and dissertation research projects.

3 EXAMPLE PROJECTS

3.1 *Slope stability – gravity switch on*

The slope stability experiment is probably the most appealing of all centrifuge experiments because students can visually confirm the development of a failure surface driven by self-weight alone. Consider the case of a saturated clay slope of height (H) having a slope angle (α). The stability of the slope is dependent on the undrained shear strength of the soil (c_u), the slope height (H) and the unit weight (γ); such that it can be related to a dimensionless group referred to as the stability coefficient (N_s) (after Taylor, 1937) as shown in Equation 1:

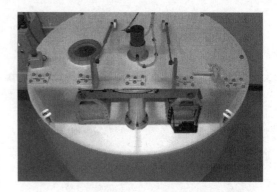

Figure 1. University of Sheffield educational centrifuge.

$$N_s = \left(\frac{c_u}{\gamma g H} \right) \quad (1)$$

Block samples were prepared by consolidating Kaolin clay slurry mixed with de-aired water at 1.5 times the Liquid Limit (LL). Consolidation pressures were ramped up to 200 kN/m² to produce consolidated homogeneous blocks of clay from which model slopes would be prepared. The clay blocks are trimmed to the correct geometry with the aid of side templates and flocked with texture for digital image analysis. Model slopes tested in the centrifuge at N times the earth's gravitational field fail by increased self-weight forces; hence, gravity switch on allowed simple simulation of slope instability without the requirement for complex actuation.

As part of a complementary self-directed laboratory activity, students are required to evaluate the undrained shear strength of the soil block from which the model would be generated. The undrained shear strength for the samples was determined by triaxial compression to be approximately 20 kN/m². Using this data and design input parameters for the slope geometry, students are tasked with predicting the gravity at which the slope will fail based on the derived shear strength. The model slopes are taken to failure by increasing gravitational acceleration until collapse occurs. Real time observations of the deforming slope and shear plane are captured by the onboard cameras and post-processed using image analysis. Comparisons in the actual test performance with the pre-test predictions are considered in conjunction with back analysis of the failed slope to determine a revised estimate for the actual shear strength.

Figure 2. Slope failure by gravity switch on.

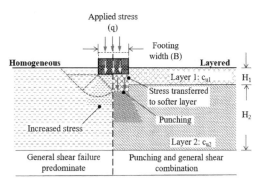

Figure 4. Bearing capacity of layered soil.

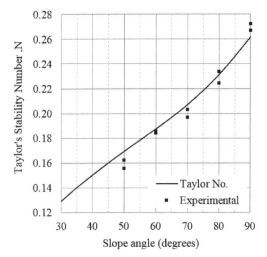

Figure 3. Theoretical Taylor stability curve compared to centrifuge experimental test data by gravity switch on.

By way of example, a failed model slope is shown in Figure 2. The model slope was 60 mm (0.06 m) in height with a slope angle of 90° (i.e. a vertical cut). The undrained strength was determined by the laboratory triaxial tests to be 22 kN/m². The saturated unit weight of soil was determined to be 17.2 kN/m³. According to Taylor's stability chart, the stability coefficient (N_s) for this slope configuration is 0.26. The g-level (N) at which the model slope was expected to fail in the centrifuge was predicted using Equation 2 as follows:

$$0.26 = \left(\frac{22kN/m^2}{(N \times 17.2kN/m^3 \times 0.06m)}\right) \Rightarrow N \approx 81g \quad (2)$$

The slope was expected to fail at 81 g, whereas it failed instantaneously at 79 g. In addition, Figure 3 presents the Taylor stability for a number of test case slopes whereby it is clear that the experimental results are in good agreement with theory; although noting that due to the reduced size of the payload, larger errors may exist than if using larger scale centrifuge systems due to boundary restrictions.

3.2 Shallow bearing capacity

Ultimate bearing capacity of strip footings resting on a single layer of homogeneous clay, described by Terzaghi (1943). Reality however is rarely this simple; soils are often non-uniform, layered and have varying strength/stiffness properties. Increased complexity such as layering, described by Davis & Booker (1973), is a significant departure from basic bearing capacity theory taught in undergraduate programmes and presents a significant challenge to students when faced with this uncertainty. A two layer, firm overlying soft soil, bearing capacity problem is considered (Fig. 4) that enabled students to evaluate aspects such as the impact on bearing capacity factor (N_c) and mode of failure.

Samples were prepared by consolidating Kaolin clay slurry mixed with de-aired water at 1.5 times the Liquid Limit. Consolidation pressures were ramped up to 200 kN/m² and 400 kN/m² to produce consolidated homogeneous blocks of clay having undrained strength of 20 and 40 kN/m² respectively.

Layers of varying thickness of soft and firm soil were required to make composite specimens. Side cutting templates where placed alongside virgin soil blocks and the sample trimmed using a wire saw. Once configured the combined sample was then placed back into the consolidation press under a nominal 100 kN/m² for 24 hours to ensure 'knitting' of the interface boundary between the upper and lower layer. Centrifuge tests were conducted at 50 g and considered footing tests on a homogeneous and layered combinations as detailed in Table 2.

The upper and lower layer properties are noted with the relevant subscript indicator, i.e. undrained shear strength of upper and lower layer are c_{u1} and c_{u2} respectively. The four upper layer thicknesses considered (10, 15, 20 and 40 mm) provided normalised thickness ratios, H_1/B, of 0.5, 0.75, 1 and 2 respectively, where B is the footing width (B = 20 mm).

Figure 5 presents the bearing capacity against normalised settlement (s/B) response for the 20 mm wide strip footing at an accelerated gravity of 50 g. Significant variation in the bearing resistance response is observed between the homogeneous soil bed

Table 2. Layered footing tests.

Test No.	Layer 1 H_1 mm [*m]	c_{u1} kN/m²	Layer 2 H_2 mm [*m]	c_{u2} kN/m²
1	80 [4.0]	40	0	N/A
2	40 [2.0]	40	40 [2.0]	20
3	20 [1.0]	40	60 [3.0]	20
4	15 [0.75]	40	65 [3.25]	20
5	10 [0.5]	40	70 [3.5]	20

*square brackets denotes prototype at N = 50 g

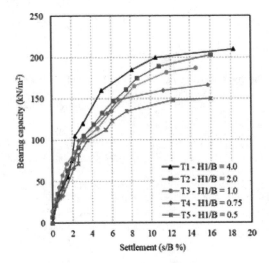

Figure 5. Experimental footing load-displacement.

($H_1/B = 4.0$) and thinnest firm layer ($H_1/B = 0.5$) case.

In the case of $H_1/B = 4.0$ the maximum bearing capacity was 210 kN/m² at the point of failure compared to that of 150 kN/m² for $H_1/B = 0.5$. For $H_1/B = 2.0$ a similar maximum bearing capacity was recorded as that in the uniform bed, albeit with a slightly reduced stiffness response over the full displacement range. Tests $H_1/B = 1.0$ and 0.75 exhibit consistent responses up to s/B = 6% at which point the bearing capacity of the latter reduces quickly as the footing penetration advances. These results clearly demonstrate the complexities that exist in bearing capacity for layered soil configurations, emphasising the challenges faced by students in adapting their basic rudimentary understanding of bearing capacity on homogeneous soils to more diverse complex conditions.

In the absence of surcharge pressure, the ultimate bearing capacity (q_u) of a strip footing on an infinite uniform purely cohesive soil can be expressed as Equation 3:

$$q_u = N_c \times c_u \qquad (3)$$

where c_u is the undrained shear strength and N_c is the bearing capacity factor.

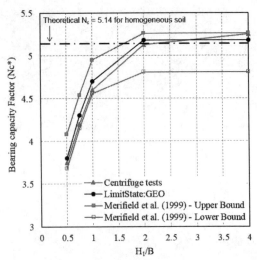

Figure 6. Bearing capacity factor from experiments compared with numerical limit analysis.

Equation 3 is valid for a homogeneous soil conditions; however, in practice non-homogeneous layered soil conditions are frequently encountered. Several authors have postulated modified bearing capacity factors to evaluate this more complex bearing problem (Merifield et al. 1999). A simplified modified bearing capacity approximation for N_c by Merifield et al. (1999), referred to as N_c^*, was approached as the undrained shear strength divided by the strength of the soil in immediate contact with the footing (i.e. the upper soil layer). Using this approximation the bearing capacity factor for the current centrifuge model tests were determined and are presented in Figure 6. It is evident that the bearing capacity factor is influenced by the depth of the upper layer and its' relative thickness to the width of the footing.

These values are also correlated with upper and lower solutions by Merifield et al. (1999) and yield good agreement. In addition, Test 1 ($H_1/B = 4.0$) represents a uniform soil strength sample and thus should conform to the classical theoretical bearing capacity factor $\pi + 2$ (Terzaghi, 1943). The bearing capacity factor in Test 1 was determined to be 5.25, approximately 2% over this theoretical value which could be due to: (i) some residual interface friction at the soil-window boundary or; (ii) increased resistance being mobilised in the soil as the penetration advances due to increased self-weight stresses, as reported by Davis and Booker (1973).

Complementary numerical analysis was carried out using LimitState:GEO (Smith and Gilbert 2007) which uses linear programming to minimise internal energy dissipated along a potential slip planes to yield an upper bound solution and critical failure mechanism. The problem was modelled at prototype to represent the test configurations outlined previously using soil strength properties determined by triaxial tests. Numerical results for the reference test case (homogeneous soil) yielded a bearing capacity factor of 5.18, which compares favourably with theory. The bearing capacity results for the numerical study in Figure 6 and shows good agreement with the upper and lower bound solutions of Merifield et al. (1999) and the centrifuge test data. These observations serve to reinforce to students the importance of determining a suitable bearing capacity factor for complex layered soil conditions as failing to do so would have catastrophic consequences on the foundation stability if the classical value $\pi + 2$ were inappropriately used.

3.3 Tunnelling

All civil engineering works generate disturbance of the ground and great care should be exercised especially when developments are in a densely populated urban environment. As large cities continue to expand, interference of adjacent structures is unavoidable and hence the impact of tunnel-structure interactions must be fully considered and understood. The work reported here pertains to preliminary investigation conducted by Song & Black (2016) to assess the viability of the small-scale centrifuge environment to suitability model a tunnel interaction problem for undergraduate research studies.

The prediction of surface settlement in 'greenfield' conditions was first reported by Peck (1969), who presented a Gaussian based settlement equation (Eq. 4) which has been shown to provide good correlation with field measurement data.

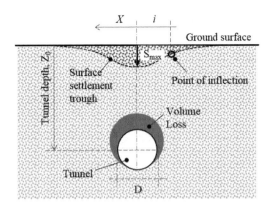

Figure 7. Test overview summary for tunnel experiments.

$$S_v = S_{max} \cdot \exp\left(\frac{-y^2}{2i^2}\right) \quad (4)$$

where vertical settlement is S_v, and the S_{max} is the maximum vertical settlement, occurring above the tunnel centre line. Horizontal offset distance from the tunnel centre line is X, and i is the location of the inflection point.

This approach forms the underlying principle of current design and key aspects are summarised in Figure 7 which indicates the maximum settlement (S_{max}), point of inflection (i) and the extent of the volume loss settlement trough.

Ground disturbances were simulated at 100 g using the conventional approach of tunnel volume loss by reducing the internal pressure of a thin latex membrane. The tunnel had a diameter of 19.05 mm, representative of an approximate 2 m prototype. Model tests were prepared from dry sand of D_{50} of 160 μm, pluviated to 73% relative density. Three C/D ratios of 1.0, 1.6 and 2.0 were considered. Soil displacement measurement and quantification of interaction performance was achieved using image correlation methods. During ramp-up the pressure within the tunnel was balanced against the increased ground stress using the pressure volume controller system.

Figure 8 presents settlement for C/D = 1.6 at a volume loss of 2.1% and 3.8% where the largest settlement displacements occur along the vertical centre line of the tunnel, diminishing with increased horizontal distance. Good agreement is observed with the classical Gaussian formulation of Peck (1969) and subsequent analytical solutions published by Jacobsz (2003) and Vorster (2006). Observations include: (i) increased levels of maximum settlement and; (ii) a changing point of inflection of the Gaussian settlement curve occur with increased volume loss. While only a preliminary study, the successful outcome of the tests to theoretical predictions confirm the potential impact to undergraduate research activities that extend beyond the scope of classic lecture design examples.

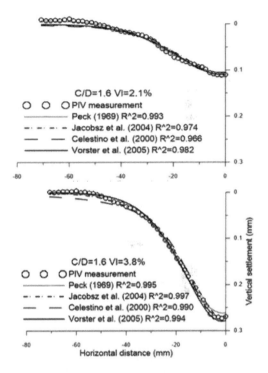

Figure 8. Surface settlement for a tunnel C/D = 1.6 at volume loss 2.1% and 3.8%.

4 CONCLUSIONS

The purpose of this paper was to provide a summary overview of 5 years working with a small-scale centrifuge and demonstrate the impact it has provided to the student learning in the undergraduate curriculum. Classic slope stability by gravity switch on has been demonstrated and correlated stability theory. Greater test complexity involving in-flight actuation for simulating bearing capacity failure of layer soil systems demonstrates the broader range of functionality that the centrifuge offers. Finally the use of the small-scale centrifuge environment is demonstrated with a focus on undergraduate research projects. A tunnel example is presented that enabled the student to achieve a high quality parametric data set for investigation. In all cases good agreement with relevant design theory has been achieved confirming the success of the modelling techniques adopted. The impact on the undergraduate learning experience is unquestionable and the authors advocate there is an immediate need for greater adoption of experimental based observation/demonstration, either conducted at 1 g or N g, to be embedded within the geotechnical undergraduate curriculum to enrich and deepen the student learning experience.

ACKNOWLEDGEMENTS

The experiments reported in this paper were completed using the UoS2gT teaching centrifuge that was developed through funding by the National Higher Education STEM Programme. Continued support by Thomas Broadbent and Son Ltd. to the Centre for Energy & Infrastructure operation and large 4 m diameter centrifuge facility is gratefully acknowledged. The contribution by the Department of Civil & Structural Engineering technical staff Dr Paul Bentley and the post-doctoral staff within CEIGR is also acknowledged. Final acknowledgments go to the CIV4501 classes of 2012–2017 for their contribution in making this educational journey a success and the test data they have contributed to this paper.

REFERENCES

Black, J.A. Baker, N. & Ainsworth, A. 2014. Development of a small-scale teaching centrifuge. Physical Modelling in Geotechnics - *Proceedings of the 8th International Conference on Physical Modelling in Geotechnics 2014*, ICPMG 2014, 181–186. Perth, Australia.

Burland, J. 2006. Interaction between Structural and Geotechnical engineers. *The Structural Engineer*, 84, 29–37.

Caicedo B. 2000. Geotechnical Centrifuge Applications to Foundations Engineering Teaching. *First International Conference on Geotechnical Engineering Education and Training Geotechnical Engineering Education and Training* (ISBN 9058091546).

Collins, B., Znidarcic, D. & Goddery, T. 1997. A New Instructional Geotechnical Centrifuge. Geotechnical News.

Craig, W.H. 1989. The Use of a Centrifuge in Geotechnical Engineering Education. *ASTM, Geotechnical Testing Journal*, 12, (4). 288–291.

Davis, E.H. & Booker, J.R. 1973. The effect of increasing strength with depth on the bearing capacity of clays. *Geotechnique*, 23, (4), 551–563.

Dewoolkar, M.M., Goddery, T. & Znidarcic, D. 2003. Centrifuge Modeling for Undergraduate Geotechnical Engineering Instruction. *ASTM, Geotechnical Testing Journal*, 26, (2), 201–209.

Jacobsz, S.W. 2003. The effects of tunnelling on piled foundations. PhD thesis, University of Cambridge Kolb, D. A. 1984. Experiential learning: Experience as the source of learning and development, Prentice-Hall, Englewood Cliffs, N.J.

Merifield, R.S., Sloan, S.W. & Yu, H.S. 1999. Rigorous plasticity solutions for the bearing capacity of two-layered clays. *Geotechnique*, 49, (4), 471–490.

Mitchell, R.J. 1994. Centrifuge Modeling as a Teaching Tool, 12(3), September.

Newson, T.A., Bransby, M.F. & Kainourgiaki, G. 2002. The use of small centrifuges for geotechnical education. *Proc., International Conference on Physical Modelling in Geotechnics*: ICPMG '02, St. Johns, Canada. 215–220.

Peck, R.B. 1969. Deep excavations and tunnelling in soft ground. In *Proc. 7th int. conf. on SMFE*, 225–290.

Smith, C. & Gilbert, M. 2007. Application of discontinuity layout optimization to plane plasticity problems. *Proceedings Royal Society A*, 463(2086), 2461–2484.

Song, G. & Black, J.A. 2016. Soil Displacement due to tunnelling using small-scale centrifuge technology. *Proceedings of the 2nd European Conference on Physical Modelling in Geotechnics 2016*, ECPMG 2016. Nantes, France.

Taylor, D.W. 1937. Stability of earth slopes. *J. Boston Soc. Civil Eng.*, 243.

Terzaghi, K. 1943. *Theoretical soil mechanics*. New York: Wiley.

Vorster, T.E.B. 2006. The effects of tunnelling on buried pipes. PhD thesis, University of Cambridge.

Wartman, J. 2006. Geotechnical Physical Modelling for Education: Learning Theory Approach. ASCE, *Journal of Professional Issues in Engineering Education and Practice*. 288–296.

Geotechnical centrifuge facility for teaching at City, University of London

S. Divall, S.E. Stallebrass, R.J. Goodey, R.N. Taylor & A.M. McNamara
City, University of London, London, UK

ABSTRACT: The London Geotechnical Centrifuge facility is located at City, University of London. The centrifuge has been used for research since it was established in 1987. However, this powerful research tool is also used to support teaching in the civil engineering undergraduate programme. Final year students are enrolled on a module entitled 'Geotechnical Analysis', which covers advanced methods for investigating non-standard geotechnical structures including physical modelling. Students are required to determine the stability of a composite structure made up of a wall or foundation and a tunnel using theorems of plastic collapse, numerical analysis and physical modelling. The students perform a centrifuge test to explore the use of this technique and to compare results from the test to analytical and numerical models. The technical details and learning objectives of this exercise are described within this paper.

1 INTRODUCTION

1.1 Background

As part of the MEng Civil Engineering programmes at City, University of London, students are introduced to advanced methods for investigating the behaviour of geotechnical structures that do not fit standard solutions for walls, foundations and slopes in the module 'Geotechnical Analysis'. This module focusses on predicting failure and looks at three key approaches i) theorems of plastic collapse, ii) numerical analysis and iii) physical modelling. As part of the assessment students use all three methods to assess the failure of a composite geotechnical structure, comprising a tunnel and wall or foundation. Typical composite structures that have been assessed are given in Figure 1.

The physical modelling aspect is explored through centrifuge modelling using City, University of London's geotechnical centrifuge facility, which was first established in 1987 and subsequently upgraded and refurbished in 2014. A laboratory experiment is organised in which undergraduate students participate in the model-making, data acquisition and analysis associated with this form of investigation. This has been an important aspect of the teaching at City since the early 1990's providing students with hands on experience of the challenges and benefits of this method of investigation.

1.2 Learning objectives

Clear learning objectives have been developed for this laboratory experiment to ensure a successful experience for students and lecturer. These objectives are to ensure students:

- Experience typical model making techniques and methods of instrumenting models.
- Apply the basic principles underlying the application of centrifuge testing techniques.
- Use centrifuge testing to assess the stability of a novel geotechnical structure.
- Interpret the observed behaviour to examine the collapse mechanism and support pressure at failure of the composite structure.
- Compare results with analytical and numerical methods

These objectives are achieved by:

- Choosing an appropriate geotechnical structure to analyse.
- Involving the students in model making and calculating the model dimensions.
- Ensuring the students follow and understand the test procedure.

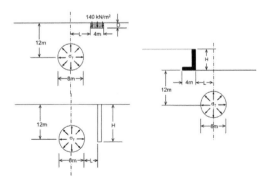

Figure 1. Prototype model geometries.

- Providing the students with data that they can analyse, interpret and compare with the other analysis methods.

1.3 Choice of prototype

The prototype modelled needed to provide a boundary value problem that could be addressed rigorously both by finite element analysis and using theorems of plastic collapse. This was most straightforward to achieve if the prototype was constructed in a clay soil that could be tested undrained. A tunnel supported by air pressure which is reduced to reach failure provides a geotechnical event that is well controlled and requires few approximations when simulated by the two analysis methods. Foundations and walls can be added to create a composite geotechnical structure for which there is no standard solution (as illustrated in Figure 1).

2 SET-UP AND RESOURCING

Clay samples are prepared prior to the laboratory. These samples are created using Speswhite kaolin clay powder and distilled water to form a slurry with a water content of 120%. This slurry is placed inside a strong-box and one-dimensionally compressed in a hydraulic press to 350 kN/m². A period of swelling to 250 kN/m² follows this initial loading. This results in a sample with an undrained shear strength of between 40 and 50 kN/m² ensuring that failure will occur after a reasonable reduction in tunnel support pressure and that the clay can be cut with precision.

It is important that the students see centrifuge modelling as a credible technique for the investigation of geotechnical problems in construction practice. Consequently, the set-up of the model should be as repeatable as possible given inexperienced modellers. To facilitate this a series of jigs (shelf, wall and tunnel) and cutters (square aluminium cutter, 50 mm diameter seamless stainless steel thin walled circular cutter and mild steel cutting bar) are specially fabricated before the laboratory. A 50 mm diameter (1 mm thick) latex bag, tunnel bag union and the required L-shaped aluminium wall, rectangular wall or foundation are also made before the test day.

The instrumentation comprises a pressure transducer to measure the support pressure inside the tunnel and 9 Linear Variable Differential Transformers (LVDTs). These are placed within manifolds and gantries, respectively, ready for calibration by the students during the laboratory.

The size of the cohort influences the number of samples and apparatus that need to be resourced before the test day. Each laboratory centrifuge experiment could be effectively run with a maximum of 15 students and 3 demonstrators. Usually, the numbers of students in the final year cohort are approximately 20–25. However, the test day has been run with as many as 45 students which required three samples to be tested during a working day.

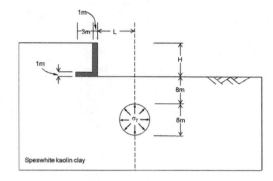

Figure 2. Typical prototype dimensions of model.

Table 1. Table for entering the calculated tunnel support pressure.

Acceleration, Ng	Support pressure, σ_T, (kPa)
1	$\sigma_T = \gamma \cdot z_0$
2	...
10	...
160	...

3 A TYPICAL LABORATORY CLASS

Each laboratory group is asked to co-operate to undertake the experiment and produce a model for testing on the centrifuge. Figure 2 shows the prototype dimensions of one of the composite geotechnical problems, which will be used as an example to illustrate the laboratory procedure.

At the beginning of the laboratory session each student is handed a booklet containing prototype dimensions (i.e. Figure 2) and simple tables (e.g. Table 1) to complete in 'Section A'. 'Section B' pertains to post-test data acquisition and analysis. The group of students is split up into three different sub groups each of which is allocated a key activity that needs to be completed prior to the test.

3.1 Sub Group 1: Calculations

This group of students calculate the required dimensions of the scale model, the effective radius, acceleration level and tunnel support pressure based on a unique value for a key dimension such as L. For this model H was kept constant and equal to 6 m (37.5 mm in the model).

The students are also given the full version of Table 1 to fill in calculating the tunnel support pressure at the spring-line of the tunnel that corresponds to different g levels.

This information will be given to the centrifuge operator during centrifuge acceleration. It is therefore imperative that the students calculate these values correctly to prevent early collapse, which is made very clear.

The students know that to perform the test at the correct stress level the effective radius and speed required to ensure 160g is acting at the appropriate height in the model needs to be computed. This is done using the model dimensions and Equations 1 and 2 below:

$$R_e = R_t + H_m/3 \qquad (1)$$

$$\omega = \sqrt{ng/R_e} \qquad (2)$$

where R_t = radius to top of model, H_m = height of model, R_e = effective radius, g = gravitational constant and ω = speed of centrifuge (in rads/sec).

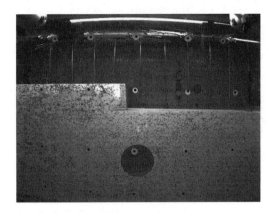

Figure 3. Image from camera on centrifuge swing.

3.2 Sub Group 2: Model-making

These students are tasked with creating the model from the sample once it has been removed from the consolidation press. They require information on the model dimensions from Sub Group 1.

Once the front wall has been removed, the shelf jig is bolted to the front of the strong box and the excess clay trimmed with the box cutter. The wall jig is then positioned onto the shelf jig to remove the clay for the horizontal element of the L-shaped wall with the mild steel cutting bar. This wall is then carefully placed.

To excavate the tunnel cavity, the tunnel jig is bolted to the front of the strongbox and the seamless stainless-steel cutter is used to excavate the soil. The latex bag and tunnel union are then placed within this newly created circular cavity and the union connected to the air supply plumbing. At this stage texture is applied to the front surface to aid the subsurface image analysis post-test. This texture is either plastic marker beads or a fine layer of Fraction B Leighton Buzzard Sand. The Poly(methyl methacrylate) or PMMA observation window and LVDT gantry are then secured to the strongbox. It is now ready for transfer to the centrifuge swing. The model is weighed before being placed on the swing and its weight is fed into the calculations from Sub Group 1. Finally, the air feed and electronic equipment are connected.

3.3 Sub Group 3: Data acquisition and sensors

The final sub group are responsible for calibrating apparatus. The data acquisition system used is a National Instruments PXIe – 1071 with a solid state hard drive and signal conditions running LabVIEW 6.1 Graphical programming software.

The tunnel pressure transducer is calibrated using a Digital Pressure Indicator (DPI) and pressurised air supply through a manostat. An LVDT is calibrated using a micrometre modified to hold the instrument's body. The calibration constants and offsets (where applicable) are then inserted into the data logging programme to convert from volts to engineering units.

4 THE CENTRIFUGE TEST

During centrifuge acceleration the supply of compressed air inside the latex bag supporting the tunnel cavity is manually increased to balance the self-weight of the soil. Once the model reaches 160g, the pressure within the cavity is gradually decreased to zero to simulate tunnel collapse. During the reduction of this tunnel support pressure, the output of the instrumentation and subsurface images are recorded at approximately one record per second. Figure 3 shows a live feed from the USB camera which is fed to a projector on the laboratory floor (outside the control room) to enable all students to watch the tunnel collapse.

When the tunnel has collapsed the test is halted and hand shear vane readings are taken by Sub Group 3, once safe access to the model is possible. The model is then removed from the swing and whilst half the group dismantle the model, the remaining students ensure that all relevant data has been logged. There is a final discussion session for all students to review their observations during the test and make a preliminary assessment of whether the results were as expected.

5 STUDENT ANALYSIS AND REPORTING

Data from the instrumentation and images taken during the test are emailed to each group. These data are intended to support the structured questions outlined in 'Section B' of the student's laboratory booklet requiring the students to report on the collapse pressure and mechanism observed.

5.1 Collapse pressure

Students plot displacement (at the tunnel centreline) against support pressure to determine a collapse pressure (see Fig. 4). They are encouraged to decide on a definition of collapse and usually use one of two approaches to arrive at a value. They either draw lines which are asymptotic to the two gradients of the curve

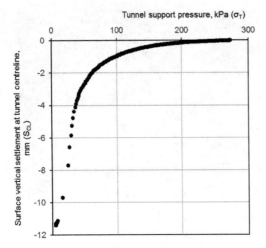

Figure 4. Tunnel support pressure against surface vertical settlement to identify collapse pressure.

or identify a magnitude of settlement that they consider to define collapse. Values obtained usually lie within a range of about 20 kPa.

5.2 Displacements

During the model-making stage the position of the LVDTs relative to the model centreline are recorded by the students. These are then used to plot a surface settlement profile at a given tunnel support pressure. Although it is not possible from the LVDTs to show rotation or horizontal displacement, only vertical displacement, students are often able to describe phenomena such as the movement of the L-shaped wall during the gradual collapse of the tunnel.

5.3 Collapse mechanism

The image analysis, using geoPIV_RG (Stanier et al. 2016), was undertaken by the laboratory demonstrator. Data sets related to specific tunnel support pressures were given to the groups of students to perform a sub-surface displacement analysis. The students are asked to identify which stages of the test would be most useful. Using the last image, often taken immediately after the tunnel collapse, the students identify the collapse mechanism and compare this with their assumed plasticity solutions (Fig. 5). Some of the more able students found it possible to calculate whether their plasticity solutions could provide representative displacements and soil strength or tunnel collapse pressure. These comparisons form the basis of their report conclusions.

5.4 Plasticity solutions vs physical modelling

Students are asked to comment on any discrepancies between the predicted and experimental values of tunnel collapse pressure. The predicted value could be

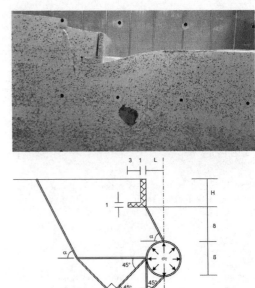

Figure 5. Collapse mechanism of model compared with plasticity solutions.

calculated using the stability number, N, or the results of an upper bound plasticity calculation.

$$N = \frac{n\gamma z_0 - \sigma_T}{S_u} \qquad (3)$$

where n = gravity scaling factor, γ = bulk unit weight of soil, z_0 = depth to tunnel axis level, σ_T = tunnel support pressure and S_u = undrained shear strength.

In general, the plasticity solutions assumed an undrained shear strength of 40 kPa. This was combined with a mechanism such as the one given above to arrive at an upper bound tunnel support pressure. Students had already undertaken this calculations for their particular geometry, which is defined by a unique combination of values for L and H and needed to adapt this solution to make a direct comparison with the centrifuge model test. Using a value for the stability number, N, of 4 (Mair 1979), they could also obtain the support pressure at collapse from equation (3) by choosing an appropriate value for z_0. In this problem, the definition of z_0 is not straightforward due to the change in ground level.

Students are also asked to use their experience of the centrifuge test and the idealisations required to generate an upper bound solution to determine how these may be expected to differ from each other. They would be expected to point out at least some of the following issues:

- The undrained shear strength of the soil will vary with depth in the model (established by the shear vane measurements)
- The failure mechanism assumed in the plasticity solution may not be the most critical failure mechanism (compared earlier in their reports)

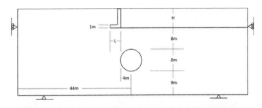
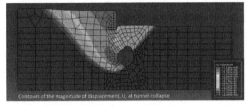

Figure 6. Finite element simulation of centrifuge model and typical pattern of deformation near failure.

- The boundary conditions to the centrifuge model may have constrained the general applicability of the results.

Students, after participating in the experience, are much more able to articulate the shortcomings of physical modelling for tunnelling and its application in design.

5.5 Finite element analysis vs physical modelling

The final approach to solving geotechnical problems that is addressed in the Geotechnical Analysis module is finite element analysis. To limit the extent of the mesh and facilitate comparisons with the centrifuge model test, the finite element analysis simulates the prototype of the centrifuge model as illustrated in Figure 6.

The soil is modelled as elastic perfectly plastic with a Tresca failure criterion to be consistent with the Upper Bound solution. Students are again asked to compare the finite element simulation with the real centrifuge event and to consider the effect of any assumptions, such as:

- Perfect contact at the soil structure interfaces
- Constant undrained shear strength with depth
- An initial *in situ* earth pressure coefficient of 1 and the effect on this of removing the soil adjacent to the wall.
- Local drainage in the centrifuge test.

Students also compare the failure mechanisms observed in the centrifuge and generated by the finite element analysis.

6 CONCLUSION

Geotechnical centrifuge modelling has been used to support teaching on the civil engineering undergraduate programme since the early 1990s. The details of a laboratory experiment conducted using the geotechnical centrifuge during the final year of the undergraduate programme have been described. Groups of 20–25 students are guided through model-making, operational calculations, data acquisition techniques, testing and evaluation of a centrifuge model of a composite geotechnical structure.

Experiential learning practices are a well-established technique for ensuring effective and enjoyable learning experiences (Kolb 1984). Moreover, Luckin et al. (2012) showed learning with specialist technology can only be effective when combined with other key elements including – learning from experts, learning with others, learning through making, learning through exploring and learning through assessment. Element testing is routinely used for demonstrating soil mechanics theories and similarly students gain immensely from undertaking experimental work using a geotechnical centrifuge facility. In particular, they are able to understand the challenges of accurate model making, the data that can be extracted from instrumentation and imaging and the effect of any simplifications that are required.

City, University of London's students are surveyed to ensure quality of teaching. All questions are rated 1 to 5. In the most recent academic years (2015/16 and 2016/17), feedback for the question "staff are enthusiastic about what they are teaching?" has resulted in scores of 4.6 and 4.7, respectively. For the question "the teaching helped my understanding of the subject?" students scored 4.5 and 4.6. The majority (85%) of free-text feedback on the positive aspects of the module commented on the benefits of the laboratory experience.

REFERENCES

Kolb, D.A. 1984. *Experimental Learning*. Prentice-Hall, Englewood Cliffs, New Jersey

Luckin, R., Bligh, B., Manches, A., Ainsworth, S., Crook, C. & Noss, R. 2012. *Decoding learning: the proof, promise and potential of digital education*. Nesta

Mair, R.J. 1979. Centrifugal Modelling of Tunnel Construction in Soft Clay. PhD thesis, Cambridge University, Cambridge, UK.

Stanier, S.A., Blaber, J., Take, W.A. & White, D.J. 2016. Improved image-based deformation measurement for geotechnical applications. *Canadian Geotechnical Journal* 53(5): 727–739.

Development of a teaching centrifuge learning environment using mechanically stabilized earth walls

A.F. Tessari
University at Buffalo, Buffalo, New York, USA

J.A. Black
Centre for Energy and Infrastructure Ground Research, University of Sheffield, Sheffield, UK

ABSTRACT: This paper discusses the development and implementation of an educational module, which incorporates a state-of-the-art teaching centrifuge into open-house demonstrations and the undergraduate educational pedagogy. The students' activities within the developed module are centred on building a model consisting of a mechanically stabilized earth (MSE) retaining wall. The students are exposed to basic educational theory to calculate the capacity of an MSE wall system which is verified through a design and build experimental modelling programme. Results obtained are compared with theoretical predictions and numerical limit equilibrium solutions for a variety of design inputs. These include different numbers of reinforcement layers, thicknesses and widths of reinforcement, and the length of reinforcement versus the surface surcharge position to consolidate their understanding of MSE design variables.

1 INTRODUCTION

1.1 Teaching centrifuge motivation

Laboratory-based exercises are a typical component of the undergraduate civil engineering pedagogy. Soil mechanics is augmented primarily with laboratory index testing that does not engage students or promote critical thinking. While index tests, such as Atterberg limits, are an important component of any geotechnical education and analysis, they do not meet the expectations of the contemporary undergraduate population. Student course reviews often identify them as uninteresting and tedious (Abdoun et al. 2013). Element testing, like direct shear, helps to bridge this gap in expectations and is rated higher by the students. However, it does not provide the level of engagement and perspective necessary to upscale into system level behaviour.

Geotechnical centrifuges have been used in research over the past few decades to advance fundamental knowledge and practice in geomechanics and geotechnical engineering. Undergraduate educational thrusts have been performed using these research-grade devices with positive results (Abdoun et al. 2013). However, due to the cost and fixed location of the apparatus, testing was performed for large groups of students spanning multiple universities and institutions. Reviews from the students that both designed and built the experimental model showed higher satisfaction and internalization of the mechanics tested when compared to the students participating via teleconference.

Low-cost small-scale teaching centrifuges fill this experimental void and provide individual educational programs with the ability to perform a wide variety of engaging and insightful experiments. There have been several unique teaching centrifuges produced, however, the lack of a commercially available model has prevented their widespread adoption due to the substantial initial financial and temporal investments. The University of Sheffield (UoS) and Thomas Broadbent and Sons Ltd (TBS) have produced a turn-key teaching centrifuge to address this need and an in-depth analysis of the history of teaching centrifuges and the device utilized hereafter is presented in Black (2014). A teaching and demonstration module utilizing a mechanically stabilized earth wall has been developed to actively engage students and reinforce concepts and theory presented in the classroom. This paper focuses on one thrust of the experimental tests.

1.2 Centrifuge specifications

The centrifuge used herein is a beam-type 1.0 m diameter with a payload rating of 20kg at 100g. It is a symmetrical design with two identical baskets, where one acts as a counter mass. It has a maximum payload size of 160 mm width by 125 mm height by 80 mm depth. Hydraulic pressure is supplied by a two port 10bar hydraulic union and electrical connections through a 4 way 24A slip ring. The centrifuge is shown in Figure 1 and is fully described in Black (2014). A 5 years review of recent operations is reported in an accompanying paper in these proceedings, see Black et al. (2018).

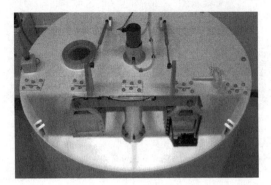

Figure 1. Overview of UoS -2 gT/1.0 teaching centrifuge.

Table 1. Properties of UoS fraction C sand.

d_{10}	0.355 mm
d_{50}	0.450 mm
d_{60}	0.470 mm
e_{min}	0.508
e_{max}	0.756
G_s	2.67

2 CENTRIFUGE-BASED LEARNING EXPERIENCE

2.1 Theory and knowledge reinforcement

The behaviour of soil is largely dependent on the magnitude and method of which it is confined. Realistic representations of global soil failure in the laboratory are difficult to achieve on a scale that is economical for both simple demonstration and teaching reinforcement. The teaching centrifuge overcomes this technical challenge by providing a low-cost method of producing valid stress-strain conditions for a variety of soil, foundation, and superstructure scenarios. Since the payload volume of the teaching centrifuge is relatively small, students can construct and perform parametric and sensitivity analyses. This approach is a core component of Kolb's learning cycle, wherein the fundamental theory presented in lecture is reinforced through experiential learning that emphasizes observation, reflection, and active experimentation (Kolb 1984). The active experimentation element leads to additional engagement of the students and further reflection of the observed results, as they the students feel personally responsible for the models they create. This reflection accompanies Bloom's taxonomy of learning, which assesses the student's level of understanding in the learning process, wherein the lowest level corresponds to knowledge obtained through instruction and independent study and building to the highest echelon, which is based on evaluation (Bloom et al. 1956).

2.2 Scaling laws and retaining wall stability

The concept of scaling laws and their interpretation can be used to reinforce core concepts of geotechnics, such as the confining stress dependent behaviour of granular soil. While this may be traditionally examined through the lens of a triaxial device, an experiment provides a single data point for one depth and does not provide an immediate link to a real-world application.

The theory of retaining wall stability and design can be utilized to explore geotechnical concepts at varying levels of complexity, producing a wide range of course and demonstration modules. This paper focuses on a

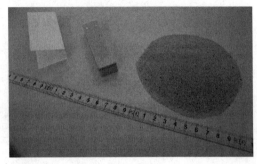

Figure 2. Paper reinforcement (left), aluminium foundation (centre), and faction C sand (right).

series of centrifuge experiments where the location of the surcharge load behind a mechanically stabilized earth (MSE) wall is varied and focuses on the global stress-strain failure behaviour of reinforced soil retaining walls. Additional parameters can be investigated, such as the length of reinforcement, geometry of reinforcement, type of reinforcement, soil type retained, and many others.

3 MSE EXPERIMENTAL MODULE

3.1 Motivation

The overall goal of this module is to provide students with the data, tools, and knowledge set to predict behaviour based on theory. The module is designed to be easily constructible as well as engaging, focusing on the global failure behaviour of a MSE retaining wall, such that it has appeal as an exhibition as well as a laboratory exercise. Individual student groups test a number of MSE configurations which contribute to a larger database of results made available for all students to access and review.

3.2 Experimental configuration

The stabilized earth wall is constructed using soil that is a fraction C sand, with properties listed in Table 1 and works well with particle image velocimetry (PIV) analysis. The soil reinforcement consists of continuous strips of typical letter paper, with wrap arounds that forms the face of the wall (Fig. 2). The reinforcement spans the full width of the wall to both improve construction time of the models as well as repeatability. Students may perform additional tests

Figure 3. Loading system (top left), payload (top right), load cell (mid left), LVDT (middle), and in-flight camera (bottom).

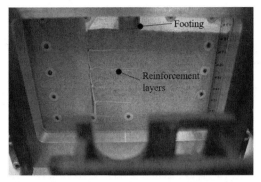

Figure 5. Completed MSE model with temporary foam block (removed during testing) and aluminium strip footing installed.

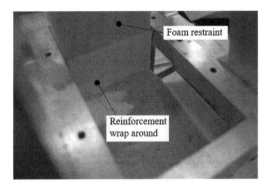

Figure 4. Construction of MSE wall showing the creation of a lift with temporary support, wall face, and wrap-around visible.

with non-continuous grid-like media or cut the paper into strips and set the spacing as a variable. Additional components to establish the model are a 1d stress actuator, load cell and image capture system (Fig. 3).

The model is constructed by placing the sand into the container in lifts, with the initial lift corresponding to the base sand layer. After the base layer has been installed, a foam block is inserted into the container to act as a temporary wall support during the construction sequence. Reinforcement is placed at the top of the base layer, with a portion of the reinforcement serving as both the wall face and wrap around. This is held flush against the foam while placing soil. Various target densities are achieved by altering the compaction effort. The wrap around tab is placed on top of the newly deposited lift prior to installing the reinforcement for the next lift (Figs 4 and 5).

Once the final lift has been installed, the 20 mm [0.6 m prototype] strip load foundation is placed on the top of the MSE wall and the temporary foam support is removed. The footing strip is comprised of aluminium and spans the full width of the container. The displacement at the foundation location is captured using an LVDT. The use of focused physical instrumentation provides key data points for observation and analysis. PIV analysis of the reinforced sand model is used to capture the 2D displacement and strain of the sand, reinforcement, and wall face.

The construction sequence is easily taught to students or demonstrated to an audience. Testing materials and consumables are inexpensive and readily available. Most participants will have a tangible feel for the flexibility and strength of typical printer paper. Since the soil is a dry sand, the lifts can be installed quickly and very little time is necessary to reach equilibrium in the testing phase.

3.3 Testing

Three models are constructed as previously described and tested using the following sequence. The parameter varied is the location of the surcharge load behind the wall face, as represented by the variable X in Figure 6. The minimum distance corresponded to the footing edge being coincident with the wall face, and the further being over the end point of the paper reinforcement. Testing is performed at $30g$, so this distance corresponds to footing centre locations from the wall face of 300 mm, 1125 mm and 1725 mm at the prototype scale.

The loading system (Fig. 7) is installed on the test basket and the point of loading is aligned with the aluminium strip footing. The loading system can be operated in force or displacement control. For the purposes of this experiment, load control is used to impart a constant load during each loading stage. The MSE wall was loaded gradually by increasing the applied vertical force on the footing incrementally until approximately 500 N, or sooner if failure was observed. The in-flight camera is installed and this is used to monitor the test as well as capture images for PIV analysis. Baseline initialization of the data acquisition is performed prior to spinning.

The centrifuge is spun up to $30g$ and the instruments are monitored until steady state conditions are achieved. Load is applied to the model incrementally and allowed to reach stable conditions. Photographs of the state of the MSE wall are taken at this point for later analysis. Students are encouraged to observe the

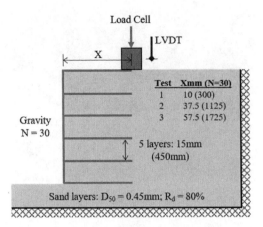

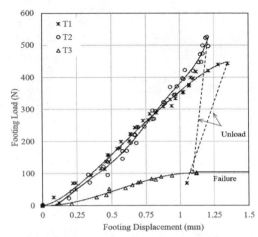

Figure 6. MSE wall schematic showing reinforcement dimensions in model (prototype) units and instrumentation locations.

Figure 8. Footing load-displacement response.

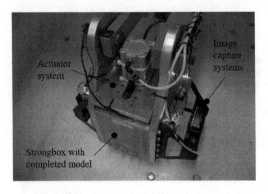

Figure 7. Container with loading system installed and ready to spin up to 30g.

displacement and behaviour of the reinforced system to appraise the real time model performance.

4 TEST RESULTS AND OBSERVATIONS

4.1 MSE system performance – load deflection

The results of the three trials highlight distinct differences in the global behaviour of the system in response to the location of applied load and effectiveness of the reinforcement. As the load both drives failure and increases the effectiveness of the reinforcement, students need to recognise this conflict and make predictions prior to the experiment. During an exhibition, where predictions are not available, this can be incorporated into the explanation of the observed results.

The load-deflection performance of the footing, as seen in Figure 8, offers some insight as to the overall stability of the MSE system in each test case. It is observed in T1 and T2 that the initial stiffness and performance is similar. This is not surprising as the footing is bearing directly on the tensile reinforcement (paper) and thus there is improved internal stability due to the increased friction at the reinforcement-soil interface. As the load exceeds 300 N the performance of T1 and T2 digress; T2 exhibiting increasing stiffness as the applied surface stress continues to generate greater internal resistance; whereas a loss in performance/softening in T1 is observed. Regarding the latter, while the reinforcement has offered capacity that otherwise would not be possible, clearly stability is compromised due to the close proximity of the footing to the edge of the MSE wall. In T3, where the footing was positioned at the end of the reinforcement, considerably lower stiffness was observed and premature failure occurred. Indeed, during the experiment T3 failed catastrophically. Collectively T1 and T3 observations serve to reinforce the need for structural ductility to prevent rapid uncontrolled failure of geotechnical systems, notwithstanding the importance of correct design of MSE systems.

4.2 MSE system performance – image analysis

The PIV technique developed by White et al. (2003) is used with the program GeoPIV to offer students greater insight of the MSE system performance observed in Figure 8. The displacement vector trajectories for T1 at a load of 400 N are shown in Figure 9 (magnified ×4) plotted over the final image captured. For clarity the start and end position of the footing and wall are indicated. It can be clearly seen from the vectors that only a relatively small volume of soil is affected which is concentrated in a local area directly beneath the footing in close proximity to the wall. Some inclination of the wall manifested as a consequence of horizontal thrust forces generated by the translation of vertical footing stress into active horizontal stresses. Evidence of these forces is clearly witnessed by the local bulging in the reinforcement facing at the mid-height of successive layers; however, they were not large enough to exceed the available tensile capacity and rupture of the reinforcement did not occur.

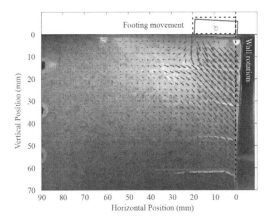

Figure 9. Observed wall displacement and footing movement in T1 at model scale. Note that vectors are magnified 4×.

The observations from the vector plot in Figure 9 are further reinforced in Figure 10 which presents the horizontal displacement contours for each test case at the maximum applied load. Note the contour plots are presented at prototype scale in metres. The localised stress concentrations leading to bulging of the reinforcement face previously described above is now easily quantifiably in T1 (Fig. 10a). Local layer budging is less prevalent in the other two cases as the stresses are dissipated more uniformly within the reinforced area as the footings are away from the wall face. Focusing on the other extreme condition (T3), large magnitudes of lateral soil movement are exhibited in the upper 2 layers (0.025–0.04 m), representative of 1.7% strain with respect to the height of the wall (2.25 m) and a rotation in excess of 1°. The impact of the footing being placed at the boundary of the reinforced material is distinct and unmistakable.

Lateral soil movement occurs throughout the full height of the wall forming a clear cut-off intercept along the end of the reinforcement strips. Although not presented herein, extensive review of the evolving displacement fields in T3 with applied increased load provides an important episodic account of the event history; and provides conclusive evidence for the catastrophic failure observed in this case. This serves as a momentous and historic lesson to aspiring young engineers in the formative years of their engineering instruction as to the dangers of ill-informed or poor design; and how if designed correctly (i.e. T2), structures will deliver satisfactory performance.

4.3 Verification with numerical and limit state analyses

The students perform a numerical analysis to predict the system performance. These predictions are used to focus on key areas and offer insight of the experimental observations. Further, at the conclusion of the experiment, the students refine their models

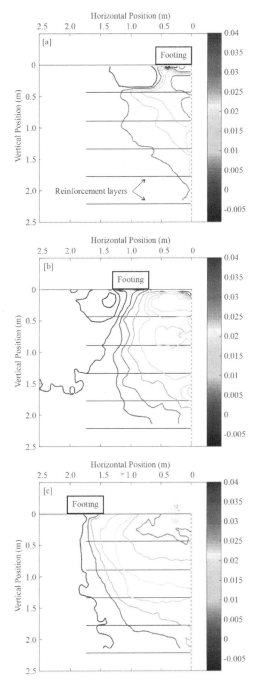

Figure 10. Horizontal MSE wall displacement contour at prototype for (a) T1: footing located at x = 0.3 m (i.e. edge of wall) (b) T2: footing located at x = 1.125 m and (c) T3: footing at x = 1.725 m.

using the experimental data/observations as calibration points. The use of simulation software provides them with chance to develop their ability to predict, verify, and calibrate. The analysis was performed using

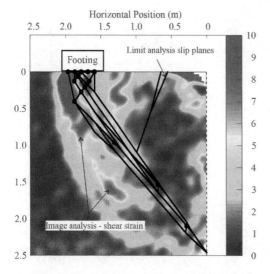

Figure 11. Numerical limit analysis of failure planes with shear strain (%) observed in T3.

LimitState:GEO, though any limit state software package could be used for the prediction. The blind prediction and resultant contours obtained through PIV from the experiments are plotted together in Figure 11. Reasonably good agreement between the experimental results and numerical analysis is observed for the max shear and location of the slip planes.

5 CONCLUSIONS

Mechanically stabilized earth walls provide a versatile case study that can be used in both demonstrations as well as laboratory experiments at varying levels of academic preparation. The key parameters can be modified to suit the course needs and to reinforce different fundamentals. In the described cases, the location of the applied surcharge load fundamentally alters 1) the location and magnitude of the soil displacement, 2) the orientation of the resultant displacement vectors, and 3) the observed displacement at the wall face. This exercise visually reinforces the concept of soil, and especially sand, being a material that develops strength based on the method in which it is confined. This may be linked with Mohr-Coloumb failure criterion in the classroom and simplified during demonstrations, where the general audience may not be familiar with stress tensors.

The walls are inexpensive to construct and may be done so rapidly, facilitating a large volume of student groups. Since the materials and consumables are ubiquitous, the year to year operating costs are relatively low. The upfront acquisition cost of a teaching centrifuge can be compared to other typical soils laboratory equipment. However, devices like direct shear and triaxial apparatuses are limited to a set of focused scope tests. The teaching centrifuge can be used to reinforce many course content topics, provided that appropriate model analogues can be constructed and tested within the geometry and boundary conditions of the system.

ACKNOWLEDGEMENTS

The authors would like to acknowledge the contributions of the laboratory staff at the Centre for Energy & Infrastructure Ground Research at the University of Sheffield, graduate students in the department of Civil and Structural Engineering, and Thomas Broadbent and Sons Ltd for their valuable support and assistance.

REFERENCES

Abdoun, T., El-Shamy, U., Tessari, A., Bennett, V., & Lawler, J. 2013. Multi-institutional physical modeling learning environment for geotechnical engineering education. *2013 ASEE Annual Conference*, Atlanta, Georgia.

Black, J.A. 2014. Development of a small scale teaching centrifuge. *Proceedings of 8th International Conference on Physical Modelling in Geotechnics*, Perth, Australia, pp. 187–192.

Black, J.A., Bayton, S.M., Cargill, A. & Tatari, A. 2018. Centrifuge modelling in the undergraduate curriculum – a 5 year reflection. *Proceedings of 9th International Conference on Physical Modelling in Geotechnics*, London.

Bloom, B. S., Englehart, M. D., Furst, E. J., Hill, W. H, & Krathwohl, D. 1956. Taxonomy of educational objectives: the classification of educational goals. Handbook I: cognitive domain, David McKay, New York.

White, D. J., Take, W. A., & Bolton, M. D. 2003. Soil deformation measurement using particle image velocimetry (PIV) and photogrammetry *Géotechnique* 53(7): 619–632.

9. Offshore

Development of a series of 2D backfill ploughing physical models for pipelines and cables

T. Bizzotto, M.J. Brown & A.J. Brennan
School of Science & Engineering, University of Dundee, Dundee, UK

T. Powell
Subsea7, Aberdeen, UK

H. Chandler
School of Engineering, University of Aberdeen, Aberdeen, UK

ABSTRACT: To investigate the process of pipeline/cable burial a pair of models has been developed to simulate a two-dimensional trench backfilling process at 1g, at two different length scaling ratios, 1:30 and 1:7.5. The aim is to investigate the influence of the velocity of the plough and the weight of the pipe or cable on its tendency to move during backfilling operations. Accelerometers attached to the model pipe ensure tracking of its position during the free motion of the pipe. The shape and velocity of the soil flow, down the slope of the trench, is monitored via particle image velocimetry (PIV) using the images from a MIRO R310 high-speed camera. The outcome of the tests will be to develop a framework to assess and quantify the risk of pipeline uplift and to improve design practice and certainty of meeting burial specifications.

1 INTRODUCTION

All offshore engineering is reliant on fast, safe, dependable and economically convenient connection between offshore and onshore facilities. Pipelines and cables usually require burial in order to protect them from mechanical damage and to improve the thermal insulation (Finch and Machin, 2001) for pipelines that transport high temperature product.

To minimize fabrication costs and ease material handling there is an obvious requirement to employ lighter materials with the restriction that the mechanical and thermal properties are met. Typically for pipelines the insulation is provided by a series of multi-layered coatings incorporating a layer of polymer foam (i.e. polypropylene, polyvinyl chloride, polyurethane) (Palmer and King, 2008), but if mechanical resistance is required to avoid the foam crushing due to hydrostatic pressure, a pipe in pipe scheme may be adopted. In such a solution the external pipe is designed to carry the mechanical and hydrostatics forces and the inner pipe carries the fluid. Both techniques rely on an increased diameter of pipe with much lighter materials, this leads to an overall decrease of the equivalent unit weight of the pipe (or specific gravity S.G.) because of the increase in cross-sectional area offset by a relatively minor increase in weight due to lower unit weight of the insulating material.

Burial of offshore pipelines and cables may be achieved through either jet trenching (Bizzotto, et al.,

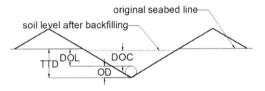

Figure 1. 2D schematic representation of a section of a post-lay ploughed trench, where TTD is target trench depth, DOL is depth of lowering, OD is the pipeline outer diameter and DOC is depth of cover after backfilling.

2017a; Bizzotto, et al., 2017b) or using a multiple phase trenching technique that involves the use of different subsea tools. For instance, mechanical trenching ploughs may excavate a trench around the pipelines after they have been laid on the seabed (Lauder, et al., 2013; Lauder and Brown, 2014). Once the trench has been excavated (Figure 1), a backfill plough pushes the excavated soil back into the trench on top of the pipeline.

During the process of backfilling, pipelines or cables could move upward or unbury themselves by uplift due to the force generated by the soil flow impacting on the pipe or by buoyant flotation (Cathie, et al., 1998; Cathie, et al., 1996). To avoid overdesign of the unit weight of the pipeline/cable (in order to mitigate or prevent these issues), it is desirable to understand the forces acting on the objects and the processes that lead to upward movement of the pipelines

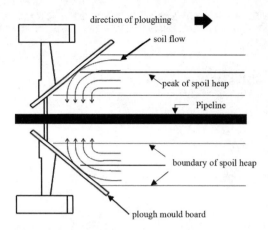

Figure 2. Schematic plan view of backfill ploughing process (redrawn from Cathie, et al. (1998)).

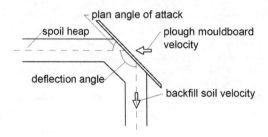

Figure 3. Schematic plan of one-blade backfilling for a conservation of momentum approach with a blade angle of attack of 45 degrees (redrawn from Kaku (1979)).

and cables. The investigation of these type of processes must consider several phenomena, and a scaled series of models has been developed to take into account scaling effects. The aim of this study is to develop a test method to investigate the effect of the backfilling plough velocity on pipe/cables of different unit weights and monitor pipe displacements and imposed forces.

2 BACKFILL PLOUGHING 2D REPRESENTATION

According to Cathie, et al. (1998) the process may be represented as shown in Figure 2. If full spoil recovery is achieved the plough velocity should convert into soil flow velocity down into the trench as represented in Figure 2.

To simplify the modelling and allow a pipe/cable representation as large as possible it was decided to model the process as a 2D plane strain representation. The advantage of designing a 2D backfilling rig is that the same blade velocity in the model may be converted to different prototype plough velocities depending on the angle of attack of the blades (Fig. 3) using an approach based on the conservation of momentum as proposed by Kaku (1979). Once the process is represented in 2D, a plane strain model can be realised

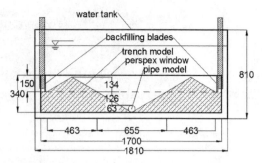

Figure 4. Schematic of large tank, scale 1:7.5 (All dimensions in mm, sliders and actuation not shown for clarity).

and the pipeline/cable can be represented by a cylindrical object with a stiffness that for the dimension of the model can be considered infinite. Therefore, any movement of the monitored pipe is due to rigid body motion of the object and not due to its deformation. This is important when considering the quantity of different pipe/cable products that are available (varying longitudinal bending stiffness), and that a 3D characterisation of the process would be intrinsically dependent on the longitudinal bending stiffness of the model pipe/cable. In our case a 2D plane strain cylindrical object will be free to move and its movements monitored through the use of MEMs accelerometers.

3 PHYSICAL MODELLING OF SUBSEA BACKFILL PLOUGHING

To investigate the backfilling process, the forces acting on the pipeline/cables and eventual scaling issues, a pair of plough backfilling rigs have been designed and fabricated. The two approximate geometric scales are 1:30 (small scale) and 1:7.5 (large scale). The dimension of the tanks are 0.98 m × 0.48 m × 0.44 m (L × W × H) and 1.89 m × 1.05 m × 0.90 m (L × W × H) (Fig. 4) for the small and large scales respectively. Both the models have two plate sliders at the top of the tank to which the simple backfilling blades are attached (Fig. 5). The plates move on 4 linear bearings that move on two parallel hardened steel shafts (an adaptation of a model first developed at University of Dundee by Taylor (2011)). The small backfill model blades are actuated by a rotary motor which winds a steel wire on a spool mounted on the axis of the motor itself. The wire is directly attached to the first slider which carries the first blade, and the linear motion is reversed for the second blade through a pulley mounted on the other side of the box. The creation of the 1:7.5 model was a non-trivial problem, due to physical size and the large frictional and inertial forces that need to be overcome to backfill at higher velocities, hence the necessity of an adequate structural performance of the rig and actuator capacity. The larger scale model motion is provided by a hydraulic piston with a maximum capacity of 69kN. As with the smaller model, the actuator pushes one of the sliders

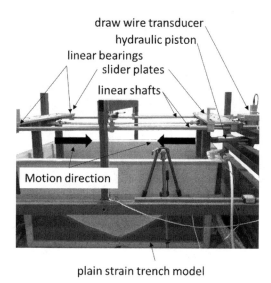

Figure 5. Model 1:7.5 2D backfilling rig actuation system.

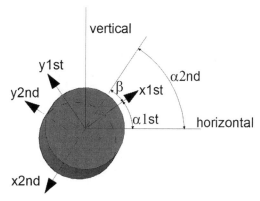

Figure 6. Accelerometer axis representation. 1st represents the axis of the accelerometer near the front Perspex window and the 2nd is the accelerometer near the back wall. β is the out of phase angle between the two x axes of the front and back accelerometers.

Table 1. Pipe diameter and cover depth of soil over the pipe.

	Scale	Pipe diameter mm	Depth of cover Diameters (mm)
Small scale test	1:30	16	1.5D (24)
Large scale test	1.75	63	1.5D (95)*

*parenthesis shows actual depth in mm

(fig. 5) with the other connected by a wire that passes through a pulley to reverse the motion.

The maximum blades velocities that each scaled model can achieve are $v_{max\text{-}small} = 0.16$ m/s and $v_{max\text{-}large} = 0.095$ m/s for the small and the large scale models respectively. At present only one pipe diameter has been investigated for each scaled model (reported in Table 1) to be buried at a depth of cover (DOC) of 1.5 diameters. The pipes can be ballasted and resealed to achieve different pipe unit weights similarly to that used by Bizzotto, et al. (2017b) in fluidised clay to investigate the influence of pipeline/cable unit weight on flotation potential.

3.1 Water depth

The minimum depth of water required on top of the trench model that does not affect the flow of the soil is calculated based on the results for gravity currents from Gonzalez-Juez, et al. (2009). Assuming that the flow of soil does not exceed the initial height of the spoil heap a minimum height of water over the spoil heaps of at least 2 times the spoil heap height will be required. In the model, the height of water used was at least 3 times the spoil heap height for the large and small-scale models.

3.2 Instrumentation

The two models have been instrumented to measure the displacement of the blades using a 150 mm linear variable displacement transducer (LVDT) from 'RDP LDC Series' and a 5 m draw-wire transducer WS17KT-5000 from 'ASM Sensors', for the small and large models respectively. The free movement of the pipe is monitored via two dual-axes MEMs accelerometers placed at both ends of each pipe. The axes of the accelerometers are oriented to capture the acceleration on the 2D plane investigated in these set of tests. Figure 6 shows the orientations of the axis of the two accelerometers on a representative cylinder. The accelerometers are ADXL203CE micro electromechanical systems (MEMS). This kind of sensors use a combination of mechanical response of a two degree of freedom lumped-mass system and the sensing capacity due to the variable parallel plate capacitors that correlate the change in the capacity (C) and the displacement of the inertial mass as schematised in Figure 7.

The ADXL203CE has a ± 1.7 g range of accelerations on both axes. The pipe β angle (Fig. 6) and the accelerometers must be calibrated every time the accelerometers are removed from the pipes and re-attached. To achieve this the pipe is rolled around its longitudinal axis in a smooth and slow pure rotational motion, in this way only the gravitational acceleration is measured, and its decomposition on the accelerometer axes gives the angle to the horizontal α 1st and α 2nd (Fig. 6). From the difference of α 1st and α 2nd we get β.

In Figure 8 and Figure 9 the accelerations during calibration are shown against time and against the back-calculated angle.

Both model tanks have a frontal Perspex window from which the model can be observed during backfilling. During the tests, the flow of the soil is recorded with a high-speed camera (Phantom MIRO R310). The

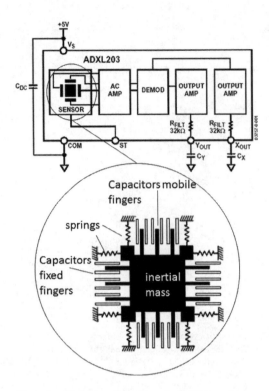

Figure 7. Internal configuration of a MEMS accelerometer ADXL203CE, electronic and mechanical configuration, from Analog Devices (2006).

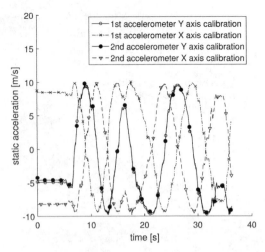

Figure 8. Accelerations measured during calibration by the front (1st) and the back (2nd) accelerometers.

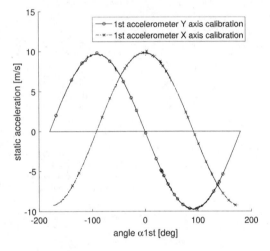

Figure 9. Accelerations from the 1st accelerometer plotted against the back-calculated angle α1st.

camera can record a video of 3200 fps (frames per seconds) at the maximum resolution of the camera, that is 1280 × 800 pixels (1 Mpx), but this limits the time duration of the video that can be saved in the volatile buffer memory of the camera. For the testing the acquisition rate was set between 350 fps and 700 fps depending on the blade velocity. Once the event to be recorded ends, the image sequence can be saved in the embedded permanent memory of the camera or through an ethernet TCP/IP connection to a personal computer provided with the Phantom dedicated software. Various file formats are available for the image recordings, here a 12-bit TIFF file that allows for a lossless image storage was adopted.

The camera trigger is provided via a limit switch, connected to a pull up resistor, that is released once the blades are set in motion, this provides a falling edge signal as a differential TTL standard to the camera. The camera trigger is recorded along with the other signals, in this way the image sequence can be synchronised with the data acquisition during the analysis of the data. Two set of lenses are used depending on the scale of the model, a Zeiss 25 mm ZF.2 DISTAGON T* for the 1:7.5 tests and a Zeiss 200mm ZF.2 MAKRO Planar for the 1:30 tests. For the large test model is only possible to record the video of only one slope of the trench (Fig.10a), for the small scale test the whole trench is recorded (Fig.10b). The data acquisition is performed with a national instrument DAQ USB 6211 with an acquisition rate of 20 kHz at 16bit.

4 DATA ANALYSIS

Both the accelerations and the video recordings need to be analysed. The accelerations require special treatment since the pipe is free to move vertically and horizontally, and it can rotate as well. If the rotation of the pipe is not considered and assuming the angle α1st remain steady the acceleration analysed may be biased, i.e. accounting for an increase in vertical acceleration when the pipe has just rotated. In this case the x axis of the 1st accelerometer (front) would start to measure a proportion of the real horizontal acceleration as well as a decreased horizontal acceleration, but this is not a linear process and it would not auto-compensate. For this

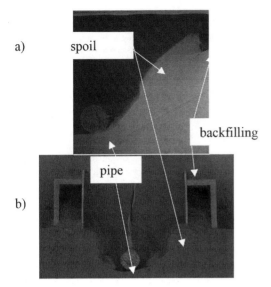

Figure 10. a) Test image from 1:7.5 model; b) Test image from 1:30 model. The soil used for the tests is a HST95 sand as in Lauder, et al. (2013).

reason, it has been chosen to have 2 accelerometers, one on each side of the pipe. Accounting for the rotation of the two reference systems of the accelerometers (Equation 1) and resolving the system allow calculation of the angle $\alpha 1st$ and $\alpha 2nd$ during the test, when the pipe may be subjected to translational and rotational rigid body motion.

$$\begin{cases} accx_{pipe} = accx_{1st} \cos \alpha_{1st} - accy_{1st} \sin \alpha_{1st} \\ accx_{pipe} = -accx_{2nd} \cos \alpha_{2nd} + accy_{2nd} \sin \alpha_{2nd} \\ \alpha_{2nd} = \alpha_{1st} + \beta \end{cases} \quad (1)$$

The images recorded during the tests must be analysed as well, this can be done with the aid of a particle image velocimetry software, the one chosen is PivLab (Thielicke & Stamhuis, 2014) provided as an add-on to MATLAB software.

The soil particles act naturally as flow tracers and the images can be analysed directly as shown in Figure 11. The results of the PIV analysis can be used to infer the velocity of the soil impacting on the model pipe, on the strain rate of the soil while flowing down the trench and on possible scaling laws.

In Figure 11a series of images with the PIV velocity vectors superimposed can be seen. The results of the tests can be compared between the two geometric scaling ratios and at the different backfilling velocities.

5 CONCLUSIONS

This article describes the background to recently developed backfilling simulation equipment at the University of Dundee. Two different scale systems

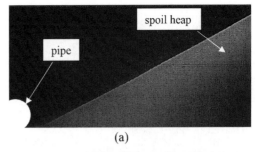

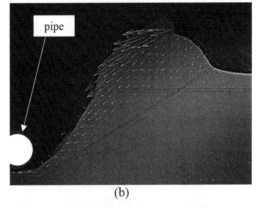

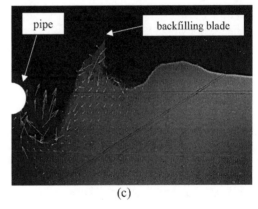

Figure 11. PIV results superimposed on the original images of a test with the large-scale model. The sequence shows the displaced pipe. a) $t = 0$ s: initial frame for reference; b) $t = 6.7$ s: the spoil heap is completely mobilised and the flow is starting to move downward; c) $t = 9.8$ s: the soil flow has impacted on the pipe and is dragging it upward. The soil used for the tests is a HST95 sand as in Lauder, et al. (2013).

have been developed at 1:7.5 and 1:30 for 2D representation of the pipeline/cable backfilling process.

The two scaled models are both provided with actuators capable of backfilling the trench at different velocities. This allows monitoring of the effect of the backfilling rate on the forces acting on the pipe. The monitoring of blade displacements in the models is carried out with LVDT and DWT.

The monitoring of the forces acting on the pipe is achieved indirectly, measuring the acceleration of the pipe with known mass. The calibration of the accelerometer on the pipe and the data analysis has been proven to be of particular importance, especially when the pipe can rotate and translate, both horizontally and vertically during the test. An additional form of data capture is provided via a high-speed camera that records the tests, the images are subsequently analysed to measure the velocity and the strain rate of the soil particles.

The maximum backfilling velocities for the small and large models are $v_{max\text{-}small} = 0.16$ m/s and $v_{max\text{-}large} = 0.095$ m/s, and the test plan include a parametric study on the velocity to understand the effect that it has on the forces exerted on the pipes and allow the consideration of scaling effects.

This research project aims to provide better insights into the industrial backfilling process, which is needed to improve the design process and optimise pipe/cable design and the speeds of installation. The design of lighter and cheaper pipelines and cables, is a key factor for faster development and cost reduction for both renewables energy and the oil & gas industries.

ACKNOWLEDGMENTS

The authors would like to acknowledge Subsea7 and the ETP Scotland (Energy Technology Partnership) for financial support and the European Regional Development Fund SMART lab at the University of Dundee for the facilities. John Anderson and Cameron Anderson that kindly provided technical knowledge and manufacturing insights.

REFERENCES

Analog Devices, 2006. ADXL103/ADXL203 Datasheet. Precision±1.7g Single/Dual Axis Accelerometer.

Bizzotto, T., Brown, M.J., Powell, T., Brennan, A.J. & Chandler, H. 2017a. Modelling of pipeline and cable flotation conditions. In Offshore Site Investigation and Geotechnics (OSIG). London, UK: Society for underwater technology, pp. 865–871.

Bizzotto, T., Brown, M.J., Powell, T., Brennan, A.J. & Chandler, H. 2017b. Toward Less Conservative Flotation Criteria for Lightweight Cables and Pipelines. In Twenty-seventh (2017) International Ocean and Polar Engineering Conference. San Francisco, CA, USA, June 25-30, 2017: International Society of Offshore and Polar Engineers (ISOPE), pp. 535–540.

Cathie, D.N., Barras, S. & Machin, J. 1998. Backfilling pipelines?: State of the Art. Offshore Pipeline Technology Conference, IBC, p. 29.

Cathie, D.N., Machin, J.B. & Overy, R.F. 1996. Engineering appraisal of pipeline floatation during backfilling. In Offshore Technology Conference. pp. 197–206.

Finch, M. & Machin, J.B. 2001. Meeting the Challenges of Deepwater Cable and Pipeline Burial. ID. OTC-13141-MS. Offshore Technology Conference.

Gonzalez-Juez, E., Meiburg, E. & Constantinescu, G. 2009. Gravity currents impinging on bottom-mounted square cylinders: Flow fields and associated forces. Journal of Fluid Mechanics, 631, pp. 65–102.

Kaku, T. 1979. A study on the resistance of snowplowing and the running stability of a snow removal truck. National Research Council (U.S.). Transportation Research Board., (185).

Lauder, K.D., Brown, M.J., Bransby, M.F. & Boyes, S. 2013. The influence of incorporating a forecutter on the performance of offshore pipeline ploughs. Applied Ocean Research, 39, pp. 121–130.

Lauder, K.L. & Brown, M.J. 2014. Scaling effects in the 1 g modelling of offshore pipeline ploughs. In 8th Int. Conf. On Physical Modelling in Geotechnics ICPMG'14. 14-17th Jan 2014. Perth, Western Australia: CRC Press, pp. 377–383.

Palmer, A.C. & King, R.A. 2008. Subsea Pipeline Engineering, PennWell Books.

Taylor, R., 2011. Backfill – Pipeline Interaction. MEng thesis. University of Dundee, UK.

Thielicke, W. & Stamhuis, E. 2014. PIVlab–towards user-friendly, affordable and accurate digital particle image velocimetry in MATLAB. Journal of Open Research Software, 2(1), pp. 1–10.

Capacity of vertical and horizontal plate anchors in sand under normal and shear loading

S.H. Chow, J. Le, M. Forsyth & C.D. O'Loughlin
Centre for Offshore Foundation Systems, University of Western Australia, Perth, Australia

ABSTRACT: Plate anchors could be a solution for mooring offshore renewable energy devices in sand. Previous work has focused on the capacity of plate anchors when loaded normal to the plate. However, for field applications the anchor is mostly likely to be installed vertically, but then rotated from a vertical to non-vertical orientation as the mooring line is tensioned. This 'keying' process will result in a progressive change in the loading angle at the anchor padeye, such that the plate is subjected to a combination of vertical, horizontal and moment loading. In order to examine the resulting change in capacity due to the changing load inclination, laboratory scale 1g plate anchor tests were conducted in dry loose silica sand to quantify the normal capacity (normal to the plate) and the shear capacity (parallel to the plate). The capacity of the pre-embedded anchor (width, $B = 20$ mm and length, $L = 40$ mm) was investigated at four embedment ratios (3, 4, 5 and 6 times the plate width), two plate orientations (horizontal and vertical) and two loading angles (parallel and normal to the plate). The data represent a first step in establishing combined loading failure envelopes for buried plate anchors in sand.

1 INTRODUCTION

Floating renewable energy devices are expected to be located in water depths less than 100 m, where the seabeds are typically coarse-grained (O'Loughlin et al. 2018). Given their high capacity to weight ratios, plate anchors are an attractive anchoring solution for offshore floating renewable energy applications (Diaz et al. 2016). Plate anchors that are considered feasible for installation in sand include fixed fluke drag embedded anchors (Neubecker & Randolph 1996), helical anchors (Byrne & Houlsby 2015) and the recently developed dynamically installed anchor described in Chow et al. (2017).

Most plate anchors involve a post-installation phase during which the plate is rotated or 'keyed' from its as-installed orientation to an 'operational' orientation. The depth of the anchor may change during this keying phase – either rising to a shallower depth or diving to a deeper depth – until an ultimate equilibrium state is reached. Hence it is crucial to be able to predict the ultimate embedment depth as this will govern the anchor's capacity. Predicting the installation process and ultimate embedment depth of plate anchors in clay is relatively well established (e.g. Tian et al. 2015) as clay is the dominant soil type in deep-water environments where plate anchors are currently used. However, there has been relatively little research on installation of plate anchors in shallow water, sandy seabeds. Adopting a similar strategy as for plate anchors in clay, the ultimate embedment depth of plate anchors in sand could be determined analytically by considering a plasticity force-resultant model that accounts for load inclination by solving the Neubecker & Randolph (1995) chain solution (e.g. Tian et al. 2015).

The plasticity force-resultant model would require a suitable yield surface and the selection of an appropriate flow rule. However, there are limited studies on establishing the yield surface for plate anchors in sand, which considers load combinations in normal, shear and rotational directions. Previous studies of plate anchor capacity in sand have mainly considered uni-directional loading, where the loading is normal to the plate (e.g. Murray & Geddes 1987, Dickin 1988, Merifield & Sloan 2006), although attempts to consider loading directions other than normal to the plate are emerging (e.g. O'Loughlin & Barron 2012, Chow et al. 2018).

As a first step in establishing the combined loading failure envelope, this study investigates the ultimate capacities of a rectangular plate anchor subjected to either pure normal loading or pure shear loading (sliding). These capacities were established experimentally through laboratory scale 1g plate anchor tests conducted in dry silica sand considering different plate orientations, loading angles and embedment ratios. The experimental results indicate that the highest uni-directional capacity is when the plate is oriented horizontally and loaded normal to the plate at the deepest embedment depth.

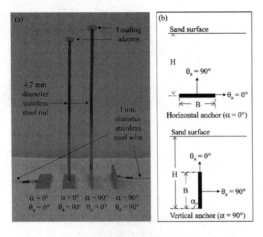

Figure 1. (a) Model plate anchors and loading configuration; (b) notation for horizontal and vertical anchors.

2 EXPERIMENTAL SETUP

2.1 Model plate anchors and mooring line

The rectangular model plate anchors were made of stainless steel and had a length, L = 40 mm, breadth, B = 20 mm and thickness, t = 4 mm. Four model anchors with different loading points were fabricated in order to investigate the effect of plate orientation, α and loading angle, θ_a (see anchor notation in Fig. 1). The load attachment points were selected such that the loading passed through the centre of mass of the anchor (see Fig. 1b) to avoid the development of a moment load that would promote anchor rotation. The surface of each plate anchor was bead-blasted to create a surface with an average normalised interface roughness, $R_n = R_{max}/d_{50} = 0.103$ (measured using a surface profilometer), indicating that the plate interface is fully rough (Garnier, 2002).

Depending on the test configuration (Fig. 1a), the anchors were loaded using either a stainless steel wire or rod. The stainless steel wire has a diameter of 1 mm and is rated to a minimum breaking load of 1.1 kN. The stainless steel rod is 4.7 mm in diameter and attaches to the plate anchors by way of a threaded connection. A loading adaptor at the top of the rod (Fig. 1a) facilitates the loading using an electrical actuator.

2.2 Soil sample properties and preparation

The soil samples were prepared using a commercially available fine sub-angular silica sand with index properties of the sand as summarised in Table 1. The sand samples were prepared in sample containers with internal dimensions of 650 × 390 × 325 mm (length × width × depth) by air pluviation to give final sample heights of between 160 and 220 mm depending on the anchor embedment depth in each sample. Global measurements of the sample mass and volume indicated an average dry density, $\rho_d = 1574$ kg/cm^3 (relative density, $I_D = 31\%$). The samples were not saturated.

Table 1. Properties of silica sand.

Specific gravity, G_s	2.67
Particle size, d_{10}, d_{50}, d_{60}	0.12, 0.18, 0.19 mm
Minimum dry density, ρ_{min}	1497 kg/m^3
Maximum dry density, ρ_{max}	1774 kg/m^3
Critical state friction angle, ϕ'_{cs}	31.6° (triaxial)
Critical state parameters, M, Γ, λ	1.27, 0.917, 0.029

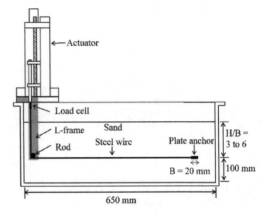

Figure 2. Test setup for horizontal anchor under shear loading.

Consideration was given to the effect of the low stress level associated with the 1g testing. The constitutive response of the sand was scaled from 1g to field conditions by adopting the equivalent state parameter approach (Altaee & Fellenius 1994). Based on the critical state parameter, $\lambda = 0.029$ (determined through triaxial tests, see Table 1), the loose samples ($I_D = 31\%$) at 1g represent the behaviour of a medium dense sand ($I_D = 76\%$) at field scale (at a mean stress level, p' of approximately 100 kPa).

2.3 Experimental setup and procedure

The test procedure involved first preparing the dry silica sand sample by air pluviation to a height of 100 mm. The four model plate anchors were then placed at the specified orientation and location with a spacing of at least 5B from rigid boundaries (sample floor and walls) and 3B between adjacent plates.

For plates that were loaded horizontally (using the steel wire), the wire was looped around a smooth horizontal rod located at the base of an inverted L-shaped frame, such that the wire was horizontal from the anchor to the rod and vertical from the rod to the surface of the sample (see Fig. 2). This allowed the loading to be applied using the vertical axis of an electrical actuator located on the surface of the sample container. For plates that were loaded vertically using the steel rod, the plates were suspended using fishing line attached to the top of the sample container to ensure that the position of the plate was maintained during the remainder of the sand pulviation.

Table 2. Test programme and summary of test results.

Sample	Test ID	Ultimate capacity, q_u (kPa)	Capacity factor, $N_\gamma = q_u/\gamma H$	Normalised displacement at ultimate capacity, δ_u/B
S3	H3N	4.0	4.4	0.06
	H3S	5.4	5.8	1.14
	V3N	17.7	19.1	1.23
	V3S	0.9	1.0	0.02
S4	H4N	6.7	5.4	0.06
	H4S	9.9	8.0	1.68
	V4N	30.1	24.4	0.90
	V4S	2.2	1.8	0.17
S5	H5N	12.5	8.1	0.10
	H5S	13.1	8.5	1.57
	V5N	42.8	27.7	1.2
	V5S	3.1	2.0	0.20
S6	H6N	19.1	10.3	0.28
	H6S	19.7	10.7	1.86
	V6N	67.6	36.5	1.49
	V6S	5.6	3.0	0.35

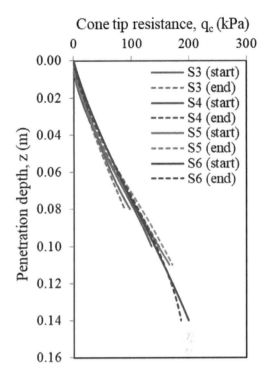

Figure 3. Cone penetrometer test results.

Each sample was characterised by cone penetrometer tests (CPTs) that were performed using a model-scale cone penetrometer penetrated at 1 mm/s. The CPTs were located outside the zone of the plate anchor tests to minimise sample disturbance. The plate anchor tests were then performed sequentially by displacing the vertical axis of the actuator at 0.5 mm/s. Following the plate anchor tests, additional CPTs were performed to ensure that the sample density has not changed significantly. The plate anchor capacity was measured using an S-shaped load cell with a capacity of 300 N and the anchor displacement was measured by the encoder on the motor of the vertical axis of the actuator.

2.4 Test programme

The anchor test programme is summarised in Table 2. The samples are identified based on the anchor embedment ratio, H/B. For instance, sample S3 refers to the sample involving all anchor tests at H/B = 3. The anchor tests are identified as PnL, where 'P' denotes the plate orientation, that is either 'H' for horizontal ($\alpha = 0°$) or 'V' for vertical ($\alpha = 90°$); 'n' denotes the anchor embedment ratio, H/B that is either 3, 4, 5 or 6; and 'L' denotes the loading angle that is either 'N' for normal ($\theta_a = 90°$) or 'S' for shear ($\theta_a = 0°$). For instance, V3S refers to the anchor test involving $\alpha = 90°$, H/B = 3 and $\theta_a = 0°$.

3 TEST RESULTS

The plate anchor test results are summarised in Table 2. Although not included in Table 2, the tests in samples S3 and S6 were repeated and good repeatability in the test results was obtained, with less than 10% difference in the measured anchor capacities.

The CPT results are provided in Figure 3, which shows that profiles of the cone tip resistance, q_c are consistent between the start and end of testing, and across samples, enabling direct comparison of test results at different embedment ratios.

The anchor test results summarised in Table 2 include the ultimate capacity, q_u, and the corresponding anchor displacement normalised by the anchor breadth, δ_u/B, and the dimensionless anchor capacity factor, N_γ, determined as

$$N_\gamma = \frac{q_u}{\gamma H} = \frac{F_u - F_f}{A\gamma H} \quad (1)$$

where γ = sand unit weight, F_u = measured ultimate load, F_f = measured wire or rod friction, A = plate loading area (BL for normal loading and 2BL for shear loading) and H = initial depth to the centroid or tip of the anchor (see definition of H in Fig. 1b). The wire or rod friction, F_f is determined experimentally by pulling the rod or wire without the plate in the sample.

3.1 Load-displacement behaviour

Figure 4 shows the dimensionless load-displacement curves of all horizontal and vertical anchor tests. The anchor response is clearly different between the horizontal and vertical anchor, and also between normal and shear loading. For the horizontal anchor under normal loading ($\alpha = 0°$, $\theta_a = 90°$) in Figure 4a, the initial load response is very stiff with the ultimate load

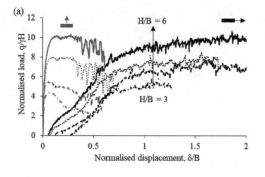

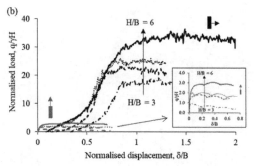

Figure 4. Load-displacement curves of (a) horizontal anchors ($\alpha = 0°$); (b) vertical anchors ($\alpha = 90°$).

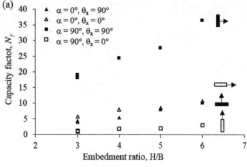

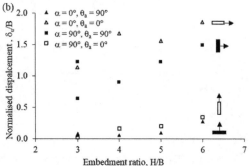

Figure 5. (a) Capacity factors; and (b) normalised displacement at ultimate capacity for anchors at different embedment ratios.

achieved at very low displacement ($\delta_u/B < 0.3$). Upon reaching the ultimate load, the load softens followed by load oscillations, which become more pronounced at higher H/B. These oscillations are attributed to the progressive infilling of the void that forms behind the plate as the anchor is lifted, as previously observed in other studies involving vertical loading of horizontal plate anchors and pipelines embedded in sand (e.g. Dickin 1994, White et al. 2008). In contrast the load response for the horizontal anchors under shear loading is much softer, with a gradual climb to a capacity that approximately plateaus, with ultimate capacity occurring at relatively high anchor displacement. The displacement at the ultimate load is in the range $\delta_u/B = 1.14$ to 1.86, approximately 6.7 to 30.5 times higher than for the horizontal anchors under normal loading.

For the vertical anchor under normal loading ($\alpha = 90°$, $\theta_a = 90°$) in Figure 4b, the load response is initially soft before stiffening to reach the ultimate capacity. In contrast, the load response for the vertical anchor under shear loading is stiff (see inset in Fig. 4b), although the ultimate capacities are significantly lower than for normal loading, and occur at very low displacements; $\delta_u/B = 0.02$ to 0.35 for shear loading compared with $\delta_u/B = 0.64$ to 1.49 for normal loading.

The experimental anchor capacity factors, N_γ and the normalised displacements, δ_u/B are summarised for all anchors at various embedment ratios in Figure 5. Further examination of the influence of embedment ratio, plate orientation and loading angle on the anchor capacity is discussed in subsequent sections.

3.2 Anchor capacity factors

As shown in Table 2 and Figure 5, the anchor capacity factors, N_γ range between 1 and 36.5. The vertical anchors under normal loading ($\alpha = 90°$, $\theta_a = 90°$) provide the highest N_γ, followed by horizontal anchors under shear ($\alpha = 0°$, $\theta_a = 0°$), horizontal anchors under normal loading ($\alpha = 0°$, $\theta_a = 90°$), and lastly vertical anchors under shear ($\alpha = 90°$, $\theta_a = 0°$). The measured N_γ values are compared with exiting theoretical solutions in Figure 6. Comparisons are provided for the normally loaded anchors only as there are no existing solutions available for anchors under shear loading. Potential failure mechanisms for all test configurations are illustrated in Figure 7 and are referred to in the following discussion.

For the normally loaded horizontal anchors ($\alpha = 0°$, $\theta_a = 90°$), the measured N_γ values are on average 56% lower than the upper bound solution proposed by Murray & Geddes (1987) for a rectangular plate anchor. The Murray & Geddes (1987) solution assumes the failure surface extends as straight lines from the edge of the plate inclined at an angle, θ, to the vertical (Fig. 7a):

$$N_\gamma = 1 + \frac{H}{B}\tan\phi\left(1 + \frac{B}{L} + \frac{\pi}{3}\frac{H}{L}\tan\phi\right) \quad (2)$$

where ϕ = peak friction angle estimated from Bolton (1986)'s correlations. The over-prediction of N_γ using the limit analysis is not surprising given the associative flow rule assumption, as duly noted by

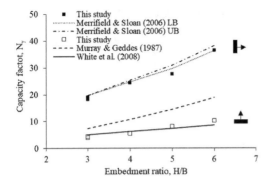

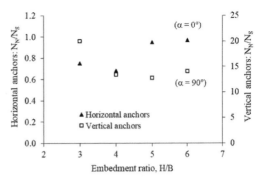

Figure 6. Comparison of experimental and theoretical anchor capacity factors.

Figure 8. Ratio of normal to shear anchor capacity, N_N/N_S.

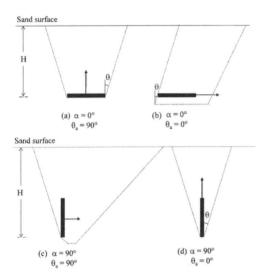

Figure 7. Possible failure mechanism for vertical and horizontal anchors undergoing normal and shear loading.

Murray & Geddes (1987). As shown in Figure 6, better agreement can be achieved using the limit equilibrium solution (White et al. 2008) developed for strip anchor, but applying a shape factor, S_f to account for the rectangular shape of the horizontal anchors (Dickin 1988):

$$N_\gamma = \left(1 + F_{up}\frac{H}{B}\right)S_f \quad (3)$$

$$F_{up} = \tan\psi + (\tan\phi - \tan\psi)\left[\frac{1+K_o}{2} - \frac{(1-K_o)\cos 2\psi}{2}\right] \quad (4)$$

$$S_f = 1 - \frac{aB}{3LH}(6B - 7H) \quad (5)$$

where ψ = dilation angle, K_o = at-rest earth pressure coefficient ($K_o = 1 - \sin\phi'_{cs}$) and the constant, a accounts for sand density, and is taken here as $a = 1$ to reflect dense sand at prototype scale.

For the normally loaded vertical anchors ($\alpha = 90°$, $\theta_a = 90°$), the measured N_γ values are in good agreement with the numerical solutions reported by Merifield & Sloan (2006) for a strip anchor when adjusted by a shape factor, S_f, to account for the rectangular shape of the vertical anchors (Ovesen & Stonmann 1972):

$$S_f = 0.42\frac{\left(\frac{H}{B}+1\right)}{L/B} + 1 \quad (6)$$

The reasonably good agreement suggests that the failure mechanism observed in the Merifield & Sloan (2006) finite element limit analyses for a normally loaded vertical anchor is broadly appropriate (see Fig 7c), despite the assumption of associated flow in their analyses assumption.

No existing comparison could be made for the horizontal and vertical anchors under shear loading. However, for the horizontal anchors under shear loading ($\alpha = 0°$, $\theta_a = 0°$), an initial estimation of the 'sliding' capacity can be made as $N_\gamma = 2\tan\delta$ by considering the friction caused by the self-weight of the sand block above the plate. However, the estimated shear capacity factor, $N_\gamma = 1.31$ is at most 23% of the measured $N_\gamma = 5.8$ to 10.7. In addition, the measured N_γ is 1 to 1.5 times higher than that for the horizontal anchors under normal loading (Fig. 8). This indicates that the zone of soil mobilised under shear loading of horizontal anchors is at least greater than that for the normally loaded horizontal anchor (Fig. 7b).

For the vertical anchor under shear loading ($\alpha = 90°$, $\theta_a = 0°$), trial finite element analyses (Roy pers. comm.) indicate that the failure mechanism is expected to be similar to that for a normally loaded horizontal plate, but with failure planes that extend from the lower most edges of the plate as shown in Figure 7d. Hence existing solutions developed for the normally loaded horizontal anchor could be adopted by simply removing the soil block above the horizontal plate. This allows the White et al. (2008) limit equilibrium solution (Equation (3)) to be simplified to $N_\gamma = F_{up}H/B$ and applied without adjustment by a shape factor as the aspect ratio becomes $L/t = 10$ rather than $L/B = 2$. The computed capacity factors are in the range $N_\gamma = 1.3$ to 2.5, which are in good agreement with the measured values, $N_\gamma = 1$ to 3 (Table 2).

3.2.1 *Effect of embedment ratio*

Consistent with the published literature, N_γ and δ_u/B increase with embedment ratio (Fig. 5a). The increase in N_γ with depth is more significant for vertical anchors under shear loading ($\alpha = 90°$, $\theta a = 0°$) with increases in N_γ of up to 237% as H/B increases from 3 to 6. Figure 5a does not indicate any clear demarcation in the rate of increase, suggesting that the failure mechanism does not change significantly as the embedment depth increases to H/B = 6.

3.2.2 *Effect of loading angle*

The effect of loading angle can be examined by considering the ratio of normal capacity to shear capacity, N_N/N_S. This is shown for all embedment ratios in Figure 8. For horizontal anchors, N_N/N_S ranges between 0.7 and 1 indicating a higher shear capacity than normal capacity, and appears to increase with increasing H/B. In contrast, for vertical anchors N_N/N_S ranges between 13 and 20, which reflects the much lower zone of soil involved in the failure mechanism for vertical anchors under shear loading (see Fig. 7d). N_N/N_S for vertical anchors also appears to be relatively independent on H/B, particularly when H/B > 3.

4 CONCLUSIONS

This experimental study has shown that the highest capacity can be achieved when the anchor is in vertical orientation, loaded normal to the plate and at its deepest depth. The test data from this study represent a first step in establishing combined loading failure envelopes for buried plate anchors in sand. The measured normal and sliding capacities, N_N and N_S represent the uniaxial bearing capacity factors which form the basis of establishing the failure envelope shape and size. In contrast to plate anchors in clay, anchor capacity in sand is more affected by plate orientation, loading angle and stress level (embedment ratio). There is a clear need to establish a comprehensive database either experimentally or numerically in order to establish the complete failure envelope. This would include investigating the anchor capacity under inclined plate orientations, loading angles, deeper embedment ratios, different sand densities, in addition to moment loading. Future experimental work should consider centrifuge tests to validate the stress scaling in these 1g tests, and also to allow for investigation of loose sand at prototype conditions which is difficult to model at 1g. In addition, future experimental studies with observation of the anchor failure mechanism using the particle image velocimetry (PIV) technique would aid the development of appropriate theoretical solutions.

ACKNOWLEDGEMENTS

This work forms part of the activities of the Centre for Offshore Foundation Systems (COFS), currently supported as a node of the Australian Research Council Centre of Excellence for Geotechnical Science and Engineering and as a Centre of Excellence by the Lloyd's Register Foundation.

REFERENCES

Altaee, A. & Fellenius, B.H. 1994. Physical modeling in sand. *Canadian Geotechnical Journal*, 31: 420–431.

Bolton, M.D. 1986. The strength and dilatancy of sands. *Géotechnique*, 36(1): 65–78.

Byrne, B.W. & Houlsby, G.T. 2015. Helical piles: an innovative foundation design option for offshore wind turbines. *Phil. Trans. R. Soc. A* 373: 20140081.

Chow, S.H., O'Loughlin, C.D., Gaudin, C., Knappett, J.A., Brown, M.J. & Lieng, J. T. 2017. An experimental study of the embedment of a dynamically installed anchor in sand. *Proc. 8th Intern. Conf. Offshore Site Investigation & Geotechnics*, London, UK, pp 1019–1025.

Chow, S. H., O'Loughlin, C. D., Gaudin, C. & Lieng, J.T. 2018 Monotonic and cyclic capacity of a dynamically installed anchor in sand. *Ocean Engineering*, 148: 588–601.

Diaz, B. D., Rasulo, M., Aubeny, C. P., Fontana, C. M., Arwade, S. R., Degroot, D. J., & Landon, M. 2016. Multiline anchors for floating offshore wind towers. *Proc. OCEANS 2016 MTS/IEEE Monterey*.

Dickin, E.A. 1988. Uplift Behavior of Horizontal Anchor Plates in Sand. *J. of Geotech. Eng.*, 114(11): 1300–1317.

Dickin, E.A. 1994. Uplift resistance of buried pipelines in sand. *Soils and Foundations*, 34(2): 41–48.

Garnier, J. 2002. Size effects in shear interfaces. In *Constitutive and Centrifuge Modelling: Two Extremes*. 335–345.

Merifield, R. & Sloan, S. 2006. The ultimate pullout capacity of anchors in frictional soils. *Canadian Geotechnical Journal*, 43: 852–868.

Murray, E.F. & Geddes, J.D. 1987. Uplift of anchor plates in sand. *ASCE Journal of Geotech. Eng.*, 113(3): 202–215.

Neubecker, S. & Randolph, M. 1995. Profile and frictional capacity of embedded anchor chains. *Journal of Geotechnical Engineering*, 121(11): 797–803.

Neubecker, S. & Randolph, M. 1996. The kinematic behaviour of drag anchors in sand. *Canadian Geotechnical Journal*, 33(4): 584–594.

O'Loughlin, C.D. & Barron, B. 2012. Capacity and keying response of plate anchors in sand. *Proc. 7th Intern. Conf. Offshore Site Investigation & Geotechnics*. pp. 649–655.

O'Loughlin, C.D., Neubecker, S.R. & Gaudin, C. 2018. Anchoring systems: anchor types, installation and design. *Encyclopaedia of Marine and Offshore Engineering*, Wiley, doi: 10.1002/9781118476406.emoe534.

Ovesen, N.K. & Stromann, H. 1972. Design Method for Vertical Anchor Slabs in Sand. *Proc. of the Speciality Conf. on Performance of Earth and Earth-Supported Structures*, V1-2, pp. 1481-1500.

Tian, Y.A., Randolph, M.F. & Cassidy, M.J. 2015. Analytical solution for ultimate embedment depth and potential holding capacity of plate anchors. *Geotechnique*, 65(6): 517–530.

White, D.J., Cheuk, C.Y. & Bolton, M.D. 2008. The uplift resistance of pipes and plate anchors buried in sand. *Géotechnique*, 58(10): 771–779.

A novel experimental-numerical approach to model buried pipes subjected to reverse faulting

R.Y. Khaksar
Faculty of Civil Engineering, Sadjad University of Technology, Mashhad, Iran

M. Moradi & A. Ghalandarzadeh
School of Civil Engineering, University College of Engineering, University of Tehran, Tehran, Iran

ABSTRACT: In this study, a combined approach of experimental-numerical modelling has been proposed to overcome the buried pipelines physical modelling dimensional limitation, subjected to reverse faulting. Initially, two reverse faulting centrifuge tests including fixed end pipelines have been conducted and the results have been employed for the calibration of a developed numerical model. Then, the calibrated numerical model has been extended longitudinally and the results have been employed to develop a novel "spring-like" end connection (SLEC) system. The developed end connection system was supposed to represent the behaviour of the omitted fault-affected length of pipeline in centrifuge tests. Moreover, a verifying centrifuge test has been conducted, employing the newly developed end connection system. For all the physical tests, the model results have also been presented.

1 INTRODUCTION

Pipeline damages have been reported all around the world due to permanent ground deformations (PGD) and especially the faulting ruptures. Among numerous researches conducted on the pipeline behaviour due to PGDs, Vazouras et al. (2010, 2012) and Xie et al. (2011, 2013) have done a series of comprehensive finite element analysis on the response of HDPE and steel pipelines subjected to strike-slip faulting. Also, the large and small scale modeling of pipelines subjected to faulting were carried out by O'Rourke et al. (2003, 2005), Choo et al. (2007), Ha et al. (2008a, 2008b, 2010). Moreover, Rojhani et al. (2012a, 2012b) and Moradi et al. (2013) conducted a series of centrifuge modelings on pipelines subjected to normal and reverse faulting. More recently, Oliveira et al. (2016) investigated the response of pipelines due to soil mass movements and Saiyar et al. (2016) studied the flexural response of pipelines to normal faulting.

Most of the mentioned experimental researches would have faced modeling geometrical limitations, resulting in the interfering effect of pipe end connections on the response of model pipe. Therefore, a combined approach of physical and numerical modeling was developed to overcome such limitation in the current study. In the proposed approach, two centrifuge tests were conducted and the results were employed to develop a calibrated finite element simulation of the prototype model. Then, a pipeline end connection system has been developed for reverse faulting mechanism, based on the calibrated numerical model results. The newly developed spring-like end connection (SLEC) system is planned to be employed in forthcoming centrifuge tests to study the effects of other factors while the model length limitation has been compensated. Moreover, the performance of this SLEC system has been compared with a centrifuge test with matching results..

2 MODELING PROCEDURE

2.1 *Step 1: Physical modeling*

In this study, the fault simulator, introduced by Rojhani et al. (2012b), has been employed to conduct the 40 g centrifuge tests for reverse faulting in the physical modeling laboratory of University of Tehran (Table 1).

Here, the available 304 grade stainless steel fine tubes were adopted as small scale model pipelines and covered by the polymer tapes to simulate the real conditions in industry. In order to get pipeline response data, the pipeline instrumentation system was planned and implemented. The soil was cut to the depth providing the designed burial depth and the pipeline was placed down into the trench. Then, the soil was returned into the trench. The Standard Firoozkouh 191 sand was adopted as the surrounding soil, complying with the size criterion proposed by Ovesen (1981) and Dickin and Leuoy (1983) for centrifuge modeling. Subsequently, a set of laboratory direct shear tests were conducted to determine the soil properties (Table 2). The soil was compacted in 4 cm-thick layers

Table 1. Summary of fixed end centrifuge models.

Test No.	Fault angle, (°)	Peak Offset, U (m)	Pipe diameter, D (m)	Pipe wall thickness, t (mm)	Buried depth, H (m)
1	60°	1.15	1.00	20.0	1.60
2	60°	1.18	0.64	16.0	1.60

* All dimensions are in prototype scale.
** The buried depth is measured to the top fiber of the pipeline.

Table 2. Standard Firoozkouh-191 sand mechanical properties.

Sand Type	Density, γ (kg/m³)	Friction Angle, (φ)	Cohesion, C (KPa)
Firoozkouh 191	1550	35°	10

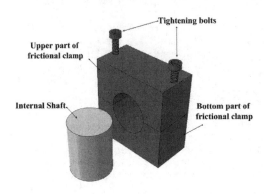

Figure 1. The frictional clamp: Schematic picture of the clamp.

with the relative density of 85% and the water content of 4.5% ~ 5.5%.

2.1.1 End connections

Since the affected length of pipe due to the faulting is generally more than the length of model pipe in the simulator box, fault affected pipelines in physical modeling inevitably face the model geometrical limitation. In the tests reported by Rojhani et al. (2012a) and Moradi et al. (2013), the pipe ends were considered fixed while they showed a semi-fixed behaviour due to the end connection welded zones. Therefore, a frictional clamp was designed in this study to eliminate the interfering effect of welded zone (Fig. 1). Such end connection was employed in the tests of Table 1 (Fig. 2).

2.1.2 Data acquisition

In the conducted tests, data acquisition of model response has been done through three methods: (1) strain gauges, installed on top and bottom fibers of the pipes; (2) displacement transducers (LVDT), monitoring the pipe and soil surface displacements and

(a)

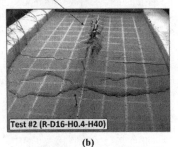

(b)

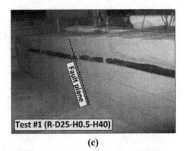

(c)

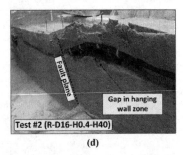

(d)

Figure 2. Soil disturbance in surface (a, b) and section view (c, d) due to reverse faulting.

(3) image processing technique (used solely in test of Table 3), employing two cameras to record the displacements of the SLEC systems during the faulting tests.

Table 3. The SLEC system test specification.

Test No.	Fault angle, (°)	Peak Offset, U (m)	Pipe diameter, D (m)	Pipe wall thickness, t (mm)	Buried depth, H (m)
3	60°	1.52	1.00	20.0	1.60

* All dimensions are in prototype scale.

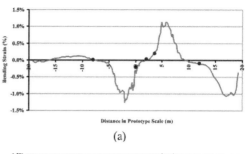

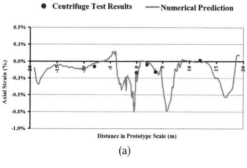

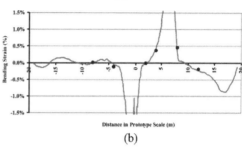

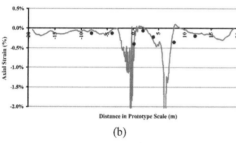

Figure 3. Axial strain comparison of centrifuge and numerical models: (a) test 1; (b) test 2.

Figure 4. Bending strain comparison of centrifuge and numerical models: (a) test 1; (b) test 2.

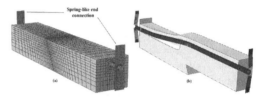

Figure 5. (a) The numerical model including SLEC system, before (a) and after (b) being subjected to the reverse faulting.

2.2 Step 2: Numerical modeling

In numerical modeling process, initially, a finite element model was developed using the finite element program, ABAQUS (2011) and calibrated then, based on the specifications and the results of centrifuge tests of Table 1 and according to the details reported by Khaksar et al. (2017). In the numerical model, the prismatic domain of soil, corresponding to the reference prototype model of centrifuge tests, was considered. Also, the inelastic behaviour of the materials and large deformation effects of faulting were accounted for. The element types used for soil continuum and pipeline were eight-node reduced integration solid (C3D8R) and four-node reduced integration shell (S4R), respectively and the soil-pipeline interaction was introduced by a contact model. Moreover, the numerical analysis was performed in two steps: (1) geostatic step in which the gravity loading was applied and (2) faulting step that the fault rupture was imposed in a displacement-controlled approach.

The diagrams depicted in Figs. 3 and 4 demonstrate the comparison of axial and bending strains of numerical model with the results of the reference centrifuge tests (Table 1). As shown, the axial and bending strain diagrams demonstrate good agreement (less than 5% diversion) with those of centrifuge tests.

Considering the good agreement between experimental and numerical results, the calibration process would be confirmed. Technically, no strain gauge can be stuck on the pipe in the critical strain location due to pipeline probable wrinkling, so no data for such points could be obtained. Therefore, such points are excluded from the calibration process.

Then, the calibrated numerical model was longitudinally extended in order to cover the whole fault-affected length of pipe. It was extended from the faulting plane to the length that stresses and strains on the pipe due to faulting would dissipate (approximately 400D each side of the faulting plane). In such a case, the numerical model would represent almost the real behavior of the pipeline subjected to the faulting. And, the displacements of the pipeline in the sections corresponding to the centrifuge model pipe end connections could be determined. In the final step of the numerical modeling, a model including the SLEC system was developed (Fig. 5). The designed SLEC was supposed to work as a semi-fixed end connection system with

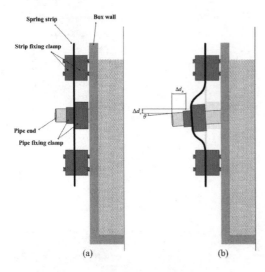

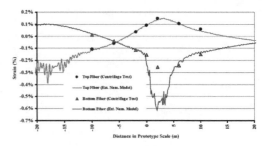

Figure 7. Strain distribution comparison for experimental and numerical reverse faulting test.

Table 4. Numerical-experimental displacement* comparison in pipe sections.

Section location on pipe	Exp. (m)	Num. (m)	Deviation** (%)
Mid Foot Wall Zone	0.307	0.333	**8.5%**
Fault Plane Intersection	0.436	0.451	**3.4%**
Mid Hanging Wall Zone	0.224	0.243	**8.5%**

* All displacement magnitudes are offered in prototype scale.
** Deviation is indicating the percentage ratio of the numerical result to the centrifuge test deviation.

Figure 6. Schematic pattern for SLEC systems: before (a) and after (b) reverse faulting, outside the soil box.

variable longitudinal, transverse and rotational stiffnesses, representing the real behaviour of the missing pipe length (Fig. 6). The data of pipe end connection displacements were employed to design the proper configuration of the connection systems by varying the free span and spring strip thickness through a numerical parametric study. Assuming the elastic behavior of the spring, the parameters were also evaluated by the analytical approach.

2.3 Step 3: Physical modeling

According to the numerical model results, the semi-fixed end connection (SLEC) system with variable stiffness was physically developed is installed outside the soil box in order to avoid the soil pressure interference (Fig. 6). Such system was expected to represent the behaviour of the fault-affected length of pipe which was inevitably omitted from the physical model due to the modeling length limitation. Although simply assembled, the system gave the chance to obtain desired stiffness by varying the free span, thickness and material (CK75 steel strips were employed here) of the spring strips and also to use it both for reverse and normal faulting mechanisms by a slight change.

One individual test (Table 3), employing the SLEC system, has been conducted on reverse faulting mechanisms in order to verify the designed SLEC systems.

The centrifuge results were compared with those of extended numerical models. The strain gauges, installed on top and bottom fibers of pipes, logged magnitudes complying with numerical model results (Fig. 7) except the peak compressive node on bottom fiber which malfunctioned due to the adhesion defect caused by wrinkling initiation.

Moreover, the pipeline displacements have been logged in three sections. The numerical model results had less than 9% diversion from the centrifuge model

Figure 8. The end connections, after reverse faulting.

results (Table 4), which could be a good approximation for the approach.

Also, the pipe displacements in the SLEC system has been captured by the aid of a simple image processing technique at 40g during the experiment through the before-after pictures taken form the pipeline end connection status due to faulting movements (Fig. 8). According to the results comparison (Table 5), there was a good agreement (around 10% diversion) between the numerical results with the physical end connection movements.

All the verification results certify that the method proposed in this study could be used to get a full scale-like response of pipelines with the centrifuge small scale modeling approach. In other words, the centrifuge modeling dimension limitation has been compensated in current study with the aid of developed SLEC system.

Table 5. Numerical-experimental displacement* comparison in end connection systems.

End connection		Foot wall	Hanging wall
Longitudinal Displacement (Δd_h)	Exp. (m)	0.423	0.264
	Num. (m)	0.469	0.238
	Deviation** (%)	**10.8%**	**9.9%**
Vertical Transverse Displacement (Δd_v)	Exp. (m)	0.025	0.066
	Num. (m)	0.026	0.072
	Deviation (%)	**4.3%**	**9.4%**
Rotation Angle, (θ)	Exp. (°)	1.026	~ 0.0
	Num. (°)	0.935	0.002
	Deviation (%)	**8.9%**	–

* All displacement magnitudes are offered in prototype scale.
** Deviation is indicating the percentage ratio of the numerical result to the centrifuge test deviation.

3 CONCLUSION

The novel physical-numerical approach has been proposed to model pipelines subjected to reverse faulting. This study focused on omitting the interfering effect of the pipe end connections on the model response which would have inevitably existed in most of physical experiments. Initially two centrifuge tests were performed with fixed end connections. Then, a calibrated numerical model was developed based on the conducted centrifuge test results followed by the numerical simulation of the novel "spring-like" end connections (SLEC). Determining the SLEC specification through the numerical parametric studies, it was made in small scale and employed in another centrifuge test to verify the performance of the developed system. The comparison between the physical and numerical results demonstrated good agreement. Overcoming the length limitation of physical modeling, the presented approach was employed in the following novel centrifuge investigations on new innovative mitigation methods (not presented here). Applying such approach may also give the chance for further investigations on pipelines subjected to permanent ground deformations.

REFERENCES

ABAQUS. 2011. Users' Manual. Dassault Systemes Simulia, Providence, RI, USA.

Choo, Y.W., Abdoun, T.H., O'Rourke, M.J. & Ha, D. 2007. Remediation for buried pipeline systems under permanent ground deformation. *Soil Dynamics and Earthquake Engineering* 27(12): 1043–1055.

Dickin, E.A. & Leuoy, C.F. 1983. Centrifuge Model Tests on Vertical Anchor Plates. *ASCE Journal of Geotechnical Engineering* 109(12): 1503–1525.

Ha, D., Abdoun, T., O'Rourke, M., Symans, M., O'Rourke, T., Palmer, M. & Stewart, H. 2008a. Centrifuge modeling of permanent ground deformation effect on buried HDPE pipelines. *Journal of Geotechnical and Geoenvironmental Engineering* 134(10): 1501–1515.

Ha, D., Abdoun, T.H., O'Rourke, M.J., Symans, M.D., O'Rourke, T.D., Palmer, M.C. & Stewart, H.E. 2008b. Buried high-density polyethylene pipelines subjected to normal and strike-slip faulting-a centrifuge investigation. *Canadian Geotechnical Journal* 45(12): 1733–1742.

Ha, D., Abdoun, T.H., O'Rourke, M.J., Symans, M.D., O'Rourke, T.D., Palmer, M.C. & Stewart, H.E. 2010. Earthquake Faulting Effects on Buried Pipelines–Case History and Centrifuge Study. *Journal of earthquake engineering* 14(5): 646–669.

Khaksar, R.Y., Moradi, M. & Ghalandarzadeh, A. 2017. Response of buried oil and gas pipelines subjected to reverse faulting: A novel centrifuge-finite element approach. *Scientia Iranica*. Article in press.

Moradi, M., Rojhani, M., Galandarzadeh, A. & Takada, S. 2013. Centrifuge modeling of buried continuous pipelines subjected to normal faulting. *Earthquake Engineering and Engineering Vibration.* 12(1): 155–164.

O'Rourke, M., Gadicherla, V. & Abdoun, T. 2003. Centrifuge modeling of buried pipelines. *Advancing Mitigation Technologies and Disaster Response for Lifeline Systems (*ASCE), pp. 757–768.

O'Rourke, M. & Gadicherla, V. & Abdoun, T. 2005. Centrifuge modeling of PGD response of buried pipe. *Earthquake Engineering and Engineering Vibration.* 4(1): 69–73.

Oliveira, J.R.M., Rammah, K.I., Trejo, P.C., Almeida, M.S. & Almeida, M.C. 2016. Modelling of a pipeline subjected to soil mass movements. *International Journal of Physical Modelling in Geotechnics*, 1–11 (Ahead of Print).

Rojhani, M., Moradi, M., Galandarzadeh, A. & Takada, S. 2012a. Centrifuge modeling of buried continuous pipelines subjected to reverse faulting. *Canadian Geotechnical Journal.* 49(6): 659–670.

Rojhani, M., Moradi, M., Ebrahimi, M.H., Galandarzadeh, A. & Takada, S. 2012b. Recent Developments in Faulting Simulators for Geotechnical Centrifuges. *ASTM geotechnical testing journal* 35(6): 924–934.

Saiyar, M., Ni, P., Take, W.A. & Moore, I.D. 2015. Response of pipelines of differing flexural stiffness to normal faulting. *Géotechnique* 66(4): 275–286.

Vazouras, P., Karamanos, S.A. & Dakoulas, P. 2010. Finite element analysis of buried steel pipelines under strike-slip fault displacements. *Soil Dynamics and Earthquake Engineering* 30(11): 1361–1376.

Vazouras, P., Karamanos, S.A. & Dakoulas, P. 2012. Mechanical behaviour of buried steel pipes crossing active strike-slip faults. *Soil Dynamics and Earthquake Engineering* 41: 164–180.

Xie, X., Symans, M.D., O'Rourke, M.J., Abdoun, T.H., O'Rourke, T.D., Palmer, M. C. & Stewart, H. E. 2011. Numerical modeling of buried HDPE Pipelines subjected to strike-slip faulting. *Journal of Earthquake Engineering* 15(8), 1273–1296.

Xie, X., Symans, M. D., O'Rourke, M. J., Abdoun, T. H., O'Rourke, T. D., Palmer, M. C. & Stewart, H. E. 2013. Numerical modeling of buried HDPE pipelines subjected to normal faulting: a case study. *Earthquake Spectra*, 29(2), 609–632.

Wave-induced liquefaction and floatation of pipeline buried in sand beds

J. Miyamoto & K. Tsurugasaki
Technical Research Institute, Naruo, Toyo Construction, Nishinomiya, Japan

S. Sassa
Port and Airport Research Institute, National Institute of Maritime, Port and Aviation Technology, Yokosuka, Japan

ABSTRACT: This paper discusses the experimental findings pertaining to the flotation of buried pipeline in association with wave-induced liquefaction in sand beds. Centrifuge wave testing with viscous scaling was applied to reproducing the build-up of residual pore pressures in sand beds leading up to liquefaction. The emphasis of the experiments was placed on investigating the relation between the occurrence/propagation of liquefaction and the start of floatation of pipeline to the soil surface. A series of wave tests using a drum centrifuge was performed under a centrifugal acceleration of 70 gravities. The characteristics of liquefaction under irregular waves were investigated as well as under regular waves. It was found that the buried pipe started moving upward upon the onset of liquefaction and reached to the soil surface in association with the significant liquefied soil motion under progressive liquefaction. Under an irregular-waves condition, liquefaction occurred and propagated in the sand bed similar to regular wave tests, and pipeline floated to the bed surface.

1 INTRODUCTION

Wave-induced liquefaction in seabed deposits or loosely packed backfills in coastal areas has received considerable attention in relation to the stability of submarine pipelines, cables or other such facilities. Some flume tests were performed under 1 g conditions and discussed the relation between the soil liquefaction and floatation/sinking of pipeline (Sumer et al. 1999, Teh et al. 2003, Sumer et al. 2006). However, scaling laws on wave-soil interaction such as liquefaction have not yet been established for 1 g wave tests. As a result, the soil-pipeline responses observed in these tests cannot be properly extrapolated to field conditions.

Characteristics of wave-induced liquefaction of sands were discussed using centrifuge wave testing with viscous scaling (Sassa & Sekiguchi 1999), and it was found that the loose sand bed subjected to severe wave conditions underwent liquefaction and the liquefaction zone propagated in the course of wave loading. It would be necessary to understand in detail the relation between the progressive nature of liquefaction and the stability of submarine structures. Recently, applications of centrifuge testing to the problems of fluid-soil-structure interactions as well as wave-induced liquefaction have been increasingly important in coastal engineering as well as in geotechnical engineering (Sumer 2014, Sassa et al. 2016).

This paper examines the flotation of buried pipeline in association with wave-induced liquefaction in sand beds in a set of centrifuge wave tests. Centrifuge wave testing on a drum channel and the experimental conditions will first be described. Then, the soil liquefaction and movement of pipe under regular waves will be presented. This will be followed by a discussion of the pipe behaviour in sand bed that was subjected to irregular wave loading.

2 EXPERIMENTATION

2.1 Wave tests in a drum centrifuge

A water channel installed in a drum centrifuge (Toyo Construction) having an effective diameter 2.2 m is shown in Figure 1. Waves were generated by using a piston type wave maker that was driven by an AC servo-motor. The wave-generating system firstly equipped in the drum channel was plunger type permitting a soil bed to be subjected to regular waves of constant and moderate amplitude (Baba et al. 2002). However, waves of varying amplitude or irregular waves were unable to be generated and severity of wave generated was below a critical value above which liquefaction occurs. In this study, these limitations have been overcome by introducing a piston type wave maker and a feedback system where the date from a laser-type displacement transducer was utilised to continuously vary the stroke of the wave paddle (Figs 1a,b). This allowed the reproduction of a sequence of storm waves staring from relatively small

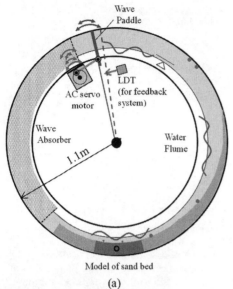

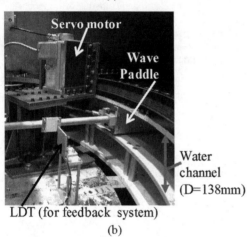

Figure 1. (a) Cross section of a water channel installed in the drum centrifuge and (b) wave-generating system developed.

Table 1. Test cases at 70 g.

	Model	Waves
Case 1	Sand bed without pipe	Regular
Case 2	Sand bed without pipe	Irregular
Case 3	Sand bed with pipe	Regular
Case 4	Sand bed with pipe	irregular

waves and increasing to very intense ones in which liquefaction occurs as well as of irregular waves supposed to be real storm wave.

Four centrifuge wave tests were performed on loosely packed fresh deposits of fine sand (Table 1). All tests were carried out under a centrifuge acceleration of 70 g. In the experiment of regular waves, a sinusoidal wave whose amplitude gradually increased

Table 2. Wave conditions (proto type scale).

Wave type	
Regular wave	Period T: 0.1 s (7 s)
	Wave height H: increasing, max 59 mm (3.4 m)
Irregular wave	Period of significant wave: T(1/3): 0.1 s (7 s)
	Significant wave height H(1/3): 30 mm (2 m)

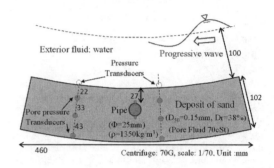

Figure 2. Cross section of sand deposit model with a buried pipe.

was generated at frequency of 10 Hz. In the experiment of irregular waves where realistic storm waves were assumed, irregular wave forms were generated based on the standard frequency spectra (modified Bretschneider-Mitsuyasu equation). These wave conditions are shown in Table 2. In this paper, experimental results from Case 3, 4 are discussed.

2.2 Experimental model

The sand deposit model with a buried pipe is shown in Figure 2. Water was used as exterior fluid and Metolose with a viscosity of 70cSt was used as the pore fluid, in order to match the time scaling laws of soil consolidation and fluid-wave propagation (Sassa & Sekiguchi 1999). The sand used was silica sand No. 7 (Gs = 2.66, $e_{max} = 1.16$, $e_{min} = 0.70$, $D_{50} = 0.15$ mm). The sand beds formed had relative densities Dr of 35–38% and the initial soil depth D of 102 mm corresponding to a 7 m thick sand bed on a prototype scale. The fluid depth, H, was kept at 100 mm.

The model of the pipeline is made of aluminium pipe (diameter of 25 mm) and urethane foam. Specific gravity of the pipe model is 1.35 which is smaller than that of liquefied soil (=1.83 (=γ_{sat}/γ_w, where γ_{sat}: Saturated unit weight of soil and γ_w: unit weight of fluid)). The burial depth of the pipe is 27 mm corresponding to a 1.9 m soil depth on a prototype scale.

The setup method for the pipe may be outlined as follows. The sand bed was formed by pouring sand into a sea of Metolose. At the time when the thickness of the sand was 50 mm, the pipe model was placed on the sand surface. Continuing to pour sand, finally the sand bed with thickness 102 mm was formed. Silicone grease was applied to the pipe ends in order to make the friction between the pipe and the wall of the water channel sufficiently small.

The wave pressures acting on the soil surface were measured by pressure transducers. Wave-induced pore pressure changes in the soil bed were measured in two vertical arrays. The movements of the pipe and soil surface were observed using High-speed camera at an image rate of 250 frames/s.

3 FLOATATION OF A PIPE DUE TO LIQUEFACTION INDUCED BY REGULAR WAVES

A typical set of experimental results from test Case3 are discussed here.

3.1 Liquefaction process

The measured time histories of the excess pore pressures at three different soil depths are shown in Figure 3, together with the wave pressure at the soil surface. Let us first look at the response at a shallow soil depth (z = 22 mm) as shown in Figure 3b. No significant build-up of the residual pore pressure occurred during the moderate level of wave loading (t = 3.0–3.6 s). The increase of wave severity brought about the build-up of the residual pore pressure. When the wave pressure at the soil surface became 8.1 kPa, the residual pore pressure reached the level of the initial vertical effective stress σ_{vo}', indicating the occurrence of liquefaction at this soil depth. After the occurrence of liquefaction at the shallow soil depth, liquefaction occurred at middle soil depth and then at the bottom, which indicate the downward advance of the liquefaction zone (Figs 3c,d). The process of the progress of liquefied zone is illustrated in Figure 4 more clearly.

3.2 Observation of floatation of pipe

The floatation process of the buried pipe under severe wave loading was observed using the high-speed CCD camera. The snapshots of the floatation at four different times are shown in Figure 5. After the occurrence of liquefaction in sand bed, the pipe moved upward gradually. In the course of wave loading the pipe reached the soil surface. A closer observation of the movement of the pipe is shown in Figure 5e. The trajectory of the pipe movement is in a shape of ellipse, showing that the pipe advanced upward in the significant fluidised motion of liquefied sand under severe waves. Hence, this behaviour of the pipe was essentially caused by the dynamic interaction between waves and liquefied soil.

3.3 Relation between liquefaction and floatation of pipe

The floatation of the buried pipe to the soil surface is a consequence of liquefaction. This aspect is illustrated

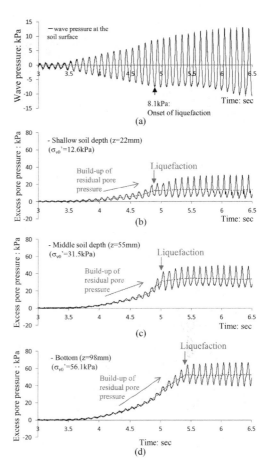

Figure 3. Time histories of (a) wave pressure acting on the soil surface and (b), (c) and (d) excess pore pressures at three different depths (Case 3).

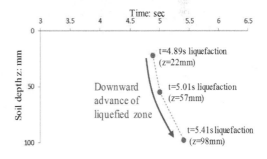

Figure 4. Measured propagation of liquefaction in the course of regular wave loading.

in Figure 6. This figure shows the vertical movement of the pipe top during the wave loading based on the observation from high speed CCD camera. No significant movement of the pipe occurred before the soil bed underwent liquefaction. Upon the occurrence of liquefaction at shallow soil depth, z = 22 mm, the buried pipe started moving upward. The motion of the pipe increased markedly in association with the progress

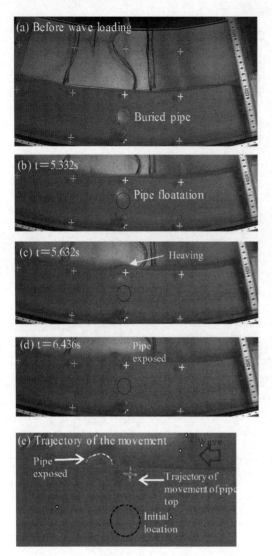

of liquefaction zone in the course of continued wave loading, and finally the pipe reached the soil surface. It is noted that the period of the ellipse motion of the pipe corresponded to that of input wave, indicating that the vibrate motion of the pipe was brought about by the significant vibration of liquefied sand due to wave loading.

4 LIQUEFACTION UNDER IRREGULAR WAVE LOADING AND FLOATATION OF PIPE

A typical set of experimental results from test Case 4 are discussed here. The measured time histories of the excess pore pressures at three different soil depths under irregular wave loading are shown in Figure 7, together with the wave pressure at the soil surface. Note here that in the case of regular wave loading, case3, liquefaction took place at the wave pressure of

Figure 5. Observed pipe floatation under regular wave loading: successive pictures of pipe floatation (a)–(d) and trajectory of the movement of pipe (e).

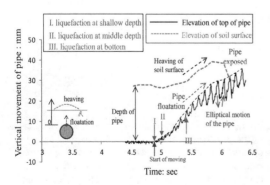

Figure 6. Measured time history of vertical movement of the pipe in association with propagation of soil liquefaction during regular wave loading.

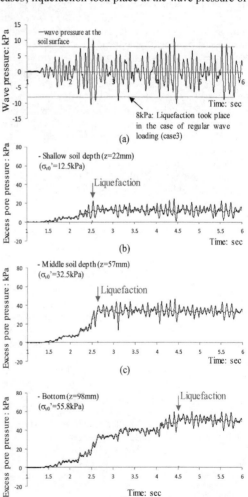

Figure 7. Time histories of (a) wave pressure acting on the soil surface and (b), (c) and (d) excess pore pressures at three different depths (Case 4).

8 kPa. When the severe waves exceeding this pressure acted on the soil surface, the residual pore pressures in the sand bed built up significantly. In fact, liquefaction took place at shallow soil depth at t = 2.5 s, and then liquefaction zone propagated downwards, leading up to liquefaction in the entire sand bed. The process of the downward propagation of the liquefaction zone during the irregular wave loading can be more clearly seen in Figure 8. It is noted that the rate of such progress of liquefaction is far less constant than that of regular wave case (see Fig. 4).

The vertical movement of the pipe top during the irregular wave loading is shown in Figure 9. No significant movement occurred before liquefaction took place. Upon the onset of liquefaction at shallow soil depth, the pipe started moving upward. In the course of downward advance of the liquefied zone, the vibrate motion of the pipe became gradually more significant. Note here that the vibration of the pipe represented the elliptical motion, whose period corresponded to that of waves. The increase in such elliptical motion of the pipe was brought about by the increase in vibratory soil motion in association with the downward propagation of the liquefaction zone. Indeed, Sassa et al. (2001) distinctly observed the relation between such vertical movement of liquefied sand and the advance of liquefied zone (Fig. 10). Namely, the soil surface started vibrating upon the onset of liquefaction and the amplitude of the vibratory soil motion increased markedly in the course of downward progress of liquefaction. Thus, it is clear that the behaviour of the pipe was closely related to the progress of liquefaction.

5 CONCLUSIONS

A set of centrifuge wave tests in a drum channel were performed to investigate the wave-induced liquefaction and floatation of buried pipe in sand bed. The principal experimental results obtained may be summarised as follows:

1. A piston type wave generating system has been developed in a drum centrifuge channel. This system can reproduce irregular waves supposed to be real storm wave.
2. Under a severe wave loading, the loosely packed fresh deposit of sand underwent liquefaction. Liquefied zone propagated downwards in the course of wave loading.
3. The floatation of the buried pipe to the soil surface was a consequence of progressive liquefaction. Indeed, in the regular waves test, upon the occurrence of liquefaction at shallow soil depth, the buried pipe started moving upward, and the motion of the pipe increased markedly in association with

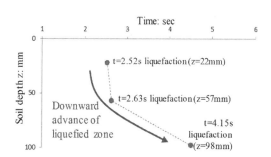

Figure 8. Measured propagation of liquefaction in the course of irregular wave loading.

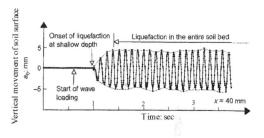

Figure 10. Relation between the increase in the vertical movement of liquefied sand and the advance of liquefied zone (centrifuge acceleration of 30 g) (adapted from Sassa et al. 2001).

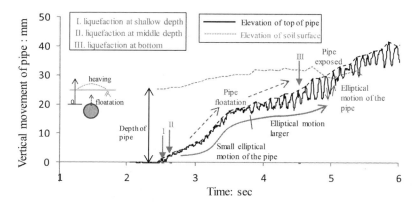

Figure 9. Measured time history of vertical movement of the pipe in association with propagation of soil liquefaction during irregular wave loading.

the progress of liquefaction in the course of wave loading.
4. The onset and spread of the liquefaction under the irregular-waves condition were essentially the same as those observed during regular wave condition. The pipe floated to the soil surface with the significant elliptical motion which took place owing to the vibratory soil motion in association with the progress of liquefaction.

REFERENCES

Baba, S., Miyake, M., Tsurugasaki, K. & Kim, H. 2002. Development of wave generation system in a drum centrifuge, Physical Modelling in Geotechnics: *ICPMG '02*, 265–270.

Sassa, S. & Sekiguchi, H. 1999. Wave-induced liquefaction of beds of sand in a centrifuge, *Géotechnique,* Vol. 49, No. 5: 621–638.

Sassa, S., Sekiguchi, H. & Miyamoto, J. 2001. Analysis of progressive liquefaction as a moving-boundary problem, *Géotechnique*, Vol. 51, No. 10: 847–857.

Sassa, S., Takahashi, H., Morikawa, Y. & Takano, D. 2016. Effect of overflow and seepage coupling on tsunami-induced instability of caisson breakwaters, *Coastal Engineering*, Vol. 117: 157–165.

Sumer, B.M. 2014. *Liquefaction around marine structures*, Advanced Ser. Ocean Eng., Vol. 39, World Scientific.

Sumer, B.M., Fredsøe, J., Christensen, S. & Lind, M. T. 1999. Sinking/floatation of pipelines and other objects in liquefied soil under waves. *Coastal Engineering,* Vol. 38: 53–90.

Sumer, B. M., Hatipoglu, F., Fredsøe, J. & Hansen, N-E. O. 2006. Critical floatation density of pipelines in soils liquefied by waves and density of liquefied soils, Journal of Waterway, *Port, Coastal and Ocean Engineering*, Vol. 132, No. 4: 252–265.

Teh, T.C., Palmer, A. & Damgaard, J. 2003. Experimental study of marine pipelines on unstable and liquefied seabed, *Coastal Engineering*, Vol. 50, No. 1–2: 1–17.

Surface pipeline buckling on clay: Demonstration

R. Phillips, J. Barrett & G. Piercey
C-CORE, St. John's, Newfoundland, Canada

ABSTRACT: Design guidelines have been recently developed to allow surface laid offshore HP/HT pipelines to buckle in a controlled manner. These guidelines need validation against physical measurements of well-defined full pipe-line system lateral buckling events. A centrifuge model testing system at C-CORE was adapted to provide such measurements. Lateral buckling of the equivalent of a 0.8 m diameter 270 m long pipeline partially embedded in soft clay due to axial displacement was demonstrated.

1 INTRODUCTION

The lateral buckling of deep-water pipelines is an important design consideration, Figure 1. The pipelines are installed on the seafloor where repeated cycles of high pressure/and high temperature (HP/HT) result in pipeline expansion and contraction.

The overall pipe-soil interaction is complex and depends on several factors that are difficult to quantify with just analytical approaches. The axial soil resistance along the pipeline provides a reaction that contributes to an effective compression within the pipe which can result in a horizontal buckle. The higher the axial friction the less pipe is available to feed into the buckle, limiting the potential for lateral buckle growth. The lateral soil resistance along the buckle has a significant influence on the buckle shape, restricting the lateral pipe movement and defining the bending stresses within the pipe.

Work has been directed at understanding the influence of the seafloor soils on lateral pipeline buckling (e.g. White et al. 2017). These studies have shown that the axial resistance alone depends on, for example, the pipe weight, soil stress, consolidation history, rate of loading, displacement level and pore pressure development. The lateral resistance depends on the same factors that impact axial resistance as well as the influence of the soil berm, which forms in front of the moving pipe and may grow in size with every operating cycle. Even with these numerous studies the prediction of the location of potential buckles and their severity is complex and standard practice attempts to mitigate this uncertainty by designing zones where the pipeline is likely to buckle. Use of buckle initiators is a common way to achieve this, such as sleepers (structural members placed perpendicular and underneath the pipe) or buoyancy sections placed at intermittent distances along the pipe. Both options are intended to reduce lateral pipe resistance and reliably initiate buckles at locations where the response can be adequately engineered.

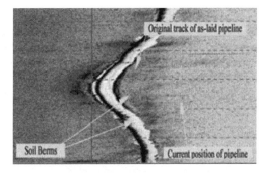

Figure 1. Typical pipeline lateral buckle (Bruton & Carr, 2011).

Despite the use of buckle initiators and the extensive studies regarding the influence of the axial and lateral soil response on buckle behaviour, much uncertainty remains and documented field cases suggest that unexpected buckles still occur. This uncertainty has dictated the use of potentially excessive and costly buckle initiators along pipelines. To date the work on the axial and lateral response has been limited to short sections of pipe tested by movements in either the axial or lateral direction (element tests).

However, these types of tests do not capture the interaction between the axial and lateral soil resistance and the 3D structural response of the pipe. For example, heavy pipe element tests have indicated that the pipe may continue to embed as the berm is pushed laterally. In the field, this may not occur due to the pipe flexural stiffness in the out of plane direction. Less cyclic embedment would reduce the lateral resistance, potentially reducing the maximum stresses within the pipe and the need for additional buckle initiators.

The 3D structural response of a buckle is not well understood and so current finite-element analysis models used in design, greatly simplify the true response. For example, much of the berm resistance is due to the pipe having to climb upwards as it meets the berm, which is not reflected in design modelling.

In addition, the lateral cyclic displacement is prescribed in simple element tests, whereas in a real buckle, the displacement varies along its length. By modelling the full length of the buckle, axial feed in is controlled, allowing variable lateral and upward displacement along the buckle to occur in a more realistic manner. There are no test or operational data available to help understand this potential response.

A joint industry project proposes to more fully evaluate three dimensional (3D) aspects of pipe-soil interaction around lateral pipeline buckles. This paper demonstrates the potential of such an evaluation.

A similar approach (Elliot et al. 2014) successfully investigated the impact of the soil response in the touchdown point region on fatigue assessments of Steel Catenary Risers. The results were used to develop a non-linear soil model that matches the measured fatigue with a high degree of accuracy, both near and away from the critical fatigue point.

Initial pipeline buckling tests (Barret et al. 2016) on dry sand have already been performed using this adapted approach. The first long pipe upheaval buckling tests in a centrifuge from thermal expansion were by Moradi & Craig (1998) in sand.

2 TEST DESIGN

2.1 *Pipeline*

A preliminary design compromise based on Hobbs (1984) showed a 1/100th scale model of a 0.8 m diameter 270 m long pipeline could be tested within the 3 m long riser test box, Figure 2. The model pipeline, comprising of an existing ¼" by 0.032" stainless steel tube with a yield stress of 285 MPa, was still wide enough to be strain gauged.

The model tube was coated with a seamless transparent heat shrink sleeve to protect 13 strain gauge bridges mounted along the tubing length. There were 2 bridges configured to measure axial load, 3 to measure vertical moment and 7 to measure lateral moment for this demonstration. The sleeved model pipeline had an outside diameter of 8.3 mm (D) and a submerged weight of 68.9 N/m.

2.2 *Clay testbed*

Alwhite kaolin slurry was overconsolidated to minimise inflight consolidation times and surface settlements to provide a clay bed depth of 8 pipe diameters, 60 mm. The clay strength profile was measured in-flight using a 2.4 mm O-bar of 35 mm diameter, specifically designed to accurately measure surface strength.

The surface clay strength (Su) was about 4 kPa, giving a normalised pipe weight of 2.1, which is at transition between light and heavy pipe behaviour, Wang et al. (2010).

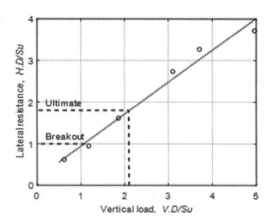

Figure 2. Ultimate lateral resistance, after Wang et al. (2010).

The pipe was initially embedded 0.5D in the mudline in a cut semi-circular trench with no berms. This provided an initial normalised breakout resistance, of 1.0, *ibid*. This resistance should ultimately increase to about 1.8, Figure 3. The increased resistance from breakout to ultimate was expected primarily from berm development, rather than increased pipe embedment, as the pipe displaced. The buckle formation should therefore be controllable rather than snap-through.

2.3 *Pipe profile*

Finite element analyses were used to define the initial out-of-straightness (OOS) of the pipeline profile. The pipe was laid around 3 props at the ¼, ½ and ¾ points, Figure 2 with lateral offsets of +12, −12 and +12 mm respectively. The ballasted wooden props at the ¼ and ¾ points along the length were removed inflight using pneumatic actuators prior to axial pipe displacement due to uncertainty of the asymmetric failure mode shape of the buckled pipeline. The apparent (undrained) lateral friction factor at 100 g of 0.48 (1.0/2.1) was sufficient to hold the initial OOS profile before pipe axial displacement.

2.4 *Equipment*

The 3 m long strongbox used by Elliot et al. (2014) was checked for operation under 100 g rather than 50 g. A reaction frame was added between the strongbox and the 1.4 m long centrifuge swinging platform to carry the increased aerodynamic buffeting loads.

An existing stepper gear motor driven, 20 mm stroke 1-kN lead-screw seesaw actuator was adapted to axially displace the upslope end of the pipeline, Figure 2. The strongbox was inclined at 2.4° inflight as the test package centre of mass was offset. This inclination was reduced by lead counterweights on the platform. The inclination was measured by 2 surface water pressure transducers at either end of the box. An inline load cell was placed between the pipeline and the actuator. A submersible LVDT was used to directly measure the

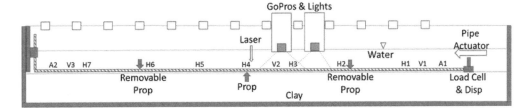

Figure 3. Model test layout.

pipe displacement with respect to the stiff strong box wall.

The other end of the pipeline was connected to the strongbox using the fixed-end arrangement used by Elliot et al. (2014). A second stepper gear motor driven lead-screw linear actuator was used to displace the O-bar penetrometer.

2.5 Instrumentation

Surface measurements of the clay and the pipe were made by two GoPro cameras, a midspan laser profiler and post flight by a coordinate measurement machine and photographs.

Apparent lateral movements of the pipe and clay surface were measured using Geo-PIV (White et al., 2003) from two 12 MP GoPro Hero3+ images taken at 2 second intervals. The vertical camera alignment and 100 g operation precluded the use of higher definition cameras such as DSLRs or mirrorless cameras. The Go-Pros were fitted with close-up filters and a ring adapter as the cameras had to be positioned closer than the minimum focus distance. The camera position dictated the depth of free water above the clay surface.

Control markers on aluminium angles, Figure 4 were placed in the field of view to correct for the distortion at the edges of the image. The marker coordinates were assessed in the plane of the clay surface. Out of plane (vertical) movements introduced error into the lateral measurements.

The white clay surface was textured using fine black sand to allow Geo-PIV measurements, Kai et al. (2016). Similarly, black permanent marker stripes on top of the pipe under the transparent heat shrink provided photographic contrast.

The cameras were placed in aluminium square sections sealed and capped by 10 mm thick acrylic windows beneath the water surface to allow the cameras to see the clay. The cameras were placed on pads directly on the acrylic windows. Care was taken during the test to prevent air bubbles sticking to the window. The surface water was added inflight to prevent swelling of the clay under low effective stress, and surface erosion during swing up. Best practice for window clarity was to raise the water level slowly under 10 g past the clean, static-free windows to minimise surface waves.

Three 2 W LED puck lights were placed either side of the two cameras to illuminate the clay surface. These lights were placed in 1 litre clear Nalgene cylindrical

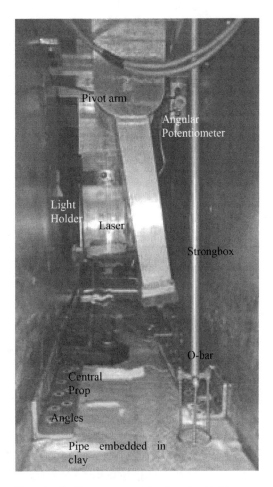

Figure 4. Surface measurement systems (post-test).

drinking water bottles, Figure 4 which were supported by the water pressure inflight.

Figure 5 shows images from the two Go-Pros inflight after 12 mm of axial displacement from well outside the left side, Figure 2. The surface texture on the clay and pipe is evident. The surface sand dusting was removed at the left and right hand ends by lateral pipe movement and berm formation.

A midspan laser profiling system, Figure 2 was developed to measure near centreline pipe and soil elevation profiles. The profiler comprised a Baumer red laser 30–130 mm distance sensor (OADM

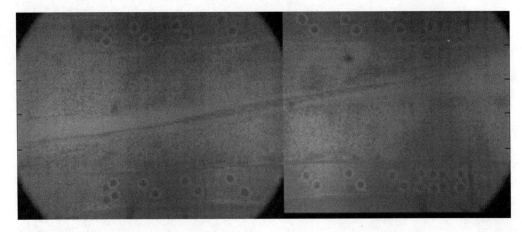

Figure 5. Go-Pro images in flight.

2OI4460/S14C) at the base of an aluminium square section sealed and capped by a 10 mm thick acrylic window, Figure 4. This laser housing was suspended from a top pivot arm and counterbalanced by a mass over a pulley. The housing was rotated through a second cable using a DC-motor driven capstan. The angle of the housing was measured by an angular potentiometer, Figure 4. The laser was calibrated through water. The laser profiler successfully measured the radial distance through water to the surface during rotation of the laser head.

The clay settlement was measured from two LVDTs.

3 DEMONSTRATION TEST

The clay was consolidated from a slurry to a uniform preconsolidation stress, with an underlying drainage sand layer to provide the necessary clay surface elevation. The clay rebound was monitored, when unloaded, to compensate for inflight elastic settlement.

The model pipeline was placed on the clay surface around the props, and its outline traced. The 0.5D deep trench was cut along this outline. The pipe was placed in the trench around the props with its ends at the correct elevation.

The test package was accelerated to 100 g with surface water added slowly at 10 g. The clay consolidation was monitored by the surface LVDTs, and then the O-bar test conducted. The water level was periodically topped up to compensate for surface evaporation. The ¼ and ¾ props were removed. The pipe, sensors and cameras were monitored throughout.

The pipe was cyclically axially displaced, with care not to cause a buckle on first loading. Further displacement cycle sets were conducted (back to initial axial load) with increasing axial displacement until buckling occurred (seen as significant reduction in axial load).

The centrifuge was decelerated and surface water drained to reduce surface erosion. Remaining water was removed and the final pipe and seabed profile

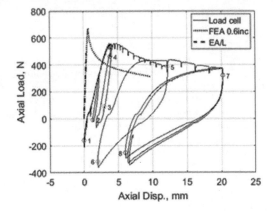

Figure 6. Axial pipe end load – displacement response.

measured at 1 g using a coordinate measurement machine.

The end of pipe was displaced only axially to a maximum of 20 mm over a 100 minute period. This pipe loading rate was considered drained. Both pipe ends were restrained from movement or rotation in other directions. The pipe end load and displacement response is shown in Figure 6.

There is a distinct change in stiffness every time the pipe is loaded between compression and tension, as some of the pipe lifts off the clay surface. The peak axial load of 560 N after 4 mm of displacement is lower than 690 N predicted by a finite element analysis (FEA) of the displaced pipe assuming an isotropic friction coefficient of 0.6 on an inclined surface. The elastic axial stiffness, EA/L of a straight pipe is much higher than the initial measured stiffness, but is comparable to the initial unload stiffness especially on the four pre-peak cycles.

The initial pipe axial displacement profiles before peak load were assessed by Geo-PIV from the left images in Figure 5. The resulting axial pipe stress – strain response indicates plastic compression of the pipe consistent with the post yield behaviour measured

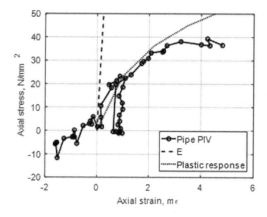

Figure 7. Axial strain response from pipe Geo-PIV.

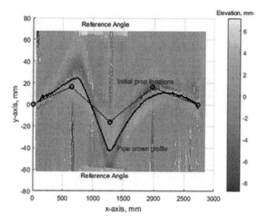

Figure 9. Post flight surface profiles.

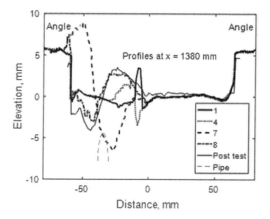

Figure 8. Midspan surface laser profiles.

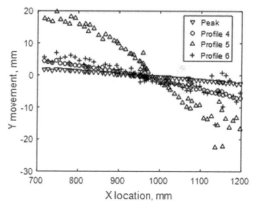

Figure 10. Lateral pipe movement profiles.

in tensile coupon tests, Figure 7, probably due to residual stresses in the drawn tubing. This highlights the importance of preloading structures to ensure their elastic response over appropriate load ranges. Such preloading was not done for this initial demonstration. The plateau above 0.2% strain is due to onset of lateral pipe movements evident in Figure 5.

The axial pipe soil friction was estimated by the difference in axial load between the motion and fixed pipe ends, Figure 2. The global axial shear stress was less than 2 kPa. The mobilised axial friction factor was assessed from the pipe weight, with an offset of 0.02 to allow for the 2.4° mudline inclination. The peak breakout factor of 0.55 was consistent with DnV-GL (2017) value of 0.54 assuming a pipe-soil roughness factor of unity. The measured residual factor was about 0.3.

Eight sets of laser surface midspan profiles were taken during the course of the pipe loading, as indicated by the open circles in Figure 6. Four of the numbered profiles were interpreted to give surface elevation profiles, Figure 8. Profile 1 is prior to pipe movement and shows the edges of the two 1/4" thick angles (Figure 4) present near midspan at −60 and +60 mm. The pipe crown elevation at +4 mm confirms the initial midspan 0.5D embedment.

The pipe crown is evident in profile 4, but is at or below the clay surface in profiles 7 & 8 due to self-embedment. Profiles 4 to 8 show the development of the clay berm ahead of the displaced pipe. The berm at profile 7 is partly on the angle, and collapses back as the pipe is unloaded to profile 8. The depression between the angle and the berm in profile 8 is due to collapse of the trench in profile 7, by comparing change in berm heights from profile 7 to 8 between −60: −40 and −40: −5 mm. The post-test surface profile (with embedded pipe location) from coordinate measurement machine, Figure 9 validates profile 8 measured in flight at the end of pipe test. The pipe was displaced from the left hand end.

The lateral pipe movements during axial displacement are shown in Figure 10 from Geo-PIV of the converted GoPro image pairs, with reference to the 8 laser profiles indicated in Figure 6. There was little lateral movement prior to the peak load. The pipe progressively self-embedded after profile 4 from midspan as evident post-test in Figure 4. The associated lateral berm movements are shown in Figure 11 near the pipe midspan.

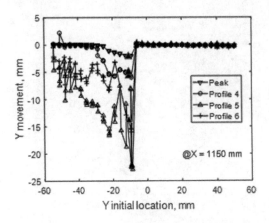

Figure 11. Lateral berm movements near midspan.

4 CONCLUSIONS

A centrifuge model testing system has been adapted to validate guidelines to control surface laid pipeline lateral buckling. A demonstration test buckled the equivalent of a 0.8 m diameter 270 m long pipeline partially initially embedded 0.5D into soft clay due to cyclic axial displacement. The results of the asymmetric pipeline buckle and berm formation are now being compared to finite element analysis predictions.

Geo-PIV provided accurate lateral surface measurements from vertically orientated GoPro cameras under 100 g looking through water. The measurement quality indicates the imaging system may be used for a 3D assessment after Le et al. (2017).

A laser profiling elevation system was also developed. This system can be extended to obtain 3D elevation profiles using a line rather than a spot laser.

ACKNOWLEDGEMENTS

The authors are grateful to their C-CORE geotechnical engineering colleagues for assistance in the execution of this demonstration test. Thanks are also extended to David Bruton of Crondall Energy, Ed Clukey of Jukes Group, and their colleagues for input into the design and ongoing analysis of this test.

REFERENCES

Barrett, J., Phillips, R. & Piercey, G. 2016. 3D Modelling of lateral behaviour of surface laid pipelines,' *Canadian Geotechnical Conf.*, Vancouver.

Bruton, D. & Carr, M. 2011. Overview of the SAFEBUCK JIP. *Offshore Tech. Conf.*, (OTC-21671), Houston.

DNV-GL. 2017. DNVGL-RP-F114: Pipe-soil interaction for submarine pipelines.

Elliott, B., Phillips, R., Macneill, A. & Piercey, G. 2014. Physical modelling of SCR in the touch-down zone under three axis motions. *Int. Conf. Phy. Model. Geotechnics*, CRC Press, Florida.

Hobbs R.E., 1984. In-service buckling of heated pipelines, *ASCE Jnl. Transportation Eng.* 110(2).

Tan, K.Q., Tung, Q.E., Lee, F.H. & Goh, S.H. 2016. Enhancing images for particle image velocimetry in centrifuge models. *15th Asian Conf. Soil Mech. & Geotechnical Engineering*

Le, B.T., Nadimi, S.S, Goodey, R.J. & Taylor, R.N. 2016. System to measure three-dimensional movements in physical models. *Géotechnique Letters* 6(4) 256–262.

Moradi, M. & Craig, W.H. 1998. Observations of upheaval buckling of buried pipelines. *Centrifuge '98*. Rotterdam: Balkema.

Wang, D, White, D.J. & Randolph, M.F. 2010. Large-deformation finite element analysis of pipe penetration and large-amplitude lateral displacement. *Canadian Geotechnical Journal* 47: 842–856.

White, D., Clukey, E.C., Randolph, M.F., Boylan, N.P., Bransby, M.F., Zakeri, A., Hill, A.J. & Jaeck, C. 2017. The state of knowledge of pipe-soil interaction for on-bottom pipeline design. *Offshore Tech. Conf.*, (OTC-27623), Houston.

White D.J., Take W.A. & Bolton M.D. 2003. Soil deformation measurement using particle image velocimetry (PIV) and photogrammetry. *Géotechnique* 53 7: 619–631.

Centrifuge modelling for lateral pile-soil pressure on passive part of pile group with platform

G.F. Ren, G.M. Xu, X.W. Gu, Z.Y. Cai & B.X. Shi
Department of Geotechnical Engineering, Nanjing Hydraulic Research Institute, Nanjing, China
State Key Laboratory of Hydrology-Water Resources and Hydraulic Engineering, Nanjing, China

A.Z. Chen
Liyuan Electronic Equipment Factory, Liyang, China

ABSTRACT: Groups of vertical cast-in-place piles with a platform have been successfully introduced to deep-water large-tonnage sheet-pile wharfs to withhold the lateral load due to horizontal soil movement towards the front wall. The work mechanism of these vertical piles is similar to that of passive stabilizing piles. However, the degree of the pile-soil interaction is far below the limit state. In order to get a knowledge of the distribution of lateral pile-soil pressure and distinguish the pile's passive part from active part, large geotechnical centrifuge model tests of two design schemes were conducted to simulate the 200,000 tonnage sheet-pile wharfs with load-relief platform in homogeneous fine sand. Some useful results are gained and presented.

1 INTRODUCTION

Sheet pile wharf structures have been widely used in the construction of port terminals located in many foreign countries in the form of its simple structure, low cost, short construction period and other unique advantages (Liu, 2014). Due to the lack of high quality steel and inclined pile construction technology for the construction of sheet piles, the structure type and construction of sheet pile wharf are greatly restricted in China, which are only used for small and medium-sized port projects. Since 2000, the structure design level and construction technology of reinforced concrete diaphragm wall pile wharf have been greatly improved in China. The curtain board pier structure and unloading sheet pile wharf structure have been successfully developed, which have been applied to dozens of large deep-water berth construction in Bohai bay. The core of these two kinds of new-type sheet pile structures with independent innovation is to exert the lateral bearing function of the vertical pile group foundation. In contrast with the traditional sheet pile wharf structure, the pile group foundation with platform above the landside soil of the front wall in the structure of the deep-water unloading sheet pile wharf is aimed to share a part of the lateral load on the front wall, so that the force and displacement deformation of the front wall can still be within a reasonable range, which can ensure the safety and stability of the whole pile wharf structure. Figure 1 shows the schematic diagram of the pile wharf structure. It consists of the front wall, pile group foundation with platform, the anchor wall and the tensile rod. The pile group foundation

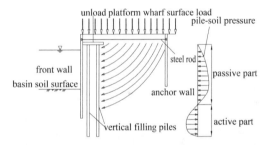

Figure 1. Sketch of sheet-pile wharfs with load-relief platform and lateral pile-soil pressure with depth.

with platform is articulated by the multi-row vertical cast-in-place piles and the unloading deck. It is well known that the unloading sheet pile wharf is composed of the structure of single anchor sheet pile wharf and the structure of the pile foundation with platform. Therefore, the pile foundation with platform and the front wall bear the lateral load together because of lateral soil movement, while the stress behaviour of the structure of the pile group foundation with platform is mainly reflected in the stress characteristics of the vertical cast-in-place piles.

As shown in Figure 1, the front wall will move to the sea side effected by the lateral unload and the vertical wharf surface load because of the basin excavation, which will also cause the soil surrounding the cast-in-place piles of the pile group foundation with platform, especially the soil above the mud surface line lateral displacement to move to the sea side, resulting in the soil pressure of the landside of the pile greater than

Figure 2. NHRI 400 g-ton large geotechnical centrifuge.

that of the seaside of the pile. Finally, a clear pressure difference between both sides exists. The lateral pressure of the pile pointing to the seaside is formed as shown in Figure 1, and the horizontal load is applied to the pile. According to the definition of the passive pile (De, 1977), this part is a real passive pile called the pile's passive part. For the pile below the mud surface line, the surrounding soil is always in a stable state without the trend of lateral movement, however, due to the deformation of the pile, the stable soil has a resistance to the pile, which belongs to the active one called pile's active part. Accordingly, the key to figure out the distribution of lateral pile-soil pressure is to distinguish the passive part from the active part of the pile.

To solve the problem, large-scale geotechnical centrifugal model tests have been carried out, based on the T-shape diaphragm wall and composite steel pipe pile scheme of a 20,000-ton unloading sheet pile wharf structure. During the test, the soil pressure test pile has been set in every row of cast-in-place piles of these two kinds of pile group foundation with platform, with the miniature earth pressure cell embedded in the seaside and landside, in order to test the distribution of lateral pile-soil pressure under the lateral unloading and the vertical wharf surface load. How to distinguish the passive part from the active part of the pile is also discussed further.

2 THE EXPERIMENTAL SET-UP

2.1 Experimental facilities

Figure 2 shows the NHRI 400 g-ton large geotechnical centrifuge with the capacity of 400 g-ton, the rotation radius of 5.5 m, maximum centrifugal acceleration of 200 g, and the maximum load of 2000 kg. Considering the model layout as the plane strain problem, a large plane model box has been used, whose inner dimensions are 1200 mm long, 400 mm wide and 800 mm high respectively.

2.2 Model design

In view of the effect range of the prototype wharf and the size of the model box, the model similarity scale

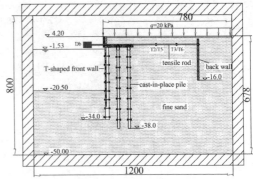

(a) sectional drawing

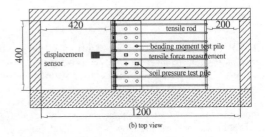

(b) top view

Figure 3. Model setup (front wall of T-shaped diaphragm) (all dimensions in mm but elevations in m).

was set to 80. After determining the model similarity scale, the similarity of the other physical quantities of the model was then settled (Xu et al. 2010).

2.3 Model layout

The aluminium alloy material has been used to make front wall, anchor wall, unloading platform and filling pile of the model. Since the elastic modulus between the model material and the prototype material is different, it is necessary to revise it according to the mechanical characteristics of the components (Xu et al. 2010). The model layout of the two schemes is shown in Figure 3 and Figure 4 respectively. The front wall, the anchor wall and the platform plate of the model have the same width as the model box. In addition, the anchor wall is 10.1 mm thick, and the platform plate is 10 mm thick. In the T-shape diaphragm wall model, the web's and flange's thickness of the front wall are 23 mm and 7.4 mm respectively. In the combined steel sheet pile scheme model, the thickness of the front wall is 7.7 mm, and the other dimensions are reduced in proportion to the size of the prototype.

The model pile size is designed according to the bending stiffness similarity principle, and the model filling piles of the two schemes are simulated by the hollow aluminium tube with inner diameter of 13 mm, outer diameter of 15mm and inner diameter of 14 mm and outer diameter of 16 mm. The length of the model pile is scaled down at the same proportion according to the length of the prototype pile. The articulation mechanism of the model unloading platform and the model piles is shown in the reference (Xu & Li, 2016).

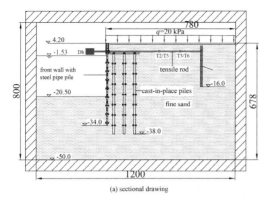

Table 1. Properties of model ground.

Indicator	ρ_{max}	ρ_{min}	ρ_d	$\varphi°$	α
Maximum dry density	1.60 g/cm³				
Minimum dry density		1.33 g/cm³			
Dry density			1.56 g/cm³		
Internal friction angle				36.2°	
Angle of repose					34.0°

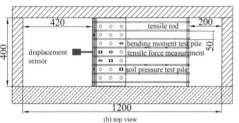

Figure 4. Model setup (front wall of composite steel pipe pile) (all dimensions in mm but elevations in m).

The steel bar is used to simulate the prototype steel rod according to the tensile stiffness similarity principle. In order to simplify the model layout, a single model rod is used to simulate three prototype rods. In this case, eight steel rods with a width of 7.2 mm and a thickness of 0.4 mm have been set within the 400 mm range of the model box.

The homogeneous fine sand foundation of the model is formed by the way of the vibration, and the fine sand is obtained from the Caofeidian port district. The specific gravity of this fine sand is 2.67. The gradation is that the particle size is greater than 0.25 mm (1.9%), the particle size is between 0.25 mm to 0.075 mm (92%), and the particle size is greater than 0.075 mm (6.1%). The relative density of the model soil is 0.87. Main physical and mechanical properties of model ground are shown in Table 1.

2.4 Model measurement

The soil pressure test pile has a rectangular section with a width of 15 mm and a thickness of 20 mm. A total of 10 earth pressure cells are arranged along the length of the pile shown in Figure 5. This kind of miniature earth pressure cell is specially developed for the geo-centrifugal model test (Xu G. M. et al. 2007). It has the outer diameter of 12 mm, the thickness of 4.2 mm, and the measurement range of 200, 400, 600 and 800 kPa.

2.5 Test procedure

Firstly, the centrifugal acceleration was raised up at a constant rate to simulate the lateral unloading effect

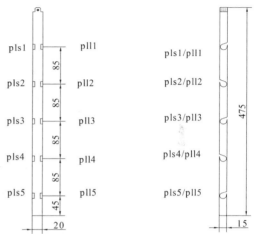

Figure 5. Model pile instrumented with total earth pressure cells (all dimensions in mm).

due to the harbour basin excavation approximatively after the soil excavation was finished under 1g. Then the model was placed on the platform of the centrifuge. Secondly, the water was poured into the preset box to simulate the surface load of 20 kPa when the centrifugal acceleration reached to the design acceleration 80 g. Figure 6 shows the duration of the test, it cost forty one minutes from 1 g to 80 g. After twenty three minutes, the wharf surface load was applied. The centrifuge was stopped as another twenty four minutes passed. It is necessary to note that the overflow tube has been used to keep the water level of the model basin unchanged during the test process.

2.6 Experimental program

Combined two kinds of schemes about 200,000 ton unloading sheet pile wharf structure, a total of 5 sets of models have been arranged. The first series of three models, M3, M4 and M5 simulated T-shape diaphragm wall, which were parallel and repeated tests. The second series of two models M6 and M7 simulated composite steel pipe pile, which were also parallel and repeated tests.

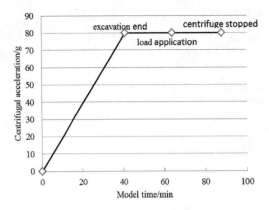

Figure 6. Simulation process of test.

3 RESULTS AND ANALYSIS

As mentioned above, the effect by the lateral unloading due to the harbour basin excavation and the vertical wharf surface load is mainly reflected in the pile-soil pressure reaction of the filling piles. Figure 7 shows the total earth pressures on two sides of seaside piles for T-shape diaphragm scheme. Figure 8 shows the lateral pile-soil pressures of seaside piles for T-shape diaphragm scheme. As shown in Figure 7, the three earth pressure distribution curves on the same side of the three parallel models are not completely coincident, but the same regularity. The change law is that the soil pressure of the landside is greater than that of the seaside above the elevation −27.6 m which is roughly from the first measurement point to the fourth measurement point of the pile body. Figure 8 shows the lateral pile-soil pressure of the pile is plus. On the contrary, the soil pressure of the landside is smaller than that of the seaside below the elevation −27.6 m. Figure 8 shows the lateral pile-soil pressure of the pile is minus. The intersection point of the soil pressure distribution curve on both sides in Figure 7 is exactly the zero point in the pile-soil pressure distribution curve in Figure 8. The position elevation of the three zero points is very close (Fig. 8), which is located near the fourth measurement point of the pile, and their average elevation is about −28.2 m. It is known that the mud surface line elevation is −20.5 m. So the zero point of the lateral pile-soil pressure of the seaside pile is located about 7.7 m below the mud surface line.

We also know that when the lateral pile-soil pressure of the pile is plus, the direction of the lateral load on the pile points to the seaside, which is in accordance with the soil movement. On the other hand, when the lateral pile-soil pressure of the pile is minus, the direction of the lateral load on the pile points to the landside, which is opposite to the soil movement. So it is called negative lateral pile-soil pressure as the pile side resistance. As shown in Figure 8, the pile body above the zero point of the lateral pile-soil pressure of the seaside pile can be used as a passive part based on the design of the anti-slide pile, and it is subjected to the lateral load in the same direction as the soil movement. For the pile's

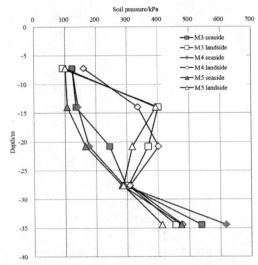

Figure 7. Total earth pressures on two sides of seaside piles for T-shape diaphragm scheme.

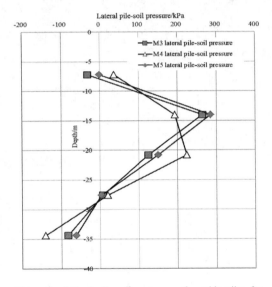

Figure 8. Lateral pile-soil pressures of seaside piles for T-shape diaphragm scheme.

active part, it withstands the lateral load that is opposite to the soil movement. It is absolutely important that the active and passive part of the filling pile have been divided according to the zero point of the lateral pile-soil pressure got from the centrifuge model tests above. Figure 9 shows the lateral pile-soil pressures of landside piles for T-shape diaphragm scheme. The lateral pile-soil pressure distribution of the landside pile is consistent with that of the seaside pile in Figure 8.

Figure 10 shows the lateral pile-soil pressures of seaside and landside piles for composite steel pipe pile scheme. The distribution law is similar to that of the first series of model tests. The lateral pile-soil pressure of the upper pile is plus and the lateral pile-soil

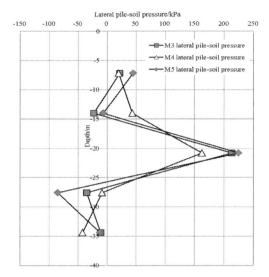

Figure 9. Lateral pile-soil pressures of landside piles for T-shape diaphragm scheme.

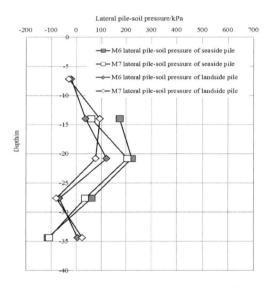

Figure 10. Lateral pile-soil pressures of seaside and landside piles for composite steel pipe pile scheme.

pressure of the lower pile is minus. The zero position elevation of the average lateral pile-soil pressure distribution curve is about −23.5 m and −24.7 m respectively, which is below the mud surface line 3.0 m and 4.2 m respectively. In general, the zero point of lateral pile-soil pressure is the demarcation point of the upper passive part and the lower active part of the seaside and the landside piles for the unloading pile wharf structure, whose elevation is below the mud surface line about 3.0 m∼7.7 m.

4 CONCLUSIONS

Two series of five large geo-centrifugal models are carried out to simulate two kinds of design schemes of unloading sheet pile wharf structure in homogeneous fine sand foundation. The lateral pile-soil pressure distribution of the vertical filling pile is mainly studied under the lateral unloading and the vertical wharf surface load. The division of passive part and active part of the filling pile is also discussed. Some useful results are obtained as follows:

1. The distribution law of the lateral pile-soil pressure is truly revealed: the lateral pile-soil pressure distribution of the pile body is characterized by the force property of anti-slip passive pile under the influence of lateral soil movement, namely, the lateral pile-soil pressure of the upper pile is plus and the direction of the action is consistent with that of the soil movement. While the lateral pile-soil pressure of the lower part of the pile is minus, the direction of the action is opposite to the soil movement.
2. The pile can be divided into the upper passive part and the lower active part according to the zero point of the lateral pile-soil pressure. Since the zero point of the lateral pile-soil pressure is always located below the mud surface, it is unreasonable to distinguish pile's passive part from active part based on the mud surface only.

ACKNOWLEDGEMENTS

The centrifuge tests were sponsored by a research program under Contract 2012AA112510 from 863 Project of China and the National Natural Science Foundation of China (No. 51679149). Grateful acknowledgement is made of the technical support by the technicians of Liyuan Electronic Equipment Factory for their assistance in the design of the miniature earth pressure cell.

REFERENCES

De. B. E. E. 1977. Piles subjected to static lateral loads, State-of the Art Report. *Proc. 9th ICSMFE, Specialty Session 10*, Tokyo. 1–14.

Liu. Y. X. 2014. Design theories and methods for sheet-pile bulkhead. *China Communications Press Co. Ltd.* 1–303.

Xu. G. M. et al. 2010. Centrifuge modeling for an innovative sheet-pile bulkhead of diaphragm. *Chinese Journal of Rock and Soil Mechanics* 31(S1): 48–52.

Xu. G. M. & Li. S. L. 2016. Experimental study of head fixity conditions of pile group in sheet-pile bulkhead. *Chinese Journal of Rock Mechanics and Engineering* (S1): 3365–3371.

Xu. G. M. et al. 2007. Measurement of boundary total stress in a multi-gravity environment. *Chinese Journal of Rock and Soil Mechanics* 28(12): 2671–2674.

Centrifuge model tests and circular slip analyses to evaluate reinforced composite-type breakwater stability against tsunami

H. Takahashi, S. Sassa & Y. Morikawa
Port and Airport Research Institute, MPAT, Yokosuka, Japan

K. Maruyama
Geodesign, Tokyo, Japan

ABSTRACT: Composite-type breakwaters made of concrete caisson and rubble mound are widely employed in Japan. The powerful tsunami of 2011 severely damaged these breakwaters. Hence, there is an urgent need to reinforce them to prevent such extensive damage in another tsunami. One of the possible reinforcement methods involves piling stones behind the caisson. A stability assessment method has been proposed, but it has not been completely verified. Thus, centrifuge modelling was implemented to examine the stabilities of the rubble mound and piled stones under the condition of a significant tsunami-induced water-level difference. The effect of reinforcement was investigated by altering the sectional shape of the piled stones. Consequently, a bearing capacity failure was observed in the cases with none and insufficient amount of piled stones. The stability observed in the centrifuge was analysed using the proposed stability assessment method with the help of the circular slip analysis. This method was verified to be able to simulate the model tests.

1 INTRODUCTION

Composite-type breakwaters made of concrete caisson and rubble mound are widely employed in Japan. The powerful tsunami due to the Great East Japan Earthquake of 2011 severely damaged these breakwaters. Hence, there is an urgent need to reinforce them to prevent such extensive damage in another tsunami. One of the possible reinforcement methods involves piling stones behind the caisson (see Fig. 1). The reactive force generated by the piled stones helps in enhancing the sliding stability. Moreover, the weight and shear resistance of the piled stones help in increasing the bearing capacity of the mound.

The method of piling stones behind the caisson has been proposed in the past to reinforce breakwaters to withstand high wind waves. The Japanese design standards for port and harbour facilities (OCDI 2009a) demonstrate the tentative stability assessment method with regard to sliding failure. Kikuchi et al. (1998) conducted large-scale horizontal loading tests to investigate the reinforcement effect on bearing capacity; moreover, numerical analyses were performed using a circular slip analysis. After the tsunami of 2011, Arikawa et al. (2013) conducted hydraulic model tests wherein reinforced composite-type breakwaters were exposed to tsunami. It confirmed the effect of piled stones placed behind the caisson. In the same period, Shinsha et al. (2014) and Takahashi et al. (2014a) examined the stability properties of reinforced breakwaters by conducting horizontal loading tests using a centrifuge. They showed that the stability could be

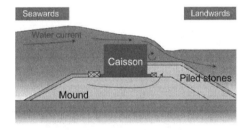

Figure 1. Composite-type breakwaters with piled stones.

successfully assessed by performing a circular slip analysis. Accordingly, Takahashi et al. (2015) proposed a stability assessment method for breakwaters reinforced with piled stones; Sato et al. (2017) considered design conditions using a circular slip analysis. However, this method was established based on the static horizontal loading tests and has not been verified under the condition of a tsunami-induced water-level difference. It might induce unexpected failure under such a condition. Additionally, it is known that the seepage force due to a water-level difference leads to a reduction in the bearing capacity of the rubble mound and piled stones (Takahashi et al. 2014b, 2015). Hence, it is important to validate the stability assessment method under the condition of a water-level difference.

In this study, a series of centrifuge model tests were conducted by changing the sectional shape of the piled stones to observe the stability of the breakwater. The centrifuge was used to reproduce a water-level difference in front and behind a caisson, thereby

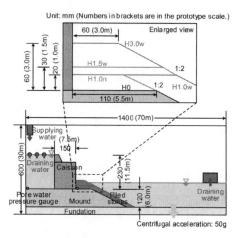

Figure 2. Schematic view of models.

Figure 3. Crushed stones used in tests.

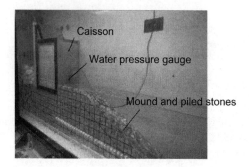

Figure 4. Example of prepared model.

simulating a tsunami. The centrifugal acceleration produced using the centrifuge helped in generating a seepage force equivalent to that produced in a prototype-scale model. The stability of the test cases was then assessed by performing a circular slip analysis. Subsequently, the accuracy of the proposed assessment method was verified.

2 CENTRIFUGE MODEL TESTS

2.1 Test conditions

Centrifuge modelling was implemented to examine the stability of the rubble mound and piled stones under the condition of a water-level difference. The centrifugal acceleration helps in generating the stress, pore water pressure, and even seepage force equivalent to those produced in a prototype-scale model. The details of the similitude ratios regarding the seepage are summarised in the paper written by Takahashi et al. (2014b).

A model of the breakwater was prepared in a steel specimen container with a width of 1400 mm, a height of 600 mm, and a depth of 80 mm. Figure 2 shows the schematic of the model. Although small stones were laid on the bottom of the container, they were covered using an acrylic board to prevent the seepage flow from scouring the stones in the mound. Silica sand was affixed to the surface of the acrylic board by using glue to increase the friction against the mound. The mound was built above the board using crushed stones with a size in the range of 5–9 mm (see Fig. 3). The height of the mound was 120 mm, which corresponds to a height of 6 m in the prototype scale. The stones used in the model were round with a relatively low shear strength. However, the crushed stones used in the real field are usually angular with a high shear strength. In the experiments conducted in this study, round stones were employed because of the following two reasons. First, Takahashi et al. (2017) confirmed that round stones were employed to successfully simulate the horizontal-loading test in the circular slip analysis by employing the shear strength ($c = 20$ kN/m^2 and $\phi = 35°$) presented in the Japanese design standard (OCDI 2009b). The shear strength in the design standard has safety margin, making it difficult to directly compare the results of a model test using angular stones with those of a simulated analysis. Second, the round stones helped in observing a clear slip surface. The round stones could easily slip between the surfaces of the stones, thus helping to demonstrate a slip surface, whereas angular stones could not be used to demonstrate a slip surface. The density of the mound produced using the soft hand-tapping method was moderate, and the submerged unit weight was 10.8 kN/m^3. The model caisson, which was made of acrylic boards, sand, and small lead balls, was laid on the mound. The weight of the caisson per unit depth was 1869 kN/m in the prototype scale. The same crushed stones as those in the mound were piled behind the caisson. Water was then gently poured into the model. Figure 4 shows the prepared model.

The centrifuge test comprised five cases wherein different sectional shapes of the pile stones were employed (see the enlarged view in Figure 2). The centrifuge machine developed by Port and Airport Research Institute was used in the tests, as shown in Figure 5. The details of the machine can be obtained in the report by Kitazume & Miyajima (1995). The manufactured models were fixed on the platform of the centrifuge; a centrifugal acceleration of 50g was applied to the models. Under the centrifugal acceleration, water was rapidly poured onto the front side of the caisson from the water tanks installed on top of the

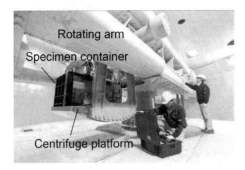

Figure 5. Centrifuge machine PARI Mark2.

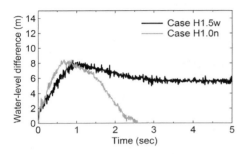

Figure 6. Time histories of water-level difference.

specimen container. This rapid water pouring helped in inducing a significant water-level difference in front and behind the caisson. To maintain the water levels for a certain period, outlets were made on both sides of the caisson in case of overflow. A g-proof camera was installed to capture the movement of the breakwater. The deformation and failure of the mound and piled stones were examined by analysing the images captured using the camera.

2.2 Test results

2.2.1 Water-level difference and horizontal forces

The movement of the breakwater was observed under the condition of a water-level difference. Here, the water-level difference generated in the tests was analysed, and the movement of the breakwater is presented in the next section. Figure 6 shows the time histories of the water-level differences in the cases H1.5w, wherein the caissons remained stationary, and H1.0n, wherein the caissons moved. The water-level difference and time histories are demonstrated in the prototype and model scales, respectively. In case H1.5w, the water-level difference reached a peak of approximately 8 m after 1 s, and subsequently, reached a steady state after 3 s. Assuming a hydrostatic distribution, the horizontal force per unit depth acting on the caisson was 552 kN/m at the peak in the prototype scale. A similar force was observed in case H1.0w, whereas a larger force of 612 kN/m was observed in case H3.0w because the amount of water supplied was considerable only in this case.

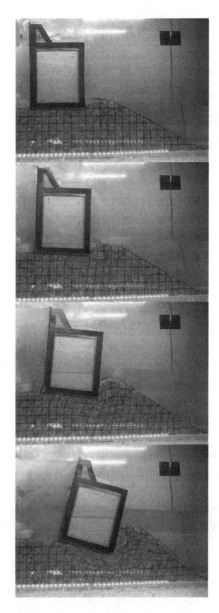

Figure 7. Sequential images at failure (Case H1.0n).

In case H1.0n, the caisson dramatically moved after the water level in front of the caisson reached a peak of approximately 8 m. Thus, the water-level difference decreased after the peak. However, the peak value of the water-level difference was similar to those of cases H1.5w and H1.0w. The maximum horizontal force was 542 kN/m in the prototype scale. In case H0, which clearly failed, the time histories of the water-level difference and maximum horizontal force were largely the same as that in case H1.0n.

2.2.2 Movement of breakwaters

The observation of the breakwaters' movement showed that only two cases of H1.0n and H0, which had

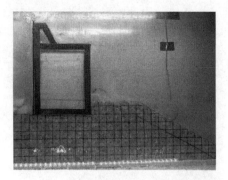

Figure 8. Breakwater during test (Case H1.5w).

insufficiently piled stones and no piled stones, respectively, completely failed in the bearing capacity failure mode. In other cases, the breakwaters remained stable, though the caisson moved slightly. These results indicated that the more the number of piled stones, the better the stability of the breakwater. Additionally, the comparison of cases H1.0n and H1.0w showed that not only the height of the piled stones but the width contributed to the stability.

Figure 7 shows the sequential images, which help in understanding the failure behaviour in case H1.0n. The mound and piled stones could not sustain the caisson pushed by the water pressure on the side surface, and the caisson dramatically inclined and moved landwards. The slip surface was observed from a point under the caisson to the deeper area of the mound when the water level reached the peak. A line in the figure indicates the slip surface, which was visually determined using the images. A broken line shows the slip surface generated when the caisson considerably moved. The type of failure observed in this case was the typical bearing capacity failure. However, in case H1.5w, shown in Figure 8, the caisson hardly moved even when the water-level difference was at its peak. Cases H3.0w, H1.0w, and H1.5w exhibited similar behaviour. It should be noted that the predominant failure mode depends on the conditions of sectional shape and ground. The present breakwaters failed in the bearing capacity failure mode.

The images were analysed to examine the failure behaviours of cases H1.0n and H0 in more detail. The particle tracking velocimetry (PTV) method was used for the analysis of the sequence images captured in the tests. A real particle such as a stone was not tracked by the PTV used in this study, whereas the centre of balance of a virtual particle generated was tracked by binarising the image. The tracking algorithm itself corresponded to the particle image velocimetry (PIV) method. Although the number of sampling points was less, a precise displacement could be obtained using this PTV method, compared with the general PIV method. Figure 9 shows the results of the image analyses. A curved slip surface, represented using broken lines, could be observed in the mound and piled stones in both cases. Although the slip surface was not perfectly a circular arc, the limit equilibrium method such

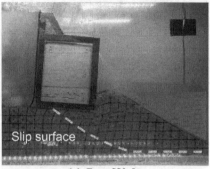

(a) Case H1.0n

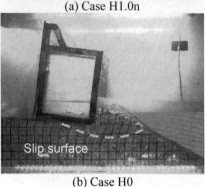

(b) Case H0

Figure 9. Tracks of particles.

as a slip analysis could be employed to assess the stability. Another interesting point was that the broken line of case H1.0n was deeper than that of case H0. The number of piled stones in case H1.0n was limited; however, the test results showed that even a small number of stones could change the failure characteristic.

3 STABILITY ASSESSMENT

3.1 Assessment method

The stability assessment method is summarised in this section with regard to sliding and overturning of the caisson and bearing capacity failure. Figure 10 illustrates the concepts of sliding and bearing capacity failure modes, which are described in the following subsections. Takahashi et al. (2015) and Sato et al. (2017) described the stability assessment method in detail. Table 1 lists the parameters used in the circular slip analyses. The values in the table were based on the data in the model tests. The horizontal force per unit depth was 612 kN/m in case H3.0w. An average horizontal force of 553 kN/m was used in other cases.

– *Assessment of sliding:* In the sliding mode, the following two successive slips were assumed: one under a caisson and a circular one beginning at the endpoint of the caisson. The stability of sliding was assessed by calculating the ratio of action and resistance forces along the slip surface. A circular slip surface could be determined by obtaining

Table 1. Parameters used in circular slip analyses.

	γ kN/m³	c kN/m²	ϕ	Friction coeff	Side face friction angle
Mound	10.8	20	35°	0.45	15°
Piled stones	10.8	20	35°	0.45	15°

Note: the load used in H3.0w was 612 kN/m whereas 553 kN/m was used in all other cases.

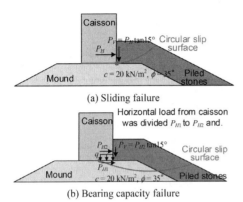

Figure 10. Sliding and bearing capacity failure modes.

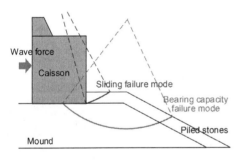

Figure 11. Slip surfaces (Case H1.5w).

the least stability. The simplified Bishop's method was employed in the circular slip analyses, wherein the shear strength of the stones was assumed to be $c = 20$ kN/m² and $\phi = 35°$. These values were obtained from the Japanese design standard (OCDI, 2009b). Moreover, Takahashi et al. (2017) confirmed that these values could be employed in the circular slip analysis to successfully simulate the horizontal loading test results as mentioned previously.

– *Assessment of overturning:* The stability of overturning of the caisson was determined via the moment equilibrium of forces acting on the caisson, including the self-weight and wave force. However, the reactive force exerted by the piled stones was excluded in the equation. This was because the point of action was low and the reactive force would be small because of the low strain when the caisson was overturned.

– *Assessment of bearing capacity failure:* The bearing capacity failure was considered to be induced by the circular slip surface that had the least stability. The circular slip surface started at the point on the bottom surface of the caisson. The stability of the bearing capacity failure was assessed using the stability calculated by performing the circular slip analysis with the help of the simplified Bishop's method. The starting point of the circular slip was obtained using the equilibrium of moments acting on the caisson. The shear strength of the stones was the same as that used in the assessment of sliding. Additionally, it should be noted that neither the friction force acting on the bottom surface of the caisson nor the reactive force exerted by the piled stones would reach the limit value. In that case, an adequate ratio of friction and reactive forces must be adopted in the analysis.

3.2 Results of assessment

The stability with respect to the three failure modes mentioned in the previous section was assessed for the centrifuge test cases. Figure 11 shows an example of the slip surfaces in the sliding and bearing capacity failure modes. These were the surfaces based on the simulation of case H1.5w. The slip surface comprised one under the caisson and circular one in the piled stones. As shown in the figure, the circular slip surface behind the caisson went through to the top surface of the piled stones. However, the slip surface will go through to the slope surface of the mound when increasing the height or narrowing the width of the piled stones behind the caisson. The circular slip surface due to the bearing capacity failure went through to the slope surface.

Table 2 lists the ratios of action and resistance forces for the three failure modes. The table includes test results confirming whether the breakwaters failed. With regard to the sliding failure, the ratio is divided into two ratios on the slips: one under a caisson and circular one in the piled stones. As presented in the table, the ratios on sliding and overturning were much higher than those of bearing capacity failure, indicating that the stability on those modes were high. It corresponded to the results of the model tests representing no sliding and overturning. On the other hand, the ratios of the bearing capacity failure of cases H1.0n and H0 were largely equal to 1.0, which were lower than those of the other cases. This implies that the stability on the bearing capacity failure mode in those cases were relatively low. In the model tests, only those cases failed in the bearing capacity failure mode. The stability assessment method was largely accurate.

3.3 Considering effect of seepage

A tsunami-induced water-level difference induced between in front and behind the caisson leads to a seepage force acting on the mound and piled stones. Takahashi et al. (2014b) verified that the seepage force

Table 2. Ratios action and resistance forces for three failure modes.

Case		H3.0w	H1.5w	H1.0w	H1.0n	H0
Sizes of piled stones	Height	3.0 m	1.5 m	1.0 m	1.0 m	–
	Top width	3.0 m	6.0 m	7.0 m	3.5 m	–
Ratio of action and resistance						
Sliding	Friction force	0.91	1.10	1.11	1.11	1.13
	Reactive force	0.50	0.33	0.20	0.20	–
	Total	1.41	1.42	1.31	1.31	1.13
Overturning		1.29	1.47	1.47	1.47	1.47
Bearing capacity		1.21 (1.02)	1.29 (1.08)	1.26 (1.06)	1.09 (0.91)	1.06 (0.89)
Centrifuge model test result		Stable	Stable	Stable	Failure	Failure

could become a dragging force for ground failure. Moreover, it could help in reducing the confining pressure and shear strength of the stones. In the model tests, a water-level difference was reproduced; this indicated that the seepage force acted on the stones. In contrast, the ratios of action and resistance forces described in the previous section did not include the effect of the seepage force. Thus, in this section, the reduction in the stability due to the seepage force is discussed.

The experimental results obtained by Takahashi et al. (2014b, 2015) showed that the effect of the seepage force could be expected to be approximately 10 and 20% at most under the water-level differences of 5 and 10 m, respectively. Thus, a reduction of 16% on the bearing capacity was assumed, considering the water-level difference of approximately 8 m induced in the model tests. The ratios of action and resistance forces were converted to the values shown in parentheses in Table 2. Consequently, the ratios in the cases H1.0n and H0 became less than 1.0, whereas those of the other cases were still more than 1.0. This indicated that only cases H1.0n and H0 failed in the bearing capacity failure mode. It was the same result as the model tests; it could be said that the accuracy of the stability assessment method was verified.

4 CONCLUSIONS

In this study, the movement of breakwaters under the condition of a water-level difference was observed to change the sectional shape of piled stones behind a caisson. Consequently, a curved slip surface was observed in the cases with none and insufficient amount of piled stones. This indicated that the limit equilibrium method such as a slip analysis could be employed to assess the stability. The stabilities of the three failure modes including sliding, overturning, and bearing capacity failure, were assessed for the centrifuge test cases. The ratios of action and resistance forces on sliding and overturning were much higher than that of the bearing capacity failure, and those of cases with none and insufficient amount of piled stones were largely equal to 1.0. Considering the effect of the seepage force on the stability of the bearing capacity failure, the ratios became less than 1.0. The analyses and model tests were in good agreement. Thus, it was verified that the stability assessment method of the rubble mound and piled stones could be used to accurately simulate the model tests.

REFERENCES

Arikawa, T., Sato, M., Shimosako, K., Tomita, T., Yeom, G.S. & Niwa, T. 2013. Effects of the back-filling to the stability of a caisson. *Technical Note of the Port and Airport Research Institute* (1269): 37. (in Japanese)

Kikuchi, Y., Shinsha, H. & Eguchi, S. 1998. Effects of the back-filling to the stability of a caisson. *Report of Port and Harbour Research Institute* 37(2): 29–58. (in Japanese)

Kitazume, M. & Miyajima, S. 1995. Development of PHRI Mark II geotechnical centrifuge. *Technical Note of the Port and Harbour Research Institute* (812): 35.

OCDI. 2009a. *Technical Standards and Commentaries for Port and Harbour Facilities in Japan*. OCDI: 606–607.

OCDI. 2009b. *Technical Standards and Commentaries for Port and Harbour Facilities in Japan*. OCDI: 431–432.

Sato, T., Miyata, M., Takahashi, H., Takenobu, M., Shimosako, K. & Suzuki, K. 2017. A Proposal on design method for wave force of breakwater with reinforcing embankment. *Technical Note of National Institute for Land and Infrastructure Management* (954): 47. (in Japanese)

Shinsha, H., Unno, T., Kikuchi, Y. & Morikawa, Y. 2014. Evaluation of the stability of a caisson with back-filling on sandy ground. *Japanese Geotechnical Journal* 9(2): 103–117. (in Japanese)

Takahashi, H., Sassa, S., Morikawa & Maruyama, K. 2017. Load bearing capacity of reinforcing embankment behind breakwaters under caisson sliding mode. *Proceeding of Annual Conference of Japanese Geotechnical Society*: 1017–1018. (in Japanese)

Takahashi, H., Sassa, S., Morikawa, Y., Takano, D., Aoki, R. & Maruyama, K. 2014a. Centrifuge tests on resistance force of breakwaters reinforced by embankment. *Journal of Japan Society of Civil Engineers, Ser. B3 (Ocean Engineering)* 70(2): 870–875. (in Japanese)

Takahashi, H., Sassa, S., Morikawa, Y., Takano, D. & Maruyama, K. 2014b. Stability of caisson-type breakwater foundation under tsunami-induced seepage. *Soils and Foundations* 54(4): 789–805.

Takahashi, H., Sassa, S., Morikawa, Y., Watabe, Y. & Takano, D. 2015. Stability of caisson-type breakwater's mound and reinforcing embankment against tsunami. *Report of the Port and Airport Research Institute* 54(2): 21–50. (in Japanese)

10. Offshore – shallow foundations

Centrifuge tests on the influence of vacuum on wave impact on a caisson

D.A. de Lange & A. Bezuijen
Deltares, Delft, The Netherlands

T. Tobita
Disaster Prevention Research Institute, Kyoto University, Japan

ABSTRACT: Combined failure mechanisms resulting in total failure of a breakwater during a tsunami wave impact were investigated through geotechnical centrifuge modelling in DPRI. In these tests, a large amount of gas bubbles was visible during the simulation of a tsunami wave. Therefore, questions remain over the test results. It was proposed to perform the same tests under vacuum conditions in the geotechnical centrifuge of Deltares in order to investigate the influence of air intrusions on the tests results. The underlying idea was that no or a negligible amount of air intrusions will be created when the wave impact is simulated under vacuum conditions as can be created in the Deltares centrifuge. The tests under vacuum however, showed no significant differences with the tests performed under normal air pressure conditions. It is thought that vapour bubbles are created by cavitation. The paper will describe the test set-up and a comparison between the tests performed by DPRI and Deltares.

1 INTRODUCTION

The great Tohoku, Japan, earthquake (Mw 9.0) of March 11, 2011 struck Japan between Tohoku (in the northeast) and the Kanto Plain. The rupture area was estimated to be approximately 450 km × 200 km off the Pacific coast of Tohoku, Japan, and generated a tsunami that struck the area from Hokkaido to Kyushu. The major cause of death was direct action of the tsunami with most fatalities occurring in the Tohoku region. Along the east-coast of the Tohoku district, the tsunami took 15,883 lives with 2,761 still missing as of June 6, 2013 (National Police Agency 2013). The number of totally and partially damaged buildings was 128,530 and 230,332 respectively, and more than 200 bridges collapsed. The estimated cost of direct damages was ¥16.9 trillion (US$211 billion). A large number of geotechnical structures were damaged, including embankments and dams, and large ground deformations due to liquefaction were observed (Tobita & Iai 2014).

One of the destroyed structures was the breakwater in Kamaishi city. The failure was caused by the wave impact and the loss of bearing capacity of the foundation. The combined failure mechanisms were investigated through geotechnical centrifuge modelling at the Disaster Prevention Research Institute, Kyoto University, Japan (Tobita & Iai 2014). However, large amounts of gas bubbles were visible in the water during the simulation of a tsunami wave. Therefore, questions were raised over the results of the centrifuge tests. "Among the others, aeration is generally acknowledged to be one of the most important causes of variability in wave impact maxima, either in the form of trapped pockets, expelled air or trapped bubbles." (Cuomo et al. 2011). It was proposed to perform the same tests under vacuum conditions in the geotechnical centrifuge of Deltares (the Netherlands) in order to investigate the influence of air intrusions in the water on the tests results (impact forces on the breakwater). The underlying idea was that no or a negligible amount of air intrusions will arise when the wave impact is simulated under vacuum conditions (air pressure close to zero). Moutzouris (1979) states that ambient pressure influences considerably the shock pressure: small air bubbles in the water jet front have a cushion effect on the shock pressure which is higher under vacuum based on experiments under normal atmospheric conditions and near vacuum conditions. Further, he mentions that the duration of existence of the air pocket is much higher under atmospheric pressure than under vacuum conditions. Furthermore, fewer bubbles were created under vacuum conditions.

2 TEST SET-UP

Centrifuge tests were conducted using a 1/200 scaled model under 25 g by applying the generalized scaling law ($\mu = 8$ and $\eta = 25$) (Iai et al. 2005). A strongbox with water tank and movable gate was installed and used to generate a tsunami in the scaled model, see Figure 1. As shown in Figure 1, the moveable gate, which was a small vertical plate that closed the water entry hole located at the bottom of the water tank, could be opened by a signal sent from the control PC to allow water to be released from the tank. A metal sheet with 3 mm punched holes was attached at the gate to control

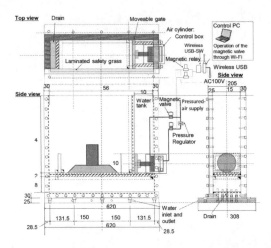
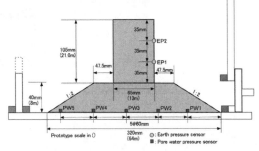

Figure 2. Cross-section of the model (model dimensions).

Figure 1. Schematic view of test set-up (Tobita & Iai 2014), that was also used in the tests in the Geo-Centrifuge at Deltares (the wireless USB connection was replaced by a wired connection for the latter tests).

the water flow. An accelerometer was installed at the moveable gate in order to register at which moment the gate has been opened. Once the water was released from the tank, it flowed into the strongbox in which the model structure was placed; it was then captured in the drainage tank attached at the base of the box to prevent a reflected wave from being generated by the opposite wall of the container. Different configurations (combination of initial water depth and water level difference) could be tested. The tests performed at Kyoto are described by Tobita & Iai (2014). The tests performed in Delft used the same set-up as the Kyoto tests. The model container and model breakwater were transported from Kyoto to Delft. Some minor adaptions were necessary to use the set-up in the Geo-Centrifuge. The wireless USB connection was replaced by a wired connection.

Model dimensions, sensor locations and a photograph of the model caisson are shown in Figure 2. The instrumentation was almost the same in both the Kyoto and in the Delft tests. Two pore water pressure transducers were installed at the base of the model to measure water height (only PW1 and PW5) and two pressure transducers were placed on the caisson (without filter stones) in order to measure the wave impact (EP1 and EP2). The pressure sensors used appeared not to be reliable under vacuum conditions. Therefore, the transducers on the caisson breakwater were changed. In order to be able to place these transducers, some minor adjustments had to be made to the model caisson. Within the limited time available for these tests, it was not possible to change the pore pressure transducers in the gravel berm too. Images were captured with a Go-Pro camera at 120 images per second in front of the model and on top of the model.

The caisson breakwater was modelled using an aluminium block with the dimensions shown in Figure 2 and a model weight of 2475 gr. A raster was drawn on the sidewall of the block next to the glass wall of the

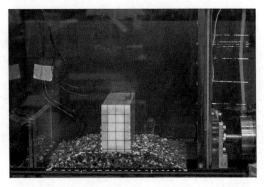

Figure 3. Photograph of model caisson and the rubble mound in the strongbox.

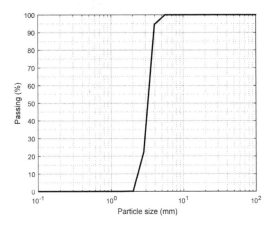

Figure 4. Grain size distribution of No. 1 silica.

model container, see Figure 3. The dimensions of this raster are 2 cm by 2 cm.

The material used in the rubble mound was No. 1 silica ($d_{50} = 3$ mm in model scale). The grain size distribution is given in Figure 4. In reality, the diameter of the rubble mound was 200 mm to 800 mm, which corresponds to 1 mm to 4 mm in model scale (=1/200 in this study). For turbulent flow the permeability scales with d. The rubble mound was built up having a relative density of 55.5 % ($n_{min} = 32.3\%$ and $n_{max} = 41.5\%$).

In order to be able to compare the tests performed in Japan with the tests performed in the Netherlands,

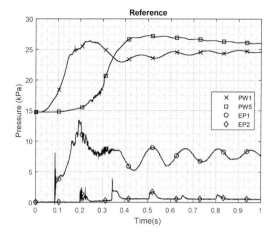

Figure 5. Pressure transducer results after opening of the gate.

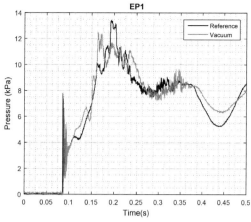

Figure 7. Reading of wave pressure on container (EP1) after opening of the gate.

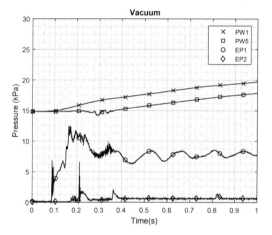

Figure 6. Pressure transducer results after opening of the gate (PW 1 and PW5 are considered to be not reliable under near vacuum conditions).

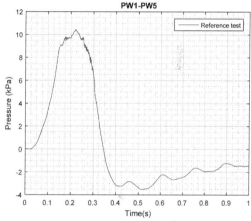

Figure 8. Difference in water pressure before and behind the caisson after opening of the gate.

one test configuration (combination of initial water heights and rubble mound material) that has been applied in Japan was repeated under normal atmospheric pressure in Delft. The same test configuration is also tested under near vacuum conditions (7.2 kPa instead of 100 kPa air pressure). De-aired water has been applied for both tests.

3 TEST RESULTS

The results of two tests are presented in this paper: a reference test (under atmospheric conditions) and a test under near vacuum conditions. The test configuration was the same for both tests: an initial water depth of 62.5 mm and a water level difference of 144 mm have been applied (the initial water level in the storage tank was equal to 206.5 mm).

The construction failed in both tests as result of the generated wave. Figures 5–8 give the results of the pressure transducers on the caisson (EP1 and EP2) and in the rubble mound (PW1 and PW5) as a function of time. At $t = 0$ s the first movement of the gate is observed. The results are presented in test scale. In both tests a pressure drop over of the model caisson was observed. PW1 and PW5 are considered not reliable under near vacuum conditions (see Fig. 6). Therefore, only the pressure difference between PW1 and PW5 of the reference test are given in Figure 8. As can be concluded from Figure 7, no significant differences in wave loading between the test under vacuum conditions and under normal atmospheric conditions are observed. The results are very similar.

As can be observed in Figure 9, which show the collapse sequence of the model caisson, white bubbles appeared in front of the caisson during both tests. These images were taken by the GoPro camera. The time in between the images is equal to 83.3 ms in this figure. The tsunami height was not sufficient to cross over the caisson. The model caisson may be rotated

Figure 9a. Reference test: Collapse sequences of the model caisson with time in between the images is equal to 83.3 ms.

Figure 9b. Vacuum test: Collapse sequences of the model caisson with time in between the images is equal to 83.3 ms.

not only by the thrust force from the tsunami, but also deformation of the rubble mound which is also loaded by the horizontal hydraulic gradient induced by the wave and has a reduced shear strength due to a decrease in the effective confining pressure. The breakwater is not damaged by the first pressure peak that hits the breakwater, but more by the massive amount of water that creates a water level difference.

Although de-aired water was applied for the tests, air bubbles did escape from the water during lowering the ambient pressure. The time in between the moment that vacuum conditions were created and the valve was opened was not sufficient to get rid of all these bubbles. It also has been observed that more time after the wave impact was needed to get rid of the white bubbles under near vacuum conditions. Therefore, it has been concluded that more bubbles have been created under near vacuum conditions.

4 DISCUSSION

Under near vacuum conditions, bubbles were visible and no differences in wave impact load were measured compared with the reference test. It is thought that the air bubbles are a result of cavitation. The cavitation number depends on the ambient pressure:

$$Ca = \frac{P_{atm} - P_{vapour}}{\frac{1}{2}\rho v^2} \quad (1)$$

Since it has been observed that more time was needed under near vacuum conditions to get rid of the "white water", it has been concluded that more gas intrusions has been created relative to the reference tests under atmospheric pressure. Therefore, it is assumed that the "white water" is a result of the manner of creating the tsunami wave, which triggers cavitation.

A cavitation bubble looks like an air bubble, but is something different, because it contains no air, just vapour. On impact these cavitation bubbles have to disappear. The fact that the impact pressures looked the same with and without vacuum could mean that the influence of air or vapor is only limited.

5 CONCLUSIONS

Vacuum conditions had no significant influence on the impact forces on the caisson during the simulation of a tsunami wave compared with the reference test (atmospheric pressure). In both tests, large amounts of air/vapour bubbles were clearly visible during the simulation of the wave impact on the breakwater. It is thought that cavitation takes place during the acceleration of the water mass (shape of the test set-up, manner of simulating a tsunami wave, obstacles) and this will happen even easier under vacuum conditions. It is concluded that creating vacuum conditions is not a solution in order to reduce the amount of air/vapour intrusions. Other measures have to be taken. However, the nature of the bubbles must be different under atmospheric or near vacuum conditions. Yet the loading and deformation in both tests are quite comparable, this indicates that the influence of the bubbles may be only limited. This is confirmed by the damage pattern observed. The breakwater is not damaged by the first pressure peak that hits the breakwater, which may be influenced by the amount of bubbles, but more by the massive amount of water that creates a water level difference.

ACKNOWLEDGEMENTS

The authors gratefully acknowledge the technicians Frans Kop, Thijs van Dijk, Ferry Schenkeveld, John Langstraat and Rob Zwaan from Deltares for their efforts to perform to the tests successfully.

REFERENCES

Cuomo G., Piscopia R. & Allsop W. 2011. Evaluation of wave impact loads on caisson breakwaters based on joint probability of impact maxima and rise times. *Coastal Engineering* 58 (2011) 9–27.

Iai, S., Tobita, T. & Nakahara, T. 2005. Generalized scaling relations for dynamic centrifuge tests. *Géotechnique* 55(5): 355–362.

Moutzouris C. 1979. *Influence of ambient air pressure on impact pressure caused by breaking waves*. Delft University of Technology, Department of Civil Engineering, Fluid Mechanics Group, Internal report no. 10–79.

Tobita T. & Iai S. 2014. Combined failure mechanisms of geotechnical structures. In C. Gaudin & D. White (Eds.), *8th International Conference on Physical modelling in Geotechnics, Proceedings* (Vol. 1, pp. 99–111). Presented at the 8th International Conference on Physical modelling in Geotechnics, Leiden, The Netherlands: CRC Press - Taylor and Francis Group.

Physical modelling of active suction for offshore renewables

N. Fiumana, C. Gaudin, Y. Tian & C.D. O'Loughlin
Centre for Offshore Foundation Systems, The University of Western Australia, Perth, Australia

ABSTRACT: Active suction is investigated as a strategy to temporarily increase the tensile capacity of bucket foundation for offshore floating renewables, which are subjected to high extreme loads. A centrifuge testing campaign was carried out in the new 10 m diameter beam centrifuge at UWA, using a purpose built suction application device. Preliminary results are presented, demonstrating the performance of the testing setup and the potential of the active suction concept to increase tensile capacity by up to 70%.

1 INTRODUCTION

Floating renewable energy devices, such as wave energy converters and floating wind turbines, experience very high extreme loads during storm events. Designing anchoring systems to resist these extreme loads appears to be extremely costly and inefficient, especially in sandy sea beds, and strategies to avoid or reduce extreme loads are sought to significantly reduce the size (and cost) of anchoring systems (Gaudin et al. 2017).

One such strategy is the application of active suction on bucket foundations (also called suction caissons). The concept of active suction consists of pumping water from the inside of the caisson to actively apply a differential pressure across the lid. The additional resistance due to this differential pressure increases the tensile capacity beyond that mobilised by friction at the soil-skirt interface (under drained loading).

From a design point of view, this potentially results in a foundation design that relies on friction at soil-skirt interface to withstand operational loads and on the temporary additional tensile capacity generated from active suction, when extreme loads are expected (Figure 1).

The increase in tensile capacity through active suction was first investigated in the 1970s through laboratory tests (e.g. Helfrich et al. 1976, Wang et al. 1975), as a means to provide a more efficient short term (2–3 days) anchoring solution for deep water buoys, coring platforms, or submersibles. More recently, a centrifuge testing campaign was conducted to investigate active suction applied to caisson foundations for the stabilisation of a vessel equipped with cranes (Allersma et al. 2003). In each case, a linear increase of the tensile capacity with differential pressure applied was observed. However, no insights on the generated physical mechanism, soil conditions and flow regime obtained with active suction are available in the literature and significant uncertainties still govern the quantification of the capacity increase.

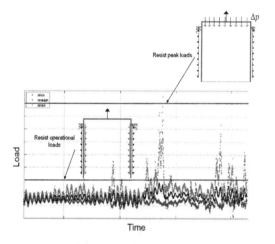

Figure 1. Load history of one mooring line of a WEC (after Herduin 2014). The active suction concept is represented as a means to resist the extreme peak loads, which can be almost three times larger than the operational loads.

The increase in tensile capacity generated by active suction is expected to be function of the differential pressure generated, the seepage regime within and around the caisson and its effect on the soil effective stresses (Fiumana et al. 2017).

To further investigate the key mechanisms governing active suction, an experimental campaign was carried out in the new 10 m diameter centrifuge at the University of Western Australia (UWA).

A new suction generation device was designed for this study, whereby suction was applied by lowering a water tank hydraulically connected to the inside of the caisson, hence creating a constant differential head pressure at a variable flow rate. This system differs from the syringe pump system used in UWA's 3.6 m diameter centrifuge for which a constant flow rate is generated at a varying suction pressure (House 2002). The constant differential head pressure is considered more appropriate for the investigation of active suction as the increase in uplift capacity can be directly associated to the constant differential pressure applied.

The testing programme presented in this paper was aimed at (i) quantifying the increase in foundation uplift capacity with respect to the differential pressure applied and the associated seepage regime in the soil, (ii) evaluating the time over which active suction can be sustained, and (iii) understanding the influence of the seepage regime on the soil behaviour, notably with respect to potential plug liquefaction and cavitation that may hamper the efficiency of active suction.

The paper presents the testing setup, the experimental procedure and some preliminary results that provide insights into the feasibility of active suction to anchor floating renewable energy devices.

2 EXPERIMENTAL SET UP

All the tests were carried out at a level of 40 g in water-saturated silica sand.

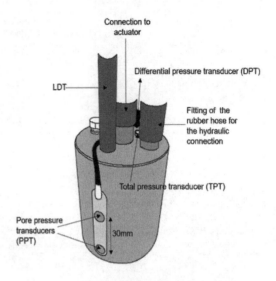

Figure 2. Scheme of the model caisson and instrumentation.

2.1 Sample preparation

The soil sample was contained in a strong box with a square base of 1000×1000 mm and a height of 500 mm.

The tests were conducted in UWA silica sand, which has been used extensively in centrifuge modelling at UWA. The properties of the sand are reported widely, e.g. Senders (2008) and Tran (2005). The sample was prepared by pluviating the silica sand over a 50 mm thick drainage layer of coarse sand covered by a layer of geotextile. The sample was subsequently levelled by vacuuming the excess sand from the surface in order to achieve a constant sample height of 200 mm. The sample was then saturated with water from the base of the strongbox, until the head of water reached 100 mm above the sand surface. The reconstitution process resulted in an effective unit weight for the sand of $\gamma' = 10.5$ kN/m^3 and a relative density $D_r = 50\%$.

Soil characterisation was performed in-flight using a miniature cone penetrometer, 10 mm in diameter. Results indicated a linear increase of the penetration resistance, with a value of $q_c/\sigma'_{v0} = 150$, where q_c is the tip resistance and σ'_{v0} is the initial vertical effective stress in the soil at a depth of 70 mm (corresponding to the tip of the installed caisson). Tests performed in multiple locations in the sample showed very little variability, indicating good sample homogeneity.

2.2 Foundation model and instrumentation

The foundation model was manufactured from aluminium, with a diameter D equal to the skirt length L of 70 mm (aspect ratio $L/D = 1$), and a top plate 10 mm thick. A schematic of the model caisson and the instrumentation is presented in Figure 2.

The wall thickness, w_t, of the skirt is 3.5 mm ($t/D = 0.05$). This value is significantly larger than typical values for field scale caissons, but was necessary to accommodate pore pressure transducers on the inside and outside face of the skirt. As discussed later in the paper, the caisson was installed at 1g, modelling a wished-in place installation, so the thickness of the skirt does not interfere with the foundation uplift capacity as (i) soil disturbance due to penetration is limited and (ii) the inverse bearing capacity potentially mobilised at the skirt tip is a negligible component of the whole capacity, which is mainly developed from the friction mobilised along the skirt (e.g. Bye et al. 1995, Feld 2001, Houlsby & Byrne 2005a).

A differential pressure transducers (DPT) was fitted on the caisson lid to obtain a direct measure of the differential pressure generated by the active suction. The lid was also instrumented with an absolute total pressure transducer (TPT) that was used to measure the initial level of the water table, and to control it during the whole testing procedure. Both sensors have a capacity of 700 kPa.

Along the inside and the outside of the caisson skirt, two pairs of pore pressure transducers (PPTs) with a capacity of 500kPa, were installed, as shown in Figure 2. The top sensors (PPT_OT (outside) and PPT_IT (inside)) were placed 40 mm from the skirt tip and the bottom sensors (PPT_OB and PPT_IB) were located 10 mm from the tip. All the transducers were calibrated by placing the caisson in a pressure chamber and gradually increasing the ambient pressure to 250 kPa.

The caisson lid was fabricated with three fittings to allow connection to the actuator, the hydraulic connection for the suction application and a linear differential transducer (LDT). The LDT was used to obtain information on the plug heave during the 1g installation phase and during active suction uplift.

The actuator adopted for the caisson extraction was instrumented with a 2 kN axial load cell. The hydraulic connection for the suction application was a rubber tube 10 mm in diameter.

2.3 Suction application device

Suction installation of bucket foundations in centrifuge tests at UWA may be achieved using two modelling techniques as described below.

A syringe pump can be used, which applies suction by moving a cylindrical piston 50 mm in diameter at a maximum velocity of 3 mm/s. This system is able to generate a maximum flow rate of 5840 mm^3/s for a total volume of 384,945 mm^3 when completing a full stroke of 196 mm (Senders 2008). Although simple to implement and closer to actual prototype conditions, this system has three shortcomings:

1. The fixed volume of the syringe pump limits the duration of suction that can be generated,
2. A constant suction cannot be applied, unless the velocity of the syringe is linked to the suction value via a feedback loop and,
3. The maximum flow rate is not sufficient to generate suction in high permeability sand, resulting in the need to use a saturation fluid more viscous than water (such as silicon oil) to compensate (Senders, 2008).

To enable a better control of the suction applied as required by the present investigation, a gravity flow system was developed. Suction application by gravity flow was successfully implemented in 1g experiments by Tran (2005), by creating a head difference between the water level in the strongbox and the outlet of the hydraulic connection attached to the caisson. A measure of the flow rate was made by measuring the volume of water pumped, inferred from the change (over time) of measurements from pressure transducers placed on the bottom of a series of interconnected reservoirs collecting the water vacuumed from the inside of the caisson.

Taking advantage of the large dimensions of the strongbox, a gravity flow setup was developed for this project. The setup was designed to (i) enable a better control of the magnitude of the differential pressure applied across the lid of the caisson, (ii) sustain the differential pressure applied for a long period of time, and (iii) measure the flow rate generated.

The setup consists of a water tank with a volume of 5000 mm^3 connected to the inside of the caisson by a 10 mm diameter reinforced rubber tube and assembled on a linear actuator fitted on the side of the strongbox. By ensuring a saturated hydraulic connection with the caisson, equilibrium of the system is achieved when the water-filled tank is levelled with the water table in the sample. By lowering the tank to a targeted elevation, a constant head difference is created and a flow from the inside of the caisson to the water tank generated. By lowering the water tank continuously at a constant velocity, an increasing differential pressure can be generated, which can be useful to model suction installation. By bringing the tank above the water table in the centrifuge sample, the seepage can be reversed to generate reverse pumping and extraction of the caisson.

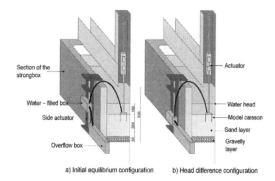

Figure 3. Schematic representation of the suction application device.

The maximum allowable differential pressure Δp is dependent on the acceleration level N:

$$\Delta p = \Delta h \cdot N g \cdot \gamma_w \qquad (1)$$

where Δh is the maximum head difference that can be generated by adjustment of the water tank height using the linear actuator (200 mm) and γ_w is the unit weight of water. At 40 g, this results in a maximum head pressure of 80 kPa.

In order for the differential pressure to be generated, the flow created by the head pressure difference must be higher than the seepage flow established in the soil under the differential pressure created across the lid (Fiumana et al. 2017). In sand, very high seepage is expected, and the setup must feature an appropriate hydraulic connection enabling a high flow rate between the caisson and the water tank. In the present case, a 10 mm rubber tube was used following some preliminary tests that showed that no suction was generated with lower diameter tubes.

The water flowing from the caisson to the water tank is allowed to overflow, through a hose of equal diameter, into a second tank placed at the base of the actuator and instrumented with a pressure transducer. The measure of pressure with time in the overflow tank provides an estimation of the flow rate generated by the active suction. The overflow tank has a length of 250 mm, a height of 100 mm and a width of 50 mm, such that the maximum volume capacity is 1250000 mm^3.

A schematic representation of the gravity flow system is presented in Figure 3.

2.4 Experimental procedure and programme

Considering the significant skirt thickness, installation was performed by simply penetrating the caisson using the electrical actuator at 1 g until the desired skirt penetration was reached. A gap of approximately 10 mm was left between the caisson lid and the soil surface, as informed by the LDT measurements. The purpose of this gap was to ensure a uniform differential pressure distribution when the active suction was applied.

Table 1. Testing programme and parameters.

Test name	Extraction type	Uplift velocity mm/s	Elevation below the water table mm	Active suction generated kPa
D	Drained	0.1	–	–
PU	Partially undrained	10	–	–
AS1	Active suction	0.1	100	28
AS2	Active suction	0.1	75	18

The centrifuge was subsequently spun to 40 g prior to the caisson uplift tests. No movement of the LDT was detected during this phase, suggesting that no further settlement of the caisson occurred while increasing the g-level. For uplift tests without active suction, which were performed as baseline reference cases, the caisson was pulled out at constant velocity. For tests where active suction was applied, tests were performed in two stages. Firstly a targeted active suction was applied by lowering the water tank to the required elevation. Secondly, after a steady flow regime was established and constant differential pressure across the caisson lid achieved, the caisson was uplifted at a constant velocity. Approximately 4 s was needed for the pore pressures to stabilise.

Four tests from a comprehensive testing program are reported in this paper. The first two tests, D and PU, were undertaken without active suction at an uplift velocity of 0.1 and 10 mm/s respectively (covering three orders of magnitude of uplift velocity), aiming to replicate drained and partially undrained conditions. In the former, the caisson was fully vented, such that the uplift resistance is solely generated by the internal and external friction along the skirt. The 0.1 mm/s uplift rate resulted in drained extraction. In the latter (10 mm/s uplift rate), the caisson was sealed, such that some passive suction was generated at the lid invert, resulting in partially undrained conditions. It is acknowledged that fully undrained conditions could not be achieved with the current setup considering the high permeability of the sand and the limited maximum extraction velocity.

The remaining two tests featured active suction (AS1 and AS2), with varying magnitude of suction applied and two different uplift velocities, as presented in Table 1.

3 RESULTS AND DISCUSSION

3.1 Tests without active suction

Results obtained for tests performed without active suction are shown in Figure 4. The extraction resistance V is reported in terms of the pressure V/A as a function of the extraction depth z. Note that $p = 0$ mm indicates full extraction of the caisson (i.e. the tip of the skirt is at the soil surface).

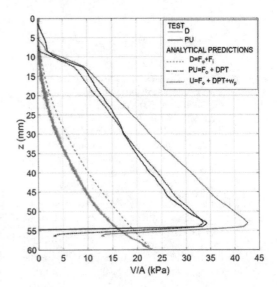

Figure 4. Extraction capacity for tests without active suction.

As evident from Figure 4, the drained uplift capacity of the caisson (Test D at $v = 0.1$ mm/s, Table 1) is correctly described by the sum of the friction mobilised along the inside and outside of the skirt. The frictional terms were computed as (Bye et al. 1995, Feld 2001, Houlsby & Byrne 2005a):

$$F_{i(o)} = \gamma' \cdot (z^2/2) \cdot (K \tan\delta) \pi D_{i(o)} \quad (2)$$

The term $K \tan \delta$ is the product of the earth pressure coefficient K and the interface friction angle, δ. A value of $K \tan \delta = 0.28$ was adopted, which is considered appropriate for the smooth stainless steel caisson skirt.

The response of the partially undrained tests undertaken at an uplift velocity $v = 10$ mm/s (Test PU, Table 1), is also reported in Figure 4. Results show a 70% increase in maximum uplift resistance, which is associated with the generation of negative excess pore pressure during extraction at the lid invert (Figure 5). A reasonable approximation of the partially undraiend uplift resistance is obtained by adding the excessive negative pore pressure measured druing the test to the external frictional resistance, computed from (2) (Figure 4). This is in agreement with existing solutions for the tensile capacity of caisson foundations under rapid loads presented by Houlsby et al. (2005). No contributions to the capacity arises from the plug weight, indicating that fully undrained conditions couldn't be achieved at this uplift velocity.

The theoretical undrained response is presented for comparison in Figure 4, computing the plug weight w_p as:

$$w_p = \gamma' \pi (D_i/2)^2 z \quad (3)$$

The plug weight adds to the external friction F_o and excessive negative pore pressure generated during

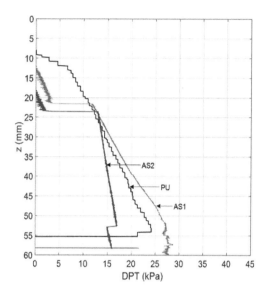

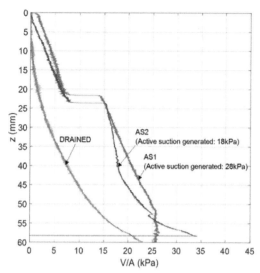

Figure 5. Differential pressures generated by active suction and compared with excess negative pore pressure generated from rapid uplift.

Figure 6. Extraction capacity for tests with active suction.

extraction. In case of a full reverse end bearing mechanism, with a gap forming at the base of the soil plug, only partial seepage might occur, with the excess negative pore pressure tranferred to the base (e.g. Senders 2008, Fiumana et al. 2017). This would result in lower magnitudes of differential pressure across the caisson lid with respect to partially undrained conditions. However, the differential pressure recorded during the test PU was considered for this prediction, giving an upper bound solution. In summary, Figure 4 indicates that the fully undrained capacity is about twice the fully drained capacity (42 kPa and 23 kPa, respectively), and the partially undrained capacity mobilised at $v = 10$ mm/s is about 70% higher than the fully drained capacity.

3.2 Tests with active suction

Tests performed with active suction, AS1 and AS2, also showed a capacity increase with respect to the drained capacity (Figure 6). The different target elevation assigned to the water tank prior to extraction, resulted in a different value of differential pressure generated (Figure 5), affecting the extraction response.

In test AS1 an initial differential pressure of 28 kPa developed, which remained constant for about 10 mm of extraction (Figure 5). This trend was reflected in the capacity mobilised during extraction, which was also held constant for the same amount of upward displacement (Figure 6) reaching a 60% increase after 5 mm extraction, with respect to the drained capacity.

For test AS2 the maximum differential pressure generated was around 18 kPa (Figure 5). However, it resulted in the initial highest capacity with a peak uplift resistance of 35 kPa, about 70% high er than the drained capacity (Figure 6). The uplift resistance of AS2 decreased almost in parallel with the drained capacity for 20 mm, despite the differential pressure decreasing with a lower gradient with respect to AS1, as shown in Figure 5.

These results suggest that the magnitude of the differential pressure governs the seepage in and around the caisson and the uplift mechanism generated during uplift and hence the uplift capacity. Interestingly, the highest active suction may not result in the highest uplift resistance.

4 CONCLUSIONS

A centrifuge testing campaign was performed in the new 10 m diameter beam centrifuge at UWA to investigate the effect of active suction on the uplift capacity of caisson foundations. The instrumentation provided insights into the increase in uplift resistance due to active suction.

A new gravity based suction application device was developed, which enabled constant active suction to be applied with varying flow rate. A preliminary testing campaign was performed, mainly aiming to validate the concept of active suction.

Provided that an adequate head pressure is applied, active suction can be generated and maintained, resulting in an increase in uplift capacity.

An increase in capacity of up to 35 kPa was observed upon application of an active suction of about 18 kPa, corresponding to 70% of the drained capacity. Interestingly, the uplift capacity upon application of active suction is close to that measured under partially undrained conditions upon rapid uplift.

However, a different response was observed between partially undrained tests and active suction tests, as a consequence of the two different magnitudes of differential pressure generated. For active suction tests, the seepage process established in the soil around the caisson seems to have a significant influence on the mechanism governing the uplift resistance of the caisson.

ACKNOWLEDGEMENTS

The research presented is supported by the Australian Research Council Discovery Programme DP150102449 and it is hosted by the Centre for Offshore Foundation Systems (COFS).

REFERENCES

Allersma, G.B. 2003. Centrifuge Tests on Uplift Capacity of Suction Caissons with Active Suction. In *Proceedings of the Thirteenth International Offshore and Polar Engineering Conference, Honolulu, Hawaii, USA*.

Bye, A., Erbrich, C. Rognlien, B. & Tjelta, T.I. 1995. Geotechnical design of bucket foundations. In *Offshore Technology Conference, Houston, TX*. OTC 7793.

Feld, T. 2001 *Suction Buckets: a new innovation concept, applied to offshore wind turbines*. Aalborg Universitetsforlag.

Fiumana, N., Gaudin, C. & Tian, Y. 2017. Active suction anchors for floating renewable energy. In *Proceeding of the 12th European Wave and Tidal Energy Conference, Cork, Ireland*.

Gaudin, C., O'Loughlin, C.D., Duong, M.T., Herduin, M., Fiumana, N., Draper, S., Wolgamot, H., Zaho, L. & Cassidy, M.J. 2017. New anchoring paradigms for floating renewables. In *Proceeding of the 12th European Wave and Tidal Energy Conference, Cork, Ireland*.

Helfrich, S. C., Brazil, R. L., Richards, A. F. & Lehigh U. 1976. Pullout Characteristics of a Suction Anchor in Sand. In *Offshore Technology Conference, Houston, TX*. OTC 2469.

Herduin M. 2014. *Dynamic behaviour of mooring systems for WEC's application*. Master thesis, University of Exeter.

Houlsby. G.T., Kelly, R.B. & Byrne, B.W. 2005. The tensile capacity of suction caissons in sand under rapid loading. In *Proceedings of the 1st International Symposium on Frontiers in Offshore Geotechnics*. 1: 405-410.

Houlsby, G.T. & Byrne, B.W. 2005a. Design procedures for installation of suction caissons in sand. *Geotechnical Engineering*. 158(3): 135-144.

House, A. 2002. *Suction caissons foundations for buoyant offshore facilities*. PhD Thesis, The University of Western Australia.

Senders, M. 2008. *Suction Caissons in Sand as Tripod Foundations for Offshore Wind*. PhD Thesis, University of Western Australia.

Tran, M.N. 2005.*Installation of Suction Caissons in Dense Sand and the Influence of Silt and Cemented Layers*. PhD Thesis, The University of Sydney.

Wang, M.C., Nacci, V.A. & Demars, K.R. 1975. Behavior of underwater suction anchors in soil. *Ocean Engineering*. 3: 47-62.

Cyclic behaviour of unit bucket for tripod foundation system under various loading characteristics via centrifuge

Y.H. Jeong, H.J. Park & D.S. Kim
Korea Advanced Institute of Science and Technology, Daejeon, Korea

J.H. Kim
Korea Institute of Civil Engineering and Building Technology, Gyeonggi-do, Korea

ABSTRACT: Offshore wind energy has been growing up as a promising renewable energy source. In recent, tripod suction bucket foundation, which is consisted of three buckets, is rapidly expanding as a foundation system supporting the offshore wind turbine. In offshore environment, wind turbine foundation structures should be designed considering cyclic loading which can lead to permanent deformation of structure, tilting problem and overall degradation of soil stiffness. However, it is technically difficult to predict the cyclic behaviour of the tripod accurately, but the good thing is that the cyclic behaviours of the tripod bucket can be inferred from vertical pullout and compression behaviours of each single bucket elements. In this paper, a series of centrifuge tests was performed by applying cyclic vertical compression and extension loadings to a single bucket element that is one element of the tripod foundation. Loading directions and level were controlled for investigating of cyclic behaviour of tripod foundation. On the basis of testing results, the permanent deformation and cyclic stiffness response of tripod suction caisson were discussed. Based on the test results, it was confirmed that the cyclic behaviour of the single bucket is affected by the load level. In addition, the behaviour showed quite different trends with the loading directions: compression; pullout; and two-way.

1 INTRODUCTION

The offshore wind energy is very rapidly expanding as a reusable energy. Recently, suction bucket foundations have been considered as a highly competitive alternative to conventional offshore wind turbine foundations due to their unique features suitable for offshore construction environments such as convenient installation, no heavy equipment for penetration, and no hammering noise from driving. Especially, tripod foundations have especially been used to improve the overturning resistance of structure by changing the horizontal load to compression and extension loads at each bucket element. With this recognition, tripod bucket foundations have been actively investigated for use as supporting systems of offshore wind turbine towers (Houlsby et al. 2005; Zhu et al. 2013).

The monotonic ultimate capacity and the installation procedure of the bucket foundation have been widely investigated and successfully put into practice (LeBlanc, 2009). The dynamic response has been studied by Andersen et al. (2009). However, forces applied to the offshore structure in open sea area are cyclic conditions. However, a lot of researches related to behaviour of foundation have mostly focused on the static or dynamic loading. Therefore, studies for understanding the behaviour of tripod suction foundation under cyclic loading are greatly needed to design safe and economical offshore foundation. When a horizontal cyclic loading is applied to the tripod suction bucket foundation, it will transform to a vertical cyclic loading at each single bucket elements as shown schematically in Figure 1. In this case, the bearing capacity of the tripod bucket foundation is conveniently evaluated by multiplying the group efficiency factor and the bearing capacity to the single bucket element

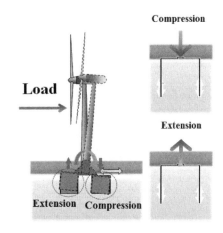

Figure 1. Behaviour of the tripod suction bucket foundation under cyclic loading.

Table 1. Specification of KAIST centrifuge.

Item	Specification
Platform radius	5.0 m
Max. capacity	240 g-tons
Max. acceleration	130 g with 1,300 kg payload
Max. model payload	2,400 kg up to 100 g

(Kim et al. 2014a; Zhan et al. 2010). Thus, it is required to evaluate the cyclic behaviour of a single bucket element by performing several parametric studies (Kim et al. 2014b).

The main goal of this study is to evaluate the stiffness and permanent displacement of the single bucket system under various stress-controlled cyclic loadings. A series of centrifuge model tests were performed under different loading characteristics: loading level and direction. On the basis of the testing results, permanent displacement of the bucket and stiffness response were assessed.

2 TESTING EQUIPMENT AND SOIL PREPARATION

2.1 KAIST geo-centrifuge

In order to simulate a prototype behaviour properly in a model test, it is necessary to replicate the in-situ soil stress condition.

In this research, a beam-type centrifuge of model C72-2 was used at KOCED Geotechnical Centrifuge Testing Centre at KAIST. This model is a 5 m radius and 240 g-tons asymmetric beam centrifuge. The general specification is listed in Table 1 (Kim et al. 2013).

2.2 Modelling of suction bucket foundation

A suction bucket was fabricated by steel to ensure its stability. The caisson diameter (D) was 143 mm, the wall height (L) was 200 mm, and the wall thickness (t) was 1.5 mm, representing 10 m × 14 m × 30 mm, (diameter × wall height × wall thickness) at 1:70 scale model. Usually, the ratio of the wall thickness to the diameter of a caisson installed on sand range from 0.3% to 0.6%. Because of the limitation in fabrication, the tip thickness ratio of the model was 1% in this study. Six holes were pierced above the bucket to prevent the pore pressure accumulation. A photograph of the suction bucket element is given in Figure 2.

2.3 Soil preparation

The soil used in model tests is Saemangeum sand, which is natural sand with high fine contests, sampled from a reclaimed site, Saemangeum, on the western coast of Korea. The Saemangeum sand was classified to SM in unified soil classification (USCS). A predetermined amount of dry soil sample per sublayers was prepared and mixed with water to obtain

Figure 2. Suction bucket element model caisson.

Table 2. Basic soil properties of the tested Saemanguem sand.

Item	Properties
Specific gravity, G_S	2.67
Fine contents (passing #200)	47%
Max. dry density (t/m^3)	1.65
Min. dry density (t/m^3)	1.2
Median grain size diameter (D_{50}, mm)	0.08
Uniformity coefficient, C_U	2.11

18.1% water contents, the optimal moisture contents. The Saemangeum sand specimens were prepared using the moist compaction method at the optimal moisture contents with diving into seven layers. In order to achieve a loose state, the relative density was approximately 40%, which represents a dry unit weight of 12.86 kN/m^3. Basic properties of the Saemanguem sand are presented in Table 2.

3 TESTING PROGRAMME AND PROCEDURE

In order to evaluate the cyclic behaviour of the single bucket element of tripod suction bucket foundation, total eight centrifuge model tests were performed. All of testing variables are tabulated in Table 3. The test program consists of monotonic and cyclic tests.

Two monotonic loading tests were carried out to assess a bearing capacity of the single bucket element foundation. One test was a compression loading test and the other test was a tension loading test. Six cyclic loading tests were performed to assess the cyclic stiffness response and permanent displacement of the single bucket element under the cyclic loads. The cyclic loading tests program consisted of the compression, extension, two-way loading tests.

Throughout the paper V represents the actual vertical force applied to the foundation. LeBlanc et al. (2010) defined a parameter characterising the applied cyclic load, $\xi_b = V_{max}/V_U$, where V_{max} = maximum force in the load cycle, V_U = static ultimate capacity, and w = vertical displacement of foundation.

Table 3. Centrifuge tests programme.

Test no.	Loads (kN)	Frequencies (Hz)	Direction	Cycles N
M-1	Monotonic	Monotonic	Compression	1
M-2	Monotonic	Monotonic	Tensile	1
C–1	$0.3\,V_u$ (FLS)	0.08	Compression	50
C–2	$0.5\,V_u$ (SLS)	0.08	Compression	50
E–1	$-0.3\,V_u$ (FLS)	0.08	Extension	50
E–2	$-0.5\,V_u$ (SLS)	0.08	Extension	50
T–1	$0.3\,V_u/-0.1\,V_u$ (FLS)	0.08	Two-way	50
T–2	$0.3\,V_u/-0.15\,V_u$ (SLS)	0.08	Two-way	50

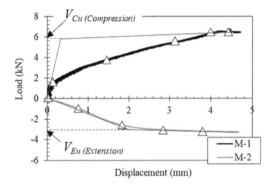

Figure 3. Monotonic test result (M-1).

After the soil preparation, water was trickled to saturate the model. 1-D vertical actuator was used to apply cyclic loading. The laser sensor was installed to measure the displacement of the caisson. The load cell was also set up to quantify the load applied to the caisson as shown in Figure 4. The data measured by the transducers were gathered at a sampling rate of 10 Hz. The bucket was installed by jacking method to the soil prior to applying cyclic loading.

4 TEST RESULTS AND DISCUSSIONS

To define the point of the static bearing capacity of bucket foundation, test M1-1 was performed by the method described by Villalobos (2006). In Figure 3, straight lines are fitted to the initial stiff elastic section and the more compliant plastic section. The intersection is used to define a yield load, which is denoted V_U. Six series of cyclic load tests were performed by changing the loading level and direction as listed in Table 3.

4.1 Comparison of cyclic behaviours in different loading directions

Figure 5 shows typical load-displacement relationships and vertical displacements of the bucket at three different load directions. The positive direction of the x and y axes in the graph shows displacement and load in the compression direction. The loads used in

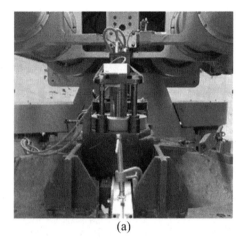

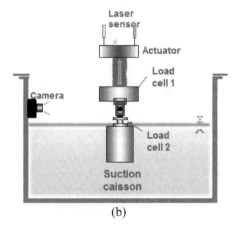

Figure 4. Centrifuge testing set up for cyclic loading tests (a) Photo of testing set up and (b) Schematic diagram of testing set up.

the tests are the SLS (50% of Ultimate limit state) loads. To illustrate the effect of cyclic load direction, Figure 5 (a) compares the load-displacement curves obtained by the compression (C-2); extension (E-2); and two-way (T-2) loading tests.

In case of the two-way loading test (T-2), the typical trend was quite different from the compression and extension tests (C-2, E-2). Despite the SLS load

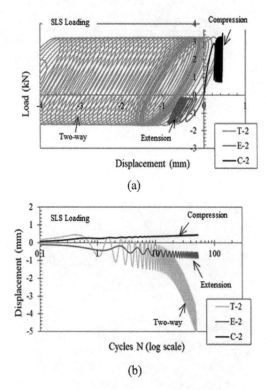

Figure 5. Load-displacement curves and vertical displacement with loading direction (a) Load-Displacement curves with direction and (b) Vertical displacements of the bucket with direction.

is only 50% of the ULS load, the slope of load-displacement curve decreases abruptly as the number of cycle increases. In the beginning of repeated loading, the slope of the graph is stiff even though some plastic deformation occurs. However, as the number of cycle increases, the slope becomes significantly flatten and the area of loop which represents a hysteretic damping increases. It leads to a nonlinear plastic behaviour, indicating the decrease in the rigidity and increase in displacement of the ground.

Generally, the two-way loadings cause a greater decrease in a contact frictional resistance between the sand and the steel compared to the one-way loads, so it is considered that permanent displacement was occurred severely (Uesugi et al. 1989).

However, in case of the compression and extension loadings, the trends of the load-displacement curve appear quite different from the two-way tests. Despite of the increase in the number of cycles, as shown in Figure 5 (a), the slope of the curve is almost unchanged and the hysteretic damping keeps small showing pseudo linear elastic behaviour even though little permanent deformation occurs. Also vertical displacements of the bucket are quite small compared to the two-way loading tests as shown in Figure 5 (b). The similar results were described in the cyclic loading tests for the tripod suction bucket foundation.

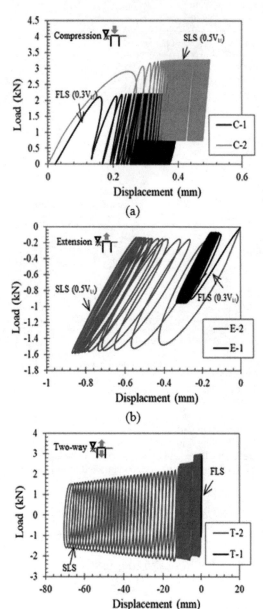

Figure 6. Load-displacement curves with loading level (a) Load-Displacement curves (Compression), (b) Load-Displacement curves (Extension) and (c) Load-Displacement curves (Two-way).

According to Kim et al. (2014), the vertical accumulated deformation of single bucket hardly occurs in the small amount of loading levels less than ultimate loading.

4.2 *Effect of loading level on cyclic behaviour*

Figure 6 and Figure 7 show load-displacement relationships and stiffness variance curves respectively for

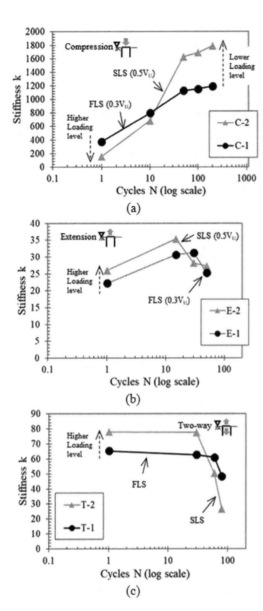

Figure 7. Cyclic stiffness with loading level (a) Stiffness with loading level (Compression), (b) Stiffness with loading level (Extension) and (c) Stiffness with loading level (Two-way).

the directions of compression, pullout, and two-way with varying loading level.

Figure 6 (a) shows that the trend of the compressional cyclic behaviour is similar for two tests, and as the loading level increases, the vertical displacement increases. Moreover, the permanent displacement increases with cycle. These trends are similar with the experiments for cyclic behaviour of monopod. Zhu et al. (2012) described that the rotational displacement of monopod is proportional to the loading level and increases in the linear log scale to the number of cycles. The initial stiffness decreases as the magnitude of load increases as shown in Figure 7 (a). This result is considered to be caused by the difference in the loading speed of cyclic loading. Also, it can be seen that as the number of cycles increases, the cyclic stiffness of soil-foundation system increases due to the soil hardening. The hardening phenomenon occurs more severely as increase in the loading magnitude.

However, it was found that the repetitive behaviours caused by pullout loads are quite different. Figure 6(b) and Figure 7(b) are graphs showing the load-displacement curves and variance in stiffness under pullout loading, respectively. As the magnitude of the load increases, the permanent displacement of the bucket increases dramatically. In the case of the variance in stiffness of the soil-foundation system, the initial stiffness tended to increase as the loading level increased due to the loading rate effects under the pullout loading. However, when the number of repetitions increased, the stiffness showed a tendency to decrease, and it appeared severely as the load level increases. These trends are quite opposite to the compression test results. This result is the same as the rotation behaviour of a monopod. In the case of the repeated loading test results of the monopod, the rigidity of the foundation decreases as the loading level increases (Vilallobos 2006). The following trends are similar in the two-way loading tests as shown in Figure 6 (c) and Figure 7 (c).

5 CONCLUSION

In this paper, a series of centrifuge model tests for evaluating the cyclic behaviours of the tripod suction bucket foundation focused on the single bucket element were performed. The stiffness and permanent displacement of the single bucket system were assessed by various cyclic loading tests. These tests were carried out by using the centrifuge facility at KAIST; the model scale suction bucket foundation; the 1D-vertical actuator and the testing sensors. These tests were performed according to the various loading characteristics: loading direction and loading magnitude. Based on these test results, the cyclic behaviours of the single bucket element were assessed.

The cyclic behaviours of the foundation were quite different depending on the difference in load direction. When the same loading state was applied repetitively, the cyclic stiffness is significantly reduced and the permanent displacement occurred more in case of the two-way loadings compared with compression and pullout loadings.

When the loading level increases, the stiffness increases in the compressive direction, and soil becomes more stable. However, in case of the pull out direction, the initial stiffness of the foundation showed the opposite tendency according to the direction of load.

Consequently, it was confirmed that the cyclic behaviour of the single bucket is affected by the load level. In addition, the behaviour showed quite opposite trends with the loading directions: compression; pullout; two-way. Therefore, when designing the highly safe wind turbine foundation, it is necessary to carry out design with consideration of the loading characteristics: cyclic loading direction and loading level.

ACKNOWLEDGEMENTS

This work was supported by the National Research Foundation of Korea (NRF) grant funded by the Korea government (MSIT) (No. 2017R1A5A1014883).

REFERENCES

Andersen, K.H. 2009. Bearing capacity under cyclic loading-offshore, along the coast, and on land, *Canadian Geotechnical Journal*, Vol. 46, pp. 513–535.

Bolton, M.D. 1986. The strength and dilatancy of sands. *Geotechnique*, 36(1), 65e78.

Houlsby, G.T., Ibsen, L.B. & Byrne, B.W. 2005a. Suction caissons for wind turbines. *1st Int. Symp. on Frontiers in Offshore Geotechnics:* ISFOG 2005, S.M. Gourvenec & M. J. Cassidy, eds., Taylor & Francis, London, 75–94.

Kim, D. J. Choo, Y.W., Kim, J.H., Kim, S. & Kim, D.S. 2014a. Investigation of monotonic and cyclic behavior of tripod suction bucket foundations for offshore wind towers using centrifuge modeling. *Journal of Geotechnical and Geoenvironmental Engineering* 140.5: 04014008.

Kim, D.S., Kim, N.R., Choo, Y.W. & Cho, G.C. 2013. A newly developed state-of-the-art geotechnical centrifuge in Korea. *KSCE Journal of Civil Engineering* 17(1): 77–84.

Kim, S.R., Hung, L.C. & Oh, M. 2014b. Group effect on bearing capacities of tripod bucket foundations in undrained clay, *Ocean Engineering*, 79.

LeBlanc, C. 2009. Design of Offshore Wind Turbine Support Structures: Selected topics in the field of geotechnical engineering, PhD Thesis, Aalborg University.

LeBlanc, C., Houlsby, G.T. & Byrne, B.W. 2010. Response of stiff piles in sand to long term cyclic loading. *Geotechnique*, 60(2), 79–90.

M. F. Randolph, Lee, K.K. & Cassidy, M. 2013. Bearing capacity on sand overlying clay soils: Experimental and finite-element investigation of potential punch-through failure. *Géotechnique* 63(15): 1271–1284.

Senders, M. 2008. Suction caissons in sand as tripod foundations for offshore wind turbines. Ph.D. thesis, University of Western Australia, Perth, Australia.

Uesugi, M., Kishida, H. & Tsubakihara, Y. 1989. Friction between sand and steel under repeated loading. *Soils and Foundations*; 29(3): 127–137.

Villalobos, F.A. 2006. Model testing of foundations for offshore wind turbines. D.Phil. thesis, University of Oxford, Oxford, UK.

Zhan, Y.G. & Liu, F.C. Numerical analysis of bearing capacity of suction bucket foundation for offshore wind turbines. *Electronic Journal of Geotechnical Engineering* 2010; 15: 633–644.

Zhu, B.M., Byrne B.W. & Houlsby, G.T. 2012. Long-Term Lateral Cyclic Response of Suction Caisson Foundations in Sand, *Journal of Geotechnical and Geoenvironmental Engineering*, ASCE, 73–83.

Physical modelling of reinstallation of a novel spudcan nearby existing footprint

M.J. Jun, Y.H. Kim, M.S. Hossain & M.J. Cassidy
Centre for Offshore Foundation Systems, The University of Western Australia, Perth, Australia

Y. Hu
School of Civil, Environmental and Mining Engineering, The University of Western Australia, Perth, Australia

S.G. Park
Daewoo Shipbuilding & Marine Engineering Co. Ltd., Seoul, Korea

ABSTRACT: The interaction between a spudcan and an existing footprint is one of the major concerns during jack-up rig installation. When a spudcan is located on or nearby an adjacent footprint slope, there is a tendency for the spudcan to slide towards the centre of the footprint, inducing excessive lateral forces and bending moments to the rig. Adverse spudcan displacement could result in an inability to install the jack-up in the required position, leg splay, structural damage to the leg, and at worst, bumping or collapsing into the neighbouring operating platform. This paper reports the results from a series of 1g and centrifuge model tests investigating spudcan-footprint interactions in clay. Three footprint shapes were tested varying depth and diameter. The footprints were generated using a cutting tool, and for the deep vertical walled footprints in the centrifuge, a plastic collar was pioneered to ensure footprint stability before testing. Spudcan loads during reinstallation were measured using a vertical load cell and two bending gauges. A novel spudcan, which was developed to minimise the interaction effects, was also tested and compared the responses with those of a generic spudcan. The results from this study indicated that the novel spudcan has potential to ease spudcan-footprint interactions.

1 INTRODUCTION

Most offshore drilling in shallow to moderate water depths (<150 m) is performed from self-elevating jack-up rigs due to their proven flexibility, mobility and cost-effectiveness (CLAROM 1993; Randolph et al. 2005). Today's jack-ups typically consist of a buoyant triangular platform supported by three independent truss legs, each attached to a large 10 to 20 m diameter spudcan. After the completion of the task, the legs are retracted from the seabed, leaving depressions at the site (referred to as a 'footprint').

Jack-ups often return to sites where previous operations have left footprints in the seabed. This is, for example, to drill additional wells or service existing wells, repair a platform or to install new structures such as jackets or wind turbines (Killalea 2002; Osborne and Paisley 2002; InSafeJIP 2010). When a spudcan is located on or nearby an adjacent footprint slope, there is a tendency for the spudcan to slide towards the centre of the footprint, inducing excessive lateral forces and bending moments to the rig (see Fig. 1). Hazards include an inability to install the jack-up in the required position, leg splay, structural damage to the leg, and at worst, bumping or collapsing into the neighbouring operating platform. The frequency of

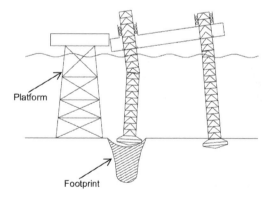

Figure 1. Spudcan reinstallation near footprint (after Kong et al. 2013).

offshore incidents during installation near footprints has increased by a factor of four between the period 1979~88 and 1996~06 (Osborne 2005) and at an even higher rate over 2005~2012 (Jack et al. 2013), with examples of offshore incidents also documented by Hunt & Marsh (2004), Brennan et al. (2006) and Handidjaja et al. (2009).

Penetration of spudcan foundations next to footprints has been addressed by a number of researchers, with each concentrating on a particular geometric aspect of the problem, such as the offset of reinstallation or the influence of the testing fixity conditions during reinstallation (e.g. Gaudin et al. 2007; Leung et al. 2007; Gan et al. 2008 & 2012; Cassidy et al. 2009; Kong et al. 2013). However, a minimum attention has been paid to mitigating spudcan-footprint interaction issues. In the field, stomping and successive leg repositioning (Brennan et al. 2006) and water jetting with the spudcan preloading (Handidjaja et al. 2009) have been used. Alternatively this study focuses on the shape of the spudcan itself and whether the shape can be manipulated to mitigate the effects of reinstallation adjacent to a footprint. A novel spudcan shape, developed to ease the footprint interactions through a series of large deformation finite element analysis (Jun et al. 2017a & b, 2018), is compared experimentally against the response of a generic spudcan at 1g and 200g conditions. Footprint geometries were varied by changing the diameter and depth, following the technique manually cutting a footprint into the sample (of Kong et al. 2013) and extending it to allow deep footprints with vertical walls to be tested.

2 MODEL TEST

2.1 Experimental program and sample preparation

The experimental programme comprised 1g and 200g modelling of penetration of spudcans nearby footprints. The spudcan tests were performed on three boxes of single layer samples of kaolin clay, with the geotechnical properties given in Table 1. A homogeneous slurry was prepared by mixing commercially available kaolin clay powder with water at 120% water content (twice the liquid limit). The slurry was then poured in beam rectangular strongboxes, which has internal dimensions of 650 (length) × 390 (width) × 325 (depth) mm. The soil samples were then pre-consolidated gradually on the high capacity press under a final pressure of 100, 200 and 550 kPa (see Table 2).

Soil characterisation tests were carried out using a T-bar penetrometer, of diameter 5 mm and length 20 mm. Typical shear strength (s_u) profiles for the three samples are shown in Figure 2, based on a T-bar deep bearing capacity factor of $N_{T\text{-bar}} = 10.5$ (where z is the

Table 1. Main properties of kaolin clay (after Stewart 1992).

Properties of kaolin clay	Value
Specific gravity, G_s	2.6
Angle of internal friction, ϕ' (degree)	23
Critical state frictional constant, M	0.92
Liquid limit, LL (%)	61
Plastic limit, PL (%)	27
Plasticity index, I_p (%)	34

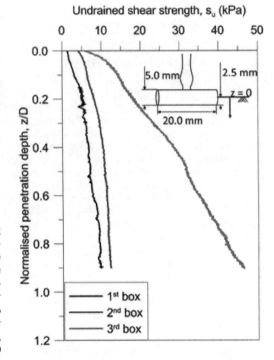

Figure 2. Undrained shear strength profile measured by T-bar penetrometer.

penetration depth of the T-bar mid-diameter and D is the spudcan diameter).

2.2 Model spudcan

Two different spudcan models of 60 mm diameter; (a) generic spudcan and (b) novel spudcan; were used,

Table 2. Summary of experimental tests.

Box	Tests	Test g level	Spudcan shape	Footprint shape D_F	z_F	Pre-consolidation pressure (kPa)	Offset distance, β	Note
1	Test 1	1g	Generic	1.0D	0.42D	100	0.50D	Deep footprint
	Test 2		Novel					
2	Test 3	200g	Generic	2.0D	0.33D	200	0.55D	Shallow footprint
	Test 4		Novel					
3	Test 5	200g	Generic	1.0D	0.50D	550	0.50D	Deep footprint
	Test 6		Novel					

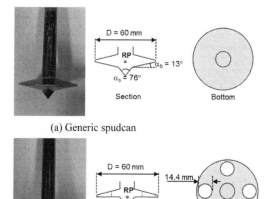

(a) Generic spudcan

(b) Novel spudcan

Figure 3. Spudcan shapes: (a) Generic spudcan; (b) Novel spudcan.

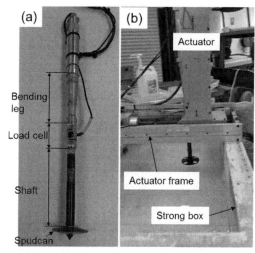

Figure 4. Experimental setup to measure the responses of footprint-spudcan interactions (a) Spudcan, shaft and model leg; (b) Experimental setup.

with their shape shown in Figure 3. The generic spudcan shape was chosen similar to the spudcans of the 'Marathon LeTourneau Design, Class 82-SDC' jack-up rig, as illustrated by Menzies & Roper (2008). A novel spudcan to minimise the horizontal force and moment induced by the interaction was developed. The basic shape of the novel spudcan is similar to the generic spudcan. However, the underside profile has been flattened and four evenly spaced holes (each of 14.4 mm diameter) inserted vertically between the spudcan top and base (see Fig. 3b). The role of each geometric factor was explored through large deformation finite element (LDFE) analyses by Jun et al. (2017a & b, 2018), and compared with the experimental results in a latter section of this paper.

Figure 4a shows the experimental set-up, including the spudcan, vertical load cell and bending leg used. The spudcan was rigidly connected to a model leg made from duraluminium. The leg was instrumented by two sets of bending strain gauges and one load cell to record the horizontal force, bending moment and vertical resistance generated during testing. The model leg was mounted on the strongbox with an actuator (see Fig. 4b). Note, in the field, the bending response of the truss leg is dictated by its stiffnesses and the connection of the leg to the hull. In this study, the tests were performed considering a fixed head boundary condition at the top level of leg, leading to a condition with very high leg stiffnesses, which is somewhat consistent with ISO (2012).

The T-bar and spudcan tests were carried out at a rate of 1 mm/s and 0.19 mm/s, respectively, targeting a non-dimensional velocity index of $V_p = vD_e/c_v > 30$ (where v is the penetration velocity, De is the area equivalent diameter) to ensure undrained conditions in the kaolin clay with $c_v \approx 2.6$ m²/year (Finnie & Randolph, 1994; Cassidy, 2012).

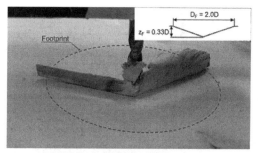

(a) Shallow footprint

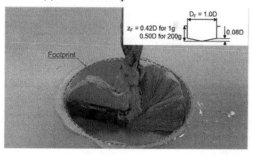

(b) Deep footprint

Figure 5. Artificial footprint generation by using two types of cutting tools: (a) Shallow footprint; (b) Deep footprint.

2.3 *Artificial footprint generation*

Artificial footprints were generated on the surface of the pre-consolidated samples. A cutting tool comprised of a mounting frame and cutting blades was developed to create the footprint cavity (see Fig. 5). The frame was first mounted on the strongbox and the cutting blade was then rotated and cut into the soil forming a cavity of ideal conical shape until the desired footprint depth was reached. With this cutting method, the disturbance of the adjacent soil around and beneath the artificial footprint was minimal. This was shown by Kong (2012) and Kong et al. (2013) who used a similar experimental cutting technique".

Depending on their sensitivity, natural fine grained soil deposits experience a degree of remoulding through the spudcan penetration-extraction event. This disturbance is recovered gradually with time through dissipation of excess pore pressure. For the kaolin clay used in the centrifuge tests in the vicinity of the footprint, it took 1~1.5 years to recovery its full strength, though of course this is a function of the soil's permeability (Leung et al. 2007). The strength recovery of the clay along and adjacent to a footprint was shown as a function of elapse time after the spudcan extraction by plotting strength contours (Gan et al. 2012) measured in the centrifuge. Because of the complexity and variety of possible soil strength gradients around the footprint (see contours of Gan et al. 2012), an artificial footprint was created in this study, where the soil strength along and adjacent to the footprint was identical to the intact strength profile. This simplified soil strength profile allows a consistent evaluation of the benefits of the spudcan shape.

From the results of a series of centrifuge tests, Hossain and Dong (2014) concluded that a conical footprint of depth $z_F = 0.22 \sim 0.33D$ was formed in soft clay whilst a cylindrical flat-based footprint of depth $z_F = 0.5 \sim 0.66D$ was formed in stiff clay. These findings are consistent with footprints measured in Gan et al. (2008), Teh et al. (2010), and Erbrich et al. (2015). In this study, the depth of the toe of the footprint z_F was varied 0.33D to 0.5D, and the width of the footprint D_F as 1~2D. The details of the investigated footprint geometries are summarised in Table 2.

2.4 Experimental cases

The spudcan was penetrated near the artificial footprint. The offset distance (β) from the footprint centre to the spudcan centre was 0.55D for the shallow footprint and 0.50D for the deep footprint. This offset was tested as it was identified as producing the largest induced loads on the spudcan by Kong et al. (2013) and Zhang et al. (2015). The experimental cases are summarised in Table 2.

Table 3. Example of normalisation of H_{max} with s_u vs $s_{u,0.5zF}$.

	Generic	Novel	Reduction using novel spudcan
H_{max}(N)	13.2	7.8	40.3%
$H_{max} / As_{u,0.5zF}$ *	0.87	0.52	40.3%
$H_{max} / As_u^{\#}$	1.53	0.47	68.9%

* $As_{u,0.5zF}^{\$} = 15.1$ N.
\# $As_u = 8.6$ N for generic spudcan; 16.5 N for novel spudcan.
\$ $s_{u,0.5zF} = 5.3$ kPa for Test 1 and 2.
   7.8 kPa for Test 3 and 4.
   20.3 kPa for Test 5 and 6.

3 RESULTS AND DISCUSSION

The effect of the novel spudcan shape (see Fig. 3b) at mitigating spudcan-footprint interactions has been evaluated through comparison with the performance of the generic spudcan shape. For direct comparison, avoiding the influence of the various undrained shear strength (s_u) profiles, all the response profiles were normalised using the measured soil strength (s_u), spudcan full plan area A (= 0.0028 m^2 for 1g test and 113.1 m^2 for 200g tests) and spudcan diameter (D). To maintain the form of the response profiles, s_u at a fixed depth was used for normalising horizontal force (H) and moment (M), while full s_u profile was used for normalising vertical force (V). s_u at a fixed depth of half of the footprint depth (0.5z_F) was selected to represent the soil strength over the footprint depth. Table 3

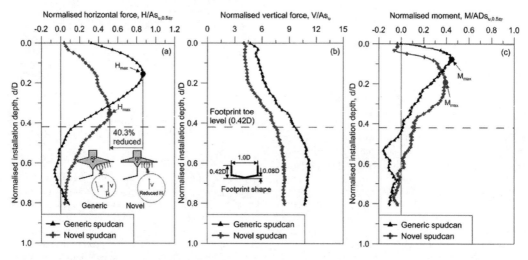

Figure 6. Responses from deep footprint shape at 1g (Test 1 and Test 2; $D_F = 1.0D$, $z_F = 0.42D$, $\beta = 0.50D$; Table 2): (a) Normalised horizontal force; (b) Normalised vertical force; (c) Normalised moment.

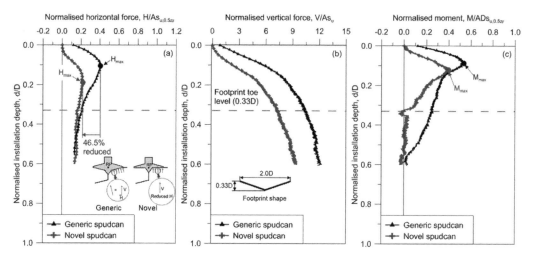

Figure 7. Responses from shallow footprint shape at 200g (Test 3 and Test 4; $D_F = 2.0D$, $z_F = 0.33D$, $\beta = 0.55D$; Table 2): (a) Normalised horizontal force; (b) Normalised vertical force; (c) Normalised moment.

shows an example of the difference in normalisation using s_u full profile and s_u at the fixed depth of $0.5z_F$.

3.1 1g test with deep footprint

Figure 6 shows the results from Test 1 and Test 2 ($D_F = 1.0D$, $z_F = 0.42D$, $\beta = 0.50D$; Table 2) conducted at 1g on the laboratory floor. The normalised horizontal force ($H/As_{u,0.5zF}$), vertical force (V/As_u) and moment ($M/As_{u,0.5zF}$) at the reference point RP of the spudcans (defined in Fig. 3) are presented as a function of the normalised penetration depth, d/D (where d is the penetration depth of the largest cross section of the spudcan; see Fig. 3). By using the novel spudcan, the maximum normalised horizontal force was reduced by 40.3% of that of the generic spudcan (see Fig. 6a). This is because of the bottom profile. The flat bottom profile of the novel spudcan minimised the horizontal contact area and hence the reduced horizontal force, compared to the sloped bottom profile of the generic spudcan (see inset figures in Fig. 6a). These findings are consistent with the numerical results presented by Jun et al. (2017a).

Within the tested depths, the normalised vertical response of the novel spudcan is lower than that of the generic spudcan partly because of the lower embedded volume of the foundation at d = 0 (see Fig. 3) and partly because the reduction in the recorded vertical load by the holes (while the full plan area A was used for normalization, although the net area was 23% lower owing to the holes). However, with the progress of penetration below the footprint toe, this gap will be diminished gradually as the soil flowing through the holes will be stopped by the pressure of the backfilled soil above the spudcan. This means the holes will be blocked and the spudcan will penetrate further as if a spudcan without holes.

The moment about the RP mainly resulted from the resultant vertical force and its eccentricity from the RP as the resultant horizontal force passes nearly through

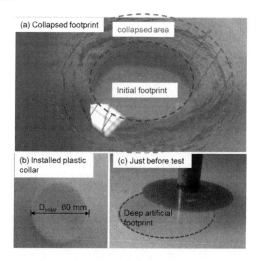

Figure 8. Deep footprint preparation: (a) Collapsed footprint; (b) Installed plastic collar (c) Just before test.

the RP (Kong et al. 2013; Zhang et al. 2015). As such, the reduced maximum moment by the novel spudcan was mainly caused by the reduced vertical force.

3.2 Centrifuge test with shallow footprint-200g

Figure 7 shows the results from Test 3 and Test 4 ($D_F = 2.0D$, $z_F = 0.33D$, $\beta = 0.55D$; Table 2) conducted at 200g. Similar to the results from tests at 1g test (Fig. 6), the novel spudcan shows a benefit in reducing horizontal force. The reduction in the maximum normalised horizontal force can be calculated as 46.5% (Fig. 7). After passing the footprint toe level (0.33D), the difference in horizontal force and moment between the two spudcans diminished.

3.3 Centrifuge test with deep footprint-200g

To investigate the effect of the footprint depth z_F at identical testing conditions, Test 5 and Test 6

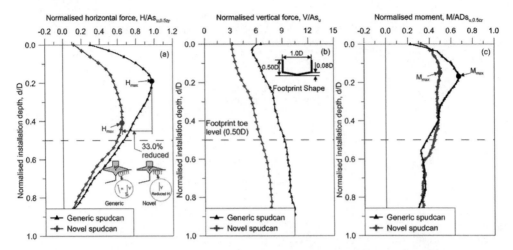

Figure 9. Responses for footprint-spudcan interaction in a deep footprint 200g (Test 5 and Test 6; $D_F = 1.0D$, $z_F = 0.50D$, $\beta = 0.50D$; Table 2): (a) Normalised horizontal force; (b) Normalised vertical force; (c) Normalised moment.

($D_F = 1.0D$, $z_F = 0.50D$, $\beta = 0.50D$; Table 2) were carried out at 200g. For the deep footing, Test 1 and Test 2 were undertaken at 1g and the craters were stable during the testing even though the undrained shear strength of the soil was low ($s_u = \sim 10$ kPa). This is because the soil overburden pressure over the footprint depth of 0.42D i.e. 25.2 mm at 1g was very low. However, in a first attempt in the centrifuge, even with clay with higher s_u of ~ 55 kPa, the edge of the crater cracked and collapsed during spinning up of the centrifuge although the crater was generated with care and precision (see Fig. 8a). This was the case even though cavity collapse was not predicted for the higher overburden stress (i.e. footprint depth 0.50D is a prototype 6 m at 200g) using Rankine's pressure theory (Meyerhof, 1972) and upper-bound plasticity analysis (Britto & Kusakabe, 1982, 1983). The main reason is believed to be that after cutting the footprint there is further swelling of the surface of the pre-consolidated sample by the sucking ambient water during centrifuge spinning. This reduced the strength of the soil close to the surface.

The lesson learnt during this initial attempt was employed in Test 5 and Test 6. The potential cavity collapse was prevented by using a very light weight plastic collar with 60 (outer diameter, D_{collar}) × 35 (height, h_{collar}) mm (see Fig. 8b). The collar weight was minimised to ensure it did not sink into the clay whilst spinning up to 200g. The collar was removed just before the penetration test. As shown in Figure 8c, no crack or collapse of the edge of the footprint occurred during Test 5 and Test 6.

Figure 9 shows the normalised results from the Tests 5 and 6. The results show that the H and M of the novel spudcan were also reduced at reinstallation with the deep footprint. By comparing with the results in Figure 7, both horizontal force and moment are higher for the deeper footprint because the balance between left and right side of a spudcan is delayed. The normalised horizontal force along with footprint

Table 4. Summary of normalised horizontal forces.

	$(H/As_{u,0.5zF})_{max}$		
Footprint depth	Generic	Novel	Reduction
Shallow footprint (200g)	0.41	0.22	40.3%
Deep footprint (1g)	0.87	0.52	46.5%
Deep footprint (200g)	0.98	0.64	33.0%

depths are summarised in Table 4. The results confirm that the novel spudcan is effective at mitigating spudcan-footprint interactions.

4 CONCLUSION

This paper presented the results from 1g and 200g centrifuge tests on a generic and a novel spudcans penetrating adjacent to an existing footprint. The artificial shallow and deep footprints were created. The spudcans were penetrated at an offset of 0.50D or 0.55D from the footprint centre. The results showed that the novel spudcan provides an effective measure at reducing horizontal force and easing the spudcan interaction with footprint. More extensive investigations are being carried out through numerical analyses and centrifuge model tests, with the objective for realistic conditions – global jack-up rig effect, remolded soil strength with penetration and extraction of a spudcan.

Collapse of footprint/cavity at elevated gravity spinning and corresponding prevention using a light weight plastic collar may be helpful for future centrifuge modellers.

ACKNOWLEDGMENTS

The research presented herein was undertaken with support from the Australian Research Council (ARC) through the Linkage Project LP140100066. The work forms part of the activities of the Centre for Offshore

Foundation Systems (COFS), currently supported as a node of the Australian Research Council Centre of Excellence for Geotechnical Science and Engineering and as a Centre of Excellence by the Lloyd's Register Foundation. This support is gratefully acknowledged.

REFERENCES

Brennan, R., Diana, H., Stonor, R.W.P., Hoyle, M.J.R., Cheng, C.P., Martin, D. & Roper, R. 2006. Installing jackups in punch through-sensitive clays. *Proc. Offshore Technology Conference, Houston, 1-4 May 2006*: OTC 18268.

Britto, A.M. & Kusakabe, O. 1982. Stability of unsupported axisymmetric excavations in soft clay. *Géotechnique*, 32(3): 261–270.

Britto, A.M. & Kusakabe, O. 1983. Stability of axisymmetric excavations in clays. *Journal of Geotechnical and Geoenvironmental Engineering*, ASCE 109(5): 666–681.

Cassidy, M.J., Quah, C.K. & Foo, K.S. 2009. Experimental investigation of the reinstallation of spudcan footings close to existing footprints. *Journal of Geotechnical and Geoenvironmental Engineering*, ASCE 135(4): 474–486.

Cassidy, M.J. 2012. Experimental observations of the penetration of spudcan footings in silt. *Géotechnique*, 62(8): 727–732.

CLAROM. 1993. Design guidance for offshore structures. *Club pour les Actions de Recherches sur les Ouvrages en Mer, Eds: Le Tirant, P. & Pérol, C., Paris*.

Erbrich, C.T., Amodio, A., Krisdani, H., Lam, S.Y., Xu, X. & Tho, K.K. 2015. Re-visiting Yolla- new insight on spudcan penetration. *Proc.15th International Conference on the Jack-up Platform Design, Construction and Operation, London*.

Finnie, I.M.S. & Randolph, M.F. 1994. Punch-through and liquefaction induced failure of shallow foundations on calcareous sediments. *Proc. International Conference on Behaviour of Offshore Structures, Boston, 12–15 July*: 217–230.

Gan, C.T., Leung, C.F. & Chow, Y.K. 2008. A study on spudcan footprint interaction. *Proc. of the 2nd BGA International Conference on Foundations, Dundee, 24–27 June 2008*: 861–872.

Gan, C.T., Leung, C.F., Cassidy, M.J., Gaudin, C., Chow, Y.K. 2012. Effect of time on spudcan-footprint interaction in clay. *Géotechnique*, 62(5): 401–413.

Gaudin, C., Cassidy, M.J., Donovan, T., Grammatikopoulou, A., Jardine, R.J., Kovacevic, N., Potts, D.M., Hoyle, M.J.R. & Hampson, K.M. 2007. Spudcan reinstallation near existing footprints. *Proc. 6th International Conference of Offshore Site Investigation and Geotechinics, London, 1 Feb. 2007*: 285–292.

Handidjaja, P., Gan, C.T., Leung, C.F. & Chow, Y.K. 2009. Jack-up foundation performance over spudcan footprints analysis of a case history. *Proc. 12th International Conference on the Jack-up Platform Design, Construction and Operation, London, 15–16 Sept. 2009*.

Hossain, M.S. & Dong, X. 2014. Extraction of spudcan foundations in single and multilayer soils. *Journal of Geotechnical and Geoenvironmental Engineering*, ASCE 140(1), 170–184.

Hunt, J.R. & Marsh, D.P. 2004. Opportunities to improve the operational and technical management of jack-up deployments. *Marine Structures*, 17: 261–273.

InSafeJIP. 2010. Improved guidelines for the prediction of geotechnical performance of spudcan foundations during installation and removal of jack-up units. *Joint Industry Funded Project*.

ISO. 2012. Petroleum and natural gas industries – Site specific assessment of mobile offshore units. *Part 1: Jackups. International Organization for Standardization*, ISO 19905-1.

Jack, R.L., Hoyle, M.J.R., Smith, N.P. & Hunt, R.J. 2013. Jack-up accident statistics – a further update. *Proc. 14th International Conference on the Jack-up Platform Design, Construction and Operation, London, 17–18 Sept*. 2013.

Jun, M.J., Kim, Y.H., Hossain, M.S., Cassidy, M.J., Hu, Y. & Sim, J.U. 2017a. Optimising spudcan shape for mitigating spudcan-footprint interaction. *Canadian Geotechnical Journal, submitted 24 July 2017*.

Jun, M.J., Kim, Y.H., Hossain, M.S., Cassidy, M.J., Hu, Y., Shim, J.U. 2017b. Physical and numerical modelling of novel spudcans for easing footprint-spudcan interaction issues. *Proc. 19th International Conference on Soil Mechanics and Geotechnical Engineering, Seoul, 17–21 Sept. 2017*: 2297–2300.

Jun, M.J., Kim, Y.H., Hossain, M.S., Cassidy, M.J., Hu, Y., Shim, J.U. 2018. Numerical investigation of novel spudcan shapes for easing spudcan-footprint interactions. *Journal of Geotechnical and Geoenvironmental Engineering, American Society of Civil Engineers, Accepted*.

Killalea, M. 2002. Jackup footprint, punch through studies underway. *Drilling Contractor*: 43.

Kong, V.W. 2012. Jack-up reinstallation near existing footprints. *Ph.D. thesis, University of Western Australia, Australia*.

Kong, V.W., Cassidy, M.J. & Gaudin, C. 2013. Experimental study of the effect of geometry on reinstallation of jack-up next to footprint. *Canadian Geotechnical Journal*, 50(5): 557–573.

Leung, C.F., Gan, C.T. & Chow, Y.K. 2007. Shear strength changes within jack-up spudcan footprint. *Proc. 17th International Offshore and Polar Engineering Conference, Lisbon, 1–6 July 2007*: 1504–1509.

Menzies, D. & Roper, R. 2008. Comparison of jackup rig spudcan penetration methods in clay. *Proc. 40th Offshore Technology Conference, Houston, 5–8 May 2008*: OTC 19545.

Meyerhof, G.G. 1972. Stability of slurry trench cuts in saturated clay. *Proc. Speciality Conf. on Performance of Earth and Earth Supported Structures*, 1: 1451–1466.

Osborne, J.J. & Paisley, J.M. 2002. S E Asia jack-up punchthrough: the way forward? *Proc. Offshore Site Investigation and Geotechnics – Diversity and Sustainability, London, 26–28 Nov. 2002*: 301–306.

Osborne, J.J. 2005. Are we good or are we lucky? *Presentation Slides for OGP/CORE Workshop: The Jackup Drilling Option-Ingredients for Success, Singapore, 27–28 Oct. 2005*.

Randolph, M.F., Cassidy, M.J., Gourvenec, S. & Erbrich, C.J. 2005. Challenges of offshore geotechnical engineering. *State of the Art paper, Proc. International Conference on the 16th Soil Mechanics Soil Mechanics and Geotechnical Engineering, Osaka, 12–16 Sept. 2005*, 1: 123–176.

Stewart, D.P. 1992. Lateral loading of piled bridge abutments due to embankment construction. *Ph.D. thesis, University of Western Australia, Perth, Australia*.

Teh, K.L., Handidjaja, P., Leung, C.F. & Chow, Y.K. 2010. Leg penetration analysis of jack-up rig installation over existing footprints. *Proc. 20th International Offshore and Polar Engineering Conference, Beijing*, 2, 427–433.

Zhang, W., Cassidy, M.J., and Tian, Y. 2015. 3D large deformation finite element analyses o jack-up reinstallations near idealised footprints. *Proc. International Conference on the 15th Jack-up Platform Design, Construction and Operation, London*.

Reduction in soil penetration resistance for suction-assisted installation of bucket foundation in sand

A.K. Koteras & L.B. Ibsen
Department of Civil Engineering, Aalborg University, Aalborg, Denmark

ABSTRACT: Bucket foundation is recently consider as a cost-effective solution for offshore wind turbines. However, the concept requires still a better understanding and a full design method that can be approved by standards. 1G laboratory tests for the installation of medium-scale bucket foundation have been performed in the laboratory of Aalborg University, Denmark. The main purpose of those tests is to investigate the interaction between the soil and the bucket skirt during the jacking and the suction installation process. The most often proposed method for the penetration resistance during installation of bucket foundation is a CPT-based method. The calculation requires information about empirical coefficients k_p and k_f, relating the skirt tip resistance and the skirt friction to the cone resistance measured during Cone Penetration Test, CPT. What is more, the suction-assisted installation in permeable soil adds additional parameters into the design resulting from the induced seepage flow around the bucket skirt. β-parameters introduced into the design describe a reduction in soil penetration resistance due to this flow. As an effect of tests results analysis, empirical coefficients and β-parameters are proposed. The use of those values leads to a reasonable fit between the applied force and the calculated soil resistance based on CPT, therefore, brings closer to the full design method for the suction installation of bucket foundation.

1 INTRODUCTION AND BACKGROUND

1.1 Motivation

The growth in the offshore wind energy market results in a large demand for more innovative and cost-effective solutions. As the foundation constitutes 20-30% of the total wind turbine budget, the cost-cuts found in the design of these sub-structures seems to be a way to go. Such a solution can be seen in the suction mono-bucket that has been proven to be feasible for challenging offshore conditions in oil and gas industry (Hogevorst 1980), (Tjelta 1995), (Bye et al. 1995). It is seen as a considerable cheap solution due to a simplicity in the geometry and a suction-assisted installation, which moves aside the requirements for heavy and expensive drilling equipment. Throughout the past few years the suction foundations have been successfully used also in the wind energy industry (Ibsen 2008).

1.2 Concept of suction bucket foundation

A bucket foundation is a skirted structure that consists of a skirt penetrated into the soil, closed with a lid element. The installation in frictional soils is accomplished in two stages. The first part is a self-weight penetration and the second part is achieved by pumping out the water from the cavity under the bucket lid. The first part is performed in order to ensure an appropriate hydraulic seal between the bucket skirt and the soil. The seal is required in order to achieve a suction pressure under the bucket lid. The second part of the installation is called a suction installation. For cohesive soil the available pump capacity is sufficient, as the penetration resistance is considerable low. Nevertheless, the installation is also possible for sandy soils. The initially high resistance is significantly reduced by the suction process. As an effect of applied suction under the bucket lid, a seepage flow is induced around the bucket skirt. This results in the inevitable gradients in the soil surrounding the skirt. The expectation for the flow is that it progress upward for the inside soil plug and downward in the soil outside of the skirt. The upward hydraulic gradient decreases the soil penetration resistance, whereas the downward gradient might give some increase. The significant decrease of resistance takes place also under the skirt tip due to a sudden change of the gradients. Studies confirm that a total penetration resistance of sand is substantially reduced and the increase of the resistance on the outside wall can be neglected (Lian et al. 2004), (Chen et al. 2016).

1.3 Effort of presented work

The knowledge about the suction installation of bucket foundation based on the recent research is still insufficient. There are no reliable standards that include all aspects of the installation. Therefore, a campaign of 10 laboratory tests for installation of medium-scale model of bucket foundation is conducted. A jacking installation tests and a suction installation tests are

Figure 1. Laboratory installation test of a bucket model.

performed in order to derive coefficients and factors required for the design. A laboratory set-up during the jacking installation test is presented in Figure 1.

2 METHODOLOGY

2.1 1-g modeling

Installation tests at 1-g have been performed in the laboratory of Aalborg University, Denmark. There are two groups of tests performed in dense sand. First group includes the jacking installed bucket (test no. 06, 07, 08 and 10) and in the second group, the bucket model is installed with the assistance of suction (test no. 01, 02, 03, 04, 05 and 09). Two different installation manners are conducted to prove the significant reduction in the soil resistance due to the seepage flow. What is more, from the jacking installation tests, the empirical coefficients k_p and k_f can be derived. Suggested values of those coefficients are next used when analyzing the suction installation tests. From those tests the β-factors can be found. All parameters can be introduced into the design method for the soil penetration resistance of suction-assisted installation of bucket foundation.

Following subsections describe shortly the model and the procedure. More detailed report about the installation tests is given by Koteras (2017).

2.1.1 Bucket foundation model

The dimensions of bucket foundation and attached elements are presented in Figure 2. The skirt thickness is 3 mm. This model is in scale 1 : 10 of a prototype size. The model is made of steel and the self-weight is equal to 201 kg. The bucket lid is equipped with 4 intake valves (1), 6 pressure transducers (2) and 1 displacement transducer (3). Measurements of pore pressure are taken from 6 different localizations on the skirt (PP1-PP6).

2.1.2 Soil preparation

A sand box of a 2.5 m diameter and a height of 1.7 m is equipped with a drainage system for a saturation. The sand box is filled with gravel til the height of 0.3 m, and

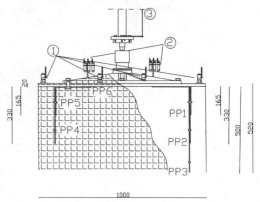

Figure 2. Model of bucket foundation.

Table 1. Soil properties.

e_{max}	e_{min}	d_{50}	C_u	d_s
[–]	[–]	[–]	mm	g/cm^3
0.854	0.549	0.14	1.78	2.64

next with sand til the height of 1.5 m. The properties of the sand are given in Table 1. Before each of the tests the soil is fully saturated and prepared for a dense, uniform condition with a mean value of a relative density, I_D, equal 89.73 (the standard deviation: $\sigma = 1, 02$). The preparation includes: first, a loosening of soil by application of a hydraulic gradient and next, a mechanical vibration. The sand condition is analyzed by cone penetration test, CPT. A cone of the device has a 15 mm diameter and it is with a 30° inclination. This test is performed before each installation and straight after each installation. The comparison allow for assessing whether the soil penetration resistance is reduced due to the installation manner.

2.1.3 Test procedure

The jacking installation is executed with a hydraulic piston as a displacement controlled operation. Open valves at the bucket lid ensure that the excess pore pressure of soil inside the bucket is not accumulated. Therefore, the soil penetration resistance is assessed to be at its initial, unreduced value. The measurements of the penetration depth, h, and the installation load, $F_{install}$, are monitored during the installation. Additionally, the pore pressure of soil around the bucket skirt and under the bucket lid are measured.

The suction installation includes two stages as mentioned before. The self-weight installation is induced by the hydraulic piston working as a force controlled. The force of a magnitude equal to the self-weight of the model is applied. When the hydraulic seal is realized, the suction pressure is applied by a vacuum system through the valves attached to the bucket lid. The measurements of the penetration depth, the applied suction, u, and the pore pressure around the soil skirt are monitored during the installation.

Table 2. Recommended values of k_p and k_f from DNV.

Empirical coefficients	Lowest expected	Highest expected
k_p	0.3	0.6
k_f	0.001	0.003

Both installations are performed in saturated soil condition, with a water level situated around 10cm above the soil level.

2.2 CPT-based method

Fairly simple design based on CPT is described by DNV (1992), where a soil penetration resistance of thin skirt structures, R_{soil}, is proposed. Both, the tip resistance of the skirt, Q_{tip}, and the friction on the inside and outside wall, F_{in} and F_{out}, are related to the cone resistance, q_c, through the empirical coefficients k_p and k_f, see Equation (1). The range of parameters is proposed for Sand and Clay in North Sea conditions. The proposal is based on the in-situ testing supported by the laboratory results, see Table 2. The method, even though it is often used, does not include the reduction in resistance due to applied suction and generated flow through the soil. A_{tip} is an area of the bucket skirt.

$$R_{soil} = F_{in} + F_{out} + Q_{tip} \qquad (1)$$

where

$$F_{in} = \pi D_i k_f \int_0^h q_c(h) dh \qquad (2)$$

$$F_{out} = \pi D_o k_f \int_0^h q_c(h) dh \qquad (3)$$

$$Q_{tip} = A_{tip} k_p q_c(h) \qquad (4)$$

Empirical coefficients relating the cone resistance to the soil penetration resistance have been studied by others as well. One example is given by Lehane et al. (2005). The solution proposed is for an open-ended pile and it emerged from the real installation data. The open-ended pile in its construction and installation method is similar to the bucket concept and therefore, this case is cited in the article. The solution is based on the difference between inside and outside diameter, D_i and D_o, of the foundation, see Equation (5). Value of C is estimated here as 0.021 and the interface friction angle, δ, lies around 30°. Given a typical bucket foundation with $\left(\frac{D_i}{D_o}\right) = 0.98$, the value of k_f would be above the values proposed in (DNV 1992).

$$k_f = C\left[1 - \left(\frac{D_i}{D_o}\right)^{0.2}\right]^{0.3} tan\delta \qquad (5)$$

Later on Senders & Randolph (2009) proposed the solution for the soil penetration resistance of suction bucket foundation based on CPT. The approach is similar to the solution proposed by DNV (1992), but the method includes the reduction in resistance due to the applied suction and the induced seepage flow. The inside friction and the tip resistance is linearly reduced from its maximum value when suction is not applied yet, to zero when the suction reaches its critical value (the critical suction is explained in the following section). For the analysis of centrifuge tests, coefficient k_f is chosen to be based on (Lehane et al. 2005), however, the value of C is adjusted in order to give values of k_f in the range of DNV values. In the research of Senders & Randolph (2009), the value $C = 0.012$ is chosen. Coefficient k_p is chosen to be equal 0.2 to represents a very dense sand condition. A very good agreement between the calculated soil resistance and the measurements of the jacking installations are obtained. Andersen (2008) have described the results of laboratory, prototype and field tests. For laboratory tests the bucket with $D = 0.557$m, $h = 0.32$m was penetrated into the soil situated in the tank of $D = 1.6$m. Conditions are closed to the test conditions described in this paper. The proposed values of k_f for laboratory tests was back-calculated to be 0.0053.

The results obtained in this research are analyzed with the CPT-based method, however, a different approach is used when including the effects of seepage flow. β-parameters are used instead of a linear change proposed by Senders & Randolph (2009), see following equations.

$$F_{in} = \beta_{in} \pi D_i k_f \int_0^h q_c(h) dh \qquad (6)$$

$$F_{out} = \beta_{out} \pi D_o k_f \int_0^h q_c(h) dh \qquad (7)$$

$$Q_{tip} = \beta_{tip} A_{tip} k_p q_c(h) \qquad (8)$$

Proposal for β-factors is given later on in the article.

2.3 Importance of the numerical analysis

The failure of the installation can happen due to the excessive applied pressure under the bucket lid. The limit gives the pressure that creates a piping channels between the bucket skirt and the soil trapped inside. In such a state, the hydraulic seal is broken and the suction process cannot proceed. Therefore, the critical value should be taken into the design. What is more, the value of critical pressure, u_{crit}, is also used for the calculations of the soil resistance against the installation. The suction initiates the loosening process for the soil and, thus, changes the resistance. As the suction pressure approaches the critical value, soil resistance drops from its maximum value to zero.

An approach based on this theory is presented by Houlsby & Byrne (2005) where the soil resistance is reduced according to the hydraulic gradients appearing in the soil. The gradients describe the changes in water pressure inside the soil, which is closely related to the applied pressure. Similarly, the same approach is given in (Senders & Randolph 2009) where the soil resistance drops linearly to zero while the applied pressure approaches its critical value.

The critical pressure against the piping is studies numerically (Erbrich & Tjelta 1999), (Senders & Randolph 2009), (Ibsen & Thilsted 2011), (Koteras et al. 2016). The most often proposed method is to relate the critical pressure to the seepage length, s, based on the definition of the critical gradient, see Equations 9 and 10. The exit hydraulic gradient, i_{exit} is obtained from the numerical simulations.

$$s = \frac{u}{i_{exit} \cdot \gamma_w} \quad (9)$$

$$u_{crit} = \left(\frac{h}{D}\right) \cdot \left(\frac{s}{h}\right) \cdot \gamma' \cdot D \quad (10)$$

Different proposition of $\left(\frac{s}{h}\right)$ can be found in the latest state of art, and all of the solutions are characterized with similar values for varying $\left(\frac{h}{D}\right)$. Interestingly, critical pressure calculated based on those values is exceeded in all of the tests in the campaign, and in none of them the installation failure was observed.

3 RESULTS AND DISCUSSION

3.1 Empirical coefficients from jacking installation

There are 4 jacking installation tests included in the analysis of the empirical coefficients. For each of those tests, the soil penetration resistance is calculated based on CPT results and then, it is related to the jacking installation force. The soil penetration resistance is calculated with Equation (1). It can only be solved when one from two of the empirical coefficients is chosen before the calculations. Value of k_f is assumed to be the same for both, the inside and the outside friction. Based on the previous research, the values of k_f are chosen. The values of k_p are derived by fitting the results of installation to the results of soil resistance from the CPT. The chosen values of k_f are as following:

- $k_f = 0.004$ based on (Lehane et al. 2005),
- $k_f = 0.0023$ based on (Senders and Randolph 2009) and
- $k_f = 0.0053$ based on (Andersen 2008).

The optimization is done for all 3 values of k_f in 4 jacking installation tests. The coefficient of determination, R^2, is found for each case. The choice is made based on this coefficient, see Table 3. As seen in the table, the best fit is obtained with $k_f = 0.0023$. The obtained values of k_p are in the range proposed by DNV (1992) −0.3 to 0.6 for installation in sand.

As an example, the results of installation load and calculated soil resistance based on CPT are presented in Figure 3 for test no. 06. The figure presents resistance calculated with 3 different values of k_f.

3.2 Critical pressure against piping

The solution for the normalized seepage length, $\left(\frac{s}{h}\right)$, is based on the numerical analysis presented by Koteras et al. (2016). However, the dimensions in the numerical model are adjusted, because it was realized that there

Table 3. Chosen values of empirical coefficients k_f and k_p.

Test no.	k_f	k_p	R^2
06	0.023	0.38	0.991
07	0.023	0.36	0.998
08	0.023	0.39	0.998
10	0.023	0.33	0.994

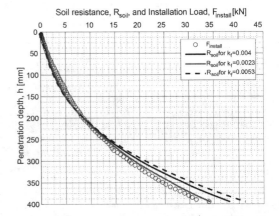

Figure 3. Soil resistance compared with installation load.

are some boundaries effects on the development of the seepage flow. The excess pore pressure at the boundary of sand box has been measured during the installation in each test. Results show considerable values of the excess pore pressure at the boundary. Therefore, the numerical model used for this paper is of the same characteristics as the laboratory model, what includes the correct diameter of the bucket model and also the boundary conditions. The results of the applied suction in each test were compared with the calculated critical pressure and it was found that the critical pressure is exceeded in each of the suction installation tests. Nevertheless, there were no piping problem observed, which suggests that the chosen solution for the seepage length is not appropriate.

Due to the observation of laboratory results the permeability of the soil inside the bucket has been increased. The results of CPT performed before and after the installation reveal the significant decrease in the relative soil density, I_D. As the soil becomes looser inside, the permeability of the soil increases. Appropriate values of the hydraulic conductivity, k, were applied for the inside and the outside soil elements of the numerical model before executing the simulations. Proposed solution for exit normalized seepage length from the numerical model that includes the increased value of k is given below.

$$\left(\frac{s}{h}\right)_{exit} = 1.25 \cdot \left(\pi - \text{atan}\left(2.5 \cdot \left(\frac{h}{D}\right)^{0.74}\right)\right)$$

$$\left(2 - \frac{1.8}{\pi}\right) \quad (11)$$

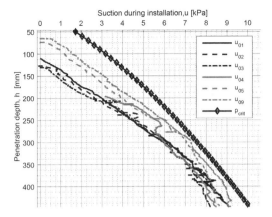

Figure 4. Applied pressure during suction installation tests.

With the proposed solution, the applied pressure in each test does not exceed the critical value, see Figure 4.

3.3 β-factors from the suction installation

The soil resistance for the suction installation is calculated by summing the inside and outside friction on the skirt and the tip resistance based on Equations (6), (7) and (8). The empirical coefficients are chosen based on the jacking installation tests as a value of 0.023 for k_f and a range of values between 0.33 to 0.39 for k_p. The proposed method includes different values of β-factors for each separate part of the total soil resistance.

$$\beta_{in} = 1 - r_{in} \cdot exp\left(\frac{u}{u_{crit}}\right) \quad (12)$$

$$\beta_{tip} = 1 - r_{tip} \cdot exp\left(\frac{u}{u_{crit}}\right) \quad (13)$$

$$\beta_{out} = 1 \quad (14)$$

Unknowns r_{in} and r_{out} are found for the minimum and the maximum value of k_p coefficient by non-linear least squares fitting. They are found as constants, see Table 4. The coefficients of determination given in table prove that the fitting process gave a fairly good estimation of the reduced soil penetration resistance for the bucket foundation. Moreover, the value of r_{tip} is closer to value 0.1 for $k_p = 0.33$, and closer to value 0.2 for $k_p = 0.39$.

As an example, the installation load calculated from the applied suction and the reduced soil penetration resistance is presented in Figure 5 for test no. 01.

4 SUMMARY AND FURTHER WORK

The suction bucket foundation has lately become a beneficial solution in the offshore wind market. However, a design method for the installation leaves a lot to be desired. First and foremost, there is still no reliable way of including the reduction in soil penetration resistance due to the applied suction. To improve the

Table 4. Proposed constants for β-factors and coefficient of determination, R^2.

Test no.	for $k_p = 0.33$			for $k_p = 0.39$		
	r_{in}	r_{tip}	R^2	r_{in}	r_{tip}	R^2
01	1	0.11	0.97	1	0.16	0.95
02	1	0.14	0.85	1	0.19	0.74
03	1	0.15	0.78	1	0.19	0.72
04	1	0.15	0.89	1	0.19	0.90
05	1	0.1	0.88	1	0.14	0.89
09	1	0.09	0.86	1	0.13	0.88

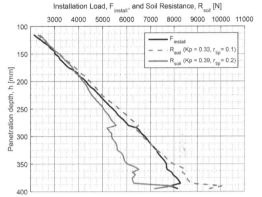

Figure 5. Soil resistance with proposed β-factors.

method, installation tests of bucket foundation have been performed at the laboratory facility on a medium scale model. The method used in the analysis is a CPT-based method, where a cone resistance is related to the soil penetration resistance by empirical coefficient. Even though, values of those coefficients can be found in literature, they vary between different sources. From the results of jacking installation, values of $k_f = 0.0023$ and $k_p = 0.33 - 0.39$ are proposed. From the results of suction installation, β-factors are proposed. β-factors cover the reduction of soil resistance due to the seepage flow, where the soil resistance is reduced as an exponential function of the ratio u/u_{crit}. Finally, calculated soil resistance is compared with the load applied during installation, and a reasonable fit is obtained.

Nevertheless, the tests results covers only medium scale testing. It is strongly advised that the proposed method is used on a prototype models and on the full-scale foundation. Then, and only then, the method can be considered as reliable.

ACKNOWLEDGMENT

The research described in this paper is a part of a PhD Study funded by "EUDP Project" and by EU7 as a part of the "INNWIND – Innovative Wind Conversion Systems (19-20MW) for Offshore Applications" project. The support is greatly appreciated.

REFERENCES

Andersen, K.H., J. H. D. R. (2008). Penetration resistance of offshore skirted foundations and anchors in dense sand. *J. Geotech. and Geoenv. Eng. 134*(1), 106–116.

Bye, A., C. Erbrich, B. Rognlien, & T. Tjelta (1995). Geotechnical design of bucket foundations. In *Proc., Offshore Technology Conference*, Houston, Texas. OTC 7793.

Chen, F., J. Lian, H. Wang, F. Liu, H. Wang, & Y. Zhao (2016). Large-scale experimental investigation of the installation of suction caissons in silt sand. *J. Applied Ocean Research 60*, 109–120.

DNV (1992). *Foundations, classification notes No. 30.4* (134 ed.). Hvik, Norway: Det Norske Veritas.

Erbrich, C. & T. Tjelta (1999). Installation of bucket foundations and suction caissons in sand geotechnical performance. In *Proc., 31st Annual Offshore Tech. Conf.*, Volume 1, Houston, Texas, pp. 725–736.

Hogevorst, J. (1980). Field trials with large diameter suction piles. In *Proc., Offshore Technology Conference*, Houston, Texas. OTC 3817.

Houlsby, G. & B. Byrne (2005). Design procedure for installation of suction caissons in sand. In *Proc., Institution of Civil Engineers*, Volume 158, pp. 135–144. Geotech. Eng.

Ibsen, L. (2008). Implementation of a new foundations concept for offshore wind farms. In *Proc., Nordisk Geoteknikermte Nr. 15: NGM 2008*, Sandefjord, Norway, pp. 19–33. Norsk Geoteknisk Forening.

Ibsen, L. & C. Thilsted (2011). Numerical study of piping limits for suction installation of offshore skirted foundations and anchors in layered sand. In *Proc., Int. Symp. on Frontiers in Offshore Geotechnics (ISFOG)*, Perth, Australia, pp. 421–426.

Koteras, A., L. Ibsen, & J. Clausen (2016). Seepage study for suction installation of bucket foundation in different soil combinations. In *Proc., 26th Int. Ocean and Polar Eng. Conf., 26 June-2 July*, Rhodos, Greece, pp. 697–704. Int. Society of Offshore and Polar Engineers.

Koteras, A. K. (2017). Set-up and test procedure for suction installation and uninstallation of bucket foundation. Dce technical report, Department of Civil Engineering, Aalborg Univ., Denmark.

Lehane, B., J. Schneider, & X. Xu (2005). The uwa-05 method for prediction of axial capacity of driven piles in sand. In *Proc., Int. Symp. on Frontiers in Offshore Geotechnics (ISFOG)*, Perth, Australia, pp. 19–21.

Lian, J., F. Chen, & H. Wang (2004). Laboratory tests on soil-skirt interaction and penetration resistance of suction caissons during installation in sand. *J. Ocean Eng. 84*, 1–13.

Senders, M. & M. Randolph (2009). Cpt-based method for the installation of suction caissons in sand. *J. Geotech. and Geoenv. Eng. 135*(1), 14–25.

Tjelta, T. (1995). Geotechnical experience from the installation of the europipe jacket with bucket foundations. In *Proc., Offshore Technology Conference*, Houston, Texas. OTC 7795.

Evaluation of seismic coefficient for gravity quay wall via centrifuge modelling

M.G. Lee, J.G. Ha, H.J. Park & D.S. Kim
Department of Civil & Environmental Engineering – KAIST, Daejeon, Republic of Korea

S.B. Jo
K-water Institute, Infrastructure Research Centre, Daejeon, Republic of Korea

ABSTRACT: Pseudo-static approach has been conventionally applied for the design of gravity type quay walls. In this method, the decision to select an appropriate seismic coefficient (k_h) is an important one, since k_h is a key variable for computing an equivalent pseudo-static inertia force. Nonetheless, there is no unified standard for definition of k_h. In Korea, there are conflicts between seismic design code and seismic performance evaluation code of port structures regarding the reference point of peak ground acceleration (PGA) used in determination of k_h according to the wall height. In this research, three dynamic centrifuge tests were performed for gravity walls of different heights (5 m, 10 m, 15 m) to clarify the reference point of PGA used in determination of k_h according to the wall height. The tests were carried out on reduced-scale models of gravity quay walls designed ($k_h = 0.13$) in dry cohesionless sand. Results from dynamic centrifuge experiments showed that in cases of 5 m and 10 m wall, the design k_h (0.13) value is similar with half of PGA from the base of the wall. On the other hand, in case of 15 m wall, the design k_h value is comparable with half of PGA from the surface of backfill soil.

1 INTRODUCTION

The conventional procedures for both seismic design and seismic stability evaluation of quay walls rely on the pseudo-static approaches. In the pseudo-static analysis, the earthquake load on the active thrust is generally considered through the well-known Mononobe-Okabe (M-O) method (Okabe 1926; Mononobe & Matsuo 1929).

In the M-O method, k_h, expressed in terms of the acceleration of gravity, is used to convert complex actual dynamic behavior to an equivalent pseudo-static inertia force for use in analysis and evaluation. Therefore, the decision to select an appropriate k_h for reducing a difference from the real dynamic behavior is a significant process in pseudo-static analysis. Nonetheless, there is no unified standard for defining the seismic coefficient. Likewise, port structure designers in Korea have a difficulty in choosing a suitable k_h definition, as there are conflicts in how k_h is defined between seismic design code of port structures (MOF 2014) and seismic performance evaluation code of harbour (MLTM 2012). The disagreement between definitions can be related to the reference point of peak ground acceleration (PGA) used in determination of k_h according to the wall height and whether it considers the application of correction factor for k_h.

In this study, three dynamic centrifuge tests were performed with different wall heights (5 m, 10 m, 15 m at prototype scale) to evaluate reference point and application correction factor of PGA according to wall height used in k_h calculation. The k_h of the gravity quay wall models were designed to be 0.13 by applying that the factor of safety against sliding equals 1. With the designed model structures, the trigger points of sliding induced by earthquake motions and the PGAs at each depth in the backfill were measured. From these results, it is confirmed that the PGA used in k_h calculation should be selected considering the wall height and that the appropriate correction factors should be applied to the PGA.

2 REVIEW OF SEISMIC COEFFICIENT DEFINITIONS OF KOREA

2.1 Ports and fishing harbours design code (MOF, 2014)

The definition of k_h suggested by MOF (2014) in seismic design for port structures in Korea is as follows:

$$k_h = \frac{1}{2} \cdot \frac{a_{max}}{g} \tag{1}$$

where, in case of walls smaller than 10 m, the PGA at the base of the backfill (base PGA) is adopted a_{max}.

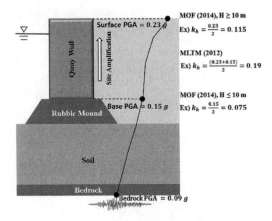

Figure 1. Locations of reference peak ground acceleration used in seismic coefficient (after Lee et al. 2017a).

On the other hand, in case of walls bigger than 10 m, the PGA at the surface of backfill (surface PGA) is used to a_{max}.

2.2 Seismic performance evaluation and improvement revision of existing structures (MLTM 2012)

The definition of k_h suggested by MLTM (2012) in the evaluation of seismic performance for port structures in Korea is as follows:

$$k_h = \frac{a_{average}}{g} \quad (2)$$

where, $a_{average}$ is the average value of the PGA along the wall height at backfill (average PGA).

An example of the k_h definitions of Korea is shown in Figure 1.

2.3 Review of k_h definitions

Lee et al. (2017b) reviewed the existing codes and guidelines for port structures to identify the current status of k_h definitions. The following three considerations for calculating k_h were deduced from the review.

– Application of the correction factor on PGA for k_h definition
– Consideration of the influence of the quay wall height
– Consideration of the allowable displacement of the quay wall

An overview of the reviewed codes and guidelines reflecting the aforementioned considerations is summarised in Table 1.

The k_h definition in MOF (2014) reflects all of the aforementioned concerns. In order to consider the effect of wall height on the k_h value, the reference point of PGA is specified based on the wall height of 10 m. Meanwhile, MLTM (2012) defines k_h as the average value of the PGA along the wall height at backfill (average PGA), regardless of the wall height and allowable displacement of the quay wall.

Table 1. Overview of reviewed codes and guidelines (after Lee et al. 2017a).

Codes/guidelines	Correction factor	Wall height	Allowable displ.
MOF (2014)	o	o	o
MLTM (2012)	N/A	N/A	N/A
CEN (2004)	o	o	o
PIANC (2001)	o	N/A	N/A
OCDI (2009)	o	o	o

3 EXPERIMENTAL PROCEDURE

In order to verify the effect of wall height on PGA used in k_h calculation and the correction factor values used in k_h definitions of Korea, three dynamic centrifuge tests were carried out on caisson gravity quay wall models for different wall heights (5 m, 10 m, 15 m) based on 10 m wall height, which is the criterion for dividing the reference point of PGA in MOF (2014).

The experiments were conducted at the KOCED Geo-Centrifuge Testing Centre at KAIST. The earthquake simulator is mounted on the centrifuge and the earthquake loadings are applied using an in-flight earthquake simulator. This setup can generate random earthquake excitations lasting up to 1 s with a model frequency ranging from 30 to 300 Hz (Kim et al., 2013a, b). In this study, the models were constructed in an Equivalent Shear Beam (ESB) model container (Lee et al., 2013), and the centrifugal accelerations used were 40 g for 5 m and 10 m walls, and 60 g for 15 m wall. All results are presented in prototype units unless otherwise stated.

3.1 Model structures and instrumentation

The caisson gravity quay wall models were designed with a k_h value of 0.13 for evaluating the reviewed k_h definitions. The structure was a simplified model based on an example of quay wall design provided by MOF (2014), and it was designed according to the method introduced in PIANC (2001). The models used in the experiments were made of aluminium alloy (T6061). Among the three seismic performance evaluation criteria (sliding, overturning, and bearing capacity failure), sliding was set to be the first failure mode to yield a safety factor of unity for this experiments. The factor of safety of 1 and other experimental conditions were then substituted into the equation for sliding safety evaluation to derive the weight of the walls corresponding to the k_h value of 0.13. The required weight of walls were obtained by adjusting

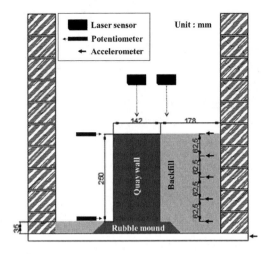

Figure 2. Schematic illustration of test instrumentation: 10 and 15 m wall height (after Lee et al. 2017a).

Table 2. Silica sand properties (after Lee et al. 2017a).

Properties	Silica sand (Rubble mound)	Silica sand (Backfill)
USCS	SP	SP
D_{50}	0.80 mm	0.22 mm
C_C	0.98	1.11
C_U	1.54	1.96
G_S	2.65	2.65
PI	NP	NP
Max. unit weight:	14.19 kN/m³	16.45 kN/m³
Min. unit weight:	12.87 kN/m³	12.44 kN/m³
D_r	86 %	80 %

the thickness of aluminium. Consequently, the height of the walls were 5 m, 10 m and 15 m, and the corresponding weight of walls were 106 ton, 424 ton and 1436 ton, respectively. The ratio of breadth (B) to height (H) in all the models was 0.57, and all the walls spanned the length (L) of the container (i.e., 484 mm at model scale, which is 2 mm shorter than length of the container to eliminate friction). Plastic sheets were attached to prevent sand from passing through those boundaries.

The instrumentation layout is presented in Figure 2. PGA values at each depth in the backfill during earthquake excitation were obtained by accelerometers buried in the soil. Residual horizontal displacements of the wall induced by earthquake (D_h) were measured by potentiometers. To measure the subsidence of backfill and wall, two laser sensors were used. Acceleration and D_h were positive for the active directions (outward from the backfill), and the settlement was positive for downward direction from the surface.

3.2 Test model preparation

The rubble mound and backfill of dynamic centrifuge test models were respectively constructed by two types of dry silica sand with different mean diameter. The physical properties of the sand are presented in Table 2. The rubble mound was densified with a relative density of 86% by compaction so as to prevent overturning and bearing capacity failure. In case of backfill, the soil was prepared using pluviation.

The sequence of model construction is as follows. (1) A 35 mm thick rubble mound was prepared underneath the structures by compaction. (2) The quay wall was placed at its designed location. (3) Sand was pluviated up to the surface height again. (4) Afterward, the front soil of the wall was removed by a vacuum cleaner. (5) Finally, all instruments were placed at pre-designed locations during pluviation. Industrial grease was covered between the model wall and the side walls of the container to reduce friction at the wall–container boundaries. Plastic sheets were attached to prevent sand from passing through those boundaries.

3.3 Earthquake input motion

Ofunato earthquake motion recorded at Miyagi-Ken Oki, Japan (Date: 1978/06/12, Magnitude = 7.9) was used as an input earthquake. The Ofunato record is characterised by a short-period-dominated spectral region, which is a common feature based on the seismicity and shallow bedrock conditions in Korea (Manandhar et al. 2017). The input earthquakes were applied incrementally, beginning with a weak acceleration amplitude, in the range of 0.01–0.43 g at the base of the ESB container, which represents the bedrock motion.

4 CENTRIFUGE TEST RESULTS

4.1 Moment of sliding failure

The results of 5 m, 10 m and 15 m wall experiments are shown in Figures 3, 4 and 5. (a) and (b) in each figure show the D_h of the wall and the settlements of the wall and backfill, respectively, based on the k_h definition proposed by MOF (2014). (c) and (d) show the results based on the k_h definitions from MLTM (2012).

The solid line in the figure is the sliding failure point of the wall calculated on the basis of MOF (2014) (i.e., wall height ≤10 m: base PGA = 2·k_h, 0.26 g; wall height > 10 m: surface PGA = 2·k_h, 0.26 g) and the dotted line means the sliding failure point of the wall calculated on the basis of the MLTM, (2012) (i.e., average PGA = k_h, 0.13 g). The shaded line means the initial point of sliding failure measured in this experiments.

In this study, the initial point for sliding failure was defined based on the following assumptions of the

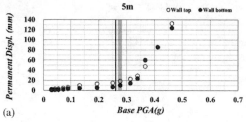

(a) Permanent displacements by base PGA Ports and fishing harbours design code (MOF 2014)

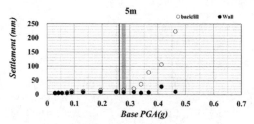

(b) Permanent settlements by base PGA Ports and fishing harbours design code (MOF 2014)

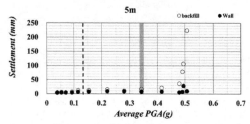

(c) Permanent displacements by average PGA (Harbours)Seismic performance evaluation & improvement revision of existing structures (MLTM 2012)

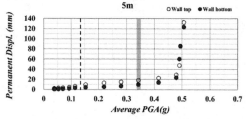

(d) Permanent settlements by average PGA (Harbours)Seismic performance evaluation & improvement revision of existing structures (MLTM 2012)

Figure 3. Determination of starting point of sliding failure (Wall height: 5 m).

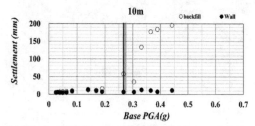

(a) Permanent displacements by base PGA Ports and fishing harbours design code (MOF 2014)

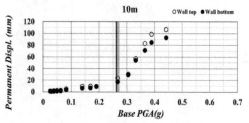

(b) Permanent settlements by base PGA Ports and fishing harbours design code (MOF 2014)

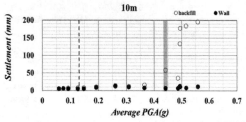

(c) Permanent displacements by average PGA (Harbours)Seismic performance evaluation & improvement revision of existing structures (MLTM 2012)

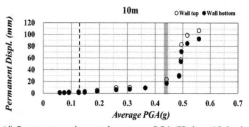

(d) Permanent settlements by average PGA (Harbours)Seismic performance evaluation & improvement revision of existing structures (MLTM 2012)

Figure 4. Determination of starting point of sliding failure (Wall height: 10 m).

M-O method, which are used conventionally in stability evaluation of sliding failure:

- The backfill is in a state of plastic equilibrium to produce minimum active pressure
- When the minimum active pressure is attained, a soil wedge behind the wall is at the point of incipient failure and the maximum shear strength is mobilised along the potential sliding surface

Firstly, upon considering first assumption, the point at which the D_h starts to increase after the backfill attains the state of minimum active pressure based on Terzaghi's experiments (Terzaghi 1941) is defined as the initial point of sliding failure. The corresponding D_h values are 5, 10 and 15 mm for 5, 10 and 15 m wall height, respectively, based on the D_h/H value of 0.001 applicable for dense sand.

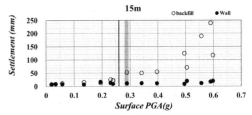

(a) Permanent displacements by surface PGA Ports and fishing harbours design code (MOF 2014)

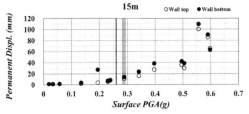

(b) Permanent settlements by surface PGA Ports and fishing harbours design code (MOF 2014)

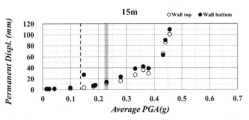

(c) Permanent displacements by average PGA (Harbours) Seismic performance evaluation & improvement revision of existing structures (MLTM 2012)

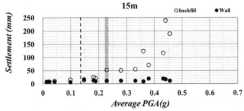

(d) Permanent settlements by average PGA (Harbours) Seismic performance evaluation & improvement revision of existing structures (MLTM 2012)

Figure 5. Determination of starting point of sliding failure (Wall height: 15 m).

Secondly, based on second assumption, the point of a soil wedge failure (i.e., the point when the relative settlement of the wall and backfill occurred) was defined as the initial point of sliding failure.

4.2 Evaluation of seismic coefficient definitions

From Figure 3, 4 and 5, it is evident that in cases of 5 m and 10 m wall, the design k_h (0.13) value is similar with that obtained using half of base PGA as suggested by MOF (2014). On the other hand, the k_h definition of MLTM (2012) showed a big difference with the design k_h. In case of 15 m wall, the design k_h value is comparable with k_h obtained from half of surface PGA as proposed by MOF (2014).

Additionally, the correction factor of 2.0 used in MOF (2014), which consider wall height, provides better estimate of design k_h than those used in MLTM (2012), which does not consider wall height. From these results, it can be stated that the PGA used in k_h calculation should be selected considering the wall height and that the appropriate correction factors derived for different cases of wall height should be applied to the PGA.

5 CONCLUSIONS

In this study, existing codes and guidelines for port structures were reviewed and a series of dynamic centrifuge tests were performed to evaluate k_h definitions of MOF (2014) and MLTM (2012). It was found that the correction factor according to seismic performance grade and the effect of wall height should be considered on PGA used in k_h calculation. Further, correction factors for the PGA used in k_h definition should be appropriately derived for different wall heights. This research has provided an important direction for arriving at a unified definition of k_h to be used in seismic codes and guidelines for simplified analysis of gravity quay walls.

ACKNOWLEDGMENTS

This research was a part of the project titled 'Development of performance-based seismic design technologies for advancement in design codes for port structures', funded by the Ministry of Oceans and Fisheries, Korea.

REFERENCES

CEN. 2004. Eurocode 8: Design of structures for earthquake resistance. Part 5: Foundations, retaining structures and geotechnical aspects. CEN, Brussels, EN 1998-5.

Kim, D.S., Kim, N.R., Choo, Y.W. & Cho, G.C. 2013a. A newly developed state-of-the-art geotechnical centrifuge in Korea, J. Civ. Eng., KSCE 17(1): 77–84.

Kim, D.S., Lee, S.H., Choo, Y.W. & Perdriat, J. 2013b. Self-balanced earthquake simulator on centrifuge and dynamic performance verification. J. Civ. Eng., KSCE 17(4): 651–661.

Lee, M.G., Jo, S.B., Cho, H.I., Park, H.J. & Kim, D.S. 2017a. A Discussion on the Definitions of Seismic Coefficient for Gravity Quay Wall in Korea. Journal of the Earthquake Engineering Society of Korea, 21(2): 77–85 (in Korean).

Lee, M.G., Ha, J.G., Jo, S.B., Park, H.J. & Kim, D.S. 2017b. Assessment of horizontal seismic coefficient for gravity quay walls by centrifuge tests. Géotechnique Letters 7(2): 211–217.

Lee, S.H., Choo, Y.W. & Kim, D.S. 2013. Performance of an equivalent shear beam (ESB) model container for dynamic geotechnical centrifuge tests. *Soil Dyn. Earthq. Eng* 44: 102–114.

Manandhar, S., Cho, H.I., & Kim, D.S. 2017. Site Classification System and Site Coefficients for Shallow Bedrock Sites in Korea. *J. Earthq. Eng.* 1–26.

Ministry of Land, Transport and Maritime Affairs (MLTM). 2012. Seismic Performance evaluation & improvement revision of existing structures (Harbours). *Korea Infrastructures Safety and Technology Corporation* (in Korean).

Ministry of Oceans and Fisheries (MOF). 2014. Ports and fishing harbours design code. *Ministry of Oceans and Fisheries* (in Korean).

Ministry of Transport (MOT). 1999. Technical standards for ports and harbour facilities in Japan. *Japan Port and Harbour Association* (in Japanese).

Mononobe, N. & Matsuo, H. 1929. On the determination of earth pressures during earthquakes. *Proceedings of the World Engineering Congress*, Tokyo, 177–185.

Noda, S., Uwabe, T. & Chiba, T. 1975. Relation between seismic coefficient and ground acceleration for gravity wall. *Report of the Port and Harbour Research Institute, Japan* 14(4): 67–111 (in Japanese).

OCDI. 2009. Technical standards and commentaries for port and harbour facilities in Japan. Japan: Overseas Coastal Area Development Institute.

Okabe, S. 1926. General theory of earth pressures. *J. Jpn. Soc. Civ. Eng.*, JSCE 12(1).

Terzaghi, K. 1941. General wedge theory of earth pressure. *ASCE Trans*. 106: 68–97.

Sleeve effect on the post-consolidation extraction resistance of spudcan foundation in overconsolidated clay

Y.P. Li & J.Y. Shi
Key Laboratory of Ministry of Education for Geomechanics and Embankment Engineering,
Geotechnical Engineering Research Institute, Hohai University, Nanjing, China

ABSTRACT: Previous studies have shown that the upward sleeve helps to enhance the spudcan bearing capacity in clay deposits. However, it is also concerned that this novel leg design may cause other problems such as extraction difficulty, the extraction resistance is demonstrated to increase with the increase of spudcan operation period. To address this problem, centrifuge model tests are performed herein to identify the upward sleeve effect on the post-consolidation extraction resistances of spudcan foundation in lightly over-consolidated clay. The centrifuge results show that compared with spudcan without sleeve, spudcan with square sleeve tends to yield a larger penetration resistance but lower maximum extraction resistance. This may be partially attributed to that the backfill soil during spudcan penetration is restricted outside the sleeve and the extraction resistance arising from the soil weight above the spudcan footing is hence reduced. This finding may demonstrate that including a sleeve at the lower part of the jack-up lattice leg with the same leg cross section area is feasible for offshore jack-up rigs, which will help to reduce the leg penetration without causing further extraction difficulty.

1 INTRODUCTION

Upon the completion of the drilling job in one location, the offshore jack-up rig is moved to other sites. The jack-up leg retrieval is usually time confusing and may encounter extraction difficulties (Purwana et al. 2009). Up to now, centrifuge model tests (Purwana et al. 2005; Hossain & Dong 2014) and numerical modelling (Zhou et al. 2009) have been conducted to study the soil failure mechanism and load-displacement response during spudcan extraction, which incorporating the effects of soil property, operation period and operation load ratio. Methods were proposed by Purwana et al. (2009) and Kohan et al. (2014) to predict the maximum spudcan extraction force, hereafter termed as breakout force. As found by Purwana et al. (2005), the breakout force required to extract the spudcan increases with the duration of the spducan operation period, and is mainly attributed to the increase of suction developed at the spudcan base. Following this, efforts were made by Gaudin et al. (2011) attempting to break the suction at the spudcan base using bottom jetting method.

As reported by Li et al. (2017), the spudcan bearing capacity is significantly enhanced by the presence of sleeve, and it is likely to reduce the spudcan penetration by including a sleeve to the lower part of the existing jack-up leg. However, it is also suspected that the including of the upward sleeve may cause other problems, such as the extraction difficulty as concerned above, especially for the case with long operation period, the extraction resistance may increase with the operation period. To address this problem, centrifuge model tests are performed herein to examine the post-consolidation breakout force for spudcans without and with square sleeve in the lightly over-consolidated clays.

2 CENTRIFUGE MODEL TEST

2.1 Centrifuge model set-up and testing procedures

The two centrifuge model tests were carried out at $100g$ using the beam Centrifuge in National University of Singapore. Reconstituted Malaysia kaolin clay was used here to prepare the lightly over-consolidated (LOC) clay beds, the soil properties are shown in Table 1. The mixed and de-aired clay slurry was firstly preloaded at $1g$ step by step, attempting to reach a final preloading of $150\,kPa$. However, as shown in Table 2, the final preloading for sample OC2 was slightly over-preloaded to $170\,kPa$. After the completion of preloading, the clay bed was moved into centrifuge to commerce self-weight consolidation and testing at $100g$. The final consolidated clay bed is approximately 27 m in prototype scale.

Figure 1 shows the model spudcans without and with sleeve used in the centrifuge model tests. A prototype spudcan diameter of 12 m was used (Figure 1(a)). Over the years, the spudcan footing has experienced rapid evolution involving both sizes and shapes to comply with the specific site conditions. The model spudcan used in this study is supposed to have a typical shape and moderate size in current mobile jack-up rigs. The upward square sleeve is extended from the top of spudcan, with a cross sectional-area of 61% of the maximum spudcan footprint. This sleeve arrangement has been studied by Li et al. (2017) with respect to its effect on spudcan penetration resistance. The same sleeve is herein used to examine its effect on spudcan extraction resistance.

A closed-loop servo-control hydraulic loading system was used to actuate the spudcan model and T-bar penetrometer, with a choice of either displacement or load control mode. Prior to the spudcan penetration, the initial soil strength was measured using a T-bar penetrometer at a constant penetration velocity of 3.2 mm/s to ensure undrained condition (Finnie & Randolph, 1994). The measured soil undrained shear strengths s_u with soil depth z are illustrated in Figure 2.

The centrifuge model testing procedure comprises four stages "penetration-unloading-consolidation-extraction" as below:

1. The spudcan was penetrated using displacement control mode with a constant velocity of 0.62 mm/s to ensure undrained condition (Finnie & Randolph, 1994) to a target depth of about 1.5D, D refers to the spudcan diameter at its largest cross sectional area.
2. The displacement control mode was switched to load control mode to conduct unloading, about 30% maximum penetration resistance was removed to simulate preload removal.
3. The remaining load was maintained for a long operation period as summarized in Table 2, which would allow the dissipation of the residual excess pore pressure caused by spudcan penetration.

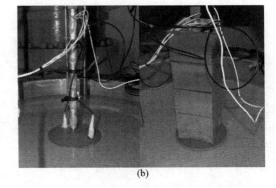

Figure 1. Model spudcan: (a) dimensions of model spudcan (unit: mm); (b) spudcan without and with square sleeve.

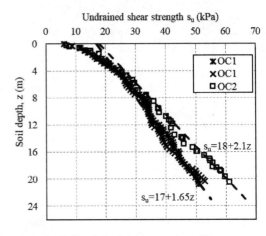

Figure 2. Soil undrained shear strength profiles.

Table 1. Soil properties of Malaysia kaolin clay.

Properties	Values
Liquid limit (LL): %	79.8
Plastic limit (PL): %	35.1
Specific gravity (G_s)	2.60
Coefficient of permeability (at 100 kPa) (k): m/s	2.0×10^{-8}
Effective unit weight (γ'): kN/m^3	6
Vertical coefficient of consolidation at different pressures C_v: (m^2/year)	
50 kPa	30
100 kPa	40
200 kPa	48
Internal friction angle at the critical state φ'	23°

Table 2. Centrifuge testing program.

Test No.	Leg type	Preloading (kPa)	Consolidation time t_c (days)	Consolidation load ratio[1]	Ratio of breakout force over maximum penetration resistance λ	Penetration depth d/D	Extraction distance to reach breakout force $\Delta d/D$
OC1	/	150	787	0.71	0.74	1.53	0.16
OC2	Square sleeve	170	2083	0.68	0.54	1.55	0.10

[1] Consolidation load ratio refers to the ratio of operation load during consolidation over the maximum installation.

Upon the completion of consolidation, the load-control mode was switched back to displacement control mode to conduct spudcan extraction, the spudcan was extracted at the same velocity of 0.62 mm/s.

2.2 Result analysis

Figure 3 shows the entire response of the spudcan load-displacement, in which d is the spudcan penetration depth, determined from the lowest point of the maximum cross section area of the spudcan footing. Take test OC1 for instance, the spudcan is installed to a depth of $1.53D$ with a penetration resistance of 54.1 MN being reached (A-B), which is followed by unloading to simulating operation loading (B-C), about 30% penetration resistance is removed. After 787 days consolidation period (C-D), the spudcan is extracted (D-E-F) out of the soil surface. As could be found in Figure 3, a breakout force of 40.2 MN is reached after an extraction distance of $0.16D$, which is about 74% of the maximum penetration resistance. This ratio value is much lower than those reported by Purwana et al. (2005) for spudcan extraction in normally consolidated (NC) Malaysia kaolin clays, in which about 97% and 110% penetration resistances are reached after a prototype operation period of 423 days and 843 days, respectively.

As could be observed from Figure 3, for spudcan with square sleeve (OC2), its breakout force is smaller than that of spudcan without sleeve (OC1) though it has a relatively larger soil strength (Fig. 2) and yields a larger penetration resistance. And the ratio of its breakout force over the maximum penetration resistance is about 54%, being reached after an extraction distance of about $0.1D$. This ratio is smaller than that of spudcan without sleeve as well. This larger maximum penetration resistance but lower breakout force maybe attributed to the presence of the sleeve. As already demonstrated by Li et al. (2017), the transient spudcan penetration resistance is enhanced by the sleeve mainly in two ways, firstly, the sleeve helps to restrict the soil backfill to the region of spudcan footprint; secondly, a larger influence zone surrounding spudcan footing is mobilized; the sleeve shaft resistance is insignificant as the back flow soil is largely remoulded and soften. Similarly, the sleeve may influence the spudcan extraction resistance in the same way. According to Hossain & Dong's (2014) PIV test result, the soil movement above the spudcan footing should be upward at the beginning of extraction for a relatively shallow embedded spudcan (less than $3D$). In this respect, the soil weight is supposed to contribute to the breakout force. Compared with spudcan without sleeve, a large amount of soil is prohibited outside the sleeve, thereafter reducing the soil weight above the spudcan footing and hence reducing the extraction resistance. As reported by Purwana et al. (2005), for spudcan without sleeve, the gain of soil strength above spudcan footing after a long operation period is minimal, thereafter is not likely to induce significant shaft resistance for spudcan with square sleeve.

In this paper, the interaction between the upward sleeve and the soil flow behaviour during spudcan extraction is not discussed, which however may influence the suction behaviour and hence the spudcan extraction resistance. This issue will be further addressed using numerical approach to shield light on the sleeve effect.

3 CONCLUSIONS

The forgoing discussion examines the post-consolidation breakout force of spudcan foundation in over-consolidated clay using centrifuge model test, which incorporates the effect of upward sleeve. The results show that the sleeve helps to enhance the spudcan penetration resistance and reduce the breakout force. This finding may demonstrate that including a sleeve at the lower part of the jack-up lattice leg with the same leg cross section area is feasible for offshore jack-up rigs, which will help to reduce the leg penetration without causing extraction difficulty.

ACKNOWLEDGMENTS

The authors would like to acknowledge the research funding provided by the National Natural Science Foundation of China (No. 41702290 and No. 41530637). The first author would also like to acknowledge the scholarship support from National University of Singapore.

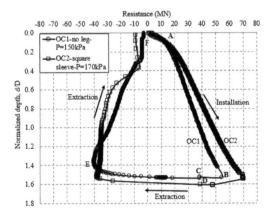

Figure 3. Load-displacement curves for spudcan penetration and extraction.

REFERENCES

Finnie, I.M.S. & Randolph, M.F. 1994. Punch-through and liquefaction induced failure of shallow foundations on calcareous sediments. *International conference on behaviour of offshore structures*, 217–230.

Gaudin, C., Bienen, B. & Cassidy, M.J. 2011. Investigation of the potential of bottom water jetting to ease spudcan extraction in soft clay. *Geotechnque*, 61(12): 1043–4054.

Hossain, S.H. & Dong, X. 2014. Extraction of Spudcan Foundations in Single and Multilayer Soils. *Journal of Geotechnical and Geoenvironmental Engineering*, 140(1): 170–184.

Li, Y.P., Yi, J.T., Lee, F.H., Goh, S.H. & Hu, J. 2018. Effect of lattice leg and sleeve on the transient vertical bearing capacity of deeply penetrated spudcans in clay. *Journal of Geotechnical and Geoenvironmental Engineering* 144(5), doi: 10.1061/(ASCE)GT.1943-5606.0001870.

Purwana, O.A., Leung, C.F., Chow, Y.K. & Foo, K.S. 2005. Influence of base suction on extraction of jack-up spudcans. *Géotechnique*, 55(10): 741–753.

Purwana, O.A., Quah, M., Foo, K.S., Nowak, S. & Handidjaja, P. 2009. Leg Extraction / Pullout Resistance – Theoretical and Practical Perspectives. *In. Proc. 12th Jack up Conf.*, London.

Zhou, X.X., Chow, Y.K. & Leung, C.F. 2009. Numerical modelling of extraction of spducans. *Geotechnique*, 59(1): 29–39.

Measuring the behaviour of dual row retaining walls in dry sands using centrifuge tests

S.S.C. Madabhushi & S.K. Haigh
Schofield Centre, Department of Engineering, University of Cambridge, UK

ABSTRACT: Protecting coastal populations requires innovative Civil Engineering solutions, as highlighted by the 2011 Tōhoku earthquake and tsunami. To this end, dual row retaining walls are being deployed along the Kochi coastline in Japan. Mechanically, how the soil-structure system resists seismic lateral loading is not well understood however. The baseline case of walls founded in dry sand is considered in this work.

Dynamic centrifuge tests of dual row walls were conducted and the performance monitored using multiple techniques. The PIV technique gives wall and soil displacements and bending moments were obtained from strain gauges. Further, the horizontal dynamic earth pressures were obtained from tactile pressure sensors.

The combination of techniques allows assessment of the internal measurement consistency and together advance understanding of the system behaviour. Decomposing the temporal variations rationalises the pre and post-earthquake results. This allows the overall structural behaviour to be linked to measured changes in the soil behaviour.

1 INTRODUCTION

In response to the catastrophic destruction of coastal communities caused by the recent combined Earthquake and Tsunami events, a number of Civil Engineering solutions are being pursued. Dual row retaining walls, which may be installed into existing Levees, can enhance the capacity of coastal sea walls. Such structures are being implemented in Kochi, Japan by Giken Ltd.

Figure 1 illustrates the dual row retaining wall concept. It is evident that the lateral capacity arises due to the combined properties of the wall and infill soil. Though primarily intended to break moderate waves and Tsunamis, it is necessary that such systems can withstand the initial earthquake loading. Thus establishing the dynamic performance of the dual row system is an important design criteria.

Particularly relevant for this application of dual row retaining walls may be the post earthquake capacity of the system, i.e. at the arrival time of a Tsunami wave. It is often observed both experimentally and in the field that seismic loading can lead to the locking in of soil stresses. This is ascribed to the plastic straining that is induced in the soil during the earthquake loading event.

In this paper results from a dynamic centrifuge test on a model dual row retaining wall are discussed. To form a baseline for future study, the case of a dual row wall in dry sand is considered first. A variety of measurement techniques are employed to elucidate the behaviour of the model. The wall displacements are measured using the Particle Image Velocimetry

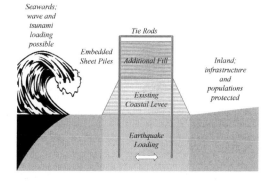

Figure 1. Dual row retaining wall, following from Madabhushi and Haigh (2016).

(PIV) technique (White et al. 2003). The curvature induced in the model walls was obtained from strain gauges arranged in full Wheatstone bridges. The lateral stresses were obtained through the use of tactile pressure sensors, following the processing technique described in Madabhushi and Haigh (2018). The examination of the wall displacement, bending moment and earth pressures before and after an applied earthquake are used to provide insight into locking in phenomena.

2 CENTRIFUGE MODEL

Schofield (1980) describes the importance of a geotechnical centrifuge in correctly capturing the

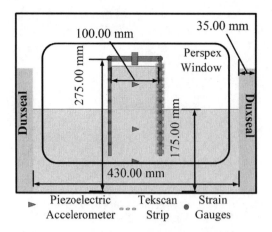

Figure 2. Simplified centrifuge model schematic.

Table 1. Details of wall system at prototype scale.

Variable	Value
Total Wall Height (*m*)	12
Embedment Depth (*m*)	6
Wall Thickness (*m*)	0.18
Bending Stiffness (MNm^2/m)	34
Tie Length (*m*)	6

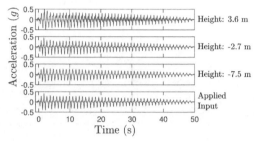

Figure 3. Input and infill accelerations.

stress strain behaviour of soils in small scale models, and the details of the 10 m Turner beam centrifuge used in these experiments. The centrifuge was used to apply an acceleration of 60 g at a point $\frac{1}{3}$ the height of the dual row models. Earthquake motions were imparted using the SAM actuator (Madabhushi et al. 1998). A simplified schematic of the centrifuge model is presented in figure 2. To allow cross-sectional views of the soil a rigid model box with a perspex front was used. Duxseal inserts were used between the soil and model container boundary to minimise the boundary effects. The degree to which these can absorb the stress wave reflections in dynamic tests has been reviewed and discussed by Campbell et al. (1991) and Coe et al. (1985).

The walls themselves are installed at 1g with S28 Hostun sand poured to a medium density in and around them utilising an automatic dry pluviation technique (Madabhushi et al. 2006). Installing the walls prior to the increase in centrifugal acceleration applied to the soil neglects the construction process of the walls and thus the static distribution of stresses developed in the model may not reflect the field case. Nevertheless, application of large earthquakes is likely to reduce the discrepancy in stress states between the model and field as the soil tends to limit states in both cases. The details of the prototype wall system tested is summarised in table 2. The walls are considerably more flexible than those previously tested and reported by Madabhushi and Haigh (2016).

Unless stated otherwise, the values reported are at the prototype scale and the sign convention that leftward displacements are positive is adopted.

3 MEASUREMENT TECHNIQUES

3.1 *Applied input motion and accelerations*

Piezoelectric accelerometers were used to measure the accelerations of the soil. An example of the recorded traces is plotted in figure 3, which shows the variation at increasing heights in the soil infill given the applied input motion. The desired input motion was a tone burst of 1 Hz with a magnitude of 0.35 g and the shaking duration was prolonged to facilitate examination of the locked in stresses. The accelerations recorded are discussed further in Section 5.

3.2 *Measuring displacements*

The measurement of wall displacements utilised the PIV technique. Discrete points were printed on the thin walls and tracked during the swing up, earthquake loading and swing down of the model. All displacements are relative to the configuration of the walls at 1g. As the soil stresses are increased one would expect the wall system to bow out as the infill soil exerts larger lateral stresses. In figure 4 the tracked displacements are shown at the prototype scale, and confirm this to be the case. Between the left and right walls the deflections are reasonably symmetric. The displacements indicate that the walls bow out above the ground level, reaching maximum deflection at approximately 2m height and reduce on either side of this value.

The post earthquake wall displacements are also shown in figure 4. The left wall has clearly accumulated a large leftwards displacement and the right wall has followed it near the tie level. However, a large majority of the right wall has displaced rightwards following the earthquake loading. This suggests that overall the soil infill has settled and moved outwards.

3.3 *Measuring wall flexure*

The deflection of the wall is clearly associated with an increase in curvature and thus internal bending moment in the wall. These are measured using a series

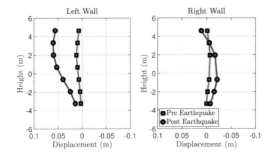

Figure 4. Wall displacements from PIV technique.

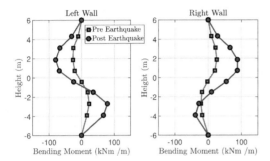

Figure 5. Wall flexure from gauges

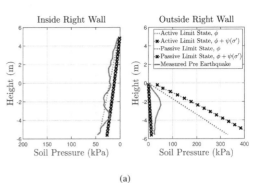

(a)

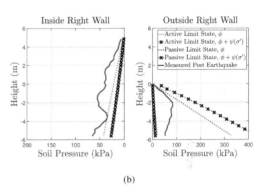

(b)

Figure 6. Earth pressures from Tekscan (a) pre-earthquake (b) post earthquake.

of full bridge strain gauges affixed to the inside and outside of the wall. The bending moment values are calculated relative to the 1g readings. Though there is not 'zero bending' in the wall at this state, the error introduced is negligibly small relative to the offset obtained from the calibration of the walls at 1g.

Figure 5 shows the bending developed in the wall following swing up and the application of the earthquake. Consistent with the wall deflections, the pre-earthquake values are quite symmetric. The bending moment profile shows that the walls bend outwards above the ground level. The magnitude of the bending moment reversal that occurs below the ground is similar to the moment above it. This is consistent with the small values of rotation that may be inferred from the pre-earthquake wall displacements. The reversal of bending moment also implies that the shear force on the wall has reversed. This is likely due to the lateral stress exerted by the soil outside the walls.

The post earthquake values are asymmetric similar to the wall displacements. The larger permanent displacements have also resulted in larger bending moments, particularly in the outward direction. Again, this suggests the bulk of the infill soil has settled and moved outwards against the dual wall system.

3.4 Measuring soil pressure

The calibration, measurement and post processing of the dynamic soil pressure data recorded by the Tekscan sheets has been discussed previously, for example by Paikowsky and Hajduk (1997), Dashti et al. (2012) and Madabhushi and Haigh (2018). The processing technique described by the latter was utilised in this work, with the resulting pre and post earthquake pressure distributions around the right wall plotted in figure 6.

The variation of the earth pressures inside and outside the wall rationalise the observed bending moments. Before the earthquake the fairly uniform driving pressure inside the right wall is balanced by a combination of the tie force and the external soil pressure. The variation in external soil pressure reaches a peak close to the soil surface and then decreases. Thus, the wall is driven outwards between the tie and the ground level, below which the restoring soil pressure causes it to inflect backwards.

Following the earthquake there is clearly an increase of earth pressures both inside and outside the right wall. The previous results have shown increased bending moment and deflection of the wall, and for the right wall the final bending moment features larger outward than inward bending. This suggests the net pressure experienced by the wall has changed, i.e. the increase in the internal and external pressures does not cancel.

It is evident from figure 6 that the external earth pressures are not readily described by a triangular distribution which may be used for 'yielding walls' or an inverted triangular distribution for 'non yielding walls' (where 'yielding' refers to the stress state of the soil). Nevertheless, comparison with limit state values may prove useful, and thus figure 6 also features envelopes

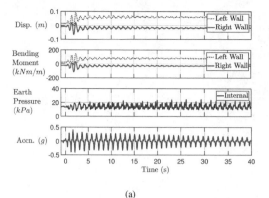

(a)

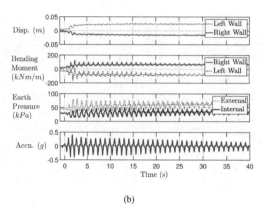

(b)

Figure 7. Total time traces (a) 2 m (b) −2 m.

based on the Coulomb earth pressure theory (Clayton et al. 1986). Because the soil was poured to a medium density, envelopes based on the critical state value of 33° are not the limiting values for all strains.

A simplified method to account for the stress dependent dilation is to follow the relations proposed by Bolton (1986), which correlate the peak dilation angle ϕ_{peak} with a relative density index I_R. This in turn accounts for the relative density I_d, confining stress p' and grain strength p'_c. Because the confining stress depends on the limit state being predicted, this is achieved via an iterative calculation outlined by equations 1–3. The comparison between the measured earth pressure and the predictions of the limting earth pressure coefficient K_{limit} in figure 6 are reasonably good, and the importance of accounting for the stress dependent dilation, even if only by using simple relations, is exemplified.

$$\phi_{peak} = \phi_{crit} + 5I_R \quad (1)$$

$$I_R = I_D I_C - 1 = I_D ln(\frac{p'_c}{p'}) - 1 \quad (2)$$

$$p' \propto K_{limit}(\phi_{peak}) \quad (3)$$

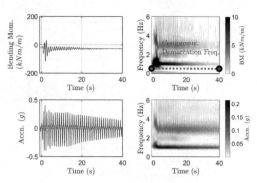

Figure 8. Using wavelet transforms to divide components of time traces.

4 TIME DEPENDENT VARIATION

To better understand the locking in phenomena the recorded variations from selected wall heights are plotted in figure 7. The wall displacements, bending moments and earth pressures around the right wall are shown. From examination of the variation at either height the variations may be divided into the immediate response to the transient loading and the overall change in the average value. The applied acceleration was continued for many cycles but neither the displacements, bending moments or pressures show continual changes in the mean value.

It is desirable to consider separately the components of the response in terms of the changing mean value and the more rapid dynamic oscillations. Wavelet transforms were taken of the responses in order to identify demarcation frequencies appropriate for all times. An illustration of this method is provided by figure 8.

Figure 9 shows the individually reconstructed parts of the signals from inverting the divided wavelet components. The traces which slowly vary are intended to reflect the best guess of the residual values that would persevere if the earthquake was stopped at that instant. From figure 9 it is interesting to see that during the initial cycles the largest mean displacement and bending moments are generated. At this instant the smallest residual passive pressure is observed at −2 m. The continued applied shaking results in reduced residual values. The dynamic variations are also largest during the initial cycling, and reduce following this. Comparison between the two wall heights shows larger dynamic variation at depth. Notably, the variation of the pressures inside and outside the wall are of similar magnitude and in phase.

5 STRESS-STRAIN BEHAVIOUR OF SOIL

From the previous section it is evident that following the initial cycles the dynamic behaviour of the system alters. Given the magnitude of the induced bending moments, the walls will behave elastically, which implies that it is the soil plastically straining that is responsible for the overall change in the system

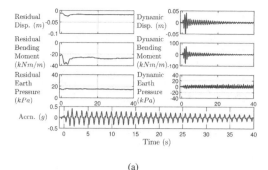

(a)

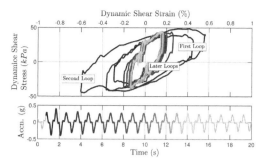

Figure 10. Shear stress-strain loops from accelerometers.

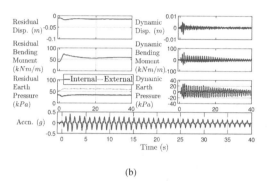

(b)

Figure 9. Components of temporal variations for Right Wall (a) 2 m (b) −2 m

response. The cyclic stress strain response of the soil is thus desired to investigate this phenomenon.

The soil shear stress strain behaviour may be inferred from the measured accelerations (Brennan et al. 2005). The relative acceleration with height and soil mass are converted to equivalent inertial loads to calculate the shear stress. The double integral of the acceleration with respect to time is linked to the relative displacement and thus shear strain between measurement points. The values are thus approximations to the average behaviour across the instrumented soil column. Further, because the accelerations must be bandpass filtered to remove miscellaneous displacement drifts (Brennan et al. 2005), only the dynamic components are obtained. Figure 10 shows the inferred loops from the infill accelerations recorded at the locations shown in figure 2. Brennan et al. (2005) discuss the importance of not over zealously filtering the accelerations to produce 'cleaner' loops which mischaracterise the real behaviour. Thus the lower and upper bandpass limits used were 0.4% and 90% of the Nyquist frequency respectively.

Figure 10 is interesting in that there is a significant change in the overall cyclic behaviour. Between early and late cycles it is evident that the area enclosed by the loops significantly decreases whilst the stiffness increases; i.e. the soil system absorbs less energy at latter cycles. As a general observation, this is consistent with the observations from figure 9; the variation of the displacement, bending moment and earth pressure only lasted a few cycles. After this, the 'residual values' are fairly constant.

It is reasonable that for approximately constant input accelerations the shear stress cycles do not diminish. Thus, the reduction in the area of the curves arises predominantly from a decrease in the cyclic shear strains that develop. From figure 10 it is evident that the soil stiffness has increased and thus the dynamic displacements, bending moments and earth pressures reduce. The soil is said to have hardened due to the stress increase; and this change is often an important parameter in soil constitutive models. However, that this hardening is achieved at the cost of permanent displacements and locked in bending moments and earth pressures is not apparent from only the cyclic variations.

To investigate this aspect of the locking in phenomenon, an alternative means of examining the soil behaviour utilising the PIV technique is explored. The dynamic displacements of the soil infill were obtained using PIV and the spatial derivatives calculated. The average strain values from discrete regions between the accelerometers in the infill were calculated and are compared with the overall average value predicted by the accelerometers in figure 11. The shear strains calculated from the PIV displacements were decomposed into the monotonic and dynamic components.

As a sense check on the calculated shear strains, if the top of a 12 m wall deflects by 0.1 m relative to its base an average shear strain of 1.6% would be expected. From this point of view, the magnitude of shear strains from either method appear quite reasonable. Both methods also show a dramatic reduction in the dynamic shear strain that corresponds with the largest variations observed in the wall bending moments and earth pressures as measured by the gauges and Tekscan respectively. The variations from the PIV are slightly coarse due to the displacement sampling rate (c.f. 900 fps with 6 kHz of the accelerometers) However, the value of examining the dynamic shear strains more locally is revealed by the variation that exists between the heights. Close to the soil surface much larger dynamic shear strains are induced than at depth.

The second advantage of calculating the dynamic shear strains from the measured soil displacements

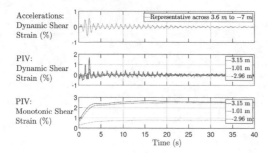

Figure 11. Comparison of strains measured using independent methods.

is that the monotonic shear strain may be examined. From figure 11 it is made clear that the reduction in the dynamic oscillations of the shear strain are coincident with the plateau in the monotonic shear strains at all depths. It is fundamentally these permanent shear strains that can explain previous observations of the soil hardening in the infill, increased residual earth pressures and hence the permanent displacement and bending of the wall.

6 CONCLUDING REMARKS

It has been illustrated how a variety of measurement techniques can be combined to further the insight into the model behaviour of a dual row retaining wall. The observation that permanent wall displacements arise from earthquake loading is attributed to locked in soil stresses. The total variation of the measured quantities are decomposed depending on the rate of variation. Changes in the rapid dynamic oscillations are rationalised using the cyclic shear stress strain behaviour of the soil. The shear strains calculated from the measured soil displacements confirmed this and also elucidated the role of the monotonically increasing component. Though the system absorbs less energy at later cycles, the dynamic response is attenuated due to the soil hardening arising from the locked in stresses.

From a practical viewpoint, the wall system is able to utilise the soils plastic capacity in the face of large and prolonged ground motions. This results in greater resistance to accumulating large permanent displacements and thus indicative of both an efficient and robust design option.

REFERENCES

Bolton, M. D. (1986). The strength and dilatancy of sands. *Géeotechnique 36*(1), 65–78.

Brennan, A. J., N. I. Thusyanthan, & S. P. G. Madabhushi (2005). Evaluation of Shear Modulus and Damping in Dynamic Centrifuge Tests. *Journal of Geotechnical and Geoenvironmental Engineering 131*(12), 1488–1497.

Campbell, D. J., J. A. Cheney, & B. L. Kutter (1991). Boundary effects in dynamic centrifuge model tests. In *Centrifuge 1991: proceedings of the International Conference Centrifuge 1991*, Boulder, Colorado, pp. 441–448. A.A. Balkema.

Clayton, C. R., R. I. Woods, A. J. Bond, & J. Milititsky (1986). *Earth Pressure and Earth-Retaining Structures* (3rd ed.). CRC Press.

Coe, C. J., J. H. Prevost, & R. H. Scanlan (1985). Dynamic stress wave reflections/attenuation: Earthquake simulation in centrifuge soil models. *Earthquake Engineering & Structural Dynamics 13*(1), 109–128.

Dashti, S., K. Gillis, & M. Ghayoomi (2012). Sensing of Lateral Seismic Earth Pressures in Geotechnical Centrifuge Models. In *15th World Conference on Earthquake Engineering, Lisbon Portugal*, pp. 1–10. Sociedade Portuguesa de Engenharia Sismica (SPES).

Madabhushi, S. P. G., N. E. Houghton, & S. K. Haigh (2006). A new automatic sand pourer for model preparation at University of Cambridge. In *Proceedings of the Sixth International Conference on Physical Modelling in Geotechnics*, Hong Kong, pp. 217–222. Taylor & Francis.

Madabhushi, S. P. G., A. N. Schofield, & S. Lesley (1998). A new Stored Angular Momentum (SAM) based earthquake actuator. In *Proceedings of The International Conference Centrifuge '98, Tokyo*, Tokyo, Japan, pp. 111–116.

Madabhushi, S. S. C. & S. K. Haigh (2016). The Influence of Embedment on the Seismic Performance of Dual Row Retaining Walls. In *6th International Conference on Recent Advances in Geotechnical Earthquake Engineering and Soil Dynamics*, New Delhi, India. Missouri University of Science and Technology.

Madabhushi, S. S. C. & S. K. Haigh (2018). Using Tactile Pressure Sensors to Measure Dynamic Earth Pressures around Dual Row Walls. *International Journal of Physical Modelling in Geotechnics*, 1–30.

Paikowsky, S. G. & E. L. Hajduk (1997). Calibration Use of Grid-Based Tactile Pressure Sensors in Granular Material. *Geotechnical Testing Journal 20*(2), 218–241.

Schofield, A. N. (1980). Cambridge Geotechnical Centrifuge Operations. *Géeotechnique 30*(3), 227–268.

White, D. J., W. A. Take, & M. D. Bolton (2003). Soil deformation measurement using particle image velocimetry (PIV) and photogrammetry. *Géeotechnique 53*(7), 619–631.

Verification of improvement plan for seismic retrofits of existing quay wall in small scale fishing port

K. Mikasa & K. Okabayashi
National Institute of Technology, Kochi College, Japan

ABSTRACT: In the Pacific coastal area around Shikoku Island especially Kochi prefecture, Japan, serious damage from the Nankai Trough earthquake and resulting tsunami is expected. Eighty-eight fishing ports around the Kochi coastline are particularly at risk. Therefore, seismic retrofits are necessary because ports function not only as a workplace for fishermen, but also as bases for rescue and restoration following disasters. Furthermore, according to the 2011 Great East Japan Earthquake damage investigation report, fishing ports functioned as important bases for transportation of emergency supplies. In this study, effectiveness of seismic retrofits using double wall sheet piling and precast concrete slabs proposed by the authors were verified by seismic response analysis applying the dynamic effective stress method and experimental study using a dynamic centrifuge representing Kaminokae Fishing Port in southwestern Kochi Prefecture as a case model.

1 INTRODUCTION

The Pacific coastal area in Tohoku suffered serious damage from the 2011 Great East Japan Earthquake.

Similarly, serious damage in the Pacific coastal area of Shikoku, Japan from the Nankai Trough earthquake and tsunami is expected. Eighty-eight fishing ports in Kochi including four primary disaster management base ports, eight secondary disaster management base ports are at risk. Quay walls at these ports require additional seismic retrofits to ensure port function during disaster restoration efforts. Some fishing villages could be isolated if roads to these villages are destroyed by Nankai Trough earthquake and tsunami. Therefore, seismic retrofits to these ports are essential for disaster mitigation. Furthermore, seismic retrofits to these ports is urgent task and they should be economical and effective because of the large number of ports.

Figure 1. Kaminokae area.

As a case model, Kaminokae Fishing Port, a small scale port of older design, was selected due to the likelihood of this village being cut off from road access after a major Nankai event.

In this study, there are two main purposes, first, to develop an efficient construction method for seismic retrofits of existing quay walls for rescue and restoration activities immediately after a disaster, second, to promptly disseminate this construction method

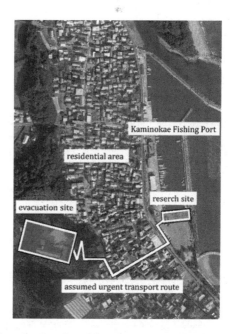

Figure 2. Location of research site and evacuation site in Kaminokae area.

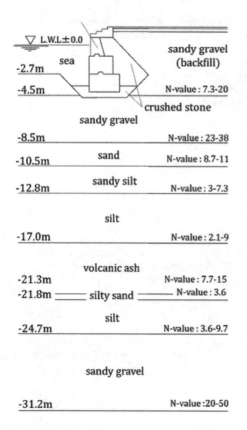

Figure 3. Geological profile of Kaminokae Fishing Port.

for seismic improvement of small scale and older ports like Kaminokae.

2 MODEL DESIGN

2.1 Test model

The test model consists of the gravity retaining quay wall constructed on sandy ground, referred to the boring survey data and the quay wall design data of Kaminokae fishing port. (Kochi Prefecture Oceanic Service Fishing Port Section 2005) (Fig. 3) This model was tested in two patterns. Case 1 is existing model before reinforcing seismic retrofits. Case 2 is reinforced seismic retrofits by double wall sheet piling. In 2011 Great East Japan Earthquake, double wall sheet pile structure showed strength against disaster.

In Case 2, the space in the behind of quay wall is about 7 m to use this port as a rescue and restoration base at the time of the disaster. This space shall not be destroyed by the earthquake. (Fig. 4)

2.2 Outline of the numerical analysis

The ground characteristics used for numerical analysis are shown in Fig.3. Test model was analyzed in two patterns. Model 1 is same condition as Case 1. Model 2 is same conditions as Case 2. (Tokuhisa 2016) Model 1 and Model 2 were analyzed by the dynamic effective stress analysis program LIQCA (Oka 2015, LIQCA Liquefaction Geo-Research Institute 2015). In Model 2, the depth of the sheet pile is 20 m. (Okabayashi, Tagaya & Hayashi 2008) The spacing of sheet piles in Model 2 is 6.3 m and connected each other by tie rods.

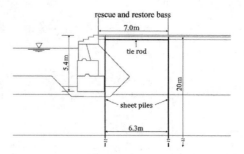

Figure 4. Rescue and restoration base plan.

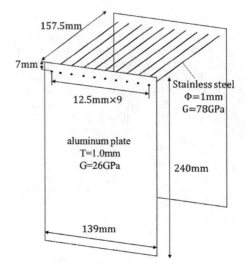

Figure 5. Details of model double wall sheet pile.

2.3 Outline of the centrifuge test

Case 1 and Case 2 were experimented by dynamic centrifuge. The dynamic centrifuge used in this study has an electronical servo-controlled shaking table. (Hashida 2016) This test was conducted at a centrifuge acceleration of 40 g. The model container is 45 cm long (18 m in prototype) and 13.9 cm wide (5.56 m in prototype). Figure 6 shows Case 2. This model was modelled by Toyoura sand ($D_r = 70\%$). The quay wall block was made with mortar. Crushed stone under the quay wall used gravel that is about $2 \sim 5$ mm grain size. The liquid used for saturation was methylcellulose solution with a viscosity 40 mpa · s. In Case 2, this model is double walled cofferdam. The model sheet piles used for Case 2 were made of aluminum plates (Fig. 5). The positions of pore water pressuremeter and accelerometer are shown in Figure 6. (Case 1 is without sheet piles in Fig. 6).

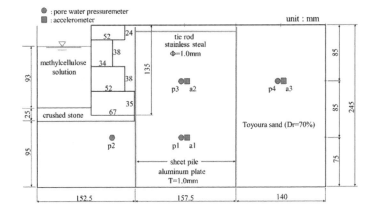

Figure 6. Model set up for centrifuge test.

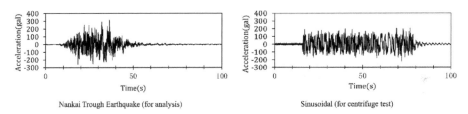

Figure 7. Input acceleration.

2.4 Earthquake wave

In the tests, two different horizontal accelerations, shown in Figure 7. In the numerical analysis, the seismic wave of the Nankai Trough Earthquake at Kaminokae Fishing Port created by Kochi Prefecture was applied. In the dynamic centrifuge experiment, a 15 Hz sinusoidal wave (20 cycle, amplitude = 3 mm) was applied, because shaking table does not have enough performance to shake the seismic wave of the Nankai Trough Earthquake at Kaminokae Fishing Port.

3 TEST RESULTS

3.1 Analysis results

Analysis results of displacement of Model 1 are shown in Figure 8. During shaking (40.5 s), the quay wall was settled and inclined towards the sea side, the backfill was also moved and settled. After shaking (100 s), finally, the quay wall has inclined about 115 cm towards the sea side and settled about 25 cm and be destroyed, the backfill has settled about 45.7 cm and moved towards quay wall. Therefore, function of port is lost.

Analysis results of displacement in Model 2 are shown in Figure 9. During shaking (40.5 s), the quay wall was settled a little, however it was not destroyed or inclined like Model 1. In the space of reinforced seismic retrofits by double wall sheet piling, the backfill

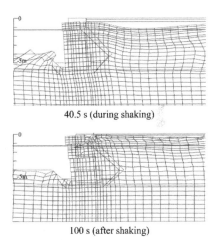

Figure 8. Displacement model 1.

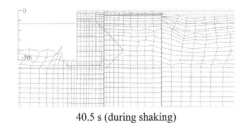

Figure 9. Displacement of model.

was settled a little. Moreover, in the behind of double wall sheet piling, the backfill was more settled.

Model 2 diverged at 45.8 seconds. Therefore, the displacement of after shaking (100 s) could not be confirmed. However, the effectiveness of seismic retrofits by double wall sheet piling can be confirmed at 40.5 s.

3.2 Centrifuge test results

3.2.1 Pore water pressure response

Time histories of excess pore water pressure ratios are shown in Figure 10. From results, increasing of excess pore water pressure were suppressed by double wall sheet piling. At all measuring point, excess pore water pressure did not increase to effective overburden pressure level. Therefore, all measuring point did not liquefy. However, p3 and p4 in Case 2 showed notable suppressing the increase of pore water pressure.

3.3 Acceleration response

Time histories of acceleration responses are shown in Figure 11. Acceleration response of Case 1 and Case 2 became almost same. Acceleration of a1 was not much different from the input acceleration. In Case 1, accelerations of a2 and a3 has amplified from the input acceleration.

3.4 Movement of quay wall and ground

Case 1: Observed movement in Case 1 are shown in Figure 12. The measured movements of Case 1 are shown in Table 1. From results, the quay wall has inclined towards the sea side and settled, the backfill has settled and been lateral spreading. Therefore, the quay wall model in Case 1 lost the function due to the increase of pore water pressure.

Case 2: Observed movement in Case 2 are shown in Figure 13. The measured movements of Case 2 are shown in Table 2. From results, the damage due to increase of porewater pressure in Case 2 was suppressed by double wall sheet piling. However, local settling and settling behind the double wall sheet pile model due to the increase of pore water pressure was confirmed between the quay wall and the sheet pile.

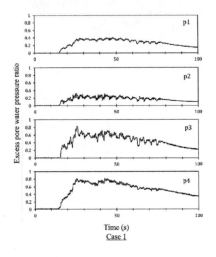

Figure 10. Excess pore water ratio.

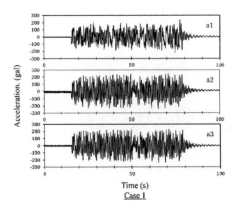 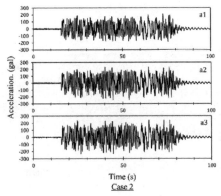

Figure 11. Acceleration response.

3.5 Comparing numerical analysis results with centrifuge test results

The ground characteristics of numerical analysis differs from centrifuge test. However, in Case 1 (existing model before reinforcing seismic retrofits), movements of the model common to numerical analysis result (Model 1) and centrifuge test result (Case 1) was confirmed.

In Case 2 (the model is reinforced seismic retrofits by double wall sheet piling), numerical analysis (Model 2) diverged. Therefore, movements of the model common to numerical analysis result (Model 2) and centrifuge test result (Case 2) could not be confirmed. Nevertheless, the effectiveness of double wall sheet piling was confirmed in each test.

4 THE PROPOSAL EXAMPLE OF SEISMIC RETROFIT PLAN

Damage of inclining the quay wall was reduced and in the space of between sheet piles, the increase of pore water pressure and displacement of ground were suppressed by double wall sheet piling. That denotes

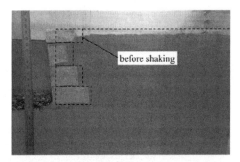
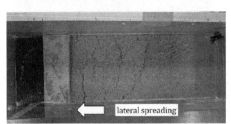

Figure 12. Observed movements in Case 1.

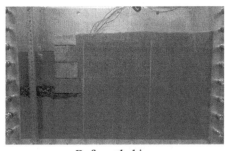
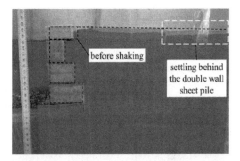
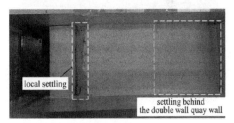

Figure 13. Observed movements in Case 2.

Table 1. Measured movements of Case 1.

	Model	Prototype
Maximum settlement of the quay wall	5 mm	24 cm
Maximum displacement of the quay wall	15 mm	60 cm
Maximum settlement of the Fig. 12 Observed movement in Case 1 backfill	12 mm	48 cm

Table 2. Measured movements of Case 1.

	Model	Prototype
Maximum settlement of the quay wall	1 mm	4 cm
Maximum displacement of the quay wall	5 mm	20 cm
Maximum settlement of the backfill	5 mm	20 cm

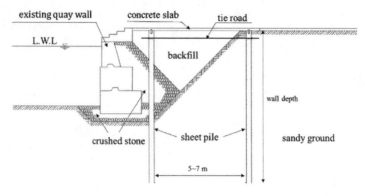

Figure 14. A proposal example of seismic retrofits.

the effectiveness of seismic retrofits by double wall sheet piling.

In Figure 14 the proposal example of seismic retrofit plan is shown. In the plan, the precast concrete slab is fixed on sheet piles. The space on the precast concrete slabs are used as rescue and restore base for unloading the relief supplies from ships. This plan is more economical than constructing the new quay wall and should not take much time to carry out construction work.

5 CONCLUSIONS

The following main conclusions can be drawn based on the results of analysis and centrifuge tests;

1. In Case1, the movements of observed in numerical analysis and centrifuge test were similar.
2. In Case 2, analysis result after shaking could not be obtained. However, from the results of the analysis and centrifuge test, the effectiveness of seismic retrofits by double wall sheet piling could be confirmed.
3. In centrifuge test of Case 1, excess pore water pressure did not increase to effective overburden pressure level. However, from movements of the quay wall and the backfill, effect of increasing of pore water pressure was confirmed.
4. In centrifuge test of Case 2, at the behind of sheet pile (p4), excess pore water pressure did not increase and the backfill did not much settle. However, in analysis of Case 2, during shaking (40.5 s) at the behind of sheet pile, backfill was settling. It was caused by inevitable side wall effect in centrifuge test.
5. In this study, effectiveness of seismic retrofits by double wall sheet piling could be confirmed. Therefore, seismic performance can be obtained by reinforcing seismic retrofits for existing quay wall instead of building new quay wall with seismic retrofits. It can be economical with construction cost and construction period and can disseminate reinforcing seismic retrofits for small scale and old type port like Kaminokae Fishing port.
6. From results we showed the proposal example of seismic retrofit plan. However, wall depth of sheet pile is not decided from this study. Because wall depth involves in the economic efficiency. Therefore, further studies are necessary to decide the wall of depth.

REFERENCES

Hashida, K. 2016. Development of Dynamic Centrifuge and Study on Vibration Earth Pressure and Embankment experiment, Thesis Research Reports, Advanced Course Kochi National College of Technology No.15 March 2016: 25–32.

Kochi Prefecture Oceanic Service Fishing Port Section. 2005. Tsunami countermeasure basic policy in fishing village.

LIQCA Liquefaction Geo-Research Institute. 2015 LIQCA 2D15 · LIQCA3D15II-2-II-38.

Oka, F. 2015. Earthquake response analysis of ground considering liquefaction using LIQCA, The Geotechnical Engineering Magazine, 693: 12–15.

Okabayashi, K., Tagaya, K. & Hayashi, Y. 2008. Study on Earthquake Resistant Reinforcement of Existing Quay in Fishing Port to Nankai Earthquake, Proceedings of The Eighteenth ISOPE Conference, 39–44.

Tokuhisa, K. 2016. A study on the liquefaction countermeasures of wharf considering the Nankai Trough earthquake, Thesis Research Reports, Advanced Course Kochi National College of Technology No. 15 March 2016: 9–16.

Visualisation of mechanisms governing suction bucket installation in dense sand

R. Ragni, B. Bienen, S.A. Stanier, M.J. Cassidy & C.D. O'Loughlin
Centre for Offshore Foundation Systems, The University of Western Australia, Perth, Australia

ABSTRACT: Suction buckets are an increasingly considered foundation option for offshore wind turbines. Although the required suction can be predicted well using existing methods, uncertainty remains around some input parameters, because the effects of suction installation on the soil state are not understood in detail. This paper visualises the mechanisms governing both initial self-weight penetration and following suction-assisted installation in dense sand. Pioneering particle image velocimetry measurements in a centrifuge environment underpinned the investigation, with details of the experimental apparatus offered in the paper. Changes in the deformation mechanisms governing the installation process and in the soil properties are revealed. The findings have an impact on the understanding of the formation of internal plug heave – the cause of premature refusal – and the prediction of the installation response. Revealing changes in void ratio and permeability also present implications on the accumulated displacements under the metocean loading, which may conflict with serviceability requirements.

1 INTRODUCTION

The suction bucket technology is nowadays regarded with interest for the foundations of offshore wind farms (Houlsby et al. 2005, Houlsby 2016), in relatively shallow waters. Associated with this application are more varied and coarser grained seabeds, requiring much stubbier suction buckets, and very different loading conditions when compared with traditional oil and gas applications (Tjelta 2015).

Suction buckets comprise a large diameter steel cylinder, open-ended at the bottom and closed at the top. In shallow water applications, typical aspect ratios are $L/D \leq 1$, where L is the skirt length and D is the bucket diameter, and $D/t = 100 \sim 250$, where t is the skirt thickness.

The first part of the installation is achieved under the self-weight of the bucket and substructure, with free flow of water through the vented bucket lid. When self-weight penetration is completed and a seal between the bucket and soil is established, the venting valve in the lid is closed and water is pumped from inside the bucket. This generates a differential pressure and a consequent additional downwards driving force in excess of the suction bucket submerged weight. In sand, the introduction of a suction-generated seepage field causes a reduction of the vertical effective stresses around the skirt tip, which reduces the vertical force required to install the bucket.

The initial self-weight penetration resembles other penetration processes, such as jacked installation of large diameter piles, although neglecting the vertical stress enhancement through the action of the downward friction has been flagged as non-conservative for the prediction of suction bucket installation (Houlsby & Byrne 2005). However, no guidance is provided on the size of this down-drag enhancement.

Suction-assisted penetration further induces changes in the soil state that are yet to be fully characterised, particularly in terms of variation in void ratio and permeability. Sound understanding, however, is crucial, as changes to the sand state considerably impact predictions. Loosening of the soil plug may result in premature refusal during suction bucket installation and adversely affect foundation performance in service.

This work presents particle image velocimetry measurements of suction bucket installation in dense sand, obtained in a centrifuge environment at enhanced gravity level. Insights into the self-weight and suction-assisted stages are provided, revealing changes in the deformation mechanisms and advancing fundamental understanding of the change of state of the sand plug.

2 EXPERIMENTAL SET-UP

2.1 Centrifuge apparatus and arrangement

The experiments presented in this paper were carried out in the geotechnical beam centrifuge at The University of Western Australia (Randolph et al. 1991), at an acceleration of 40 g. Particle Image Velocimetry (PIV) techniques (Stanier & White 2013, Stanier et al. 2015) were used to reveal the failure mechanism associated with the suction bucket installation, where a half model of the foundation was penetrated against a Perspex window, and photographs of the event were captured.

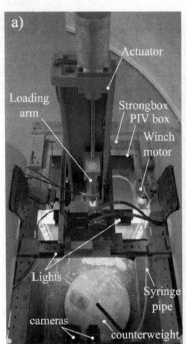
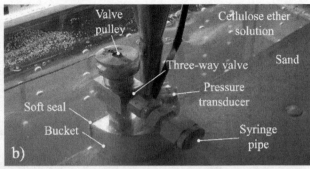

Figure 1. Illustration of the set-up for centrifuge testing.

An overview of the apparatus is presented in Figure 1a. A PIV box (inner dimensions 335×225×300 mm) containing the soil sample was housed in a larger sample container, termed 'strongbox'. The latter housed the remaining equipment; in particular, two cameras were placed opposite the PIV box, along with a 20 kg counterweight to limit the unbalance of the strongbox as it swung into position at 40 g. The bucket was connected to an electrically driven actuator using a steel shaft, and a 2 kN axial load cell was located between the bucket and the shaft. The actuator controlling the bucket motion was located on top of the strongbox. The actuator has two degrees of freedom: vertical, which was used to initially penetrate the bucket and control the bucket self-weight (via the axial load cell), and horizontal, which was used to position the bucket against the Perspex window. Two LED light panels provided sufficient lighting, such that the exposed face of the model was evenly and well illuminated. Finally, a winch and a syringe pump were fitted on the strongbox to control the fluid flow from the bucket (as described later).

The aluminium bucket model was designed with a skirt length $L = 50$ mm and outer diameter $D = 50$ mm (Fig. 2). This gives an aspect ratio $L/D = 1$, which is within the typical range and allowed PIV investigation of suction-assisted penetration over a significant depth. The skirt thickness t was 2 mm, such that $D/t = 25$, which although higher than employed for field scale buckets, was required to create an effective seal between the bucket and the window. Nonetheless, Tran (2005) demonstrated the skirt thickness to have a marginal effect on the penetration resistance during

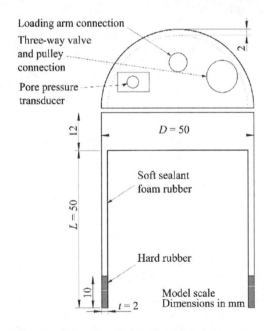

Figure 2. Schematic of the suction bucket model.

suction installation. A combination of soft sealant foam tape and hard rubber at the skirt tips was used at the bucket-window interface (Figs. 1b, 2), which was lubricated with petroleum jelly.

This provided: (i) adequate sealing of the bucket interior from the external fluid; (ii) sufficient resistance against peeling of the foam tape during installation (enhanced by the hard rubber strips); and (iii) minimal friction generation during installation. The testing procedure required the fluid flow through the bucket lid to be: (i) directed to ambient; (ii) connected to the syringe pump; and (iii) stopped. This was achieved using a three-way valve placed on the lid top with a winch manipulating the valve configuration (Zhu et al. 2018). The valve was connected to the syringe pump via a flexible pipe. The bucket was equipped with a differential total pressure transducer at the lid invert, which measured the suction pressure generated during installation.

Two 5-megapixel cameras mounted on a rigid crossbar captured images for the PIV analyses (Fig. 1c). The master macro-camera (8 mm lens) provided a view of the entire window, whereas the slave microcamera (43 mm lens) focused on the region around the bucket tips, providing highly detailed images (at ∼20 times the effective resolution of the macro camera) of this region of interest. The cameras captured images at a rate of 2 frame/s throughout the test. The thickness of the transparent window (50 mm) was estimated to limit lateral deflection of the window to less than 0.16 mm, which is sufficient to maintain approximately K_0 conditions following ramp-up to 40 g (Haigh & Madabhushi, 2014). An array of regularly spaced, pre-installed control markers inside the window was also required for the photogrammetric correction of the measurements. Further technical details regarding the dual camera arrangement and the lighting system can be found in Teng et al. (2017).

2.2 Soil sample

Commercially available fine silica sand ($d_{50}=0.18$ mm) was used to prepare the sample. The sand properties are reported in Table 1 (after Chow et al. 2018). A fraction of the sand was dyed black in order to optimise the particle contrast as recommended in Stanier & White (2013) and then thoroughly mixed with the natural, white portion (75% natural, 25% black). A 175 mm high dense sample was achieved through dry pluviation of sand (including a more permeable 10-mm bottom layer of coarse sand separated by a layer of geotextile), following the procedure described in Xu (2007). Measurements of mass and volume of the sample confirmed a value of relative density $D_r = 93\%$. The sample was then saturated from the base on the laboratory floor at 1 g with a solution of water and 1.1% cellulose ether content, to obtain a viscosity of 40cSt (DOV 2002). Matching the viscosity with the enhanced gravity level was necessary to correctly scale the drainage properties of the soil (Tan & Scott 1985, Taylor 1987, Dewoolkar et al. 1999). Throughout the centrifuge testing, a fluid head above the sample of 100 mm was maintained.

Prior to the suction bucket tests, three cone penetration tests (CPTs) were carried out using a model scale cone penetrometer with a diameter of 10 mm at the testing acceleration of 40 g. These CPTs provided a basis for characterising each sample and to quantify potential density variations in and across the samples.

Table 1. Silica sand properties.

Specific gravity, G_s		2.67
Mean particle size, d_{50}	mm	0.18
Minimum dry density, ρ_{min}	kg/m^3	1497
Maximum dry density, ρ_{max}	kg/m^3	1774
Critical state friction angle, ϕ'_{cv}	[°]	31.6

2.3 Testing procedure

The initial self-weight installation was modelled by penetrating the bucket vertically in displacement control, rather than under the self-weight of the foundation. This was necessary to ensure that the suction-assisted penetration initiated at the same bucket embedment depth in each test. A target normalised depth limited to $z/L = 0.2$ guaranteed a large portion of the skirt penetration to be investigated in suction-assisted regime. A constant rate of 0.057 mm/s was selected, as it is similar to the penetration rate measured in rigorous load-controlled self-weight penetration for full suction bucket tests reported in Bienen et al. (2018). At $z/L = 0.2$, the penetration resistance corresponded to a vertical force $V = 82.5$ N (at model scale), which was maintained constant during the subsequent suction-assisted penetration stage. The self-weight penetration was performed with the valve on the bucket lid set to 'vented', to allow fluid flow from the inside to the outside of the bucket as the foundation advanced into the soil.

After the bucket reached the targeted embedment depth, $z/L = 0.2$, the actuator was switched to load control to hold the vertical load previously achieved, and the winch motor was activated to connect the inner chamber of the suction bucket with the syringe pump. The syringe pump was then activated using a piston displacement rate of 0.06 mm/s, which resulted in a constant fluid flow of 117.8 mm^3/s being extracted from the bucket (at model scale). This is similar to the full suction bucket tests at the medium pumping flow rate in Bienen et al. (2018), which is typical of field conditions. This translates to a bucket penetration rate that reduces with time, since the seepage occurring below the skirt tips reduces the development of suction pressure. The pumping flow rate was maintained constant until full installation of the bucket was reached (penetration was arrested immediately before the lid touched the surface to avoid sand plugging the fluid access port on the total pressure sensor or the three-way valve). The generation of suction was continuously monitored via the differential total pressure transducer located in the bucket lid.

3 PIV ANALYSIS

The PIV results presented in Figures 3-4 analysed a bucket penetration of $\Delta z = 1$ mm at different depths. For the generation of the contour plots, the measured displacement fields were normalised by the vertical displacement of the bucket $\Delta z = 1$ mm.

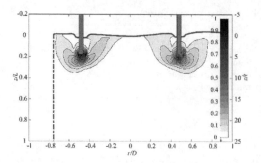

Figure 3. Normalised resultant velocity contours at the end of self-weight installation ($z/L = 0.2$).

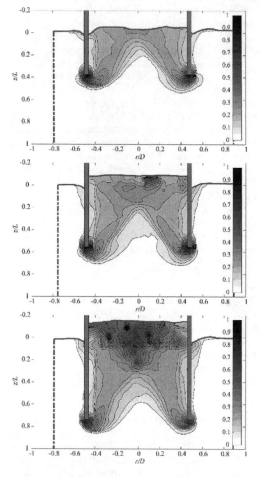

Figure 4. Normalised resultant velocity contours at (a) z/L = 0.18; (b) z/L = 0.36; (c) z/L = 0.54 outside, which is simply generated by the skin friction.

In contrast to the self-weight installation phase (Fig. 3), which was carried out at a constant penetration rate, the suction assisted stage (Fig. 4) involved a bucket penetration rate $\dot{z}$ that reduced with depth, as a steady pumping flow rate was maintained. The image acquisition rate was also kept constant at all times (2 frame/s).

GeoPIV-RG (Stanier et al. 2015) was used to analyse the image sequences recorded during the tests. The module relies on an array of subsets to analyse the soil domain, with the subset size and spacing of the subsets being user-defined. A subset size of 50 pixels was adopted here, equally spaced every 50 pixels (i.e. no subset overlap). The automatic reference image updating scheme was utilised in this work to track the soil displacements throughout the whole image series automatically whilst ensuring maximal correlations.

In general, modern image-based deformation algorithms can tolerate relatively large deformations (∼10% strain), so long as the deformations are not overly localised (Stanier et al. 2016). In this work problems occurred at the surface of the sand plug due to localised piping, where the minimum correlation coefficient (CCZNCC) for a given image pair was less than the default criteria for GeoPIV-RG recommended by Stanier et al. (2015) (recommended CCZNCC ≥ 0.75; where 0 = no correlation and 1 = perfect correlation). Where these problems occurred, the analyses were concluded by relaxing the correlation coefficient criteria from the recommended values (from CCZNCC = 0.75 to 0.1 < CCZNCC < 0.5) before manually filtering all subsets from the analyses with a correlation coefficient < 0.75. This typically resulted in automated deletion of ∼2% of the subsets analysed in a highly localised zone at the top of the sand plug.

4 RESULTS

The results presented here are part of a larger campaign investigating the effects of the variation of several key parameters. This paper focuses mainly on the experimental apparatus and the PIV analyses, reporting the results of a representative test only.

4.1 Self-weight installation

The resultant velocity contours for self-weight installation at $z/L = 0.195$ (PIV analysis of the preceding 1 mm penetration) are shown in Figure 3 (the dashed lines delimit the area captured by the micro-camera). The development of two independent mechanisms generated by the penetrating tips can be observed. The absence of overlapping at any point shows the lack of mutual influence.

Analogies with a penetrating strip footing (Houlsby & Byrne 2005) or cone (Senders & Randolph 2009) can be found. Bolton et al. (1999), defined a critical depth ($z/t \sim 10$) for a centrifuge scale CPT in sand, which is the depth at which the behaviour transitions from shallow to deep. In this sense, the response displayed in Figure 3 aligns with the prediction of shallow foundation behaviour expected at $z/t \sim 5$.

However, asymmetry of the failure mechanism under each skirt is apparent. In order to understand the reason behind this, changes in soil stresses generated by skirt friction are examined. Houlsby & Byrne (2005) suggested that the penetration-generated friction increases the vertical effective stress around the skirt, and consequently also the tip resistance. This down-drag effect is expected to be more pronounced on the inside of the bucket compared to the outside, leading to a shift in vertical stresses at the skirt tip from a symmetric (under equal conditions inside and outside of the bucket) towards an asymmetric stress distribution, with consequences for the symmetry of the mechanism. Similar changes in soil stresses have been reported for jacked piles in Henke & Bienen (2014). The effect of changes in soil stress induced by the advancing skirts is evident in Figure 3, where the mechanism under the skirt tip has shifted outside the bucket, and now consequently extends to the soil surface at different gradients inside compared to outside of the bucket.

4.2 Suction-assisted installation

Figure 4 shows the normalised resultant velocity contours at increasing normalised depths: $z/L = 0.38$; 0.56; 0.74. In contrast to Figure 3, overlapping of the mechanisms originating from each tip is now apparent. In fact, a single mechanism now characterises the process. The mechanism shows marked changes compared to that under self-weight (jacked) penetration. The mechanism under the skirt tip has shifted to concentrate inside the bucket. The prevalent inwards and upwards flow of soil observed inside the bucket contrasts with the extremely limited downwards motion.

The generation of suction is responsible for the seepage field established inside and outside the bucket, with seepage patterns that have been extensively investigated numerically (Erbrich & Tjelta 1999, Tran 2005, Senders 2008). The high velocity observed around the bucket tips in Figure 4, along with the upwards movement of the internal soil, is consistent with steep upwards hydraulic gradients generated by the suction. Furthermore, they are consistent with the theory that resistance in suction installation is reduced through reduction of effective vertical stresses around the skirt tip. Further, the lack of any soil movement outside the bucket is consistent with the theory of low hydraulic gradients in this region (Senders & Randolph 2009).

As the bucket advances into the soil as a consequence of the increasing suction, Figure 4 shows that the shape of the failure mechanism remains largely unaffected; it simply moves with the penetrating bucket. The rate at which the plug moves upwards appears to be a constant fraction of the bucket penetration rate (as a reducing penetration rate $\dot{z}$ also involves the plug moving more slowly). The prevalent upwards mobilisation of the soil can be associated with the inner seepage flow. The gradual formation of soil plug heave led to a final recorded value of $\sim 0.15\,L$ above the original soil surface, which is close to the volume of soil displaced by the skirts (quantified in $0.136\,L$). According to Figure 4 (and to the vector plots not included here) it is reasonable to assume that all the soil displaced by the penetrating skirts entered the bucket. It can then be estimated that the skirt penetration is the main cause of the plug heave observed ($\sim 90\%$), with seepage-triggered sand dilation only partially responsible ($\sim 10\%$). Dilation caused by shearing of the dense sand along the skirts (Tran et al. 2005) does not appear to have a particular influence on the plug heave formation.

Critical hydraulic gradients are thought to have occurred around the tips and within the soil plug. However, the consequent changes in soil state are not uniform across the plug, as localised singularities of large soil displacement can be observed in its uppermost part in Figure 4c. In contrast to the tips, the sand close to the surface is not confined, so that localised phenomena of piping, associated with negative pressures above the critical value and localised high-velocity regions were captured. It should be clarified that, as piping occurs, it becomes increasingly complex for PIV analysis to track the particle displacement (see Section 3), for this is associated with extremely rapid velocities and highly localised deformations. The finding that plug heave formation is mainly a consequence of skirt penetration, along with the observation of these localisations only towards the final stage of suction installation, would suggest that only a minor increase in void ratio, and thus soil permeability, took place in the soil plug. However, further study on the measured volumetric strain is required to accurately determine the changes in void ratio.

5 CONCLUDING REMARKS

The Particle Image Velocimetry (PIV) technique was successfully applied to the problem of suction bucket installation in a centrifuge environment. Velocity contours revealed substantial differences between the mechanisms governing the initial self-weight installation and the subsequent suction stage. The effect of down-drag enhancement of soil stresses due to friction on the suction bucket skirt was observed during the self-weight penetration phase. The analyses revealed the mechanism during suction-assisted penetration to consist of shearing, and confined within the bucket. Non-uniform positive volume change within the soil plug was associated with events of localised piping. Minimal soil displacement was generated by the friction along the external skirt face.

The PIV centrifuge experiment discussed in this paper forms part of a larger testing programme, which aims to explore the effects of the bucket diameter (and hence confining conditions), the initial sand relative density and the pumping flow rate on the soil response.

ACKNOWLEDGEMENTS

This work forms part of the activities of the Centre for Offshore Foundation Systems (COFS), which is currently supported as one of the primary nodes of the Australian Research Council (ARC) Centre of Excellence for Geotechnical Science and Engineering and as a Centre of Excellence by the Lloyd's Register Foundation. The third author is supported by an ARC DECRA Fellowship. This support is gratefully acknowledged.

REFERENCES

Andersen, K.H., Jostad, H.P. & Dyvik, R. 2008. Penetration resistance of offshore skirted foundations and anchors in dense sand. *Journal of geotechnical and geoenvironmental engineering* 134(1): 106-116.

Bienen, B., Klinkvort, R.T., O'Loughlin, C.D., Zhu, F. & Byrne, B.W. 2018. Suction buckets in dense sand: Installation, limiting capacity and drainage. *Géotechnique* https://doi.org/10.1680/jgeot.16.p.281.

Bolton, M.D., Gui, M.W., Garnier, J., Corte, J.F., Bagge, G., Laue, J. & Renzi, R. 1999. Centrifuge cone penetration tests in sand. *Géotechnique* 49(4): 543-552.

Chow, S.H., O'Loughlin, C.D., Gaudin, C. & Lieng, J.T. 2018. Drained monotonic and cyclic capacity of a dynamically installed plate anchor in sand. *Ocean Engineering* 148: 588-601.

Dewoolkar, M.M., Ko, H.Y., Stadler, A.T. & Astaneh, S.M.F. 1999. A substitute pore fluid for seismic centrifuge modeling. *Geotechnical Testing Journal* 22(3): 196-210.

DOV. 2002. Methocel Cellulose ethers. *Technical Handbook*.

Haigh, S.K. & Madabhushi, S.P.G. 2014. Discussion of "Performance of a transparent flexible shear beam container for geotechnical centrifuge modelling of dynamic problems by Ghayoomi, Dashti and McCartney". *Soil Dynamics and Earthquake Engineering* 67: 359-362.

Henke, S. & Bienen, B. 2014. Investigation of the influence of the installation method on the soil plugging behaviour of a tabular pile. *In Proc. International Conference on Physical Modelling in Geotechnics, Perth, Australia*: 14-19.

Houlsby, G.T. & Byrne, B.W. 2005. Design procedures for installation of suction buckets in sand. *In Proc. of the Institution of Civil Engineers-Geotechnical Engineering* 158(3): 135-144.

Houlsby, G.T., Ibsen, L.B. & Byrne, B.W. 2005. Suction buckets for wind turbines. *In Proc. International Symposium Frontiers in Offshore Geotechnics, Perth, Australia*: 75-93.

Houlsby, G.T. 2016. Interactions in offshore foundation design. *Géotechnique* 66(10): 791-825.

Randolph, M.F., Jewell, R.J., Stone, K.J.L. & Brown, T.A. 1991. Establishing a new centrifuge facility. *In Proc. International Conference Centrifuge, Boulder, USA*: 3–9.

Senders, M. 2008. Suction buckets in sand as tripod foundations for offshore wind turbines. *PhD thesis, The University of Western Australia, Australia*.

Senders, M. & Randolph, M.F. 2009. CPT-based method for the installation of suction buckets in sand. *Journal of Geotechnical and Geoenvironmental engineering* 135(1): 14-25.

Stanier, S.A. & White, D.J. 2013. Improved Image-Based Deformation Measurement in the Centrifuge Environment. *Geotechnical Testing Journal* 36(6): 915-928

Stanier, S.A., Blaber, J., Take, W.A. & White, D.J. 2015. Improved Image-Based Deformation Measurement for Geotechnical Applications. *Canadian Geotechnical Journal* 53(5): 727-739.

Stanier, S.A., Dijkstra, J., Leśniewska, D., Hambleton, J., White, D.J. & Wood, D.M. 2016. Vermiculate artefacts in image analysis of granular materials. *Computers and Geotechnics* 72: 100-113.

Tan, T.S. & Scott, R.F. 1985. Centrifuge scaling considerations for fluid-particle systems. *Géotechnique*, 35(4): 461-470.

Taylor, R.N. 1987. Discussion of "Tan, T.S. and Scott, R.F. (1985). Centrifuge scaling considerations for fluid-particle systems." *Géotechnique* 37(1): 131-133.

Teng, Y., Stanier, S.A. & Gourvenec, S.M. 2017. Synchronised multi-scale image analysis of soil deformations. *International Journal of Physical Modelling in Geotechnics* 17(1): 53-71.

Tjelta, T.I. 1990. The Skirt piled Gullfaks C installation. *In Proc. Offshore Technology Conference, Houston, USA*.

Tjelta, T.I. 2015. The suction foundation technology. *In Proc. International Symposium Frontiers in Offshore Geotechnics, Oslo, Norway*: 85-93.

Tran, M.N. 2005. Installation of suction buckets in dense sand and the influence of silt and cemented layers. *PhD thesis, The University of Sydney, Australia*.

Tran, M.N., Randolph, M.F. & Airey, D.W. 2005. Study of sand heave formation in suction buckets using Particle Image Velocimetry (PIV). *In Proc. International Symposium Frontiers in Offshore Geotechnics, Perth, Australia*: 259-265.

Tran, M.N. & Randolph, M.F. 2008. Variation of suction pressure during bucket installation in sand. *Géotechnique*, 58(1): 1-11.

Xu, X. 2007. Investigation of the end bearing performance of displacement piles in sand. *PhD thesis, The University of Western Australia, Australia*.

Zhu, F., Bienen, B., O'Loughlin, C.D., Cassidy, M.J. & Morgan, N. 2018. Suction caisson foundations for offshore wind energy: installation and cyclic response in sand and sand over clay. *Géotechnique, under review*.

Recent advances in tsunami-seabed-structure interaction from geotechnical and hydrodynamic perspectives: Role of overflow/seepage coupling

S. Sassa
Port and Airport Research Institute, National Institute of Maritime, Port and Aviation Technology, Yokosuka, Japan

ABSTRACT: The paper reports some recent research advances on tsunami-seabed-structure interaction following the 2011 Tohoku Earthquake Tsunami, Japan. It presents and discusses a comparison of utilising a geotechnical centrifuge and a large-scale hydro flume for the modelling of tsunami-seabed-structure interaction. The similitudes for laminar and turbulent seepage flows in a geo-centrifuge are discussed, showing that the viscous scaling for studying wave-soil interaction problems can be replaced by the similitude for turbulent seepage flow in studying the stability of rubble foundations under tsunami. Notably, both geo-centrifuge and large-scale flume experiments proved to be consistent with each other, based on the role of tsunami-induced seepage. The paper also highlights the results from our recent experiments in which a new tsunami overflow-seepage-coupled centrifuge system developed was applied to investigate the concurrent processes of the instability involving the scour, flow of the foundation, and the failure of caisson breakwaters, elucidating the role of overflow and seepage coupling.

1 INTRODUCTION

The 2011 Tohoku earthquake tsunami devastated the eastern part of Japan and caused significant damage and destruction of breakwaters. The key factors contributing to the damage involved the scour of the mound in the vicinity of breakwaters due to tsunami-induced overflow above the caissons (Arikawa et al. 2012, Arikawa & Shimosako 2013). The seepage flow in mounds, stemming from the water level difference between offshore and onshore sides of the caissons, also affected the stability of mounds (Takahashi et al. 2014, Sassa, 2014). However, the concrete mechanism at work in the instability of the breakwater foundation under the concurrent actions of the overflow and seepage due to a tsunami remained unclear.

Here, some recent research advances in tsunami-seabed-structure interaction from geotechnical and hydrodynamic perspectives are highlighted. Also, the paper presents and discusses a comparison of utilising a geotechnical centrifuge and a large-scale hydro flume for the modelling of tsunami-seabed-structure interaction. For this purpose, it first presents the similitudes for tsunami overflow and seepage in light of those used for wave-soil interaction problems in a centrifuge. This is followed by a review of the effect of tsunami-induced seepage on piping/boiling, seepage erosion and bearing capacity decrease and failure of caisson breakwaters. Then, a large scale hydro flume study is described and discussed in view of the stability assessment for the design of tsunami-resistant structures. The role of overflow and seepage coupling in tsunami-seabed-structure interaction is summarised with conclusions.

2 SIMILITUDES FOR TSUNAMI OVERFLOW AND SEEPAGE IN A GEO-CENTRIFUGE

It is essential to reproduce a prototype-scale stress field in the mounds that support breakwaters in order to clarify the instability of the breakwater foundations in the presence of a tsunami. A geo-centrifuge makes this possible and has proved effective in studying fluid–soil interaction problems (Sassa & Sekiguchi 1999). Indeed, the role and importance of centrifuge testing in the field of coastal and ocean engineering has recently been emphasised (Sumer & Fredsøe 2002 Sumer 2014). The theoretical background and similitudes for the tsunami overflow–seepage coupled centrifuge experiment are described below.

Suppose that a centrifuge tsunami test with a 1/N scale model is performed under a constant centrifugal acceleration of Ng, where N is the scale factor, and g represents the acceleration due to the Earth's gravity. According to the Depuit-Forchheimer approximation (Dupuit 1863, Forchheimer 1901, Bear 1972, Lage 1998, Bordier & Zimmer 2000) the relationship between the hydraulic gradient i and mean seepage flow velocity $\bar{v}$ can be expressed as

$$i = a\bar{v} + b\bar{v}^2 \tag{1}$$

Table 1. Similitudes for tsunami overflow and seepage.

	Prototype	Present study *	Viscous scaling **
g	1	N	N
Size	1	$1/N$	$1/N$
Grain size	1	$1/N$	1
Dynamic viscosity	1	1	N
Time	1	$1/N$	$1/N$
Pressure, stress	1	1	1
Hydraulic gradient	1	1	1
a* or b** in Eq (1)	1	1	1
Mean seepage flow velocity	1	1	1
Mean overflow velocity	1	1	NA

*Turbulent flow, ** Laminar flow

$$a = \alpha_0 \frac{\mu}{Ng} \frac{(1-n)^2}{n^3 d_{15}^2} \quad b = \beta_0 \frac{1}{Ng} \frac{1-n}{n^3 d_{15}} \quad (2)$$

The forms of equation (2) represent those in the widely used formulas (Bear 1972, Macdonald et al. 1979, Lage 1998, among others). Here, α_0 and β_0 are coefficients, μ is the dynamic viscosity of the pore fluid, n is the porosity of the ground, and d_{15} is the grain size passing 15% finer by weight. The hydraulic gradients in the horizontal and vertical directions can be written as

$$i_x = \frac{1}{\gamma_w} \frac{\partial \Delta u}{\partial x} = \frac{1}{\rho Ng} \frac{\partial \Delta u}{\partial x} \quad (3)$$

$$i_y = \frac{1}{\gamma_w} \frac{\partial \Delta u}{\partial y} = \frac{1}{\rho Ng} \frac{\partial \Delta u}{\partial y} \quad (4)$$

where Δu is the excess pore water pressure, γ_w is the unit weight of the fluid, and ρ is the mass density of the fluid.

In the laminar flow regime, the viscous term, i.e., the first term on the right-hand side of Equation (1), becomes dominant. For such a Darcy flow, the associated similitude can be fulfilled by using viscous fluid whose dynamic viscosity is N-times that of water, with a hydraulic gradient and mean flow velocity equivalent to those of the corresponding prototype (Eqs. (2)-(4), Table 1). With this technique, called viscous scaling, one can satisfy the time-scaling laws for fluid wave propagation and the consolidation of the soil (Sassa & Sekiguchi 1999). By contrast, in the turbulent flow regime that manifests itself in rubble mounds under tsunami (Takahashi et al. 2014, Sassa et al. 2014), the inertia term, i.e., the second term on the right-hand side of Equation (1), becomes dominant, for which one can satisfy the associated similitude by using the 1/N scale grain diameter, with a hydraulic gradient and mean flow velocity equivalent to those of the corresponding prototype (Eqs. (2)-(4), Table 1) for a given packing state. As a consequence, the time in the centrifuge model becomes a 1/N scale of the time in the prototype, in other words, the 1/N time scale is necessary to satisfy the similitude for the time, as shown in Table 1. The Froude law is satisfied naturally for the same overflow velocity in a 1/N scale model under Ng, which fulfills the mechanical similarity between the model and the prototype.

3 REVIEW OF THE EFFECT OF TSUNAMI-INDUCED SEEPAGE

The effect of tsunami-induced turbulent seepage in piping/boiling, erosion and bearing capacity decrease and failure of caisson breakwaters will be briefly summarised below, based on our experimental and numerical work described in Takahashi et al. (2014).

3.1 Piping/boiling

The occurrence of piping and boiling was tested against models of actual breakwaters, Omaezaki and Kamaishi in Japan. The tsunami-induced water level differences were set at 8 m on prototype scale for the Omaezaki model: caisson height 12 m and width 10 m, and at 16 m for the Kamaishi model: caisson height 20 m and width 16 m. Both conditions correspond to severest possible conditions in the field. The experimental results showed that the piping and boiling did not take place. This fact suggests that piping/boiling is not a matter of concern, however, situations distinctly changed under the coupling actions of overflow and seepage as described later in this paper.

3.2 Seepage erosion

Tsunami-induced seepage can induce erosion in a sandy ground underneath the rubble mound of caisson breakwaters. Experiments were performed at 50 g under conditions where no sliding and/or overturning of the caisson was allowed to occur in order to focus on the erosion process of the sandy ground. The results on the Omaezaki model show that such seepage erosion progressed with time at the vicinity of the rubble/sand interface (Fig. 1). The eroded mass deposited in the onshore side of the mound in accordance with the onshore direction of the seepage flow. Accordingly, the caisson settled as a consequence of the tsunami-induced seepage erosion.

3.3 Bearing capacity decrease and failure

The bearing capacity decrease and failure characteristics of the rubble mounds of caissons were examined under tsunami-induced seepage. The centrifuge test (50 g) results showed that the bearing capacity of the mound decreased significantly due to the seepage effect. In order to confirm this experimental finding, finite element analyses were performed concerning the typical breakwater dimensions such as caisson width 10m and height 12 m. Although the seepage flow in the mound was turbulent, the analysis assumed

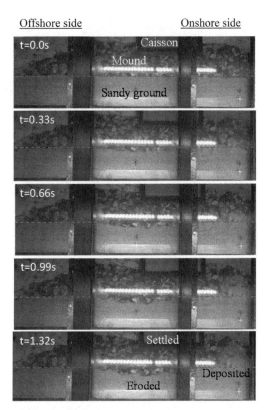

Figure 1. Progress of seepage erosion. Descriptions are added to Fig. 7 of Takahashi et al. (2014). Settlement of the caisson was observed to occur with seepage erosion.

Darcy flow adopting the linear relationship between the seepage flow velocity and the hydraulic gradient. As such, the material parameters involving the permeability coefficient were selected so as to reproduce the measured hydraulic gradient ($i = 0.4$) in the top shoulder of the mounds, whose region proved to be important for the stability of the caisson. The numerical results show that the seepage effect became more pronounced with increasing water pressure (level) differences (Fig. 2). Indeed, a water level difference of 10 m, i.e. water pressure difference 98 kN/m^2 gave rise to the decrease of the bearing capacity by 20% owing to the tsunami-induced seepage. The influence of placing a surcharge embankment as countermeasures was also examined. The embankment as thick as 2 m had a certain effect on improving the bearing capacity of the caisson breakwaters (Fig. 2).

4 LARGE SCALE HYDRO FLUME STUDY AND DISCUSSION

This section discusses a large-scale hydro flume study performed by Arikawa et al. (2013), in comparison to the above geo-centrifuge study. Specifically, it focuses on the bearing capacity failure of rubble mounds of caisson breakwaters. The hydro flume used was 184m

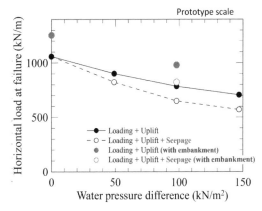

Figure 2. Effect of tsunami-induced seepage on bearing capacity of the mound: numerical results. Figs. 23 and 26 of Takahashi et al. (2014) are combined and descriptions are added. Water pressure difference corresponds to water level difference between inside and outside the caisson under tsunami.

long and 3.5 m wide and 12 m deep. Arikawa et al. (2013) modelled the Omaezaki breakwater with a 1/7.5 scaled model that was 40.98 kN/m and 1.5 m thick. With reference to Fig. 3, dummy caissons which were 10% heavier than the model caisson were installed at both ends of the channel. Water was discharged from the offshore side gate and taken in the onshore side gate of the flume, creating a water level difference between outside and inside of the caisson. The modelled water level difference was 4.5 m on prototype scale. In order to prevent the occurrence of erosion of the sandy ground, a thin concrete plate was installed just underneath the rubble mound. The experimental results show that the bearing capacity failure took place accompanying flow deformations of the mound.

In the design of caisson breakwaters, the safety against sliding, overturning and bearing capacity failure needs to be clarified and assured in the Technical Standard for Port and Harbour Facilities, Japan. Conventional slip-circle Bishop analysis has been used to calculate the safety factor for the bearing capacity of the mound. It is important here to remark that the safety factor for the bearing capacity calculated by Arikawa et al. (2013) for the above hydro flume experiment was higher than unity, whose discrepancy with the actual caisson failure was of the order of 10%. This appreciable discrepancy can be explained as follows. Namely, the geo-centrifuge study combined with the numerical analysis has shown that the bearing capacity decreased essentially linearly with increasing water level difference due to the effect of the tsunami-induced seepage. This means that the water level difference of 4.5 m as modelled in the large-scale hydro flume experiment, caused the bearing capacity decrease of approximately 10% on the basis of Figure 2. This quantitative agreement is noteworthy considering the two distinctly different approaches from both geotechnical and hydrodynamic perspectives.

Figure 3. Large-scale hydro flume experiment of Arikawa & Shimosako (2013), showing views before and after the experiment. Caisson failed accompanying flow deformations of the rubble mound under tsunami.

5 ROLE OF OVERFLOW AND SEEPAGE COUPLING

On the basis of the above results, this section highlights the role of overflow and seepage coupling in the tsunami-seabed-structure interaction. For this purpose, a series of experiments were performed by utilising a new tsunami overflow-seepage-coupled centrifuge experimental system developed (Sassa et al. 2016). With reference to Section 2 of this paper, the similitudes for tsunami overflow as well as turbulent seepage flows in the mounds were satisfied at 50g in addition to the mechanical similarity between the model and the prototype.

The concurrent processes of the instability involving the scour of the mound/sandy seabed, bearing capacity failure and flow of the foundation and the failure of caisson breakwaters under the coupled overflow and seepage are elucidated here, as shown in Figure 4. Namely, the experimental results first demonstrated that the coupled overflow-seepage actions promoted the development of the mound scour significantly (Fig. 5). Notably, the scour front developed in the form of progressive slip failure at the vicinity of caissons, regardless of the mound thickness. This stems from the fact that the development of the mound scour shortened the seepage path around the shoulder area of the mound, enhancing the coupling effect of the overflow and seepage. Indeed, the scour stopped far from the caisson toe in the absence of seepage. By contrast, the scour development due to the coupled overflow-seepage approached the caisson toe and brought about bearing capacity failure of the mound, resulting in the total failure of the caisson breakwater, which otherwise remained stable without the coupling effect (Fig. 5). The velocity vectors obtained from the high-resolution image analysis illustrated the series of such concurrent scour/bearing-capacity-failure/flow processes leading to the instability of the breakwater (Fig. 4). Further, the experiments that constrained caisson movement showed that the scour front reached the toe of the remaining caisson due to the coupled overflow and seepage (Fig. 5), resulting in washout and cavity formation as observed in the field (Sassa et al. 2016). This contrasts sharply with their absence solely under the effect of the seepage as described in Section 3.1 of this paper.

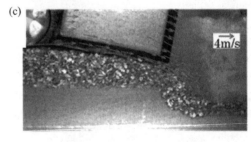

Figure 4. Tsunami overflow and seepage coupling induced (a) scour of the mound/seabed ground, (b) bearing capacity failure, and (c) flow of the mound in the concurrent processes of the caisson instability, shown with velocity vectors. The velocity vectors correspond to those at (a) $t = 15$-15.25s, (b) $t = 38$-44.5s, (c) $t = 55$-55.25s respectively after commencement of overflow on the prototype scale.

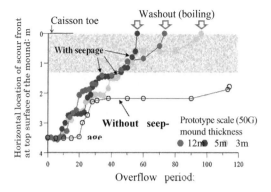

Figure 5. Tsunami overflow and seepage coupling induced mound scour. The shaded zone indicates the region where the caisson became unstable.

6 CONCLUSIONS

The paper has summarised some recent research advances on tsunami-seabed-structure interaction through the use of geo-centrifuge and large-scale hydro flume. The similitudes for laminar and turbulent seepage flows in a geo-centrifuge were discussed, showing that the viscous scaling, which has successfully been used for studying wave-soil interaction problems, can be replaced by the similitude for turbulent seepage flow in studying the stability of rubble mound foundations under tsunami. Notably, both geo-centrifuge and large-scale hydro flume experiments proved to be consistent with each other, based on the role of the tsunami-induced seepage in the bearing capacity failure of caisson breakwaters. This provides support for both approaches to tsunami-seabed-structure interaction problems. Accordingly, the role of such tsunami-induced seepage should be taken into account in the design practice in order to facilitate the rational stability assessment of tsunami-resistant caisson breakwaters. The paper also highlighted the importance, via a new tsunami overflow-seepage-coupled centrifuge experimental system, of the concurrent processes of the instability involving the scour, flow of the foundation, and the failure of caisson breakwaters. These findings elucidate the crucial role of overflow/seepage coupling in tsunami-seabed-structure interaction, warranting an enhanced disaster resilience.

REFERENCES

Arikawa, T., Sato, S., Shimosako, K., Tomita, T., Tatsumi, D., Ren, Y. & Takahashi, K. 2012. Investigation of the failure mechanism of Kamaishi breakwaters due to tsunami-Initial report focusing hydraulic characteristics-. *Technical Note of the Port and Airport Research Institute* 1251: 52p. (in Japanese).

Arikawa, T. & Shimosako, K. 2013. Failure mechanism of breakwaters due to tsunami: a consideration to the resiliency. *Proc. 6th Civil Engineering Conference in the Asian Region*, Jakarta, Indonesia: 1–8.

Bear, J. 1972. *Dynamics of Fluids in Porous Media*, Elsevier, New York.

Bordier, C. & Zimmer, D. 2000. Drainage equations and non-Darcian modelling in coarse porous media or geosynthetic materials. *Journal of Hydrology* 228(3-4): 174–187.

Dupuit, J. 1863. *Etudes Théoriques et Pratiques sur le Mouvement des Eaux*, Dunod, Paris.

Forchheimer, P. 1901. Wasserbewegung durch Boden, *Z. Ver. Dtsch. Ing.* 45: 1736–1741, 1781–1788.

Lage, J. L. 1998. The fundamental theory of flow through permeable media from Darcy to turbulence. In D. B. Ingham & I. Pop (eds), *Transport Phenomena in Porous Media*: 1–30. Oxford: Elsevier Science.

Macdonald, I. F., El-Sayed, M. S., Mow, K. & Dullien, F. A. L. 1979. Flow through porous media-the Ergun equation revisited. *Industrial & Engineering Chemistry Fundamentals* 18(3): 199–208.

Sassa, S. & Sekiguchi, H. 1999. Wave-induced liquefaction of beds of sand in a centrifuge. *Géotechnique* 49(5): 621–638.

Sassa, S. 2014. Tsunami-Seabed-Structure Interaction from Geotechnical and Hydrodynamic Perspectives. *Geotechnical Engineering Journal* 45(4): 102–107. Special Issue on Offshore and Coastal Geotechnics.

Sassa, S., Takahashi, H., Morikawa, Y. & Takano, D. 2016. Effect of overflow and seepage coupling on tsunami-induced instability of caisson breakwaters. *Coastal Engineering* 117: 157–165.

Sumer B. M. & Fredsøe J. 2002. The mechanics of scour in the marine environment. *Advanced Series on Ocean Engineering* 17: 536p. World Scientific.

Sumer B. M. 2014. Liquefaction around marine structures. *Advanced Series on Ocean Engineering* 39: 453p. World Scientific.

Takahashi, T., Sassa, S., Morikawa, Y., Takano, D. & Maruyama, K. 2014. Stability of caisson-type breakwater foundation under tsunami-induced seepage. *Soils and Foundations* 54(4): 789–805.

Evaluation of seismic behaviour of reinforced earth wall based on design practices and centrifuge model tests

Y. Sawamura, T. Shibata & M. Kimura
Kyoto University, Kyoto, Japan

ABSTRACT: Although it is well known that reinforced earth walls have strong earthquake stability, the mechanisms by which they show strong earthquake performance is not clear. When determining the reinforcement length, the earth pressure acting on the wall and the tensile force of reinforcement in the resistance zone are assumed to be at equilibrium with each other to simplify the internal stability calculation. In the external stability calculation, the whole reinforced earth wall is considered as one rigid body. We have focused on these calculations, and aimed to clarify the mechanical role of reinforcement laid in the active failure zone through dynamic centrifuge model tests. From the experimental results, it has been confirmed that the reinforcement laid in the active failure zone contributes to the stability of the reinforced earth wall, and that the deformation of the reinforced earth wall is smaller than that of a rigid body during strong shaking.

1 INTRODUCTION

The reinforced earth wall (REW) method (Figure 1) is a method for supporting road embankments on vertical gradients that involves constructing a retaining wall using a concrete wall panel and laying steel strip at predetermined intervals. This method was developed by H. Vidal in 1963 and adopted for road embankments in France in 1964. It was first applied to highway embankments in Japan in 1972, and has also seen wide application in other countries.

Dynamic model experiments (Richardson et al. 1975) and large-scale experiments (Richardson et al. 1977) have been conducted to investigate the seismic behavior of REW. More recently, dynamic centrifugal model experiments using the ground motions measured in the 1995 Hanshin Awaji Great Earthquake demonstrated that the REW exhibited high ductility (Siddharthan et al. 2004). Recent studies have attempted to accurately reproduce the tensile force generated in reinforcement by statically examining the interaction between the ground and the reinforcement using numerical analysis (Yan et al, 2015). In addition, it has been proposed to design an appropriate reinforcement length from the relationship between the reinforcement length and the deformation of the REW after the earthquake (Yazdandoust 2015). However, since the interactions between the wall, reinforcement and embankment are complicated, the earthquake behaviour and reinforcement mechanisms of the REW are still unclear.

In current design practices, the inside of the area where the reinforcement tensile force is maximised is assumed as the active failure zone (AFZ), and internal stability calculation is conducted. In the internal

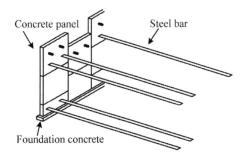

Figure 1. Reinforced earth wall.

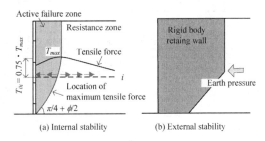

Figure 2. Concept of internal stability and external stability in current design method.

stability calculation, the length of the reinforcement is determined based on the limit balance of the earth pressure acting on the wall and the tensile force generated in the reinforcement in the resistance zone (Figure 2(a)). Subsequently, external reinforcement is examined by evaluating the whole REW as a rigid body retaining wall (Figure 2(b)). In the above calculations, two assumptions are made: (1) In the internal stability calculation, the reinforcement laid in the AFZ is negligible. (2) In the external stability calculation, the REW follows the deformation of the surrounding ground and foundation, and the entire REW behaves as a rigid body during an earthquake.

Therefore, in this study, centrifuge model tests were conducted to elucidate the dynamic behaviour and the reinforcing mechanism of the REW by clarifying the mechanical role of the reinforcement in the AFZ and considering the rigidity of the REW.

2 EXPERIMENTAL OUTLINE

2.1 Experimental object

Centrifuge model tests were performed under a gravitational acceleration of 20 g. A soil chamber 630 mm long, 500 mm deep, and 150 mm wide, with a transparent front window was used for the tests. The experimental model represented a REW with a vertical wall 8.25 m in height.

2.2 Model ground and boundary condition of soil chamber

In this experiment, both the foundation and embankment were made from dry Toyoura sand. Table 1 shows the material properties of Toyoura sand. The internal friction angle of Toyoura sand was determined by direct shear test. Direct shear tests were also performed to determine the friction angle between the reinforcement and Toyoura sand using a stainless steel plate with Toyoura sand adhered to it by double-sided tape. Figure 3 shows the results of the tests.

The sand was pluviated into the chamber, with the falling height adjusted such that a relative density of 80% was achieved. However, in the experimental case using a rigid body retaining wall with an oblique angle, the embankment was constructed by compaction because pluviation was not feasible. Gel sheets 3 mm in thickness were inserted as cushioning between the soil and the chamber in the direction of the shaking to suppress reflected waves from the boundary of the soil chamber. To reduce friction between the reinforced wall and the soil chamber in the orthogonal direction of shaking, a sponge of 5 mm thickness with a Teflon sheet was attached to the wall.

2.3 Experimental cases

In this study, to elucidate the seismic behaviour and reinforcing mechanism of REW, the experimental cases were determined based on the internal and external stability calculations in the current Japanese design method. Figure 4 shows the schematic diagram of the experimental cases.

Case-1: Basic case designed using the current design manual with a factor of safety $F_{sE} = 1.2$ for pull-out of the reinforcement during an earthquake.

Case-2: Case minimising the role of reinforcement in the AFZ by using a stainless steel wire ($\phi = 0.45$ mm) in the AFZ to connect the wall surface and the reinforcement in the resistance zone. In this case, the same factor of safety for the internal stability calculation ($F_{sE} = 1.2$) is satisfied.

Case-3: AFZ is substituted by a rigid body block, and the reinforcements are arranged such that the factor of safety F_{sE} becomes 1.2. The aim of Case-3 is to verify the behaviours of REW in the case where no AFZ exists.

Case-4: In compliance with the external stability calculation, the whole REW is substituted by a rigid

Table 1. Material properties of Toyoura sand.

Material property	Toyoura sand
Specific gravity G_s	2.64
Average diameter D_{50} [mm]	0.20
Internal friction angle ϕ [deg]	40.6
Cohesion c [kPa]	0.0
Maximum void ratio e_{max}	0.975
Minimum void ratio e_{min}	0.585

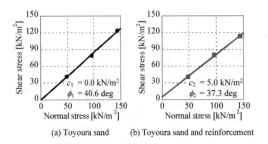

Figure 3. The results of direct shear tests.

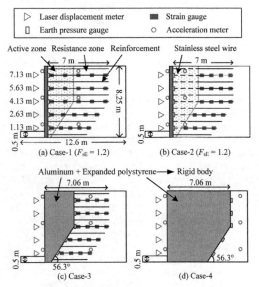

Figure 4. Experimental cases.

block to compare the behaviour of the REW and rigid body retaining wall during earthquake.

2.4 Modelling of wall and rigid retaining wall

In a REW, the vertical wall is kept stable by the frictional force acting on steel reinforcement strips attached to the back of the wall member. In the actual wall, crossed concave panels (thickness 0.14 m) 1.5 m × 1.5 m are connected vertically by steel bars. Hence, it is thought that the wall is flexible. To simulate the bending stiffness of an actual wall, the model wall was made using an aluminium plate ($E = 7.03 \times 10^7$ kN/m^2) with a thickness of 3 mm. Figure 5 shows the size of the model wall and the rigid body retaining wall. The model wall is 412.5 mm in height, 140 mm in width 3 mm in thickness. For the rigid body retaining wall, aluminium and styrene were used to make a model with the same volume and mass as the reinforced region. In addition, by adhering double-sided tape and attaching Toyoura sand to the back side of the wall surface, the continuity between the wall and sand was improved.

2.5 Modelling of reinforcing material

The standard reinforcement in actual REWs is 60 mm wide and 4 mm thick. If the reinforcement were reduced to the model scale, the reinforcement would be a width of 3 mm and a thickness of 0.2 mm. Because it is extremely difficult to make such a small reinforcement model, stainless steel strips with a width of 6 mm and a thickness of 0.5 mm were used for the reinforcement. The reinforcement intervals were set to 1.5 m horizontally and 0.75 m vertically in the prototype scale. Here, the reinforcement density, defined as the strip width per unit area of the wall, is approximately 60 mm × 4/(1.5 m × 1.5 m) = 107 mm/m^2 for the actual structure. The reinforcement density of the experimental model is 120 mm × 22/(8.25 m × 3.0 m) = 107 mm/m^2 in prototype scale, which is consistent with the actual structure. The reinforcement was attached to the model wall using an L-shaped angle piece. In order to measure the tensile force acting on the reinforcement, strain gauges were attached to the second, fourth, sixth, eighth and tenth reinforcements from the bottom of the wall. Dry Toyoura sand was adhered to the surface of the reinforcement using double-sided tape to increase the friction with the embankment.

2.6 Input waves

Tapered sinusoidal waves with a frequency of 2 Hz were used as input waves, and the duration of the waves was 10 seconds. The waves were input 8 times, from Step 1 to 8, with a gradual increase in acceleration of 1.0 m/s^2 per step. Figure 6 shows the input acceleration waveform measured at the shaking table in Step 4.

3 EXPERIMENTAL RESULTS

3.1 Earthquake behaviour of each case

Figure 7 shows the relationship between the input acceleration and the maximum displacement of the

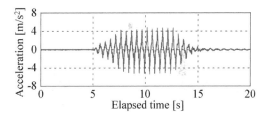

Figure 6. Input acceleration waveform measured at the shaking table in Step 4.

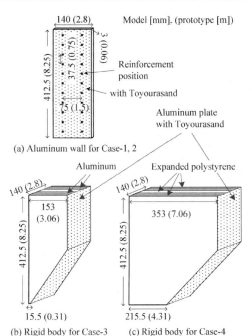

Figure 5. Size and structure of aluminium wall and rigid body retaining wall.

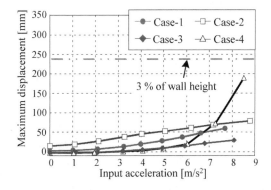

Figure 7. Relationship between the input acceleration and the maximum displacement of the wall.

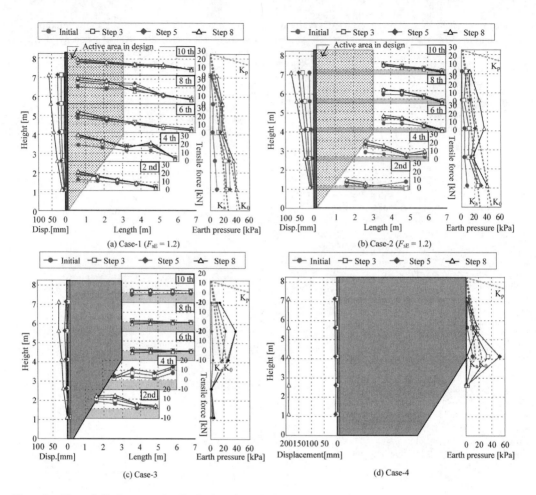

Figure 8. The wall displacement, the distribution of tensile force generated in reinforcements, and distribution of earth pressure acting of the wall.

wall. In Figure 7, 3% of the wall height (maximum allowable displacement of reinforced earth wall) is also noted. Figure 8 shows the wall displacement, the distribution of tensile force generated in reinforcement, and the distribution of earth pressure acting on the wall. Figure 9 shows the state before and after shaking of each case.

3.1.1 Seismic behaviour of Case-1

In Case-1, reinforcement was laid in the AFZ in the manner of real structures. From Step 0 (centrifugal acceleration up to 20 G) to Step 2, the maximum displacement was observed at the centre of the wall ($H = 4.13$ m). From Step 3 to Step 8, the deformation mode changed and the displacement at the upper end of the wall ($H = 7.13$ m) became the maximum. This is because the confining stress acting on the reinforcement at the upper part is smaller than at the centre and lower part, so the reinforcement is more easily pulled out when the vibration level increases. As shown in Figure 7 the amount of deformation was accumulated every excitation step and never sharply increased. The maximum wall displacement after the final excitation was 60 mm, and no slip surface was observed in the embankment (Figure 9). The tensile force generated in the reinforcement was at its maximum near the wall ($L = 0.6$ m) at any height, and the distribution of the tensile forces did not draw a bowed shape. This point was different from the current design concept. Additionally, the earth pressure acting on the wall increased with every excitation.

3.1.2 Seismic behaviour of Case-2

In Case-2, the pullout resistance of reinforcement in the AFZ was minimised to mimic the design internal stability calculation. Rather, the tensile force of the reinforcement laid in the resistance zone is expected to stabilise the whole REW.

Although 15 mm of displacement occurred at the top of the wall at Step 0, the amount of deformation accumulated across each excitation as seen in Case-1 and never increased sharply. The maximum displacement after the final excitation was 79 mm, and no slip surface was observed in the embankment (Figure 9). Despite the lack of reinforcement in the AFZ, the whole structure remained stable even after Step 8.

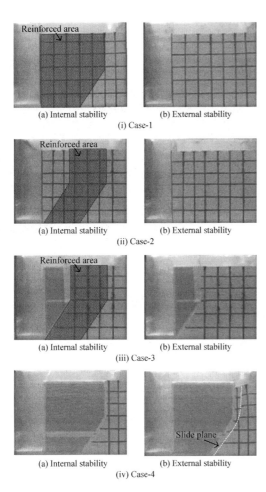

Figure 9. The state before and after excitation.

At Step 0, the rigid body retaining wall was in an upright state. When the reinforcement is not attached to the retaining wall, the retaining wall may collapse into the embankment and large deformation may occur at the lower part because the centre of gravity is pushed backwards into the embankment. However, from the reinforcement axial force distributions shown in Figure 8, it is visible that the reinforcement in the upper portions was under compression, while the lower portions were under tension. As a result, it is thought that the upper part is resisting collapse into the embankment and the lower part is resisting outward displacement, keeping the whole retaining wall upright.

3.1.4 Seismic behaviour of Case-4

In Case-4, the whole REW was replaced with a rigid body retaining wall. From Step 0 to Step 6, the deformation pattern was the same as Case-3. However, the wall displacement sharply increased from Step 7, and the residual displacement after the final excitation was 190 mm. As the retaining wall slid, a slip surface was confirmed inside the embankment in Step 7, and the embankment greatly deformed in Step 8 (Figure 9). In Case-4, the difference in rigidity between the rigid body retaining wall and the embankment was large, and the continuity of the rigid body retaining wall and the embankment was lost during earthquake, resulting in brittle collapse.

3.2 Mechanical role of reinforcing material laid in the active failure zone

The mechanical role of the reinforcement in the AFZ is visible by comparing Cases 1, 2 and 3. From Figure 7, the maximum wall displacement for all input accelerations was in the order of Case-2 > Case-1 > Case-3. The AFZ contained strip reinforcement for Case-1, stainless steel wires ($\phi = 0.45$ mm) for Case-2, and a rigid body retaining wall for Case-3. Hence, the stiffness of the AFZ is largest in Case-3 and smallest in Case-2. Accordingly, due to the relationship between the rigidity of the AFZ and the wall displacement, an inverse correlation was observed for the wall displacement.

In Case-1, the deformation mode changed from Step 3 to an overturning pattern, whereas in Case-2 an overturning deformation mode was seen from Step 0. By restraining the embankment in the upper part of the embankment where the confining stress is small and the reinforcement is likely to be pulled out, the reinforcement in the AFZ plays a role of suppressing the deformation of the whole embankment.

3.3 Rigid body behaviour and seismic performance of reinforced earth wall

By comparing the seismic behaviour of Case-1, 3 and 4, the rigidity of the REW is examined.

This is because (1) the reinforcement laid in the resistance zone inhibited the deformation of the embankment inside the AFZ, and (2) the stainless wire transfers the tensile force of the reinforcement to the wall, sandwiching the AFZ between the wall and the resistance zone and suppressing deformation.

An overturning deformation mode was observed for all steps of shaking. This is thought to be because the reinforcement is likely to be pulled out at the upper part due to the small confining stress and because of the lack of reinforcement close to the wall binding it to the sand.

3.1.3 Seismic behaviour of Case-3

In Case-3, the area corresponding to the AFZ in design was replaced with a rigid body retaining wall. From Step 0 to Step 3, almost no deformation occurred. The deformation began to accumulate from Step 4, but the maximum wall displacement after the final excitation was only 30 mm. This is because installation of reinforcement in the resistance zone added ductility to an effectively rigid AFZ.

From Figure 7, the maximum wall displacement until Step 6 was Case-1 > Case-3 ≒ Case-4. In Case-3 and Case-4, the maximum wall displacement was almost the same until Step 6. From these results, we observed that if a part or the whole of the REW is handled as a rigid body, the displacement becomes smaller than that of the actual structure.

The wall displacement after the final excitation was in the order of Case-4 > Case-1 > Case-3. The wall displacement sharply increased for Case-4, and became larger than that of Case-1, resulting in a different pattern than seen in the first steps. On the other hand, in Case-1 and Case-3 with reinforcement, ductile failure was observed without any sudden increase in wall displacement. This is because the reinforcement is laid in the embankment, reducing the difference in rigidity and maintaining continuity between the structure and surrounding soil.

From the above, when the input acceleration is small, it is considered that the whole REW can be handled as an integrated rigid retaining wall. However, when the input acceleration becomes large, treating the REW as an integrated rigid retaining wall neglects the ductility of the REW and the continuity with the surrounding soil, causing different behaviour during earthquakes.

4 CONCLUSIONS

In this study, centrifuge model tests were conducted to elucidate the dynamic behaviour and the reinforcing mechanism of the reinforced earth wall by clarifying the mechanical role of the reinforcement in the active failure zone and considering the rigidity of the reinforced earth wall. The findings obtained by this research are as follows.

A.: Role of reinforcement laid in the active failure zone

1) Rigidity of the embankment is increased by laying reinforcement in the active failure zone, reducing the amount of deformation during earthquakes.
2) In the upper part of the embankment where confining stress is small and deformation is likely to occur, reinforcement restrains the overturning motion around the wall.

B.: Rigid body behaviour and seismic performance of reinforced earth wall

3) When the earthquake ground motion is relatively small, treating the entire reinforced earth wall as a rigid body results in similar deformation to that of an actual reinforced soil wall.
4) 4. Reinforced earth walls demonstrate toughness because the reinforcement is laid in the embankment, reducing the difference in rigidity from the surrounding soil and allowing the reinforcing earth wall and the surrounding ground to behave continuously during an earthquake.

ACKNOWLEDGEMENTS

This research was supported by the Konoike Scholarship Foundation.

REFERENCES

Richardson, G.N. 1975. Seismic design of reinforced earth walls. *Journal of Geotechnical and Geoenvironmental Engineering*, 101(ASCE# 11143 Proc Paper).

Richardson, G.N., Lee, K.L., Fong, A.C. & Feger, D.L. 1977. Seismic testing of reinforced earth walls. *Journal of the Soil Mechanics and Foundations Division*, 103(1): 1–17.

Siddharthan, R.V., Ganeshwara, V., Kutter, B.L., El-Desouky, M. & Whitman, R.V. 2004. Seismic deformation of bar mat mechanically stabilized earth walls. II: A multiblock model. *Journal of Geotechnical and Geoenvironmental Engineering*, 130(1): 26–35.

Yu, Y., Bathurst, R.J., & Miyata, Y. 2015. Numerical analysis of a mechanically stabilized earth wall reinforced with steel strips. *Soils and Foundations*, 55(3): 536–547.

Yazdandoust, M. 2015. Study on Seismic Performance of Reinforced Soil Walls to Modify the Pseudo Static Method. World Academy of Science, Engineering and Technology, International *Journal of Civil, Environmental, Structural, Construction and Architectural Engineering*, 9(9): 1248–1259.

Centrifuge tests investigating the effect of suction caisson installation in dense sand on the state of the soil plug

M. Stapelfeldt
Institute of Geotechnical Engineering and Construction Management, Hamburg University of Technology, Hamburg, Germany

B. Bienen
Centre for Offshore Foundation Systems, University of Western Australia, Perth, Australia

J. Grabe
Institute of Geotechnical Engineering and Construction Management, Hamburg University of Technology, Hamburg, Germany

ABSTRACT: Suction caissons are a promising foundation system for offshore wind farms. However, uncertainties remain regarding the effect of the suction installation in dense sand on the load bearing behaviour. This subject is addressed here through a series of centrifuge tests, in which model suction caissons were installed in saturated fine silica sand. Following the installation, and without stopping the centrifuge, a cone penetration test was conducted inside the caisson. Comparison of the results allows relative evaluation of the effects of pumping flow rate and lid contact. Cone test results from undisturbed free field sites further provide an indication of the altered stress state inside and below the caisson. The centrifuge results indicate that the suction installation related effects on the load bearing behaviour appear to be influenced by the role of the lid contact.

1 INTRODUCTION

1.1 Caissons for wind farms in the North Sea

More than a decade ago offshore wind energy was established to become a sustainable, alternative electric power supply in Northern Europe. Since then, a growing number of offshore wind farm projects is commissioned every year – especially in the North Sea. Since offshore wind energy is on the rise, suction caissons are considered a suitable foundation system for bottom-fixed wind turbines (e. g. Houlsby & Byrne 2005, Ibsen et al. 2005 and Tjelta 2015). The sea bed in the North Sea often consists of dense sand in areas of offshore wind development, which provides appropriate conditions for large diameter structures with relatively short skirts, i. e. caissons with an aspect ratio of $\frac{L}{D} \approx 0.5$. These skirted shallow foundations provide considerable capacities for moments, horizontal and tensile loads. Since lid contact is crucial for caisson shallow foundations, grout beneath the lid might be required to ensure full contact and hence load transferral into the underlying soil. Recent efforts and ongoing developments prepared the caisson technology for applications in offshore wind farm projects at commercial scale (Tjelta 2015).

1.2 Suction caisson installation in dense sand

A suction caisson is installed in two stages: The self-weight penetration and the suction installation. In the first stage, the caisson is lowered to the mud line and the skirt penetrates into the seabed due to the self-weight of the caisson and the wind turbine substructure. The actual penetration depth results from the submerged self-weight of the structure and the bearing capacity of the skirt tip; mainly depending on the wall thickness and the *in situ* soil characteristics (e. g. Tran et al. 2004, Houlsby & Byrne 2005 and Andersen et al. 2008). When an equilibrium is reached, the self-weight penetration is completed and the suction installation can commence. Water trapped inside the caisson is pumped out and a differential pressure to ambient (i. e. suction) is created, which results in seepage flow around the skirt into the caisson. Consequently, the effective stresses inside the caisson and at the skirt tip are reduced. The combination of suction pressure, seepage flow and self-weight, enables the installation of a caisson foundation – even in dense or very dense sand (e. g. Tjelta 1995, Houlsby & Byrne 2005 and Bienen et al. 2018).

Alongside the seepage flowing upwards inside the caisson, the effective stress state and the plug volume and therefore the void ratio and the permeability change during the suction installation (Tran 2005, Houlsby & Byrne 2005). This soil state changing process depends on the actual hydraulic conditions inside the caisson during the suction installation (Tran et al. 2004, Tran & Randolph 2008). These conditions govern the characteristics of the soil plug until the suction installation is completed. Possible scenarios for the soil plug state at this stage are:

1. If the caisson reaches the target penetration depth and contact is established between the lid and the soil, only insignificant plug heave and therefore minimal uncertainties concerning the in-service performance are expected as the plug is unlikely to have loosened substantially.
2. If the caisson stops before the target penetration depth is reached but the lid touches down, significant plug heave has occurred, which raises uncertainties concerning the in-service performance. This is due to the partial skirt embedment and likely soil plug loosening.
3. If the caisson stops penetrating before the lid touches down, uncertainties concerning the in-service performance arise from the partial skirt embedment and the unknown state of the soil plug. The caisson lid can not be relied on to transfer loads into the soil. Hence, in-service loads might exceed allowable settlements or perhaps the bearing capacity.

In each of those cases, the actual conditions of the soil plug is crucial to the bearing capacity and the expected settlements of the caisson foundation system. This necessitates the targeted research and provides the background of this series of centrifuge tests.

2 HYPOTHESIS

Since the seepage flow into the caisson and skirt penetration occur simultaneously, the majority of the soil located underneath the tip is expected to travel into the caisson. The reason for this assumption is the occurrence of momentum exchange between the flowing pore water and the soil grains, which forces the sand to preferably displace alongside the direction of the seepage flow, i.e. heading into the caisson. Consequently the volume of the soil plug increases. In addition to the shearing-induced dilatancy, the momentum exchange inside the caisson lifts the soil grains and therefore decreases the effective stresses, the plug density and increases its volume and permeability. Both effects, the inflow of sand and loosening, lead to a considerably altered stress state inside the caisson during the suction installation. Apart from the soil properties, the degree of disturbance primarily depends on the seepage flow velocity and therefore on the applied suction pressure, which is related to the pumping flow rate.

Just before the end of the suction installation, i.e. at the moment the target penetration depth is reached, the plug volume and the void ratio reach their respective maximum. Subsequently, the pumping flow rate reduces to zero and the suction pressure dissipates. The seepage flow and the momentum exchange between liquid and solid phase stops. Consequently, the plug has the opportunity to settle due to its self-weight to regain density. This is more likely if the lid touches down and compressive stresses consolidate the plug. Similar to a specimen in an oedometer cell the soil plug is confined and drains at the bottom. Apart from the sand that was displaced form the skirt tip into the caisson, plug heave may be reversible through this mechanism.

This hypothesis requires experimental verification. Therefore, a centrifuge testing program was designed that allowed *in situ* investigations through a cone penetration test (CPT) of the soil plug inside an installed suction caisson.

3 EXPERIMENTAL SET-UP

The experiments were performed at $100\,g$ in the Acutronic Model 661 centrifuge (Randolph et al. 1991) at the University of Western Australia (UWA) to ensure a realistic stress state during the model scale experiments. These had the following requirements:

- The caisson needed to be suspended just above the sand surface until the centrifuge had reached the target acceleration, to ensure penetration at stress levels representative of the prototype.
- The self-weight penetration and the suction installation were to be done in controlled manner.
- The CPT inside the caisson was to be performed without stoppage of the centrifuge to retain the soil stress state following the suction installation.

Additional CPTs at free field sites were to be performed to characterise the soil sample itself.

3.1 *Characteristics of the saturated soil sample*

The soil sample was prepared in three steps: First a 20 mm filter layer containing of coarse sand was constructed at the base of the strong box, which was covered with a geotextile before pluviating the fine silica sand. The dry sand was rained from a height of approximately 1.2 m by means of an automatic pluviator. After reaching a minimum total sample height of 200 mm, the surface was vacuum levelled. The unit weight was determined from weight and volume measurements. The relative density D_r for these samples was approximately 79%.

Subsequently, the sand sample was saturated from the base. Taking scaling laws for $100\,g$ conditions into account, the viscosity η of the pore fluid was increased from approximately 1 cSt for water to 100 cSt by means of the addition of Methocel (DOW 2002), targeting a permeable soil, i.e. $k_f = 1 \cdot 10^{-4}$ m/s at $100\,g$. Since the concentration of Methocel powder diluted in water defines the fluid viscosity, the permeability of the saturated soil can be adjusted. Increasing the viscosity of the fluid up to 300 cSt, a prototype permeability coefficient of $k_f = 3.3 \cdot 10^{-5}$ m/s was achieved. This allows investigations of the effect of uncertainty in the *in situ* permeability, with target values ensuring drainage behaviour that represents typical North Sea soil characteristics (Taylor 1995).

3.2 *Caisson and CPT model for ng-tests*

The model caisson had a diameter of $D = 80$ mm and a skirt length of $L = 40$ mm or $L = 80$ mm – representing a diameter $D = 8$ m and a skirt length of $L = 4$ m or $L = 8$ m in prototype scale. The cone had diameter $D_{cpt} = 6.3$ mm. The cone was in place, with the tip retracted inside the caisson lid during the installation. The opening in the caisson lid was sealed with two levels of o-rings. The clearance between the cone tip and the caisson skirt was more than three times the cone diameter to minimise interaction.

Controlled suction caisson and CPT penetration were facilitated with an actuator. Figure 1 shows the experimental set-up. The caisson was connected to the actuator with a steel wire only with the dual purpose of suspending the foundation prior to testing and allowing self-weight penetration to take place at a controlled rate. The CPT was rigidly connected to the actuator. The caisson and the CPT were lowered simultaneously, i.e. without relative movement, and at a constant rate. The movement was controlled through a feedback loop, maintaining constant the measurement of a linear displacement transducer (LDT), which was attached to the actuator at one end and rested on the caisson lid at the other.

Verticality of the caisson installation was ensured by a guide rod arrangement (see Figure 1). The self-weight of the steel rod ensured a total self-weight of 350 N. This represents a prototype structure with a self-weight of 3.5 MN per caisson.

The three way valve, presented in Figure 1 vented to ambient during the self-weight installation, connected the caisson to the syringe pump through the hose during suction installation and was sealed during the CPT. The valve was operated by a cord connected to an electric winch, which was located next to the actuator. An eyelet below the actuator set the pulling direction of the cord in line with the CPT and the caisson centreline to avoid rotation of the caisson.

Between each test site and the boundaries of the box, there was a space of at least $2D$. Furthermore, the sample height was more than $2L$ to prevent boundary effects (Tran & Randolph 2008).

3.3 *Testing procedure and measurements*

The caisson was suspended above the soil surface until the centrifuge had reached $100\,g$ and any pore pressures in the sand had dissipated. The self-weight installation was performed by controlled lowering at a penetration rate of $\dot{z}_{caisson} = 0.1$ mm/s. After the caisson stopped penetrating under self-weight, the steel wire went slack and caisson self-weight and the

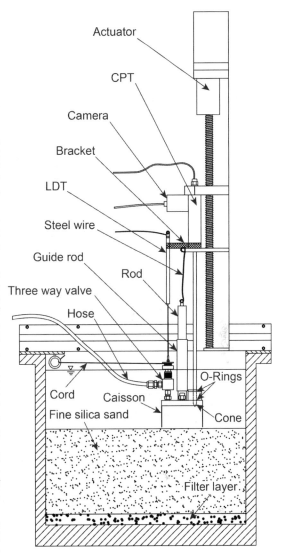

Figure 1. Centrifuge test set-up for suction caisson installation including CPTs through the lid.

bearing capacity were in equilibrium. At this point, the three way valve was switched to connect the caisson to the syringe pump, which was mounted on the centrifuge. Fluid was then evacuated from the caisson through the hose due to suction provided by the pump, which was operated at constant pumping flow rate of 250π mm^3/s or 1000π mm^3/s in model scale. After the suction installation was terminated, the three way valve was sealed, a wait period of 60 s was observed to allow suction pressures to dissipate and the CPT was conducted at a rate of 1 mm/s. Figure 2 provides an overview of the testing programme.

Total pressures were measured with total pressure transducers (TPTs) at the top and the bottom of the caisson lid and inside the syringe pump. Pore pressures were measured at the bottom of the lid, i.e. inside the caisson, by means of a pore pressure transducer (PPT).

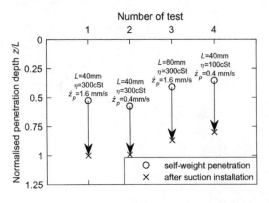

Figure 2. Normalised penetration depths of all installation tests.

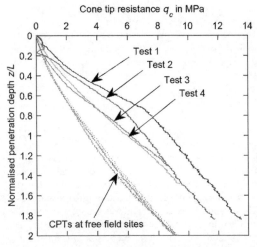

Figure 3. CPTs before and after caisson installation.

This instrumentation allowed further insight into the plug conditions during the whole test. Furthermore, the fluid temperature was measured, since the viscosity of the Methocel-modified water depends significantly on the temperature. A constant fluid temperature was ensured by a fixed air conditioned temperature of $T = 20\,°C$.

4 RESULTS

A total of four CPTs and four caisson installations with accompanying CPTs through the lid are discussed here. The CPTs at free field sites, i. e. tests in virgin soil, were penetrated at a constant rate of $\dot{z}_{cpt} = 1$ mm/s, which is expected to result in drained conditions (Finnie & Randolph 1994). Those CPTs were penetrated to a depth of at least $2L$ or $19D_{cpt}$, respectively. The CPTs inside the caissons were penetrated at the same rate to a maximum depth of $1.8L$ – restricted by the clearance between guide rod and actuator (see Figure 1 and 3).

4.1 Cone tip resistance profiles in the sand samples

The CPTs were performed at different locations close to the caisson installation sites to check the uniformity of the saturated soil sample and provide reference profiles for the CPTs inside the caisson. All four cone resistance curves from the free field sites presented in Figure 3 show values within a narrow range, confirming that the sample densities were uniform.

4.2 Cone tip resistance profiles inside the caissons

As presented in Figure 2, the first two caissons were completely installed and the second two caissons were partially installed. Test 3 was conducted using a caisson with a skirt length of $L = 80$ mm to evoke incomplete installation. However, the penetration depth is normalised by $L = 40$ mm.

Figure 3 shows, that the cone penetration resistance profile inside each caisson is higher than it is at the free field sites. This indicates that the stress level is considerably higher inside the caisson. This effect continues beyond the maximum skirt penetration depth of $\frac{z}{L} = 1$.

The four q_c profiles through the soil plugs of the installed caissons show two different characteristic behaviours: 1) two gently sloping curves (Tests 3 and 4) and 2) two curves with a distinct change in gradient at approximately $\frac{z}{L} = 0.8$ (Tests 1 and 2), with the latter also exhibiting higher stiffness at shallow penetration.

Comparing these two groups, the only systematic difference is complete installation in Tests 1 and 2 ($\frac{z}{L} \approx 1$) and incomplete installation in Tests 3 and 4 ($\frac{z}{L} < 0.9$). The significantly higher cone resistance at shallow depth in Tests 1 and 2 compared to Tests 3 and 4 is due to the surcharge pressure provide by the lid, which changes the stress state inside and below the caisson. Therefore it is concluded that the plug is reconsolidated, which leads to higher effective stresses and explains the increased cone resistance compared to the partially installed caisson in Tests 3 and 4. The increased cone resistance profile inside the partially installed caissons is due to penetration of the cone within the confinement of the soil plug inside the caisson skirt.

4.3 Influence of the pumping flow rates

The influence of the pumping flow rate on the soil plug state is investigated through comparison of the tests with different penetration rates (see Figure 4). The penetration depths of each test is normalised by $L = 40$ mm. The penetration rate is normalised by the reference piston velocity $\dot{z}_{p,ref.} = 1.6$ mm/s.

Figure 4 shows that the penetration rates of the slowly installed caisson in Tests 2 and 4 were almost constant, indicating that no significant seepage flow occurred during the installation. Hence, no seepage flow induced changes of the plug condition are expected. The pumping action was transferred to

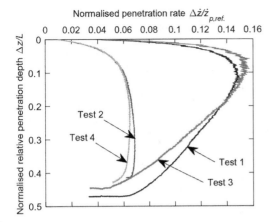

Figure 4. History of normalised caisson penetration rates during suction installation.

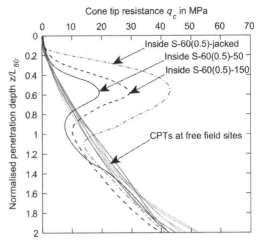

Figure 5. CPTs before and after installation from centrifuge tests referred by Tran (2005).

caisson penetration throughout this process. However, the initially high penetration rates in Tests 1 and 3 deceased significantly over the course the installation. This behaviour indicates that the seepage flow and hence the permeability of the soil plug increased significantly. Therefore, the relative density of the soil plug decreased during the suction installation. However, this is not reflected in the respective cone resistance profiles presented in Figure 3.

Consequently, the caisson penetration rate, which is determined by the pumping flow rate, has no significant influence on the cone resistance once the installation is completed. It is concluded that the stress distribution through the lid and the confined state of the plug inside the caisson govern the stress state at this stage. This is consistent with the hypothesis in section 2 and the findings presented in section 4.2.

Test 3 and 4 also feature different penetration rates in accordance to Figure 4. Analogue to Test 1 and 2, the penetration rate of the fast installed caisson decreases significantly during the suction installation. Again, the cone resistances shown in Figure 3 are almost identical, which supports the previous findings. Furthermore, it is concluded that the effective stress state of the soil is influenced by the dissipating suction pressure. Since the lid did not touch down in Test 3 and 4, it is additionally assumed that the plug re-consolidates under the submerged self-weight of the sand particles only, which results in softer response at shallow depth (see Figure 3).

Furthermore, the *in situ* permeability and changes during the suction installation can be crucial to the installation process (Tran et al. 2004, Houlsby & Byrne 2005). Therefore, these parameters are investigated. It must be noted that the effective permeability in Test 4 is higher, compared to Test 3, due to a lower viscosity of the pore fluid – i.e. $\eta = 100$ cSt instead of $\eta = 300$ cSt. Since the cone resistance in both tests is almost identical and the influence of the caisson penetration rate is considered to be insignificant here, it is found that a variation in the permeability by a factor of three, indicative of the uncertainty associated with permeability measurements, has no considerable influence on the cone resistance and therefore on the stress state of the soil plug for the investigated cases.

5 COMPARISON TO PREVIOUS TESTS

Comparable investigations were published by Tran (2005) and Tran et al. (2005). Several aspects, e.g. the centrifuge used, the caisson dimensions, the soil characteristics and the acceleration level are similar. Furthermore, Tran et al. (2005) also performed CPTs inside suction installed caissons, However, the results differ considerably.

The original data from two graphs referred in Tran (2005) is summarised in Figure 5. The penetration depth z is normalised by $L_{60} = 60$ mm to facilitate visual comparison with Figure 3.

A significant difference between both series of tests is the caisson self-weight. Tran (2005) referred applying 0.5 MN in prototype scale instead of 3.5 MN as mentioned in section 3. Therefore, the self-weight penetration depths obviously differ considerably, but the effect on the suction installation is proven to be insignificant (Tran & Randolph 2008). A comparison between Figure 3 and 5 supports the conclusion that the self-weight and therefore the overburden pressure determines the stress state of the soil plug.

A further significant difference between the testing series is that the soil stress state was maintained in the present tests, whereas the centrifuge was stopped between suction caisson installation and CPT in Tran (2005). This may have affected the soil stress state.

However, the curves in Figure 5 approach the values obtained from CPTs outside the caisson again at a normalised penetration depth of approximately 1.6 – in contrast to the curves presented in Figure 3. This is consistent, since the self-weight applied to the caisson to obtain the results presented Figure 3 is seven times higher than for Figure 5. So it is reasonable that the the area of increased effective stresses in the results reported by Tran (2005) is limited to the soil plug inside the skirt. It is assumed that this area expands under higher vertical loading, so that it exceeds the embedded depth of the caisson.

6 CONCLUDING REMARKS

The centrifuge test results presented in this paper were obtained to gain insight into the condition of the soil plug within a caisson installed by suction in dense sand. Hence, CPTs inside the caisson were conducted to investigate the actual stress state of the plug and the underlying soil after the installation. The main findings are summarised below:

- Data measured during the installation indicates that significant soil plug loosening most probably occurred during suction installation with high pumping flow rates. However, neither the achieved penetration depth nor the measured cone resistance indicate a permanent impact, with cone resistance profiles similar to more slowly installed caissons. This holds for both: partially and completely installed caissons.
- Since the cone resistance for within fully installed caissons is considerably higher than within partially installed caissons, in particular at shallow depth, it is concluded that the pressure applied through the lid dominates the stress state of the plug and the underlying soil.
- Variations in the installation properties – e.g. changing the pumping flow rate or the soil permeability – induced only minor differences regarding the soil plug conditions.

The hypotheses of the lid transferring stresses to the soil plug and confinement of the soil plug within the caisson governing the *in situ* stress and soil state is confirmed.

ACKNOWLEDGEMENTS

We acknowledge the support from the Deutsche Forschungsgemeinschaft (DFG) for our research project GR 1024/26-1 "Sauginstallation Maritimer Strukturen" (SIMS).

This work forms part of the activities of the Centre for Offshore Foundation Systems (COFS), which is currently supported as one of the primary nodes of the Australian Research Council (ARC) Centre of Excellence for Geotechnical Science and Engineering and as a Centre of Excellence by the Lloyds Register Foundation. Lloyds Register Foundation helps to protect life and property by supporting engineering-related education, public engagement and the application of research. This support is gratefully acknowledged.

REFERENCES

Andersen, K. H., H. P. Jostad, & R. Dyvik 2008. Penetration resistance of offshore skirted foundations and anchors in dense sand. *Journal of Geotechnical and Geoenvironmental Engineering 134*(1), 106–116.

Bienen, B., R. T. Klinkvort, C. O'Loughlin, F. Zhu, & B. W. Byrne 2018. Suction caissons in dense sand, Part I: Installation, limiting capacity and drainage. *Géotechnique 0*(0), 1–47.

DOW 2002. *Methocel cellulose ethers – Technical handbook*. United States of America: DOW.

Finnie, I. M. S. & M. F. Randolph 1994. Punch-through and liquefaction induced failure of shallow foundations on calcareous sediments. In *Seventh International Conference on the Behaviour of Offshore Structures*, Volume 1, United Kingdom, pp. 217–230. Pergamon.

Houlsby, G. T. & B. W. Byrne 2005. Design procedures for installation of suction caissons in sand. *Geotechnical Engineering 158*(3), 135–144.

Ibsen, L. B., M. Liingaard, & S. A. Nielsen 2005. Bucket foundation, a status.

Randolph, M. F., R. J. Jewell, K. J. L. Stone, & T. A. Brown 1991. Establishing a new centrifuge facility. In *Proc. of Intern. Con. on Centrifuge Modelling*, pp. 3–9.

Taylor, R. N. 1995. *Geotechnical centrifuge technology*. Glasgow: Blackie Academic and Professional.

Tjelta, T. I. 1995. Geotechnical experience from the installation of the europipe jacket with bucket foundations.

Tjelta, T. I. 2015. The suction caisson foundation technonlogy. In V. Meyer (Ed.), *Frontiers in Offshore Geotechnics III*, Volume III, pp. 85–93. London: Taylor & Francis Group and CRC Press/Balkema.

Tran, M. N. 2005. *Installation of suction caissons in dense Sand and the influence of silt and cemented layers*. Phd-thesis, The University of Sydney, Sydney.

Tran, M. N., D. W. Airey, & M. F. Randolph 2005. Study of seepage flow and sand plug loosening in installation of suction caissons in sand. In International Society of Offshore and Polar Engineers (Ed.), *15th International Offshore and Polar Engineering Conference (ISOPE)*, pp. 516–521.

Tran, M. N. & M. F. Randolph 2008. Variation of suction pressure during caisson installation in sand. *Géotechnique 58*(1), 1–11.

Tran, M. N., M. F. Randolph, & D. W. Airey 2004. Experimental study of suction installation of caissons in dense sand. In *Proc. of the 23rd Intern. Conf. on Offshore Mechanics and Arctic Engineering*, New York, NY, pp. 105–112. ASME.

Centrifuge model tests on stabilisation countermeasures of a composite breakwater under tsunami actions

K. Tsurugasaki, J. Miyamoto & R. Hem
Toyo Construction Co., Ltd, Nishinomiya, Japan

T. Iwamoto & H. Nakase
Tokyo Electric Power Services Co., Ltd., Tokyo, Japan

ABSTRACT: In this study, tsunami overflow experiments of composite breakwaters with different types of surcharge embankments behind the caisson were conducted with a drum centrifuge. In order to stabilise the foundation rubble mound against seepage flow generated by tsunami, it is important to keep pressing the mound just under the caisson toe with an embankment. The embankment by gravel gradually becomes ineffective due to continuation of the scouring. On the other hand, the embankment using the mesh bag units can resiliently hold down the mound just under the caisson toe, so it was confirmed effective even with a small amounts of mesh bag units against scouring and seepage flow in the mound.

1 INTRODUCTION

In the 2011 Pacific Coast of Tohoku Earthquake, many composite breakwaters were severely damaged by the huge tsunami in the Tohoku region, Japan. As the influence of the tsunami on breakwater, it is necessary to consider the initial tsunami wave impact, tsunami overflow, and backrush wave in order. The process leading to the overflow including the initial tsunami wave impact was shown by the drum centrifuge experiments with dam-break wave generation method by Miyamoto et al. (2015). The breakwater was confirmed stable against the initial tsunami wave impact until the overflow, but it collapsed immediately after the overflow. When the tsunami overflows the breakwater, the foundation rubble mound behind the caisson is scoured, and eventually the caisson collapses. In this case, it has been pointed out that the mound becomes unstable not only by the scouring but also by the seepage flow induced by the differential pressure between the front and back of breakwater, so that the breakwater failure (Imase et al. 2012, Takahashi et al. 2014, Sassa 2014). One of the countermeasures for stabilising the mound against the seepage flow is installing a surcharge embankment behind the caisson. The embankment is expected to improve the stability of the caisson and mound bearing capacity and to reduce the scouring of mound.

The purpose of this study is to investigate the stability of breakwater with several types of surcharge embankments by drum centrifuge and to compare the resilience of each case against to the continuous tsunami overflow. In this study, the 'resilience' was quantified by the length of time from the start of the overflow to the failure of the foundation mound.

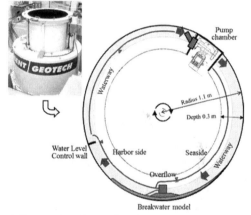

a) Cross section of water channel

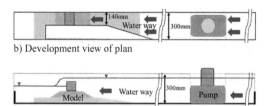

b) Development view of plan

c) Development view of cross section

Figure 1. Cross section of typical waterway.

2 TSUNAMI TESTS BY DRUM CENTRIFUGE

2.1 *Tsunami tests method in centrifugal force field*

A series of tsunami tests were conducted by drum centrifuge apparatus (channel radius = 1.1 m) owned by

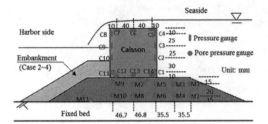

Figure 2. Composite breakwater sectional view (with embankment) and measuring position.

Toyo Construction Co., Ltd. Figure 1 shows a typical centrifugal water channel cross section including a composite breakwater. Examples of waterway experiments in centrifugal force fields are described in the literature (Miyake et al. 2002, Baba et al. 2002). In tsunami tests with a drum centrifuge, waves have been generated by a dam break system or a piston system (Miyake et al. 2009, Sawada et al. 2013, Miyamoto et al. 2015). In this study, a circulation pump was installed to reproduce a continuous tsunami overflow for a long time (Miyamoto et al. 2015, Tsurugasaki et al. 2016). Tests were conducted at 70G in centrifugal force field (modelling scale by 1/70). Table 1 shows the scaling laws related to the tests. The scaling laws of tsunami and seepage flow in the mound were referred to Takahashi et al. (2014). Accordingly, by assuming that the fluid flow in the mound is turbulent, the time scaling rate of tsunami in the Froude law and that of the seepage flow in the mound are the same.

Table 1. Scaling laws referred to Takahashi et al. (2013).

Parameter	Prototype	Model
Acceleration	1	N
Dimension	1	$1/N$
Grain size	1	$1/N$
Density	1	1
Pore water pressure	1	1
Tsunami (external force) – Froude Law		
Time	1	$1/N$
Pressure	1	1
Velocity	1	1
Seepage within mould		
Pore water pressure	1	1
Stress	1	1
Hydraylic gradient	1	1
Turbulent		
Mean flow velocity	1	1
Seepage time	1	$1/N$

Table 2. Test conditions.

Case	Countermeasure	Water level at front and back of caisson
1	Without embankment	44 kPa
2-1	Embankment by gravel (overall)	48 kPa
2-2		52 kPa
3	Embankment by unit (overall)	52 kPa
4	Embankment by unit (partial)	55 kPa

2.2 Breakwater model

Figure 2 shows the composite breakwater model and measurement locations. The breakwater model was composed of gravel mound and box type aluminium caisson filled with sand. Both the caisson's height and width were 100 mm (equal to 7 m in prototype size). The pressure at the bottom of caisson was 92 kPa (fully submerged) in 70 G centrifuge force field. The mound was composed of the gravels with the size of 4.75–9.5 mm (equal to 300–700 mm in prototype size) and its porosity was 45%. The breakwater model was set up on PVC fixed bed in the experimental channel. The friction coefficients, which were obtained from the direct shear tests, were 0.61, between aluminium caisson bottom and gravel mound and 0.76, between gravel mound and PVC bed. The sides of the breakwater model were coated with silicone grease to minimise the model side frictions. Wave height measurement transducers and piezometers for water level monitoring were set in the experimental channel, and piezometers and pressure meters were set in the mound and around the caisson. The experiments were observed with multiple high-speed video cameras. Drum centrifuge tests were held sideways in the channel. All models were set sideway, too. Breakwater models were prepared in another container and frozen after being saturated by water. Thereafter, water was poured into the channel while it rotates, and the model was thawed.

2.3 Test cases

Table 2 shows test conditions and Figure 3 shows a schematic diagram cases and unit model. In the test cases, there were no-measure case (Case 1) and embankment cases (Case 2-4). The embankment cases were divided into the normal case using gravel (Case 2) and the mesh-bag unit containing gravel (Cases 3 and 4). In the case of the unit, two cases were carried out: the case covering the mound on the back of the caisson in its entirety (Case 3) and the case holding only the mound just under the back of the caisson (Case 4) in consideration of economy. Units were prepared by filling a mesh bag with gravel of particle size of 2.0–4.75 mm (equal to the real size of 150–300 mm). One unit was set to have the weight of 15 g, which corresponds to the actual weight of 5 t.

3 TEST RESULTS

3.1 The instability of mound due to seepage flow and the effect of embankment

Figure 4a shows the time history (from the start of circulation pump) of the differential water pressure between the front and back of breakwater and the hydraulic gradient in the mound of Case 1 (no countermeasure).

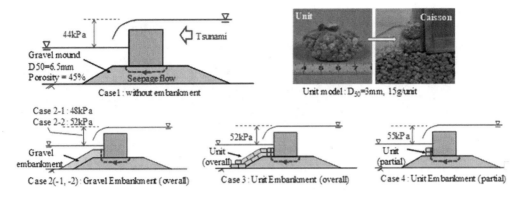

Figure 3. Embankment model and unit model.

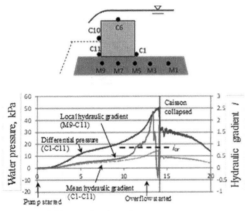

(a) Case 1 (without embankment)

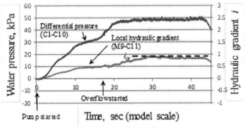

(b) Case 2-1 (embankment by gravel)

Figure 4. Water pressure differential between front and back of caisson and hydraulic gradient in mound.

Two kinds of hydraulic gradient in the mound are shown, i.e. the mean gradient (C1-C11) and the local gradient just under the caisson toe (M9-C11). Hydraulic gradient was calculated from pore water pressures and the distance between pressure gauges as presented in Figure 2. The hydraulic gradient of the seepage flow in the mound increased with the rise in the differential water pressure between front and back of breakwater. The local hydraulic gradient just under the caisson toe increased more than the mean hydraulic gradient and showed instability beyond the critical hydraulic gradient i_{cr}. At this point, the mound just under the caisson toe was unstable due to seepage flow, indicating that the caisson collapsed. Here, the critical hydraulic gradient i_{cr} was calculated by $i_{cr} = \gamma'/\gamma_w$, where γ' is the submerged unit weight of the mound material and γ_w is the unit weight of water.

Figure 4b shows the result of Case 2-1 with embankment. While the differential water pressure between the front and back of the breakwater increased and remained constant, the hydraulic gradient just under the caisson toe also had a constant value and the mound kept stable. It is revealed that the stability of the mound against the seepage flow was improved by the embankment.

3.2 *Behaviour of embankment cases*

The behaviour when the composite breakwater with an embankment receives a large tsunami are as follows. Figure 5 shows the behaviour of Case 2-2 which was subjected to the embankment by gravel. After starting the overflow, the embankment was scoured, and at the same time the amount of gravel at the back of the caisson was greatly reduced (Figure 5b). As the scour hole grew larger, the gravel slipped down to the hole. With the continuation of overflow, (while repeating this situation,) the gravel at the back of the caisson disappeared (Figs 5c,d). Finally, the embankment disappeared, the breakwater immediately collapsed (Fig. 5e). Next, the behaviour of Case 3 in which the embankment constituted by mesh-bag units is described (Figure 6). From the start of the overflow, the units in the area where the overflow water hit directly were blown off (Figure 6b), and the scour hole developed. However, unlike the case of gravel (Case 2-2), the units did not fall into the scour hole and some units continued to stay at the back of the caisson. Furthermore, although the overflow continued, the size of the scour hole did not develop and the units kept staying. Even with the continuation of the subsequent overflow, the scour hole did not expand (Figure 6d). The reason is as shown in Figure 6d that the blown units protected the mound by holding down the leg of the mound. In this case, the mound and the caisson remained stable within the experimental observation time (Figure 6e).

Figure 5. Behaviour under overflow. (Case 2-2).

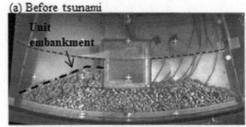
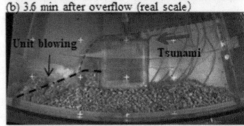

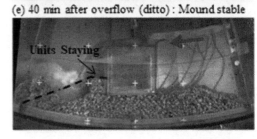

Figure 6. Behaviour under overflow. (Case 3).

From Case 3, it is important that the units hold down the just the caisson toe, so the effect of placing the units only on the back of the caisson (Case 4) was investigated. The mound was scoured by the overflow, but because the embankment was composed of units, it did not fail like a gravel embankment and kept staying for a while (Figs 7b,c). Finally, with the continuation of the overflow, however, the scouring of the mound had gradually progressed, accompanied by the steep slope of the mound (Fig. 7d), which resulted in reducing the bearing capacity of the mound, and led to the breakwater failure (Fig. 7e).

3.3 Mound instability by seepage flow

The change in stability caused by seepage flow in the mound of the embankments cases are shown in Figure 8. As the seepage distance increases by the embankment, the hydraulic gradient of the seepage

(a) Before tsunami

(b) 2.3 min after overflow (real scale)

(c) 3.5 min after overflow (ditto)

(d) Just before mound failure (7.3min ditto)

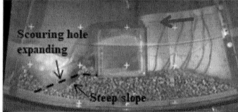

(e) Foundation mound failure

Figure 7. Behaviour under overflow (Case 4).

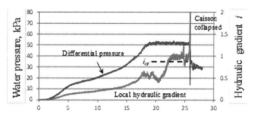

(a) Case 2-2 (embankment by gravel)

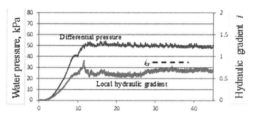

(b) Case 3 (embankment by units)

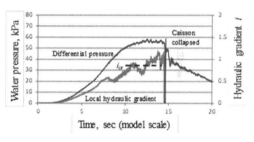

(c) Case 4 (embankment by partial units)

Figure 8. Water pressure differential between front & back of caisson and hydraulic gradient in the mound.

flow decreases and the mound stabilises. However, when the embankment by gravel is scoured as in Case 2-2, the seepage distance gradually decreases, so the hydraulic gradient of the seepage flow increases, and finally the critical hydraulic gradient i_{cr} may be exceeded. Figure 8a shows the change in the hydraulic gradient of the seepage flow in the mound of Case 2-2, taking into account the decrease in seepage distance. The hydraulic gradient of this figure was calculated by reading the change of the gravel layer thickness on the back of the caisson from the video image. As the thickness of the gravel layer decreased due to the overflow scouring, the hydraulic gradient increased and exceeded the critical hydraulic gradient i_{cr}. At this time, it shows that the mound's bearing capacity was decreasing. Next, in Case 3, as shown in Figure 8b, there was no change in the seepage distance because the units stayed on the back of the caisson.

While the overflow continued, the hydraulic gradient in the mound was almost constant, indicating that the mound was stable. For Case 4 where the units were placed only on the back of the caisson, the hydraulic gradient of the seepage flow in the mound is shown in Figure 8c. During overflow, the hydraulic gradient exceeds i_{cr}, and it shows that the mound's bearing capacity was decreasing.

4 COMPARISON OF EMBANKMENT RESILIENCE

Figure 9 shows the comparison of the time from the start of overflow until the foundation mound failure. Displacement of caisson toe was read using high

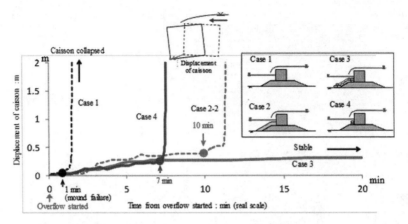

Figure 9. Caisson displacement-time history: comparison of embankment resilience.

speed video images. When using the units to cover the entire mound, the mound was most stable against overflow (Case 3). In the case of embankment with gravel (Case 2-2), it endured about 10 minutes (actual time) from the start of the overflow, whereas even if the units were placed only on the back of the caisson (Case 4), it endured about 7 minutes. In Case 4, there was an effect of 70% (=7 minutes/10 minutes) compared to Case 2 with whole embankment with gravel despite a small countermeasure amount (20% of the total embankment volume).

5 CONCLUSIONS

In this study, tsunami overflow tests of composite breakwaters with different types of surcharge embankments behind the caisson were conducted with a drum centrifuge. The main conclusions are as follows.

1. At the time of the tsunami overflow, the local hydraulic gradient just under the caisson toe was larger than the critical hydraulic gradient, and the foundation rubble mound lost its bearing capacity. On the other hand, by setting the surcharge embankments behind the caisson, the seepage distance increased and the local hydraulic gradient decreased, so the foundation mound was stable.
2. The embankments by the gravel gradually lost their effect by scouring. On the other hand, in the case of embankments by mesh-bag units, gravel particles were not scattered and they kept suppressing the mound on the back of the caisson. Therefore, placing the units only on the back of the caisson showed resiliency. In this study, even with a measure amount of only 20% of the total embankment using gravel, it showed 70% resiliency as an effect.
3. Furthermore, by placing the units up to the leg of the mound, the progress of scouring was suppressed, and the mound was found to be most stable against the tsunami overflow.

REFERENCES

Baba, S., Miyake, M., Tsurugasaki, K., & Kim, H. 2002. Development of wave generation system in a drum centrifuge, *Proc. Int. Conf. Phys. Modelling in Geotech.* 265–270.

Imase, T., Maeda, K., Miyake, M., Sawada, Y., Sumida, H., & Tsurugasaki, K. 2012. Destabilization of a caisson-type breakwater by scouring and seepage failure of the seabed due to a tsunami, *Proc. Int. Conf. Scour and Erosion.* 128–135.

Miyake, M., Yanagihata, S., Baba, S. & Tsurugasaki, K. 2002. A large-scale drum type centrifuge facilities and its application, *Proc. Int. Conf. Phys. Modelling in Geotech.* 43–48.

Miyake, M., Sumida, H., Maeda, K., Sakai, H. & Imase, T. 2009. Development of centrifuge modeling for tsunami and its application stability of a caisson type breakwater, *Annual Journal of Japan Society of Civil Engineers*, 25, 87–92 (in Japanese).

Miyamoto, J., Miyake, M., Tsurugasaki, K. & Sumida, H. 2015. Instability of breakwater foundation during tsunami in a drum centrifuge. *Proc. Twenty-fifth (2015) Int. Ocean and Polar Engineering Conference*, 846–850.

Sassa, S. 2014. Tsunami-Seabed Structure Interaction from Geotechnical and Hydrodynamic Perspectives, *Geotechnical Engineering Journal of the SEAGS & AGSSEA*, 45, No. 4, 102–107.

Sawada, Y., Miyake, M., Sumida, H., Tsurugasaki, K., Maeda, K. & Imase, T. 2013. Redundancy and ductility for the breakwater head subjected to tsunami. *Journal of Japan Society of Civil Engineers*, Ser. B3 (Ocean Engineering) 69, I_461-I_466 (in Japanese).

Takahashi, H., Sassa, S., Morikawa, Y., Takano, D. & Maruyama, K. 2014. Stability of caisson-type breakwater foundation under tsunami-induced seepage, *Soils and Foundations*, 54(4), 789–805.

Tsurugasaki, K., Miyamoto, J., Hem, R., Nakase, H. & Iwamoto, T. 2016. Collapse mechanism of composite breakwater under continuous tsunami overflow and its countermeasure. *Proc. Twenty-sixth (2016) Int. Ocean and Polar Engineering Conference*, 754–760.

Interaction between jack-up spudcan and adjacent piles with non-perfect pile cap

Y. Xie
Fugro AG, Perth, Australia

C.F. Leung & Y.K. Chow
Department of Civil and Environmental Engineering, National University of Singapore, Singapore

ABSTRACT: With ageing, older jacket platforms may have less than perfect pile cap conditions. A centrifuge model study has been conducted to evaluate the spudcan-pile interaction under free and fixed pile head conditions during spudcan penetration in soft clay. Owing to the weakened restraint at the pile head, the pile lateral deflection increases significantly whereas the bending moment reduces by a large extent. The bending pattern of a free-head pile is different from that of a fixed-head pile, and the pile bends towards the spudcan at shallow spudcan penetration depths and subsequently bends away from the spudcan when the spudcan penetrates further. An analytical approach is then conducted to evaluate the pile responses with partially fixed pile cap condition.

1 INTRODUCTION

Owing to its mobile nature, a jack-up rig may need to move in and out of a drilling site in offshore waters. During the installation of jack-up spudcan foundation of 10 m to 25 m diameter, the large soil movements may induce severe stresses on the piles (typically 1-m to 3-m diameter steel pipe piles) supporting the adjacent jacket platform. The results of extensive centrifuge model tests on spudcan-pile interaction have been reported by Xie et al. (2012, 2017) on fully fixed head piles with no deflection and no rotation allowed.

Although this fully fixed-head condition provides a conservative design in terms of pile bending stresses, the pile head is typically partially restrained by the tubular jacket members and hence labelled as "partially fixed" (Tomlinson & Woodward 2014). As the piles are usually grouted or welded to the jacket leg (Gerwick 2000), a number of jacket platforms have non-grouted jacket legs (Flick & Green 1983, Chakrabarti 2005, Joseph 2009). Owing to the annular gap between the pile and the ungrouted jacket leg, the pile head constraints provided by the jacket could be reduced due to ageing.

Jack-ups are also commonly employed for wind-turbine installation, repair and maintenance (The Crown Estate 2014) in offshore wind farms. While most wind-turbine foundations are partially fixed (e.g. tripod piles, jacket piles), Bhattacharya (2014) reported that another type of foundation called "mono-pile" has typically free-head condition. In view of the above, a study on free-head jacket platform pile would provide insights of mono-pile behaviours. Further centrifuge model tests have been conducted at the National University of Singapore (NUS) to investigate spudcan-pile interaction under weak pile cap conditions. Tests were conducted on free head piles which denote the worst pile cap condition (free deflection and free rotation). The results are then compared with those of fixed-head piles and distinct differences in the pile bending moment and deflection profiles are noted for these two extreme pile head conditions.

A two-stage analytical study was then carried out to provide insights on spudcan-pile interaction with pile cap having restraint in between the free and fixed pile head conditions. This would enable offshore geotechnical engineers to better equip in analysing spudcan-pile interaction issues for different types of jacket platforms with mobile jack-up rig activities.

2 EXPERIMENTAL SETUP AND PROCEDURE

2.1 *Centrifuge model setup*

All the centrifuge model tests in the present study were conducted at 100 g. Figures 1 and 2 show a sketch and a photograph of the model set-up, respectively. The stainless steel model container has a diameter of 500 mm and a height of 400 mm. The model spudcan has a diameter of 100 mm with an 11° base angle and an 80° conical tip, simulating a 10-m diameter prototype spudcan at 100 g.

A model aluminium pile of 12.6-mm width (1.26-m in prototype scale) was employed. Preliminary tests revealed that the lateral pile response is more severe as compared to the axial pile response. Hence the pile flexural rigidity EI is deemed critical with the model

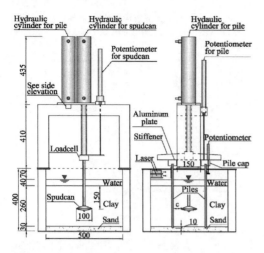

Figure 1. Centrifuge model setup for free-head piles (all dimensions in model scale in mm).

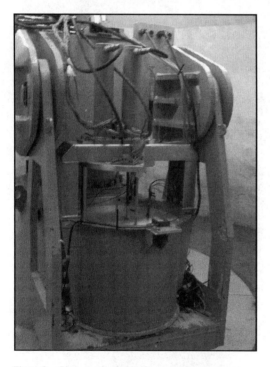

Figure 2. Photograph of centrifuge model setup.

pile EI converted to prototype scale having the same EI of the prototype steel pile of the same width and certain wall thickness. Half-bridge strain gauge circuits were placed on the pile shafts to measure the pile bending moments. The circuits were evenly distributed along the pile shaft and protected by a thin layer of epoxy. A pile hook assembly device was employed to hang the pile in the air before installation and subsequently facilitate the jacking of the pile into the soil. Above the hook assembly is a thick aluminium plate which locates and fixes the assembly at half spudcan diameter from the edge of the spudcan.

A closed-loop hydraulic servo-valve control system was employed to jack the spudcan and the pile, respectively, using two hydraulic cylinders that were fixed onto the loading frame attached to the model container. Potentiometers were employed to monitor the penetration depth of the spudcan and the movement of the piles, while a load cell was attached to the top of the spudcan to measure the spudcan load during its penetration, operation and extraction.

2.2 *Soil preparation*

The saturated soil bed consists of a 260-mm thick kaolin clay underlain by a 30-mm thick sand, as shown in Figure 1. The kaolin clay has a specific gravity G_s of 2.60, liquid limit of 80% and plastic limit of 35%. The silica sand has a mean grain size d_{50} of 0.5 mm and a special gravity G_s of 2.65.

Firstly, the sand was rained into the model container to achieve the bearing and bottom drainage layer of 30 mm thick. The dense sand has a dry density of 15.10 kN/m³, corresponding to a relative density of 85%. Water was then slowly sucked in by vacuum through a hole at the bottom of the container to saturate the sand. Subsequently, kaolin slurry with 120% water content was poured on the top of the sand and subjected to a consolidation pressure of 20 kPa at 1 g. Thereafter, the sample was consolidated under self-weight at 100 g in the centrifuge. T-bar penetrometers were employed to measure the clay strength profile at 100 g. The undrained shear strength profile indicates slight over-consolidation for the upper 40 mm (4 m in prototype scale) followed by an approximately linear increase in shear strength of 1.56 kPa/m depth (prototype scale).

2.3 *Experimental procedure*

After the self-weight consolidation of the clay has been completed, the piles were jacked in at a constant displacement rate of 5 mm/s under a displacement control mode until the embedment length of 270 mm (27 m in prototype scale) with the pile tip socketed 10 mm (1 m in prototype scale) into the underlying sand layer. Afterwards, there was a 2-hour waiting period (over 2 years in prototype scale) to allow the excess pore pressure to dissipate.

Thereafter, the spudcan was jacked into the soil at a constant model loading rate of 2 kPa/s. This load rate resulted in a displacement rate of over 0.45 mm/s. According to Finnie (1993), undrained condition is preserved when the dimensionless velocity group factor vD/c_v is greater than 30, where v is the displacement rate, D is the spudcan diameter and c_v is the coefficient of consolidation of soil (=40 m²/year). Based on the present clay properties, the spudcan penetration was deemed to be conducted under undrained conditions.

The target spudcan penetration depth was 150 mm, while the final spudcan penetration depth may vary slightly due to the difficulties in load control. The spudcan was then unloaded to 50% of the maximum installation load. After the operation period, the spudcan was extracted with a constant displacement rate of 1 mm/s also deemed to be under undrained condition.

3 EXPERIMENTAL RESULTS

3.1 Lateral responses of free-head pile

The measured pile bending moment and lateral deflection profiles for a free-head pile at 5 different spudcan penetration depths are shown in Figures 3 and 4, respectively. Negative bending moment denotes the pile bending towards spudcan, and vice versa. It should be noted that due to the interval of 2.5 m (in prototype) between two adjacent pairs of strain gauges along the pile, the measured recorded maximum bending moment may deviate slightly from the actual maximum bending moment.

The lateral pile shaft deflection profile was obtained by integration from the bending moment distribution, using the calculated pile head deflection and rotation as the two boundary conditions. The pile head deflection at the ground surface is extrapolated from the readings obtained from the two laser transducers placed at different heights on the exposed part of the pile head above the ground surface.

At a relatively shallow spudcan penetration depth (i.e. before 6 m), much of the soil movements due to spudcan penetration also occur at shallow depths as established in Xie et al. (2010). These outward soil movements would push the pile top to deflect away from the spudcan but there would be a kick back in the lower pile section resulting in pile bending towards the spudcan. This is denoted by the negative bending moments at 3 m and 6 m spudcan penetration depths shown in Figure 3. The pile lateral deflection increases rapidly and the pile shaft exhibits a convex shape shown in Figure 4.

As the spudcan penetrates deeper (e.g. at 12 m and 15 m depth), the soil movement extends deeper and the greatest soil movement takes place at some depths beneath the ground surface (Xie et al. 2010). This significant soil movement would push the mid-section of the pile and cause it to bend away from the spudcan, as represented by the positive bending moments in the right hand side of Figure 3. The pile bending is in the opposite direction to that at 3 m and 6 m spudcan penetration depths. The pile head deflection increases at a much slower rate and meanwhile the pile shaft changes to a concave shape.

It can be established from the present study that at shallow spudcan penetration depths (i.e. before 6 m or 0.6D depth), the soil movement and soil pressure concentrate at the upper part of the pile, and causes the pile to bend like a cantilever wall. For deep penetration (i.e. beneath 12 m or 1.2 D penetration depth), the soil movement and soil pressure mainly focus at the mid-pile section and the pile behaves like a hinged-girder.

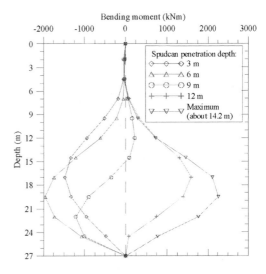

Figure 3. Induced bending moments of free-head pile during spudcan penetration.

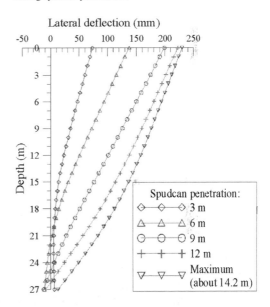

Figure 4. Induced pile lateral deflection of free-head pile during spudcan penetration.

3.2 Comparison with a fixed-head pile

The results from Xie et al. (2012) on a fixed-head pile are adopted for the comparison. It is noted that for the fixed-head pile test, the spudcan and pile dimensions are different from those of the free-head pile study. For example, the spudcan diameter is 12 m instead of 10 m, and the fixed-head pile has a circular cross-section with a diameter of 1.26 m and embedded 27 m in the soft clay only. More details on the fixed-head pile can be found in Xie et al. (2012). The test program for the free-head and fixed-head pile tests is compared in Figure 5. Despite the differences, this section aims to compare mainly the pile bending patterns rather

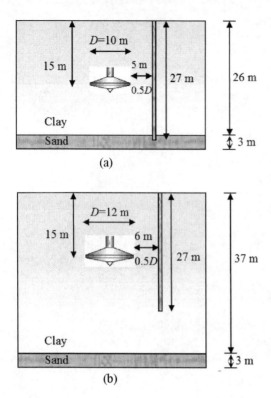

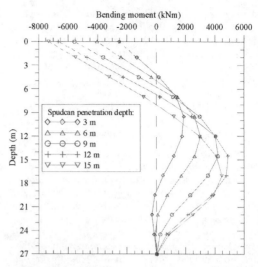

Figure 6. Induced bending moments of fixed-head pile during spudcan penetration (test results reported in Xie et al., 2012).

Figure 5. Test program for (a) free-head piles, and (b) fixed-head piles.

than the magnitudes of pile responses, hence providing revelations on the effect of pile cap conditions.

The measured bending moment profiles of the fixed-pile at 5 different spudcan penetration depths are shown in Figure 6. The gauge readings along the upper pile shaft reveal that it is reasonable to obtain the pile head moments by linearly extrapolating the top 2 or 3 strain gauge readings below the pile head. When subjected to lateral outward soil movements induced by spudcan penetration, fairly large negative bending moments are developed at the pile head due to restraint by the fixed pile cap. On the other hand, positive bending moments with smaller magnitudes are achieved along the lower pile shaft.

As the spudcan penetrates deeper, the maximum negative bending moment at the pile head increases. The maximum positive bending moment near the mid-depth of the pile also increases except after 12 m spudcan penetration depth, and the elevation of the maximum positive bending moment shifts downwards. Beneath 9 m spudcan penetration depth, the rate of increase in bending moment at the pile head decreases. This is because the majority of outward soil movements shift towards the mid-pile; meanwhile back-flow induced inward soil movements start to influence the pile portion near the pile head.

It is evident that the bending moment profile of a free-head pile is different from that of a fixed-head pile (Fig. 3 versus Fig. 6). A large bending moment is developed at the fixed pile head as it is fully restrained from deflection and rotation. By contrast, the allowance for free motion at the pile head for a free-headed pile causes a large lateral pile deflection (see Fig. 4) which greatly buffers the soil pressure on the pile and subsequently reduces the pile bending moment. In addition, the bending shape along a fixed-head pile is consistent throughout the spudcan penetration, with the maximum negative and positive bending moments developed at the pile head and along the pile shaft, respectively. On the contrary, the bending shape along a free-head pile changes from bending towards spudcan at a shallow spudcan penetration to bending away from the spudcan when the spudcan penetrates deeply in clay.

For fixed head piles, the pile moment profile is more important than the pile deflection profile and hence only the former is presented in this paper due to paper length limit. The differences between fixed-head and free-head piles are evident since they simulate essentially the two extremes of pile head condition. It is expected that the bending moment in a partially fixed-head pile should be smaller than a fully fixed-head pile, but greater than a free-head pile. This hypothesis is consistent with the numerical study on the effect of pile head condition on the lateral pile responses by Chow & Yong (1996), who concluded that the boundary conditions at the pile head have a significant effect on the pile performance.

4 ANALYTICAL ANALYSIS

Under centrifuge test environment, it would be difficult to achieve a precise partially fixed pile head condition as a small model pile head movement would translate to a large prototype pile head movement.

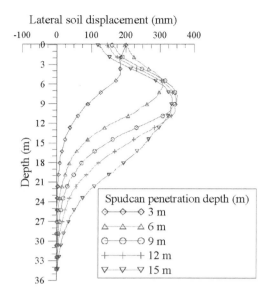

Figure 7. Induced cumulative lateral free-field soil movements at 0.5D from spudcan edge during spudcan penetration.

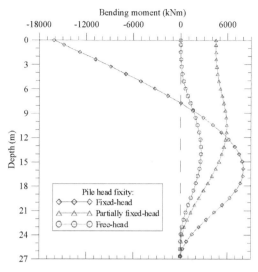

Figure 8. Comparison of pile bending moment profiles at 15 m spudcan penetration depth from analytical analysis.

An analytical study is hence conducted to evaluate the partially fixed pile head condition.

4.1 Methodology & inputs

This analytical study consists of two stages. The first stage involves estimating the lateral free-field soil movement profile at 0.5D from spudcan edge during spudcan penetration. This soil movement profile was obtained from the interpretation of the half-spudcan centrifuge test data by using the Particle Velocimetry Technology (PIV) as detailed in Xie (2009). The second stage involves applying lateral free-field soil movements in a pile–soil interaction program (Chow & Yong 1996) with elastic-perfectly plastic soil springs. Different pile cap conditions can be specified in this program. Details of this approach are reported in Leung & Xie (2017).

In the present study, the adopted cumulative lateral free-field soil movements at 0.5D from spudcan edge are taken from Xie (2009) and repeated in Figure 7, which correspond to the test configuration shown in Figure 5b. The progressive developments of the soil movement patterns during spudcan penetration are consistent with the earlier finding.

In the pile-soil interaction program, the main input soil parameters are the lateral soil stiffness K_h (or modulus of subgrade reaction), and the limiting soil pressure p_y. K_h was taken to be the soil undrained elastic modulus E_u. The E_u/c_u ratio was taken to be 150 for soft clay. The limiting soil pressure p_y follows Broms (1964), where p_y increases from $2s_u B$ (B is the pile diameter) at the ground surface, to $9s_u B$ at a depth of $3B$ and remains $9s_u B$ at a greater depth.

The pile has a diameter of 1.26 m and a length of 27 m, to simulate the model test shown in Figure 5b.

The following notations for the boundary conditions at the pile head are adopted:

1. fixed: no deflection and no rotation;
2. partially fixed: free deflection and no rotation;
3. free: free deflection and free rotation.

4.2 Results

Owing to the space constraint for this article, only the predicted lateral pile responses at 15 m spudcan penetration depth are compared. The corresponding bending moment and lateral deflection profiles are shown in Figures 8 and 9, respectively.

It can be seen that the boundary conditions at the pile head have a significant effect on the bending moment and lateral deflection profiles. Consistent with the experimental results, the fixed-head pile has the largest bending moment, while the free-head pile has the greatest lateral pile deflection. Once the restraint at the pile head is weakened, the bending moments reduce significantly as the pile head is allowed to deflect and rotate.

The lateral responses of a partially fixed-head pile are in-between fixed-head and free-head piles. The maximum bending moment of a partially fixed-head pile is about 60% less than that of a fixed-head pile, but more than twice of that of a free-head pile. The lateral deflection at the top of a partially fixed-head pile is 30% less than a free-head pile, but the difference reduces with depth and becomes negligible below a depth of 14 m.

Comparing bending moments for a fixed-head pile at 15 m spudcan penetration depth between Figures 6 and 8, the predicted maximum pile moments are more than twice of that from the centrifuge model test due to simplification of the analytical approach. In reality, the spudcan displaces the soil gradually and the adjacent

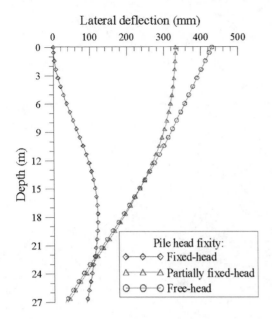

Figure 9. Comparison of pile lateral deflection profiles at 15 m spudcan penetration depth from analytical analysis.

pile is progressively "loaded" by the incremental soil displacements. Hence the current analytical approach can be improved to simulate the progressive loading on the adjacent pile. Considering the main purpose of this study is to address the effect of pile head fixity, the adopted analytical approach is considered sufficient.

5 CONCLUSIONS

A centrifuge model study has been carried out to evaluate the spudcan-pile interaction under free and fixed pile head conditions during spudcan penetration in soft clay. Owing to the weakened restraint at the pile head, the pile lateral deflection increases significantly whereas the bending moment significantly. The free-head pile bending pattern is different from that of a fixed-head pile, and the pile bends towards the spudcan at a shallow spudcan penetration depth and subsequently bends away from the spudcan when the spudcan penetrates deeper.

A parallel analytical study has been performed to further evaluate the effect of pile head conditions. The partially fixed-head pile can be estimated through this analytical approach, and the lateral pile responses are found in-between those of fixed-head and free-head piles. It is hence noted that it is of great importance to adopt an appropriate pile head condition for analysis and design.

ACKNOWLEDGEMENT

The study is conducted under A*STAR SERC Singapore-Korea Joint Research Program: Interaction of Jack-up Spudcans with Adjacent Piles in Stiff over Soft Soil (SERC grant number 1628200007 and NUS number R-261-502-035-305).

REFERENCES

Bhattacharya, S. 2014. Challenges in design of foundations for offshore wind turbines. *J. Eng. Technol. Ref.* 1(1): 1–9, doi: 10.1049/etr.2014.0041.

Broms, B.B. 1964. Lateral resistance of piles in cohesive soils. *J. Soil Mech. Found. Div.*, ASCE, 90(SM2): 27–63.

Chakrabarti, S. 2005. *Handbook of offshore engineering*. 1st Ed., Amsterdam: Elsevier.

Chow, Y.K. & Yong, K.Y. 1996. Analysis of piles subject to lateral soil movements. *J. Inst. Eng. Singapore* 36(2): 43–49.

Finnie, I.W.S. 1993. *Performance of shallow foundations in calcareous soils*. Ph.D Thesis, University of Western Australia.

Flick, L.D. & Green, D.J. 1983. Lateral stability of piles in ungrouted jacket legs. *Offshore Technology Conference*. OTC 4645.

Gerwick, B.C. 2007. *Construction of marine and offshore structures*. 3rd Ed., San Francisco: CRC Press.

Joseph, T. 2009. *Assessment of kinematic effects on offshore piled foundations*. Master Thesis, Università degli Studi di Pavia.

Leung, C.F. & Xie, Y. 2017. Effect of lateral soil movements on piles: an analysis prospective. *Proc. 27th Int. Offshore and Polar Eng. Conf.*: 461–466. San Francisco: Int. Society of Offshore and Polar Engineers.

The Crown Estate. 2014. *Jack-up vessel optimisation*. London: The Crown Estate.

Tomlinson, M. & Woodward, J. 2014. *Pile design and construction practice*. 6th Ed., New York: CRC Press.

Xie, Y. 2009. *Centrifuge model study on spudcan-pile interaction*. Ph.D Thesis, National University of Singapore.

Xie, Y., Leung, C.F. & Chow, Y.K. 2010. Study of soil movements around a penetrating spudcan. *Proc. 7th Int. Conf. Physical Modeling in Geotechnics*: 1075–1080. Zurich: Taylor & Francis Group.

Xie, Y., Leung, C.F. & Chow, Y.K. 2012. Centrifuge modelling of spudcan-pile interaction in soft clay. *Géotechnique* 62(9): 799–810.

Xie, Y., Leung, C.F. & Chow, Y.K. 2017. Centrifuge modelling of spudcan-pile interaction in soft clay overlying sand. *Géotechnique* 67(1): 69–77.

11. Offshore – deep foundations

Centrifuge modelling of long term cyclic lateral loading on monopiles

S.M. Bayton & J.A. Black
Centre for Energy and Infrastructure Ground Research, University of Sheffield, Sheffield, UK

R.T. Klinkvort
Norwegian Geotechnical Institute, Oslo, Norway

ABSTRACT: A series of centrifuge cyclic monopile lateral loading experiments in dry sand are presented. Model foundation tests were performed at 100 gravities (100g) of a prototype pile 5 m in diameter with an embedment depth of $L/D = 5$. Observations indicate that permanent rotational failure criteria (θ at mudline equal to 0.25°) may not be reached for load magnitudes 40% or less of the monotonic failure load for 10^7 cycles. A cyclic degradation model for monopile rotation accumulation is also presented. Results from the cyclic tests are used to plot contour lines of predicted rotations based on load magnitude and number of cycles applied. This model is then used to predict cyclic rotational accumulation of a load magnitude ramp test and appears to perform well.

1 INTRODUCTION

Over the past decade, there has been a gradual increase in the deployment of both on and offshore wind turbine structures for generation of electrical energy. The United Kingdom (UK) alone has increased its proportion of electrical energy from wind to over 11% in 2016 (Carbon Brief 2017), overtaking coal power generation for the first time in history. This represents a milestone, given the UK's historic reliance on coal. Of this, the offshore market makes up one third and is increasing year on year.

The monopile remains the foundation of choice for the offshore wind industry, owing to a number of factors; it's simple, suitability for mass production and has a previous successful track record. As the offshore industry seek to minimise the cost of energy production, current constraints include water depth and subsequent pile diameters, are repeatedly being increased presenting potential design challenges, both in terms of ultimate limit state (ULS) and long term fatigue limit state (FLS).

This paper presents the results from a suite of centrifuge cyclically loaded model scale monopile tests. The aim of the current research is to expand on the limited available datasets as well as to present a cyclic accumulation model adapted for monopile cyclic design.

2 CURRENT CYCLIC KNOWLEDGE

2.1 Design codes and developments

For ULS loading, offshore monopile foundations are today often designed by modelling the structure as a beam supported by soil medium represented as a series of uncoupled non-linear elastic springs (API 2011, recommended in DNV GL (2016) but with further validation). This method has been successfully adopted in the oil and gas industry for many years; however, these piles are particularly slender in comparison to today's rigid monopiles.

Much research has subsequently been carried out to assess the transferability of this method for wind turbine monopile design (e.g. Klinkvort 2012, Byrne et al. 2015). When it comes to the prediction of monopile cyclic degradation, the recommended design code procedure is to simply apply a one-off reduction factor to the capacity of the non-linear elastic springs. This is widely accepted as not being adequate for the offshore wind turbine design given the criticality of its cyclic performance.

As a result, many authors have attempted to develop cyclic predictive models, through full scale (Long & Vanneste 1994, Lin & Liao 1999) and model scale cyclic experiments (LeBlanc et al. 2010 at 1g, Klinkvort 2012, Kirkwood 2015 at Ng), and numerical methods (Lesny & Hinz 2007, Achmus & Albiker 2014). The issue with these, however, is the lack of suitable experimental data to validate the various models.

2.2 Advancement in monopile experimental datasets

There exists relatively little available literature with results of cyclic lateral load tests. There are no accessible published field scale datasets on monopiles and thus we rely on scaled physical model tests. Table 1 presents a selection of the publically available data,

Table 1. Cyclic datasets performed on rigid, large diameter monopiles.

Author(s)	Scale	Max cycles
LeBlanc et al. (2010)	1g	65000
Peralta & Achmus (2010)	1g	10000
Cuéllar (2011)	1g	5×10^6
Klinkvort (2012)	75g	10000
Rudolph et al. (2014)	200g	13000
Truong & Lehane (2015)	40g	46
Kirkwood (2015)	100g	1000

Table 2. HST95 sand properties.

Property	Value
Particle size, d_{50}	0.20 mm
Specific gravity, G_s	2.65
Maximum void ratio, e_{max}	0.827
Minimum void ratio, e_{min}	0.514
Peak angle of shear, φ'_{peak}	37.5° (at $R_d = 80\%$)

the experimental method by which they were achieved, and the maximum number of cycles performed. Note that during the lifetime of typical wind turbine structures, the foundations are expected to be subject to over 10^7 cycles. As shown, the maximum number of cycles previously performed in a geotechnical centrifuge at replicative stresses is of the order of magnitude of 10^4. This needs to be addressed in order to be confident of an entire lifetime of adequate performance.

3 CENTRIFUGE EXPERIMENTAL SETUP

Given the complex behaviour of the pile-sand interaction, prototype stress conditions are required to adequately capture the small-strain stiffness and dilation responses. This is secured when testing in a geotechnical centrifuge.

A series of centrifuge model tests, at a centrifugal acceleration of 100 times Earth's gravity (100g at 2/3rd pile embedment depth) were conducted using the University of Sheffield's 4 m diameter 50g-tonne geotechnical beam centrifuge (Black et al. 2014). The increase in acceleration was achieved having a radius to the sand surface of 1.65 m and an angular velocity rotational frequency of 23.0 rad/s. A fine grained sand commercially known as HST95 sand (see Table 2) was pluviated dry around an aluminium pipe section, embedded to a depth of 5D. This simulates that the pile is wished in-place removing any installation effects. The pile had an outer diameter of 50 mm [5 m at prototype in dry sand] and a wall thickness of 2.8 mm [84 mm for equivalent steel section]. Ten pairs of strain gauges were positioned along the pile extremity at 50 mm intervals from the base (Bayton & Black 2016). The model chamber had an internal diameter and height of both 500 mm, and provided a rigid boundary condition.

The load actuation system was capable of providing both cyclic and monotonic loading that replicated prototype wind and wave conditions. As illustrated in Figure 1, pneumatic actuators were employed for both loading configurations, with cyclic loading applied at an eccentricity (e_1) of approximately 5D from the sand surface, and the monotonic loading applied at an eccentricity of (e_2) 11D. Continuous 'wave' cyclic loading was provided using a dual acting SMC low friction 25 mm diameter bore piston together with

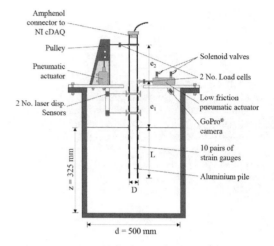

Figure 1. Centrifuge experimental setup.

two opposing solenoid valves. This generated a load capacity of 375 N at 7 Bar of input air pressure at a maximum frequency of 10 Hz. Individual test load magnitudes and frequency were controlled by the piston input pressure and the solenoid digital square wave frequency respectively. Two Baumer (OADM 12 type) laser sensors were directed perpendicular to the pile at two different positions to measure pile deflection (at 40 mm and 140 mm from the sand surface). These had the advantage over traditional LVDTs given the non-contact measurement and the capability of setting a specific measuring range to maximise signal to noise ratio.

4 CYCLIC DEGRADATION MODEL

The Norwegian Geotechnical Institute (NGI) cyclic degradation approach is a powerful tool used in cyclic design of geotechnical structures (Andersen 2009). It was originally proposed for the evaluation of undrained cyclic loading conditions in order to evaluate the cyclic soil strength, which was subsequently used in bearing capacity and serviceability calculations (Sturm et al. 2012). The 3D contour models are based on cyclic element laboratory tests, both triaxial and direct simple shear. The capability of the approach is such that the stress and cyclic state in a cyclic laboratory tests is related to a point in the three dimensional strain space. For greater, more in-depth information, Andersen (2009) presents a comprehensive history and

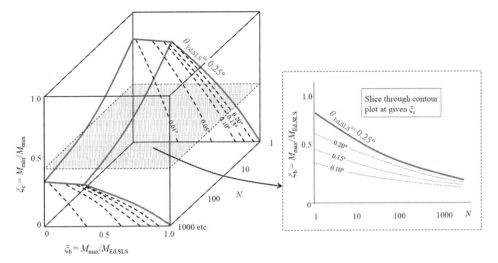

Figure 2. Illustrative cyclic accumulation contour plot.

application of the NGI cyclic accumulation approach with a number of examples.

It is proposed here to develop a NGI style cyclic degradation approach specific to the monopile design scenario. The three-dimensional contour diagram would allow the user to specify any given cyclic loading scenario and the associated applied cycles to predict the monopile displacements, or more suitably rotation response. It is to be noted that this model would be for a fully drained condition since the tests here are performed in dry sand.

While in the NGI model, a specific value of element shear strain is specified, here it is chosen to select a value of pile rotation at the mudline. DNV GL standard recommends permanent rotations of only 0.25° degrees of rotation at the sand surface as its serviceability limit state (SLS). This is in addition to the allowable 0.25° of rotation tolerance at installation. The permanent 0.25° limit at zero load is therefore selected and used forthwith. With this in mind, Figure 2 presents an illustrative failure contour plot where the x axis is defined by ξ_b ($M_{max}/M_{Ed,SLS}$), the y axis by ξ_c (M_{min}/M_{max}) and the z axis by the number of cycles applied. The ξ_b and ξ_c non-dimensionalised parameters are universally adopted for the description of applied overturning moments at the mudline on a monopile foundation.

5 EXPERIMENTAL RESULTS

In order to construct a cyclic degradation contour plot, it is first necessary to determine the monotonic moment capacity, $M_{Ed,SLS}$, at the point of mudline permanent rotation equal to 0.25° (i.e. on unload) upon which will form the basis of the model. Figure 3 shows the plot of moment against rotation for a monotonic test and the dimensionless design moment of 32.7 was used throughout this study.

Table 3. Test matrix (N.B. $M_{min} = 0$ and $\xi_c = 0$ for all tests).

Test (#)	R_d (%)	M_{max} ($M/\gamma D^4$)	ξ_b (–)	Number of Cycles (N)
1	80.2	32.7	1.0	1 (Monotonic)
2	81.0	18.9	0.58	76
3	79.5	12.9	0.39	1063
4	81.8	6.5	0.20	65000
5	80.7	4.5	0.14	96550 followed by
		(6.5)	0.20	49000)
6	80.4	2.2	0.07	91600 followed by
		(4.5)	0.14	50000)
7	80.6	Cyclic ramp (see Table 4 for details)		

A series of cyclic experiments was then performed to populate the contour plot. Table 3 presents the test matrix. As can be seen, all tests were performed at ξ_c equal to zero and therefore this dataset represents one horizontal slice through the three dimensional contour plot.

The first cycle for each of these tests is also added to Figure 3 to illustrate the precision of the sample preparation since each curve coincides well with the monotonic response. In order to fully populate the entire 3D contour space, it will be necessary in the future to perform experiments at various values of ξ_c, for both one-way and two-way loading scenarios.

Firstly, Figure 4 presents the complete moment-rotation response of Test 1 ($\xi_b = 0.58$). It can be seen that the monopile has reached serviceability permanent rotational failure after approximately 20 cycles. The rotation accumulates cycle by cycle and in addition, there has been an increase in secant stiffness of the soil-pile system with cycles. This has increased the secant stiffness by a factor of 1.8 from the first to the 76th cycle. This can be attributed to localised changes in sand grain distribution and density, and therefore new soil-structure interaction regimes on each cycle.

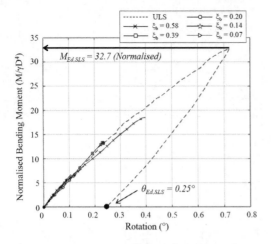

Figure 3. Monotonic 'backbone' failure curve and definition of $M_{Ed,SLS}$ at $\theta = 0.25°$ of permanent rotation.

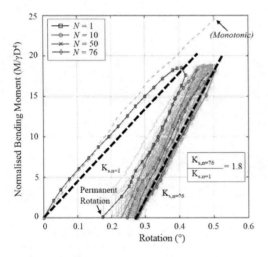

Figure 4. Moment-rotation evolution for Test 1 ($\xi_b = 0.58$).

Ultimately, increased soil-structure stiffness has implications on the structure's natural frequency, an important design criterion. The range of allowable structure natural frequencies is small and minor changes to this can have implications on structural resonance response and turbine efficiency; although this is beyond the scope of the current work. It is also noted that this effect has been observed in similar experiments previously when monopiles are tested in dry sand (Klinkvort 2012, Kirkwood 2015).

For the cyclic tests the permanent rotation was stored and is presented in Figure 5 as a function of number of cycles. The accumulation of rotations plot linearly on the logarithmic plot, confirming the tendencies seen in other studies (LeBlanc et al. 2010). It is also seen that larger values of ξ_b generate steeper accumulation plots.

According to LeBlanc et al. (2010) the progression of cyclic rotation accumulation can be predicted using

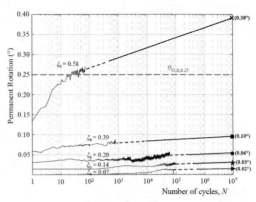

Figure 5. Increase in rotation with cycles. Solid markers at 10^6 show extrapolated predictions.

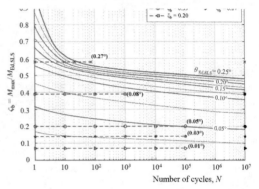

Figure 6. Cyclic accumulation contour diagram at $\xi_c = 0$. Results from individual cyclic tests are overlaid alongside their final value of rotation at the mudline after N cycles (in brackets). Solid markers at 10^7 cycles show extrapolated predictions.

predictive best fit curves. With this in mind, the data has been extrapolated to reach 10^7 cycles and these predictions are also shown in Figure 5.

The cyclic results were then used to populate the cyclic degradation contour diagram in Figure 6. To use Test 1 as an example (marked with crosses), the line is traced from cycle 1 to cycle 76 and incremental markers of mudline rotation (at exponents of number of cycles) are located and form the basis of the contour lines. It can be seen that the permanent rotational failure contour line of $0.25°$ is reached after the 20 cycles. Given only two experimental results cross this $0.25°$ contour, a degree of estimation of curve fitting has been applied based on expected behaviour. The remaining mudline rotation contours (0.20, 0.15° and 0.10° etc.) are positioned accordingly based on the full test matrix results and interpolations. Extrapolated values (solid markers) at 10^7 cycles have been added and appear to fit well with the contour lines. Note that Figure 6 makes up one slice of the three-dimensional contour plot.

Despite only an initial dataset of 5 baseline cyclic tests being performed, a reasonable contour fit has been achieved. In order to have further confidence in the accurate location of the individual contours, cyclic experiments of different magnitudes which result in cross-over of rotation at different numbers of applied cycles would have been preferable. Instead here the results from the monotonic test (taken as cycle 1 for each load increment) have been used to predict contour positioning.

In principle, this plot can then be used to predict the mudline rotation for any given cyclic load ratio and number of cycles (e.g. for a value of $\xi_b = 0.5$, a permanent rotation of 0.25° is expected to be reached after approximately 10^5 cycles for this test setup at this relative density). Interestingly, from Figure 6, it can be suggested that SLS rotation failure will never be reached for a lifetime of load cycles at a certain load level. Further experimental tests with much greater number of cycles will need to be performed in order to evaluate this level with confidence.

6 CYCLIC ACCUMULATION MODEL

The real potential of the cyclic degradation model arises when wanting to analyse a load sequence with many different load magnitudes and cycles. This resembles the expected loading for a prototype monopile where magnitudes and frequencies are highly variable and dependant on climatic conditions. The methodology of application is described below.

In practice, the individual load packages are first sorted from a full load sequence (by means of the rainfall count method or similar) and ordered in separate load packages alongside their respective number of cycles. The load packages are applied to the contour diagram with the given load magnitude and the number of cycles applied. This will determine the accumulated rotation at the end of load package, e.g. at $\xi_b = 0.30$, a line towards $N = 10^5$ cycles is drawn. At this point, the value of rotation is read from the contour (this is the first marker and reads 0.075°). The rotation is then used as a state parameter and the corresponding number of cycles for the next load level for the given rotation is then found by tracing back up the rotation contour line. A correction for the change load level is applied before the next load package number of cycles is added to the current cycle number at this load level to obtain a new rotation at the mudline. If upon tracing back up the contour line, the subsequent load package magnitude is not reached before cycle 1, the new load magnitude then starts at cycle 1. This signifies that the rotation obtained from the cycles at the previous load package has no future effect on the rotation from the subsequent load package. The process is continued until all the load packages are analysed and an equivalent number of cycles at one specific load is obtained.

To test this accumulative model, one cyclic ramp experiment was performed (Test 7). Seven load packages (LP) of 5000 cycles each were applied, equally 35000 cycles in total. Table 4 presents the magnitude

Table 4. Experimental cyclic ramp 'load packages'.

LP (#)	M_{max} ($M/\gamma D^4$)	ξ_b (−)	N (−)	Permanent θ, end of package (°)
1	1.4	0.04	5000	0.005
2	2.6	0.08	5000	0.010
3	3.6	0.11	5000	0.018
4	4.7	0.14	5000	0.027
5	5.9	0.18	5000	0.045
6	7.1	0.22	5000	0.055
7	8.3	0.25	5000	0.065

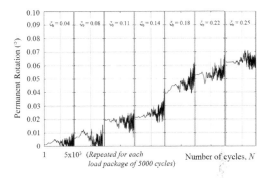

Figure 7. Evolution of rotation with each cyclic ramp package. Each load package is 5000 cycles.

of the loads and Figure 7 shows the evolution of the rotation for each load package. On the x axis, individual logarithmic scales have been specified for each load package to ease the viewing of the data.

As can be seen, with the increase in load magnitude for each package, the rotation increases, this is of course intuitive and to be expected. For the most part, on the first cycle of application of a new load package, the majority of the rotation at that load magnitude is achieved.

The cyclic packages from the experimental ramp results can be traced onto the contour plot. Figure 8 shows this process of accumulation. As shown, the first 3 load packages appear to make no impact on the behaviour of the subsequent loadings. It is only when LP4 is reached that this has implications on the subsequent LP5, LP6 and LP7 and its final rotation. However as can be seen, the trace back up the contour from the end of LP6 reaches approximately 100 cycles equivalently applied of load magnitude from LP7. This represents a minor addition in comparison to the further 5000 cycles to be applied, and in reality this will have little effect on the accumulative rotation.

Finally, within this plot, the rotation points at the end of each load package which were observed experimentally have been added (marked by crosses). It can be seen that these coincide well with the cyclic predictions and drawn contours lines. This provides a different approach of validation of the method.

In order to further validate the positioning of the accumulation model rotation contour lines, multiple additional cyclic ramp experiments are ongoing to

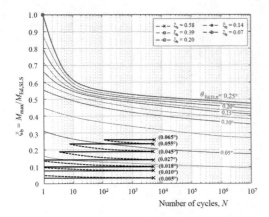

Figure 8. Cyclic accumulation contour diagram with cyclic ramp test model overlaid. Values in brackets at crossed markers represent obtained experimental results as per Figure 7.

populate the plot. From observations of the contour plot, the most suitable load levels for these appear to be at larger values ($\xi_b > 0.4$) in order to minimise the number of cycles required to achieve noticeable interactions between successive load packages.

7 CONCLUSION

A series of centrifuge cyclic monopile lateral loading experiments at different load magnitudes have been performed at the University of Sheffield. The tests add to the available literature with centrifuge experiments with 1000s of cycles. Results show an increase in rotation with cycles, however for magnitudes of ξ_b equal to or less than 0.4, it would appear that SLS will never be reached for 10^7 cycles, suggesting potential significant monopile overdesign for actual offshore load magnitudes.

A cyclic degradation model for the prediction of monopile rotation accumulation based on the NGI element cyclic degradation model has been presented. Results from the cyclic tests have been used to plot contour lines of predicted mudline rotations based on cyclic load magnitude and number of cycles applied. This model was then used to predict cyclic rotational accumulation for a load ramp test. The model appears to perform well.

Additional strategically selected cyclic experiments are required in order to further validate this method; these are ongoing at CEIGR in Sheffield.

REFERENCES

Achmus, M. & Albiker, J. 2014. Prediction of accumulated deformations of cyclic laterally loaded piles in sand. Numerical Methods in Geotechnical Engineering, London:1225–30.

American Petroleum Institute (API). 2011. RP 2A-WSD Recommended practice for planning, designing and constructing fixed offshore platforms – working stress design, 21st Edition. Washington, USA.

Andersen, K.H. 2009. Bearing capacity under cyclic loading – offshore, along the coast, and on land. The 21st Bjerrum Lecture presented in Oslo, 23 November 2007. Canadian Geotechnical Journal, 46:512–535.

Bayton, S. & Black, J. 2016. The effect of soil density on offshore wind turbine monopile foundation performance. Proc. of the 3rd European conference on physical modelling in geotechnics 1: 245–251.

Black, J., Baker, N. & Ainsworth, A. 2014. Establishing a 50 g-ton geotechnical centrifuge at the University of Sheffield. Proceedings of the 8th International Conference on Physical Modelling in Geotechnics. Perth, Aus., 14–18 January 2014.

Byrne, B.W., McAdam, R., Burd, H.J., Houlsby, G.T., Martin, C.M., Zdravković, L., Taborda, D.M.G., Potts, D.M., Jardine, R.J., Sideri, M., Schroeder, F.C., Gavin, K., Doherty, P., Igoe, D., Muir Wood, A., Kallehave, D. & Skov Gretlund, J., 2015. New design methods for large diameter piles under lateral loading for offshore wind applications. 3rd ISFOG, 2015, Oslo, Norway.

Carbon Brief. 2017. Analysis: UK Wind generated more electricity than coal in 2016. January 2017.

Cuéllar, P. 2011. Pile Foundations for offshore wind turbines: numerical and experimental investigations on the behaviour under short- and long-term cyclic loading. PhD Thesis, Technical University of Berlin

DNVGL ST-0126 2016. Support structures for wind turbines. Edition April.

Kirkwood, P. 2015. Cyclic lateral loading of monopile foundations in sand. PhD Thesis, University of Cambridge.

Klinkvort, R.T. 2012. Centrifuge modelling of drained lateral pile – soil response. Application for offshore wind turbine support structures. PhD Thesis, DTU Report R-271.

LeBlanc, C., Houlsby, G.T. & Byrne, B.W. 2010. Response of stiff piles in sand to long-term cyclic lateral loading. Géotechnique, 60(2): 79–90.

Lesny, K. & Hinz, P. 2007. Investigation of monopile behaviour under cyclic lateral loading. Proc. 6th Int. Conf. on Offshore Site Investigation and Geotechnics, London:383–390.

Lin, S. & Liao, J. 1999. Permanent strains of piles in sand due to cyclic lateral loads. Journal of Geotechnical and Geoenvironmental Engineering, 125:798–802.

Long, J.H. & Vanneste, G. 1994. Effects of cyclic lateral loads on piles in sand. Journal of the Geotechnical Engineering Division, 120:225–244.

Peralta, P. & Achmus, M. 2010. An Experimental investigation of piles in sand subjected to lateral cyclic loads. Physical Modelling in Geotechnics:985–990.

Rudolph, C., Grabe, J. & Bienen, B. 2014. Response of monopiles under cyclic lateral loading with a varying loading direction. Physical Modelling in Geotechnics:453–458.

Sturm, H., Andersen, K.H., Langford, T. & Saue, M. 2012. An Introduction to the NGI cyclic accumulation approach in the foundation design of OWTS.

Truong, P. & Lehane, B.M. 2015. Experimental trends from lateral cyclic tests of piles in sand. International Symposium on Frontiers in Offshore Geotechnics III 1:747–752.

Centrifuge modelling of screw piles for offshore wind energy foundations

C. Davidson, T. Al-Baghdadi, M.J. Brown, A. Brennan & J.A. Knappett
University of Dundee, UK

C. Augarde, W. Coombs & L. Wang
Durham University, UK

D.J. Richards & A. Blake
University of Southampton, UK

J. Ball
Roger Bullivant Limited, UK

ABSTRACT: Screw piles (helical piles) can provide a viable, cost-effective and low-noise installation alternative to increasing the size of existing foundation solutions (e.g. monopiles) to meet the demand for the advancement of offshore wind energy into deeper water. Significant upscaling of widely used onshore screw pile geometries will be required to meet the loading conditions of a jacket supported offshore wind turbine. This increase in size will lead to greater installation force and torque. This paper presents preliminary results from centrifuge tests investigating the requirements to install screw piles designed for an offshore wind energy application using specially developed equipment. Results indicate that the equipment is suitable to investigate these screw pile requirements and that significant force is required for such upscaled screw piles, with 19 MN vertical force and 7 MNm torque for the standard design. Optimisation of the screw pile geometry, reduced these forces by 29 and 11% for the vertical and rotational forces respectively.

1 INTRODUCTION

Recently, the development and installation of offshore wind energy has increased greatly and this trend is expected to continue for the foreseeable future. Overall in the UK, wind turbines are currently installed in water depths up to 40 m, with 81% of all turbines supported by monopiles and 6.6% by jackets (Wind Europe 2017). As companies look to further develop wind energy resources, wind farms will have to be sited in deeper water. Utilising monopiles in deep water, of around 25 m, raises concerns over their in-service performance and ability to manufacture very large monopiles (Golightly 2014).

An alternative solution is to expand the use of jacket structures. Typically, jackets are founded on long steel piles, driven with pile driving hammers, causing significant levels of noise above (Marine Management Organisation 2016) and below the water surface (Bruns et al. 2014). This has led German authorities to impose strict hydro sound limits, as underwater noise levels during piling operations are considered disruptive to many marine mammal species. Mitigation of the noise through techniques and equipment such as bubble curtains and sound dampers (Bruns et al. 2014) and in extreme cases large physical barriers is possible but expensive.

Screw piles have been proposed as a replacement for conventional straight-shafted piles for jacket supported wind turbine foundations (Byrne & Houlsby 2015, Spagnoli & Gavin 2015), as they offer several advantages. Firstly, installation is achieved through continuous simultaneous application of vertical crowd and rotational forces. This eliminates problems associated with pile driving hammers as the rotational installation method is significantly quieter (Byrne & Houlsby 2015, Knappett et al. 2014). Secondly, the superior axial capacity, compared to straight-shafted piles, generated by the helical plates and the soil trapped (soil-soil shear) between them is a significant advantage in resisting the substantial environmental loads acting on the jacket and turbine. Thirdly, screw piles can potentially reduce the amount of steel required for the foundations due to improved efficiency of the load carrying mechanisms, leading to cost savings for a wind farm project.

Rotating a screw pile into the soil generates large amounts of torque from the frictional resistance on the soil-steel interface. This is an important factor in the development of screw piles for offshore wind applications and is an active area of research (Schiavon et al. 2016, Spagnoli et al. 2016, Tsuha 2016). Numerous analytical and empirical methods (Hoyt & Clemence 1989, Ghaly & Hanna 1991) are available to predict the installation torque, which are derived from observations on small onshore piles and may not be applicable to the larger screw piles required for offshore projects. Methods correlating torque to Cone Penetration Test (CPT) data are also available (Spagnoli 2016, Al-Baghdadi et al. 2017).

With offshore screw pile installation torque predictions of 5 MNm (Byrne & Houlsby 2015) or greater, the need for reliable and accurate design methods is paramount in anticipating and potentially minimising the required torque. Additionally, optimisation of the screw pile design should be undertaken to reduce the torque and material requirements for a given pile capacity where possible. There is an absence of literature on the design and performance of large-scale screw piles with variable shaft geometry. To address this issue, a series of centrifuge tests were conducted using novel equipment to explore the optimisation of a screw pile designed for deep-water offshore wind energy foundations.

The work presented in this paper was conducted as part of the EPSRC project EP/N006054/1: Supergen Wind Hub Grand Challenges Project: Screw piles for wind energy foundations. This project is developing understanding of large scale screw piles through a combination of physical modelling (University of Dundee), field testing (University of Southampton) and advanced numerical modelling using the material point method (Wang et al. 2017) to simulate the effects of screw pile installation, ahead of performance analysis with traditional finite element techniques (Durham University).

2 MODEL SCREW PILE DESIGN

To investigate the optimisation of a realistic screw pile design, which could be used for an offshore wind energy development, in terms of the installation torque and force (crowd) required, two screw piles were designed for a theoretical loading scenario.

2.1 Loading scenario

It is expected from an economic perspective that jacket supported turbines will be deployed in water depths of 45 to 80 m, between the proven capability of monopiles and future floating structures. Using a worst-case design approach, loads were calculated for an 8 MW turbine on a jacket in 80 m of water on homogenous sand with the properties discussed below.

Environmental loads were calculated using DNV (2007) methods and the parameters in Table 2, representing conditions with a 1% exceedance level. Loads were assumed to act in unison diagonally across the jacket (from corner to corner) to calculate the upwind tensile and downwind compressive loads. A pinned jacket-pile connection was assumed. A steel density of

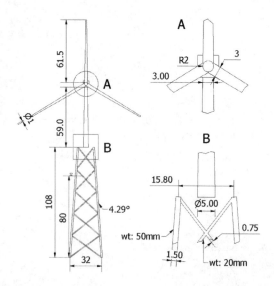

Figure 1. Schematic of conceptual 8 MW wind turbine and steel jacket in 80 m water depth (dimensions in metres).

Table 1. Loads acting on screw pile (negative value indicates tensile load).

Load direction	Upwind	Downwind
Horizontal (MN)	6.28	6.28
Vertical (MN)	−26.14	34.85

Table 2. Aerodynamic and hydrodynamic properties for loading calculations.

Parameter	Units	Value
Reference wind speed	m/s	32.7
Reference wind speed elevation	M	10
Air temperature	°C	5
Kinematic viscosity of air at 5°	m^2/s	13.60×10^{-6}
Density of air at 5°C	kg/m^3	1.226
Significant wave height	M	11.5
Wave period	S	15
Storm duration	Hours	3
Sea water temperature	°C	5
Kinematic viscosity of water at 5°C	m^2/s	1.56×10^{-6}
Density of sea water at 5°C	kg/m^3	1027.6
Sea water current speed	m/s	0.6

7700 kg/m^3 was applied for the dead weight calculations. The turbine and jacket design and the calculated loads, with a factor of safety of 1.35, are summarised in Figure 1 and Table 1 respectively.

2.2 Screw pile designs

Two screw piles were designed to suit the loads and soil properties previously defined, using a single pile at each corner of the jacket, to provide a worst-case design. However, it is likely that a multiple screw pile template would be used at each corner of the jacket to reduce installation requirements.

Figure 2. Photograph of optimised (upper) and standard uniform (lower) screw piles tested in the centrifuge.

Table 3. Screw pile dimensions in metres at prototype scale (mm at model scale of 1/80th).

Parameter		Uniform screw pile	Optimised screw pile
Length, L		13 (162.5)	
Core diameter, d_c	Upper	0.88 (11)	
	Lower	0.88 (11)	0.60 (7.5)
Helix diameter, D_h	Upper	1.70 (21.25)	
	Lower	1.70 (21.25)	1.34 (16.75)
Pitch, p	Upper	0.56 (7)	0.56 (7.5)
	Lower		
Thickness, t	Upper	0.11 (1.4)	0.11 (1.4)
	Lower		
Helix spacing ratio, S/D_h		2	2

The following approaches to calculate the appropriate capacities were used: tensile resistance from the multi-helix method in Das & Shukla (2013); cylindrical shear method in Perko (2009) for compressive capacity; analytical methods in Fleming et al. (2008) for the lateral capacity, with contributions from the helices to the lateral capacity neglected (Perko 2009). The first design had uniform core and helix dimensions, while the second, optimised design, had reduced core and bottom helix diameters (Fig. 2 and 3). Morais & Tsuha (2014) demonstrated a substantial reduction in installation torque of small diameter onshore screw piles with a section of reduced core diameter compared to designs with a uniform core diameter.

Reduction of the core and helix diameters in the optimised version was possible because of the substantial axial bearing capacity generated in compression by the lower surface of the helix and solid pile base. In the uniform design, the axial compressive capacity far exceeds the required amount from Table 1. Thus, the diameter was reduced to match the loading conditions. As the screw pile was calculated to fail with a 'long pile' mechanism under lateral loading, the core diameter was reduced below the hinge point, while maintaining the necessary capacities and torsional resistance. It is acknowledged that a more robust and practical design could be achieved by introducing the step change in pile core diameter at the position of the upper helix rather than above it as adopted here. Alternatively, a transitional change in diameter could be used rather than a stepped approach. The large core diameter was required to deal with the significant bending moments associated with the jacket structure loading regime (lateral loading) and demonstrates one of the significant changes in geometry required for offshore deployment compared with onshore, where maximum core diameters are typically 406 mm (Tappenden & Sego 2007). Al-Baghdadi et al. (2015) investigated placing helices close to the top of the screw pile (and seabed surface) but found that although increased lateral capacity could be generated (up to 22%), the helix was most efficient just below the seabed surface making it prone to being uncovered due to scour.

3 CENTRIFUGE TESTING EQUIPMENT

3.1 Installation and loading system

Previous investigations of the effects of installing model piles at 1g before subsequent testing at high-g levels observed significant differences in the response of the model pile compared to in-flight installation and testing (Klinkvort et al. 2013). Therefore, equipment was developed at the University of Dundee which is capable of installing and testing screw piles in one continuous centrifuge flight (Al-Baghdadi et al. 2016), at up to 60g (Figure 3).

The testing equipment allows for very precise bi-directional axial displacement (up to 300 mm) through a geared belt-driven ball screw system, powered by a AKM54H servo motor (Motor 1 in Figure 3). Rotation was provided by a second (slave) servo motor (AKM53H) connected to the screw pile via a 4:1 ratio gearbox to increase available torque from 10 Nm to 41 Nm. A custom F310-Z combined torque transducer and axial loadcell was mounted to the gearbox of the slave motor to measure the installation torque and crowd force as well as the force during tension/compression tests. The loadcell can measure torque to 30 Nm and axial force to ± 20 kN.

As the loadcell was mounted directly above the screw pile and rotated during the installation phase, this created a challenge regarding handling of the loadcell cable. The initial process in Al-Baghdadi et al. (2016) to control the cable under high-g conditions during a screw pile installation was to allow the cable, which was pre-wound around the loadcell, to unwind as the loadcell rotated and collect at the base of the rig. This approach was quickly found to be flawed, as on several occasions the cable became entangled with the screw pile because of the increased self-weight under high-g conditions (leading to aborted tests).

An alternative approach was adopted by Al-Baghdadi et al. (2016) which reversed the initial method by winding the cable on to the loadcell, from a pre-coiled position, through a large plastic funnel and cable sleeve to control the position of the cable. A large diameter plate added below the loadcell created a spool for the cable to wind on to.

Although this method worked for the initial testing program, a third system was implemented for this project which was considered more robust and

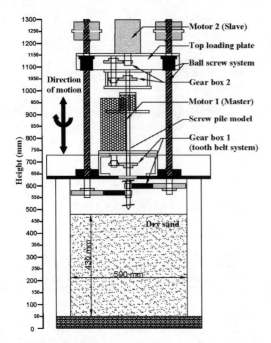

Figure 3. Schematic diagram of the screw pile centrifuge testing equipment mounted on a model container (800 mm long) (Al-Baghdadi et al. 2016).

Table 4. HST95 sand material properties (Al-Defae, 2013).

Property	Units	Value
Grading description	–	Fine
Effective particle size, D_{10}	Mm	0.09
Average particle size, D_{50}	Mm	0.14
Critical state friction angle, ϕ'_{crit}	(°)	32
Typical interface friction angle, δ'_{crit}	(°)	24
Angle of dilation*, ψ	(°)	16
Maximum dry density, ρ_{max}	kN/m³	17.58
Minimum dry density, ρ_{min}	kN/m³	14.59

*As measured at 80% relative density (Al-Defae 2013).

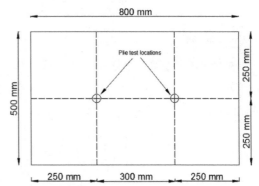

Figure 4. Plan locations of the centrifuge pile tests in the sand container.

eliminated any concerns over the handling of the load-cell cable without affecting the quality of the data. Although other researchers have used different designs (Tsuha et al. 2007), a SR002 8 channel slip ring was installed on a shaft extension above the loadcell. This slip ring was designed for low voltage, signal carrying applications and has particularly low electrical noise of less than 10 mΩ at 5 rpm.

Axial displacement data was measured with a WPS-500-MK30 draw-wire sensor, with back-up recording provided by the encoded servo-motors.

Data from the loadcell and draw wire were acquired with a Micro Analog 2 modular instrument data acquisition system (DAQ), with amplification factors of 1 for the draw wire and 200 for the loadcell. The servo-motors were supplied with a 3-phase alternating current which can introduce electrostatic and magnetic noise, from capacitive coupling and the flow of electric current respectively, into a strain gauge based device such as the F310-Z loadcell. The DAQ system has very effective screening of such noise and was able to consistently record high quality, low-noise data. Furthermore, shielded cable was used for all instruments to reduce electrical noise.

Labview 2013 software was used to provide a graphical user interface and control system for the servo-motors. Live data from the DAQ was monitored for the duration of the centrifuge test.

3.2 Soil properties

All tests used dry HST95 fine-grained quartz sand. The physical properties of this sand, summarised in Table 4, have been characterised at the University of Dundee (Al-Defae 2013, Jeffrey et al. 2016).

A container with internal dimensions of 500 × 800 × 550 mm (Figure 3) was used for both tests. Sand was prepared to a depth of 430 mm and relative density of 81% using an air pluviation system described by Jeffrey et al. (2016). Such dense to very dense sands are commonly encountered offshore Europe and represent a worst-case condition for installation forces. The container size allows for two tests in a single test bed (by moving the actuator between separate centrifuge flights) without creating any boundary effects. The shortest distance to the container boundaries from the test locations (Figure 4) was greater than 10 times the largest helix diameter (Phillips & Valsangkar 1987, Bolton et al. 1999).

Grain size effects were also considered, with a minimum value of 53 satisfying the smallest pile diameter (7.5 mm) to d_{50} particle size ratio of less than 50 recommended by Garnier et al. (2007).

3.3 Model container and sand bed preparation

See Figure 4.

3.4 Screw pile installation

The parameters used to install screw piles are critical to their in-service performance, as over-rotating (augering) during installation can reduce tensile

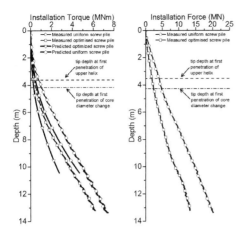

Figure 5. a) Vertical installation force and b) installation torque for uniform and optimised screw pile designs.

capacity (Perko 2009). Therefore, screw piles should be installed to penetrate the soil by an amount equal to at least 80% of the helix pitch per revolution (Perko 2009). This was achieved by using displacement controlled installation such that sufficient vertical force was applied to install the screw pile by 7 mm per revolution at a vertical rate of 21 mm/min.

4 RESULTS AND DISCUSSION

Two centrifuge tests were performed to measure the installation torque and crowd force of the standard and optimised screw piles designs at 50g. The results of both tests are shown in Figure 5 at prototype scale. To compensate for the dry soil conditions, the test data was scaled by a factor of 80 to simulate the design soil conditions of saturated soil behaving in a fully drained manner (Li et al. 2010).

The crowd and torque of the uniform and optimised screw pile designs were successfully measured in the centrifuge tests, indicating that the equipment is suitable for further investigating the magnitude of the full-scale screw pile installation requirements. The equipment also allows for the investigation of pile geometry optimisation and its effect on torque and crowd force. It is apparent from Figure 5 that the magnitude of both the torque and crowd force during installation are significant ($V_{max} = 19$ MN, $T_{max} = 7$ MNm) for the standard uniform pile. To put this in context, an onshore screw piling system based on an excavator mounted torque head, may be capable of only generating a torque of 250 kNm and vertical force of 257 kN, although casing rotators are capable of supplying torque of around 5 MNm and crowd force of 1.2 MN. The solid screw pile core design, reflecting a worst-case scenario, also influenced the large installation forces, but it is expected that full-scale, hollow core, screw piles would behave in an unplugged manner during installation, thus reducing the torque and crowd force.

The results from this preliminary testing (Figure 5) of the optimised design show that this approach was effective in reducing the final crowd force and torque values by 29% and 11% respectively. Furthermore, optimisation of the screw pile design reduced the surface area and volume by 11% and 18% respectively, which could present a substantial saving of material and costs across a large array of turbines, while still achieving the in-service structural requirements.

These findings are comparable to those of Morais & Tsuha (2014) who observed a 33% reduction in installation torque for a screw pile with a 28% smaller surface area, during field scale tests of screw piles with uniform and reduced core diameters.

Figure 5 also includes a torque prediction for each test using the CPT correlation by Al-Baghdadi et al. (2017) with the rotation reduction factors equal to one and q_b equal to q_c. For the depth covered by the CPT data, the torque prediction matches the observed torque reasonable closely, although there is a slight under-prediction in this case, it is suggested that this method is appropriate for estimating the installation torque of large scale screw piles in sand.

5 CONCLUSIONS

As offshore wind energy progresses into deeper water, new foundation solutions are required. Steel jackets founded on screw piles are one such solution, with screw piles offering quiet installation and potentially superior capacity compared to straight shafted piles.

Centrifuge testing of model scale screw piles offers an alternative to large scale field testing, of which there is very limited existing research, especially for large or novel geometries. Centrifuge tests of two screw pile designs, in dense sand, were conducted to investigate the full-scale installation force and torque requirements of screw piles designed to sustain the loads of a jacket supported turbine.

The results show that equipment specially developed at the University of Dundee can successfully capture the installation data of model scale screw piles along with the subtle changes to the geometry, intended to optimise the design. It is apparent that the torques and vertical forces generated during installation are very significant and one or two orders of magnitude greater than those used onshore. Armed with this information, offshore contractors will have insights into the requirements for installation equipment development or installation strategies (screw pile groups) to minimize installation requirements. Further centrifuge tests of additional designs in various sand densities are planned to continue the investigation into the behaviour of large screw piles.

ACKNOWLEDGEMENTS

This work is funded by the EPSRC project EP/N006054/1: Supergen windhub grand challenges

Project: Screw piles for wind energy foundations. The authors would also like to acknowledge the support of the European Regional Development Fund SMART Centre at the University of Dundee.

REFERENCES

Al-Baghdadi, T., Davidson, C., Brown, M., Knappett, J., Brennan, A., Augarde, C., Coombs, W., Wang, L., Richards, D. & Blake, A. 2017. CPT-based design procedure for installation torque prediction for screw piles installed in sand. *8th Int. Conf. on Site Investigation and Geotechnics*. London, UK, 12–14 September. Society for Underwater Technology (SUT OSIG).

Al-Baghdadi, T.A., Brown, M.J. & Knappett, J.A. 2016. Development of an inflight centrifuge screw pile installation and loading system. In Thorel, L., Bretschneider, A., Blanc, M. & Escoffier, S. (eds.), *3rd European Conf. on Physical Modelling in Geotechnics (EUROFUGE 2016)*. IFSTTAR Nantes Centre, France, 1–3 June.

Al-Baghdadi, T.A., Brown, M.J., Knappett, J.A. & Ishikura, R. 2015. Modelling of laterally loaded screw piles with large helical plates in sand. *Frontiers in Offshore Geotechnics III*. CRC Press.

Al-Defae, A.H.H. 2013. *Seismic performance of pile-reinforced slopes*. Ph D. University of Dundee

Bolton, M.D., Gui, M.-W., Garnier, J., Corte, J.F., Bagge, G., Laue, J. & Renzi, R. 1999. Centrifuge cone penetration tests in sand. *Géotechnique*, 49: 543–552.

Bruns, B., Stein, P., Kuhn, C., Sychla, H. & Gattermann, J. 2014. Hydro sound measurements during the installation of large diameter offshore piles using combinations of independent noise mitigation systems. *INTER-NOISE and NOISE-CON Congress and Conf. Proc.* Melbourne, Australia: Institute of Noise Control Engineering.

Byrne, B.W. & Houlsby, G.T. 2015. Helical piles: an innovative foundation design option for offshore wind turbines. *Phil. Trans. R. Soc. A*, 373.

Das, B.M. & Shukla, S.K. 2013. *Earth anchors*, USA: J. Ross Publishing.

DNV. 2007. *Recommended Practice DNV-RP-C205: Environmental Conditions and Environmental Loads*, Høvik, Norway: DNV.

Fleming, K., Weltman, A., Randolph, M. & Elson, K. 2008. *Piling engineering*, Abingdon, UK: Taylor and Francis.

Garnier, J., Gaudin, C., Springman, S., Culligan, P., Goodings, D., Konig, D., Kutter, B., Phillips, R., Randolph, M. & Thorel, L. 2007. Catalogue of scaling laws and similitude questions in geotechnical centrifuge modelling. *Int. Journal of Physical Modelling in Geotechnics*, 7: 01–23.

Ghaly, A. & Hanna, A. 1991. Experimental and theoretical studies on installation torque of screw anchors. *Canadian Geotechnical Journal*, 28: 353–364.

Golightly, C. 2014. Technical Paper: Tilting of monopiles Long, heavy and stiff; pushed beyond their limits. *Ground Engineering*, January 2014: 20–23.

Hoyt, R.M. & Clemence, S.P. 1989. Uplift capacity of helical anchors in soil. *Proc. of the 12th Int. Conf. on Soil Mechanics and Foundation Engineering*. Rio de Janeiro, Brazil.

Jeffrey, J.R., Brown, M.J., Knappett, J.A., Ball, J.D. & Caucis, K. 2016. CHD pile performance: part I–physical modelling. *Proc. of the Institution of Civil Engineers-Geotechnical Engineering*, 169: 421–435.

Klinkvort, R.T., Hededal, O. & Springman, S.M. 2013. Scaling issues in centrifuge modelling of monopiles. *Int. Journal of Physical Modelling in Geotechnics*, 13: 38–49.

Knappett, J.A., Brown, M.J., Andrew, J.B. & Hamilton, L. 2014. Optimising the compressive behaviour of screw piles in sand for marine renewable energy applications. *DFI/EFFC 11th Int. Conf. On Piling & Deep Foundations*. Stockholm, Sweden, 21–23 May. Deep Foundations Institute.

Li, Z., Haigh, S.K. & Bolton, M.D. 2010. Centrifuge modelling of mono-pile under cyclic lateral loads. *Physical Modelling in Geotechnics*: 965–970.

Marine Management Organisation. 2016. Noise and wind-farm construction off the Sussex coast. https://www.gov.uk/government/news/noise-and-windfarm-construction-off-the-sussex-coast: GOV.UK.

Morais, T.d.S.O. & Tsuha, C.d.H.C. 2014. A new experimental procedure to investigate the torque correlation factor of helical anchors. *Electronic Journal of Geotechnical Engineering*, 19: 3851–3864.

Perko, H.A. 2009. *Helical piles: a practical guide to design and installation*, Hoboken, New Jersey: John Wiley & Sons.

Phillips, R. & Valsangkar, A. 1987. *An experimental investigation of factors affecting penetration resistance in granular soils in centrifuge modelling*, Cambridge, UK: University of Cambridge.

Schiavon, J.A., Tsuha, C.H.C., Neel, A. & Thorel, L. 2016. Physical modelling of a single-helix anchor in sand under cyclic loading. In Thorel, L., Bretschneider, A., Blanc, M. & Escoffier, S. (eds.), *3rd European Conf. on Physical Modelling in Geotechnics (EUROFUGE 2016)*. IFSTTAR Nantes Centre, France, 1–3 June.

Spagnoli, G. 2016. A CPT-based model to predict the installation torque of helical piles in sand. *Marine Georesources & Geotechnology*: 1–8.

Spagnoli, G. & Gavin, K. 2015. Helical piles as a novel foundation system for offshore piled facilities. *Abu Dhabi Int. Petroleum Exhibition and Conf.* Abu Dhabi, UAE, 9–12 November. Society of Petroleum Engineers.

Spagnoli, G., Jalilvand, S. & Gavin, K. 2016. Installation torque measurements of helical piles in dry sand for offshore foundation systems. In Zekkos, D., Farid, A., De, A., Reddy, K. R. & Yesiller, N. (eds.), *Geo-Chicago 2016*. Chicago, Illinois, 14–18 August 2016. ASCE.

Tappenden, K.M. & Sego, D.C. 2007. Predicting the axial capacity of screw piles installed in Canadian soils. *OttawaGeo2007*. Ottawa, Canada: Canadian Geotechnical Society.

Tsuha, C.H.C. 2016. Physical Modelling of the Behaviour of Helical Anchors. In Thorel, L., Bretschneider, A., Blanc, M. & Escoffier, S. (eds.), *3rd European Conf. on Physical Modelling in Geotechnics (EUROFUGE 2016)*. IFSTTAR Nantes Centre, France, 1st-3rd June.

Tsuha, C.H.C., Aoki, N., Rault, G., Thorel, L. & Garnier, J. 2007. Physical modelling of helical pile anchors. *Int. Journal of Physical Modelling in Geotechnics*, 7: 1–12.

Wang, L., Coombs, W., Augarde, C., Brown, M., Knappett, J., Brennan, A., Humaish, A., Richards, D. & Blake, A. 2017. Modelling screwpile installation using the MPM. *1st Int. Conf on the Material Point Method, MPM 2017*. Delft, The Netherlands. Procedia Engineering.

Wind Europe. 2017. The European offshore wind industry: Key trends and statistics 2016.

General study on the axial capacity of piles of offshore wind turbines jacked in sand

I. El Haffar, M. Blanc & L. Thorel
IFSTTAR, GERS-GMG, Bouguenais, France

ABSTRACT: A general study has been realised on the axial capacity of the foundation of the offshore wind turbines with the use of the centrifuge modelling. Piles with different diameters have been jacked at $100 \times g$ in dense Fontainebleau sand. The effect of the saturation, pile diameter and the type of the pile tip (open or close-ended) on the axial compression of the piles is studied. The plug creation during installation is clearly impacted by the saturation of the dense sand. Plug formation show also to be improved with the decrease of the diameter of open-ended piles. A linear relationship is obtained between the jacking load and the embedded volume of the tested piles with different diameters.

1 INTRODUCTION

The project of many countries to increase its use of renewable energy, especially wind energy, makes onshore wind turbines insufficient. Thus, the pace of development of offshore wind turbine projects and design methods is increasing drastically. Until now, most of the already installed offshore wind turbines are constructed in relatively shallow water. The use of gravity bases and large monopiles, therefore, suits their installation best. The construction of wind turbines in deeper water supported by jacket structures, however, is increasing. Such structure consists of a three or four-legged steel lattice frame founded on a single pile placed below each leg. This new demand for jacked structure built over deep driven open-ended piles and the need for higher and deeper offshore wind turbines make their piles exposed to more severe weather impacts such as tide and wind. So in order to design the present piles of the jacked structure the API (American Petroleum Institute) standard developed by the oil and gas American industry is used. Also in the commentary of the new 22nd edition of the API RP 2A recommendations (2007) four CPT-based methods developed by researcher to be used in the design of offshore piles were included (ICP-05 (Jardine et al. 2005), Fugro-05 (Kolk et al. 2005), NGI-05 (Clausen et al. 2005), UWA-05 (Lehane et al. 2005)). But the use of the API and the CPT-based methods contain a good number of uncertainty and bias.

The objective of this study is to decrease the uncertainty related to the design of offshore piles using centrifuge tests. Tests will be performed in dense and medium dense sand saturated and dry to study the effect of the density and the effect of sand saturation. In addition, the test will be realised using open- and close-ended piles with different diameter to have clearer ideas about the effect of plugging on the behaviour of the piles and the effects of pile diameter on the phenomena of pile plugging.

2 METHODOLOGY

2.1 Centrifuge modelling

The centrifuge modelling consists of the use of small scale model installed in a high field of gravity to permit the replication of the stress state that exists in the prototype. The use of the 5.5 m diameter beam centrifuge in the study of deep foundations is important in order to have a negligible gradient of g between the top and the bottom of the model pile.

Tests presented in this study have been performed at acceleration level of 100 times the earth gravity ($100 \times g$) on piles scaled by 1/100.

2.2 Model pile

All the model piles used in this study have a thickness of 1 mm and an embedded depth of 250 mm which give a thickness of 0.1 m and a depth of 25 m in the prototype scale. The other properties and the diameter of the piles are presented in (Table 1). A 25 kN load sensor (FN3070 from *FGP*) located between the pile head and the hydraulic jack (Fig. 1) measures the total bearing capacity of the pile. The pile displacement can be determined using a magnetostrictif displacement sensor (1/3000350S010–1E01 from *TWK*) which controls the displacement of the hydraulic jack.

2.3 Model soil

The model soil is the Fontainebleau NE34 poorly graded sand (Table 2). A total of 4 rectangular strongboxes are prepared for this study with two relative densities of 58% and 99% obtained by filling the

Table 1. Experimental campaign.

Description	Test name	Open pile	Closed pile	Ø* mm	Øp* mm
C1 Dry sand with relative density of 99%	C1O18	×		18	1.8
	C1C18		×	18	1.8
	C1O16	×		16	1.6
	C1C16		×	16	1.6
	C1O14	×		14	1.4
	C1C14		×	14	1.4
C Dry sand with relative density of 58%	C2O18	×		18	1.8
	C2C18		×	18	1.8
	C2O16	×		16	1.6
	C2C16		×	16	1.6
	C2O14	×		14	1.4
	C2C14		×	14	1.4
	C2O12	×		12	1.2
	C2C12		×	12	1.2
C3 Saturated sand with relative density of 99%	C3O18	×		18	1.8
	C3C18		×	18	1.8
	C3O16	×		16	1.6
	C3C16		×	16	1.6
	C3O14	×		14	1.4
	C3C14		×	14	1.4
C4 Saturated sand with relative density of 58%	C4O18	×		18	1.8
	C4C18		×	18	1.8
	C4O16	×		16	1.6
	C4C16		×	16	1.6
	C4O14	×		14	1.4
	C4C14		×	14	1.4
	C4O12	×		12	1.2
	C4C12		×	12	1.2

* Where Ø is the model and prototype pile diameter

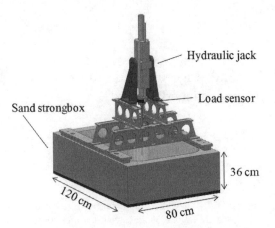

Figure 1. Experimental set up (inside dimensions).

Table 2. Characteristics of used sand Fontainbleau NE34.

Sand	$U_C = d_{60}/d_{10}$	d_{50} μm	γ_{min} g/cm^3	γ_{max} g/cm^3
Fontainebleau NE34	1.53	210	1.46	1.71

Where U_c is coefficient of uniformity (Matias Silva 2014); d_x is grain size at which x% of particles by weight, respectively, are smaller (Matias Silva 2014); γ_{min}, γ_{max} is minimum and maximum, respectively, dry unit weigh tested in the lab according to the standard (NF P 94-059)

strongboxes with sand using the air pluviation technique. At the end two strongboxes from each density are prepared.

One box from each density has been connected by the bottom to a water tank until the total saturation is reached.

2.4 Piles installation and experimental campaign

The piles are jacked in flight at 100×g to represent the installation effects and the displacement of the soil that develop during the installation of displacement piles. All the piles have been jacked to the desired embedded depth of 250 mm without stopping the centrifuge.

In Table 1, a total of 28 load tests realised in the experimental campaign are presented. The name of each test contains this information in the order: the number of the container in which the pile is jacked, the type of the tip (open or close ended), the pile diameter. For example, C1O18 is 18-mm diameter open pile jacked in container 1.

3 EXPERIMENTAL TEST ANALYSIS

3.1 Determination of the ultimate capacity

The final jacking force is considered as the ultimate compression capacity of the pile, this method is in good accordance with the results obtained in Deeks et al. (2005) where they found that the failure of the piles was reached at a load equal to the installation force.

3.2 Bearing capacity tests

The effect of pile plugging, saturation and sand density on the behaviour of sand is examined using the results from the 28 loading tests are presented below.

Figure 2 shows the jacking tests performed in the strongboxes C1(dry dense sand) and C3(saturated dense sand). A first comparison between the piles C3O18 and C3C18 shows that at the beginning of the loading and until the depth of 3 m. is reached, the behaviour and initial stiffness of both piles is different. As the piles are driven deeper into the soil, their behaviour becomes similar and approximately the same capacity is obtained at the end of the jacking phase. This observation may be accounted for by plug formation. On the other hand, in dry dense sand, C1O18 and C1C18 curves, in spite of different initial stiffness, are also different at the beginning of the jacking phase. Both piles behave differently until the depth of 13 m where the C1O18 curve starts to move upward to converge towards that of C1C18. Plug formation, here, does not start until the depth of 13 m.

Similar behaviour has been found also for the piles with 1.6 m of diameter. Concerning the 1.4 m diameter piles, C1O14 and C1C14 jacked in dry sand show

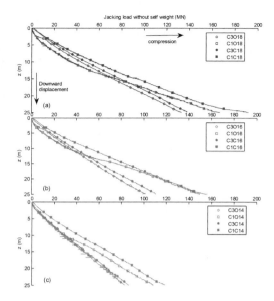

Figure 2. Jacking load of piles installed in saturated and dry dense sand.

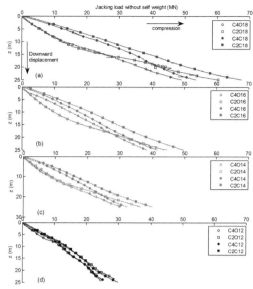

Figure 3. Jacking load of piles installed in saturated and dry medium dense sand.

similar behaviour of the piles C1O18 and C1C18 but the behaviour of C3O14 and C3C14 is equal and no difference at early depth is found.

These results show important information concerning the behaviour of open and closed piles and how it is affected by the sand saturation. The behaviour of piles jacked in dry or saturated sands demonstrates the formation of plugging as the pile depth increases. This phenomenon seems to be clearly influenced by the saturation of dense sand. As explained in the previous section C3O18 show different behaviour of the closed pile C3C18 put in similar conditions only in early depth and then C3O18 formed a rigid plug and show similar behaviour of C3C18. On the other hand in the case of C1O18 the plug formation is only happening after 13 m of installation depth when the behaviour of C1O18 begins to approach that of the closed pile C1C18. It is clear that the saturation of the sand speed up the phenomenon of the plug formation in comparison with dry sand as the plug was created in earlier depth in the case of saturated sand than in the case of dry one.

The plug formation is affected also by the pile diameter. As the pile diameter decrease from 1.8 m to 1.4 m the difference between the open and closed piles is decreasing until the two piles perform the same way like the case of C3O14 and C3C14.

Figure 3 show the results of piles tested in the medium dense dry and saturated sand. Although in this density the open-ended piles jacked in dry or saturated sands develop plugging at the same time of their respective test, the plug was formed at higher depth. It is even more visible in the case of saturated sand, C4O18 (medium dense) formed a plug beyond 16 m on the other hand C3O18 (dense) developed a plug from 3 m. Moreover, pile diameter effect on pile behaviour is still present. The difference between the performance of open-ended and close-ended piles decreases with the decrease in diameter until merging with a diameter of 1.2 m. This confirms the previous results, for which the decrease in pile diameter improves pile plugging causing open-ended piles to behave as close-ended ones.

These results are in accordance with the work of Kumara et al. (2016b) where they found that relatively smaller diameter open-ended piles produce higher degree of soil plugging under both dense and loose sand. Kumara et al. (2016a) found also that a fully-plugged open-ended pile behaves similar to a closed-ended pile. Lee et al. (2003) found also that at depth equal to 17 times the pile diameter, the base resistance of open-ended piles was approximately the same as that of the closed-ended piles under similar conditions. This ratio between pile depth and pile diameter is close to the ratio of studied piles presented earlier.

The ultimate compression capacity of all tested piles is presented in Figure 4. This figure highlights the result found previously that the ultimate capacity of open-ended piles is comparable to the closed ones. Two significant findings of this figure are: (1) the ultimate compression capacities of piles jacked in dry sand is clearly higher of the capacities of piles jacked in saturated sand. This results from that in saturated sand, the capacity of the pile results from the effective stresses and not the total one like in dry sand. (2) The ultimate compression capacity in dense sand is higher than the capacity in medium dense sand in both dry and saturated sand. This is consistent with the work of Paik and Salgado (2003) where they found that the ultimate unit base resistance increases significantly with increasing relative density and increasing horizontal stress.

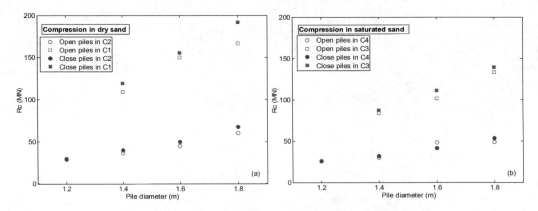

Figure 4. Ultimate capacity in compression (Rc) : (a) Rc for dry sand – (b) Rc for saturated sand.

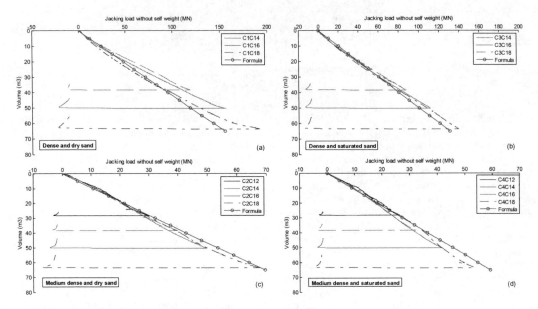

Figure 5. Closed piles embedded volume in function of the jacking load: (a) for dense and dry sand – (b) for dense and saturated sand – (c) for medium dense and dry sand – (d) for medium dense and saturated sand.

3.3 Embedded volume in function of the jacking load

An interesting find in this study is presented in Figure 5 where a linear relationship has been found between the embedded volume of the close-ended piles and their jacking loads. Despite that no certified scientific explanation is found to be able to describe the existence of such linear relationship, it is more likely that this result is in relation with the quantity of the displaced sand. This came from that the used piles were installed in a displacement method known to lead to high level of displacement of the sand with the advancement of the jacking. This relation exists for any sand density and saturation. This relationship shows also that for a wished jacking load, the designer can make the choice between the use of deeper pile with small diameter or a larger one but with shallower depth.

A survey of the literature reveals that little research is focusing on this subject. Still, some similar results are discussed in Robinsky et al. (1964). The authors present the results obtained from the study conducted on two straight sided piles of different diameters and one tapered pile. According to them, each pile presents a straight-line relationship between its embedded volume and its respective capacity. However, in this case, slopes are different contrary to the results obtained in the present work.

The same study has been done for the open-ended piles and presented in Figure 6. The majority of these piles are shown also to have a straight-line relationship between their embedded volume and jacking load. However, the slopes are different contrary to close-ended piles. The only exception is found in the results of the open-ended piles installed in the dense and

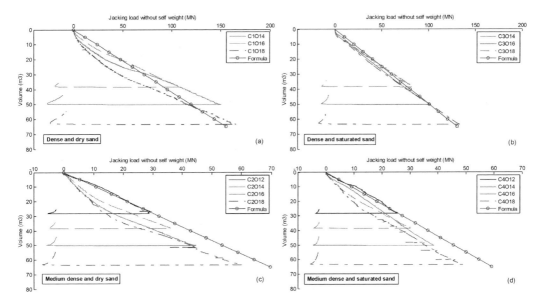

Figure 6. Open piles embedded volume in function of the jacking load: (a) for dense and dry sand – (b) for dense and saturated sand – (c) for medium and dry sand – (d) for medium and saturated sand.

saturated sand (C3) where the lines of the relationships of the piles have the same slope. As discussed above in the first section, this can be accounted for by the fact that plug formation is initiated at a very shallow depth in this high density and saturation strongbox and the open-ended piles then behave similarly to close-ended piles.

The fact that all the curves of the close-ended piles tested in each strongbox have the same slope whatever their diameter points to the fact that the slope can be expressed using a formula, which depends on the sand density and the friction angle of contact between the pile and sand and is independent from the pile diameter. With this aim in view, the formula describing the curve initial slope of the close-ended piles in each strongbox takes the form:

$$k = V/v = \frac{7 D_r^2}{6.75 \tan \alpha} \left[\frac{\gamma'}{\gamma} \right]^{0.35} \quad (1)$$

where V is the jacking load without self-weight in MN, v is the Volume in m³, D_r is the relative density of the soil, α is the friction angle in degrees, γ is the dry density of the sand and γ' is the effective density of the sand.

For each strongbox in Figure 5, a line having k as the slope has been plotted to test the performance of this formula against the experimental results. The same lines that have been used in Figure 5 have been plotted also in Figure 6. The theoretical line plotted in Figure 6-b verify the conclusion presented above that the piles in this strongbox have plugged from the beginning of the installation phase and performed similarly to the close-ended piles.

The line in the dry strongbox C2 shows similar slope to the piles C2O12 which was plugged from the beginning of the jacking phase. Concerning the other piles in C2, the initial slopes of these piles are well below the theoretical line, but with the increase of the embedded volume and as these piles was creating plugs the slope of the curves change and become more rigid. It seems that with the initial slope of these piles, they can be considered as the unplugged slope. So if the piles didn't plug it is more likely that they were going to continue with this same slope. The plug formation let to a change in the behaviour of these piles and they become more similar to the closed ones. Similar results have been found in C1 and C4.

4 CONCLUSIONS

Different piles with 25 mm of embedded depth has been installed in the centrifuge at 100×g and studied in order to have a clear idea about the conditions that can influence the behaviour of the piles for jacket structures of offshore wind turbines.

Firstly, the saturation of the dense sand proves to have an important influence on the plug creation during the installation. The decrease in the pile diameters are shown to improve the potential for plug formation of the open-ended piles. Piles installed in dry sand are shown to have higher capacity than that installed in saturated sand in compression.

An important result has been also found of the existence of a linear relationship between the jacking load and the embedded volume of the tested piles.

ACKNOWLEDGEMENTS

The authors wish to thank the IFSTTAR and the Region Pays de Loire for their financial support to the thesis grants, within the context of which this study has been conducted. Special thanks to the IFSTTAR Centrifuge team for its technical support and assistance during the centrifuge experimental campaign.

REFERENCES

API RP 2GEO. 2011. Geotechnical and foundation design considerations. 120.

API RP2A. 2007. Recommended Practice for planning, Designing and Constructing Fixed Offshore Platforms, 22nd edn. American Petroleum Institute, Washington, DC, USA, API RP2A.

Clausen, C.J.F., Aas, P.M. & Karlsrud, K. 2005. Bearing capacity of driven piles in sand, the NGI approach. Proc., Int. Symp. On Frontiers in Offshore Geomechanics, ISFOG, Taylor & Francis, London, 677-681.

Deeks, A.D., White, D.J. & Bolton, M.D. 2005. A comparison of jacked, driven and bored piles in sand. The 16th International Conference on Soil Mechanics and Geotechnical Engineering (16ICSMGE, Osaka, Japan)

Jardine, R., Chow, F., Overy, R. & Standing, J. 2005. ICP design methods for driven piles in sands and clays. Thomas Telford, London, 105.

Kolk, H.J., Baaijiens, A.E. & Senders, M. 2005. Design criteria for pipes piles in silica sands. Proc., Int. Symp. On Frontiers in Offshore Geomechanics, ISFOG, Taylor & Francis, London: 711-716.

Kumara, J.J., Kikuchi, Y., Kurashina, T. & Yajima, T. 2016a. Effects of inner sleeves on the inner frictional resistance of open-ended piles driven into sand. Frontiers of Structural and Civil Engineering 10(4): 499-505.

Kumara, J.J., Kikuchi, Y. & Kurashina. 2016b. The effectiveness of thickened wall at the pile base of open-ended piles in increasing soil plugging. Japanese Geotechnical Society Special Publication 4(6): 138-143.

Lee, J. & Salgado, R. 2003. Estimation of Load Capacity of Pipe Piles in Sand Based on Cone Penetration Test Results. *J. Geotech. Geoenviron. Engng.* 129(5): 391-403.

Lehane, B.M., Schneider, J.A. & Xu, X. 2005. The UWA-05 method for prediction of axial capacity of driven piles in sand. Proc., Int. Symp. On Frontiers in Offshore Geomechanics, ISFOG, Taylor & Francis, London, 683-689.

Paik, K. & Salgado, R. 2003. Determination of Bearing Capacity of Open – Ended piles in Sand. *J. Geotech. Geoenviron. Engng.* 129(1): 46-57.

Robinsky, E.I., Sagar, W.L. & Morrison, C.F. 1964. Effect of shape and volume of the capacity of model piles in sand. *Canad. Geotech. J.* 1(4).

Silva, M. 2014. Experimental study of ageing and axial cyclic loading effect on shaft friction along driven piles in sands. Ph. D. thesis. Université de Grenoble.

Dynamic load tests on large diameter open-ended piles in sand performed in the centrifuge

E. Heins
Institute of Geotechnical Engineering and Construction Management, Hamburg University of Technology, Germany

B. Bienen & M.F. Randolph
Centre for Offshore Foundation Systems, The University of Western Australia, Australia

J. Grabe
Institute of Geotechnical Engineering and Construction Management, Hamburg University of Technology, Germany

ABSTRACT: Offshore infrastructure is commonly founded on piles. In accordance with design guidelines, piling contracts usually include load tests to verify the bearing capacity following pile installation. Offshore, these are generally performed as dynamic load tests (DLT) rather than static load tests often used onshore, for economical and practical considerations. Substantial experience exists in evaluating DLT on closed-ended piles. However, DLT on open-ended profiles have not been investigated sufficiently to date, particularly for piles with low length to diameter ratio. In order to derive realistic static axial pile bearing capacities, the processes during DLT taking place in the open-ended pile and soil need to be understood. Centrifuge tests provide the possibility of model tests with defined and reproducible conditions under consideration of realistic stress states. This paper discusses centrifuge test results of static and dynamic load tests on monopiles investigating the effects of pile installation method. The findings of this study provide a basis towards improved understanding of DLT on open-ended piles.

1 MOTIVATION

Dynamic load tests (DLT) are necessary for offshore pile foundations because common design guidelines such as Eurocode 7-1 (EC7 2014) and the design standard of the Federal Maritime and Hydrographic Agency (BSH) (2015) require verification of the pile bearing capacity under compression load after installation of the piles. Open-ended piles are commonly used for foundations of offshore structures such as oil rigs and offshore wind turbines. Even though the axial pile capacity could be evaluated directly from static load tests (SLT), such tests are rarely performed offshore due to economic and practical reasons. DLTs are an alternative for SLTs with extensive experience gained for closed-ended piles.

During a DLT dynamic impact is applied at the pile head. This causes a wave to propagate through the pile, which as a result of the difference in density between pile and soil, is reflected at the pile tip. The wave propagating through the pile is measured as strain and acceleration at the pile head. The measured pile head signals are evaluated to deduce the static axial pile bearing capacity. Evaluation techniques have been developed for closed-ended piles, based on one dimensional wave theory. Although differences are expected, this technique has been transferred to the interpretation of open-ended piles also, for lack of alternative guidance. The pile-soil interaction due to the soil column inside the open-ended pile and possible plugging may lead to significant differences compared to closed-ended piles. Additionally, the wave propagation cannot be simplified as one dimensional (Paikowsky & Chernauskaus 2008). The key parameters of influence of DLT on open-ended piles have not been investigated comprehensibly yet such that pile head signals obtained from DLT on open-ended piles are not understood in sufficient detail to date. The fact that DLT are already conducted on open-ended piles highlights the importance of accurate evaluation of axial capacities from DLT. The processes taking place in the pile and soil during DLT therefore have to be understood. This is particularly important for monopiles of low length to diameter ratio such as are used in the offshore wind industry.

In this study, investigations of DLTs on open-ended piles have been carried out by means of centrifuge tests. For a better understanding of the evaluation

Table 1. Mechanical properties of very fine silica sand.

Parameter	Value	Description (Unit)
d_{50}	0.19	mean particle size (mm)
U	1.9	index of uniformity
φ_c	30	critical state friction angle (°)
$\rho_{d,min}$	14.9	minimum dry density (kN/m³)
$\rho_{d,max}$	18.0	maximum dry density (kN/m³)

of DLT to deduce the static bearing capacity, DLT and SLT have been performed on a large diameter monopile installed into loose, dry sand at 100 g. This allowed pile head signals to be measured and compared to load-displacement curves from SLT. Furthermore, the effect of pile jacking and impact driving, the latter being the most common installation technique offshore, on the comparison of DLT and SLT was investigated in the physical model tests.

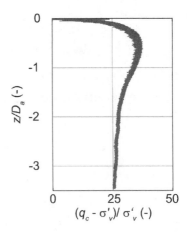

Figure 1. Typical result of the cone penetration test in loose, dry sand.

2 CENTRIFUGE MODELLING

A fixed beam geotechnical centrifuge of type Acutronic 661 with a platform radius of 1.8 m situated at the Centre for Offshore Foundation Systems at the University of Western Australia was used to perform the ng-physical model tests at 100 g.

2.1 Soil sample

A very fine silica sand with the mechanical properties as taken from Pucker, Bienen, & Henke (2013) listed in Table 1 was chosen. The soil sample was prepared within an internally 0.39 m wide, 0.65 m long and 0.425 m high strongbox. A sample height of 0.31 m was targeted. A loose soil sample was aimed for to ensure sufficient penetration depth of the pile during pile installation within the centrifuge tests, within the capabilities of the actuator. A loose, dry sand sample with an initial relative density of $I_D = 0.34$ was achieved by manually and carefully pouring the sand into the strongbox with minimum falling height.

The soil sample was characterized by four cone penetration tests (CPT) performed in the corners of the strong box with a minimum distance from the side of the strongbox of $7.5 D_{cone}$. The penetrometer has a diameter of $D_{cone} = 10$ mm (1 m prototype). The results of the CPTs into dry, loose sand are illustrated in Figure 1. Within this figure the cone resistance q_c is normalized by the effective vertical stress σ'_v and is plotted over the normalized distance below ground surface. The penetration depth below ground surface z is normalized by the outer pile diameter D_a (5 m prototype).

2.2 Model pile and instrumentation

The prototype large-diameter tubular steel pile considered for these centrifuge tests has an outer diameter of $D_a = 5.0$ m. Maintaining geometric similarity, the outer diameter was scaled to $D_a = 0.05$ m for the model pile. An embedment ratio typical for offshore piles of $D_a/L_e = 5$ is considered for the performed tests. Wave propagation through the pile and pile movement during this process are characterized by the axial stiffness of the pile EA. Ensuring similarity between prototype and model scale for EA an aluminium tube with a wall thickness of 1.6 mm is chosen as the model pile. The wall thickness is slightly larger than that of a typical offshore monopile which may influence the ratio between tip and shaft friction. However, this is the case for all performed tests enabling comparison between the scenarios. Furthermore, pile plugging, assumed to be a key aspect in pile response, is not influenced.

The pile response during installation, DLT and SLT was captured as acceleration (or velocity), strain and load at the pile head. As illustrated in Figure 2 the corresponding sensors are located $1.6 D_a$ from load application. This meets the requirements by EA-Pfähle (2012) according to which the measurement devices have to be located at least $1.5 D_a$ below load application in order to ensure a plain wave propagation at measurement level during DLT.

The strain is measured by strain gauges attached directly to the pile. Measurement of acceleration at small scale is difficult. Therefore, two different systems to capture the acceleration of the wave travelling through the pile during DLT are applied. A MEMS 3501A1260KG accelerometer with a measuring range of 60 000 g was mounted to the pile. Further a Hopkinson bar as described in Bruno & Randolph (1999) was manufactured. The Hopkinson bar is a polyvinyl chloride (PVC) strip to which two adjacent strain gauges are attached. The PVC strip used here is 220 mm long, 3 mm wide and 1.6 mm thick. Wave propagation through the pile will initiate a wave propagation through the PVC strip which is captured by the strain gauges of the Hopkinson bar. The wave speed within the PVC strip is significantly lower than within the pile. Hence, the returning wave from pile tip is measured before the first wave induced in the Hopkinson

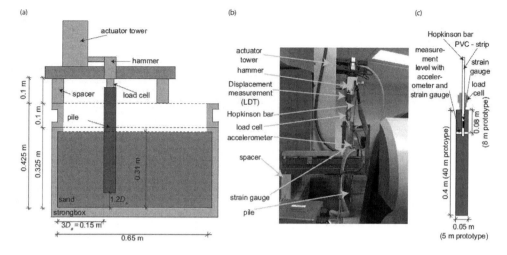

Figure 2. (a) Schematic testing set-up and (b) photograph of the set-ups.

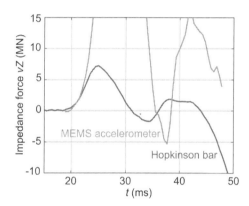

Figure 3. Comparison of a impedance force derived from measurements of the MEMS accelerometer and measurements of the Hopkinson bar.

bar returns from the end of the PVC strip. Consequently, the wave inside the pile can be captured by the strain gauges of the Hopkinson bar without any other effects. In order to transfer the measured strain of the PVC strip into velocity of the wave inside the pile $v(t)$ this measured strain $\varepsilon_{PVC}(t)$ is multiplied by the wave speed of the Hopkinson bar c_{PVC}

$$v(t) = \varepsilon_{PVC}(t) \cdot c_{PVC}. \qquad (1)$$

Instead of measuring the acceleration and integrating this to velocity as for the MEMS accelerometer, the Hopkinson bar measures the velocity directly. The measurements during centrifuge testing reveal that measurement of acceleration at small scale is influenced by oscillation and resonance effects. The first rise within the signals might be reliable but hold uncertainty. An example comparison of the acceleration measurement of the MEMS accelerometer and the measured velocity by the Hopkinson bar is illustrated in Figure 3. The equality within the first rise is obvious. Afterwards a large drift and increase within the measurements of the accelerometer is present. The uncertainty of the measurement of the acceleration is reflected by the large impedance forces which do not correlate with the Hopkinson bar or strain gauge measurements. Measurement of velocity over time for the whole DLT is necessary for evaluation of the response. Additionally, the measured acceleration with the MEMS accelerometer remained well below 60 000 g. To verify the acceleration measurements and the precision of the measurement, the accelerometer was moved vertically through air within the g-field of the operating centrifuge. Instead of gaining the same acceleration with depth every time, the acceleration over estimated the vertical distance within the acceleration field and showed hysteresis. From this test a precision of measurement of approximately 60 g is deduced. Hence, the small obtained accelerations might also be influenced by an insufficient precision of measurements for the performed tests. As a consequence, the data from the Hopkinson bar instead of the accelerometer are used for the evaluation of the performed centrifuge tests.

2.3 *Testing set-up*

As depicted in Figure 4 the pile is attached to an anvil which in return is fixed to a miniature driving hammer as developed by De Nichola & Randolph (1994) and also used in Henke & Bienen (2014). The drop weight of the model scale hammer is 50 g corresponding to a weight of 50 t of a prototype hammer. A drop height of 17 mm in model scale was maintained during the performed tests. The miniature pile driving hammer was mounted on the vertical lead screw of the dual axis servo controlled actuator (compare Figure 4). The vertical movement of the pile measured by a displacement transducer is integrated into a feed back loop causing the actuator and hence the miniature pile driving hammer to follow the penetration of the pile. The

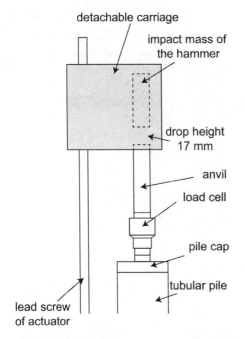

Figure 4. Schematic drawing of the pile-hammer connection.

horizontal movement of the pile was restricted by the actuator ensuring the positioning of the pile within the strongbox. The pile was situated $3D_a$ from the end of the strongbox (compare Figure 2a) allowing two tests to be performed within the same soil sample. At the target embedment depth of $5D_a$ soil of a distance of $1.2D_a$ remained below the pile. Due to the limited size of a strongbox influences from wave reflection at the walls of the strongbox cannot be prevented during pile installation. However, the spacings between the sides of the strongbox and the pile are sufficiently large to neglect influences from wave reflection during DLT, ensuring valid results.

2.4 *Testing procedure*

The centrifuge tests were performed with the aim of studying the response during SLT and DLT of a large-diameter monopile within a loose, dry sand analysing the influence of different pile installation techniques on the results. After acceleration of the testing set-up to $100\,g$ acting at a distance of $1/3L_e$ below ground surface, the pile installation was conducted. For the static pile installation of jacking the pile was pushed into the soil with a constant velocity of 0.1 mm/s by the actuator. The jacking process was stopped when the resistance approaches the load limit of the actuator leaving sufficient capacity for the SLT. The impact driven pile installation was performed with continuous operation of the pile driving hammer at a frequency of 5 Hz (0.05 Hz prototype). The driving process was stopped when no further pile penetration was reached with the set characteristics. Both the jacked and the impact driven set-up were performed within one soil sample to ensure that comparable characteristics of the soil sample are present.

After pile installation a waiting time of at least 120 s was observed before performing a DLT to eliminate effects of wave propagation on the load test results. The DLT was performed by applying one single hammer blow at the pile head. During DLT the data were sampled with a sampling rate of 200 kHz. After another immobilization time of approximately 600 s a SLT was conducted by pushing the pile further into the soil with a constant velocity of 0.1 mm/s until a displacement of $0.1D_a$ or until the load limit of the actuator was reached.

The jacked pile reached a final embedment and therefore testing depth of DLT and SLT of $2D_a$. The impact driven pile was installed to a final embedment length of $4D_a$, where a pair of DLT and SLT were performed. To ensure comparability between the results of the DLT an extra DLT was performed for the impact driven pile at an embedment depth of $2.8D_a$.

3 RESULTS

3.1 *Dynamic Load Tests (DLTs)*

The measurements of velocity $v(t)$ by the Hopkinson bar and strain $\varepsilon(t)$ by the strain gauges on the pile are converted to impedance force vZ by multiplying the velocity with the impedance of the open-ended pile Z and with the Young's modulus E and pile metal cross-sectional area A to force F by

$$F = \varepsilon(t) \cdot E \cdot A. \qquad (2)$$

Force F and impedance force vZ are plotted in prototype scale for the DLT in Figure 5. The first peak within both forces describes the time ($\approx$25 ms) when the first wave travels downward through the pile. After the return time of around 12.2 ms, at approximately 37.2 ms the return wave from the pile tip is captured by the sensors. At 45 to 50 ms the return wave inside the Hopkinson bar reaches the strain gauges on the PVC strip. This explains the strong decrease of impedance force vZ at this time.

The results for force F and impedance force vZ are very similar for both pile installation methods considered. The first slope of both forces is the same. The value of the first peak within the force F signal differs marginally, showing a slightly larger force F for the impact driven pile. The slopes of the decrease after the first peak are very similar again. The returning wave from pile tip is a tension wave for both piles at an embedment depth of $2.8D_a$, with some difference at a penetration depth of $4D_a$. A tension wave characterizes a pile tip with low resistance. The values of the force F for the impact driven pile at $2.8D_a$ are slightly less within the return wave but no significant difference can be observed. The impedance force vZ of impact and jacked pile is quite similar at $2.8D_a$, although there is a more significant difference during the return wave

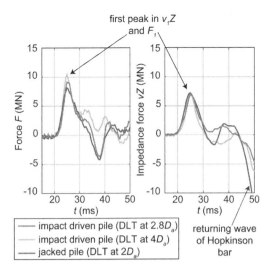

Figure 5. Dynamic load test results of an impact driven and a jacked pile in loose, dry sand.

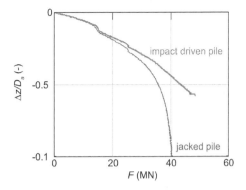

Figure 6. Obtained SLT results for an impact driven and a jacked pile in loose, dry sand.

as the impedance force for the impact driven pile does not go negative (implying smaller soil resistance mobilized). Overall, this comparison reveals that no distinct or significant difference is observed between the DLT results of the jacked and the impact driven pile.

Although the pile was successfully driven into the soil by the miniature pile driving hammer, the maximum impact force was quite low and the blowcounts were high, with over 7300 blows required for the penetration of $4D_a$. At the penetration of $2.8D_a$ the set per blow was only 2.6 mm (just under 100 blows per 250 mm), while by the end of penetration the set was down to around 0.5 mm/blow.

3.2 Static Load Tests (SLTs)

For the SLTs performed the load and pile displacement are measured directly. The results of the SLTs are plotted in Figure 6 in prototype scale. The measured force F is depicted over the pile penetration during the SLT normalized by the outer pile diameter D_a. The typical criterion of $0.1D_a$ was employed to define the pile capacity. The slight unevenness in both measured curves at approximately 15 MN and 25 MN results from seating within the hammer-pile connection.

The results of the SLTs reveal the expected trend of increasing force with increasing penetration. Both measured load-displacement curves commence very similar for both piles up to a force of approximately 15 MN. This implies that both piles have the same initial stiffness. Greater static pile displacement leads to an increasing discrepancy between the SLT results of the jacked and the impact driven pile. However, the impact driven pile behaves significantly stiffer during the SLT. The load limit of the actuator was reached before the impact driven pile achieved the desired displacement of $0.1D_a$. Therefore, the ultimate capacity of the impact driven pile within this centrifuge test is not known. However, the curve suggests it to be significantly larger than that of the jacked pile. The jacked pile shows an ultimate capacity of approximately 40 MN.

In contrast to the experimental data, higher bearing capacity for the jacked pile would be expected because of the influence of the jacking in comparison to impact driving on the state of the soil. Jacked piles usually have a greater tendency towards plugging (Randolph & Gourvenec 2011). This goes along with jacking causing higher radial stresses within the soil while resulting in the least change in void ratio distribution (Mahutka, König, & Grabe 2006). In these centrifuge tests, the jacked pile was pushed into the soil until the load limit of the actuator was reached. Therefore, no plugging or only a partially plugged state is expected to have occurred, as evidenced by the low pile tip resistance deduced from the DLTs. It appears, though, that conditions for plugging during the jacking process might not have developed completely up to this embedment depth. The impact driven pile on the other hand was subsequently installed until no further penetration was achieved. The DLT at final penetration reflects greater mobilisation of soil resistance. Due to the low initial relative density densification of the soil during impact driving is assumed. Due to dilatancy effects plugging is possible. For pile jacking no significant effect from dilatancy is expected making plugging unlikely to occur. Hence, the measured results for the SLTs are plausible.

3.3 Comparison of DLT and SLT

The measured results from DLTs and SLTs on a jacked and an impact driven pile lead to different conclusions. While the pile head signals of the DLTs are similar and hence it is assumed that an evaluation would lead to similar axial capacities, the load displacement curves of the SLTs are similar only at the beginning, i.e at the small displacement generated during the DLT. Afterwards the capacities of the impact driven and jacked pile show significant differences. As a consequence, the open-ended piles have different bearing capacities. The static axial bearing capacities of the tested jacked and the tested impact driven piles are not the same but vary greatly despite similar DLTs.

The DLT involves very small pile displacements. Considering the comparison of the SLT results with the same initial response of both piles, the stiffness of both piles might be similar during DLTs. Hence, the similarity of the DLT results might be a reason of the condition governing small strain response. However, this is an assumption and explanation of the similarity between DLTs and the difference in SLTs is, at this stage, somewhat speculative. This highlights that the evaluation of a DLT on an open-ended pile is not fully understood yet. In order to derive the static axial capacity, therefore, an extensive centrifuge testing program is necessary to study the key parameters of influence of pile head signals and to evaluate the impact of different variations on pile head signals from DLTs and load-displacement curves from SLTs.

A more extensive set of centrifuge tests has subsequently been planned with the installation method varied between no installation (wished-in-place pile), static pile installation (jacking) and dynamic pile installation (impact driving). Dry samples, water-saturated samples and samples saturated with a pore fluid with scaled viscosity are to be investigated. Additionally to the loose sample prepared for the tests considered here, a dense sample aiming at offshore conditions typical for the North Sea, for instance, will be modelled. For all of these variations the pile will be installed and tested dynamically and statically.

4 CONCLUSIONS

Centrifuge tests at $100\,g$ were performed to gain a better understanding for deducing the static axial pile bearing capacity from dynamic load test (DLT) on open-ended piles. For this purpose a jacked and an impact driven pile installation were performed on a large diameter tubular pile in loose, dry sand. The installation was followed by a pair of DLT and SLT.

The results of the DLTs of both open-ended piles are similar and only a slight difference within the returning wave is observed. In contrast, the load-displacement curves obtained from SLT of jacked and impact driven piles are similar only at the very beginning. The impact driven pile shows a stiffer behaviour while the jacked pile softens to an ultimate load. Therefore, a significant difference between the DLT and SLT results obtained on these open-ended piles exists and the derived ultimate capacities will also be different.

In order to understand DLTs on open-ended piles comprehensively and to deduce realistic static axial bearing capacities further investigations are necessary. These will be undertaken as part of an extensive centrifuge testing program.

REFERENCES

EC7 (2014). Handbuch Eurocode 7-1 Entwurf, Berechnung und Bemessung in der Geotechnik – Teil 1: Allgemeine Regeln, Deutsche Fassung EN 1997-1:2005 + AC:2009 + A1:2013.

BSH (2015). Standard design – minimum requirements concerning the constructive design of offshore structures within the exclusive economic zone (eez): Bsh-no. 7005.

Bruno, D. & M. F. Randolph (1999). Dynamic and static load testing of model piles driven into dense sand. *Journal of Geotechnical and Geoenvironmental Engineering 125*(11), 988–998.

De Nichola, A. & M. F. Randolph (1994). Development of a miniature pile driving actuator. In C. F. Leung, F. H. Lee, and T. S. Tan (Eds.), *Proceedings of International Conference on Centrifuge Modelling 1994*, pp. 473–478.

EA-Pfähle (2012). *Empfehlungen des Arbeitskreises "Pfähle"– EA-Pfähle der Deutschen Geschellschaft für Geotechnik (DGGT)* (2 ed.). Ernst & Sohn.

Henke, S. & B. Bienen (2014). Investigation of the influence of the installation method on the soil plugging behaviour of a tubular pile. In *Proceedings of International Conference on Physical Modelling in Geotechnics (ICPMG) 2014 in Perth/Australia*, Volume 2, pp. 1–8.

Mahutka, K.-P., F. König, & J. Grabe (2006). Numerical modelling of pile jacking, driving and vibratory driving. In R. Th. Triantafyllidis, Balkema (Ed.), *Proceedings of International Conference on Numerical Simulation of Construction Processes in Geotechnical Engineering for Urban Enviroment (NSC06), Bochum*, pp. 235–246.

Paikowsky, S. & L. R. Chernauskaus (2008). Dynamic analysis of open ended pipe piles. In J. A. Santos (Ed.), *Proceedings of 8th International Conference on the Application of Stress Wave Theory to Piles in Lisbon, Portugal*, pp. 59–76. IOS Press.

Pucker, T., B. Bienen, & S. Henke (2013). CPT based prediction of foundation penetration in siliceous sand. *Applied Ocean Research 41*, 9–18.

Randolph, M. F. & S. Gourvenec (2011). *Offshore Geotechnical Engineering*. Spon press.

Centrifuge model tests on holding capacity of suction anchors in sandy deposits

K. Kita
Tokai University, Shizuoka, Japan

T. Utsunomiya
Kyushu University, Fukuoka, Japan

K. Sekita
Marine River Technology Engineering Inc., Tokyo, Japan

ABSTRACT: Holding capacity of suction anchor was studied by centrifuge model tests, in view of their application to the mooring system for floating wind power generation in offshore Japan. Capped stainless steel pipes placed in loose sandy deposits were pulled at about mid depth of the anchor embedment in nearly horizontal direction (initial angle of elevation of 12.5 to 23.7 degrees) under 20G centrifugal conditions. Measured holding capacities were compared with capacities estimated from a limit equilibrium model considering Reese type failure mechanism, as well as from a practical method proposed by Deng and Carter, both of which were associated with friction angles evaluated from centrifugal miniature cone penetration tests carried out prior to the anchor placement. Capacities predicated by Deng and Carter method underestimated measured ones.

1 INTRODUCTION

Offshore wind power generation is considered to be a promising renewable energy production in Japan because of the geographical feature. Floating wind turbines were constructed in offshores of Nagasaki and Fukushima prefectures, and drag type anchors were adopted for the mooring systems in these cases. Suction anchors have advantages over drag anchors in precise placement and applicability of geotechnical design framework, and have been employed for offshore oil/gas production platforms in deep sea areas (Andersen et al., 2005). In case of the use in mooring foundations of floating wind power generation in Japan, anchors would be installed at sites with shallower water depths where the seabed might be composed of silty or sandy deposits.

A pilot project has been conducted since 2015 to introduce a suction anchor for one of three mooring lines for a floating mast in offshore of Goto city, Nagasaki prefecture, Japan, where the water depth is about 100 meters and the seabed is formed mainly by silt and sand grains (Utsunomiya et al., 2017). Diameter and embedment of the anchor are 3.8 and 3.0 meters respectively, and the catenary mooring chain is connected at the mid-depth of the embedment with predicted elevation angle of 13.4 degrees.

This paper describes the results of centrifuge model tests on holding capacity of suction anchor placed in sandy deposits. In view of the aforementioned project in Goto, anchor with the diameter of 76 mm and the embedment of 56 to 63 mm was pulled almost horizontally under 20G centrifugal conditions, therefore the prototype sizes of the anchor in the experiments were geometrically about 1/2.5 of the anchor used in the project. Holding capacities measured in the experiments were compared with capacities estimated from a limit equilibrium model considering Reese (1962) type failure mechanism, as well as from a practical method proposed by Deng and Carter (2000).

2 EXPERIMENTATION

2.1 Experimental setup

Experimental setup for the pullout test of suction anchor is shown in Figure 1.

Horizontal loose deposits are formed by raining air-dried fine sand grains into water which is reserved in a rectangular strong box of 510 mm long and 200 mm wide. After de-aeration of the ground by suction, the ground is consolidated by self-weight under 20G condition in a centrifuge with the effective spinning radius of 2.5 meters. This is followed by the miniature cone penetration test under 20G condition, with the rod diameter of 10 mm and the penetration rate of 1.1 mm/s. Model anchor with mooring wire line is then fixed to a vertically slidable guide rod. The anchor is firstly inserted into the ground by the self-weight

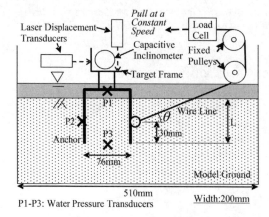

Figure 1. Experimental setup for pullout tests.

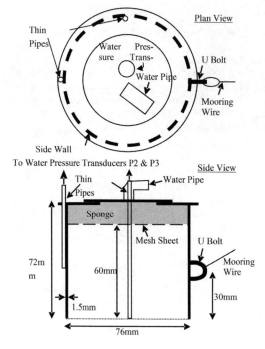

Figure 2. Model anchor.

under 1G condition and then under 20G centrifugal condition. This is followed by pushing the guide rod downward by hand under 1G condition to attain the prescribed anchor embedment. After the guide rod is removed, a target frame is placed above the anchor top. The mooring wire is linked through pulleys to the actuator and is pulled at a constant speed under 20G condition. During the pullout test, wire tension is measured by a load cell and water pressures are observed at three locations around the anchor (P1 to P3 in Fig. 1). Anchor position is monitored by laser displacement transducers and the capacitive inclinometer on the target frame.

The model anchor is schematically drawn in Figure 2. The anchor is made by stainless steel pipe with

Table 1. Physical properties of soils.

Soil	A	B	C
Density of soil particles (kg/m^3)	2637	2637	2688
Maximum void ratio	1.429	1.291	NA
Minimum void ratio	0.669	0.763	NA

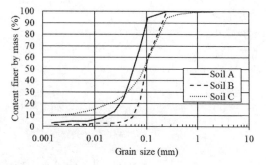

Figure 3. Grain size distribution of soils.

a plate soldered at the upper end. The outer diameter, the thickness of the side wall and the anchor height are 76 mm, 1.5 mm and 72 mm respectively. Inside of the anchor is separated to two parts by mesh sheet. Sponge is inserted in the upper room to prevent soil grains inside the anchor from flowing out during the anchor installation stage. The lower room is 60 mm high and is open at the lower end. A U bolt is attached to the front wall for connecting the mooring wire. Two thin pipes are soldered vertically along the side wall; one is on the back and is open-ended at almost the same height as the wire connection point (P2 in Fig. 1), the other is on the inner left wall and is open-ended at the anchor tip (P3 in Fig. 1). The upper ends of the thin pipes are connected by flexible polyurethane tubes to water pressure transducers. A water pressure transducer and a water pipe are equipped to the lid on the anchor top. The water pipe is open during anchor installation, and then closed during the pullout test.

2.2 Physical properties of soils

Three kinds of sandy soil were used in this study. Physical properties of these soils are shown in Table 1 and their grain size distribution curves are depicted in Figure 3. Soils A and B are silica sands, while soil C was obtained from the seabed surface at offshore of Goto city.

2.3 Test cases

Five pullout tests were carried out as summarized in Table 2, in view of the aforementioned pilot project in offshore Goto, where the suction anchor was installed in silty to sandy deposits and the mooring line was connected with low elevation angle to the anchor. The model ground was formed in loose state considering quiet sedimentation at seabed with water depth of

Table 2. Summary of test cases.

Case	1	2	3	4	5
Soil used	A	A	A	B	C
Void ratio of ground, e^*	1.03	1.00	1.04	1.15	1.77
Ground thickness, H^* (mm)	176	163	176	182	142
Anchor embedment, L (mm)	63	60	61	56	59
Internal friction angle by miniature CPT, ϕ (degree)	34.3	35.0	34.1	35.0	32.2
Initial pullout angle of elevation, θ (degree)	14.8	17.8	23.7	12.5	15.1
Wire pullout speed, v (mm/s)	0.84	0.73	0.75	0.55	0.49
Measured maximum wire tension, T_{max} (N)	256	254	211	222	184
Calculated holding capacities					
P_R (N)**	186	179	174	145	105
P_{DC} (N)***	176	159	152	138	112

* Values measured after miniature cone penetration tests
** Holding capacity calculated by the limit equilibrium method
*** Holding capacity calculated by Deng & Carter method

around 100 meters. Silica sand with higher fines content (soil A) was used as ground material in cases 1 to 3, where the holding capacities were investigated by changing the initial pullout angle of elevation θ from 14.8 to 23.7 degrees. Here the 'initial' pullout angle is defined as the elevation angle of the straight line tangent to the pulley on the container wall, originating from the front point of the U bolt on the anchor wall. Coarser silica sand (soil B) was employed in case 4 while the seabed soil of offshore Goto (Soil C) was used for Case 5, in both cases the anchors were dragged in nearly horizontal directions.

3 RESULTS AND DISCUSSION

3.1 Experimental results

Variation of wire tension with anchor displacement is shown in Figure 4a for case 1, in which the initial pullout angle of elevation is 14.8 degrees. Displacement at wire connection point is calculated from measured displacements and inclination at the target frame. The tensile force increases instantly at very small displacement, and then grows gradually to the maximum value of 256 N at 25 mm displacement. Water pressures measured at three locations inside and outside the anchor (P1 to P3 in Fig. 1) are plotted against displacement in Figure 4b. One may observe negative water pressures to generate at all points, while the negative pressure at the back wall (P2) dissipates rapidly. The effective overburden pressure at the initial anchor tip level is approximately 10 kPa for this case. Anchor position

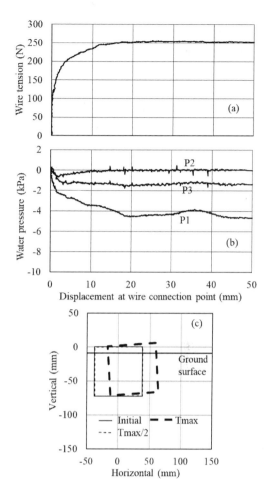

Figure 4. Results of case 1.

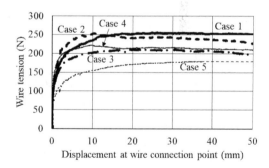

Figure 5. Variation of wire tension with displacement at wire connection point.

when the wire tension reached maximum (T_{max}), is drawn by thick broken lines in Figure 4c, together with the initial position (thin solid lines) and the position when the wire tension was $T_{max}/2$ (thin broken lines). The anchor displaced horizontally and inclined slightly backward at the moment of maximum wire tension.

Measured responses of wire tension are shown in Figures 5 for all cases, while Figure 6 shows anchor

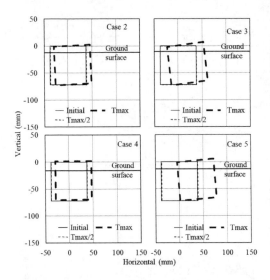

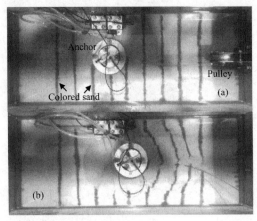

Photograph 1. Plan view of the ground surface (a) before and (b) after the pullout test for case 1.

Figure 6. Anchor positions at three phases (initial, wire tension equal to $T_{max}/2$ and T_{max}) for cases 2–5.

positions at three particular phases (initial, wire tension equal to $T_{max}/2$ and T_{max}) for cases 2 to 5. Measured peak wire tension is noted in Table 2. It may be seen that larger the initial pullout angle of elevation, smaller the maximum wire tension for cases 1 to 3. From Figures 4c and 6, anchors displace very little till the wire tensions reach $T_{max}/2$ for all cases. The anchor displacement at T_{max} varies substantially among cases, while the anchor rotation angle at T_{max} tends to be larger for larger initial pullout angle in these experiments.

3.2 Discussion on holding capacity

Holding capacities of anchors evaluated from a limit equilibrium method and Deng & Carter method are compared with observed maximum wire tension in this section.

In the limit equilibrium method, passive soil resistance, which acts on the vertical plane area projected from embedded part of anchor, is calculated by considering the failure mechanism proposed by Reese (1962) for horizontal movement of a pile. This mechanism incorporates widening failure region in front of the pile (anchor) observed in experiments (see Photo. 1) in a simple way. Assumed failure blocks are depicted in Figure 7, and the soil resistance P is derived as equation (1):

$$P = \frac{\gamma' L^2}{\sin(\xi - \phi - \delta)} \left\{ \frac{D}{2} \tan\xi \cos(\xi - \phi) \right. $$
$$+ \frac{L}{3}(\tan\xi)^2 \tan\varepsilon \tan(\xi - \phi) \quad (1)$$
$$\left. + \frac{KL}{3}\tan\xi(\sin\phi - \tan\varepsilon \sin(\xi - \phi)) \right\}$$

where γ' = submerged unit weight of soil; ϕ = internal friction angle of soil; δ = friction angle at wall;

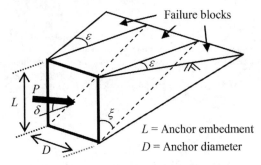

Figure 7. Failure blocks for the limit equilibrium method.

K = earth pressure coefficient; L = anchor embedment; D = anchor diameter; and ξ, ε = angles to determine the shape of failure regions (see Fig.7). In this study, friction angle δ is set as $\delta = \phi/3$ and the earth pressure coefficient K is given by the active earth pressure coefficient. Angle ε is given by $\varepsilon = \phi/2$ after Reese (1962). The minimum value of P is searched by changing the angle ξ, and then P_R is calculated as the horizontal component of minimum P.

Deng & Carter (2000) conducted parametric finite element analyses, and developed a workable method to estimate holding capacity of suction anchor in dense sand under fully drained condition. In spite of these assumptions, this method is examined for loose ground cases in this study because it could provide a practical design framework. Because the ground was formed in loose state in the experiments, dilatancy angle of 5 degrees is employed here as the minimum value considered in Deng & Carter method. The coefficient of earth pressure at rest is set 0.5.

For calculation of holding capacities by these two methods, internal friction angles of soils were evaluated from miniature CPT results associated with the conversion method by Lunne & Christoffersen (1983). Cone tip resistance is obtained by subtracting the side friction from the total soil resistance during penetration. The side friction is not measured directly during penetration but evaluated from pullout resistance of the penetrometer. The side friction was 7 to 11% of total penetration resistance. Estimated internal friction angles are given in Table 2.

Holding capacities obtained by the limit equilibrium method (P_R) and Deng & Carter method (P_{DC}) are listed in Table 2. As seen in measured capacities, the calculated capacity tends to decrease with increasing pullout angle of elevation for cases 1 to 3. This is consistent with analytical pullout capacities of suction anchors calculated by Bang et al. (2011). Ratios of measured to calculated capacities are between 1.21 and 1.75 in case for the prediction by the limit equilibrium method, while the ratios are between 1.39 and 1.64 for Deng & Carter method. These methods therefore underestimated the measured holding capacities by 17 to 43%.

4 CONCLUSION

Holding capacity of suction anchor in sandy ground were investigated by centrifuge model tests, in view of the use in catenary mooring system for floating wind turbines. Measured capacities were compared with those calculated by a limit equilibrium method and practical Deng & Carter method. Calculated capacities underestimated the measured ones by 17 to 43% in this study where the anchors were dragged in nearly horizontal directions.

ACKNOWLEDGMENTS

The Authors sincerely acknowledge the funding support from Low Carbon Technology Research, Development and Demonstration Program (2015–2017), Ministry of the Environment, Government of Japan. We also thank Mr. Masato Ito of Tokai University for his cooperation in experiments.

REFERENCES

Andersen, K.H., Murff, J.D., Randolph, M.F., Clukey, C.T., Erbrich, C.T., Jostad, H.P., Hansen, B., Aubeny, C., Sharma, P. & Supachawarote, C. 2005. Suction anchors for deepwater application. In S. Gourvenec & M. Cassidy (eds), *Frontiers in Offshore Geotechnics*: 3–30, Boca Raton: CRC Press.

Bang, S., Jones, K.D., Kim, K.O., Kim, Y.S. & Cho, Y. 2011. Inclined loading capacity of suction piles in sand. *Ocean Engineering* 38: 915–924.

Deng, W. & Carter, J.P. 2000. Inclined Uplift Capacity of Suction Caissons in Sand. *Proc. Offshore Technology Conference*: OTC12196.

Lunne, T. & Christoffersen, H. P. 1983. Interpretation of Cone Penetrometer for Offshore Sands. Proc. *Offshore Technology Conference*: OTC4464.

Reese, L.C. 1962. Ultimate resistance against a rigid cylinder moving laterally in a cohesionless soil. *Society of Petroleum Engineers Journal* 2(4): 355–359.

Utsunomiya, T., Sekita, K., Kita, K. & Sato, I. 2017. Demonstration test for using suction anchor and polyester rope in floating offshore wind turbine. Proc. 36th international conf. Ocean, Offshore and Arctic Engineering: OMAE2017-62197.

A review of modelling effects in centrifuge monopile testing in sand

R.T. Klinkvort
Norwegian Geotechnical Institute, Oslo, Norway

J.A. Black & S.M. Bayton
Centre for Energy and Infrastructure Ground Research, University of Sheffield, Sheffield, UK

S.K. Haigh & G.S.P. Madabhushi
Department of Engineering, University of Cambridge, Cambridge, UK

M. Blanc & L. Thorel
IFSTTAR – GERS Department, Geomaterials & Modelling in Geotechnics Laboratory, Bouguenais, France

V. Zania
Technical University of Denmark, Lyngby, Denmark

B. Bienen & C. Gaudin
Centre for Offshore Foundation Systems, Perth, Australia

ABSTRACT: Monopile supports are to this day the preferred foundation solution for offshore wind turbines. There is still an empirical gap, however, between the in-situ conditions for these monopiles and the test conditions the current design methods are based on. This relates to both pile geometry and load conditions. The gap introduces uncertainties and in order to optimise design, this gap needs to be minimised. Scaled modelling in an increased acceleration field, i.e. testing in a geotechnical centrifuge, can be an effective way of understanding behaviour and obtaining empirical evidence. Reliable testing is underpinned by thorough consideration of scaling laws. To address fundamental understanding of modelling effects in centrifuge testing of laterally loaded monopiles in sand, a review of the latest relevant research for performing state of the art centrifuge testing of monopiles in sand has been carried out. Based on this review, modelling effects that introduce some uncertainties in the scaling of the results are identified, and based on that, a coordinated centrifuge-testing program is presented. The test program will be performed across five centrifuge facilities and aims at minimising the identified modelling uncertainties.

1 INTRODUCTION

The monopile is still the preferred foundation solution for offshore wind turbines. The simple geometry (a hollow steel pile) together with the experience gathered over years, still makes it the most attractive foundation concept for offshore wind turbines. Wind turbines are continuously being installed in deeper waters, with larger turbines and bigger monopiles in the search of minimising the cost of energy. Today the monopiles have a diameter of ∼8 m and penetration depths between 4–6 diameters, and will most probably continue to increase. Even though the monopiles are so widely used today, the design guidelines used are still based on tests of slender piles primarily used for oil and gas platforms.

In the last 10 years, large attention has been given to the lateral response of monopiles. For example, Cuéllar et al. (2009), LeBlanc et al. (2010) and Peralta & Achmus (2010) all performed scaled testing at 1 g on a laterally cyclic loaded monopile. In these tests, the focus was on the accumulation of displacement/rotations of the monopile when subjected to cyclic loading. Cyclic loading of laterally loaded slender piles in the centrifuge was the focus in Verdure et al. (2003) and Rosquet et al. (2007). More dedicated monopile centrifuge tests were done in Li et al. (2010), Klinkvort (2012), Rudolph et al. (2014), Kirkwood (2015), Truong & Lehane (2015) and Choo & Kim (2016). Also here the focus was on cyclic behaviour.

All of these investigations focused on the cyclic response, which is logical, considering the nature of the offshore environmental loading condition. However, there is still important information to gain from monotonic testing. For example, failure mechanism and forces acting on the pile are yet to be clarified.

In Klinkvort (2012), Choo and Kim (2016) and Bayton & Black (2016) a more fundamental investigation of the monotonic response was performed in order to understand the behaviour of a stiff rigid pile.

Even though monopiles have been a research focus over the last 10 years, there is still a lack of fundamental understanding of monopile behaviour, especially in sand and layered soils.

Centrifuge modelling can contribute to narrowing the knowledge gap providing modelling effects are carefully accounted for. This paper presents modelling considerations for centrifuge monopile testing in sand, it highlights where additional data is needed and presents a coordinate test program involving five different centrifuges facilities around the world targeting these gaps.

2 METHODOLOGY

When performing scaled experiments of any given boundary value problem and especially of monopiles, it is important to understand the fundamental soil behaviour. Sand has a non-linear stiffness and strength behaviour. This is well established and is here illustrated showing the empirical relationship between mean effective pressure (p'), critical state angle of friction (φ'_{cr}) and mobilised triaxial angle of friction (φ'_{max}) proposed by Bolton (1986).

$$\varphi'_{max} - \varphi'_{cr} = 3 \cdot (D_r(Q - \ln p') - R) \quad (1)$$

where, $D_r(e)$ = relative density; Q = grain coefficient; and R = crushing material coefficient.

In the same way, the shear stiffness stress dependency for sand can be written as shown in Equation 2, from Yang & Liu (2016).

$$G_0 = A \cdot F \left(\frac{p'}{p_a}\right)^n \quad (2)$$

where, G_0 = initial shear stiffness; p' = mean effective stress; p'_a = atmospheric pressure; A & n = best fit parameters; and $F(e)$ = function of the void ratio, e.

Equations 1 and 2 provide empirical evidence that the sand behaviour is not only a function of the void ratio but also depends on the effective stress. In experiments of monopiles in sand it is therefore important to ensure a stress field as close as possible to the prototype stress field. This ensures soil behaviour similar to the prototype. Stresses and strains are identical which allows the use of well-established linear scaling laws for other quantities.

For a scaled model, the stress similarity between model and prototype can be obtained by an increased stress field. The increase in stress field can be achieved by placing the model in a centrifuge in an increased acceleration field. The increase in acceleration is calculated as shown in Equation 3, Taylor (2003).

$$N = \frac{\omega^2 \cdot R}{g} \quad (3)$$

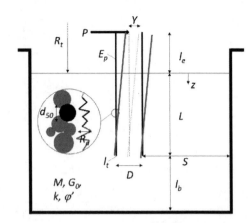

Figure 1. Sketch of centrifuge container with a lateral loaded model monopile installed in sand.

where, N = increase in acceleration; ω = rotational frequency; R = radius and g = Earth's gravity.

Using Equation 3 together with the specific unit weight of the sand (ρ) allows the vertical stress field in the model sand sample to be calculated.

$$\sigma_v = \rho \cdot \omega^2 \cdot z \cdot \left(R_t + \frac{z}{2}\right) \quad (4)$$

where, z = depth in soil sample and R_t = radius to the top of the sand sample.

In a drum centrifuge, Equation 4 is exact, whereas in a beam centrifuge, where the swing is not completely horizontal, the difference is insignificant.

It is, though, not sufficient to control relative density and effective stress in order to achieve full similarity between model and prototype and the paper explores the consequences of this.

Figure 1 illustrates the different parameters controlling the lateral response. Dimensional analysis of the monopile response shows that the pile resistance to overall loading is a function of several parameters that needs to be constant between model and prototype, shown in Equation 5.

$$\frac{P}{\gamma' D^3} = f\left(\frac{Y}{D}, \frac{L}{D}, \frac{l_e}{D}, \frac{I_p}{D^4}, \frac{E_p}{\gamma' D}, \frac{G_0}{\gamma' D}, \frac{Mkt}{\gamma_w D^2}, \varphi', \frac{l_t}{d_{50}}, \frac{D}{d_{50}}, \frac{R_a}{d_{50}}\right) \quad (5)$$

where, P = applied load; Y = pile head displacement; D = pile diameter; γ' = effective unit weight; L = pile length; l_e = load eccentricity; I_p = pile moment of inertia; E_p = pile stiffness; M = modulus stiffness; k = permeability; t = time; γ_w = effective unit weight of water; φ' = effective angle of friction; l_t = thickness of the pile wall; d_{50} = average sand grain size and R_a = roughness factor.

3 LATERAL RESPONSE

As an initial attempt to review the monotonic response, pile head load displacement curves have been collected

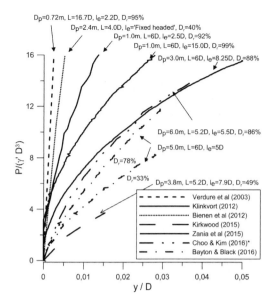

Figure 2. Normalised load – displacement response at sand surface. Deflection given at sand surface. D_p = prototype diameter.

from different publications, and are shown in Figure 2. All results are normalised using the normalisation proposed in Equation 5. There is a clear trend in the data and the response from the model monopiles groups well. For example smaller load eccentricities leads to stiffer response, higher relative density of the sand leads to a stiffer response and piles with approximately same prototype dimensions gives approximately same responses. There is a range in the results and it is reasonably to believe that this is because of the different test conditions, as also highlighted in Figure 2. It is this uncertainty we want to narrow down. This is done by first a review of different modelling effects and afterwards a dedicated coordinated test program exploring the most critical effects.

4 TESTING CONSIDERATIONS

Several factors need to be addressed in an appropriately designed centrifuge experiment. Garnier et al. (2007) provides a catalogue of scaling laws and similitude consideration. It is a review of centrifuge modelling in general. This catalogue gives some insights of laterally loaded piles but does not address monopiles specifically. The following section will present considerations relevant to monopile testing in sand. We will the use the catalogue as a basis and will aim to update with newer findings.

4.1 Sand reconstitution effects

The first point to address is to create a representative sand sample. This means that the sample density needs to be well controlled to achieve the target void ratio, and comparable with real sand conditions. The preparation of the sand samples may seem as a trivial task but is maybe one of the most important tasks in order to obtain reliable results.

4.1.1 Pluviation
The preparation of the sand in the test container can be done in several ways. Garnier (2001) suggests that using the appropriate equipment and techniques, repeatable sand samples can be obtained within $+/-0.1\,\mathrm{kN/m^3}$ of the dry unit weight. Madabhushi et al. (2006) presented an automated spot pouring hopper that is found to be the best-suited way of sand preparation. It is here noted that the effect of a difference of $0.2\,\mathrm{kN/m^3}$ for a dense sand leads to a change in relative density of 76% to 83%.

4.1.2 Aging of sand
Another effect of the sand reconstitution technique that also needs to be acknowledged is the effect of sand aging. The effect of aging is widely documented. Schertmann (1991) provides a review of the aging effect on sand. Aging can be associated with two different processes, chemical bonding and stress rearrangement, both as a function of time. This time effect does not scale in the centrifuge and the same time is therefore needed in the centrifuge in order to create similar effects. A sample with a true aging effect is therefore not practical to create in the centrifuge. The effect of aging will lead to a stiffer and stronger pile response. This is important to remember if centrifuge test results are compared with field tests. It is though not a problem if the centrifuge test is used to compare with numerical analysis where element testing of the un-aged sand are used as calibration of the constitutive models.

4.2 Scale effects

In centrifuge modelling of monopiles, it is often chosen to have an angular rotational frequency of the centrifuge, which ensures stress similarity at a depth 2/3 of the pile length. The increase in acceleration at this point is often called the scaling factor. All length dimensions in the model are therefore N times smaller than in the prototype. The sand grain-size is usually kept constant. The reason for this is that decreasing the grain-sizes of the sand will fundamentally change the soil behaviour and that is not wanted. This introduces a series of scaling effects which will examined further.

Figure 3 shows the influence of the sand grain-size on the soil-pile interface. It is illustrated here how the grain-size of the sand may influence the different parts of the response. For a pile that moves vertically through the soil (under installation or axial loading), the soil loads the pile with resultant stresses on the pile tip (tip pressure) and along the interface of the pile (shear resistance). For a pile that moves horizontally through the sand, similar analogy can be made. The soil loads the pile with resultant stresses in front of the pile

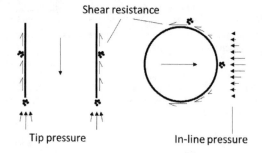

Figure 3. Sketch of stresses action on a pile moving vertically (side view) and horizontally through the soil (top view).

(in-line pressure) and along the interface of the pile (two components of shear resistance, circumferential and longitudinal). These 3 (tip, shear, in-line) responses are all a function of displacement and grain-size.

4.2.1 *Effect on pile tip resistance (l_t/d_{50})*

When the pile wall thickness is scaled down, the ratio between pile wall thickness and grain-sizes reduces such that it may affect the tip response. It is often not possible to have a sufficiently large ratio to avoid scaling issues. It is therefore not possible to obtain full similarity between model and prototype and only general observation and behaviors can be studied in the centrifuge. It is also important to note that this affects the stresses after installation and this may affect the lateral response. Klinkvort (2012) showed no clear difference in the response of a hollow and a solid pile (where l_t/d_{50} should be sufficiently large). This can be explained if the hollow pile plug during installation. On the other hand, the lateral capacity and ultimate lateral resistance of a closed ended hollow pile was found smaller than the same pile when it was coring (Zania et al., 2015). There exist therefore still some uncertainties on how to model this, as the previously mentioned indications should be verified.

4.2.2 *Effects on the in-line pressure (D/d_{50})*

For a full-scale monopile the ratio D/d_{50} is very large and the sand in front of the pile behaves as a continuum. When the pile diameter is scaled down, the ratio between the sand grains and the pile diameter is reduced. Below a threshold value of D/d_{50}, the sand will not act as a representative continuum. This ratio can be found by experimental testing using the modelling of models technique. Nunez et al. (1988) and Remaud (1999) have studied the effect of the D/d_{50} ratio for laterally loaded slender piles. For D/d_{50} ratios larger than 40 or 60, respectively, no grain-size effect was observed. Klinkvort (2012) demonstrated that a ratio of 88 was necessary to avoid grain-size effect in tests on monopiles. Therefore, there are some uncertainties related to this effect. The interface roughness may also play a role here. This requires further research.

4.2.3 *Effects on the shear resistance (R_a/d_{50})*

The resistance contributes to the pile response both during vertical and horizontal loading. The resistance comes through the interface between sand and pile. This interface behaviour is a function of displacement and roughness of the pile. Some research has been done in this field, mainly on plates, e.g. Garnier & Kønig (1998), Foray et al. (1998), Fioravante (2002), Lings & Dietz (2005). In general, the normalised roughness ($R_n = R_a/d_{50}$, Kishida & Uesugi, 1987) should be the same in the model and in the prototype. However, if the interface is perfectly rough or perfectly smooth, the value of normalised roughness has no effect on the mobilisation of shear strength. It was found that if $R_n < 0.01$ the interface is smooth and if $R_n > 1$ then the interface is rough.

Offshore monopiles have rough interfaces; it is though easier to keep a model pile smooth. If the centrifuge tests is used for calibration of numerical models where the effect of pile roughness can be addressed afterwards this is acceptable.

4.3 *Boundary value effects*

There are a series of other effects that need to be considered when designing an appropriate test program. Care needs to be taken when test results are interpreted. Often the effects can be taken into account by applying the appropriate boundary effects when test results are analysed. In the following sub-sections, we will comment on some of the most important ones.

4.3.1 *Container boundary effects (S/D & l_b/D)*

The size of the test container can affect the test results if the container is too small. In CPT testing, it was found that the ratio between CPT diameter (D) and the distance to the container wall (S) should be larger than 10 for sand (Bolton et al. 1999). The effect of S/D is density dependent and can be higher for very dense samples. The distance from the pile tip to the container floor (l_b/D) also needs attention, but further research is needed to quantify this effect.

4.3.2 *Stress level effects (R_t/L)*

The distance from the rotation centre to a point in the sand sample together with the rotational frequency defines the acceleration, see Equation 3. The acceleration increases with depth in the sand sample and the vertical effective stress is therefore non-linear. The non-linear stress distribution is well known and the effect of this is small if the ratio between rotation radius and height of soil model R_t/L is large. Klinkvort (2012) showed an effect of the non-linear stress distribution in a series of modelling of models with R_t/L ratios between 9 and 23. The effect was seen when comparing the normalised load displacement response. The non-linear stress distribution can be calculated using Equation 4 and should be considered when interpreting the results.

4.3.3 Installation effects

Several studies have shown the effect of installation on the lateral response, e.g. Dyson and Randolph (2001) and Klinkvort (2012). The effect of installation therefore also needs to be considered when performing centrifuge test of monopiles. Offshore monopiles are installed by driving and the driving effect of the soil and pore fluid response must be considered if full similarity to the prototype is the goal. There exist still several uncertainties on how to model the driving process correctly.

4.3.4 Effect of load eccentricity (l_e/D)

For an offshore wind turbine, the ratio between the applied bending moment (M) and horizontal force (P), also known as the load eccentricity ($l_e = M/P$), is not constant. The complex dynamic load situation from both wind and waves constantly changes and the load eccentricity is therefore not constant. This change in load eccentricity is extremely difficult to model and often a constant load eccentricity is chosen when testing monopiles. To investigate the effect of load eccentricity, Klinkvort (2012) performed a test series of laterally loaded monopiles with different load eccentricities ($l_e/D = 2.50$–17.25). From this test series, it was seen that increasing the load eccentricity decreased the bearing capacity as expected. However, it also indicated that pile-soil response (often modelled as p-y curves) did not change with load eccentricity. This shows that the general pile-soil interaction is similar no matter the eccentricity. No "correct" load eccentricity exists for monopile testing though it needs to be sufficient to ensure a rotational failure mechanism.

4.3.5 Loading rate ($Mkt/(\gamma_w D^2)$)

Monopiles supporting offshore wind turbines are installed in saturated sand. The effect of flow of water in the soil can therefore have an effect on the response of the monopile. Most testing on monopiles has been performed in dry sand or at load rates sufficiently slow that only the drained response is investigated. With loading periods around 3 seconds and increasing diameters, the effect pore pressure is increasingly important to understand. Both for monotonic and especially for cyclic loading. Because the grain-size distribution normally is not scaled, the permeability in a saturated sample is the same in prototype and model scale. This introduces a modelling problem that needs to be addressed in order to retain similarity. The problem is solved by scaling the load rate and the viscosity of the fluid. The scaling of the fluid is often done by either oil or water mixed with cellulose ethers additive. This is a routine exercise in modelling of earthquake related problems in the centrifuge, (Madabhushi, 2014). There is though some issues that need to be considered when changing the pore fluid. The changes of water with a different fluid may introduce a different soil behaviour. The effect of the new fluid can be examined in triaxial testing prior to centrifuge testing. The sample needs to be fully saturated. This is probably best achieved by flooding the sand sample with CO2 before saturation starts from the bottom of the test container. The last comment is here made to the system that applies the load to the pile. This load control system needs to be of high quality in order to achieve the required loads and loading rates. This is especially important for cyclic loading.

5 DISCUSSION

There are several effects that need to be considered when performing monopile testing in sand in the centrifuge. We have here listed the most important issues that need to be addressed. Some of the effects are fairly well understood and can be addressed if care is taken. However, several issues that can introduce scaling problems need to be resolved, including uncertainties related to the ratio of grain-size to diameter, interface and pile wall thickness. The grain-size also affects the permeability. Together with the loading rate and the pore fluid response, the role of which is poorly understood.

5.1 Proposed test programme

In order to explore these scaling issues in greater details we propose the following test programme. The program will be conducted in five different beam centrifuges as a joint research project between NGI, CEIGR, DTU, COFS, CUED and IFSTTAR. The basic outline of the proposed test program is as follows: First, a common benchmark test, modelling the same prototype is carried out at all five centrifuge centres to demonstrate repeatability of the results:

1. Monotonic benchmark test for all centrifuges. Lateral loading of a monopile, with the following prototype properties: $D_r = 80\%$ sand, $D = 2$m, $L = 5D$, $l_e = 4D$.

Each centrifuge centre will afterwards explore one of the following points:

1. Modelling of models tests, with different D/d_{50} and R_t/L ratios.
2. Tests addressing the shear (interface) effect, with different R_a/d_{50} ratios.
3. Tests addressing the effect of plugging/coring, at different stress levels and with different t/d_{50} ratios.
4. Tests addressing the effect of installation method, with constant t/d_{50} ratio.
5. Tests addressing the effect of loading rate with different $Mkt/(\gamma_w D^2)$ ratios.

The idea of the test program is that all centrifuges first model the same prototype monopile test. This allows for a benchmark of the test results. Different sands and model monopiles will be used, relying on existing apparatus, which places constraints on achievable D, L and l_e. The effect of these parameters can be investigated economically through numerical analysis, once confidence in the model has been established through validation against physical data. With two types of sand being used at two testing facilities each, this allows both repeatability within the same sand and among the different silica sands to be assessed. Afterwards, each centrifuge facility will investigate one of the effects listed above by performing 4–5 tests. Before the centrifuge test-program commerce, all test-sands will be tested at NGI for a comparison on element level.

6 CONCLUSION

A review of modelling effects in centrifuge modelling of monopiles in sand has been given. The review takes its starting point from the scaling catalogue of scaling laws (Garnier et al., 2007). The relevant scaling considerations given in the catalogue relevant for monopiles derived from tests on slender piles. We have included new monopile research in our review and identified uncertainties relevant for monopiles centrifuge testing are present. Based on this review, we have proposed a joint research program to address the most important scaling issues for monopiles. The program will be based on monotonic tests but the findings should be equally important for cyclic testing.

REFERENCES

Bayton, S. & Black, J. 2016. The effect of soil density on offshore wind turbine monopile foundation performance. *Proc. In the 3rd Eurofuge '16*: 245–251.

Bolton, M.D. 1986. The strength and dilatancy of sands. *Géotechnique* 36(1): 65–79.

Bolton, M.D., Gui, M. W., Garnier, J., Corte, J. F., Bagge, G., Laue, J. & Renzi, R. 1999. Centrifuge cone penetration test in sand. Geotechnique, 49(4): 543–552.

Choo, Y.W & Kim, D. 2016. Experimental development of the p-y relationship for large diameter offshore monopiles in sands: Centrifuge tests. *Journal of Geotechnical and Geoenvironmental Engineering* 142(1).

Cuéllar, P., Bassler, M & Rücker, W. 2009. Racheting convective cells of sand grains around offshore piles under cyclic lateral loads. *Granular Matter*, 11(6):379–390.

Dyson, G.J. & Randolph, M.F. 2001. Monotonic lateral loading of piles in calcareous sand. *Journal of Geotechnical and Geoenvironmental Engineering* 127(4): 346–352.

Fioravante, V. 2002. On the shaft friction modelling of non-displacement piles in sand. Soils and Foundations, *Japanese Geotechnical Society*. 42(2): 23–33.

Foray, P., Balachowski, L. & Rault, G. 1998. Scale effect in shaft friction due to the localisation of deformations. *Proc. International Conference Centrifuge 98*, 1: 211–216.

Garnier, J. 2001. Physical models in geotechnics: state of the art and recent advances, 1st Coulomb lecture, Paris

Garnier, J., Gaudin, C., Springman, S.M., Culligan, P.J., Goodings, D., Konig, D., Kutter, B. Phillips, R., Randolph, M.F. & Thorel, L. 2007. Catalogue of scaling laws and similitude questions in geotechnical centrifuge modelling. ICPMG 3: 01–23.

Garnier, J. & Konig, D. 1998. Scale effects in piles and nail loading tests in sand. Proc. *International Conference Centrifuge 98*, Tokyo, T. Kimura, O. Kusakabe, J. Takemura (eds), Balkema, Rotterdam, 1: 205–210.

Kirkwood, P. 2015. Cyclic lateral loading of monopile foundations in sand. PhD Thesis, University of Cambridge.

Kishida, H. & Uesugi, M., 1987 Tests of the interface between sand and steel in the simple shear apparatus. *Géotechnique* 37(1): 45–52.

Klinkvort, R.T. 2012. Centrifuge modelling of drained lateral pile – soil response. Application for offshore wind turbine support structures. PhD Thesis, DTU Report R-271.

LeBlanc, C., Houlsby, G.T. & Byrne, B.W. 2010. Response of stiff piles in sand to long-term cyclic lateral loading. *Géotechnique*, 60(2): 79–90.

Li, Z., Haigh, S. K. & Bolton, M. D. 2010. Centrifuge modelling of mono-pile under cyclic lateral loads. *ICPMG 2010*, 1: 965–970.

Lings, M. L. & Dietz, M. S. 2005. The peak strength of sand-steel interfaces and the role of dilation. Soils and Foundations, Vol. 45(6): 1–14.

Madabhushi, S. P. G., Houghton, N. E., & Haigh, S. K. 2006. A new automatic sand pourer for model preparation at University of Cambridge, *Proc. 6th International Conference on Physical Modelling in Geotechnics*, 1: 217–222.

Madabhushi, S. P. G. 2014. Centrifuge Modelling for Civil Engineers. CRC Press ISBN 9780415668248

Nunez, I.L., Philips, R., Randolph, M.F. & Wesselink BD 1988. Modelling laterally loaded piles in calcareous sand. *In Proc. of Int. Conf. Centrifuge*, 1: 353–362.

Peralta, P. & Achmus, M. 2010. An Experimental Investigation of Piles in Sand Subjected to Lateral Cyclic Loads. *ICPMG 2010*.

Remaud, D. 1999. Pieux sous Charges Laterales: Etude Experimentale de l'Effet de Groupe. PhD thesis, LCPC, France.

Rosquoët, F., Thorel, L., Garnier, J. & Canepa, Y. 2007. Lateral cyclic loading of sand-installed piles, *Soils and Foundations*, 47(5): 821–832.

Rudolph, C., Bienen, B. & Grabe, J. 2014. Effect of variation of the loading direction on the displacement accumulation of large-diameter piles under cyclic lateral loading in sand. *Canadian Geotechnical Journal* 51:1196–1206

Truong, P. & Lehane, B.M. 2015. Experimental trends from lateral cyclic tests of piles in sand. ISFOG 1:747–752.

Schmertmann, J.H. 1991. The mechanical aging of soils, *Journal of Geotechnical Engineering*, 117(9): 1288–1330.

Verdure, L., Garnier, J. & Levacher, D. 2003. Lateral cyclic loading of single piles in sand, *International Journal of Physical Modelling in Geotechnics*, 3 (3): 17–28.

Yang, J. & Liu, X. 2016. Shear wave velocity and stiffness of sand: the role of non-plastic fines. *Géotechnique* 66(6): 500–514.

Yang, S.X., Jardine, R.J., Zhu, B.T., Foray, P. & Tsuha, C.H.C. 2010. Sand grain crushing and interface shearing during displacement pile installation in sand. *Géotechnique* 60(6): 469–482.

Zania, V., Hededal, O. & Klinkvort, R.T. 2015. Effect of relative pile's stiffness on the lateral pile response under loading of large eccentricity. *ISFOG 1*:753–758.

Experimental modelling of the effects of scour on offshore wind turbine monopile foundations

R.O. Mayall, R.A. McAdam, B.W. Byrne, H.J. Burd & B.B. Sheil
Department of Engineering Science, University of Oxford, Oxford, UK

P. Cassie
E.ON Climate & Renewables, Coventry, UK

R.J.S. Whitehouse
HR Wallingford, Wallingford, UK

ABSTRACT: Local and global scour around offshore wind turbine monopile foundations can lead to a reduction in system stiffness, and a consequential drop in the natural frequency of the combined monopile-tower-nacelle structure. If unchecked this could lead to operational problems such as accelerated fatigue damage and de-rating or decommissioning of the turbine. Research exploring the interaction between scour, foundation stiffness, and structural dynamic behaviour is therefore critical if scour formation is to be properly accounted for in predictions of structural performance, and to guide the implementation of scour remediation strategies. This paper describes experimental work that explores these interactions, conducted on a 1:20 scale driven monopile foundation and tower-nacelle superstructure in a prepared sand test-bed at HR Wallingford's Fast Flow Facility. The flume allowed realistic scour geometries to be developed, providing a means to explore the effectiveness of different remediation strategies. Measured acceleration and strain caused by harmonic lateral loading are interpreted to deduce changes in structural performance as scour develops.

1 INTRODUCTION

1.1 Background

Scour of seabed sediments can occur around offshore structures due to the interaction between the structure and the water flow around it. The resulting reduction in strength and stiffness of the structure's foundation can present operational and structural problems. For monopile wind turbine structures this can be of particular concern due to the consequential reduction in natural frequency, and potential for increased fatigue damage.

Ongoing research at the University of Oxford involves development of a numerical modelling framework for analysis of the effects of scour and scour protection on monopile offshore wind turbines. The numerical framework involves structural dynamic analysis using a 1-dimensional (1D) finite element model of the monopile-tower system, calibrated using three-dimensional (3D) finite element analyses and scale model testing, and further validated using experimental and field data.

This paper describes a programme of experiments conducted using the Fast Flow Facility (FFF) at HR Wallingford. The experiments were designed to provide validation cases for the numerical framework and to inform predictions of the effects of scour and scour protection on the structural dynamic response of field scale monopile offshore wind turbines.

1.2 Scour around offshore piles

Scour around offshore foundations can take the form of a combination of global scour (S_G) and local scour (S_L), as illustrated in Figure 1.

Global scour, also termed general scour, is a global lowering of the seabed, caused by overall seabed movement, migrating sand waves and sandbanks, or migrating channel features.

Local scour is the formation of a scour pit around a foundation due to the effects of horseshoe vortices that develop around the pile in a flow (Sumer & Fredsøe, 2002). The scour pit will typically reach an equilibrium depth for a particular set of current and wave conditions and no global scour.

Empirical calculation methods for predicting the local scour depth have been developed by several authors (e.g. Breusers et al., 1977; Høgedal & Hald, 2005), and are typically normalized by pile diameter, D. Commonly cited is the experimental data of Breusers et al. (1977), with a mean equilibrium local scour depth of $S_L/D = 1.5$ (or 2 to be on the safe side

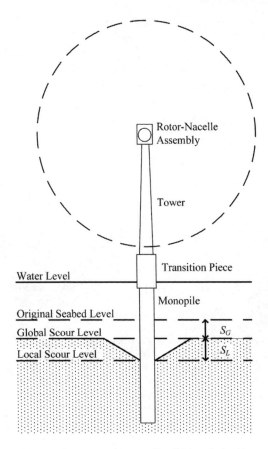

Figure 1. Scour around a monopile offshore wind turbine.

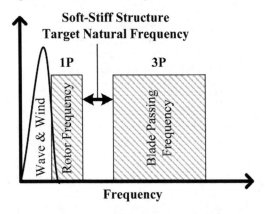

Figure 2. Offshore wind turbine loading frequencies.

for design) or $S_L/D = 1.3$ with a standard deviation of 0.7 (Sumer & Fredsøe, 2002). Design standards from the petroleum industry (API, 2011; ISO, 2007) recommend a design $S_L/D = 1.5$ whilst DNVGL (2016) recommends $S_L/D = 1.3$.

1.3 Natural frequencies of offshore wind turbines

Wind turbine structures are designed such that their fundamental natural frequency avoids the loading frequencies of the rotor (1P), blade passing (3P) and wind and wave frequencies, to minimise the dynamic amplification of loads (Kallehave et al., 2015). Monopile wind turbines are soft-stiff structures with the design natural frequency placed between the 1P and 3P loading bands (Fig. 2), and typically above the wind and wave frequencies.

With the development of scour the embedded pile length, L, is reduced, and the foundation stiffness and natural frequency drifts downwards. A turbine may fall out of warranty or have to be decommissioned if the natural frequency of the structure becomes too close to the 1P frequency; this leads to a loss of revenue unless the natural frequency can be restored by remediation.

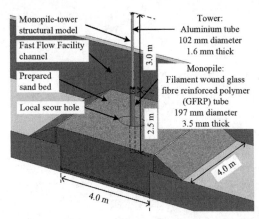

Figure 3. Fast Flow Facility experiment layout.

2 EXPERIMENTAL SETUP

2.1 Scope of experiments

An experimental programme was conducted to investigate the influence of scour and scour protection on the natural frequencies of an instrumented model monopile-tower system. The tests were conducted at 1:20 scale in the Fast Flow Facility (FFF) at HR Wallingford. The experiment elements include a prepared sand test bed, the monopile-tower model, and development of scour using the flume (Fig. 3).

Six separate tests were conducted, each of which included the following sequence of activities:

1. Soil bed preparation and saturation
2. Filling the flume to the required water level
3. Pile driving to target initial embedment L/D
4. Initial structural dynamics testing
5. Developing local scour
6. Installing scour protection (if any)
7. Developing global scour
8. Draining the flume
9. Excavating pile and disturbed areas of soil

Table 1 presents the variations in the initial embedment, target local scour before installing scour protection, scour protection type, and target global scour

Table 1. Summary of target scour conditions for the Fast Flow Facility experiments.

Test Number	Initial Pile Embedment L/D	Local Scour S_L/D	Scour Protection	Global Scour S_G/D
1	4.5	1.5	None	1.5
2	4.5	1.5	Tyre-filled nets	1.5
3	4.5	1.5	Rock armour	1.5
4	4.5	0.75	Rock armour	1.5
5	4.5	0.0	Rock armour	1.5
6	5.5	1.5	Rock armour	2.5

Figure 4. Pit section of the Fast Flow Facility prior to installation of the experiments.

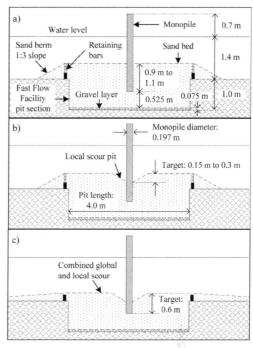

Figure 5. Schematic of scouring process, tower not shown. a) Condition at end of bed preparation and pile driving. b) Condition after development of local scour. c) Condition after removing retaining bars and flow segments for global scour.

depth for the tests. The details of the scour protection models falls outside the scope of this paper.

2.2 Fast Flow Facility (FFF)

The FFF is a research flume capable of producing flows up to 4.9 m³/s and generating wave heights of up to 1 m. The FFF channel has a cross section 4.0 m wide × 2.5 m deep and length of 57 m between sediment traps. The experiments described in this paper were prepared in a deeper 4.0 m long × 3.5 m deep pit section of the FFF (Fig. 4).

Scour development in the FFF tests comprised local and global scouring stages (Fig. 5). Scour was developed using currents, including cycling of direction; wave generation was not used.

The scour development was scaled using conventional approaches to produce a live bed scour regime. To ensure that all sediment was mobile the ratio of current speed to threshold current speed for sand motion was selected to be greater than 1.0 (Whitehouse, 1998). This regime generates realistic local scour depths scaled on the pile diameter, in line with empirical predictions.

The extent of scour is controlled by the sand properties and by vortex shedding downstream of the pile. Cycling the flow direction simulates a tidal situation and enables the scour extent to grow in both directions.

Typically 8.25 hours of flow was used to develop local scour.

Global scour was developed by incrementally removing temporary retaining bars, and running further flow at higher velocity. After each 0.1 m deep bar was removed the flow was again run for 8.25 hours before the next bar was removed.

The same water level (1.8 m above pile toe) was used throughout the testing, as a result the water depth varied as global scour developed. The soil bed bathymetry was measured using an underwater laser scanner between flow cycles, and with a terrestrial laser scanner at test completion.

2.3 Soil bed preparation

Scour phenomena are particularly prevalent at wind farm sites where superficial sand layers exist. A fine-grained silica sand with a mean particle size of 0.16 mm was therefore used for the experiments.

After initial filling of the pit with sand, the central area of the pit was re-prepared for each test. The preparation consisted of excavating a hole of at least 1.5 m width to a depth of 50 mm to 100 mm below the target pile toe depth. Sand was then added in 50 mm thick compacted layers (using a 240 V plate compactor, Fig. 6) up to the target bed level.

Following compaction of the final soil layer the soil bed was saturated, prior to filling the FFF to the required water level. To aid the saturation process a

Figure 6. Compaction of sand layers using plate compactor.

Figure 7. Perforated pipes installed in the pit section of the Fast Flow Facility.

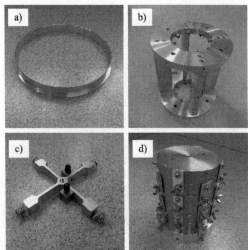

Figure 8. Structure model components. a) Pile shoe. b) Transition piece. c) Mounted accelerometer pair. d) Variable top mass.

set of perforated pipes (Fig. 7) were installed in the base of the pit within a 75 mm thick gravel bed, and overlain by a geotextile membrane. This allowed saturation of the sand using an upwards hydraulic gradient by connection to a header tank.

2.4 Monopile-tower structural model

The model monopile-tower structure consisted of a pile, transition piece, tower, top masses, and top stiffener. The model structure dimensions and materials (Fig. 3) were specified to achieve geometry and stiffness that scale appropriately to a full-sized structure, within practical limitations. The relative lengths and diameters of the pile and tower are similar to field scale structures.

The monopile geometry and material was selected such that the non-dimensional group $E_p I_p / E_s L^4$ (Poulos & Davis, 1980) was similar to field scale, where $E_p I_p$ is the flexural stiffness of the pile and E_s is an estimate of the soil stiffness. The tower geometry and material was selected using the non-dimensional group $E_t I_t / E_p I_p$ (Arany et al., 2016) where $E_t I_t$ is the flexural stiffness of the tower. For practical reasons the moment of inertia of the rotor-nacelle was

not replicated; this is anticipated to have a negligible influence and does not restrict the use of the results in validating the numerical framework.

An annular aluminium driving shoe was bonded to the pile toe to reduce risk of damage during installation (Fig. 8a). The pile and tower were attached by bolted connections using an aluminium transition piece (Fig. 8b). The model also included accelerometers mounted within the tower (Fig. 8c), and a variable top mass to mimic a rotor-nacelle assembly (Fig. 8d).

2.5 Pile driving

The monopile-tower model was installed by pile driving. Fig. 9 presents the set-up of the pile driving equipment used in the experiments. Driving took place with the structure fully assembled, with the tower attached to the pile. A pair of pulleys were used to manually raise the driving hammer to a set elevation, with the tower acting as a guide for the hammer. A polyethylene driving cushion was used, underlain by a rubber layer on top of the pile annulus to reduce the risk of damaging the pile. The hammer had a mass of 7.75 kg and the maximum stroke used in any test was 1 m.

A set of guides with inwards-facing rubber wheels were used to ensure pile verticality during driving. These guides were fixed to a portal-layout scaffold frame constructed over the FFF.

2.6 Structural dynamics testing

2.6.1 Structural dynamics testing overview

Structural dynamics measurements were repeated at each scour depth, to track changes of response with scour condition. Tests were conducted with different amounts of mass attached to the top of the tower

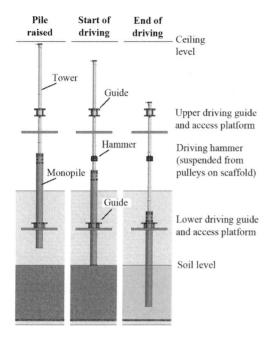

Figure 9. Illustration of pile driving process.

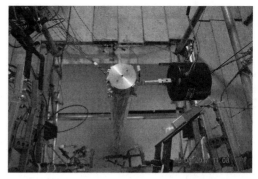

Figure 10. Modal shaker attached to the top of the tower.

Figure 11. Local scour pit formed around the model monopile. From Test 6 in the Fast Flow Facility.

(Fig. 8d). Increasing the top mass reduces the natural frequency and also changes the shape of the higher order bending modes. Changing the top mass therefore provides valuable additional information on the foundation response for validating and/or calibrating the numerical models.

The structural natural frequencies and mode shapes were measured using accelerometer pairs mounted within the tower structure (Fig. 8c) at 2.33 m, 2.83 m, 3.18 m, 3.53 m, 4.13 m and 5.38 m above pile tip (APT). This instrumentation was mounted before pile driving; accelerometers were therefore required with a high shock rating to withstand pile driving. The accelerometer axes were aligned to measure the response in the streamwise and spanwise directions.

Additional instrumentation included: bending gauges on the pile in the streamwise direction at 1.61 m APT, a temposonics displacement sensor mounted to the pile in the streamwise direction at 1.95 m APT, and force sensors attached to the top of the tower in the streamwise and spanwise directions. All instruments were sampled and logged at 3 kHz.

Measurements of natural frequencies were made using impulse loading, and also by measuring the response with a modal shaker attached, as described below.

2.6.2 *Impulse loading*

The impulse loading tests were used to measure the natural frequencies for the first three bending modes. The impulses were applied as a light impact to the force sensor attached at the top of the tower; the magnitude of the impulse could therefore be recorded. The natural frequencies were identified from the power spectral density (PSD) of the measured accelerations.

2.6.3 *Modal shaker tests*

Measurements of natural frequencies were also made using a modal shaker (Fig. 10). The modal shaker was suspended via lightweight cables, and attached to the required force sensor in the streamwise or spanwise direction. Two types of test were performed using the modal shaker:

1. Frequency sweeps, generating sinusoidal signals at a series of discrete frequencies around a target frequency.
2. White noise tests, in which white noise vibrations were sent to the modal shaker.

In the frequency sweeps the magnitude of accelerations and force provide information on the structural dynamics. The natural frequencies are determined as the frequencies at which the phase difference between the measured input force signal and acceleration signal is 90°.

3 EXAMPLE RESULTS

An example of the scour pit developing during the FFF experiments is shown in Figure 11, from Test 6 (see Table 1). The image in Figure 11 was taken part-way through the full duration of flow for local scour, with an approximate depth of $S_L/D = 0.75$. This test continued until the target $S_L/D = 1.5$ was achieved.

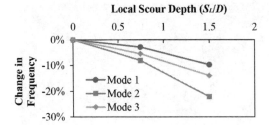

Figure 12. Measured change in natural frequency of model wind turbine with local scour (Test 1, no top mass).

In Test 1 (see Table 1) structural dynamics testing was performed at approximate local scour depths $S_L/D = 0.75$ and 1.5. With no top mass, the natural frequencies (interpreted from peaks of the PSD) were 6.5 Hz, 18.4 Hz and 47.6 Hz for the first three modes. Figure 12 presents the changes in the natural frequency of the first three modes for Test 1 as the local scour develops. These example results show natural frequencies reducing as local scour develops, consistent with the loss of foundation stiffness caused by the scouring process. All three frequencies show increasing sensitivity to scour as the scour depth increases. An interesting observation is the mode 2 and mode 3 frequencies are more strongly affected by local scour than mode 1.

4 CONCLUSIONS

This paper has described experiments to measure the response of a monopile wind turbine structure to scour developed in a flume. The experiments were successful in providing a measurable change in natural frequency as the scour developed.

The observation that mode 2 and 3 natural frequencies are more strongly affected by scour than mode 1 may provide useful insight for structural health monitoring of offshore wind turbine structures to inform detection of scour without bathymetry surveys.

Further results to be extracted from the experiments include: measured damping, mode shapes, effects of global scour on the structural dynamics, and a comparison of the natural frequency measurement methods employed in the tests. The trends in the natural frequency measurements will be used to validate numerical models of scour on monopile wind turbine structures.

ACKNOWLEDGEMENTS

This research project is supported through funding from E.ON and HR Wallingford, and by grant EP/L016303/1 for Cranfield University and the University of Oxford, Centre for Doctoral Training in Renewable Energy Marine Structures – REMS (http://www.rems-cdt.ac.uk/) from the UK Engineering and Physical Sciences Research Council (EPSRC).

The authors would like to thank the individuals that assisted during the experimental campaign; from E.ON: Ben Holland, Patrick Rainey, and Steve Heald; from HR Wallingford: David Todd, Esther Gomes, Joe Mitchell, Amelia Astley, James Sutherland, John Harris, Andrew Huckstep, plus the electronics lab, build team, and workshop staff; and from the University of Oxford: Iona Richards, Toby Balaam, and Teodor Totev.

REFERENCES

API 2011. RP 2GEO. *Recommended Practice for Geotechnical Foundation Design Consideration.*

Arany, L., Bhattacharya, S., Macdonald, J.H. & Hogan, S.J. 2016. Closed form solution of Eigen frequency of monopile supported offshore wind turbines in deeper waters incorporating stiffness of substructure and SSI. *Soil Dynamics and Earthquake Engineering*, 83, pp. 18–32.

Breusers, H.N.C., Nicollet, G. & Shen, H.W. 1977. Local scour around cylindrical piers. *Journal of Hydraulic Research*, 15(3), pp. 211–252.

DNVGL 2016. *Standard DNVGL-ST-0126, Support structures for wind turbines*, Edition April 2016.

Høgedal, M. & Hald, T. 2005. Scour assessment and design for scour for monopile foundations for offshore wind turbines. In: *Proceedings of the Copenhagen Offshore Wind*. Copenhagen, Denmark.

International Organization for Standardization (ISO) 2007. *ISO 19902 Petroleum and natural gas industries – Fixed steel offshore structures*. Geneva: ISO.

Kallehave, D., Byrne, B.W., LeBlanc Thilsted, C. & Mikkelsen, K.K. 2015. Optimization of monopiles for offshore wind turbines. Phil.Trans.R.Soc.A 373: 20140100. http://dx.doi.org/10.1098/rsta.2014.0100

Poulos, H. G. & Davis, E. H., 1980. *Pile foundation analysis and design*, John Wiley & Sons, New York.

Sumer, B.M. & Fredsøe, J. 2002. *The mechanics of scour in the marine environment*. World Scientific, Singapore.

Whitehouse, R., 1998. *Scour at Marine Structures*. Thomas Telford, London.

Centrifuge tests on the response of piles under cyclic lateral 1-way and 2-way loading

C. Niemann & O. Reul
Department of Geotechnical Engineering, University of Kassel, Germany

Y. Tian, C.D. O'Loughlin & M.J. Cassidy
Centre for Offshore Foundation Systems, University of Western Australia, Perth, Australia

ABSTRACT: As part of a research project to develop a framework for quantifying cyclic effects on piles, centrifuge model tests on single piles and pile groups in dry silica sand have been carried out at the Centre for Offshore Foundation Systems (COFS) in Perth, Australia. The piles were subjected to lateral 1-way and 2-way cyclic loading, investigating the vertical and horizontal load-displacement behaviour. Strain gauges located along the pile shaft provided information about the bending behaviour. This paper presents the tests on single piles and highlights the influence of load type, cyclic load amplitude and number of cycles on the observed response. As a characteristic value for the general bearing behaviour, the cyclic performance of the modulus of subgrade reaction is calculated additionally.

1 INTRODUCTION

Understanding the response of pile foundations to cyclic loading is of importance in civil engineering design including offshore and onshore wind turbines (BSH 2007) and structures for transport infrastructure (Berger et al. 2003). Cyclic loading may be caused by wind and waves (e.g. offshore wind turbines) or temperature induced constraints (e.g. integral bridges without joints and bearings).

The behaviour of both fine and coarse grained soil, under cyclic loading is complex (e.g. Wichtmann 2016). For example, lateral cyclic loading on piles in sand has been shown to lead to both densification and loosening of the surrounding soil depending on the relative density (e.g. Reese & van Impe 2011). However, the design of pile foundations under cyclic loading is typically based on modifications to the more simplistic monotonic loading conditions. A better understanding of pile-soil interaction during lateral cyclic loading is therefore required to establish more efficient design strategies for these types of foundations.

As part of a research project to develop a framework for quantifying cyclic effects on piles and pile groups, centrifuge model tests on single piles and freestanding pile groups in dry silica sand have been carried out at the Centre for Offshore Foundation Systems (COFS) in Perth, Australia. The aims of the single pile tests were to identify specific factors influencing the lateral capacity of single piles, establish phenomenological correlations and provide a reference for the tests on pile groups.

While the results achieved on pile groups will be the subject of future publications, parts of the results for single piles subjected to 1-way loading have been published in Niemann et al. (2017).

2 CENTRIFUGE TESTS

2.1 Test design

The centrifuge tests described below were carried out at an acceleration of 100g using the 40 g-tonne, 1.8 m radius beam centrifuge located at The University of Western Australia (Randolph et al. 1991).

The test program comprised monotonic and cyclic lateral loading on single piles and cyclic lateral loading on pile groups. In Figure 1, the load characteristics according to the German technical guideline for piles and pile groups EA-P (DGGT 2012) are defined. Cyclic loading can be characterised as either 1-way or 2-way, where 1-way cyclic loading is uni-directional with no load reversal and 2-way cyclic loading is bi-directional with load reversals in each cycle.

2.2 Experimental setup

The pile was connected to the vertical axis of a 2D electrical actuator via a loading leg and axial load cell as shown in Figure 2. The actuator allows motion along both the vertical and horizontal axes; the horizontal axis was used in displacement control in the monotonic tests and in load control in the cyclic tests, whereas the

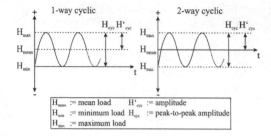

Figure 1. Cyclic loading characteristics.

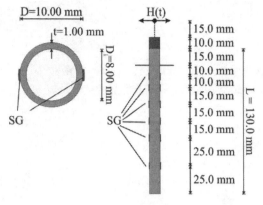

Figure 3. Pile geometry and instrumentation (not to scale).

Table 1. Properties of the model sand (Bienen et al. 2012).

Parameter		Units	
Mean grain size	d_{50}	mm	0.19
Maximum void ratio	e_{max}	–	0.79
Minimum void ratio	e_{min}	–	0.49
Critical state friction angle	φ_c	(°)	30°
Grain density	ρ_s	g/cm³	2.65
Minimum unit weight	γ_{min}	kN/m³	14.9
Maximum unit weight	γ_{max}	kN/m³	18.0
Coefficient of uniformity	U	–	1.9

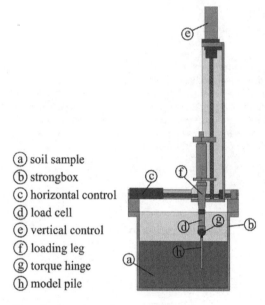

Figure 2. Experimental setup (not to scale).

vertical axis was used in load control to maintain a constant vertical load on the pile. Horizontal loading was applied via a hinge located on top of the pile, 40 mm above the sample surface. The hinge ensured that the top of the pile was free to rotate in the direction of loading, such that the moment at this load application point was zero.

Horizontal displacements were measured using the optical encoder on the horizontal axis of the actuator, and the vertical load was measured using an 'S-type' axial load cell. The horizontal load was calculated from bending measurements from strain gauges located on the loading leg (Figure 3), assuming linear moment distribution between the strain gauges. From the strain measurements taken during the tests, the deflection curves and subsequently bending moments, shear forces and lateral pressures have been derived assuming beam theory (e.g. Reese & van Impe 2011).

The centrifuge tests used circular model piles that were fabricated from stainless steel to have a diameter, D = 10 mm, a length, L = 130 mm and a wall thickness, t = 1 mm (see Figure 3). A cap at the base of the pile prevented soil entering the interior of the pile during installation. The surface of the piles were sand-blasted, such that the pile interface was, comparable to bored piles in-situ. The embedment length of the piles was $L_p = 115$ mm.

2.3 Test soil and sample preparation

The experiments were conducted in a fine to medium sub-angular silica sand, with properties as summarised in Table 1.

All samples were prepared using the air pluviation technique to achieve repeatable medium dense sand samples. The surface of the samples was vacuum levelled in order to achieve a level surface and to facilitate estimation of the sample density from measurements of the sample mass and volume. These measurements resulted in a relative density, $I_D = 45\%$.

Cone penetration tests (CPTs) were conducted to characterise each sample using a 7 mm diameter model cone penetrometer. Typical profiles of penetration resistance q_c with depth are shown in Figure 4 in prototype scale. The two profiles correspond to different locations within the sample and are in good agreement, particularly over the upper 8 m.

Schneider & Lehane (2006) derived a correlation for (centrifuge scale) cone penetration resistance q_c and relative density I_D (Eq. 1).

$$I_D = \sqrt{\frac{q_c}{250 \cdot \sigma'_{v0}}} \qquad (1)$$

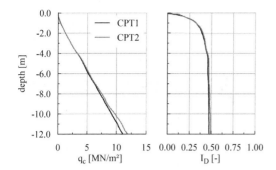

Figure 4. Profile of cone penetration resistance qc and relative density ID (prototype scale).

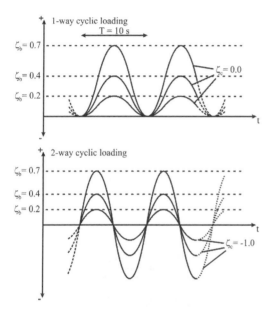

Figure 5. Load application and definition.

Table 2. Test program for cyclic loading of single piles.

Parameter		
No. of cycles	N	500
Magnitude of cyclic loading	ζ_b	0.2; 0.4; 0.7*
Load characteristics	ζ_c	0.0; −1.0

*The test with $\zeta_c = -1.0$ and $\zeta_b = 0.7$ was executed, but failed within the first 5 cycles due to a high rate of upward pile displacement

A total of six cyclic single pile tests were conducted, with varying amplitudes for 1-way and 2-way cyclic loading (Figure 5, Table 2). The cyclic tests were carried out under load control.

To categorise cyclic loads relative to the ultimate lateral capacity of a single pile H_{ult}, LeBlanc et al. (2010) introduce the cyclic loading parameters ζ_b and ζ_c for single piles (Eq. 2).

$$\zeta_b = \frac{H_{max}}{H_{ult}} \text{ and } \zeta_c = \frac{H_{min}}{H_{max}} \quad (2)$$

where H_{min} and H_{max} are the minimum and maximum load during one cycle (see Figure 1).

The piles were jacked into the soil at 1g rather than in flight, as the in-flight installation resistance would have exceeded the 7 kN vertical capacity of the actuator. According to Li et al. (2010) the behaviour of 'pre-jacked' model piles is comparable to that of filed scale bored piles. After pile installation, the centrifuge was accelerated to 100 g and the pile was loaded.

A loading frequency of 0.1 Hz was selected to ensure that the load control was not compromised by limitations on the feedback control loop or the maximum actuator speed. The data were acquired at a sampling frequency of 10 Hz. After completion of a test, the centrifuge was stopped and the pile was relocated within the sample.

where q_c = cone penetration resistance and σ'_{v0} = initial effective vertical stress. Figure 4 shows the profile of I_D derived using Eq. 1.

After approximately six cone diameters (4 m) Eq. 1 results in $I_D = 0.47$ to 0.49, which is in reasonable agreement with $I_D = 0.45$ as established from laboratory measurements of the sample mass and volume before spinning in the centrifuge.

2.4 Test program for single piles

Prior to the cyclic tests, a monotonic load test was conducted to determine the ultimate capacity of a single pile. The monotonic test was conducted by loading the pile head in displacement control at a velocity of 0.2 mm/s. The test was carried out to derive the ultimate lateral load H_{ult} at the pile head, from which the load amplitude in the cyclic tests was scaled.

3 TEST RESULTS

Results from the displacement controlled monotonic load test are provided in Figure 6, which shows the load-displacement response at the pile head (2.5 m above the sample surface). In this paper the ultimate lateral load was defined as $H_{ult} = 2$ MN, which is within the range defined by several analytical approaches for the given soil properties and pile geometry (Figure 6). These results – and all results presented throughout the remainder of the paper – are in prototype scale unless specified otherwise.

The remainder of the paper focuses on the tests featuring 1-way and 2-way cyclic loading. Example load displacement responses during cycling loading are shown in Figure 7. The load displacement responses exhibit the hysteretic behaviour that is characteristic of cyclic loading, and also confirm that the load control during the tests was of high quality.

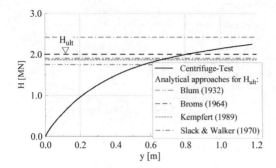

Figure 6. Monotonic loading: Pile head load vs. displacement.

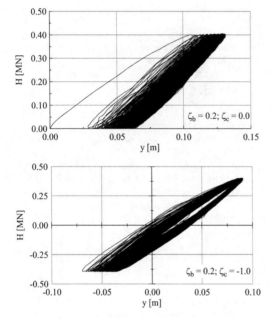

Figure 7. Typical cyclic load-displacement response for 1-way and 2-way loading.

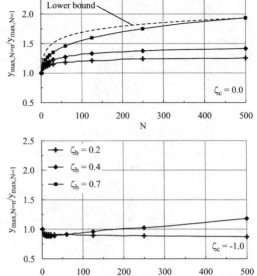

Figure 8. Evolution of horizontal displacement ratio with cycle number.

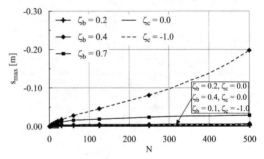

Figure 9. Evolution of vertical displacement with cycle number.

The results are compared by expressing the measured quantities using the following normalised ratios:

$$\frac{y_{max,N=n}}{y_{max,N=1}}; \frac{M_{max,N=n}}{M_{max,N=1}}; \frac{\overline{k}_{s,N=n}}{\overline{k}_{s,N=1}} \qquad (3)$$

where $y_{max,N=1}$ and $y_{max,N=n}$ are the maximum lateral displacement at the 1st and nth cycle, $M_{max,N=1}$ and $M_{max,N=n}$ are the maximum bending moment during the 1st and nth cycle, and $k_{s,N=1}$ and $k_{s,N=n}$ are the mean modulus of subgrade reaction at the peak of the 1st and nth cycle, respectively.

The development of pile head displacement with number of cycles N during 1-way and 2-way loading is shown in Figure 8, which highlights the influence of the load characteristics on the pile head displacement. For 1-way loading ($\zeta_c = 0.0$) and at small load amplitudes ($\zeta_b = 0.2$) the displacement ratio does not increase significantly after approximately 200 cycles, indicating shakedown (e.g. Poulos 1982). However, the tests with higher load amplitudes reveal a different behaviour, with higher displacement-ratios and increasing cyclic soil degradation (evident by pile head deformation accumulating with cycle number, see Poulos 1982).

In contrast to the results derived from 1-way loading, the maximum pile head displacement under 2-way loading ($\zeta_c = -1.0$) reduces within the first 150 cycles for the two investigated cases. After 150 cycles for $\zeta_b = 0.4$ the ratio $y_{max,N=n}/y_{max,N=1}$ increases almost linearly, which can be attributed to the increasing uplift of the pile and the resulting reduction of the embedment length, where Figure 9 shows the vertical displacement of the pile. It can be assumed that this effect depends on the magnitude of the vertical load. For higher vertical loads this effect will probably only occur in combination with larger peak-to-peak amplitudes. Generally, the qualitative behaviour for the two

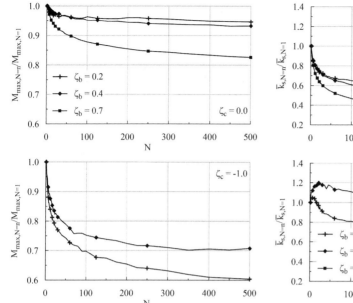
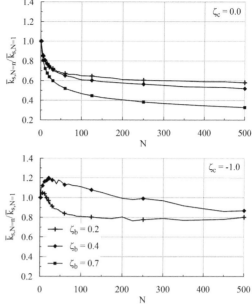

*Measurements taken in a depth of 6.5 m below ground surface.

Figure 10. Evolution of bending moment ratio with cycle number.

Figure 11. Evolution of ratio of modulus of subgrade reaction with cycle number.

investigated load cases is in a good agreement with the results from small-scale model tests derived by Hettler (1981).

The distribution of bending moment along the pile length was derived from the measured strains at the locations indicated by Figure 3. Maximum bending for all tests occurred at the strain gauge located 6.50 m below the sample surface. At this depth, the maximum values of the bending moments during each cycle were evaluated and are shown in Figure 10.

Zhu et al. (2016) link decreases in the maximum bending moment to densification of the surrounding soil and increases in the bending moment to loosening of the surrounding soil. Figure 10 shows that (in these tests) the maximum bending moment decreases over the 500 applied cycles. The extent of this reduction (quantified as $M_{max,N=n}/M_{max,N=1}$) appears to be controlled by the peak-to-peak cyclic load amplitude, characterised by a combination of ζ_b and ζ_c. Furthermore, the behaviour of the bending moment depends on the load type. In case of $\zeta_c = 0.0$, a larger decrease can be found for the highest amplitude (about 15% for $\zeta_b = 0.7$ after N = 500). In contrast, a larger amplitude at 2-way loading seems to cause a smaller reduction of the maximum bending moment within the 500 investigated cycles. The smaller decrease at a higher amplitude for the test with $\zeta_c = -1.0$ and $\zeta_b = 0.4$ can be attributed to the uplift of the pile, as shown in Figure 9. Overall, the reduction of M_{max} due to 2-way cyclic loading is higher (~30% to 40%) compared to 1-way cyclic loading.

Frequently, pile design for monotonic loading is based on the distribution of the modulus of subgrade reaction $k_s(z)$ along the pile shaft. To analyse changes in the modulus of subgrade reaction during cyclic loading, the distribution was established for each load cycle from the lateral pressure $p_{max,N=n}(z)$ and the lateral deflection of the pile $y_{max,N=n}(z)$, see Eq. (4).

$$k_{s,N=n}(z) = \frac{p_{max,N=n}(z)}{y_{max,N=n}(z)} \quad (4)$$

In order to show the general evolution with cycle number, weighted mean values of the modulus of subgrade reaction $k_{s,N=n}$ at each measurement point were calculated.

The 1-way loading test ($\zeta_c = 0.0$) show a significant reduction of the modulus of subgrade reaction during cyclic loading, by up to 70% (Figure 11). The reduction increases as the load amplitude increases, due to the larger displacements associated with higher magnitude loading.

A different behaviour was observed for 2-way lateral loading. For the small load amplitude ($\zeta_b = 0.2$), $k_{s,N=n}/k_{s,N=1}$ initially rises over the first 10 cycles, but then reduces to $k_{s,N=n}/k_{s,N=1} = 0.8$ by cycle number 150, which is maintained to N = 500. A similar pattern is observed for $\zeta_b = 0.4$, albeit that the initial increase is approximately 20% (and over 10 cycles), and the subsequent decrease is steady over the 500 cycles, reaching $k_{s,N=n}/k_{s,N=1} = 0.85$. The reason for the reduction is considered to be a combination of increasing lateral displacements and the reduction of the general lateral capacity due to the uplift of the pile.

4 CONCLUSIONS

This paper examines the response of single piles in sand to drained lateral 1-way and 2-way cyclic loading through a series of centrifuge model tests at 100g. Differences between different modes of loading (1-way and 2-way) are highlighted.

In general, the cyclic load tests reveal significant changes in the lateral bearing behaviour due to cyclic loading. The results for 1-way and 2-way cyclic loading show the influence of the peak-to-peak load amplitude, on the accumulation of pile head displacements as well as on the changes in bending moments and the reduction of subgrade reaction. The data indicate that 1-way loading, irrespective of the size of the peak-to-peak load amplitude, has a more substantial influence on the lateral bearing behaviour than 2-way loading. In particular, the changes in the modulus of subgrade reaction are more distinctive under 1-way cyclic loading than under 2-way cyclic loading.

For the pile head displacements and bending moment, the tests show results comparable to the data presented by Rosquoet et al. (2007) and Hettler (1981).

ACKNOWLEDGEMENTS

The co-operation between COFS and the University of Kassel was financially supported by the "Group of Eight Australia – Germany Joint Research Co-operation Scheme" sponsored by the Group of Eight (Go8) and the German Academic Exchange Service (DAAD).

The overall project is funded by the Deutsche Forschungsgemeinschaft (DFG, German Research Foundation) – RE 3881/1-1.

REFERENCES

Berger, D., Graubner, C.A., Pelke, E. & Zink, M. 2003. Entwurfshilfen für integrale Straßenbrücken. *Schriftenreihe der Hessischen Straßen- und Verkehrsverwaltung (50).*

Bienen, B., Dührkop, J., Grabe, J., Randolph, M. F. & White, D. 2012. Response of Piles with Wings to Monotonic and Cyclic Lateral Loading in Sand. *Journal of Geotechnical and Geoenvironmental Engineering* 138 (3): 364–375.

Blum, H. 1932. Wirtschaftliche Dalbenformen und deren Berechnung. *Bautechnik* 10: 50–55.

Broms, B.B. 1964. Lateral resistance of piles in cohesionless soils. *Journal of the soil mechanics and foundations division.* 90 (3): 123–156.

Bundesamt für Seeschifffahrt und Hydrographie (BSH). 2007. *Standard – Konstruktive Ausführung von Offshore-Windenergieanlagen.*

Deutsche Gesellschaft für Geotechnik (DGGT). 2012. *Empfehlungen des Arbeitskreises Pfähle – EA Pfähle (EA-P), 2nd Edition.* Berlin: Ernst & Sohn.

Hettler, A. 1981. Verschiebungen starrer und elastischer Gründungskörper in Sand bei monotoner und zyklischer Belastung. *Veröffentlichungen des Institutes für Bodenmechanik und Felsmechanik der Universität Fridericiana in Karlsruhe, Heft 90.*

Kempfert, H. G. 1989. Dimensionierung kurzer, horizontal belasteter Pfähle. *Bauingenieur* 64 (5): 201–207.

LeBlanc, C., Houlsby, G.T. & Byrne, B.W. 2010. Response of stiff piles in sand to long term cyclic loading. *Géotechnique* 60 (2): 79–90.

Li, Z., Haigh, S.K. & Bolton, M. D. 2010. The response of pile groups under cyclic lateral loads. *International Journal of Physical Modelling in Geotechnics* 10 (2): 47–57.

Niemann, C., Tian, Y., O'Loughlin, C., Cassidy. M. & Reul, O. 2017. Response of piles under cyclic lateral loading – centrifuge tests. *Proceedings of the 19th International Conference on Soil Mechanics and Geotechnical Engineering, Seoul* (accepted for publication).

Poulos, H.G. 1982. Single Pile Response to Cyclic Lateral Load. *Journal of the Geotechnical Engineering Division.* 108 (3): 355–375.

Randolph, M.F., Jewell, R.J., Stone, K.J.L. & Brown, T.A. 1991. Establishing a new centrifuge facility. *Proc. Int. Conf. on Centrifuge Modelling, Centrifuge 91, Boulder, Colorado*: 3–9.

Reese, L.C. & Van Impe, W.F. 2011. *Single Piles and Pile Groups Under Lateral Loading, 2nd Edition.* Boca Raton: CRC Press.

Rosquoet, F., Thorel, L., Garnier, J. & Canepa, Y. 2007. Lateral cyclic loading of sand-installed piles. *Soils and Foundations* 47 (5): 821–832.

Schneider, J.A. & Lehane, B.M. 2006. Effects of width for square centrifuge displacement piles in sand. *Proc. 6th Int. Conf. Physical Modelling in Geotechnics, Hong Kong*: 867–873.

Slack, D.C. & Walker, J.N. 1970. Deflections of shallow pier foundations. *Journal of the Soil Mechanics and Foundations Division* 96 (4): 1143–1157.

Wichtmann, T. 2016. Soil behaviour under cyclic loading – experimental observations, constitutive descriptions and applications. *Veröffentlichungen des Institutes für Bodenmechanik und Felsmechanik am Karlsruher Institut für Technologie (KIT), Heft 181.*

Zhu, B., Li, T., Xiong, G. & Liu, J.C. 2016. Centrifuge model tests on laterally loaded piles in sand. *International Journal of Physical Modelling in Geotechnics* 16 (4): 160–172.

Physical modelling of monopile foundations under variable cyclic lateral loading

I.A. Richards, B.W. Byrne & G.T. Houlsby
Department of Engineering Science, University of Oxford, Oxford, UK

ABSTRACT: Monopile foundations supporting offshore wind turbines experience cyclic lateral loading due to wind, waves and turbine operation. This loading varies in amplitude, frequency and direction. Understanding and predicting the monopile's response to cyclic lateral loading is necessary for optimised design, as it can lead to permanent foundation rotation and evolution of the foundation's dynamic response. This paper describes new laboratory apparatus designed to perform computer-controlled cyclic load tests on model-scale monopiles. Two perpendicular electric actuators facilitate multi-directional loading, with the apparatus capable of applying loads that continuously vary in amplitude, frequency and direction. Resolution of the instrumentation is also sufficient to capture the response, including accumulation of ratcheting rotation, at low loads. Results in cohesionless soil demonstrate the capabilities of the new apparatus, begin to reveal the monopile's response to variable cyclic loading, and will allow validation of a recently developed design method for monopiles under cyclic lateral loading.

1 INTRODUCTION

1.1 Application

Monopile foundations dominate the offshore wind industry, supporting 80% of turbines in EU waters (EWEA 2015). As the industry has developed, these stiff, open-ended piles have increased in size to support larger turbines in deeper waters. Modern 'large-diameter' monopiles may be around 40 m in length (L) with diameters (D) in excess of 6 m.

Compared to conventional piles, monopiles have very large diameters and small aspect ratios ($\eta = L/D$), impacting the lateral failure mechanism; they also experience atypical loading, with vertical loads small relative to the lateral cyclic loads.

Offshore wind turbine structures are typically designed to experience up to 10^7 cycles of lateral loading throughout their 20–25 year lifetime (Leblanc et al. 2010). This cyclic loading is caused by wind, waves and turbine operation, and continuously varies in amplitude, frequency and direction. Repeated cyclic loading has been shown to cause permanent accumulated pile rotation (ratcheting) and evolution of the foundation's dynamic response with cycle number (e.g. Leblanc et al. (2010), Abadie (2015)).

Permanent pile rotation can impact turbine operation. Strict tolerances are typically defined by turbine manufacturers and DNV GL (2016) suggest permanent rotations under SLS should be kept below 0.25°. Evolution of the foundation's dynamic response is also a concern, as it can impact the whole structure's natural frequency, potentially leading to dynamic amplification and increased structural fatigue damage.

1.2 Background

Many model-scale experimental studies have explored the monopile's response to long-term cyclic lateral loading in cohesionless soil at various densities. Li et al. (2010) and Klinkvort (2012) performed centrifuge testing, while laboratory floor tests have been performed by Leblanc et al. (2010), Peralta (2010), Cuéllar (2011), Nicolai and Ibsen (2014), Abadie (2015), Arshad and O'Kelly (2016) and others.

Informed by these studies, various empirical relationships have been suggested for global permanent pile rotation as a function of cycle number. Both power-law (e.g. Nicolai and Ibsen (2014), Arshad and O'Kelly (2016)) and logarithmic relations (e.g. Li et al. (2010)) have been suggested, including various coefficients to account for soil properties and load characteristics. These relations give insight into the impact of cyclic loading on the pile response, and may be useful for initial design. However, such empirical expressions are not well adapted for predicting the monopile's response to complex lifetime loading in a range of soil conditions. Moreover, there is uncertainty in how empirical parameters derived from small-scale tests apply to full-scale design.

A new theoretical model for a mechanical system under cyclic loading is presented by Houlsby et al. (2017). This Hyperplastic Accelerated Ratcheting Model (HARM) was developed to capture the key features of the monopile's cyclic response, as observed by Abadie (2015). HARM can capture both ratcheting and evolution of the dynamic response, and its rigorous theoretical basis gives confidence for application to a range of loading and soil conditions. However, further validation and potential development is needed to move towards a model that may be used for practical monopile design.

1.3 Objectives

Previous experimental studies have mostly used electro-mechanical systems to apply cyclic loading to model monopiles. Such systems can reliably apply long-term cyclic loading, but typically have a fixed frequency and are restricted to uni-directional or basic multi-directional loading (e.g. Peralta (2010), Rudolph et al. (2014)).

Additionally, limited data is available for the pile response at low cyclic load amplitudes. It is important to understand the response at low loads and to extend design models into this region, as much of the cyclic loading experienced by monopiles in the field occurs at low amplitudes.

To explore the monopile's response to more diverse and realistic loading, and to inform further development of HARM, new computer-controlled laboratory apparatus has been developed. Focus has been placed on achieving highly controlled multi-directional and continuously time-varying loading, and on capturing the pile's response at low load amplitudes.

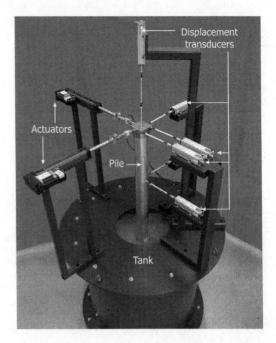

Figure 1. Laboratory apparatus.

2 LABORATORY APPARATUS

2.1 Apparatus design

Although physical modelling of this problem could be performed in either the centrifuge or at laboratory scale, laboratory testing has been pursued as it enables more sophisticated apparatus to be developed and longer-term tests to be performed.

Two perpendicular Zaber BAR electric actuators facilitate planar multi-directional loading while six MTS Temposonics magnetostrictive displacement transducers allow the pile's position in 3D space to be determined. A removable annulus attaches to the cylindrical tank's top flange, on which brackets are affixed to support the actuators and transducers, shown in Figure 1.

Dührkop & Grabe (2008) also developed computer-controlled multi-directional apparatus, using a combined stage device with two axes. Instead, this design uses two separate actuators to allow application of vertical load and inclusion of flexible couplings.

The rigid aluminium pile has diameter 80 mm, scaled approximately 1:100 in relation to 'large-diameter' monopiles. The model pile properties are summarised in Table 1. The tank has an internal diameter of 800 mm, chosen to minimise boundary effects (Achmus et al. 2007).

Figure 2 shows the specially designed 'pile cap' and connecting components. The pile cap transfers loads applied by the actuators to the pile and attaches to four of the six displacement transducer linkages. The remaining two transducer linkages attach directly onto the pile 330 mm below. The transducer linkages

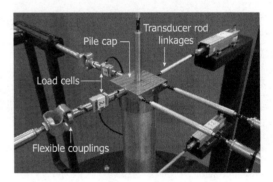

Figure 2. Detail of laboratory apparatus.

Table 1. Model pile properties.

Diameter	D	80	mm
Embedded length	L	320	mm
Aspect ratio	$\eta = L/D$	4	–
Wall thickness	t_W	5	mm

connect the transducer rods (which move along a single axis) to the pile, with miniature universal joints at either end to accommodate rotation and settlement of the pile.

The tests presented in this paper have been conducted without additional vertical loading. However, the pile cap has been designed to accommodate ring-shaped masses to apply static vertical loading if required.

Combined moment and horizontal loading is applied by the actuators, located 800 mm above the soil surface. This results in a dimensionless eccentricity $\tilde{e}$ (see Section 2.3) within the expected range for monopiles in the field. In-line with each actuator, flexible polymer couplings introduce significant compliance to the system to simplify load control, while miniature S-beam load cells allow the x and y load components to be resolved. Universal joints are again necessary to accommodate pile rotation and settlement.

Pile driving apparatus has also been developed to allow controlled, repeatable pile installation, but is not described further here.

2.2 Data acquisition and load control

During testing, the following operations must be performed simultaneously: calculation of the pile's position, resolution of the x and y load components and control of the actuators to apply the demanded load. An application developed in LabVIEW (National Instruments 2015) performs these operations, alongside test set-up and data logging tasks. Associated hardware include a CompactDAQ system for data acquisition and signal conditioning and a Zaber X-MCB2 controller which communicates with both actuators via USB.

Central to the LabVIEW application are two PID controllers, which simultaneously adjust the actuator's speeds dependent on the error between the actual and demanded load components. To resolve the load components the pile's 3D position (pose) must be known.

The pile's pose is calculated from the current transducer measurements and the transducer's initial coordinates using forward kinematics calculations, which require a numerical solution. The calculations used by Byrne and Houlsby (2005) for pivoting devices have been adapted for this system with fixed transducers and pivoting linkages.

Each PID controller is called at a rate of 50 Hz, allowing accurate load control for cyclic load frequencies up to 0.2 Hz. The flexible couplings also play a key role in accurate load control: introducing a significant compliance, they ensure any changes to the pile-soil system's stiffness remain negligible, so the optimum controller variables are constant even as the pile-soil stiffness changes.

Load demand files are prepared as .csv files, and read during each call to the PID controller. The LabVIEW application also logs measurement data and computed outputs at a rate appropriate for each test.

2.3 Scaling

If model-scale results are to appropriately inform full-scale design, rigorous consideration of scaling is essential. Dimensionless frameworks allow model-scale results to be related to full-scale, and vice-versa. However, it is not usually possible to scale all aspects of the problem. Ideally, the dimensionless framework is validated with comparison to large-scale results (e.g. Kelly et al. (2006)).

Table 2. Key dimensionless parameters (Leblanc et al. 2010).

Horizontal force	$\tilde{H} = \dfrac{H}{L^2 D \gamma'}$
Moment	$\tilde{M} = \dfrac{M}{L^3 D \gamma'}$
Rotation	$\tilde{\theta} = \theta \sqrt{\dfrac{p_a}{L\gamma'}}$
Eccentricity	$\tilde{e} = \dfrac{M}{HL}$

Table 3. Leighton Buzzard 14/25 sand characteristics. Critical friction angle from Schnaid (1990), other values measured.

Particle sizes	$D_{10}, D_{30},$	0.56, 0.69,	mm
	D_{50}, D_{60}	0.81, 0.87	
Maximum unit weight	γ_{max}	17.64	kN/m³
Minimum unit weight	γ_{min}	14.43	kN/m³
Critical friction angle	ϕ_{cr}	34.3	degrees

Leblanc et al. (2010) derive a dimensionless framework for monopiles under lateral loading, starting from an expression for shear modulus as a function of stress level. Key parameters from this dimensionless framework are summarised in Table 2. Pile geometry, the soil's effective unit weight (γ') and atmospheric pressure (p_a) are required inputs.

The effect of stress level on friction angle must also be considered when performing model-scale tests. The empirical relationships presented by Bolton (1986) show how friction angle (ϕ') relates to relative density (R_D) and mean effective stress (p'). At model-scale (lower stress level) the soil sample's relative density must be appropriately reduced to match the friction angle and represent dilatancy at full scale.

Following Bolton's work, Schnaid (1990) performed a triaxial test program to determine the best-fitting empirical coefficients for Leighton Buzzard 14/25 sand, leading to:

$$\phi' = 34.3° + 3(R_D(9.9 - \ln p')) \qquad (1)$$

2.4 Soil properties

The tests presented in this paper were conducted using Leighton Buzzard 14/25 sand, whose characteristics are summarised in Table 3. Dry samples were prepared by pouring from a low drop height to achieve a density of 1475.7 ± 6.6 kg/m³ or $R_D \approx 2\%$, which corresponds to $R_D \approx 4\%$ at full scale, employing Equation 1 to match ϕ'.

Very low density samples are simple to prepare in the laboratory, but even when considering stress level effects, such low densities would not be found at real sites. Tests at higher densities, representative of typical offshore sites, are planned.

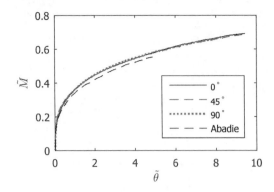

Figure 3. Monotonic response in three directions, compared to the results of Abadie 2015.

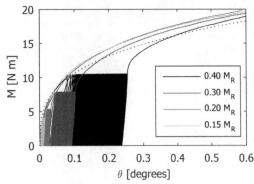

Figure 4. Response to 1000 load cycles at various amplitudes, followed by monotonic loading. Typical monotonic response shown dotted.

2.5 Commissioning

A series of monotonic and cyclic commissioning tests were performed to check the operation of the new apparatus and LabVIEW application. Figure 3 presents the results of three monotonic tests, each conducted in a different loading direction. These results demonstrate good repeatability and invariance to loading direction, which is essential for investigations into multi-directional loading.

Figure 3 is plotted in $\tilde{M}$–$\tilde{\theta}$ space to allow comparison to a typical monotonic test performed by Abadie (2015) using mechanical loading apparatus. The results compare well, giving confidence in the new system's operation.

The pile's average reference moment capacity M_R was also obtained from these monotonic tests. With M_R defined at 2° rotation at model scale, $M_R = 26$ N m.

3 EXAMPLE RESULTS

3.1 Uni-directional, constant amplitude cyclic loading

Uni-directional, constant amplitude cyclic loading has been the focus of many previous experimental studies, and is a sensible place to begin exploring the pile's response.

Figure 4 presents the response to 1000 cycles of 1-way loading at various load amplitudes, followed by monotonic re-loading. The average virgin monotonic curve or backbone curve is also plotted (dotted). The results demonstrate the high resolution of the instrumentation and accuracy of the load control system. Results are consistent down to a cyclic load amplitude of $0.15M_R$ or 3.9 N m, where the relative rotation across one cycle is less than 0.005°. These results also indicate an increase in capacity following cyclic loading, as observed by Nicolai et al. (2017).

The data in Figure 4 is re-cast in Figure 5 to show the accumulation of rotation $\Delta\theta$ with cycle number N. $\Delta\theta = \theta_N - \theta_0$, neglecting the rotation contribution from initial monotonic loading θ_0. The results suggest a power-law relationship between $\Delta\theta$ and N is

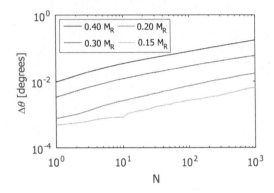

Figure 5. Accumulated rotation with cycle number at various amplitudes.

appropriate, in agreement with Leblanc et al. (2010), Arshad and O'Kelly (2016) and others. The response at $0.40M_R$ has also been validated against results from Abadie (2015).

3.2 Multi-directional loading

Figures 6 and 7 plot the pile's response to two multi-directional load paths: a 2-way spiral with 5 cycles, and a 1-way spiral with 8 cycles, both up to a maximum amplitude of $0.8M_R$. The top left quadrant plots the applied moment in x–y space, with the resulting rotation plotted on the bottom right; the x and y direction responses are plotted accordingly.

These unusual and unrealistic load paths, with continuously varying x and y load components and no dominant load direction, robustly test the new system's ability to apply multi-directional loading. The applied load deviates slightly from the load demand as the pile's direction of movement changes—visible as small spikes on the moment plots. This error is caused by small backlash in the universal joints in-line with the actuators, and may be reduced with adaptive tuning of the PID controllers.

HARM has been theoretically extended to two-dimensions to capture multi-directional loading

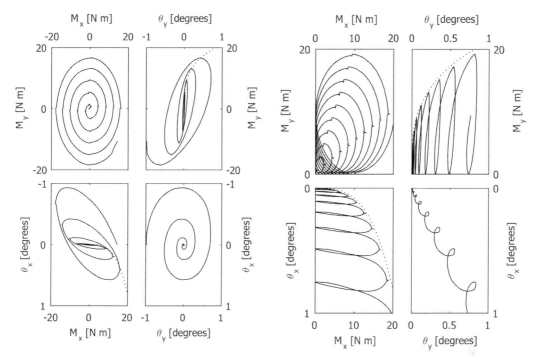

Figure 6. Response to multi-directional two-way spiral load path. Typical monotonic response shown dotted.

Figure 7. Response to multi-directional one-way spiral load path. Typical monotonic response shown dotted.

(Houlsby et al. 2017), but has yet to be validated. Such multi-directional data will allow validation, and inform potential development, of two-dimensional HARM.

If HARM is able to capture such unusual multi-directional responses as shown in Figures 6 and 7, it would give confidence in its applicability to realistic multi-directional loading. Nonetheless, future tests will explore the response to more realistic loading, for example, applying cyclic loading in a fan-shape about a dominant loading direction as investigated by Rudolph et al. (2014) and (Nanda et al. 2017).

3.3 Uni-directional, time-varying loading

Figure 8(a) shows a continuously time-varying moment applied to the model pile. This signal is based on scaled signals generated in Bladed (Garrad Hassan & Partners 2014), using a turbulent wind field superimposed by gusts of variable amplitude and period. The turbulent wind field gives rise to very small scale load fluctuations with period ≈ 10 s.

The resulting rotation response is plotted in Figure 8(b) and in $M - \theta$ space in Figure 9, where the average backbone curve is also plotted (dotted). The response is dominated by the large load events, and there also appears to be some increase in capacity compared to the backbone response, as observed in the constant amplitude cyclic tests (Section 3.1).

The natural next step is to apply realistic multi-directional time-varying cyclic loading to the model monopile. As well as giving insight into the monopile's

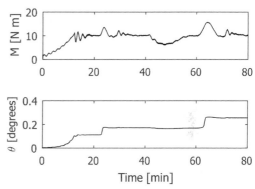

Figure 8. Time-varying cyclic loading (a) applied moment, (b) resulting rotation response.

behaviour, this would provide valuable data for model validation.

4 CONCLUSIONS

This paper describes Pile geometry, the soil's new laboratory apparatus developed to perform a range of lateral cyclic loading tests on model monopile foundations. Two computer-controlled electric actuators facilitate multi-directional loading, and six transducers allow the pile's position in 3D space to be determined, even at very small displacements. The critical role that the LabVIEW application plays in performing tests is also discussed.

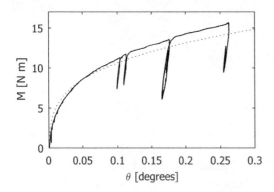

Figure 9. Response to time-varying cyclic loading. Typical monotonic response shown dotted.

Example results in loose, cohesionless soil demonstrate the capabilities of the apparatus and begin to reveal the monopile's response to low amplitude, multi-directional and continuously time-varying cyclic loading.

Future test series are identified, including tests in higher density soil samples more representative of offshore sites. Future tests will allow validation and potential development of HARM under diverse and more realistic loading, ultimately moving towards a robust model that may be used for monopile design.

REFERENCES

Abadie, C. N. (2015). *Cyclic lateral loading of monopile foundations in cohesionless soils*. DPhil thesis, University of Oxford.
Achmus, M., K. Abdel-Rahman, & Y. S. Kuo (2007). Numerical Modelling of Large Diameter Steel Piles under Monotonic and Cyclic Horizontal Loading. In *Proceedings of the Tenth International Symposium on Numerical Models in Geomechanics (NUMOG X)*, Hannover, Germany.
Arshad, M. & B. O'Kelly (2016). Model Studies on Monopile Behavior under Long-Term Repeated Lateral Loading. *International Journal of Geomechanics 17*(1).
Bolton, M. D. (1986). The strength and dilatancy of sands. *Géotechnique 36*(1), 65–78.
Byrne, B. W. & G. T. Houlsby (2005). Investigating 6 degree-of-freedom loading on shallow foundations. In *Proceedings of the International Symposium on Frontiers in Offshore Geotechnics*, Perth, Australia, pp. 477–482.
Cuéllar, V. P. (2011). *Pile Foundations for Offshore Wind Turbines : Numerical and Experimental Investigations on the Behaviour under Short-Term and Long-Term Cyclic Loading*. Ph. D. thesis, Technischen Universitat Berlin.
DNV GL (2016). *Support structures for wind turbines DNVGL-ST-0126* (April ed.).
Dührkop, J. & J. Grabe (2008). Monopilegründungen von Offshore-windenergieanlagen – Zum Einfluss Einer Veränderlichen Zyklischen Lastangriffsrichtung. *Bautechnik 85*(5), 317–321.
EWEA (2015). The European offshore wind industry – key trends and statistics 2015. Technical report.
Garrad Hassan & Partners (2014). Bladed 4.6.
Houlsby, G. T., C. N. Abadie, W. J. A. P. Beuckelaers, & B. W. Byrne (2017). A model for nonlinear hysteretic and ratcheting behaviour. *International Journal of Solids and Structures 120*.
Kelly, R. B., B. W. Byrne, & G. T. Houlsby (2006). A comparison of field and laboratory tests of caisson foundations in sand and clay. *Géotechnique 56*(9), 617–626.
Klinkvort, R. T. (2012). *Centrifuge modelling of drained lateral pile-soil response*. Ph. D. thesis, DTU.
Leblanc, C., G. T. Houlsby, & B. W. Byrne (2010). Response of stiff piles in sand to long-term cyclic lateral loading. *Géotechnique 60*(2), 79–90.
Li, Z., S. K. Haigh, & M. D. Bolton (2010). Centrifuge modelling of mono-pile under cyclic lateral loads. In *Proceedings of the 7th International Conference on Physical Modelling in Geotechnics*, Volume 2, Zurich, pp. 965–970.
Nanda, S., I. Arthur, V. Sivakumar, S. Donohue, A. Bradshaw, R. Keltai, K. Gavin, P. Mackinnon, B. Rankine, & D. Glynn (2017). Monopiles subjected to uni- and multilateral cyclic loading. In *Proceedings of the Institute of Civil Engineers – Geotechnical Engineering (Ahead of print, online)*.
National Instruments (2015). LabVIEW 2015.
Nicolai, G. & L. B. Ibsen (2014). Small-Scale Testing of Cyclic Laterally Loaded Monopiles in Dense Saturated Sand. *Journal of Ocean and Wind Energy 1*(4), 240–245.
Nicolai, G., L. B. Ibsen, C. D. O'Loughlin, & D. J. White (2017). Quantifying the increase in lateral capacity of monopiles in sand due to cyclic loading. *Géotechnique Letters 7*(3), 1–8.
Peralta, P. (2010). *Investigations on the Behaviour of Large Diameter Piles under Long-Term Lateral Cyclic Loading in Cohesionless Soil*. Ph. D. thesis, Leibniz Universitat Hannover.
Rudolph, C., B. Bienen, & J. Grabe (2014). Effect of variation of the loading direction on the displacement accumulation of large-diameter piles under cyclic lateral loading in sand. *Canadian Geotechnical Journal 51*.
Schnaid, F. (1990). *A Study of the cone-pressuremeter test in sand*. DPhil thesis, University of Oxford.

Centrifuge model testing of fin piles in sand

S. Sayles
Balfour Beatty, UK

K.J.L. Stone & M. Diakoumi
School of Environment and Technology, University of Brighton, Brighton, UK

D.J. Richards
Faculty of Engineering and the Environment, University of Southampton, UK

ABSTRACT: This paper presents the results of a series of centrifuge model tests undertaken to evaluate the lateral response of circular piles fitted with fins. The geometry of the fins in terms of their length and orientation with regard to the direction of loading (i.e. projected area), and depth of embedment were investigated. The study was conducted at an enhanced acceleration level of 50 gravities in medium to dense uniformly graded dry sand. Only monotonic lateral loading was applied, and the results are presented through load versus displacement plots, from which lateral capacities are derived. The data set obtained from the tests is compared and appraised with data from other researchers and the influence of the fin geometry on the development of lateral capacity is discussed.

1 INTRODUCTION

Monopiles are the most popular form of foundation for large offshore wind turbines. They have to withstand not just large vertical forces from the weight of the turbine but also significant lateral loads from wind, wave, current and tide, which can exceed one-third of the gravity load (Peng, 2006). As such, lateral bearing capacity is often the principal design concern. As wind farms use increasingly larger turbines, and move into deeper water higher lateral capacity is required. For example, steel piles up to 4 m in diameter have been installed in the North Sea (Randolph & Gourvenec, 2011) and monopiles are now close to the maximum fabrication diameter that can be supported by UK manufacturers.

1.1 Fin piles

Early studies on the performance of a 40 ft long octagonal "rocket-shaped" pile in soft clay under cyclic and lateral loading was reported by Lee and Gilbert in 1980. More recently the Fin Pile concept has received significant attention due to the increase in demand for more efficient offshore wind turbine foundations.

Single gravity tests have demonstrated that the addition of fins can significantly enhance the lateral resistance of single monopiles (Peng, 2006, DTI, 2006). These studies also investigated the cyclic response and the potential optimisation of fin location and orientation on the pile shaft. It was demonstrated that fins located at the top and middle of the pile have a significant impact on the lateral capacity. Although it is noted that the study piles were short and rigid and fins located in the middle may be less effective for long flexible piles. The lateral capacity was found to be very similar for piles loaded at 45° and 90°.

Duhrkope & Grabe (2008) investigated laterally loaded piles with bulge, where the bulge refers to "the application of vertical steel plates somewhere near the ground surface to improve the lateral bearing capacity". A series of 1g tests were carried out in dry dense sand for a range of fin shapes. It was found that a ratio $h1/L$, where $h1$ is depth to top of fins and L is the total embedded depth of the pile, of about 0.3, maximises the positive effect of the bulge for all investigated soil types. For strongly cohesive soils or layered ground with weak soils on top, the optimal ratio is less.

Duhrkop (2010) report a series of centrifuge tests, with the aim of further exploring the validity of the "winged pile" concept. Short, solid piles were modelled to represent a 2.4 m diameter (D), 9.7 m embedment length prototype pile when tested at 200 g. Wings were 1.3D long, 2/3D wide and 1 mm thick. The model piles were connected stiffly to the actuator and the tests were therefore not representative of the more typical field conditions of a free-headed pile. All 8 tests were done in the same medium dense ($\gamma = 15.9\,kN/m^3$) dry sand sample. It was recorded that starting and stopping the centrifuge led to densification of the soil (proved by in-flight CPTs – relative density increased from 0.4 to 0.49). Performance of the piles was investigated in both the installation and lateral loading phases. Piles were installed in-flight and the centrifuge then stopped to set up the lateral loading system. Piles were monotonically pushed to a lateral displacement of at least 10% of the diameter of the reference pile (*D*) at a model rate of 0.01 mm/s. All were pushed to a maximum of 0.2D. In 6 of the experiments, a cyclic loading stage preceded a lateral monotonic capacity test.

The study demonstrated that piles with wings provided a considerably higher lateral resistance and stiffer behaviour. However, due to the low maximum displacement the piles were tested to, there was no evidence of a limiting load being reached in any of the tests. The soil was also observed to strain-harden. A very stiff behaviour was initially exhibited by piles with a previous cyclic loading history as a result of soil densification during the cyclic loading phase, although the ultimate resistance of these piles was similar. The lateral capacity of a winged pile was found to be 40% higher than that of a monopile at the same pile head displacement.

Bienen (2012) carried out cyclic tests with a fixed headed pile using the piles from Duhrkop & Grabe (2008) and another pair of longer piles. The longer winged piles were found, as with the shorter winged piles, to exhibit significantly higher stiffness than the monopile, which was sustained for serviceability pile head displacements. This higher stiffness led to smaller pile head deflections under cyclic loading, although the relative rate of displacement accumulation was found to be similar to the monopile. It was also found that head deflections were reduced by approximately 50% compared with regular monopiles for the same load level. Also, winged piles with 4*D* embedment were found to have a similar stiffness to a monopile embedded to a depth of 5*D*.

The effect of wings under cyclic loading was investigated in a field test (Rudolph and Grabe 2013). Two piles – one of them equipped with wings – are installed in a harbour basin in Bremerhaven, Germany. The piles had a diameter of $D = 0.8$ m and an embedment length of 8 m, 5 m of which was load-bearing. The tests, contrary to expectations, found that the finned pile did not perform as well as the monopile. Most likely this was due to the piles having been installed in inhomogeneous soil with variable silt layer thickness and also because the vibratory driving installation disturbed a larger area with the finned pile. However, the finpile did manage to suppress the effect of a changing cyclic loading direction. Nasr (2014) carried out model tests in sand of different relative densities, the results of which were compared to results of 3D finite element analysis using PLAXIS 3D. Model piles were 21 mm diameter, 315 mm and 777 mm long to represent short rigid and long flexible piles. Fins were assumed to be sufficiently rigid for fin deformation to have a negligible effect.

2 EXPERIMENTAL PROGRAMME

A total of 8 tests are reported herein as summarised in Table 1 below. The tests were designed to investigate the influence of the fin geometry, specifically the fin length, on the performance of a single monopile. The pile embedment depth remained the same for all the tests, and for all but one of the tests the fins were embedded to be flush with the soil surface.

Table 1. Summary of model tests undertaken. (FP – finned pile; MP-monopile).

Test	Fin length mm	Depth to back of fin mm	Loading direction to fin axis
MP	0	0	–
FP10	10	0	90°
FP10X	10	0	45°
FP20	20	0	90°
FP20X	20	0	45°
FP30	30	0	90°
FP30X	30	0	45°
FP10E*	10	20	90°

*10 mm embedment depth to back of fin.

2.1 Material and model preparation

The sand used for the model tests was uniformly graded Fraction D (150–300 micron) silica sand supplied by David Ball Limited. The sand has maximum and minimum void ratios of 1.06 and 0.61 respectively, with a critical state angle of friction φ_{crit} of 32°. The models were prepared though a combination of dry pluviation and vibration using a vibrating table. This method produced consistent soil specimens with a bulk density of 1700 kg/m^3.

2.2 Centrifuge test package

All the tests reported were carried out on the balanced beam centrifuge at the University of Brighton. This machine is manufactured by Thomas Broadbent & Sons Ltd. and is 6 g-tonne machine (20 kg payload to 300 g). The tests were conducted in circular steel tub with an internal diameter of 272 mm and sample height of 114.6 mm. The steel tub was placed in a cradle with is then mounted on the centrifuge arms. A single degree of freedom actuator is then mounted on the cradle, and through the use of a wire and pulley mechanism is able to apply a horizontal load to the model finpile, refer to Figure 1. The steel wire is attached to a load cell fixed to the actuator traveller plate. The applied horizontal load is assumed to be equal to the tension measured in the steel wire (i.e. friction effects are neglected). The horizontal displacement at the point of loading is determined from the vertical displacement of the actuator, assuming no stretch of the steel cable occurs.

The model piles were driven into the soil approximately 55 mm from the edge of the container with light blows from a soft hammer. Great care was taken during driving to ensure verticality of the pile through the

Figure 1. Model package assembled and loaded on UoB centrifuge.

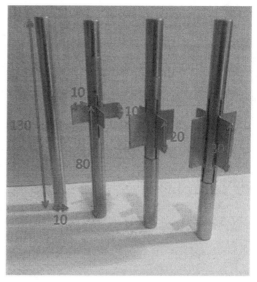

Figure 2. Assembled model piles, from left to right: MP, FP10, FP20 and FP30 (dimensions in mm).

use of a guide. Since the pile is displaced away from the tub walls and towards the centre of the sample, boundary effects are not considered significant.

The model finpiles were constructed from solid 10 mm diameter aluminium rod. The finpiles were composed of three sections with a middle section consisting of solid rod with the fins glued into slots cut into the rod. A section of pile was then fixed using male-female couplers either side of the finned section to create the complete finpile (Figure 2). Fins were 1 mm thick, 10 mm wide, and 10 mm, 20 mm and 30 mm long.

The shaft of the pile was loaded via a collar attached at 80 mm above the model surface. This height was selected primarily for ease of connection with the pulley and actuator arrangement such that loading could be applied horizontally.

2.3 Test procedure

The model and actuator is set-up on the laboratory bench and then loaded onto the centrifuge. The actuator is connected to the power and control system and a digital video camera is mounted on the top of the cradle to observe the model during the test. All the tests were conducted at and acceleration level of 50 g.

Once the model had achieved the test acceleration the actuator was run at a velocity of 5 mm/minute and data from the load cell and actuator encoders recorded continuously until the end of the test. Each test configuration was repeated three times.

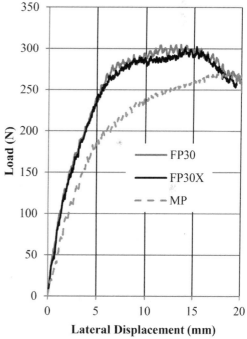

Figure 3. Load-displacement responses for monopile (MP) and finpiles, loaded at 0 (FP30) and 45 degrees (FP30X) to the fin orientation.

3 EXPERIMENTAL RESULTS

Figure 3 presents a typical load-displacement response for a laterally loaded finpile, loaded at both 0 and 45° to the fin orientation. The response of the monopile is presented on the same plot for comparison. It is noted

745

Figure 4. Post-test photograph of finpile loaded at 45 degrees to fin axis showing extent and shape of passive failure zone.

that the loading direction does not have any significant influence on the finpile response. In this test the increase in ultimate lateral capacity due to the fins is approximately 20%.

Figure 4 shows the corresponding post-test photograph of the model where the shape of the passive soil mound is clearly visible. It is interesting to note that the passive mound has an oval shape in plan which reflects the complex geometry of the passive 'wedge' associated with the soil dilatancy.

4 ANALYSIS AND DISCUSSION

As is apparent in Figure 3, a significant displacement is required to mobilise the ultimate lateral capacities of the model piles. It is thus of interest to compare the results of the models at a more conventional displacement, in this case taken at 0.1D where D is the pile diameter. This value is often used to define the lateral capacity for design purposes. Using this definition for the lateral capacity of the model pile, the effect of the fins can be evaluated in relation to developing a potential design methodology. Table 2 below summarises the results of the model piles in terms of their lateral load capacity at a displacement of 0.1D.

From the data presented in Table 2 above, and Figures 3 and 5, it is apparent that the presence of a fin can significantly enhance the lateral capacity and lateral stiffness of a monopile.

A simple approach to design is to consider the lateral load development as the sum of the lateral capacity of the pile plus an additional contribution from the fin. The extent of this contribution will be a function of the fin geometry and its location on the pile. In the tests reported here only one test was undertaken where the fin was embedded below the soil surface (Test FP10E).

Table 2. Summary of lateral capacity developed at 0.1D (1 mm).

Pile Test	Lateral load at 0.1D N	% increase
MP	50	–
FP10/10X	61.7	23
FP20/20X	76.0	52
FP30/30X	83.4	67
FP10 E	61.7	23

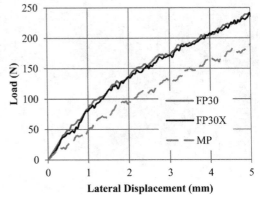

Figure 5. Lateral load displacement plots for monopile with 30 mm fins loaded at 90 degrees (FP30) and 45 degrees (FP30X) and monopile (MP).

Interestingly for this single test the lateral response was very similar to the response of the non-embedded fin, implying that for this geometry the effect of fin burial was not significant.

Using the data presented in Table 2 a simple linear expression can be derived to estimate the lateral capacity of the finpile as a function of the fin length as follows,

$$P_{FP} = P_{MP}\left(1 + 1.8743\left(\frac{l_F}{L_P}\right)\right) \quad (1)$$

where P_{FP} and P_{MP} are the lateral load developed by the finpile and monopile respectively and L_F and L_P are the respective lengths of the pile (embedment depth) and fin.

A comparison to the results from Peng (2006) and Nasr (2014) static loading tests on short rigid piles in sand together with the results from this study are presented in Figure 6. The findings from this study are very similar to Nasr's 2014 study, but the results reported by Peng (2006) are significantly lower. It is also noted that Peng's study appears to indicate a linear relationship between lateral load and fin length, especially for long fins, whereas this study and that of Nasr's, tends to suggest that linearity reduces as fin length increases.

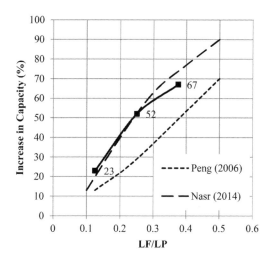

Figure 6. Comparison of lateral capacity with previous studies, this study shown as solid line.

5 CONCLUSIONS

This paper has presented a centrifuge model study on the effect of fin geometry on the performance of monopiles under monotonic lateral loading in dry sand. There are many variables that would contribute to the response of the system and this study has essentially focused on only the length of the fin. In this respect the tests reported are in reasonable agreement to some earlier studies, in that they indicate that a linear relationship between lateral capacity and fin length is a reasonable approximation. However, it is emphasised that the actual relationship would appear to be non-linear and further research is required to better define this relationship. In particular, it should be noted that an assumed linear relationship is likely to provide optimistic lateral capacities for finned piles, most notably for those with longer fins.

REFERENCES

Bienen, B. Dührkop, J. Grabe, J., Randolph, M.F. & White, D.J. 2012. Response of Piles with Wings to Monotonic and Cyclic Lateral Loading in Sand. *Journal of Geotechnical and Geoenvironmental Engineering.* 138 (3), pp. 364–375.

Department of Trade and Industry (DTI). 2006. Finpile Project: Final Report. London: Department of Trade and Industry.

Duhrkope, J. & Grabe, J. 2008. Laterally Loaded Piles with Bulge. *Journal of Offshore Mechanics and Arctic Engineering*, Vol. 130(4), 041602 (2008).

Duhrkop, J., Grabe, J., Bienen, B, White, D. & Randolph, M. 2010. Centrifuge experiments on laterally loaded piles with wings. *Proc. 7th Int. Conf. on Physical Modelling in Geotechnics (ICPMG), Zurich, Switzerland, Vol. 2.* Rotterdam: CRC Press/Balkema, pp. 919–924.

Nasr, A. 2014. Experimental and theoretical studies of laterally loaded finned piles in sand. *Canadian Geotechnical Journal*, 10.1139/cgj-2013-0012, 381–393.

Peng, J. 2006. Behaviour of Finned Piles in Sand under Lateral Loading. Unpublished PhD thesis. University of Newcastle upon Tyne.

Peng, J., Rouainia, M. & Clarke, B. G. 2010. Finite element analysis of laterally loaded fin piles. *Computers and Structures.* 88 (21-22), pp. 1239–1247.

Randolph, M.F. & Gourvenec, S. 2017 Offshore Geotechnical Engineering, CRS Press, ISBN 9781138074729.

Rudolph, C. & Grabe, J. 2013. Laterally loaded piles with wings: In situ testing with cyclic loading from varying directions. *ASME 2013 32nd International Conference on Ocean, Offshore and Arctic Engineering*, Nantes, France.

Dynamic behaviour evaluation of offshore wind turbine using geotechnical centrifuge tests

J.T. Seong, J.H. Kim & D.S. Kim
Korea Advanced Institute of Science and Technology, Daejeon, Korea

ABSTRACT: Understanding of dynamic response of offshore wind turbine is important to evade structural hazards by dynamic loadings such as resonance vibration and earthquake. Although dynamic characteristics of the offshore wind turbine are affected by soil stiffness and foundation type, there have been only a few experimental studies on this subject to supplement and calibrate the analytical studies considering soil-foundation-structure interaction (SFSI). Geotechnical centrifuge experiment can provide reliable experiment result as it can reproduce field stress condition of the soil with a scaled model test. The goal of this research is to evaluate the natural frequency and seismic behaviour of the offshore wind turbine for monopile, monopod, and tripod foundation considering SFSI by geotechnical centrifuge test. Natural frequency was evaluated in fixed-based scaled model test and SFSI geotechnical centrifuge test to evaluate the period lengthening due to SFSI. Results showed that if the same amount of material was invested, Tripod foundation showed most stiff behaviour. Also, monopile and monopod test results were compared with existing analytical evaluation method. Results showed less than 6% difference in period lengthening ratio (PLR). Seismic behavior was evaluated by applying series of seismic loadings and measuring the structural acceleration, permanent rotational displacement. The result showed that difference in seismic behaviour by foundation types should be considered.

1 INTRODUCTION

The size of the offshore wind turbine has been continuously increased since its first introduction. As the size and cost of offshore wind turbine increases, securing the structural stability of wind turbine against dynamic loading becomes one of the important issues. Wind turbine's thin, flexible structure is vulnerable to dynamic loading and resonance vibrations could reduce the efficiency of the generator, even afflicts fatigue damage to the structure (Haritos, 2007). With the introduction of the variable speed wind turbine, the frequency bandwidth of the structural load widens, and the permissible natural frequency bandwidth of the offshore wind turbine becomes narrow. Due to this reason, it is important to estimate the exact natural frequency of the wind turbine (Adhikari & Bhattacharya, 2011). Also, consideration of offshore wind turbine's behaviour against seismic loading is needed as a large number of offshore wind turbines are planned to be installed at active seismic areas around Asia.

It is important to consider the effect of soil-foundation-structure interaction (SFSI) for more precise estimation of the dynamic behaviour of the system. Bazeos et al. (2002) evaluated natural frequency reduction due to SFSI in elastic foundation condition. Hongwang (2012) suggested a simplified spring-damper model for a shallow foundation to evaluate dynamic behaviour considering SFSI. Prowell et al. (2011) and Kjørlaug et al. (2015) performed finite element analysis to evaluate dynamic response of offshore wind tower.

In the analytical study, a simplified spring system is widely used to consider SFSI. Although it is a convenient way to assess it, experimental approach is needed to verify and calibrate the analytical study. As geotechnical centrifuge can replicate field stress condition in model scale, it can produce reliable experimental result while easy to repeat compared to field test (Seong et al. 2017).

This research aims to evaluate the natural frequency and seismic behaviour of offshore wind turbine with monopile, monopod, and tripod foundations using the geotechnical centrifuge test. Research targeted three offshore foundation types; monopile, monopod, and tripod. The natural frequency of offshore wind turbine model was measured in fixed-based condition and SFSI condition to evaluate the effect of SFSI to the natural frequency. In addition, the model was exposed to series of seismic loadings with different intensities to evaluate the seismic behaviour of the offshore wind turbine during the earthquake, and permanent rotational displacement after the earthquake.

Table 1. Centrifuge scaling law.

Parameter	Scaling factor
Mass	$1/N^3$
Length	$1/N$
Acceleration	N
Time	$1/N$
Frequency	N

Table 2. Offshore wind turbine Structural parameters in model and prototype scale.

	Value	
Parameter	In model scale (mm)	In Prototype scale (m)
Tower height	500	21.9
Tower diameter	36	1.58
Tower wall thickness	3	0.13
Head height	50	2.19
Head diameter	60	2.63

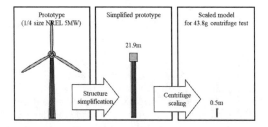

Figure 1. Scaled model production procedure.

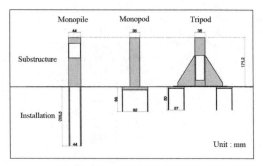

Figure 2. Foundation schematics in model scale.

2 CENTRIFUGE EXPERIMENT OVERVIEW

In a geotechnical experiment, it is important to reproduce the correct field stress condition as the mechanical behaviour of soil depends on the effective stress level. Centrifuge experiment can raise the level of confinement to real field condition for the scaled model by applying centrifugal force, thereby reproducing the actual field behaviour of the soil.

The result of model scale centrifuge experiment can be converted into the prototype scale by using centrifuge scaling laws (Schofield, 1980). The scaling laws that correlate Ng scaled model to 1 g field model (prototype) are described in Table 1.

3 SCALED MODEL PRODUCTION

A 1/4 size NREL 5 MW reference wind turbine (Jonkman et al. 2009) was used as a prototype for scaled model production. The prototype was simplified into a model of lumped mass on hollow cylindrical shape tower, then reduced into scaled model following the centrifuge scaling laws in Figure 1.

The scaled model represents 1/4 size NREL 5 MW offshore wind turbine when converted to prototype from 43.8 g centrifuge experiment. Dimensions of the scaled model are presented in Table 2.

Scaled model of monopile, monopod, and tripod foundation was produced for the experiment. Two types of models (fixed-based foundation and SFSI foundation) were produced for each foundation types.

The fixed-based model consists of substructure and bolt-fixed base. This model was produced to evaluate wind turbine's natural frequency without considering SFSI. Substructure of monopod and tripod foundation was designed to have the same diameter with the tower. The substructure diameter of monopile was designed to have the same diameter with installation to match up with the general shape of each offshore foundation types. All three substructures were designed to have same mass and height.

SFSI model consists of substructure and installation. All three installations were designed to have the same mass. Details of foundation models are described in Figure 2.

4 TEST PROCEDURE

4.1 Fixed-based condition test

1 g scaled model test in fixed-based condition was conducted to evaluate wind turbine's natural frequency without considering SFSI. Each fixed-based model was bolted to aluminium base plate to reproduce fixed-based condition.

Natural frequency was evaluated by impact hammer test. The impact was applied to the base plate as shown in Figure 3. An accelerometer attached to the tower head measured time-acceleration history.

Obtained time-acceleration history was converted into frequency-acceleration history by using fast Fourier transformation (FFT) method. Natural frequency was evaluated by picking the highest point in the frequency-acceleration history. Setting and experiment procedure was identical for all three foundations.

 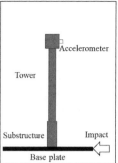  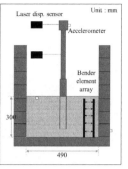

Figure 3. Picture and schematics of monopod fixed-based test.

Figure 4. Picture and schematics of monopile SFSI test.

4.2 SFSI condition test

43.8 g geotechnical centrifuge test in SFSI condition was conducted to evaluate the natural frequency and seismic behaviour considering SFSI.

Equivalent shear beam (ESB) box with the inner dimension of 490 mm (length) × 490 mm (width) × 630 mm (height) was used for containing the model. ESB box is consisted of multiple frames with rubber layers in between to move alongside with the soil. This structure reduces the boundary effect when dynamic loading is applied to the box (Lee et al. 2013).

300 mm depth dry silica soil layer with a relative density of 85% was produced by using sand raining. Accelerometers were installed on the bottom of the model box and the soil surface to check the input acceleration and the surface acceleration. Five sets of bender element arrays were installed in each 50mm depth of soil layer to measure shear wave velocity. Bender elements were fixed in parallel aluminium bars (Kim & Kim, 2010).

Accelerometers were attached to the tower head to measure wind turbine's time-acceleration history, and two laser displacement sensors were installed to measure wind turbine's displacement during the test. Soil condition and sensor installation locations were identical for all three tests.

Foundations were installed in the middle of the ESB box by applying vertical weight to the foundation. Pictures and schematics for SFSI condition centrifuge test are described in Figure 4.

Figure 5 shows soil's Young's modulus by depth. Young's modulus was evaluated from the shear wave velocity profile acquired from bender element test. Shear wave velocity was converted to shear modulus (G_{max}) by using equation (1), then converted to Young's modulus (E_{max}) by using equation (2). The density of silica soil layer was 1530 kg/m³ and Poisson's ratio (ν) was assumed 0.25.

$$G_{max} = \rho V_s^2 \qquad (1)$$

$$E_{max} = G_{max} * 2(1+\nu) \qquad (2)$$

Geotechnical centrifuge test was conducted in two phases. First, the natural frequency was evaluated by

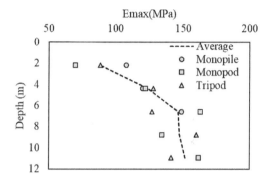

Figure 5. Variation of Young's modulus of soil with depth.

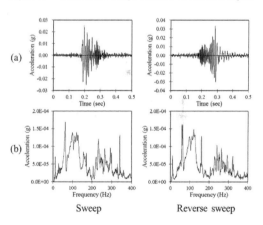

Figure 6. (a) Time, (b) frequency history of input signal.

using centrifuge mounted shaking table (Kim et al. 2013). Two types of frequency sweeps, 1) sweep signal, 30~300 Hz low-to-high frequency, 2) reverse sweep signal, 300~303 Hz high-to-low frequency were applied. Signals were applied from the bottom of the model box. Time and frequency history of input signals are shown in Figure 6. Time-acceleration history of the wind turbine was measured by the accelerometer attached to the tower head. The result was converted into frequency-acceleration history by FFT. Natural frequency was evaluated by picking the highest point in the frequency response.

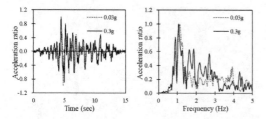

Figure 7. Time and frequency history of Hachinohe earthquake signal at 0.03 g and 0.3 g intensity.

Second, seismic behaviour was evaluated by applying Hachinohe earthquake signal with various intensities from 0.03 g up to 0.3 g. Time and frequency history of Hachinohe earthquake signal for 0.03 g input dynamic loading is shown in Figure 7. Results were converted into the prototype using the scaling law. Acceleration intensities were normalised by the ratio to the peak to show the agreement of signal shapes in different intensities.

5 TEST RESULT

5.1 Natural frequency

Figure 8 shows the natural frequency evaluation result for fixed-based scaled model test and SFSI geotechnical centrifuge test. Both results were converted into the prototype using the scaling law. Acceleration intensities were normalised by the ratio to the peak for more obvious comparison.

Tripod foundation showed 21%, monopile showed 28%, and monopod showed 42% natural frequency reduction due to the effect of SFSI. As three foundations have same mass and substructure height, it can be considered that if the amount of material used and condition of site is identical, monopod foundation would showed much more flexible behaviour then other foundation types.

Monopile and Monopod test results were compared with analytical evaluation of period lengthening ratio (PLR) suggested in FEMA 440 (FEMA, 2005) as shown in equation (3)

$$\text{PLR} = [1+(k^*_{fix}/k_x)+(k^*_{fix}h^2/k_\theta)]^{0.5} \quad (3)$$

Fixed structural stiffness (k^*_{fix}) was evaluated by fixed-based added mass test (Kim et al. 2014), effective height (h) was evaluated by calculating the centre of mass (FEMA, 2005).

Previous researchers suggested Horizontal (k_x) and rotational (k_θ) foundation stiffness for monopile and monopod foundations. Arany et al. (2017) collected stiffness evaluation methods for monopile foundation, Gelagoti et al. (2015) and Jalbi et al. (2018) suggested stiffness evaluation methods for monopod foundation. Also, stiffness evaluation methods for embedded foundation suggested in DNV-OS-J101 (Veritas, 2010) can be applied to estimate monopod foundation stiffness (Kjørlaug et al. 2015).

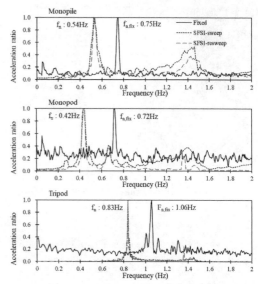

Figure 8. Natural frequency comparison between fixed-based and SFSI condition

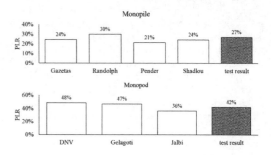

Figure 9. Comparison of PLR between analytical calculations and experiment result.

Figure 9 shows the comparison between analytical and experimental PLR. Both monopile and monopod showed promising match between analytical and experimental evaluation result with less than 6% difference in PLR, verified the reliability of conventional evaluation method.

5.2 Seismic behaviour

Figure 10 shows the comparison between spectral acceleration evaluated from surface response spectrum and peak acceleration recorded at the tower head. Results were converted into the prototype by using the scaling law. The results showed that conventional seismic evaluation methods underestimates the seismic behaviour of the offshore wind turbine. Especially in case of monopile and monopod, design acceleration evaluation considered that offshore wind turbine is mostly unaffected by the earthquake. However, recorded peak acceleration showed notable seismic behaviour.

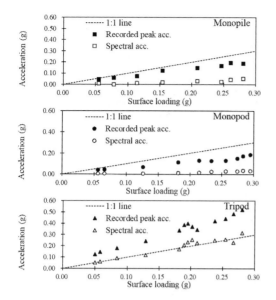
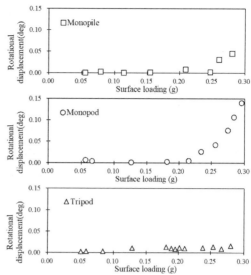

Figure 10. Spectral acceleration and recorded peak acceleration of offshore wind turbine.

Figure 11. Permanent rotational displacement of offshore wind turbine.

DNV-OS-J101 (Veritas 2010) suggested that the limit of the permanent accumulated rotation is 0.25-degree for Monopile foundation. Figure 11 shows permanent rotational displacement measured after each earthquake was applied. Monopile test result showed recognizable permanent rotational displacement over 0.25 g surface loading and showed 0.06-degree total rotational displacement after the series of earthquakes were applied. Monopod test result showed recognizable permanent rotational displacement over 0.21 g surface loading and showed 0.14-degree total rotational displacement after the series of earthquakes were applied. Tripod test result showed little to no rotational displacement through the series of earthquake and showed only 0.02-degree total rotational displacement.

Considering both the peak acceleration and the permanent rotational displacement, tripod foundation is favorable for reducing permanent deformations against possible strong earthquake, while monopod foundation is favorable for reducing peak acceleration suffered by the wind turbine structure against weak and frequent earthquake.

6 CONCLUSION

In this research, natural frequency and seismic behaviour of offshore wind turbine considering SFSI was evaluated by using geotechnical centrifuge test.

Natural frequency evaluation result showed that if the amount of material used and condition of site is identical, monopod foundation showed much more flexible behaviour then other foundation types. Also, the comparison between analytical evaluation and geotechnical centrifuge test result verifies the reliability of conventional PLR evaluation methods based on a simplified spring system for monopile and monopod foundation. Further study is needed for evaluating stiffness for a complex foundation type such as a tripod.

Seismic behaviour evaluation result showed that conventional seismic evaluation methods underestimates the seismic behaviour of the offshore wind turbine. Also comparison of seismic behaviour between foundation types showed that that foundation types need to be considered based on the earthquake tendency of the target site.

Overall, these results verified the effect of SFSI toward the natural frequency and seismic behaviour through geotechnical centrifuge test. Still, further experiments considering actual field soil conditions are needed to provide results that are more precise. Especially effect of liquefaction should be considered in the further experiment as it can affect both dynamic behaviour and permanent deformation of soil condition.

ACKNOWLEDGEMENT

This work was supported by the National Research Foundation of Korea (NRF) grant funded by the Korea government (MSIT) (No. 2017R1A5A1014883).

REFERENCES

Adhikari, S., Bhattacharya, S. 2011. Vibrations of wind-turbines considering soil-structure interaction. *Wind and Structures,* 14(2): 85–112.

Arany, L., Bhattacharya, S., Macdonald, J., & Hogan, S. J. 2017. Design of monopiles for offshore wind turbines in

10 steps. *Soil Dynamics and Earthquake Engineering*, 92, 126–152.

Bazeos, N., Hatzigeorgiou, G., Hondros, I., Karamaneas, H., Karablis, D., & Beskos, D. 2002. Static, Seismic and Stability Analyses of a Prototype Wind Turbine Steel Tower. *Engineering Structures*, 24(8): 1015–1025

FEMA, A. 2005. 440, Improvement of nonlinear static seismic analysis procedures. FEMA-440, Redwood City.

Gelagoti, F. M., Lekkakis, P. P., Kourkoulis, R. S., & Gazetas, G. 2015. Estimation of elastic and non-linear stiffness coefficients for suction caisson foundations. *Proceedings of the XVI ECSMGE*

Haritos, N. 2007. Introduction to the analysis and design of offshore structures–an overview. *Electronic Journal of Structural Engineering*, 7 (Special Issue: Loading on Structures): 55–65.

Hongwang, M. 2012. Seismic analysis for wind turbines including soil-structure interaction combining vertical and horizontal earthquake. *15th World Conference on Earthquake Engineering, Lisbon, Portugal.*

Jalbi, S., Shadlou, M., & Bhattacharya, S. 2018. Impedance functions for rigid skirted caissons supporting offshore wind turbines. *Ocean Engineering*, 150, 21–35.

Jonkman, J., Butterfield, S., Musial, W., & Scott, G. 2009. Definition of a 5-MW reference wind turbine for offshore system development (No. NREL/TP-500-38060). National Renewable Energy Laboratory (NREL), Golden, CO.

Kim, D. K., Lee, S. H., Kim, D. S., Choo, Y. W., & Park, H. G. 2014. Rocking effect of a mat foundation on the earthquake response of structures. *Journal of Geotechnical and Geoenvironmental Engineering*, 141(1), 04014085.

Kim, D. S. & Kim, N. R. 2010. A shear wave velocity tomography system for geotechnical centrifuge testing. *Geotechnical Testing Journal*, 33(6), 1–11.

Kim, D. S., Lee, S. H., Choo, Y. W., & Perdriat, J. 2013. Self-balanced earthquake simulator on centrifuge and dynamic performance verification. *KSCE Journal of Civil Engineering*, 17(4), 651–661.

Kjørlaug, R.A. & Kaynia, A.M. 2015. Vertical earthquake response of megawatt-sized wind turbine with soil-structure interaction effects. *Earthquake Engng Struct. Dyn.* 44:2341–2358.

Lee, S.H., Choo, Y.W., & Kim, D.S. 2013. Performance of an equivalent shear beam (ESB) model container for dynamic geotechnical centrifuge tests. *Soil Dynamics and Earthquake Engineering*, 44: 102–114.

Prowell, I., Elgamal, A., & Jonkman, J. FAST Simulation of Wind Turbine Seismic Response. Technical Report No.NREL/CP-500-46225, National Renewable Energy Laboratory March 2010.

Schofield, AN. 1980. Cambridge geotechnical centrifuge operations. *Géotechnique*, 30(3): 227–268.

Seong, J. T., Ha, J. G., Kim, J. H., Park, H. J., & Kim, D. S. 2017. Centrifuge modeling to evaluate natural frequency and seismic behavior of offshore wind turbine considering SFSI. *Wind Energy*, 20(10), 1787–1800.

Veritas, D. D. N. 2010. DNV-OS-J101 offshore standard. Design of offshore wind turbine structures.

An investigation on the performance of a self-installing monopiled GBS structure under lateral loading

K.J.L. Stone & A. Tillman
School of Environment and Technology, University of Brighton, Brighton, UK

M. Vaziri
Ramboll, UK

ABSTRACT: This paper presents a physical model study of the feasibility and performance of a monopiled gravity base structure (MGBS) for offshore foundation applications. The research builds on the current interest in hybrid foundation systems such as monopiled footings. The proposed MGBS is essentially a conventional GBS structure with a projecting monopile or caisson. The system relies on the self-weight of the GBS to drive the projecting monopile into the seabed. Once installed additional ballast can be added to the GBS, and if necessary installation of the projecting monopile can be enhanced through the use of suction. The test programme has concentrated on the response of the system once installed, and has investigated a range of geometries for circular monopiles and cruciform shaped piles. The results clearly demonstrate a significant enhancement to lateral resistance offered by the addition of a monopile to a GBS.

1 INTRODUCTION

To date monopiled foundations have been used extensively and very effectively for foundation solutions for the majority of offshore, and to a lesser extent onshore, wind farm installations. However, to date many offshore installations have been located where water depths are such that the installation and performance of the monopile makes it the most economically viable solution. However, the development of deep water sites is not favourable to the monopile solution due to difficulties in fabricating large enough monopiles, and undertaking their installation in deep water locations.

In view of the limitations to the monopile solution, alternative designs such as floating and/or tension leg semi-submersible platforms, and large gravity base units are proposed for deep water locations.

This paper presents a study to investigate the concept of a monopiled gravity base structure (MGBS). The proposed MGBS is essentially a conventional GBS structure with a projecting monopile or caisson. The system relies on the self-weight of the GBS to drive the projecting monopile into the seabed. Once installed additional ballast can be added to the GBS, and if necessary installation of the projecting monopile/caisson can be enhanced through the use of suction.

1.1 Hybrid foundations

In recent years much research has been reported on the behaviour and viability of 'hybrid' foundation systems.
Dixon (2005) proposed the use of a bearing plate in conjunction with a monopile to increase the lateral capacity of the pile. This hybrid monopiled-footing concept is not dissimilar to that of a retaining wall with a stabilising base, see for example Carder & Brookes (1993), Carder et al. (1999) and Powrie & Daly (2007).

Model tests are reported for the case where the bearing plate is rigidly connected 'coupled' to the pile, see for example, Stone et al. (2007), Stone et al. (2010), Lehane et al. (2014), Arshi (2016), or where it is allowed to slide 'decoupled' on the pile shaft (Arshi & Stone 2015, Anastasopoulos & Theofilou 2015, Arshi 2016). In both arrangements the addition of the bearing plate is seen to enhance the lateral capacity and stiffness of the monopile. In the decoupled configuration little or no moment is transferred between the pile and the plate and enhanced lateral resistance is provided by the shear resistance between the plate and the underlying soil. In the coupled arrangement the presence of the bearing plate provides a degree of both lateral and moment fixity at the mudline leading to enhanced lateral resistance from both the shear resistance and moment restraint provided by the plate.

This study is an extension of the coupled arrangement for potential deep water deployment, where the self-weight of the structure is utilised to install the pile.

It is noted that studies on the coupled monopiled-footing in cohesionless soil have shown that for the system to be effective it is necessary that the bearing plate develops a positive contact with the underlying soil, and hence applied vertical loads should be in excess of the vertical pile capacity. If the self-weight

Table 1. Summary of model tests undertaken.

Test	Pile type	Pile diameter or width mm (m)*	Pile embedment mm (m)*
GBS-A/B/C	n/a	n/a	–
MP6-A/B/C	Circular	6.35 (0.32)	30 (1.5)
MP12-A/B/C	Circular	12.7 (0.64)	30 (1.5)
MPX-A/B/C	Cruciform	12.7 (0.64)	30 (1.5)
MP60-A/B/C	Circular	12.7 (0.64)	60 (3.0)

*The corresponding prototype dimensions are given in the brackets.

of the GBS is used to install the monopile, then this requirement will be met. However, it is noted that in a cohesionless material the sliding resistance of the GBS is directly proportional to the vertical load. Hence any reduction in the vertical load by the presence of the pile will result in a reduction of the sliding resistance of the GBS, which may or may not be offset by the lateral resistance provided by the pile.

In a cohesive soil, GBS sliding resistance is not affected by the presence of the pile and depends only on the undrained shear strength of the soil. The research reported here presents a very preliminary assessment of the behaviour of a monopiled GBS in dry sand under monotonic loading conditions, further tests using cohesive soils are planned.

2 EXPERIMENTAL PROGRAMME

A total of 15 tests are reported using four different model pile designs. Each model pile was tested with three different GBS weights of 0.256, 0.376 and 0.496 kg designated A, B and C respectively. The applied weight consists of the self-weight of the GBS and loading mast (0.256 kg), plus any additional weight added to the model GBS. The lateral performance of the GBS alone, for each vertical load, was also evaluated. The test programme is summarised below in Table 1.

3 MATERIALS AND TEST EQUIPMENT

3.1 Materials and equipment

The sand used for the model tests was uniformly graded Fraction D silica sand supplied by David Ball Limited. The sand has maximum and minimum void ratios of 1.06 and 0.61 respectively, with a critical state angle of friction φ_{crit} of 32°.

All the tests reported were carried out on the balanced beam centrifuge at the University of Brighton. This machine is manufactured by Thomas Broadbent & Sons Ltd. and is 6 g-tonne machine (20 kg payload to 300 g) and a radius of 760 mm. The tests were conducted in a circular steel tub with an

(a) (b)

Figure 1. a) Assembled model GBS and loading mast and b) assortment of model pile attachments.

internal diameter of 272 mm and sample height of 114.6 mm.

The model GBS and piles were fabricated from aluminium. Two solid circular piles of 6.35 and 12.7 mm diameter were fabricated together with a cruciform shaped pile with an overall width of 20 mm. All these piles projected 30 mm below the base of the GBS. A further 12.7 mm diameter 60 mm long pile was also fabricated. The GBS is a flat-bottomed cylinder of 75 mm external diameter with a wall thickness of 10 mm. A threaded hole is formed in the centre of the GBS base and the pile attachments are bolted through this hole and into an aluminium mast which projects above the GBS unit to provide the loading attachment for a wire and pulley arrangement. Figure 1 shows the model GBS and the piles used in the test programme.

4 MODEL PREPARATION AND TEST PROCEDURES

The soil models were prepared though a combination of dry pluviation and vibration using a vibrating table. This method produced consistent soil specimens with a bulk density of 1700 kg/m³.

The fully assembled models, including any additional weights were installed by lightly pushing the structure into the soil until firm contact with the sand surface was achieved. The steel tub was placed in a cradle which is then mounted on the centrifuge arm. A single degree of freedom actuator is then mounted on the cradle, and through the use of a wire and pulley mechanism is able to apply a horizontal load to the mast. The steel wire is attached to a load cell fixed to the actuator traveller plate. The applied horizontal load is assumed to be equal to the tension measured in the steel wire (i.e. friction effects are neglected).

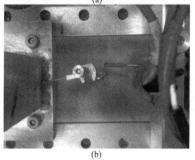

Figure 2. (a) Post-test photograph of model and (b) detail of lateral load and displacement attachment.

The horizontal displacement at the point of loading is measured by a linear potentiometer connected via a 'sheep's crook' attachment to the loading mast, refer to Figure 2b.

The loading wire was attached to the mast at a height of 120 mm above the model surface.

4.1 Test procedure

After mounting the test package on the centrifuge, the actuator is connected to the power and control system, and a digital video camera is mounted on the top of the cradle to observe the model during the test. All the tests were conducted at an acceleration level of 50 g.

Once the model had achieved the test acceleration the actuator was run at a velocity of 5 mm/minute and data from the load cell and displacement transducers are recorded continuously until the end of the test. Figure 2a shows a typical model after testing.

5 EXPERIMENTAL RESULTS

Figures 3a to 3c present load-displacement responses for the GBS alone, and with the 12.7 mm diameter pile (MP12) and cruciform pile (MPX) for the three applied vertical loads (A = 0.265 kg, B = 0.376 kg and C = 0.496 kg).

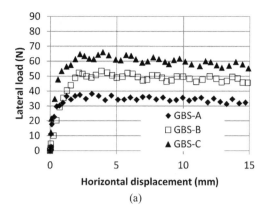

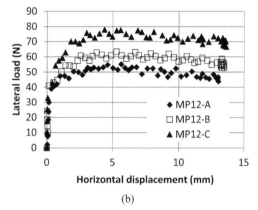

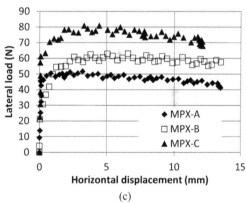

Figure 3. Lateral load displacement plots for (a) GBS, (b) GBS with 12.7 mm dia, 30 mm long pile and (c) 30 mm long cruciform pile (blades at 45 deg to load).

Whilst the data presented above is unfortunately somewhat noisy as a result of worn signal slip rings, it is still apparent that the addition of the piles to the GBS enhances the lateral capacity of the system by approximately 20–25%. It is also interesting to note that the initial lateral stiffness of the GBS appears to be slightly enhanced by the presence of the 30 mm long piles. For all the tests the load displacement curves are seen to plateau at about 2–2.5 mm of lateral displacement

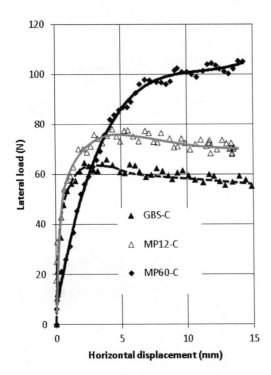

Figure 4. Lateral load displacement plots for GBS compared with response for GBS with 30 mm and 60 mm long 12.7 mm diameter piles.

at the point of loading, which corresponds to about 1.2 degrees of rotation as estimated from in-flight photographs.

Figure 4 above compares the lateral response of the heaviest GBS-C with and without 12.7 mm diameter piles of 30 and 60 mm in length. It is interesting to note that the increase in length of the pile from 30 to 60 mm results in a significant increase in lateral capacity, and also a less distinct development of a load plateau, coupled with a less stiff lateral response.

6 DISCUSSION AND CONCLUSIONS

The tests reported here are preliminary in nature, and due to difficulties with instrumentation and logging the quality of the data is poor. Nevertheless the tests do provide some useful insights into the potential performance of a GBS unit with a monopile extending from its base. A range of circular piles of various lengths and diameters were tested, together with a cruciform shaped pile.

It is suggested that for deep water sites where a GBS type of installation is proposed, then the combination of the GBS with a monopile would provide a more efficient solution. The use of the GBS to install the pile would negate the requirement to drive the piles, either from a barge or from an autonomous sea-bed system.

The tests undertaken in this study have focused on the response of the system post-installation, where it is essentially similar to a hybrid monopiled-footing arrangement. For the current arrangement the pile was rigidly fixed to the base of the GBS. However it is also possible that the GBS's weight can be used to install a monopile which is not fixed to the GBS unit. In this arrangement the pile remains decoupled from the GBS such that the GBS is restrained laterally by the pile but is also able to move vertically relative to the pile.

From the selected results presented in this paper the following initial conclusions can be made.

- The lateral capacity of the GBS can be enhanced by the presence of a monopile extending beneath its base.
- The response of the system is similar to a monopiled footing.
- A cruciform shaped pile is effective.
- Increasing the length of the pile increases the lateral capacity but may reduce the lateral stiffness.

It is clear that there is a lot of further study required to evaluate the full potential of a self-installing monopiled-GBS. Issues such as the performance of the system in clay, cyclic loading, optimisation of geometries for self-installation with satisfactory lateral performance, and the nature of the pile-GBS connection, are just a few of the areas that need to be addressed.

ACKNOWLEDGEMENTS

The authors are grateful for the assistance of David Harker in the preparation and operation of the centrifuge testing equipment at the University of Brighton.

REFERENCES

Anastasopoulos, I. & Theofilou, M. 2015. On the development of a hybrid foundation for offshore wind turbines. *Proceedings of the 3rd International Symposium on Frontiers in Offshore Geotechnics* (Mayer ed.), Norway, ISBN:978-1-138-02848-7.

Arshi, H.S. 2016. Physical and numerical modelling of hybrid monopile-footing foundation systems. PhD dissertation, University of Brighton.

Arshi, H.S. & Stone, K.J.L. 2015. Improving the lateral resistance of offshore pile foundations for deep water application. Proceedings of the 3rd International Symposium on *Frontiers in Offshore Geotechnics* (Mayer ed.), Norway, ISBN:978-1-138-02848-7.

Carder, D.R., Watson, G.V.R., Chandler R.J. & Powrie, W. 1999. Long term performance of an embedded retaining wall with a stabilizing base. *Proc. Instn. Civ. Engrs Geotech. Engng* 137, No. 2, 63–74.

Carder, D.R & Brookes, N.J. 1993. Discussion. In Retaining Structures (ed. C.R.I Clayton), London, Thomas Telford, 498-501.Dixon RK (2005) Marine Foundations. WO (Patent application) 2005/038146.

Dixon, R.K. 2005. Marine Foundations. WO (Patent application) 2005/038146.

Lehane, B.M, Pedram, B, Doherty, J.A. & Powrie, W. 2014. Improved performance of monopiles when combined with footings for tower foundations in sand. *Journal of Geotechnical and Geoenvironmental Engineering* 140(7):04014027

Powrie, W, & Daly, MP. (2007). Centrifuge modelling of embedded retaining wall with stabilising bases. Geotechnique. 57(6), 485–497.

Stone, K.J.L, Newson, T.A. & Sandon, J. 2007. An investigation of the performance of a 'hybrid' monopole-footing foundation for offshore structures. *Proceedings of 6th International on Offshore Site Investigation and Geotechnics*. London: SUT, 391–396.

Stone, K.J.L, Newson, T.A. & El Marassi, M. 2010. An investigation of a monopiled-footing foundation. *International Conference on Physical Modelling in Geotechnics, ICPMG2010*. Rotterdam: Balkema, 829–833.

Model tests on the lateral cyclic responses of a caisson-piles foundation under scour

C.R. Zhang, H.W. Tang & M.S. Huang
Department of Geotechnical Engineering, Tongji University, Shanghai, China

ABSTRACT: As a new type of deep foundation, the caisson-piles composite foundation is a selection for the Qiongzhou Straits bridges. It can greatly reduce the height of the caisson by connecting the lower pile groups to the upper caisson, and improves its mechanical behaviour under lateral load. A series of model tests are conducted at 1g to investigate the scour effects on the responses of the caisson-piles composite foundation under long-term lateral cyclic load, in which two scour depths and two cyclic load amplitudes are applied. The cyclic load pattern is single cycle sinusoidal curve with 10,000 cycles. Based on the test results, the accumulated residual rotations are found to be dependent on the scour depth and the applied cyclic load amplitude. An empirical method is developed to predict the accumulated rotation due to long-term cyclic loading, in which the effect of scour is considered by a revised dimensionless cyclic load level, relating the cyclic load magnitude to the post-scour lateral capacity of the composite foundation.

1 INTRODUCTION

The strategic planning of a 26.3-km-long cross Qiongzhou strait passage project linking Hainan Island to mainland is under discussion in China. In the construction site, the bed rock is underlain by a 300-meter-thick soil stratum. The cross-sea bridge passes through a water depth of more than 50m for a length of 8 km. Considering the complex geological conditions, the caisson-piles composite foundation, composed of an upper caisson and a lower pile group, is brought up (see Fig. 1) to reduce the embedment depth of the caisson for construction. Under the marine environment, the foundations sustain long-term horizontal cyclic load from wind, wave, current and so on. Besides, scouring induced soil erosion is also a challenge, for the local scour hole can reduce the lateral resistance of the foundation and leads to an increase of excessive deformation (Qi et al. 2016). The cyclic behaviour of monopile in sand is widely investigated based on results of field tests (Li et al. 2015, Lin & Liao 1999, Long & Vanneste 1994), centrifuge tests (Zhu et al. 2016) and laboratory tests (Chen et al. 2015, Cuéllar et al. 2012, Leblanc et al. 2010). However, the effects of both cyclic load and scouring on the responses of piles are seldom considered. Let alone the long-term cyclic responses of the caisson-piles composite foundation under scouring.

In this paper, laboratory tests are carried out to simulate a caisson-piles composite foundation in dry sand subjected to 10,000 load cycles of one way sinusoidal cyclic loads. Meanwhile, the effects of scour on the responses of the composite foundation are analysed

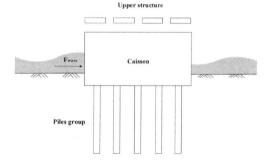

Figure 1. Schematic plot of a typical caisson-piles composite foundation.

by considering the scour-hole geometry around the foundation with two different scour hole depths. This paper aims to study the cyclic interaction between the caisson-piles composite foundation and the soil in dry sand under scouring.

2 TEST SETUP AND PROGRAMME

2.1 Test preparation and loading device

The sand used in the test is taken from a river in the city of Shanghai, the properties of which are shown in Table 1. By mass controlling, the soil was poured into the container layer by layer. Each layer was compacted to a thickness of 0.03 m with a grinding wheel. Having a sample box buried in the soil bed, the measured

Table 1. Properties of the river sand.

Property		Value
Particle sizes: $D_{10}, D_{30}, D_{50}, D_{60}$	mm	0.18/0.3/ 0.37/0.42
Specific gravity, G_S	–	2.65
Minimum dry unit weight, γ_{min}	kN/m³	1.322
Maximum dry unit weight, γ_{max}	kN/m³	1.573
Relative density, D_r	%	60
Effective internal friction angle, φ	(°)	30

Table 2. Properties of the model caisson-pile foundation.

Property	Value
Caisson diameter, mm	300
Caisson length, mm	300
Caisson wall thickness, mm	10
Caisson embedment depth, mm	130
Number of piles	8
Pile diameter, mm	10
Pile length, mm	300
Pile wall thickness, mm	1

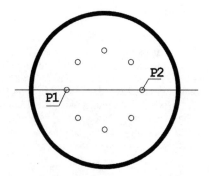

Figure 2. The pile group configuration.

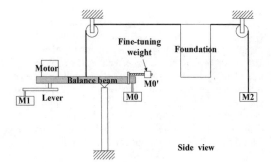

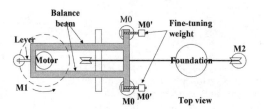

Figure 3. Schematic plot of the cyclic loading device.

relative density of the river sand is approximately 60%, which represents a dry unit weight of 1.462 kN/m³.

For the scheme design of the cross Qiongzhou strait passage project, the caisson-piles foundation is designed to sustain a cable-stayed bridge with two 1500 m spans. The diameter and length of the caisson were 88.4 m and 90 m with an embedment depth of 40 m. The pile group is composed of 92 steel pipe piles with 1.8 m in diameter, 90 m in length. As the limited size of the soil container (1.3 m × 1 m × 0.7 m), the dimension of the model caisson and the length of model pile are scaled to approximately 1:300. Since the shallow caisson-piles composite foundation are brought out to replace a deep caisson foundation having a length of 120 m and embedment depth of 70 m, it is thought that the piles groups in scheme design are too conservative. The number of model piles is reduced to 8 for simplicity with the plan view as a circle (See Fig. 2). Based on the same embedment area in the soil stratum for the deep caisson and the shallow-piles composite foundation, the model pile groups represents a pile group with 32 piles and each pile with a diameter of 0.92m. The properties of the caisson and the pile is listed in Table 2.

The aluminium alloy made composite foundation is placed in the centre of the container when preparing the soil sample. As the stress in the laboratory test at 1 g is much smaller than in the field, the stress history of sand due to general scouring is not taken into account. To consider the effects of local scour, a scour hole is created around the caisson in the shape of an inverted truncated cone. A scour hole slope of 30° and two different scour depth, 40 mm and 80 mm are adopted. The scour hole is excavated manually with a shovel.

A simple mechanical load rig has been used in previous studies (Leblanc et al. 2010, Zhu et al. 2013) to supply a stable and long-term cyclic load for model tests, which is later improved by Chen (Chen et al. 2015). Based on Chen's work, some improvements are done as below in Figure 3. The double balance beams can reduce the unbalanced force at the pivot and the lengthened beam increases the magnitude of the cyclic load. Two additional weight hanger are designed to move along the screw, in order to fine tune the initial balance of the loading device. As the motor drives the mass M_1 to move in a circle, the loading system generates a sinusoidal cyclic force, with the size and type adjusted by the weight M_1 and M_2. The composite foundation is fixed horizontally during the balance adjustment. The horizontal deflection and rotation of the foundation are measured by two LVDTs positioned at two levels on the caisson. The physical photograph of the model test is shown in Figure 4.

2.2 *Test program*

The testing program comprises monotonic and cyclic tests with and without scour. After the cyclic load is

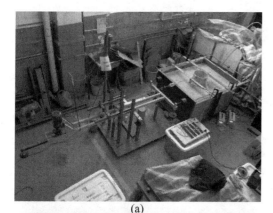

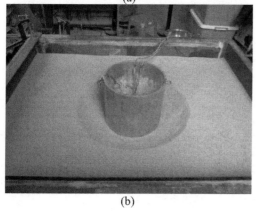

Figure 4. Photographs of the (a) loading device (b) scour hole used in the laboratory tests.

Table 3. Test programme.

Test name	Scour depth S_d mm	Scour angle θ (°)	Buried depth H mm	Cyclic amplitude N	Cyclic number
A1	–	–	430	–	–
A2	–	–	430	40	10,000
A3	–	–	430	80	10,000
B1	40	30°	390	–	–
B2	40	30°	390	40	10,000
B3	40	30°	390	80	10,000
C1	80	30°	350	–	–
C2	80	30°	350	40	10,000
C3	80	30°	350	80	10,000

completed, a monotonic load test was applied to obtain the post-cyclic performance of the composite foundation. Two scour hole depth and the one-way cyclic load with two amplitudes are conducted. The details about the scour hole and the cyclic load are given in Table 3.

The frequency of the cyclic load is 0.1 Hz, low enough that only the effect of repeated load is important. Based on the Leblanc (2010) definition, two independent parameters to characteristic the sinusoidal load at the pile head, are

$$\zeta_b = \frac{F_{max}}{F_u}, \quad \zeta_b = \frac{F_{max}}{F_{min}}, \qquad (1)$$

in which F_{max} and F_{min} are maximum and minimum load in the load cycle, F_u is the static capacity, ζ_b is a measure of the size of the cyclic load and ζ_c represents the type of the cyclic load, (0 for one-way cyclic load, −1 for symmetric two-way cyclic load). For the model tests of the composite foundation subjected to long-term cyclic load, the time consuming is a big challenge. Since we focus on the accumulated rotations of the composite foundation combined with the effect of scouring, only the one-way cyclic load is applied, which means $F_{min} = 0$, F_{max} = the amplitude of the cyclic load, relating to $\zeta_c = 0$. Based on the static load-rotation curve without scouring in Figure 5(a), the capacity of the horizontal load is obtained as 160 N, which is defined by a rotation of 0.5°. The value of ζ_b is 0.25 and 0.5 corresponding to the cyclic load amplitude as 40 N and 80 N.

3 TEST RESULTS

3.1 Load displacement responses

The representative load-rotation curves for the monotonic test and post-cyclic tests are given in Figure 5(a) and (b), in relation to the cyclic load amplitude of 40 N and the scour hole depth of 40 mm respectively. A work-hardening behaviour of the measured load-displacement is obtained, which makes the identification of a distinct failure corresponding to the turning point difficult. Therefore, the capacity of the model foundation F_u is taken as the load at a rotation of 0.5°, and the results for the 9 tests are shown in Figure 5c. The figure indicates that the stiffness of the composite foundation decreases with scour depth, at an increasing rate, while the effect of the previous cyclic load is opposite, which induces an increases of stiffness, and at a decreasing rate with cyclic load amplitude.

3.2 Accumulated displacements

In Leblanc's (2010) work, it is explained that a function of log(N) will underestimate the accumulation for large cyclic numbers ($N > 500$) while an exponential function with N gives a good fit with the measured values. This may relates to the two distinct phases of the accumulated deformation, an initial stage with rapid increase of accumulated rotation under small cyclic numbers and a following steady stage with limited rotation accumulation which is too small in each cycle but has a profound influence on the long-term behaviour of the foundation with respect to the fatigue limit state as approximately 10^7 cyclic numbers.

As plotted in Figure 6 (the dotted line), a good fit of the accumulated residual rotations of the composite

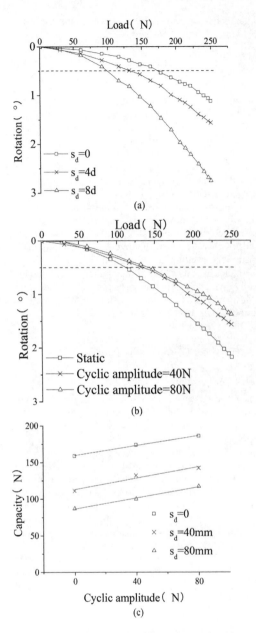

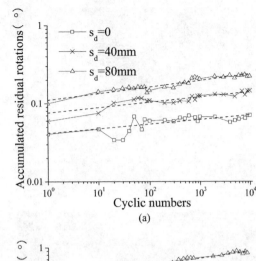

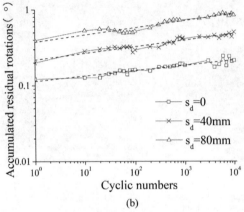

Figure 6. Accumulated residual rotations with cyclic numbers in double logarithmic axis (a) cyclic amplitude = 40N; (b) cyclic amplitude = 80N.

Figure 5. Load displacement curves of a composite foundation (a) cyclic amplitude = 40 N; (b) scour depth = 40 mm; (c) capacity.

foundation subjected to one-way cyclic load by an exponential function is given as

$$\theta_N = \theta_1 N^\alpha \qquad (2)$$

where the value of the exponent α is a constant as 0.09. It is noted $\alpha = 0.31$ and 0.39 respectively for the laboratory tests of a cyclically loaded monopile and a cyclically loaded suction caisson by Leblanc et al. (2010) and Zhu et al. (2013).

θ_1 in Equation (2) is the residual rotation for the first cyclic load, which is determined from the point of interaction with the vertical axis when N = 1. As only one-way cyclic load with $\zeta_c = 0$ is applied in the laboratory tests, the relation of θ_1 is only discussed with respect to the cyclic load size $\zeta_b = F_{max}/F_u$. Figure 7(a) shows that when plotted in terms of ζ_b, the value of θ_1, depends on three scour depths, in which θ_1 is obtained as the intercept of the dotted line of the six tests data in Fig. 6. In order to simplify the calculation, the modification to consider the scour effect is given below. By reviewing the definition of $\zeta_b = F_{max}/F_u$ in Leblanc et al. (2010), it is found that the quantification of the cyclic load amplitude is expressed as a dimensionless parameter in relation to the static capacity of the foundation.

The comparison of the results of the monotonic tests with and without scouring in Figure 5 reveals that scour has a disadvantage effect on lateral responses of the foundation, such as a reduced stiffness and a smaller capacity. Therefor a modified definition of the size of the one-way cyclic load ζ_b' is used to consider the

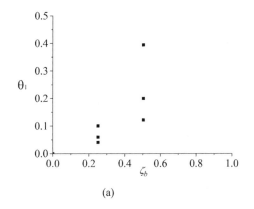

(a)

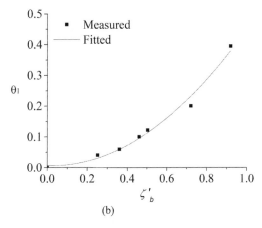

(b)

Figure 7. Relationship between θ_1 and characteristic cyclic load level: (a) with ζ_b (b) with ζ_b'.

effects of scour hole geometry on the residual rotation accumulation θ_1, as

$$\zeta_b' = \frac{F_{max}}{F_{u,s}} \quad (3)$$

where $F_{u,s}$ is the static lateral capacity of the foundation affected by scouring. The θ_1 is replotted in Figure 7(b) as a function of the updated cyclic load size ζ_b', where the behavior of the function $\theta_1(\zeta_b')$ is apparent and the curve is easily fitted.

In the rest part, we discuss the effect of scour hole depth to the static lateral capacity of the composite foundation $F_{u,s}$, which is an important parameter to determine the dimensionless cyclic amplitude ζ_b'. Based on the results of the wave flume model test, Liang et al. (2016) advised a 50% discount of the maximum scour depth by USA specifications HEC-18 (Richardson & Davis 2001), for the caisson with a diameter large than 15 m. According to Liang's suggestion, the maximum scour depth for the caisson in this paper is 16cm and the corresponding dimensionless scour depth $\tilde{s}_d$ is 0, 0.25, 0.5, respectively. In Liang's work, it is explained that the calculated scour depth is not a conservative value and a careful adjustment should be done when the caisson diameter larger than

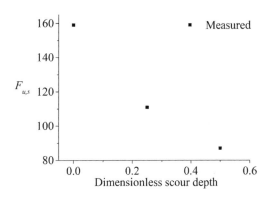

Figure 8. Values of $F_{u,s}$ in relation to dimensionless scour depth.

30 m. It may be the reason that 16 cm here is larger than the embedment depth of 13 cm for the caisson in the model test. As only three test results are available as shown in Figure 8, the post-scour capacity of the composite foundation $F_{u,s}$ can just be interpolated or extrapolated to obtain the value for other scour depths. It is noted that the maximum dimensionless scour depth in Figure 8 should be smaller than 0.8125. The measure in calculating $F_{u,s}$ is doubtful, while for a preliminary study, the aim of the work in this paper is to provide a feasible way in evaluating the long-term cyclic responses of the caisson-piles composite foundation considering the scour effects, both of which are crucial for the foundations under marine environments.

4 CONCLUSIONS

For the application of the caisson-piles composite foundation to cross Qiongzhou strait passage project, a series of laboratory tests is carried out on the model foundation subjected to lateral cyclic load and under scouring in sand. The main conclusions are drawn as below:

1. The one-way cyclic loading increases the stiffness and capacity of the post-cyclic behaviour of the composite foundation subjected to monotonic load. And the scouring has a disadvantageous effect on the behaviour of the composite foundation with a decreased capacity.
2. The caisson-piles composite foundation shows a good performance under long-term cyclic load since the accumulation rate of the residual rotation at the head of the composite foundation is small.
3. An exponential expression based on the model test results is presented to predict the accumulated residual rotations with large cyclic numbers. To consider the effects of scour on the long-term responses of the foundation, a modified dimensionless cyclic load size in relation to the static foundation capacity under scouring should be used.

ACKNOWLEDGEMENTS

The authors acknowledge the National Basic Research Program of China (973 Program) (Grant No. 2013CB036304) and the National Natural Science Funds of China (Grant No. 51378392, No. 51779175) for the financial support.

REFERENCES

Chen, R.P., Sun, Y.X., Zhu, B. & Guo, W.D. 2015. Lateral cyclic pile–soil interaction studies on a rigid model monopile. *Geotechnical Engineering* 168(2): 120–130.

Cuéllar, P., Georgi, S., Baeßler, M. & Rücker, W. 2012. On the quasi-static granular convective flow and sand densification around pile foundations under cyclic lateral loading. Granular Matter 14(1): 11–25.

Leblanc, C., Houlsby, G.T. & Byrne, B.N. 2010. Response of stiff piles in sand to long-term cyclic lateral loading. *Géotechnique* 60(2): 79–90.

Li, W.C., Igoe, D. & Gavin, K. 2015. Field tests to investigate the cyclic response of monopiles in sand. *Geotechnical engineering* 168(5): 407–421.

Liang, F.Y., Wang, C. & Huang, M.S. 2016. Scale effects on local scour configurations and around caisson foundations and dynamic evolutions. *China Journal of highway and transport* 29(9): 59–65.

Lin, S.S. & Liao, J.C. 1999. Permanent strains of piles in sand due to cyclic lateral loads. *Journal of Geotechnical and Geoenvironmental Engineering* 125(9): 798–802.

Long, J.H. & Vanneste, G. 1994. Effects of cyclic lateral loads on piles in sand. *Journal of Geotechnical Engineering* 120(1): 225–244.

Qi, W.G., Gao, F.P., Randolph, M.F. & Lehane, B.M. 2016. Scour effects on p–y curves for shallowly embedded piles in sand. *Géotechnique* 66(8): 1–13.

Richardson, E.V. & Davis, S.R. 2001. Evaluating scour at bridges. *Washionton DC: FHWA*.

Zhu, B. Byrne, B. & Houlsby, G.T. 2013. Long-Term Lateral Cyclic Response of Suction Caisson Foundations in Sand. *Journal of Geotechnical & Geoenvironmental Engineering* 139(1): 73–83.

Zhu, B., Li, T., Xiong, G. & Liu, J.C. 2016. Centrifuge model tests on laterally loaded piles in sand. *International Journal of Physical Modelling in Geotechnics* 16(4): 1–13.

Comparison of centrifuge model tests of tetrapod piled jacket foundation in saturated sand and clay

B. Zhu, K. Wen, L.J. Wang & Y.M. Chen
Institute of Geotechnical Engineering, Zhejiang University, Hangzhou, China

ABSTRACT: Tetrapod piled jacket (TPJ) foundations have been widely used in offshore wind farms due to their advantages of rigidity and influence by wave action. This paper compares centrifuge model tests of TPJ foundation in saturated sand and in normally consolidated clay when subjected to lateral monotonic loads. It is found that the depth of maximum bending moment and inflection point in sand are obviously shallower than that in soft clay. Based on the distribution curve of lateral pile displacement derived from the bending moment, limited pile deflection in sand can be observed below a depth of approximately 7D (D denotes pile diameter), while this value is about 12D in clay. The soil resistance around back row piles is about 40% of the front row piles in sand since the increase of axial force in piles could change the effective stress around the piles, while that difference is not significant for the undrained soft clay. The failure of TPJ foundation under lateral static loads is due to the back row piles pulling out both in sand and clay, as observed from the vertical load-displacement curves of the piles.

1 INTRODUCTION

Offshore wind power has been paid more attention worldwide due to its high generation efficiency and impact on the surrounding environment. With the increase of offshore wind farm distance and water depth, the development of tetrapod piled jacket (TPJ) foundations has been a priority in most offshore wind farms. The TPJ foundations for offshore wind turbine are mainly subjected to lateral loads induced by the wave, wind, earthquake, etc. during its service cycle, which can be facilitated by a lateral load applying to a certain height (Byrne & Houlsby 2003). Therefore, the study on the lateral bearing capacity of TPJ foundation for offshore wind turbines is of great importance.

A number of researchers have investigated jacket foundation for offshore wind turbines (e.g. Elshafey et al. 2009; Shi et al. 2011; Zhao et al. 2017), while most of them focus on the overall response of the foundation without examining the detailed behaviour of the piles. These researches mainly adopt numerical methods to study the lateral loading behaviour of TPJ foundations. Mao et al. (2015) established a 1/10 scale experimental model to study the influence of foundation degradation on the TPJ foundation. However, 1g model tests can't duplicate the non-linearity and stress-dependent behaviour of the soil and the loading response of TPJ foundation subjected to lateral loads.

Centrifuge modelling offers an effective and convenient method to understand the influence of lateral monotonic load on TPJ foundation. In centrifuge tests, small-scale model loading tests are conducted in acceleration fields many times greater than that of the earth's gravity to ensure the highly non-linear soil behaviour and stress can also be replicated realistically (Li et al. 2010). Thereby, Zhu et al. (2015) performed a set of centrifuge tests to understand the loading behaviours of TPJ foundation in sand when subjected to lateral monotonic and cyclic loading.

Based on the previous centrifuge tests reported by Zhu et al. (2015), the objective of this paper is to compare the loading behaviour of TPJ foundations in saturated sand with that in soft clay, including the bending moment, lateral deflection and pile-soil interaction of the TPJ foundation.

2 METHODOLOGY

2.1 Test apparatus

The centrifuge tests of TPJ foundation were carried out at 100 g on the centrifuge ZJU-400 at Zhejiang University. The payload capacity and maximum acceleration of centrifuge are 400 g-ton and 150 g respectively, and more details about the centrifuge can be found in Chen et al. (2010).

2.2 Model TPJ foundation

The prototype dimension of TPJ foundation was chosen to be the same as that of a typical 3 MW offshore wind turbine in Guishan offshore wind farm in China. The acceleration (n g) of the test was 100 g, and the sizes of the testing model were chosen according to the similarity principle of centrifuge modelling (e.g. scaling the diameter and length as 1/n, and axial

Table 1. Properties of aluminium alloy pipe.

Properties	Values
Ultimate tensile strength	124 MPa
Yield strength	55.2 MPa
Elongation	25.0%
Elasticity modulus	68.9 GPa
Poisson ratio	0.33
Fatigue strength	62.1 MPa

Table 2. Parameters of the components of the model TPJ foundation.

Components	Outer diameter × Wall thickness (m × m)	Number × Length (m)
Model pile	0.0259 × 0.00295	4 × 0.55
M1	0.048 × 0.0025	1 × 0.07
M2	0.012 × 0.02	4 × 0.09
M3	0.016 × 0.0015	4 × 0.173
M4	0.008 × 0.0005	8 × 0.136, 8 × 0.16

stiffness as $1/n^2$). The model jacket and piles were made of 6061 aluminium alloy pipe, the property of which can been found in Table 1. The outer diameter (D), bending stiffness, and tensile stiffness of the model piles, after being instrumented and coated, were 0.0259 m, 286.51 N.m² and 5.11 MN, respectively. The sizes of the model jacket and piles can be found in Table 2. More details of centrifuge model tests of TPJ foundation can be found in Zhu et al. (2015).

Figure 1 presents the schematic diagram of model TPJ foundation and the test set-up in the centrifuge. The lateral load was applied by load motor and the loading position was 31.5 m above the mud surface. Three laser displacement transducers were setup to measure the lateral displacement at top of piles (Including front row piles and back row piles) and loading position. Two linear variable differential transformers (LVDT) were installed at the top of the front row piles and the back row piles, as illustrated in Figure 1, to measure the vertical displacement. Moreover, fourteen pairs of full-bridge strain gauges were attached on pile evenly to measure the bending moment at various depth during the loading process.

The soil container has an internal dimension of 0.85-m long, 0.7-m wide, and 0.75-m deep. The separation between the pile base and the bottom box is 100 mm, which is around 4D (D denotes pile diameter). Besides, that lateral distance from pile to strongbox is 350 mm. which is larger than 7D. On this condition, the boundary effect has little influence on the testing results according to the reports by Xu & Zhang (1996).

2.3 Property of soils

In the present tests, centrifuge model of TPJ foundation was subjected to lateral load in saturated sand and soft clay, respectively.

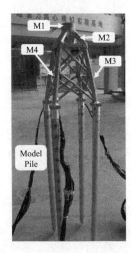

(a) Photo of the model TPJ foundation (Zhu et al. 2015)

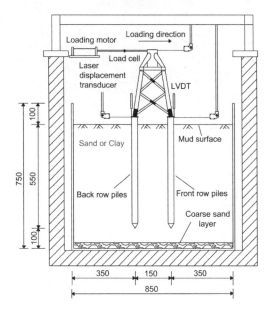

(b) Centrifuge test arrangement (Unit: mm)

Figure 1. Schematic diagram of centrifuge model test.

The sand used was commercially available Fujian standard sand, which has a mean particle diameter $D_{50} = 0.17$ mm, uniformity coefficient $C_u = 1.542$, specific gravity $G_s = 2.643$ and the critical state friction angle of 35°. The maximum and minimum dry unit weight are 1.645 kN/m³ and 1.335 kN/m³ respectively. At first, a coarse sand layer with a thickness of 0.03 m was rained at the bottom of the soil container, providing base drainage. Then, a piece of non-woven geotextile was placed on the coarse sand layer. The dry sand was prepared by the raining method using a spot-pouring hopper to a thickness of 0.60 m and a relative density of 60%.

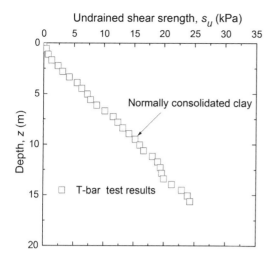

Figure 2. Undrained soil shear strength from T-bar test.

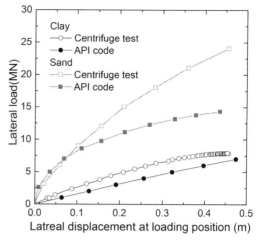

Figure 3. Lateral load–displacement curves in sand and clay.

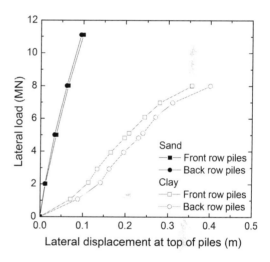

Figure 4. Lateral load-displacement curves at top of piles.

For another set of centrifuge tests, the normally consolidated clay was prepared by thoroughly mixing dry kaolin powder in a soil-mixing machine at a water content of 173% (two times its liquid limit). The slurry then subjected to 8 h of self-weight consolidation at 100 g (over 9 years in prototype time) under double drainage. Subsequently, the distribution of the undrained soil strength (s_u) along depth was obtained from T-bar penetrometer tests using a resistance factor of 10.5 (Stewart & Randolph, 1991), as shown in Figure 2.

3 TEST RESULTS

3.1 Load-displacement behaviour

Figure 3 shows the lateral load-displacement curves at loading position both in sand and in soft clay. The static load-displacement behaviour is of a work hardening type and it can be found that the lateral bearing capacity of the TPJ foundation in the soft clay is considerably lower than that in saturated sand. Also the calculated lateral load-displacement curves using API's static p–y responses are plotted in Figure 3 and the results underestimate the bearing capacity of TPJ foundations in both kind of soils, indicating the API code (2012) is likely to give conservative predictions. Similar observations can be found in Jeanjean (2009).

Figure 4 shows the lateral load-displacement curves at the top of piles during the loading process. It is found that the load-displacement curves at the top of each pile are of work hardening type as well and the lateral displacement of piles in clay is significantly larger than that in sand when subject the same lateral load. Moreover, the lateral displacement of front row piles is larger than that of back row piles in both two soil types. Of interest is that the difference between the front- and back-row piles in the soft clay is larger than that in sand. That is due to the upper jacket is inclined with the increase of lateral load and the inclination of jacket in the soft clay is greater than that in sand.

To investigate the vertical response of TPJ foundation when subjected to lateral load, load-vertical displacement curves at the top of piles are presented in Figure 4. The vertical displacement of the back row piles (Uplift piles) is larger than that of the front row piles (Pushdown piles) in both kinds of soil, indicating the failure mode of the TPJ foundation under the lateral static load is due to the back row piles pulling out. API code (2012) suggests that the bearing capacity of piles can be determined based on the load-displacement curves. In the present study, the ultimate bearing capacity of the TPJ foundation is taken as the load corresponding to an inflection point at the lateral load-vertical displacement curves of back row piles. As illustrated in Figure 5, for TPJ foundations in clay, the ultimate bearing capacity of the back row piles is

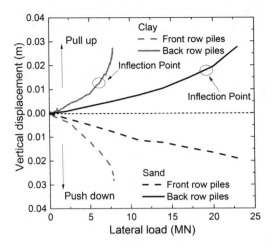

Figure 5. Lateral load-vertical displacement behaviour of pile shafts in two kinds of soil.

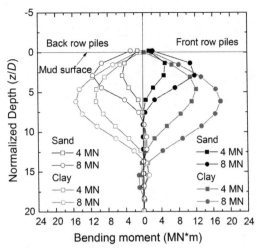

Figure 6. Distribution of bending moment along depth in different soil (H = 4 MN, 8 MN).

about 7 MN, while it is about 20 MN in sand, confirming that ultimate bearing capacity of TPJ foundation in clay is lower than sand, corresponding with the trend observed in Figure 3.

3.2 Bending moment of pile shafts

In centrifuge tests, the bending moments of the piles were measured by fourteen pairs of strain gauges placed on the surface of pile shafts (Zhu et al. 2015). Figure 4 presents the distribution of the bending moment along the piles when the lateral load (H) is 4 MN and 8 MN, respectively. Of interest is that with the increase of lateral load, the position of the maximum bending moment and inflection point in two kinds of soil move downwards gradually due to the expansion of plastic zone in front of the pile shafts. It can also be found that in saturated sand, the maximum bending moment and inflection point is considerably shallower than those in clay, as illustrated in Figure 6. Moreover, the maximum bending moment of pile shafts in clay are larger than that in sand when subject the same lateral load. The difference of bending moment between front row piles and back row piles in sand is more significant than that in clay. The best explanation is that the axial force of the pile shafts in sand has apparent effect on the effective stress (Zhu et al, 2015), but not in undrained soft clay.

3.3 Lateral deflection of pile shafts

Since the embedment length of piles is much larger than its diameter, the deflection and rotation at pile toe are considered to be zero. Based on these two boundary conditions, the lateral deflection along front row piles and back row piles can be obtained. Figure 7 presents the distribution of lateral deflection along pile under two different loading cases (H = 4 MN, 8 MN). It can be seen that the difference of lateral deflection between front row piles and back row piles in the soft clay is

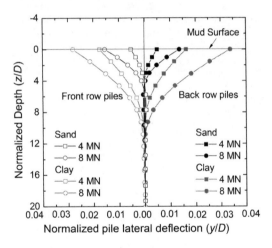

Figure 7. Distribution of lateral displacement along depth in different soil (H = 4 MN, 8 MN).

obviously larger than that in sand. In addition, limited pile deflection in sand can be observed below a depth of approximately 7D (D denotes pile diameter), while this value is 12D in clay. This illustrates that the TPJ foundation has a deeper effect zone in clay than that in sand when subjected to the lateral load.

3.4 P-y curves

The p-y curve proposed by Matlock (1970) firstly can comprehensively reflect the non-linearity of pile-soil interaction and the relative stiffness of pile and soil. Based on the centrifugal model test of the TPJ foundation, the p-y curve can be obtained by measuring the bending moments at various depth firstly and then calculating based on the reciprocal relationship between the bending moment, lateral pile deformation and soil resistance (Jeanjean 2009).

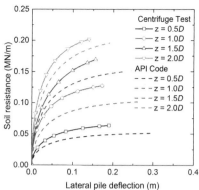

(a) P-y curves of front row piles in clay

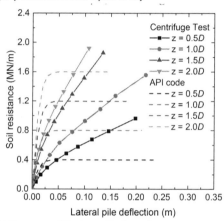

(b) P-y curves of front row piles in sand

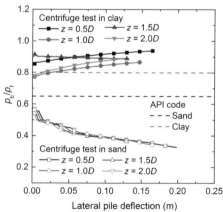

(c) Ratio of resistance between back row piles and front row piles

Figure 8. Comparison of p-y curves between centrifuge test and API code in two kinds of soil.

Figure 8 presents the comparison of p-y curves between centrifuge test and API code in two kinds of soil. It could be found that the soil resistance around piles in sand is obviously greater than that in clay when at the same depth, conforming that the lateral bearing capacity of TPJ foundations in sand is larger than in clay (see Fig. 3). Compared with the results predicted by API code, the soil resistance of front row piles obtained from centrifuge test is much larger, indicating API code is likely to be more conservative. When comparing the soil resistance of back row piles with that of front row piles, as illustrated in Figure 8(c), the former within the depth range of 2D was about 40% of the latter in sand. Besides, it is interesting to be found that the soil resistance of back row piles in sand predicted by API code is larger than that in centrifuge test, which means the axial force has obvious effect on the pile group effect. However, the difference of soil resistance in clay between front row piles and back row piles is not significant (the ratio was about 90%). The best explanation is that the static loading rate is so fast in the centrifugal tests that the clay is approximately in undrained condition during the loading process (Stewart & Randolph 1991), so the axial force of the pile does not significantly change effective stress of soil. Additionally, the spacing of TPJ foundation in this test is 5.8D and pile group effect is not obvious (Ilyas et al. 2004). These two factors mentioned above lead to a small difference in the p-y curve between front row piles and back row piles of TPJ foundation in soft clay.

4 CONCLUSIONS

A series of centrifuge model tests of tetrapod jacket (TPJ) foundation have been carried out in two kinds of soil type (including saturated sand and normally consolidated clay) to investigate static pile-soil interaction and deformation of TPJ foundation. The main findings are summarized as follows:

1. With the increase of the lateral load, the position of maximum bending moment and inflection point gradually move downwards both in clay and sand. Additionally, when subjected to the same lateral load, the depth of maximum bending moment and inflection point in sand are significantly shallower than that in soft clay. Based on the distribution of lateral pile deformation, little pile deflection can be observed in sand below 7D, while this value is 12D in clay.
2. The lateral load applied to TPJ foundation will induce large axial force at the piles. According to the lateral load-vertical displacement curves of pile shafts, the failure mechanism of TPJ foundation under the lateral static load is due to the back row piles being pulled out.
3. The soil resistance of back row piles is about 40% of the front piles in sand, since the change in the axial force of piles affects the effective stress in sand. However, the gap of p-y curves in soft clay between the front row piles and back row piles is not significant. That is due to the clay is in undrained condition during the loading process and the change of axial force only has no effect on the shear strength of soil around piles.

ACKNOWLEDGEMENTS

The authors acknowledge the National Natural Science Foundation of China (research grant: 51679211, 51708494), Guangdong Electric Power Design Institute (research grant: 2011-CGSQ-264), China postdoctoral Science Foundation (Grant: 2017M611994) and particularly Zhejiang Electric Power Design Institute, for their financial support.

REFERENCES

API Code. 2012. Recommended Practice for Planning, Designing and Constructing Fixed Offshore Platforms-Working Stress Design, API Recommended Practice 2A-WSD, 21 edn, American Petroleum Institute, Washington, D.C.

Byrne, B.W. & Houlsby, G.T. 2003. Foundations for offshore wind turbines. *Philosophical Transactions*, 361(1813), 2909.

Chen, Y.M., Kong, L.G., Zhou, Y.G. Jiang J.Q., Tang, X.W., Niu, B. & Lin, M. 2010. Development of a large geotechnical centrifuge at Zhejiang University. *In Proceedings of 7th International Conference on Physical Modeling in Geotechnics*, Zurich, 28 June–1 July, 223–228.

Elshafey, A.A., Haddara, M.R. & Marzouk, H. 2009. Dynamic response of offshore jacket structures under random loads. *Marine Structures*, 22(3): 504–521.

Ilyas, T., Leung, C.F., Chow, Y.K., & Budi, S.S. 2004. Centrifuge model study of laterally loaded pile groups in clay. *Journal of Geotechnical and Geoenvironmental Engineering*, 130(3), 274–283.

Jeanjean, P. 2009. Re-assessment of p-y curves for soft clays from centrifuge testing and finite element modeling. *Proc. Offshore Technology Conf.*, Paper OTC20158, Houston.

Li, T. 2015. Study on lateral loading behaviours of jacket foundation for offshore wind turbines, master dissertation, Zhejiang University. (In Chinese)

Li, Z., Haigh, S.K., & Bolton, M.D. 2010. The response of pile groups under cyclic lateral loads. *International Journal of Physical Modelling in Geotechnics*, 10(2): 47–57.

Matlock, H. 1970. Correlation for Design of Laterally Loaded Piles in Soft Clay. Off*shore Technology in Civil Engineering*, 1: 77–94.

Mao, D., Zhong, C., Zhang, L. & Chu, G. 2015. Dynamic response of offshore jacket platform including foundation degradation under cyclic loadings. *Ocean Engineering*, 100: 35–45.

Shi, W., Park, H.C., Chung, C.W., & Kim, Y.C. 2011. Comparison of dynamic response of monopile, tripod and jacket foundation system for a 5-MW wind turbine. *In The Twenty-first International Offshore and Polar Engineering Conference*. International Society of Offshore and Polar Engineers.

Stewart, D.P., & Randolph, M.F. 1991. A new site investigation tool for the centrifuge. In *Proc. Int. Conf. Centrifuge 91*: 531–538.

Xie, Y., Leung, C.F., & Chow, Y.K. 2012. Centrifuge modelling of spudcan–pile interaction in soft clay. *Géotechnique*, 62(9): 799–810.

Xu, G.M. & Zhang, W.M. 1996. A study of size effect and boundary effect in centrifugal tests. *Chinese Journal of Geotechnical Engineering*. 18(3): 80–85. (in Chinese)

Zhu, B., Li, T. & Ren, J. 2015 Centrifuge model tests on laterally monotonic and cyclic pile-soil interaction of a tetrapod jacket foundation. *In Proceeding of International Symposium on Frontiers in offshore Geotechnics III*: ISFOG, Oslo, Norway, 673–678.

Zhu, B., Li, T., Xiong, G. & Liu, J. C. 2016. Centrifuge model tests on laterally loaded piles in sand. *International Journal of Physical Modelling in Geotechnics*, 16(4): 160–172.

Author index

Abdoun, T. 293, 519, 847, 943, 949
Abrashitov, A.A. 203
Abuhajar, O.S. 221
Achmus, M. 1347
Adamidis, O. 113, 937
Adams, M.T. 1291
Aggarwal, P. 1113
Ahmed, U. 1025
Airey, D.W. 513
Ajagbe, W.O. 1353
Akoochakian, A. 1157
Akram, I. 1253
Al-Baghdadi, T. 695
Al-Defae, A.H. 241
Al-Fergani, M. 507
Alber, S. 113
Ali, U. 233
Almeida, M.C.F. 1031, 1093
Almeida, M.S.S. 1031, 1043, 1093
Alvarez Grima, M. 469
Anastasopoulos, I. 113, 1321
Apostolou, E. 1163
Arakawa, M. 1469
Archer, A. 489
Arnold, A. 119
Asaka, Y. 921
Ashtiani, M. 885
Askarinejad, A. 119, 461, 469, 987, 1119, 1149
Audrain, P. 465
Augarde, C. 695
Azúa-González, C. 469

Ball, J. 695
Barrett, J. 577
Bathurst, R.J. 1259
Bayton, S.M. 163, 533, 689, 719
Beber, R. 125
Becker, L.B. 527
Beckett, C.T.S. 823
Beddoe, R.A. 175
Beemer, R.D. 279
Beltrán-Rodriguez, L.N. 865
Bengough, A.G. 401, 425
Bezuijen, A. 285, 597, 815, 1037
Bhattacherjee, D. 337
Bienen, B. 33, 651, 669, 707, 719
Bilotta, E. 3, 955
Bisht, R.S. 1329
Bizzotto, T. 553

Black, J.A. 163, 507, 533, 545, 689, 719, 835
Blake, A. 695
Blanc, M. 465, 701, 719, 1031, 1043, 1341
Blom, C.B.M. 443
Boksmati, J. 343
Borghei, A. 247, 349
Boulanger, R.W. 21
Bowman, E.T. 377, 1075
Breen, J. 501
Brennan, A. 695
Brennan, A.J. 553, 1163, 1265
Bretschneider, A. 465
Broere, W. 317, 323, 443
Bronner, C.E. 21
Bronner, J.D. 21
Brown, M.J. 241, 553, 695
Burd, H.J. 725
Byrne, B.W. 725, 737

Cabrera, M. 1075
Cai, Z.Y. 583, 1143, 1187, 1475
Caicedo, B. 131, 155, 413, 1285, 1407
Carey, T. 293, 829
Cargill, A. 533
Caro, S. 155
Cassidy, M.J. 279, 615, 651, 731
Cassie, P. 725
Castillo, D. 155
Chalaturnyk, R.J. 1271
Chan, D.Y.K. 1421
Chandler, H. 553
Charles, J.A. 835
Chen, A.Z. 583
Chen, C.H. 873
Chen, C.H. 873
Chen, H. 909
Chen, S.S. 495
Chen, Y.M. 293, 407, 767, 929
Cheng, Y.B. 495
Chian, S.C. 355
Chow, S.H. 559
Chow, Y.K. 681
Coelho, P.A.L.F. 993
Coleman, J. 299
Colletti, J. 299
Combe, G. 841
Constantinou, M.C. 519
Coombs, W. 695
Cui, G. 137, 359

Cuira, F. 1401
Cumming-Potvin, D. 809
Cunha, R.P. 1407

da Silva, T.S. 1169
Dafni, J. 365
Darby, K.M. 21
Dasaka, S.M. 169
Dashti, S. 1005, 1199
Dave, T.N. 169
Davidson, C. 695
de Boorder, M. 1119
de Jager, R.R. 987
de Lange, D.A. 597
De, A. 779
DeJong, J.T. 21
DeJong, M.J. 437, 449
den Hamer, D.A. 285
Deng, C. 1427
Detert, O. 1175
Diakoumi, M. 743
Dijkstra, J. 317, 323
Dimitriadis, K. 903
Divall, S. 143, 539
Dobrisan, A. 125
Dobry, R. 519, 949
Doreau-Malioche, J. 841
Dos Santos Mendes, C. 1377

Ebeido, A. 1335
Ebizuka, H. 233
Egawa, T. 879
Ehrlich, M. 1223
Eichhorn, G.N. 785
El Haffar, I. 701, 1341
El Shafee, O. 943
El-Sekelly, W. 949
Elgamal, A. 1335
Elmrom, T. 163
Elshafie, M.Z.E.B. 791, 1169
Emeriault, F. 1377
Escobar, J. 131
Escoffier, S. 293, 1401
Exton, M.C. 371

Fagundes, D.F. 1031, 1043
Farrin, M. 797
Fasano, G. 955
Faustin, N.E. 791
Fioravante, V. 955
Fiumana, N. 603
Flora, A. 955
Forsyth, M. 559

Fourie, A.B. 823
Franza, A. 209
Frick, D. 1347
Frolovsky, Y.K. 1277
Fukutake, K. 215
Funahara, H. 961

Gaber, F. 377
Ganchits, V.V. 1277
Ganiyu, A.A. 1353
Garala, T.K. 1359
García-Torres, S. 1181
Garzón, L.X. 1285
Gaudin, C. 33, 279, 501, 603, 719
Gavras, A. 293
Geirnaert, K. 285
Gelagoti, F. 1383
Georgiou, I. 1383
Ghaaowd, I. 1365
Ghalandarzadeh, A. 565, 885
Ghayoomi, M. 247, 349
Giardina, G. 449
Giretti, D. 955
Girout, R. 1031, 1043
Gng, Z. 1149
Goodey, R.J. 43, 143, 191, 539, 853, 1395
Gorlov, A.V. 1277
Gourvenec, S. 51
Grabe, J. 669, 707, 1229, 1451
Gu, X.W. 495, 583, 1187, 1475
Gütz, P. 1347

Ha, J.G. 629, 897
Haigh, S.K. 125, 293, 639, 719, 785, 993, 1427
Hajialilue-Bonab, M. 797, 829
Halai, H. 191
Hallett, P.D. 401
Harry, S. 371
Hartmann, D.A. 1031
Hasebe, M. 921
Hasegawa, G. 1067
Hatanaka, Y. 891
Heins, E. 707
Hem, R. 431, 675
Heron, C.M. 137, 209, 359, 455
Hicks, M.A. 987
Higo, Y. 215
Hoffman, W. 299
Hong, J.Z. 495
Hori, T. 383, 1463
Horii, Y. 1371
Hossain, M.S. 329, 615
Hou, Y.J. 77, 1125
Houda, M. 1377
Houlsby, G.T. 737
Hu, L.M. 975
Hu, Y. 329, 615, 1309
Huang, B. 407
Huang, J.X. 975

Huang, M. 909, 1297
Huang, M.S. 761
Huang, Q. 519
Huang, X. 1365
Huang, Y.H. 1143
Hughes, F.E. 967
Hung, H.M. 1193
Hung, W.Y. 293, 975

Iai, S. 265, 1017
Ibsen, L.B. 623
Idinyang, S. 209
Imamura, S. 981
Indiketiya, S. 803
Iskander, M. 389, 859
Isobe, K. 879, 891
Iwamoto, T. 675
Iwasa, N. 1235

Jacobsz, S.W. 179, 185, 305, 311, 809, 1081
Jahnke, S.I. 311
Jegatheesan, P. 803
Jenck, O. 1377
Jeong, Y.H. 609
Jia, C.H. 1125
Jia, Y. 909
Jiang, Q. 395
Jo, S.B. 629, 1055
Jun, M.J. 615
Juneja, A. 1113, 1329

Kailey, P. 1075
Kamalzare, M. 1049
Kamchoom, V. 1131
Kanai, H. 1253
Kang, X. 1297
Kavand, A. 1157
Kearsley, E.P. 179, 185, 311, 809, 1081
Khaksar, R.Y. 565
Khan, I.U. 507
Kim, D.S. 293, 609, 629, 749, 897
Kim, J.H. 609, 749
Kim, N.R. 1055
Kim, Y.H. 615
Kimura, M. 663, 915, 1067
Kiriyama, T. 215
Kirkwood, P.B. 1199
Kishida, K. 915
Kita, K. 713
Kitazume, M. 1217
Klinkvort, R.T. 689, 719
Knappett, J.A. 87, 241, 401, 425, 475, 481, 695, 1011, 1265
Ko, K.W. 897
Kogure, T. 1253
Kokkali, P. 847
König, D. 1175
Koteras, A.K. 623
Kotronis, P. 1401

Koudelka, P. 1433
Kourkoulis, R. 1383
Kunasegarm, V. 1439
Kundu, S. 1205
Kutter, B. 293, 829
Kutter, B.L. 371
Kuwano, J. 1193, 1253
Kuwano, R. 233, 803, 1087, 1099
Kwa, K.A. 513

Lai, C.G. 955
Laporte, S. 175
Larrahondo, J.M. 865
Lawler, J. 943
Le Cossec, J. 903
Le, B.T. 853
Le, J. 559
Lee, F.H. 1329
Lee, M.G. 629
Lehane, B.M. 1309
Leung, A.K. 475, 481, 1131
Leung, C.F. 681, 1025
Li, J.C. 929
Li, L. 859, 1457
Li, L.F. 395
Li, Y.P. 635
Liang, F. 909
Liang, J.H. 1125
Liang, T. 401, 481
Liao, T.W. 975
Liel, A.B. 1005
Ling, D.S. 407
Ling, H.I. 1457
Linhares, R.M. 527
Liu, K. 293, 407
Liu, W. 495
Loades, K.W. 401, 425
Loli, M. 241, 1383
Louw, H. 1081
Lozada, C. 413
Lundberg, A.B. 317
Luo, F. 1137
Lutenegger, A.J. 1291

Ma, X.F. 419
Madabhushi, G.S.P. 437, 719, 937, 993, 1181, 1359
Madabhushi, S.P.G. 125, 293, 343, 967, 1421
Madabhushi, S.S.C. 125, 639
Maeda, K. 431
Maghsoudloo, A. 987
Mair, R.J. 449, 791
Mamaghanian, J. 1211
Manikumar, C.H.S.G. 1211
Manzari, M. 293
Marques, A.S.P.S. 993
Marques, F.L. 527
Marshall, A.M. 137, 209, 359, 455

Maruyama, K. 589
Mason, H.B. 371
Matsuda, S. 1217
Matsuda, T. 431
Mayall, R.O. 725
Mayanja, P. 1229
McAdam, R.A. 725
McCartney, J. 1365
McNamara, A.M. 191, 539, 1395, 1445
Meijer, G.J. 401, 425
Mikami, T. 1017
Mikasa, K. 645
Miles, S. 999
Mirmoradi, S.H. 1223
Mirshekari, M. 247
Miyamoto, J. 431, 571, 675
Miyata, Y. 1259
Miyazaki, Y. 915
Mizumoto, M. 1469
Mohan Gowda, K.T. 253
Molenkamp, F. 987
Momoi, M. 1217
Moormann, C. 197
Moradi, M. 565, 1157, 1247
Morikawa, Y. 589, 1259
Moug, D.M. 21
Movasat, M. 797
Mu, L. 1297
Muir Wood, D. 401

Nadimi, S. 853
Nagao, T. 1371
Nagula, S. 1229
Nakamoto, S. 1235
Nakase, H. 675
Néel, A. 465
Newson, T.A. 221
Ng, C.W.W. 489
Nicoll, B.C. 425
Niemann, C. 731
Nigorikawa, N. 921
Nishiura, D. 227
Nonoyama, H. 1259
Nordal, S. 1105
Norris, S. 1389

O'Loughlin, C.D. 33, 501, 559, 603, 651, 731
Ohara, Y. 1087
Okabayashi, K. 645
Okamura, M. 293
Olarte, J.C. 1005
Oliveira, F.S. 527
Oliveira, J.R.M.S. 1093
Omidvar, M. 859
Ong, D.E.L. 1025
Osman, M.H. 1353
Ota, M. 1099
Otsubo, M. 233
Ottolini, M. 323

Ovalle-Villamil, W. 259
Özcebe, A.G. 955

Panchal, J.P. 1395, 1445
Paramasivam, B. 1005
Parchment, J. 149
Park, H.J. 609, 629, 897
Park, S.G. 615
Pathmanathan, R. 803
Pearson, A. 1303
Pelekis, I. 437
Peng, R. 1125
Pérez-Herreros, J. 1401
Petryaev, A.V. 1277
Phillips, R. 577
Phoon, K.K. 1285
Piercey, G. 577
Pooresmaeili, A. 1247
Powell, T. 553
Prada-Sarmiento, L.F. 865
Pradhan, R.N. 1105
Pua, L.M. 155

Qi, S. 1011
Qin, C. 355

Ragni, R. 651
Rahadian, R. 443
Rammah, K.I. 1093
Ramos-Cañón, A.M. 865
Randolph, M.F. 707
Rashid, A.S.A. 1353
Raymond, A.J. 21
Razeghi, H.R. 1211
Ren, G.F. 495, 583, 1187, 1475
Reul, O. 731
Richards, D.J. 695, 743
Richards, I.A. 737
Ritchie, E.P. 143
Ritter, S. 449
Robinson, S. 1265
Rodríguez, E. 1407
Rosenbrand, E. 1037
Rotte, V.M. 1241
Ryan, C. 903

Sabermahani, M. 1247
Saboya, F. 465, 1365
Sakaguchi, H. 227
Sakellariadis, L. 1321
Salehi, A. 1309
Sánchez-Peralta, J.A. 865
Sánchez-Silva, M. 1285
Saran, R.K. 1061
Sasanakul, I. 259
Sassa, S. 589, 657
Sawada, K. 265
Sawamura, Y. 663, 915, 1067
Sayles, S. 743
Schanz, T. 1175
Schenkeveld, F. 461

Schiavon, J.A. 1315
Schindler, R. 1321
Schmoor, K.A. 1347
Seitz, K.-F. 1451
Seki, S. 1413, 1439
Sekita, K. 713
Seong, J.T. 749
Sera, R. 1087, 1099
Sett, K. 299
She, Y. 407
Sheil, B.B. 725
Shepley, P. 149, 1303, 1389
Shi, B.X. 583
Shi, J.Y. 635
Shibata, T. 663
Shields, L. 241
Shiraga, S. 1067
Siemens, G.A. 175
Silva, M. 841
Singla, R. 1113
Smit, M.S. 179, 185
Smith, C.C. 835
Soe, A.A. 1253
Song, G. 455
Song, L.B. 395
Sovso, S.L. 1309
Stallebrass, S.E. 143, 539, 1445
Stanier, S.A. 651
Stapelfeldt, M. 669
Stathas, D. 1457
Still, J. 999
Stone, K.J.L. 221, 743, 755, 903
Stone, N. 829
Stringer, M. 999
Sturm, A. 21
Sydrakov, A.A. 203

Taeseri, D. 1321
Takada, Y. 1017
Takahashi, H. 589
Takano, D. 1259
Take, W.A. 101
Takemura, J. 1235, 1413, 1439
Tamate, S. 383, 1463
Tang, H.W. 761
Tanghetti, G. 191
Tatari, A. 533
Taylor, R.N. 539, 853
Tehrani, F.S. 461
Tessari, A. 299, 847
Tessari, A.F. 545
Thakur, V. 1105
Thevanayagam, S. 519, 949
Thorel, L. 413, 465, 701, 719, 1031, 1043, 1315, 1341
Thusyanthan, N.I. 343
Tian, Y. 603, 731
Tibana, S. 1365
Tillman, A. 755
Tituaña-Puente, J.S. 865
Tobita, T. 271, 597

Tomita, N. 961
Toni, J.B. 841, 1377
Trejo, P.C. 1093
Trujillo-Vela, M.G. 865
Tsatsis, A. 1383
Tsuha, C.H.C. 1315
Tsurugasaki, K. 431, 571, 675
Tyagi, A. 1329

Ueda, K. 265, 293, 1017
Ueng, T.S. 873
Utsunomiya, T. 713

Valencia-Galindo, M.D. 865
van Beek, V.M. 1037
van der Woude, S. 443
van der Zon, J. 1119
van Zeben, J.C.B. 469
van 't Hof, C. 469
Vandenboer, K. 1037
Vaziri, M. 755
Vian, G. 1377
Viggiani, G. 841
Vincke, L. 285
Viswanadham, B.V.S. 253, 337, 1061, 1205, 1211, 1241
Vitali, D. 475

Wang, C. 1125
Wang, J.P. 1457
Wang, K. 1265
Wang, L. 695
Wang, L.J. 767, 929
Wang, N.X. 495
Wang, Y. 329, 1271
Wartman, J. 365
Watanabe, K. 1469
Wehr, J. 1163
Wen, K. 767
Wesseloo, J. 809
Whitehouse, R.J.S. 725
Wilschut, D. 443
Wilson, D.W. 21
Worbes, R. 197
Wu, W. 1075

Xia, P. 407
Xie, Y. 681
Xu, G.M. 495, 583, 1143, 1187, 1475
Xu, J.W. 419
Xu, L. 1457
Xu, T. 815

Yaba, O. 1377

Yamamoto, S. 227
Yamanashi, T. 879
Yao, C.F. 1413
Yeh, H. 371
Yifru, A.L. 1105

Zambrano-Narvaez, G. 1271
Zania, V. 719, 1309
Zayed, M. 1335
Zaytsev, A.A. 203, 1277
Zeghal, M. 293
Zhang, C. 1143
Zhang, C.R. 761
Zhang, G. 1137
Zhang, H. 909
Zhang, M. 395
Zhang, W. 119, 1119, 1149
Zhang, W.M. 495
Zhang, X. 481
Zhang, X.D. 1125
Zhang, Y. 1297
Zhang, Z. 355
Zhao, R. 475
Zhou, Y.G. 293, 407
Zhu, B. 767, 929
Zimmie, T.F. 779, 1049
Ziotopoulou, K. 21

Printed and bound by PG in the USA